Inference:

Confidence interval for Parameter $= Estimate \pm Critical\ value \times SE(Estimator)$

Test statistic $= \dfrac{Estimate - Parameter}{SE(Estimator)}$ [Replace SE by SD if latter is known]

Parameter	Estimator	SD (Estimator)	SE (Estimator)
p	$\hat{p}$	$\sqrt{\dfrac{pq}{n}}$	$\sqrt{\dfrac{\hat{p}\hat{q}}{n}}$
$p_1 - p_2$	$\hat{p}_1 - \hat{p}_2$	$\sqrt{\dfrac{p_1 q_1}{n_1} + \dfrac{p_2 q_2}{n_2}}$	$\sqrt{\dfrac{\hat{p}_1 \hat{q}_1}{n_1} + \dfrac{\hat{p}_2 \hat{q}_2}{n_2}}$
μ	$\bar{y}$	$\dfrac{\sigma}{\sqrt{n}}$	$\dfrac{s}{\sqrt{n}}$
$\mu_1 - \mu_2$	$\bar{y}_1 - \bar{y}_2$	$\sqrt{\dfrac{\sigma_1^2}{n_1} + \dfrac{\sigma_2^2}{n_2}}$	$\sqrt{\dfrac{s_1^2}{n_1} + \dfrac{s_2^2}{n_2}}$
μ_d	$\bar{d}$	$\dfrac{\sigma_d}{\sqrt{n}}$	$\dfrac{s_d}{\sqrt{n}}$
σ_ε	$s_e = \sqrt{\dfrac{\sum(y - \hat{y})^2}{n - 2}}$	(divide by $n - k - 1$ in multiple regression)	
β_1	b_1	(in simple regression)	$\dfrac{s_e}{s_x \sqrt{n - 1}}$
μ_ν	$\hat{y}_\nu$	(in simple regression)	$\sqrt{SE^2(b_1) \cdot (x_\nu - \bar{x})^2 + \dfrac{s_e^2}{n}}$
y_ν	$\hat{y}_\nu$	(in simple regression)	$\sqrt{SE^2(b_1) \cdot (x_\nu - \bar{x})^2 + \dfrac{s_e^2}{n} + s_e^2}$

Pooling: For testing difference between proportions: $\hat{p}_{pooled} = \dfrac{y_1 + y_2}{n_1 + n_2}$

For testing difference between means (when $\sigma_1 = \sigma_2$): $s_p = \sqrt{\dfrac{(n_1 - 1)s_1^2 + (n_2 - 1)s_2^2}{n_1 + n_2 - 2}}$

Substitute these pooled estimates in t̶ ̶ ̶ ̶ ̶ ̶ ̶ groups when assumptions and conditions are met.

Chi-square: $\chi^2 = \sum \dfrac{(Obs - Exp)^2}{Exp}$

One-way ANOVA: $SS_T = \sum \sum (\bar{y}_j - \bar{\bar{y}})^2$; $\qquad MS_T = SS_T/(k - 1)$

$SS_E = \sum \sum (\bar{y}_{ij} - \bar{y}_j)^2$; $\qquad MS_E = SS_E/(N - k)$

$F = MS_T/MS_E$ with df $= (k - 1, N - k)$

STATS
DATA AND MODELS

THIRD CANADIAN EDITION

STATS
DATA AND MODELS

RICHARD D. DE VEAUX
WILLIAMS COLLEGE

PAUL F. VELLEMAN
CORNELL UNIVERSITY

DAVID E. BOCK
CORNELL UNIVERSITY

AUGUSTIN M. VUKOV
UNIVERSITY OF TORONTO

AUGUSTINE C.M. WONG
YORK UNIVERSITY

Vice President, Editorial: Anne Williams	**Photo Permissions Research:** Integra Publishing Services
Acquisitions Editor: Mark Grzeskowiak	**Text Permissions Research:** Integra Publishing Services
Marketing Manager: Euan White	**Interior Designer:** Anthony Leung
Content Manager: Emily Dill	**Cover Designer:** Anthony Leung
Project Manager: Pippa Kennard	**Cover Image:** Rawpixel.com/Shutterstock.com, Graph is based on free material from GAPMINDER.ORG, CC-BY LICENSE
Content Developer: Paul Donnelly	**Vice-President, Cross Media and Publishing Services:** Gary Bennett
Media Developer: Kelli Cadet	
Production Services: Cenveo® Publisher Services	
Permissions Project Manager: Joanne Tang	

Pearson Canada Inc., 26 Prince Andrew Place, North York, Ontario M3C 2H4.

9780134301051

1 17

Library and Archives Canada Cataloguing in Publication

De Veaux, Richard D., author
        Stats : data and models / Richard D. De Veaux, Williams College, Paul F. Velleman, Cornell University, David E. Bock, Cornell University, Augustin M. Vukov, University of Toronto, Augustine C.M. Wong, York University.—Third Canadian edition.

Includes index.

ISBN 978-0-13-430105-1 (hardcover)

        1. Statistics—Textbooks. 2. Mathematical statistics—Textbooks. I. Velleman, Paul F., 1949-, author II. Bock, David E., author III. Vukov, Augustin M., author IV. Wong, Augustine C. M., author V. Title.

QA276.12.D42 2017   519.5   C2017-902493-0

To Sylvia, who has helped me in more ways than she'll ever know,
and to Nicholas, Scyrine, Frederick, and Alexandra,
who make me so proud in everything that they are and do

—Dick

To my sons, David and Zev, from whom I've learned so much,
and to my wife, Sue, for taking a chance on me

—Paul

To Greg and Becca, great fun as kids and great friends as adults,
and especially to my wife and best friend, Joanna, for her
understanding, encouragement, and love

—Dave

To my adopted country, Canada, its amazing citizens,
indigenous peoples, immigrants and refugees, whose stories inspire me,
and the University of Toronto without which this
dedication would never have happened.

—Gus

To Camilla and Octavia for their
understanding and encouragement

—Augustine

Meet the Authors

Richard D. De Veaux is an internationally known educator and consultant. He has taught at the Wharton School and the Princeton University School of Engineering, where he won a "Lifetime Award for Dedication and Excellence in Teaching." Since 1994, he has been Professor of Statistics at Williams College. Dick has won both the Wilcoxon and Shewell awards from the American Society for Quality. He is a fellow of the American Statistical Association (ASA). In 2008, he was named Statistician of the Year by the Boston Chapter of the ASA. Dick is also well known in industry, where for more than 25 years he has consulted for such Fortune 500 companies as American Express, Hewlett-Packard, Alcoa, DuPont, Pillsbury, General Electric, and Chemical Bank. Because he consulted with Mickey Hart on his book *Planet Drum*, he has also sometimes been called the "Official Statistician for the Grateful Dead." His real-world experiences and anecdotes illustrate many of this book's chapters.

Dick holds degrees from Princeton University in Civil Engineering (B.S.E.) and Mathematics (A.B.) and from Stanford University in Dance Education (M.A.) and Statistics (Ph.D.), where he studied dance with Inga Weiss and Statistics with Persi Diaconis. His research focuses on the analysis of large data sets and data mining in science and industry.

In his spare time, he is an avid cyclist and swimmer. He also is the founder of the "Diminished Faculty," an a cappella Doo-Wop quartet at Williams College and sings bass in the college concert choir. Dick is the father of four children.

Paul F. Velleman has an international reputation for innovative Statistics education. He is the author and designer of the multimedia Statistics program *ActivStats*, for which he was awarded the EDUCOM Medal for innovative uses of computers in teaching statistics, and the ICTCM Award for Innovation in Using Technology in College Mathematics. He also developed the award-winning statistics program, *Data Desk*, and the Internet site Data and Story Library (DASL) (lib.stat.cmu.edu/DASL/), which provides data sets for teaching Statistics. Paul's understanding of using and teaching with technology informs much of this book's approach.

Paul has taught Statistics at Cornell University since 1975. He holds an A.B. from Dartmouth College in Mathematics and Social Science, and M.S. and Ph.D. degrees in Statistics from Princeton University, where he studied with John Tukey. His research often deals with statistical graphics and data analysis methods. Paul co-authored (with David Hoaglin) *ABCs of Exploratory Data Analysis*. Paul is a Fellow of the American Statistical Association and of the American Association for the Advancement of Science. Paul is the father of two boys.

David E. Bock taught mathematics at Ithaca High School for 35 years. He has taught Statistics at Ithaca High School, Tompkins-Cortland Community College, Ithaca College, and Cornell University. Dave has won numerous teaching awards, including the MAA's Edyth May Sliffe Award for Distinguished High School Mathematics Teaching (twice), Cornell University's Outstanding Educator Award (three times), and has been a finalist for New York State Teacher of the Year.

Dave holds degrees from the University at Albany in Mathematics (B.A.) and Statistics/Education (M.S.). Dave has been a reader and table leader for the AP Statistics exam, serves as a Statistics consultant to the College Board, and leads workshops and institutes for AP Statistics teachers. He has recently served as K–12 Education and Outreach Coordinator and a senior lecturer for the Mathematics Department at Cornell University. His understanding of how students learn informs much of this book's approach.

Dave and his wife relax by biking or hiking, spending much of their free time in Canada, the Rockies, or the Blue Ridge Mountains. They have a son, a daughter, and four grandchildren.

Augustin M. Vukov has been teaching at the University of Toronto for several decades. For much of that time, he was the course coordinator for *The Practice of Statistics*—a large multi-section course designed to introduce the basic concepts and practice of Statistical Science to students from a great variety of (mostly non-mathematical) disciplines. Having taught so many students their required Stats course, he was not surprised to hear his new family doctor, his new ophthalmologist and a prospective new dentist say "Your name looks familiar. Ah yes, I used to sit in your Statistics lecture!". A big fan of Minitab™ software for use in introductory Statistics courses, he has also authored several Minitab manuals.

Augustine C.M. Wong is a professor of Statistics at York University. He completed his Ph.D. at the University of Toronto in 1990 and taught at the University of Waterloo and University of Alberta before coming to York in 1993. His research interests include asymptotic inference, computational methods in statistics, and likelihood-based methods. He is an author or co-author of over 80 research articles and 2 book chapters. At York, he teaches various statistics courses at both the undergraduate and graduate levels.

Contents

*Optional section.

*Optional section.

Part VI Assessing Associations Between Variables

*Optional section.

---

*Optional section.

Preface

We are often asked why we write Statistics texts. After all, it takes a lot of work to find new and better examples, to keep datasets current, and to make a book an enjoyable and effective learning tool. So we thought we'd address that question first.

We do it because it's fun.

Of course, we care about teaching students to think statistically; we are teachers and professional statisticians. And we want to change the way students think about the world. But Statistics can be a particularly challenging subject to teach. The student encounters many new concepts, many new methods, and many new terms. Our challenge has been to write a book that students would read, learn from, and enjoy. And we recommit to this goal with each new edition.

The book you hold is more succinct and, we hope, even more readable than previous editions. Of course, we've kept our conversational style and background anecdotes.[1] But we've tightened discussions and adjusted the order of some topics. The story we tell about Statistics moves even more quickly to interesting real-world questions.

Increasingly teachers are using examples from Statistics to provide intuitive examples of how a little bit of math can help us say a lot about the world. Students expect Statistics to be about real-world insights. This edition of *Stats: Data and Models* keeps students engaged and interested because we show Statistics in action right from the start. Students will be solving problems of the kind they're likely to encounter in real life sooner. In Chapter 4, they will be comparing groups and in Chapter 6, they'll see relationships between two quantitative variables—and, of course, always with real, modern data. By emphasizing statistical thinking, these early chapters lay the foundation for the more advanced chapters on Analysis of Variance and Multiple Regression at the end of the book, telling a consistent story throughout.

We hope our book will help students realize Statistics can be fun, useful, and empowering. There are few things more rewarding than discovering the world anew through the lens of statistical analysis. And few things more fun for authors than helping students make those discoveries.

So, What's New in This Edition?

We've scrutinized literally every sentence and paragraph throughout the book, and rewritten many sections to make them clearer and more interesting, along with introducing some new up-to-the-minute motivating examples.

We've added some new features, each with the goal of making it even easier for students to put the concepts of Statistics together into a coherent whole.

1. New and improved pedagogical tools: We've replaced the list of sections at the start of each chapter with a more informative list of *Learning Objectives*. We've also replaced *Where Are We Going* with *Connections*, while deleting the similarly named item that used to sit (we believe, almost invisibly) at the end of chapters in earlier editions. Chapter study materials at the end of chapters still include *What Have We Learned* including the *Review of Terms*. Students who understand the latter items are well prepared for future chapters (not to mention well on their way to being ready for any tests).

2. Streamlined content: Our reorganization has shortened the book from 30 to 29 chapters (Chapter 29 is only available as part of the eText in MyLab™ Statistics). We've also grouped some important concepts, in new and better presentation orders. Each chapter is still a focused discussion, and most can be taught in one lesson.

---

[1]and our footnotes.

3. Content changes: Here's how we've reorganized or changed the content:

   a. We've reorganized material in *Chapter 3*: *Displaying and Summarizing Quantitative Data*. Instead of measures of centre followed by measures of spread, we start with the more intuitive measures: median, IQR, and 5-number summary, followed later by the mean and standard deviation.

   b. We've removed the chapter *Understanding Randomness*, while moving its section on Random Numbers into the following chapter on Sample Surveys.

   c. We've tried to improve the organization of material in Parts V–VII:

   - *Inference About Means* appears earlier now, in Part V. We still lead the discussion of inference with inference for proportions (for reasons we explain in the *Test Bank and Resource Guide*), but now we turn sooner to inference for means so students can see the methods side by side. Students can then also see that the reasoning is essentially the same.

   - *Comparing Two Proportions* was moved from Part V to Part VI, so that it now follows the similar discussion for comparing two means, with the chapter on chi-square tests immediately after.

   - *Inference for Regression* was moved from Part VII to Part VI, where it more logically belongs, though in the same chapter order as before.

   d. We've added an optional section on the Regression ANOVA to the *Inference for Regression* chapter.

4. Section Exercises: A new feature of this edition is the inclusion of *Section Exercises*. In addition to the usual end-of-chapter exercises, at the end of each section, you will now find these new single-concept exercises.

5. Exercises: We've updated most exercises that use real-world data, retired some that were getting old, and added new exercises. As in previous editions, many of the exercises feature Canadian content and data with up-to-date references to Indigenous peoples, sports, health care, education, the environment, and other social and political issues.

Our Approach

Statistical Science is practiced with technology. We think a modern Statistics text should recognize that fact from the start. And so do our students. You won't find tedious calculations worked by hand. But you will find equation forms that favour intuition over calculation. You'll find extensive use of real data—even large data sets. And, most important, you'll find a focus on statistical thinking rather than calculation. The question that motivates each of our hundreds of examples is not "how do you *find* the answer?" but "how do you *think* about the answer?"

We've been guided in the choice and order of topics by several fundamental principles. First, we have tried to ensure that each new topic fits into the growing structure of understanding that we hope students will build. We know that learning a new subject like Statistics requires putting together many new ideas, and we have tried to make that process as natural as possible. That goal has led us to cover some topics in a different order than is found in other texts.

For example, we introduce inference by looking at a confidence interval for a proportion rather than the more traditional approach that starts with the one-sample mean. Everyone has seen opinion polls. Most people understand that pollsters use a sample of voters to try to predict the preferences of the entire population and that the result is an estimate with a margin of error. Showing how to construct and interpret a confidence interval in this context introduces key concepts of inference in a familiar setting.

We next examine hypothesis tests for a proportion, which use formulas and a test statistic that is now familiar, so it is easier to focus on the logic and structure of a formal hypothesis test. When we then turn our attention to the topic of small-sample inference for means, the only new issue is the t-distribution. However, should you wish to discuss the z-test and confidence interval for means (known sigma or large n) earlier, in tandem with the z-procedures for proportions, optional sections are included in those chapters.

Textbooks can be defined more by what they choose not to cover than by what they do cover. We've structured this text so that each new topic fits into a student's growing understanding. Several topic orders can support this goal. While this text contains material sufficient for two semesters, it offers good choices for those teaching a one-semester course, for which we recommend coverage of Chapters 1–21, followed by (time permitting) one or more of Chapters 22, 23, 24, and 28. Some topics in Chapters 9–10 on sampling and experimental design may be de-emphasized or omitted entirely depending on the needs of the particular audience. Likewise, the material on probability and random variables in Chapters 11–13 is considerable and portions may be de-emphasized. Optional sections, indicated by asterisks, are indeed just that and may be omitted without harm. We have also differentiated the coverage in our Canadian edition from its U.S. counterpart in some ways. We added two additional chapters (authored by A. Vukov), on nonparametric procedures (Chapter 28) and bootstrapping (Chapter 29). We also consistently put more emphasis on power & sample size issues for tests of hypothesis.

GAISE Guidelines

The Guidelines for Assessment and Instruction in Statistics Education (GAISE) report adopted by the American Statistical Association urges that Statistics education should

1. emphasize Statistical literacy and develop Statistical thinking,

2. use real data,

3. stress conceptual understanding rather than mere knowledge of procedures,

4. foster active learning,

5. use technology for developing concepts and analyzing data, and

6. make assessment a part of the learning process.

We've designed our text and supporting materials to encourage this approach to the introductory course. We urge you to think about these guidelines with each class meeting.

Our Goal: Read This Book!

The best text in the world is of little value if students don't read it. Here are some of the ways we have made this edition even more approachable:

- ***Readability.*** This book doesn't read like other Statistics texts. Our style is both colloquial and informative, engaging students to actually read the book to see what it says. We've tightened the discussions and removed digressions.

- ***Humour.*** We know that humour is the best way to promote learning. You will find quips and wry comments throughout the narrative, in margin notes, and in footnotes.

- ***Informality.*** Our informal diction doesn't mean that we treat the subject matter lightly or informally. We try to be precise and, wherever possible, we offer deeper explanations and justifications than those found in most introductory texts.

- ***Focused lessons.*** The chapters are shorter than in most other texts so instructors and students can focus on one topic at a time.

- ***Consistency.*** We try to avoid the "do what we say, not what we do" trap. Having taught the importance of plotting data and checking assumptions and conditions, we model that behavior right through the rest of the book. (Check, for example, the exercises in Chapter 23. You'll see that we still require and demonstrate the plots and checks that were introduced in the early chapters.) This consistency helps reinforce these fundamental principles.

- *The need to read.* Students who plan just to skim the book may find our presentation frustrating. The important concepts, definitions, and sample solutions don't sit in little boxes. Statistics is a consistent story about how to understand the world when we have data. The story can't be told piecemeal. This is a book that needs to be read, so we've tried to make the reading experience enjoyable.

Mathematics

Mathematics can

1. provide a concise, clear statement of important concepts.

2. describe calculations to be performed with data.

3. embody proofs of fundamental results.

Of these, we emphasize the first. Mathematics can make discussions of Statistics concepts, probability, and inference clear and concise. We don't shy away from using math where it can clarify without intimidating. But we know that some students are put off by equations, so we always provide a verbal description and a numerical example as well. Some theorems about Statistics are quite interesting, and many are important. Often, though, their proofs are not enlightening to introductory Statistics students and can distract the audience from the concepts we want them to understand. So we avoid them here.

Nor do we slide in the opposite direction and concentrate on calculation. Although Statistics calculations are generally straightforward, they are also usually tedious. And, more to the point, they are often unnecessary. Today, virtually all statistics are calculated with technology, so there is little need for students to spend time summing squared deviations by hand. We have selected the equations that do appear for their focus on illuminating concepts and methods. Although these equations may be the best way to understand the concepts, they may not be optimal for hand calculation. When that happens, we give an alternative formula, better suited for hand calculation, for those who find following the process a better way to learn about the result.

Technology and Data

To experience the real world of Statistics, use modern technology to explore real data sets.

Technology. We assume that you are using some form of technology—a statistics package, a calculator, a spreadsheet, or some combination of these—in your Statistics course. We also assume that you'll put little emphasis on calculating answers by hand, even though we often show how. However, this is not a technology-heavy book. The role of technology in this book is to get the calculations out of the way so we can focus on statistical thinking. We discuss generic computer output, but we don't adopt any particular statistics software. We do offer guidance to help students get started on eight common software platforms: Excel®, Minitab®, JMP®, SPSS®, TI-83/84 Plus graphing calculators, StatCrunch®, and R®. *StatCrunch* can be accessed through *MyLab Statistics*, available from Pearson with the text. The **On the Computer** section at the end of most chapters is specific to the methods learned in that chapter.

Data. Because we use technology for computing, we don't limit ourselves to small, artificial data sets. You'll find some small data sets, but we also base examples and exercises on real data with a moderate number of cases—usually more than you would want to enter by hand into a program or calculator. Machine-readable versions of the data, appropriate for a variety of statistical packages, are included on the book's website, **www.pearsoncanada.ca/deveaux.**

New and Continuing Features

Enhancing Understanding

Learning Objectives and Connections. Each chapter starts with a brief summary of the chapter's key learning objectives, as well as a paragraph or two that point out the kinds of questions students will learn how to answer in the chapter and how this connects with or builds on earlier material in the text.

Each chapter ends with a **What Have We Learned?** summary, which includes new learning objectives and definitions of terms introduced in the chapter. Students can think of these as study guides.

In each chapter, our innovative **What Can Go Wrong?** sections highlight the most common errors that people make and the misconceptions they have about Statistics. One of our goals is to arm students with the tools to detect statistical errors and to offer practice in debunking misuses of statistics, whether intentional or not.

Margin and in-text boxed notes. Throughout each chapter, boxed margin and in-text notes enhance and enrich the text. Boxes with essential or practical information are screened. Conversational notes that enhance the text and entertain the reader are unscreened.

Math Boxes. In many chapters we present the mathematical underpinnings of the statistical methods and concepts. By setting these proofs, derivations, and justifications apart from the narrative, we allow the student to continue to follow the logical development of the topic at hand, yet also refer to the underlying mathematics for greater depth.

By Hand. Even though we encourage the use of technology to calculate statistical quantities, we realize the pedagogical benefits of occasionally doing a calculation by hand. The By Hand boxes break apart the calculation of many simpler formulas to help the student through the calculation of a worked example.

Reality Check. We regularly remind students that Statistics is about understanding the world with data. Results that make no sense are probably wrong, no matter how carefully we think we did the calculations. Mistakes are often easy to spot with a little thought, so we often ask students to stop for a reality check before interpreting their result.

Notation Alert. Throughout this book, we emphasize the importance of clear communication, and proper notation as part of the vocabulary of Statistics. We've found that it helps students when we are clear about the letters and symbols statisticians use to mean very specific things, so we've included Notation Alerts whenever we introduce a special notation that students will see again.

Learning by Example

For Example. As we introduce each important concept, we provide a focused example applying it—often with real up-to-the-minute data. Many For Examples carry the discussion through the chapter, picking up the story and moving it forward as students learn more about the topic.

Step-by-Step Examples: Think, Show, Tell. Step-by-Step examples repeat the mantra of Think, Show, and Tell in every chapter. These longer, worked examples guide students through the process of analyzing the problem with the general explanation on the left and the worked-out problem on the right. They emphasize the importance of thinking about a Statistics question (What do we know? What do we hope to learn? Are the assumptions and conditions satisfied?) and reporting our findings (the Tell step). The Show step contains the mechanics of calculating results and conveys our belief that it is only one part of the process.

The result is a better understanding of the concept, not just number crunching. In this edition, we've updated many Think, Show, Tell Step-by-Step examples with new demonstrations and data.

Testing Understanding

Just Checking. Just Checking questions are quick checks throughout the chapter; most involve very little calculation. These questions encourage students to pause and think about what they've just read. The Just Checking answers are at the end of the exercise sets in each chapter so students can easily check themselves.

Exercises. We've added section-specific single-concept exercises at the end of almost every section so students can be sure they have a clear understanding of each important topic before they're asked to tie them all together in more comprehensive exercises at chapter end. Chapter exercises have been updated with the most recent data available. Many come from news stories; some from recent research articles. Whenever possible, the data are available on MyLab Statistics and the Companion Website for the book so students can explore them further.

Technology

Data Sources. Most of the data used in examples and exercises are from real-world sources, and whenever we can, we include references to the Internet data sources used, often in the form of URLs. The data we use are usually on MyLab Statistics and the Companion Website. If you seek the data—or an updated version of the data—on the Internet, we try to direct you to a good starting point.

On the Computer. In the real world, Statistics is practiced with computers. We prefer not to choose a particular Statistics program. Instead, at the end of most chapters, we summarize what students can find in the most common packages, often with annotated output. We then offer specific guidance for several of the most common packages (Excel®, JMP®, Minitab®, R®, SPSS®, StatCrunch®, and TI-83/84 Plus[2]) to help students get started with the software of their choice.

---

[2]For brevity, we will write TI-83/84 Plus for the TI-83 Plus and/or TI-84 Plus. Keystrokes and output remain the same for the TI-83 Plus and the TI-84 Plus, so instructions and examples serve for both calculators.

Supplements

For Instructors

Instructor resources are password protected and available for download via the Pearson online catalogue at **http://catalogue.pearsoned.ca**.

Instructor's Solutions Manual. This resource provides complete, detailed, worked-out solutions for all the exercises in the textbook and is available through the online catalogue in both PDF and Word formats.

TestGen and Test Item File. For your convenience, our test bank is available in two formats. TestGen is a computerized testbank containing a broad variety of multiple-choice, short answer, and more complex problem questions. Questions can be searched and identified by question type, level of difficulty, and skill type (computational or conceptual). Each question has been checked for accuracy and is available in the latest version of TestGen software. This software package allows instructors to custom design, save, and generate classroom tests. The test program permits instructors to edit, add, or delete questions from the test bank; edit existing graphics and create new ones; analyze test results; and organize a database of tests and student results. This software allows for greater flexibility and ease of use. It provides many options for organizing and displaying tests, along with search and sort features. The same questions can also be found in a Test Item File available in Word format. The TestGen testbank and Test Item File can be downloaded from the online catalogue.

PowerPoint® Presentations. These PowerPoint® lecture slides provide an outline to use in a lecture setting, presenting definitions, key concepts, and figures from the textbook. These lecture slides can be downloaded from the online catalogue.

Image Library. The Image Library provides access to many of the images, figures, and tables in the textbook and is available to instructors on the online catalogue.

For Students

Student Solutions Manual. This solutions manual provides complete worked-out solutions to all of the odd-numbered exercises in the book, expanding on the answers provided in the Appendix at the back of the text.

Companion Website. The Companion Website provides additional resources (and data files) for instructors and students: **www.pearsoncanada.ca/deveaux**.

Technology Resources

MyLab™ Statistics Online Course (access code required)

- MyLab Statistics is a course management system that delivers **proven results** in helping individual students succeed.

- MyLab Statistics can be successfully implemented in any environment—lab-based, hybrid, fully online, traditional—and demonstrates the quantifiable difference that integrated usage has on student retention, subsequent success, and overall achievement.

- MyLab Statistics' comprehensive online gradebook automatically tracks students' results on tests, quizzes, homework, and in the study plan. Instructors can use the gradebook to provide positive feedback or intervene if students have trouble. Gradebook data can be easily exported to a variety of spreadsheet programs, such as Microsoft Excel. You can determine which points of data you want to export, and then analyze the results to determine success.

MyLab Statistics provides **engaging experiences** that personalize, stimulate, and measure learning for each student. In addition to the following resources, each course includes a full interactive online version of the accompanying textbook.

- **Tutorial Exercises with Multimedia Learning Aids:** The homework and practice exercises in MyLab Statistics align with the exercises in the textbook, and they regenerate algorithmically to give students unlimited opportunity for practice and mastery. Exercises offer immediate helpful feedback, guided solutions, sample problems, animations, videos, and eText clips for extra help at point-of-use.

- **StatTalk Videos:** *24 Conceptual Videos to Help You Actually Understand Statistics.* Fun-loving statistician Andrew Vickers takes to the streets of Brooklyn, New York, to demonstrate important statistical concepts through interesting stories and real-life events. These fun and engaging videos will help students actually understand statistical concepts. Available with an instructor's user guide and assessment questions.

- **Getting Ready for Statistics:** A library of questions now appears within each MyLab Statistics to offer the developmental math topics students need for the course. These can be assigned as a prerequisite to other assignments, if desired.

- **Conceptual Question Library:** In addition to algorithmically regenerated questions that are aligned with your textbook, there is a library of 1000 Conceptual Questions available in the assessment manager that requires students to apply their statistical understanding.

- **StatCrunch®:** MyLab Statistics integrates the web-based statistical software, StatCrunch, within the online assessment platform so that students can easily analyze data sets from exercises and the text. In addition, MyLab Statistics includes access to **www.StatCrunch.com**, a website where users can access more than 15,000 shared data sets, conduct online surveys, perform complex analyses using the powerful statistical software, and generate compelling reports.

- **Statistical Software Support:** Knowing that students often use external statistical software, we make it easy to copy our data sets, both from the eText and the MyLab Statistics questions, into software such as StatCrunch, Minitab, Excel, and more. Students have access to a variety of support tools—Technology Tutorial Videos, Technology Study Cards, and Technology Manuals for select titles—to learn how to effectively use statistical software.

And, MyLab Statistics comes from a **trusted partner** with educational expertise and an eye on the future.

- Knowing that you are using a Pearson product means knowing that you are using quality content. That means that our eTexts are accurate and our assessment tools work. Whether you are just getting started with MyLab Statistics, or have a question along the way, we're here to help you learn about our technologies and how to incorporate them into your course.

To learn more about how MyLab Statistics combines proven learning applications with powerful assessment, visit **www. mystatlab.com** or contact your Pearson representative.

MathXL® for Statistics Online Course (access code required)

MathXL is the homework and assessment engine that runs MyLab Statistics. (MyLab Statistics is MathXL plus a learning management system.)

With MathXL for Statistics, instructors can:

- Create, edit, and assign online homework and tests using algorithmically generated exercises correlated at the objective level to the textbook.
- Create and assign their own online exercises and import TestGen tests for added flexibility.
- Maintain records of all student work, tracked in MathXL's online gradebook.

With MathXL for Statistics, students can:

- Take chapter tests in MathXL and receive personalized study plans and/or personalized homework assignments based on their test results.
- Use the study plan and/or the homework to link directly to tutorial exercises for the objectives they need to study.
- Students can also access supplemental animations and video clips directly from selected exercises.

MathXL for Statistics is available to qualified adopters. For more information, visit our website at **www.mathxl.com**, or contact your Pearson representative.

StatCrunch®

StatCrunch is powerful web-based statistical software that allows users to perform complex analyses, share data sets, and generate compelling reports of their data. The vibrant online community offers more than 15,000 data sets for students to analyze.

- **Collect.** Users can upload their own data to StatCrunch or search a large library of publicly shared data sets, spanning almost any topic of interest. Also, an online survey tool allows users to quickly collect data via web-based surveys.
- **Crunch.** A full range of numerical and graphical methods allow users to analyze and gain insights from any data set. Interactive graphics help users understand statistical concepts, and are available for export to enrich reports with visual representations of data.
- **Communicate.** Reporting options help users create a wide variety of visually-appealing representations of their data.

Full access to StatCrunch is available with MyLab Statistics, and StatCrunch is available by itself to qualified adopters. StatCrunch Mobile is now available to access from your mobile device. For more information, visit our website at **www. StatCrunch.com**, or contact your Pearson representative.

Acknowledgments

A very big thanks goes to Shivon Sue-Chee for her hard work on the exercise sections, data sets, and appendix solutions. Her hard work and conscientiousness were, as always, a marvel. Also, a big thanks to Aaron Springford and Andy Leung for their very precise work in updating the Instructor's Solutions Manual.

Thank you also to the team at Pearson who supported the revision of this text throughout the writing and production process including Paul Donnelly, Senior Content Developer; Pippa Kennard, Project Manager; Emily Dill, Content Manager; and Mark Grzeskowiak, Acquisitions Editor.

Finally, thank you to the reviewers who provided feedback on our work and helped guide our plans for this new edition, including:

James Adcock, *University of Waterloo*
Nancy Chibry, *University of Calgary*
Torben Drewes, *Trent University*
Tracy Finlay, *Fleming College*
Renzo Gomez, *Durham College*
Erin Kirk, *Fleming College*
Michele Millar, *Mount Saint Vincent University*
Julie Orsini, *Durham College*
Scott Robison, *University of Calgary*
Tim Swartz, *Simon Fraser University*
Asokan Mulayath Variyath, *Memorial University of Newfoundland*

Index of Application

Note: (E) = Exercise, (FE) = For Example, (JC) = Just Checking, (IE) = in-text example (SBS) = Step-by-Step

Agriculture

American Angus Association (FE) Ch 18, p. 560

BST (E) Ch 20, p. 664

Cloud seeding (E) Ch 4, p. 122–123

Ephedra (E) Ch 9, p. 316

Fertilizers (E) Ch 24, p. 821–822

Genetically modified corn (JC) Ch 12, p. 386

Grass species (IE) Ch 2, p. 43

Peas (E) Ch 25, p. 860

Pesticides (E) Part VII Review, p. 939

Seeds (E) Ch 14, p. 470

Sprouts (E) Ch 25, p. 855

Tiny black potato flea beetles (JC) Ch 22, p. 704

Tomatoes (E) Ch 5 p. 162–163 Ch 10 p. 323, p. 329, p. 332, p. 343

Vineyards (E) Ch 1 p. 14 Part II Review, p. 281 Part III Review, p. 352

Arts and Literature

Barbershop music (E) Part VII Review, p. 933–934

Lefties and music (E) Part VII Review, p. 935

Mozart (E) Ch 10, p. 344

Music and memory (E) Part I Review p. 173 Ch 19, p. 635

Music students (E) Ch 28, p. 967

Pottery (E) Ch 6, p. 200–201 Part VI Review, p. 777

Rap (E) Ch 19, p. 635

Rock concert accidents (E) Ch 3, p. 96

Singers (E) Part I Review p. 166

Biology

Beetles (E) Ch 10, p. 346

Bird counts (IE) Ch 3, p. 89

Cattle (E) Ch 5, p. 158

Crocodile (E) Part II Review, p. 285

Cuckoos (E) Ch 19, p. 635 Ch 28, p. 970

Eggs (E) Ch 5 p. 162 Ch 13, p. 432

Elephants and hippos (E) Ch 8, p. 271

Feeding fish (E) Part VI Review, p. 778

Fish (E) Ch 9, p. 317–318

Fruit flies (E) Ch 22, p. 717

Gators (E) Ch 7, p. 243

Irises (E) Part VI Review, p. 784

Manatees (E) Part II Review, p. 282–283

Oranges (E) Ch 13, p. 402, p. 405

Weighing bears (E) Ch 1, p. 13–14

Business

Absenteeism (E) Ch 12, p. 397

Accounting (E) Ch 9, p. 317

Ad campaign (E) Ch 19, p. 631–632

Assets (E) Ch 4, p. 128

Bank tellers (E) Ch 13, p. 435

Bookstore sales (E) Ch 7, p. 217, p. 222, p. 226

CEO (E) Ch 14, p. 471

Cosmetic marketing (E) Ch 11, p. 382 Ch 12, p. 395, p. 396

Credit card charges (E) Ch 18, p. 586

Dairy sales (E) Part VI review, p.782–783

E-commerce (E) Ch 10, p. 321

Expected value (E) Ch 13, p. 430

Fire insurance (E) Ch 11, p. 370

Fraud detection (E) Part I Review, p. 167–168

Frumpies (E) Ch 10, p. 344

Hiring (E) Ch 15, p. 500

Industrial experiment (E) Ch 4, p. 124

Industry codes (E) Ch 3, p. 91

Insurance (E) Ch 13, p. 431 Part IV Review, p. 442

Interest rates (E) Ch 8, p. 269, p. 270

Labour force (E) Part VI Review, p. 782

Loans (E) Ch 17, p. 553–554

Marketing (E) Ch 14, p. 450

Market segments (E) Ch 8, p. 250

Mutual funds (E) Ch 27, p. 931

Online insurance (E) Ch 20, p. 658–659

Pay (E) Part I Review, p. 174–175

Payroll (E) Ch 3, p. 87 Ch 5, p. 157

Profits (E) Part I Review, p. 176 Part II Review, p. 289

Quality control (E) Part IV Review, p. 440

Real estate (E) Ch 7, p. 235–236 Ch 11, p. 371

Receptionist performance (E) Ch 26, p. 888

Revenue and ticket sales (E) Ch 8, p. 250

Revenue (E) Ch 8, p. 256

Sales and Profits (E) Ch 23, p. 765, p. 766

Sex sells (E) Ch 20, p. 656

Sick days (E) Ch 3, p. 86–87

Software (E) Ch 13, p. 431

Stock market (E) Ch 13, p. 430

Stocks (E) Part IV Review, p. 442

Tips (E) Ch 10, p. 323, p. 329, p. 332 Part III Review, p. 350

Finance

Food and Drinks

Games

Genetics

Health

Law

Surveys

Technology

Telecommunication

Transport

LEARNING OBJECTIVES

1 Recognize the vital role of Statistics in answering real-world questions involving variation and uncertainty.

2 Understand that data are values with their context.

3 Identify cases and variables.

4 Identify whether a variable is being used as categorical or quantitative.

Connections

Statistics gets no respect. People say things like, "you can prove anything with Statistics." People will write off a claim based on data as "just a statistical trick." And Statistics courses don't have the reputation of being students' first choice for a fun elective.

But Statistics can be fun, though you might not guess it from some of the book covers popping up now and then at Indigo, with titles like *Statistics for the Terrified* and *Statistics without Tears*. Well, we certainly hope to calm your fears and hold back the tears while helping you learn to think clearly when dealing with data.

This is a book about understanding the world by using data. So we'd better start by understanding data. There's more to that than you might have thought.

"But where shall I begin?" asked the White Rabbit.
"Begin at the beginning," the King said gravely,
"and go on till you come to the end: then stop."

—*Lewis Carroll, Alice's Adventures in Wonderland*

1.1 What Is Statistics?

People around the world have one thing in common—they all want to figure out what's going on. You'd think with the amount of information available today this would be an easy task, but actually, as the amount of information grows, so does our need to understand what it can tell us.

At the base of all this information, on the Internet and all around us, are data. We'll talk about data in more detail in the next section, but for now, think of **data** as any collection of numbers, characters, images, or other items that provide information about something. What sense can we make of all this data? You certainly can't make a coherent picture from random pieces of information. Whenever there are data and a need for understanding the world, you'll find Statistics.[1]

This book will help you develop the skills you need to understand and communicate the knowledge that can be learned from data. By thinking clearly about the question you're trying to answer and learning the statistical tools to show what the data are saying, you'll acquire the skills to tell clearly what it all means. Our job is to help you make sense of the concepts and methods of Statistics and to turn them into a powerful, effective approach to understanding the world through data.

Data vary. Ask different people the same question and you'll get a variety of answers. Even when you ask the same person the same question at different times, you may have a variety of answers! Statistics helps us to make sense of the world by seeing past the underlying variation to find patterns and relationships. This book will teach you skills to help with this task and ways of thinking about variation that are the foundation of sound reasoning about data.

Consider the following:

- If you have a Facebook account, you have probably noticed that the ads you see online tend to match your interests and activities. Coincidence? Hardly. According to the *Wall Street Journal*,[2] much of your personal information has probably been sold to marketing or tracking companies. Why would Facebook give you a free account and let you upload as much as you want to its site? Because your data are valuable! Using your Facebook profile, a company might build a profile of your interests and activities: what movies and sports you like; your age, sex, education level, and hobbies; where you live; and, of course, who your friends are and what *they* like. From Facebook's point of view, your data are a potential gold mine. Gold ore in the ground is neither very useful nor pretty. But with skill, it can be turned into

---

[1] *Statistical Science* would be a more appropriate name. Statistics is indeed a science and integral to the scientific method. Its methodology extends far beyond the stereotype of data compilation and tabulation, and calculation of summary *statistics*. It overlaps with a rather new field of knowledge called *Data Science*.

[2] blogs.wsj.com/digits/2010/10/18/referers-how-facebook-apps-leak-user-ids/

Q: What is Statistics?
A: Statistics is a way of reasoning, along with a collection of tools and methods, designed to help us understand the world.
Q: What are statistics?
A: Statistics (plural) are particular calculations made from data.
Q: So what is data?
A: You mean, "What are data?" Data is the plural form. The singular is datum.
Q: OK, OK, so what are data?
A: Data are values along with their context.

The ads say, "Don't drink and drive; you don't want to be a statistic." But you can't be a statistic.
    We say: "Don't be a datum."

Statistics is about *variation*.
Data vary because we don't see everything, and because even what we do see and measure, we measure imperfectly.
    But Statistics helps us understand and model this variation so that we can see through it to the underlying truths and patterns.[4]

something both beautiful and valuable. What we're going to talk about in this book is how you can mine your own data and gain valuable insights about the world.

■ In December 1992, British engineer Neil Papworth (now a Montrealer) texted "Happy Christmas' to a company director. That was the first text message ever sent, followed by trillions more in ensuing years, as the number of text messages and messaging apps have grown at a ridiculous pace. Some of these messages were sent or read while the sender or the receiver was driving. How dangerous is texting while driving?

How can we study the effect of texting while driving? One way is to measure reaction times of drivers faced with an unexpected event while driving and texting. Researchers at the University of Utah tested drivers on simulators that could present emergency situations. They compared reaction times of sober drivers, drunk drivers, and texting drivers.[3] The results were striking. The texting drivers actually responded more slowly and were more dangerous than those who were above the legal limit for alcohol. In this book, you'll learn to design and analyze experiments like this.

How do we assess the risk of genetically engineered foods being considered by the Canadian Food Inspection Agency or the safety and effectiveness of new drugs submitted to Health Canada for approval? How can we predict the number of new cases of AIDS by regions of the country or the number of customers likely to respond to a sale at the mall? Or determine whether enriched early education affects later performance of school children, and whether vitamin C really prevents illness? Statistics is all about answering real-world questions such as these.

You'll learn how to interpret data and to communicate the message you see to others. You'll also learn how to spot deficiencies and weaknesses in conclusions drawn by others that you see in newspapers and on the Internet every day. Statistics can help you become a more informed citizen by giving you the tools to understand, question, and interpret data.

Statistics in a Word

It can be fun and sometimes useful to summarize a discipline in only a few words. So,

Economics is about … *Money (and why it is good).*
Psychology: *Why we think what we think (we think).*
Paleontology: *Previous Life.*
Biology: *Life.*
Religion: *After Life*
Anthropology: *Who?*
History: *What, where, and when?*
Philosophy: *Why?*
Engineering: *How?*
Accounting: *How much?*
In such a caricature, Statistics is about … *Variation.*

FRAZZ © 2003 Jef Mallett. Distributed by Universal Uclick. Reprinted with permission. All rights reserved.

---

[3]"Text Messaging During Simulated Driving," Drews, F. A. et al., Human Factors: hfs.sagepub.com/content/51/5/762
[4]Seeing the forest *through* the trees

Section 1.1 Exercises

1. **Grocery shopping** Many grocery store chains offer customers a card they can scan when they check out and offer discounts to people who do so. To get the card, customers must give information, including a mailing address and e-mail address. The actual purpose is not to reward loyal customers but to gather data. What data do these cards allow stores to gather, and why would they want that data?

2. **Online shopping** Online retailers such as Amazon.com keep data on products that customers buy, and even products they look at. What does Amazon hope to gain from such information?

1.2 Data

©Dorling Kindersley

Amazon.com opened for business in July 1995, billing itself as "Earth's Biggest Bookstore." By 1997, Amazon had a catalogue of more than 2.5 million book titles and had sold books to more than 1.5 million customers in 150 countries. In 2015, the company's sales reached $107.0 billion (a 20% increase from the previous year and more than double of 2011 sales). Amazon has sold a wide variety of merchandise, including a $400,000 necklace, yak cheese from Tibet, and the largest book in the world (see photo at left). Amazon. ca launched in June 2002, following up in April 2011 with a new shipping facility built in Mississauga, Ontario. How did Amazon become so successful and how can it keep track of so many customers and such a wide variety of products? The answer to both questions is *data*.

But what are data? Think about it for a minute. What exactly *do* we mean by "data"? Do data have to be numbers? The amount of your last purchase in dollars is numerical data. But your name and address in Amazon's database are also data, even though they are not numerical.

Let's look at some hypothetical data values that Amazon.ca might collect:

105-2686834-3759466	Quebec	Nashville	Kansas	10.99	819	N	B0000015Y6	Katherine H.
105-9318443-4200264	Nova Scotia	Orange County	Boston	16.99	902	Y	B000002BK9	Samuel P.
105-1872500-0198646	Alberta	Bad Blood	Chicago	15.98	403	N	B000068ZVQ	Chris G.
103-2628345-9238664	Ontario	Let Go	Mammals	11.99	416	N	B0000010AA	Monique D.
002-1663369-6638649	Quebec	Best of Kansas	Kansas	10.99	819	N	B002MXA7Q0	Katherine H.

Try to guess what they represent. Why is that hard? Because there is no *context*. If we don't know what attributes or characteristics are being measured and what is measured about them, the values displayed are meaningless. We can make the meaning clear if we organize the values into a **data table** such as this one:

Order Number	Name	Province	Price	Area Code	Previous Album Download	Gift?	ASIN	New Purchase Artist
105-2686834-3759466	Katherine H.	Quebec	10.99	819	Nashville	N	B0000015Y6	Kansas
105-9318443-4200264	Samuel P.	Nova Scotia	16.99	902	Orange County	Y	B000002BK9	Boston
105-1372500-0198646	Chris G.	Alberta	15.98	403	Bad Blood	N	B000068ZVQ	Chicago
103-2628345-9238664	Monique D.	Ontario	11.99	416	Let Go	N	B0000010AA	Mammals
002-1663369-6638649	Katherine H.	Quebec	10.99	819	Best of Kansas	N	B002MXA7Q0	Kansas

Now we can see that these are purchase records for album download orders from Amazon. The column titles tell *what* has been recorded. Each row is about a particular purchase.

What information would provide a **context**? Newspaper journalists know that the lead paragraph of a good story should establish the "Five W's": *who, what, when, where,* and (if possible) *why.* Often, we add *how* to the list as well. The answers to the first two questions are essential. If we don't know *what* attributes or characteristics are measured and *who* they are measured on, the values displayed are meaningless.

Who and What

In general, the rows of a data table correspond to individual **cases** about whom (or about which, if they're not people) we record some characteristics. Those characteristics (called variables—see Section 1.2) constitute the *"what"* of the study. The *"who"* or the cases go by different names, depending on the situation.

Individuals who answer a survey are called **respondents**. People on whom we experiment are **subjects** or (in an attempt to acknowledge the importance of their role in the experiment) **participants**. Animals, plants, websites, and other inanimate subjects are often called **experimental units** or **observational units**. Often we simply call *cases* what they are: for example, *customers, economic quarters,* or *companies.* In a database, rows are called **records**—in this example, purchase records. Perhaps the most generic term is *cases,* but in any event the rows represent the *who* of the data.

Look at all the columns to see exactly what each row refers to. Here the cases are different purchase records. You might have thought that each customer was a case, but notice that, for example, Katherine H. appears twice, both in the first and the last row. Each row of data is information about a single case, and that case is often identified in the leftmost column. In this example, it's the order number. If you collect the data yourself, you'll know what the cases are. But, often you'll be looking at data that someone else collected and you'll have to ask or figure that out yourself.

Often the cases are a **sample** from some larger **population** that we'd like to understand. Amazon doesn't care about just these customers; it wants to understand the buying patterns of *all* its customers, and, generalizing further, it wants to know how to attract other Internet users who may not have made a purchase from Amazon's site. To be able to generalize from the sample of cases to the larger population, we'll want the sample to be *representative* of that population—a kind of snapshot image of the larger world.

We must know *who* and *what* to analyze data. Without knowing these two, we don't have enough information to start. Of course, we'd always like to know more. The more we know about the data, the more we'll understand about the world. If possible, we'd like to know the *when* and *where* of data as well. Values recorded in 1803 may mean something different than similar values recorded last year. Values measured in Tanzania may differ in meaning from similar measurements made in Mexico. And knowing *why* the data were collected may also tell us much about its reliability and quality.

How the Data Are Collected

How the data are collected can make the difference between insight and nonsense. As we'll see later, data that come from a voluntary survey on the Internet are almost always worthless. One primary concern of Statistics, to be discussed in Part III, is the design of sound methods for collecting data. Throughout this book, whenever we introduce data, we'll provide a margin note listing the W's (and H) of the data. Identifying the W's is a habit that we recommend you adopt.

The first step of any data analysis is to know what you are trying to accomplish and what you want to know (*why?*). To help you use Statistics to understand the world and make decisions, we'll lead you through the entire process of *thinking* about the problem, *showing* what you've found, and *telling* others what you've learned. Every guided example in this book is broken into these three steps: *Think, Show,* and *Tell.* Identifying the problem and the *who* and *what* of the data is a key part of the *Think* step of any analysis. Make sure you know these before you proceed to *Show* or *Tell* anything about the data.

Data trumps intuition

Amazon monitors and evolves its website to better serve customers and maximize sales. To decide which changes to make, analysts experiment with new designs, offers, recommendations, and links. Statisticians want to know how long you'll spend browsing the site and whether you'll follow the links or purchase the suggested items. As Ronny Kohavi, former director of Data Mining and Personalization for Amazon, said, "Data trumps intuition. Instead of using our intuition, we experiment on the live site and let our customers tell us what works for them … Data is king at Amazon."

Wavebreakmedia/Shutterstock

For Example IDENTIFYING THE "WHO"

In December 2013, *Consumer Reports* published an evaluation of 46 tablets from a variety of manufacturers (http://www.consumerreports.org/cro/magazine/2013/12/best-electronics-brands/index.htm).

QUESTION: Describe the population of interest, the sample, and the *Who* of the study.

ANSWER: The magazine is interested in the performance of tablets currently offered for sale. It tested a sample of 46 tablets, which are the "Who" for these data. Each tablet selected represents all similar tablets offered by that manufacturer.

Section 1.2 Exercises

3. Super Bowl Sports announcers love to quote statistics. During the Super Bowl, they particularly love to announce when a record has been broken. They might have a list of all Super Bowl games, along with the scores of each team, total scores for the two teams, margin of victory, passing yards for the quarterbacks, and many more bits of information. Identify the *who* in this list.

4. Nobel laureates The web page at www.nobelprize.org/nobel_prizes/peace/laureates/ allows you to look up all the Nobel prizes awarded in any year. The data are not listed in a table. Rather you drag a slider to a list of prize categories and then see a list of the awardees for each category for every year. Describe the *who* in this scenario.

1.3 Variables

The characteristics (the *"what"*) recorded about each individual are called **variables**. They are usually found as the columns of a data table with a name in the header that identifies what has been recorded. In the Amazon data table we find the variables *Order Number, Name, Province, Price,* and so on.

Categorical Variables

Some variables just tell us what group or category each individual belongs to. Are you male or female? Pierced or not? We call variables like these **categorical**, or **qualitative, variables**. (You may also see them called **nominal variables** because they name categories.) Some variables are clearly categorical, like the variable *Province*. Its values are text and those values tell us what category the particular case falls into. But numerals are often used to label categories, so categorical variable values can also be numerals. For example, Amazon collects telephone area codes that *categorize* each phone number into a geographical region. So area code is considered a categorical variable even though it has numeric values. (But see the story in the following box.)

Donatas1205/Fotolia

Area codes—numbers or categories? The *What* and *Why* of area codes are not as simple as they may first seem. When area codes were first introduced, AT&T was still the source of all telephone equipment, and phones had dials.

To reduce wear and tear on the dials, the area codes with the lowest digits (for which the dial would have to spin least) were assigned to the most populous regions—those with the most phone numbers and thus the area codes most likely to be dialled. New York City was assigned 212, Chicago 312, and Los Angeles 213, while Halifax was 902, Quebec City 418, and Vancouver 604. For that reason, at one time the numerical value of an area code could be used to guess something about the population of its region. Since the advent of push-button phones, area codes have now become just categories.

Descriptive responses to questions are often categories. For example, the responses to the questions "Who is your cell phone provider?" or "What is your marital status?" yield categorical values. When Amazon considers a special offer of free shipping to customers, it might first analyze how purchases have been shipped in the recent past. Amazon might start by counting the number of purchases shipped in each category: ground transportation, second-day air, and overnight air. *Counting is a natural way to summarize and learn about a categorical variable* like *Shipping Method*. The counts though are just summaries of the actual raw categorical data, which looks like a long list of shipping methods, one shipping method for each order. Chapter 2 discusses summaries and displays of categorical variables more fully.

Quantitative Variables

When a variable contains measured numerical values with measurement *units*, we call it a **quantitative variable**. Quantitative variables typically record an amount or degree of something. For quantitative variables, its **measurement units** provide a meaning for the numbers. Even more important, units such as yen, cubits, carats, angstroms, nanoseconds, kilometres per hour, or degrees Celsius tell us the *scale* of measurement, so we know how far apart two values are. Without units, the values of a measured variable have no meaning. It does little good to be promised a raise of 5000 a year if you don't know whether it will be paid in Euros, Canadian dollars, U.S. dollars, or Chinese yuan.

Aside from the process of measurement, quantitative variables may also arise from the process of counting. Counts also measure the amounts of things. How many songs are on your iPod? How many siblings do you have? To measure these quantities, we would naturally count, instead of measure, and we'd consider the units to be "number of …" or just "counts." Amazon might be interested in the number of teenage customers visiting its site each month, in order to track customer growth and forecast sales. So, Who = months and What = # teens (whereas when Amazon counts the number of ground, second-day air, and overnight shipments during some time period, these counts are just summaries for the categorical variable *Shipping Method*, where Who = purchases and What = method). Chapter 3 discusses summaries and displays for quantitative variables.

Sometimes a variable with numeric values may be treated as categorical. Amazon could record your *Age* in years. That seems quantitative, and it would be if the company wanted to know the average age of those customers who visit their site after 3 a.m. But suppose Amazon wants to decide which album to feature on its site when you visit. Then placing your age into one of the categories *Child, Teen, Adult,* or *Senior* might be more useful. You can always convert a quantitative variable to a categorical one, by simply breaking up the range of values into several intervals. There is some loss of information, but if this facilitates a particular method of analysis, it may still be worthwhile doing.

Identifiers

For a categorical variable like *Sex*, each individual is assigned one of two possible values, say *M* or *F*. But for a variable with ID numbers, such as *Student ID*, each individual receives a unique value. A variable like this, used solely to uniquely identify each of the cases, is called an **identifier variable**. Identifiers are useful, but not typically for analysis.

Amazon wants to know who you are when you sign in again and doesn't want to confuse you with some other customer. So it assigns you a unique identifier. Amazon also wants to send you the right product, so it assigns a unique Amazon Standard Identification Number (ASIN) to each item it carries. You'll want to recognize when a variable is playing the role of an identifier so you aren't tempted to analyze it.

Identifier variables themselves don't tell us anything useful about their categories because we know there is exactly one individual in each. However, they are crucial in this era of large data sets because by uniquely identifying the cases, they make it possible to combine data from different sources, protect (or violate) privacy, and provide unique labels. Many large databases are *relational* databases. In a relational database, different data tables

Privacy and the Internet You have many identifiers: a social insurance number, a student ID number, a passport number, a health insurance number, and probably a Facebook account name. Privacy experts are worried that Internet thieves may match your identity in these different areas of your life, allowing, for example, your health, education, and financial records to be merged. Even online companies, such as Facebook and Google, are able to link your online behaviour to some of these identifiers, which carries with it both advantages and dangers. The National Strategy for Trusted Identities in Cyberspace (www.wired.com/images_blogs/threat-level/2011/04/NSTICstrategy_041511.pdf) proposes ways that we may address this challenge in the near future.

link to one another by matching identifiers. In the Amazon example, *Transaction Number* is clearly an identifier. What about a CD's ASIN? It depends on the purpose of the study. If we want to determine the best-selling CD, then its ASIN is a categorical variable of interest and we would count how often each ASIN appears in a row or transaction. But if we are examining the range of prices for CDs, the ASIN could be used as an identifier variable.

Ordinal Variables

A typical course evaluation survey asks, "How valuable do you think this course will be to you?" 1 = Worthless; 2 = Slightly; 3 = Middling; 4 = Reasonably; 5 = Invaluable. Is *Educational Value* categorical or quantitative? Often the best way to tell is to look to the *why* of the study. A teacher might just count the number of students who gave each response for her course, treating *Educational Value* as a categorical variable. When she wants to see whether the course is improving, she might treat the responses as the *amount* of perceived value—in effect, treating the variable as quantitative.

But what are the units? There is certainly an *order* of perceived worth: Higher numbers indicate higher perceived worth. A course that averages 4.5 seems more valuable than one that averages 2, but we should be careful about treating *Educational Value* as quantitative. To treat it as quantitative, the teacher will have to imagine that it has "educational value units" or some similar arbitrary construct. Because there are no natural units, any conclusions she draws should be viewed with considerable caution! Variables that report order without natural units are called **ordinal variables**. But saying "that's an ordinal variable" doesn't get you off the hook. You must still look to the *why* of your study and understand what you want to learn from the variable to decide whether to treat it as categorical or quantitative. Ordinal variables are often viewed as a subclass of categorical variables, along with nominal (no natural ordering) variables, since nominal and ordinal variables both arise through a process of categorizing your cases.

For Example IDENTIFYING "WHAT" AND "WHY" OF TABLETS

RECAP: A *Consumer Reports* article about 46 tablets lists each tablet's manufacturer, price, battery life (hrs.), operating system (iOS/Android/Windows), and overall performance score (0–100), and whether or not it has a memory card reader.

QUESTION: Are these variables categorical or quantitative? Include units where appropriate, and describe the "Why" of this investigation.

ANSWER: The variables are

- manufacturer (categorical)
- price (quantitative, $)
- battery life (quantitative, hrs.)
- operating system (categorical)
- performance score (quantitative, no units)
- memory card reader (categorical)

The magazine hopes to provide consumers with the information to choose a good tablet.

 ## Just Checking

In the 2004 Tour de France, Lance Armstrong made history by winning the race for an unprecedented sixth time. In 2005, he became the only seven-time winner and once again set a new record for the fastest average speed. And on January 17, 2013, on the Oprah Winfrey show, he made history yet again, confessing to a long history of extensive doping! You can find data on all the Tour de France races in the data set **Tour de France 2015**. Here are the first three and last eight lines of our data set. Keep in mind that the entire data set has over 100 entries. And as promised in our second edition, the Armstrong entries, like his medals, have vanished with our third edition!

1. List as many of the W's as you can for this data set.

2. Classify each variable as categorical or quantitative; if quantitative, identify the units.

Year	Winner	Country of Origin	Age	Team	Total Time (h/min/s)	Avg. Speed (km/h)	Stages	Total Distance Ridden (km)	Starting Riders	Finishing Riders
1903	Maurice Garin	France	32	La Française	94.33.00	25.7	6	2428	60	21
1904	Henri Cornet	France	20	Cycles JC	96.05.00	25.3	6	2428	88	23
1905	Louis Trousseller	France	24	Peugeot	112.18.09	27.1	11	2994	60	24
⋮										
2007	Alberto Contador	Spain	24	Discovery Channel	91.00.26	38.97	20	3547	189	141
2008	Carlos Sastre	Spain	33	CSC-Saxo Bank	87.52.52	40.50	21	3559	199	145
2009	Alberto Contador	Spain	26	Astana	85.48.35	40.32	21	3460	180	156
2010	Andy Schleck	Luxembourg	25	Saxo Bank	91.59.27	39.590	20	3642	180	170
2011	Cadel Evans	Australia	34	BMC	86.12.22	39.788	21	3430	198	167
2012	Bradley Wiggins	Great Britain	32	Sky	87.34.47	39.827	20	3488	198	153
2013	Christopher Froome	Great Britain	28	Sky	83.56.40	40.551	21	3404	198	169
2014	Vincenzo Nibali	Italy	29	Astana	89.56.06	40.735	21	3663.5	198	164
2015	Christopher Froome	Great Britain	30	Sky	84.46.14	39.64	21	3360.3	198	160

Section 1.3 Exercises

5. **Grade levels** A person's grade in school is generally identified by a number.
   a) Give an example of a *why* in which grade level is treated as categorical.
   b) Give an example of a *why* in which grade level is treated as quantitative.

6. **ZIP codes** Unlike in Canada, the U.S. Postal Service uses five-digit ZIP codes to identify locations to assist in delivering mail.
   a) In what sense are ZIP codes categorical?
   b) Is there any ordinal sense to ZIP codes? In other words, does a larger ZIP code tell you anything about a location compared to a smaller ZIP code?

7. **Affairs** A May 2015 Gallup Poll Canada question asked, "Please tell me whether you personally believe that in general it is morally acceptable, or morally wrong for married men and women having an affair?" The possible responses were "Morally acceptable," "Morally wrong,"

"Depends on situation," "Not a moral issue," and "No opinion." What kind of variable is the response?

8. **Economic outlook** An April 2016 Gallup Poll Canada asked Canadians, "What is your opinion about the local economy six months from now?" The possible responses were "Grows strong," "Remains the same," "Gets weaker," and "Don't know." What kind of variable is the response?

9. **Medicine** A pharmaceutical company conducts an experiment in which a subject takes 100 mg of a substance orally. The researchers measure how many minutes it takes for half of the substance to exit the bloodstream. What kind of variable is the company studying?

10. **Stress** A medical researcher measures the increase in heart rate of patients under a stress test. What kind of variable is the researcher studying?

There's a world of data on the Internet These days, one of the richest sources of data is the Internet. With a bit of practice, you can learn to find data on almost any subject. Many of the data sets we use in this book were found in this way. The Internet has both advantages and disadvantages as a source of data. Among the advantages are that often you'll be able to find even more current data than those we present. The disadvantage is that references to Internet addresses can "break" as sites evolve, move, and die.

Our solution to these challenges is to offer the best advice we can to help you search for the data, wherever they may be residing. We usually point you to a website. We'll sometimes suggest search terms and offer other guidance.

Some words of caution, though: Data found on Internet sites may not be formatted in the best way for use in statistics software. Although you may see a data table in standard form, an attempt to copy the data may leave you with a single column of values. You may have to work in your favourite statistics or spreadsheet program to reformat the data into variables. You will also probably want to remove commas from large numbers and extra symbols such as money indicators ($, ¥, £); few statistics packages can handle these.

WHAT CAN GO WRONG?

- **Don't label a variable as categorical or quantitative without thinking about the data and what they represent.** The same variable can sometimes take on different roles.

- **Don't assume that a variable is quantitative just because its values are numbers.** Categories are often given numerical labels. Don't let that fool you into thinking they have quantitative meaning. Look at the context.

- **Always be skeptical.** One reason to analyze data is to discover the truth. Even when you are told a context for the data, it may turn out that the truth is a bit (or even a lot) different. The context colours our interpretation of the data, so those who want to influence what you think may slant the context. A survey that seems to be about all students may in fact report just the opinions of those who visited a fan website. The question that respondents answered may be posed in a way that influences responses.

  Be very cautious (if not skeptical) when an ordinal variable is converted to a numerical scale for analysis.

What Have We Learned?

Understand that data are information (recorded as numbers or labels) in a context.

- *Who, what, why, where, when* (and *how?*)—the W's—help nail down the context of the data.

- We must know *who, what,* and *why* to be able to say anything useful based on the data. The *who* are the cases. The *what* are the variables. A variable gives information about each of the cases. The *why* helps us decide which way to treat the variables.

- Stop and identify the W's whenever you have data, and be sure you can identify the cases and the variables.

Consider the source of your data and the reasons the data were collected. That can help you understand what you might be able to learn from the data.

We treat variables in two basic ways, as *categorical* (sometimes called qualitative) or *quantitative*.

- Categorical variables identify a category for each case. Usually we think about the counts of cases that fall in each category. (An exception is an identifier variable that just names each case.)

- Quantitative variables record measurements or amounts of something; they must have units.

- Sometimes we may treat the same variable as categorical or quantitative depending on what we want to learn from it, which means some variables can't be pigeonholed as one type or the other.

Review of Terms

The key terms are in chapter order so you can use this list to review the material in the chapter.

Data	Recorded values whether numbers or labels, together with their context (p. 1).
Data table	An arrangement of data in which each row represents a case and each column represents a variable (p. 3).
Context	The context ideally tells *who* was measured, *what* was measured, *how* the data were collected, *where* the data were collected, and *when* and *why* the study was performed (p. 4).
Case	An individual element under study about whom or which we have data (p. 4).
Respondent	Someone who answers, or responds to, a survey (p. 4).
Subject	A human experimental unit. Also called a participant (p. 4).
Participant	A human experimental unit. Also called a subject (p. 4).
Experimental unit or Observational unit	An individual in a study for which or for whom data values are recorded. Human experimental units are usually called subjects or participants (p. 4).
Record	Information about an individual in a database (p. 4).
Sample	A subset of a population, examined in hope of learning about the population (p. 4).
Population	The entire group of individuals or instances about whom we hope to learn (p. 4).
Variable	A variable holds information about the same characteristic for many cases (p. 5).
Categorical (or qualitative) variable	A variable that names categories with words or numerals (p. 5).
Nominal variable	The term "nominal" is often applied to a categorical variable whose values are used only to name categories lacking some natural ordering (p. 5).
Quantitative variable	A variable in which the numbers are values of measured quantities with units (p. 6).
Measurement unit	A quantity or amount adopted as a standard of measurement, such as dollars, hours, or grams (p. 6).
Identifier variable	A categorical variable that records a unique value for each case, used to name or identify it (p. 6).
Ordinal variable	The term "ordinal" may be applied to a variable whose categorical values possess some kind of order (p. 7).

On the Computer DATA

Most often we do statistical analysis of data on a computer using a program, or *package*, designed for that purpose. There are many different statistics packages, but they all do essentially the same things.

If you understand what the computer needs to know to do what you want and what it needs to show you in return, you can figure out the specific details of most packages pretty easily.

For example, to get your data into a computer statistics package, you need to tell the computer:

- Where to find the data. This usually means directing the computer to a file previously stored on your computer or to data on a database. Or it might just mean that you have copied the data from a spreadsheet program or Internet site and it is currently on your computer's clipboard. Usually the data is in the form of a data table. Most computer statistics packages prefer the *delimiter* that marks the division between elements of a data table to be a tab character (comma is another common delimiter) and the delimiter that marks the end of a case to be a *return* character.

- Where to put the data. (Usually this is handled automatically.)

- What to call the variables. Some data tables have variable names as the first row of the data, and often statistics packages can take the variable names from the first row automatically.

- Excel is often used to help organize, manipulate and prepare data for other software packages. Many of the other packages take Excel files as inputs. Alternatively, you can copy a data table from Excel and Paste it into many packages, or export Excel spreadsheets as tab delimited (.txt) or comma delimited files (.csv), which can be easily shared and imported into other programs. All data files provided with this text are in these two formats.

EXCEL

To open a file containing data in Excel:
- Choose **File** > **Open**.
- Browse to find the file to open. Excel supports many file formats.
- Other programs can import data from a variety of file formats, but all take both tab delimited (.txt) and comma delimited (.csv) text files.
- You can also copy tables of data from other sources, such as Internet sites, and paste them into an Excel spreadsheet. Excel can recognize the format of many tables copied this way, but this method may not work for some tables.
- Excel may not recognize the format of the data. If data include dates or other special formats ($, €, ¥, etc.), identify the desired format. Select the cells or columns to reformat, right click and choose **Format** > **Cell**. Often, the General format is the best option.

*Screenshot from Microsoft Excel®. Used by permission of Microsoft Corporation.

DATA DESK

To read data into Data Desk
- **File** > **Open.** The dialog lets you specify variable names (or take them from the first row of the data), the delimiter, or how to read formatted data.
- The **File** > **Import** command works in the same way, but instead of starting a new data file, it adds the data in the file to the current data file. Data Desk can work with multiple data tables in the same file.

- If the data are already in another program, such as, for example, a spreadsheet, **Copy** the data table (including the column headings). In Data Desk choose **Edit** > **Paste** Variables. There is no need to create variables first; Data Desk does that automatically. You'll see the same dialog as for Open and Import.

JMP

To import a text file, choose:
- **File** > **Open** and select the file from the dialog. At the bottom of the dialog screen you'll see **Open As:**—be sure to change to **Data (Using Preview)**. This will allow you to specify the delimiter and make sure the variable names are correct. (**JMP** also allows various formats to be imported directly, including.xls files.)

You can also paste a data set in directly (with or without variable names) by selecting:
- **File** > **New** > **New Data Table** and then **Edit** > **Paste (or Paste with Column Names** if you copied the names of the variables as well).

Finally, you can import a data set from a URL directly by selecting:
- **File** > **Internet Open** and pasting in the address of the website. JMP will attempt to find data on the page. It may take a few tries and some edits to get the data set in correctly.

MINITAB

To import a text or Excel file, choose:

- **File > Open.** In the lower right corner, you should specify the file type, by changing **All Minitab Files** to either **Text** or **Excel Files** as appropriate, unless the file has already been saved in Minitab format.
- Browse to find and select the file.

- Choose **Open**, and if Excel or Text, you will see a preliminary view of how the data will be entered in a Minitab worksheet. If okay, click **Open** again.
- **Comments** In MINITAB EXPRESS, choose Open Worksheet. Acceptable file types do not include Text files.

R

R can import many types of files, but text files (tab or comma delimited) are easiest. If the file is tab delimited and contains the variable names in the first row, then:

> mydata = read.delim(file.choose())

will give a dialog where you can pick the file you want to import. It will then be in a data frame called mydata. If the file is comma delimted, use:

>mydata=read.csv(file.choose())

COMMENTS

(RStudio provides an interactive dialog that may be easier to use). For other options including the case that the file does not contain variable names, consult **R** help.

SPSS

To import a text file, choose:

- **File > Open > Data.** Under "Files of type," choose **Text (*.txt,*.dat)**. Select the file you want to import. Click **Open.**

- A window will open called **Text Import Wizard.** Follow the steps, depending on the type of file you want to import.

STATCRUNCH

Statcrunch offers several ways to enter data. Choose MyStatCrunch > My Data. Choose a dataset to analyze the data or edit its properties.

MY DATA

Click a data set link to analyze the data or edit its properties. Want to load a new data set?

- Select a file on my computer.
- Enter the WWW address of a file.

- Paste data into a form.
- Type or paste data into a blank data table.

For the "select a file on my computer" option, Statcrunch offers a choice of whitespace, comma, tab or semicolon delimiters. You may also choose to use the first line as the names of the variables.

After making your choices, select the **Load File** button at the bottom of the screen.

TI 83/84 PLUS

Locate the TI List on the book's URL: pearsonhighered.com/deveaux or **ActivStats DVD**.

- Link your calculator to the computer using the USB cable.

- Right-click on the file name and choose "Send to TI-Device."

Chapter Exercises

For the description of data in each of the exercises below, identify the W's (and How, if known), name the variables, specify for each variable whether its use indicates it should be treated as categorical or quantitative, and, for any quantitative variable, identify the units in which it was measured (or note that they were not provided).

11. **The news** Find a newspaper or magazine article in which some data are reported. For the data discussed in the article, answer the questions above. Include a copy of the article with your report.

12. **The Internet** Find an Internet source that reports on a study and describes the data. Print out the description and answer the questions above.

13. **Ovulation** A study conducted by a team of American and Canadian researchers found that during ovulation, a woman can tell whether a man is gay or straight by looking at his face. To explore the subject, the authors conducted three investigations, the first of which involved 40 undergraduate women who were asked to guess the sexual orientation of 80 men based on photos of their face. Half of the men were gay, and the other half were straight. All held similar expressions in the photos or were deemed to be equally attractive. None of the women were using any contraceptive drugs at the time of the test. The result: the closer a woman was to her peak ovulation, the more accurate her guess. (http://consumer.healthday.com/mental-health-information-25/behavior-health-news-56/does-ovulation-boost-a-woman-s-gaydar-654279.html)

14. **Blindness** A study begun in 2011 examines the use of stem cells in treating two forms of blindness, Stargardt's disease and dry age-related macular degeneration. Each of the 24 patients entered one of two separate trials, one for each form of blindness, in which embryonic stem cells were to be used to treat the blindness. (www.redorbit.com/news/health/2066060/stem_cell_trials_to_begin_for_blindness_treatment/)

15. **Investments** Which mutual funds to choose? An investor with some RRSP funds to invest was considering investing in China or India funds. She went to www.globeinvestor.com, did a search, and found 48 China/India/Chindia funds, where she viewed the one-month, one-year, and five-year returns for each of these funds.

16. **Biomass of trees** Because of the difficulty of determining the biomass (or volume) of trees in the wild, Canadian forestry ecology researchers chose a sample of 25 trees in a BC forest, from each of three different species, then measured each tree's diameter (at chest height), and estimated its height via a special viewing instrument (and basic Pythagorean math for triangles). Later these trees were chopped down and the biomass of each was determined by a precise method.

17. **Air travel** Transport Canada and the Canadian Transportation Agency monitor airlines for safety and customer service. For each flight, carriers must report the type of aircraft, the number of passengers, whether or not the flights departed and arrived on schedule, and any mechanical problems.

18. **Tracking sales** A start-up company is building a database of customers and sales information. For each customer, it records name, ID number, region of the country (1 = East, 2 = Central, 3 = Prairies, 4 = West), date of last purchase, amount of purchase, and item purchased.

19. **Cars** A survey of automobiles parked in student and staff lots at a large university recorded the make, country of origin, type of vehicle (car, van, SUV, etc.), and age.

20. **Stats students** An online survey of students in a large Statistics class asked them to report their height, shoe size, sex, which degree program they were in, and their birth order (1 = only child or first born). The data were used for classroom illustrations.

21. **Honesty** Coffee stations in offices often just ask users to leave money in a tray to pay for their coffee, but many people cheat. Researchers at Newcastle University alternately taped two posters over the coffee station. During one week, it was a picture of flowers; during the other, it was a pair of staring eyes. They found that the average contribution was significantly higher when the eyes poster was up than when the flowers were there. Apparently, the mere feeling of being watched—even by eyes that were not real—was enough to encourage people to behave more honestly. (Source: *NY Times,* Dec. 10, 2006)

22. **Molten iron** The Cleveland Casting Plant is a large, highly automated producer of grey and nodular iron automotive castings for Ford Motor Company. According to an article in *Quarterly Engineering* (7[1995]), Cleveland Casting is interested in keeping the pouring temperature of the molten iron (in degrees Fahrenheit) close to the specified value of 2550 degrees. Cleveland Casting measured the pouring temperature for 10 randomly selected crankshafts.

23. **Weighing bears** Because of the difficulty of weighing a bear in the woods, researchers caught and measured

54 bears, recording their weight, neck size, length, and sex. They hoped to find a way to estimate weight from the other, more easily determined quantities.

24. **Schools** The provincial or territorial ministry of education requires local school districts to keep the following records on all students: age, race or ethnicity, days absent, current grade level, standardized test scores in reading and mathematics, and any disabilities or special educational needs.

25. **Tim Hortons doughnuts** At the Tim Hortons Website (www.timhortons.com), there is a link called "Nutritional Information" that gives nutrition facts for various categories of food that it sells, such as doughnuts, sandwiches, etc. Under doughnuts, for example, we see a listing of all the doughnut types, and for each, information is provided about the number of calories and the amounts of trans fat, total fat, sodium, sugar, protein, and "% daily value" of iron and of calcium.

26. **Trudeaumania?** Even before the 2015 federal election, early signs of Justin Trudeau's appeal were becoming evident. An April 2013 Harris/Decima poll indicated that 57% of Canadians held a favourable view of new Liberal leader Justin Trudeau, while only 30% held an unfavourable impression. These survey results were based on 1006 telephone respondents, via the company's national telephone omnibus survey. For each respondent, gender, province, party preference, and age group (18–34, 35–54, 55+) was recorded, along with the answer to the favourability question.

27. **Babies** Medical researchers at a large city hospital investigating the impact of prenatal care on newborn health collected data from 882 births between 1998 and 2000. They kept track of the mother's age, the number of weeks the pregnancy lasted, the type of birth (caesarean, induced, or natural), the level of prenatal care the mother received (none, minimal, or adequate), the birth weight and sex of the baby, and whether the baby exhibited health problems (none, minor, or major).

28. **Flowers** In a study appearing in the journal *Science*, a research team reported that plants in southern England are flowering earlier in the spring. Records of the first flowering dates for 385 species over a period of 47 years show that flowering has advanced an average of 15 days per decade, an indication of climate warming, according to the authors.

29. **Herbal medicine** Scientists at a major pharmaceutical firm conducted an experiment to study the effectiveness of an herbal compound to treat the common cold. They exposed each patient to a cold virus, then gave them either the herbal compound or a sugar solution known to have no effect on colds. Several days later they assessed each patient's condition using a cold severity scale ranging from 0 to 5. They found no evidence of the benefits of the compound.

30. **Vineyards** Business analysts hoping to provide information helpful to grape growers compiled the following data about vineyards: size (acres), number of years in existence, province, varieties of grapes grown, average case price, gross sales, and percent profit.

31. **Streams** As research for an ecology class, students at Laurentian University collect data on streams each year. They record a number of biological, chemical, and physical variables, including the stream name, the substrate of the stream (limestone, shale, or mixed), the acidity of the water (pH), the temperature (°C), and the BCI (a numerical measure of biological diversity).

32. **Fuel economy** Transport Canada and Natural Resources Canada track the fuel economy of automobiles. Among the data they collect are the manufacturer (Ford, Toyota, etc.); the vehicle type (car, SUV, etc.); and the weight, horsepower, and gas mileage (L/100 km) for city and highway driving.

33. **Refrigerators** A *Consumer Reports* article evaluating refrigerators listed 41 models, giving the brand, cost, size (cu. ft.), type (such as top freezer), estimated annual energy cost, overall rating (good, excellent, etc.), and the repair history for that brand (percentage requiring repairs over the past five years).

34. **Walking in circles** People who get lost in the desert, mountains, or woods often seem to wander in circles rather than walk in straight lines. To see whether people naturally walk in circles in the absence of visual clues, researcher Andrea Axtell tested 32 people on a football field. One at a time, they stood at the centre of one goal line, were blindfolded, and then tried to walk to the other goal line. She recorded each individual's sex, height, handedness, the number of yards each was able to walk before going out of bounds, and whether each wandered off course to the left or the right. No one made it all the way to the far end of the field without crossing one of the sidelines.[5]

35. **Kentucky Derby 2015** The Kentucky Derby is a horse race that has been run every year since 1875 at Churchill Downs, Louisville, Kentucky. The race started as a 1.5-mile race, but in 1896, it was shortened to 1.25 miles because experts felt that 3-year-old horses shouldn't run such a long race that early in the season. (It has been run in May every year but one—1901—when it took place on April 29). Here are the data for the first four and some recent races.

---

[5]*Source: STATS* No. 39, Winter 2004.

Year	Winner	Jockey	Trainer	Owner	Time
2015	American Pharoah	V. Espinoza	B. Baffert	Zayat Stables	2:03.02
2014	California Chrome	V. Espinoza	A. Sherman	Coburn & Perry	2:03.66
2013	Orb	J. Rosario	S. McGaughey	Stuart Janney & Phipps Stable	2:02.89
2012	I'll Have Another	M. Gutierrez	D. O'Neill	Reddam Racing	2:01.83
2011	Animal Kingdom	J. Velazquez	H. G. Motion	Team Valor	2:02.04
2010	Super Saver	C. Borel	T. Pletcher	WinStar Farm	2:04.45
⋮					
1878	Day Star	J. Carter	L. Paul	T. J. Nichols	2:37.25
1877	Baden Baden	W. Walker	E. Brown	Daniel Swigert	2:38
1876	Vagrant	R. Swim	J. Williams	William Astor	2:38.25
1875	Aristides	O. Lewis	A. Williams	H. P. McGrath	2:37.75

Source: Data from Kentucky Derby Winners. http://horsehats.com/KentuckyDerbyWinners.html

36. Stanley Cup 2015 Lord Stanley's Cup is the oldest professional sports trophy in North America and has been awarded to the championship-winning hockey team every year since 1893. It was awarded to amateur champions in the early years, before the professional leagues had been formed. Below are some data for recent and not-so-recent Stanley Cup finals. The current best-of-seven championship format was not the case until 1939, as you can see from the earlier championship series listed (some were best of five, some best of three).

Year	Winning Team	Series (W–L)	Finalist	Time of Winning Goal	Winning Goal Scored By
1926	Montreal Maroons	3–1	Victoria Cougars	N/A	N/A
1927	Ottawa Senators	2–0	Boston Bruins	27:30	Denneny
1928	New York Rangers	3–2	Montreal Canadiens	43:35	Boucher
1929	Boston Bruins	2–0	New York Rangers	58:02	Carson
1930	Montreal Canadiens	2–0	Boston Bruins	21:00	Morenz
1931	Montreal Canadiens	3–2	Chicago Blackhawks	29:59	Gagnon
1932	Toronto Maple Leafs	3–0	New York Rangers	55:07	Bailey
1933	New York Rangers	3–1	Toronto Maple Leafs	67:34	Cook
1934	Chicago Blackhawks	3–1	Detroit Red Wings	90:05	March
1935	Montreal Maroons	3–0	Toronto Maple Leafs	36:18	Northcott
1936	Detroit Red Wings	3–1	Toronto Maple Leafs	49:45	Kely
1937	Detroit Red Wings	3–2	New York Rangers	19:22	Barry
1938	Chicago Blackhawks	3–1	Toronto Maple Leafs	36:45	Voss
1939	Boston Bruins	4–1	Toronto Maple Leafs	37:54	Conacher
1940	New York Rangers	4–2	Toronto Maple Leafs	62:07	Hextall
⋮					
2013	Chicago Blackhawks	4-2	Boston Bruins	59:01	Bolland
2014	Los Angeles Kings	4-1	New York Rangers	94:43	Martinez
2015	Chicago Blackhawks	4-2	Tampa Bay Lightning	37:13	Keith

(Note: above, 29:59 means 9:59 into the second period, since regular time periods are 20 minutes)

Just Checking ANSWERS

1. *Who*—Tour de France races; *What*—year, winner, country of origin, age, team, total time, average speed, stages, total distance ridden, starting riders, finishing riders; *How*—official statistics at race; *Where*—France (for the most part); *When*—1903 to 2014; *Why*—not specified (To see progress in speeds of cycling racing?)

2.

Variable	Type	Units
Year	Quantitative or Identifier	Years
Winner	Categorical	
Country of Origin	Categorical	
Age	Quantitative	Years
Team	Categorical	
Total Time	Quantitative	Hours/minutes/seconds
Average Speed	Quantitative	Kilometres per hour
Stages	Quantitative	Counts (stages)
Total Distance	Quantitative	Kilometres
Starting Riders	Quantitative	Counts (riders)
Finishing Riders	Quantitative	Counts (riders)

Displaying and Describing Categorical Data

LEARNING OBJECTIVES

1. Summarize a single categorical variable and display its frequency distribution, using
   - Bar charts
   - Pie charts
2. Explore the relationship between two categorical variables, using
   - Contingency tables
   - Conditional and marginal distributions
   - Bar charts

Connections

What is your field of study: Humanities, Social Sciences, or Sciences? What is your blood type? Which candidate do you plan to vote for?

Data are not just numbers—the answers to all these questions are categories. We'll study categorical variables like these in this chapter. Things get even more interesting when we look at how categorical variables work together. Are Grade 5 students more likely to experience bullying than students in Grade 9? Are homosexuals more likely to be victims of crime than heterosexuals?

You often see categorical data displayed in pie charts or bar charts, or summarized in tables, sometimes in confusing ways. But, with a little skill it's not hard to do it right.

A single death is a tragedy; a million is a statistic.

—*Joseph Stalin*

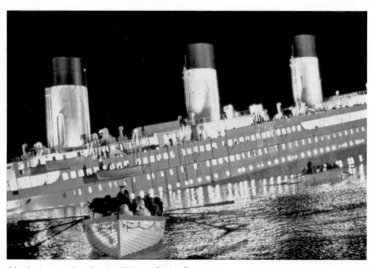

Moviestore collection Ltd/Alamy Stock Photo

What happened on the *Titanic* at 11:40 on the night of April 14, 1912, is well known. Frederick Fleet's cry of "Iceberg, right ahead" and the three accompanying pulls of the crow's nest bell signalled the beginning of a nightmare that has become legend. By 2:15 a.m., the *Titanic*, thought by many to be unsinkable, had sunk, leaving almost 1500 passengers and crewmembers on board to meet their icy fate.

Below are some data about the passengers and crew aboard the *Titanic*. Each case (row) of the data table represents a person on board the ship. The variables are the person's *Survival* status (Dead or Alive), *Age* (Adult or Child), *Sex* (Male or Female), and ticket *Class* (First, Second, Third, or Crew). (Data in the file **Titanic** at the Companion Website)

Table 2.1

Part of a data table showing four variables for nine people (one per row) aboard the *Titanic*. We omit a case identifier column (usually the first column), but we really don't need to know their names.

Survival	Age	Sex	Class
Dead	Adult	Male	Third
Dead	Adult	Male	Crew
Dead	Adult	Male	Third
Dead	Adult	Male	Crew
Dead	Adult	Male	Crew
Dead	Adult	Male	Crew
Alive	Adult	Female	First
Dead	Adult	Male	Third
Dead	Adult	Male	Crew
⋮	⋮	⋮	⋮

The problem with a data table like this—and in fact with all data tables—is that you can't see what's going on. And seeing is just what we want to do. We need ways to show the data so that we can see patterns, relationships, trends, and exceptions.

2.1   Summarizing and Displaying a Single Categorical Variable

So, what should we do with data like these? There are three things you should always do first with data:

Who	People on the *Titanic*
What	Survival status, age, sex, ticket class
When	April 14, 1912
Where	North Atlantic
How	A variety of sources and Internet sites
Why	Historical interest

The Three Rules of Data Analysis

1. **Make a picture.** A display of your data will reveal things you are not likely to see in a table of numbers and will help you to *Think* clearly about the patterns and relationships that may be hiding in your data.
2. **Make a picture.** A well-designed display will *Show* the important features and patterns in your data. A picture will also show you the things you did not expect to see: the extraordinary (possibly wrong) data values or unexpected patterns.
3. **Make a picture.** The best way to *Tell* others about your data is with a well-chosen picture.

These are the three rules of data analysis. There are pictures of data throughout the book and new kinds of visual displays are constantly being developed. These days, technology makes it easy to draw pictures of data, so there is no reason not to follow the three rules.

Figure 2.1

A Picture to Tell a Story. Are you a binge drinker? Hopefully not. How should one define binge or heavy drinking, or even measure alcohol consumption? Determining a precise numerical scale is not an easy task. Suppose we want to know how heavy drinking differs across regions of Canada. Pictures can vibrantly display relationships between variables and patterns or trends in variables over time or space (geographically). Here is a thematic graph from Statistics Canada showing the relationship between heavy drinking and geographical region. Two categorical variables used were: *Heavy Drinker? - Yes or No*, per person and *Percentage of Heavy Drinkers*, per census area. While percentage of heavy drinkers is quantitative, it often makes sense to categorize it for superior clarity in graphs. It might also be enlightening to study additional relationships such as gender differences or changes over time.

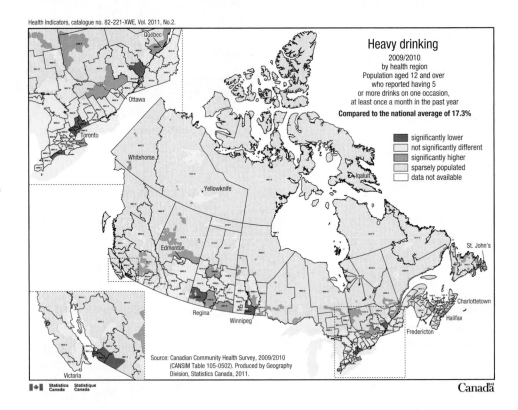

Health Indicators, catalogue no. 82-221-XWE, Vol. 2011, No.2.

Heavy drinking
2009/2010
by health region
Population aged 12 and over
who reported having 5
or more drinks on one occasion,
at least once a month in the past year
Compared to the national average of 17.3%

■ significantly lower
☐ not significantly different
▨ significantly higher
☐ sparsely populated
☐ data not available

Source: Canadian Community Health Survey, 2009/2010 (CANSIM Table 105-0502). Produced by Geography Division, Statistics Canada, 2011.

Statistics Canada / Statistique Canada

Canadä

The Area Principle

We're ready to follow the three rules of data analysis and make a picture of the data. But a bad picture can distort our understanding rather than help it. Figure 2.2 is a

Figure 2.2

How many people were in each class on the *Titanic*? From this display it looks as though the service must have been great, since most aboard were crewmembers. Although the length of each ship here corresponds to the correct number, the impression is all wrong. In fact, only about 40% were crew.

Class	Count
First	325
Second	285
Third	706
Crew	885

Table 2.2

A frequency table of the *Titanic* passengers and crew.

Class	%
First	14.77
Second	12.95
Third	32.08
Crew	40.21

Table 2.3

A relative frequency table for the same data.

colourful graph of the *Titanic* data, showing the total number of persons in each class, as well as the number of crewmembers aboard. What impression do you get about who was aboard the ship?

It looks like most of the people on the *Titanic* were crewmembers, with a few passengers along for the ride. That doesn't seem right. What's wrong? The lengths of the ships do match the number of people in each ticket class category. However, experience and psychological tests show that our eyes tend to be more impressed by the area than by other aspects of each ship image. So, even though the *length* of each ship matches up with one of the totals on the horizontal axis, it's the associated *area* in the image that we notice. There were about three times as many crew as second-class passengers, and the ship depicting the number of crew is about three times longer than the ship depicting second-class passengers. The problem is that it occupies about nine times the area.[1] That just isn't a correct impression.

The best data displays observe a fundamental principle of graphing data called the area principle. The **area principle** says that the area occupied by a part of the graph should correspond to the magnitude of the value it represents. Violations of the area principle are a common way to lie (or, since most mistakes are unintentional, we should say err) with statistics.

Frequency Tables

A display that distorts the number of cases can be misleading. But before we can follow the area principle to accurately display a categorical variable, we need to organize the number of cases associated with each category. Even when we have thousands of cases, a variable like ticket *Class*, with only a few categories, is easy to organize. We just count the number of cases corresponding to each category and put them in a table. A **frequency table** (Table 2.2) records the totals and uses the category names to label each row. For ticket *Class*, these are "First," "Second," "Third," and "Crew."

For a variable with dozens or hundreds of categories a frequency table will be much harder to read. You might want to combine categories into larger headings. For example, instead of counting the number of foreign students from each state in the U.S. who attend Montreal's prestigious McGill University, you might group the states into regions like "Northeast," "South," "Midwest," "Mountain States," and "West." If the number of cases in several categories is relatively small, you can put them together into one category labelled "Other."

Counts are useful, but sometimes we want to know the fraction or proportion of the data in each category, so we divide the counts by the total number of cases. Often we multiply by 100 to express these proportions as percentages (Table 2.3). A **relative frequency table** displays the *proportions* or *percentages*, rather than the counts, of the values in each category. Both types of tables show how the cases are distributed across the categories. In this way, they describe the **distribution** of a categorical variable because they name the possible categories and tell how frequently each occurs.

Bar Charts

Figure 2.3 on the next page shows a chart that obeys the area principle. It's not as visually entertaining as the ships, but it does give an accurate visual impression of the distribution. The height of each bar shows the count for its category. The bars are the same width, so their heights determine their areas, and the areas are proportional to the counts in each class. Now it's easy to see that the majority of people on board were

---

[1]Edward Tufte would say the Lie Factor here is 3.0, since the impression or displayed effect is three times bigger than it should be (*The Visual Display of Quantitative Information*).

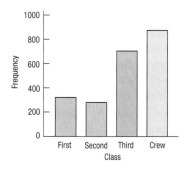

Figure 2.3

People on the Titanic ***by ticket class.***
With the area principle satisfied, we can see the true distribution more clearly.

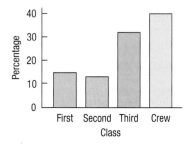

Figure 2.4

The relative frequency bar chart looks the same as the bar chart (Figure 2.3), but shows the percentage of people in each category rather than the counts. The vertical axis may also show proportions: 0, 0.10, 0.20, 0.30, 0.40.

not crew, as the ships picture led us to believe. We can also see that there were about three times as many crew as second-class passengers. And there were more than twice as many third-class passengers as either first- or second-class passengers, something you may have missed in the frequency table. Bar charts make these kinds of comparisons easy and natural.

A **bar chart** may be used to display the distribution of a categorical variable, showing the counts for each category next to each other for easy comparison. Bar charts should have small spaces between the bars to indicate that these are freestanding bars that could be rearranged into any order. The bars are lined up along a common base.

Usually they stick up like this , but sometimes they run sideways

like this 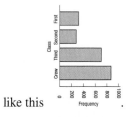 .

If we really want to draw attention to the relative proportion of passengers falling into each of these classes, we could replace the counts with percentages (or proportions) and use a **relative frequency bar chart** like Figure 2.4 in the margin.

Pie Charts

Another common display that shows how a whole group breaks into several categories is a pie chart. **Pie charts** show the whole group of cases as a circle. They slice the circle into pieces whose size is proportional to the fraction of the whole in each category.

Pie charts give a quick impression of how a whole group is partitioned into smaller groups. Because we're used to cutting up pies into 2, 4, or 8 pieces, pie charts are good for seeing relative frequencies near 1/2, 1/4, or 1/8. For example, you may be able to tell that the pink slice, representing the second-class passengers, is very close to 1/8 of the total (Figure 2.5). It's harder to see that there were about twice as many third-class as first-class passengers. Which category had the most passengers? Were there more crew or more third-class passengers? Comparisons such as these are easier in a bar chart.

Figure 2.5

Pie chart showing the number of *Titanic* crew and passengers in each class.

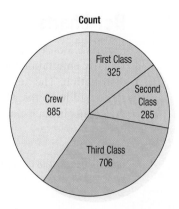

For example, Figure 2.6 below shows three pie charts that look pretty much alike along with bar charts of the same data. The bar charts show three distinctly different patterns, but it is almost impossible to see those in the pie charts.

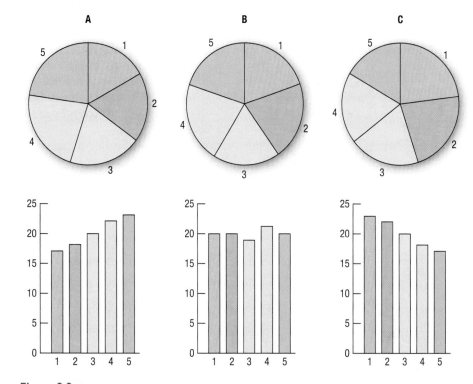

THINK BEFORE YOU DRAW

Our first rule of data analysis is *Make a picture*. But what kind of picture? We don't have a lot of options—yet. There's more to Statistics than pie charts and bar charts, and knowing when to use each type of graph is a critical first step in data analysis. That decision depends in part on what type of data we have.

If you want to make a relative frequency bar chart or a pie chart, you'll need to also make sure that the categories don't overlap, so no individual is counted twice. If the categories do overlap, you can still make a bar chart, but the percentages won't add up to 100%. For the *Titanic* data, either kind of display is appropriate because the categories don't overlap.

For any statistical analysis, a crucial decision is the choice of a proper method. Be sure to think about the situation at hand, and make sure to check that the type of analysis you plan is appropriate. The Categorical Data Condition is just the first of many such checks.

Figure 2.6
Patterns that are easy to see in the bar charts are often hard to see in the corresponding pie charts.

No matter what display you choose for your variable, make sure that the data are counts or percentages of cases. For example, a pie chart showing the price of each ticket class wouldn't make sense. Price is quantitative. So before making a bar chart or a pie chart, always check the **Categorical Data Condition** that the data are counts or percentages of individuals that were placed into categories.

Section 2.1 Exercises

1. **Automobile fatalities** The following table gives the numbers of passenger car occupants killed in accidents in 2011 by car type.

Subcompact and Mini	1351
Compact	3789
Intermediate	4050
Full	2627
Unknown	164

Source: Bureau of Transportation Statistics. Automobile fatalities. Taken from www.bts.gov/publications/national_transportation_statistics/html/table_02_19.html

Convert this table to a relative frequency table.

2. **Nonoccupant fatalities** The frequencies of traffic fatalities of nonoccupants of vehicles are shown in the following bar chart. Change this to a bar chart of relative frequencies.

3. Movie genres The Motion Picture Association of America (MPAA) rates each film to designate the appropriate audience. The ratings are G, PG, PG-13, and R. The ratings of the 20 top-grossing movies in the years 1996, 2006, 2008, and 2012 are shown in the following bar charts. The pie charts show the same data, but are unlabeled. Match each pie chart with the correct year.

i)

ii)

iii)

iv)

A)

B)

C)

D)
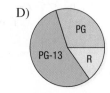

4. Marriage in decline Changing attitudes about marriage and families prompted Pew Research to ask how people felt about recent trends (pewresearch.org/pubs/1802/decline-marriage-rise-new-families). For each trend, participants were asked whether the trend "is a good thing," "is a bad thing," or "makes no difference." Some participants said they didn't know or chose not to respond. The following bar charts show the number of each response for each trend. The pie charts show the same data without the trends being identified. Match each pie chart with the correct trend and bar chart.

i)

ii)

iii)

iv)

A)

B)

C)

D)
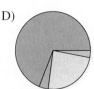

2.2 Exploring the Relationship Between Two Categorical Variables

Contingency Tables: Children and First-Class Ticket Holders First?

Only 32% of all those aboard the *Titanic* survived. Was the survival rate the same for men and women? For children and adults? For all ticket classes? It's often more interesting to ask if one variable relates to another. For example, was there a relationship between the kind of ticket a passenger held and the passenger's chances of surviving? To answer this question, we need to look at the two categorical variables *Survival* and ticket *Class* together.

To look at two categorical variables together, we often arrange the counts in a two-way table. Table 2.4 shows each person aboard the *Titanic* classified according to both their ticket *Class* and their *Survival*. Because the table shows how the individuals are distributed along each variable, contingent on the value of the other variable, such a table is called a **contingency table**.

Table 2.4

Contingency table of ticket class and survival. The bottom line of "Total" is the same as the previous frequency table.

		First	Second	Third	Crew	Total
	Alive	203	118	178	212	711
Survival	Dead	122	167	528	673	1490
	Total	325	285	706	885	2201

(Class across top; Survival down side)

Each *cell* of the table gives the count for a combination of values of the two variables. The margins of the table, on the right and at the bottom, give totals. The bottom line of the table just shows the frequency distribution of the variable ticket *Class*. The right column of the table shows the frequency distribution of the variable *Survival*. When presented like this, in the margins of a contingency table, the frequency distribution of one of the variables is called its **marginal distribution**. The marginal distribution can be expressed either as counts or percentages.

If you look down the column for second-class passengers to the first row, you'll find the cell containing the 118 second-class passengers who survived. Looking at the cell to its right we see that more third-class passengers (178) survived. But, does that mean that third-class passengers were *more likely* to survive? It's true that more third-class passengers survived, but there were many more third-class passengers on board the ship. To compare the two numbers fairly, we need to express them as percentages—but as a percentage of what?

For any cell, there are three choices of percentage. We could express the 118 second-class survivors as 5.4% of all 2201 passengers on the *Titanic* (the *overall percent*), as 16.6% of all 711 survivors (the *row percent*), or as 41.4% of all 285 second-class passengers (the *column percent*). Each of these percentages is potentially interesting.

Statistics programs generally offer all three choices. Sometimes, you may see them all together in each cell of the table. The resulting table holds lots of information, but it can be hard to understand:

Table 2.5

Another contingency table of ticket
Class. This time we see not only the
counts for each combination of *Class*
and *Survival* (in bold), but also the
percentages these counts represent.
For each count, there are three
choices for the percentage: by row, by
column, and by table total. There's
probably too much information here
for this table to be useful.

			Class				
			First	Second	Third	Crew	Total
Alive	Count		**203**	**118**	**178**	**212**	**711**
	% of Row		28.6	16.6	25.0	29.8	100
	% of Column		62.5	41.4	25.2	24.0	32.3
	% of Overall Total		9.2	5.4	8.1	9.6	32.3
Dead	Count		**122**	**167**	**528**	**673**	**1490**
	% of Row		8.2	11.2	35.4	45.2	100
	% of Column		37.5	58.6	74.8	76.0	67.7
	% of Overall Total		5.5	7.6	24.0	30.6	67.7
Total	Count		**325**	**285**	**706**	**885**	**2201**
	% of Row		14.8	12.9	32.1	40.2	100
	% of Column		100	100	100	100	100
	% of Overall Total		14.8	12.9	32.1	40.2	100

*(Survival labels the left margin of the table.)*

To simplify the table, let's first pull out the overall percent values:

Table 2.6

A contingency table of class
by survival with only the table
percentages

		Class				
		First	Second	Third	Crew	Total
Survival	Alive	9.2%	5.4%	8.1%	9.6%	32.3%
	Dead	5.5%	7.6%	24.0%	30.6%	67.7%
	Total	14.8%	12.9%	32.1%	40.2%	100%

These percentages tell us what percent of all passengers belong to each combination
of column and row category (called the **joint distribution** of the two variables). For
example, we see that although 8.1% of the people aboard the *Titanic* were surviving
third-class ticket holders, only 5.4% were surviving second-class ticket holders. Comparing these percentages, you might think that the chances of surviving were better in
third class than in second. But be careful. There were many more third-class than
second-class passengers on the *Titanic*, so we should naturally expect there to be more
third-class survivors. Comparing the overall percentages, 5.4% and 8.1%, doesn't
answer the question about whether second-class passengers were more or less likely to
survive than third-class passengers. Overall percentages can be useful, but they don't
answer questions like this, involving comparisons of percentages of one variable across
levels of the other.

PERCENT OF WHAT?

The English language can be tricky when we talk about percentages. If you're asked, "What percent *of the
survivors* were in second class?" it's pretty clear that we're interested only in survivors. So we should
restrict our attention only to the survivors and look at the number of second-class passengers among all
the survivors—in other words, the row percent.

But if you're asked, "What percent were second-class passengers who survived?" you have a different question. Be careful; now we need to consider everyone on board, so 2201 should be the denominator, and the
answer is the overall percent (5.4%).

And if you're asked, "What percent of the second-class passengers survived?" you have yet another question. Now the denominator is the 285 second-class passengers, and the answer is the column percent (41.4%).

Always be sure to ask "percent of what?" That will help you to know *Who* we're talking about and whether
we want *row*, *column*, or *overall* percentages.

For Example FINDING MARGINAL DISTRIBUTIONS

STILL LIVING WITH YOUR PARENTS, EH? Of the 4.3 million young Canadian adults aged 20-29 in 2011, 1.8 million were still living at home with their parents (Statistics Canada, census of population, 2011). Below we show some (rounded) counts, in thousands, for living arrangements and sex:

		Sex		
		Male	Female	Total
Living arrangements	Living with parents*	1020	810	1830
	Living as part of a couple	530	770	1300
	Other living arrangements	630	560	1190
	Total	2180	2140	4320

*May also be part of a couple, with or without children, or a lone parent.

QUESTION: What's the marginal distribution of *Living arrangements*?

ANSWER: To determine the percentages for the three types of living arrangements, divide the count for each living arrangement by the total number of young adults:

$$\frac{1830}{4320} = 42.4\%, \quad \frac{1300}{4320} = 30.1\%, \quad \frac{1190}{4320} = 27.5\%$$

42.3% of young adults (aged 20-29) were still living with their parents, while 30.1% were living as part of a couple and 27.5% of them had alternative living arrangements.

Conditional Distributions

Rather than looking at the overall percentages, it's more interesting to ask whether the chance of survival *depended* on ticket class. We can look at this question in two ways. First, we could ask how the distribution of ticket *Class* changes between survivors and non-survivors. To do that, we look at the *row percentages*:

Table 2.7

The conditional distribution of ticket *Class* conditioned on each value of *Survival*: alive and dead.

		Class				
		First	Second	Third	Crew	Total
Survival	Alive	203	118	178	212	711
		28.6%	16.6%	25.0%	29.8%	100%
	Dead	122	167	528	673	1490
		8.2%	11.2%	35.4%	45.2%	100%

By focusing on each row separately, we see the distribution of *Class* for each *condition* of surviving or not. The sum of the percentages in each row is 100%, and we divide that up by ticket class. First, we restrict our attention to survivors, and make a pie chart for them. Then we focus on the non-survivors and make their pie chart. These pie charts (Figure 2.7) show the distribution of ticket classes *for each row* of the table: survivors and non-survivors. The distributions we create this way are called **conditional distributions** because they show the distribution of one variable for just those cases that satisfy a condition on another variable.

Figure 2.7

Pie charts of the conditional distributions of ticket class for survivors and non-survivors, separately. Do the distributions appear to be the same pie charts or bar charts can be used to compare these two distributions.

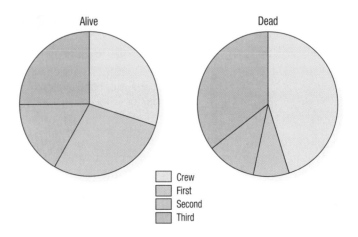

For Example   FINDING CONDITIONAL DISTRIBUTIONS

RECAP: The table below shows 2011 census counts for young adults, aged 20-29, classified by sex and living arrangements. We have grouped together various 'other living arrangements' such as living with non-relatives, living alone, or living as a lone parent.

		Sex		
		Male	**Female**	**Total**
Living arrangements	**Living with parents***	1020	810	**1830**
	Living as part of a couple	530	770	**1300**
	Other living arrangements	630	560	**1190**
	Total	**2180**	**2140**	**4320**

*May also be part of a couple, with or without children, or a lone parent.

QUESTION: Describe the conditional distribution of *Sex* for those still living at home.

ANSWER: Find the percent of males among those who are living at home, and repeat for females:

$$\frac{1020}{1830} = 55.7\%, \quad \frac{810}{1830} = 44.3\%$$

Substantially more young men (55.7%) than women (44.3%) are living at home with the parents.

But we still haven't answered our question about whether survival depends on ticket class. Instead, we can turn the conditioning around and look at the distribution of *Survival* for each category of ticket *Class*. The *column percentages* show us whether the chance of surviving was roughly the same *for each of the four classes*. Now the percentages in each column add to 100%, because we've restricted our attention, in turn, to each of the four ticket classes:

Table 2.8

A contingency table of class by survival with only counts and column percentages. Each column represents the conditional distribution of *Survival* for a given category of ticket *Class*.

			Class				
			First	**Second**	**Third**	**Crew**	**Total**
Survival	**Alive**	**Count**	203	118	178	212	**711**
		% of Column	62.5%	41.4%	25.2%	24.0%	**32.3%**
	Dead	**Count**	122	167	528	673	**1490**
		% of Column	37.5%	58.6%	74.8%	76.0%	**67.7%**
	Total	**Count**	325	285	706	885	2201
			100%	100%	100%	100%	100%

Looking at how the percentages change across each row, it sure looks like ticket class mattered for survival. To illustrate this more vividly, we could display the percentages for Alive and Dead for each *Class* in a side-by-side bar chart (Figure 2.8):

Figure 2.8

Side-by-side bar charts showing the conditional distribution of *Survival* for each category of ticket *Class*. The corresponding pie charts would have only two categories in each of four pies, so bar charts seem the better alternative.

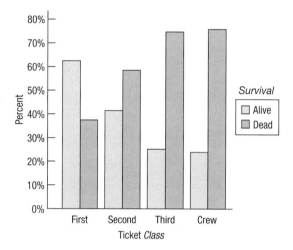

These bar charts are simple because for the variable *Survival*, we have only two alternatives: Alive and Dead. When we have only two categories, we can simplify the display even more by dropping one category (Figure 2.9). Knowing the percentage that died tells us the percentage that survived. Here are just the percentages of dying across the classes displayed in one chart:

Figure 2.9

Bar chart showing just non-survivor percentages for each value of ticket *Class*. Because we have only two values, the second bar wouldn't add any information.

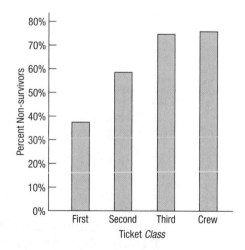

Now it's easy to compare the risks. Among first-class passengers, 37.5% perished, compared to 58.6% for second-class ticket holders, 74.8% for those in third class, and 76.0% for crewmembers.

If the risk had been about the same across the ticket classes, we would have said that survival was *independent* of class. But it's not. The differences we see among these conditional distributions suggest that survival may have depended on ticket class. You may find it useful to look at the conditional distributions of each variable in a contingency table in order to explore the dependence between them.

It is interesting to know that *Class* and *Survival* are associated. That's an important part of the *Titanic* story. And we know how important this is because the margins show us the actual numbers of people involved.

Variables can be associated in many ways and to different degrees. The best way to tell whether two variables are associated is to ask whether they are *not*.[2] In a contingency table, when the distribution of *one* variable is the same for all categories of another, we say that the variables are **independent**. That tells us there's no association between these variables. We'll see a way to formally check for independence later in the book. For now, we'll just informally compare the distributions.

For Example LOOKING FOR ASSOCIATIONS BETWEEN VARIABLES

RECAP: The table below shows 2011 census results classified by sex and living arrangements for young Canadian adults.

		Sex		
		Male	**Female**	**Total**
	Living with parents*	1020	810	1830
	Living as part of a couple	530	770	1300
	Other living arrangements	630	560	1190
	Total	2180	2140	4320

*May also be part of a couple, with or without children, or a lone parent.

QUESTION: Does there seem to be an association between type of living arrangement and a young person's sex?

ANSWER: First find the distribution of *Living Arrangements* for the men (the column percentages):

$$\frac{1020}{2180} = 46.8\%$$

$$\frac{530}{2180} = 24.3\%$$

$$\frac{630}{2180} = 28.9\%$$

Then, do the same for the women, and display the two distributions with a side-by-side bar chart:

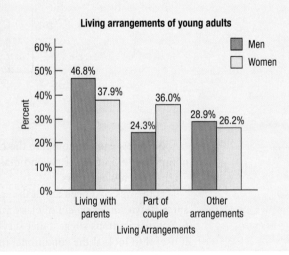

---

[2]This kind of "backward" reasoning shows up surprisingly often in science—and in Statistics. We'll see it again later on when we discuss tests of hypotheses.

It appears that women are less likely to still be living at home, but more likely to be living as part of a couple (actually the most impressive difference), compared with the men. There is not much of a difference between the sexes in the likelihood of some other living arrangement.

We could also mimic Figure 2.8 and show the conditional distribution of *Living Arrangements* for males, in blue, on one side, and the conditional distribution of *Living Arrangements* for females, in yellow, on the other side. Your software may construct it this way, rather than intermingling the two distributions as above (which does allow for more precise comparisons of percentages).

Just Checking

Below are data from the Ontario Health Supplement survey of 4888 individuals (over age 15 with no history of physical or sexual abuse) about the effect of spanking (or slapping) on future antisocial behaviours during teen or adult years.[3] Externalizing of problems—including illicit drugs and antisocial behaviour—were assessed by an interviewer-administered questionnaire.

	Never spanked	Rarely spanked	Sometimes/ often spanked	Total
Externalizing problems	75	309	240	624
No externalizing problems	921	2146	1197	4264
Total	996	2455	1437	4888

1. What percent of those rarely spanked had externalizing problems?
2. What percent of those with externalizing problems were rarely spanked?
3. What percent of those in the study were rarely spanked and had externalizing problems?
4. Determine the (marginal) distribution of *Spanking (Never, Rarely, Sometimes)*
5. Determine the conditional distribution of the variable *Spanking* for those with externalizing problems.
6. Compare the percent of the never spanked who had externalizing problems with the percent of all individuals who had externalizing problems.
7. Does it appear that *Externalizing* and *Spanking* are independent? Explain.

Segmented Bar Charts

We could display the *Titanic* information by dividing up bars rather than circles. The resulting **segmented bar chart** treats each bar as the "whole" and divides it proportionally into segments corresponding to the percentage in each group. We can clearly see (Figure 2.10) that the two conditional distributions of ticket *Class* are different, indicating again that survival was not independent of ticket *Class*.

Figure 2.10

A segmented bar chart for **Class** *by* **Survival.** Notice that although the totals for survivors and non-survivors are quite different, the bars are the same height because we have converted the numbers to percentages. Compare this display with the side-by-side pie charts of the same data in Figure 2.7.

[3]H. L. MacMillan, M. H. Boyle, M. Y. Wong, E. K. Duku, J. E. Fleming, and C. A. Walsh, 1999, "Slapping and spanking in childhood and its association with lifetime prevalence of psychiatric disorders in a general population sample." *CMAJ*, 161(7): 805–809.

Sometimes we want to examine associations among more than two categorical variables. This requires a series of two way tables. But segmented bar charts are very good for their visualization, as in the example below (Figure 2.11), showing associations among sex, age group (20–24, 25–29) and living arrangements:

Figure 2.11

Associations among three variables. We can see, for example, that women are more likely than men to be living as part of a couple at both age categories, and less likely to be living solo, with a big (over 50%) drop in both sexes still at home by their late 20's.

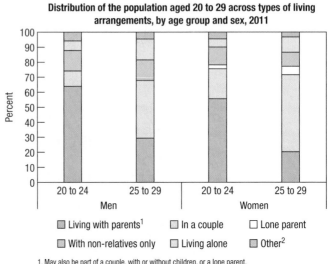

Distribution of the population aged 20 to 29 across types of living arrangements, by age group and sex, 2011

☐ Living with parents[1]   ☐ In a couple   ☐ Lone parent
☐ With non-relatives only   ☐ Living alone   ☐ Other[2]

1. May also be part of a couple, with or without children, or a lone parent.
2. Other living arrangements include living with relatives, as well as children in census families for whom the parent is not the economic family reference person.

Source: Anne Milan (2016). Insights on Canadian Society Diversity of young adults living with their parents, Statistics Canada, June 15, 2016

Step-by-Step Example   EXAMINING CONTINGENCY TABLES

Medical researchers followed 6272 Swedish men for 30 years to see if there was any association between the amount of fish in their diet and prostate cancer. Their results are summarized in this table:

We asked for a picture of a man eating fish. This is what we got.

Pieter De Pauw/Fotolia

		Prostate Cancer	
		No	**Yes**
Fish Consumption	Never/seldom	110	14
	Small part of diet	2420	201
	Moderate part	2769	209
	Large part	507	42

Question: Is there an association between fish consumption and prostate cancer?

THINK ➡ Plan State what the problem is about.

I want to know if there is an association between fish consumption and prostate cancer.

Variables Identify the variables and report the W's.

The individuals are 6272 Swedish men followed by medical researchers for 30 years. The variables record their fish consumption and whether or not they were diagnosed with prostate cancer.

Be sure to check the appropriate condition.

✓ **Categorical Data Condition**: I have counts for both fish consumption and cancer diagnosis. The categories of diet do not overlap, and the diagnoses do not overlap. It's okay to draw pie charts or bar charts.

SHOW ➡ Mechanics Check the marginal distributions first before looking at the two variables together.

| | Prostate Cancer | | |
	No	Yes	Total
Never/Seldom	110	14	**124 (2.0%)**
Small part of diet	2420	201	**2621 (41.8%)**
Moderate part	2769	209	**2978 (47.5%)**
Large part	507	42	**549 (8.8%)**
Total	**5806 (92.6%)**	**466 (7.4%)**	**6272 (100%)**

Fish Consumption (row label)

Two categories of the diet are quite small, with only 2.0% Never/Seldom eating fish and 8.8% in the "Large part" category. Overall, 7.4% of the men in this study had prostate cancer.

Then, make appropriate displays to see whether there is a difference in the relative proportions. These pie charts compare fish consumption for men who have prostate cancer to fish consumption for men who don't.

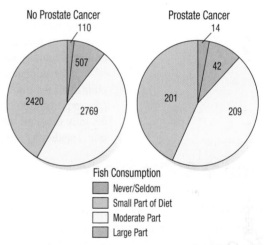

It's hard to see much difference in the pie charts. So, I made a display of the row percentages. Because there are only two alternatives, I chose to display the risk of prostate cancer for each group:

Both pie charts and bar charts can be used to compare conditional distributions. Here we compare prostate cancer rates based on differences in fish consumption.

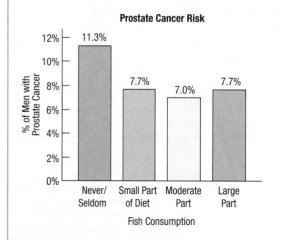

TELL ➡ **Conclusion** Interpret the patterns in the table and displays in context. If you can, discuss possible real-world consequences. Be careful not to overstate what you see. The results may not generalize to other situations.

Overall, there is a 7.4% rate of prostate cancer among men in this study. Most of the men (89.3%) ate fish either as a moderate or small part of their diet. From the pie charts, it's hard to see a difference in diet between the cancer and non-cancer groups. But in the bar chart, it looks like the cancer rate for those who never/seldom ate fish may be somewhat higher.

However, only 124 of the 6272 men in the study fell into this category, and only 14 of them developed prostate cancer. More study would probably be needed before we would recommend that men change their diets.[4]

This study is an example of looking at a sample of data to learn something about a larger population, one of the main goals of this book. We care about more than these particular 6272 Swedish men. We hope that learning about their experiences will tell us something about the value of eating fish in general. That raises several questions. What population do we think this sample might represent? Do we hope to learn about all Swedish men? About all men? How do we know that other factors besides that amount of fish they ate weren't associated with prostate cancer? Perhaps men who eat fish often have other habits that distinguish them from other Swedish men, and maybe those other habits are what actually kept their cancer rates lower.

Observational studies, like this one, often lead to contradictory results because we can't control all the other factors. In fact, a later paper, published in 2011, based on data from a cancer prevention trial on 3400 men from 1994 to 2003, showed that some fatty acids may actually increase the risk of prostate cancer.[5] We'll discuss the pros and cons of observational studies as well as experiments in which we can control the factors in Chapter 11.

Section 2.2 Exercises

5. Movies again The following table summarizes 348 films from 2010 that have been classified into a genre and have a MPAA rating.

	G	PG	PG-13	R	Total
Action/Adventure	3	18	18	14	53
Comedy	2	18	29	41	90
Documentary	4	8	8	9	29
Drama	0	16	36	65	117
Musical	0	2	2	0	4
Suspense/Thriller/Horror	0	1	13	41	55
Total	9	63	106	170	348

Source: www.the-numbers.com/

a) What percentage of these films were rated R?
b) What percentage of these films were R-rated comedies?
c) What percentage of R-rated films were comedies?
d) What percentage of comedies were rated R?

6. Labour force The following table shows the labour force status of individuals 15 years of age and older in Canada in June 2016. Counts are in hundreds of people.

Age	Employed	Unemployed	Not in Labour Force	Total
15 to 24 Years	2618	380	1389	4387
25 to 54 Years	12 016	675	1939	14 630
55 to 64 Years	3031	174	1661	4866
65 Years and Over	745	32	4912	5689
Total	18 410	1261	9901	29 572

Source: http://www5.statcan.gc.ca/cansim

---

[4]"Fatty fish consumption and risk of prostate cancer," *Lancet*, June 2001. The original study actually used pairs of twins, which enabled the researchers to discern that the risk of cancer for those who never ate fish actually was substantially greater. Using pairs is a special way of gathering data. We'll discuss such study design issues and how to analyze the data in later chapters.

[5]"Serum phospholipid fatty acids and prostate cancer risk: Results from the Prostate Cancer Prevention Trial," *American Journal of Epidemiology*, 2011.

a) What percentage of the population were unemployed?

b) What percentage of the population were unemployed and aged 25 to 54 years?

c) What percentage of 15 to 24 year olds were unemployed?

d) What percentage of employed were aged 65 years and over?

WHAT CAN GO WRONG?

- **Don't violate the area principle.** This is probably the most common mistake in a graphical display. It is often made in the cause of artistic presentation. Here, for example, are two displays of the pie chart of the *Titanic* passengers by class:

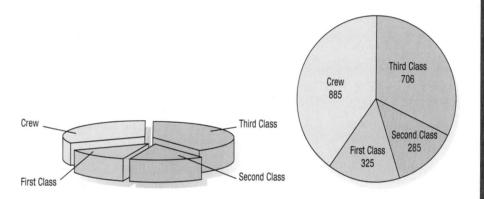

The one on the left looks pretty, doesn't it? But showing the pie on a slant violates the area principle and makes it much more difficult to compare fractions of the whole made up of each class—the principal feature that a pie chart ought to show.

- **Keep it honest.** Here's a pie chart that displays data on the percentage of high school students who engage in specified dangerous behaviours as reported by the Centers for Disease Control. What's wrong with this plot?

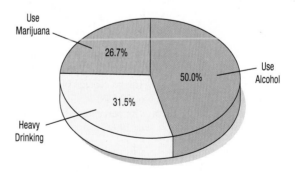

Try adding up the percentages. Or look at the 50% slice. Does it look right? Then think: What are these percentages of? Is there a "whole" that has been sliced up? In a pie chart, the proportions shown by each slice of the pie must add up to 100% and each individual must fall into *only one category*. Of course, showing the pie on a slant makes it even harder to detect the error.

Here's another one. The following chart shows the average number of texts in various time periods by American cell phone customers in the period 2010 to 2014.

(continued)

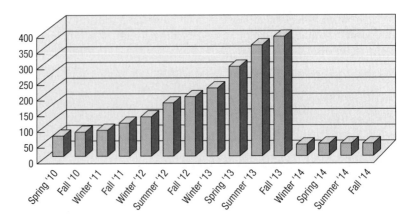

First, we should note that not every plot that uses bars is a bar chart. *Bar charts* display counts of categories. This is in fact a *time plot*, plotting a quantitative variable (average number of texts) versus time. Any symbol could have been plotted against the desired number on the y-axis.

One obvious problem can be seen by examining the x-axis. Some time periods are missing! And it looks as if text messaging decreased suddenly in 2014, which probably doesn't jibe with your experience. The worst mistake is one that can't be seen from the plot. In 2014, the company reporting the data switched from reporting the average number of texts per year (reported each quarter) to the average number of texts per month. So, the numbers in the last four bars should be multiplied by 12 to make them comparable to the rest.

Another problem is the use of three dimensional bars. Using three-dimensional figures to represent one-dimensional numbers is a waste of ink and leads to problems related to perceived depth and volume of the bars.

■ **Don't confuse similar-sounding percentages.** These percentages sound similar but are different:

			Class			
		First	Second	Third	Crew	Total
Survival	**Alive**	203	118	178	212	**711**
	Dead	122	167	528	673	**1490**
	Total	**325**	**285**	**706**	**885**	**2201**

■ The percentage of the passengers who were both in first class and survived: This would be 203/2201, or 9.2%.
■ The percentage of the first-class passengers who survived: This is 203/325, or 62.5%.
■ The percentage of the survivors who were in first class: This is 203/711, or 28.6%.

In each instance, pay attention to the *who* implicitly defined by the phrase. Often there is a restriction to a smaller group (all aboard the *Titanic*, those in first class, and those who survived, respectively) before a percentage is found. Your discussion of results must make these differences clear.

■ **Don't forget to look at the variables separately, too.** When you make a contingency table or display a conditional distribution, be sure you also examine the marginal distributions. It's important to know how many cases are in each category.

■ **Be sure to use enough individuals.** When you consider percentages, take care that they are based on a large enough number of individuals. Take care not to make a report such as this one: *We found that 66.67% of the rats improved their performance with training. The other rat died.*

Entering Centreville

Established	1793
Population	7943
Elevation	710
Average	3482

Don't use inappropriate averages. Sometimes they aren't as obvious as the Centreville sign.

Table 2.9

On-time flights by **Time of Day** *and* **Pilot.** Look at the percentages within each *Time of Day* category. Who has a better on-time record during the day? At night? Who is better overall?

- **Don't overstate your case.** Independence is an important concept, but it is rare for two variables to be *entirely* independent. We can't conclude that one variable has no effect whatsoever on another. Usually all we know is that little effect was observed in our study. Other studies of other groups under other circumstances could find different results.

Simpson's Paradox

- **Be wary of looking only at overall percentages.** It's easy to make up an example showing that looking at *overall* proportions can be misleading. Suppose there are two pilots, Pierre and Usha. Pierre argues that he's the better pilot of the two, since he managed to land 83% of his last 120 flights on time compared with Usha's 78%. But let's look at the data a little more closely. Here are the results for each of their last 120 flights, broken down by the time of day they flew:

		Time of Day		
		Day	**Night**	**Overall**
Pilot	**Pierre**	90 out of 100 90%	10 out of 20 50%	100 out of 120 83%
	Usha	19 out of 20 95%	75 out of 100 75%	94 out of 120 78%

Look at the day- and night-time flights separately. For day flights, Usha had a 95% on-time rate, and Pierre only a 90% rate. At night, Usha was on time 75% of the time, and Pierre only 50%. So Pierre is better "overall," but Usha is better both during the day and at night. How can this be?

What's going on here is a problem known as **Simpson's paradox**, named after the statistician who first described it in the 1960s. It rarely occurs in real life, but there have been several well-publicized cases of it. As we can see from the pilot example, the problem is combining percentages over different groups. Usha has mostly night flights, which are more difficult, so her *overall average* is heavily influenced by her night-time average. Pierre, on the other hand, benefits from flying mostly during the day, with its higher on-time percentage. With their very different patterns of flying conditions, taking an overall percentage is misleading. It's not a fair comparison.

The moral of Simpson's paradox is to be careful when you combine across different levels of a second variable. It's always better to compare percentages *within* each level of that other variable (*controlling* for that variable, one might say). The overall percentage may be misleading. From a practical standpoint, the issue often is whether you are unknowingly averaging or combining over an important unidentified variable buried in your data (lurking). Think about and try to identify variables that have the potential to alter your understanding of a relationship. They may not be as easy to catch as "time of day" or "graduate school" in the previous and next examples.

> **Simpson's Paradox**
>
> One famous example of Simpson's paradox arose during an investigation of admission rates for men and women at the University of California at Berkeley's graduate schools. As reported in an article in *Science*[6], about 45% of male applicants were admitted, but only about 30% of female applicants got in. It looked like a clear case of discrimination. However, when the data were broken down by school (Engineering, Law, Medicine, etc.), it turned out that within each school, the women were admitted at nearly the same or, in some cases, much *higher* rates than the men. How could this be? Women applied in large numbers to schools with very low admission rates (Law and Medicine, for example, admitted fewer than 10%). Men tended to apply to Engineering and Science. Those schools have admission rates above 50%. When the *average* was taken, the women had a much lower *overall* rate, but the average didn't really make sense.

[6]P. J. Bickel, E. A. Hammel, and J. W. O'Connell (1975). "Sex Bias in Graduate Admissions: Data from Berkeley." *Science*, 187(4175): 398–404. doi:10.1126/science.187.4175.398.

What Have We Learned?

Make and interpret a frequency table for a categorical variable.

- We can summarize categorical data by counting the number of cases in each category, sometimes expressing the resulting distribution as percentages.

Make and interpret a bar chart or pie chart.

- We display categorical data using the area principle in either a **bar chart** or a **pie chart**.

Make and interpret a contingency table.

- When we want to see how two categorical variables are related, we put the counts (and/or percentages) in a two-way table called a **contingency table**.
- Focusing on the total, row, or column percentages changes who you are interested in. Restricting the *who* of the study in different ways answers different questions.

Make and interpret bar charts and pie charts of marginal and conditional distributions.

- We may look at the **marginal distribution** of each variable (found in the margins of the table). We may also look at the **conditional distribution** of a variable within each category of the other variable.
- Comparing conditional distributions of one variable across categories of another tells us about the association between variables. If the conditional distributions of one variable are the same for every category of the other, the variables are **independent**.

Review of Terms

The key terms are in chapter order so you can use this list to review the material in the chapter.

Frequency table (Relative frequency table)	A table that lists the categories in a categorical variable and gives the count (or percentage) of observations for each category (p. 19).
Distribution	The distribution of a variable gives - the possible values of the variable, and - the relative frequency of each value (p. 19).
Area principle	In a statistical display, each data value should be represented by the same amount of area (p. 19).
Bar chart (Relative frequency bar chart)	A chart that shows bars whose heights represent the count (or percentage) of observations for each category of a categorical variable (p. 20).
Pie chart	A chart that shows how a "whole" divides into non-overlapping categories by showing a wedge of a circle whose area corresponds to the proportion in each category (p. 20).
Categorical data condition	The methods in this chapter are appropriate for displaying and describing categorical data. Be careful not to use them with quantitative data (p. 21).
Contingency table	A table that displays counts and, sometimes, percentages of individuals falling into named categories on two or more variables. The table categorizes the individuals on all variables at once to reveal possible patterns in one variable that may be contingent on the category of the other (p. 23).

Marginal distribution	In a contingency table, the distribution of either variable alone. The counts or percentages are the totals found in the margins (last row or column) of the table (p. 23).
Joint distribution	The distribution of both variables jointly in a contingency table, expressed as a percentage of the total count (p. 24).
Conditional distribution	The distribution of a variable restricting the *Who* to consider only a smaller group of individuals, such as those belonging to one category of another variable (p. 25).
Independent	Variables are said to be independent if the conditional distribution of one variable is the same for each category of the other. We'll show how to check for independence in a later chapter (p. 28).
Segmented bar chart	A chart that displays the conditional distribution of a categorical variable within each category of another variable (p. 29).
Simpson's paradox	Relationships among proportions taken within different groups or subsets can appear to contradict relationships among the overall proportions (p. 35).

On the Computer   DISPLAYING CATEGORICAL DATA

Although every package makes a slightly different bar chart, they all have similar features:

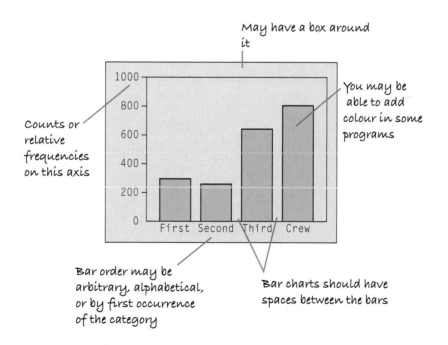

May have a box around it

You may be able to add colour in some programs

Counts or relative frequencies on this axis

Bar order may be arbitrary, alphabetical, or by first occurrence of the category

Bar charts should have spaces between the bars

Sometimes the count or a percentage is printed above or on top of each bar to give some additional information. You may find that your statistics package sorts category names in annoying orders by default. For example, many packages sort categories alphabetically or by the order the categories are seen in the data set. Often, neither of these is the best choice.

EXCEL

- First make a Pivot Table (Excel's name for a frequency table). From the **Data** menu choose **Pivot Table** and **Pivot Chart Report**.
- When you reach the Layout window, drag your variable to the row area and drag your variable again to the data area. This tells Excel to count the occurrences of each category. Once you have an Excel Pivot Table, you can construct bar charts and pie charts.
- Click inside the Pivot Table.
- Click the Pivot Table Chart Wizard button. Excel creates a bar chart.

- A longer path leads to a pie chart; see your Excel documentation.

COMMENTS

Excel uses the Pivot Table to specify the category names and find counts within each category. If you already have that information, you can proceed directly to the Chart Wizard.

JMP

JMP makes a bar chart and frequency table together.
- From the **Analyze** menu, choose **Distribution**.
- In the Distribution dialog, drag the name of the variable into the empty variable window beside the label "Y, Columns"; click **OK**.
  To make a pie chart:
- Choose **Chart** from the **Graph** menu.

- In the Chart dialog, select the variable name from the Columns list.
- Click on the button labelled "Statistics," and select "N" from the drop-down menu.
- Click the "**Categories, X, Levels**" button to assign the same variable name to the x-axis.
- Under Options, click on the **second** button—labelled **Bar Chart**—and select **Pie** from the drop-down menu.

MINITAB

To make a bar chart:
- Choose **Bar Chart** from the **Graph** menu.
- Select "Counts of unique values" in the first menu, and select "Simple" for the type of graph. Click **OK**.
- In the Chart dialog, enter the name of the variable that you wish to display in the box labelled "Categorical variables."
- Click **OK**.

COMMENTS

For summarized, rather than raw data, instead of "Counts of unique values," choose "Values from a table," or "Summarized data" if using MINITAB EXPRESS.

R

To make a bar chart or pie chart in **R**, first create the frequency table for the desired variable:
- **table(X)** will give a frequency table for a single variable X.
- **barplot(table(X))** will give a bar chart for X.
- **pie(table(X))** will give a pie chart.

COMMENTS

Stacked bar charts of two variables, X and Y, can be made using **barplot(xtabs(~X+Y))** or directly from a two-way table of counts or percentages. Legends and other options are available for all charts using various functions.

SPSS

To make a bar chart:

- Open the **Chart Builder** from the **Graphs** menu.
- Click the **Gallery** tab.
- Choose **Bar Chart** from the list of chart types.
- Drag the appropriate bar chart onto the canvas.
- Drag a categorical variable onto the x-axis drop zone.
- Click **OK**.

COMMENT

A similar path makes a pie chart by choosing **Pie chart** from the list of chart types.

STATCRUNCH

To make a bar chart or pie chart:

- Click on **Graphics**.
- Choose the type of plot ≫ **with data** or ≫ **with summary**.
- Choose the variable name from the list of **Columns**; if using summaries, also choose the counts.
- Click on **Next**.

- Choose **Frequency/Counts** or (usually) **Relative frequency/Percents**. Note that you may elect to group categories under a specified percentage as "Other."
- Click on **Create Graph**.

TI-83/84 PLUS

The TI-83 won't do displays for categorical variables.

Chapter Exercises

T indicates data is available in a saved file at the Companion Website at www.pearsoncanada.ca/deveaux

7. **Graphs in the news I** Find a bar chart of categorical data from a newspaper, a magazine, or the Internet.
   a) Is the graph clearly labelled?
   b) Does it violate the area principle?
   c) Does the accompanying article tell the W's of the variable?
   d) Do you think the article correctly interprets the data? Explain.

8. **Graphs in the news II** Find a pie chart of categorical data from a newspaper, a magazine, or the Internet.
   a) Is the graph clearly labelled?
   b) Does it violate the area principle?
   c) Does the accompanying article tell the W's of the variable?
   d) Do you think the article correctly interprets the data? Explain.

9. **Tables in the news I** Find a frequency table of categorical data from a newspaper, a magazine, or the Internet.
   a) Is it clearly labelled?
   b) Does it display percentages or counts?

   c) Does the accompanying article tell the W's of the variable?
   d) Do you think the article correctly interprets the data? Explain.

10. **Tables in the news II** Find a contingency table of categorical data from a newspaper, a magazine, or the Internet.
    a) Is it clearly labelled?
    b) Does it display percentages or counts?
    c) Does the accompanying article tell the W's of the variable?
    d) Do you think the article correctly interprets the data? Explain.

11. **Forest fires 2014** Every year, thousands of forest fires burn throughout Canada. According to the National Forestry Database[7], in 2014, a fewer-than-average 5152 forest fires were started, affecting 4 563 327 hectares of land. Of these, 2375 fires were caused by lightning,

---

[7]*Source*: National Forestry Database http://nfdp.ccfm.org

affecting 4 522 257 hectares of land; 2648 were caused by human activities, affecting 40 613 hectares of land, while the remainder were of unknown origin. Determine the relative frequency distribution of forest fire causes for 2014. Describe your findings (in a sentence or two).

12. **Forest fires 2014 by region** Of those forest fires discussed in Exercise 11, 292 were in Quebec, 303 in Ontario, 1470 in Alberta, and 1465 in British Columbia. Determine the relative frequency distribution of forest fire locations within Canada, and describe it in a sentence or two.

13. **Teen smokers** The organization Monitoring the Future (www.monitoringthefuture.org) asked 2048 U.S. Grade 8 students who said they smoked cigarettes what brands they preferred. The table below shows brand preferences for two regions of the country. Write a few sentences describing the similarities and differences in brand preferences among Grade 8 students in the two regions listed.

Brand Preference	South	West
Marlboro	58.4%	58.0%
Newport	22.5%	10.1%
Camel	3.3%	9.5%
Other (over 20 brands)	9.1%	9.5%
No Usual Brand	6.7%	12.9%

14. **Bad countries** According to the January 2008 Canada's World Poll (conducted by Environics, principal sponsor The Simon Foundation[8]), when asked to name countries that stand out as a negative force in the world today, approximately half (52%) of Canadians pointed to their nearest neighbour, the U.S. Other countries making the list were Iran (21%), Iraq (19%), China (13%), Afghanistan (11%), Pakistan (9%), North Korea (8%), and Russia (6%). Another 4% identified Israel or the Palestinian territories. Four percent named Canada. (To see who the good guys are, you'll have to read the report.)
    a) Display this information in an appropriate display. Explain clearly which displays are appropriate and which are not and why.
    b) Some could not name any country as a negative force. Can you determine what percent this was from the given information? Explain. Could it be possibly be over 30%?
    c) Is it true that 49% (21 + 19 + 9) named Iran or Iraq or Pakistan as negative forces? Explain.

15. **Oil spills 2013** Data from the International Tanker Owners Pollution Federation Limited give the cause of spillage for 459 oil tanker accidents resulting in spills of more than 700 tons of oil from 1970 to 2013 (www.itopf.com/information-services/data-and-statistics/statistics).

Causes of Oil Spillage

Causes of Spillage

Cause
- Collisions
- Groundings
- Hull or equipment failure
- Fire & Explosions
- Other/Unknown

a) Write a brief report interpreting what the displays show.
b) Is a pie chart an appropriate display for these data? Why or why not?
c) Were there more spills due to Collisions or to Groundings? Is it easier to tell that from the bar chart or the pie chart?

16. **Winter Olympics 2014** Twenty-six countries won medals in the 2014 Winter Olympics in Sochi. The table lists them along with the total number of medals each won.

Country	Medals	Country	Medals
Russia	33	Japan	8
United States	28	Slovenia	8
Norway	26	South Korea	8
Canada	25	Belarus	6
Netherlands	24	Poland	6
Germany	19	Finland	5
Austria	17	Great Britain	4
France	15	Latvia	4
Sweden	15	Australia	3
Switzerland	11	Ukraine	2
China	9	Croatia	1
Czech Republic	8	Kazakhstan	1
Italy	8	Slovakia	1

Source: https://en.wikipedia.org/wiki/2014_Winter_Olympics_medal_table

---

[8]Canada's World Poll's principal sponsor is the Simmons Foundation, and it is also sponsored by the Environics Institute, the Canadian Broadcasting Corporation (CBC), *The Globe and Mail*, and *Le Devoir*.

a) Try to make a display of these data. What problems did you encounter?

b) Can you find a way to organize the data so that the graph is more successful?

17. **Global warming** The Pew Research Center for the People and the Press (http://people-press.org) asked a representative sample of U.S. adults about global warming, repeating the question over time. In October 2010, the responses reflected an increased belief that global warming is real and due to human activity. Here's a display of the percentages of respondents choosing each of the major alternatives offered:

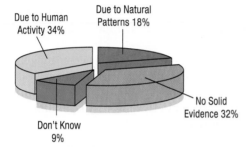

List the errors in this display.

18. **Modalities** A survey of athletic trainers[9] asked what modalities (treatment methods, such as ice, whirlpool, ultrasound, or exercise) they commonly use to treat injuries. Respondents were each asked to list three modalities. The article included the figure shown here.

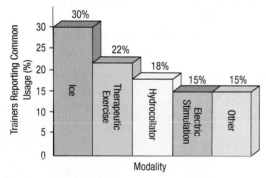

a) What problems do you see with the graph?

b) Consider the percentages for the named modalities. Do you see anything odd about them?

19. **Complications** The survey of athletic trainers reported in Exercise 18 summarizes the complications most commonly associated with cryotherapy (ice). The trainers were asked to report the most common side effects. Of those identifying cryotherapy, 86 respondents reported allergic reactions, 23 reported burns, 16 reported pain intolerance, and 6 reported frostbite.

a) Make an appropriate display of these data.

[9]*Source*: Scott F. Nadler, Michael Prybicien, Gerard A. Malanga, and Dan Sicher, 2003, "Complications from therapeutic modalities: results of a national survey of athletic trainers," *Archives of Physical Medical Rehabilitation*, 84.

b) Specify the *Who* for these data. Would the data provide the most useful information about the risks of cryotherapy?

T 20. **Indigenous languages 2011** Many Indigenous languages in Canada are endangered, and extinction for some may be only a generation away. In the 2011 Census, a vast majority of Canadians of Indigenous identity reported English as their mother tongue. Following are Census counts for Indigenous people who reported an Indigenous language as their mother tongue (excluding multiple responses, e.g., English and Indigenous language).

a) Find the relative frequency distribution of Indigenous mother tongue.

b) Portray it with two different types of graphs. Point out the most interesting or salient characteristic(s) of the distribution. Which type of graph do you prefer? Explain.

c) The first seven in the list are all *Algonquian* languages. Repeat part b) after combining the first seven languages into a category named "Algonquian languages." Included in the "Other" category are 2670 people with other Algonquian languages as their mother tongue, so adjust for this, too.

Indigenous mother tongue	Total = 200 700
Cree	78 200
Ojibway	17 625
Oji-Cree	9 835
Innu/Montagnais	10 785
Mi'kmaq	7 635
Atikamekw	5 820
Blackfoot	2 860
Inuktitut	33 500
Dene	11 215
Dakota/Stoney/Siouan	4 200
Other Indigenous languages	19 025

Source: Based on data from Statistics Canada, 2011 Census of Population, Statistics Canada catalogue no. 98-314-XCB2011016 (Canada, Code01

21. **Spatial distribution** We divided a tract of land into four equal quadrants, in the east–west direction. We randomly selected 100 longleaf pine trees from the tract of land and counted the number of trees in each quadrant, yielding the counts below:

Quadrant 1	Quadrant 2	Quadrant 3	Quadrant 4
18	21	22	39

a) Is it possible that the longleaf pine trees are randomly dispersed over the region, which would result in a uniform or even distribution over quadrants? Find the relative frequency distribution of quadrant location. Do you think the proportions are similar enough to support this theory? (We will learn how to formally test such theories or *hypotheses* in later chapters.)

b) Counts for these same quadrants were also obtained for a random sample of 50 trees of a different species selected from this tract of land:

Quadrant 1	Quadrant 2	Quadrant 3	Quadrant 4
6	12	14	18

Find the distribution of quadrant location for this species, and compare it with the distribution in part a). Are they fairly similar? Do you think the differences are within the bounds of chance or random variation in the samples? (We'll learn how to answer this question more precisely when we discuss tests of hypotheses.)

22. **Politics** Students in an Intro Stats course were asked to describe their politics as "Liberal," "Moderate," or "Conservative." Here are the results:

		Politics			
		L	**M**	**C**	**Total**
Sex	Female	35	36	6	77
	Male	50	44	21	115
	Total	**85**	**80**	**27**	**192**

a) What percent of the class is male?

b) What percent of the class considers themselves to be "Conservative"?

c) What percent of the males in the class consider themselves to be "Conservative"?

d) What percent of all students in the class are males who consider themselves to be "Conservative"?

23. **Politics revisited** Look again at the table of political views for the Intro Stats students in Exercise 22.

a) Find the conditional distributions (percentages) of political views for the females.

b) Find the conditional distributions (percentages) of political views for the males.

c) Make a graphical display that compares the two distributions.

d) Do the variables *Politics* and *Sex* appear to be independent? Explain.

24. **More politics** Look once more at the table summarizing the political views of Intro Stats students in Exercise 22.

a) Produce a graphical display comparing the conditional distributions of males and females among the three categories of politics.

b) Comment briefly on what you see from the display in part a).

T 25. **Canadian languages 2011** Statistics Canada provides, on its website, the following data on the Canadian population.

a) What percent of Canadian citizens speak only English?

b) What percent of Canadian citizens speak French?

c) What percent of Quebec residents speak French?

d) What percent of French-speaking Canadians live in Quebec?

e) Do you think that language knowledge and province of residence are independent? Explain.

	Knowledge of official languages[1]				
Geographic name	**English only**	**French only**	**Both**	**Neither**	**Total**
Canada	22 564 665	4 165 015	5 795 575	595 920	33 121 175
Newfoundland and Labrador	485 740	135	23 450	625	509 950
Prince Edward Island	120 590	130	17 005	715	138 435
Nova Scotia	814 670	875	93 435	1635	910 615
New Brunswick	426 675	66 380	245 885	955	739 900
Quebec	363 860	4 047 175	3 328 725	76 195	7 815 955
Ontario	10 984 360	42 980	1 395 805	298 920	12 722 065
Manitoba	1 074 335	1 490	103 145	14 135	1 193 095
Saskatchewan	965 925	430	46 570	5 395	1 018 310
Alberta	3 321 815	3 205	235 565	49 600	3 610 185
British Columbia	3 912 955	2 050	296 645	144 560	4 356 205
Yukon	29 050	90	4 420	95	33 655
Northwest Territories	37 045	45	3 720	235	41 040
Nunavut	27 665	35	1 200	2 860	31 765

Province (row label for the above table)

[1]Refers to the ability of the individual to conduct a conversation in English only, in French only, in both English and French, or in neither English nor French at the time of the census (May 10, 2011).

Source: Adapted from Statistics Canada, Language Highlight Tables, 2011 Census (Catalogue no.: 98-314-XWE2011002

26. Tattoos A study by the University of Texas Southwestern Medical Center examined 626 people to see if there was an increased risk of contracting hepatitis C associated with having a tattoo. If the subject had a tattoo, researchers asked whether it had been done in a commercial tattoo parlour or elsewhere. Write a brief description of the association between tattooing and hepatitis C, including an appropriate graphical display.

	Tattoo done in commercial parlour	Tattoo done elsewhere	No tattoo
Has hepatitis C	17	8	18
No hepatitis C	35	53	495

T 27. Weather forecasts Just how accurate are the weather forecasts we hear every day? The following table compares the daily forecast with a city's actual weather for a year.

		Actual Weather	
		Rain	No rain
Forecast	Rain	27	63
	No rain	7	268

a) On what percent of days did it actually rain?
b) On what percent of days was rain predicted?
c) What percent of the time was the forecast correct?
d) Do you see evidence of an association between the type of weather and the ability of forecasters to make an accurate prediction? Write a brief explanation, including an appropriate graph.

28. Attraction, repulsion In a certain geographical region, 200 one-metre-square quadrants were randomly selected. In each quadrant, it was noted whether there were any specimens of grass species *Agropyron* and whether there were any specimens of grass species *Agrostis*. We are often interested in whether species tend to attract or repel each other. The following summary counts were recorded:

	Agrostis present	Agrostis absent
Agropyron present	10	70
Agropyron absent	40	80

a) How often is *Agrostis* present in plots where *Agropyron* is present? On the other hand, how often is *Agrostis* present when *Agropyron* is absent?
b) Find the conditional distribution of *Agropyron* presence/absence when *Agrostis* is present, and repeat for

when *Agrostis* is absent. Plot these two distributions and compare. Describe the nature of the association between these two species. Do they attract or repel? (In later chapters, we will learn how to do a precise test of the association hypothesis.)

T 29. Drivers' licenses 2014 The following table shows the number of licensed Canadian drivers by age and by sex (https://www.tc.gc.ca/media/documents/roadsafety/cmvtcs2014_eng.pdf).

Age	Male Drivers	Female Drivers	Total
Under 16	29 782	25 145	54 927
16-19	601 391	553 183	1 154 574
20-24	1 015 355	940 890	1 956 245
25-34	2 173 377	2 096 856	4 270 233
35-44	2 180 884	2 123 336	4 304 219
45-54	2 456 317	2 342 910	4 799 227
55-64	2 193 457	2 068 965	4 262 421
65 and over	2 182 217	1 929 988	4 112 205
Total	12 832 778	12 081 272	24 914 051

a) What percent of total drivers are under 20?
b) What percent of total drivers are male?
c) Write a few sentences comparing the number of male and female licensed drivers in each age group.
d) Do a driver's age and sex appear to be independent? Explain.

30. Fat and Fatter Following are estimated population counts (in thousands), using data from the National Population Health Survey, on changes from one body mass index (BMI) category to another between 1995 and 2011 for the Canadian household population aged 18 to 56 (who reported their height and weight in the survey). BMI is defined as body weight in kilograms divided by height in metres squared.

	2011			
1995	Underweight	Normal	Overweight	Obese
Underweight	57	240	?	?
Normal	90	3909	2776	441
Overweight	?	521	2676	1670
Obese	?	?	243	1491

Source: Adapted from the Statistics Canada CANSIM database http://cansim2.statcan.gc.ca table104-7030. July 15, 2016

"The "?" indicates a value too small to be estimated reliably. You may plug in zero for the "?," if needed, to answer any question below."

a) What percent of normal weight Canadians in 1995 became overweight or obese by 2011?

b) What percent of normal weight Canadians in 2011 were overweight or obese in 1995?

c) What percent of overweight or obese Canadians in 1995 got their weight down to normal by 2011?

31. Anorexia Hearing anecdotal reports that some patients undergoing treatment for the eating disorder anorexia seemed to be responding positively to the antidepressant Prozac, medical researchers conducted an experiment to investigate. They found 93 women being treated for anorexia who volunteered to participate. For one year, 49 randomly selected patients were treated with Prozac and the other 44 were given an inert substance called a placebo. At the end of the year, patients were diagnosed as healthy or relapsed, as summarized in the table:

	Prozac	Placebo	Total
Healthy	35	32	67
Relapse	14	12	26
Total	49	44	93

Do these results provide evidence that Prozac might be helpful in treating anorexia? Explain.

T 32. Neighbourhood density How do people distribute themselves in our cities? From the 2001 Canadian census, we know what types of neighbourhoods people live in, for example, whether the neighbourhoods are high, medium, or low density. Low density means that 66.6% or more of the housing stock is composed of single family dwellings, semi-detached dwellings, or mobile homes, while less than 33.3% means high density, and in between is medium-density. Following is a table with neighbourhood density data for some larger cities.

a) Are these percentages column percentages, row percentages, or table percentages?

	Density		
	High	**Medium**	**Low**
Quebec City	30%	24%	46%
Montreal	47%	19%	34%
Ottawa	22%	37%	40%
Toronto	23%	31%	47%
Winnipeg	10%	26%	64%
Calgary	6%	27%	67%
Edmonton	12%	30%	58%
Vancouver	25%	38%	37%

Source: Adapted from Statistics Canada (2008). Canadian Social Trends, 11-008-XIE2008001, vol. 85, June 2008.

b) Use segmented bar charts to show and compare the different percentages of city dwellers' high-, medium-, and low-density neighbourhoods for each of the cities (i.e., the conditional distribution of density, by city). Order the cities according to percentage of high density. Do you prefer this plot or some other type?

c) Can you construct a table showing what percent of all the high-density neighbourhood dwellers live in each city, what percent of all the medium-density dwellers live in each city, and what percent of all the low-density dwellers live in each city? If not, what additional information would you need? Would row total counts suffice? Would the total sample size and marginal distribution of city suffice?

T 33. Smoking gene? In a study of Icelandic smokers, the association of smoking and a genetic variant is shown below.

a) Find the marginal distribution of genotype for these individuals.

b) Find the conditional distribution of genotype for each category of smoker.

Parameter	Genotype of rs1051730			Total n (frequency)	Frequency of T allele
	GG	**GT**	**TT**		
Cigarettes per day (SQ level)					
1 to 10 (0)	1 743	1 558	326	3 627 (0.260)	0.305
11 to 20 (1)	2 727	2 865	810	6 402 (0.459)	0.350
21 to 30 (2)	1 145	1 416	427	2 988 (0.214)	0.380
31 and more (3)	341	448	139	928 (0.067)	0.391
All levels (frequency)	5956 (0.427)	6 287 (0.451)	1 702 (0.122)	13 945 (1.000)	0.347

GENOTYPE STATUS AND SQ LEVEL OF 13 945 ICELANDIC SMOKERS

Source: Stephen J. Chanock and David J. Hunter. "Genomics: When the smoke clears." *Nature*, 452, 537–538 (3 April 2008). Reprinted by permission from Macmillan Publishers Ltd. www.nature.com/nature

c) Plot the conditional distributions and compare. Does the allele T seem to be related to smoking? If so, do you think that this proves that presence of T increases susceptibility to nicotine addiction?

34. Pet ownership The U.S. Census Bureau reports on the number of households owning various types of pets. Specifically, they keep track of dogs, cats, birds, and horses.

	INCOME DISTRIBUTION OF HOUSEHOLDS OWNING PETS (PERCENT)			
	Pet			
Income	Dog	Cat	Bird	Horse
Under $12 500	14	15	16	9
$12 500 to $24 999	20	20	21	21
$25 000 to $39 999	24	23	24	25
$40 000 to $59 999	22	22	21	22
$60 000 and over	20	20	18	23
Total	100	100	100	100

a) Do you think the income distributions of the households who own these different animals should be roughly the same? What is your guess? Explain.
b) The table shows column percentages. Are these the percentages of income levels for each type of animal owned or the percentages of type of pet owned for each income class?
c) Do the data support your initial guess? Explain.

35. Antidepressants and bone fractures. For a period of five years, physicians at McGill University Health Centre followed more than 5000 adults over the age of 50. The researchers were investigating whether people taking a certain class of antidepressants (SSRIs = Selective Serotonin Reuptake Inhibitors) might be at greater risk of bone fractures. Their observations are summarized in the following table.

	Taking SSRI	No SSRI	Total
Experienced fractures	14	244	258
No fractures	123	4627	4750
Total	137	4871	5008

Do these results suggest there's an association between taking SSRI antidepressants and experiencing bone fractures? Explain.

T 36. Blood proteins Blood samples were obtained from a random sample of 100 individuals, and blood type (A or B) along with presence or absence of a certain blood protein were determined for each individual. The following counts were obtained:

Type A, protein	Type A, no protein	Type B, protein	Type B, no protein
5	35	20	40

Examine the association between the presence of the protein and blood type by reorganizing the data into an appropriate two-way table. Give relevant conditional distributions and describe any observed association.

37. Cell phones A researcher wanted to determine the possible association between using a cell phone and car accidents. He surveyed 180 drivers and found that 78 regularly used a cell phone while driving, of which 20 had had crashes in the past year, whereas 102 drivers did not use cell phones while driving, of which 10 had had accidents in the past year.
a) Create a two-by-two table of crash versus non-crash, cell phone versus non-cell phone for these data. Find the relevant conditional distributions (condition on cell phone use), and comment on any observed association.
b) On the basis of this study, would you conclude that the use of a cell phone increases the risk of a car accident? Explain your answer.

T 38. Twins In 2000, the *Journal of the American Medical Association (JAMA)* published a study that examined pregnancies that resulted in the birth of twins.[10] Births were classified as preterm with intervention (induced labour or caesarean), preterm without procedures, or term/post-term. Researchers also classified the pregnancies by the level of prenatal medical care the mother received (inadequate, adequate, or intensive). The data, from the years 1995 to 1997, are summarized in the following table. Figures are in thousands of births.

		Twin Births 1995–1997 (in thousands)			
		Preterm (induced or caesarean)	Preterm (without procedures)	Term or post-term	Total
Level of Prenatal Care	Intensive	18	15	28	61
	Adequate	46	43	65	154
	Inadequate	12	13	38	63
	Total	76	71	131	278

a) What percent of these mothers did not receive adequate medical care during their pregnancies?
b) What percent of all twin births were preterm?

---

[10]Adapted from Dr. Michael D. Kogan et al (2000). Trends in Twin Birth Outcomes and Prenatal Care Utilization in the United States, 1981–1997. *JAMA* 284 [2000]: 335–341.

c) Among the mothers who did not receive adequate medical care, what percent of the twin births were preterm?

d) Create an appropriate graph comparing the outcomes of these pregnancies by the level of medical care the mother received.

e) Write a few sentences describing the association between these two variables.

T 39. Blood pressure A company held a blood pressure screening clinic for its employees. The results are summarized in the table below by age group and blood pressure level.

		Age		
		Under 30	30–49	Over 50
Blood Pressure	Low	27	37	31
	Normal	48	91	93
	High	23	51	73

a) Find the marginal distribution of blood pressure level.

b) Find the conditional distribution of blood pressure level within each age group.

c) Compare these distributions with a segmented bar graph.

d) Write a brief description of the association between age and blood pressure among these employees.

e) Do you think that this proves that people's blood pressure increases as they age? Explain.

T 40. Self-reported BMI Self-reported measures are often unreliable. For example, people tend to understate their weight and overstate their height, resulting in understatement of one's body mass index or BMI (weight divided by height squared). Analyses based on understated BMIs may overstate health risks of overweight status. The following table shows estimated counts in the entire Canadian population (based on a sample) where the true height, weight, and BMI were accurately measured and compared with self-reported measures for each individual. The counts are in thousands, but just ignore the three extra zeroes in any calculations. (Due to rounding to nearest 1000, column entries may not add perfectly to totals.)

Self-reported BMI category (kg/m²)	Measured BMI category (kg/m²)									
	Underweight (less than 18.5)		Normal weight (18.5 to 24.9)		Overweight (25.0 to 29.9)		Obese class I (30.0 to 34.9)		Obese class II/III (35 or more)	
	'000	%	'000	%	'000	%	'000	%	'000	%
Total	402	100	10 859	100	8 746	100	4 288	100	1 562	100
Underweight (less than 18.5)	271	67	308	3	1	0	6	0	0	0
Normal weight (18.5 to 24.9)	131	33	10 163	94	2 651	30	120	3	4	0
Overweight (25.0 to 29.9)	0	0	388	4	5 851	67	1 894	44	134	9
Obese class I (30.0 to 34.9)	0	0	0	0	244	3	2 247	52	603	39
Obese class II/III (35.0 or more)	0	0	0	0	0	0	22	1	822	53

Source: Adapted from Statistics Canada, Estimates of obesity based on self-report versus direct measures, *Health Reports*, 82-003XIE2008002, vol 19 no. 2, May, 2008. www.statcan.gc.ca/bsolc/olc-cel/olc-cel?catno=82-003-XWE&lang=eng.

a) Give the conditional distribution of self-reported BMI for each of the three measured overweight classes.

b) Give the conditional distribution of measured BMI for each of the three self-reported overweight classes.

c) Why do you know in advance that the two categorical variables in this study will not be independent? How is this clear by visual inspection of the table?

d) What have you learned about self-reported measures of overweight?

e) Name another (categorical) variable that it might be wise to incorporate into such an analysis, for

example, breaking up the data according to values of such a variable prior to analysis.

f) This analysis shows a general underestimation of overweight. How does this affect our estimates of the health risk (for diabetes, etc.) to Canadians of being overweight when self-reported data are used? Will risk of overweight to health be over- or underestimated?

T 41. Indigenous identity 2011 Following are 2011 estimated counts of the Indigenous identity population for Canada, provinces, and territories.

Geographic name	Total population ▼▲	Indigenous population[11] ▼▲	North American Indian ▼▲	Métis ▼▲	Inuit ▼▲	Non-Indigenous population ▼▲
Canada	**32 852 320**	**1 400 685**	**851 560**	**451 795**	**59 440**	**31 451 640**
Newfoundland and Labrador	507 270	35 800	19 315	7 660	6 265	471 470
Prince Edward Island	137 375	2 230	1 520	410	55	135 145
Nova Scotia	906 175	33 845	21 895	10 050	695	872 325
New Brunswick	735 835	22 620	16 120	4 850	485	713 215
Quebec	7 732 520	141 915	82 420	40 960	12 570	7 590 610
Ontario	12 651 790	301 430	201 105	86 015	3 360	12 350 365
Manitoba	1 174 345	195 895	114 225	78 835	575	978 450
Saskatchewan	1 008 760	157 740	103 210	52 450	290	851 020
Alberta	3 567 980	220 700	116 670	96 870	1 985	3 347 280
British Columbia	4 324 455	232 285	155 020	69 475	1 570	4 092 170
Yukon Territory	33 320	7 705	6 590	840	175	25 610
Northwest Territories	40 795	21 160	13 350	3 245	4 340	19 640
Nunavut	31 700	27 360	130	130	27 070	4 335

Source: Statistics Canada, 2011 National Household Survey, Statistics Canada Catalogue no. 99-011-X2011027

a) Aside from the placement of marginal totals at the top and left rather than bottom and right, the table is not exactly the standard contingency table. Why not?

b) If you re-labelled and changed the numbers in the second column of numbers, you would produce a standard contingency table. How would you recalculate and rename this column? (But don't bother to do it.)

c) What percent of Canadians are Inuit from Nunavut?

d) What percent of Indigenous Canadians are Inuit from Nunavut?

e) What percent of Indigenous Canadians are Inuit?

f) What percent of Indigenous Canadians are from Nunavut?

g) What percent of the people of Nunavut are Inuit?

h) What percent of Nunavut Indigenous Canadians are Inuit?

i) What percent of Inuits live in Nunavut?

j) What percent of Ontario Indigenous Canadians could not be simply classified as Inuit, Métis, or N.A. Indian?

k) It is no surprise that the distribution of Indigenous identity group (Inuit, etc.) differs according to region in Canada. Find the percentage of total provincial population for each Indigenous identity group (Inuit, Métis, N.A. Indian) for Newfoundland, Ontario, Saskatchewan, and Alberta; graph suitably, and compare and comment. (Note that since we are omitting the "others" in each province, these are not proper conditional distributions. But if we include "others," the graph will be harder to read, since this will be such a large and dominating percentage.)

T **42. Dim sum 2011** When in China, one of the authors would sometimes mention a favourite Chinese treat, called dim sum—and some types of dim sum like *har gow*, *cheong fun*, and *siu mei*—to people in various parts of the country. But they didn't understand! These expressions are in the Cantonese dialect and are not usually understood by Mandarin speakers. Of the many languages spoken in Canada, Chinese has grown rapidly, becoming the third most common mother tongue after English and French due to large numbers of immigrants. Following are "language spoken most often at home" Census counts for Mandarin and Cantonese (the two major dialects, but not the only dialects) in several cities:

Calgary:		
	Cantonese	Mandarin
2011	16 920	9 900
2006	12 785	5 345
Toronto:		
	Cantonese	Mandarin
2011	156 425	91 670
2006	129 925	44 990
Vancouver:		
	Cantonese	Mandarin
2011	113 610	83 825
2006	94 760	51 465

Source: Based on data from Statistics Canada, Census of Population, 2011 & 2006 Census of Population, Statistics Canada catalogue no. 97-555-XCB2006039.

[11]Includes the Indigenous groups (North American Indian, Métis, and Inuit), multiple Indigenous responses, and Indigenous responses not included elsewhere.

a) How many categorical variables do we have here? Describe each in a few words.
b) Find the conditional distribution of spoken Chinese dialect for each year in Toronto. Repeat for Calgary and again for Vancouver.
c) Can you estimate the proportion of Chinese-speaking Calgarians whose dialect was Mandarin in 2006?
d) Would you prefer to use pie charts or bar charts to display all the information captured in your part b) calculations? Explain. Create your display and tell what you learned.
e) Collapse (add) the counts across the variable "city," give the resulting two-way table by year and dialect, and find the conditional distribution of dialect for each year. What did you learn? People from Hong Kong generally speak Cantonese, while those from other parts of China more often speak Mandarin. What is likely happening in terms of immigration patterns?
f) Using just 2011 data, find the marginal distribution of dialect and the conditional distribution of dialect for each city. Which conditional distribution differs most from the marginal distribution?

43. **Hospitals** Most patients who undergo surgery make routine recoveries and are discharged as planned. Others suffer excessive bleeding, infection, or other post-surgical complications and have their discharges from the hospital delayed. Suppose your city has a large hospital and a small hospital, each performing major and minor surgeries. You collect data to see how many surgical patients have their discharges delayed by postsurgical complications, and find the results shown in the following table.

		Discharge Delayed	
		Large hospital	Small hospital
Procedure	Major surgery	120 of 800	10 of 50
	Minor surgery	10 of 200	20 of 250

a) Overall, for what percent of patients was discharge delayed?
b) Were the percentages different for major and minor surgery?
c) Overall, what were the discharge delay rates at each hospital?
d) What were the delay rates at each hospital for each kind of surgery?
e) The small hospital claims that it has a lower rate of postsurgical complications. Do you agree?
f) Explain, in your own words, why this confusion occurs.

44. **Delivery service** A company must decide which of two delivery services they will contract with. During a recent trial period they shipped numerous packages with each service and kept track of how often deliveries did not arrive on time. Here are the data:

Delivery Service	Type of Service	Number of Deliveries	Number of Late Packages
Pack Rats	Regular	400	12
	Overnight	100	16
Boxes R Us	Regular	100	2
	Overnight	400	28

a) Compare the two services' overall percentage of late deliveries.
b) Based on the results in part a), the company has decided to hire Pack Rats. Do you agree that they deliver on time more often? Why or why not? Be specific.
c) The results here are an instance of what phenomenon?

45. **Graduate admissions** A 1975 article in the magazine *Science* examined the graduate admissions process at Berkeley for evidence of sex discrimination. The following table shows the number of applicants accepted to each of four graduate programs.

		Males accepted (of applicants)	Females accepted (of applicants)
Program	1	511 of 825	89 of 108
	2	352 of 560	17 of 25
	3	137 of 407	132 of 375
	4	22 of 373	24 of 341
	Total	1022 of 2165	262 of 849

a) What percent of total applicants were admitted?
b) Overall, were a higher percentage of males or females admitted?
c) Compare the percentage of males and females admitted in each program.
d) Which of the comparisons you made do you consider to be the most valid? Why?

46. **Be a Simpson!** Can you design a Simpson's paradox? Two companies are vying for a city's "Best Local Employer" award, to be given to the company most committed to hiring local residents. While both employers hired 300 new people in the past year, Company A brags that it deserves the award because

70% of its new jobs went to local residents, compared to only 60% for Company B. Company B concedes that those percentages are correct, but points out that most of its new jobs were full-time, while most of Company A's were part-time. Not only that, says Company B, a higher percentage of its full-time jobs went to local residents than did Company A's, and the same was true for part-time jobs. Thus, Company B argues, it's a better local employer than Company A. Show how it's possible for Company B to fill a higher percentage of both full-time and part-time jobs with local residents, even though Company A hired more local residents overall. You can make up some numbers for illustrative purposes.

 ## Just Checking ANSWERS

1. 12.6%
2. 49.5%
3. 6.3%
4. 20.4% Never, 50.2% Rarely, 29.4% Sometimes
5. 12.0% Never, 49.5% Rarely, 38.5% Sometimes
6. 7.5% of the never spanked had externalizing problems compared to 12.8% of all subjects.
7. Since those never spanked appear less likely to have problems, they may not be independent.

MathXL

MyLab Statistics

Go to MathXL at **www.mathxl.com** or MyLab™ Statistics at **www.mystatlab.com**. You can practise exercises for this chapter as often as you want. The guided solutions will help you find answers step by step. You'll find a personalized study plan available to you too!

Displaying and Summarizing Quantitative Data

Connections

In the previous chapter, we examined descriptive methods for categorical variables. Quantitative variables—such as weight loss, blood pressure, survival time, Arctic sea ice volume, Raptors game attendance—can be much more interesting and also challenging to display and summarize adequately. Your first picture might well be a histogram. We've all looked at histograms, but what should we look *for*? We'll describe histograms, and we'll do more—reporting numerical summaries of the centre as well as the spread. The latter are less commonly reported, but in Statistics, they can be even more important.

Numerical quantities focus on expected values, graphical summaries on unexpected values.

—John Tukey

Toru Hanai/Reuters

On March 11, 2011, the most powerful earthquake ever recorded in Japan created a wall of water that devastated the northeast coast of Japan and left 20 000 people dead or missing. Tsunamis like this are most often caused by earthquakes beneath the sea that shift the earth's crust, displacing a large mass of water. The 2011 tsunami in Japan was caused by a 9.0 magnitude earthquake. It was particularly noted for the damage it caused to the Fukushima Daiichi nuclear power plant, bringing it perilously close to a complete meltdown and causing international concern.

As disastrous as it was, the Japan tsunami was not nearly as deadly as the tsunami of December 26, 2004, that occurred off the west coast of Sumatra which killed an estimated 297 248 people, making it the most lethal tsunami on record. The earthquake that caused it was magnitude 9.1—more than 25% more powerful than the Japanese earthquake.[1] Were these earthquakes truly extraordinary, or did they just happen at unlucky times and places? The U.S. National Geophysical Data Centre (NGDC)[2] has data on more than 5000 earthquakes dating back to 2150 B.C.E., and we have estimates of the magnitude of the underlying earthquake for the 1168 earthquakes that were known to cause tsunamis. What can we learn from these data?

[1] Earthquake magnitudes are measured on a logarithmic scale.
[2] www.ngdc.noaa.gov.

3.1 Displays for Quantitative Variables

Let's start with a picture. For categorical variables, it is easy to draw the distribution because it's natural to count the cases in each category. But for quantitative variables, there are no categories. Instead, we usually slice up all the possible values into bins and then count the number of cases that fall into each bin. The bins, together with these counts, give the **distribution** of the quantitative variable and provide the building blocks for one common type of display of the distribution, called a **histogram**. By representing the counts as bars and plotting them against the bin values, the histogram displays the distribution at a glance.

Histograms

Figure 3.1 is a histogram of the *Magnitudes* (on the Richter scale) of the 1168 earthquakes in the NGDC data:

Figure 3.1

A histogram shows the distribution of magnitudes (in Richter scale units) of tsunami-generating earthquakes.
(Data in **Tsunami Earthquakes_2013**)

Who	1168 earthquakes known to have caused tsunamis for which we have data or good estimates
What	Magnitude (Richter scale[3]), date, and location
When	From 2000 B.C.E. to the present
Where	All over the earth

The height of each bar of the histogram shows the number of cases that fall in its bin. In this histogram of earthquake magnitudes, each bin has a width of 0.5 units, so, for example, the height of the tallest bar says that there were slightly more than 300 earthquakes with magnitudes between 7.0 and 7.5.

> **DESIGNING YOUR HISTOGRAM**
>
> Different features of the distribution may appear more obvious at different bin width choices. When you use technology, it's usually easy to vary the bin width interactively so you can make sure that a feature you think you see isn't a consequence of a certain bin width choice.

Stepping back from the histogram, you can see the entire distribution of earthquake magnitudes. Does this distribution look as you expected? It is often a good idea to *imagine* what the distribution might look like before you make the display. That way you'll be less likely to be fooled by errors in the data or when you accidentally graph the wrong variable. We can see that these earthquakes typically have magnitudes around 7. Most are between 5.5 and 8.5, but one is less than 4 and a few are 9 or larger. Now we can answer the question about the Sumatra and Japan tsunamis. With values of 9.1 and 9.0, it's clear that these earthquakes were extraordinarily powerful—among the largest on record.

In a histogram, the bins slice up all the values of the quantitative variable, so that every value falls in a bin. Choosing the bin width used to be an important exercise in introductory statistics courses. Today, technology will choose a default width for you, but usually you can vary the bin width interactively. You'll probably want to have between 5 and 30 bins (unless your data set is very large), and you'll want the bin boundaries to be aesthetically pleasing (such as 7.0 and 7.5 instead of 7.098 and 7.464). You may find that

---

[3]Technically, Richter scale values are in units of log dyne-cm. But the Richter scale is so common now that usually the units are assumed.

some features of the distribution appear or disappear as you change the bin width. Figure 3.2 a histogram with a smaller bin width of 0.2 for the tsunami data:

Figure 3.2

A histogram of earthquake magnitudes with a smaller bin size highlights the surprising number of earthquakes with values at 7.0.

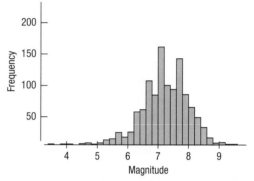

The overall distribution appears similar, as it should, but we're struck by two features that we didn't see before. There is a spike around magnitude 7.0. The 1168 earthquake magnitudes include historical values, some more than 2000 years old. Many of the values were estimated by experts and not measured by modern seismographs. Perhaps the experts rounded some of the values to 7, reasoning that this was a typical value for a moderately sized tsunami-causing earthquake. That might explain the overabundance of values of 7.0. There is also a gap in the histogram between 4.2 and 4.4. Unlike a bar chart that uses gaps just to separate categories, a **gap** in a histogram indicates that there is a bin (or bins) with no cases. There were no earthquakes among the 1168 that had values between 4.2 and 4.4. Gaps often reveal interesting features about the distribution.

Sometimes it is useful to make a **relative frequency histogram**, replacing the counts on the vertical axis with the *percentage* of the total number of cases falling in each bin. Of course, the shape of the histogram (Figure 3.3) is exactly the same; only the vertical scale is different.

Figure 3.3

A relative frequency histogram looks just like a frequency histogram except for the labels on the *y*-axis, which now shows the percentage of earthquakes in each bin. Compare this one with Figure 3.1.

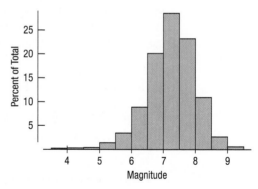

Stem-and-Leaf Displays

Histograms provide an easy-to-understand summary of the distribution of a quantitative variable, but they don't show the data values themselves. Figure 3.4 is a histogram of the pulse rates of 24 women, taken by a researcher at a health clinic:

Figure 3.4

The pulse rates of 24 women at a health clinic.

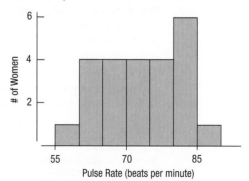

Stem-and-leaf or Stemplot? The stem-and-leaf display was devised by John W. Tukey, one of the greatest statisticians of the twentieth century. It is called a "stemplot" in some texts and computer programs, but we prefer Tukey's original name for it.

The story seems pretty clear. We can observe the entire span of the data and can easily see what a typical pulse rate might be. But is that all there is to these data?

A **stem-and-leaf display** is like a histogram, but it shows the individual values. It's also easier to make by hand. Here's a stem-and-leaf display of the same data:

```
5 | 6
6 | 0444
6 | 8888
7 | 2222
7 | 6666
8 | 000044
8 | 8
```
Pulse Rate
(5|6 means 56 beats/min)

Turn the stem-and-leaf on its side (or turn your head to the right) and squint. It should look roughly like the histogram of the same data. Does it?[4]

What does the first line of the display which says 5|6 mean? It stands for a pulse of 56 beats per minute (bpm). We've taken the tens place of the number and made that the "stem." Then we sliced off the ones place and made it a "leaf." The next line down is 6|0444. The four leaves tell us that there are four pulse rates in this row. Attach each leaf to the common stem of 6, and you get one pulse rate of 60 and three of 64 bpm.

Stem-and-leaf displays are especially useful when you make them by hand for batches of fewer than a few hundred data values. They are a quick way to display—and even to record—numbers. Because the leaves show the individual values, we can sometimes see even more in the data than the distribution's shape. Take another look at all the leaves of the pulse data. See anything unusual? At a glance, they are all even. With a bit more thought you can see that they are all multiples of four—something you couldn't possibly determine from a histogram. How do you think the nurse took these pulses? Counting beats for a full minute or counting for only 15 seconds and multiplying by four?

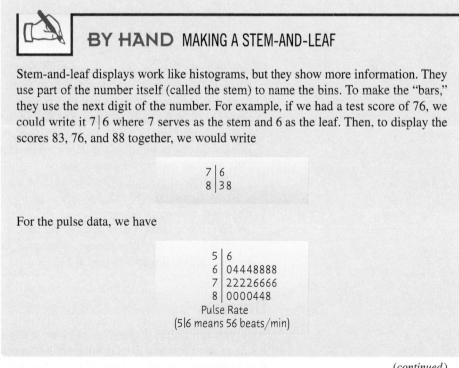

BY HAND MAKING A STEM-AND-LEAF

Stem-and-leaf displays work like histograms, but they show more information. They use part of the number itself (called the stem) to name the bins. To make the "bars," they use the next digit of the number. For example, if we had a test score of 76, we could write it 7|6 where 7 serves as the stem and 6 as the leaf. Then, to display the scores 83, 76, and 88 together, we would write

```
7 | 6
8 | 38
```

For the pulse data, we have

```
5 | 6
6 | 04448888
7 | 22226666
8 | 0000448
```
Pulse Rate
(5|6 means 56 beats/min)

(continued)

[4]You could make the stem-and-leaf with the higher values on the top. Putting the lower value at the top matches the histogram, when turned; putting the higher value at the top matches the way a vertical axis works in other displays.

This display is OK, but a little crowded. A histogram might split each line into two bars. With a stem-and-leaf, we can do the same by putting the leaves 0–4 on one line and 5–9 on another, as we saw earlier:

```
5 | 6
6 | 0444
6 | 8888
7 | 2222
7 | 6666
8 | 000044
8 | 8
```
Pulse Rate
(5|6 means 56 beats/min)

We could have stretched out the scale even more, using five lines for each distinct stem, assigning leaves 0 and 1 to the first line, 2 and 3 to the second line, etc. We need an equal number of possible leaves per stem, to maintain the integrity of our display. Here's what that would look like just for the pulses from 60 to 69:

```
6 | 0
6 |
6 | 444
6 |
6 | 8888
```
Pulse Rate
(6|4 means 64 beats/min)

That's probably too spread out to be useful. Which resolution you choose is up to you. For numbers with three or more digits, you'll need to decide on the resolution as well. Often you'll truncate (or round) the number to two digits, using the first digit as the stem and the second as the leaf. So, if you had 432, 540, 571, and 638, you might display them as

```
4 | 3
5 | 47
6 | 3
```

with an indication that 6|3 means 630 (though actually 6|3 could be any value from 630 to 639, due to the truncation process). Some stats packages will simply say "leaf unit = 10," telling you that a leaf of 3 is 3 tens, or 30. Could you also use two-digit leaves? Yes, if doing it by hand. You would have to separate them by commas or spaces—for example, 5|40,71—but this is tricky stuff for computer software, and frankly that extra digit adds little useful information to the display, while leaving little space on each row for many leaves.

But what if the data were 520, 541, 542, 547, 530 . . .? In this case, you would use two-digit stems (52, 53, 54), one digit leaves, and perhaps do some stem splitting to get a more useful picture.

When you make a stem-and-leaf by hand, make sure to give each leaf the same width in order to preserve the area principle. (That can lead to some fat 1s and thin 8s, but it makes the display honest.)

Dotplots

A **dotplot** is a simple display. It just places a dot along an axis for each case in the data. It's like a stem-and-leaf display but with dots instead of leaves. Dotplots are another great way to display a small data set. Figure 3.5 is a dotplot of the time (in seconds) that the winning horse took to win the Kentucky Derby in each race between the first Derby in 1875 and the 2014 Derby.

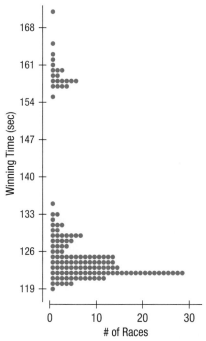

Figure 3.5

A dotplot of Kentucky Derby winning times plots each race as its own dot. We can see two distinct groups corresponding to the two different race distances (Data in **Kentucky Derby_2014**).

Dotplots show basic facts about distribution. We can find the slowest and quickest races by finding times for the topmost and bottommost dots. It's also clear that there are two clusters of points here, one just below 160 seconds and the other at about 122 seconds. Something strange happened to the Derby times. Once we know to look for it, we can find out that in 1896 the distance of the race was changed from 1.5 miles (2.4 km) to the current 1.25 miles (2.0 km). That explains the two clusters of winning times.

Some dotplots stretch out horizontally, with the counts on the vertical axis, like a histogram. Others, such as the one shown here, run vertically, like a stem-and-leaf display. Some dotplots place points next to each other when they would otherwise overlap. Others just place them on top of one another. Newspapers sometimes offer dotplots using little pictures, called pictographs, instead of dots.

Think Before You Draw

Suddenly, we face a lot more options when it's time to invoke our first rule of data analysis and make a picture. *Think* carefully about which type of graph to make. In the previous chapter, you learned to check the Categorical Data Condition before making a pie or bar chart. Now, before making a stem-and-leaf display, a histogram, or a dotplot, you need to check the

Quantitative Data Condition; The data are values of a quantitative variable whose units are known.

Although a bar chart and histogram may look somewhat similar, they're not the same display. You can't display categorical data in a histogram or quantitative data in a bar chart. Always check the condition that confirms what type of data you have before proceeding with your display.

Step back from a histogram or stem-and-leaf display. What can you say about the distribution? Every histogram can tell its own story, but you need to develop a vocabulary to help you explain it to others. When you describe a distribution, you should always discuss its *shape*, *centre*, and *spread*, as well as any unusual features or departures from the overall pattern.

Section 3.1 Exercises

1. Explain
   a) Explain the differences between a bar chart and a histogram.
   b) Explain the differences between a histogram and a stem-and-leaf plot

2. Stem-and-leaf Consider the following data set:
   18 28 28 28 37 31 24 29 8 27 24 12 21 32 27
   24 23 33 31 29

   a) Display the stem-and-leaf plot without splitting the "stems".
   b) Redo part (a) but put the leaves 0-4 on one line and 5-9 on another.
   c) Which stem-and-leaf plot do you prefer? Explain.

3.2 Shape

1. *Does the histogram have a single, central hump or several separated humps?* These humps are called **modes**.[5] The earthquake magnitudes have a single mode at about 7. A histogram with one peak, such as the earthquake magnitudes, is dubbed **unimodal**; histograms with two peaks are **bimodal**; and those with three or more are called **multimodal**.[6] Figure 3.6 below shows a bimodal histogram.

---

[5]Well, technically, the value on the horizontal axis of the histogram is the mode, but anyone asked to point to the mode would point to the hump.

[6]Apparently, statisticians don't like to count past two.

Figure 3.6

A bimodal histogram has two apparent peaks. Always ask: Are there two distinguishable groups present?

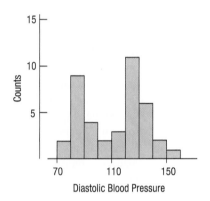

A histogram (such as Figure 3.7) that doesn't appear to have any mode and in which all the bars are approximately the same height is called **uniform**.

Figure 3.7

In this histogram, the bars are all about the same height. The histogram doesn't appear to have a mode and is called uniform.

2. *Is the histogram* **symmetric***?* Can you fold it (Figure 3.8) along a vertical line through the middle and have the edges match pretty closely, or are more of the values on one side?

Figure 3.8

A symmetric histogram can fold in the middle so that the two sides almost match.

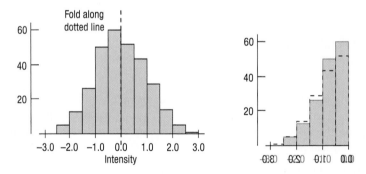

The (usually) thinner ends of a distribution are called the **tails**. If one tail stretches out farther than the other (Figure 3.9), the histogram is said to be **skewed** to the side of the longer tail.

Figure 3.9

Two skewed histograms showing data on two variables for all female heart attack patients in New York State in one year. The blue one (age in years) is skewed to the left (or negatively skewed). The purple one (charges in $) is skewed to the right (or positively skewed).

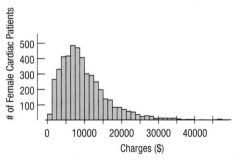

Pie à la mode? You've heard of pie à la mode. Is there a connection between pie and the mode of a distribution? Actually, there is! The mode of a distribution is a *popular* value near which a lot of the data values gather. And "à la mode" means "in style"—*not* "with ice cream." That just happened to be a *popular* way to have pie in Paris around 1900.

Olivier Juneau/Shutterstock

Figure 3.10
A histogram with outliers. There are three cities in the leftmost bar. We wonder why they might be different.

WHAT IS THE MODE?

The *mode* is sometimes defined as the single value that appears most often. That definition is fine for categorical variables because all we need to do is count the number of cases for each category. For quantitative variables, the mode is more ambiguous. What is the mode of the Kentucky Derby times? Well, seven races were timed at 122.2 seconds—more than any other race time. Should that be the mode? Probably not. For quantitative data, it makes more sense to use the term "mode" in the more general sense of the peak of the histogram rather than as a single summary value. In this sense, the important feature of the Kentucky Derby races is that there are two distinct modes, representing the two different versions of the race and warning us to consider those two versions separately.

---

3. *Do any unusual features stick out?* Often such features tell us something interesting or exciting about the data. You should always mention any stragglers, or **outliers**, that stand off away from the body of the distribution. For example, if you're collecting data on nose lengths and Pinocchio is in the group, you'd probably notice him, and you'd certainly want to mention it.

Outliers can affect almost every method we discuss in this course, so we'll always be on the lookout for them. An outlier can be the most informative part of your data. Or it might just be an error (fix it if you can). But don't throw it away without comment. Sometimes we point it out, try to explain it, and set it aside, rather than have it distort our analysis and understanding of the true story present in the rest of the data. The histogram in Figure 3.10 has three clear outlying cities, and we should investigate what makes those three cities different from the rest. Be sure to look for outliers—always—though there is no general rule for deciding whether an observation is a genuine outlier. It's always a judgment call. Even the shape of the distribution matters. If very skewed, then the long tail that continues that skewness is part of the overall pattern, so points would need to be farther away to be considered likely outliers.

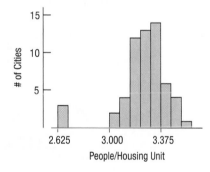

Are there any *gaps* in the distribution? The Kentucky Derby data we saw in the dotplot (Figure 3.5) on page 55 has a large gap between two groups of times, one near 120 seconds and one near 160. Gaps help us see multiple modes and encourage us to notice when the data may come from different sources or contain more than one group.

For Example DESCRIBING HISTOGRAMS

A credit card company wants to see how much customers in a particular segment of their market use their credit card. They have provided you with data[7] on the amount spent by 500 selected customers during a three-month period and have asked you to summarize the expenditures. Of course, you begin by making a histogram.

---

[7]These data are real, but cannot be further identified for privacy reasons.

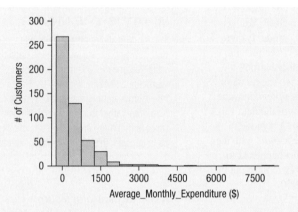

QUESTION: Describe the shape of this distribution. (Data in **Credit_Card_Exp**)

ANSWER: The distribution of expenditures is unimodal and skewed to the high end. The largest value is almost $8000, and some of the expenditures are negative.

Why do I get the feeling we're not in math class anymore?   When we summarize data, our goal is usually more than just developing detailed knowledge of the data we have at hand. We want to know what the data say about the world, so we'd like to know whether the patterns we see in histograms and summary statistics generalize to other individuals and situations. Scientists generally don't care about the particular guinea pigs in their experiment, but rather about what their reaction to different treatments says about how other animals (and, perhaps, humans) might respond.

Because we want to see broad patterns rather than focus on the details of the data set we're looking at, many of the most important concepts in Statistics are not defined as precisely as concepts in Mathematics. Whether a histogram is symmetric or skewed, whether it has one or more modes, whether a case is far enough from the rest of the data to be considered an outlier—these are all somewhat vague concepts. They all require judgment.

You may be used to finding a single correct and precise answer, but in Statistics, there may be more than one interpretation. That may make you a little uncomfortable at first, but soon you'll see that leaving room for judgment brings you both power and responsibility. It means that your own knowledge about the world and your judgment matter. You'll use them, along with the statistical evidence, to draw conclusions and make decisions about the world.

Just Checking

It's often a good idea to think about what the distribution of a data set might look like before we collect the data. What do you think the distribution of the following data sets will look like? Be sure to discuss shape. Where do you think the centre might be? How spread out do you think the values will be?

1. Number of kilometres run by Saturday morning joggers at a park.

2. Hours spent by Canadian adults last Saturday watching CBC's *Hockey Night in Canada*.[8]

3. Amount of winnings of all people playing Lotto 6/49 last week.

4. Ages of the faculty members at your school.

5. Last digit of phone numbers on your campus.

---

[8]Saturday night hockey is broadcast on CBC television. It is one of the highest-rated Canadian television programs, and is the world's oldest television sports program still on the air.

Section 3.2 Exercises

3. Incomes in Canada What do you expect the distribution of individual incomes in Canada to look like? Will it be uniform, unimodal, or bimodal? Will it be symmetric or skewed? Explain why.

4. Incarcerated This histogram displays the number of incarcerated prisoners in each U.S. state as of 2012.

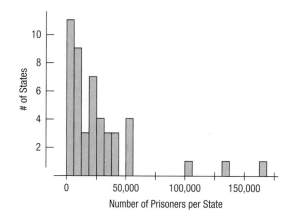

Describe the distribution of the number of incarcerated prisoners per state.

3.3 Centre

Let's return to the tsunami earthquakes. But this time, let's look at just 25 years of data: 221 earthquakes that occurred from 1989 through 2013 (selected from **Tsunami Earthquakes**). See Figure 3.11. These should be more accurately measured than prehistoric quakes because seismographs were in wide use. Try to put your finger at the value on the histogram you think is typical. (Read the value from the horizontal axis and remember it.) When we think of a typical value, we usually look for the **centre** of the distribution. Where do you think the centre of this distribution is? For a unimodal, symmetric distribution such as the earthquake data, it's easy. We'd all agree on the centre of symmetry, where we would fold the histogram to match the two sides. But when the distribution is skewed or possibly multimodal, it's not immediately clear what we even mean by the centre.

The Median

One natural choice of typical value is the data value that is literally in the middle, with half the values below it and half above it.

Figure 3.11

Tsunami-causing earthquakes (1989–2013). The median splits the histogram into two halves of equal area. It appears to be about 7.2 here.

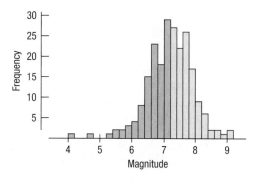

■ **NOTATION ALERT**

We always use *n* to indicate the number of values. Some people even say, "How big is the *n*?" when they mean the number of data values.

Histograms follow the area principle, and each half of the data has about 110 earthquakes, so each coloured region has the same area in the display. The middle value that divides the histogram into two equal areas is called the **median**. You can see it is slightly above 7 here, or about 7.2.

BY HAND FINDING THE MEDIAN

Finding the median of a batch of n numbers is easy as long as you remember to order the values first. If n is odd, the median is the middle value. Counting in from

the ends, we find this value in the $\dfrac{n+1}{2}$ position.

When n is even, there are two middle values. So, in this case, the median is the average of the two values in positions $\dfrac{n}{2}$ and $\dfrac{n}{2} + 1$.

Here are two examples:
Suppose the batch has these values: 14.1, 3.2, 25.3, 2.8, −17.5, 13.9, 45.8.
First we order the values: −17.5, 2.8, 3.2, 13.9, 14.1, 25.3, 45.8.
Since there are 7 values, the median is the $(7+1)/2 = 4th$ value, counting from the top or bottom: 13.9. Notice that 3 values are lower, 3 higher.
Suppose we had the same batch with another value at 35.7. Then the ordered values are −17.5, 2.8, 3.2, 13.9, 14.1, 25.3, 35.7, 45.8.
The median is the average of the 8/2 or 4th, and the $(8/2) + 1$ or 5th, values. So the median is $(13.9 + 14.1)/2 = 14.0$. Four data values are lower, and four higher.

The median has the same units as the data. Be sure to include the units whenever you discuss the median.

For the recent tsunamis, there are 221 earthquakes, so the median is found at the $(221 + 1)/2 = 111th$ place in the sorted data. The median earthquake magnitude is 7.2.

The median is one way to find the centre of the data. But there are many others. We'll look at an even more important measure later in this chapter.

Knowing the median, we could say that in recent years, a typical tsunami-causing earthquake has been about 7.2 on the Richter scale. How much does that really say? How well does the median describe the data? After all, not every earthquake has a Richter scale value of 7.2. Whenever we find the centre of data, the next step is always to ask how well it actually summarizes the data.

Section 3.3 Exercises

5. Blue Jays The numbers of season wins that the Toronto Blue Jays have since they won the World Series in 1993 are

| 55 | 56 | 74 | 76 | 88 | 84 | 83 | 80 | 78 | 86 | 67 |
| 80 | 87 | 83 | 86 | 75 | 85 | 81 | 73 | 74 | 83 | 93 |

What is the median number of wins by the Blue Jays since they last won the World Series?

(*Source:* http://toronto.bluejays.mlb.com/tor/history/year_by_year_results.jsp)

6. Divorce Rate The divorce rates (per 100 000 residents per year) in Canada from 1989 to 2008 are

| 296.9 | 283.4 | 274.7 | 278.6 | 272.7 | 272.0 | 264.9 | 241.6 | 225.4 | 229.1 |
| 233.2 | 231.8 | 229.2 | 223.8 | 223.9 | 218.0 | 221.0 | 229.3 | 222.2 | 210.8 |

What is the median divorce rate (per 100 000 residents per year) in Canada from 1989 to 2008?

(*Source:* https://en.wikipedia.org/wiki/Marriage_in_Canada)

3.4 Spread

If every earthquake that caused a tsunami registered 7.2 on the Richter scale, then knowing the median would tell us everything about the distribution of earthquake magnitudes. The more the data vary, however, the less the median alone can tell us. So we need to measure the **spread**—or how much the data values vary around the centre. In other words, how spread out are they? When we describe a distribution numerically, we always report a measure of its spread along with its centre. Spread, like centre and shape, is a general description of the distribution, not a precise quantity. There are several different measures of spread. Some are simple, some more complex, and, as we'll see, each has its own advantages and disadvantages.

> **WHAT WE DON'T KNOW**
>
> Statistics pays close attention to what we *don't* know as well as what we do know. Understanding how spread out the data are is a first step in understanding what a summary *cannot* tell us about the data. It's the beginning of telling us what we don't know.

The Range

How should we measure the spread? We could simply look at the extent of the data. How far apart are the two extremes? The **range** of the data is defined as the *difference* between the maximum and minimum values:

$$Range = max - min.$$

Notice that the range is a *single number, not* an interval of values, as you might think from its use in common speech. The maximum magnitude of these earthquakes is 9.1 and the minimum is 4.0, so the *range* is 9.1 − 4.0 = 5.1.

The range has the disadvantage that a single extreme value can make it very large, giving a value that doesn't really represent the data overall.

The Interquartile Range

A better way to describe the spread of a variable might be to ignore the extremes and concentrate on the middle of the data. We could, for example, find the range of just the middle half of the data. What do we mean by the middle half? Divide the data in half at the median. Now divide both halves in half again, cutting the data into four quarters. We call these new dividing points **quartiles**. One quarter of the data lies below the **lower quartile**, and one quarter of the data lies above the **upper quartile**, so half the data lies between them. The quartiles border the middle half of the data. For any percentage of the data, there is a corresponding **percentile**—the value that leaves that percentage of the data below it. The lower and upper quartiles are also known as the 25th and 75th percentiles of the data, respectively, since 25% of the data falls below the lower quartile and 75% of the data falls below the upper quartile. The median is the 50th percentile.

 BY HAND FINDING QUARTILES

A simple way to find the quartiles is to start by splitting the batch into two halves at the median. (When *n* is odd, some statisticians include the median in both halves; others omit it. We will omit it.[9]) The lower quartile is the median of the lower half, and the upper quartile is the median of the upper half.

(continued)

---

[9]Unless otherwise noted, we will use the method of excluding the median in both halves when calculating the quartiles in examples and answers. Your answer may vary slightly.

Here are our two examples again.

The ordered values of the first batch were −17.5, 2.8, 3.2, 13.9, 14.1, 25.3, and 45.8, with a median of 13.9. Notice that 7 is odd; we'll exclude the median in both halves to get −17.5, 2.8, 3.2, and 14.1, 25.3, 45.8.

Each half has 3 values, so the median of each is the 2nd value. So, the lower quartile is 2.8 and the upper quartile is 25.3.

The second batch of data had the ordered values −17.5, 2.8, 3.2, 13.9, 14.1, 25.3, 35.7, and 45.8 with a median of 14. Here *n* is even, so the two halves of 4 values are −17.5, 2.8, 3.2, 13.9 and 14.1, 25.3, 35.7, 45.8. Now the lower quartile is $(2.8 + 3.2)/2 = 3.0$ and the upper quartile is $(25.3 + 35.7)/2 = 30.5$.

The difference between the quartiles tells us how much territory the middle half of the data covers and is called the **interquartile range**. It's commonly abbreviated **IQR** (and pronounced "eye-cue-are"):

$$IQR = upper\,quartile - lower\,quartile.$$

For the earthquakes, there are 110 values below the median and 110 values above the median. The midpoint of the lower half is the average of the 55th and 56th values in the ordered data; that turns out to be 6.7. In the upper half, we find the average of the 55th and 56th values, standing above the median, to be a magnitude of 7.6, which is the third quartile. The *difference* between the quartiles gives the IQR:

$$IQR = 7.6 - 6.7 = 0.9.$$

Now we know that the middle half of the earthquake magnitudes extends across a (interquartile) range of 0.9 Richter scale units. This seems like a reasonable summary of the spread of the distribution, as we can see from Figure 3.12 below:

Why the quartiles?
Could we use other percentiles besides the quartiles to measure the spread? Sure, we could, but the IQR is the most commonly used percentile difference and the one you're most likely to see in practice.

Figure 3.12
The quartiles bound the middle 50% of the values of the distribution. This gives a visual indication of the spread of the data. Here we see that the IQR is 0.9 Richter scale units.

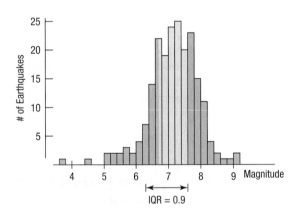

Even if the distribution itself is skewed or has some outliers, the IQR should provide useful information. The upper and lower quartiles should also be reported along with the IQR, for skewed data, so that we can see the *difference in spread* on both sides of the median. And we will see in the next chapter how useful the IQR can be for comparing the spread of two or more distributions (with side-by-side boxplots). The one exception is when the data are strongly bimodal. For example, remember the dotplot of winning times in the Kentucky Derby (page 55)? Because the race distance was changed, we really have data on two different races, and they shouldn't be summarized together.

SO, WHAT IS A QUARTILE ANYWAY?

Finding the quartiles sounds easy, but surprisingly, the quartiles are not well-defined. It's not always clear how to find a value such that exactly one quarter of the data lies above or below that value. We offered a simple rule for finding quartiles in the box on page 61: Find the median of each half of the data split by the median. When n is odd, we exclude the median with each of the halves. Some other texts include the median from each half before finding the quartiles. Both methods are commonly used. If you are willing to do a bit more calculating, there are several other methods that locate a quartile somewhere between adjacent data values. We know of at least six different rules for finding quartiles. Remarkably, each one is in use in some software package or calculator.

So don't worry too much about getting the "exact" value for a quartile. All of the methods agree pretty closely when the data set is large. When the data set is small, different rules will disagree more, but in that case there's little need to summarize the data anyway.

Remember, Statistics is about understanding the world, not about calculating the right number. The "answer" to a statistical question is a sentence about the issue raised in the question.

Section 3.4 Exercises

7. Blue Jays The numbers of wins that the Toronto Blue Jays have since they won World Series are recorded in question 5. Compute the IQR.

8. Divorce Rate The divorce rates (per 100,000 residents per year) in Canada from 1989 to 2008 are recorded in question 6. Compute the IQR.

3.5 Boxplots and 5-Number Summaries

The **5-number summary** of a distribution reports its median, quartiles, and extremes (maximum and minimum). The 5-number summary for the recent tsunami earthquake *Magnitudes* looks like this:

■ **NOTATION ALERT**

We always use Q1 to label the lower (25%) quartile and Q3 to label the upper (75%) quartile. We skip the number 2 because the median would, by this system, naturally be labeled Q2—but we don't usually call it that.

Max	9.1
Q3	7.6
Median	7.2
Q1	6.7
Min	4.0

It's a good idea to report the number of data values and the identity of the cases (the *Who*). Here there are 221 earthquakes.

The 5-number summary provides a good overview of the distribution of magnitudes of these tsunami-causing earthquakes. For a start, we can see that the median magnitude is 7.2. Because the IQR is only $7.6 - 6.7 = 0.9$, we see that many quakes are close to the median magnitude. Indeed, the quartiles show us that the middle half of these earthquakes had magnitudes between 6.7 and 7.6. One quarter of the earthquakes had magnitudes above 7.6, although one tsunami was caused by a quake measuring only 4.0 on the Richter scale.

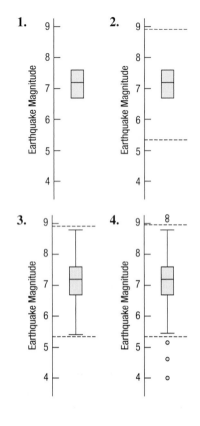

Boxplots

Once we have a 5-number summary of a (quantitative) variable, we can display that information in a **boxplot**. To make a boxplot of the earthquake magnitudes, follow these steps:

1. Draw a single vertical axis spanning the extent of the data.[10] Draw short horizontal lines at the lower and upper quartiles and at the median. Then connect them with vertical lines to form a box. The box can have any width that looks OK.[11]

2. To construct the boxplot, erect "fences" around the main part of the data. Place the upper fence 1.5 IQRs above the upper quartile and the lower fence 1.5 IQRs below the lower quartile. For the earthquake magnitude data, we compute

$$Upper fence = Q3 + 1.5\ IQR = 7.6 + 1.5 \times 0.9 = 8.95$$

and

$$Lower fence = Q1 - 1.5\ IQR = 6.7 - 1.5 \times 0.9 = 5.35.$$

The fences are just to help with construction and are not part of the display. We show them here with dotted lines for illustration. You should never include them in your boxplot.

3. We use the fences to grow "whiskers." Draw lines from the ends of the box up and down to *the most extreme data values found within the fences*. If a data value falls outside one of the fences, we do *not* connect it with a whisker.

4. Finally, we display any data values lying beyond the fences with special symbols. These are worthy of consideration as possible outliers. Let's call them *suspect outliers*.

A boxplot highlights several features of the distribution. The central box shows the middle half of the data, between the quartiles. The height of the box (or width, if we turn it sideways) is equal to the IQR. If the median is roughly centred between the quartiles, then the middle half of the data is roughly symmetric. If the median is not centred, the distribution is skewed. The whiskers show skewness as well if they are not roughly the same length. Any suspect outliers are displayed individually, both to keep them out of the way for judging skewness and to encourage you to give them special attention. They may be mistakes, or they may be the most interesting cases in your data.

For the recent tsunami earthquake data, the central box (Figure 3.13) contains all the earthquakes whose magnitudes are between 6.7 and 7.6 on the Richter scale. From the shape of the box, it looks like the central part of the distribution of earthquakes is roughly symmetric, but the longer lower whisker indicates that the distribution stretches out slightly at the lower end. We also see a few very small magnitude earthquakes and the two large earthquakes we've discussed. Boxplots are particularly good at pointing out possible outliers.

Why 1.5 IQRs? One of the authors asked the prominent statistician, John W. Tukey, the originator of the boxplot, why the outlier nomination rule cut at 1.5 IQRs beyond each quartile. He answered that the reason was that 1 IQR would be too small and 2 IQRs would be too large. That works for us.

Figure 3.13

By turning the boxplot and putting it on the same scale as the histogram, we can compare both displays of the earthquake magnitudes and see how each represents the distribution.

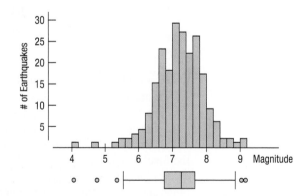

---

[10]The axis could also run horizontally.

[11]Some computer programs draw wider boxes for larger data sets. That can be useful when comparing groups.

Step-by-Step Example SHAPE, CENTRE, AND SPREAD: FLIGHT CANCELLATIONS

MO_SES Premium/Shutterstock

The U.S. Bureau of Transportation Statistics (www.bts.gov) reports data on airline flights. Let's look at data giving the percentage of flights cancelled each month between 1994 and 2013. (Data in **Pct Cancelled 2013**)

Question: How often are flights cancelled?

Who   Months
What   Percentage of flights cancelled at U.S. airports
When   1994–2013
Where   United States

THINK ➡ **Variable** Identify the *variable*, and decide how you wish to display it. To identify a variable, report the W's. Select an appropriate display based on the nature of the data and what you want to know.	I want to learn about the monthly percentage of flight cancellations at U.S airports. I have data from the U.S. Bureau of Transportation Statistics giving the percentage of flights cancelled at U.S. airports each month between 1994 and 2013. ✓ **Quantitative Data Condition**: Percentages are quantitative. A histogram and numerical summaries would be appropriate.
SHOW ➡ **Mechanics** We usually make histograms with a computer or graphing calculator.	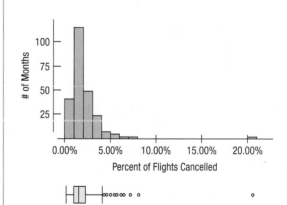 The histogram shows a distribution skewed to the high end and one extreme outlier, a month in which more than 20% of flights were cancelled. In most months, fewer than 5% of flights are cancelled and usually only about 2% or 3%. That seems reasonable.
REALITY CHECK It's always a good idea to think about what you expect to see so that you can check whether the histogram looks like what you expected.	

With 238 months, we probably have more data than you'd choose to work with by hand. The results given here are from technology.

Some summary numbers:

Count	238
Max	20.24
Q3	2.39
Median	1.68
Q1	1.16
Min	0.38
IQR	1.23

TELL ➡ **Interpretation** Describe the shape, centre, and spread of the distribution. Report on the symmetry, number of modes, and any gaps or outliers. You should also mention any concerns you may have about the data.

The distribution of cancellations is skewed to the right, and this makes sense: The values can't fall below 0%, but can increase almost arbitrarily due to bad weather or other events.

The median is 1.68% and the IQR is 1.23%. The low IQR indicates that in most months the cancellation rate is close to the median. In fact, it's between 1.16% and 2.39% in the middle 50% of all months, and in only 1/4 of the months were more than 2.39% of flights cancelled.

There is one extraordinary value: 20.2%. Looking it up, I find that the extraordinary month was September 2001. The attacks of September 11 shut down air travel for several days, accounting for this outlier.

Section 3.5 Exercises

9. **Temperature** The day time high temperatures (in Celsius) for Toronto in June 2016 are:

20	30	25	26	25	25	20
14	20	23	30	21	19	21
24	26	29	31	33	34	26
26	23	27	28	33	33	22
26	27					

Obtain the 5-number summary and the boxplot for this data set.

(*Source*: http://www.timeanddate.com/weather/canada/toronto/historic?month=6&year=2016)

10. **Bicycle Deaths** The total number of bicyclists killed in crashes in the U.S. from 1994 to 2013 are:

796	828	761	811	757	750	689	729	663	626
722	784	769	699	716	628	616	680	730	747

Obtain the 5-number summary and the boxplot for this data set.

(*Source*: http://www.bhsi.org/stats.htm)

3.6 The Centre of Symmetric Distributions: The Mean

Medians do a good job of summarizing the centre of a distribution, even when the shape is skewed or when there is an outlier, as with the flight cancellations. But when we have symmetric data, there's another alternative. You probably already know how to average values. In fact, to find the median when *n* is even, we said you should average the two middle values, and you didn't even flinch.

The earthquake magnitudes are pretty close to symmetric, so we can also summarize their centre with a mean. The mean recent tsunami earthquake magnitude is 7.15—about what we might expect from the histogram. You already know how to average values, but this is a good place to introduce notation that we'll use throughout the book. We use the Greek capital letter sigma, Σ, to mean "sum" (sigma is "S" in Greek), and we'll write

$$\bar{y} = \frac{Total}{n} = \frac{\sum y}{n}.$$

The formula says that to find the **mean**, add up all the values of the variable and divide that sum by the number of data values, *n*—just as you've always done.[12]

Once we've averaged the data, you'd expect the result to be called the *average*, but that would be too easy. Informally, we speak of the "average person" but we don't add up people and divide by the number of people. A median is also a kind of average. To make this distinction, the value we calculated is called the mean, $\bar{y}$ and pronounced "*y*-bar."

The mean feels like the centre because it is the point where the histogram (Figure 3.14) balances:

Figure 3.14

The mean is located at the *balancing point* of the histogram. For the recent earthquake magnitudes, $\bar{y} = 7.15$.

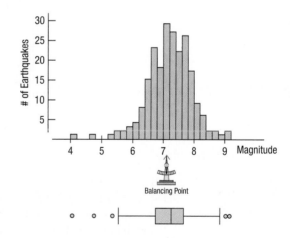

Mean or Median?

Using the centre of balance makes sense when the data are symmetric. But data are not always this well behaved. If the distribution is skewed or has outliers, the centre is not so well defined and the mean may not be what we want.

---

[12]You may also see the variable called x and the equation written $\bar{x} = \dfrac{Total}{n} = \dfrac{\sum x}{n}$. Don't let that throw you.

You are free to name the variable anything you want, but we'll generally use *y* for variables like this that we want to summarize, model, or predict. (Later we'll talk about variables that are used to explain, model, or predict *y*. We'll call them *x*.)

For example, the mean salary for the Toronto Maple Leaf players in a recent season was $1 996 143. Is this a typical player's salary? Figure 3.15 is a histogram of their salaries. Let's balance the histogram:

Figure 3.15

Toronto Maple Leaf players salaries, 2012–2013.

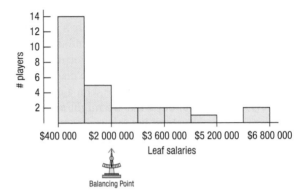

The mean is about $2 million, but 19 of 28 players had salaries below that, so the mean doesn't feel like a good overall summary. Why is the balancing point so high? It's easy to see that the very big salaries on the right are pulling the balancing point in that direction. So, the mean here is being heavily influenced by some super-star salaries, whereas the median salary is only $1 150 000, which seems like a more typical summary. The median splits the data into two halves, so we can estimate it on the histogram by splitting it up into two regions of equal area (blue and pink in Figure 3.16):

Figure 3.16

The median splits the area of the histogram in half at $1 150 000, at the upper edge of the blue rectangle. Because the distribution is skewed to the right, the mean is higher than the median. The high salaries at the right have pulled the mean toward them, away from the median.

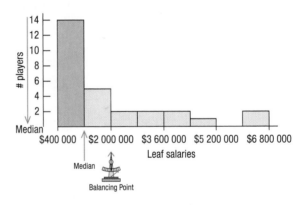

Because the median considers only the order of the values, it is **resistant** to values that are extraordinarily large or small; it simply notes that they are one of the "big ones" or the "small ones" and ignores their distance from the centre.

How should we choose between the mean and the median? In the salary data, it certainly seems more sensible to use the median salary, or salary of the player who stands right in the middle. The distribution is skewed to the right, and even one very big salary will pull the mean away from the centre. Like a child who sits way out on a see-saw, one extreme value can pull the balancing point toward it. For a distribution like this, the median is usually a better descriptor.

For the recent tsunami-causing earthquakes, it doesn't seem to make much difference—the mean is 7.15 and the median is 7.20. When the data are symmetric, the mean and median will be close, but when the data are skewed, the median is likely to be a better choice. So why not just use the median? Well, for one, the median can go overboard. It's not just resistant to occasional outliers, but can be unaffected by changes in most of the data values. By contrast, the mean includes input from each data value and gives each one equal weight. It's also easier to work with, so when the distribution is unimodal and fairly symmetric, we'll use the mean.

A **resistant measure** is one that is not affected too much by changes, big or small, to a small portion of the data. Such a measure will pay little attention to outliers. If recording errors are discovered in a small subset of your data, resistant measures may not even need to be recalculated, as they will barely move. The median is one example of a resistant measure.

Of course, to choose between mean and median as a descriptor of centre, we'll start by looking at the data. If the histogram is roughly symmetric and there are no outliers, we'll prefer the mean. However, if the histogram is skewed or has outliers, we're usually better off with the median. If you're not sure, report both and discuss why they might differ. If symmetric, except for one or two clear outliers, you might also consider reporting the outliers separately, and the mean just for the rest of the data.

For Example DESCRIBING CENTRE

RECAP: You want to summarize the expenditures of 500 credit card company customers, and have looked at a histogram.

QUESTION: You have found the mean expenditure to be $478.19 and the median to be $216.28. Which is the more appropriate measure of centre, and why?

ANSWER: Because the distribution of expenditures is skewed, the median is a very appropriate measure of centre. Unlike the mean, it's not affected by the large outlying value or by the skewness. Half of these credit card customers had average monthly expenditures less than $216.48 and half more.

WHEN TO EXPECT SKEWNESS

Even without making a histogram, we can expect some variables to be skewed. When values of a quantitative variable are bounded on one side but not the other, the distribution may be skewed. For example, incomes and waiting or survival times can't be less than zero, so they are often skewed to the right. Amounts of things (dollars, employees) are often skewed to the right for the same reason. If a test is too easy, the distribution will be skewed to the left because many scores will bump against 100%. And combinations of things are often skewed. In the case of a hockey team's salaries, often a small group of superstars will have very inflated . . . er, impressive salaries . . . so including them along with the more typical players and the minimum salary players can lead to a very skewed (or even bimodal) distribution.

But there are also times when the purpose or intended use of the measure is key, rather than just trying to provide a useful descriptor of the data. For example, the players on a hockey team might be keenly interested in the median salary, as this is what the "average" player is making, while the team owner would be more interested in the average or mean salary, since this is how much is being paid out per player and directly related to total payroll (mean × number of players = total payroll). Many sports teams have highly right-skewed salary distributions due to the presence of a few star or superstar players, hence a much bigger mean than median salary.

Or if you are interested in your potential long-term returns from some type of investment, the mean of past returns would interest you more than the median, since the longer you hold the investment, the closer your average return will be to the long-term mean (assuming future returns will be similar to past returns). This is a consequence of the Law of Large Numbers—to be discussed in later chapters.

Some other possible measures of centre include:

■ *Mode(s) or modal bin(s).* See the note on page 57 about these peaks or high-frequency values in a distribution. The modes of a bimodal distribution, for example, would be far more informative than a mean or median. And if I could win a lottery by guessing the approximate price of the next resale condo sold in Toronto, please tell me the mode—not the mean or median—of the recent distribution of sale prices, since more houses sell around the modal price than any other price.

■ *Trimmed mean.* We can make the ordinary mean more *resistant* by trimming off a small percent (like 5% or 10%) of the biggest and smallest values before averaging, creating what is called a trimmed mean. See exercise #49 at the end of this chapter for more details.

Section 3.6 Exercises

11. Adoptions II Here is a histogram showing the total number of adoptions in each of the 50 states and the District of Columbia.

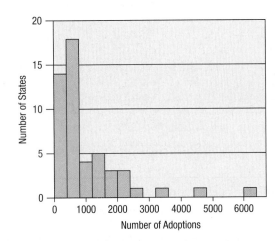

Would you expect the mean number of adoptions or the median number of adoptions to be higher? Why?

12. Test score centres After entering the test scores from her Statistics class of 25 students, the instructor calculated the mean and the median of the scores. Upon checking, she discovered that she had entered the top score as 46, but it should have been 56. When she corrects this score, how will the mean and median be affected?

3.7 The Spread of Symmetric Distributions: The Standard Deviation

The IQR is usually a reasonable summary of spread, but because it uses only the two quartiles of the data, it ignores much of the information about how individual values vary. A more powerful approach uses the standard deviation, which takes into account how far *each* value is from the mean. Like the mean, the standard deviation is appropriate only for reasonably symmetrical data.

One way to think about spread is to examine how far each data value is from the mean. This difference is called a *deviation*. For example, consider the two samples of data in Figure 3.17, one in orange, on top of the horizontal axis, and one in orange, below the axis.

Figure 3.17

Deviations from the mean reflect spread. The sample on top of the horizontal axis has bigger deviations, on average, than the sample on the bottom, though the ranges are identical.

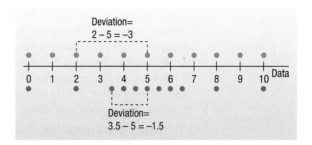

Both samples have the same central value (5.0) and range (10.0), but in the blue sample, the data tend to cluster more tightly around the centre. This is captured in the size of the deviations or distances from the centre. The third smallest data point is 1.5 units below the centre in the blue sample whereas the third smallest data point is 3 units below the centre in the orange sample Similarly, if you run through all the corresponding data points, left to right, in each graph, you will see that each data point is closer to the middle in the blue sample, except for the maximum (10.0), the minimum (0.0), and the mean (5.0), which have similar deviations of +5, −5, and 0 respectively, in each case.

One single deviation or even a few of them won't be very useful. We want to use all the information about data spread—use all the deviations together—combining them all into one useful summary number. Our first thought might be to just average the deviations, but the positive and negative differences exactly cancel each other out! The average deviation is always zero—not very helpful.

To keep them from cancelling out, one possible approach would be to take absolute values of the deviations, but instead we will elect to square them. Statisticians rarely see a deviation not worth squaring! There are very good reasons for squaring, having to do with a certain underlying right triangle geometry and mathematics—a discussion quite beyond the level of this course though—sorry, just take it on faith! Squaring always gives a positive or zero value, so the sum cannot be negative. (Can it be zero? When?) That's great. Squaring also emphasizes larger differences (for example, a deviation of 3 is thrice a deviation of 1, but after squaring, the former becomes $3^2 = 9$ times bigger than $1^2 = 1$)—a feature that turns out to be both good and bad.

When we add up these squared deviations and find their average (almost), we call the result the sample **variance**:

$$s^2 = \frac{\Sigma(y - \bar{y})^2}{n - 1}.$$

Why almost? It *would* be a mean if we divided the sum by n. Instead, we divide by $n - 1$. Why? Well, you might think it's just to drive you crazy, but actually s^2 would be too small if we divided by n. Why $n - 1$ fixes it is an amazing fact, but showing it is beyond the scope of this course. Nevertheless, we'll say more in a later chapter.

The variance will play an important role later in this book, but it has a problem as a measure of spread. Whatever the units of the original data are, the variance is in *squared* units. We want measures of spread to have the same units as the data. And we probably don't want to talk about squared dollars or $1/100$ km^2. So, to get back to the original units, we take the square root of s^2. The result, s, is the sample **standard deviation**:

$$s = \sqrt{\frac{\Sigma(y - \bar{y})^2}{n - 1}}.$$

■ NOTATION ALERT

s^2 always means the variance of a set of data, and s always denotes the standard deviation.

You will almost always rely on a calculator or computer to do the calculating. For the recent 221 tsunami-causing earthquakes, we found the standard deviation of the magnitudes to be 0.70 (in Richter scale units).

Does this number "0.70" have any simple or intuitive interpretation? Is it an average absolute deviation? Not quite, since squaring, averaging, and square-rooting absolute deviations does not somehow cancel the first and last operation to produce a simple average. But it is in the same ballpark, so one can think of it as a somewhat average or typical deviation from the mean. And for distributions that are roughly symmetrical and bell-shaped in appearance, with no extreme outliers, we can say more, based on the *Empirical Rule* first discovered some 300 years ago:

  i. About 68% of the data will lie within 1 standard deviation of the mean.

 ii. About 95% of the data will lie within 2 standard deviations of the mean.

iii. Virtually all of the data will lie within 3 standard deviations of the mean.

For example, many studies report a mean and standard deviation for various subject characteristics upon entrance to the study. If a study reported the mean weight of 50 subjects to be 70 kg with standard deviation of 10 kg (and weights generally are somewhat bell-shaped), we would know that about two-thirds (30–35 subjects) weighed between 60 kg and 80 kg, with just a few subjects less than 50 kg or more than 90 kg. Likely no one weighed less than 40 kg or more than 100 kg (since $100 = 70 + 3 \times 10$, and $40 = 70 - 3 \times 10$). Now, let's check this out for the tsunami-causing earthquake data set. Calculate the mean $\pm$ two standard deviations: $7.15 - 2 \times 0.70 = 5.75$ and $7.15 + 2 \times 0.70 = 8.55$. Locate these two points on the histogram, and you can see that about 95% of tsunami-causing earthquake magnitudes fall between these numbers.

The empirical rule gives reasonable approximations even for mildly skewed mound-shaped distributions.[13] Just keep in mind that it is rather common to find data in the

BY HAND HOW TO CALCULATE A STANDARD DEVIATION

To find the standard deviation, start with the mean, $\bar{y}$. Then find the *deviations* by taking $\bar{y}$ from each value: $(y - \bar{y})$. Square each deviation: $(y - \bar{y})^2$.

Now just add up the squared deviations and divide by $n - 1$. That gives you the variance, s^2. To find the standard deviation, s, take the square root. Here we go:

Suppose the batch of values is 14, 13, 20, 22, 18, 19, and 13.

The mean is $\bar{y} = 17$. So the deviations are found by subtracting 17 from each value:

Original Values	Deviations	Squared Deviations
14	$14 - 17 = -3$	$(-3)^2 = 9$
13	$13 - 17 = -4$	$(-4)^2 = 16$
20	$20 - 17 = 3$	9
22	$22 - 17 = 5$	25
18	$18 - 17 = 1$	1
19	$19 - 17 = 2$	4
13	$13 - 17 = -4$	16

Add up the squared deviations: $9 + 16 + 9 + 25 + 1 + 4 + 16 = 80$

Now divide by $n - 1$: $\qquad\qquad\qquad\qquad\qquad\qquad\qquad 80/6 = 13.33$

Finally, take the square root: $\qquad\qquad\qquad\qquad\qquad\qquad s = \sqrt{13.33} = 3.65$

---

[13] To make similar types of statements using means and standard deviations about an arbitrary distribution of any shape, we would have to invoke Chebyshev's Rule, which, due to its complete generality, has more theoretical than practical value, so we won't discuss it here (but see The Worse Case margin note on page 145 in Chapter 5).

vicinity of 1 standard deviation from the mean, but rather uncommon to find data more than 2 standard deviations from the mean for such distributions. One particularly important bell-shaped distribution is called the Normal model, for which the mean and standard deviation together provide not just a good but a complete description. But you'll have to wait until Chapter 5 to ring that bell.

The bad news about the standard deviation (as with its partner in crime, the mean) is its lack of resistance to changes to small portions of the data set, and a single outlier can hugely inflate its value (and that outlier's big deviation also gets squared!).

Thinking About Variation

Statistics is about variation, so spread is an important fundamental concept in Statistics. Measures of spread help us to be precise about what we *don't* know. If many data values are scattered far from the centre, the IQR and the standard deviation will be large. If the data values are close to the centre, then these measures of spread will be small. If all our data values were exactly the same, we'd have no question about summarizing the centre, and all measures of spread would be zero—and we wouldn't need Statistics. You might think this would be a big plus, but it would make for a boring world. Fortunately (at least for Statistics), data do vary.

Measures of spread tell how well other summaries describe the data. That's why we always (always!) report a spread along with any summary of the centre.

For Example   DESCRIBING SPREAD

RECAP: The histogram has shown that the distribution of credit card expenditures is skewed, and you have used the median to describe the centre. The quartiles are $73.84 and $624.80.

QUESTION: What is the IQR and why is it a suitable measure of spread?

ANSWER: For these data, the interquartile range (IQR) is $624.80 − $73.84 = $550.96. Like the median, the IQR is not affected by the outlying value or by the skewness of the distribution, so it is an appropriate measure of spread (more so than the standard deviation) for describing the given expenditures.

 ## Just Checking

6. Statistics Canada reports the median family income in its summary of census data. Why do you suppose they use the median instead of the mean? What might be the disadvantages of reporting the mean?

7. You've just bought a new car that claims to get a highway fuel efficiency of 9.8 litres per 100 kilometres. Of course, your mileage will "vary." If you had to guess, would you expect the IQR of gas mileage attained by all cars like yours to be 10, 1, or 0.1 L/100 km? Why?

8. A company selling a new MP3 player advertises that the player has a mean lifetime of five years. If you were in charge of quality control at the factory, would you prefer that the standard deviation of lifespans of the players you produce be two years or two months? Why?

Section 3.7 Exercises

13. Test score spreads The professor in Exercise 12 had also calculated the IQR and standard deviation of the test scores before realizing her mistake. (She had entered the top score as 46 but it was actually 56.) What effect will correcting the error have on the IQR and the standard deviation?

14. Fuel economy The boxplot shows the fuel economy (miles per gallon) ratings for 67 model year 2012 sub-compact cars. Some summary statistics are also provided. The extreme outlier is the **Mitsubishi i-MiEV**, an electric car whose electricity usage is *equivalent* to 112 mpg.

Combined Fuel Economy

Mean	sd	Min	Q1	Med	Q3	Max	n
23.76	11.87	14	20	22	25	112	67

If that electric car is removed from the data set, how will the standard deviation be affected? The IQR?

3.8 Summary—What to *Tell* About a Quantitative Variable

What should you *Tell* about a quantitative variable?

- Start by making a histogram or stem-and-leaf display, and discuss the shape of the distribution.
- Next, discuss the centre and spread.
  - Always pair the median with the IQR and the mean with the standard deviation. It's not useful to report one without the other. Reporting a centre without a spread is dangerous. You may think you know more than you do about the distribution. Reporting only the spread leaves us wondering where we are.
  - If the shape is skewed, report the 5-number summary and be sure to discuss the median and IQR. You may want to include the mean and standard deviation as well, but you should point out why the mean and median differ.
  - If the shape is symmetric, report the mean and standard deviation and possibly the median and IQR as well. For unimodal symmetric data, the IQR is usually a bit larger than the standard deviation. If that's not true of your data set, look again to make sure that the distribution isn't skewed and there are no outliers.

- Also, discuss any unusual features.
  - If there are multiple modes, try to understand why. If you can identify a reason for separate modes (for example, women and men typically have heart attacks at different ages), it may be a good idea to split the data into separate groups.
  - If there are any clear outliers, you should point them out. If you are reporting the mean and standard deviation, report them with the outliers present and with the outliers omitted. The differences may be revealing. (Of course, the median and IQR won't be affected very much by the outliers.)

> **HOW "ACCURATE" SHOULD WE BE?**
>
> Don't think you should report means and standard deviations to a zillion decimal places; such implied accuracy is really meaningless. Although there is no ironclad rule, statisticians commonly report summary statistics to one or two decimal places more than the original data have.

For Example   CHOOSING SUMMARY STATISTICS

RECAP: You have provided the credit card company's board of directors with a histogram of customer expenditures, and you have summarized the centre and spread with the median and IQR. Knowing a little Statistics, the directors now insist on having the mean and standard deviation as summaries of the spending data.

QUESTION: Although you know that the mean is $478.19 and the standard deviation is $741.87, you need to explain to them why these are not suitable summary statistics for these expenditures data. What would you give as reasons?

ANSWER: The high outlier at $7000 pulls the mean up substantially and inflates the standard deviation. Locating the mean value on the histogram shows that it is not a typical value at all, and the standard deviation suggests that expenditures vary much more than they do. The median and IQR are more resistant to the presence of skewness and outliers, giving more realistic descriptions of centre and spread.

Step-by-Step Example SUMMARIZING A DISTRIBUTION

One of the authors owned a Nissan Maxima for 8 years. Being a statistician, he recorded the car's fuel efficiency (in mpg[14]) each time he filled the tank. He wanted to know what fuel efficiency to expect as "ordinary" for his car. (Hey, he's a statistician. What would you expect?[15]) Knowing this, he was able to predict when he'd need to fill the tank again and to notice if the fuel efficiency suddenly got worse, which could be a sign of trouble.

Question: How would you describe the distribution of *Fuel efficiency* for this car?

Dmitriy Ystuyjanin/Fotolia

THINK ➡ **Plan** State what you want to find out.		I want to summarize the distribution of Nissan Maxima fuel efficiency.
Variable Identify the variable and report the W's.		The data are the fuel efficiency values in miles per gallon for the first 100 fill-ups of a Nissan Maxima.
Be sure to check the appropriate condition.		✓ **Quantitative Data Condition**: The fuel efficiencies are quantitative with units of miles per gallon. Histograms are an appropriate way to display the distribution. Numerical summaries are appropriate as well.

SHOW ➡ **Mechanics** Make a histogram. Based on the shape, choose appropriate numerical summaries.

A histogram of the data shows a fairly symmetric distribution with a low outlier.

REALITY CHECK A value of 22 mpg seems reasonable for such a car. The spread is reasonable, although the range looks a bit large.

Count	100
Mean	22.4 mpg
StdDev	2.45
Q1	20.8
Median	22.0
Q3	24.0
IQR	3.2

---

[14]Data collected by one of the authors in the U.S., where they use miles per gallon to measure fuel efficiency.

[15]He also recorded the time of day, temperature, price of gas, and phase of the moon. (OK, maybe not phase of the moon.) (Data in **Nissan**)

	The mean and median are close, so the outlier doesn't seem to be a problem. I can use the mean and standard deviation.
TELL ➡ **Conclusion** Summarize and interpret your findings in context. Be sure to discuss the distribution's shape, centre, spread, and unusual features (if any).	The distribution of mileage is unimodal and roughly symmetric with a mean of 22.4 mpg. There is a low outlier that should be investigated, but it does not influence the mean very much. The standard deviation suggests that from tankful to tankful, I can expect the car's fuel economy to differ from the mean on average by about 2.45 mpg.

I GOT A DIFFERENT ANSWER: DID I MESS UP?

When you calculate a mean, the computation is clear: You sum all the values and divide by the sample size. You may round your answer less or more than someone else (we recommend one more decimal place than the data), but all books and technologies agree on how to find the mean. Some statistics, however, are more problematic. For example, we've already pointed out that methods of finding quartiles differ.

Differences in numeric results can also arise from decisions in the middle of calculations. For example, if you round off your value for the mean before you calculate the sum of squared deviations, your standard deviation probably won't agree with a computer program that calculates using many decimal places. (We do recommend that you use as many digits as you can during the calculation and round only when you are done.)

Don't be overly concerned with these discrepancies, especially if the differences are small. They don't mean that your answer is "wrong," and they usually won't change any conclusion you might draw about the data. Sometimes (in footnotes and in the answers in the back of the book) we'll note alternative results, but we could never list all the possible values, so we'll rely on your common sense to focus on the meaning rather than on the digits. Remember: Answers are sentences—not single numbers!

WHAT CAN GO WRONG?

A data display should tell a story about the data. To do that it must speak in a clear language, making plain what variable is displayed, what any axis shows, and what the values of the data are. And it must be consistent in those decisions.

A display of quantitative data can go wrong in many ways. The most common failures arise from only a few basic errors:

■ **Don't make a histogram of a categorical variable.** Just because the variable contains numbers doesn't mean it's quantitative. Figure 3.18 below is a histogram of the insurance policy numbers of some workers. It's not very informative because the policy numbers are just labels. A histogram or stem-and-leaf display of a categorical variable makes no sense. A bar or pie chart would be more appropriate.

Figure 3.18
It's not appropriate to display these data with a histogram.

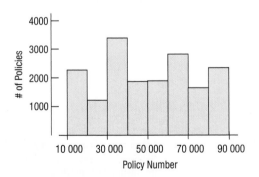

- **Don't look for shape, centre, and spread of a bar chart.** A bar chart showing the sizes of the piles displays the distribution of a categorical variable, but the bars could be arranged in any order from left to right. Concepts like symmetry, centre, and spread make sense only for quantitative variables.

- **Don't use bars in every display—save them for histograms and bar charts.** In a bar chart, the bars indicate how many cases of a categorical variable are piled in each category. Bars in a histogram indicate the number of cases piled in each interval of a quantitative variable. In both bar charts and histograms, the bars represent counts of data values. Some people create other displays that use bars to represent individual data values. Beware: Such graphs are neither bar charts nor histograms. For example, a student was asked to make a histogram from data showing the number of juvenile bald eagles seen during each of the 13 weeks in the winter at a site in Rock Island, IL. Instead, he made this plot (Figure 3.19):

Figure 3.19

This isn't a histogram or a bar chart. It's an ill-conceived graph that uses bars to represent individual data values (number of eagles sighted) week by week.

Look carefully. That's not a histogram. A histogram shows *What* we've measured along the horizontal axis and counts of the associated *Who* represented as bar heights. This student has it backwards: He used bars to show counts of birds for each week.[16] We need counts of weeks. A correct histogram (Figure 3.20) should have a tall bar at "0" to show there were many weeks when no eagles were seen, like this:

Figure 3.20

A histogram of the eagle-sighting data shows the number of weeks in which different counts of eagles occurred. This display shows the distribution of juvenile-eagle sightings.

(continued)

---

[16]Edward Tufte, in his book *The Visual Display of Quantitative Information*, proposes that graphs should have a high data-to-ink ratio. That is, we shouldn't waste a lot of ink to display a single number when a dot would do the job. Actually, replace the bars by dots in Figure 3.19, and we get a timeplot (discussed in the next chapter).

■ **Choose a bin width appropriate to the data.** Computer programs usually do a pretty good job of choosing histogram bin widths. Often there's an easy way to adjust the width, sometimes interactively. Here are tsunami earthquake magnitudes with two (rather extreme) choices for the bin size:

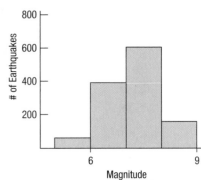

For the usual amounts of data we display, between 5 and 30 bins show the features of a distribution best. Too few bins tend to hide too much and too many bins make unimportant features too prominent. The task of summarizing a quantitative variable is relatively simple, and there is a simple path to follow. However, you need to watch out for certain features of the data that make summarizing them with a number dangerous. Here's some advice:

■ **Don't forget to do a reality check.** Don't let the computer or calculator do your thinking for you. Make sure the calculated summaries make sense. For example, does the mean look like it is in the centre of the histogram? Think about the spread: An IQR of 10 L/100 km would clearly be wrong for gas mileage. And no measure of spread can be negative. The standard deviation can take the value 0, but only in the very unusual case that all the data values equal the same number. If you see an IQR or standard deviation equal to 0, it's probably a sign that something's wrong with the data.

■ **Don't forget to sort the values before finding the median or percentiles.** It seems obvious, but when you work by hand, it's easy to forget to sort the data first before counting in to find medians, quartiles, or other percentiles. Don't report that the median of the five values 194, 5, 1, 17, and 893 is 1 just because 1 is the middle number.

■ **Don't worry about small differences when using different methods.** Finding the 10th percentile or the lower quartile in a data set sounds easy enough, but it turns out that the definitions are not exactly clear. If you compare different statistics packages or calculators, you may find that they give slightly different answers for the same data. These differences, however, are unlikely to be important in interpreting the data, the quartiles, or the IQR, so don't let them worry you.

■ **Don't compute numerical summaries of a categorical variable.** Neither the mean (5-digit) U.S. ZIP (postal) code nor the standard deviation of Canadian social insurance numbers is meaningful. If the variable is categorical, you should instead report summaries such as percentages of individuals in each category. It is easy to make this mistake when using technology to do the summaries for you. After all, the computer doesn't care what the numbers mean.

■ **Don't report too many decimal places.** Statistical programs and calculators often report a ridiculous number of digits. A general rule for numerical summaries when you report them is to use one or two more digits than the data. For example, earlier we saw a dotplot of Kentucky Derby race times. The mean and standard deviation of those times could be reported as

$$\bar{y} = 129.56779 \text{ sec}$$

$$s = 13.05657 \text{ sec}$$

Gold Card Customers—Regions National Banks		
Month	**April 2007**	**May 2007**
Average ZIP Code	45 034.34	38 743.34

but we knew the race times only to the nearest quarter-second, so the extra digits are meaningless.

- **Don't round in the middle of a calculation.** Don't *report* too many decimal places, but it's best not to do *any* rounding until the end of calculation. Even though you might report the mean of the earthquakes as 7.05, it's really 7.05281. Use the more precise number in your calculations if you're finding the standard deviation by hand—or be prepared to see small differences in your final result.

- **Watch out for multiple modes.** The summaries of the Kentucky Derby times are meaningless for another reason. As we saw in the dotplot, the Derby was initially a longer race. It would make much more sense to report that the old 1.5 mile Derby had a mean time of 159.6 seconds, while the current Derby has a mean time of 124.3 seconds. If the distribution has multiple modes, consider separating the data into different groups and summarizing each group separately.

- **Beware of outliers.** The median and IQR are resistant to outliers, but the mean and standard deviation are not. To help spot outliers . . .

- **Don't forget to make a picture (make a picture, make a picture).** The sensitivity of the mean and standard deviation to outliers is one reason why you should always make a picture of the data. Summarizing a variable with its mean and standard deviation when you have not looked at a histogram or dotplot to check for outliers or skewness invites disaster. You may find yourself making absurd or dangerously wrong conclusions about the data. And, of course, you should demand no less of others. Don't accept a mean and standard deviation blindly without some evidence that the variable they summarize is unimodal, reasonably symmetric, and free of outliers.

What Have We Learned?

Make and interpret displays of the distribution of a variable.

- Display the distribution of a quantitative variable with a histogram, stem-and-leaf display, or dotplot.
- Understand distributions in terms of their shape, centre, and spread.

Describe the shape of a distribution.

- A **symmetric** distribution has roughly the same shape reflected around the centre.
- A **skewed** distribution extends farther on one side than on the other.
- A **unimodal** distribution has a single major hump or mode; a bimodal distribution has two; multimodal distributions have more.
- **Outliers** are values that lie far from the rest of the data.
- Report any other **unusual feature** of the distribution, such as gaps.

Describe the centre of a distribution by computing the mean and median, and know when it is best to use each.

- The **mean** is the sum of the values divided by the count. It is a suitable measure of centre for unimodal, symmetric distributions.
- The **median** is the middle value; half the values are above and half are below the median. It is generally a better measure of centre when the distribution is skewed or has outliers.

Compute the standard deviation and interquartile range (IQR), and know when it is best to use each to describe the spread.

- The **standard deviation** is roughly the square root of the average squared difference between each data value and the mean. It is the measure of choice for the spread of unimodal, symmetric distributions.

- The **IQR** is the difference between the quartiles. It often provides a better measure of spread for skewed distributions or data with outliers.
- When the distribution is skewed, we'll find the 5-number summary and use it to report the median, the IQR, and to examine the difference in spread on the two sides of the distribution. If it's symmetric, we'll summarize the distribution with the mean and standard deviation (and possibly the 5-number summary as well). Always pair the median with the IQR and the mean with the standard deviation.

Recognize that the methods of this chapter assume that the data are quantitative.

- The **Quantitative Data Condition** serves as a reminder to check that the data are quantitative. A good way to be sure is to report the measurement units.

Find the 5-number summary and using it, make a boxplot. Use the boxplot's suspect outlier nomination rule to identify cases that may deserve special attention.

- A **5-number summary** consists of the median, the quartiles, and the extremes of the data.
- A **boxplot** shows the quartiles as the upper and lower ends of a central box, the median as a line across the box, and "whiskers" that extend to the most extreme values that are not nominated as possible outliers.
- Boxplots display separately any case that is more than 1.5 IQRs beyond each quartile. These cases may be considered as possible outliers.

Review of Terms

Distribution	The distribution of a variable is a representation of its possible values along with their corresponding frequencies or relative frequencies. (p. 51).
Histogram (relative frequency histogram)	A histogram slices up all the possible values of the variable into (usually) equal-width bins. Over each bin is a rectangle whose height represents the frequency (or relative frequency) of values falling into that bin. (p. 51).
Gap	A region of the distribution where there are no values (p. 52).
Stem-and-leaf display	A display that organizes quantitative data by breaking up each value into a stem and a leaf, and then attaching all the leaves to their proper (ordered) stems. It's best described in detail by example (p. 53).
Dotplot	graphs a dot for each case against a single axis (p. 54).
Quantitative Data Condition	specifies that the data are values of a quantitative variable whose units are known (p. 55).
Shape	To describe the shape of a distribution, look for (p. 55)

- single versus multiple modes
- symmetry versus skewness
- outliers and gaps (p. 57).

Mode	A hump or local high point in the shape of the distribution of a variable. The apparent location of modes can change as the binning of a histogram is changed (p. 56).
Unimodal (Bimodal)	Having one mode. This is a useful term for describing the shape of a histogram when it's generally mound-shaped. Distributions with two modes are called **bimodal**. Those with more than two are **multimodal** (p. 56).
Uniform	A distribution is said to be uniform distribution in shape when the histogram is rather flat and doesn't appear to have any clear mode. (p. 56).
Symmetric	A distribution is symmetric if the two halves on either side of the centre look approximately like mirror images of each other (p. 57).

Tails	The parts of a distribution that typically trail off on either side. Distributions can be characterized as having long tails (if they straggle off for some distance) or short tails (if they don't) (p. 57).
Skewed	A distribution is skewed if it's not symmetric and one tail stretches out farther than the other. Distributions are said to be **skewed left** (or negatively) when the longer tail stretches to the left (in the negative direction), and **skewed right** (or positively) when it goes to the right (in the positive direction) (p. 57).
Outliers	Extreme values that don't appear to belong with the rest of the data. They may be unusual values that deserve further investigation or just mistakes; there may be no obvious way to tell. Don't delete outliers automatically—you have to think about them. Outliers can affect many statistical analyses, so you should always be alert for them. Boxplots display points more than 1.5 IQR from either end of the box individually, but this is just a rule-of-thumb and not a definition of what is an outlier. (p. 57).
Centre	The place in the distribution of a variable that you'd point to if you want to attempt the impossible by summarizing the entire distribution with a single number. Measures of centre include the mean and median (p. 59).
Median	The middle value, with half of the data above and half below it. If *n* is even, it is the average of the two middle values. It is usually paired with the IQR (p. 60).
Spread	A numerical measure of how tightly the values are clustered around the centre. Measures of spread include the standard deviation and IQR (p. 61).
Range	The difference between the lowest and highest values in a data set. *Range = max − min* (p. 61).
Percentile	The *i*th percentile is the number that falls above *i* percent of the data (p. 62).
Quartile	**The lower quartile** (Q1) is the value with a quarter of the data below it. **The upper quartile** (Q3) has three quarters of the data below it. The median and quartiles divide data into four parts of equal numbers of data values (p. 62).
Interquartile range (IQR)	The difference between the first and third quartiles. *IQR = Q3 − Q1*. It is usually reported along with the median and 5-number summary (p. 62).
5-number summary	A useful summary of data consisting of minimum, Q1, median, Q3, and maximum (p. 64).
Boxplot	displays the 5-number summary as a central box with whiskers that extend to the most extreme values not designated as possible outliers. Boxplots are particularly effective for comparing groups and for spotting possible outliers (p. 64).
Mean	found by summing all the data values an d dividing by the count: $$\bar{y} = \frac{Total}{n} - \frac{\sum y}{n}$$ It is usually paired with the standard deviation (p. 67).
Resistant measure	A calculated summary measure is said to be resistant if it is affected only a limited amount by any small bunch of cases, such as outliers (p. 69).
Variance	The sum of squared deviations from the mean, divided by the count minus one (p. 71). $$s^2 = \frac{\sum (y - \bar{y})^2}{n - 1}$$
Standard deviation	The square root of the variance. $$s = \sqrt{\frac{\sum (y - \bar{y})^2}{n - 1}}$$ It is usually reported along with the mean (p. 71).

On the Computer   DISPLAYING AND SUMMARIZING QUANTITATIVE VARIABLES

Almost any program that displays data can make a histogram, but some will do a better job of determining where the bars should start and how they should partition the span of the data.

The vertical scale may be counts or proportions. Sometimes it isn't clear which. But the shape of the histogram is the same either way.

Most packages choose the number of bars for you automatically. Often you can adjust that choice.

The axis should be clearly labelled so you can tell what "pile" each bar represents. You should be able to tell the lower and upper bounds of each bar.

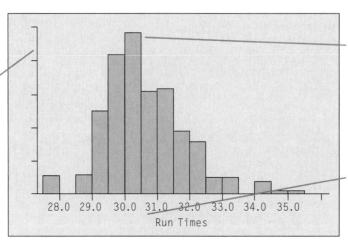

Run Times

Many statistics packages offer a pre-packaged collection of summary measures. The result might look like this:

```
Variable: Weight
N = 234
Mean = 143.3          Median = 139
St. Dev. = 11.1       Q1 = 134.0
                      Q3 = 148.0
```

Alternatively, a package might make a table for several variables and summary measures.

Variable	N	mean	median	stdev	Q1	Q3
Weight	234	143.3	139	11.1	134.0	148.0
Height	234	68.3	68.1	4.3	66.0	71.0
Score	234	86	88	9	86	91

It is usually easy to read the results and identify each computed summary. You should be able to read the summary statistics produced by any computer package.

Packages often provide many more summary statistics than you need. Of course, some of these may not be appropriate when the data are skewed or have outliers. It is your responsibility to check a histogram or stem-and-leaf display and decide which summary statistics to use.

It is common for packages to report summary statistics to many decimal places of "accuracy." Of course, it is rare that data have such accuracy in the original measurements. Just because a package calculates to six or seven digits beyond the decimal point doesn't mean that those digits have any meaning. Generally, it's a good idea to round these values, allowing perhaps one more digit of precision than was given in the original data.

Displays and summaries of quantitative variables are among the simplest things you can do in most statistics packages.

EXCEL

Excel cannot make histograms or dotplots without a third party add-in.

To calculate summaries, for example, the mean:

- Click on an empty cell.
- Go to the Formulas tab in the Ribbon. Click on the drop down arrow next to AutoSum and choose **Average**.
- Enter the data range in the formula displayed in the empty box you selected earlier.
- Press **Enter**. This computes the mean for the values in that range.

To compute the standard deviation:

- Click on an empty cell.
- Go to the Formulas tab in the Ribbon and click the drop down arrow next to "AutoSum" and select **More functions** . . .
- In the dialog window that opens, select **STDEV** from the list of functions and click **OK**. A new dialog window opens. Enter a range of fields into the text fields and click **OK**.

Excel computes the standard deviation for the values in that range and places it in the specified cell of the spreadsheet.

JMP

To make a histogram and find summary statistics:

- Choose **Distribution** from the **Analyze** menu.
- In the **Distribution** dialogue, drag the name of the variable that you wish to analyze into the empty window beside the label **Y, Columns**.

- Click **OK**. JMP computes standard summary statistics along with displays of the variables.

MINITAB

To make a histogram:

- Choose **Histogram** from the **Graph** menu.
- Select "Simple" for the type of graph and click **OK**.
- Enter the name of the quantitative variable you wish to display in the box labelled "Graph variables." Click **OK**.

To calculate summary statistics:

- Choose **Basic statistics** from the **Stat** menu. From the **Basic Statistics** submenu, choose **Display Descriptive Statistics**.

- Assign variables from the variable list box to the Variables box. MINITAB makes a Descriptive Statistics table.

COMMENTS

In MINITAB EXPRESS, choose Statistics > Describe > Descriptive Statistics.

R

For a quantitative variable X:

- **summary(X)** gives a 5-number summary and the mean.
- **mean(X)** gives the mean and **sd(X)** gives the standard deviation.
- **hist(X)** produces a histogram and **boxplot(X)** makes a boxplot.

COMMENTS

Many other summaries are available, including **min( )**, **max ( )**, **quantile(X,prob=p)** (where p is a probability between 0 and 1), and **median( )**. Both hist and boxplot have many options.

SPSS

To make a histogram in SPSS open the Chart Builder from the Graphs menu:

- Click the **Gallery** tab.
- Choose **Histogram** from the list of chart types.
- Drag the histogram onto the canvas.
- Drag a scale variable to the *y*-axis drop zone.
- Click **OK**.

To calculate summary statistics:

- Choose **Explore** from the **Descriptive Statistics** sub-menu of the **Analyze** menu. In the Explore dialog, assign one or more variables from the source list to the Dependent List and click the **OK** button.

STATCRUNCH

To make a histogram, dotplot, or stem-and-leaf plot:

- Click on **Graphics**.
- Choose the type of plot.
- Choose the variable name from the list of **Columns**.
- Click on **Next**.
- (For a histogram) Choose **Frequency** or (usually) **Relative frequency**, and (if desired) set the axis scale by entering the **Start** value and **Bin width**.
- Click on **Create Graph**.

To calculate summaries:

- Click on **Stat**.
- Choose **Summary Stats » Columns**.
- Choose the variable name from the list of **Columns**.
- Click on **Calculate**.

COMMENTS

- You may need to hold down the Ctrl or command key to choose more than one variable to summarize.
- Before calculating, click on **Next** to choose additional summary statistics.

TI-83/84 PLUS

To make a histogram:

- Turn a **STATPLOT** on.
- Choose the **histogram icon** and specify the List where the data are stored.
- **ZoomStat**, then adjust the **WINDOW** appropriately.

To calculate summary statistics:

- Choose **1-VarStats** from the **STAT CALC** menu and specify the List where the data are stored. You must scroll down to see the 5-number summary.

- To make a boxplot, set up a **STAT PLOT** using the box-plot icon.

COMMENTS

If the data are stored as a frequency table (say, with data values in L1 and frequencies in L2), set up the Plot with Xlist: L1 and Freq: L2.

Chapter Exercises

T indicates data is available in a saved file at the Companion Website at www.pearsoncanada.ca/deveaux

15. **Histogram** Find a histogram that shows the distribution of a variable in a newspaper, a magazine, or on the Internet.
   a) Does the article identify the *W*'s?
   b) Discuss whether the display is appropriate for the data.
   c) Discuss what the display reveals about the variable and its distribution.
   d) Does the article accurately describe and interpret the data? Explain.

16. Not a histogram Find a graph other than a histogram that shows the distribution of a quantitative variable in a newspaper, a magazine, or on the Internet.
a) Does the article identify the *W*'s?
b) Discuss whether the display is appropriate for the data.
c) Discuss what the display reveals about the variable and its distribution.
d) Does the article accurately describe and interpret the data? Explain.

17. In the news Find an article in a newspaper, a magazine, or on the Internet that discusses an "average."
a) Does the article discuss the *W*'s for the data?
b) What are the units for the variable?
c) Is the average used the median or the mean? How can you tell?
d) Is the choice of median or mean appropriate for the situation? Explain.

18. In the news II Find an article in a newspaper, a magazine, or on the Internet that discusses a measure of spread.
a) Does the article discuss the *W*'s for the data?
b) What are the units for the variable?
c) Does the article use the range, IQR, or standard deviation?
d) Is the choice of measure of spread appropriate for the situation? Explain.

19. Thinking about shape Would you expect distributions of these variables to be uniform, unimodal, or bimodal? Symmetric or skewed? Explain.
a) The number of speeding tickets each student in their final year of university has ever had.
b) Players' scores (number of strokes) at the RBC Canadian Open golf tournament in a given year.
c) Weights of female babies born in a particular hospital over the course of a year.
d) The length of the average hair on the heads of students in a large class.

20. More shapes Would you expect distributions of these variables to be uniform, unimodal, or bimodal? Symmetric or skewed? Explain.
a) Ages of people at a pee-wee hockey game.
b) Number of siblings of people in your class.
c) Pulse rates of college- or university-age males.
d) Number of times each face of a die shows in 100 tosses.

Ⓣ 21. Sugar in cereals The histogram displays the sugar content (as a percent of weight) of 49 brands of breakfast cereals.

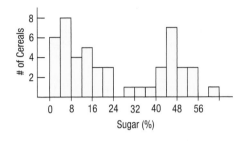

a) Describe this distribution.
b) What do you think might account for this shape?

22. Chorus heights The display shows the heights of some of the singers in a chorus collected, so that the singers could be positioned on stage with shorter ones in front and taller ones in back.

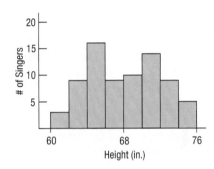

a) Describe the distribution.
b) Can you account for the features you see here?

23. Test scores Below is a histogram of 110 students' test scores on a one-hour test in STA220 at the University of Toronto during summer 2007. The test was out of 50, not 100.

a) Approximately what percentage of students got A grades? An A was 40 or higher, since the test was out of 50 (some scored above 50 due to bonus marks).
b) What percent got C or D grades? (i.e., between 50% and 69%, or 25 and 35 in the graph since any scores of 35 went into the higher bin)
c) Write a brief description of this distribution (shape, centre, spread, and unusual features). Can you account for any of the features you see here?

24. Run times One of the authors collected the times (in minutes) it took him to run four miles on various courses during a 10-year period. Here is a histogram of the times.

Describe the distribution and summarize the important features. What is it about running that might account for the shape you see?

T 25. Election 2011 Below is a histogram of percentages of rejected ballots for each of the 308 federal electoral districts in the 2011 Canadian federal election.

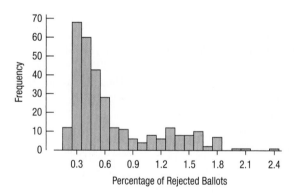

a) From the histogram, would you expect the mean or median to be larger? Explain. What can you say about the actual numerical value of the median from the graph?
b) Write a few sentences describing this distribution (shape, centre, spread, unusual features). Precise numerical calculations are not necessary (though some numbers are). Is there anything of possible practical importance to be learned from this graph? Any further investigation you'd suggest?

T 26. Emails A university professor saved every email received from students in a large Introductory Statistics class during an entire term. He then counted, for each student who had sent him at least one email, how many emails each student had sent.

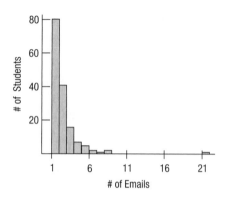

a) From the histogram, would you expect the mean or the median to be larger? Explain. What can you say about the actual value of the median from this graph?
b) Write a few sentences describing this distribution (shape, centre, spread, and unusual features).

27. Summaries Here are costs of 10 electric smoothtop ranges rated very good or excellent by *Consumer Reports* on their website www.consumerreports.org.

$850, 900, 1 400, 1 200, 1 050, 1 000, 750, 1 250, 1 050, 565

Find these statistics *by hand* (no calculator!):
a) Mean
b) Median and quartiles
c) Range and IQR

28. Tornadoes 2011 Here are the annual numbers of deaths from tornadoes in the United States from 1998 through 2011 (*Source:* NOAA):

130, 94, 40, 40, 555, 54, 35, 38, 67, 81, 125, 21, 45, 544

Find these statistics *by hand* (no calculator!):
a) Mean
b) Median and quartiles
c) Range and IQR

29. Mistake A clerk entering salary data into a company spreadsheet accidentally put an extra "0" in the boss's salary, listing it as $2 000 000 instead of $200 000. Explain how this error will affect these summary statistics for the company payroll:
a) Measures of centre: median and mean
b) Measures of spread: range, IQR, and standard deviation

30. Sick days During contract negotiations, a company seeks to change the number of sick days employees may take, saying that the annual "average" is seven days of absence per employee. The union negotiators counter that the "average" employee misses only three days of work each year. Explain how both sides might be correct, identifying the measure of centre you think each side is using and why the difference might exist.

31. Standard deviation I For each lettered part, a) through c), examine the two given sets of numbers. Without doing any calculations, decide which set has the larger standard deviation and explain why. Then check by finding the standard deviations *by hand*.

	Set 1	Set 2
a)	3, 5, 6, 7, 9	2, 4, 6, 8, 10
b)	10, 14, 15, 16, 20	10, 11, 15, 19, 20
c)	2, 6, 6, 9, 11, 14	82, 86, 86, 89, 91, 94

32. Standard deviation II For each lettered part, a) through c), examine the two given sets of numbers. Without doing any calculations, decide which set has the larger standard deviation and explain why. Then check by finding the standard deviations *by hand*.

	Set 1	Set 2
a)	4, 7, 7, 7, 10	4, 6, 7, 8, 10
b)	100, 140, 150, 160, 200	10, 50, 60, 70, 110
c)	10, 16, 18, 20, 22, 28	48, 56, 58, 60, 62, 70

33. Payroll A small warehouse employs a supervisor at $1200 a week, an inventory manager at $700 a week, six stock boys at $400 a week each, and four drivers at $500 a week each.
a) Find the mean and median wage.
b) How many employees earn more than the mean wage?
c) Which measure of centre best describes a typical wage at this company, the mean or the median?
d) Which measure of spread would best describe the payroll, the range, the IQR, or the standard deviation? Why?

T 34. Singers The frequency table shows the heights (in inches) of 130 members of a choir.

Height	Count	Height	Count
60	2	69	5
61	6	70	11
62	9	71	8
63	7	72	9
64	5	73	4
65	20	74	2
66	18	75	4
67	7	76	1
68	12		

a) Find the 5-number summary and IQR.
b) Find the mean and standard deviation.
c) Display these data with a histogram.
d) Write a few sentences describing the distribution of heights.

T 35. Alberta casinos 2013 Below are the numbers of gaming machines in each of Alberta's 24 casinos:

Apex (Gold Dust) Casino	330
Baccarat Casino	330
Boomtown Casino	415
Camrose Casino	200
Cash Casino – Calgary	600
Cash Casino – Red Deer	299
Casino by Vanshaw	230
Casino Calgary	862
Casino Dene	170
Casino Edmonton	860
Casino Lethbridge	437
Casino Yellowhead	783
Century Casino – Calgary	525
Century Casino – Edmonton	758
Cowboys (Stampede) Casino	400
Deerfoot Inn & Casino	767
Eagle River Casino & Travel Plaza	250
Elbow River Casino	600
Great Northern Casino	419
Grey Eagle Casino & Bingo	600
Jackpot Casino	204
Palace Casino	704
River Cree Resort & Casino	1000
Stoney Nakoda Resort & Casino	299

Source: gamblinginstitute.ualberta.ca/
LibraryResources/ReferenceSources/AlbertaCasinos

Construct a stem-and-leaf plot and a dotplot for these data. Describe the distribution (including a 5-number summary).

T 36. He shoots, he scores During his 20 seasons in the NHL, the "Great One", Wayne Gretzky, scored 50% more points than anyone who ever played professional hockey. He accomplished this amazing feat while playing in 280 fewer games than Gordie Howe, the previous record holder. Here are the total number of points scored by Gretzky during each season, in order, from the 1979–80 season to the 1998–99 season:

137, 164, 212, 196, 205, 208, 215, 183, 149, 168, 142, 163, 121, 65, 130, 48, 102, 97, 90, 62

a) Create a stem-and-leaf display for these data with 7 to 10 bins (rows). You will need to use split stems, and to either truncate (i.e., chop off last digit) or round the values. Another alternative is to carry two-digit leaves separated by commas.
b) Describe the shape and any unusual features of the distribution. What might explain it?
c) Give the most appropriate and useful numerical summary measures for this distribution, and explain your choice.

T 37. How tall? Students in a large class were asked to estimate the teacher's height in centimetres. Here's a histogram of their estimates:

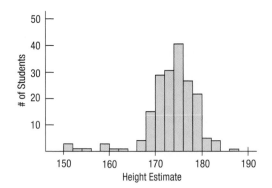

You want a good guess of the teacher's true height (or at least a good guess of how tall the students think the teacher is). How would you summarize these data? Why? What might explain guesses as low as 150 centimetres?

T 38. He shoots, he scores again In Exercise 36, you examined the total number of points scored by hockey great Wayne Gretzky during his 20-year career in the NHL.
a) Would you use the median or the mean to describe the centre of this distribution? Why?
b) Find the median.
c) Without actually finding the mean, would you expect it to be higher or lower than the median? Explain.

T 39. World Series champs On Oct 23, 1993, Joe Carter hit the 9th inning game-winning home run (with the Jays trailing 6–5) to make the Toronto Blue Jays World Series champions for a second straight year. Here is a stem-and-leaf display of the number of home runs hit by Joe Carter during the 1983–98 seasons. (*Source:* www.base-ballreference.com/players/c/cartejo01.shtml). Describe the distribution, mentioning its shape and any unusual features.

```
0 | 0
0 |
1 | 3
1 | 58
2 | 14
2 | 5779
3 | 02334
3 | 5
(3|2 means 32)
```

T 40. Bird species 2013 The Cornell Lab of Ornithology holds an annual Christmas Bird Count (www.birdsource.org), in which bird watchers at various locations around the country see how many different species of birds they can spot. Here are the number of species counted from the 20 sites with the most species in 2013:

184	98	101	126	150
166	82	136	124	118
133	83	86	101	105
97	88	131	128	106

a) Create a stem-and-leaf display of these data.
b) Write a brief description of the distribution. Be sure to discuss the overall shape as well as any unusual features.

T 41. Hurricanes 2015 The following data give the number of hurricanes classified as major hurricanes in the Atlantic Ocean each year from 1944 through 2015, as reported by NOAA (*Source:* www.nhc.noaa.gov):

3, 3, 1, 2, 4, 3, 8, 5, 3, 4, 2, 6, 2, 2, 4, 2, 2, 7, 1, 2, 6, 1, 3, 1, 0, 5, 2, 1, 0, 1, 2, 3, 2, 1, 2, 2, 2, 3, 1, 1, 1, 3, 0, 1, 3, 2, 1, 2, 1, 1, 0, 5, 6, 1, 3, 5, 3, 4, 2, 3, 6, 7, 2, 2, 5, 2, 5, 4, 2, 0, 2, 2

a) Create a dotplot of these data.
b) Describe the distribution.

T 42. Horsepower Create a stem-and-leaf display for these horsepowers of autos reviewed by *Consumer Reports* one year, and describe the distribution:

155	103	130	80	65
142	125	129	71	69
125	115	138	68	78
150	133	135	90	97
68	105	88	115	110
95	85	109	115	71
97	110	65	90	
75	120	80	70	

T 43. World Series Champs again Students were asked to make a histogram of the number of home runs hit by Joe Carter from 1983 to 1998 (see Exercise 39). One student submitted the following display (he noted that Carter had hit no home runs in 1983 while playing in only 23 games):

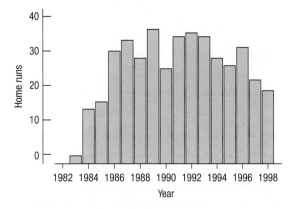

a) Comment on this graph.
b) Create your own histogram of the data.

44. Return of the birds 2013 Students were given the assignment to make a histogram of the data on bird counts reported in Exercise 40. One student submitted the following display:

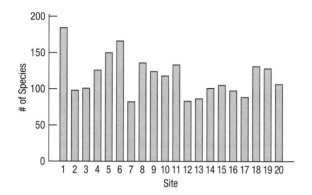

a) Comment on this graph.
b) Create your own histogram of the data.

45. Acid rain Two researchers measured the pH (a scale on which a value of 7 is neutral and values below 7 are acidic) of water collected from rain and snow over a six-month period in Allegheny County, Pennsylvania. Describe their data with a graph and a few sentences.

4.57 5.62 4.12 5.29 4.64 4.31 4.30 4.39 4.45

5.67 4.39 4.52 4.26 4.26 4.40 5.78 4.73 4.56

5.08 4.41 4.12 5.51 4.82 4.63 4.29 4.60

46. Sip size Researchers in Cornell University's Food Sciences department study how people experience foods. One study considered how much liquid people typically take into their mouths in one "sip" (in millilitres). The researchers also recorded the height (metres) and weight (kilograms) of the participants. Here are histograms of three of the variables from that study:

a) For which of the variables depicted in the histograms would you be most satisfied to summarize the centre with a mean? Explain.
b) For which of the variables depicted in the histograms would you most strenuously insist on using an IQR rather than a standard deviation to describe spread? Explain.

T 47. Housing price boom House prices were booming in some Canadian cities during 2015–16. Below are percentage increases in new home prices in major cities:

Metropolitan Area	Percentage increase in new housing prices between March 2015 and March 2016
St. John's	0.3
Halifax	−0.3
Charlottetown	−0.3
Saint John, Fredericton and Moncton	0.1
Québec	−0.6
Montréal	1.4
Ottawa–Gatineau	0.4
Toronto and Oshawa	9
Hamilton	3.7
St. Catharines–Niagara	6.8
Kitchener	2.3
London	2.4
Windsor	1.8
Greater Sudbury and Thunder Bay	−0.1
Winnipeg	2.3
Regina	−0.4
Saskatoon	−3.5
Calgary	−1.9
Edmonton	−0.3
Vancouver	7.7
Victoria	3.8

Source: "Adapted from Statistics Canada, New Housing Price Index, The Daily, May 12, 2016."

a) Give a stem-and-leaf plot for these data after rounding (not truncating) the values to the nearest 1.0. Split your stems, writing each one twice, once for the five lowest value leaves and once for the five highest value leaves.
b) Calculate the 5-number summary using the stem-and-leaf plot.
c) Find the mean. Did you expect it to be larger or smaller than the median? Why?
d) Give a useful brief report about annual metropolitan area house price increases (including any relevant summary measures).
e) Data taken over time or space can show a dependence on time or location. We can easily plot data versus time (see next chapter), but showing dependence on location is trickier. Do you think that the data above show any apparent or identifiable spatial (geographical) patterns? If so explain. (Optional: Find a map of Canada and put bars at each city listed above, with height equal to new housing percentage increase.)

T 48. Getting high 2011 The European School Survey Project on Alcohol and other Drugs issued in its 2011 report: *Substance Use Among Students in 36 European Countries,* (www.espad.org). Among other issues, the survey investigated the percentages of 15–16 year olds who had used alcohol together with pills to get high. Here are the results for the 36 European countries. Create an appropriate graph of these data, and describe the distribution.

Country	%Booze+Pills	Country	%Booze+Pills
Albania	2	Latvia	6
Belgium	2	Liechtenstein	7
Bosnia & Herz.	1	Lithuania	4
Bulgaria	4	Malta	8
Croatia	10	Moldava	1
Cyprus	4	Monaco	5
Czech Republic	16	Montenegro	2
Denmark	3	Norway	2
Estonia	4	Poland	5
Faroe Islands	3	Portugal	3
Finland	10	Romania	3
France	7	Russian Fed.	3
Germany	8	Serbia	2
Greece	4	Slovak Republic	8
Hungary	10	Slovenia	4
Iceland	2	Sweden	4
Ireland	5	Ukraine	2
Italy	3	United Kingdom	5

49. Trimmed mean We can make the ordinary mean a more resistant measure by trimming off a certain percent of the biggest values and the smallest values. Let TrMean(x%) represent the mean of the data after placing the data in order from smallest to biggest and deleting x% at each end. So the 10% trimmed mean, for example, is the average of the middle 80% of the data. If you can't trim off exactly x%, trim off the percentage that is closest to x% (a computer would interpolate).
a) Find the mean, median and 10% trimmed mean of the following data: 6, 8, 8, 8, 1, 5, 7, 7, 6, 8.
b) Find the mean, median and 5% trimmed mean of the data from Exercise 48.
c) Explain the relative order of the three measures in each of the above data sets. Note that the ordinary mean is just a 0% trimmed mean and the median is essentially a 50% trimmed mean. A 5% or 10% trimmed mean can be a nice compromise between these two measures, more resistant than the mean, while using more of the data than the median.

T 50. Raptors 2011 Below are the lengths of all the Toronto Raptor loss streaks in the 2011–2012 season:

3, 2, 8, 1, 2, 3, 4, 1, 1, 1, 2, 1, 3, 2, 1, 4, 4
a) Construct a stem-and-leaf plot for these data.
b) Calculate the median, the 5% trimmed mean (see previous exercise), and the ordinary mean. Place these three measures in order from smallest to largest, and explain why you would expect them to be in this order for data such as these.

c) What is the mode of these data (highest frequency value)?
d) Can you think of another variable (categorical) that perhaps we should take into account in order to improve an analysis of loss streaks (rather than just ignore such a variable and let it lurk)?

51. Final grades A professor (of something other than Statistics!) distributed the following histogram to show the distribution of grades on his 200-point final exam. Comment on the display.

52. Final grades revisited After receiving many complaints about his final-grade histogram from students currently taking a Statistics course, the professor from Exercise 51 distributed the following revised histogram:

a) Comment on this display.
b) Describe the distribution of grades.

53. Zip codes Holes-R-Us, an Internet company that sells piercing jewellery, keeps transaction records on its sales. At a recent sales meeting, one of the staff presented a histogram and summary statistics for the U.S. zip codes of the last 500 U.S. customers so that they might understand where sales are coming from. Comment on the usefulness and appropriateness of the display and summary statistics.

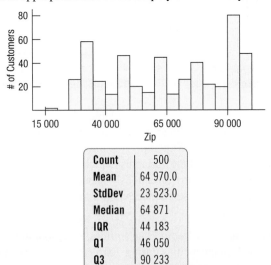

Count	500
Mean	64 970.0
StdDev	23 523.0
Median	64 871
IQR	44 183
Q1	46 050
Q3	90 233

54. Industry codes An investment analyst has been assigned to analyze the companies that make up the Fortune 800, a collection of the largest U.S. companies. He codes each company by the type of industry to which it belongs. Here is a table showing some of the industry types and the corresponding codes he uses.

Industry	Industry Code
Financial services	1
Food/drink/tobacco	2
Health	3
Insurance	4
Retailing	6
Forest products	9
Aerospace/defense	11
Energy	12
Capital goods	14
Computers/communications	16
Entertainment/information	17
Consumer nondurables	18
Electric utilities	19

The analyst produces the following histogram of the industry codes for the companies:

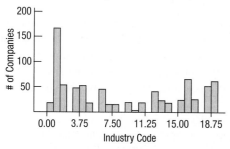

a) What might account for the gaps seen in the histogram?
b) Is the histogram unimodal? While the mode is 1, the median is 6. Interpret these numbers.
c) What advice might you give the analyst about the appropriateness of this display?

55. Golf courses One measure of the difficulty of a golf course is its length: the total distance (in yards) from tee to hole for all 18 holes. Here are the histogram and summary statistics for the lengths of all the golf courses in Vermont.

Count	45
Mean	5 892.91 yd
StdDev	386.59
Min	5 185
Q1	5 585.75
Median	5 928
Q3	6 131
Max	6 796

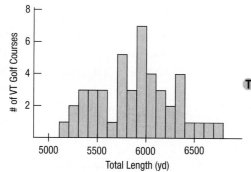

a) By hand, construct a boxplot directly below the histogram, using the same *x*-axis.
b) Between what lengths do the central 50% of these courses lie?
c) What summary statistics would you use to describe these data?
d) Write a brief description of these data (shape, centre, and spread).
e) Compute the intervals $(\bar{y} \pm s)$ and $(\bar{y} \pm 2s)$. Count, approximately, using the histogram, the number of golf courses with lengths within each interval, and convert to percents. Are the percents close to what you expected?

56. Women's Olympic alpine 2010 The women's super combined alpine event consists of a downhill and a slalom. Here are the total times (adding the two race times) for the 2010 Vancouver Olympics.

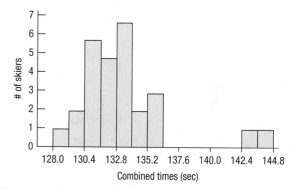

Count	28
Mean	133.43
StdDev	3.40
Min	129.14
Q1	131.09
Median	132.87
Q3	134.11
Max	144.15

a) By hand, construct a boxplot directly below the histogram, using the same *x*-axis (assume data points fall at midpoints of bins when helpful).
b) Between what lengths do the central 50% of these times lie?
c) Using appropriate summary statistics, write a brief description of these data (shape, centre, and spread).

T 57. Math scores 2009 The Programme for International Student Assessment (PISA) reported 2009 average mathematics performance scores for 15-year-olds in 34 Organisation for Economic Co-operation and Development (OECD) nations. Korea led the group, with an average score of 546, while Mexico had the lowest average of 419. (Non-OECD China topped everyone, with a 600 average). The average scores for each nation are given below:[17]

Country	Score	Country	Score
Australia	514	Japan	529
Austria	496	Korea	546
Belgium	515	Luxembourg	489
Canada	527	Mexico	419
Chile	421	Netherlands	526
Czech Republic	493	New Zealand	519
Denmark	503	Norway	498
Estonia	512	Poland	495
Finland	541	Portugal	487
France	497	Slovakia	497
Germany	513	Slovenia	501
Greece	466	Spain	483
Hungary	490	Sweden	494
Iceland	507	Switzerland	534
Ireland	487	Turkey	445
Israel	447	United Kingdom	492
Italy	483	United States	487

a) Find the 5-number summary and IQR, the mean, and the standard deviation of these national averages. Can you explain why the mean is a bit smaller than the median?

b) Which summary of the data would you report? Why?

c) Write a brief summary of the performance of 15-year-olds in OECD countries. Be sure to comment on the performance of Canada and the United States.

d) In Canada, the standard deviation of the individual students' scores was 88 and the distribution was fairly bell-shaped. About 68% of students scored between ___and ___(fill in the blanks). Only about 5% scored less than ___or more than ___(fill in the blanks). Only a real math genius could have scored above ___(fill in the blank).

T 58. Incarceration rates 2013 Below is a portion of the data set you can find online at http://www.pearsoncanada.ca/deveaux showing incarceration rates (prisoners per 100 000 of the nation's population) for 221 nations of the world, according to the World Prison Brief at the International Centre for Prison Studies:

Country	Rate
United States of America	716
St. Kitts and Nevis	649
Seychelles	641
Virgin Islands (USA)	539
Rwanda	527
Cuba	510
Russian Federation	490
: :	
: :	
: :	
Central African Republic	19
Comoros	19
Faroe Islands	17

a) Graph these data.

b) Calculate appropriate summary statistics.

c) Write a few sentences about these data. (Include shape, centre, spread, and unusual features.) Discuss where Canada and the U.S. stand respectively in this distribution.

T 59. First Nations 2010 Below are the first few entries from a data file you can find online at www.pearsoncanada.ca/deveaux showing the sizes of First Nations registry groups in Ontario. Registry groups are usually the same as bands, except that occasionally a band may consist of more than one registry group, like the Six Nations of Grand River in Ontario, which consists of 13 registry groups denoted in the complete data file. Hence in Ontario, there are 126 bands but 138 registry groups. http://www.aadnc-aandc.gc.ca/DAM/DAM-INTER-HQ/STAGING/texte-text/ai_rs_pubs_sts_ni_rip_rip10_rip10_1309289046808_eng.pdf

Registry Group	Size
Aundeck-Omni-Kaning	752
Batchewana_First_Nation	2460
Dokis	988
Garden_River_First_Nation	2396
Henvey_Inlet_First_Nation	654
Magnetawan	233
M'Chigeeng_First_Nation	2329
Mississauga	1104
Nipissing_First_Nation	2245
. . .	
. . .	

[17]Based on The National Assessment of Educational Progress, NCES.

a) Using appropriate graphical displays and summary statistics, write a report on the distribution of registry group size in Ontario.

b) Suppose we were to analyze bands rather than registry groups. Explain how each of the following would change if you analyzed bands rather than registry groups: mean, median, standard deviation, IQR, and histogram. Use your logic here, not new calculations.

T 60. **Election 2011 Maritimes** Below are results from the 2011 Canadian federal election for the Maritime provinces, showing the riding name, winning candidate, and percentage of rejected ballots:

Electoral district	Percentage of Ballots Rejected	Elected candidate
Newfoundland and Labrador		
Avalon	0.5	Andrews, Scott Liberal
Bonavista—Gander—Grand Falls—Windsor	0.5	Simms, Scott Liberal
Humber—St. Barbe—Baie Verte	0.3	Byrne, Gerry Liberal
Labrador	0.5	Penashue, Peter Conservative
Random—Burin—St. George's	0.5	Foote, Judy Liberal
St. John's East	0.3	Harris, Jack NDP
St. John's South—Mount Pearl	0.3	Cleary, Ryan NDP
Prince Edward Island		
Cardigan	0.5	MacAulay, Lawrence Liberal
Charlottetown	0.6	Casey, Sean Liberal
Egmont	0.8	Shea, Gail Conservative
Malpeque	0.4	Easter, Wayne Liberal
Nova Scotia		
Cape Breton—Canso	0.9	Cuzner, Rodger Liberal
Central Nova	0.6	MacKay, Peter G. Conservative
Cumberland—Colchester—Musquodoboit Valley	0.6	Armstrong, Scott Conservative
Dartmouth—Cole Harbour	0.6	Chisholm, Robert NDP
Halifax	0.5	Leslie, Megan NDP
Halifax West	0.5	Regan, Geoff Liberal
Kings—Hants	0.5	Brison, Scott Liberal
Sackville—Eastern Shore	0.6	Stoffer, Peter NDP
South Shore—St. Margaret's	0.7	Keddy, Gerald Conservative
Sydney—Victoria	0.7	Eyking, Mark Liberal
West Nova	0.8	Kerr, Greg Conservative

New Brunswick		
Acadie—Bathurst	1.3	Godin, Yvon NDP
Beauséjour	1.2	LeBlanc, Dominic Liberal
Fredericton	0.5	Ashfield, Keith Conservative
Fundy Royal	0.6	Moore, Rob Conservative
Madawaska—Restigouche	1.6	Valcourt, Bernard Conservative
Miramichi	1.1	O'Neill Gordon, Tilly Conservative
Moncton—Riverview—Dieppe	0.7	Goguen, Robert Conservative
New Brunswick Southwest	0.6	Williamson, John Conservative
Saint John	0.5	Weston, Rodney Conservative
Tobique—Mactaquac	0.8	Allen, Mike Conservative

Source: Elections Canada

a) Make a suitable display of the percentages of rejected ballots.

b) Find the mean and standard deviation.

c) Report the 5-number summary.

d) Why do the mean and median differ here?

e) Which of b) and c) above does a better job of summarizing the distribution of percentage rejected ballots? Why?

f) Suppose the true percentage for Egmont was 1.8% and not 0.8%. How would you expect the mean, median, standard deviation, and IQR to change? Explain your expectations for each (no computations, please).

g) Write a brief report about rejected ballots in the Maritimes.

T 61. **Bi-lingual Ni-lingual 2011** Canada is officially bilingual, according equal treatment of English and French languages, but individual Canadians who are fluent in English and French are uncommon in many regions of the country. And how many Canadians speak neither English nor French? Below are data from the 2011 Census for major metropolitan areas in Canada. Compute the proportion of the total bilingual population in each city, and the proportion knowledgeable in neither language.

a) Examine the distribution of the proportion of city residents who are bilingual, using suitable graphical and numerical methods. Summarize and report your findings.

| Population by knowledge of official language, by census metropolitan area (2011 Census) | | | | |
City	Total number	English only	French only	Both English and French	Neither English nor French
Abbotsford-Mission (BC)	167 815	151 740	45	7 830	8 200
Barrie (ON)	185 345	172 680	150	11 775	740
Brantford (ON)	133 915	127 665	35	5 620	600
Calgary (AB)	1 205 175	1 088 785	1 040	89 345	26 000
Edmonton (AB)	1 146 600	1 045 145	1 310	83 640	16 500
Greater Sudbury/Grand Sudbury (ON)					
	159 200	95 390	1 665	61 795	350
Guelph (ON)	140 410	127 285	55	11 620	1 450
Halifax (NS)	386 440	338 470	380	46 405	1 190
Hamilton (ON)	712 580	656 790	480	45 060	10 250
Kelowna (BC)	177 615	165 370	65	11 515	660
Kingston (ON)	155 405	135 645	315	18 860	580
Kitchener-Cambridge-Waterloo (ON)					
	472 095	434 105	220	30 275	7 490
London (ON)	469 010	432 600	240	31 095	5 065
Moncton (NB)	136 145	68 060	4 650	63 245	195
Montréal (QC)	3 785 915	280 785	1 401 455	2 039 035	64 635
Oshawa (ON)	353 460	326 965	245	24 640	1 605
Ottawa-Gatineau (ON)	1 222 760	556 405	105 280	547 625	13 450
Peterborough (ON)	117 595	109 420	30	7 895	250
Québec (QC)	756 405	1 550	479 895	273 335	1 620
Regina (SK)	208 085	194 785	140	11 700	1 455
Saguenay (QC)	156 585	150	125 335	31 050	45
Saint John (NB)	126 280	107 145	160	18 695	280
Saskatoon (SK)	256 900	239 330	125	15 350	2 105
Sherbrooke (QC)	199 190	3 060	111 455	83 855	815
St. Catharines-Niagara (ON)	386 525	353 445	490	29 910	2 680
St. John's (NL)	194 935	180 240	40	14 345	305
Thunder Bay (ON)	119 925	110 485	135	8 755	550
Toronto (ON)	5 541 880	4 872 010	5 000	424 265	240 600
Trois-Rivières (QC)	149 710	190	106 775	42 535	210
Vancouver (BC)	2 292 115	1 997 605	1 265	164 785	128 460
Victoria (BC)	339 725	303 850	185	33 060	2 635
Windsor (ON)	316 515	279 865	410	31 650	4 600
Winnipeg (MB)	721 115	635 575	1 050	75 715	8 775

Source: Adapted from Statistics Canada. (2012). Language Highlight Tables. 2011 Census. Statistics Canada Catalogue no. 98-314-XWE2011002

b) Make changes to your graph above (by hand is fine) so that the cities in Quebec or New Brunswick have different symbols (in shape or colour) than the other cities. Does that group of cities appear to differ from the rest in bilingualism?

c) Examine the distribution of the proportion of city residents who speak neither English nor French, using suitable graphical and numerical methods. Summarize and report your findings.

T 62. First Nations 2010 on reserves Open up the data file First Nations 2010 available online at http://www.pearsoncanada.ca/deveaux. For each band, you see both the band size as well as the number of individuals living on reserve. Let's analyze the latter variable.

a) Construct a histogram for band counts of those living on reserve. Describe the distribution as best you can without doing any computations. How does it differ from the distribution of band sizes?

b) Instead of the actual count of those living on reserve, we might be interested in the proportion of the band that lives on reserve. Do you think its distribution should be similar to the distribution of on-reserve counts? Compute this proportion for each band by dividing the reserve count by the band size. Examine the distribution graphically and describe it as best you can without doing precise calculations. From your graph, guess the mean and standard deviation (this will certainly be an easier task after you have read Chapter 5).

c) Compute the mean and standard deviation for the on-reserve proportions. Are they close to your guesses?

d) For the on-reserve proportions, do you expect the mean and median to be similar? Why? For the living on reserve counts, do you expect the mean and median to be similar? Why?

e) Are the bands lying far from the middle of the distribution of proportions likely to be larger or smaller bands? Why?

f) Is the mean of the proportions the same as the overall proportion of the (band registered) First Nations population living on reserve?

T **63. Election 2011 again** Below are a graph and some calculations describing the distribution of voter turnout percentage for all federal ridings in the 2011 federal election in Canada.

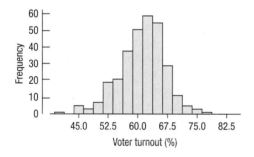

Descriptive Statistics: Voter turnout percentage			
N:	308	Q1:	57.700
Mean:	61.068	Median:	61.700
St.Dev.:	5.772	Q3:	65.075
Minimum:	40.300	Maximum:	77.000

a) What does the relationship between the mean and median suggest?

b) Give a brief report about percentage of voter turnout in the 2011 election.

c) Work out the two numbers: "$\bar{y} + 2s$" and "$\bar{y} - 2s$" and count the number of observations that lie in between (it is easier to count the number that lie outside, then subtract). Is it close to 95%? Did you expect it to be? To count, either use the actual data

that are available online, or estimate from the graph above, assuming that data are evenly spaced or spread out across any bin.

d) Since the mean of the percentages is 61.07, we can assume that the percentage of eligible voters in all these ridings combined who went to vote was 61.07%—true or not? Explain.

64. Family income 2011 According to Statistics Canada, Income Statistics Division, the mean after-tax family income in 2011 was $63 000, while the median was $50 700. In constant 2011 dollars, the mean and median in 1980 were $54 000 and $49 500 respectively.

a) Why did you expect to find a higher mean than median in both years? Draw a very rough plot of what the histogram for 2011 incomes might look like.

b) How much higher was the mean compared with the median in 2011, in percentage terms? How much higher in 1980? What does this suggest about how the income distribution has changed in Canada? Draw two very rough histograms to represent what the two distributions might look like in comparison to each other. Are the rich getting richer, and the poor poorer? Explain your answer.

c) Suppose that in 2030, the mean and median in actual 2030 dollars are $80 000 and $60 000 respectively. Have the rich gotten richer, and the poor poorer? Say what you can about what appears to be going on with incomes, based on this limited information.

65. Run times continued For the times provided in Exercise 24, explain how you would expect the mean, median, standard deviation, range, and IQR to change if we were to:

a) Change all times above 32.0 to 32.0.

b) Add 50 new run times, all between 30.0 and 31.0.

c) Add 30 new run times around 32.7 and another 30 times around 28.7 (the median and mean are 30.5 and 30.7 respectively).

d) Subtract 1 minute from each time.

e) Add 5 times around 29.5 and another 5 times around 35.0.

f) Change 10 times of 32.0 to 32.5 and 10 times of 29.4 to 28.9.

66. Test scores continued For the data provided in Exercise 23, explain how you would expect the mean, median, standard deviation, range, and IQR to change (with brief explanation), if you were to:

a) Delete all the test scores between 0 and 5.

b) Change all the test scores above 50 to scores of exactly 50.

c) Correct some data entry errors: One score recorded as 35 was actually 3.5 and another score recorded as 45 was actually 4.5.

d) Delete half the test scores between 30 and 35.

e) Add 30 additional graded tests with scores all between 30 and 35.

f) Add 20 additional test papers, where 10 scored 50 and the other 10 scored 15.

g) Adjust all the test scores upward by +5.

h) Double all the raw test scores so that you get percentage scores out of 100.

67. Grouped data Suppose that the data on heights provided in Exercise 34 had been organized into only six bins, as follows:

Height	Count
60–62	17
63–65	32
66–68	37
69–71	24
72–74	15
75–77	5

a) If you no longer had the actual data, but just this grouped data, could you still calculate the mean and standard deviation? Try the following: Pretend that each value in a bin lies exactly at the midpoint (i.e., 17 members stand 61 inches, 32 stand 64 inches, etc.). Now proceed exactly as you did before for the ungrouped data, and find the mean and standard deviation. Did you get good approximations to the actual values? Why do you think this method works as well as it does?

b) Write down a formula for the mean as you calculated it in part a), using "m_i" for midpoint of the ith bin and "f_i" for the frequency (count) of the ith bin. Do likewise for the variance.

ⓣ 68. Grouped data II Look back at the points scored per season by Wayne Gretzky in Exercise 36.

a) Compute his mean points per season.

b) Group the data into bins: 40–60, 60–80, etc., and find the counts for each bin. Multiply each bin midpoint (50, 70, . . .) by its respective bin count. Add all of these products. Divide by 20, the number of seasons. This is your 'grouped mean' calculation. We are pretending that each data point lies right at the midpoint, for this calculation (obviously they don't, but errors high and low may just balance out nicely in certain calculations). How close is this approximation to the actual mean you calculated in part (a)?

69. Unequal bin widths Look back at the histogram of times in Exercise 24.

a) Construct by hand another histogram, for these data (estimate counts as best you can), where the bins are 28.5–29.0, 29.0–29.5, 29.5–30.0, 30.0–30.5, 30.5–31.0, 31.0–31.5, 31.5–32.0, 32.0–32.5, 32.5–35.5.

Plot as usual frequency for the height of each bin's rectangle. Any problems? If so, explain.

b) Redo your histogram but for the last bin, plot the frequency divided by 6. Do you prefer this histogram or the one in part a)? How does the area of the rectangle over 32.5–35.5 compare with the area of all the rectangles starting from 32.5 in the original histogram? What basic principle (mentioned in Chapter 2) was violated in part a) and corrected in part b)? (The vertical axis is now a measure of "frequency density".)

70. Unequal bin widths II We emphasize using equal widths for the bins in a histogram. What's the problem with using unequal bin widths?

a) Draw a histogram for data containing five measurements between 10 and 11 cm, four measurements between 11 and 12 cm, three measurements between 12 and 13 cm, two measurements between 13 and 14 cm, and one measurement between 14 and 15 cm. Plot frequency (count) on the y-axis, as usual.

b) Now combine the last two bins into one, so that the final bin is 13 to 15 cm, which hence contains three measurements. Draw a histogram now, the usual way. What's wrong? Was any basic principle (mentioned in Chapter 2) violated?

c) Correct the histogram by plotting frequency divided by bin width on the y-axis. Why does this correct the problem?

71. Rock concert accidents Crowd Management Strategies (www.crowdsafe.com) monitors accidents at rock concerts. In their database, they list the names and other variables of victims whose deaths were attributed to "crowd crush" at rock concerts. Here are the histogram and boxplot of the victims' ages for data from a recent one-year period:

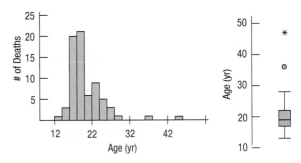

a) What features of the distribution can you see in both the histogram and the boxplot?

b) What features of the distribution can you see in the histogram that you could not see in the boxplot?

c) What summary statistic would you choose to summarize the centre of this distribution? Why?

d) What summary statistic would you choose to summarize the spread of this distribution? Why?

72. Slalom times 2010 The Men's Giant Slalom skiing event consists of two runs whose times are added together for a final score. Two displays of the giant slalom times in the 2010 Winter Olympics are shown below.

Slalom Times (sec)

Slalom Times

a) What features of the distribution can you see in both the histogram and the boxplot?
b) What summary statistic would you choose to summarize the centre of this distribution? Why?
c) What summary statistic would you choose to summarize the spread of this distribution? Why?

 Just Checking ANSWERS

(Thoughts will vary.)

1. Roughly symmetric, slightly skewed to the right. Centre around 5 kilometres, few over 15 kilometres.

2. Bimodal or trimodal, with peaks around zero, three, and maybe six hours, since many people watch no hockey, and others watch most of one or two games. Spread from zero hours to about six hours.

3. Strongly skewed to the right, with almost everyone at $0; a few small prizes, with the winner an outlier.

4. Fairly symmetric, somewhat uniform, perhaps slightly skewed to the right. Centre in the 40s? Few ages below 25 or above 70.

5. Uniform, symmetric. Centre near 5. Roughly equal counts for each digit 0–9.

6. Incomes are probably skewed to the right and not symmetric, making the median the more appropriate measure of centre. The mean will be influenced by the high end of family incomes and not reflect the "typical" family income as well as the median would. It will give the impression that the typical income is higher than it is.

7. An IQR of 10 L/km would mean that only 50% of the cars get gas mileages in an interval 10 L/km wide. Fuel economy doesn't vary that much. 1 L/km is reasonable. It seems plausible that 50% of the cars will be within about 1 L/km of each other. An IQR of 0.1 L/km would mean that the gas mileage of half the cars varies little from the estimate. It's unlikely that cars, drivers, and driving conditions are that consistent.

8. We'd prefer a standard deviation of two months. Making a consistent product is important for quality. Customers want to be able to count on the MP3 player lasting somewhere close to five years, and a standard deviation of two years would mean that lifespans were highly variable.

MathXL

MyLab Statistics

Go to MathXL at **www.mathxl.com** or MyLab™ Statistics at **www.mystatlab.com**. You can practise exercises for this chapter as often as you want. The guided solutions will help you find answers step by step. You'll find a personalized study plan available to you too!

Understanding and Comparing Distributions

Connections

The most interesting questions are usually not about variables in isolation, but rather about relationships between variables or about patterns and changes in a variable over time or space. Do people who drink green tea live longer? Does the presence of some gene allele affect one's health? Are tsunami-creating earthquakes bigger in certain geographic regions? Have they gotten bigger in recent times? In Chapter 2 we examined relationships between two categorical variables. In this chapter, we'll examine the relationship between a quantitative variable and a categorical variable, by comparing groups (for two quantitative variables, see Chapter 6) as well as trends over time.

I'm not an outlier; I just haven't found my distribution yet!

—Ronan M. Conroy

The Hopkins Memorial Forest is a 2500-acre preserve in Massachusetts, New York, and Vermont managed by the Williams College Center for Environmental Studies (CES). As part of their mission, CES monitors forest resources and conditions over the long term.

One of the variables measured in the forest is wind speed. Each day, remote sensors record the minimum, maximum, and average wind speed (in miles per hour). Wind is caused as air flows from areas of high pressure to areas of low pressure. Centres of low pressure often accompany storms, so both high winds and low pressure are associated with some of the fiercest storms. Wind speeds can vary greatly during a day and from day to day, but if we step back a bit farther, we can see patterns. By modelling these patterns, we can understand things about *Average Wind Speed* and the insights it provides about weather that we may not have known.

In Chapter 2, we looked at the association between two categorical variables using contingency tables and displays. Here we'll look at different ways to examine the relationship between two variables when one is quantitative and the other indicates groups to compare. We are given wind speed averages for each day of the year. But we can gather the days together into groups and compare the wind speeds among them. If we partition *Time* in this way, we'll gain enormous flexibility. When we change the size of the groups, from the entire year, to seasons, to months, and finally to days, different patterns emerge that together increase our understanding of the entire pattern.

Let's start with the "big picture." Here's a histogram and 5-number summary (Figure 4.1) of the daily *Average Wind Speed* for every day in 2011.

Figure 4.1

A histogram of daily *Average Wind Speed* for 2011. It is unimodal and skewed to the right. The boxplot below the histogram suggests some possible outliers that may warrant our attention.

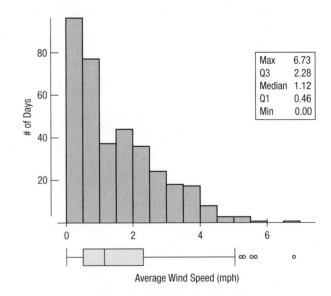

Max	6.73
Q3	2.28
Median	1.12
Q1	0.46
Min	0.00

We can see that the distribution is unimodal and skewed to the right, so we'll focus on the 5-number summary rather than the mean and standard deviation. Median daily wind speed is about 1.12 mph. On half of the days, the average wind speed was between 0.46 and 2.28 mph, because the quartiles frame the middle half of the data. We also see some possible outliers, in particular one rather windy 6.73 mph day (the other *suspect outliers* don't look much like outliers, given the long tail). Was this an unusual weather event or just the windiest day of the year? To answer that, we'll need to work with the summaries a bit more.

4.1 Comparing Groups

It is almost always more interesting to compare groups. Is it windier in the winter or the summer? Are any months particularly windy? Are weekends a special problem? Let's split the year into two groups: April through September (Summer) and October through March (Winter). To compare the groups, we create two histograms, being careful to use the same scale. Figure 4.2 displays average daily wind speeds for Summer and for Winter:

Figure 4.2

Histograms of *Average Wind Speed* for days in Summer (left) and Winter (right) show very different patterns.

Who	Days during 2011
What	Average daily wind speed (miles per hour), average barometric pressure (mb), average daily temperature (degrees Celsius)
When	2011
Where	Hopkins Forest, in western Massachusetts
Why	Long-term observations to study ecology and climate (Data in **Hopkins Forest_2011**)

The distribution of wind speeds for the summer months is unimodal and skewed to the right. By contrast, the distribution for the winter months is less strongly skewed and more nearly uniform. During the summer, the median wind speed is less than 1 mph. In winter, the wind speed is higher on average, and there is at least one noticeably high value. There is also more variability in the winter (IQR = 1.82 mph) than in summer (IQR = 1.27 mph)

Summaries for *Average Wind Speed* by Season								
Group	Mean	StdDev	Median	IQR	Q1	Q3	Minimum	Maximum
Winter	1.90	1.29	1.72	1.82	0.84	2.66	0.02	6.73
Summer	1.11	1.10	0.71	1.27	0.34	1.62	0.00	5.47

For Example COMPARING CRIME IN ONTARIO AND THE WEST WITH STEM-AND-LEAF

Statistics Canada records police-reported crime for various offences, by census metropolitan area (CMA: an urban area with over 100 000 people, with at least 50 000 living in the urban core). Below are comparisons of homicide rates and also of motor vehicle theft rates for a number of CMAs in Ontario and Western Canada (Data in

CrimeByCMA_2014). Rates are incidents per 100 000 of population. We have data for 14 Ontario CMAs and 9 Western Canada CMAs. There are only 23 values, so a *back-to-back stem-and-leaf plot* is an effective display, and easy (and necessary) to make by hand (since I know no software that can put them back to back). In these displays, the stem runs down the middle, with the leaves for the two regions to the left or right. For values on the left side, note that e.g. 34|1 corresponds to 13 and 14, not 31 and 41.

QUESTION: How do these regions compare?

Motor Vehicle Theft Rates
(per 100 000)

Ontario		Western Canada
67789	0	
011234	1	1
9	1	
3	2	
	2	
	3	011
6	3	6789
	4	1

(3|5|6 means 530 per 100 000
for an Ontario CMA and 560 per
100 000 for a Western CMA)

Homicide Rates (per 100 000)

Ontario		Western Canada
0045567899	0	
235	1	158
	2	0126
	3	02
	4	
	5	
	6	
	7	
	8	
0	9	

(3|5|6 means 5.3 per 100 000 for an Ontario CMA
and 5.6 per 100 000 for a Western CMA)

ANSWER: All Western CMAs have higher motor vehicle theft rates than Ontario CMAs (generally under about 200 per 100 000 in Ontario but over 300 per 100 000 in the West), but for a high Ontario outlier (Brantford with 630 thefts per 100 000) and a low Western outlier (Victoria with only 110 per 100 000—seems the Garden City is not only exceptional for its gardens!).

Regarding homicide rates, there is a bit more overlap here, but again the Western cities clearly tend to have the higher rates—mostly under 1.0 per 100 000 for Ontario and always over 1.0 per 100 000 for the West. In fact, 10 of 14 Ontario cities have lower homicide rates than all the Western cities. There is an Ontario and overall outlier, Thunder Bay with 9.0 homicides per 100 000! (This more than tripled their rate in 2013, roughly tripled the next biggest rate, for Winnipeg, and enabled the city to steal the title of Murder Capital of Canada from 2013 champ Regina. City officials had no simple explanation).

And note that we have let one outlier unduly affect the detail in this plot. We would be better off to remove the outlier, report it separately, and redo the stem-and-leaf plot, stretching it out by splitting each stem, as discussed in Chapter 3.

Source: Statistics Canada. Canadian Centre for Justice Statistics, Uniform Crime Reporting Survey (2014). Retrieved 9 July, 2017, from http://www.statcan.gc.ca/pub/85-002-x/2015001/article/14211/tbl/tbl07-eng.htm

Section 4.1 Exercises

T **1. Load factors, 2013** The Research and Innovative Technology Administration of the Bureau of Transportation Statistics (www.TranStats.bts.gov/Data_Elements.aspx?Data=2) reports load factors (passenger-miles as a percentage of available seat-miles) for commercial airlines for every month from January 2000 through September 2013. Here are histograms comparing the domestic and international load factors for this time period:

	Mean	Median	Std Dev	IQR	n	Missing
Domestic	77.36715	78.02	6.45481	9.20	165	0
International	77.75976	77.98	5.22681	6.78	165	0

Compare and contrast the distributions.

T **2. Load factors, 2013 by season** We can also compare domestic load factors for September through March versus those for April through August:

Compare and contrast what these histograms show.

4.2 Comparing Boxplots

Are some months windier than others? Even residents may not have a good idea which parts of the year are the most windy. (Do you know for your hometown?) And what about the spreads? Are wind speeds equally variable from month to month, or do some months show more variation?

Histograms or stem-and-leaf plots are a fine way to look at one distribution or two, but it would be hard to see patterns by comparing 12 histograms. Boxplots offer an ideal balance of information and simplicity, hiding the details while displaying the overall summary information. We often plot them side by side for groups or categories we wish to compare.

By placing boxplots side by side, we can easily see which groups have higher medians, which have greater IQRs, where the central 50% of the data is located in each group, and which have the greater overall range. And, when the boxes are in some natural order, we can get a general idea of patterns in both the centres and the spreads. Equally important, we can see past any outliers in making these comparisons because boxplots display them separately.

Figure 4.3 shows boxplots of the *Average Daily Wind Speed* by month:

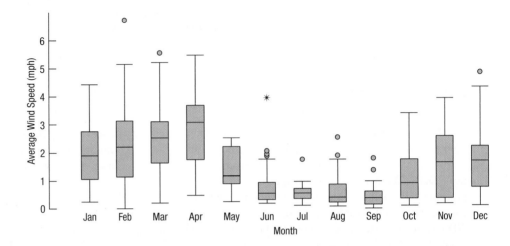

Here we see that wind speeds tend to decrease in the summer. The months in which the winds are both strongest and most variable are October through April.

When we looked at a boxplot for wind speeds of the entire year (Figure 4.1), there were only five suspect outliers. But the monthly boxplots show different suspect outliers than before because some days that seemed ordinary when placed against the entire year's data look like outliers for the month that they're in. Outliers are context dependent. That windiest day in August (the outlier at about 2.5 mph) certainly wouldn't stand out in November or December, but for August it was a remarkable day—as we'll soon see.

nra/Shutterstock

For Example COMPARING DISTRIBUTIONS

Roller coaster[1] riders want a coaster that goes fast. There are two main types of roller coasters: those with wooden tracks and those with steel tracks. Do they typically run at different speeds? Below are their boxplots (Data in **Coasters 2014**):

QUESTION: Compare the speeds of wood and steel roller coasters.

ANSWER: The median speed for the wood track coasters is a bit higher than for the steel coasters but overall, their typical speeds are about the same. However, the speeds of the steel track coasters vary much more and include both high and low outliers.

---

[1]See the roller coaster database at www.rcdb.com.

Step-by-Step Example COMPARING GROUPS

23d/Shutterstock

Most scientific studies compare two or more groups. It is almost always a good idea to start an analysis of data from such studies by comparing boxplots for the groups. Here's an example.

For her class project, a student compared the efficiency of various coffee mugs. She compared four different mugs, testing each of them eight different times. Each time, she heated water to 180°F[2], poured it into a mug, and sealed it. After 30 minutes, she measured the temperature again and recorded the difference in temperature. Because these are temperature differences, smaller differences mean that the liquid stayed hot—just what we would want in a coffee mug. (A century ago, Sir Ronald Fisher described the basic principles of statistical experiments, such as this. We'll have much more to say about experimental design later in this book.)

Question: What can we say about the effectiveness of these four mugs? (Data in **Mugs**)

THINK ➡ Plan State what you want to find out. **Variables** Identify the *variables* and report the *W*'s. Be sure to check the appropriate condition.	I want to compare the effectiveness of the four different mugs in maintaining temperature. I have eight measurements of Temperature Change for each of the mugs. ✓ **Quantitative Data Condition**: The temperature changes are quantitative, with units of °F. Boxplots and summary statistics are appropriate displays for comparing the groups.

SHOW ➡ Mechanics Report the 5-number summaries of the four groups. Include the IQR as a check.

	Min	Q1	Median	Q3	Max	IQR
CUPPS	6°F	6	8.25	14.25	18.50	8.25
Nissan	0	1	2	4.50	7	3.50
SIGG	9	11.50	14.25	21.75	24.50	10.25
Starbucks	6	6.50	8.50	14.25	17.50	7.75

Make a picture. Because we want to compare the distributions for four groups, boxplots are a great choice.

---

[2]This experiment was conducted in the U.S., where, like in the Bahamas, Belize, Palau, Cayman Islands—and likely on your oven's temperature readings—the Fahrenheit scale still prevails. To convert to Celsius, the formula is °C = (°F − 32)/1.8. So, 180°F = 82°C (and this was much more pleasant for your Canadian authors than converting the data and all text references to °C!)

TELL ➡ **Conclusion** Interpret what the boxplots and summaries say about the ability of these mugs to maintain heat. Compare the shapes, centres, and spreads, and note any outliers.

The individual distributions are all slightly skewed to the high end. The Nissan mug does the best job of keeping liquids hot, with a median loss of only 2 °F, and the SIGG mug does the worst job, typically losing 14 °F. The difference is large enough to be important; a coffee drinker would be likely to notice a 14°F (8°C) drop in temperature. And the mugs are clearly different; 75% of the Nissan tests showed less heat loss than any of the other mugs in the study. The IQR for the Nissan results is also the smallest of all, indicating that it is a consistent performer.

Just Checking

Spoiled ballots are a real threat to democracy. Below are displays of data from Elections Canada showing percentage of spoiled ballots (number of spoiled ballots divided by total number of ballots cast) for all 338 electoral ridings in the 2015 Canadian federal election, and another display comparing results by region. (Data in **FederalElection2015**. See Exercise 38 for some rather disheartening results from the 2000 federal election.)

1. Describe what the histogram says about spoiled ballots.

2. What does the histogram suggest that you can't see in the boxplot?

3. Are there many outliers present in the nationwide data? Explain.

4. Describe the regional variation in spoiled ballots.

5. Why is one of the regional boxplots missing whiskers?

6. Could it possibly be true that three quarters of results in one region were worse than all the individual riding results in all other regions? Explain.

7. Which region(s) had the worst result? The best result?

8. How does the last display above clarify what is seen in the histogram?

Section 4.2 Exercises

T **3. Load factors 2013 by month** Here is a display of the international load factors by month for the period from 2000 to 2013.

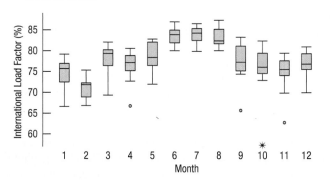

Discuss the patterns you see in this display.

T **4. Load factors 2013 by year** Here is a display of the domestic load factors by year.

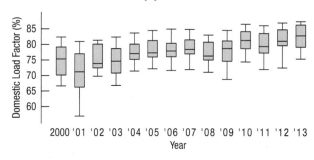

Discuss the patterns you see in this display.

4.3 Outliers

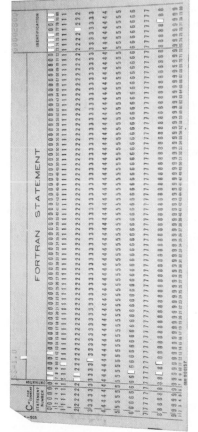

1950s computer card

Richard D. De Veaux

In the boxplots for the daily *Average Wind Speed by Month*, several days are nominated by the boxplots as possible outliers. Cases that stand out from the rest of the data almost always deserve attention. An outlier is a value that doesn't fit with the rest of the data, but exactly how different it should be to receive special treatment is a judgment call. Boxplots provide a rule of thumb to highlight possibly unusual cases, but that rule is only a rough guide, and it doesn't tell you what to do with them. So, what *should* you do with outliers?

Outliers arise for many reasons. They *may* be the most important values in the data set, pointing out an exceptional case or illuminating a pattern by being the exception to the rule. They may be values that just happen to lie above the limits *suggested* by the box plot rule. Or they may be errors. A decimal point may have been misplaced, digits transposed, or digits repeated or omitted. Some outliers are obviously wrong. For example, if a class survey includes a student who claims to be 170 inches tall (about 14 feet), you can be sure that's an error. But, maybe the units were wrong. Did the student mean 170 centimetres (about 65 inches)? There are many ways errors can creep into data sets and such errors occur remarkably often. If you can identify the correct value, then you should certainly use it. One important reason to look into outliers is to correct errors in your data.

14-year-old widowers? Two researchers, Ansley Coale and Fred Stephan, looking at data from the 1950 census, noticed that the number of widowed 14-year-old boys had increased from 85 in 1940 to a whopping 1600 in 1950. The number of divorced 14-year-old boys had increased, too, from 85 to 1240. Oddly, the number of teenaged widowers and divorcés decreased for every age group after 14, from ages 15 to 19. When Coale and Stephan also noticed a large increase in the number of young Native Americans in the Northeast United States, they began to look for the cause of these problems. Data in the 1950 census were recorded on computer cards. Cards are hard to read and mistakes are easy to make. It turned out that data punches had been shifted to the right by one column on hundreds of cards. Because each card column meant something different, the shift turned 43-year-old widowed males into 14-year-olds, 42-year-old divorcés into 14-year-olds, and children of white parents into Native Americans. Not all outliers have such a colourful (or famous) story, but it is always worthwhile to investigate them. And, as in this case, the explanation may well be surprising. (Source: A. Coale and F. Stephan, "The case of the Indians and the teen-age widows," *J. Am. Stat. Assoc.* 57 [Jun 1962]: 338–347.)

Values that are large (or small) in one context may not be remarkable in another. The boxplots of *Average Wind Speed by Month* show possible outliers in several of the months. The windiest day in February was an outlier not only in that windy month, but for the entire year as well. The windiest day in June lies more than 3 IQRs from the upper quartile for that month, but it wouldn't have been unusual for a winter day. And, the windiest day in August seemed much windier than the days around it, even though it was relatively calm.

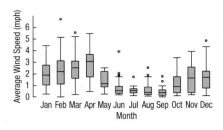

Many outliers are not wrong; they're just different. And most repay the effort to understand them. You can sometimes learn more from the extraordinary cases than from summaries of the entire data set.

That blustery day in February turned out to be a blast that brought four days of extreme cold ($-27°C$) to the region. John Quinlan of the National Weather Service predicted at the time: "The Berkshires have definitely had some of the highest gusts with this wind event. This is definitely the worst we'll be seeing."

What about that June day? It was June 2. A search for weather events on the Internet for June 2, 2011, finds a rare tornado in western Massachusetts. *The Berkshire Eagle* reported: "It probably never occurred to most Berkshire residents horrified by the photos and footage of recent tornado damage in Alabama and Missouri that tornadoes would soon strike close to home. They did, of course, on Wednesday, sowing destruction and in some cases costing lives in Springfield, West Springfield, Monson and other towns near enough to us that the destruction had a sobering effect."

Finally, let's look at the more extreme outlier in August. It wouldn't be remarkable in most other months. It turns out that that was Hurricane Irene, whose eye passed right over the Hopkins Forest. According to *The New York Times*, "it was probably the greatest number of people ever threatened by a single storm in the United States."

Not all outliers are as dramatic as Hurricane Irene, but all deserve attention. If you can correct an error, you should do so (and note the correction). If you can't correct it, or if you confirm that it is correct, you can simply note its existence and leave it in the data set. But the safest path is to report summaries and analyses with *and* without the outlier so that readers can judge the influence of the outlier for themselves.

There are two things you should *never* do with outliers. You should not leave an outlier in place and proceed as if nothing were unusual; analyses of data with outliers are very likely to be wrong. On the other hand, you should never omit an outlier from the analysis without comment. If you want to exclude an outlier, you must announce your decision and, to the extent you can, justify it. But a case lying just over the fence suggested by the boxplot may just be the largest (or smallest) value at the end of a stretched-out tail. A histogram is often a better way to see more detail about how the outlier fits in (or doesn't) with the rest of the data, and how large the gap is between it and the rest of the data.

For Example CHECKING OUT THE OUTLIERS

RECAP: We looked at the speeds of roller coasters earlier. There is one quite extraordinarily high value (plotted with a slightly different symbol, as is done by some software for the more extreme values).

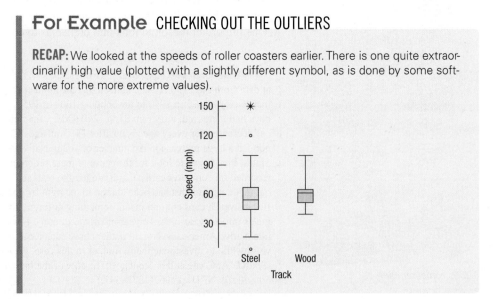

QUESTION: The fastest coaster in this collection turns out to be "Formula Rossa" at Ferrari world in Abu Dhabi. What might make this roller coaster unusual? You'll have to do some research, but that's often what happens with outliers.

ANSWER: Formula Rossa is easy to find in an Internet search. We learn that it is a "hydraulic launch" coaster. That is, it doesn't get its remarkable speed just from gravity, but rather from a kick-start by a hydraulic piston. That could make it different from the other roller coasters. It accelerates to 240 km/h in 4.8 seconds. Riders must wear safety goggles.

Section 4.3 Exercises

T **5. Extraordinary months** Look at the boxplots by month in Exercise 3. The three outliers in September, October, and November are all from the year 2001 and reflect the terrorist attacks of 9/11. Do you think the data for these months should be set aside in any overall analysis of load factor patterns by month? Explain.

T **6. Extraordinary months again** Here's a boxplot of all of the domestic load factors. It shows a single outlier. The boxplots of Exercise 4 show the same data, but have no outliers. Why do you think this is?

4.4 Timeplots: Order, Please!

The Hopkins Forest wind speeds are reported as daily averages.[3] Previously, we grouped the days into months or seasons, but we could look at the wind speed values day by day. Whenever we have data measured over time, it is a good idea to look for patterns by plotting the data in time order. Figure 4.4 plots the daily average wind speeds over time:

Figure 4.4

A timeplot of *Average Wind Speed* shows the overall pattern and changes in variation.

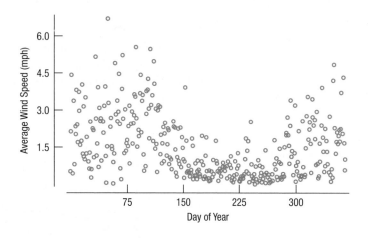

[3]Here's another warning that the word "average" is used to mean different things. Many "daily average wind speeds" are computed as the midrange—the average of the highest and lowest speeds seen during the day. But the Hopkins Forest values are actually measured continuously all day and averaged over all those observations. When you see the word "average," it is wise to check for the details.

A display of values against time is called a *time series plot* or more simply, **timeplot**. This timeplot reflects the pattern that we saw when we plotted the wind speeds by month. But without the arbitrary divisions between months, we can see a calm period during the summer, starting around day 150 (the beginning of June), when the wind is relatively mild and doesn't vary greatly from day to day. We can also see that the wind becomes both more variable and stronger during the late fall and winter months.

*Smoothing Timeplots

Timeplots often show a great deal of point-to-point variation, as evident in Figure 4.4. We usually want to see past this variation to understand any underlying smooth trends, and then also think about how the values vary around that trend—the timeplot version of centre and spread. You could take your pencil, squint at the plot a bit to blur things, and sketch a smooth curve. And you'd probably do a pretty good job. But we can also ask a computer to do it for us.

There are many ways for computers to run a smooth trace through a timeplot. Some provide an equation that gives a typical value for any given time point, others just offer a smooth trace. For our purposes here, it doesn't matter all that much. You can try the methods offered by your favourite program and pick the one that looks best for the timeplot you are looking at and the questions you'd like to answer.

You'll often see timeplots drawn with all the points connected, especially in financial publications. This trace through the data falls at one extreme, following every bump. For the wind speeds, the daily fluctuations are so large that connecting the points (Figure 4.5) doesn't help us a great deal:

Figure 4.5

The *Average Wind Speeds* of Figure 4.4, drawn connecting all the points. Sometimes this can help you see the underlying pattern, but here there is too much daily variation for this to be very useful.

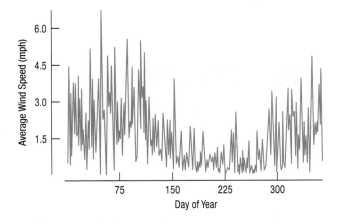

A smooth trace can highlight long-term patterns and help you see them through the more local variation. Figure 4.6 shows the daily average wind speed values with a smooth trace found by a method called *lowess*, available in many statistics programs.

Figure 4.6

The *Average Wind Speeds* of Figure 4.4, with a smooth trace added to help your eye see the long-term pattern.

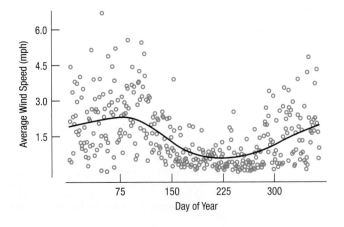

With the smooth trace, it's a bit easier to see a pattern. The trace helps your eye follow the main trend and alerts you to points that don't fit the overall pattern.

For Example   TIMEPLOTS AND SMOOTHING

RECAP: We have looked at the world's fastest roller coasters. Have coaster tracks been getting shorter? Our data include coasters going back to 1909. Here's a time-plot with a *lowess* smooth trace:

QUESTION: What does the timeplot say about the trend in roller coaster track lengths?

ANSWER: Roller coaster tracks for these coasters seem to be getting shorter, although there are still some with very long tracks. The trend to shorter tracks seems to have begun after the 1970s and is continuing to this day.

How do smoothers work? If you planned to go camping in the Hopkins Forest around July 15 and wanted to know how windy it was likely to be, you'd probably look at a typical value for mid-July. Maybe you'd consider all the wind speeds for July, or maybe only those from July 10th to July 20th. That's just the kind of thing we do when we smooth a timeplot.

One simple way to smooth is with a **moving average**. To find a smooth value for a particular time, we average the values around that point in an interval called the "window." To find the value for the next point, we *move* the window by one point in time and take the new *average*. The size of the window you choose affects how smooth the resulting trace will be. For the Hopkins Forest winds, we might use a five-day moving average. Stock analysts often use a 50- or 200-day moving average to help them (attempt) to see the underlying pattern in stock price movements.

Can we use smoothing to predict the future? We have only the values of a stock in the past, but (unfortunately) none of the future values. We could use the recent past as our window and take the simple average of those values. A more sophisticated method, **exponential smoothing**, gives more weight to the recent past values and less and less weight to values as they recede into history.

Looking into the Future

People have always wanted to peer into the future, and we predict that they will continue to do so! Sometimes that makes sense. Most likely, the Hopkins Forest climate follows regular seasonal patterns. It's probably safe to predict a less windy June than December next year. But we certainly wouldn't predict another hurricane on August 28.

Other patterns are riskier to extend into the future. If a stock has been rising, will it continue to go up? No stock has ever increased in value indefinitely, and no stock analyst has consistently been able to forecast when a stock's value will turn around. Stock prices, unemployment rates, and other economic, social, or psychological concepts are much harder to predict than physical quantities. The path a ball will follow when thrown from a certain height at a given speed and direction is well understood. Even some dogs can predict the flight of a frisbee well enough to run to where it will be within reach and catch it. The path interest rates will take is much less clear. Unless you have strong (nonstatistical) reasons for doing otherwise, you should resist the temptation to think that any trend you see will continue, even into the near future.

Statistical models often tempt those who use them to think beyond the data. We'll pay close attention later in this book to understanding when, how, and how much we can justify doing that.

Section 4.4 Exercises

T **7. Load factors 2013 over time** Here is a timeplot of each monthly load factor for domestic flights for 2000 to 2013 along with a lowess smooth.

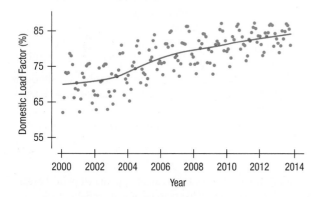

a. Describe the patterns you see in this plot.
b. Do you expect the overall pattern to continue for another decade? Why or why not?

T **8. Load factors 2013 over time, a second look** Here is a timeplot of the same domestic load factors as in Exercise 7, but this time with a different smoother (called a median smoother) applied.

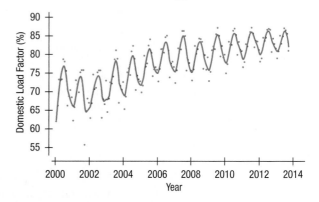

a. What patterns do you see in this plot?
b. Do you expect the patterns you see here to continue for the next decade? Why or why not?

4.5 Re-expressing Data: A First Look

Re-expressing to Improve Symmetry

When data are skewed, it can be hard to summarize them simply with a centre and spread, and hard to decide whether the most extreme values are outliers or just part of the stretched-out tail. How can we say anything useful about such data? The secret is to *re-express* the data by applying a simple function to each value.

Many relationships and "laws" in the sciences and social sciences include functions such as logarithms, square roots, and reciprocals. Similar relationships often show up in data.

Consider this example: In 1980, large companies' chief executive officers (CEOs) made, on average, about 42 times what workers earned. In the next two decades, CEO compensation soared compared to the average worker. By 2008, that multiple had jumped to 344. What does the distribution of the compensation of Fortune 500 companies' CEOs look like? Figure 4.7 shows a histogram and boxplot for 2012 compensation (Data in **CEO salaries 2012**)[4]:

---
[4]www.forbes.com/lists/2012/12/ceo-compensation-12_land.html

Figure 4.7

CEOs' compensation for the Forbes 500 companies in 2012.

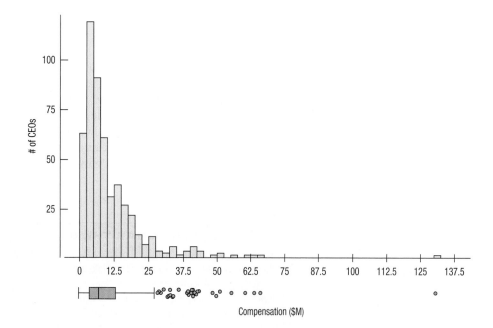

We have 500 CEOs and nearly 60 possible histogram bins, about half of which are empty—and many of which contain only one or two CEOs. The boxplot indicates that some CEOs received extraordinarily high pay, while the majority received relatively "little"—but look at the values of the bins. The first four bins together include about two-thirds of the CEOs, and cover compensations from $0 up to $10 000 000. The long stretch of this graph minimizes the detail here, where the CEO salaries are most densely concentrated.

The histogram and boxplot clearly show a distribution that is highly skewed to the right. It can be hard to decide what we mean by the "centre" of a skewed distribution, so it's difficult to pick a typical value to summarize the distribution. What would you say was a typical CEO total compensation? The mean value is $10 475 830 while the median is "only" $6 967 500. Each tells us something different about the data.

One approach is to **re-express**, or **transform**, the data by applying a simple function to make the skewed distribution more symmetric. For example, we could take the square root or logarithm of each pay value. Taking logs works pretty well for the CEO compensations, as you can see below (Figure 4.8):[5]

Figure 4.8

The logarithms (base 10) of 2012 CEO compensations are much more nearly symmetric.

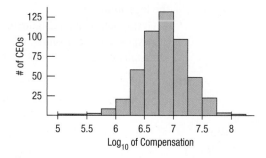

The histogram of the logs of the total CEO compensations is much more nearly symmetric, so we can see that the log compensations are between 5.2, which corresponds to $160 000, and 8.1, corresponding to about $125 000. And it's easier to talk about a typical value for the logs. Both the mean and median log compensations are about 6.84. (That's about $6 900 000 but who's counting?)

---

[5]Three CEOs, including Larry Page of Google—accepted no pay. When we took logs, their pay dropped out because we couldn't take log(0).

Against the background of a generally symmetric main body of data, it's easier to decide whether the largest or smallest pay values are outliers. In fact, the most highly compensated CEO is identified as an outlier by the boxplot rule of thumb even after this re-expression. It's impressive to be an outlier CEO in annual compensation. It's even more impressive to be an outlier in the log scale! (On the other hand, four CEOs earned less than $500,000, making them low outliers, on the log scale)

Variables that are skewed to the right often benefit from a re-expression by square roots, logs, or reciprocals. Those skewed to the left may benefit from squaring the data or using the exponential function. Because computers and calculators can do the calculating, re-expressing data is quite easy. Consider re-expression as a helpful tool whenever you have skewed data.

DEALING WITH LOGARITHMS

You have probably learned about logs in math courses and seen them in psychology or science classes. In this book, we use them only for making data behave better. Base 10 logs are the easiest to understand, but natural logs are often used as well. (Either one is fine.) You can think of base 10 logs as roughly one less than the number of digits you need to write the number. So 100, which is the smallest number to require three digits, has a $\log_{10}$ of 2. And 1000 has a $\log_{10}$ of 3. The $\log_{10}$ of 500 is between 2 and 3, but you'd need a calculator to find that it's approximately 2.7. All salaries of "six figures" have $\log_{10}$ between 5 and 6. Logs are incredibly useful for making skewed data more symmetric. But don't worry—nobody does logs without technology and neither should you. Often, remaking a histogram or other display of the data is as easy as pushing another button.

Re-expressing to Equalize Spread Across Groups

Researchers measured the concentration (nanograms per millilitre) of cotinine in the blood of three groups of people: nonsmokers who have not been exposed to smoke, nonsmokers who have been exposed to smoke (ETS), and smokers. Cotinine is left in the blood when the body metabolizes nicotine, so this measure gives a direct measurement of the effect of passive smoke exposure. The boxplots of the cotinine levels of the three groups (Figure 4.9) tell us that the smokers have higher cotinine levels, but if we want to compare the levels of the passive smokers (ETS) to those of the nonsmokers, we're in trouble, because on this scale, the cotinine levels for both nonsmoking groups are too low to be seen.

Re-expressing can help alleviate the problem of comparing groups that have very different spreads. For measurements like the cotinine data, whose values can't be negative and whose distributions are skewed to the high end, a good first guess at a re-expression is the logarithm.

After taking logs (Figure 4.10), we can compare the groups and see that the nonsmokers exposed to environmental smoke (ETS) do show increased levels of (log) cotinine, although not the high levels found in the blood of smokers.

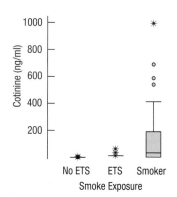

Figure 4.9

Cotinine levels (nanograms per millilitre) for three groups with different exposures to tobacco smoke. Can you compare the ETS (exposed to smoke) and No ETS groups?

Figure 4.10

Blood cotinine levels after taking logs. What a difference a log makes!

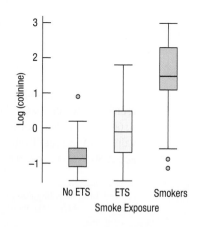

Notice that the same re-expression has also improved the symmetry of the cotinine distribution for smokers and pulled in most of the apparent outliers in all of the groups. It is not unusual for a re-expression that improves one aspect of data to improve others as well. We'll talk about other ways to re-express data as the need arises throughout the book.

Section 4.5 Exercises

9. Exoplanets Discoveries of planets beyond our solar system have grown rapidly. Here is a histogram showing the distance (in light-years) from earth to stars having known planets (as of 2014).

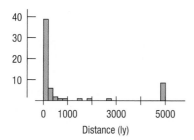

Explain why it might be beneficial to re-express these distances.

10. Exoplanets re-expressed Here are the exoplanet distances of Exercise 9, re-expressed to the log scale.

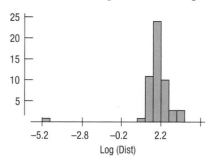

a) Is this a better scale to understand these distances?

b) The low outlier is "sol"– that is, it is the distance of the sun from the earth. Do you think it belongs with these data?

WHAT CAN GO WRONG?

■ **Avoid inconsistent scales.** Parts of displays should be mutually consistent—no fair changing scales in the middle or plotting two variables on different scales but on the same display. When comparing two groups, be sure to compare them on the same scale.

■ **Label clearly.** Variables should be identified clearly and axes labelled so a reader knows what the plot displays.

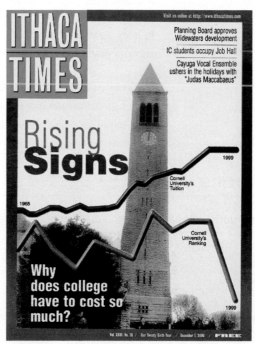

Ithaca Times

(continued)

Here's a remarkable example of a plot gone wrong. It illustrated a news story about rising college costs. It uses timeplots, but it gives a misleading impression. First think about the story you're being told by this display. Then try to figure out what has gone wrong.

What's wrong? Just about everything.

● The horizontal scales are inconsistent. Both lines show trends over time, but exactly for what years? The tuition sequence starts in 1965, but rankings are graphed from 1989. Plotting them on the same (invisible) scale makes it seem that they're for the same years.

● The vertical axis isn't labelled. That hides the fact that there are actually two distinct implicit vertical axes: one for tuition dollars and one for Cornell ranking.

This display violates three of the rules. And even worse: It violates a rule that we didn't even bother to mention.

● The two inconsistent scales for the vertical axis don't point in the same direction! The line for Cornell's rank shows that it has "plummeted" from 15th place to 6th place in academic rank. Most of us think that's an *improvement,* but that's not the message of this graph.

■ **Beware of outliers.** If the data have outliers and you can correct them, you should do so. If they are clearly wrong or impossible, you should remove them and report on them. Otherwise, consider summarizing the data both with and without the outliers.

What Have We Learned?

Choose the right tool for comparing distributions.

■ Compare the distributions of two groups with histograms or back-to-back stemplots.

■ Compare several groups with boxplots, which make it easy to compare centres and spreads and spot outliers, but hide much of the detail of distribution shape.

Treat outliers with attention and care.

■ Outliers may be "local" or global. Different cases may emerge as outliers when we group data in different ways.

■ Track down the background for outliers—it may be informative.

Timeplots graph individual values versus time, allowing us to see trends or patterns over time.

Re-express data to make it easier to work with.

■ Re-expression can make skewed distributions more nearly symmetric.

■ Re-expression can make the spreads of different groups more nearly comparable.

Review of Terms

Comparing distributions   When comparing the distributions of several groups using histograms or stem-and-leaf displays, consider their:

■ Shape

■ Centre

■ Spread (p. 102)

Comparing boxplots   When comparing groups with boxplots:

■ Compare the shapes. Do the boxes look symmetric or skewed? Are there differences between groups?

- Compare the medians. Which group has the higher centre? Is there any pattern to the medians?
- Compare the IQRs. Which group is more spread out? Is there any pattern to how the IQRs change?
- Using the IQRs as a background measure of variation, do the medians seem to be different, or do they just vary in the way that you'd expect from the overall variation?
- Check for possible outliers. Identify them if you can and discuss why they might be unusual. Of course, correct them if you find that they are errors (p. 101).

Outliers   Values that are large or small in one context may not be remarkable in another. Be sure to consider outliers in their appropriate context (p. 105).

Timeplot   A timeplot (often called a *time series plot*) displays data that change over time. Often, successive values are connected with lines to show trends more clearly. Sometimes a smooth curve is added to the plot to help show long-term patterns and trends (p. 108).

Re-expression (transformation)   In order to facilitate analysis, we often re-express data by taking the logarithm, the square root, the reciprocal, or some other mathematical operation of all values of a variable (p. 111).

On the Computer   COMPARING DISTRIBUTIONS

Most programs for displaying and analyzing data can display plots to compare the distributions of different groups. Typically, these are boxplots displayed side by side.

Side-by-side boxplots should be on the same y-axis scale so they can be compared.

Some programs offer a graphical way to assess how much the medians differ by drawing a band around the median or by "notching" the boxes.

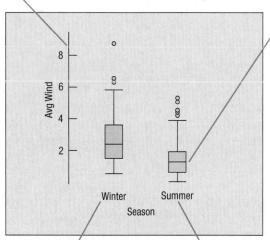

Boxes are typically labelled with a group name. Often they are placed in alphabetical order by group name—not the most useful order.

There are two ways to organize data when we want to compare groups. Each group can be in its own variable (or list, on a calculator). In this form, the experiment comparing mugs would have four variables, one for each type of mug:

CUPPS	Nissan	SIGG	Starbucks
6	12	2	13
6	16	1.5	7
6	9	2	7
18.5	23	3	17.5
10	11	0	10
17.5	20.5	7	15.5
11	12.5	0.5	6
6.5	24.5	6	6

But there's another way to think about and organize the data. What is the variable of interest (the *What*) in this experiment? It's the number of degrees lost by the water in each mug. And the *Who* is each time she tested a mug. We could gather all of the temperature values into one variable and put the names of the mugs in a second variable listing the individual results, one on each row. Now, the *Who* is clearer—it's an experimental run, one row of the table. Most statistics packages prefer data on groups organized in this way.

Mug	Temperature Difference (°F)	Mug	Temperature Difference (°F)
CUPPS	6	SIGG	12
CUPPS	6	SIGG	16
CUPPS	6	SIGG	9
CUPPS	18.5	SIGG	23
CUPPS	10	SIGG	21
CUPPS	17.5	SIGG	20.5
CUPPS	11	SIGG	12.5
CUPPS	6.5	SIGG	24.5
Nissan	2	Starbucks	13
Nissan	1.5	Starbucks	7
Nissan	2	Starbucks	7
Nissan	3	Starbucks	17.5
Nissan	0	Starbucks	10
Nissan	7	Starbucks	15.5
Nissan	0.5	Starbucks	6
Nissan	6	Starbucks	6

That's actually the way we've thought about the wind speed data in this chapter, treating wind speeds as one variable and the groups (whether seasons, months, or days) as a second variable.

EXCEL

Excel cannot make boxplots.

JMP

- Choose **Fit y by x**.
- Assign a continuous response variable to **Y, Response** and a nominal group variable holding the group names to **X, Factor**, and click **OK**. JMP will offer (among other things) dotplots of the data.

- Click the red triangle and, under **Display Options**, select Boxplots.

Note: If the variables are of the wrong type, the display options might not offer boxplots.

MINITAB

- Choose **Boxplot** . . . from the **Graph** menu.

If your data are in the form of one quantitative variable and one group variable:

- Choose **One Y** and **with Groups.**

If your data are in separate columns of the worksheet:

- Choose **Multiple Ys.**

R

If the data for each group are in separate variables, X1, X2, X3 . . . then:

- **boxplot(X1,X2,X3. . .)** will produce side-by-side boxplots.

If the quantitative values are in variable X and the grouping values are in categorical variable Y:

- **boxplot(X~Y)** will produce side-by-side boxplots—one for each level of Y.

SPSS

To make a boxplot in SPSS, open the **Chart Builder** from the Graphs menu:

- Click the **Gallery** tab.
- Choose **Boxplot** from the list of chart types.
- Drag a single or 2-D (side-by-side) boxplot onto the canvas.

- Drag a scale variable to the *y*-axis drop zone.
- To make side-by-side boxplots, drag a categorical variable to the *x*-axis drop zone.
- Click **OK**.

STATCRUNCH

To make a boxplot:

- Click on **Graphics**.
- Choose **Boxplot**.

- Choose the variable name from the list of **Columns**.
- Click on **Next**.

TI-83/84 PLUS

To make a boxplot, set up a **STAT PLOT** by using the boxplot icon.

To compare groups with boxplots, enter the data in lists, for example in L1 and L2.

Set up STATPLOT's Plot1 to make a boxplot of the L1 data:

- Turn the plot On.
- Choose the first boxplot icon (so the plot will indicate outliers).

- Specify Xlist:L1 and Freq:1, and select the Mark you want the calculator to use for displaying any outliers.
- Use ZoomStat to display the boxplot for L1. You can now TRACE to see the statistics in the 5-number summary.
- Then set up Plot2 to display the L2 data. This time when you use ZoomStat with both plots turned on, the display shows the boxplots in parallel.

Chapter Exercises

T indicates data is available in a saved file at the Companion Website at www.pearsoncanada.ca/deveaux

11. In the news Find an article in a newspaper or a magazine or online that compares two or more groups of quantitative data.
a) Does the article discuss the *W*'s?
b) Is the choice of graphics appropriate for the situation? Explain.
c) Discuss what the display reveals about the groups.
d) Does the article accurately describe and interpret the data? Explain.

12. In the news again Find an article in a newspaper or a magazine or online that shows a timeplot.
a) Does the article discuss the *W*'s?
b) Is the timeplot appropriate for the data? Explain.
c) Discuss what the timeplot reveals about the variable.
d) Does the article accurately describe and interpret the data? Explain.

13. Population in Ontario The table below gives the percentage of the Ontario population aged 65 years and older according to the Canadian Census from 1956 to 2011.

Year	Percentage of Ontario population aged 65 years and older
1956	8.4
1961	8.1
1966	8.2
1971	8.4
1976	8.9
1981	10.1
1986	10.9
1991	11.7
1996	12.4
2001	12.9
2006	13.6
2011	14.6

Make an appropriate display and discuss what it shows.

(*Source:* https://www12.statcan.gc.ca/census-recensement/2011/as-sa/98-311-x/2011001/fig/fig7-eng.cfm and also from Business Statistics by Sharpe, De Veaux, Velleman and Write, 2CE Ch5 Q 27)

14. Population in Canada by age The table below gives the age distribution of the Canadian population in 2015 according to the data published by Statistics Canada.

Age (in years)	Male (in 1000)	Female (in 1000)
Under 5	990.9	942.5
5 to 9	999.6	952.2
10 to 14	957.5	906.7
15 to 19	1079.5	1019.9
20 to 24	1263.3	1202.3
25 to 29	1239.9	1227.7
30 to 34	1247.3	1254.6
35 to 39	1197.6	1205.6
40 to 44	1173.5	1176.8
45 to 49	1219.0	1214.1
50 to 54	1387.2	1376.5
55 to 59	1298.4	1304.0
60 to 64	1103.5	1130.6
65 to 69	931.7	978.1
70 to 74	652.2	719.1
75 to 79	456.2	542.2
80 to 84	318.9	424.0
85 to 89	180.8	297.4
90 and over	79.7	200.5

Make an appropriate display to compare the groups and interpret what you find.

(*Source:* Statistics Canada. Table 051-0001 - Estimates of population, by age group and sex for July 1, Canada, provinces and territories, annual (persons unless otherwise noted), CANSIM (database). (accessed: 2016-08-08)

15. Graduation? A survey of major universities asked what percentage of incoming freshmen usually graduate "on time" in four years. Use the summary statistics given to answer these questions.

	% on Time
Count	48
Mean	68.35
Median	69.90
StdDev	10.20
Min	43.20
Max	87.40
Range	44.20
25th %tile	59.15
75th %tile	74.75

a) Would you describe this distribution as symmetric or skewed? Explain.

b) Are there any outliers? Explain.

c) Create a boxplot of these data.

d) Write a few sentences about the graduation rates.

16. Vineyards Here are summary statistics for the sizes (in acres) of Finger Lakes vineyards.

Count	36
Mean	46.50 acres
StdDev	47.76
Median	33.50
IQR	36.50
Min	6
Q1	18.50
Q3	55
Max	250

a) Would you describe this distribution as symmetric or skewed? Explain.

b) Are there any outliers? Explain.

c) Create a boxplot of these data.

d) Write a few sentences about the sizes of the vineyards.

17. Dots to boxes For the blood pressure measurements: 110 113 116 118 120 122 124 126 132 134 144 164, create a dotplot (by hand or by computer). Take the dotplot and superimpose upon it a boxplot, following the steps below:

i. Divide the data into two halves, spot the median, and draw a bar there.

ii. Divide the upper portion of the data into two halves, spot the upper quartile, and draw a bar there.

iii. Divide the lower portion of the data into two halves, spot the lower quartile, and draw a bar there.

iv. Finish drawing the box.

v. Draw a whisker on each side extending out from the box through the data points, one data point at a time. Stop when extending to the next data point would be too far to travel, i.e., further than 150% of the width of the box.

vi. Draw an asterisk or star on any data point(s) beyond the whiskers that you drew.

18. Dots to boxes again For the blood pressure measurements: 102 115 124 128 129 130 132 133 135 137 142 160, create a dotplot (by hand or by computer). Take the dotplot and superimpose upon it a boxplot, following the steps below:

i. Divide the data into two halves, spot the median, and draw a bar there.

ii. Divide the upper portion of the data into two halves, spot the upper quartile, and draw a bar there.

iii. Divide the lower portion of the data into two halves, spot the lower quartile, and draw a bar there.

iv. Finish drawing the box.

v. Draw a whisker on each side extending out from the box through the data points, one data point at a time. Stop when extending to the next data point would be too far to travel, i.e., further than 150% of the width of the box.

vi. Draw an asterisk or star on any data point(s) beyond the whiskers that you drew.

19. Hospital stays The U.S. National Center for Health Statistics compiles data on the length of stay by patients in short-term hospitals and publishes its finding in *Vital and Health Statistics*. Data from a sample of 39 male patients and 35 female patients on length of stay (in days) are displayed in the histograms below.

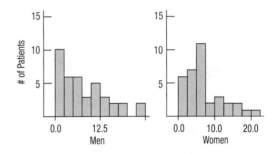

a) What would you suggest be changed about these histograms to make them easier to compare?

b) Describe these distributions by writing a few sentences comparing the duration of hospitalization for men and women.

c) Can you suggest a reason for the peak in women's length of stay?

20. Deaths 2011 A National Vital Statistics Report (www.cdc.gov/nchs/) indicated that nearly 290 000 black Americans died in 2011, compared with just over 2 million white Americans. Below are histograms displaying the distributions of their ages at death.

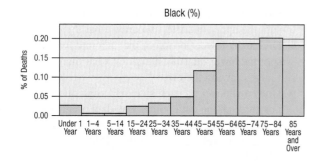

Black (%)

a) Describe the overall shapes of these distributions.
b) How do the distributions differ?
c) Look carefully at the bar definitions. Where do these plots violate the rules for statistical graphs?

T 21. Women's basketball Here are boxplots of the points scored during the first 10 games of the season for both Alexandra and Scyrine.

a) Summarize the similarities and differences in their performance so far.
b) The coach can take only one player to the provincial championship. Which one should she take? Why?

22. Wines The boxplots display case prices (in dollars) of varieties of wines produced by vineyards along three of the Finger Lakes in New York State.

a) Which lake region produces the most expensive wine?

b) Which lake region produces the cheapest wine?
c) In which region are the wines generally more expensive?
d) Write a few sentences describing these wine prices.

23. Test scores Three Statistics classes all took the same test. Histograms and boxplots of the scores for each class are shown below. Match each class with the corresponding boxplot.

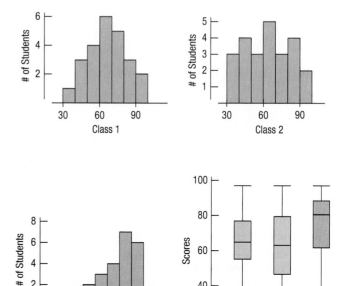

24. Cereals Sugar is a major ingredient in many breakfast cereals. The histogram displays the sugar content as a percentage of weight for 49 brands of cereal. The boxplots compare sugar content for adult and children's cereals.

a) What is the range of the sugar contents of these cereals?
b) Describe the shape of the distribution.

c) What aspect of breakfast cereals might account for this shape?

d) Are all children's cereals higher in sugar than adult cereals?

e) Which group of cereals varies more in sugar content? Explain.

25. Caffeine A student study of the effects of caffeine asked volunteers to take a memory test two hours after drinking soda. Some drank caffeine-free cola, some drank regular cola (with caffeine), and others drank a mixture of the two (getting a half-dose of caffeine). Here are the 5-number summaries for each group's scores (number of items recalled correctly) on the memory test:

	n	Min	Q1	Median	Q3	Max
No caffeine	15	16	20	21	24	26
Low caffeine	15	16	18	21	24	27
High caffeine	15	12	17	19	22	24

a) Describe the *W*'s for these data.

b) Name the variables and classify each as categorical or quantitative.

c) Create parallel boxplots to display these results as best you can with this information.

d) Write a few sentences comparing the performances of the three groups.

T 26. PISA math scores 2009 Below are some summary statistics available at www.pisa.oecd.org showing mean, standard deviation, 5th, 10th, 25th, 75th, 90th, and 95th percentiles for 2009 PISA math scores of 15-year-old students in various Organisation for Economic Co-operation and Development (OECD) countries, and also China.

a) Since you do not have sufficient information to construct standard boxplots, construct modified boxplots instead, showing whiskers extending to 5th and 95th percentiles on each side of the box, with the box constructed as usual, but marking off the mean rather than the median inside the box. Use such plots to compare the countries and then summarize the noticeable differences among countries.

b) Do the distributions appear to be close to symmetric or very skewed? Explain. Are the mean and standard deviation useful summary numbers for these data?

c) If the distributions are mound-shaped and not very skewed, then the empirical rule (see Chapter 3) should hold. Calculate the mean + two standard deviations and the mean − two standard deviations for Canada and the U.S. Can you check if more than 90% of scores lie between these two numbers

for each country? Explain. Can you confirm that between 50% and 80% of the data fall within the mean ± one standard deviation for Canada and for the U.S.? Explain.

Country	Mean	StDev	5th	10th	25th	75th	90th	95th
Australia	514	94	357	392	451	580	634	665
Canada	527	88	379	413	468	588	638	665
Finland	541	82	399	431	487	599	644	669
France	497	101	321	361	429	570	622	652
Germany	513	98	347	380	443	585	638	666
Japan	529	94	370	407	468	595	648	677
Korea	546	89	397	430	486	609	659	689
Mexico	419	79	289	318	366	472	520	547
UK	492	87	348	380	434	552	606	635
USA	487	91	337	368	425	551	607	637
China	600	103	421	462	531	674	726	757

Source: Based on data from OECD PISA 2009 online database https://www.acer.org/ozpisa/publications-and-data#2009, accessed on Feb 22, 2013.

T 27. Population growth 2010 Following is a "back-to-back" stem-and-leaf display that shows two data sets at once—one going to the left, one to the right. The display compares the percent change in population for two regions of the United States (based on census figures for 2000 and 2010). The fastest growing state was Nevada at 35%. To show the distributions better, this display breaks each stem into two lines, putting leaves 0–4 on one stem and leaves 5–9 on the other.

NE/MW States		S/W States
4433332200	0	134
987777666555	0	7799
2	1	0023344
5	1	57889
	2	1
	2	5
	3	4
	3	5

Population Growth rate
(2|1|0 means 12% for a NE/NW state
and 10% for a S/W state)

a) Use the data displayed in the stem-and-leaf display to construct comparative boxplots.

b) Write a few sentences describing the difference in growth rates for the two regions of the United States.

28. Reading An educator believes that new reading activities for elementary schoolchildren will improve reading comprehension scores. She randomly assigns Grade 3 students to an eight-week program in which some will use these activities and others will experience traditional teaching methods. At the end of the experiment, both groups take a reading comprehension exam. Here are their scores:

Control: 17 28 37 48 43 42 42 55 62 85 60 55 10 53 42 42 42 41 33 37 20 26

New Activities: 49 54 57 58 43 24 33 71 67 62 52 53 56 59 43 43 46 61

a) Construct a back-to-back stem-and-leaf display, similar to the one in the previous exercise (no need to split stems).
b) Describe the difference in scores for the two groups. Do these results suggest that the new activities might be better?

29. Derby speeds 2014 (Data in **Kentucky Derby 2014**) How fast do horses run? Kentucky Derby winners top 30 miles per hour, as shown in this graph. The graph shows the percentage of Derby winners that have run *slower* than each given speed. Note that few have won running less than 33 miles per hour, but about 86% of the winning horses have run less than 37 miles per hour. (A cumulative frequency graph like this is called an "ogive.")

Suppose you had only the ogive (and no access to the data):
a) Estimate the median winning speed.
b) Estimate the quartiles.
c) Estimate the range and the IQR.
d) Create a boxplot of these speeds.
e) Write a few sentences about the speeds of the Kentucky Derby winners.

30. Framingham The Framingham Heart Study recorded the cholesterol levels of more than 1400 men. Here is an ogive of the distribution of these cholesterol measures. (An ogive shows the percentage of cases at or below a certain value.) Create an approximate boxplot for these data and write a few sentences describing the distribution.

31. Reading scores A class of Grade 4 students took a diagnostic reading test and the scores were reported by reading grade-level. The 5-number summaries for the 14 boys and 11 girls are shown:

	Min	Q1	Median	Q3	Max
Boys:	2.0	3.9	4.3	4.9	6.0
Girls:	2.8	3.8	4.5	5.2	5.9

a) Which group had the highest score?
b) Which group had the greatest range?
c) Which group had the greatest interquartile range?
d) Which group's scores appear to be more skewed? Explain.
e) Which group generally did better on the test? Explain.
f) If the mean reading level for boys was 4.2 and for girls was 4.6, what is the overall mean for the class?

32. Cloud seeding In an experiment to determine whether seeding clouds with silver iodide increases rainfall, 52 clouds were randomly assigned to be seeded or not. The amount of rain they generated was then measured (in acre-feet).

	n	Mean	Median	SD	IQR	Q1	Q3
Unseeded	26	164.59	44.20	278.43	138.60	24.40	163
Seeded	26	441.98	221.60	650.79	337.60	92.40	430

a) Which of the summary statistics are most appropriate for describing these distributions? Why?
b) Do you see any evidence that seeding clouds may be effective? Explain.

T **33. PISA Math by gender 2009** At the PISA Website, www.pisa.oecd.org, there are summary statistics for math scores of 15-year-old males and females separately. For each OECD country below, we show the mean math scores for males, the mean math scores for females, and also the difference of the two means, followed by graphs of the data.

Country	Males	Females	Difference
Australia	519	509	10
Austria	506	486	19
Belgium	526	504	22
Canada	533	521	12
Chile	431	410	21
Czech Republic	495	490	5
Denmark	511	495	16
Estonia	516	508	9
Finland	542	539	3
France	505	489	16
Germany	520	505	16
Greece	473	459	14
Hungary	496	484	12
Iceland	508	505	3
Ireland	491	483	8
Israel	451	443	8
Italy	490	475	15
Japan	534	524	9
Korea	548	544	3
Luxembourg	499	479	19
Mexico	425	412	14
Netherlands	534	517	17
New Zealand	523	515	8
Norway	500	495	5
Poland	497	493	3
Portugal	493	481	12
Slovakia	498	495	3
Slovenia	502	501	1

Spain	493	474	19
Sweden	493	495	−2
Switzerland	544	524	20
Turkey	451	440	11
United Kingdom	503	482	20
United States	497	477	20

Source: Based on data from OECD PISA 2009 online database http://pisa2009.acer.edu.au, accessed on Feb 22, 2013.

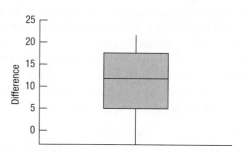

a) Use the graph(s) to compare performance of male and female students.
b) Locate Canada in all plots above and place an "x" to mark the spot. Repeat for the U.S. and for Mexico (using different symbols). For each plot, indicate which countries are outliers and what makes them unusual.
c) Give the 5-number summaries for male scores, for female scores, and for differences. Confirm that Mexico is a suspect outlier by the 1.5 IQR rule in the distribution of female scores.
d) What do you learn from the top graph that is not apparent in the other one? What do you learn from the bottom graph that is not apparent in the other one? Which graph is more useful for learning about gender differences in OECD countries? Why?

T **34. Cholesterol** A study examining the health risks of smoking measured the cholesterol levels of people who had smoked for at least 25 years and people of similar ages who had smoked for no more than five years and then stopped. Create appropriate graphical displays for both groups and write a brief report comparing their cholesterol levels.

Smokers				Ex-Smokers		
225	211	209	284	250	134	300
258	216	196	288	249	213	310
250	200	209	280	175	174	328
225	256	243	200	160	188	321
213	246	225	237	213	257	292
232	267	232	216	200	271	227
216	243	200	155	238	163	263
216	271	230	309	192	242	249
183	280	217	305	242	267	243
287	217	246	351	217	267	218
200	280	209		217	183	228

T 35. Industrial experiment Engineers at a computer production plant tested two methods for accuracy in drilling holes into a PC board. They tested how fast they could set the drilling machine by running 10 boards at each of two different speeds. To assess the results, they measured the distance (in inches) from the centre of a target on the board to the centre of the hole. The data and summary statistics are shown in the table:

	Distance (in.)	Speed	Distance (in.)	Speed
	0.000101	Fast	0.000098	Slow
	0.000102	Fast	0.000096	Slow
	0.000100	Fast	0.000097	Slow
	0.000102	Fast	0.000095	Slow
	0.000101	Fast	0.000094	Slow
	0.000103	Fast	0.000098	Slow
	0.000104	Fast	0.000096	Slow
	0.000102	Fast	0.975600	Slow
	0.000102	Fast	0.000097	Slow
	0.000100	Fast	0.000096	Slow
Mean	0.000102		**Mean**	0.097647
StdDev	0.000001		**StdDev**	0.308481

Write a report summarizing the findings of the experiment. Include appropriate visual and verbal displays of the distributions, and make a recommendation to the engineers if they are most interested in the accuracy of the method.

T 36. Women researchers 2009 Examine the display of data from Eurostat showing the percentage of female researchers in each of several employment categories for 35 countries, mostly EU members, but not exclusively (e.g., Japan is included).

a) Graph the data with four side-by-side boxplots (one for each employment category). Point out any outliers, indicating the country and its particular distinction.

b) Describe the distribution of these nations' percentages of women researchers, across all sectors, with words and numbers.

c) Compare the distributions for the three different sectors (shape, centre, spread, and unusual features).

d) Which countries are in the upper quarter for women researchers in the business enterprise sector?

e) In higher education, 75% of countries have at least _____% women researchers (fill in the blank).

Country	All sectors	Govt.	Higher Education	Business Enterprise
Belgium	32.7	32.5	39	24
Bulgaria	47.6	54	43.2	43.4
Czech Republic	28.9	37.2	35.4	15.7
Denmark	31.7	36.2	41	23.9
Germany	24.9	32.4	34.7	12.7
Estonia	42.5	61.4	46	27.5
Ireland	33.3	38.9	38.7	25.8
Spain	38.1	48.5	39.8	28.8
France	26.9	35.1	34.4	19.9
Italy	33.8	43.8	37.8	20.7
Cyprus	35.6	46.3	36.5	27.3
Latvia	52.4	53.3	52.1	53.4
Lithuania	51	54.1	53.3	31.2
Luxembourg	21.2	35.5	35.8	11.4
Hungary	32.1	40	36.1	21
Malta	29.4	44	29.5	26.6
Netherlands	25.9	30.4	36.9	14.2
Austria	28.4	43.1	37.8	16.3
Poland	39.5	41.2	42.1	22.1
Portugal	45.8	60.4	49.6	30.2
Romania	44.7	49.5	45.6	37.6
Slovenia	35.7	45.5	40.7	23.4
Slovakia	42.5	44.6	44.6	21.8
Finland	31.4	42.4	46.6	17
Sweden	35.7	38.9	44.5	25.5
United Kingdom	37.9	35.3	43.7	19.1

Country	All sectors	Govt.	Higher Education	Business Enterprise
Iceland	41	46.8	43.8	31.8
Norway	35.2	42	44.1	22.1
Switzerland	30.2	33.5	34.4	18.7
Croatia	46.4	51.8	45.4	40.7
Macedonia	51.3	50.6	49.5	80
Turkey	36.3	29.2	40.6	23.6
Russia	41.9	45.2	40.7	39.9
Japan	13.6	14.6	23.9	7.6
South Korea	15.8	18.4	25	11.4

Source: Researchers, by institutional sector, Eurostat.

T 37. Crime comparisons Below are crime rates per 100 000 residents for 18 of the biggest Canadian and U.S. cities. Rates are shown for six classes of criminal activity.
a) For each type of crime, graphically compare Canadian cities with U.S. cities. Point out crimes where rates tend to be higher in the U.S. or higher in Canada or similar in both countries.
b) Which Canadian city could pass for American in aggravated assault? Which U.S. city could pass for Canadian in homicide rate? Which Canadian city is an outlier in one of the last three listed crime categories?
c) What extraneous (i.e., lurking, or not included) quantitative variable might be important and worth taking into account, in some appropriate fashion, in order to make these Canada–U.S. comparisons as honest as possible?

City	Homicide	Aggravated Assault	Robbery	Break and Enter	Motor Vehicle Theft	Theft	Country
Toronto	1.7	157	107	553	365	1 692	Canada
Montreal	2.1	152	173	1 195	800	2 068	Canada
Vancouver	2	164	187	1 430	1 058	4 415	Canada
Calgary	1.7	132	105	814	580	2 616	Canada
Edmonton	2	180	134	986	539	2 559	Canada
Ottawa	1	123	96	690	558	1 835	Canada
Quebec	1.7	76	70	925	230	1 771	Canada
Winnipeg	2.5	276	251	1 228	1 425	2 779	Canada
Hamilton	1.3	144	86	815	698	1 832	Canada
LA	10.6	607	298	636	674	1 726	USA
NYC	7.8	474	372	453	428	1 785	USA
Philadelphia	8.1	350	270	507	492	2 199	USA
Washington	7.4	265	171	452	484	2 223	USA
Detroit	10.6	472	229	735	919	2 280	USA
Houston	7.7	433	242	960	645	2 724	USA
Boston	2.1	N/A	106	408	426	1 640	USA
Dallas	8.5	377	252	1 081	747	3 146	USA
Phoenix	7.2	354	170	1 111	1 010	3 254	USA

Source: Adapted from Statistics Canada, Juristat. 85-002-XIE2001011 Vol. 21 no. 11. Released December 18, 2001.

T 38. Election 2000 and spoiled ballots Open the Election 2000 data set available online at www.pearsoncanada.ca/deveaux, a small part of which is shown below.

District & Elected Member	Province	% Rejected Ballots
Abitibi—Baie-James—Nunavik [Guy St-Julien (Lib.)]	Quebec	2.28
Acadie—Bathurst [Yvon Godin (N.D.P.)]	NB	1.22
Ahuntsic [Eleni Bakopanos (Lib.)]	Quebec	2.37
Algoma—Manitoulin [Brent St. Denis (Lib.)]	Ontario	0.43
:	:	:

a) Use boxplots to compare the percent of rejected ballots by province (ignore territories) and point out the most noticeable differences among the provinces. Comments?
b) Are there any outliers for the western provinces? What would you suggest or recommend to Canadian officials?
c) Are there any outliers for the province of Quebec? If so, which party had the winning candidate for each outlier? Which outlier(s) are a particularly grievous smirch upon the democratic process?
d) Is the mean percentage of rejected ballots per riding bigger or smaller than the median for Manitoba? For B.C.? Explain without using any calculations. From

the boxplots, determine which province is most likely to have a smaller mean than median.

T 39. Election 2011 Open the 2011 Election data set available online at www.pearsoncanada.ca/deveaux, a small part of which is shown below.

Electoral district	Province	Voter turnout (%)	Elected candidate	Party
Avalon	NL	56.8	Andrews, Scott	Liberal
Bonavista—Gander— Grand Falls—Windsor	NL	44.8	Simms, Scott	Liberal
Humber—St. Barbe—Baie Verte	NL	50.9	Byrne, Gerry	Liberal
Labrador	NL	52.9	Penashue, Peter	Conservative
⋮	⋮	⋮	⋮	⋮

a) Use boxplots to compare the percentage of voter turnout by province, and point out the more (voting) motivated provinces and the less motivated provinces.

b) Do you consider the ridings designated as suspect outliers in the boxplots to be genuine outliers? Explain.

c) For Canadian ridings, overall, are there any suspect outliers (by the boxplot rule)? If so, can you explain the cause or nature of any of the suspect outliers?

d) Is the mean or median smaller for Manitoba? For B.C.? No calculations, but explain your answer.

T 40. Election 2006 Open up the Election 2006 data set available online at www.pearsoncanada.ca/deveaux, a small part of which is displayed below.

Prov. Voter	Electoral District	Region	% Rejected Ballots	% Turnout
NL	Avalon	Atlantic	1.65	59.6
NL	Bonavista— Gander— Grand Falls— Windsor	Atlantic	0.50	54.2
NL	Humber— St. Barbe— Baie Verte	Atlantic	0.58	54.8
NL	Labrador	Atlantic	0.45	58
⋮	⋮	⋮	⋮	⋮

a) Use boxplots to compare the percentage of voter turnout among the regions (ignore the territories), and point out the differences among regions. Be sure to comment on more (voting) motivated and less motivated regions.

b) Which ridings are considered to be suspect outliers within their respective regions by the 1.5 IQR rule? Are those ridings also suspect outliers for Canada as a whole?

c) Which regional distribution is least symmetric (ignore territories and the Atlantic region due to small numbers)? Use any appropriate graphs.

d) Find the mean and standard deviation for turnout percentages in all Quebec ridings. Compute "$\bar{y} + (2 \times s)$" and "$\bar{y} - (2 \times s)$". Predict the percent of ridings whose turnout percentages will fall between these two numbers (see Empirical Rule, Chapter 3). Go through the Quebec data and count how many (and what percent) actually do fall between these two numbers.

e) If you averaged the turnout percentages for all ridings in Ontario, would you get the percentage turnout for the province?

T 41. Homicides 2011 Examine the displayed data on homicides from Statistics Canada. The rate shown is the average homicide rate per 100 000 population over the 11 years from 2001 to 2011. Cities were classified into three size categories. We also classified them by geographic region below.

a) Use graphs to examine the relationship between crime rate and city size (category). Describe any patterns you see.

b) What type of variable is city size, if determined precisely by a count of residents? What type is it in our analysis above? Is this change a good thing or a bad thing? If you wanted to use the actual city sizes, can you think of another type of graph to use to explore the relationship between crime rate and city size? If so, produce this graph. (More on this in Chapter 6.)

c) Use graphs to examine the relationship between crime rate and geographical region. Describe any geographical patterns you see.

d) We have used the average homicide rate over 10 years. Is there another relationship we should examine to more fully understand homicide patterns in Canada? If so, explain what type of data plot(s) you would like to request from StatsCan.

City	Size	Population	Region	Homicide Rate
Toronto	500 000+	5 841 126	Ontario	1.78
Montreal	500 000+	3 917 875	Quebec	1.53
Vancouver	500 000+	2 426 200	West	2.43
Calgary	500 000+	1 268 704	West	1.98

Edmonton	500 000+	1 196 704	West	3.07
Ottawa	500 000+	945 489	Ontario	1.20
Winnipeg	500 000+	765 770	West	3.51
Quebec	500 000+	760 961	Quebec	0.62
Hamilton	500 000+	749 722	Ontario	1.43
Kitchener	250 000–500 000	498 876	Ontario	0.90
London	250 000–500 000	496 156	Ontario	1.30
Halifax	250 000–500 000	409 662	East	1.92
St. Cath-Niagara	250 000–500 000	405 216	Ontario	1.42
Oshawa	250 000–500 000	370 448	Ontario	1.18
Victoria	250 000–500 000	361 685	West	1.19
Windsor	250 000–500 000	332 539	Ontario	1.48
Gatineau	250 000–500 000	310 449	Quebec	1.15
Saskatoon	250 000–500 000	272 771	West	2.88
Regina	100 000–250 000	219 286	West	3.67
Sherbrooke	100 000–250 000	200 695	Quebec	0.36
St. John's	100 000–250 000	197 524	East	0.67
Abbotsford-Mission	100 000–250 000	177 064	West	2.98
Sudbury	100 000–250 000	164 196	Ontario	1.60
Kingston	100 000–250 000	164 089	Ontario	1.67
Saguenay	100 000–250 000	151 394	Quebec	0.74
Trois-Rivieres	100 000–250 000	147 657	Quebec	0.82
Saint John	100 000–250 000	128 534	East	0.92
Thunder Bay	100 000–250 000	127 049	Ontario	1.86

Source: Adapted from 'Homicide in Canada' - Statistics Canada, Catalogue no. 85-002X, Juristat, ISSN 1209-6393

42. Eye and hair colour A survey of 1021 school-age children was conducted by randomly selecting children from several large urban elementary schools. Two of the questions concerned eye and hair colour. In the survey, the following codes were used:

Hair Colour	Eye Colour
1 = Blond	1 = Blue
2 = Brown	2 = Green
3 = Black	3 = Brown
4 = Red	4 = Grey
5 = Other	5 = Other

The Statistics students analyzing the data were asked to study the relationship between eye and hair colour. They produced this plot:

Is their graph appropriate? If so, summarize the findings. If not, explain why not.

T 43. He shoots, he scores the hat-trick Look once more at the data on total points scored each season by Wayne Gretzky, seen in Exercise 36 Chapter 3.
a) Find the 5-number summary.
b) Construct a boxplot.
c) Explain why the season when Gretzky scored only 48 points would not be deemed a suspect outlier in a boxplot.
d) Plot total points versus year. What do you learn from this plot that is masked in the boxplot?

T 44. Drunk Driving 2014 Accidents involving drunk drivers account for about 40% of all deaths on the nation's highways. The following table tracks the number of alcohol-related fatalities for 33 years.

Year	Deaths (in 1000)	Year	Deaths (in 1000)
1982	26.2	1999	16.6
1983	24.6	2000	17.4
1984	24.8	2001	17.4
1985	23.2	2002	17.5
1986	25.0	2003	17.1
1987	24.1	2004	16.9
1988	23.8	2005	16.9
1989	22.4	2006	15.8
1990	22.6	2007	15.4
1991	20.2	2008	13.8
1992	18.3	2009	10.8
1993	17.9	2010	10.2
1994	17.3	2011	9.9
1995	17.7	2012	10.3
1996	17.7	2013	10.1
1997	16.7	2014	10.0
1998	16.7		

(*Source:* www.madd.org/drunk-driving/about/history.html)

a) Create and a stem-and-leaf display or a histogram of these data.
b) Create a timeplot.
c) Using features apparent in the stem-and-leaf display (or histogram) and the timeplot, write a few sentences about deaths caused by drunk driving.

T **45. Hurricanes 2015 again** Here are the number of Atlantic hurricanes classified as major for the years 1944 to 2015:

3, 3, 1, 2, 4, 3, 8, 5, 3, 4, 2, 6, 2, 2, 4, 2, 2, 7, 1, 2, 6, 1, 3,

1, 0, 5, 2, 1, 0, 1, 2, 3, 2, 1, 2, 2, 2, 3, 1, 1, 1, 3, 0, 1, 3, 2,

1, 2, 1, 1, 0, 5, 6, 1, 3, 5, 3, 4, 2, 3, 6, 7, 2, 2, 5, 2, 5 , 4, 2, 0, 2, 2

You examined these data in the previous chapter's exercises. Now create a timeplot of the data and tell if there is anything new that you have learned. (Use a smoother if your software has one available.)

46. World Series champs again What type of plot was the graph of Joe Carter's home runs shown in Exercise 43 in Chapter 3? Summarize what you see.

T **47. Tsunamis 2013** Look again at the magnitudes of tsunami generating earthquakes (updated in the Tsunamis_2013 data file to early 2013). Delete all data from 1950 and earlier, so that only relatively recent data, from 1951 to 2013, will be examined. Use computer software to answer the following questions about these more recent tsunami-creating earthquakes:
a) Examine graphically and describe the distribution of tsunami-creating earthquake magnitudes since 1951.
b) Plot magnitude versus year and apply a smoother to the plot. Do you learn anything more from this plot that is not observable from your plot(s) in part a)?
c) Now examine how magnitude depends on country with side-by-side boxplots. Any problems? Any suggestions?
d) Delete data for all countries that have suffered 10 or fewer tsunami-creating earthquakes. Give side-by-side boxplots again. Where did quakes tend to be biggest? Smallest? Which three countries had the biggest quakes (and how big?) Where and when was the smallest quake?

T **48. Assets** Here is a histogram of the assets (in millions of dollars) of 79 companies chosen from the *Forbes* list of top corporations.

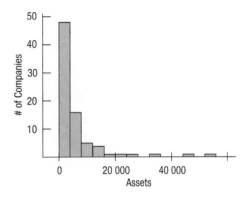

a) What aspect of this distribution makes it difficult to summarize or discuss centre and spread?
b) What would you suggest doing with these data if we want to understand them better?

T **49. Assets again** Here are the same data you saw in Exercise 48 after re-expressions as the square root of assets and the logarithm of assets.

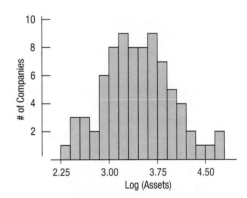

a) Which re-expression do you prefer? Why?
b) In the square root re-expression, what does the value 50 actually indicate about the company's assets?
c) In the logarithm re-expression, what does the value 3 actually indicate about the company's assets?

T **50. Rainmakers** The table below lists the amount of rainfall (in acre-feet) from the 26 clouds seeded with silver iodide discussed in Exercise 32:

2745	703	302	242	119	40	7
1697	489	274	200	118	32	4
1656	430	274	198	115	31	
978	334	255	129	92	17	

a) Why is acre-feet a good way to measure the amount of precipitation produced by cloud seeding?

b) Plot these data, and describe the distribution.

c) Create a re-expression of these data that produces a more advantageous distribution.

d) Explain what your re-expressed scale means.

T 51. Stereograms Stereograms appear to be composed entirely of random dots. However, they contain separate images that a viewer can "fuse" into a three-dimensional (3D) image by staring at the dots while defocusing the eyes. An experiment was performed to determine whether knowledge of the embedded image affected the time required for subjects to fuse the images. One group of subjects (group NV) received no information or just verbal information about the shape of the embedded object. A second group (group VV) received both verbal information and visual information (specifically, a drawing of the object). The experimenters measured how many seconds it took for the subject to report that he or she saw the 3D image.

a) What two variables are discussed in this description?

b) For each variable, is it quantitative or categorical? If quantitative, what are the units?

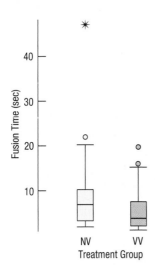

c) The boxplots above compare the fusion times for the two treatment groups. Write a few sentences comparing these distributions. What does the experiment show?

T 52. Stereograms revisited Because of the skewness of the distributions of fusion times, we might consider a re-expression. Here are the boxplots of the log of fusion times. Is it better to analyze the original fusion times or the log fusion times? Explain.

T 53. First Nations 2010 by province Let's compare sizes of First Nations bands in Ontario and Saskatchewan.

a) Using the boxplots, what is the approximate size of the largest band in Ontario? Do band sizes tend to be bigger in one province than in the other? From the plot, what can you say about the difference in median band size? In mean band size?

b) How could we improve readability of the boxplots above?

c) Describe the shapes of the two distributions. Do you consider most of the suspect outliers to be genuine outliers? Explain.

d) If we were to exclude the biggest Ontario band, the Six Nations of the Grand River, which measures of centre and spread would hardly need the trouble of recalculation, and why?

e) Why is the mean band size bigger than the median band size in both provinces?

f) Seventy-five percent of Saskatchewan bands are at least what size? Half of Saskatchewan bands are between _____ and _____ in size (fill in the blanks). Just estimate from the boxplots as best you can.

g) Calculate the logarithms (any base) of all band sizes in both provinces, and produce side-by-side boxplots. Describe the major effects of this transformation.

h) After this transformation, does the Six Nations still appear to be such an extreme outlier? Are there new outliers (if so, which band is the farthest outlying)?

✓ Just Checking ANSWERS

1. The distribution is very right-right skewed and somewhat bimodal with one big group centred around 0.3% and another smaller group centred around 1.5%.

2. The boxplot doesn't show the bimodality well.

3. There are some suspect outliers in the boxplot, but the bimodality in the histogram suggests that they may just be members of a distinct subgroup.

4. Quebec results are poorest, with more than 3/4 of ridings exceeding 1% spoiled ballots, and also showing the largest variability in results. Excluding a few outliers, western provinces did best at limiting spoiled ballots, with most ridings around 0.3%.

5. Only three territories for the far-right boxplot.

6. Yes, we see that the lower quartile of Quebec's boxplot (lower edge of the box) exceeds the highest results from all regions.

7. Worst result is in Quebec. Best are in the West and Atlantic provinces, at about 0.2%.

8. We see that the unusually poor results in Quebec created the bimodality in the histogram.

MathXL **MyLab Statistics**	Go to MathXL at **www.mathxl.com** or MyLab™ Statistics at **www.mystatlab.com**. You can practise exercises for this chapter as often as you want. The guided solutions will help you find answers step by step. You'll find a personalized study plan available to you too!

The Standard Deviation as a Ruler and the Normal Model

LEARNING OBJECTIVES

1 Standardize data and use *z*-scores for comparisons

2 Know how shifting and scaling of data affect shape, centre and spread

3 Understand how density curves, and in particular the Normal Model, work

4 Apply the 68–95–99.7 Rule

5 Find other percentiles of the Normal Model

6 Use Normal Probability Plots to assess the adequacy of the Normal Model

Connections

Let's expand our discussion of quantitative variables. We have looked at histograms and tried to describe the shape of the distribution. Now we will *model* that shape with a curve, such as the well-known *Normal curve*, and we'll see how this conceptualization can help us.

But first we'll discuss some common transformations of variables—e.g. if we convert temperature measurements in Celsius to Fahrenheit, how would summary measures change?

What if we have two graduate school candidates from overseas, and we want to compare their TOEFFL scores but one wrote the Internet-based exam while the other wrote the paper-based exam? One solution is to convert their scores to *standardized scores*, which measure how far each score is from the mean on that test—but measured by counting *standard deviations*.

All models are wrong—but some are useful.

—*George Box, famous statistician*

The women's heptathlon in the Olympics consists of seven track-and-field events: the 200 m and 800 m runs, 100 m high hurdles, shot put, javelin, high jump, and long jump. In the 2012 Olympics, Tatanya Chernova of Russia posted a long jump of 6.54 metres—about half a metre farther than the mean distance for all contestants.

Jessica Ennis of Great Britian won the 200 m run with a time of 22.83 seconds—more than two seconds faster than the average. Each contestant is awarded points for each event based on her performance. So, which performance deserves more points? It's not clear how to compare them. They aren't measured in the same units, or even in the same *direction* (longer jumps are better but shorter times are better).

We want to compare individual values, not whole groups as we did earlier. To see which value is more extraordinary, we need a way to judge them against the background of data they come from. The tool we'll use is one you've already seen: the standard deviation. We'll judge how unusual a value is by how far, in standard deviation units, it lies from the mean. Statisticians use the standard deviation as a ruler throughout Statistics, and we'll do that for the rest of the book.

We'll start with stem-and-leaf displays (Figure 5.1) because they show individual values, and because it is easy to orient them either high-to-low or low-to-high so the best performances can be at the top of both displays.

Figure 5.1

Stem-and-leaf displays for the 200 m race and the long jump in the 2012 Olympic Heptathlon. Jessica Ennis (green scores) won the 200 m, and Tatanya Chernova (red scores) won the long jump. The stems for the 200 m race run from faster to slower and the stems for the long jump from longer to shorter so that the best scores are at the top of each display. (Data in **Womens Heptathalon 2012**)

200 m Run		Long Jump	
Stem	Leaf	Stem	Leaf
22	88	6	55
23	3	6	0000111111122233334
23	6677779	6 5	77888889999
24	12333444	5	4
24	556778	4	
25	012334444	4	1
25	679	3	7
22	8=22.8 seconds	6 5=6.5 metres	

Which of the two winning scores is the better one? Chernova's 6.54 m long jump was 0.63 m longer than the mean jump of 5.91 m (summary statistics are in the margin below). The standard deviation for this event was 0.56 m, so her jump was a little more than one SD better (longer) than the mean. More precisely, it was $(6.54 - 5.91)/0.56 = 1.1$ standard deviations better than the mean, so we could give her 1.1 points. For the 200 m run,

	Long Jump	200 m
Mean (all contestants)	5.91 m	24.48 s
SD	0.56 m	0.80 s
n	35	36
Chernova	6.54 m	23.67 s
Ennis	6.48 m	22.83 s

Ennis's winning time of 22.83 was more impressive because it was just over 2 SD's better (faster) than the mean. She would get 2.1 points for that.

We could turn each result into a score like this, and simply add them up over the 7 events to determine the winner. If we did this, we'd find that Ennis got a total of 8.4 (and won the gold). Chernova got a 4.3 to take the bronze. Olympic judges actually use a point system based on similar calculations (but based on performances from many more competitions). In those points, Ennis got 6955 and Chernova 6628. However, the correlation between our system and the Olympic point system is 0.99! (More about *correlations* in the next chapter.)

5.1 Standardizing with *z*-Scores

Expressing a distance from the mean in standard deviations *standardizes* the performances. To **standardize** a value, we subtract the mean and then divide this difference by the standard deviation:

$$z = \frac{y - \bar{y}}{s}$$

The values are called **standardized values**, and are commonly denoted with the letter *z*. Usually we just call them **z-scores**.

z-scores measure the distance of a value from the mean in standard deviations. A *z*-score of 2 says that a data value is two standard deviations above the mean. It doesn't matter whether the original variable was measured in fathoms, dollars, or carats; those units don't apply to *z*-scores. Data values below the mean have negative *z*-scores, so a *z*-score of −1.6 means that the data value was 1.6 standard deviations below the mean. Of course, regardless of the direction, the farther a data value is from the mean, the more unusual it is, so a *z*-score of −1.3 is more extraordinary than a *z*-score of 1.2.

By using the standard deviation as a ruler to measure statistical distance from the mean, we can compare values from the same data set, or even two values measured on different variables or with different scales. However, the comparison must still make sense. The *z*-score simply tells us how far a value is from its mean in terms of standard deviations.

We can even combine individual *z*-scores for the heptathlon to get an overall score (our approach, not the Olympic system). In the table below we combine the two events, the long jump and the 200 m run. Notice that for the running events we would naturally reverse the sign of the *z*-score for those events because a time *below* the mean is better.

		Event	
		Long Jump	**200 m Run**
	Mean **SD**	5.91 m 0.56 m	24.48 s 0.80 s
Chernova	**Performance** ***z*-score** **Total for two events**	6.54 m (6.54 − 5.91)/0.56 = 1.13 1.13 + 1.01 = 2.14	23.67 s (23.67 − 24.48)/0.80 = −1.01
Ennis	**Performance** ***z*-score** **Total for two events**	6.48 m (6.48 − 5.91)/0.56 = 1.02 1.02 + 2.06 = 3.08	22.83 s (22.83 − 24.48)/0.80 = −2.06

Ruslan Kudrin/Fotolia

For Example STANDARDIZING SKIING TIMES

The men's super-combined skiing event debuted in the 2010 Winter Olympics in Vancouver. It consists of two races: a downhill and a slalom. Times for the two events are added together, and the skier with the lowest total time wins. At Vancouver, the mean slalom time was 52.67 seconds with a standard deviation of 1.614 seconds. The mean downhill time was 116.26 seconds with a standard deviation of 1.914 seconds. Bode Miller of the United States, who won the gold medal with a combined time of 164.92 seconds, skied the slalom in 51.01 seconds and the downhill in 113.91 seconds.

QUESTION: On which race did he do better compared to the competition?

ANSWER: Standardize each of his times by subtracting the mean and dividing by the standard deviation, as follows:

$$z_{Slalom} = \frac{51.01 - 52.67}{1.614} = -1.03$$

$$z_{Downhill} = \frac{113.91 - 116.26}{1.914} = -1.23$$

Keeping in mind that faster times are *below* the mean, Miller's downhill time of 1.23 SDs below the mean is even more remarkable than his slalom time, which was 1.03 SDs below the mean.

 # Just Checking

1. Your Statistics teacher has announced that the lower of your two tests will be dropped. You got a 90 on test 1 and an 80 on test 2. You're all set to drop the 80 until she announces that she grades "on a curve." She standardized the scores in order to decide which is the lower one. If the mean on the first test was 88 with a standard deviation of 4 and the mean on the second was 75 with a standard deviation of 5

   a) Which one will be dropped?
   b) Does this seem "fair"?

Section 5.1 Exercises

1. Stats test The mean score on the Stats exam was 75 points with a standard deviation of 5 points, and Gregor's z-score was −2. How many points did he score?

2. Mensa People with z-scores above 2.5 on an IQ test are sometimes classified as geniuses. If IQ scores have a mean of 100 and a standard deviation of 16 points, what IQ score do you need to be considered a genius?

5.2 Shifting and Scaling

There are two steps to finding a z-score. First, the data are *shifted* by subtracting the mean. Then, they are *rescaled* by dividing by the standard deviation. How do these operations work?

Who	80 male participants of the NHANES survey between the ages of 19 and 24 who measured between 68 and 70 inches tall
What	Their weights
Unit	Kilograms
When	2001–2002
Where	United States
Why	To study nutrition and health issues and trends
How	National survey

Figure 5.2

Histogram and boxplot for the men's weights. The shape is skewed to the right with several high outliers.

Shifting to Adjust the Centre

Since the 1960s, the Centers for Disease Control and Prevention's National Center for Health Statistics has been collecting health and nutritional information on people of all ages and backgrounds. The most recent survey, the National Health Interview Survey (NHIS), collected data on over 75 000 individuals, based on interviews with nearly 40 000 families. An earlier National Health And Nutrition Examination Survey (NHANES) 2001–2002[1] measured a wide variety of variables, including body measurements, cardiovascular fitness, blood chemistry, and demographic information on more than 11 000 individuals.

Included in this group were 80 men between 19 and 24 years old of average height (between 5′8″ and 5′10″ tall). (Data in **NHANES**) Figure 5.2 displays a histogram and boxplot of their weights:

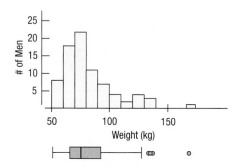

Their mean weight is 82.36 kg. For this age and height group, the National Institutes of Health recommends a maximum healthy weight of 74 kg, but we can see that some of the men are heavier than the recommended weight. To compare their weights to the recommended maximum, we could subtract 74 kg from each of their weights. What would that do to the centre, shape, and spread of the histogram? Figure 5.3 gives us the new picture:

Figure 5.3

Subtracting 74 kilograms shifts the entire histogram down, but leaves the spread and the shape exactly the same.

On average, they weigh 82.36 kg, so on average they're 8.36 kg overweight. And, after subtracting 74 from each weight, the mean of the new distribution is 82.36 − 74 = 8.36 kg. In fact, when we **shift** the data by adding (or subtracting) a constant to each value, all measures of position (centre, percentiles, min, max) will increase (or decrease) by the same constant.

What about the spread? What does adding or subtracting a constant value do to the spread of the distribution? Look at the two histograms again. Adding or subtracting a constant changes each data value equally, so the entire distribution just shifts. Its shape doesn't

---

[1]www.cdc.gov/nchs/nhanes.htm

change and neither does the spread. None of the measures of spread we've discussed—not the range, not the IQR, not the standard deviation—changes.

> *Adding (or subtracting) a constant to every data value adds (or subtracts) the same constant to measures of position, but leaves measures of spread unchanged.*

Rescaling to Adjust the Scale

Not everyone thinks naturally in metric units. Suppose we want to look at the weights in pounds instead. We'd have to rescale the data. Because there are about 2.2 pounds in every kilogram, we'd convert the weights by multiplying each value by 2.2. Multiplying or dividing each value by a constant—**rescaling** the data—changes the measurement units. Figure 5.4 shows histograms of the two weight distributions, plotted on the same scale, so you can see the effect of multiplying:

Figure 5.4

Men's weights in both kilograms and pounds. How do the distributions and numerical summaries change?

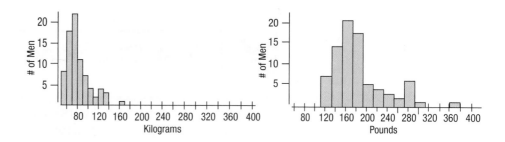

What happens to the shape of the distribution? Although the histograms don't look exactly alike, we see that the shape really hasn't changed: Both are unimodal and skewed to the right.

What happens to the mean? Not too surprisingly, it gets multiplied by 2.2 as well. The men weigh 82.36 kg on average, which is 181.19 pounds. As the boxplots and 5-number summaries below show, all measures of position act the same way. They all get multiplied by this same constant.

What happens to the spread? Take a look at the boxplots (Figure 5.5). The spread in pounds (on the right) is larger. How much larger? If you guessed 2.2 times, you've figured out how measures of spread get rescaled.

Figure 5.5

The boxplots (drawn on the same scale) show the weights measured in kilograms (on the left) and pounds (on the right). Because 1 kg is 2.2 lb, all the points in the right box are 2.2 times larger than the corresponding points in the left box. So each measure of position and spread is 2.2 times as large when measured in pounds rather than kilograms.

	Weight (kg)	Weight (lb)
Min	54.3	119.46
Q1	67.3	148.06
Median	76.85	169.07
Q3	92.3	203.06
Max	161.5	355.30
IQR	25	55
SD	22.27	48.99

> *When we multiply (or divide) all the data values by any constant, all measures of position (such as the mean, median, and percentiles) and measures of spread (such as the range, the IQR, and the standard deviation) are multiplied (or divided) by that same constant.* If the constant is negative, though, make any measures of spread positive. A variance gets multiplied by the constant squared.

For Example RESCALING THE MEN'S COMBINED TIMES

RECAP: The times in the men's combined event at the winter Olympics are reported in minutes and seconds. The mean and standard deviation of the 34 final super combined times at the 2010 Olympics were 168.93 seconds and 2.90 seconds, respectively.

QUESTION: Suppose instead that we had reported the times in minutes—that is, that each individual time was divided by 60. What would the resulting mean and standard deviation be?

ANSWER: Dividing all the times by 60 would divide both the mean and the standard deviation by 60:

Mean = 168.93/60 = 2.816 minutes   SD = 2.90/60 = 0.048 minutes.

 # Just Checking

2. SAT test scores are required of applicants to many U.S. universities. In 1995, the Educational Testing Service (ETS) adjusted the scores of SAT tests. Before ETS recentred the SAT Verbal test, the mean of all test scores was 450.
   a) How would adding 50 points to each score affect the mean?
   b) The standard deviation was 100 points. What would the standard deviation be after adding 50 points?
   c) Suppose we drew boxplots of test takers' scores a year before and a year after the '+50' recentring. How would the boxplots of the two years differ?

3. A company manufactures wheels for in-line skates. The diameter of the wheels has a mean of 3 inches and a standard deviation of 0.1 inches. Because so many of its customers use the metric system, the company decides to report their production statistics in millimetres (1 inch = 25.4 mm). They report that the standard deviation is now 2.54 mm. A corporate executive is worried about this increase in variation. Should he be concerned? Explain.

Non-linear versus linear transformations

Shifting and rescaling are examples of linear transformations (or re-expressions) of data. Sometimes we use *non-linear transformations*, such as log, square root, etc., to re-express data in a useful manner—perhaps to change the shape to something close to Normal. Linear transformations do not change the shape of the distribution, and measures of centre and spread change in an easily predictable manner. On the other hand, non-linear transformations will change the shape of the distribution, and measures of centre and spread do not change in a predictable manner. You will have to re-compute them after transformation. So, for example, in the earlier example, the mean weight of 80 NHANES participants was 82.36 kg. If we take the logs of all 80 weights, the new distribution of the log-weights will be less skewed to the right, and the mean of this distribution will *not* equal the log of 82.36. Re-compute the mean and standard deviation or other relevant summary measures using the log-transformed data.

The distribution of *z*-scores

Standardizing data into *z*-scores is just shifting them by the mean and rescaling them by the standard deviation. Now we can see how standardizing affects the distribution. When we subtract the mean of the data from every data value, we shift the mean to zero. As we have seen, such a shift doesn't change the standard deviation.

When we *divide* each of these shifted values by *s*, however, the standard deviation should be divided by *s* as well. Since the standard deviation was *s* to start with, the new standard deviation becomes 1.

How, then, does standardizing affect the distribution of a variable? Let's consider three aspects of a distribution: the shape, centre, and spread.

> *z*-scores have mean 0 and standard deviation 1.

■ Standardizing into *z*-scores does not change the *shape* of the distribution of a variable.

■ Standardizing into *z*-scores changes the *centre* by making the mean 0.

■ Standardizing into *z*-scores changes the *spread* by making the standard deviation 1.

Step-by-Step Example WORKING WITH STANDARDIZED VARIABLES

The School of Graduate Studies (SGS) at the University of Toronto requires non-Canadian applicants to demonstrate facility in the English language if English is not the applicant's primary language nor language of instruction. TOEFL test scores may be used as proof. SGS requires a minimum score of 580 on the paper-based TOEFL test (PBT). In a recent year, the mean and standard deviation for all paper-based TOEFL writers, who were graduate school applicants, were 548 and 57, respectively. An alternative test is the Internet-based TOEFL test (iBT), which had a mean of 82 and standard deviation of 23 for graduate school writers.[2]

Question: If SGS wanted to require a similar standing on the Internet-based test for applicants, what minimal score would you recommend that they accept?

THINK ➡ Plan State what you want to find out. **Variables** Identify the variables and report the *W*'s (if known). Check the appropriate conditions.	I want to know what iBT score corresponds approximately to the required PBT score, based on having knowledge only of the mean and standard deviation for both tests for graduate-level test writers. ✓ **Quantitative Data Condition:** Scores for both tests are quantitative where the units are points.
SHOW ➡ Mechanics Standardize the variables. The raw score is '*z*' standard deviations above the mean.	The PBT minimal score is 580, which standardized gives a *z*-score of $$z = \frac{580 - 548}{57} = 0.56$$ So a PBT score of 580 is 0.56 standard deviations above the mean of all test takers. For the iBT, 0.56 standard deviations above the mean is $82 + 0.56(23) = 94.9$.
TELL ➡ Conclusion Interpret your results in context.	I'd recommend accepting an IBT score of 95 in place of the PBT score of 580. [Caution: Identical *z*-scores does not necessarily mean identical percentile standing—unless the two distributions have similar shapes.]

---

[2]Test statistics from Educational Testing Service (www.ets.org).

Section 5.2 Exercises

3. Shipments A company selling clothing on the Internet reports that the packages it ships have a median weight of 68 ounces and an IQR of 40 ounces.
   a) The company plans to include a sales flyer weighing 4 ounces in each package. What will the new median and IQR be?
   b) If the company recorded the shipping weights of these new packages in pounds instead of ounces, what would the median and IQR be? (1 lb. = 16 oz.)

4. Hotline A company's customer service hotline handles many calls relating to orders, refunds, and other issues.

The company's records indicate that the median length of calls to the hotline is 4.4 minutes with an IQR of 2.3 minutes.
   a) If the company were to describe the duration of these calls in seconds instead of minutes, what would the median and IQR be?
   b) In an effort to speed up the customer service process, the company decides to streamline the series of push-button menus customers must navigate, cutting the time by 24 seconds. What will the median and IQR of the length of hotline calls become?

5.3 Density Curves and the Normal Model

A z-score gives us an indication of how unusual a value is because it tells us how far it is from the mean. If the data value sits right at the mean, it's not very far at all and its z-score is 0. A z-score of 1 tells us the data value is 1 standard deviation above the mean, while a z-score of -1 tells us that the value is 1 standard deviation below the mean. How far from 0 does a z-score have to be to be interesting or unusual? We know that 50% of the data lie between the quartiles. For symmetric data, the standard deviation is usually a bit smaller than the IQR, and it's not uncommon for at least half of the data to have z-scores between -1 and 1. But no matter what the shape of the distribution, a z-score of 3 (plus or minus) or more is rare, and a z-score of 6 or 7 shouts out for attention.

To say more about how big we expect a data value or its z-score to be, we need to *model* the data's distribution. A model will let us say much more precisely how often we'd be likely to see data values of different sizes. Of course, like all models of the real world, the model will be wrong—wrong in the sense that it can't match reality exactly. But it can still be useful. Like a physical model, it's something we can look at and manipulate in order to learn more about the real world.

Models help our understanding in many ways. Just as a model of an airplane in a wind tunnel can give insights even though it doesn't show every rivet,[3] models of data give us summaries that we can learn from and use, even though they don't fit each data value exactly. Without models, what we can learn about the world at large is limited to only what we can say about the data we have at hand.

What do we mean by a statistical model? To model the frequency distribution of a quantitative variable, start by imagining a histogram for a variable based on lots of data; for example, think about the weights, in grams, of at-term newborns. If we observed, say, 5000 births, we might get a histogram resembling the one in Figure 5.6a.

We can easily imagine fitting a curve going through the tops of the rectangles. But what if we observed 5000 more births? The heights of the bars would roughly double, changing the fitted curve. We don't want our model to depend on the number of observations. What we are really interested in modelling is the relative frequencies or proportions of births in certain ranges. So let's switch to relative frequencies, which removes the dependence on the amount of data. But there is still a problem: If we halve the bin width, the relative frequencies all decrease by about 50%, changing the picture again. How can we stabilize our picture of the distribution? What we'll do is set the *area of a*

---

[3]In fact, the model is useful *because* it doesn't have every rivet. It is because models offer a simpler view of reality that they are so useful as we try to understand reality.

Figure 5.6a

Simulated weights of 5000 newborns

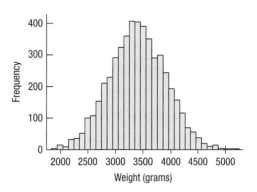

rectangle equal to the *proportion* or *relative frequency* of individuals falling into that bin:

Area of Rectangle = Width × Height = Relative Frequency

Now solve the above for *Height,* and we get:

Height = Relative Frequency of Bin ÷ Bin Width (also known as *frequency density*)

The histogram above becomes (Figure 5.6b):

Figure 5.6b

Histogram of newborn weights, with Area = Relative Frequency or Proportion

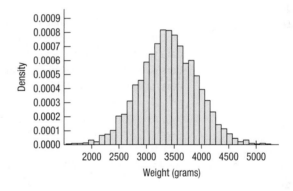

So area equals relative frequency, and if you add up the relative frequencies or areas of all the rectangles now you have to get 1.00. Decrease or increase the bin width and things change a bit as you trace a path across the tops of the rectangles, but the overall pattern stays about the same and the total area stays put at 1.00. What would happen if you could keep adding data (while gradually decreasing the bin width)? You would get a smoother and smoother shape as you get closer and closer to a true unclouded picture of the underlying distribution.

Now you have to decide what type of smooth curve to fit to the data. Such curves are called **density curves**. In Figure 5.6c, we fitted a *Normal* density curve, which will be discussed in more detail shortly.

Of interest are the proportions of data in specified ranges. If we want to estimate the proportion of at-term newborns weighing between 2450 and 2750 grams, we could add up the areas of the three corresponding rectangles, shown in red in Figure 5.6c, or approximate it by calculating the area under the fitted curve between 2450 and 2750 (Figure 5.6c–second diagram). If we are interested in the proportion weighing between 2500 and 2700, we take the area under the curve between 2500 and 2700, even though they are not cutpoints, or boundaries, of bins. Once we use the smooth curve model, there no longer are any bins or cutpoints; we can cut anywhere on the weight-axis. Finding areas under a curve actually requires the use of calculus, but for the most common models, we can use tables or computer software.

Figure 5.6c

Matching up areas under the histogram and under the density curve

So, if we believe that we can adequately model a distribution with some particular density curve, we use areas under that density curve to approximate percentages of the data falling in ranges of interest. The area under the density curve between any two numbers *a* and *b* will represent the proportion of the data that lie between those same two numbers, as shown below (Figure 5.6d):

Figure 5.6d

Area under the density curve between *a* and *b* represents the proportion of the data between *a* and *b*.

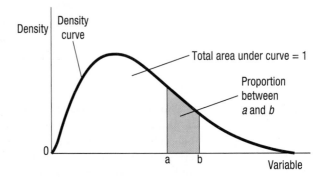

There are many possible curves that might serve as useful models, but any density curve has to satisfy the following conditions:
1. Always positive or zero (or we'd have negative percentages)
2. Total area under the curve above the *x*-axis equal to 1.00 (or we would not cover exactly 100% of the observations)

One density curve stands out in particular, due to its central role in statistical theory (which we will delve into further in Chapter 14): the Normal curve. This is a bell-shaped curve (not the only bell-shaped curve used in Statistics), and you could draw it—if you had to—using the following formula:

$$y = \frac{1}{\sigma\sqrt{2\pi}}e^{-\frac{1}{2}\left(\frac{x-\mu}{\sigma}\right)^2} \text{ for } -\infty < x < +\infty$$

You cannot draw this curve until you choose values for the two *parameters* appearing in the equation, μ and σ, which control the centre and dispersion of the distribution, respectively. For example, Figure 5.7 shows two different Normal curves for different choices of these parameters, one modelling perhaps a set of rather average-looking blood pressure measurements, the other modelling another possible set of blood pressure measurements with a higher centre and greater dispersion:

Figure 5.7

Two Normal curves with different choices for parameters.

	μ	σ
——	120	5
- - -	140	10

Is the Normal normal?
Don't be misled. The name "Normal" doesn't mean that these are the *usual* shapes for histograms. The name follows a tradition of positive thinking in Mathematics and Statistics in which functions, equations, and relationships that are easy to work with or have other nice properties are called "normal," "common," "regular," "natural," or similar terms. It's as if by calling them ordinary, we could make them actually occur more often and simplify our lives.

■ NOTATION ALERT

$N(\mu, \sigma)$ always denotes a Normal model. The μ, pronounced "mew," is the Greek letter for "m," and always represents the mean in a model. The σ, sigma, is the lowercase Greek letter for "s" and always represents the standard deviation in a model.

Note that the curve extends indefinitely in both directions, though after a while, the remaining area is close enough to zero to just be ignored. The curve with the bigger σ is not as tall as the other, because the total area has to be maintained at 1.00.

The parameters μ and σ in the Normal curve equation are indeed just the mean and standard deviation of an idealized Normal frequency distribution. We write $N(\mu, \sigma)$ to represent a **Normal model** with a mean of μ and standard deviation of σ. Why the Greek? Well, *this* mean and standard deviation are not numerical summaries of data. They are part of the model—numbers that we choose to help specify the model. Such numbers are called **parameters** of the model.

We don't want to confuse the model parameters with summaries of the data such as $\bar{y}$ and s, so we use special symbols. In Statistics, we almost always use Greek letters for parameters. By contrast, summaries of data are called **statistics** and are usually written with Latin letters. If you're thinking that statistics calculated from data might be used to estimate (unknown) model parameters—you're quite right! But this big and important topic—called *statistical inference*—will be deferred to later chapters in this text.

If we model data with a Normal model and standardize values using some presumed μ and σ, we still call the standardized value a **z-score**, and we write

$$z = \frac{y - \mu}{\sigma}$$

Usually it's easier to standardize data first (using its mean and standard deviation). Then we need only the model $N(0, 1)$. The Normal model with mean 0 and standard deviation 1 is called the **standard Normal model** (or the **standard Normal distribution**).

But be careful. You shouldn't use a Normal model for just any data set. Remember that standardizing won't change the shape of the distribution. If the distribution is not unimodal and symmetric to begin with, standardizing won't make it Normal.

When we use the Normal model, we assume that the distribution of the data is, well, Normal. Practically speaking, there's no way to check whether this **Normality Assumption** is true. In fact, it is almost certainly not perfectly true. Real data don't behave like mathematical models. Models are idealized; real data are real. The good news, however, is that to use a Normal model, it's usually sufficient to check the following condition:

> **Nearly Normal Condition.** The shape of the data's distribution is unimodal and fairly symmetric. Check this by making a histogram (or a Normal probability plot, which we'll explain later).

Don't model data with a Normal model without checking whether this condition is satisfied. All models make *assumptions*. Whenever we model—and we'll do that often—we'll be careful to point out the assumptions that we're making. And, what's even more important, we'll check the associated *conditions* in the data to make sure that those assumptions are reasonable.

IS THE STANDARD NORMAL A STANDARD?

Yes. We call it the "standard Normal" because it models standardized values. It is also a "standard" because this is the particular Normal model that we almost always work with.

> **One in a million.** These magic 68, 95, 99.7 values come from the Normal model. As a model, it can give us corresponding values for any z-score. For example, it tells us that fewer than 1 out of a million values have z-scores smaller than -5.0 or larger than $+5.0$. So if someone tells you you're "one in a million," they must really admire your z-score.

Figure 5.8

Reaching out one, two, and three standard deviations on a Normal model gives the 68–95–99.7 Rule.

The 68–95–99.7 Rule for Normal Models

Normal models give us an idea of how extreme a value is by telling us how likely it is to find one that far from the mean. We'll soon show how to find these numbers precisely—but one simple rule is often all we need.

It turns out that in a Normal model, approximately 68% of the values fall within one standard deviation of the mean, approximately 95% of the values fall within two standard deviations of the mean, and approximately 99.7%—almost all—of the values fall within three standard deviations of the mean. (see Figure 5.8.) These facts are summarized in a rule that we call (let's see . . .) the **68–95–99.7 Rule**.[4]

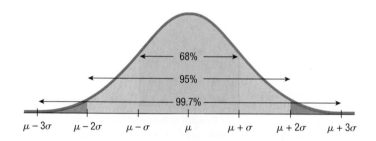

For Example THE 68–95–99.7 RULE AND CLARA HUGHES

QUESTION: In the Vancouver 2010 Olympic Games, Canada's female speed skaters excelled, winning four medals in the six events (only the Germans won as many). Perhaps Canada's most amazing female athlete, Clara Hughes (two cycling Olympic medals in 1996 summer games and four speed skating medals over three winter games) had a time of 6:55.73 minutes in the 5000 metre race, 4.81 seconds behind the gold medalist Martina Sablikova. Her time was slightly faster than one standard deviation below the mean. Where do you think she finished, out of 15 competitors? Though there were only 15 race times, their distribution was in fact nearly Normal.

ANSWER: From the 68–95–99.7 Rule, we expect 68% of the skaters to be within one standard deviation of the mean. Of the remaining 32%, we expect half on the high end and half on the low end. 16% of 15 is 2.4, so we'd expect about two or maybe three racers to have skated faster than Hughes. Did she win the bronze medal? (Yes!)

 # Just Checking

4. As a group, the Dutch are among the tallest people in the world. The average Dutch man is 184 cm tall (and the average Dutch woman is 170.8 cm tall). If a Normal model is appropriate and the standard deviation for men is about 8 cm, what percentage of all Dutch men will be over 2 metres tall?

5. Suppose it takes you 20 minutes, on average, to drive to school, with a standard deviation of 2 minutes. Suppose a Normal model is appropriate for the distributions of driving times.
   a) How often will you arrive at school in less than 22 minutes?
   b) How often will it take you more than 24 minutes?
   c) Do you think the distribution of your driving times is unimodal and symmetric?
   d) What does this say about the accuracy of your predictions? Explain.

---

[4]This rule is just the observation-based "empirical rule" of Chapter 3, made exact for the Normal model. The empirical rule was first published by Abraham De Moivre in 1733, 75 years before the Normal model was discovered.

The First Three Rules for Working with Normal Models

1. Make a picture.
2. Make a picture.
3. Make a picture.

Although we're thinking about models, not histograms of data, the three rules don't change. To help you think clearly, a simple hand-drawn sketch is all you need. Even experienced statisticians sketch pictures to help them think about Normal models. You should too.

Of course, when we have data, we'll also need to make a histogram to check the **Nearly Normal Condition** to be sure we can use the Normal model to model the data's distribution. Other times, we may be told that a Normal model is appropriate based on prior knowledge of the situation or on theoretical considerations.

How to sketch a Normal curve that looks Normal To sketch a good Normal curve, you need to remember only three things:

- The Normal curve is bell-shaped and symmetric around its mean. Start at the middle, and sketch to the right and left from there.

- Even though the Normal model extends forever on either side, you need to draw it only for 3 standard deviations. After that, there's so little left that it isn't worth sketching.

- The place where the bell shape changes from curving downward to curving back up—the *inflection point*—is located exactly one standard deviation away from the mean (and there is another inflection point symmetrically located on the other side).

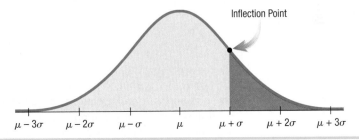

Step-by-Step Example WORKING WITH THE 68–95–99.7 RULE

TOEFL scores had a distribution that was roughly unimodal and symmetric with an overall mean of about 540 and a standard deviation of 60 for all test takers, including graduate students, undergraduates, and professionals.

Question: Suppose you earned a 600 on this TOEFL exam. Where do you stand among all students who took that test?

You could calculate your z-score and find out that it is $z = (600 - 540)/60 = 1.0$, but what does that tell you about your percentile? You'll need the Normal model and the 68–95–99.7 Rule to answer that question.

THINK ➡ **Plan** State what you want to know.	I want to see how my score compares with all other students. To do that, I'll need to model the distribution.
Variables Identify the variable and report the W's.	Let y = my TOEFL score. Scores are quantitative but have no meaningful units other than points.
Be sure to check the appropriate conditions.	✓ **Nearly Normal Condition:** If I had data, I would check the histogram. I have no data, but I am told that the scores are roughly unimodal and symmetric.
Specify the parameters of your model.	I will model TOEFL score with a $N(540,60)$ model.

SHOW ➡ **Mechanics** Make a picture of this Normal model (a simple sketch is all you need).

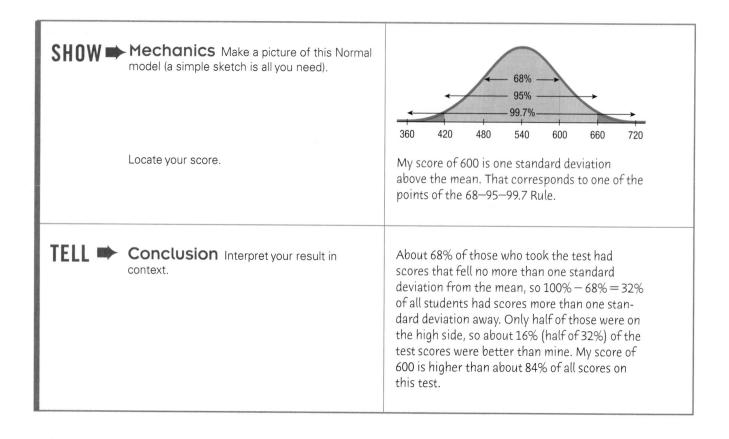

Locate your score.

My score of 600 is one standard deviation above the mean. That corresponds to one of the points of the 68–95–99.7 Rule.

TELL ➡ **Conclusion** Interpret your result in context.

About 68% of those who took the test had scores that fell no more than one standard deviation from the mean, so 100% − 68% = 32% of all students had scores more than one standard deviation away. Only half of those were on the high side, so about 16% (half of 32%) of the test scores were better than mine. My score of 600 is higher than about 84% of all scores on this test.

The upper bound of TOEFL scoring was actually 677, a score that sits between two and three standard deviations above the mean of 540. Hence the upper tail of the distribution is very slightly compressed as compared with a perfect Normal. If we extended the scoring a bit, we'd only get more information about less than (100 − 95)/2 = 2.5% of students—students who clearly have superlative English skills, so it would hardly be worthwhile.

Section 5.3 Exercises

5. Guzzlers? Environmental Protection Agency (EPA) fuel economy estimates for automobile models tested recently predicted a mean of 24.8 mpg and a standard deviation of 6.2 mpg for highway driving. Assume that a Normal model can be applied.
a) Draw the model for auto fuel economy. Clearly label it, showing what the 68–95–99.7 Rule predicts.
b) In what interval would you expect the central 68% of autos to be found?
c) About what percent of autos should get more than 31 mpg?
d) About what percent of cars should get between 31 and 37.2 mpg?
e) Describe the gas mileage of the worst 2.5% of all cars.

6. IQ Some IQ tests are standardized to a Normal model, with a mean of 100 and a standard deviation of 16.
a) Draw the model for these IQ scores. Clearly label it, showing what the 68–95–99.7 Rule predicts.
b) In what interval would you expect the central 95% of IQ scores to be found?
c) About what percent of people should have IQ scores above 116?
d) About what percent of people should have IQ scores between 68 and 84?
e) About what percent of people should have IQ scores above 132?

5.4 Finding Normal Percentiles

A TOEFL score of 600 is easy to assess, because we can think of it as one standard deviation above the mean. If your score was 648, though, where do you stand among the rest of
the people tested? Your *z*-score is 1.80, so you're somewhere between one and two standard deviations above the mean. We figured out that no more than 16% of people scored
better than 600. By the same logic, no more than 2.5% of people scored better than 660.
Can we be more specific than "between 16% and 2.5%"?

When the value doesn't fall exactly one, two, or three standard deviations from the
mean, we can look it up in a table of **Normal percentiles**[5] or use technology. We convert
our data to *z*-scores in order to use the table. Your TOEFL score of 648 has a *z*-score of
$(648 - 540)/60 = 1.80$.

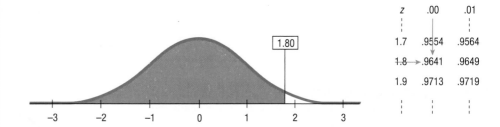

Figure 5.9
A table of Normal percentiles (Table Z in Appendix C) lets us find the percentage of individuals in a
standard Normal distribution falling below any specified *z*-score value.

In the piece of the table shown (Figure 5.9), you can find the *z*-score by looking down
the left column for the first two digits, 1.8, and across the top row for the third digit, 0. The
table gives the percentile as 0.9641. That means that 96.4% of the *z*-scores are less than
1.80. Only 3.6% of people, then, scored better than 648 on this TOEFL.

These days, finding percentiles from a Normal probability table is a "desert island"
method—something we might do if we desperately needed a Normal percentile and were
stranded miles from the mainland with only a Normal probability table. (Of course, you
might feel just that way during a Statistics exam, so it's a good idea to know how to do it.)
Fortunately, most of the time, we can use a calculator, a computer, a smart phone or tablet
app, or the Internet. Most technology tools that find Normal percentiles offer to draw the
picture as well. With modern technology tools, it's easy to answer a question like what
percentage of test takers scored between 550 and 650 on the paper-based TOEFL, once
you specify the Normal model and its parameters.

Other Models

Of course, the Normal is not the only model for data. There are models for skewed data
and we'll see models for variables that can take on only a few values (Binomial and
Poisson, to name two). But the Normal will return later on in an important and surprising way.

---

[5]See Table Z in Appendix C. But most statistics computer packages do this, too—and more easily!

Step-by-Step Example PRACTICE—WORKING WITH NORMAL MODELS PART I

The Normal model is our first model for data. It's the first in a series of modelling situations where we step away from the data at hand to make more general statements about the world. We'll become more practised in thinking about and learning the details of models as we progress through the book. To give you some practice in thinking about the Normal model, here are several problems that ask you to find percentiles in detail.

Question: What proportion of the TOEFL scores fell between 510 and 600?

THINK ➡ Plan State the problem.	I want to know the proportion of TOEFL scores between 510 and 600.
Variables Name the variable.	Let y = TOEFL score.
Check the appropriate conditions and specify which Normal model to use.	✓ **Nearly Normal Condition:** We are told that TOEFL scores are nearly Normal.
	I'll model TOEFL scores with a $N(540,60)$ model, using the mean and standard deviation specified for them.
SHOW ➡ Mechanics Make a picture of this Normal model. Locate the desired values and shade the region of interest.	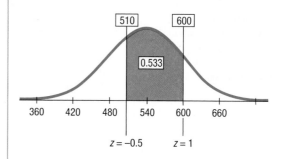
The easiest way is to use software that allows you to input the raw data values directly, after specifying the Normal model and its parameters (μ, σ). Alternatively, you can convert 510 and 600 to z-scores in order to use the table.	Standardizing the two scores, I find $$z = \frac{(y - \mu)}{\sigma} = \frac{(600 - 540)}{60} = 1.00$$ and $$z = \frac{(510 - 540)}{60} = -0.50$$
Using the table, find the two left-side areas. Then subtract to find the area *between* the two values.	So, Area $(510 < y < 600)$ = Area $(-0.5 < z < 1.0)$ = 0.5328. (From Table Z, Area $(z < 1.0)$ = 0.8413 and Area $(z < -0.5)$ = 0.3085, so the proportion of z-scores between them is 0.8413 − 0.3085 = 0.5328, or 53.28%.)
TELL ➡ Conclusion Interpret your result in context.	If the Normal model is accurate, we can estimate that about 53.3% of TOEFL scores fell between 510 and 600.

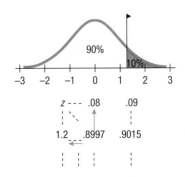

From Percentiles to Scores

Finding areas from *z*-scores is one way that we sometimes work with the Normal model. But at other times we start with areas and are asked to work backward to find the corresponding *z*-score and original data value. For instance, what *z*-score cuts off the top 10% in a Normal model?

Make a picture like the one shown, shading the rightmost 10% of the area. Notice that this is the 90th percentile. Look in Table Z for an area of 0.900. The exact area is not there, but 0.8997 is pretty close. That shows up in the table with 1.2 in the left margin and .08 in the top margin. The *z*-score for the 90th percentile, then, is approximately *z* = 1.28. And once you have the *z* score, you can easily transform it into the corresponding raw score or data value.

Of course, computers and calculators will determine the cut point more precisely (and more easily).

Step-by-Step Example   WORKING WITH NORMAL MODELS PART II

Question: Shuang is hoping to score among the top 10% on the TOEFL. How high a score does she need?

THINK ➡ **Plan** State the problem.	How high a TOEFL score does Shuang need to be in the top 10% of all test takers?
Variable Define the variable.	Let *y* = Shuang's score.
Check to see if a Normal model is appropriate, and specify which Normal model to use.	✓ **Nearly Normal Condition**: I am told that TOEFL scores are nearly Normal. I'll model them with N(540,60).
SHOW ➡ **Mechanics** Make a picture of this Normal model. Locate the desired percentile approximately by shading the rightmost 10% of the area. The cutoff score is the 90th percentile in order for Shuang to make the top 10%. Find the corresponding *z*-score using Table Z as shown earlier.	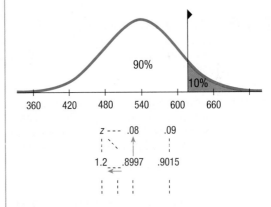 The cut point, when standardized, is *z* = 1.28.
Convert the *z*-score back to the original units. If you want a formula, it is: Original score = μ + *z* × σ	A *z*-score of 1.28 is 1.28 standard deviations above the mean. Since the SD is 60, that's 1.28 × 60 = 76.8 points above the mean of 540, so the cutoff is 540 + 77 = 617.
TELL ➡ **Conclusion** Interpret your results in the proper context.	To place in the top 10%, Shuang needs a score of 617.

Step-by-Step Example MORE WORKING WITH NORMAL MODELS

Working with Normal percentiles can be a little tricky, depending on how the problem is stated. Here are a few more worked examples of the kind you're likely to see.

A cereal manufacturer has a machine that fills the boxes. Boxes are labelled "16 ounces," so the company wants to have that much cereal in each box, but since no packaging process is perfect, there will be minor variations. If the machine is set at exactly 16 ounces and the Normal model applies (or at least the distribution is roughly symmetric), then about half of the boxes will be underweight, making consumers unhappy and exposing the company to bad publicity and possible lawsuits. To prevent underweight boxes, the manufacturer has to set the mean a little higher than 16.0 ounces.

Based on their experience with the packaging machine, the company believes that the amount of cereal in the boxes fits a Normal model with a standard deviation of 0.2 ounces. Let's use that model to answer a series of questions about these cereal boxes.

Question 1: If the manufacturer decides to set the machine to put an average of 16.3 ounces in each box, what fraction of the boxes will be underweight?

THINK ➡ Plan State the problem.	What proportion of boxes weigh less than 16 ounces?
Variable Name the variable.	Let y = weight of cereal in a box.
Check to see if a Normal model is appropriate.	✓ **Nearly Normal Condition**: I have no data, so I cannot make a histogram, but I am told that the company believes the distribution of weights from the machine is Normal.
Specify which Normal model to use.	I'll use a $N(16.3, 0.2)$ model.
SHOW ➡ Mechanics Make a picture of this Normal model. Locate the value you're interested in on the picture, label it, and shade the appropriate region.	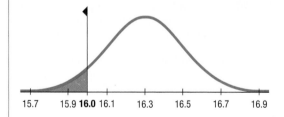
REALITY CHECK Estimate from the picture the percentage of boxes that are underweight. (This will be useful later to check that your answer makes sense.) It looks like a low percentage. Likely less than 20%.	I want to know what fraction of the boxes will weigh less than 16 ounces.
Convert your cutoff value into a z-score if using the table of standard Normal areas.	$z = \dfrac{y - \mu}{\sigma} = \dfrac{16 - 16.3}{0.2} = -1.50$
Find the area with your software or use the Normal table.	Area $(y < 16)$ = Area $(z < -1.50)$ = 0.0668
TELL ➡ Conclusion State your conclusion, and check that it's consistent with your earlier guess. It's below 20%—seems okay.	I estimate that approximately 6.7% of the boxes will contain less than 16 ounces of cereal.

Question 2: The company's lawyers say that 6.7% is too high. They insist that no more than 4% of the boxes can be underweight. So the company needs to set the machine to put a little more cereal in each box. What mean setting do they need?

THINK ➡ **Plan** State the problem.	What mean weight will reduce the proportion of underweight boxes to 4%?
Variable Name the variable.	Let y = weight of cereal in a box.
Check to see if a Normal model is appropriate.	✓ **Nearly Normal Condition**: I am told that a Normal model applies.
Specify which Normal model to use. This time you are not given a value for the mean!	I don't know μ, the mean amount of cereal. The standard deviation for this machine is 0.2 ounces. The model is $N(\mu, 0.2)$.
REALITY CHECK We found out earlier that setting the machine to $\mu = 16.3$ ounces made 6.7% of the boxes too light. We'll need to raise the mean a bit to reduce this fraction.	No more than 4% of the boxes can be below 16 ounces.

SHOW ➡ **Mechanics** Make a picture of this Normal model. Centre is at μ (since you don't know the mean). Shade the region below 16 ounces.	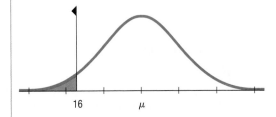
Using software or the Normal table, find the z-score that cuts off the lowest 4%.	The z-score that has a 0.04 area to the left of it is $z = -1.75$.
Use this information to find μ. It's located 1.75 standard deviations to the right of 16. Since σ is 0.2, that's 1.75×0.2, or 0.35 ounces more than 16. (Or set $(16 - \mu)/0.2 = -1.75$ and solve)	For 16 to be 1.75 standard deviations below the mean, the mean must be $$16 + 1.75(0.2) = 16.35 \text{ ounces.}$$

TELL ➡ **Conclusion** Interpret your result in context. This makes sense; we knew it would have to be just a bit higher than 16.3.	The company must set the machine to average 16.35 ounces of cereal per box.

Question 3: The company president vetoes that plan, saying the company should give away less free cereal, not more. Her goal is to set the machine no higher than 16.2 ounces and still have only 4% underweight boxes. The only way to accomplish this is to reduce the standard deviation. What standard deviation must the company achieve, and what does that mean about the machine?

THINK ➡ **Plan** State the problem.	What standard deviation will allow the mean to be 16.2 ounces and still have only 4% of boxes underweight?
Variable Name the variable.	Let y = weight of cereal in a box.
Check conditions to be sure that a Normal model is appropriate. Specify which Normal model to use. This time you don't know σ.	✓ **Nearly Normal Condition**: The company believes that the weights are described by a Normal model.
REALITY CHECK We know the new standard deviation must be less than 0.2 ounces.	I know the mean, but not the standard deviation, so my model is $N(16.2, \sigma)$.

SHOW ➡ **Mechanics** Make a picture of this Normal model. Centre it at 16.2, and shade the area you're interested in. We want 4% of the area to the left of 16 ounces.

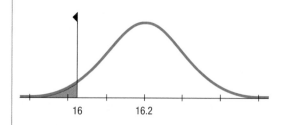

Find the z-score that cuts off the lowest 4%.

I know that the z-score with 4% below it is $z = -1.75$.

We need 16 standardized to equal −1.75. Write this down and solve for σ. (Or: We need 16 to be 1.75 σ's below 16.2, so 1.75 σ must be 0.2 ounce. You could start with that equation.)

$$z = \frac{y - \mu}{\sigma}$$

$$-1.75 = \frac{16 - 16.2}{\sigma}$$

$$1.75\sigma = 0.2$$

$$\sigma = 0.114$$

TELL ➡ **Conclusion** Interpret your result in context.

As we expected, the standard deviation is lower than before—actually, quite a bit lower.

The company must get the machine to box cereal with a standard deviation of only 0.114 ounces. This means the machine must be more consistent (by nearly a factor of two) in filling the boxes.

Section 5.4 Exercises

7. Normal cattle The Virginia Cooperative Extension reports that the mean weight of yearling Angus steers is 1152 pounds. Suppose that weights of all such animals can be described by a Normal model with a standard deviation of 84 pounds. What percent of steers weigh
a) over 1250 pounds?
b) under 1200 pounds?
c) between 1000 and 1100 pounds?

8. IQs revisited Based on the Normal model $N(100, 16)$ describing IQ scores, what percent of people's IQs would you expect to be
a) over 80?
b) under 90?
c) between 112 and 132?

5.5 Normal Probability Plots

In the examples we've worked through, we've assumed that the underlying data distribution was roughly unimodal and symmetric, so using a Normal model makes sense. When you have data, you must *check* to see whether a Normal model is reasonable. How? Make a picture, of course! Drawing a histogram of the data and looking at the shape is one reasonable way to see if a Normal model might work.

There's a more specialized graphical display that can help you to decide whether the Normal model is appropriate: the **Normal probability plot (NPP)**.[6] If the distribution of

---

[6]Also called a *Normal quantile plot*.

the data is roughly Normal, the plot is roughly a diagonal straight line. Systematic deviations from a straight line indicate that the distribution is not Normal. This plot is usually able to show deviations from Normality more clearly than the corresponding histogram, but it's usually easier to understand *how* a distribution fails to be Normal by looking at its histogram.

The NPP plots the data values on one axis versus corresponding *Normal scores* on the other axis. More about Normal scores later, but suffice to say for now, that a histogram of the Normal scores will look perfectly Normal. The NPP matches up the pattern in the data with the pattern of Normal data by plotting one set of values versus the other set (each ordered low to high).

Some data on a car's fuel efficiency provide an example of data that are nearly Normal. The overall pattern of the Normal probability plot (Figure 5.10) is straight. The two trailing low values correspond to the values in the histogram that trail off the low end. They're not quite in line with the rest of the data set. The Normal probability plot shows us that they're a bit lower than we'd expect of the lowest two values in a Normal model.

Figure 5.10

Histogram and Normal probability plot for gas mileage (mpg or miles per gallon) recorded by one of the authors over the eight years he owned a Nissan Maxima. The vertical axes are the same, so each dot on the probability plot would fall into the bar on the histogram immediately to its left. On the other hand, if you project the dots straight down onto the horizontal axis, and make a histogram, it will look perfectly Normal.

By contrast, the Normal probability plot of the men's *weights* from the NHANES Study (Figure 5.11) is far from straight. The weights are skewed to the high end, and the plot is curved. We'd conclude from these pictures that approximations using the 68–95–99.7 Rule for these data would not be very accurate.

Figure 5.11

Histogram and Normal probability plot for men's weights. Note how a skewed distribution corresponds to a bent probability plot.

How Does a Normal Probability Plot Work?

Why does the Normal probability plot work like that? We looked at 100 fuel efficiency measures for the author's Nissan car. The smallest of these was 14.7 mpg which has a z-score of −3.16. The Normal model can tell us what value to expect for the smallest z-score in a batch of 100 if a Normal model were appropriate. That turns out to be −2.58. So our first data value is smaller than we would expect from the Normal. In the margin NPP, you can see that we plotted the data value, 14.7 mpg, versus its corresponding Normal score, −2.58.

We can continue this and ask a similar question for each value. For example, the 14th smallest fuel efficiency has a z-score of almost exactly -1, and that's just what we should expect (well, -1.1 to be exact). A Normal probability plot takes each data value and plots it against the z-score you'd expect that point to have if the distribution were perfectly Normal.[7]

When the values match up well, the line is straight. If one or two points are surprising from the Normal's point of view, they don't line up. When the entire distribution is skewed or different from the Normal in some other way, the values don't match up very well and the plot will bend or show some type of curvature.

It turns out to be tricky to find these average Normal z-scores and you can't easily look them up in some tables. That's why probability plots are best made with technology and not by hand. Don't confuse these Normal z-scores with the z-scores of your data. If you plot your data versus their z-scores, guess what, the patterns match up perfectly and you just get a straight line! Better to call them *Normal scores* rather than z-scores to emphasize what matters about them—their Normality, not the fact that they also happen to be standardized or $N(0,1)$.

The best advice on using Normal probability plots is to see whether they are essentially straight, aside from very random-looking squiggles. If so, then your data look like data from a Normal model. If not—and particularly if you see clear curving patterns or points way out of line with the rest—make a histogram to understand how they differ from the model.

Section 5.5 Exercises

9. Music library Corey has 4929 songs in his computer's music library. The lengths of the songs have a mean of 242.4 seconds and standard deviation of 114.51 seconds. A Normal probability plot of the song lengths looks like this:

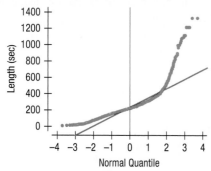

a) Do you think the distribution is Normal? Explain.
b) If it isn't Normal, how does it differ from a Normal model?

10. Wisconsin ACT math The histogram shows the distribution of mean ACT mathematics scores for all Wisconsin public schools in 2011. The vertical lines show the mean and one standard deviation above and below the mean. 78.8% of the data points are between the two outer lines.

a) Give two reasons why a Normal model is not appropriate for these data.

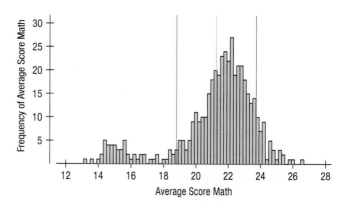

b) The Normal probability plot on the left shows the distribution of these scores. The plot on the right shows the same data with the Milwaukee area schools (mostly in the low mode) removed. What do these plots tell you about the shape of the distributions?

---

[7]A Normal probability plot is also often drawn with the two axes switched, putting the data on the x-axis and the standard Normal scores on the y-axis.

WHAT CAN GO WRONG?

■ **Don't use a Normal model when the distribution is not unimodal and fairly symmetric.**
Normal models are so easy and useful that it is tempting to use them even when they don't
describe the data very well. That can lead to wrong conclusions. Don't use a Normal model
without first checking the **Nearly Normal Condition**. Look at a picture of the data to
check that it is unimodal and symmetric. A histogram, or a Normal probability plot, can
help you tell whether a Normal model is appropriate.

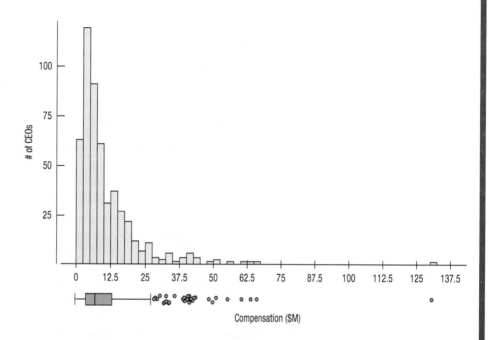

Annual compensations for Fortune 500 CEOs (first discussed in Chapter 4) are displayed
in the histogram above. In that year, they had a mean total compensation of $10.48M with
a standard deviation of $11.46M. Using the Normal model rule, we should expect about
68% of the CEOs to have had compensations between −$0.98M and $21.94M. In fact,
more than 90% of the CEOs had annual compensations in this range. What went wrong?
The distribution is extremely skewed, not symmetric. Using the 68–95–99.7 Rule for data
like these will lead to silly results.

■ **Don't use the mean and standard deviation when outliers are present.** Both means and
standard deviations can be distorted by outliers, and no model based on distorted values
will do a good job. A z-score calculated from a distribution with outliers may be misleading. It's always a good idea to check for outliers. How? Make a picture.

■ **Don't round your results in the middle of a calculation.** We *reported* the mean of the
heptathletes' 200 m run as 24.48 seconds. More precisely, it was 24.4844444444444456
seconds.

You should use all the precision available in the data for all the intermediate steps of a
calculation. Using the more precise value for the mean (and also carrying 15 digits for the
SD), the z-score calculation for Ennis's run comes out to

$$z = \frac{22.83 - 24.4844444444444456}{0.80034437032511907} = -2.06716571739284216$$

We'd likely report that as −2.067, as opposed to the rounded-off value of 2.1 we got earlier
from the table. So, . . .

■ **Do what we say, not what we do.** When we showed the z-score calculations for Ennis, we
rounded the mean to 24.48 seconds and the SD to 0.80 seconds. Then to make the story
clearer we used *those values* in the displayed calculation.

(continued)

We'll continue to show simplified calculations in the book to make the story simpler. When you calculate with full precision, your results may differ slightly from ours. So, we also advise . . .

■ **Don't worry about minor differences in results.** Because various calculators and programs may carry different precision in calculations, your answers may differ slightly from those we show in the text and Step-By-Steps, or even from the values given in the answers in the back of the book. Those differences aren't anything to worry about. They're not the main story Statistics tries to tell.

What Have We Learned?

Understand how z-scores facilitate comparisons by standardizing variables to have zero mean and unit standard deviation. But recognize that standardizing data values does not change the shape of the histogram.

Shifting and scaling of data, like standardizing, are linear transformations of data, hence only affect centre and spread, in a simple predictable manner, but do not change the shape of the distribution (unlike non-linear transformations of data, such as logs).

Central to Statistics is the use of models such as density curve models, including the Normal model, whose importance will only later be fully appreciated:

■ The Normal model uses areas under a density curve to model proportions of values falling into various intervals.

■ Before using a Normal model, always check that the data are plausibly from a normally distributed population. Make a histogram and check whether it is unimodal, roughly symmetric and bell-shaped.

■ A Normal probability plot provides evidence that the data are Normally distributed if it is roughly linear. It is usually a better tool than a histogram for such judgments.

Understand how to use the Normal model.

■ Standardize values to make z-scores. Then refer to a standard Normal distribution (using software or tables) to determine the proportion of values in some range, or to find percentile points.

■ Use the 68–95–99.7 Rule as a rule-of-thumb to judge how unusual a value is.

Review of Terms

Standardize   We standardize to eliminate units. Standardized values can be compared and combined even if the original variables had different units and magnitudes (p. 132).

Standardized value   A value found by subtracting the mean and dividing by the standard deviation (p. 132).

z-score   A z-score tells how many standard deviations a value is from the mean; z-scores have a mean of zero and a standard deviation of one. When working with data, use the statistics $\bar{y}$ and s:

$$z = \frac{y - \bar{y}}{s}$$ (p. 132).

When working with models, use the parameters μ and σ:

$$z = \frac{y - \mu}{\sigma}$$ (p. 141).

Shifting   Adding a constant to each data value adds the same constant to the mean, the median, and the quartiles, but does not change the standard deviation or IQR (p. 134).

Rescaling   Multiplying each data value by a constant multiplies both the measures of position (mean, median, and quartiles) and the measures of spread (standard deviation and IQR) by that constant (p. 135).

Density curve   A model for the frequency distribution of data using areas under the curve to represent relative frequencies (p. 139).

Parameter	A numerically valued attribute of a model. For example, the values of μ and σ in a $N(\mu, \sigma)$ model are parameters (p. 141).
Statistic	A value calculated from data to summarize aspects of the data. For example, the mean, $\bar{y}$, and standard deviation, s, are statistics (p. 141).
Normal model	A useful family of density curve models for unimodal, symmetric distributions, indexed by two parameters, one to centre the curve and one controlling the spread. (p. 141).
Standard Normal model	A Normal model, $N(\mu, \sigma)$, with mean $\mu = 0$ and standard deviation $\sigma = 1$. Also called the **standard Normal distribution** (p. 141).
Nearly Normal Condition	A distribution is nearly Normal if it is unimodal and fairly symmetric. Check by looking at a histogram or a Normal probability plot (p. 141).
68–95–99.7 Rule	In a Normal model, approximately 68% of values fall within one standard deviation of the mean, approximately 95% fall within two standard deviations of the mean, and approximately 99.7% fall within three standard deviations of the mean (p. 142).
Normal percentile	The Normal percentile corresponding to a z-score gives the percentage of values in a standard Normal distribution found at that z-score or below (p. 145).
Normal probability plot (NPP)	A display to help assess whether a distribution of data is approximately Normal. If the plot is nearly straight, the data satisfy the **Nearly Normal Condition** (p. 150).

On the Computer NORMAL PROBABILITY PLOTS

The best way to tell whether your data can be modeled well by a Normal model is to make a picture or two. We've already talked about making histograms. Normal probability plots are almost never made by hand because the values of the Normal scores are tricky to find. But most statistics software make Normal plots, though various packages call the same plot by different names (such as 'quantile plot') and array the information differently.

EXCEL

Excel offers a "Normal probability plot" as part of the Regression command in the Data Analysis extension, but (as of this writing) it is not a correct Normal probability plot and should not be used.

JMP

To make a "Normal Quantile Plot" in JMP:
- Make a histogram using **Distributions** from the **Analyze** menu.
- Click on the drop-down menu next to the variable name.
- Choose **Normal Quantile Plot** from the drop-down menu.
- JMP opens the plot next to the histogram.

COMMENTS

JMP places the ordered data on the vertical axis and the Normal scores on the horizontal axis. The vertical axis aligns with the histogram's axis, a useful feature.

MINITAB

To make a "Normal Probability Plot" in MINITAB:
- Choose **Probability Plot** from the **Graph** menu.
- Select "Single" for the type of plot. Click **OK**.
- Enter the name of the variable in the "Graph variables" box. Click **OK**.

COMMENTS

MINITAB places the ordered data on the horizontal axis and in place of the Normal scores, shows 'Percent' on the vertical axis. The interpretation is exactly the same, since each Normal score is just replaced by its percentile standing in the Normal distribution. You can edit the y-axis to show Normal scores instead.

R

To make a Normal probability (Q-Q) plot for x:
- **qqnorm(X)** will produce the plot.

To standardize a variable *x*:
- **z = (X−mean (X))/sd(X)** will create a standardized variable **Z**.

COMMENTS

By default, R places the ordered data on the vertical axis and the Normal scores on the horizontal axis, but that can be reversed by setting **datax =TRUE** inside qqnorm.

SPSS

To make a Normal "P-P plot" in SPSS:
- Choose **P-P** from the **Graphs** menu.
- Select the variable to be displayed in the source list.
- Click the arrow button to move the variable into the target list.
- Click the **OK** button.

COMMENTS

SPSS places the ordered data on the horizontal axis and the Normal scores on the vertical axis. You may safely ignore the options in the P-P dialog.

STATCRUNCH

To make a Normal probability plot:
- Click on **Graphics**.
- Choose **QQ Plot**.
- Choose the variable name from the list of **Columns**.
- Click on **Create Graph**.

To work with Normal percentiles:
- Click on **Stat**.
- Choose **Calculators » Normal**.
- Choose a lower tail ($\leq$) or upper tail ($\geq$) region.
- Enter the *z*-score cutoff, and then click on **Compute** to find the probability.

OR

Enter the desired probability, and then click on **Compute** to find the *z*-score cutoff.

TI-83/84 PLUS

To make a Normal Probability plot:
- Turn a STATPLOT On.
- Tell it to make a Normal probability plot by choosing the last of the icons.
- Specify your datalist and which axis you want the data on. (Use Y to make the plot look like those here.)

- Specify the Mark you want the plot to use.
- Now ZoomStat does the rest.

Although most people wouldn't open a statistics package just to find a Normal model value they could find in a table, you *would* use a calculator for that function.

- So, to find what percent of a Normal model lies between two z-scores, choose **normalcdf** from the **DISTRibutions** menu and enter the command **normalcdf(zLeft, zRight)**.
- To find the z-score that corresponds to a given percentile in a Normal model, choose **invNorm** from the **DISTRibutions** menu and enter the command **invNorm(percentile)**.

COMMENTS

We often want to find Normal percentages from a certain z-score to infinity. On the calculator, indicate "infinity" as a very large z-score, say, 99. For example, the percentage of a Normal model over 2 standard deviations above the mean can be evaluated with **normalcdf(2, 99)**.

Chapter Exercises

T indicates data is available in a saved file at the Companion Website at www.pearsoncanada.ca/deveaux

11. Payroll Here are the summary statistics for the weekly payroll of a small company: lowest salary = $300, mean salary = $700, median = $500, range = $1200, IQR = $600, first quartile = $350, standard deviation = $400.
   a) Do you think the distribution of salaries is symmetric, skewed to the left, or skewed to the right? Explain why.
   b) Between what two values are the middle 50% of the salaries found?
   c) Suppose business has been good and the company gives every employee a $50 raise. Determine the new value of each of the summary statistics.
   d) Instead, suppose the company gives each employee a 10% raise. Determine the new value of each of the summary statistics.

12. Hams A specialty foods company sells gourmet hams by mail order. The hams vary in size from 4.15 to 7.45 pounds, with a mean weight of 6 pounds and standard deviation of 0.65 pounds. The quartiles and median weights are 5.6, 6.2, and 6.55 pounds.
   a) Find the range and the IQR of the weights.
   b) Do you think the distribution of the weights is symmetric or skewed? If skewed, which way? Why?
   c) If these weights were expressed in kilograms (1 kg = 2.2 pounds) what would the mean, standard deviation, quartiles, median, IQR, and range be?
   d) When the company ships these hams, the box and packing materials add 900 grams. What are the mean, standard deviation, quartiles, median, IQR, and range of weights of boxes shipped (in kg)?
   e) One customer made a special order of a 5 kg ham. Which of the summary statistics of part d) might *not* change if that data value were added to the distribution?

13. SAT or ACT? Each year thousands of high-school students in the U.S. take either the SAT or the ACT, standardized tests used in the college admissions process.

Combined SAT Math and Verbal scores go as high as 1600, while the maximum ACT composite score is 36. Since the two exams use very different scales, comparisons of performance are difficult. A convenient rule of thumb is $SAT = 40 \times ACT + 150$; that is, multiply an ACT score by 40 and add 150 points to estimate the equivalent SAT score. An admissions officer reported the following statistics about the ACT scores of 2355 students who applied to her college one year. Find the summaries of equivalent SAT scores.

Lowest score = 19 Mean = 27 Standard deviation = 3
Q3 = 30   Median = 28 IQR = 6

14. Cold U? A high school senior in the U.S. uses the Internet to get information on December temperatures in the Canadian town where he'll be going to college. He finds a website with some statistics, but they are given in degrees Celsius. The conversion formula is $°F = \frac{9}{5}°C + 32$. Determine the Fahrenheit equivalents for the summary information below.

Maximum temperature = 11°C Range = 33° Mean = 1°
Standard deviation = 7°   Median = 2° IQR = 16°

15. Temperatures A town's January high temperatures average 2°C with a standard deviation of 6°, while in July the mean high temperature is 24° and the standard deviation is 5°. In which month is it more unusual to have a day with a high temperature of 13°? Explain.

16. Placement exams An incoming freshman took her college's placement exams in French and Mathematics. In French, she scored 82, and in Math, 86. The overall results on the French exam had a mean of 72 and a standard deviation of 8, while the mean Math score was 68, with a standard deviation of 12. On which exam did she do better compared with the other freshmen?

17. Final exams Anna, a language major, took final exams in both French and Spanish and scored 83 on both. Her roommate Megan, also taking both courses, scored 77 on the French exam and 95 on the Spanish exam. Overall, student scores on the French exam had a mean of 81 and a standard deviation of 5, and the Spanish scores had a mean of 74 and a standard deviation of 15.
   a) To qualify for language honours, a major must maintain at least an 85 average for all language courses taken. So far, which student qualifies?
   b) Which student's overall performance was better?

18. MP3s Two companies market new batteries targeted at owners of personal music players. DuraTunes claims a mean battery life of 11 hours, while RockReady advertises 12 hours.
   a) Explain why you would also like to know the standard deviations of the battery lifespans before deciding which brand to buy.
   b) Suppose those standard deviations are 2 hours for DuraTunes and 1.5 hours for RockReady. You are headed for 8 hours at the beach. Which battery is most likely to last all day? Explain.
   c) If your beach trip is all weekend, and you probably will have the music on for 16 hours, which battery is most likely to last? Explain.

19. Cattle The Virginia Cooperative Extension reports that the mean weight of yearling Angus steers is 1152 pounds. Suppose that weights of all such animals can be described by a Normal model with a standard deviation of 84 pounds.
   a) How many standard deviations from the mean would a steer weighing 1000 pounds be?
   b) Which would be more unusual, a steer weighing 1000 pounds, or one weighing 1250 pounds?

T 20. Car speeds John Beale of Stanford, CA, recorded the speeds of cars driving past his house, where the speed limit is 20 mph (miles per hour). The mean of 100 readings was 23.84 mph, with a standard deviation of 3.56 mph. (He actually recorded every car for a two-month period. These are 100 representative readings.)
   a) How many standard deviations from the mean would a car going under the speed limit be?
   b) Which would be more unusual, a car travelling 34 mph or one going 10 mph?

21. More cattle Recall that the beef cattle described in Exercise 19 had a mean weight of 1152 pounds, with a standard deviation of 84 pounds.
   a) Cattle buyers hope that yearling Angus steers will weigh at least 1000 pounds. To see how much over (or under) that goal the cattle are, we could subtract 1000 pounds from all the weights. What would the new mean and standard deviation be?

   b) Suppose such cattle sell at auction for 40 cents a pound. Find the mean and standard deviation of the sale prices (in dollars) for all the steers.

T 22. Car speeds again For the car speed data in Exercise 20, recall that the mean speed recorded was 23.84 mph, with a standard deviation of 3.56 mph. To see how many cars are speeding, John subtracts 20 mph from all speeds.
   a) What is the mean speed now? What is the new standard deviation?
   b) His friend in Inuvik, NWT, wants to study the speeds, so John converts all the original miles per hour readings to kilometres per hour by multiplying all speeds by 1.609 (km per mile). What is the mean now? What is the new standard deviation?

23. Cattle, part III Suppose the auctioneer in Exercise 21 sold a herd of cattle whose minimum weight was 980 pounds, median was 1140 pounds, standard deviation was 84 pounds, and IQR was 102 pounds. They sold for 40 cents a pound, and the auctioneer took a $20 commission on each animal. Then, for example, a steer weighing 1100 pounds would net the owner $0.40(1100) - 20 = \$420$. Find the minimum, median, standard deviation, and IQR of the net sale prices.

24. Caught speeding Suppose police set up radar surveillance on the Stanford street described in Exercise 20. They handed out a large number of tickets to drivers going a mean of 28 mph, with a standard deviation of 2.4 mph, a maximum of 33 mph, and an IQR of 3.2 mph. Local law prescribes fines of $100 plus $10 per mile per hour over the 20 mph speed limit. For example, a driver convicted of going 25 mph would be fined $100 + 10(5) = \$150$. Find the mean, maximum, standard deviation, and IQR of all the potential fines.

25. Professors A friend tells you about a recent study dealing with the number of years of teaching experience among current college professors. He remembers the mean but can't recall whether the standard deviation was 6 months, 6 years, or 16 years. Tell him which one it must have been, and why.

26. Rock concerts A popular band on tour played a series of concerts in large venues. They always drew a large crowd, averaging 21 359 fans. While the band did not announce (and probably never calculated) the standard deviation, which of these values do you think is most likely to be correct: 20, 200, 2000, or 20 000 fans? Explain your choice.

27. Trees A forester measured 27 of the trees in a large wooded area that is up for sale. He found a mean diameter of 26 cm and a standard deviation of 12 cm. Suppose these trees provide an accurate description of the whole forest and that a Normal model applies.
   a) Draw the Normal model for tree diameters.

b) In what size range would you expect the central 95% of all trees to be?

c) About what percent of the trees should be less than 2 cm in diameter?

d) About what percent of the trees should be between 14 and 26 cm in diameter?

e) About what percent of the trees should be over 38 cm in diameter?

28. Rivets A company that manufactures rivets believes the shear strength (in kg) is modelled by $N(400, 25)$.

a) Draw and label the Normal model.

b) Would it be safe to use these rivets in a situation requiring a shear strength of 375 kg? Explain.

c) About what percent of these rivets would you expect to fall below 450 kg?

d) Rivets are used in a variety of applications with varying shear strength requirements. What is the maximum shear strength for which you would feel comfortable approving this company's rivets? Explain your reasoning.

29. Trees, part II Later on, the forester in Exercise 27 shows you a histogram of the tree diameters he used in analyzing the wooded area that was for sale. Do you think he was justified in using a Normal model? Explain, citing some specific concerns.

T 30. Car speeds, the picture Here is the histogram, box-plot, and Normal probability plot of the 100 readings for the car speed data from Exercise 20. Do you think it is appropriate to apply a Normal model here? Explain.

T 31. Winter Olympics 2010 downhill Fifty-nine men completed the men's alpine downhill race in Vancouver. The gold medal winner finished in 114.3 seconds. Here are the times (in seconds) for all competitors (*Source:* Data from 2010 Winter Olympics Results - Alpine Skiing from http://www.espn.com/olympics/winter/2010/results/_/sport/1/event/2:

114.3	115.0	115.7	116.7	118.6	119.8
114.4	115.2	115.8	116.7	118.7	120.0
114.4	115.2	116.0	117.0	118.8	120.1
114.5	115.3	116.1	117.2	118.9	120.6
114.6	115.3	116.2	117.2	119.2	121.7
114.7	115.4	116.2	117.4	119.5	121.7
114.8	115.4	116.3	117.7	119.6	122.6
114.8	115.5	116.4	117.9	119.7	123.4
114.9	115.6	116.6	118.1	119.8	124.4
114.9	115.7	116.7	118.4	119.8	

a) The mean time was 117.34 seconds, with a standard deviation of 2.465 seconds. If the Normal model is appropriate, what percent of times will be less than 114.875 seconds?

b) What is the actual percent of times less than 114.875 seconds?

c) Why do you think the two percentages don't agree?

d) Make a histogram of these times. What do you see?

32. Check the model Recall that the mean of the 100 car speeds in Exercises 20 and 30 was 23.84 mph, with a standard deviation of 3.56 mph.

a) Using a Normal model, what values should border the middle 95% of all car speeds?

b) Here are some summary statistics.

Percentile		Speed
100%	**Max**	34.060
97.5%		30.976
90.0%		28.978
75.0%	**Q3**	25.785
50.0%	**Median**	23.525
25.0%	**Q1**	21.547
10.0%		19.163
2.5%		16.638
0.0%	**Min**	16.270

From your answer in part a), how well does the model do in predicting those percentiles? Are you surprised? Explain.

T 33. TV watching A survey of 200 university students showed the following distribution of the number of hours of television watched per week.

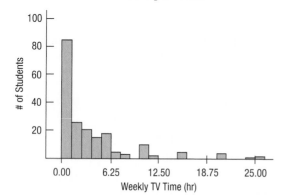

The mean is 3.66 hours, with a standard deviation of 4.93 hours.

a) According to the Normal model, what percent of students will watch fewer than one standard deviation below the mean number of hours?
b) For these data, what does that mean? Explain.
c) Explain the problem in using the Normal model for these data.

T 34. Receivers 2013 NFL data from the 2013 football season reported the number of yards gained by each of the league's 181 wide receivers:

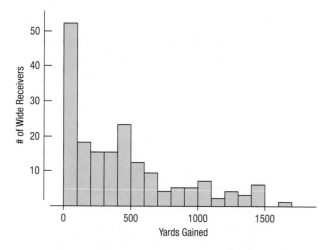

The mean is 426.98 yards, with a standard deviation of 408.34 yards.

a) According to the Normal model, what percent of receivers would you expect to gain more yards than 2 standard deviations above the mean number of yards?
b) For these data, what does that mean?
c) Explain the problem in using a Normal model here.

35. Normal models What percent of a standard Normal model is found in each region? Be sure to draw a picture first.
a) $z > 1.5$
b) $z < 2.25$

c) $-1 < z < 1.15$
d) $|z| < 0.5$

36. Normal models, again What percent of a standard Normal model is found in each region? Draw a picture first.
a) $z > -2.05$
b) $z < -0.33$
c) $1.2 < z < 1.8$
d) $|z| < 1.28$

37. More Normal models In a standard Normal model, what value(s) of z cut(s) off the region described? Don't forget to draw a picture.
a) the highest 20%
b) the highest 75%
c) the lowest 3%
d) the middle 90%

38. Yet another Normal model In a standard Normal model, what value(s) of z cut(s) off the region described? Remember to draw a picture first.
a) the lowest 12%
b) the highest 30%
c) the highest 7%
d) the middle 50%

39. Human Gestation The length of pregnancy is assumed to be normally distributed with a mean of 266 days and a standard deviation of 16 days. What percentage of pregnancies last
a) over 285 days?
b) under 275 days?
c) between 237 and 256 days?

40. More Human Gestation The length of pregnancy is assumed to be normally distributed with a mean of 266 days and a standard deviation of 16 days. What are the cutoff values for
a) the longest 10% of the length of pregnancies?
b) the shortest 20% of the length of pregnancies?
c) the middle 40% of the length of pregnancies?

41. Human Gestation, finis The length of pregnancy is assumed to be normally distributed with a mean of 266 days and a standard deviation of 16 days.
a) What length of pregnancies represent the 40th percentile?
b) What length of pregnancies represent the 99th percentile?
c) What's the IQR of the length of pregnancies?

42. TOEFL scores, Internet test The mean and standard deviation for all 2007 writers of the Internet-based TOEFL were 78 and 24, respectively.[8] The test results are a bit left or negatively skewed.
a) The maximum score possible is 120. How does this suggest that a Normal model might not be a very accurate approximation?

8www.ets.org/Media/Research/pdf/71943_web.pdf

b) The middle 68% of scores were between 52 and 104. If you used a Normal model, what proportion would you expect to find between these two scores?

c) The middle 95% of scores were between 29 and 115. If you used a Normal model, what proportion would you expect to find between these two scores? What scores does the Normal model predict for an interval to capture 95% of the data?

43. Cholesterol Assume the cholesterol levels of adult women can be described by a Normal model with a mean of 188 mg/dL and a standard deviation of 24.

a) Draw and label the Normal model.

b) What percent of adult women do you expect to have cholesterol levels over 200 mg/dL?

c) What percent of adult women do you expect to have cholesterol levels between 150 and 170 mg/dL?

d) Estimate the interquartile range of the cholesterol levels.

e) Above what value are the highest 15% of women's cholesterol levels?

44. Tires A tire manufacturer believes that the treadlife of its snow tires can be described by a Normal model with a mean of 52 000 km and standard deviation of 2500 km.

a) If you buy a set of these tires, would it be reasonable for you to hope they'll last 60 000 km? Explain.

b) Approximately what fraction of these tires can be expected to last less than 50 000 km?

c) Approximately what fraction of these tires can be expected to last between 50 000 and 55 000 km?

d) Estimate the IQR of the treadlifes.

e) In planning a marketing strategy, a local tire dealer wants to offer a refund to any customer whose tires fail to last a certain number of kilometres. However, the dealer does not want to take too big a risk. If the dealer is willing to give refunds to no more than 1 of every 25 customers, for how many kilometres can he guarantee these tires to last?

45. Kindergarten Companies who design furniture for elementary school classrooms produce a variety of sizes for kids of different ages. Suppose the heights of kindergarten children can be described by a Normal model with a mean of 97.0 cm and standard deviation of 4.9 cm.

a) What fraction of kindergarten kids should the company expect to be less than 91 cm tall?

b) In what height interval should the company expect to find the middle 80% of kindergarteners?

c) At least how tall are the biggest 10% of kindergarteners?

46. Body temperatures Most people think that the "Normal" adult body temperature is 37.0°C. That figure, based on a nineteenth-century study, has recently been challenged. In a 1992 article in the *Journal of the American Medical Association,* researchers reported that a more accurate figure may be 36.8°C. Furthermore, the standard deviation appeared to be around 0.4°C. Assume that a Normal model is appropriate.

a) In what interval would you expect most people's body temperatures to be? Explain.

b) What fraction of people would be expected to have body temperatures above 37.0°C?

c) Below what body temperature are the coolest 20% of all people?

47. Undercover? We learned in this chapter that the average Dutch man is 184 cm tall. The standard deviation of Caucasian adult male heights is about 7 cm. The average Greek 18-year-old in Athens is 167.8 cm tall. How easily could the average Dutch man hide in Athens? (Let's assume he dyes his hair, if necessary.) That is, would his height make him sufficiently extraordinary that he'd stand out easily? Assume heights are nearly Normal.

48. Big mouth! A Cornell University researcher measured the mouth volumes of 31 men and 30 women. She found a mean of 66 cc for men (SD = 17 cc) and a mean of 54 cc for women (SD = 14.5 cc). The man with the largest mouth had a mouth volume of 111.2 cc. The woman with the largest mouth had a mouth volume of 95.8 cc.

a) Which had the more extraordinarily large mouth?

b) If the distribution of mouth volumes is nearly Normal, what percentage of men and of women should have even larger mouths than these?

49. Helmet sizes Adult women's head circumferences are approximately Normally distributed with a mean of 56.0 cm and standard deviation of 1.8 cm. You are manufacturing a new type of helmet for female recruits in the Canadian Armed Forces, and are planning on producing three sizes: small, medium, large. Each size will fit a range of head circumferences. Assume female recruits have similar head circumferences to the general population.

a) If we wanted to adequately fit 99.7% of recruits, the small size should fit what head size at the smallest, and the large size should fit what head size at the biggest?

b) If medium is to fit about 60% of recruits, it should fit what head circumferences?

c) If small fits those with head circumferences between 51 cm and 54 cm, what percent of recruits will need this size?

d) If the large size is to fit the biggest 15% of heads, what is the smallest head circumference suitable for a large helmet?

e) If you decide to make a special production run of a few extra small helmets, suitable for the smallest 2% of recruit circumferences, what circumference range will it fit?

50. **Newborns** At-term newborns in Canada vary in weight according to, approximately, a Normal distribution, with a mean of 3500 grams and standard deviation of 500 grams.
    a) Heavy birth weight (HBW) babies are those weighing over 4500 grams. Approximately how many at-term newborns among the next 10 000 will be HBW babies?
    b) Low birth weight (LBW) babies are those weighing less than 2500 grams. Approximately how many at-term newborns among the next 10 000 will be LBW babies? How can you answer this using what you learned in part a), with no new calculation?
    c) Approximately how many at-term newborns among the next 10 000 will be babies weighing between 3300 and 4300 grams?
    d) A very low birth weight (VLBW) is sometimes defined as less than 1500 grams. If we wanted to set a VLBW limit at the 0.1th percentile of the Canadian distribution, i.e., at a weight such that only 1 in 1000 babies weigh less, what would this VLBW limit equal?
    e) One in ten at-term babies weighs more than _____. Fill in the blank.

51. **TOEFL scores again** TOEFL test scores are commonly used to assess English language ability. The paper-based test scores are close to Normally distributed, though the top scores are slightly more tightly compressed than the bottom scores, perhaps due to bumping up against a maximum score of 677. The mean and standard deviation for all writers of the paper-based test in 2007 were approximately 540 and 60, respectively.[9]
    a) Some schools require applicants for admission to obtain a TOEFL score of at least 600. Approximately what proportion of those writing the test actually obtained such a score?
    b) The 10th and 90th percentiles were reported as 460 and 620, respectively. See if the Normal model gives you very similar values.
    c) The IQR was actually 84. What would you expect using the Normal model?
    d) That maximum of 677 seems to cheat the top students; i.e., if a true Normal distribution applied, with mean and standard deviation as indicated above, what percent of test-takers get their grade reduced to 677 from something potentially higher? (Find the probability of scoring higher than 677.)

52. **Blackjack** If you play the card game of Blackjack using the well-known *basic strategy*, you can expect to lose an average 3 cents for every 10-dollar bet, in the long run (learn to count cards and you can do even better). Suppose that thousands of people played Blackjack using

the basic strategy, each playing 10 000 times and betting $10 each time. A histogram of the final net gains (or losses) for those thousands of players would closely resemble a Normal distribution with a mean of approximately −$30.00 and a standard deviation of approximately $1140.00.[10] How can we make such a prediction without actually observing thousands of players? Such predictability follows from the laws of probability, which we will study later in this book.
    a) What percent of the players lost money (have a negative outcome)?
    b) What percent of the players won over $1000.00?
    c) The luckiest 10% of players won over _____ dollars. (Fill in the blank.)
    d) The unluckiest 5% of players lost more than _____ dollars. (Fill in the blank.)

53. **Eggs** Hens usually begin laying eggs when they are about six months old. Young hens tend to lay smaller eggs, often weighing less than the desired minimum weight of 54 grams.
    a) The average weight of the eggs produced by the young hens is 50.9 grams, and only 28% of their eggs exceed the desired minimum weight. If a Normal model is appropriate, what would the standard deviation of the egg weights be?
    b) By the time these hens have reached the age of one year, the eggs they produce average 67.1 grams, and 98% of them are above the minimum weight. What is the standard deviation for the appropriate Normal model for these older hens?
    c) Are egg sizes more consistent for the younger hens or the older ones? Explain.
    d) A certain farmer finds that 8% of his eggs are underweight and that 12% weigh over 70 grams. Estimate the mean and standard deviation of his eggs (Challenging).

54. **Tomatoes** Agricultural scientists are working on developing an improved variety of Roma tomatoes. Marketing research indicates that customers are likely to bypass Romas that weigh less than 70 grams. The current variety of Roma plants produces fruit that average 74 grams, but 11% of the tomatoes are too small. It is reasonable to assume that a Normal model applies.
    a) What is the standard deviation of the weights of Romas now being grown?
    b) Scientists hope to reduce the frequency of undersized tomatoes to no more than 4%. One way to accomplish this is to raise the average size of the fruit. If the standard deviation remains the same, what target mean should they have as a goal?
    c) The researchers produce a new variety with a mean weight of 75 grams, which meets the 4% goal. What

[9]www.ets.org/Media/Research/pdf/71943_web.pdf

[10]Courtesy of http://wizardofodds.com/blackjack

is the standard deviation of the weights of these new Romas?

d) Based on their standard deviations, compare the tomatoes produced by the two varieties.

55. **Flight Departure Delays** The flights delay time (in minutes) data from Vancouver to Toronto are collected with a mean of 11.22 minutes and a standard deviation of 31.25 minutes. The Normal probability plot shows the following:

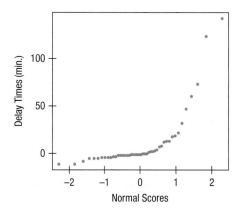

a) Do you think the distribution of delay times is Normal? Explain.

b) If it isn't Normal, how does it differ from a Normal model?

56. **Ages of Oscar Winners** The histogram below shows the distribution of ages of actors and actresses at the time that they won Oscars for the categories of Best Actor and Best Actress. The vertical lines show the mean and one standard deviation above and below the mean. 66.46% of the data points are between the two outer lines.

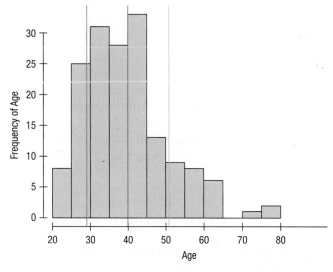

a) Give a reason that a Normal model is not appropriate for these data.

b) The Normal probability plots show the distribution of the age of actors and actresses. What does the plot tell you about the shape of the distribution?

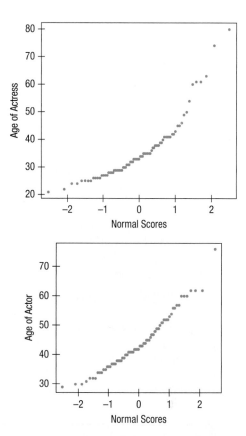

Actor:
https://en.wikipedia.org/wiki/List_of_Academy_Award_Best_Actor_winners_by_age

Actress:
https://en.wikipedia.org/wiki/List_of_Academy_Award_Best_Actress_winners_by_age

T 57. **Tsunamis 2013 again** We discussed tsumani-causing earthquakes in the previous chapter. Plot the magnitudes of these earthquakes (updated to early 2013 in our data file) in a histogram, boxplot, and NPP. Does the distribution appear roughly Normal? If not, how does it appear to depart from Normality, and how do you see possible Normality or lack thereof in each graph?

T 58. **First Nations bands 2010 again** Plot the data on sizes of the 138 First Nations registry groups in Ontario from Exercise 59 in Chapter 3 in a histogram, boxplot, and NPP. Does the distribution appear roughly Normal? If not, how does it appear to depart from Normality, and how do you see possible Normality or lack thereof in each graph?

59. **Simulations** Use your statistical software to draw a random sample of 30 observations from a standard Normal distribution (look for menu items like *Random* and *Normal* and plug in *0* and *1* for the parameters). Plot the data in a histogram and in a Normal probability plot.

a) Compare the two plots. Is it easier to judge *bell-shapedness* or straightness?

b) Repeat four more times, and answer the same question. Is it easier to be fooled by the purely random squiggles in the data when viewing the NPPs or the histograms? Or put another way, in which type of plot is it easier to see the essential pattern through the noise or random jitter?

60. More simulations Use your statistical software to draw a random sample of 30 observations from a *uniform distribution* extending from 0 to 1 (look for menu items like *Random* and *Uniform* and plug in *0* and *1* for the parameters). Plot the data in a histogram and in a Normal probability plot.
   a) Compare the two plots. Is the shape of the histogram roughly uniform or flat? What shape do you see in the NPP?
   b) Repeat four more times and answer the same questions. Try to explain why the NPP has such a shape here. In which type of plot is it easier to see through the noise or random squiggles to the underlying pattern?

61. Some more simulations Use your statistical software to draw a random sample of 100 observations from a standard Normal distribution (look for menu items like *Random* and *Normal* and plug in *0* and *1* for the parameters). Plot the data in a histogram and in a Normal probability plot.
   a) Is the NPP nearly perfectly straight? Is the histogram nearly perfectly bell-shaped? Double the original

number of bins in the histogram. Can the choice of number of bins change one's impression of the bell-shapedness? Is it easier to judge bell-shapedness of the histogram or straightness of the NPP?
   b) Repeat four more times, and answer the same questions. Is it more difficult to see through the random variation or noise in your data to the underlying pattern with a histogram or NPP?

62. And more simulations Use your statistical software to draw a random sample of 100 observations from a uniform distribution extending from 0 to 1 (look for menu items like *Random* and *Uniform* and plug in *0* and *1* for the parameters). Plot the data in a histogram and in a Normal probability plot.
   a) Is the histogram nearly perfectly flat? Change the bin cut points or endpoints to 0, 0.1, 0.2 . . . 0.9, 1.0. Does your impression of the flatness change at all? What is the shape of the NPP? Try to explain why the NPP has such a shape here.
   b) Repeat four more times, and answer the same questions. Is it more difficult to see through the random variation or noise in your data to the underlying pattern with a histogram or NPP?

Just Checking ANSWERS

1. a) On the first test, the mean is 88 and the SD is 4, so $z = (90 - 88)/4 = 0.5$. On the second test, the mean is 75 and the SD is 5, so $z = (80 - 75)/5 = 1.0$. The first test has the lower z-score, so it is the one that will be dropped.
   b) The second test is one standard deviation above the mean, farther away than the first test, so it's the better score relative to the class.

2. a) The mean would increase to 500.
   b) The standard deviation is still 100 points.
   c) The two boxplots would look nearly identical (the shape of the distribution would remain the same), but the later one would be shifted 50 points higher.

3. The standard deviation is now 2.54 millimetres, which is the same as 0.1 inches. Nothing has changed. The standard deviation has "increased" only because we're reporting it in millimetres now, not inches.

4. The mean is 184 cm, with a standard deviation of 8 cm. Two metres is 200 cm, which is two standard deviations above the mean. We expect 5% of the men to be more than two standard deviations below or above the mean, so half of those, 2.5%, are likely to be above 2 metres.

5. a) We know that 68% of the time, we'll be within one standard deviation (two minimum) of 20. So 32% of the time we'll arrive in less than 18 or more than 22 minutes. Half of those times (16%) will be greater than 22 minutes, so 84% will be less than 22 minutes.
   b) 24 minutes is two standard deviations above the mean. Because of the 95% rule, we know 2.5% of the times will be more than 24 minutes.
   c) Traffic incidents may occasionally increase the time it takes to get to school, so the driving times may be skewed to the right, and there may be outliers.
   d) If so, the Normal model would not be appropriate and the percentages we predict would not be accurate.

MathXL

MyLab Statistics

Go to MathXL at **www.mathxl.com** or MyLab™ Statistics at **www.mystatlab.com**. You can practise exercises for this chapter as often as you want. The guided solutions will help you find answers step by step. You'll find a personalized study plan available to you too!

Review

Exploring and Understanding Data

Quick Review

It's time to put it all together. Real data don't come tagged with instructions for use. So let's step back and look at how the key concepts and skills we've seen work together. This brief list and the review exercises that follow should help you check your understanding of Statistics so far.

- We treat data two ways: categorical and quantitative.

- To describe categorical data:
  - Make a picture. Bar graphs work well for comparing counts in categories.
  - Summarize the distribution with a table of counts or relative frequencies (percents) in each category.
  - Pie charts and segmented bar charts display divisions of a whole.
  - Compare distributions with plots side by side.
  - Look for associations between variables by comparing conditional distributions.

- To describe quantitative data:
  - Make a picture. Use histograms, boxplots, stem-and-leaf displays, or dotplots. Stem-and-leafs are great when working by hand and good for small data sets. Histograms are a good way to see the distribution. Boxplots are best for comparing several distributions.
  - Describe distributions in terms of their shape, centre, and spread, and note any unusual features, such as gaps or outliers.
  - The shape of most distributions you'll see will likely be uniform, unimodal, or bimodal. It may even be multimodal. If it is unimodal, then it may be symmetric or skewed.
  - A 5-number summary provides a good numerical description of a distribution: minimum, Q1, median, Q3, and maximum. It gives information about shape, centre, and spread, and is particularly appropriate for describing skewed distributions. From the 5-number summary, you can obtain the median and IQR ($Q_3 - Q_1$) which are *resistant* measures of centre and spread.

- A distribution that is severely skewed may benefit from re-expressing the data. If it is skewed to the high end, taking logs often works well.

- If the distribution is unimodal and symmetric, describe its centre and spread with the mean and standard deviation.

- Use the standard deviation as a ruler to tell how unusual an observed value may be, or to compare or combine measurements made on different scales.

- Shifting a distribution by adding or subtracting a constant affects measures of position but not measures of spread. Rescaling by multiplying or dividing by a constant affects both.

- When a distribution is roughly unimodal and symmetric, a Normal model may be useful. For Normal models, the 68–95–99.7 Rule is a good rule of thumb.

- If the Normal model fits well (check a histogram or Normal probability plot), then Normal percentile tables or functions found in most statistics technology can provide more detailed values.

Need more help with some of this? It never hurts to reread sections of the chapters! And in the following pages we offer you more opportunities[1] to review these concepts and skills.

The exercises that follow use the concepts and skills you've learned in the first five chapters. To be more realistic and more useful for your review, they don't tell you which of the concepts or methods you need. But neither will your exam.

---

[1] If you doubted that we are teachers, this should convince you. Only a teacher would call additional homework exercises "opportunities."

Review Exercises

T 1. Homicide across Canada 2011 Below are annual homicide rates (per 100 000 population) in Canadian cities bigger than 100 000 in population, averaged over the period 2001–11.

City	Homicide Rate
Regina	3.67
Winnipeg	3.51
Edmonton	3.07
Abbotsford	2.98
Saskatoon	2.88
Vancouver	2.43
Kelowna	2.06
Calgary	1.98
Halifax	1.92
Thunder Bay	1.86
Toronto	1.78
Kingston	1.67
Greater Sudbury	1.60
Montreal	1.53
Windsor	1.48
Hamilton	1.43
St. Catharines–Niagara	1.42
London	1.30
Ottawa	1.20
Moncton	1.20
Victoria	1.19
Oshawa	1.18
Brantford	1.16
Gatineau	1.15
Barrie	1.03
Saint John	0.92
Kitchener	0.90
Peterborough	0.83
Trois-Rivières	0.82
Saguenay	0.74
St. John's	0.67
Guelph	0.66
Quebec City	0.62
Sherbrooke	0.36

Source: Based on data from Statistics Canada (2011). Canadian Centre for Justice Statistics, Homicide Survey.

a) Display the data with an appropriate graph, and describe the distribution as best you can. Include any appropriate summary statistics.

b) The graph in a) misses many important factors, such as changes over time and over space (geography). We could plot homicide rates versus year to assess the effect of time. Instead, let's look at geographical effects. Group cities by region—for example Atlantic Canada, Quebec, Ontario, Prairies, Western Canada—and assess if there are any geographical trends.

2. Prenatal care Results of a 1996 American Medical Association report about the infant mortality rate for twins carried for the full term of a normal pregnancy are shown below, broken down by the level of prenatal care the mother had received.

Full-Term Pregnancies, Level of Prenatal Care	Infant Mortality Rate Among Twins (deaths per thousand live births)
Intensive	5.4
Adequate	3.9
Inadequate	6.1
Overall	5.1

a) Is the overall rate the average of the other three rates? Should it be? Explain.

b) Do these results indicate that adequate prenatal care is important for pregnant women? Explain.

c) Do these results suggest that a woman pregnant with twins should be wary of seeking too much medical care? Explain.

3. Singers The boxplots shown display the heights (in inches) of 130 members of a choir.

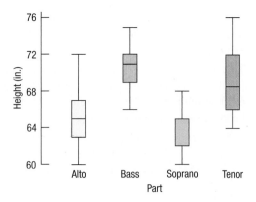

a) It appears that the median height for sopranos is missing, but actually the median and the upper quartile are equal. How could that happen?

b) Write a few sentences describing what you see.

4. Dialysis In a study of dialysis, researchers found that "of the three patients who were currently on dialysis, 67% had developed blindness and 33% had their toes amputated." What kind of display might be appropriate for these data? Explain.

5. Beanstalks Beanstalk Clubs are social clubs for very tall people. To join, a man must be over 6'2" tall, and a woman over 5'10". Heights of adults are approximately Normally distributed, with mean heights of 69.1" for men and 64.0" for women. The respective standard deviations are 2.8" and 2.5".
   a) You are probably not surprised to learn that men are generally taller than women, but what does the greater standard deviation for men's heights indicate?
   b) Who are more likely to qualify for Beanstalk membership, men or women? Explain.

6. Bread Tobermory Bakery is trying to predict how many loaves to bake. In the past 100 days, they have sold between 95 and 140 loaves per day. Here is a histogram of the number of loaves they sold for the past 100 days.

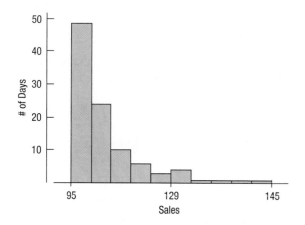

   a) Describe the distribution.
   b) Which should be larger, the mean number of sales or the median? Explain.
   c) Here are the summary statistics for Tobermory Bakery's bread sales. Use these statistics and the histogram above to create a boxplot. You may approximate the values of any outliers.

Summary of Sales	
Median	100
Min	95
Max	140
25th %tile	97
75th %tile	105.5

   d) For these data, the mean was 103 loaves sold per day, with a standard deviation of 9 loaves. Do these statistics suggest that Tobermory Bakery should expect to sell between 94 and 112 loaves on about 68% of days? Explain.

7. University survey Public relations staff at a Canadian university phoned 850 local residents. After identifying themselves, the callers asked the survey participants their ages, whether they had attended college or university, and whether they had a favourable opinion of the university. The official report to the university's directors claimed that, in general, people had very favourable opinions about the university.
   a) Identify the *W*'s of these data.
   b) Identify the variables, classify each as categorical or quantitative, and specify units if relevant.
   c) Are you confident about the report's conclusion? Explain.

8. Acid rain Based on long-term investigation, researchers have suggested that the acidity (pH) of rainfall in Shenandoah Mountains storms can be described by the Normal model $N(4.9, 0.6)$.
   a) Draw and carefully label the model.
   b) What percent of storms produce rainfall with pH over 6?
   c) What percent of storms produce rainfall with pH under 4?
   d) The lower the pH, the more acidic the rain. What is the pH level for the most acidic 20% of all storms?
   e) What is the pH level for the least acidic 5% of all storms?
   f) What is the IQR for the pH of rainfall?

9. Fraud detection A credit card bank is investigating the incidence of fraudulent card use. The bank suspects that the type of product bought may provide clues to the fraud. To examine this situation, the bank looks at the Standard Industrial Code (SIC) of the business related to the transaction. This is a code that was used by Statistics Canada and the U.S. Census Bureau to identify the type of every registered business in North America.[2] For example, 1011 designates Meat and Meat Products (except Poultry), 1012 is Poultry Products, 1021 is Fish Products, 1031 is Canned and Preserved Fruits and Vegetables, and 1032 is Frozen Fruits and Vegetables.

---

[2]Since 1997, the SIC has been replaced by the NAICS, a code of six letters.

A company intern produces the following histogram of the SIC codes for 1536 transactions:

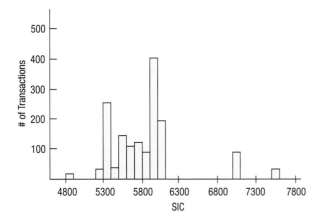

He also reports that the mean SIC is 5823.13 with a standard deviation of 488.17.

a) Comment on any problems you see with the use of the mean and standard deviation as summary statistics.

b) How well do you think the Normal model will work on these data? Explain.

T 10. Streams As part of the course work, a class at a northern BC college collects data on streams each year. Students record a number of biological, chemical, and physical variables, including the stream name, the substrate of the stream (*limestone*, *shale*, or *mixed*), the pH, the temperature (°C), and the BCI, a measure of biological diversity.

Group	Count	%
Limestone	77	44.8
Mixed	26	15.1
Shale	69	40.1

a) Name each variable, indicate whether it is categorical or quantitative, and give the units if available.

b) These streams have been classified according to their substrate—the composition of soil and rock over which they flow—as summarized in the table. What kind of graph might be used to display these data?

T 11. Cramming One Thursday, researchers gave students enrolled in a section of basic French a set of 50 new vocabulary words to memorize. On Friday, the students took a vocabulary test. When they returned to class the following Monday, they were retested—without advance warning. Both test scores (side by side) for the 25 students are shown in the table.

Fri	Mon	Fri	Mon
42	36	50	47
44	44	34	34
45	46	38	31
48	38	43	40
44	40	39	41
43	38	46	32
41	37	37	36
35	31	40	31
43	32	41	32
48	37	48	39
43	41	37	31
45	32	36	41
47	44		

a) Create a graphical display to compare the two distributions of scores.

b) Write a few sentences about the scores reported on Friday and Monday.

c) Create a graphical display showing the distribution of the *changes* in student scores.

d) Describe the distribution of changes.

12. Band sizes 2011 Below is a chart of First Nations registry groups (similar to bands) in Canada by size, on December 31, 2010[3]:

Size of band	Number of bands
<100	20
100–249	75
250–499	112
500–999	167
1 000–1 999	139
2 000+	118

Source: Based on data Registered Indian Population by Sex and Residence 2010. From https://www.aadnc-aandc.gc.ca/DAM/DAM-INTER-HQ/STAGING/texte-text/ai_rs_pubs_sts_ni_rip_rip10_rip10_1309289046808_eng.pdf

a) Which types of graphs are appropriate for displaying these data? Name them all.

b) Present the data in an appropriate graph.

[3]*Registered Indian Population by Sex and Residence 2010.* Ottawa: Indian and Northern Affairs Canada, 2008. http://www.aadnc-aandc.gc.ca/DAM/DAM-INTER-HQ/STAGING/texte-text/ai_rs_pubs_sts_ni_rip_rip10_rip10_1309289046808_eng.pdf. Reproduced with the permission of the Minster of Public Works and Government Services Canada, 2010.

c) The size of a band is actually what type of variable? How is it that we can use a pie chart for such a variable?

d) Is it possible to convert a categorical variable to a quantitative one? Or a quantitative variable to a categorical one? If so, what are the possible advantages and disadvantages?

13. Off or on reserve 2011 Below is a table with information about age and residence status of the registered First Nations population of Canada as of December 31, 2010[4]:

Age Group	0–24	25–34	35–64	65+
Residence Status				
On reserve	49.5%	14.6%	30.2%	5.7%
Off reserve	38.7%	15.1%	38.4%	7.7%

a) Compare the two conditional age distributions, using pie charts. Describe the difference.

b) The total living on-reserve was 459 159, and the total living off-reserve was 365 182. Compare the four conditional residence distributions, using an appropriate bar chart(s), and describe the relationship between residence status and age.

c) Are the variables of interest here quantitative or categorical? Explain clearly.

d) Would it be possible for the average age of First Nations males to be less than the average age of First Nations females on reserve as well as off reserve, but for the average age of all First Nations males to be greater than the average age of all First Nations females? Explain as best you can (you can make up some hypothetical numbers).

T 14. Accidents Progressive Insurance asked customers who had been involved in auto accidents how far they were from home when the accident happened. The data are summarized in the table.

Miles from Home	% of Accidents
Less than 1	23
1 to 5	29
6 to 10	17
11 to 15	8
16 to 20	6
Over 20	17

[4]Ibid.

a) Create an appropriate graph of these data.

b) Do these data indicate that driving near home is particularly dangerous? Explain.

T 15. Hard water In an investigation of environmental causes of disease, data were collected on the annual mortality rate (deaths per 100 000) for males in 61 large towns in England and Wales. In addition, the water hardness was recorded as the calcium concentration (parts per million, ppm) in the drinking water.

a) What are the variables in this study? Indicate whether each variable is quantitative or categorical and what the units are.

b) Here are histograms of calcium concentration and mortality. Describe the distributions of the two variables.

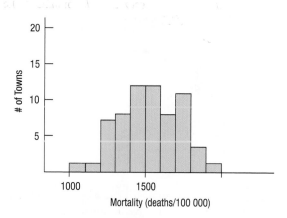

T 16. Hard water II The data set from England and Wales also notes for each town whether it was south or north of Derby. Here are some summary statistics and a comparative boxplot for the two regions.

Summary of Mortality				
Group	Count	Mean	Median	StdDev
North	34	1631.59	1631	138.470
South	27	1388.85	1369	151.114

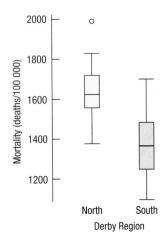

a) What is the overall mean mortality rate for the two regions?

b) Do you see evidence of a difference in mortality rates? Explain.

17. Seasons The histograms below show the average daily temperatures in January and in July for 60 large U.S. cities.

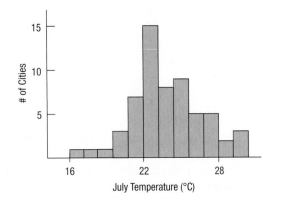

a) What aspect of these histograms makes it difficult to compare the distributions?

b) What differences do you see between the distributions of January and July average temperatures?

c) Differences in temperatures (July–January) for each of the cities are displayed in the boxplot below. Write a few sentences describing what you see.

Cities

T 18. Old Faithful It is a common belief that Yellowstone's most famous geyser erupts once an hour at very predictable intervals. The histogram below shows the time gaps (in minutes) between 222 successive eruptions. Describe this distribution.

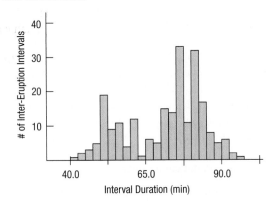

T 19. Old Faithful? Does the duration of an eruption have an effect on the length of time that elapses before the next eruption?

a) The histogram below shows the duration (in minutes) of those 222 eruptions. Describe this distribution.

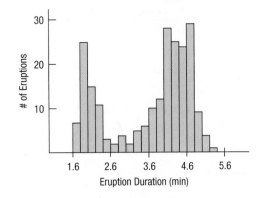

b) Which summary statistics would you choose to describe this distribution? Explain your choices.

c) Let's classify the eruptions as "long" or "short," depending on whether they last at least three minutes. Describe what you see in the comparative boxplots.

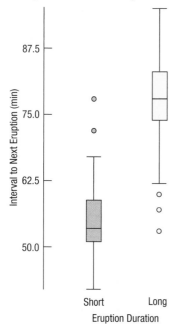

20. Very, very rich According to the World Wealth Report 2008[5], we have the following distributions of High Net Worth Individuals (HNWIs; $1 million or more in assets) and Ultra High Net Worth Individuals (Ultra-HNWIs; $30 million or more in assets):

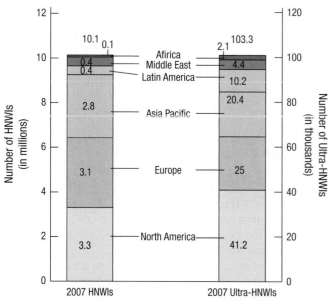

a) What is this type of display called? To make it look similar to displays in this text, what would you have to change about the vertical axis in this graph?

[5]Capgemini/Merril Lynch, "World Wealth Report 2008," www.ml.com/media/100502.pdf.

b) Answer the following or explain why you can't:
   (i) What percent of all Ultra-HNWIs are in Asia-Pacific?
   (ii) What percent of Latin Americans are Ultra-HNWIs?
   (iii) What percent of HNWIs in North America are Ultra-HNWIs?

c) Explain the major differences between the two distributions displayed.

d) If you entered the numbers above into a 2×6 table of counts, would you then have a proper contingency table? If not, how would you convert it to one?

T 21. Liberty's nose Is the Statue of Liberty's nose too long? Her nose measures 4'6" (4.5 feet), but she is a large statue, after all. Her arm is 42 feet long. That means her arm is $42/4.5 = 9.3$ times as long as her nose. Is that a reasonable ratio? Shown in the table are arm and nose lengths of 18 women in a Statistics class, and the ratio of arm-to-nose length for each.

Arm (cm)	Nose (cm)	Arm/Nose Ratio
73.8	5.0	14.8
74.0	4.5	16.4
69.5	4.5	15.4
62.5	4.7	13.3
68.6	4.4	15.6
64.5	4.8	13.4
68.2	4.8	14.2
63.5	4.4	14.4
63.5	5.4	11.8
67.0	4.6	14.6
67.4	4.4	15.3
70.7	4.3	16.4
69.4	4.1	16.9
71.7	4.5	15.9
69.0	4.4	15.7
69.8	4.5	15.5
71.0	4.8	14.8
71.3	4.7	15.2

a) Make an appropriate plot and describe the distribution of the ratios.

b) Summarize the ratios numerically, choosing appropriate measures of centre and spread.

c) Is the ratio of 9.3 for the Statue of Liberty unrealistically low? Explain.

T 22. Winter Olympics 2010 speed skating The top 34 women's 500-m speed skating times are listed in the following table.

Nation	Athlete	Result
Korea	Sang-Hwa Lee	38.249
Germany	Jenny Wolf	38.307
China	Beixing Wang	38.487
Netherlands	Margot Boer	38.511
China	Shuang Zhang	38.530
Japan	Sayuri Yoshii	38.566
Russian Federation	Yulia Nemaya	38.594
China	Peiyu Jin	38.686
United States	Heather Richardson	38.698
Germany	Monique Angermuller	38.761
China	Aihua Xing	38.792
Japan	Nao Kodaira	38.835
Canada	Christine Nesbitt	38.881
Netherlands	Thijsje Oenema	38.892
DPR Korea	Hyon-Suk Ko	38.893
Japan	Shihomi Shinya	38.964
Japan	Tomomi Okazaki	38.971
United States	Elli Ochowicz	39.002
Kazakhstan	Yekaterina Aydova	39.024
United States	Jennifer Rodriguez	39.182
Netherlands	Laurine van Riessen	39.302
Canada	Shannon Rempel	39.351
Germany	Judith Hesse	39.357
Russian Federation	Olga Fatkulina	39.359
Czech Republic	Karolina Erbanova	39.365
Korea	Bo-Ra Lee	39.396
Russian Federation	Svetlana Kaykan	39.422
Italy	Chiara Simionato	39.480
United States	Lauren Cholewinski	39.514
Korea	Jee-Min Ahn	39.595
Australia	Sophie Muir	39.649
Russian Federation	Yekaterina Malysheva	39.782
Korea	Min-Jee Oh	39.816
Canada	Anastasia Bucsis	39.879
Belarus	Svetlana Radkevich	39.899
Netherlands	Annette Gerritsen	97.952

a) The mean finishing time was 40.72 seconds, with a standard deviation of 9.82 seconds. If the Normal model is appropriate, what percent of the times should be within 0.5 second of 40.72?

b) What percent of the times actually fall within this interval?

c) Explain the discrepancy between parts a) and b).

23. **Simpson's paradox** A study in South Africa focusing on the impact of health insurance identified 1590 children at birth and then sought to conduct follow-up health studies five years later. Only 416 of the original group participated in the five-year follow-up study. This made researchers concerned that the follow-up group might not accurately resemble the total group in terms of health insurance. The table below summarizes the two groups by race and by presence of medical insurance when the child was born. Carefully explain how this study demonstrates Simpson's paradox.[6]

		Number (%) Insured	
		Follow-up	Not traced
Race	Black	36 of 404 (8.9%)	91 of 1048 (8.7%)
	White	10 of 12 (83.3%)	104 of 126 (82.5%)
	Overall	46 of 416 (11.1%)	195 of 1174 (16.6%)

T 24. **Sluggers** Babe Ruth was the first great "slugger" in baseball. His record of 60 home runs in one season held for 34 years until Roger Maris hit 61 in 1961. Mark McGwire, with the aid of steroids, shocked the baseball world with 70 home runs in 1998. Listed below are the home run totals for each season McGwire played. Also listed are Babe Ruth's home run totals (starting in 1920 when he joined the N.Y. Yankees and stopped pitching).

McGwire: 3*, 49, 32, 33, 39, 22, 42, 9*, 9*, 39, 52, 58, 70, 65, 32*, 29*

Ruth: 54, 59, 35, 41, 46, 25, 47, 60, 54, 46, 49, 46, 41, 34, 22

a) Find the 5-number summary for McGwire's career.

b) Do any of his seasons appear to be outliers? Explain.

c) McGwire played in only 18 games at the end of his first big league season, and missed major portions of some other seasons because of injuries to his back and knees. Those seasons are marked with asterisks in the list above. Omit these values and make parallel boxplots comparing McGwire's career to Babe Ruth's.

d) Write a few sentences comparing the two sluggers.

e) Create a back-to-back stem-and-leaf display comparing the careers of the two players (Omit those injured seasons).

f) What aspects of the distributions are apparent in the stem-and-leaf displays that the boxplots did not clearly show?

25. **Be quick!** Avoiding an accident when driving can depend on reaction time. That time, measured from the moment the driver first sees the danger until he or she

_____

[6]*Birth to Ten Study*, Medical Research Council, South Africa.

gets a foot on the brake pedal, is thought to follow a Normal model with a mean of 1.5 seconds and a standard deviation of 0.18 seconds.

a) Use the 68–95–99.7 Rule to draw the Normal model.

b) Write a few sentences describing driver reaction times.

c) What percent of drivers have a reaction time less than 1.25 seconds?

d) What percent of drivers have reaction times between 1.6 and 1.8 seconds?

e) What is the interquartile range of reaction times?

f) Describe the reaction times of the slowest 1/3 of all drivers.

26. **Music and memory** Is it a good idea to listen to music when studying for a big test? In a study conducted by some Statistics students, 62 people were randomly assigned to listen to rap music, Mozart, or no music while attempting to memorize objects pictured on a page. They were then asked to list all the objects they could remember. Here are the 5-number summaries for each group:

	n	Min	Q1	Median	Q3	Max
Rap	29	5	8	10	12	25
Mozart	20	4	7	10	12	27
None	13	8	9.5	13	17	24

a) Describe the *W*'s for these data: *Who, What, Where, Why, When, How.*

b) Name the variables and classify each as categorical or quantitative.

c) Create parallel boxplots as best you can from the summary statistics to display these results.

d) Write a few sentences comparing the performances of the three groups.

T 27. **Mail** Here are the number of pieces of mail received at a school office for 36 days:

123	70	90
80	78	72
52	103	138
112	92	93
118	118	106
95	131	59
151	115	97
100	128	130
66	135	76
143	100	88
110	75	60
115	105	85

a) Plot these data.

b) Find appropriate summary statistics.

c) Write a brief description of the school's mail deliveries.

d) What percent of the days actually lie within one standard deviation of the mean? Comment.

T 28. **Birth order** Is your birth order related to your choice of program of study? A Statistics professor at a large university polled his students to find out what their specialist programs were and what position they held in the family birth order. The results are summarized in the table.

a) What percent of these students are oldest or only children?

b) What percent of Humanities specialists are oldest children?

c) What percent of oldest children are Humanities students?

d) What percent of the students are oldest children specializing in the Humanities?

		Birth Order*				
		1	**2**	**3**	**4+**	**Total**
Field	Math/Science	34	14	6	3	**57**
	Agriculture	52	27	5	9	**93**
	Humanities	15	17	8	3	**43**
	Other	12	11	1	6	**30**
	Total	**113**	**69**	**20**	**21**	**223**

* 1 = oldest or only child

29. **Herbal medicine** Researchers for the Herbal Medicine Council collected information on people's experiences with a new herbal remedy for colds. They went to a store that sold natural health products. There they asked 100 customers whether they had taken the cold remedy and, if so, to rate its effectiveness (on a scale from 1 to 10) in curing their symptoms. The Council concluded that this product was highly effective in treating the common cold.

a) Identify the *W*'s of these data.

b) Identify the variables, classify each as categorical or quantitative, and specify units, if relevant.

c) Are you confident about the Council's conclusion? Explain.

T 30. **Birth order revisited** Consider again the data on birth order and university programs of study in Exercise 28.

a) What is the marginal distribution of field?

b) What is the conditional distribution of field for the oldest children?

c) What is the conditional distribution of field for the children born second?

d) Do you think that program of study appears to be independent of birth order? Explain.

31. **Engines** One measure of the size of an automobile engine is its "displacement," the total volume (in litres or cubic inches) of its cylinders. Summary statistics for

several models of new cars are shown below. These displacements were measured in cubic inches.

Summary of Displacement	
Count	38
Mean	177.29
Median	148.5
StdDev	88.88
Range	275
25th %tile	105
75th %tile	231

a) How many cars were measured?
b) Why might the mean be so much larger than the median?
c) Describe the centre and spread of this distribution with appropriate statistics.
d) Your neighbour is bragging about the 227-cubic-inch engine he bought in his new car. Is that engine unusually large? Explain.
e) Are there any engines in this data set that you would consider to be outliers? Explain.
f) Is it reasonable to expect that about 68% of car engines measure between 88 and 266 cubic inches? (That's 177.29 ± 88.88.) Explain.
g) We can convert all the data from cubic inches to cubic centimetres (cc) by multiplying by 16.4. For example, a 200-cubic-inch engine has a displacement of 3280 cc. How would such a conversion affect each of the summary statistics?

32. Engines, again Horsepower is another measure commonly used to describe auto engines. Below are the summary statistics and histogram displaying horsepowers of the same group of 38 cars.

Summary of Horsepower	
Count	38
Mean	101.7
Median	100
StdDev	26.4
Range	90
25th %ile	78
75th %ile	125

a) Describe the shape, centre, and spread of this distribution.
b) What is the interquartile range?
c) Are any of these engines outliers in terms of horsepower? Explain.
d) Do you think the 68–95–99.7 Rule applies to the horsepower of auto engines? Explain.
e) From the histogram, make a rough estimate of the percentage of these engines whose horsepower is within one standard deviation of the mean.
f) A fuel additive boasts in its advertising that it can "add 10 horsepower to any car." Assuming that is true, what would happen to each of these summary statistics if this additive were used in all the cars?

T 33. Toronto students Below is information about region of birth of Toronto public school students (2005). Make a pie chart with two categories: Canada and Outside of Canada. Next, make both a pie chart and a bar chart showing the breakdown for those born outside of Canada. In the pie chart, do some logical grouping, since there are so many categories (10 categories at most).

Region of Birth	Number of Students
Canada	192 956
US	2 297
English-speaking Caribbean/region	4 853
Non-English speaking Caribbean	260
Central & South America & Mexico	4 767
Central Africa	376
Eastern Africa	3 240
Northern Africa	727
Southern Africa	199
Western Africa	1 098
UK & Republic of Ireland	752
South & Western Europe	2 700
Eastern Europe	9 163
Central Asia	400
Eastern Asia	19 335
Southeast Asia	3 002
South Asia	22 927
Western Asia	9 605
Oceania	253

34. Pay According to the U.S. Bureau of Labor Statistics, the mean hourly wage for chief executives in 2009 was

$80.43 and the median hourly wage was $77.27. By contrast, for general and operations managers, the mean hourly wage was $53.15 and the median was $44.55. Are these wage distributions likely to be symmetric, skewed left, or skewed right? Explain.

T 35. Toronto teams Below are salaries for the 2008 season for two popular Toronto sports teams.

Maple Leaf Salaries in $	Blue Jay Salaries in $
2 150 000	12 000 000
2 500 000	11 625 000
4 500 000	10 000 000
525 000	6 750 000
1 400 000	4 835 000
600 000	3 750 000
3 500 000	3 687 500
850 000	2 250 000
3 000 000	1 900 000
525 000	1 550 000
700 000	1 125 000
4 250 000	700 000
5 000 000	640 000
1 400 000	410 000
900 000	404 100
2 240 000	402 500
1 750 000	400 300
1 700 000	395 800
4 000 000	395 500
3 350 000	392 200
850 000	13 200 000
500 000	

Source: Based on data from Data Courtesy Sports-Reference.com

a) Using appropriate methods, compare the two distributions with graphs and numbers.
b) Which team would you rather play for if you are a true superstar? A new guy starting out? An average player? (Just think financially, please.)
c) If you were the owner, which salary structure would you prefer to see, and why?

T 36. Men with brooms 2013 Canadian men will do the sweeping, as long as it is on ice. Canada excels not only in hockey, but also in the sport of curling. Below is a portion of the complete data set (available online at

http://www.pearsoncanada.ca/deveaux) showing scores in games between Canada and the U.S. in the men's world championships since 1968. The tournaments were played in Canada in the following years: 1968, 1973, 1978, 1980, 1981, 1983, 1986, 1987, 1991, 1995, 1996, 1998, 1999, 2003, 2005, 2007, 2009, 2011, and 2013.

Year	Round	Canada	U.S.	Score
2013	8	Brad Jacobs	Brady Clark	7–2
2012	5	Glenn Howard	Heath McCormick	8–7
2011	8	Jeff Stoughton	Pete Fenson	5–3
2010	2	Kevin Koe	Pete Fenson	6–3
2009	13	Kevin Martin	John Shuster	9–6
⋮				
1970	6	Don Duguid	Art Tallackson	10–8
1969	F	Ron Northcott	Raymond "Bud" Somerville	9–6
1969	5	Ron Northcott	Raymond "Bud" Somerville	10–12
1968	SF	Ron Northcott	Raymond "Bud" Somerville	12–2
1968	2	Ron Northcott	Raymond "Bud" Somerville	10–6

Source: Based on data from World Curling Federation (WCF), http://results.worldcurling.org/Associations.aspx.

a) Clearly, Canada usually wins. Let's look at the score differential. Calculate the difference between the Canadian and the American score for each game. Examine the distribution with an appropriate plot. Describe the distribution as best you can (shape, centre, spread).
b) Check for possible Normality of the score differentials via a Normal probability plot, and comment on what the shape of the plot suggests.
c) Maybe the analysis above is missing some important factors, such as time or location. Examine possible dependence on time, graphically, and comment on what you see. How does this plot help you better understand the distribution examined in parts a) and b)?
d) Examine the possible home country effect (Does playing in Canada improve the point differential?) with appropriate graphs, and comment.

37. Some assembly required (Challenging) A company that markets build-it-yourself furniture sells a computer desk that is advertised with the claim "less than an hour to assemble." However, through post-purchase surveys the company has learned that only 25% of its customers succeeded in building the desk in under an hour; 5% said it took them over two hours. The company assumes that consumer assembly time follows a Normal model.

a) Find the mean and standard deviation of the assembly time model.

b) One way the company could solve this problem would be to change the advertising claim. What assembly time should the company quote so that 60% of customers succeed in finishing the desk in that time?

c) Wishing to maintain the "less than an hour" claim, the company hopes that revising the instructions and labelling the parts more clearly can improve the 1-hour success rate to 60%. If the standard deviation stays the same, what new lower mean time does the company need to achieve?

d) Months later, another post-purchase survey shows that new instructions and part labelling did lower the mean assembly time, but only to 55 minutes. None-theless, the company did achieve the 60%-in-an-hour goal, too. How was that possible?

T **38. Profits** The following stem-and-leaf display shows profits as a percent of sales for 29 of the Forbes 500 largest U.S. corporations. The stems are split; each stem represents a span of 5%, from a loss of 9% to a profit of 25%.

a) Find the 5-number summary.

b) Draw a boxplot for these data.

c) Find the mean and standard deviation.

d) Describe the distribution of profits for these corporations.

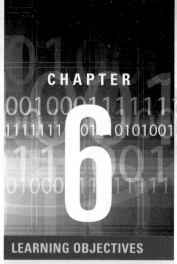
Scatterplots, Association, and Correlation

"The cause of lightning" Alice said very decidedly, for she felt quite sure about this, "is the thunder—no, no" she hastily corrected herself, "I meant it the other way."

"It's too late to correct it," said the Red Queen, "When you've once said a thing, that fixes it, and you must take the consequences."

—From Alice's Adventures in Wonderland by Lewis Carroll

Connections

In previous chapters, we dealt with a single quantitative variable—displaying, summarizing, modelling its distribution. We also examined the relationship between a quantitative variable and a categorical variable (comparing boxplots, etc.).

Now we turn our attention to associations between two quantitative variables.

Is the price of running shoes related to how long they last? Is your alertness in class related to how much (or little) sleep you got the night before?

We'll start by looking at scatterplots and describing what we see—the direction, form, and strength of the association. Then, as we did for histograms, we'll find a quantitative summary (the *correlation*) of what we learn from the display.

Humans have fairly big and heavy brains. Most other species have comparatively smaller brains. What affects the brain size for a particular species? Why are ours so large? For bigger brains, perhaps more time is needed in the womb, which is called the gestation period. Let's explore this relationship for a number of placental mammal species. Figure 6.1 plots one point for each of 23 different species. On the *x*-axis, we show the average gestation period, in days, for that species, and on the *y*-axis, we show the average neonatal brain weight (grams). When we plot one quantitative variable versus another quantitative variable, the resultant graph is called a **scatterplot**.

Figure 6.1

Scatterplot of neonatal brain weight versus gestation period for a number of placental mammal species. Is the general pattern roughly straight or curved? Are any species somewhat unusual compared with the rest? Are there distinct groups or clusters? (Do you recognize any of the plotted species?)

Scatterplot of Brain Weight vs Gestation Period

So what do you see? Brain weight clearly increases, in general, as gestation time increases, which we call a *positive association*. The general pattern though does not closely resemble a straight line but instead curves up. For example, if you look at longer gestation periods—say over 200 days—the brain weight "dividend," or increase, for every extra day in the womb seems to be greater than when gestation periods are shorter (like an acceleration in brain-growth rate).

If it were a straight-line type pattern, this would not be the case, and every extra day of gestation would produce a roughly similar brain weight dividend anywhere in the plot. There may be other reasonable ways to measure brain size—for example, taking the logarithm of each brain weight—that might produce a more straight-line like or linear pattern, but let's postpone the topic of data re-expression until later in this chapter.

Who	Placental mammal species
What	Gestation period and neonatal brain weight
Units	Days; Grams
When	1974
Where	All over the Earth
How	Published in *The American Naturalist*
Why	To learn about vertebrate growth

Are any species unusual? It is hard to say with such a curved pattern, but when you get to around 270 days in the plot, one species' brain weight seems unusually high, at about 340 g. And what about the species with a brain weight of 130 g? Does this appear unusually low to you? Hey, did you notice that one of those species is you guys (i.e., humans[1])?

Are there distinct groups or clusters present? Look very closely at the eight species to the left. Could this be a separate group whose pattern is curving upward very sharply? This is not at all clear from the plot, so let's find out more about the actual species included. It turns out that seven of those eight species on the left are carnivores, while all the remaining species are primates. Figure 6.2 is another plot of the same data, distinguishing the carnivores from the primates:

Figure 6.2

Introducing a categorical variable into the scatterplot of brain weight versus gestation, in order to separate the primates from the carnivores

The answer is pretty clear now. There are two distinct subgroups, with different relationships between brain weight and gestation for each group. The relationship between gestation and brain weight now appears to curve upward even more sharply than before, particularly for carnivores. It might be best to separate the data according to subgroup and analyze primates and carnivores separately. We have discovered the existence of an important *lurking variable* in our initial analysis, namely carnivore/primate status. Lurking variables are hidden variables that lurk or stand behind an analysis but may have an important influence on the relationship being studied.

Sometimes we can see separate subgroups by careful inspection of the scatterplot, but in other cases we have to use our knowledge of categorical variables that may make a difference, introduce the categorical variable into the plot, and see if the relationship between the two quantitative variables is similar or not for the different subgroups.

Scatterplots may be the most common displays for data. Just looking at them, you can see patterns, trends, relationships, and even the occasional extraordinary value sitting apart from the others. As the great philosopher Yogi Berra[2] once said, "You can observe a lot by watching."[3] Scatterplots are the best way to start observing the relationship between two *quantitative* variables.

Relationships between variables are often at the heart of what we'd like to learn from data:

- Are grades higher now than they used to be?
- Do people tend to reach puberty at a younger age than in previous generations?
- Does applying magnets to parts of the body relieve pain? If so, are stronger magnets more effective?
- Do students learn better when they get more sleep?

> **Where did the origin go?** Scatterplots usually don't—and shouldn't—show the origin, because often one or both of the variables do not have values near 0. The display should focus on the part of the coordinate plane that actually contains the data. In our opening example, no species had gestation times anywhere near 0, so the computer drew the *x*-axis starting around 50 days, and the drawn axes do not intersect at (0,0).

---

[1]I confess, I'm Klingon.

[2]Hall of Fame catcher and manager of the New York Mets and Yankees.

[3]But then he also said, "I really didn't say everything I said." So we can't really be sure.

Questions such as these relate two quantitative variables and ask whether there is an **association** between them. Scatterplots are the ideal way to *picture* such associations.

You've seen a scatterplot earlier in this book—the plot of average wind speed versus day of year—in Chapter 4. When the *x*-variable is time, the scatterplot is usually called a timeplot. Timeplots can show quite complex patterns, such as cyclical and seasonal patterns, often requiring what are called *time series methods of analysis*, which we won't be discussing in this text.

6.1 Scatterplots

Almost everyone has seen a scatterplot. But if asked, many people would find it hard to say what to look for in a scatterplot. What are the main features or aspects of the plot that we should focus on and discuss in order to accurately report the story being told by the scatterplot?

The **direction** of the association is important. A pattern like this that runs from the upper left to the lower right   is said to be **negative**. A pattern running the other way

> Look for **Direction**: What's my sign—positive, negative, or neither?

is called **positive**. In our example, there was a positive association between brain weight and gestation period.

The second thing to look for in a scatterplot is its **form**. If there is a straight-line relationship, it will appear as a cloud or swarm of points stretched out in a generally consistent, straight form. In this case, we describe the association as **linear**.

If the relationship isn't straight but curves gently, while still increasing or decreasing

> Look for **Form**: straight, curved, something exotic, or no pattern?

steadily like in the chapter opening example   , we can often find ways to make it more nearly straight (see the last section of this chapter). But if it curves sharply—up

and then down, for example like this: , we may well need to resort to the multiple regression methods of Chapter 26.

The third feature to look for in a scatterplot is the **strength** of the relationship. At one

> Look for **Strength**: How much scatter?

extreme, do the points appear tightly clustered in a single stream   (whether straight, curved, or bending all over the place)? Or, at the other extreme, does the swarm of points seem to form a vague cloud through which we can barely discern any trend or

pattern?   For example, the brain weight and gestation plot shows only a moderate degree of scatter, indicating a moderately strong relationship between a species' gestation period and its neonatal brain weight.

Finally, always look for the unexpected. Often the most interesting thing to see in a scatterplot is something you never thought to look for. One example of such a surprise is

> Look for **Unusual Features**: Are there outliers or subgroups?

an **outlier** standing away from the overall pattern of the scatterplot. Such a point is almost always interesting and always deserves special attention. Is the highest point in our plot an outlier (humans)? Possibly, but hard to say, given the curved pattern. *Clusters or subgroups* that stand away from the rest of the plot or show a trend in a different direction than the rest of the plot should raise questions about why they are different. They may be a clue that you should split the data into subgroups (like primates and carnivores in our example) instead of looking at all the cases together.

For Example COMPARING PRICES WORLDWIDE

If you travel overseas, you know that what's really important is not the amount in your wallet but the amount it can buy. UBS (one of the largest banks in the world) prepared a report comparing prices, wages, and other economic conditions in cities around the world for their international clients. Some of the variables they measured in 73 cities are *Cost of Living, Food Costs, Average Hourly Wage*, average number of *Working Hours* per Year, average number of *Vacation Days*, hours of work (at the average wage) needed to buy an *iPhone*, minutes of work needed to buy a *Big Mac*, and *Women's Clothing Costs*.[4] For your burger fix, you might want to live in Tokyo where it takes only about 9 minutes of work to afford a Big Mac. In Nairobi, you'd have to work almost an hour and a half.

Of course, these variables are associated, but do they consistently reflect costs of living? Plotting pairs of variables can reveal how and even if they are associated. The variety of these associations illustrate different directions and kinds of association patterns you might see in other scatterplots.

QUESTION: Describe the patterns shown by each of these plots.

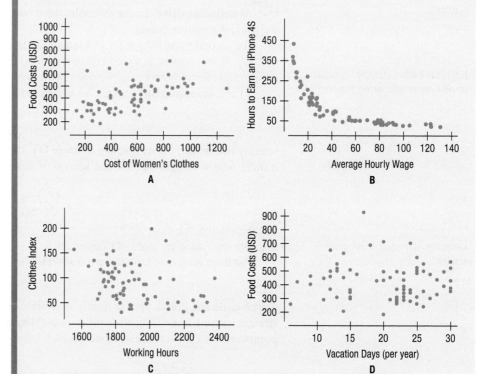

ANSWER: In Plot A, the association between *Food Costs* and *Cost of Women's Clothes* is positive and straight with a few high outliers, and moderately strong. In Plot B, the association between *Hours to Earn an iPhone4S* and *Average Hourly Wage* is negative and strong, but the form is not straight. In Plot C, the association between the *Clothes Index* and *Working Hours* is weak, negative, and generally straight with one or two possible high outliers. In Plot D, there does not appear to be any association between *Food Costs* and *Vacation Days*.

---

[4]Detail of the methodology can be found in the report *Prices and Earning: A comparison of purchasing power around the globe/2012 edition*. www.economist.com/node/14288808?story_id=14288808 (Data in **Prices and Earnings 2012**)

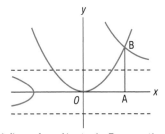

A figure from Newton's *Enumeration of lines of the third order*[5] labels the axes in the modern way.

Scatterplot history The idea of using two axes at right angles to define a field on which to display values can be traced back to the philosopher, scientist, and mathematician René Descartes (1596–1650). The playing field he defined in this way is formally called a Cartesian plane in his honour. Sir Isaac Newton (1642–1727) may have been the first to use the now-standard convention of denoting the vertical axis *y* and the horizontal axis *x*.

Both Descartes and Newton used the Cartesian plane to plot functions, as you have probably done yourself. Plotting individual data values works in a similar way. Each case is displayed as a dot or symbol on a scatterplot at a position that corresponds to its values on two variables—one measured along the *x*-axis and the other along the *y*-axis. When you work with data, you should label the axes with the names of the variables rather than just using *y* and *x*.

Roles for Variables

Which variable should go on the *x*-axis and which on the *y*-axis? What we want to know about the relationship can tell us how to make the plot. We often have questions such as:

■ Do older houses sell for less than newer ones of comparable size and quality?
■ Do people who drink more coffee tend to have faster pulse rates?
■ How accurately can we estimate a person's percent body fat by measuring their waist or wrist size?
■ Do baseball teams that score more runs sell more tickets to their games?

In each of these examples, the two variables play different roles. We'll call the variable of interest the **response variable** and the other the **explanatory** or **predictor variable**. It's common practice to place the response variable on the *y*-axis and the explanatory variable on the *x*-axis. Sometimes, we may even just call them the *x*- and *y*-variables. When you make a scatterplot, you can assume that those who view it will think this way, so carefully choose which variables to assign to which axes.

The roles we choose for variables are more about how we *think* about them than about the variables themselves. Just placing a variable on the *x*-axis doesn't necessarily mean that it explains *anything*. And the response variable may not actually *respond* to it in any way. We plotted neonatal brain weight on the *y*-axis against gestation period on the

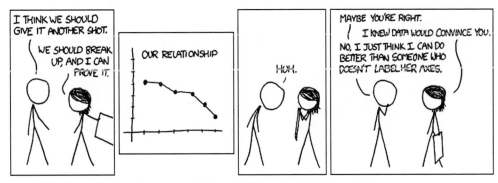

x-axis because we were interested in seeing how brain weight *depends* on gestation period, and perhaps *predicting* brain weight from gestation period. Could we have plotted them the other way? If we were interested in estimating how much time in gestation is typically needed to allow the neonatal brain to reach some particular size, we might have

[5]Sir Isaac Newton's *Enumeration of lines of the third order, generation of curves by shadows, organic description of curves, and construction of equations by curves.* p. 86. Translated from the Latin. With notes and examples, by C.R.M. Talbot. Available at quod.lib.umich.edu/u/umhistmath/ABQ9451.0001.001/81?rgn=full+text; view=pdf

plotted the variables the other way around. For some scatterplots, it can make sense to use either choice, so you have to think about how the choice of role helps to answer the question you have.

Older textbooks and disciplines other than Statistics sometimes refer to the *x*- and *y*-variables as the *independent* and *dependent* variables, respectively. The idea was that the *y*-variable depended on the *x*-variable and the *x*-variable acted independently to make *y* respond. These names, however, conflict with other uses of the same terms in Statistics.

Section 6.1 Exercises

1. **Association** Suppose you were to collect data for each pair of variables given below. You want to make a scatterplot. Which variable would you use as the explanatory variable and which as the response variable? Why? What would you expect to see in the scatterplot? Discuss the likely direction, form, and strength.
   a) Apples: weight in grams, weight in ounces
   b) Apples: circumference (cm), weight (g)
   c) University freshmen: shoe size, grade point average
   d) Gasoline: number of kilometres you drove since filling up, litres remaining in your tank

2. **Association II** Suppose you were to collect data for each pair of variables given below. You want to make a scatterplot. Which variable would you use as the explanatory variable and which as the response variable? Why? What would you expect to see in the scatterplot? Discuss the likely direction, form, and strength.
   a) T-shirts at a store: price of each, number sold
   b) Scuba diving: depth, water pressure
   c) Scuba diving: depth, visibility
   d) All elementary-school students: weight, score on a reading test

3. **Bookstore sales** Consider the following data from a small bookstore.

Number of Sales People Working	Sales (in $1000)
2	10
3	11
7	13
9	14
10	18
10	20
12	20
15	22
16	22
20	26
$\bar{x} = 10.4$	$\bar{y} = 17.6$
$S_x = 5.64$	$S_y = 5.34$

   a) Prepare a scatterplot of *Sales* against *Number of Sales People Working*.
   b) What can you say about the direction of the association?
   c) What can you say about the form of the relationship?
   d) What can you say about the strength of the relationship?
   e) Does the scatterplot show any outliers?

4. **Disk drives 2014** Disk drives have been getting larger. Their capacity is now often given in *terabytes* (TB) where 1 TB = 1000 gigabytes, or about a trillion bytes. A search of prices for external disk drives on Amazon.com in early 2014 found the following data:

Capacity (in TB)	Price (in $)
0.50	59.99
1.0	79.99
2.0	111.97
3.0	109.99
4.0	149.99
6.0	423.34
8.0	596.11
12.0	1079.99
32.0	4461.00

   a) Prepare a scatterplot of *Price* against *Capacity*.
   b) What can you say about the direction of the association?
   c) What can you say about the form of the relationship?
   d) What can you say about the strength of the relationship?
   e) Does the scatterplot show any outliers?

6.2 Correlation

Data collected from students in a Statistics class included their *Height* (in inches) and *Weight* (in pounds), and are displayed in Figure 6.3. It's no great surprise that there is a positive association between the two. As you might suspect, taller students tend to weigh more. (If we had reversed the roles and chosen weight as the explanatory variable, we might say that heavier students tend to be taller.)[6] And the form of the scatterplot is fairly straight as well, although there appears to be a high outlier.

Figure 6.3

Weight (lbs) versus *Height (in)* from a Statistics class.

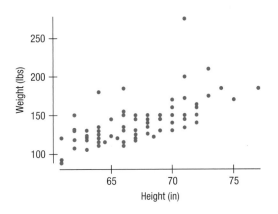

Who	Students
What	Height (inches), Weight (pounds)
Where	Ithaca, NY
Why	Data for class
How	Survey

The pattern in the scatterplot looks straight, and the positive association is clear, but how strong is that association? If you had to put a number (say, between 0 and 1) on the strength, what would you say? The strength of the association shouldn't depend on the units we choose for *x* or *y*. For example, if we had measured weight in *kg* instead of *lbs* and height in *cm* instead of *inches,* the plot (Figure 6.4) would look the same.

Figure 6.4

Changing the units (from pounds to kilograms for *Weight* and from inches to centimetres for *Height*) did not change the direction, form, or strength of the association between the variables.

If we subtract the mean from each variable, that shouldn't change the strength of the association either. We would just move the means of both variables to zero. This makes it easier to see what each point contributes to the strength of the association because we can tell which points are above and which are below the means of each variable. Figure 6.5 colours them to make this even clearer. The green points in the upper right (first quadrant) and lower left (third quadrant) are consistent with a positive association, while the red points in the other two quadrants are consistent with a negative association. Points lying on the *x*- or *y*-axis don't really add any information; they are coloured blue.

---

[6]The young son of one of the authors, when told (as he often was) that he was tall for his age, used to point out that, actually, he was young for his height.

Figure 6.5

In this scatterplot, points are coloured according to how they affect the association: green for positive, red for negative, and blue for neutral.

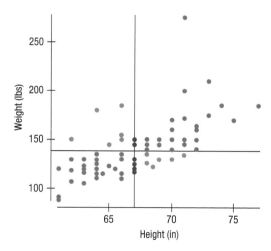

We said that the units shouldn't affect our measure of strength. Now that we've subtracted means, a natural way to remove the units is to standardize each variable and work instead with the z-scores. Recall that to find z-scores for any variable, we subtract its mean from each value and divide by its standard deviation. So instead of plotting (x, y), we plot

$$(z_x, z_y) = \left(\frac{x - \bar{x}}{s_x}, \frac{y - \bar{y}}{s_y} \right)$$

and these z-scores do just what we want. They centre the plotted points around the origin and they remove the original units, replacing them with standard deviations. As we've seen, measuring differences in standard deviations is a fundamental idea in Statistics.

Figure 6.6

Standardized *Weight* and *Height*. We use the standard deviation to judge the distance of each point from the centre of the plot and the influence it should have.

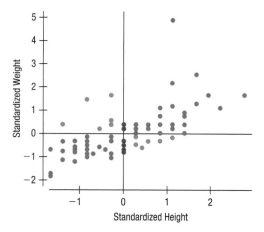

For the green points in Figure 6.6, both z-scores are positive, or both are negative. In either case, their product, $z_x z_y$, is positive. But the red points have opposite signs, so their products are negative. We can measure the strength of the association by adding up these products:

$$\sum z_x z_y$$

This summarizes both the direction and strength of the association. Points farther from the origin have larger z-scores, so they'll contribute more to the sum. But the sum can keep growing as we add more points, so we convert it to an average, but as before, instead of dividing by the obvious choice n, we go with the *statistical* choice of $n - 1$.[7]

The result is the famous **correlation coefficient**:

$$r = \frac{\sum z_x z_y}{n - 1}$$

---

[7]Yes, the same $n - 1$ as in the standard deviation calculation. And we offer the same promise to explain it better later.

Dividing the sum by $n - 1$ serves two purposes. It adjusts the strength for the number of points and it makes the correlation lie between values of -1 and $+1$. For the students' heights and weights, the correlation is 0.644. Because it is based on z-scores, which have no units, the correlation has no units either. It will stay the same if you change measurement units from inches to centimetres, fathoms, or Ångstroms.

There is an alternative formula for the correlation coefficient, using x and y in their original units, that is also faster to compute. The correlation can be expressed as follows:

$$r = \frac{s_{xy}}{s_x \times s_y}$$

where

$$s_{xy} = \frac{\sum (x - \bar{x})(y - \bar{y})}{(n - 1)}$$

is called the *covariance* of x and y. Plug in x for y in the covariance formula above, and you will see that the covariance of x and x is just the variance of x. Covariance generalizes the concept of variance. And the covariance of x standardized and y standardized produces the correlation. For a convenient computational formula with just the x and y values, we can plug in formulas for the two standard deviations in the denominator, and for the covariance in the numerator, and cancel out all the "$n - 1$" terms. We get $r = \dfrac{\sum (x - \bar{x})(y - \bar{y})}{\sqrt{\sum (x - \bar{x})^2 \sum (y - \bar{y})^2}}$

 BY HAND FINDING THE CORRELATION COEFFICIENT

Start with the summary statistics for both variables: $\bar{x}$, $\bar{y}$, s_x, and s_y. Then find the deviations from the means as we did for the standard deviation, but now for both variables: $(x - \bar{x})$ and $(y - \bar{y})$. Form the product for each data pair, and add up these products for all data pairs. Divide the sum by $(n - 1)$ to get the covariance, then divide again by the product of the two standard deviations $s_x \times s_y$, to get the correlation coefficient.

Here's an example:

Suppose the data pairs are:

x	6	10	14	19	21
y	5	3	7	8	12

We calculate $\bar{x} = 14$, $\bar{y} = 7$, $s_x = 6.20$, and $s_y = 3.39$.

Deviations in x	Deviations in y	Product
$6 - 14 = -8$	$5 - 7 = -2$	$-8 \times -2 = 16$
$10 - 14 = -4$	$3 - 7 = -4$	16
$14 - 14 = 0$	$7 - 7 = 0$	0
$19 - 14 = 5$	$8 - 7 = 1$	5
$21 - 14 = 7$	$12 - 7 = 5$	35

Add up the products: $16 + 16 + 0 + 5 + 35 = 72$.
Average the products by dividing by $(n - 1) = (5 - 1)$ to get a covariance of 18.00.
Divide by $s_x \times s_y = 6.20 \times 3.39 = 21.02$ to remove the scale dependence, producing the correlation coefficient:

$$r = 18/21.02 = 0.856$$

Assumptions and Conditions for Correlation

Correlation measures the strength of *linear* association between two *quantitative* variables. To interpret a correlation, we must assume that there is a true underlying linear relationship. Of course, in general, we can't know that. But we *can* check whether that assumption is plausible by looking at the data we do have. To distinguish between what we *assume* about variables from what we can *check* by looking at the data, we'll call the things we check *conditions*. Checking the conditions almost always means either thinking about how the data were collected or looking at a graph—but if you follow the first three rules, you'll have already made the plot, so that takes no extra effort.

There are three conditions to check before you use a correlation:[8]

- **Quantitative Variables Condition** Don't make the common error of calling an association involving a categorical variable a correlation. Correlation is only about quantitative variables.
- **Straight Enough Condition** The best check for the assumption that the variables are truly linearly related is to look at the scatterplot to see whether it looks reasonably straight. That's a judgment call, but not a difficult one.
- **No Outliers Condition** Outliers can distort the correlation dramatically, making a weak association look strong or a strong one look weak. Outliers can even change the sign of the correlation. But it's easy to see outliers in the scatterplot, so to check this condition, just look.

Correlations are often reported without supporting data or plots. Nevertheless, you can still think about the conditions. Be cautious in interpreting (or accepting others' interpretations of) a correlation when you can't check the conditions.

For Example   CORRELATIONS FOR SCATTERPLOT PATTERNS

Look back at the scatterplots of the economic variables in cities around the world (p. 180). The correlations for those plots are (a to d) 0.614, −0.791, −0.388, and −0.040, respectively.

QUESTION: Check the conditions for using correlation. If you feel they are satisfied, interpret the correlation.

ANSWER: All of the variables examined are quantitative and none of the plots shows an outlier. However, the relationship between *Hours to Earn an iPhone* and *Average Wage* is not straight, so the correlation coefficient isn't an appropriate summary. For the others:

A correlation of 0.614 between *Food Costs* and *Women's Clothing Costs* indicates a moderately strong positive association.

A correlation of −0.388 between *Clothes Index* and *Working Hours* indicates a weak negative association.

The small correlation value of −0.040 between *Food Costs* and *Vacation Days* suggests essentially no linear association between them.

 Just Checking

Your Statistics teacher tells you that the correlation between the scores (points out of 50) on Exam 1 and Exam 2 was 0.75.
1. Before answering any questions about the correlation, what would you like to see? Why?

---

[8]Quite generally, the terms 'correlation' or 'correlation coefficient' refer to the particular measure we have defined in this chapter, also known as *Pearson's product moment correlation*. However, there are some other measures of association where the word 'correlation' may appear as part of the name (e.g. Spearman's rank correlation, discussed in Chapter 28), and they may have very different *conditions*.

2. If she adds 10 points to each Exam 1 score, how will this change the correlation?

3. If she standardizes scores on each exam, how will this affect the correlation?

4. In general, if someone did poorly on Exam 1, are they likely to have done poorly or well on Exam 2? Explain.

5. If someone did poorly on Exam 1, can you be sure that the student also did poorly on Exam 2? Explain.

Step-by-Step Example   LOOKING AT ASSOCIATION

When your blood pressure is measured, it is reported as two values, systolic blood pressure and diastolic blood pressure.

Questions: How are these variables related to each other? Do they tend to be both high or both low? How strongly associated are they?

THINK ➡ Plan State what you are trying to investigate.	I'll examine the relationship between two measures of blood pressure. (Data in **Framingham**)
Variables Identify the two quantitative variables whose relationship we wish to examine. Report the *W*'s and be sure both variables are recorded for the same individuals.	The variables are systolic and diastolic blood pressure (SBP and DBP) recorded in millimetres of mercury (mm Hg) for each of 1406 participants in the Framingham Heart Study, a long-running health study in Framingham, MA.[9]
Plot Make the scatterplot. Use a computer program or graphing calculator if possible.	
Check the conditions.	✓ **Quantitative Variables Condition**: Both variables are quantitative with units of mm Hg. ✓ **Straight Enough Condition**: The scatterplot is quite straight. ✓ **Outliers?** There are a few cases with moderately high SBP (compared with their DBP), but not extreme enough to matter, particularly in light of the large number of cases.
REALITY CHECK Looks like a strong positive linear association. We shouldn't be surprised if the correlation coefficient is positive and fairly large.	I have two quantitative variables that satisfy the conditions, so correlation is a suitable measure of association.

---

[9]http://www.framinghamheartstudy.org/

| SHOW ➡ **Mechanics** We usually calculate correlations with technology. Here we have 1406 cases, so we'd never try it by hand. | The correlation coefficient is $r = 0.792$. |
| TELL ➡ **Conclusion** Describe the direction, form, and strength you see in the plot, along with any unusual points or features. Be sure to state your interpretations in the proper context. | The scatterplot shows a positive association, with higher SBP going with higher DBP. The plot is generally straight with only a moderate amount of scatter. The correlation of 0.792 is consistent with what I saw in the scatterplot. |

Correlation Properties

Here's a useful list of facts about the correlation coefficient:

- The sign of a correlation coefficient gives the direction of the association.
- Correlation is always between −1 and +1. Correlation *can* be exactly equal to −1.0 or +1.0, but these values are unusual in real data because they mean that all the data points fall *exactly* on a single straight line.
- Correlation treats x and y symmetrically. The correlation of x with y is the same as the correlation of y with x.
- Correlation has no units. This fact can be especially appropriate when the data's units are somewhat vague to begin with (IQ score, personality index, socialization, and so on). Correlation is sometimes given as a percentage, but you really shouldn't do that because it suggests a percentage of *something*—and correlation, lacking units, has no "something" of which to be a percentage.
- Correlation is not affected by changes in the centre or scale of either variable. Changing the units or baseline of either variable has no effect on the correlation coefficient. Correlation depends only on the z-scores, and they are unaffected by changes in centre or scale.
- Correlation measures the strength of the *linear* association between the two variables. Variables can be strongly associated but still have a small correlation if the association isn't linear.
- Correlation is sensitive to outliers. A single outlying value can make a small correlation large or make a large one small.

UNITS DON'T MATTER

We could have measured the students' weights in stones. A stone is a measure in the pre-metric UK system of measures equal to 14 pounds. And we could have measured heights in hands. Hands are still commonly used to measure the heights of horses. A hand is 4 inches. But no matter what units we use to measure the two variables, *the correlation* stays the same.

How strong is strong? You'll often see correlations characterized as "weak," "moderate," or "strong," but be careful. There's no agreement on what those terms mean. The same numerical correlation might be strong in one context and weak in another. You might be thrilled to discover a correlation of 0.7 between the new summary of the economy you've come up with and stock market prices, but you'd consider it a design failure if you found a correlation of "only" 0.7 between two tests intended to measure the same skill. Deliberately vague terms like "weak," "moderate," or "strong" that describe a linear association can be useful additions to the numerical summary that correlation provides. But be sure to include the correlation and show a scatterplot so others can judge for themselves.

For Example CHANGING SCALES

RECAP: Several measures of prices and wages in cities around the world show a variety of relationships, some of which we can summarize with correlations.

QUESTION: Suppose that, instead of measuring prices in U.S. dollars and recording work time in hours, we had used Canadian dollars and minutes. How would those changes affect the conditions, the values of correlation, or our interpretation of the relationships involving those variables?

ANSWER: Not at all! Correlation has no units, so the conditions, value of r, and interpretation are all unaffected by changes to the units of the individual variables.

Correlation Tables

It is common in some fields to compute the correlations between every pair of variables in a collection of variables and arrange these correlations in a table (see Table 6.1). The rows and columns of the table name the variables, and the cells hold the correlations.

Correlation tables are compact and give a lot of summary information at a glance. They can be an efficient way to start to look at a large data set, but a dangerous one. By presenting all of these correlations without any checks for linearity and outliers, the correlation table risks showing truly small correlations that have been inflated by outliers, truly large correlations that are hidden by outliers, and correlations of any size that may be meaningless because the underlying form is not linear.

The diagonal cells of a correlation table (if they appear) always show correlations of exactly 1. (Can you see why?) Correlation tables are commonly offered by computer statistics packages. These same packages often offer simple ways to make all the scatterplots that go with these correlations.[10]

Table 6.1

A correlation table of data reported by *Forbes* magazine on several financial measures for large companies. From this table, can you be sure that the variables are linearly associated and free from outliers?

	Assets	Sales	Market Value	Profits	Cash Flow	Employees
Assets	1.000					
Sales	0.746	1.000				
Market Value	0.682	0.879	1.000			
Profits	0.602	0.814	0.968	1.000		
Cash Flow	0.641	0.855	0.970	0.989	1.000	
Employees	0.594	0.924	0.818	0.762	0.787	1.000

Section 6.2 Exercises

5. Correlation facts If we assume that the conditions for correlation are met, which of the following are true? If false, explain briefly.
a) A correlation of -0.98 indicates a strong, negative association.
b) Multiplying every value of x by 2 will double the correlation.
c) The units of the correlation are the same as the units of y.

6. Correlation facts II If we assume that the conditions for correlation are met, which of the following are true? If false, explain briefly.
a) A correlation of 0.02 indicates a strong positive association.
b) Standardizing the variables will make the correlation 0.
c) Adding an outlier can dramatically change the correlation.

6.3 Warning: Correlation ≠ Causation

Whenever we have a strong correlation, it's tempting to try to explain it by imagining that the predictor variable has *caused* the response to change. Humans are like that; we tend to see causes and effects in everything.

Sometimes this tendency can be amusing. A scatterplot (Figure 6.7) of the human population (y) of Oldenburg, Germany, in the beginning of the 1930s plotted against the number of storks nesting in the town (x) shows a tempting pattern.

---

[10]A table of scatterplots arranged just like a correlation table is sometimes called a *scatterplot matrix*.

Figure 6.7

A scatterplot of the number of storks in Oldenburg, Germany, plotted against the population of the town for seven years in the 1930s. The association is clear. How about the causation? (Data in **Storks**)

(*Source*: Based on data from Statistics for Experimenters, Box, Hunter and Hunter from Ornithologische Monatsberichte, 44, no. 2)

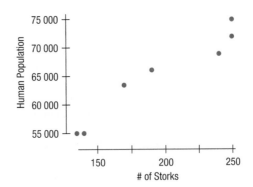

The two variables are obviously related to each other (the correlation is 0.97!), but that doesn't prove that storks bring babies. It turns out that storks nest on house chimneys. More people means more babies, but also more houses, more nesting sites, and so more storks. The causation is actually in the *opposite* direction, but you can't tell from the scatterplot or correlation. You need additional information—not just the data—to determine the real mechanism.

> **Does cancer cause smoking?** Even if the correlation of two variables is due to a causal relationship, the correlation itself cannot tell us what causes what.
>
> Sir Ronald Aylmer Fisher (1890–1962) was one of the greatest statisticians of the twentieth century. Fisher testified in court (paid by the tobacco companies) that a causal relationship might underlie the correlation of smoking and cancer:
>
> > "Is it possible, then, that lung cancer . . . is one of the causes of smoking cigarettes? I don't think it can be excluded . . . the pre-cancerous condition is one involving a certain amount of slight chronic inflammation . . .
> >
> > A slight cause of irritation . . . is commonly accompanied by pulling out a cigarette, and getting a little compensation for life's minor ills in that way.
> > And . . . is not unlikely to be associated with smoking more frequently."
>
> Ironically, the proof that smoking indeed is the cause of many cancers came from experiments conducted following the principles of experiment design and analysis that Fisher himself developed—and that we'll see in Chapter 10.
>
> *Source: Cigarettes, Cancer, And Statistics Centennial Review, 2: 151-166 (1958)*

Scatterplots and correlation coefficients *never* prove causation. That's one reason it took so long for warning labels to appear on cigarettes. Although there was plenty of evidence that increased smoking was *associated* with increased levels of lung cancer, it took years to provide evidence that smoking actually *causes* lung cancer.

A scatterplot of the damage (in dollars) caused to a house by fire would show a strong correlation with the number of firefighters at the scene. Surely the financial damage doesn't cause firefighters. And firefighters do seem to cause damage, spraying water all around and chopping holes. Does that mean we shouldn't call the fire department? Of course not. There is an underlying variable that leads to both more damage and more firefighters: the size of the blaze. We may say that damage and firefighters are exhibiting a **common response** to the size of the blaze (Figure 6.8).

Figure 6.8

Amount of damage and number of firefighters exhibit a common response to the size of the blaze.

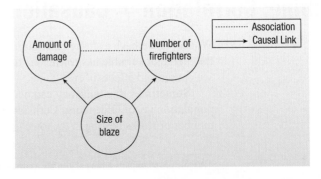

A hidden variable that stands behind a relationship but that may affect or change our understanding of the relationship is called a **lurking variable**. You can often debunk cause-and-effect claims made about data by finding a variable lurking behind the scenes.

In the example of cancer and smoking, while it is now well established that smoking causes cancer, what might have been some other possible explanations for the observed association? Maybe cancer causes increased smoking (as suggested by Fisher). Maybe both cancer and smoking are a common response to some gene(s), making one genetically predisposed to both developing cancer and nicotine addiction. Or—one more theory—maybe smokers differ from nonsmokers in certain lifestyle characteristics (such as alcohol consumption, diet, exercise, stress level, etc.) and perhaps it is some of those lifestyle differences that cause the cancer, not the smoking. Figure 6.9 shows such a *confounding* of effects.

Figure 6.9

The effect of smoking on cancer is confounded with the effect of certain lifestyle differences.

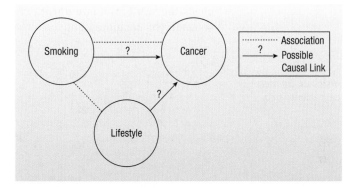

The effects of *Lifestyle* and *Smoking*, on cancer, are **confounded**, making it difficult to determine which (if either) of the two confounded variables is the causal culprit. Maybe both have a causal link to cancer, maybe neither. Maybe some other undiscovered lurking variable is the real culprit, its hidden effects confounded with smoking habits. The causal connection between smoking and lung cancer was eventually established through great effort over many years, via many different types of corroborative investigations. Proving causation in the absence of controlled experiments is exceedingly difficult.

Section 6.3 Exercises

7. **Bookstore sales again** A larger firm is considering acquiring the bookstore of Exercise 3. An analyst for the firm, noting the relationship seen in Exercise 3, suggests that when they acquire the store they should hire more people because that will drive higher sales. Is his conclusion justified? What alternative explanations can you offer? Use appropriate statistics terminology.

8. **Blizzards** A study finds that during blizzards, online sales are highly associated with the number of snow plows on the road; the more plows, the more online purchases. The director of an association of online merchants suggests that the organization should encourage municipalities to send out more plows whenever it snows because, he says, that will increase business. Comment.

*6.4 Straightening Scatterplots

Correlation is a suitable measure of strength for straight relationships only. When a scatterplot bends but shows a steadily increasing pattern or a steadily decreasing pattern, we can often straighten the form of the plot by re-expressing (aka *transforming*) one or both variables.

Some camera lenses have an adjustable aperture, the hole that lets the light in. The size of the aperture is expressed by a mysterious number called the f/stop. Each increase of one f/stop number corresponds to a halving of the light that is allowed to come through. The f/stops of one digital camera are

f/stop:	2.8	4	5.6	8	11	16	22	32

When you halve the shutter speed, you cut down the light, so you have to open the aperture one notch. We could experiment to find the best f/stop value for each shutter speed. A table of recommended shutter speeds and f/stops for a camera lists the relationship like this:

Shutter speed:	1/1000	1/500	1/250	1/125	1/60	1/30	1/15	1/8
f/stop:	2.8	4	5.6	8	11	16	22	32

The correlation of these shutter speeds and f/stops is 0.979. That sounds pretty high. But a high correlation doesn't necessarily mean a *linear* relationship. When we check the scatterplot (we *always* check the scatterplot) in Figure 6.10, it shows that something is not quite right:

Figure 6.10

A scatterplot of *f/stop* vs. *Shutter Speed* shows a bent relationship.

We can see that the f/stop is not *linearly* related to the shutter speed. Can we find a transformation of *f/stop* that straightens out the line?

In Chapter 4, we saw that **re-expressing** data by raising each value to a power or taking the logarithm couldmake the shape of the histogram of the data more nearly symmetric and could make the spreads of different groups more nearly equal. Now we hope to make a curved relationship straighter. Remarkably, these three goals are closely related, so we can use the same methods to achieve all of them.

The Ladder of Powers

One common approach is to select a re-expression from a simple collection of functions— the powers and the logarithm. We can raise each data value in a quantitative variable to the same power. It is important to recall a few facts about powers:

- The ½ power is the same as taking the square root.
- The −1 power is the reciprocal: $y^{-1} = 1/y$
- Putting those together, $y^{-1/2} = 1/\sqrt{y}$

In addition, we'll use the logarithm in place of the "0" power. Although it doesn't matter for this purpose, the base–10 logarithm is usually easier to think about.[11]

Where to start? It turns out that certain kinds of data are more likely to be helped by particular re-expressions. Knowing that gives you a good place to start your search for a re-expression.

> **Re-expressions in scientific laws** Scientific laws often include simple re-expressions. For example, in Psychology, Fechner's Law states that sensation increases as the logarithm of stimulus intensity ($S = k \log R$).

---

[11]You may have learned facts about calculating with logarithms. Don't worry; you won't need them here.

Power	Name	Comment
2	The square of the data values, y^2.	Try this for unimodal distributions that are skewed to the left or to re-express y in a scatterplot that bends downward.
1	The raw data—no change at all. This is "home base." The farther you step from here up or down the ladder, the greater the effect.	Data that can take on both positive and negative values with no bounds are less likely to benefit from re-expression.
1/2	The square root of the data values, $\sqrt{y}$.	Counts often benefit from a square root re-expression. For counted data, start here.
"0"	Although mathematicians define the "0-th" power differently,[12] for us the place is held by the logarithm. You may feel uneasy about logarithms. Don't worry; the computer or calculator does the work.[13]	Measurements that cannot be negative, and especially values that grow by percentage increases such as salaries or population sizes, often benefit from a log re-expression. When in doubt, start here. If your data have zeros, try adding a small constant to each value before finding the logs.
−1/2	The reciprocal square root, $1/\sqrt{y}$.	This is an uncommon re-expression, but sometimes useful. Change the sign and take the *negative* of the reciprocal square root if you want to preserve the direction of relationships.
−1	The reciprocal, $1/y$.	Ratios of two quantities (kilometres per hour, for example) often benefit from a reciprocal. (You have about a 50–50 chance that the original ratio was taken in the "wrong" order for simple statistical analysis and would benefit from re-expression.) Often, the reciprocal will have simple units (hours per km). Put a *negative* sign in front if you want to preserve the direction of relationships. If your data have zeros, try adding a small constant to all values before finding the reciprocal.

The powers and logarithm have an important advantage as re-expressions: They are *invertible*. That is, you can get back to the original values easily. For example, if we re-express y by the square root, we can get back to the original values by squaring the re-expressed values. To undo a logarithm, make the value a power of 10.[14]

The positive powers and the logarithm preserve the order of values—the larger of two values is still the larger one after re-expression. Negative powers reverse the order (4 > 2 but ¼ < ½) so we sometimes negate the values (−¼ > −½) to preserve the order.

Most important, the *effect* that any of these re-expressions has on straightening a scatterplot is ordered. Choose a power. If it doesn't straighten the plot by enough, then try the next power. If it overshoots so the plot bends the other way, then step back. That insight makes it easy to find an effective re-expression, and it leads to referring to these functions as the **ladder of powers**. With modern technology, finding a suitable re-expression is no harder than the push of a button.

f/Stops Again

Following the advice in the table, we can try taking the logarithm of the *f/stop* values:

Figure 6.11

Log(f/stop) values vs. *Shutter Speed*.

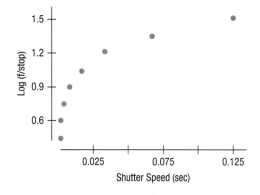

---

[12]You may remember that for any nonzero number y, $y^0 = 1$ This is not a very exciting re-expression for data; every data value would be the same. We use the logarithm in its place.

[13]Your calculator or software package probably gives you a choice between "base 10" logarithms and "natural (base e)" logarithms. Don't worry about that. It doesn't matter which you use; they have exactly the same effect on the data. If you want to choose, base 10 logarithms can be a bit easier to interpret.

[14]Your computer or calculator knows how.

But, as Figure 6.11 above shows, the logarithm of *f/stop* values is even *more* curved vs. *Shutter Speed*. The ladder of powers says that we'd do better if we move in the opposite direction. That would take us to the square.

That scatterplot looks much straighter Figure 6.12.[15] It's appropriate to summarize the strength of this relationship with a correlation, which turns out to be 0.998.[16] (If the square was in the right direction but not strong enough, you might next try the stronger exponential function, though not in our ladder.)

Figure 6.12

(f/stop)² vs. *Shutter Speed*.

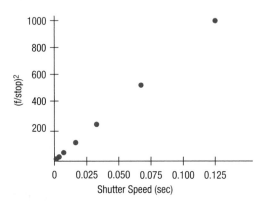

Could we also try re-expressing the *x*-variable? Sure, or even both variables. In fact here, taking the square root of *Shutter Speed* works just as well! And surprisingly often, taking logs of both variables can improve a scatterplot.

But watch out for scatterplots that turn around. Re-expression can straighten many bent relationships, but not those that go up and then down or down and then up. You should refuse to analyze such data with methods that require a roughly linear form.

Section 6.4 Exercises

*9. **Salaries and logs** For an analysis of the salaries of your company, you plot the salaries of all employees against the number of years they have worked for the company. You find that plotting the base-10 logarithm of salary makes the plot much straighter. A part-time shipping clerk, who has worked at the company for one year earns $10 000. A manager earns $100 000 after 15 years with the firm. The CEO, who founded the company 30 years ago, receives $1 000 000. What are the values you will plot? Will the plot of these three points be straight enough?

*10. **Dexterity scores** Scores on a test of dexterity are recorded by timing a subject who is inverting pegs by picking them up with one hand, manipulating them to turn them over, and then placing them back in a frame. A typical 4-year-old needs about 2.5 seconds to invert a peg, but a 9-year-old takes 2 seconds and a 12-year-old can do it in 1.5 seconds. Plot these three points. Now try the reciprocal re-expression $(1/y)$ vs. age. Which version is straighter?

---

[15]Sometimes, we can do a "reality check" on our choice of re-expression. In this case, a bit of research reveals that f/stops are related to the diameter of the open shutter. Since the amount of light that enters is determined by the *area* of the open shutter, which is related to the diameter by squaring, the square re-expression seems reasonable. Not all re-expressions have such nice explanations, but it's a good idea to think about them.

[16]That's a high correlation, but it does not confirm that we have the best re-expression. After all, the correlation of the original relationship was 0.979—high enough to satisfy almost anyone. To judge a re-expression, look at the scatterplot. Your eye is a better judge of straightness than any statistic.

WHAT CAN GO WRONG?

- **Don't say "correlation" when you mean "association."** How often have you heard the word "correlation"? Chances are pretty good that when you've heard the term, it's been misused. It's one of the most widely misused Statistics terms, and given how often statistics are misused, that's saying a lot. One of the problems is that many people use the specific term correlation when they really mean the more general term *association*.

    "Association" is a deliberately vague term describing the relationship between two variables. "Correlation" is a precise term that measures the strength and direction of the linear relationship between quantitative variables.

- **Don't correlate categorical variables.** Did you know that there's a strong correlation between playing an instrument and drinking coffee? No? One reason might be that the statement doesn't make sense. People who misuse the term "correlation" to mean "association" often fail to notice whether the variables under discussion are quantitative. But correlation is valid only for *quantitative* variables. Be sure to check this condition.

- **Don't confuse correlation with causation.** One of the most common mistakes people make in interpreting statistics occurs when they observe a high correlation between two variables and jump to the perhaps tempting conclusion that one thing must be causing the other. Scatterplots and correlations *never* demonstrate causation. At best, these statistical tools can only reveal an association between variables, and that's a far cry from establishing cause and effect.

- **Make sure the association is linear.** Not all associations between quantitative variables are linear. Correlation can miss even a strong nonlinear association. A student project evaluating the quality of brownies baked at different temperatures reports a correlation of −0.05 between judges' scores and baking temperature. That seems to say there is no relationship until we look at the scatterplot:

The relationship between brownie taste *Score and Baking Temperature is* strong, but not at all linear.

There is a strong association, but the relationship is not linear. Don't forget to check the **Straight Enough Condition**.

- **Don't assume the relationship is linear just because the correlation coefficient is high.** The correlation of f/stops and shutter speeds is 0.979 and yet the relationship is clearly not straight. A high correlation is no guarantee of straightness. Nor it is safe to use just the correlation to judge the best re-expression. It's always important to look at the scatterplot.

A scatterplot of *f/stop* vs. *Shutter Speed* shows a bent relationship even though the correlation is $r = 0.979$.

(*continued*)

AP Images

- **Beware of outliers.** You can't interpret a correlation coefficient safely without a background check for outliers (and subgroups). Here's a silly example:

The relationship between IQ and shoe size among comedians shows a surprisingly strong positive correlation of 0.50. To check assumptions, we look at the scatterplot.

What is the relationship? The correlation is 0.50, but it is all due to the outlier (the green x) in the upper right corner. Who does that point represent?

The outlier is Bozo the Clown, known for his large shoes and widely acknowledged to be a comic "genius." Without Bozo the correlation is near zero.

Even a single outlier can dominate the correlation value because its z-scores will be large and their product even larger. That's why you need to check the scatterplot for outliers.

What Have We Learned?

Make a scatterplot to display the relationship between two quantitative variables.

- Look at the direction, form, and strength of the relationship, and for any outliers that stand away from the overall pattern. Look for distinct subgroups.

Provided the form of the relationship is straight, summarize its strength with a correlation, r.

- The sign of the correlation gives the direction of the relationship.
- $-1 \leq r \leq 1$; A correlation of 1 or -1 is a perfect linear relationship. A correlation of 0 indicates that there is no linear relationship.
- Correlation has no units, so shifting or scaling the data, standardizing, or even swapping the variables has no effect on the numerical value.

A large correlation is not a sign of a causal relationship.

Recognize nonlinearity, and straighten plots when possible.

- If a scatterplot of y vs. x isn't straight, correlation isn't appropriate.
- Re-expressing (transforming) one or both variables can often improve the straightness of the relationship.
- The powers, roots, and the logarithm provide an ordered collection of re-expressions so you can search up and down the *ladder of powers* to find an appropriate one.

Review of Terms

Scatterplot   A graph plotting one variable versus another, that shows the relationship between two quantitative variables measured on the same cases (p. 177).

Association
- **Direction:** A **positive** direction or association means that, in general, as one variable increases, so does the other. When increases in one variable generally correspond to decreases in the other, the association is **negative** (p. 179).

- **Form:** The simplest form is straight (**linear**), but you should describe any other patterns you see in scatterplots (p. 179).
- **Strength:** A scatterplot is said to show a strong association if there is little scatter around the underlying relationship (p. 179).

Outlier   A point that does not fit the overall pattern seen in the scatterplot (p. 179).

Response variable, explanatory variable, *y*-variable, *x*-variable   In a scatterplot, you may need to choose a role for each variable. Assign the response to the *y*-axis, the variable that you hope to predict or explain. Assign to the *x*-axis the explanatory or predictor variable that accounts for, explains, predicts, or is otherwise responsible for the *y*-variable (p. 181).

Correlation coefficient   A numerical measure of the direction and strength of a linear association (p. 185).

$$r = \frac{\sum z_x z_y}{n - 1}$$

Common response   Changes in both *x* and *y* are caused by a lurking variable (p. 190).

Lurking variable   A variable not present in our analysis that may influence our understanding of the relationship between *x* and *y* (p. 191).

Confounded   Variables whose effects on the response variable, *y*, are entangled and difficult to distinguish are said to be *confounded*. (p. 191).

Re-expression   We re-express data by taking the logarithm, the square root, the reciprocal, or some other mathematical operation of all values of a variable (p. 192).

Ladder of Powers   Places many re-expressions in order according to the magnitude of their effects on the data (p. 193).

On the Computer SCATTERPLOTS AND CORRELATION

Statistics packages generally make it easy to look at a scatterplot to check whether the correlation is appropriate. Some packages make this easier than others.

Many packages allow you to modify or enhance a scatterplot, altering the axis labels, the axis numbering, the plot symbols, or the colours used. Some options, such as colour and symbol choice, can be used to display additional information on the scatterplot.

EXCEL

To make a scatterplot in Excel:

- Select the columns of data to use in the scatterplot. You can select more than one column by holding down the control key while clicking.
- In the Insert tab, click on the **Scatter** button and select the **Scatter with only Markers** chart from the menu.

Unfortunately, the plot this creates is often statistically useless. To make the plot useful, we need to change the display:

- With the chart selected, click on the **Gridlines** button in the Layout tab to cause the Chart Tools tab to appear.
- Within Primary Horizontal Gridlines, select **None**. This will remove the gridlines from the scatterplot.

- To change the axis scaling, click on the numbers of each axis of the chart, and click on the **Format Selection** button in the Layout tab.
- Select the **Fixed** option instead of the Auto option, and type a value more suited for the scatterplot. You can use the pop-up dialog window as a straightedge to approximate the appropriate values.

Excel automatically places the leftmost of the two columns you select on the *x*-axis, and the rightmost one on the *y*-axis. If that's not what you'd prefer for your plot, you'll want to switch them.

(continued)

To switch the X- and Y-variables:

- Click the chart to access the **Chart Tools** tabs.
- Click on the **Select Data** button in the Design tab.
- In the pop-up window's Legend Entries box, click on **Edit**.
- Highlight and delete everything in the Series X Values line, and select new data from the spreadsheet. (Note that selecting the column would inadvertently select the title of the column, which would not work well here.)
- Do the same with the Series Y Values line.
- Press **OK**, then press **OK** again.

To find the correlation in Excel:

- Click on a blank cell in the spreadsheet.

- Go to the **Formulas** tab in the Ribbon and click **More Functions** —>**Statistical**.
- Choose the **CORREL** function from the drop-down menu of functions.
- In the dialogue that pops up, enter the range of one of the variables in the space provided.
- Enter the range of the other variable in the space provided.
- Click OK

COMMENTS

The correlation is computed in the selected cell. Correlations computed this way will update if any of the data values are changed.

JMP

To make a scatterplot and compute correlation:

- Choose **Fit Y by X** from the **Analyze** menu.
- In the Fit Y by X dialogue, drag the *y* variable into the "Y, Response" box, and drag the *x* variable into the "X, Factor" box.
- Click the **OK** button.

Once JMP has made the scatterplot, click on the red triangle next to the plot title to reveal a menu of options:

- Select **Density Ellipse** and select 0.95. JMP draws an ellipse around the data and reveals the **Correlation** tab.
- Click the blue triangle next to Correlation to reveal a table containing the correlation coefficient.

MINITAB

To make a scatterplot:

- Choose **Scatterplot** from the **Graph** menu.
- Choose "Simple" for the type of graph. Click **OK**.
- Enter variable names for the Y variable and X variable into the table. Click **OK**.

To compute a correlation coefficient:

- Choose **Basic Statistics** from the **Stat** menu.

- From the Basic Statistics submenu, choose **Correlation**. Specify the names of at least two quantitative variables in the "Variables" box.
- Click **OK** to compute the correlation table.

COMMENTS

In MINITAB EXPRESS, use Statistics > Correlation

R

To make a scatterplot of two variables, X and Y:

- **plot(X,Y)** or equivalently **plot(Y~X)** will produce the plot.
- **cor(X,Y)** finds the correlation.

COMMENTS

Your variables X and Y may be variables in a data frame. If so, and DATA is the name of the data frame, then you will need to attach the data frame, or use

with(DATA,plot(Y~X)) or **plot(Y~X,data=DATA)**.

SPSS

To make a scatterplot in SPSS:

- Open the Chart Builder from the **Graphs** menu.
- Click the **Gallery** tab.
- Choose Scatterplot from the list of chart types.
- Drag the scatterplot onto the canvas.
- Drag a scale variable you want as the response variable to the *y*-axis drop zone.
- Drag a scale variable you want as the factor or predictor to the *x*-axis drop zone.
- Click **OK**.

To compute a correlation coefficient:

- Choose **Correlate** from the **Analyze** menu.
- From the Correlate submenu, choose **Bivariate**.
- In the Bivariate Correlations dialogue, use the arrow button to move variables between the source and target lists.

Make sure the **Pearson** option is selected in the Correlation Coefficients field.

STATCRUNCH

To make a scatterplot:

- Click on **Graphics**.
- Choose **Scatter Plot**.
- Choose **X** and **Y** variable names from the list of **Columns**.
- Click on **Create Graph**.

To find a correlation:

- Click on **Stat**.
- Choose **Summary Stats » Correlation**.
- Choose two variable names from the list of **Columns**. (You may need to hold down the ctrl or command key to choose the second one.)
- Click on **Calculate**.

TI-83/84 PLUS

To create a scatterplot:

- Set up the **STAT PLOT** by choosing the **scatterplot** icon (the first option). Specify the lists where the data are stored as Xlist and Ylist.
- Set the graphing **WINDOW** to the appropriate scale and **GRAPH** (or take the easy way out and just ZoomStat!).

To find the correlation:

- Go to **STAT CALC** menu and select **8: LinReg(a + bx)**.
- Then specify the lists where the data are stored. The final command you will enter should look like **LinReg(a + bx) L1, L2**.

COMMENTS

Notice that if you **TRACE** the scatterplot, the calculator will tell you the *x*- and *y*-value at each point.

If the calculator does not tell you the correlation after you enter a LinReg command, try this: Hit **2nd CATALOG**. You now see a list of everything the calculator knows how to do. Scroll down until you find **DiagnosticOn**. Hit **ENTER** twice. (It should say Done.) Now and forevermore (or until you change batteries), you can find a correlation by using your calculator.

Chapter Exercises

T indicates data is available in a saved file at the Companion Website at www.pearsoncanada.ca/deveaux

11. **Association III** Suppose you were to collect data for each pair of variables given below. You want to make a scatterplot. Which variable would you use as the explanatory variable and which as the response variable? Why? What would you expect to see in the scatterplot? Discuss the likely direction, form, and strength.
    a) When climbing a mountains: altitude, temperature
    b) For each week: ice cream cone sales, air-conditioner sales

    c) People: age, grip strength
    d) Drivers: blood alcohol level, reaction time

12. **Association IV** Suppose you were to collect data for each pair of variables given below. You want to make a scatterplot. Which variable would you use as the explanatory variable and which as the response variable? Why? What would you expect to see in the scatterplot? Discuss the likely direction, form, and strength.
    a) Long-distance calls: time (minutes), cost

b) Lightning strikes: distance from lightning, time delay of the thunder

c) A streetlight: its apparent brightness, your distance from it

d) Cars: weight of car, age of owner

13. Scatterplots Which of the scatterplots below show

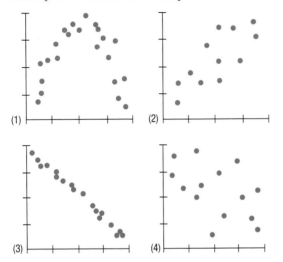

(1)   (2)   (3)   (4)

a) little or no association?
b) a negative association?
c) a linear association?
d) a moderately strong association?
e) a very strong association?

14. Scatterplots II Which of the scatterplots below show

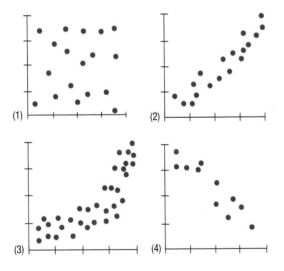

(1)   (2)   (3)   (4)

a) little or no association?
b) a negative association?
c) a linear association?
d) a moderately strong association?
e) a very strong association?

T 15. Performance IQ scores versus brain size A study examined brain size (measured as pixels counted in a digitized magnetic resonance image [MRI] of a cross-section of the brain) and IQ (four performance scales of the Weschler IQ test) for university students. The scatterplot shows the performance IQ scores versus brain size. Comment on the association between brain size and IQ as seen in this scatterplot.

T 16. Kentucky Derby 2014 The fastest horse in Kentucky Derby history was Secretariat in 1973. The scatterplot shows speed (in miles per hour) of the winning horses each year.

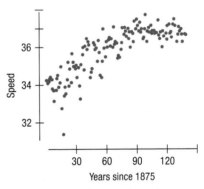

What do you see? In most sporting events, performances have improved and continue to improve, so surely we anticipate a positive direction. But what of the form? Has the performance increased at the same rate throughout the past 140 years?

17. Firing pottery A ceramics factory can fire eight large batches of pottery a day. Sometimes in the process a few of the pieces break. To better understand the problem, the factory records the number of broken pieces in each batch for three days and then creates the scatterplot shown.

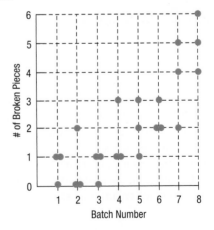

a) Make a histogram showing the distribution of the number of broken pieces in the 24 batches of pottery examined.

b) Describe the distribution as shown in the histogram. What feature of the problem is more apparent in the histogram than in the scatterplot?

c) What aspect of the company's problem is more apparent in the scatterplot?

d) In the histogram, "batch" was a "lurking variable," lurking behind the data display, and not visible. Plotting the number of broken pieces against batch helped to explain the variation in breakage. Can you think of any other lurking variables to investigate here?

18. Coffee sales Owners of a new coffee shop tracked sales for the first 20 days and displayed the data in a scatterplot (by day):

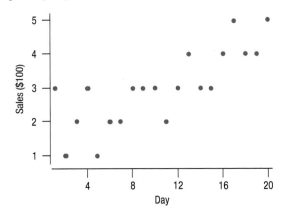

a) Make a histogram of the daily sales since the shop has been in business.

b) State one fact that is obvious from the scatterplot but not from the histogram.

c) State one fact that is obvious from the histogram but not from the scatterplot.

19. Matching Here are several scatterplots. The calculated correlations are −0.923, −0.487, 0.006, and 0.777. Which is which?

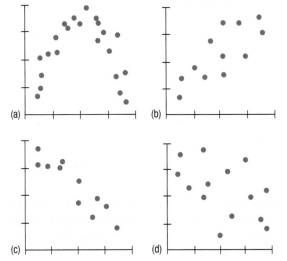

20. Matching II Here are several scatterplots. The calculated correlations are −0.977, −0.021, 0.736, and 0.951. Which is which?

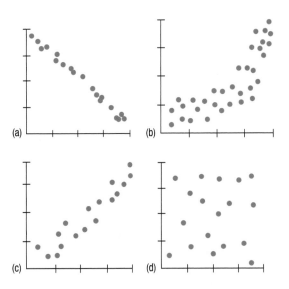

T 21. Roller coasters 2014 Most roller coasters get their speed by dropping down a steep initial incline, so it makes sense that the height of that drop might be related to the speed of the coaster. Here's a scatterplot of top *Speed* and largest *Drop* for 108 roller coasters around the world.

a) Does the scatterplot indicate that it is appropriate to calculate the correlation? Explain.

b) In fact, the correlation of *Speed* and *Drop* is 0.937. Describe the association.

T 22. Antidepressants A study compared the effectiveness of several antidepressants by examining the experiments in which they had passed the FDA requirements. Each of those experiments compared the active drug to a placebo, an inert pill given to some of the subjects. In each experiment, some patients treated with the placebo had improved, a phenomenon called the *placebo effect*. Patients' depression levels were evaluated on the Hamilton Depression Rating Scale, where larger numbers indicate greater improvement. The scatterplot shown compares mean improvement levels for the antidepressants and placebos for several experiments.

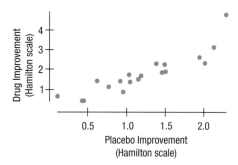

a) Is it appropriate to calculate the correlation? Explain.
b) The correlation is 0.898. Explain what you have learned about the results of these experiments.

T 23. Lunchtime Does how long children remain at the lunch table help predict how much they eat? The table gives data on 20 toddlers observed over several months at a nursery school. "Time" is the average number of minutes a child spent at the table when lunch was served. "Calories" is the average number of calories the child consumed during lunch, calculated from careful observation of what the child ate each day.

Calories	Time
472	21.4
498	30.8
465	37.7
456	33.5
423	32.8
437	39.5
508	22.8
431	34.1
479	33.9
454	43.8
450	42.4
410	43.1
504	29.2
437	31.3
489	28.6
436	32.9
480	30.6
439	35.1
444	33.0
408	43.7

a) Find the correlation for these data.
b) Suppose we were to record time at the table in hours rather than in minutes. How would the correlation change? Why?

c) Write a sentence or two explaining what this correlation means for these data. Remember to write about food consumption by toddlers rather than about correlation coefficients.
d) One analyst concluded, "It is clear from this correlation that toddlers who spend more time at the table eat less. Evidently something about being at the table causes them to lose their appetites." Explain why this explanation is not an appropriate conclusion from what we know about the data.

T 24. Vehicle weights The Minnesota Department of Transportation hoped that they could measure the weights of big trucks without actually stopping the vehicles by using a newly developed "weight-in-motion" scale. To see if the new device was accurate, they conducted a calibration test. They weighed several trucks when stopped (*Static Weight*), assuming that this weight was correct. Then they weighed the trucks again while they were moving to see how well the new scale could estimate the actual weight. Their data are given in the table.

Weight of a Truck (thousands of pounds)	
Weight-in-Motion	**Static Weight**
26.0	27.9
29.9	29.1
39.5	38.0
25.1	27.0
31.6	30.3
36.2	34.5
25.1	27.8
31.0	29.6
35.6	33.1
40.2	35.5

a) Make a scatterplot for these data.
b) Describe the direction, form, and strength of the plot.
c) Write a few sentences telling what the plot says about the data. (*Note*: The sentences should be about weighing trucks, not about scatterplots.)
d) Find the correlation.
e) If the trucks were weighed in kilograms, how would this change the correlation?
(1 kilogram = 2.2 pounds)
f) Do any points deviate from the overall pattern? What does the plot say about a possible recalibration of the weight-in-motion scale?

T 25. Fuel economy 2014 Here are advertised engine size (in litres) and gas mileage (estimated combined city and highway, in miles per gallon) for some 2014 vehicles. (www.kbb.com)

Manufacturer	Model	Eng Displ	Comb MPG
Audi	A4 quattro	2	26
LEXUS	IS 250	2.5	24
Mercedes- Benz	CL 63	5.5	18
Chrysler	Mini Cooper Paceman	1.6	31
BMW	ActiveHybrid 5	3	26
Nissan	SENTRA	1.8	30
Ford	TAURUS AWD FFV	3.5	21
Porsche	Panamera GTS	4.8	19
Subaru	IMPREZA WAGON AWD	2	28
GMC	TERRAIN FWD	2.4	26
Jeep	Cherokee FWD	2.4	25
BMW	X3 xDrive35i	3	21
Chevrolet	EQUINOX AWD	3.6	19
Volvo	XC60 AWD	3.2	20
Chevrolet	C10 TAHOE 2WD	5.3	17
Porsche	Boxster S	3.4	23
BMW	650i Coupe	4.4	20
Chevrolet	SONIC	1.4	31
Volkswagen	BEETLE	2	26
Chevrolet	SONIC 5	1.4	33
Kia	OPTIMA	2.4	27
Mercedes-Benz	E 350	3.5	24
Toyota	AVALON HYBRID	2.5	40
Chevrolet	IMPALA	2.4	29
Chrysler	Charger AWD	3.6	21
Porsche	Panamera Turbo Executive	4.8	18
Volkswagen	JETTA SPORTWAGEN	2	34
Mercedes-Benz	E 350 4Matic (Wagon)	3.5	21
Nissan	TITAN 4WD	5.6	14
GMC	H1500 SAVANA AWD PASS	5.3	14
Nissan	QX60 AWD	3.5	21
Ford	MKT AWD	3.5	18
Nissan	QX60 AWD	2.5	26
Jeep	Grand Cherokee SRT8	6.4	15
Toyota	GX 460	4.6	17

a) Make a scatterplot for these data.
b) Describe the direction, form, and strength of the plot.
c) Find the correlation between engine size and miles per gallon.
d) Write a few sentences telling what the plot says about fuel economy.

T **26. Drug abuse** A survey was conducted in the United States and 10 countries of Western Europe to determine the per-centage of teenagers who had used marijuana and other drugs. The results are summarized in the following table.

Country	Percent Who Have Used	
	Marijuana	Other Drugs
Czech Rep.	22	4
Denmark	17	3
England	40	21
Finland	5	1
Ireland	37	16
Italy	19	8
No. Ireland	23	14
Norway	6	3
Portugal	7	3
Scotland	53	31
United States	34	24

a) Create a scatterplot.
b) What is the correlation between the percent of teens who have used marijuana and the percent who have used other drugs?
c) Write a brief description of the association.
d) Do these results confirm that marijuana is a "gateway drug"—that is, that marijuana use leads to the use of other drugs? Explain.

T **27. Burgers** Fast food is often unhealthy because much of it is high in both fat and sodium. But are the two related? Here are the fat and sodium contents of several brands of burgers. Analyze the association between fat content and sodium.

Fat (g)	19	31	34	35	39	39	43
Sodium (mg)	920	1500	1310	860	1180	940	1260

T **28. Burgers II** In the previous exercise, you analyzed the association between the amounts of fat and sodium in fast food hamburgers. What about fat and calories? Here are data for the same burgers. Analyze the association using correlation and scatterplots.

Fat (g)	19	31	34	35	39	39	43
Calories	410	580	590	570	640	680	660

T **29. Attendance 2013** American League baseball games are played under the designated hitter rule, meaning that pitchers, often weak hitters, do not come to bat. Baseball owners believe that the designated hitter rule means more runs scored, which in turn means higher atten-dance. Is there evidence that more fans attend games if the teams score more runs? Data collected from American League games during the 2013 season indicate a correlation of 0.278 between runs scored and the number of people at the game. (espn.go.com/mlb/attendance)

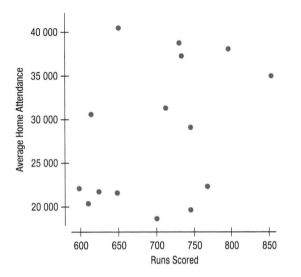

a) Does the scatterplot indicate that it's appropriate to calculate a correlation? Explain.

b) Describe the association between attendance and runs scored.

c) Does this association prove that the owners are right that more fans will come to games if the teams score more runs?

T 30. **Second inning 2013** Perhaps fans are just more interested in teams that win. The displays below are based on American League teams for the 2013 season. (espn.go.com) (Use the data set Attendance_2013).

	Wins	Runs	Attendance
Wins	1.000		
Runs	0.723	1.000	
Attendance	0.296	0.278	1.000

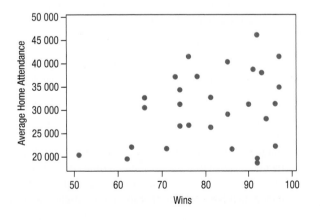

a) Do winning teams generally enjoy greater attendance at their home games? Describe the association.

b) Is attendance more strongly associated with winning or scoring runs? Explain.

c) How strongly is scoring runs associated with winning games?

31. **Politics** A candidate for office claims that "there is a correlation between television watching and crime." Criticize this statement in statistical terms.

32. **Association V** A researcher investigating the association between two variables collected some data and was surprised when he calculated the correlation. He had expected to find a fairly strong association, yet the correlation was near 0. Discouraged, he didn't bother making a scatterplot. Explain to him how the scatterplot could still reveal the strong association he anticipated.

33. **Height and reading** A researcher studies children in elementary school and finds a strong positive linear association between height and reading scores.

a) Does this mean that taller children are generally better readers?

b) What might explain the strong correlation?

34. **Smart phones and life expectancy** A survey of the world's nations in 2010 shows a strong positive correlation between percentage of the country using smart phones and life expectancy in years at birth.

a) Does this mean that smart phones are good for your health?

b) What might explain the strong correlation?

T 35. **Hard water** In a study of streams in the Adirondack Mountains, the following relationship was found between the pH of the water and the water's hardness (measured in grains):

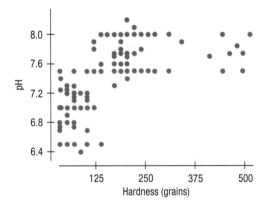

Is it appropriate to summarize the strength of association with a correlation? Explain.

36. Traffic headaches A study of traffic delays in 68 U.S. cities found the following relationship between total delays (in total hours lost) and mean highway speed:

Is it appropriate to summarize the strength of association with a correlation? Explain.

37. Correlation errors Your Economics instructor asks the class to investigate factors associated with the gross domestic product (GDP) of nations. Each student examines a different factor (life expectancy, literacy rate, etc.) for a few countries and reports to the class. Apparently, some of your classmates do not understand statistics very well because you know several of their conclusions are incorrect. Explain the mistakes in their statements below.
a) "My correlation of −0.772 shows that there is almost no association between GDP and infant mortality rate."
b) "There was a correlation of 0.44 between GDP and continent."

38. More correlation errors Students in the Economics class discussed in Exercise 37 also wrote these conclusions. Explain the mistakes they made.
a) "There was a very strong correlation of 1.22 between life expectancy and GDP."
b) "The correlation between literacy rate and GDP was 0.83. This shows that countries wanting to increase their standard of living should invest heavily in education."

39. Baldness and heart disease Medical researchers followed 1435 middle-aged men for a period of five years, measuring the amount of baldness present (none = 1, little = 2, some = 3, much = 4, extreme = 5) and presence of heart disease (no = 0, yes = 1). They found a correlation of 0.089 between the two variables. Comment on their conclusion that this shows that baldness is not a possible cause of heart disease.

40. Sample survey A California polling organization is checking its database to see if the three data sources they used sampled the same zip codes (five-digit U.S. postal codes). The variable *Datasource* = 1 if the data source is MetroMedia, 2 if the data source is DataQwest, and 3 if it's RollingPoll. The organization finds that the correlation between *Zip Code* and *Datasource* is −0.0229. It concludes that the correlation is low enough to state that there is no dependency between zip code and the source of data. Comment.

T 41. Thrills 2011 Since 1994, the Best Roller Coaster Poll (www.ushsho.com/bestrollercoasterpoll.htm) has been ranking the world's best roller coasters. In 2011, Bizarro earned the top steel coaster rank for the sixth straight year. Here are data on the top 10 steel coasters from this poll:

Rank	Roller Coaster	Park	Location	Initial Drop (ft.)	Duration (sec)	Height (ft.)	Max Speed (mph)	Max Vert Angle (degrees)	Length (ft.)
1	Bizarro	Six Flags New England	MA	221	155	208	77	72	5400
2	Expedition GeForce	Holiday Park	DE	184	75	188	74.6	82	4003
3	Intimidator 305	Kings Dominion	VA	300	98	305	93	85	5100
4	Kawasemi	Tobu Zoo	JP		60	108	54	67.4	2454
5	Nemesis	Alton Towers	UK	104	80	43	50	40	2349
6	Piraten	Djurs Sommerland	DK	100	61	105	56	70	2477
7	Goliath	Walibi World	NL	152	92	155	66.9	70	3984
8	Millennium Force	Cedar Point	OH	300	120	310	93	80	6595
9	Katun	Mirabilandia	IT	148	142	169	65		3937
10	iSpeed	Mirabilandia	IT	181	60	180	74.6	90	3281

What do these data indicate about the *Length* of the track and the *Duration* of the ride you can expect?

42. Thrills II For the roller coaster data in Exercise 41:
a) Examine the relationship between *Initial Drop* and *Max Speed*.
b) Examine the relationship between *Initial Drop* and *Height*.
c) What conclusions can you safely draw about the initial drop of a roller coaster? Is *Initial Drop* strongly correlated with other variables as well?

43. Thrills III For the roller coaster data in Exercise 41:
a) Explain why in looking for a variable that explains rank, you will be hoping for a negative correlation.
b) Do any of the provided variables provide a strong predictor for roller coaster rank?
c) What other (unaccounted for) variables might help explain the rank?

44. Ontario migration 2015 Below are data showing the net interprovincial migration each year for Ontario (number arriving minus number leaving):

Year	Ontario Net Migration	Year	Ontario Net Migration	Year	Ontario Net Migration
1979	−4 325	1992	−11 045	2005	−11 172
1980	−22 362	1993	−14 189	2006	−17 501
1981	−33 247	1994	−9 420	2007	−20 047
1982	−5 665	1995	−2 841	2008	−14 750
1983	23 585	1996	−2 822	2009	−15 601
1984	36 400	1997	1 977	2010	−4 662
1985	33 885	1998	9 231	2011	−4 007
1986	33 562	1999	16 706	2012	−10 611
1987	42 601	2000	22 369	2013	−13 901
1988	35 215	2001	18 623	2014	−14 565
1989	9 739	2002	5 354	2015	−8 763
1990	−5 961	2003	637		
1991	−11 627	2004	−6 935		

Source: Adapted from Statistics Canada. CANSIM table 051-0004, March 31, 2013

a) Calculate the correlation between Ontario net migration and year.
b) A reporter claimed that the low correlation between year and migration shows that Ontario migration has remained fairly steady over this period of time. Do you agree with this interpretation? Explain.

45. Bigger and bigger According to the UK National Sizing Survey, women's average body-size measurements have increased over time as follows:

Year	Bust size (in)	Waist size (in)	Hips size (in)
1920	32	20	32
1940	33	21	33
1960	34	24	33
1980	35	26	37
2000	36	28	38

Source: P. Treleaven, (September 2007), "How to fit into your clothes," *Significance* 4(3), 113–117.

a) Calculate all possible correlations among the four variables.
b) Plot the body-size measurement sizes against each other and against time. Describe the relationships you see. Which relationship is the strongest?
c) Offer some plausible explanation(s) for these high correlations.

46. Obvious correlations Looking at the largest 152 urban areas in Canada, the correlation between the number of renter-occupied dwellings and number of owner-occupied dwellings per urban area is 0.955, while the correlation between the percentage of renter-occupied dwellings and percentage of owner-occupied dwellings is −1.0. Explain why these two correlation values are just what you would expect.

47. Census at school The International *CensusAtSchool* Project collects data on primary and secondary school students from various countries, including Canada. We selected a random sample of 111 Canadian secondary school students age 14 and over. Below are the first four rows of the 111-row worksheet of data, available at http://www.pearsoncanada.ca/deveaux. All measurements are in centimetres and were taken in the classroom. Middlefinger is distance from wrist to tip of middle finger. Wristbone is distance from wrist bone to elbow. Armspan is distance from tip of left hand to tip of right hand with arms spread wide. Foot is length of right foot.

Gender	Age	Height	Arm-span	Wrist-bone	Middle-finger	Foot
Boy	14	166	169	25	19	26
Boy	14	160	170	27	19	28
Boy	14	167	163	26	20	26
Girl	16	165	165	24	17	22
⋮	⋮	⋮	⋮	⋮	⋮	⋮

Source: Based on Data from The Royal Statistical Society Centre for Statistical Education, CensusAtSchool project.

a) Find all the pair-wise correlations. Which two variables possess the strongest linear correlation?
b) Suppose we are interested in the relationship between armspan and height. Below is a plot of armspan versus height:

Scatterplot of Armspan vs Height

Describe the nature of the relationship between armspan and height.
c) In the plot above, we also included information on another variable, gender. Does gender appear to be an important variable in understanding the relationship between armspan and height? If we omitted the information about gender in the plot above, it would become a _____ variable. Fill in the blank.

48. **Interprovincial migration 2015** Below are data showing the net interprovincial migration each year for Ontario and Alberta (the number arriving minus the number leaving). Note that each year actually stands for a non-calendar year period ending in that particular year.

Year	Ontario	Alberta
1978/1979	−4 325	33 426
1979/1980	−22 362	41 435
1980/1981	−33 247	44 250
1981/1982	−5 665	36 562
1982/1983	23 585	−11 650
1983/1984	36 400	−31 986
1984/1985	33 885	−20 771
1985/1986	33 562	−3 831
1986/1987	42 601	−29 998
1987/1988	35 215	−23 223
1988/1989	9 739	−1 528
1989/1990	−5 961	5 593
1990/1991	−11 627	8 983
1991/1992	−11 045	2 983
1992/1993	−14 189	−1 181
1993/1994	−9 420	−1 630
1994/1995	−2 841	−556
1995/1996	−2 822	7 656
1996/1997	1 977	26 282
1997/1998	9 231	43 089
1998/1999	16 706	25 191
1999/2000	22 369	22 674
2000/2001	18 623	20 457
2001/2002	5 354	26 235
2002/2003	637	11 903
2003/2004	−6 935	10 606
2004/2005	−11 172	34 423
2005/2006	−17 501	45 795
2006/2007	−20 047	33 809
2007/2008	−14 750	15 317
2008/2009	−15 601	13 184
2009/2010	−4 662	−3 271
2010/2011	−4 007	8 443
2011/2012	−10 611	27 652
2012/2013	−13 901	38 598
2013/2014	−14 564	35 382
2014/2015	−8 763	28 921

Source: "Adapted from Statistics Canada (2015). CANSIM table 051-0004, March 31, 2015"

a) Plot Alberta net migration versus Ontario net migration for each year.
b) Describe the nature of the association.
c) Calculate the correlation.
d) Why would you expect a negative association? Is it possible to have a positive correlation here? Why (or why not)? If there were only two provinces in Canada, what would the correlation have to equal, and why?

49. Planets (more or less) On August 24, 2006, the International Astronomical Union voted that Pluto is not a planet. Some members of the public have been reluctant to accept that decision. Let's look at some of the data. Is there any pattern to the locations of the planets? The table shows the average distance of each of the traditional nine planets from the Sun.

Planet	Position Number	Distance from Sun (million miles)
Mercury	1	36
Venus	2	67
Earth	3	93
Mars	4	142
Jupiter	5	484
Saturn	6	887
Uranus	7	1784
Neptune	8	2796
Pluto	9	3666

a) Make a scatterplot and describe the association. (Remember: direction, form, and strength!)
b) Why would you not want to talk about the correlation between a planet's *Position Number* and *Distance from the sun*?
c) Make a scatterplot showing the logarithm (any base) of *Distance* vs. *Position Number*. What is better about this scatterplot?

50. Human Development Index 2010 The United Nations Development Programme (UNDP) uses the Human Development Index (HDI) in an attempt to summarize in one number the progress in health, education, and economics of a country. In 2010, the HDI was as high as 0.938 for Norway and as low as 0.14 for Zimbabwe. The gross domestic product per capita (GDPPC) is used to summarize the overall economic strength of a country. Here is a plot of GDP per capita against HDI for 169 countries throughout the world:

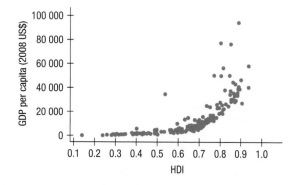

a) GDPPC is measured in dollars. Incomes and other economic measures tend to be highly right-

skewed. Taking logs often makes the distribution more unimodal and symmetric. Compare the histogram of GDPPC to the histogram of log(GDPPC).
b) Use the log re-expression to make a scatterplot of log(GDPPC) against HDI. Comment on the effects of the re-expression.
c) The HDR report classifies countries into high, medium, and low development based in part on their HDI. Look at the scatterplot of log(GDPPC) against HDI for only the 127 medium- and high-development countries (currently viewed as the highest 75%, so HDI > 0.48 here). Does this relationship appear stronger or weaker than when we include all 169 countries?
d) Find the correlation of log(GDPPC) with HDI for all 169 countries. Find the correlation for only the 127 medium- and high-development countries. Which is higher? (Restricting the range of a variable usually reduces the correlation, but not always.)

51. Mandarin or Cantonese 2011 Below are data on Canadian cities showing mother-tongue counts from the 2011 census:

Urban Centre	Cantonese	Mandarin
Abbotsford (BC)	310	400
Barrie (ON)	215	295
Brantford (ON)	125	90
Calgary (AB)	20 175	11 380
Edmonton (AB)	13 595	7 295
Guelph (ON)	770	650
Halifax (NS)	460	805
Hamilton (ON)	1 975	2 215
Kelowna (BC)	235	210
Kingston (ON)	335	395
Kitchener-Cambridge-Waterloo (ON)	1 390	3 060
London (ON)	1 165	1 415
Moncton (NB)	70	75
Montréal (QC)	10 515	10 180
Oshawa (ON)	730	645
Ottawa-Gatineau (ON-QC)	6 570	7 275
Peterborough (ON)	90	110
Québec (QC)	130	270
Regina (SK)	560	505
Saguenay (QC)	0	15
Saint John (NB)	45	125
Saskatoon (SK)	865	790
Sherbrooke (QC)	20	65
St. Catharines-Niagara (ON)	640	555
St. John's (NL)	105	165

Sudbury (ON)	140	90
Thunder Bay (ON)	75	75
Toronto (ON)	170 485	100 045
Trois-Rivières (QC)	10	20
Vancouver (BC)	128 110	90 190
Victoria (BC)	2 930	1 945
Windsor (ON)	1 200	1 085
Winnipeg (MB)	3 220	2 080

Source: "Based on data from Statistics Canada, 2011 Census of Population and Statistics Canada, catalogue no. 98-314-XCB"

a) Find the correlation between these two mother-tongue counts. Are you surprised that the correlation is so high? Explain.
b) Plot Cantonese speaker count versus Mandarin speaker count. Describe the relationship. Is a correlation a useful and appropriate summary measure here?
c) Take the log (base 10) of both columns of numbers using your computer or calculator. Replot the data. Have things improved in any way? Calculate the correlation. Is the correlation more appropriate now?
d) What is one obvious reason (for example, a lurking variable) for such a high correlation between the two log counts here? Suggest another useful way

of defining the two variables that would produce a smaller correlation. (In real-life studies, there are always issues about how to best define and measure the relevant variables.)

T 52. HDI and phones 2010 In Exercise 50, we examined the relationship between GDPPC and the Human Development Index for a large number of countries. The number of phone subscribers per 100 people is also associated with economic progress in a country. Here's a scatterplot of phone subscribers per 100 people against HDI:

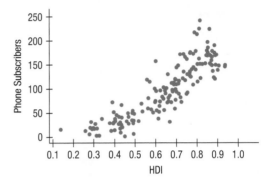

a) The square root transformation is often useful for counts. Examine the histogram of *Phone Subscribers* and the histogram of square-rooted *Phone Subscribers*. Comment.
b) Use the re-expression in part a) for the scatterplot against HDI. Comment.

 Just Checking ANSWERS

1. We know the scores are quantitative. We should check to see if the Straight Enough Condition and the No Outliers Condition are satisfied by looking at a scatterplot of the two scores.
2. It won't change.
3. It won't change.
4. They are likely to have done poorly. The positive correlation means that low scores on Exam 1 are associated with low scores on Exam 2 (and similarly for high scores).
5. No. The general association is positive, but the correlation is not 1.0, so individual performances may vary considerably.

MathXL

MyLab Statistics

Go to MathXL at **www.mathxl.com** or MyLab™ Statistics at **www.mystatlab.com**. You can practise exercises for this chapter as often as you want. The guided solutions will help you find answers step by step. You'll find a personalized study plan available to you too!

Linear Regression

LEARNING OBJECTIVES

1. The linear model
2. Use the least-squares criterion to find the line of 'best fit'
3. Understand the differential roles of the response and explanatory variables
4. Compute, graph and assess the residuals
5. Summarize model utility with $R^2 =$ the variation explained by the model
6. Know the assumptions and check the necessary conditions for regression. Use scatterplots and residual plots

All Mathematics would suggest,

A steady straight line as the best,

But left and right alternately,

Is consonant with History.

—W. H. Auden

Connections

We know that the sales price of a downtown condo is correlated with its size, but how much should you expect to pay for a 750-square foot condo? Hubble discovered that galaxies farther away from us are moving away faster than nearer galaxies, but what velocity would you predict for a galaxy that is 5 million light-years away? We can calculate a correlation between the two quantitative variables in each case, but it won't tell us the answer.

We need to set up and fit a model to the data (using a procedure called 'least-squares') in order to better understand the relationship and to predict one variable from the other.

The Whopper™ has been Burger King's signature sandwich since 1957. One Triple Whopper with cheese provides 71 grams of protein—all the protein you need in a day. It also supplies 1230 calories and 82 grams of fat. The Daily Value (based on a 2000-calorie diet) for fat is 65 grams. So after a Triple Whopper you'll want the rest of your calories that day to be fat-free.[1]

Of course, the Whopper isn't the only item Burger King sells. How are fat and protein related on the entire BK menu? The Figure 7.1 scatterplot of the *Fat* (in grams) versus the *Protein* (in grams) for foods sold at Burger King shows a positive, moderately strong linear relationship.

Figure 7.1

Total *Fat* versus *Protein* for 122 items on the BK menu. The Triple Whopper is in the upper right corner. It's extreme, but is it out of line? (Data in **Burger King Items**)

7.1 Least Squares: The Line of "Best Fit"

If there are 25 grams of protein in your lunch, how much fat should you expect to consume at Burger King? The correlation between *Fat* and *Protein* is 0.76, which tells us that the linear association seen in the scatterplot is fairly strong. But *strength* of the relationship is only part of the picture. The correlation says, "The linear association between these two variables is fairly strong," but it doesn't tell us *how to predict one variable from the other*.

Of course, the points in the scatterplot don't all line up perfectly. But anyone looking at the plot can see that the points move from the lower left to the upper right in a consistent

---

[1]Sorry about the fries.

way. You could probably sketch a straight line that summarizes that relationship pretty well. We'll take that one step further and find a **linear model** that gives an equation of a straight line through the data. We want to predict the fat content of any Burger King menu item given its protein, so fat content is our response variable and protein is the explanatory variable or predictor. This is an important distinction and will affect the line that we fit to the data.

No straight line can go through all the points, but a linear model can summarize the general pattern with only a couple of parameters. Like all **models** of the real world, the line will be wrong—wrong in the sense that it can't match reality *exactly*. But it can help us understand how the variables are associated. Without models, our understanding of the world is limited to only the data we have at hand.

Not only can't we draw a line through all the points, the best line might not even hit any of the points. Then how can it be the "best" line? We want to find the line that somehow comes closer to all the points than any other line. Some of the points will be above the line and some below. For example, the line might suggest that a BK Tendergrill® chicken sandwich on ciabata with 37 grams of protein[2] should have 42.2 grams of fat when, in fact, it actually has only 18. We call the estimate made from a model the **predicted value** (or fitted value), and write it as $\hat{y}$ (called *y-hat*) to distinguish it from the observed response, *y*. The difference between the observed response and its associated predicted value is called the **residual**. The residual value tells us how far off the model's prediction is for that case or point. The Tendergrill chicken sandwich residual would be $y - \hat{y} = 18 - 42.2 = -24.2$ g of fat.

<table>
<tr><td>**Who**</td><td>Items on the Burger King menu</td></tr>
<tr><td>**What**</td><td>Protein content and total fat content</td></tr>
<tr><td>**Unit**</td><td>Grams of protein, Grams of fat</td></tr>
<tr><td>**How**</td><td>Supplied by BK on request or at their Website</td></tr>
</table>

Figure 7.2

The residual is the difference between the actual Fat (our response variable) and the predicted Fat (for some value of Protein). So it is a difference in the vertical direction in our plot.

To find the residuals, we always subtract the predicted value from the observed one.

Residual = Observed response − Predicted response

The negative residual tells us that the actual fat content of the Tendergrill chicken sandwich is about 24.2 grams *less* than the model predicts for a typical Burger King menu item with 37 grams of protein. We often use residuals in this way. The line tells us the average fat content we'd expect for a given amount of protein. Items with negative residuals have less fat than we'd expect, while those with positive residuals have more.

Our challenge now is how to find the right line.

The size of the residuals tells us how well our line fits the data. A line that fits well will have small residuals. But we can't just add up all the residuals to see how well we've done, because some of the residuals are negative. We faced the same issue when we calculated a standard deviation to measure spread. And we deal with it the same way here: by squaring the residuals (and as before, using absolute values is just a poor choice from a mathematical standpoint). Now that they're all positive, adding them up makes sense. Also as before, note that squaring emphasizes the large residuals. When we add all the squared residuals together, that sum indicates how well the line we drew fits the data—the smaller the sum, the better the fit. A different line will produce a different sum, maybe bigger,

> **RESIDUALS**
>
> A positive residual means the actual value is larger than the predicted value—an underestimate. A negative residual shows that the model made an overestimate.

---

[2]The sandwich comes with mayo unless you ask for it without. That adds an extra 24 grams of fat, which is more than the original sandwich contained.

maybe smaller. The **line of best fit** is the line for which the sum of the squared residuals is smallest, hence called the **least squares** line.

You might think that finding this line would be pretty hard. Surprisingly, it's not, although it was an exciting mathematical discovery when Legendre published it in 1805.

> **Who's on first**   In 1805, the French mathematician Adrien-Marie Legendre was the first to publish the *least squares* solution to the problem of fitting a line to data when the points don't all fall exactly on the line. The main challenge was how to distribute the errors "fairly." After considerable thought, he decided to minimize the sum of the squares of what we now call the residuals. When Legendre published his paper, however, the German mathematician Carl Friedrich Gauss claimed he had been using the method since 1795. Gauss later referred to the *least squares* solution as "*our* method" (*principium nostrum*), which certainly didn't help his relationship with Legendre.

Section 7.1 Exercises

1. **True or false** If false, explain briefly.
   a) We choose the linear model that passes through the most data points on the scatterplot.
   b) The residuals are the observed *y*-values minus the *y*-values predicted by the linear model.
   c) Least squares means that the square of the largest residual is as small as it could possibly be.

2. **True or false II** If false, explain briefly.
   a) Some of the residuals from a least squares linear model will be positive and some will be negative.
   b) Least Squares means that some of the squares of the residuals are minimized.
   c) We write $\hat{y}$ to denote the predicted values and *y* to denote the observed values.

7.2 The Linear Model

■ NOTATION ALERT

"Putting a hat on it" is standard Statistics notation to indicate that something has been predicted by a model. Whenever you see a hat over a variable name or symbol, you can assume it is the predicted version of that variable or symbol (and look around for the model).

In a linear model, we use b_1 for the slope and b_0 for the *y*-intercept.

You may remember from Algebra that a straight line can be written as

$$y = mx + b$$

where *m* is the slope and *b* is the *y*-intercept. We'll use this form for our linear model, but in Statistics we use slightly different notation:

$$\hat{y} = b_0 + b_1 x$$

We write $\hat{y}$ (*y*-hat) to emphasize that the points that satisfy this equation are just our *predicted* values, not the actual data values (which scatter around the line). If the model is a good one, the data values will scatter closely around it.

We write b_0 and b_1 for the intercept and slope of the line. The *b*'s are also called the *coefficients* of the linear model. The coefficient b_1 is the **slope**, which tells how rapidly $\hat{y}$ changes with respect to *x*. The coefficient b_0 is the **y-intercept**, which tells where the line hits (intercepts) the *y*-axis.[3]

For the Burger King menu items, the line of best fit is

$$\widehat{Fat} = 8.4 + 0.91\,Protein.$$

What does this mean? The **slope**, 0.91, says that a Burger King item with one more gram of protein can be expected, on average, to have 0.91 more grams of fat. Less formally, we might say that, on average, Burger King foods pack about 0.91 grams of fat per additional gram of protein. Slopes are always expressed in *y*-units per *x*-unit. They tell how the *y*-variable changes (in its units) for a one-unit change in the *x*-variable. When you see a phrase like "students per teacher" or "kilobytes per second," think slope.

---

[3]We change from $mx + b$ to $b_0 + b_1 x$ for a reason: Eventually we'll want to add more *x*-variables (predictors) to the model to make it more realistic, and we don't want to use up the entire alphabet. What would we use after *m*? The next letter is *n* and that one's already taken. *o*? See our point? Sometimes subscripts are the best approach.

What about the **intercept**, 8.4? Algebraically, that's the value the line takes when x is zero. Here, our model predicts that even a BK item with no protein would have, on average, about 8.4 grams of fat. Is that reasonable? Well, there are two items on the menu with 0 grams of protein (apple fries and applesauce). Neither has any fat either (they are essentially pure carbohydrates), though we could imagine a protein-less item with 8.4 grams of fat. But often 0 is not a plausible value for x (predicted price of a 0 square foot condo?). Then the intercept serves only as a value that positions the line properly in the vertical direction, and we don't interpret it as a meaningful predicted value.

NASA

For Example A LINEAR MODEL FOR HURRICANES

The barometric pressure at the centre of a hurricane is often used to measure the strength of the hurricane because it can predict the maximum wind speed of the storm. A scatterplot shows that the relationship is straight, strong, and negative. It has a correlation of −0.896. (Data in **Hurricanes 2014**)

Using technology to fit the straight line, we find

$$\widehat{MaxWindSpeed} = 1024.464 - 0.968\ CentralPressure.$$

QUESTION: Interpret this model. What does the slope mean in this context? Does the intercept have a meaningful interpretation?

ANSWER: The negative slope says that lower *CentralPressure* goes with higher *MaxWindSpeed*. This is consistent with the way hurricanes work: Low central pressure pulls in moist air, driving the rotation and the resulting destructive winds. The slope's value says that, on average, the maximum wind speed is about 0.968 knots higher for every 1-millibar drop in central pressure.

It's not meaningful, however, to interpret the intercept as the wind speed predicted for a central pressure of 0—that would be a vacuum.

Section 7.2 Exercises

3. Least squares interpretations A least squares regression line was calculated to relate the length (cm) of newborn boys to their weight in kg. The line is $\widehat{weight} = -5.94 + 0.1875\ length$. Explain in words what this model means. Should new parents (who tend to worry) be concerned if their newborn's length and weight don't fit this equation?

4. Residual interpretations The newborn grandson of one of the authors was 48 cm long and weighed 3 kg. According to the regression model of Exercise 3, what was his residual? What does that say about him?

Paul Velleman

7.3 Finding the Least Squares Line

How do we find the slope and intercept of the least squares line? The formulas are built from the summary statistics you already know: the correlation (to tell us the strength of the linear association), the standard deviations (to give us the units), and the means (to tell us where to put the line). Remarkably, the correlation and standard deviations are all you need to find the slope[4] of the line:

$$b_1 = r\frac{s_y}{s_x}$$

> **SLOPE**
>
> $$b_1 = r\frac{s_y}{s_x}$$

We've already seen that the correlation tells us the *sign* as well as the strength of the relationship, so it's no surprise to see that the slope inherits this sign as well.

Correlations don't have units, but slopes do. How x and y are measured—what units they have—doesn't affect their correlation, but can change their slope. If children grow an average of 3 inches per year, that's the same as 0.21 millimetres per day. For the slope, it matters whether you express age in days or years and whether you measure height in inches or millimetres. Changing the units of x and y affects their standard deviations directly. And that's how the slope gets its units—from the ratio of the two standard deviations. So the units of the slope are a ratio, too. The units of the slope are always the units of y *per* unit of x.

> **UNITS OF y PER UNIT OF x**
>
> Get into the habit of identifying the units by writing down "y-units per x-unit," with the unit names put in place. You'll find it'll really help you to *Tell* about the line in context.

What about the intercept? If you had to predict the y-value for a data point whose x-value was the average, $\bar{x}$, you'd probably pick $\bar{y}$, and that's just what the least squares line does. Putting that into our equation, we find

$$\bar{y} = b_0 + b_1\bar{x}$$

Or, rearranging the terms, we get an equation for the intercept:

$$b_0 = \bar{y} - b_1\bar{x}$$

> **INTERCEPT**
>
> $$b_0 = \bar{y} - b_1\bar{x}$$

For Example   FINDING THE REGRESSION EQUATION

Let's try out the slope and intercept formulas on the Burger King data and see how to find the line. We checked the conditions when we calculated the correlation. The summary statistics are as follows:

Protein	Fat
$\bar{x} = 18.0\,g$	$\bar{y} = 24.8\,g$
$s_x = 13.5\,g$	$s_y = 16.2\,g$
	$r = 0.76$

So, the slope is

$$b_1 = r\frac{s_y}{s_x} = 0.76 \times \frac{16.2}{13.5} = 0.91 \text{ grams of fat per gram of protein.}$$

The intercept is

$$b_0 = \bar{y} - b_1\bar{x} = 24.8 - (0.91 \times 18.0) = 8.4\,g \text{ fat}$$

Putting these results back into the equation gives

$$\widehat{Fat} = 8.4 + 0.91\,Protein.$$

In summary, the predicted fat of a Burger King food item in grams is $8.4 + 0.91 \times$ the protein content (in grams).

---

[4]The slope may also be written: $b_1 = \dfrac{s_{xy}}{s_x^2} = \dfrac{covariance\ of\ x\ and\ y}{variance\ of\ x}$

With an estimated linear model, it's easy to predict fat content for any menu item we want. For example, for a BK Tendercrisp chicken sandwich w/o mayo with 31 grams of protein, we can plug in 31 grams for the amount of protein and see that the predicted fat content is $8.4 + 0.91(31) = 36.6$ grams of fat (Figure 7.3). (Actually, rounding has caught up with us. Using full precision not shown here, the predicted value to one decimal place should be 36.7. Always use full precision until you report your results). Because the BK Tendercrisp chicken sandwich actually has 22 grams of fat, its residual is

$$Fat - \widehat{Fat} = 22 - 36.7 = -14.7\,\text{g}$$

Least squares lines are commonly called regression lines. In a few pages, we'll see that this name is an accident of history. For now, you just need to know that *regression* almost always means *the model fit by least squares*. And we say that we are *regressing the response variable on the explanatory variable*.

To use a regression model, we should check the same **Conditions** for regression as we did for correlation, making sure that the variables are **Quantitative**, that the relationship is **Straight Enough**, and that there are **No Outliers**.

Figure 7.3

Burger King menu items with the regression line. The regression model lets us make predictions. The predicted fat value for an item with 31 grams of protein is 36.6 g.

Step-by-Step Example CALCULATING A REGRESSION EQUATION

Lawrence Sawyer/Getty Images

During the evening rush hour of August 1, 2007, an eight-lane steel truss bridge spanning the Mississippi River in Minneapolis, Minnesota, collapsed without warning, sending cars plummeting into the river, killing 13 and injuring 145. Although similar events had brought attention to North America's aging infrastructure, this disaster put the spotlight on the problem and raised the awareness of the general public.

How can we tell which bridges are safe?

Most states conduct regular safety checks, giving a bridge a structural deficiency score on various scales. The New York State Department of Transportation uses a scale that runs from 1 to 7, with a score of 5 or less indicating "deficient." Many factors contribute to the deterioration of a bridge, including amount of traffic, material used in the bridge, weather, and bridge design.

New York has more than 17 000 bridges. We have available data on the 194 bridges of Tompkins County[5]. (Data in **Tompkins County Bridges 2014**)

One natural concern is the age of a bridge. A model that relates age to safety score might help the DOT to focus inspectors' efforts where they are most needed.

Question: Is there a relationship between the age of a bridge and its safety rating?

---

[5]Coincidentally, the home county of two of the authors.

THINK ➡ **Plan** State the problem.

Variables Identify the variables and report the *W*'s.

Just as we did for correlation, check the conditions for a regression by making a picture. Never fit a regression without looking at the scatterplot first.

I want to know whether there is a relationship between the age of a bridge in Tompkins County, NY, and its safety rating.

I have data giving the Safety Score and Age for 194 bridges constructed since 1887.

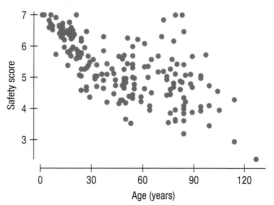

Conditions:

✓ **Quantitative Variables**: We'll assume the Safety Scores are determined on a reasonably precise quantitative scale. Age is in years.

✓ **Straight Enough**: Scatterplot looks straight.

✓ **No Outliers**: None are evident.

It is okay to use a linear regression to model this relationship.

SHOW ➡ **Mechanics** Find the equation of the regression line. Summary statistics give the building blocks of the calculation.

(We generally report summary statistics to one more digit of accuracy than the data. We do the same for intercept and predicted values, but for slopes we usually report an additional digit. Remember, though, not to round until you finish computing an answer.)[6]

Find the slope, b_1.

Find the intercept, b_0.

Write the equation of the model, using meaningful variable names.

Age
$\bar{x} = 48.52$
$s_x = 28.84$

Safety Score
$\bar{y} = 5.27$
$s_y = 0.959$

Correlation
$r = -0.659$

$b_1 = r\dfrac{s_y}{s_x} = -0.659\dfrac{0.959}{28.84}$

$\quad = -0.0219$ *Safety Score points per year*

$b_0 = \bar{y} - b_1\bar{x} = 5.27 - (-0.0219)48.52$

$\quad = 6.33$

The least squares line is

$\hat{y} = 6.33 - 0.0219x$

or

$\widehat{\text{Safety Score}} = 6.33 - 0.0219\,\text{Age}$

---

[6]We warned you that we'll round in the intermediate steps of a calculation to show the steps more clearly, and we've done that here. If you repeat these calculations yourself on a calculator or statistics program, you may get somewhat different results.

TELL ➡	**Conclusion** Interpret what you have found in the context of the question. Discuss in terms of the variables and their units.	The condition of the bridges in Tompkins County, NY, decreases with the age of the bridges at the time of inspection. Bridge condition declines by about 0.022 points on the 1 to 7 scale for each year of age. The model uses a base or intercept of 6.3, which is quite reasonable because a new bridge (0 years of age) should have a safety score near the maximum of 7.
		Because I have only data from one county, I can't tell from these data whether this model would apply to bridges in other counties or locations.

Section 7.3 Exercises

5. Bookstore sales revisited Recall the data we saw in Chapter 6, Exercise 3 for a bookstore. The manager wants to predict *Sales* from *Number of Sales People Working*.

Number of Sales People Working	Sales (in $1000)
2	10
3	11
7	13
9	14
10	18
10	20
12	20
15	22
16	22
20	26
$\bar{x} = 10.4$	$\bar{y} = 17.6$
$SD(x) = 5.64$	$SD(y) = 5.34$
$r = 0.965$	

a) Find the slope estimate, b_1.
b) What does it mean, in this context?
c) Find the intercept, b_0.
d) What does it mean, in this context? Is it meaningful?
e) Write down the equation that predicts *Sales* from *Number of Sales People Working*.
f) If 18 people are working, what *Sales* do you predict?
g) If sales are actually $25 000, what is the value of the residual?
h) Have we overestimated or underestimated the sales?

6. Disk drives 2014 again Recall the data on disk drives we saw in Chapter 6, Exercise 4. Suppose we want to predict *Price* from *Capacity*.

Capacity (in TB)	Price (in $)
0.50	59.99
1.0	79.99
2.0	111.97
3.0	109.99
4.0	149.99
6.0	423.34
8.0	596.11
12.0	1079.99
32.0	4461.00
$\bar{x} = 7.6111$	$\bar{y} = 785.819$
$SD(x) = 9.855$	$SD(y) = 1418.67$
$r = 0.9876$	

a) Find the slope estimate, b_1.
b) What does it mean, in this context?
c) Find the intercept, b_0.
d) What does it mean, in this context? Is it meaningful?
e) Write down the equation that predicts *Price* from *Capacity*.
f) What would you predict for the price of a 20 TB drive?
g) A 20 TB drive on Amazon.com was listed at $2017.86. According to the model, does this seem like a good buy? How much would you save compared to what you expected to pay?
h) Does the model overestimate or underestimate the price?
i) The correlation is very high. Does this mean that the model is accurate? Explain. (Hint: Revisit the scatterplot from Chapter 6, Exercise 4.)

7.4 Regression to the Mean

Suppose you were told that a new male student was about to join the class, and you were asked to guess his height in inches. What would be your guess? Since you have no other information, a good guess would be the mean height of all male students. (If you thought about this in z-scores, you'd say $\hat{z}_{Height} = 0$). Now suppose you are also told that this student has a grade point average (*GPA*) of 3.9—about two SDs above the mean *GPA*. Would that change your guess? Probably not. The correlation between *GPA* and *Height* is near 0, so knowing something about *GPA* wouldn't tell you anything about *Height*.

On the other hand, if you were told that, measured in centimetres, the student's height was two SDs above the mean, what could you say about his height in inches? There's a perfect correlation between *Height* in inches and *Height* in centimetres, so you'd predict that he's two SDs above the mean height in inches as well. Or, writing this as z-scores, you could say: $\hat{z}_{Htin} = z_{Htcm}$.

What if you're told that the student is two SDs above the mean in his shoe size? Would you still guess that he's of average height? You might guess that he's taller than average, since there's a positive correlation between *Height* and *Shoe Size*. But would you guess that he's two SDs above the mean? When there was no correlation, we didn't move away from the mean at all. With a perfect correlation, we moved our guess the full two SDs. Any correlation between these extremes should lead us to guess somewhere between zero and two SDs above the mean. As we'll see in a minute, if we think in z-scores, the relationship is remarkably simple. When the correlation between x and y is r, and x has a z-score of z_x, the predicted value of y is just $\hat{z}_y = rz_x$. For example, if our student was two SDs above the mean shoe size, the formula tells us to guess $r \times 2$ standard deviations above the mean for his height.

How did we get that? We know from the least squares line that $\hat{y} = b_0 + b_1x$. We also know that $b_0 = \bar{y} - b_1\bar{x}$. If we standardize x and y, both their means will be 0, and so for standardized variables, $b_0 = 0$. And the slope $b_1 = r\dfrac{s_y}{s_x}$. For standardized variables, both s_y and s_x are 1, so $b_1 = r$. In other words, $\hat{z}_y = rz_x$. Figure 7.4 illustrates this for the BK data.

Why is correlation "r"? In his original paper, Galton standardized the variables and so the slope was the same as the correlation. He used r to stand for this coefficient from the (standardized) regression.

Figure 7.4

Standardized Fat vs. *Standardized Protein* with the regression line. Each one-standard-deviation change in *Protein* results in a predicted change of r standard deviations in *Fat*.

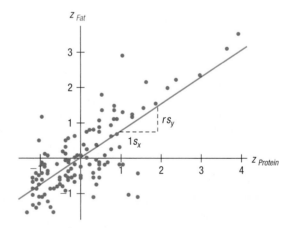

Even more important than deriving the formula is understanding what it says about regression. While we rarely standardize variables to do a regression, we can think about what the z-score regression means. It says that when the correlation is r, and an x-value lies k standard deviations from its mean, our prediction for y is $r \times k$ standard deviations from its mean.

The z-score equation has another profound consequence. It says that for linear relationships, you can never predict that y will be *farther* from its mean than x was from its mean. That's because r is always between -1 and 1. So, each predicted y tends to be closer to its mean (in standard deviations) than its corresponding x was. Sir Francis Galton discovered this property when trying to predict heights of offspring from their parents. What he found was that even though tall parents tend to have tall children, their children are closer to the mean of all offspring than their parents were. He called this phenomenon "regression to mediocrity." He had hoped that tall parents would have children whose heights were even more extraordinary, but that's not what he found. Often when people relate essentially the same variable in two different groups, or at two different times, they see this same phenomenon—the tendency of the response or predicted variable to be closer to the mean than the predictor. Unfortunately, people try to interpret this by thinking that the performance of those far from the mean is deteriorating, but it's just a mathematical fact about correlation. So, today we try to be less judgmental about this phenomenon and we call it **regression to the mean**. We managed to get rid of the term "mediocrity," but the name regression stuck as a name for the whole least squares fitting procedure—and that's where we get the term **regression line**.[7]

Just Checking

A scatterplot of house price (in dollars) versus house size (in square feet) for houses sold recently in your small Maritime city shows a relationship that is straight, with only moderate scatter and no outliers. The correlation between house price and house size is 0.84 and the equation of the regression model is

$$\widehat{Price} = 90\,564 + 182.74\,Size$$

1. What does the slope of 182.74 mean?
2. What are the units of the slope?
3. Your house is 2000 sq. ft. bigger than your neighbour's house. How much more do you expect it to be worth?
4. You read an ad about a house that is 2 standard deviations above the mean in size. What can you say about its price?
5. The mean price of all these houses is $338 000, with standard deviation of $83 790. What is the predicted price for a house that is 2 SDs above the mean in size?

Section 7.4 Exercises

7. **Sophomore slump?** A CEO complains that the winners of his "rookie junior executive of the year" award often turn out to have less impressive performance the following year. He wonders whether the award actually encourages them to slack off. Can you offer a better explanation?

8. **Sophomore slump again?** An online investment blogger advises investing in mutual funds that have performed badly the past year because "regression to the mean tells us that they will do well next year." Is he correct?

7.5 Examining the Residuals

■ **NOTATION ALERT**

Galton used *r* for correlation, so we use *e* for "error." But it's not a mistake. Statisticians often refer to variability not explained by a model as error.

The linear model we are using assumes that the relationship between the two variables is a straight line. The residuals are the part of the data that *hasn't* been modelled. Conceptually, we think

$$Data = Model + Residual$$

or, equivalently,

$$Residual = Data - Model$$

---

[7]Half way through the 2016 baseball season (the time of this edit), Toronto Blue Jay slugger Edwin Encarnacion has a league-leading 71 rbi's (runs batted in) in 81 games. Would you predict more or fewer rbi's in the second half of the season? (Will all the chance factors line up again so well? Or will he *regress to the mean*?)

More precisely, the actual individual residuals are defined as:

$$\text{Residual} = \text{Observed response} - \text{Predicted response}$$

or, in symbols,

$$e = y - \hat{y}$$

When we want to know how well the model fits, we can ask instead what the model missed. No model is perfect, so it's important to know how and where it fails. To see that, we look at the residuals.

For Example KATRINA'S RESIDUAL

RECAP: The linear model relating hurricanes' maximum wind speeds to their central pressures was

$$\widehat{MaxWindSpeed} = 1024.464 - 0.968 \, CentralPressure$$

Let's use this model to make predictions and see how those predictions do.

QUESTION: Hurricane Katrina had a central pressure measured at 920 millibars. What does our regression model predict for her maximum wind speed? How good is that prediction, given that Katrina's actual maximum wind speed was measured at 150 knots?

ANSWER: Substituting 920 for the central pressure in the regression model equation gives

$$\widehat{MaxWindSpeed} = 1024.464 - 0.968(920) = 133.90$$

The regression model predicts a maximum wind speed of 133.90 knots for Hurricane Katrina.

The residual for this prediction is the observed value minus the predicted value:

$$150 - 133.90 = 16.1 \, kts$$

In the case of Hurricane Katrina, the model predicts a wind speed 16.1 knots lower than was actually observed.

Residuals help us to see whether the model makes sense. When a regression model is appropriate, it should model the underlying relationship. Nothing interesting should be left behind. So after we fit a regression model, we usually plot the residuals in the hope of finding . . . nothing interesting. Figure 7.5 plots the residuals versus *Protein* for the BK menu regression.

Figure 7.5

The residuals for the BK menu regression look appropriately boring. There are no obvious patterns, although there are a few points with large residuals. The negative ones turn out to be grilled chicken items, which are among the lowest fat (per protein) items on the menu. The two high outliers contain hash browns, which are high fat per protein items.

A scatterplot of the residuals versus the *x*-values should be the most boring scatterplot you've ever seen. It shouldn't have any interesting features, like a direction or shape. It should stretch horizontally, with about the same amount of scatter throughout. It should show no bends and it should have no outliers. If you see any of these features, investigate them if possible, and find out what the regression model missed. For example, in Figure 7.6 below, we see a curving scatterplot and the corresponding bent residual plot (note how the bend is accentuated in the residual plot):

Figure 7.6

A case of the bends.

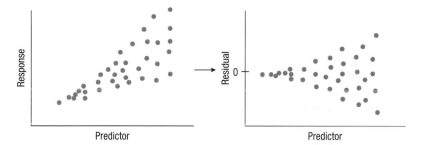

Most computer statistics packages plot the residuals against the predicted values $\hat{y}$, rather than against x.[8] When the slope is positive, they're virtually identical except for the axis labels. When the slope is negative, the two versions are mirror images. Since all we care about is the patterns (or, better, lack of patterns) in the plot, it really doesn't matter which way we plot the residuals.

If the residuals show no interesting pattern when we plot them against x, we can look at how big they are. Since their mean is always zero, though, it's only sensible to look at how much they vary (from zero). The standard deviation of the residuals, s_e, will give us a measure of how much the points spread around the regression line, but for this summary to make sense, the residuals should all share the same underlying spread. That's why we check to make sure that the residual plot has about the same amount of scatter throughout the plot.

This gives us a new assumption: the **Equal Variance Assumption**. The associated condition to check is the **Does the Plot Thicken? Condition**. We check to make sure that the spread is about the same throughout. We can check that either in the original scatterplot of y against x or in the scatterplot of residuals.

For example, the variability of measurements on the response variable often increases as the measurements get bigger, such as illustrated below (Figure 7.7):

Figure 7.7

A steady trend in dispersion. The plot thickens (as we go from left to right).

This is not the only way the problem of uneven dispersion can arise, but it is the easiest to deal with. Often a non-linear transformation, like a logarithm, will fix it (sometimes even fixing other problems as well, like non-linearity). This problem may also arise when we mix together different subgroups in our analysis—again with a simple solution—separate the groups. But at other times, there may be no simple or easy fix for big differences in response variability across the scatterplot.

The Residual Standard Deviation

We estimate the **standard deviation of the residuals** in almost the way you'd expect:

$$s_e = \sqrt{\frac{\Sigma e^2}{n - 2}}$$

Why $n - 2$ rather than $n - 1$? We used $n - 1$ for s when we estimated the mean. Now we're estimating both a slope and an intercept. Looks like a pattern—and it is. We subtract one more for each parameter we estimate in our fitted model.

[8]They have a good reason for this choice. When regression models use more than one x-variable, there can be many x's to plot against, but there's still only one $\hat{y}$.

We don't need to subtract the mean because the mean of the residuals $\bar{e} = 0$. The $n - 2$, like the $n - 1$ in the denominator of the variance, is a statistical adjustment we'll discuss in more detail in a later chapter.

For the Burger King menu items, the standard deviation of the residuals is 10.6 grams of fat. That looks about right in the scatterplot of residuals, as a fairly typical distance from the mean of zero. The residual for the BK Tendergrill chicken was –24.2 grams, a bit over 2 standard deviations in size, making it a stand-out in low fat content.

It's a good idea to make a histogram of the residuals. If we see a unimodal, symmetric histogram, then we can apply the 68–95–99.7 Rule to see how well the regression model describes the data. In particular, we would know that about 95% of the residuals should be no farther from their mean of zero than $2s_e$, hence no bigger in size than $2s_e$. The Burger King residuals look like (Figure 7.8):

Figure 7.8

The histogram of residuals is symmetric and unimodal, centred at 0, with a standard deviation of 10.6. Only a few values lie outside of 2 standard deviations. The low ones are chicken items. The high ones contain hash browns.

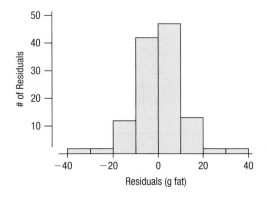

Sure enough, most are less than 2 × 10.6, or 21.2 grams of fat in size. In other words, *about 95% of the fat measurements lie within 21.2 grams of their predicted values.*

Section 7.5 Exercises

9. Bookstore sales once more Here are the residuals for a regression of *Sales* on *Number of Sales People Working* for the bookstore of Exercise 5:

Number of Sales People Working	Residual
2	0.07
3	0.16
7	–1.49
9	–2.32
10	0.77
10	2.77
12	0.94
15	0.20
16	–0.72
20	–0.37

a) What are the units of the residuals?
b) Which residual contributes the most to the sum that was minimized according to the Least Squares Criterion to find this regression?
c) Which residual contributes least to that sum?

10. Disk drives 2014, residuals Here are the residuals for a regression of *Price* on *Capacity* for the hard drives of Exercise 6. (Based on the hand-computed coefficients.)

Capacity	Residual
0.50	285.16
1.0	234.07
2.0	123.88
3.0	–20.27
4.0	–122.44
6.0	–133.43
8.0	–245.00
12.0	–329.80
32.0	207.81

a) Which residual contributes the most to the sum that is minimized by the Least Squares criterion?
b) Five of the residuals are negative. What does that mean about those drives? Be specific and use the correct units.

7.6 R^2—The Variation Accounted For by the Model

The variation in the residuals is the key to assessing how well the model fits the data. Let's compare the variation of the response variable with the variation of the residuals. The *Fat* content has a standard deviation of 16.2 grams. The standard deviation of the residuals is 10.6 grams. If the correlation were 1.0 and the model predicted the *Fat* values perfectly, the residuals would all be zero and have no variation. We couldn't possibly do any better than that.

On the other hand, if the correlation were zero, the model would simply predict 24.8 grams of fat (the mean) for all menu items (Figure 7.9 on the left), and the residuals would just be the observed *Fat* values minus their mean. These residuals would have the same variability as the original data because, as we know, just subtracting the mean doesn't change the spread.

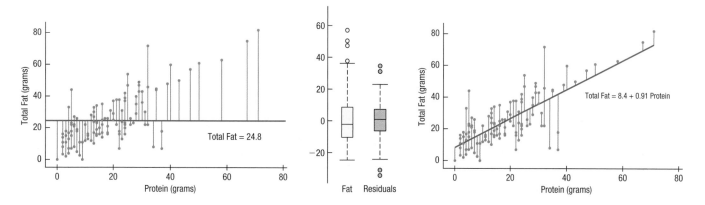

Figure 7.9

Compare the variability of the *Fat* content measurements with the variability of the residuals from the regression. In the boxplots, we compare the deviations of the individual *Fat* values from their mean of 24.8 (on the left), with the deviations or residuals from the line of best fit (on the right). The variation left in the residuals is unaccounted for by the model, but it's less than the variation in the original data (more precisely compared in the boxplots).

How well does the BK regression model do? Look at the boxplots in Figure 7.9. The variation in the residuals (on the right) is smaller than variation in the data, but certainly bigger than zero. That's nice to know, but how much of the variation in the data is still left in the residuals? If you had to put a number between 0% and 100% on the fraction of the variation left in the residuals, what would you say?

All regression models fall somewhere between the two extremes of zero correlation and perfect correlation. We'd like to gauge where our model falls. Can we use the correlation to do that? Well, a regression model with correlation −0.5 is doing as well as one with correlation +0.5. They just have different directions. If we *square* the correlation coefficient, we'll get a value between 0 and 1, and the direction won't matter. The squared correlation, r^2, gives the fraction of the data's variation accounted for by the model, and $1 - r^2$ is the fraction of the original variation left in the residuals. For the Burger King model, $r^2 = 0.76^2 = 0.58$, and $1 - r^2$ is 0.42, so 42% of the variability in *Fat* content has been left in the residuals. How close was that to your guess? (Don't feel bad about guessing too high. We are comparing variances, not standard deviations!)

All regression analyses include this statistic, although by tradition, it is written with a capital letter, $\mathbf{R^2}$, pronounced "*R*-squared," and also may be called the

TWICE AS STRONG?

Is a correlation of 0.80 twice as strong as a correlation of 0.40? Not if you think in terms of R^2. A correlation of 0.80 means an R^2 of $0.80^2 = 64\%$. A correlation of 0.40 means an R^2 of $0.40^2 = 16\%$—only a quarter as much of the variability accounted for. A correlation of 0.80 gives an R^2 *four times as strong* as a correlation of 0.40 and accounts for four times as much of the variability.

coefficient of determination. An R^2 of 0 means that none of the variance in the data is in the model; all of it is still in the residuals . It would be hard to imagine using that model for anything. Because R^2 is a fraction of a whole, it is often given as a percentage.[9] For the Burger King data, R^2 is 58%.

When interpreting a regression model, you need to be able to explain what R^2 means. According to our linear model, 58% of the variability in the fat content of Burger King sandwiches is accounted for by variation in the protein content.

How can we see that R^2 is really the fraction of variance accounted for by the model? It's a simple calculation. The variance of the fat content of the Burger King foods is $16.2^2 = 262.44$. If we treat the residuals as data, the variance of the residuals is 110.53.[10] As a fraction, that's $110.53/262.44 = 0.42$ *or* 42%. That's the fraction of the variance that is *not* accounted for by the model. The fraction that is accounted for is $100\% - 42\% = 58\%$, just the value we got for R^2.

For Example INTERPRETING R^2

RECAP: Our regression model that predicts maximum wind speed in hurricanes based on the storm's central pressure has $R^2 = 80.3\%$.

QUESTION: What does that say about our regression model?

ANSWER: An R^2 of 80.3% indicates that 80.3% of the variation in maximum wind speed can be accounted for by the hurricane's central pressure. Other factors, such as temperature and whether the storm is over water or land, may account for some of the remaining variation.

 ## Just Checking

Back to our regression of house price (in dollars) on house size (in square feet):

$$\widehat{Price} = 90\,564 + 182.74\,Size$$

The R^2 value is reported as 70.0%, and the standard deviation of the residuals is 45 900.

6. What does the R^2 value mean about the relationship of price and size?

7. Is the correlation of price and size positive or negative? How do you know?

8. If we measured the size in square metres instead of square feet, would the R^2 value change? How about the slope?

9. You find that your house is worth $55 000 more than the regression model predicts. You are undoubtedly pleased, but is this actually a surprisingly large value?

---

[9] By contrast, we usually report correlation coefficients as decimal values between –1.0 and 1.0.

[10] This isn't quite the same as squaring the s_e that we discussed a couple of pages ago (due to different denominators, $n - 1$ vs $n - 2$), but it's very close. We'll deal with the distinction in Chapter 23.

How Big Should R^2 Be?

R^2 is always between 0% and 100%. But what's a "good" R^2 value? The answer depends on the kind of data you are analyzing and on what you want to do with it. Just as with correlation, there is no value for R^2 that automatically determines that the regression is "good." Data from scientific experiments often have R^2 in the 80% to 90% range and even higher. Data from observational studies and surveys, though, often show relatively weak associations because it's so difficult to measure responses reliably. An R^2 of 50% or 30% or even lower might be taken as evidence of a useful regression. The standard deviation of the residuals can also give us information about the usefulness of the regression by telling us how much scatter there is around the line.

As we've seen, an R^2 of 100% is a perfect fit, with no scatter around the line. The s_e would be zero. All of the variance is accounted for by the model and none is left in the residuals. This sounds great, but it's too good to be true for real data.[11]

Along with the slope and intercept for a regression, you should always report R^2 so that readers can judge for themselves how successful the regression is at fitting the data. Statistics is about variation, and R^2 measures the success of the regression model in terms of the fraction of the variation of the response variable accounted for by the regression. R^2 is the first part of a regression that many people look at because, along with the scatterplot, it tells whether the regression model is even worth thinking about.

A Tale of Two Regressions

Regression slopes may not behave exactly the way you'd expect at first. Our regression model for the Burger King menu items was $\widehat{Fat} = 8.4 + 0.91\,Protein$. That equation allows us to estimate that a menu item with say 31 grams of protein would have 36.7 grams of fat. Suppose, though, that we knew the fat content and wanted to predict the amount of protein. It might seem natural to think that by solving our equation for *Protein* we'd get a model for predicting *Protein* from *Fat*. But that doesn't work.

To see why it doesn't work, let's *make a picture*. On the left side in Figure 7.10 is a scatterplot with simulated data for two variables *x* and *y*, showing the least-squares fit for predicting *y* from *x*. Note how reasonable the line seems: i.e., at any *x*, the data points are roughly equally above and below the line, and the line nicely slices through the data (in a vertical sense) as you move from low *x* to high *x*, or from left to right.

Figure 7.10

Plot A is a scatterplot with best fit line to predict *y* from *x*.
Plot B is the same graph but rotated 90 degrees, so *x* is the vertical axis and *y* is the horizontal axis. Is the same line good for predicting *x* from *y*?

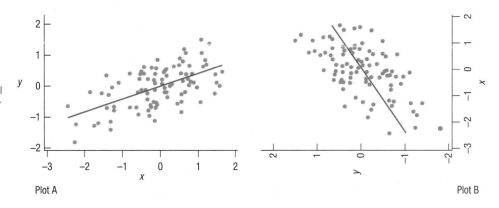

Plot A   Plot B

Now examine the plot on the right. We see a very poorly fitted straight line if predicting the vertical axis variable from the horizontal axis variable. For example, at

---

[11]If you see an R^2 of 100%, it's a good idea to figure out what happened. You may have discovered a new law of physics, but it's much more likely that you accidentally regressed two variables that measure the same thing.

values on the right side of the horizontal axis, the predicted values on the line fall way below the data points, so predictions here are very low, while on the left side of the graph the line shoots high above the actual data. For a value of 1 on the horizontal axis, the prediction is just off the graph! Yet the plot on the right is exactly the same as the plot on the left, except for rotating the plot by 90 degrees (which is why the symbol y is upside down and the positive direction on the horizontal axis is to the left, not to the right[12]). Clearly, the line's tilt is too extreme, and a good prediction model would have a less severe slope. To find the best fit, we have to reverse the roles of the two variables and refit, rather than solving the prediction equation from the graph on the left for x in terms of y, which would just give us the hideous prediction line displayed on the right side.

Our original model was $\hat{y} = b_0 + b_1 x$, but now we want to find an $\hat{x}$ based on a value of y. We don't have y in our original model, only $\hat{y}$, and that makes all the difference. Our model doesn't fit the data values perfectly, and the least squares criterion focuses on the *vertical* errors (in the y direction) the model makes in using x to predict y—not on *horizontal* errors related to prediction of x from y.

A quick look at the equations reveals why. Simply solving our original equation for x would give a new line whose slope is the reciprocal of ours. To model y in terms of x, our slope is $b_1 = r\frac{s_y}{s_x}$. To model x in terms of y, we'd need to use the slope $b_1 = r\frac{s_x}{s_y}$. And that's *not* the reciprocal of the former.

Sure, if the correlation, r, were 1.0 or –1.0, the slopes *would* be reciprocals, but that would happen only if we had a perfect fit. Real data don't follow perfect straight lines, so in the real world y and $\hat{y}$ aren't the same, r is a fraction, and the slopes of the two models are not simple reciprocals of one another.

If we want to predict x from y, we need to create that particular model from the data. For example, to predict *Protein* from *Fat* we find the slope, $b_1 = 0.76\frac{13.5}{16.2} = 0.63$ grams of protein per gram of fat. The regression model is then $\widehat{Protein} = 2.29 + 0.63\ Fat$, so we'd predict that a menu item with say 36.7 grams of fat should have 25.4 grams of protein—not the 31 grams of protein from which we predicted 36.7 grams of fat in the first paragraph of this discussion.

Protein	Fat
$\bar{x} = 18.0\,g$	$\bar{y} = 24.8\,g$
$s_x = 13.5\,g$	$s_y = 16.2\,g$
$r = 0.76$	

Moral of the story: *Think*. Decide which variable you want to use (the explanatory or x variable) to predict values for the other (the response or y variable). Then find the model that does that. If, later, it turns out that you were mistaken and actually want to make predictions in the other direction, you'll need to start over and fit the correct model from scratch.

Section 7.6 Exercises

11. Bookstore sales last time For the regression model for the bookstore of Exercise 5, what is the value of R^2 and what does it mean?

12. Disk drives encore For the hard drive data of Exercise 6, find and interpret the value of R^2.

---

[12]We could flip it over like a page in a book to correct the direction of values on the horizontal axis, but that wouldn't change the story (just make it easier to read).

7.7 Regression Assumptions and Conditions

MAKE A PICTURE (OR TWO)

You can't check the conditions just by checking boxes. You need to examine both the original scatterplot of *y* against *x* before you fit the model, and the plot of residuals afterwards. These plots can save you from making embarrassing errors and … losing points on the exam.

The linear regression model may be the most widely used model in all of Statistics. It has everything we could want in a model: two easily estimated parameters, a meaningful measure of how well the model fits the data, and the ability to predict new values. It even provides an easy way to see violations of conditions in plots of the residuals.

Like all models, though, linear models are only appropriate if some assumptions are true. We can't confirm assumptions, but we often can check related conditions.

First, be sure that both variables are quantitative. It makes no sense to perform a regression on categorical variables. After all, what could the slope possibly mean? Always check the **Quantitative Variables Condition**.

The linear model only makes sense if the relationship is linear. It is easy to check the associated **Straight Enough Condition**. Just look at the scatterplot of *y* vs. *x*. You don't need a *perfectly* straight plot, but it must be straight enough for the linear model to make sense. If you try to model a curved relationship with a straight line, you'll usually get just what you deserve. If the scatterplot is not straight enough, stop here. You can't use a linear model for *any* two variables, even if they are related. They must have a *linear* association, or the model won't mean a thing. [But if you don't want to stop here, try to find a re-expression of one or both variables that will straighten the relationship. If that doesn't work, you may need to fit a more complex model using more advanced regression techniques.]

The response variable needs to exhibit a roughly similar amount of dispersion or spread regardless of the value of *x*. Fitting the model and using it for inference both depend on this assumption. If the scatterplot of *y* vs. *x* looks equally spread out everywhere and (often more vividly) if the *residual plot* of residuals *vs.* predicted values also has a consistent spread, then the assumption is reasonable. The most common violation of that equal variance assumption is for the data or residuals to spread out more for *larger* values of *x*, so a good mnemonic for this check is the **Does the Plot Thicken? Condition**.

Outlying points can dramatically change a regression model. They can even change the sign of the slope, which could give a very misleading impression of the relationship between the variables if you only look at the fitted model. So check the **Outlier Condition**. Check the scatterplot of *y* against *x* and the residual plot to be sure there are no outliers. The residual plot often shows violations more clearly and may reveal other unexpected patterns or interesting quirks in the data. Of course, any outliers are likely to be interesting and informative, so be sure to look into why they are unusual.

To summarize:

Before starting, be sure to check the

- **Quantitative Variables Condition** If either *y* or *x* is categorical, you can't make a scatterplot and you can't perform a regression. Stop.

From the scatterplot of *y* against *x*, check the

- **Straight Enough Condition** Is the relationship between *y* and *x* straight enough to proceed with a linear regression model?
- **Outlier Condition** Are there any outliers that might dramatically influence the fit of the least squares line?
- **Does the Plot Thicken? Condition** Does the spread of the data around the generally straight relationship seem to be consistent (aside from some natural random variation) for all values of *x*?

After fitting the regression model, make a plot of residuals against the predicted values (or *x* values) and look for

- Any bends that would violate the **Straight Enough Condition**
- Any outliers that weren't clear before, and
- Any change in the spread of the residuals from one part of the plot to another

Step-by-Step Example REGRESSION

If you plan to hit the fast food joints for lunch, you should have a good breakfast. Nutritionists, concerned about "empty calories" in breakfast cereals recorded facts about 77 cereals, including the *Calories* and *Sugar* content (in grams) per serving. (Data in **Cereals**)

QUESTION: How are calories and sugar content related in breakfast cereals?

THINK ➡ **PLAN** State the problem.

I am interested in the relationship between sugar content and calories in cereals.

VARIABLES Name the variables and report the *W*'s.

I have two quantitative variables, *Calories* and *Sugar* content per serving, measured on 77 breakfast cereals. The units of measurement are calories and grams of sugar, respectively.

✓ **Quantitative Variables**: Both variables are quantitative.

Check the conditions for a regression by making a picture. Never fit a regression without looking at the scatterplot first.

I'll check the remaining conditions from the scatterplot (and later, the residual plot):

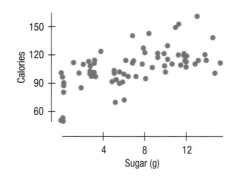

Calories are reported to the nearest 10 calories and sugar content is reported to the nearest gram, so many cereals have the same values for both variables. In order to show them better, we have added a little random noise to both variables in the scatterplot, a process called *jittering*. Of course, we use the actual values for calculation.

✓ **Outlier Condition**: There are no obvious outliers or distinct groups.

✓ **Straight Enough**: The scatterplot looks straight to me.

✓ **Does the Plot Thicken?** The spread around the line looks about the same throughout, but I'll check this and the previous condition again in a residual plot, after fitting the model.

So far so good, so I will proceed to fit a regression model to these data.

SHOW ➡ **Mechanics** If there are no clear violations of the conditions, fit a straight line model of the form $\hat{y} = b_0 + b_1 x$ to the data. Summary statistics give the building blocks of the calculation. (More often, you will just use software to find the fitted model.)

Calories

$\bar{y} = 106.88$ *calories*

$s_y = 19.48$ *calories*

Sugar

$\bar{x} = 6.94$ *grams*

$s_x = 4.42$ *grams*

Correlation

$r = 0.564$

Find the slope.

$$b_1 = r\frac{s_y}{s_x} = 0.564\frac{19.48}{4.42}$$

$$= 2.49 \text{ calories per gram of sugar}$$

Find the intercept.

$$b_0 = \bar{y} - b_1\bar{x} = 106.88 - 2.49(6.94)$$

$$= 89.6 \text{ calories}$$

So the least squares line is

$$\hat{y} = 89.6 + 2.49x,$$

Write the equation using meaningful variable names.

or $\widehat{Calories} = 89.6 + 2.49\,Sugar.$

Squaring the correlation gives

State the value of R^2.

$$R^2 = 0.564^2 = 0.318 \text{ or } 31.8\%.$$

TELL ➡ **Conclusion** Describe what the model says in words and numbers. Be sure to use the names of the variables and their units.

The scatterplot shows a positive linear relationship and no outliers. The least squares regression line fit through these data has the equation

$$\widehat{Calories} = 89.6 + 2.49\,Sugar$$

The key to interpreting a regression model is to start with the phrase "b_1 *y*-units per *x*-unit," substituting the estimated value of the slope for b_1 and the names of the respective units. The intercept may (as in this example) be meaningful, or not (when $x = 0$ is not realistic).

So, cereals have about 2.49 calories more per gram of additional sugar, on average.

And sugar-free cereals would still average about 89.6 calories.[13]

R^2 gives the fraction of the variability of y accounted for by the linear regression model.

The R^2 says that 31.8% of the variability in *Calories* is accounted for by variation in *Sugar* content.

Find the standard deviation of the residuals, s_e, and compare it to the original s_y.

$s_e = 16.2$ calories. That's smaller than the original cereal calorie count SD of 19.5, but still relatively large.

THINK ➡ CHECK AGAIN Even though we looked at the
AGAIN scatterplot *before* fitting a regression model, a plot of the residuals is essential to any regression analysis because it is the best check for additional patterns and interesting quirks in the data. (As we did earlier, the points have been jittered to see the pattern more clearly.)

The residuals show a horizontal direction, a shapeless form, and roughly equal scatter for all predicted values. The linear model appears to be appropriate.

---

[13]Predictions at $x = 0$ only makes sense if $x = 0$ is within the range of the observed data.

Reality Check: Is the Regression Reasonable?

Statistics don't come out of nowhere. They are based on data. The results of a statistical analysis should reinforce your common sense, not fly in its face. If the results are surprising, then either you've learned something new about the world or your analysis is wrong.

Whenever you perform a regression, think about the coefficients and ask whether they make sense. Is a slope of 2.5 calories per gram of sugar reasonable? That's hard to say right off. We know from the summary statistics that a typical cereal has about 100 calories and 7 grams of sugar. A gram of sugar contributes some calories (actually, 4, but you don't need to know that), so calories should go up with increasing sugar. The direction of the slope seems right.

To see if the *size* of the slope is reasonable, a useful trick is to consider its order of magnitude. Start by asking if deflating the slope by a factor of 10 seems reasonable. Is 0.25 calories per gram of sugar enough? The 7 grams of sugar found in the average cereal would contribute less than 2 calories. That seems too small.

Now let's try inflating the slope by a factor of 10. Is 25 calories per gram reasonable? Then the average cereal would have 175 calories from sugar alone. The average cereal has only 100 calories per serving, though, so that slope seems too big.

We have tried inflating the slope by a factor of 10 and deflating it by 10 and found both to be unreasonable. So, like Goldilocks, we're left with the value in the middle that's just right. And an increase of 2.5 calories per gram of sugar is certainly *plausible*.

The small effort of asking yourself whether the regression equation is plausible is repaid whenever you catch errors or avoid saying something silly or absurd about the data. It's too easy to take something that comes out of a computer at face value and assume that it makes sense.

Always be skeptical and ask yourself if the answer is reasonable.

Section 7.7 Exercises

13. Residual plots Here are residual plots (residuals plotted against predicted values) for three linear regression models. Indicate which condition appears to be violated (linearity, outlier, or equal spread) in each case.

a)

b)

c)
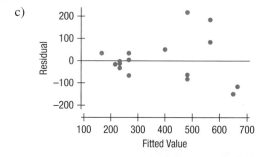

14. Disk drives 2014, residuals again Here is a scatterplot of the residuals from the regression of the hard drive prices on their sizes from Exercise 6.

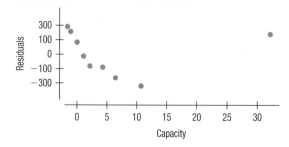

a) Are any assumptions or conditions violated? If so, which ones?

b) What would you recommend about this regression?

WHAT CAN GO WRONG?

There are many ways in which data that appear at first to be good candidates for regression analysis may be unsuitable. And there are ways that people use regression that can lead them astray. Here's an overview of the most common problems. We'll discuss them at length in the next chapter.

■ **Don't fit a straight line to a nonlinear relationship.** Linear regression is suited only to relationships that are, well, *linear*. Fortunately, we can often easily improve the linearity by using re-expression (as discussed in Chapter 6).

■ **Don't ignore outliers.** Outliers can have a serious impact on the fitted model. You should identify them and think about why they are extraordinary. If they turn out not to be obvious errors, read the next chapter for advice.

■ **Don't invert the regression.** The BK regression model was $\widehat{Fat} = 8.4 + 0.91 \, Protein$. Knowing protein content, we can predict the amount of fat. But that doesn't let us switch the regression around. We can't use this model to predict protein values from fat values. To swap the predictor–response roles of the variables in a regression (which rarely makes sense), we must fit a new regression equation (it turns out to be $\widehat{Protein} = 2.29 + 0.63 \, Fat$).

■ **Don't choose a model based on R^2 alone.** Although R^2 measures the strength of the linear association, a high R^2 does not demonstrate the *appropriateness* of the regression. The relationship may still be clearly curved or have very uneven spread about the line. A single outlier or data that separate into two groups may hugely inflate or deflate the value of R^2. Always look at the scatterplot!

What Have We Learned?

A linear model shows the response y as a linear (straight-line) function of the explanatory or x-variable.

■ The best fit model is the one that minimizes the sum of the squared residuals—the Least Squares Model.

■ Find the coefficients of the least squares line from the correlation and the summary statistics of each variable: $b_1 = r\dfrac{s_y}{s_x}$

$$b_0 = \bar{y} - b_1\bar{x}$$

■ Interpret a regression slope as the average change in y per one unit change in x.

■ Understand the correlation coefficient as the number of standard deviations by which one variable is expected to change for a one standard deviation change in the other.

Always report and interpret R^2, which reports the fraction of the variation of y accounted for by the linear model. Also known as the Coefficient of Determination.

■ R^2 is the square of the correlation coefficient (but only for the *simple linear regression* model discussed in this chapter).

Be sure to check the conditions for regression before reporting or interpreting a regression model. Use the scatterplot and the residual plot. Check that:

■ The relationship is Straight Enough.

■ A scatterplot does not "thicken" in any region.

■ There are no obvious outliers.

Review of Terms

Linear model	An equation of the form

$$\hat{y} = b_0 + b_1 x$$

where the x-variable is being used as an *explanatory variable* to help predict the *response variable y*. To interpret a linear model, we need to know the variables (along with their W's) and their units (p. 211).

Model	An equation or formula that simplifies and represents reality (p. 211).
Predicted (fitted) value	The value of $\hat{y}$ found for a given x-value in the data. A predicted value is found by substituting the x-value in the regression equation. The predicted values are the values on the fitted line; the points $(x, \hat{y})$ all lie exactly on the fitted line (p. 211).

The predicted values are found from the linear model that we fit:

$$\hat{y} = b_0 + b_1 x.$$

Residuals	The differences between the observed values of the response variable y and the corresponding values predicted by the regression model—or, more generally, values predicted by any model ($\hat{y}$) (p. 211).

$$\text{Residual} = \text{Observed } y\text{-value} - \text{Predicted } y\text{-value} = y - \hat{y}$$

Least squares	The least squares criterion specifies the unique line that minimizes the variance of the residuals or, equivalently, the sum of the squared residuals (p. 212).
Slope	The slope, b_1, gives a value in "y-units *per* x-unit." Changes of one unit in x are associated with changes of b_1 units in predicted values of y (p. 212).

The slope can be found by

$$b_1 = r \frac{s_y}{s_x}$$

Intercept	The intercept, b_0, gives the $\hat{y}$-value when x is 0. You can find the intercept from $b_0 = \bar{y} - b_1 \bar{x}$ (p. 212).
Regression line (Line of best fit)	The particular linear equation

$$\hat{y} = b_0 + b_1 x$$

that satisfies the least squares criterion is called the least squares regression line. Casually, we often just call it the regression line, or the line of best fit (p. 219).

Regression to the mean	Because the correlation is always less than 1.0 in magnitude, each predicted $\hat{y}$ tends to be fewer standard deviations from its mean than its corresponding x was from its mean (p. 219).
Standard deviation of the residuals (s_e)	The standard deviation of the residuals is found by $s_e = \sqrt{\dfrac{\Sigma e^2}{n - 2}}$. When the residuals are roughly Normally distributed (check their histogram), their sizes can be well described by using this standard deviation and the 68–95–99.7 Rule (p. 221).
Coefficient of determination R^2	The square of the correlation between y and x (p. 224).

- R^2 gives the fraction of the variability of y accounted for by the least squares linear regression on x.
- R^2 is an overall measure of how successful the regression is in linearly relating y to x.

On the Computer REGRESSION

All statistics packages make a table of results for a regression. These tables may differ slightly from one package to another, but all are essentially the same—and all include much more than we need to know for now. Every computer regression table includes a section that looks something like this:

Standard dev of residuals (s_e)

R-squared

The "dependent," response, or y-variable

```
Dependent variable is: Total Fat
R-squared = 69.0%
s = 9.277

Variable    Coefficient   SE(Coeff)   t-Ratio   P-Value
Intercept   6.83077       2.664       2.56      0.0158
Protein     0.971381      0.1209      8.04      ≤0.0001
```

The "explanatory," predictor, or x-variable

The slope

The intercept

We'll deal with all of these later in the book. You may ignore them for now.

The slope and intercept coefficient are given in a table such as this one. Usually the slope is labelled with the name of the x-variable, and the intercept is labelled "Intercept" or "Constant." So the regression equation shown here (for data different from the current chapter example) is

$$\widehat{Fat} = 6.83077 + 0.97138 \, Protein.$$

It is not unusual for statistics packages to give many more digits of the estimated slope and intercept than could possibly be estimated from the data. (The original data were reported to the nearest gram.) Ordinarily, you should round most of the reported numbers to one digit more than the precision of the data, and the slope to two. We will learn about the other numbers in the regression table later in the book. For now, all you need to be able to do is find the coefficients, the s_e, and the R^2 value.

EXCEL

Make a scatterplot of the data. You will now see in the menu, under Chart Tools > Design, a choice of various Chart Layouts. Choose Layout 9, which show the fitted line and its equation, and R^2.

COMMENTS

The computer section for Chapter 6 shows how to make a scatterplot.

JMP

- Choose Fit Y by X from the **Analyze** menu.
- Specify the y-variable in the Select Columns box and click the y, **Response** button.
- Specify the x-variable and click the **X, Factor** button.
- Click **OK** to make a scatterplot.

- In the scatterplot window, click on the red triangle beside the heading labelled "**Bivariate Fit . . .**" and choose **Fit Line**. JMP draws the least squares regression line on the scatterplot and displays the results of the regression in tables below the plot.

MINITAB

- Choose **Regression** from the **Stat** menu.
- From the Regression submenu, choose **Fitted Line Plot**.
- In the Fitted Line Plot dialog, click in the **Response Y** box, and assign the y-variable from the variable list.
- Click in the **Predictor X** box, and assign the x-variable from the Variable list. Make sure that the Type of Regression Model is set to Linear. Click the **OK** button.

COMMENTS

Alternatively, **Stat > Regression > Regression > Fit Regression Model** may be used—no graph but more output options.
In MINITAB EXPRESS, select Statistics>Regression>Simple Regression

R

- lm(Y ~ X) produces the linear model.
- summary(lm(Y ~ X)) produces more information about the model.

COMMENTS

Typically, your variables X and Y will be in a data frame. If DATA is the name of the data frame, then
- lm(Y ~ X, data = DATA) is the preferred syntax.

SPSS

- Choose **Interactive** from the **Graphs** menu.
- From the interactive Graphs submenu, choose **Scatterplot**.
- In the Create Scatterplot dialog, drag the y-variable into the **y-axis target**, and the x-variable into the **x-axis target**.

- Click on the Fit tab.
- Choose **Regression** from the **Method** popup menu. Click the **OK** button.

STATCRUNCH

- Click on **Stat**.
- Choose **Regression » Simple Linear**.
- Choose X and Y variable names from the list of columns.
- Click on **Next** (twice) to **Plot the fitted line** on the scatterplot.
- Click on **Calculate** to see the regression analysis.
- Click on **Next** to see the scatterplot.

COMMENTS

Remember to check the scatterplot to be sure a linear model is appropriate.
Note that before you **Calculate**, clicking on **Next** also allows you to:
- enter an X-value for which you want to find the predicted Y-value
- save all the fitted values
- save the residuals
- ask for a residuals plot

TI-83/84 PLUS

To find the equation of the regression line (add the line to a scatterplot), choose **LinReg(a+bx)**, tell it the list names, and then add a comma to specify a function name (from **VARS Y-Vars 1:Function**). The final command looks like

LinReg (a+bx)L1, L2, Y1

- To make a residuals plot, set up a **STATPLOT** as a scatterplot.
- Specify your explanatory data list as Xlist.
- For Ylist, import the name RESID from the **LIST NAMES** menu. ZoomStat will now create the residuals plot.

COMMENTS

Each time you execute a **LinReg** command, the calculator automatically computes the residuals and stores them in a data list named RESID. If you want to see them, go to **STAT EDIT**. Space through the names of the lists until you find a blank. Import RESID from the LIST NAMES menu. Now every time you have the calculator compute a regression analysis, it will show you the residuals.

Chapter Exercises

T indicates data is available in a saved file at the Companion Website at www.pearsoncanada.ca/deveaux

15. Regression equations Fill in the missing information in the table below.

	$\bar{x}$	s_x	$\bar{y}$	s_y	r	$\hat{y} = b_0 + b_1 x$
a)	10	2	20	3	0.5	
b)	2	0.06	7.2	1.2	−0.4	
c)	12	6			−0.8	$\hat{y} = 200 - 4x$
d)	2.5	1.2		100		$\hat{y} = -100 + 50x$

16. More regression equations Fill in the missing information in the table below.

	$\bar{x}$	s_x	$\bar{y}$	s_y	r	$\hat{y} = b_0 + b_1 x$
a)	30	4	18	6	−0.2	
b)	100	18	60	10	0.9	
c)		0.8	50	15		$\hat{y} = -10 + 15x$
d)			18	4	−0.6	$\hat{y} = 30 - 2x$

17. Residuals Tell what each of the residual plots below indicates about the appropriateness of the linear model that was fit to the data.

18. Residuals II Tell what each of the residual plots below indicates about the appropriateness of the linear model that was fit to the data.

19. Least squares Consider the four points (10, 10), (20, 50), (40, 20), and (50, 80). The least squares line is $\hat{y} = 7.0 + 1.1x$. Explain (or show) what *least squares* means using these data as a specific example.

20. Least squares II Consider the four points (200, 1950), (400, 1650), (600, 1800), and (800, 1600). The least squares line is $\hat{y} = 1975 - 0.45x$. Explain (or show) what *least squares* means using these data as a specific example.

21. Real estate The Re/Max August/Sept 2008 Market Report lists the asking price and size, in square feet, of 38 condos under 1500 square feet in downtown Toronto. A regression to predict the asking price, in thousands of dollars, from size has an R^2 of 70.2%. The plot of residuals indicates that a linear model is appropriate.

a) What are the variables and units in this regression?
b) What units does the slope have?
c) Do you think the slope is positive or negative? Explain.

22. Money and degrees The 2001 Canadian Census provides data on a number of variables for each census tract (neighbourhood). For example, we can find the average income and percent with a Bachelor's degree for each of the 237 census tracts in Ottawa-Gatineau. A regression to predict average income in dollars from percent with degree has an R^2 of 54.2%. The plot of residuals versus predictor indicates that a linear model is appropriate.

a) What are the variables and units in this regression?
b) What units does the slope have?
c) Do you think the slope is positive or negative? Explain.

23. Real estate again The regression of price on size of condos in Toronto, as described in Exercise 21, had $R^2 = 70.2\%$. Write a sentence (in context, of course) summarizing what the R^2 value says about this regression.

24. Money again The regression of average income on percent with degree for Ottawa-Gatineau census tracts, as described in Exercise 22, had $R^2 = 54.2\%$. Write a sentence (in context, of course) summarizing what the R^2 value says about this regression.

25. Real estate redux The regression of price on size of homes in Toronto, as described in Exercise 21, had $R^2 = 70.2\%$.

a) What is the correlation between size and price?
b) What would you predict about the price of a home one standard deviation above average in size?
c) What would you predict about the price of a home two standard deviations below average in size?

26. More money The regression of average income on percent with degree for census tracts in Ottawa-Gatineau, as described in Exercise 22, had $R^2 = 54.2\%$.

a) What is the correlation between average income and percent with degree?
b) What would you predict about the average income of a census tract one standard deviation above average in percent with degree?
c) What would you predict about the average income of a census tract two standard deviations below average in percent with degree?

27. More real estate Consider the Toronto downtown condo data from Exercise 21 again. The regression analysis gives the model $\widehat{Price} = 49.30 + 0.37\ Size$.

a) Explain what the slope of the line says about condo prices and condo size.

b) What asking price would you predict for a 1000-square-foot condo in this market?

c) A real estate agent shows a potential buyer a 1200-square-foot condo, saying that the asking price is $6000 less than what one would expect for a condo of this size. What is the asking price, and what is this $6000 called?

28. Lots of money Consider the Ottawa-Gatineau census data from Exercise 22 again. The regression analysis gives the model $\overline{Average\ income} = \$23\ 129 + 527.4\ Degree$.

a) Explain what the slope of the line says about average income and percent with degree.

b) What average income would you predict for a census tract with 50% degree holders?

c) One census tract with 60% degree holders had an average income of $81 000. Is this more or less than you'd expect? By how much? What's that called?

d) Another census tract with 42% degree holders had an average income of $25 000. Is this more or less than you'd expect? By how much? What's that called?

e) Make up a rough plot of what the data and fitted line might look like here, and indicate where the census tracts in parts c) and d) would lie. Note that these two tracts were rather unusual in their average incomes, as the standard deviation of the residuals was about 7000.

T 29. Cigarettes Is the nicotine content of a cigarette related to the "tars"? A collection of data (in milligrams) on 29 cigarettes produced the scatterplot, residual plot, and regression analysis shown:

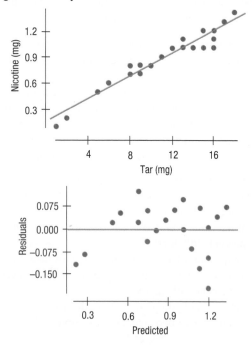

Dependent variable is: nicotine
R squared = 92.4%

Variable	Coefficient
Constant	0.154030
Tar	0.065052

a) Do you think a linear model is appropriate here? Explain.

b) Explain the meaning of R^2 in this context.

T 30. Attendance 2013, revisited In the previous chapter, you looked at the relationship between the number of wins by American League baseball teams and the average attendance at their home games for the 2013 season. Here are the scatterplot, the residuals plot, and part of the regression analysis:

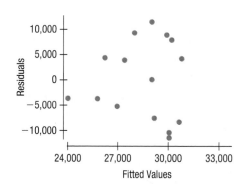

Dependent variable is Home Attendance
R squared = 6.45%

Variable	Coefficient
Intercept	16484.0
Wins	147.4

a) Do you think a linear model is appropriate here? Explain.

b) Interpret the meaning of R^2 in this context.

c) Do the residuals show any pattern worth remarking on?

d) The point at the top of the plots is the New York Yankees. What can you say about the residual for the Yankees?

T 31. Another cigarette Consider again the regression of nicotine content on tar (both in milligrams) for the cigarettes examined in Exercise 29.

a) What is the correlation between tar and nicotine?

b) What would you predict about the average nicotine content of cigarettes that are two standard deviations below average in tar content?

c) If a cigarette is one standard deviation above average in nicotine content, what do you suspect is true about its tar content?

T 32. Second inning 2010 Consider again the regression of *Average Attendance* on *Wins* for the baseball teams examined in Exercise 30.
 a) What is the correlation between *Wins* and *Average Attendance*?
 b) What would you predict about the *Average Attendance* for a team that is 2 standard deviations above average in *Wins*?
 c) If a team is 1 standard deviation below average in attendance, what would you predict about the number of games the team has won?

T 33. Last cigarette Take another look at the regression analysis of tar and nicotine content of the cigarettes in Exercise 29.
 a) Write the equation of the regression line.
 b) Estimate the nicotine content of cigarettes with 4 milligrams of tar.
 c) Interpret the meaning of the slope of the regression line in this context.
 d) What does the *y*-intercept mean?
 e) If a new brand of cigarette contains 7 milligrams of tar with a residual of −0.5 mg, what is the nicotine content?

T 34. Last inning 2010 Refer again to the regression analysis for average attendance and games won by American League baseball teams, seen in Exercise 30.
 a) Write the equation of the regression line.
 b) Estimate the *Average Attendance* for a team with 50 *Wins*.
 c) Interpret the meaning of the slope of the regression line in this context.
 d) In general, what would a negative residual mean in this context?
 e) The San Francisco Giants, the 2010 World Champions, are not included in these data because they are a National League team. During the 2010 regular season, the Giants won 92 games and averaged 41 736 fans at their home games. Calculate the residual for this team, and explain what it means.

35. What slope? If you create a regression model for predicting the weight of a car (in pounds) from its length (in feet), is the slope most likely to be 3, 30, 300, or 3000? Explain.

36. What slope again? If you create a regression model for estimating the height of a pine tree (in centimetres) based on the circumference of its trunk (in centimetres), is the slope most likely to be 0.1, 1, 10, or 100? Explain.

37. Misinterpretations A Biology student who created a regression model to use a bird's height when perched for predicting its wingspan made these two statements. Assuming the calculations were done correctly, explain what is wrong with each interpretation.
 a) My R^2 of 93% shows that this linear model is appropriate.
 b) A bird 30 cm tall will have a wingspan of 34 cm.

38. More misinterpretations A Sociology student investigated the association between a country's literacy rate and life expectancy, then drew the conclusions listed below. Explain why each statement is incorrect. (Assume that all the calculations were done properly.)
 a) The literacy rate determines 64% of the life expectancy for a country.
 b) The slope of the line shows that an increase of 5% in literacy rate will produce a two-year improvement in life expectancy.

39. ESP People who claim to "have ESP" participate in a screening test in which they have to guess which of several images someone is thinking of. You and a friend both took the test. You scored two standard deviations above the mean, and your friend scored one standard deviation below the mean. The researchers offer everyone the opportunity to take a retest.
 a) Should you choose to take this retest? Explain.
 b) Now explain to your friend what his decision should be and why.

40. SI jinx Players in any sport who are having great seasons—turning in performances that are much better than anyone might have anticipated—often are pictured on the cover of *Sports Illustrated*. Frequently, their performances then falter somewhat, leading some athletes to believe in a "*Sports Illustrated* jinx." Similarly, it is common for phenomenal rookies to have less stellar second seasons—the so-called "sophomore slump." While fans, athletes, and analysts have proposed many theories about what leads to such declines, a statistician might offer a simpler (statistical) explanation. Explain.

T 41. SAT scores The SAT is a test often used in the U.S. as part of an application to college. SAT scores are between 200 and 800. Tests are given in both math and verbal areas. Doing the SAT-Math problems also involves the ability to read and understand the questions, but can a person's verbal score be used to predict the math score? Verbal and math SAT scores of a high-school graduating class are displayed in the scatterplot, with the regression line added.

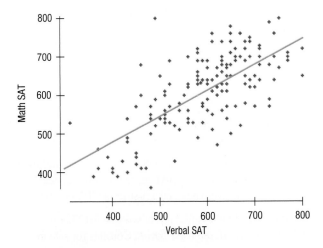

a) Describe the relationship.

b) Are there any students whose scores do not seem to fit the overall pattern?

c) For these data, $r = 0.685$. Interpret this statistic.

d) These verbal scores averaged 596.3 with a standard deviation of 99.5, and the math scores averaged 612.2 with a standard deviation of 96.1. Write the equation of the regression line.

e) Interpret the slope of this line.

f) Predict the math score of a student with a verbal score of 500.

g) Every year some student scores a perfect 1600. Based on this model, what would that student's residual be for their math score?

42. **Success in college** Many U.S. colleges use SAT scores in the admissions process because they believe these scores provide some insight into how a high-school student will perform at the college level. Suppose the entering freshmen at a certain college have mean combined SAT score of 1833, with a standard deviation of 123. In the first semester these students attained a mean GPA of 2.66, with a standard deviation of 0.56. A scatterplot showed the association to be reasonably linear, and the correlation between SAT score and GPA was 0.47.

a) Write the equation of the regression line.

b) Explain what the *y*-intercept of the regression line indicates.

c) Interpret the slope of the regression line.

d) Predict the GPA of a freshman who scored a combined 1400.

e) Based upon these statistics, how effective do you think SAT scores would be in predicting academic success during the first term of the freshman year at this college? Explain.

f) As a student, would you rather have a positive or a negative residual in this context? Explain.

T 43. **SAT, take 2** Suppose we wanted to use SAT math scores to estimate verbal scores based on the information in Exercise 41.

a) What is the correlation?

b) Write the equation of the line of regression predicting verbal scores from math scores.

c) In general, what would a positive residual mean in this context?

d) A person tells you her math score was 500. Predict her verbal score.

e) Using that predicted verbal score and the equation you created in Exercise 41, predict her math score.

f) Why doesn't the result in part e) come out to 500?

44. **Success, part 2** Based on the statistics for college freshmen given in Exercise 42, what SAT score might be expected among freshmen who attained a first-term GPA of 3.0?

T 45. **Used cars 2013** Classified ads at www.auto123.com (on July 22, 2013) offered used Toyota Corollas for sale in

central Ontario. Listed below are ages and advertised prices for some Corolla CEs with automatic transmission:

Age (years)	Price (dollars)
1	13 900
1	14 610
2	13 998
2	16 887
3	10 900
3	12 995
3	14 995
4	10 999
4	11 988
4	11 995
4	12 495
5	8 988
5	9 488
5	10 488
5	10 989
7	11 400
8	6 995
9	6 988
10	10 000
12	2 700

a) Make a scatterplot for these data.

b) Describe the association between age and price of a used Corolla.

c) Do you think a linear model is appropriate?

d) Computer software says that $R^2 = 74.8\%$. What is the correlation between age and price?

e) Explain the meaning of R^2 in this context.

f) Why doesn't this model explain 100% of the variability in the price of a used Corolla?

T 46. **Drug abuse** In the exercises of the last chapter you examined results of a survey conducted in the United States and 10 countries of Western Europe to determine the percentage of teenagers who had used marijuana and other drugs. Below is the scatterplot. Summary statistics showed that the mean percent that had used marijuana was 23.9, with a standard deviation of 15.6%. An average of 11.6% of teens had used other drugs, with a standard deviation of 10.2%.

a) Do you think a linear model is appropriate? Explain.
b) For this regression, R^2 is 87.3%. Interpret this statistic in this context.
c) Write the equation you would use to estimate the percentage of teens who use other drugs from the percentage who have used marijuana.
d) Explain in context what the slope of this line means.
e) Do these results confirm that marijuana is a "gateway drug"—that is, that marijuana use leads to the use of other drugs?

47. More used cars 2013 Use the advertised prices for Toyota Corollas given in Exercise 45 to create a linear model for the relationship between a car's age and its price.
a) Find the equation of the regression line.
b) Explain the meaning of the slope of the line.
c) Explain the meaning of the y-intercept of the line.
d) If you want to sell a 6-year-old Corolla, what price seems appropriate?
e) You have a chance to buy one of two cars. They are about the same age and appear to be in equally good condition. Would you rather buy the one with a positive residual or a negative residual? Explain.
f) You see a "For Sale" sign on a 5-year-old Corolla stating the asking price as $9000. What is the residual?
g) Would this regression model be useful in establishing a fair price for a 17-year-old car? Explain.

48. Veggie burgers Recently Burger King introduced a meat-free burger. The nutrition label is shown here.

Nutrition Facts

Calories	330
Fat	10g*
Sodium	760mg
Sugars	5g
Protein	14g
Carbohydrates	43g
Dietary Fibre	4g
Cholesterol	0

* (2 grams of saturated fat)

RECOMMENDED DAILY VALUES
(based on a 2,000-calorie/day diet)

Iron	20%
Vitamin A	10%
Vitamin C	10%
Calcium	6%

a) Use the regression model created in this chapter, $\widehat{Fat} = 8.4 + 0.91\ Protein$, to predict the fat content of this burger from its protein content.
b) What is its residual? How would you explain the residual?

c) Write a brief report about the fat and protein content of this new menu item. Be sure to talk about the variables by name and in the correct units.

49. Burgers In the last chapter's exercises, you examined the association between the amounts of fat and calories in fast-food hamburgers. Here are the data:

Fat (g)	19	31	34	35	39	39	43
Calories	410	580	590	570	640	680	660

a) Create a scatterplot of *Calories vs. Fat.*
b) Interpret the value of R^2 in this context.
c) Write the equation of the line of regression.
d) Use the residuals plot to explain whether your linear model is appropriate.
e) Explain the meaning of the y-intercept of the line.
f) Explain the meaning of the slope of the line.
g) A new burger containing 28 grams of fat is introduced. According to this model, its residual for *Calories* is 133. How many calories does the burger have?

50. Chicken Chicken sandwiches are often advertised as a healthier alternative to beef because many are lower in fat. Tests on 11 brands of fast food chicken sandwiches produced the following summary statistics and scatterplot from a graphing calculator:

	Fat (g)	Calories
Mean	20.6	472.7
St. Dev.	9.8	144.2
Correlation	0.947	

a) Do you think a linear model is appropriate in this situation?
b) Describe the strength of this association.
c) Write the equation of the regression line.
d) Explain the meaning of the slope.
e) Explain the meaning of the y-intercept.
f) What does it mean if a certain sandwich has a negative residual?
g) If a chicken sandwich and a burger each advertised 35 grams of fat, which would you expect to have more calories? (See Exercise 49.)
h) McDonald's Filet-O-Fish sandwich has 26 grams of fat and 470 calories. Does the fat/calorie relationship in this sandwich appear to be very different from that found in chicken sandwiches or in burgers (see Exercise 49)? Explain.

51. A second helping of burgers In Exercise 49, you created a model that can estimate the number of calories in a burger when the fat content is known.

a) Explain why you cannot use that model to estimate the fat content of a burger with 600 calories.

b) Using an appropriate model, estimate the fat content of a burger with 600 calories.

T 52. Cost of living 2013 Numbeo.com lists the cost of living (COL) for many cities around the world. These rankings scale New York City as 100, and express the cost of living in other cities as a percentage of the New York cost. For example, the table below shows 25 of the most expensive cities in the world, and indicates that the cost of living in Paris was 30.3% higher than New York in 2010, but only 25.0% higher in 2013.

a) Using the scatterplot below, describe the association between costs of living in 2010 and 2013.

b) The correlation is 0.714. Find and interpret the value of R^2.

c) The regression equation predicting the 2013 cost of living from the 2010 figure is $\widehat{COL\ 2013} = -38.292 + 1.319\ COL\ 2010$. Use this equation to find the residual for Oslo.

d) Explain what the residual means.

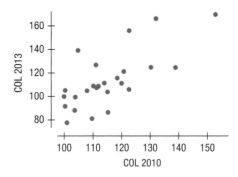

City	COL 2013	COL 2010
Oslo	170.04	152.85
Copenhagen	124.9	138.91
Zurich	166.51	132.03
Paris	125.02	130.3
Geneva	156.17	122.69
Milan	106.12	122.58
Dublin	121.5	120.79
Brussels	111.55	120
London	116.06	118.47
Berlin	86.65	115.24
Lyon	104.08	115.04
Helsinki	111.76	113.91
Tampere	109.04	111.96
Amsterdam	107.61	111.3
Stockholm	126.96	111.02
Oxford	109.15	110.15
Tampa	81.32	109.64
Munich	105.01	107.96
Sydney	139.25	104.83
Vienna	99.5	103.88
Madrid	88.27	103.67
Atlanta	77.77	100.99
Edmonton	105.24	100.36
Seattle	91.79	100.33
New York	100	100

T 53. New York bridges We saw in this chapter that in Tompkins County, NY, older bridges were in worse condition than newer ones. Tompkins is a rural area. Is this relationship true in New York City as well? Here are data on the *Safety Score* (as measured by the state Department of Transportation) and *Age at Inspection* for bridges in New York City.

Dependent variable is: Safety Score
R-squared = 2.6%
s = 0.6708

Variable	Coefficient
Intercept	4.95147
Age at Inspection	−0.00481

a) New York State defines any bridge with a safety score less than 5 as *deficient*. What does this model predict for the safety scores of New York City bridges?

b) Our earlier model found that the safety scores of bridges in Tompkins County was decreasing at about 0.023 per year. What does this model say about New York City bridges?

c) How much faith would you place in this model? Explain.

T 54. Candy 2009 The table below shows the increase in Halloween candy sales over a 6-year period as reported by the National Confectioners Association (www.candyusa.com). Using these data, estimate the amount of candy sold in 2009. Discuss the appropriateness of your model and your faith in the estimate. Then comment on the fact that NCA reported 2009 sales of $2.207 million. (Enter *Year* as 3, 4, . . .)

2003	1.993
2004	2.041
2005	2.088
2006	2.146
2007	2.202
2008	2.209

55. Climate change 2013 The earth's climate is getting warmer. The most common theory attributes the increase to an increase in atmospheric levels of carbon dioxide (CO_2), a greenhouse gas. Here is a scatterplot showing the mean annual CO_2 concentration in the atmosphere, measured in parts per million (ppm) at the top of Mauna Loa in Hawaii, and the mean annual air temperature over both land and sea across the globe, in degrees Celsius (°C) for the years 1970 to 2013.

A regression predicting *Temperature* from CO_2 produces the following output table (in part):

Dependent variable is Global Temperature (°C)
R-squared $= 73.6\%$

Variable	Coefficient
Intercept	10.64
CO_2 (ppm)	0.0103

a) What is the correlation between CO_2 and *Temperature*?
b) Explain the meaning of *R*-squared in this context.
c) Give the regression equation.
d) What is the meaning of the slope in this equation?
e) What is the meaning of the *y*-intercept of this equation?
f) Here is a scatterplot of the residuals vs. predicted values. Does this plot show evidence of the violation of any assumptions behind the regression? If so, which ones?

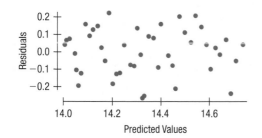

g) CO_2 levels will probably reach 400 ppm by 2020. What mean *Temperature* does the regression predict for that concentration of CO_2?
h) Does the answer in g) mean that when the CO_2 level hits 400 ppm, the *Temperature* will reach the predicted level? Explain briefly.

T 56. Birthrates 2010 The following table shows the number of live births per 1000 population in the United States, starting in 1965. (National Center for Health Statistics (2010). Data for United States in 2010.)

Year	1965	1970	1975	1980	1985	1990	1995	2000	2005	2010
Rate	19.4	18.4	14.8	15.9	15.6	16.4	14.8	14.4	14.0	13.0

a) Make a scatterplot and describe the general trend in *Birthrates*. (Enter *Year* as years since 1900: 65, 70, 75, etc.)
b) Find the equation of the regression line.
c) Check to see if the line is an appropriate model. Explain.
d) Interpret the slope of the line.
e) The table gives rates only at 5-year intervals. Estimate what the rate was in 1978.
f) In 1978, the birthrate was actually 15.0. How close did your model come?
g) Predict what the *Birthrate* will be in 2010. Comment on your faith in this prediction.
h) Predict the *Birthrate* for 2025. Comment on your faith in this prediction.

T 57. Body fat It is difficult to accurately determine a person's body fat percentage without immersing him or her in water. Researchers hoping to find ways to make a good estimate immersed 20 male subjects, then measured their waists and recorded their weights.

a) Create a model to predict *% Body Fat* from *Weight*.
b) Do you think a linear model is appropriate? Explain.
c) Interpret the slope of your model.
d) Is your model likely to make reliable estimates? Explain.
e) What is the residual for a person who weighs 190 pounds and has 21% body fat?

Waist (in.)	Weight (lb)	Body Fat (%)
32	175	6
36	181	21
38	200	15
33	159	6
39	196	22
40	192	31
41	205	32
35	173	21
38	187	25
38	188	30
33	188	10
40	240	20
36	175	22
32	168	9
44	246	38
33	160	10
41	215	27
34	159	12
34	146	10
44	219	28

T **58. Body fat, again** Would a model that uses the person's *waist* size be able to predict the *% Body Fat* more accurately than one that uses a person's weight? Using the data in Exercise 57, create and analyze that model.

T **59. Heptathlon 2012** We discussed the women's 2012 Olympic heptathlon in Chapter 5. Here are the results from the high jump (metres), 800-metre run (seconds), and long jump (metres) for the women who successfully completed all three events in the 2012 Olympics:

Athlete	Country	Points	High Jump	Long Jump	800 m
Jessica Ennis	GBR	6955	1.86	6.48	128.65
Lilli Schwarzkopf	GER	6649	1.83	6.30	130.50
Tatyana Chernova	RUS	6628	1.80	6.54	129.56
Lyudmyla Yosypenko	UKR	6618	1.83	6.31	133.28
Austra Skujytė	LTU	6599	1.92	6.25	140.59
Antoinette Nana Djimou Ida	FRA	6576	1.80	6.13	133.62
Jessica Zelinka	CAN	6480	1.68	5.91	129.15
Kristina Savitskaya	RUS	6452	1.83	6.21	132.27
Laura Ikauniece	LAT	6414	1.83	6.13	132.13
Hanna Melnychenko	UKR	6392	1.80	6.40	132.90
Brianne Theisen	CAN	6383	1.83	6.01	129.27
Dafne Schippers	NED	6324	1.80	6.28	135.52
Nadine Broersen	NED	6319	1.86	5.94	136.98
Jessica Samuelsson	SWE	6300	1.77	6.18	131.31
Katarina Johnson-Thompson	GBR	6267	1.89	6.19	130.76
Sharon Day	USA	6232	1.77	5.85	131.31
Yana Maksimava	BLR	6198	1.89	5.99	133.37
Eliška Klučinová	CZE	6109	1.80	6.13	136.08
Ellen Sprunger	SUI	6107	1.71	5.88	137.54
Olga Kurban	RUS	6084	1.80	5.83	139.82
Marisa De Aniceto	FRA	6030	1.71	5.76	136.20
Györgyi Farkas	HUN	6013	1.80	6.07	137.83
Grit Šadeiko	EST	6013	1.74	6.11	143.01
Sofia Ifadidou	GRE	5947	1.68	5.81	142.03
Ivona Dadic	AUT	5935	1.80	6.00	135.90
Sarah Cowley	NZL	5873	1.80	6.00	139.01
Louise Hazel	GBR	5856	1.59	5.77	138.78
Ida Marcussen	NOR	5846	1.68	5.82	133.62
Chantae McMillan	USA	5688	1.68	5.37	160.55

Source: Data from Olympics 2012. From http://www.espn.in/?src=com

Let's examine the association among these events. Perform a regression to predict high jump performance from the 800-metre results. Set aside Ms. McMillan in your analysis, in view of her very atypical 800 m time.

a) What is the regression equation? What does the slope mean?

b) What is the R^2 value?

c) Do good high jumpers tend to be fast runners? (Be careful—low times are good for running events and high distances are good for jumps.)

d) What does the residual plot reveal about the model?

e) Do you think this is a useful model? Would you use it to predict high jump performance? (Compare the residual standard deviation to the standard deviation of the high jumps.)

60. Heptathlon 2012 again We saw the data for the women's 2012 Olympic heptathlon in Exercise 59. Are the two jumping events associated? Perform a regression of the long jump results on the high jump results. Set aside in your analysis the last place (points) finisher, an apparent outlier.

a) What is the regression equation? What does the slope mean?

b) What is the R^2 value?

c) Do good high jumpers tend to be good long jumpers?

d) What does the residual plot reveal about the model?

e) Do you think this is a useful model? Would you use it to predict long jump performance? (Compare the residual standard deviation to the standard deviation of the long jumps.)

T **61. Hard water** In an investigation of environmental causes of disease, data were collected on the annual mortality rate (deaths per 100 000) for males in 61 large towns in England and Wales. In addition, the water hardness was recorded as the calcium concentration (parts per million, ppm) in the drinking water. The following display shows the relationship between mortality and calcium concentration for these towns:

a) Describe what you see in this scatterplot, in context.

b) Here is the regression analysis of mortality and calcium concentration. What is the regression equation?

Dependent variable is: Mortality
$R^2 = 43\%$
$s = 143.0$

Variable	Coefficient
Intercept	1676
Calcium	−3.23

c) Interpret the slope and *y*-intercept of the line, in context.

d) The largest residual, with a value of −348.6, is for the town of Exeter. Explain what this value means.

e) The hardness of Derby's municipal water is about 100 ppm of calcium. Use this equation to predict the mortality rate in Derby.

f) Explain the meaning of R^2 in this situation.

62. Gators Wildlife researchers monitor many wildlife populations by taking aerial photographs. Can they estimate the weights of alligators accurately from the air? Here is a regression analysis of the weight of alligators (in pounds) and their length (in inches) based on data collected from captured alligators.

Dependent variable is: Weight

R-squared = 83.6%

s = 54.01

Variable	Coefficient
Intercept	−393
Length	5.9

a) Did they choose the correct variables to use as the response variable and the predictor? Explain.

b) What is the correlation between an alligator's length and weight?

c) Write the regression equation.

d) Interpret the slope of the equation in this context.

e) Do you think this equation will allow the scientists to make accurate predictions about alligators? What part of the regression analysis indicates this? What additional concerns do you have?

T 63. Brakes The table below shows stopping distances in metres for a car tested three times at each of five speeds. We hope to create a model that predicts stopping distance from the speed of the car.

Speed (km/h)	Stopping Distances (m)
32	19.5, 18.9, 18.0
48	34.8, 36.0, 32.0
64	46.7, 52.1, 50.3
80	70.4, 61.9, 72.6
95	96.6, 97.9, 84.1

a) Explain why a linear model is not appropriate. Be sure to examine the residual plot.

b) Re-express the data to straighten the scatterplot, by using the square roots of the stopping distances.

c) Fit an appropriate model. Is the residual plot any better after re-expression?

d) Estimate the stopping distance for a car travelling 90 km/h.

T 64. Ontario interprovincial migration 2015 Below are data showing the net interprovincial migration each year for Ontario (number arriving minus number leaving). The data are not for calendar years, but for a one-year period ending in the displayed year.

Year	Ontario Net Migration
1979	−4 325
1980	−22 362
1981	−33 247
1982	−5 665
1983	23 585
1984	36 400
1985	33 885
1986	33 562
1987	42 601
1988	35 215
1989	9 739
1990	−5 961
1991	−11 627
1992	−11 045
1993	−14 189
1994	−9 420
1995	−2 841
1996	−2 822
1997	1 977
1998	9 231
1999	16 706
2000	22 369
2001	18 623
2002	5 354
2003	637
2004	−6 935
2005	−11 172
2006	−17 501
2007	−20 047
2008	−14 750
2009	−15 601
2010	−4 662
2011	−4 007
2012	−10 611
2013	−13 901
2014	−14 564
2015	−8 763

Source: Adapted from Statistics Canada CANSIM table 051-0004, March 31, 2015

a) Fit a least squares line to predict Ontario interprovincial migration from year.

b) Use your equation to predict Ontario's interprovincial migration in 2016.

c) Do you think your prediction is reasonable? Do you think your regression line is an appropriate model?

T 65. Immigrant commuters Recent immigrants are more likely than Canadian-born to use public transit. Below are data for selected Census Metropolitan Areas (CMA) on the percentage of individuals who commute to work by public transit, for recent immigrants (<10 years since arrival) and for Canadian-born:

CMA	Canadian Born	Recent Immigrant
Montreal	20.1	47.8
Toronto	20.0	35.4
Ottawa–Gatineau	17.8	33.5
Calgary	12.7	24.6
Winnipeg	13.6	24.3
Vancouver	10.9	20.1
Edmonton	8.8	19.3
Victoria	10.3	17.2
Hamilton	8.3	16.4
London	6.7	15.1
Windsor	3.3	10.0
Kitchener	4.4	8.8
Abbotsford	2.1	1.2

Source: Adapted from Statistics Canada trends and conditions in Census Metropolitan Areas, immigrants in Canada's Census Metropolitan Areas, 89-613-MIE2004003, www.statcan.gc.ca/bsolc/olc-cel/olc-cel?catno=89-613-MIE&lang=eng.

a) Plot the immigrant percentage versus the Canadian-born percentage. Describe the plot.
b) Fit a line to predict the immigrant percent from the Canadian-born percent and superimpose this line on the data scatterplot.
c) For every additional 1% ridership by Canadian-born commuters, how much additional ridership (as a percent) by recent immigrants do we tend to find, on average?
d) What is unusual about Montreal? Use some numbers (via the regression) to explain more precisely this unusualness.
e) Plot residuals from the regression versus the fitted (predicted) values or versus the Canadian-born percentage. Describe the pattern you see and what it suggests. Do you think Montreal has anything to do with the pattern?

T 66. Math and gender 2009 Below are mean PISA (Programme for International Student Assessment) math scores for samples of 15-year-old male and female students from a number of randomly selected schools in each of various OECD (Organization for Economic Cooperation and Development) and other industrialized nations (4000–10 000 per country).

Country	Mean Male Score	Mean Female Score
OECD		
Australia	519	509
Austria	506	486
Belgium	526	504
Canada	533	521
Chile	431	410
Czech Republic	495	490
Denmark	511	495
Estonia	516	508
Finland	542	539
France	505	489
Germany	520	505
Greece	473	459
Hungary	496	484
Iceland	508	505
Ireland	491	483
Israel	451	443
Italy	490	475
Japan	534	524
Korea	548	544
Luxembourg	499	479
Mexico	425	412
Netherlands	534	517
New Zealand	523	515
Norway	500	495
Poland	497	493
Portugal	493	481
Slovak Republic	498	495
Slovenia	502	501
Spain	493	474
Sweden	493	495
Switzerland	544	524
Turkey	451	440
United Kingdom	503	482
United States	497	477
Partners		
Albania	372	383
Argentina	394	383
Azerbaijan	435	427
Brazil	394	379
Bulgaria	426	430
Colombia	398	366
Croatia	465	454
Dubai (UAE)	454	451
Hong Kong-China	561	547
Indonesia	371	372
Jordan	386	387
Kazakhstan	405	405
Kyrgyzstan	328	334
Latvia	483	481

Country	Mean Male Score	Mean Female Score
Liechtenstein	547	523
Lithuania	474	480
Macao-China	531	520
Montenegro	408	396
Panama	362	357
Peru	374	356
Qatar	366	371
Romania	429	425
Russian Federation	469	467
Serbia	448	437
Shanghai-China	599	601
Singapore	565	559
Chinese Taipei	546	541
Thailand	421	417
Trinidad and Tobago	410	418
Tunisia	378	366
Uruguay	433	421

Source: Based on data from OECD PISA 2009 online database, http://pisa2009.acer.edu.au, accessed on Feb 22, 2013.

a) Plot female mean score versus male mean score. Describe the relationship.
b) Fit a straight line for predicting female means from male means, and find and interpret the R^2 value.
c) Plot residuals versus male means or versus predicted (fitted) values. What do you learn?
d) In which country or countries do the female students perform unusually poorly in relation to the male students, according to the regression model? Give the residual(s).

67. **Regressing** Test scores over the years in an introductory statistics course have shown a 0.6 correlation between term test and final exam. Suppose test and exam both have a mean of 70 with similar standard deviations.
a) Predict the exam score of someone who gets 85 on the test.
b) Predict the test score of someone who gets 79 on the exam.
c) Is something wrong? Explain.

Just Checking ANSWERS

1. An increase in home size of one square foot is associated with an increase of $182.74, on average, in price.
2. Units are dollars per square foot.
3. About $365 480, on average.
4. You would guess the price to be about 2 × 0.84 = 1.68 standard deviations above the average price.
5. 1.68 SDs above the mean price is 338 000 + 1.68 × 83 790 = $478 767
6. Differences in the size of houses account for about 70% of the variation in house prices.
7. It's positive. The correlation and the slope have the same sign.
8. The correlation won't change, so neither will R^2. The slope will change because the units have changed.
9. You shouldn't be surprised since the value of your house is only slightly more than one standard deviation above the regression prediction. If it were more than two standard deviations above the predicted value, you might well be pleasantly surprised.

MathXL

MyLab Statistics

Go to MathXL at www.mathxl.com or MyLab™ Statistics at www.mystatlab.com. You can practise exercises for this chapter as often as you want. The guided solutions will help you find answers step by step. You'll find a personalized study plan available to you too!

LEARNING OBJECTIVES

① Perform model diagnostics using residual plots

② Check for unusual observations such as high leverage points, and assess their influence on the fitted model

③ Exercise proper caution when extrapolating

④ Don't assume a strong predictive relationship implies causation

⑤ Exercise caution when using averaged or restricted range data

Connections

How can we tell if our regression model is appropriate and sensible? We discussed some issues in the previous chapter, and here we will extend the discussion.

Regression opens new vistas of potential problems to look for—like curvature and unequal spread—but also some that are not dissimilar from ones we've thought about before.

Our concern with unusual cases harks back to our earlier concern with outliers in histograms and boxplots—and for many of the same reasons. And again we look for multiple modes, because it is also important that our data be from a single group and not made up of disparate groups. For regression, careful examination of scatterplots and residual plots is key!

Regression Wisdom

Regression is a powerful tool for forecasting. Economists using it successfully predicted ten out of the last two recessions.

—Hershey M. Friedman

Regression is used every day throughout the world to predict customer loyalty, numbers of admissions at hospitals, sales of automobiles, and many other things. Because regression is so widely used, it's also widely abused and misinterpreted. This chapter presents examples of regressions in which things are not quite as simple as they seem at first, and shows how you can still use regression to discover what the data have to say.

8.1 Examining Residuals

No regression analysis is complete without a display of the residuals to check that the linear model is reasonable. Because the residuals are what is "left over" after the model describes a relationship, they often reveal subtleties that were not clear from a plot of the original data. Sometimes these are additional details that help confirm or refine our understanding. Sometimes they reveal violations of the regression conditions that require our attention.

For example, the fundamental assumption in working with a linear model is that the relationship you are modelling is, in fact, linear. That sounds obvious, but you can't take it for granted. It may be hard to detect nonlinearity from the scatterplot you looked at before you fit the regression model. Sometimes you can see important features such as nonlinearity more readily when you plot the residuals.

Getting the "Bends": When the Residuals Aren't Straight

Jessica Meir and Paul Ponganis study emperor penguins at the Scripps Institution of Oceanography's Center for Marine Biotechnology and Biomedicine at the University of California at San Diego. Says Jessica:[1]

> *Emperor penguins are the most accomplished divers among birds, making routine dives of 5–12 minutes, with the longest recorded dive over 27 minutes. These birds can also dive to depths of over 500 metres! Since air-breathing animals like penguins must hold their breath while submerged, the duration of any given dive depends on how much oxygen is in the bird's body at the beginning of the*

---

[1]Excerpt from Research Note on Emperor Penguins, Scripps Institution of Oceanography's Center for Marine Biotechnology and Biomedicine at the University of California at San Diego by Jessica Meir. Published by Meir, Jessica.

dive, how quickly that oxygen gets used, and the lowest level of oxygen the bird can tolerate. The rate of oxygen depletion is primarily determined by the penguin's heart rate. Consequently, studies of heart rates during dives can help us understand how these animals regulate their oxygen consumption in order to make such impressive dives.

The researchers equip emperor penguins with devices that record their heart rates during dives. Figure 8.1 is a scatterplot of the *Dive Heart Rate* (beats per minute) and the *Duration* (minutes) of dives by these high-tech penguins. (Data in **Penguins**)

Figure 8.1

The scatterplot of *Dive Heart Rate* in beats per minute (bpm) versus *Duration* (minutes) shows a strong, roughly linear, negative association.

The scatterplot has a consistent trend from the upper left to the lower right and a moderately strong negative association ($R^2 = 71.5\%$). The linear regression equation

$$\widehat{DiveHeartRate} = 96.9 - 5.47\,Duration$$

says that for longer dives, the average dive heart rate decreases by about 5.47 beats per dive minute, starting from a heart rate of 96.9 beats per minute.

The scatterplot of the residuals against *Duration* in Figure 8.2 makes things much clearer. The Linearity Assumption says we should see no pattern, but instead there's a strong bend, starting high on the left, dropping down in the middle of the plot, and rising again at the right. Graphs of residuals often reveal patterns such as this that were not readily apparent in the original scatterplot.

> **STRAIGHT ENOUGH CONDITION**
>
> We can't *know* whether the Linearity Assumption is true, but we can see if it's *plausible* by checking the Straight Enough Condition.

Figure 8.2

Plotting the residuals against *Duration* reveals a bend. It was also in the original scatterplot, but here it's easier to see.

Now looking back at the original scatterplot, we can see that the scatter of points isn't really straight. There's a slight bend to that plot, but the bend is much easier to see in the residuals. Even though it means rechecking the Straight Enough Condition *after* you run the regression, it's always a good idea to check your scatterplot of the residuals for bends that you might have overlooked in the original scatterplot. And you might also wonder—is it possible to unbend the bends? A re-expression (logarithms for example) of one or both of the variables might well help straighten out the relationship (as discussed in Section 6.4).

> **Why Not Just Use a Curve?** While it is possible to fit various curves to data using our familiar "least-squares" criterion, the mathematics and calculations can be considerably more difficult. Straight lines are easier to fit and to understand. We know how to think about the slope and the *y*-intercept, for example. Powerful and simple statistical methods are available for analyzing linear models. So, it's usually better to linearize.

Sifting the Data and Residuals for Groups

In the Step-By-Step analysis in Chapter 7 to predict *Calories* from *Sugar* content in breakfast cereals, we examined scatterplots of the data (reposted in the margin below) and of the residuals (Data in **Cereals**). Our first impression was that they had no particular structure. But let's look again.

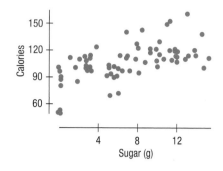

Look at Figure 8.3 below, a histogram of the residuals. How would you describe its shape? It looks like there might be small modes on both sides of the central body of the data. One group of cereals seems to stand out (on the left) as having large negative residuals, with fewer calories than our model predicts. Another stands out (on the right) with large positive residuals. Whenever we suspect multiple modes, we ask whether there might be different subgroups in the data.

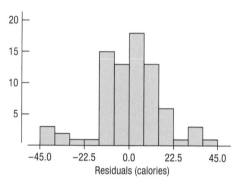

Figure 8.3

A histogram of the Chapter 7 breakfast cereal regression residuals shows small modes both above and below the central large mode. These may be worth a second look.

Figure 8.4 shows the residual scatterplot, with the cereals in those modes highlighted. We can see that those two groups stand away from the central pattern in this plot. If you look very carefully at the data scatterplot, there too you can spot these same two cereal clusters standing a bit apart from the central pattern. The high-residual cereals are Just Right Fruit & Nut; Muesli Raisins, Dates & Almonds; Peaches & Pecans; Mueslix Crispy Blend; and Nutri-Grain Almond Raisin. Do these cereals seem to have something in common? They all present themselves as "healthy." This might be surprising, but in fact "healthy" cereals often contain more fat, and therefore more calories, than we might expect from looking at their sugar content alone.

Figure 8.4

A scatterplot of the breakfast cereal regression residuals versus predicted values (points have been jittered) for the cereal regression, highlighting some possible clusters. The green "x" points are cereals whose calorie content is much higher than the linear model predicts. The red "–" points show cereals with many fewer calories than the model predicts. Is there something special about these cereals?

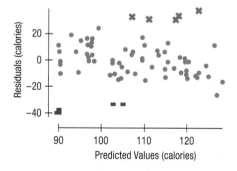

The low-residual cereals are Puffed Rice, Puffed Wheat, three bran cereals, and Golden Crisps. You might not have grouped these cereals together before. What they have in common is a low calorie count *relative to their sugar content*—even though their sugar contents are quite different.

These observations may not lead us to question the overall linear model, but they do help us understand that other factors may be part of the story. A careful examination of the data and residuals often leads us to discover groups of observations that are different from the rest.

When we discover that there is more than one group in a regression, we may decide to analyze the groups separately, using a different model for each group. Or we can stick with the original model and simply note that there are groups that are a little different. Either way, the model will be wrong, but useful, so it will improve our understanding of the data.

Subsets

Here's an important unstated condition for fitting models: **All the data must come from the same population**.

Sometimes it's not careful data inspection, but rather our special knowledge about the data that suggests the possibility of distinct subgroups or subsets. Cereal manufacturers aim cereals at different segments of the market. Supermarkets and cereal manufacturers try to attract different customers by placing different types of cereals on certain shelves. Cereals for kids tend to be on the "kid's shelf," at their eye level. Toddlers wouldn't be likely to grab a box from this shelf and beg, "Mom, can we please get this All-Bran with Extra Fibre?"

Should we take this extra information into account in our analysis? Figure 8.5 shows a scatterplot of *Calories* and *Sugar*, coloured according to the shelf on which the cereals were found and with a separate regression line fit for each. The top shelf is clearly different. We might want to report two regressions, one for the top shelf and one for the bottom two shelves.[2]

Figure 8.5

Calories and Sugar (y-values jittered) coloured according to the shelf on which the cereal was found in a supermarket, with regression lines fit for each shelf individually. Do these data appear homogeneous? That is, do all the cereals seem to be from the same population of cereals? Or are there different kinds of cereals that we might want to consider separately?

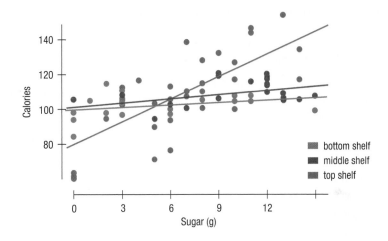

Section 8.1 Exercises

1. **Credit card spending** An analysis of spending by a sample of credit card bank cardholders shows that spending by cardholders in January (*Jan*) is related to their spending in December (*Dec*):

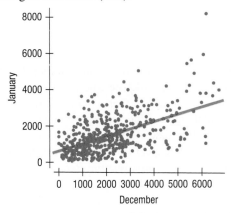

The assumptions and conditions of the linear regression seemed to be satisfied and an analyst was about to predict January spending using the model

$$\widehat{Jan} = \$612.07 + 0.403 \; Dec.$$

Another analyst worried that different types of cardholders might behave differently. She examined the spending patterns of the cardholders and placed them into five market *Segments*. When she plotted the data using different colors and symbols for the five different segments, she found the following:

Look at this plot carefully and discuss why she might be worried about the predictions from the model
$$\widehat{Jan} = \$612.07 + 0.403 \; Dec.$$

2. **Revenue and talent cost** A concert production company examined its records. The manager made the following scatterplot. The company places concerts in two venues,

---

[2]More complex models can take into account both sugar content and shelf information. This kind of *multiple regression* model is a natural extension of the model we're using here. We'll see such models in Chapters 26 and 27.

a smaller, more intimate theater (plotted with blue circles) and a larger auditorium-style venue (red x's).

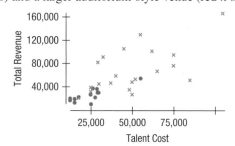

a) Describe the relationship between *Talent Cost* and *Total Revenue*. (Remember: direction, form, strength, outliers.)
b) How are the results for the two venues similar?
c) How are they different?

3. **Market segments** The analyst in Exercise 1 tried fitting the regression line to each market segment separately and found the following:

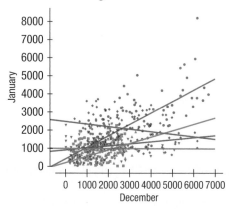

What does this say about her concern in Exercise 1? Was she justified in worrying that the overall model $\widehat{Jan} = \$612.07 + 0.403\ Dec$ might not accurately summarize the relationship? Explain briefly.

4. **Revenue and ticket sales** The concert production company of Exercise 2 made a second scatterplot, this time relating *Total Revenue* to *Ticket Sales*.

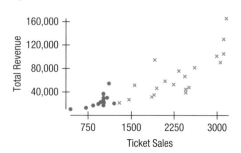

a) Describe the relationship between *Ticket Sales* and *Total Revenue.*
b) How are the results for the two venues similar?
c) How are they different?

8.2 Outliers, Leverage, and Influence

The outcome of the 2000 U.S. presidential election was determined in Florida amid much controversy. Even years later, historians continue to debate who really won the election. The main race was between George W. Bush and Al Gore, but two minor candidates played a significant role. To the political right of the main party candidates was Pat Buchanan, while to the political left was Ralph Nader. Generally, Nader earned more votes than Buchanan throughout the state. We would expect counties with larger vote totals to give more votes to each candidate. Here's a regression relating *Buchanan* vote totals by county in the state of Florida to *Nader* vote totals:

Dependent variable is: Buchanan

R-squared = 42.8%

Variable	Coefficient
Intercept	50.3
Nader	0.14

The regression model,

$$\widehat{Buchanan} = 50.3 + 0.14\ Nader,$$

Jim Cole/AP Images

says that, in each county, Buchanan received about 0.14 times (or 14% of) the vote Nader received, plus an additional 50.3 votes.

This seems like a reasonable regression, with an R^2 of almost 43%. But *we've violated all three Rules of Data Analysis* by going straight to the regression summary without making, making, making a picture.

Figure 8.6 is a scatterplot that shows the votes for Buchanan in each county of Florida plotted against votes for Nader. The striking **outlier** is Palm Beach County.

The so-called "butterfly ballot," used only in Palm Beach County, was a source of controversy. It has been claimed that the format of this ballot confused voters so that some who intended to vote for Democrat Al Gore punched the wrong hole next to his name and as a result voted for Buchanan.

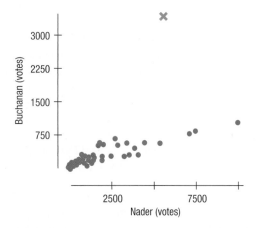

Figure 8.6

Votes received by Buchanan against votes for Nader in all Florida counties in the presidential election of 2000. The red "x" point is Palm Beach County, home of the "butterfly ballot."

The scatterplot shows a strong, positive, linear association, and one striking point. With Palm Beach removed from the regression, the R^2 jumps from 42.8% to 82.1% and the slope of the line changes to 0.1 (the blue line in Figure 8.7), suggesting that Buchanan received only about 10% of the votes that Nader received. Palm Beach County now stands out, not as a Buchanan stronghold, but rather as a clear violation of the model that begs for explanation.

Figure 8.7

The red line shows the effect that one unusual point can have on a regression.

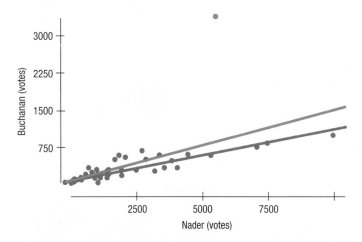

One of the great values of models is that they help us to see when and how data values are unusual. In regression, a point can stand out in two different ways. First, a data value can have a large residual, as Palm Beach County does in this example. Because they seem to be different from the other cases, points whose residuals are large always deserve special attention.

"Give me a place to stand and I will move the Earth."

—*Archimedes (287–211 BCE)*

"For whoever knows the ways of Nature will more easily notice her deviations; and, on the other hand, whoever knows her deviations will more accurately describe her ways."

—*Francis Bacon (1561–1626)*

A data point can also be unusual if its *x*-value is far from the mean of the *x*-values. Such a point is said to have high **leverage**. The physical image of a lever is exactly right. We know the line must pass through $(\bar{x}, \bar{y})$, so you can picture that point as the fulcrum of the lever. Just as sitting farther from the hinge on a see-saw gives you more leverage to pull it your way, points with values far from $\bar{x}$ pull more strongly on the regression line.

A point with high leverage has the potential to change the regression line, but it doesn't always use that potential. If the point lines up with the pattern of the other points, then including it doesn't change our estimate of the line. How can you tell if a high-leverage point actually changes the model? Just fit the linear model twice, both with and without the point in question. We say that a point is **influential** if omitting it from the analysis changes the model enough to make a meaningful difference.[3]

Influence depends on both leverage and residual; a case with high leverage whose *y*-value sits right on the line fit to the rest of the data is not influential. Removing that case won't change the slope, even if it does affect R^2. A case with modest leverage but a very large residual (such as Palm Beach County) can be influential. Of course, if a point has enough leverage, it can pull the line right to it. Then it's highly influential, but its residual is small. The only way to be sure is to fit both regressions.

Unusual points in a regression often tell us a great deal about the data and the model. We face a challenge: The best way to identify unusual cases is against the background of a model, but a model dominated by a single case is unlikely to be useful for identifying unusual cases. (That insight's at least 400 years old. See the Francis Bacon quote in the margin.) You can set aside cases and discuss what the model looks like with and without them, but arbitrarily deleting cases can give a false sense of how well the model fits the data. Your goal should be understanding the data, not just making R^2 as big as possible.

In 2000, George W. Bush won Florida (and thus the presidency) by only a few hundred votes, so Palm Beach County's residual was big enough to be meaningful. It's the rare unusual case that determines a presidency, but all are worth examining and trying to understand.

A point with so much influence that it pulls the regression line close to it can make its residual deceptively small. Influential points like that can have a shocking effect on the regression. Figure 8.8 is a plot of *IQ* against *Shoe Size*, with fitted regression line, again from the fanciful study of intelligence and foot size in comedians we saw in Chapter 6. What if Bozo's IQ had instead been very low—say, 50—as shown in Figure 8.9? The slope of the line drops from approximately +2 IQ points per shoe size to about −1.5 IQ point per shoe size.

Figure 8.8

Bozo's extraordinarily large shoes give his data point high leverage in the regression. Wherever Bozo's IQ falls, the regression line will follow.

Figure 8.9

If Bozo's IQ were low, the regression slope would change from positive to negative. A single influential point can change a regression model drastically.

---

[3]How big a change would make a difference depends on the circumstances and is a judgment call. Note that some textbooks use the term *influential point* for any observation that influences the slope, intercept, or R^2. We'll reserve the term for points that influence the slope.

No matter where Bozo's IQ is, the line tends to follow it because his shoe size, being so far from the mean shoe size, makes this a high-leverage point.

Without Bozo, there is little correlation between shoe size and IQ, and $R^2 = 0.7\%$, whereas including Bozo with the high IQ yields an $R^2 = 24.8\%$, so almost all of the variance accounted for is due to one case, namely Bozo. Even though this example is far-fetched, similar situations often occur in real life. For example, a regression of sales against floor space for hardware stores that looked primarily at small-town businesses could be dominated in a similar way if a Home Depot were included.

 Just Checking

Each of these scatterplots shows an unusual point. For each, tell whether the point is a high-leverage point, would have a large residual, or is influential.

Section 8.2 Exercises

5. **Abalone** Abalones are edible sea snails that include over 100 species. A researcher is working with a model that uses the number of rings in an Abalone's shell to predict its age. He finds an observation that he believes has been miscalculated. After deleting this outlier, he redoes the calculation. Does it appear that this outlier was exerting very much influence?

   Before:
   Dependent variable is Age
   R-squared = 67.5%

Variable	Coefficient
Intercept	1.736
Rings	0.45

   After:
   Dependent variable is Age
   R-squared = 83.9%

Variable	Coefficient
Intercept	1.56
Rings	1.13

6. **Abalone again** The researcher in Exercise 5 is content with the second regression. But he has found a number of shells that have large residuals and is considering removing them all. Is this good practice?

8.3 Extrapolation: Reaching Beyond the Data

Linear models give a predicted value for each case in the data. Put a new x-value into the equation, and it gives a predicted value, $\hat{y}$, to go with it. But when the new x-value lies far from the data we used to build the regression, how trustworthy is the prediction?

The simple answer is that the farther the new x-value is from $\bar{x}$, the less trust we should place in the predicted value. Once we venture into new x territory, such a prediction is called an **extrapolation**. Extrapolations are dubious because they require the very questionable assumption that nothing about the relationship between x and y changes even at extreme values of x and beyond.

Extrapolation is a good way to see just where the limits of our model may be. But it requires caution. When the x-variable is *Time*, extrapolation becomes an attempt to peer into the future. People have always wanted to see into the future, and it doesn't take a crystal ball to foresee that they always will.

> "Prediction is very difficult, especially about the future."
>
> —*Niels Bohr, Danish physicist*

WHEN THE DATA ARE YEARS . . .

. . . we usually don't enter them as four-digit numbers. Here we used 0 for 1970, 10 for 1980, and so on. Or we may simply enter two digits, using 82 for 1982, for instance. Rescaling years like this often makes calculations easier and equations simpler. We recommend you do it, too. But be careful: If 1982 is 82, then 2004 is 104 (not 4), right?

Figure 8.10

The scatterplot shows an average *increase* in the price of a barrel of oil of about $6.75 per year from 1971 to 1982.

Those with a more scientific outlook may use a linear model as their digital crystal ball. Some physical phenomena do exhibit a kind of "inertia" that allows us to guess that current systematic behaviour will continue, but be careful in counting on that sort of regularity in phenomena, such as stock prices, sales figures, hurricane tracks, or public opinion.

Extrapolating from current trends is so tempting that even professional forecasters sometimes expect too much from their models—and sometimes the errors are striking. In the mid-1970s, oil prices surged and long lines at gas stations were common. In 1970, oil cost about US$17 a barrel (in 2014 dollars)—about what it had cost for 20 years or so. But then, within just a few years, the price surged to over $40. In 1975, a survey of 15 top econometric forecasting models (built by groups that included Nobel prize–winning economists) found predictions for 1985 oil prices that ranged from $300 to over $700 a barrel (in 2014 dollars). How close were these forecasts?

Figure 8.10 shows a scatterplot of oil prices from 1971 to 1982 (in 2014 dollars; Data in **Historical Oil Prices 2014**).

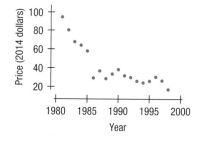

Figure 8.11

This scatterplot of oil prices from 1981 to 1998 shows an average decrease of nearly $5 per year.

The fitted regression model

$$\widehat{Price} = 12.79 + 6.75 \; Years \; Since \; 1970$$

says that prices had been going up on average $6.75 per year, or $67.50 over 10 years. If you assume that they would *keep going up*, it's not hard to imagine almost any price you want. That would result in a forecast of about $316 for the year 2015.

So how did the forecasters do? Well, Figure 8.11 shows the period from 1981 to 1998, in which oil prices didn't exactly continue that steady increase. In fact, they went down so much that by 1998, prices (adjusted for inflation) were the lowest they'd been since before World War II.

Not one of the experts' models predicted that.

Of course, these decreases clearly couldn't continue, or oil would have been free by the year 2000. The Energy Information Administration (EIA) has provided the U.S. Congress with both short- and long-term oil price forecasts every year since 1979. In that year they predicted that 1995 prices would be $87 (in today's $). As we've just seen, they were closer to $25. So, 20 years later, what did the EIA learn about forecasting? Let's see how well they did with their future forecasts made in 1998 and again in 1999. Figure 8.12 shows a timeplot of the EIA's predictions and the actual prices (in 2014 dollars):

Figure 8.12

Here are the EIA forecasts compared with the actual prices since 1981. Neither forecast predicted the sharp run-up after 2003.

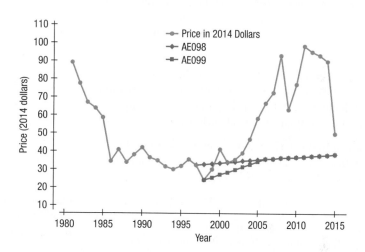

Oops! They seemed to have missed the sharp run-up in oil prices leading up to the financial crisis of 2008, and then the sharp drop in prices at the beginning of 2009 back to about $64 per barrel and the bounce back up to $106 (and the subsequent drop from late 2014, and . . .).

Where do you think oil prices will go in the next decade? *Your* guess may be as good as anyone's!

Of course, knowing that extrapolation requires thought and caution doesn't stop people. The temptation to see into the future is hard to resist. So our more realistic advice is this:

If you extrapolate into the future, at least don't believe blindly that the prediction will come true.

For Example BREAKFAST CEREALS AGAIN

In the Step-By-Step analysis in Chapter 7 (also discussed earlier in this chapter), we found the following model for predicting *Calories* from *Sugar* content (in grams) in breakfast cereals (per serving):

$$\widehat{Calories} = 89.6 + 2.49 \; Sugar$$

> **QUESTION:** Predict the calories for a cereal containing 30 grams of sugar per serving.
>
> **ANSWER:** Are you kidding? The sugar content in the observed data range from 0 to 15 grams, so predicting at 30 grams is such an extreme extrapolation as to be worthless. [You did remember to go back and look at a *picture* of the data, right?]
>
> **QUESTION:** Predict the calories for a cereal containing 10 grams of sugar that will sit on the top shelf.
>
> **ANSWER:** As we saw earlier in this chapter, the shelf matters, so going back to the data, we fit a model just for cereals on the top shelf and get: $\widehat{Calories} = 73.7 + 5.23$ *Sugar*. Residual plots show no problems, so plugging in *Sugar* = 10, we would predict 126 calories.

Section 8.3 Exercises

7. Cell phone costs Noting a recent study predicting the increase in cell phone costs, a friend remarks that by the time he's a grandfather, no one will be able to afford a cell phone. Explain where his thinking went awry.

8. Stopping times Using data from 20 compact cars, a consumer group develops a model that predicts the stopping time for a vehicle by using its weight. You consider using this model to predict the stopping time for your large SUV. Explain why this is not advisable.

9. Revenue and large venues A regression of *Total Revenue* on *Ticket Sales* by the concert production company of Exercises 2 and 4 finds the model

$$\widehat{Revenue} = -14\,228 + 36.87\ TicketSales.$$

a) Management is considering adding a stadium-style venue that would seat 10 000. What does this model predict that revenue would be if the new venue were to sell out?

b) Why would it be unwise to assume that this model accurately predicts revenue for this situation?

10. Revenue and advanced sales The production company of Exercise 9 offers advanced sales to "Frequent Buyers" through its website. Here's a relevant scatterplot:

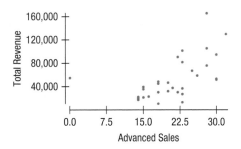

One performer refused to permit advanced sales. What effect would that point have if we were to fit a regression line to predict *Total Revenue* from *Advanced Sales*?

8.4 Cautions

Lurking Variables and Causation

> **CAUSAL INTERPRETATIONS**
>
> One common way to interpret a regression slope is to say that "a change of 1 unit in x results in a change of b_1 units in y." This way of saying things encourages causal thinking. Beware.

In Chapter 6, we tried to make it clear that no matter how strong the correlation is between two variables, it doesn't show that one variable causes the other. Putting a regression line through a cloud of points just increases the temptation to think and to say that the x-variable *causes* the y-variable. Let's repeat the point again: No matter how strong the association, no matter how large the R^2 value, no matter how straight the line, you cannot conclude from a regression alone that one variable causes the other. There's always the possibility that some third variable may help explain the association. With observational data, as opposed to data from a designed experiment, there is no way to be sure that a **lurking variable** is not responsible for the apparent association.

Here's an example: Figure 8.13 shows the *Life Expectancy* (average of men and women, in years) for each of 40 countries of the world, plotted against the square root of the number of *Doctors per person* in the country. (The square root is used here to make the relationship satisfy the Straight Enough Condition; Data in **Doctors and Life Expectancy**)

Figure 8.13

The relationship between life expectancy (in years) and availability of doctors (measured as √doctors per person) for countries of the world is strong, positive, and linear.

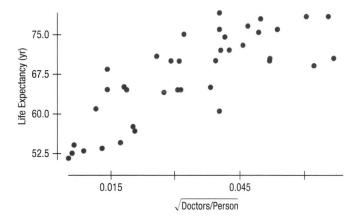

The strong positive association ($R^2 = 62.4\%$) seems to confirm our expectation that more doctors per person improves healthcare, leading to longer lifetimes and greater life expectancy. The strength of the association would *seem* to argue that we should send more doctors to developing countries to increase life expectancy.

That conclusion is about the consequences of a change. Would sending more doctors increase life expectancy? Specifically, do doctors *cause* greater life expectancy? Perhaps, but these are observed data, so there may be another explanation for the association.

In Figure 8.14, the similar-looking scatterplot's *x*-variable is the square root of the number of *Televisions per person* in each country. The positive association in this scatterplot is even *stronger* than the association in the previous plot ($R^2 = 72.5\%$). We can fit the linear model, and use the number of TVs to help predict life expectancy. Should we conclude that increasing the number of TVs actually extends lifetimes? If so, we should send TVs instead of doctors to developing countries. Not only is the correlation with life expectancy higher, but TVs are much cheaper than doctors and possibly more fun.

Figure 8.14

To increase life expectancy, don't send doctors, send TVs; they're cheaper and more fun. Or maybe that's not the right interpretation of this scatterplot of life expectancy against availability of TVs (as √TVs per person).

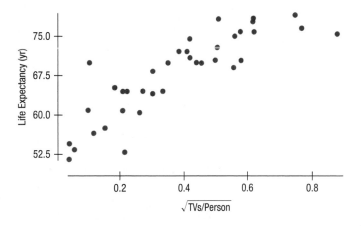

What's wrong with this reasoning? Maybe we were a bit hasty earlier when we concluded that doctors *cause* longer lives. Maybe there's a lurking variable here. Countries with higher standards of living have both longer life expectancies *and* more doctors (and more TVs). Could higher living standards cause changes in the other variables? If so, then improving living standards might be expected to prolong lives, increase the number of doctors, and increase the number of TVs—a conjecture illustrated in Figure 8.15 below:

Figure 8.15

Diagram illustrating a common response to the lurking variable *Living Standards*.

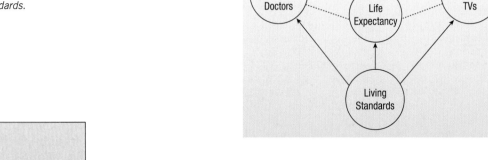

To think about lurking variables you must think "outside the box." What variables are not in your data that ought to be?[4]

From this example, you can see how easy it is to fall into the trap of mistakenly inferring causality from a regression. For all we know, doctors (or TVs!) *do* increase life expectancy. But we can't tell that from data like these, no matter how much we'd like to. Resist the temptation to conclude that *x* causes *y* from a regression, no matter how obvious that conclusion seems to you.

For Example CAUSATION AND REGRESSION

Our regression model in the previous chapter predicting hurricane wind speeds from the central pressure was reasonably successful. The negative slope indicates that in general, storms with lower central pressures have stronger winds.

QUESTION: Can we conclude that lower central barometric pressure causes the higher wind speeds in hurricanes?

ANSWER: No. While it may be true that lower pressure causes higher winds, a regression model for observed data like these cannot demonstrate causation. Perhaps higher wind speeds reduce the barometric pressure, or perhaps pressure and wind speed are both driven by some other variable we have not observed.

(As it happens, in hurricanes it is reasonable to say that the low central pressure at the eye is responsible for the high winds because it draws moist, warm air into the centre of the storm, where it swirls around, generating the winds. But as is often the case, things aren't quite that simple. The winds themselves contribute to lowering the pressure at the centre of the storm as it becomes organized into a hurricane. The lesson is that to understand causation in hurricanes, we must do more than just model the relationship of two variables; we must study the mechanism itself.)

Predicting Changes in the Response Variable

Not only it is incorrect and dangerous to interpret association with causation, but when using regression there's a more subtle danger. Never interpret a regression slope coefficient as predicting how *y* is likely to change if its *x* value in the data were changed. Here's an example: In Chapter 7, we found a regression model relating calories in breakfast cereals to their sugar content as

$$\widehat{Calories} = 89.6 + 2.49 \, Sugar.$$

---

[4]If you wonder about the graphic: this is one of the boxes you should think outside of. There are many others. Stay alert for them.

It might be tempting to interpret this slope as implying that adding 1 gram of sugar is expected to *lead to* a change of 2.49 calories. Predicting a change of b_1 in y for a 1 unit change in x might be true in some instances, but it isn't a consequence of the regression. In fact, adding a gram of sugar to a cereal increases its calorie content by 3.90 calories—that's the calorie content of a gram of sugar.

A safer interpretation of the slope is that cereals that have a gram more sugar in them tend to have about 2.49 more calories per serving. That is, the regression model describes how the cereals differ, but does not tell us how they might change if circumstances were different.

We've warned against using a regression model to predict far from the x-values in the data. You should also be cautious about using a regression model to predict what might happen to cases in the regression if they were changed. Regression models describe the data as they are, not as they might be under other circumstances.

Working with Summary Values

Scatterplots of statistics summarized over groups tend to show less variability than we would see if we measured the same variable on individuals. This is because the summary statistics themselves vary less than the data on the individuals do—a fact we will make more specific in coming chapters.

In Chapter 6, we looked at the heights and weights of individual students, portrayed again in Figure 8.16 below. We saw a correlation of 0.644, so R^2 is 41.5%.

Suppose, instead of data on individuals, we knew only the mean weight for each height value. The scatterplot of mean weight by height in Figure 8.17 shows less scatter. And the R^2 would increase to 80.1%.

Figure 8.16

Weight (lb) against Height (in.) for a sample of men. There's a strong, positive, linear association.

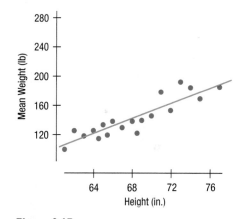

Figure 8.17

Mean Weight shows a stronger linear association with Height than do the weights of individuals. Means vary less than individual values.

Scatterplots of summary statistics show less scatter than the raw data on individuals and can give a false impression of how well a line summarizes the relationship. There's no simple correction for this phenomenon. Once we're given summary data, there's no simple way to get the original values back.

In the life expectancy example, we have no good measure of exposure to doctors or to TVs on an individual basis. But if we did, we should expect the scatterplot to show more variability and the corresponding R^2 to be smaller. The bottom line is that you should be a bit suspicious of conclusions based on regressions of summary data. They may look better than they really are. All of which also reminds us of that first step in proper interpretation, to think clearly about the *Who*—just who does each individual case in your plot actually represent (one country, one student, all students at the same height, etc.)?

Restricted Range

In the scatterplot of individuals' weights and heights above, block out all those shorter than 71 inches. What do you think the correlation is now? Or block out all those taller than 64 inches. Guess the correlation. Block out everyone except those between 64 and 68 inches. Guess the correlation again. In each case, the correlation appears to be close to zero, and certainly much less than the 0.644 correlation based on data over the full range of heights.

Correlations reported from a study may be smaller than expected if the range of observations on one or both variables has been restricted in some way. For example, we would expect the correlation between students' high school average and GPA for McGill University students to be less than it would be if all high school graduates (not just the best) were able to enrol and study at this school. If you just study the relationship for McGill students, you might be fooled into thinking high school averages are almost worthless at predicting success at university.

moodboard/Fotolia

For Example USING SEVERAL OF THESE METHODS TOGETHER

Motorcycles designed to run off-road, often known as dirt bikes, are specialized vehicles. We have data on 114 dirt bikes (Data in **Dirt Bikes 2014**). Some cost as little as $1399, while others are substantially more expensive. Let's investigate how the size and type of engine contribute to the cost of a dirt bike. As always, we start with a scatterplot.

Here's a scatterplot of the manufacturer's suggested retail price (*MSRP*) in dollars against the engine *displacement* (in cubic centimetres), along with a regression analysis:

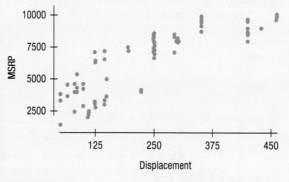

Dependent variable is MSRP
R-squared = 70.4%   s = 1335

Variable	Coefficient
Intercept	2856.84
Displacement	15.2567

QUESTION: What do you see in the scatterplot?

ANSWER: There is a strong positive association between the engine displacement of dirt bikes and the manufacturer's suggested retail price. There is some curvature and there seem to be a group of points with small displacement whose prices are substantially below the rest but show a similar slope.

Here's a scatterplot of the residuals:

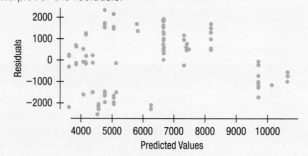

QUESTION: What do you see in the residuals plot?

ANSWER: The points at the far right and the bottom left don't fit well with the other dirt bikes. Overall, there appears to be a bend in the relationship, so a linear model may not be appropriate.

Let's try a re-expression. Here's a scatterplot showing *MSRP* against the cube root of *Displacement* to make the relationship closer to straight. (Since displacement is measured in cubic centimetres, its cube root has the simple units of centimetres.) In addition, we've coloured the plot according to the cooling method used in the bike's engine: liquid or air. Each group is shown with its own regression line, as we did for the cereals on different shelves.

QUESTION: What does this plot say about dirt bikes?

ANSWER: There appears to be a positive, linear relationship between *MSRP* and the cube root of *Displacement*. In general, the larger the engine a bike has, the higher the suggested price. Liquid-cooled dirt bikes, however, typically cost more than air-cooled bikes with comparable displacement.[5]

Section 8.4 Exercises

11. Skinned knees There is a strong correlation between the temperature and the number of skinned knees on playgrounds. Does this tell us that warm weather causes children to trip?

12. Cell phones and life expectancy The correlation between cell phone usage and life expectancy is very high. Should we buy cell phones to help people live longer?

13. Grading A team of Calculus teachers is analyzing student scores on a final exam compared to the midterm scores. One teacher proposes that they already have every teacher's class averages and they should just work with those averages. Explain why this is problematic.

14. Average GPA An athletic director proudly states that he has used the average GPAs of the university's sports teams and is predicting a high graduation rate for the teams. Why is this method unsafe?

WHAT CAN GO WRONG?

This entire chapter has held warnings about things that can go wrong in a regression analysis. So let's just recap. When you make a linear model,

■ **Make sure the relationship is straight.** Check the Straight Enough Condition. Always examine the residuals for evidence that the Linearity Assumption has failed. It's often easier to see deviations from a straight line in the residuals plot than in the scatterplot of the original data. Some variables are just not related in a way that's simple enough for a linear model to fit very well. When that happens, report the failure and stop (well, stop and think about alternative approaches, like re-expression of variables).

(continued)

---

[5]Jiang Lu, Joseph B. Kadane, and Peter Boatwright, "The dirt on bikes: An illustration of CART models for brand differentiation," provided data on 2005-model bikes, lib.stat.cmu.edu/datasets/dirtbike_aug.csv.

■ **Be on guard for different groups in your regression.** Check for evidence that the data consist of separate subsets. If you find subsets that behave differently, consider fitting a different linear model to each subset.

■ **Beware of extrapolating.** Beware of extrapolation beyond the *x*-values that were used to fit the model. Although it's common to use linear models to extrapolate, the practice is dangerous.

■ **Beware especially of extrapolating into the future!** Be especially cautious about extrapolating into the future with linear models. To predict the future, you must assume that future changes will continue at the same rate you've observed in the past. Predicting the future is particularly tempting and particularly dangerous.

■ **Look for unusual points.** Unusual points always deserve attention, and may well reveal more about your data than the rest of the points combined. Always look for them and try to understand why they stand apart. A scatterplot of the data is a good way to see high-leverage and possibly influential points. A scatterplot of the residuals against the predicted values is a good tool for finding points with large residuals (as is a histogram of the residuals).

■ **Beware of high-leverage points, and especially of those that are influential.** Influential points can alter the regression model a great deal. The resulting model may say more about one or two points than about the overall relationship.

■ **Treat unusual points honestly.** To see the impact of outliers on a regression, it's often wise to run two regressions, one with and one without the extraordinary points, and then to discuss the differences. If you remove enough carefully selected points, you will eventually get a regression with a high R^2, but it won't give you much understanding.

■ **Beware of lurking variables.** Think about lurking variables before interpreting a linear model. It's particularly tempting to explain a strong regression by thinking that the *x*-variable *causes* the *y*-variable. A linear model alone can never demonstrate such causation in part, because it cannot eliminate the chance that it is a lurking variable, and not *x*, that is causing the variation in *y*.

■ **Watch out when dealing with data that are summaries.** Be cautious in working with data values that are themselves summaries, such as means or medians. Such statistics are less variable than the data on which they are based, so they tend to inflate the impression of the strength of a relationship.

■ **Watch out when dealing with range-restricted data.** Not including data over a full range of the variables will often deflate the impression of the strength of a relationship.

What Have We Learned?

Be skeptical of regression models. Always plot and examine the residuals for unexpected behaviour. Be alert to a variety of possible violations of the standard regression assumptions and know what to do when you find them.

Be alert for subgroups in the data.

■ Often these will turn out to be separate groups that should not be analyzed together in a single analysis.

■ Often identifying subgroups can help us better understand what is going on in the data.

Be especially cautious about extrapolating beyond the data.

Look out for unusual and extraordinary observations.

■ Cases that are extreme in *x* have high leverage and can affect a regression model strongly.

■ Cases that are extreme in *y* have large residuals and may also influence a regression model.

- Running two regressions, one with and one without the unusual cases, can help you assess their influence on the model. Consider removing and reporting these cases separately if their influence is excessive or undesirable.

Interpret regression models appropriately. Don't infer causation from a regression model.

Be careful when you interpret regressions based on summarized data or with restricted data ranges. Regressions with summaries tend to look stronger than the regression based on all the individual cases, whereas restricting the ranges of variables usually makes regressions weaker.

Diagnose and treat nonlinearity.

- If a scatterplot of y vs. x isn't straight, a linear regression model isn't appropriate.
- Re-expressing one or both variables can often improve the straightness of the relationship.
- The powers, roots, and the logarithm provide an ordered collection of re-expressions so you can search up and down the "ladder of powers" (see Section 6.4) to find an appropriate one.

Review of Terms

Outlier Any data point that stands away from the others can be called an outlier. In regression, cases can be extraordinary in two ways: by standing far apart in the y-direction and having a large residual or by standing far apart in the x-direction and having high leverage (p. 251).

Leverage Data points whose x-values are far from the mean of x are said to exert leverage on a linear model. High-leverage points pull the line close to them, and so they can have a large effect on the line, sometimes completely determining the slope and intercept. With high enough leverage, their residuals can be deceptively small (p. 252).

Influential point A point (case) that, if omitted from the data, results in a very different regression model (p. 252).

Extrapolation Although linear models provide an easy way to predict values of y for a given value of x, it is unsafe to predict for values of x far from the ones used to fit the linear model. At times, such extrapolation may pretend to see into the future, but the predictions should not be trusted (p. 254).

Lurking variable A variable that is not explicitly part of a model but affects the way the variables in the model appear to be related. Because we can never be certain that observational data are not hiding a lurking variable that influences the relationship between x and y, it is never safe to conclude that a linear model demonstrates a causal relationship, no matter how strong the linear association (p. 256).

On the Computer REGRESSION DIAGNOSTICS

Most statistics technology offers simple ways to check whether your data satisfy the conditions for regression. We have already seen that these programs can make a simple scatterplot. They can also help us check the conditions by plotting residuals.

EXCEL

The Data Analysis add-in for Excel includes a Regression command. The dialog box it shows offers to make plots of residuals.

COMMENTS

Do not use the Normal probability plot offered in the regression dialog. It is not what it claims to be and is wrong.

JMP

- From the **Analyze** menu, choose **Fit Y by X**. Select **Fit Line**.
- Under Linear Fit, select **Plot Residuals**. You can also choose to **Save Residuals**.

- Subsequently, from the **Distribution** menu, choose **Normal quantile plot** or **histogram** for the residuals.

MINITAB

- From the **Stat** menu, choose **Regression**.
- From the **Regression** submenu, select **Regression** again and then select **Fit Regression Model**.
- In the Regression dialogue, enter the response variable name in the "Responses" box and the predictor variable name in the "Continuous predictor" box.
- To save residuals and predicted values, in the Regression dialog, click **Storage**.
- Check "Residuals" and "Fits." Click **OK**.

- To easily produce some residual plots, in the Regression dialog, click **Graphs**.
- Under "Residual Plots," select "Four in One" for a set of four very useful residual plots (or just above, select individual plots of interest, such as "Residuals versus fits").
- Click **OK**. Now back in the Regression dialog, click **OK**. Minitab computes the regression and the requested saved values and graphs.

R

Save the regression model object by giving it a name, such as "myreg":

- myreg = lm(Y~X) or myreg = lm(Y~X, data = DATA) where DATA is the name of the data frame.
- plot(residuals(myreg)~predict(myreg)) plots the residuals against the predicted values.

- qqnorm(residuals(myreg)) gives a normal probability plot of the residuals.
- plot(myreg) gives similar plots (but not exactly the same).

SPSS

- From the **Analyze** menu, choose **Regression**.
- From the Regression submenu, choose **Linear**.
- After assigning variables to their roles in the regression, click the "**Plots . . .**" button.

In the Plots dialog, you can specify a Normal probability plot of residuals and scatterplots of various versions of standardized residuals and predicted values.

COMMENTS

A plot of *ZRESID against *PRED will look most like the residual plots we've discussed. SPSS standardizes the residuals by dividing by their standard deviation. (There's no need to subtract their mean; it must be zero.) The standardization doesn't significantly affect the scatterplot.

STATCRUNCH

To create a residuals plot:

- Click on **Stat**.
- Choose **Regression » Simple Linear** and choose X and Y.
- Click on **Next** and click on **Next** again.
- Indicate which type of residuals plot you want.
- Click on **Calculate**.

COMMENTS

Note that before you click on **Next** for the second time you may indicate that you want to save the values of the residuals. Residuals becomes a new column, and you may use that variable to create a histogram or residuals plot.

TI-83/84 PLUS

- To make a residuals plot, set up a **STATPLOT** as a scatterplot.
- Specify your explanatory data list as **Xlist**.
- For **Ylist**, import the name **RESID** from the **LIST NAMES** menu. **ZoomStat** will now create the residuals plot.

COMMENTS

Each time you execute a **LinReg** command, the calculator automatically computes the residuals and stores them in a data list named **RESID**. If you want to see them, go to **STAT EDIT**. Space through the names of the lists until you find a blank. Import **RESID** from the **LIST NAMES** menu. Now every time you have the calculator compute a regression analysis, it will show you the residuals.

Chapter Exercises

T indicates data is available in a saved file at the Companion Website at www.pearsoncanada.ca/deveaux

T 15. Marriage age 2011 Is there evidence that the age at which women get married has changed over the past 100 years? The scatterplot shows the trend in age at first marriage for American women (Adapted from Bureau of the Census (2010). Median age at first marriage: 1890 to present. From https://www.census.gov/hhes/families/files/graphics/MS-2.pdf).

a) Is there a clear pattern? Describe the trend.
b) Is the association strong?
c) Is the correlation high? Explain.
d) Is a linear model appropriate? Explain.

T 16. Smoking 2011 The Centers for Disease Control and Prevention track cigarette smoking in the United States. How has the percentage of people who smoke changed since the danger became clear during the last half of the 20th century? The scatterplot shows percentages of smokers among men 18–24 years of age, as estimated by surveys, from 1965 through 2011 (http://www.cdc.gov/nchs/).

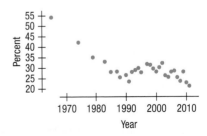

a) Is there a clear pattern? Describe the trend.
b) Is the association strong?
c) Is a linear model appropriate? Explain.

17. Gas mileage One of the important factors determining a car's fuel efficiency is its weight. Let's examine this relationship for 11 cars. In this exercise, we are using the U.S. units for fuel efficiency, miles per (U.S.) gallon.

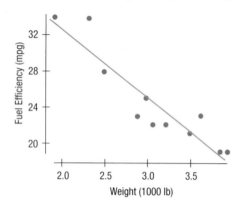

a) Describe the association between these variables shown in the scatterplot.
b) Here is the regression analysis for the linear model. What does the slope of the line say about this relationship?

Dependent variable is: Fuel Efficiency
R-squared = 85.9%

Variable	Coefficient
Intercept	47.9636
Weight	−7.65184

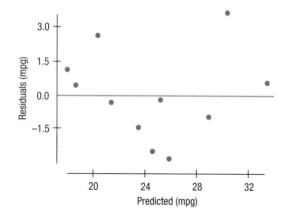

c) Do you think this linear model is appropriate? Use the residual plot to explain your decision.

18. Gas mileage revisited Let's try the re-expressed variable fuel consumption (gal/100 mi) to examine the fuel efficiency of the 11 cars in Exercise 17. Here are the revised regression analysis and residual plot:

Dependent variable is: Fuel Consumption
R-squared = 89.2%

Variable	Coefficient
Intercept	0.624932
Weight	1.17791

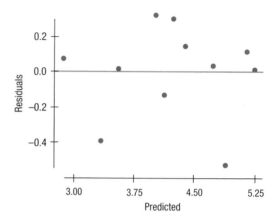

a) Explain why this model appears to be better than the model used in Exercise 17.
b) Using the regression analysis above, write an equation of the model.
c) Interpret the slope of the line.
d) Based on this model, how many miles per gallon would you expect a 3500-pound car to get?
e) Canadians might wish instead to know the number of litres per 100 km expected for a 3500-pound car. What would you estimate? (1 US gallon = 3.7854 litres; 1 mile = 1.61 km)

19. Good model? In justifying his choice of a model, a student wrote, "I know this is the correct model because $R^2 = 99.4\%$."
a) Is this reasoning correct? Explain.
b) Does this model allow the student to make accurate predictions? Explain.

20. Bad model? A student who has created a linear model is disappointed to find that her R^2 value is a very low 13%.
a) Does this mean that a linear model is not appropriate? Explain.
b) Does this model allow the student to make accurate predictions? Explain.

T 21. Movie dramas Here's a scatterplot of the production budgets (in millions of dollars) versus the running time (in minutes) for major release movies in 2005. Dramas are plotted as red x's and all other genres are plotted as blue dots. (The re-make of *King Kong* is plotted as a black "–." At the time, it was the most expensive movie ever made, and not typical of any genre.) A separate least squares regression line has been fitted to each group. For the following questions, examine the plot:

a) What are the units for the slopes of these lines?

b) In what way are dramas and other movies similar with respect to this relationship?

c) In what way are dramas different from other genres of movies with respect to this relationship?

T 22. Smoking 2011, women and men In Exercise 16, we examined the percentage of men aged 18–24 who smoked from 1965 to 2011 according to the Centers for Disease Control and Prevention. How about women? Here's a scatterplot showing the corresponding percentages for both men and women:

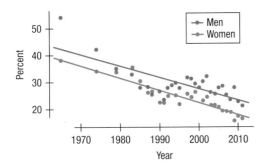

a) In what ways are the trends in smoking behaviour similar for men and women?

b) How do the smoking rates for women differ from those for men?

c) Viewed alone, the trend for men may have seemed to violate the Linearity Condition. How about the trend for women? Does the consistency of the two patterns encourage you to think that a linear model for the trend in men might be appropriate? (Note: there is no correct answer to this question; it is raised for you to think about.)

T 23. Oakland passengers 2013 The scatterplot below shows the number of passengers at Oakland (CA) airport month by month since 1997. (www.oaklandairport.com)

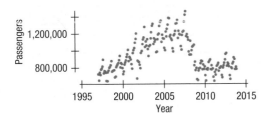

a) Describe the patterns in passengers at Oakland airport that you see in this time plot.

b) Until 2009, analysts got fairly good predictions using a linear model. Why might that not be the case now?

c) If they considered only the data from 2009 to the present, might they get reasonable predictions into the future?

T 24. Tracking hurricanes 2012 In a previous chapter, we saw data on the errors (in nautical miles) made by the National Hurricane Center in predicting the path of hurricanes. The following scatterplot shows the trend in the 24-hour tracking errors since 1970 (www.nhc.noaa.gov).

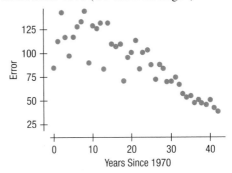

Dependent variable is Error
R-squared = 72.5% $s_e = 16.13$

Variable	Coefficient
Intercept	133.024
Years - 1970	−2.0602

a) Interpret the slope and intercept of the model.

b) Interpret s_e in this context.

c) The Center would like to achieve an average tracking error of 45 nautical miles by 2015. Will they make it? Defend your response.

d) What if their goal were an average tracking error of 25 nautical miles?

e) What cautions would you state about your conclusion?

25. Unusual points Each of these four scatterplots shows a cluster of points and one "stray" point. For each, answer these questions:

a) In what way is the point unusual? Does it have high leverage, a large residual, or both?

b) Do you think that point is an influential point?

c) If that points were removed, would the correlation become stronger or weaker? Explain.

d) If that points were removed, would the slope of the regression line increase or decrease? Explain.

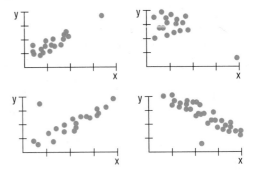

26. More unusual points Each of the following scatterplots shows a cluster of points and one "stray" point. For each, answer these questions:

a) In what way is the point unusual? Does it have high leverage, a large residual, or both?

b) Do you think that point is an influential point?

c) If that points were removed, would the correlation become stronger or weaker? Explain.

d) If that points were removed, would the slope of the regression line increase or decrease? Explain.

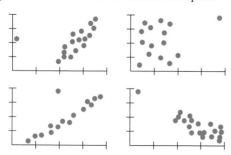

27. The extra point The scatterplot below shows five small blue data points at the left. Not surprisingly, the correlation for these points is $r = 0$. Suppose *one* additional data point is added at one of the five positions (a–e) suggested below in green. Match each point (a–e) with the correct new correlation from the list given.

a) -0.90 b) -0.40 c) 0.00 d) 0.05 e) 0.75

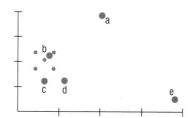

28. The extra point revisited The original five points in Exercise 27 produce a regression line with slope 0. Match each of the green points (a–e) with the slope of the line after that one point is added:

a) -0.45 b) -0.30 c) 0.00 d) 0.05 e) 0.85

29. What's the cause? Suppose a researcher studying health issues measures blood pressure and the percentage of body fat for several adult males and finds a strong positive association. Describe three different possible cause-and-effect relationships that might be present.

30. What's the effect? A researcher studying violent behaviour in elementary-school children asks the children's parents how much time each child spends playing computer games and has their teachers rate each child on the level of aggressiveness they display while playing with other children. Suppose that the researcher finds a moderately strong positive correlation. Describe three different possible cause-and-effect explanations for this relationship.

31. Reading To measure progress in reading ability, students at an elementary school take a reading comprehension test every year. Scores are measured in "grade-level" units; that is, a score of 4.2 means that a student is reading at slightly above the expected level for a Grade 4 student. The school principal prepares a report to parents that includes a graph showing the mean reading score for each grade. In his comments, he points out that the

strong positive trend demonstrates the success of the school's reading program.

a) Does this graph indicate that students are making satisfactory progress in reading? Explain.
b) What would you estimate the correlation between *Grade* and *Average Reading Level* to be?
c) If, instead of this plot showing average reading levels, the principal had produced a scatterplot of the reading levels of all the individual students, would you expect the correlation to be the same, higher, or lower? Explain.
d) Although the principal did not do a regression analysis, someone as statistically astute as you might. (But don't bother.) What value of the slope of that line would you view as demonstrating acceptable progress in reading comprehension? Explain.

32. Grades A U.S. college admissions officer, defending the college's use of SAT scores in the admissions process, produced the graph below. It shows the mean GPAs for last year's freshmen, grouped by SAT scores. How strong is the evidence that *SAT Score* is a good predictor of *GPA*? What concerns you about the graph—the statistical methodology or the conclusions reached?

33. Heating After keeping track of his heating expenses for several winters, a homeowner in the U.S. believes he can estimate the monthly cost from the average daily Fahrenheit temperature by using the model $\widehat{Cost} = 133 - 2.13\,Temp$. The residuals plot for his data is shown.

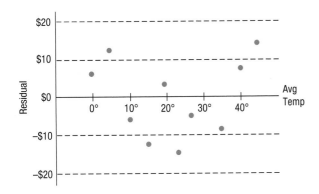

a) Interpret the slope of the line in this context.

b) Interpret the *y*-intercept of the line in this context.

c) During months when the temperature stays around freezing, would you expect cost predictions based on this model to be accurate, too low, or too high? Explain.

d) What heating cost does the model predict for a month that averages 10°F?

e) During one of the months on which the model was based, the temperature did average 10°F. What were the actual heating costs for that month?

f) Do you think the homeowner should use this model? Explain.

g) Would this model be more successful if the temperature were expressed in degrees Celsius? Explain.

34. Speed How does the speed at which you drive affect your fuel economy? To find out, researchers in the United States drove a compact car for 200 miles at speeds ranging from 35 to 75 miles per hour, and measured fuel efficiency in miles per gallon (mpg). From their data, they created the model $\widehat{Fuel\ Efficiency} = 32 - 0.1\ Speed$ and created this residual plot:

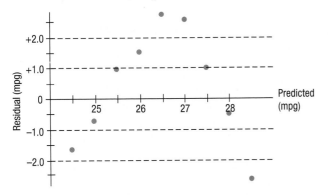

a) Interpret the slope of this line in context.

b) Explain why it's silly to attach any meaning to the *y*-intercept.

c) When this model predicts high *Fuel Efficiency,* what can you say about those predictions?

d) What *Fuel Efficiency* does the model predict when the car is driven at 50 mph?

e) What was the actual *Fuel Efficiency* when the car was driven at 45 mph?

f) Do you think there appears to be a strong association between *Speed* and *Fuel Efficiency*? Explain.

g) Do you think this is the appropriate model for that association? Explain.

h) If we use km/h and litres/100 km in our model instead, the scatterplot and residual plots should look essentially the same. True or false? Explain.

T 35. Interest rates 2014 Here's a plot showing the federal rate on 3-month Treasury bills from 1950 to 1980, and a regression model fit to the relationship between the *Rate* (in %) and *Years Since 1950* (www.gpoaccess.gov/eop/).

a) What is the correlation between *Rate* and *Year*?

b) Interpret the slope and intercept.

c) What does this model predict for the interest rate in the year 2000?

d) Would you expect this prediction to have been accurate? Explain.

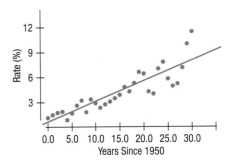

Dependent variable is Rate
R-squared = 77.6%   s = 1.232

Variable	Coefficient
Intercept	0.61149
Year - 1950	0.24788

T 36. Marriage age, 2011 The graph shows the ages of both men and women at first marriage (www.census.gov).

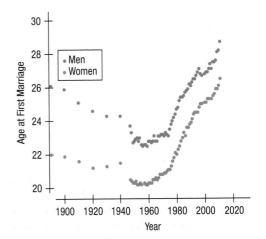

Clearly, the patterns for men and women are similar. But are the two lines getting closer together?

Here's a timeplot showing the *difference* in average age (men's age–women's age) at first marriage, the regression analysis, and the associated residuals plot.

Dependent variable is Age Difference
R-squared = 74.9%   s = 0.2346

Variable	Coefficient
Intercept	32.968
Year	−0.01549

a) What is the correlation between *Age Difference* and *Year*?
b) Interpret the slope of this line.
c) Predict the average age difference in 2015.
d) Describe reasons why you might not place much faith in that prediction.

T 37. Interest rates 2014 revisited In Exercise 35, you investigated the federal rate on 3-month Treasury bills between 1950 and 1980. The scatterplot below shows that the trend changed dramatically after 1980, so we computed a new regression model for the years 1981 to 2013.

Here's the model for the data from 1981 to 2013:

Dependent variable is Rate
R-squared = 77.0%   s = 1.626

Variable	Coefficient
Intercept	18.87316
Year - 1950	−0.30317

a) How does this model compare to the one in Exercise 35?
b) What does this model estimate the interest rate was in 2000? How does this compare to the rate you predicted in Exercise 35?
c) Do you trust this newer predicted value? Explain.
d) What does each of the two models predict the interest rate on 3-month Treasury bills will be in 2020?
e) Comment on your predictions in d)

T 38. Marriage age 2011 again Has the trend of decreasing difference in age at first marriage seen in Exercise 36 gotten stronger recently? The scatterplot and residual plot for the data from 1980 through 2011, along with a regression for just those years, are below.

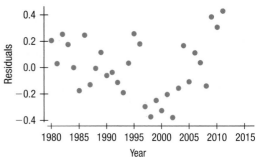

Dependent variable is Men − Women
R-squared = 49.4%   s = 0.225

Variable	Coefficient
Intercept	48.727
Year	−0.0234

a) Is this linear model appropriate for the post-1980 data? Explain.
b) What does the slope say about marriage ages?
c) Explain why it's not reasonable to interpret the *y*-intercept.

T 39. Gestation For humans, pregnancy lasts about nine months. In other species of animals, the length of time from conception to birth varies. Is there any evidence that the gestation period is related to the animal's lifespan? The first scatterplot shows *Gestation Period* (in days) vs. *Life Expectancy* (in years) for 18 species of mammals. The highlighted point at the far right represents humans.

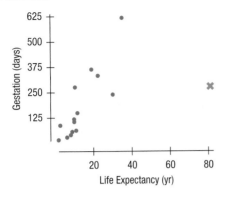

a) For these data, $r = 0.54$, not a very strong relationship. Do you think the association would be stronger or weaker if humans were removed? Explain.
b) Is there reasonable justification for removing humans from the data set? Explain.
c) Here are the scatterplot and regression analysis for the 17 nonhuman species. Comment on the strength of the association.

Dependent variable is: Gestation
R-squared = 72.2%

Variable	Coefficient
Constant	−39.5172
LifExp	15.4980

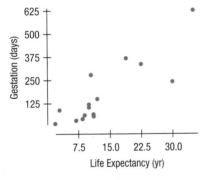

d) Interpret the slope of the line.
e) Some species of monkeys have a life expectancy of about 20 years. Estimate the expected gestation period of one of these monkeys.

T 40. Swim the lake 2013 People swam across Lake Ontario 52 times between 1974 and 2013 (www.soloswims.com). We might be interested in whether they are getting any faster or slower. Here are the regression of the crossing *Times* (minutes) against the *Year* since 1974 of the crossing and the residuals plot:

Dependent variable is Time
R-squared = 1.2%   s = 439.1

Variable	Coefficient
Intercept	1213.423
Year since 1974	3.986

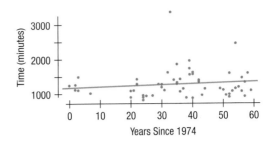

a) What does the R^2 mean for this regression?
b) Are the swimmers getting faster or slower? Explain.
c) The outlier seen in the residuals plot is a crossing by Vicki Keith in 1987 in which she swam a round trip, north to south, and then back again. Clearly, this swim doesn't belong with the others. Would removing it change the model dramatically? Explain.

41. Elephants and hippos We removed humans from the scatterplot in Exercise 39 because our species was an outlier in life expectancy. The resulting scatterplot shows two points that now may be of concern. The point in the upper right corner of this scatterplot is for elephants, and the other point at the far right is for hippos.

a) By removing one of these points, we could make the association appear to be stronger. Which point? Explain.
b) Would the slope of the line increase or decrease?
c) Should we just keep removing animals to increase the strength of the model? Explain.
d) If we remove elephants from the scatterplot, the slope of the regression line becomes 11.6 days per year. Do you think elephants were an influential point? Explain.

T 42. Another swim 2013 In Exercise 40, we saw that Vicki Keith's round-trip swim of Lake Ontario was an obvious outlier among the other one-way times. Here is the new regression after this unusual point is removed:

Dependent variable is Time
R-Squared = 4.5%   s = 322.7

Variable	Coefficient
Intercept	1139.794
Year since 1974	5.691

a) In this new model, the value of s_e is smaller. Explain what that means in this context.
b) Now would you be willing to say that the Lake Ontario swimmers are getting faster (or slower)?

43. Marriage age 2011 revisited Suppose you wanted to predict the trend in marriage age for American women into the early part of this century.
a) How could you use the data graphed in Exercise 15 to get a good prediction? Marriage ages in selected years starting in 1900 are listed below. Use all or part of these data to create an appropriate model for predicting the average age at which women will first marry in 2020.

| 1900–1950 (10-yr intervals): | 21.9, 21.6, 21.2, 21.3, 21.5, 20.3 |
| 1955–2010 (5-yr intervals): | 20.3, 20.3, 20.6, 20.8, 21.1,22.0, 23.3, 23.9, 24.5, 25.1, 25.3, 26.1 |

b) How much faith do you place in this prediction? Explain.
c) Do you think your model would produce an accurate prediction about your grandchildren, say, 50 years from now? Explain.

T 44. South American Immigrants Below are data on immigration to Canada from South America compared with total immigrants, by year:

Year	Total immigrants	South America
1986	99 343	6 699
1987	152 031	10 798
1988	161 534	7 237
1989	191 516	8 663
1990	216 424	8 986
1991	232 776	10 674
1992	254 856	10 471
1993	256 754	9 616
1994	224 395	7 936
1995	212 875	7 545
1996	226 061	6 114
1997	216 034	5 689
1998	174 184	4 971
1999	189 971	5 610
2000	227 429	6 782
2001	250 638	8 510
2002	229 049	8 825
2003	221 349	11 070
2004	235 824	12 292
2005	262 240	13 720
2006	251 644	13 423
2007	236 754	12 899

Source: Adapted from Statistics Canada, CANSIM table 051-0006, June 30, 2013.

a) Fit a model to the data to predict the annual percentage of total immigrants who are from South America in future years. You may ignore any number of earlier years, and start your fit at the year of your choice. (First you need to compute percents from the last two columns shown.)
b) Predict the percent of South American immigrants in 2012 using your model. The percent was in fact 3.8. What was the error of your prediction? Are you surprised?
c) In fact, the percents in 2008–2012 were 5.1, 5.2, 4.4, 4.1, 3.8, respectively. Add these points to your data plot, and now draw a (straight or curving) line out from 2007 or earlier to 2018 indicating where you think things are headed. What do you predict for 2018? Can you see the future? (Good luck! Check www.statcan.gc.ca to see how well you did.)

T 45. Life expectancy 2013 Data for 26 Western Hemisphere countries can be used to examine the association between life expectancy and the birth rate (number of births per 1000 population).

Country	Birth Rate (births/ 1000 population)	Life Expectancy
Argentina	17.12	76.1
Barbados	12.1	77
Belize	25.58	76.3
Canada	10.28	81.1
Chile	14.12	79.3
Colombia	16.98	73.9
Costa Rica	16.25	79.4
Cuba	9.92	79.3
Dominican Republic	19.21	73.6
Ecuador	19.23	75.8
El Salvador	17.12	72.4
Guatemala	25.99	71.4
Guyana	16.31	70.2
Haiti	23.35	62.4
Honduras	24.16	73.4
Jamaica	18.65	73.3
Mexico	18.61	77.1
Nicaragua	18.77	74.3
Panama	18.91	76.3
Paraguay	16.95	72.7
Saint Kitts and Nevis	13.79	73.3
Saint Lucia	14.19	74.8
Saint Vincent and the Grenadines	14.12	72.5
Trinidad and Tobago	14.07	70.3
Uruguay	13.28	77.2

a) Create a scatterplot relating *Life Expectancy* to *Birth Rate* and describe the association. Are the regression assumptions satisfied?

b) Is there an outlier? If so, set it aside and identify it before computing the regression.

c) Find the equation of the regression line with the outlier removed.

d) Make a plot of the residuals and comment.

e) Is the line an appropriate model?

f) Interpret the value of R^2.

g) If government leaders want to increase life expectancy, should they encourage women to have fewer children? Explain.

T 46. Tour de France 2014 We met the Tour de France data set in Chapter 1 (in Just Checking). One hundred years ago, the fastest rider finished the course at an average speed of about 25.3 kph (around 15.8 mph). By the 21st century, riders were averaging over 40 kph (nearly 25 mph).

a) Make a scatterplot of *Avg Speed* against *Year.* Describe the relationship of *Avg Speed* by *Year*, being careful to point out any unusual features in the plot.

b) Find the regression equation of *Avg Speed* on *Year.*

c) Are the conditions for regression met? Comment.

T 47. Rich getting richer 2011 Below are annual percentages of Canadian families that earned over $150 000 (in constant 2011 dollars) from 1986–2011. This percentage has gone up, but alas, we're not all getting richer! During the same period, the percentage earning under $40 000 increased from 41.3% in 1986 to 41.9% in 2011, partly attributable to an increase in part-time workers.

Year	over $150 000 (%)	Year	over $150 000 (%)
1986	3.7	1999	4.5
1987	4.1	2000	5
1988	4.5	2001	5.1
1989	4.8	2002	5.3
1990	4.5	2003	5.4
1991	4.2	2004	5.6
1992	4.1	2005	5.9
1993	3.4	2006	6.4
1994	3.7	2007	6.8
1995	3.7	2008	7.1
1996	3.9	2009	7.3
1997	4	2010	7.2
1998	4.5	2011	7.7

Source: Adapted from Statistics Canada, CANSIM table 202-0408 (accessed June 30, 2013).

a) Plot the data appropriately and describe the relationship.

b) Fit a straight line to the data. What percent of the variation in the percentage earning over $150 000 is explained by the straight line relationship?

c) For the year 1997, what is the predicted percent, using the linear model, and what is the residual for that observation? Does this model predict too high or too low for years around 1997?

d) Now use your best judgment to create a model from these data that you would consider using for prediction into some future years (you may ignore some earlier years and start your fit where you choose). What is R^2 for your model? Use it to predict the percentage earning over $150 000 in 2013 and in 2040, and comment on the reliability of your two predictions.

T 48. Second stage 2014 Look once more at the data from the Tour de France. In Exercise 46, we looked at the whole history of the race, but now let's consider just the post-World War II era.

a) Find the regression of *Avg Speed* by *Year* only for years from 1947 to the present. Are the conditions for regression met?

b) Interpret the slope.

c) In 1981, Bernard Hinault averaged 39.86 kph, while in 2013, Christopher Froome averaged 40.55 kph. Which was the more impressive performance? Why?

T 49. Gender gap 2011 Below is the difference in median income, for full-time, full-year workers, between males and females in Canada, by year (in constant 2011 dollars):

Year	Difference in Median Earnings (Male – Female)
1976	20 500
1977	20 600
1978	20 400
1979	19 500
1980	19 900
1981	18 800
1982	18 800
1983	17 500
1984	17 800
1985	18 300
1986	17 000
1987	17 600
1988	18 500
1989	17 600
1990	16 100
1991	15 200
1992	14 500
1993	14 000
1994	14 800
1995	13 400
1996	13 200

Year	Difference in Median Earnings (Male – Female)
1997	13 500
1998	13 300
1999	15 100
2000	14 500
2001	14 000
2002	13 600
2003	14 100
2004	13 700
2005	13 500
2006	12 800
2007	13 700
2008	12 800
2009	11 700
2010	12 200
2011	12 600

Source: Adapted from Statistics Canada, CANSIM table 202-0101, June 30, 2013.

a) Plot the difference in median earnings versus year. Describe the relationship.

b) Fit a straight line and assess your model by examining both R^2 and a plot of residuals versus year.

c) Predict the difference in the year 2070 using your fitted straight-line model from part b). Do you think this is likely to be high or low?

d) Do you think a quadratic curve might be a better model? Below is a scatterplot of residuals after fitting a quadratic curve by the same least squares principle:

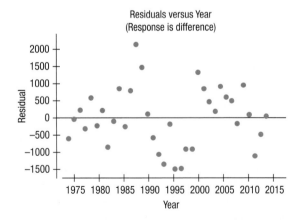

Residuals versus Year
(Response is difference)

Compare this residual plot with your plot from part b). What does it suggest about appropriate choice of model?

T 50. Interprovincial migration 2015 In Chapter 6 exercises, we examined data showing the net interprovincial migration each year for Ontario and Alberta (number arriving minus number leaving). Below are the first few and last few lines of the data file available at www.pearsoncanada. ca/deveaux.

Year	Ontario	Alberta
1978/1979	−4 325	33 426
1979/1980	−22 362	41 435
1980/1981	−33 247	44 250
⋮	⋮	⋮
⋮	⋮	⋮
2012/2013	−13 901	38 598
2013/2014	−14 564	35 382
2014/2015	−8763	28 921

Source: Adapted from Statistics Canada. CANSIM table 051-0004, September 15, 2016.

a) Plot Alberta net migration versus Ontario net migration for each year.

b) Fit a straight line model to the data for predicting Alberta migration from Ontario migration. What percent of the variation in Alberta migration numbers is explained by Ontario migration?

c) Examine the residuals. Are they approximately Normal? Is the plot of residuals versus predicted values (or the explanatory variable) fairly random-looking?

d) Plot residuals versus year. Is there a pattern? What do you learn from this plot? Does the pattern you see have anything to do with the Alberta economy and its dependence on oil prices? Do you think oil prices might be a better predictor of Alberta migration numbers?

T 51. Math and gender 2009 Below are mean PISA (Programme for International Student Assessment) math scores for samples of 15-year-old male and female students from a number of randomly selected schools in each of various OECD and other industrialized nations (4000–10 000 per country).

Country	Male mean	Female mean
OECD		
Australia	519	509
Austria	506	486
Belgium	526	504
Canada	533	521
Chile	431	410
Czech Republic	495	490
Denmark	511	495
Estonia	516	508
Finland	542	539
France	505	489
Germany	520	505
Greece	473	459
Hungary	496	484
Iceland	508	505
Ireland	491	483

Country	Male mean	Female mean
Israel	451	443
Italy	490	475
Japan	534	524
Korea	548	544
Luxembourg	499	479
Mexico	425	412
Netherlands	534	517
New Zealand	523	515
Norway	500	495
Poland	497	493
Portugal	493	481
Slovak Republic	498	495
Slovenia	502	501
Spain	493	474
Sweden	493	495
Switzerland	544	524
Turkey	451	440
United Kingdom	503	482
United States	497	477
OECD total	496	481
OECD average	501	490
Partners		
Albania	372	383
Argentina	394	383
Azerbaijan	435	427
Brazil	394	379
Bulgaria	426	430
Colombia	398	366
Croatia	465	454
Dubai (UAE)	454	451
Hong Kong-China	561	547
Indonesia	371	372
Jordan	386	387
Kazakhstan	405	405
Kyrgyzstan	328	334
Latvia	483	481
Liechtenstein	547	523
Lithuania	474	480
Macao-China	531	520
Montenegro	408	396
Panama	362	357
Peru	374	356
Qatar	366	371
Romania	429	425
Russian Federation	469	467
Serbia	448	437
Shanghai-China	599	601
Singapore	565	559
Chinese Taipei	546	541
Thailand	421	417
Trinidad and Tobago	410	418
Tunisia	378	366
Uruguay	433	421

Source: Based on data from OECD PISA 2009 online database, http://pisa2009.acer.edu.au, accessed on Feb 22, 2013.

a) Plot male mean score versus female mean score, and superimpose the least-squares regression line on the plot (for prediction of male means from female means).

b) Plot residuals from the regression versus female means or versus predicted (fitted) values. What do you learn?

c) Plot a histogram and Normal probability plot of the residuals. Are they approximately Normally distributed?

d) Plot residuals versus row number. Are the OECD countries less variable than the non-OECD countries?

e) Which two countries have the greatest leverage on the fit? Explain in simple words what this means.

f) The slope does not differ much from 1.0. What would a slope of 1.0 indicate about the nature of the relationship? If we fitted a model with the slope fixed at 1.0, what prediction equation would you expect to get? (Hint: Find the mean scores for males and females.)

g) If we examined only those countries with means over 500 for both sexes, what would happen to the correlation? Why?

Ⓣ **52. Mandarin or Cantonese 2011** Below are data on Canadian cities, showing mother-tongue counts from the 2011 Census.

Urban Centre	Cantonese	Mandarin
Abbotsford (BC)	310	400
Barrie (ON)	215	295
Brantford (ON)	125	90
Calgary (AB)	20 175	11 380
Edmonton (AB)	13 595	7 295
Guelph (ON)	770	650
Halifax (NS)	460	805
Hamilton (ON)	1 975	2 215
Kelowna (BC)	235	210
Kingston (ON)	335	395
Kitchener-Cambridge- Waterloo (ON)	1 390	3 060
London (ON)	1 165	1 415
Moncton (NB)	70	75
Montréal (QC)	10 515	10 180
Oshawa (ON)	730	645
Ottawa-Gatineau (ON-QC)	6 570	7 275
Peterborough (ON)	90	110
Québec (QC)	130	270
Regina (SK)	560	505
Saguenay (QC)	0	15
Saint John (NB)	45	125
Saskatoon (SK)	865	790
Sherbrooke (QC)	20	65
St. Catharines- Niagara (ON)	640	555
St. John's (NL)	105	165

Urban Centre	Cantonese	Mandarin
Sudbury (ON)	140	90
Thunder Bay (ON)	75	75
Toronto (ON)	170 485	100 045
Trois-Rivières (QC)	10	20
Vancouver (BC)	128 110	90 190
Victoria (BC)	2 930	1 945
Windsor (ON)	1 200	1 085
Winnipeg (MB)	3 220	2 080

Source: Statistics Canada, 2011 Census of Population and Statistics Canada, catalogue no. 98-314-XCB. Last modified: 2013-02-13.

a) Plot Cantonese speaker count versus Mandarin speaker count. Would you fit a regression line for predicting Cantonese speakers from Mandarin speakers? Explain.

b) Take the log (base 10) of both columns of numbers using your computer or calculator (why do you lose one city?). Now would you use regression for prediction purposes? What percentage of the variation in Cantonese counts is explained by the linear association with Mandarin counts?

   And stick to this log-log model for the remaining parts of this exercise.

c) Predict the Cantonese log count in a city where the Mandarin log count is 4.0. What would you predict for the actual Cantonese count?

d) What is the residual for the city of Vancouver?

e) Which city has the most surprising number of Cantonese speakers (considering the number of resident Mandarin speakers)? Give the residual.

f) Which cities have the greatest leverage on the fit?

T 53. Bigger and bigger According to the UK National Sizing Survey, women's average body size measurements have increased over time as follows:

Year	Bust size (in)	Waist size (in)	Hips size (in)
1920	32	20	32
1940	33	21	33
1960	34	24	33
1980	35	26	37
2000	36	28	38

Source: P. Treleaven, (September 2007), "How to fit into your clothes," Busts, waists, hips and the UK National Sizing Survey, *Significance* 4(3), 113–117.

a) Compute all correlations among all pairs of the four variables.

b) Why are the correlations so high?

c) Predict the average waist size for UK women in 1950 and in 2080. Comment on the reliability of your predictions.

d) Predict average waist size for UK women if their average bust size increases to 44. Comment on the reliability of your prediction.

T 54. More Mandarin 2011 Mandarin speakers in Canada are increasing rapidly. Here are 2001 and 2011 Census figures on Mandarin mother-tongue residents in some urban areas:

Urban Area	2001 Count	2011 Count
Abbotsford (BC)	105	400
Calgary (AB)	3 235	11 380
Edmonton (AB)	2 230	7 295
Guelph (ON)	290	650
Halifax (NS)	170	805
Hamilton (ON)	740	2 215
Kingston (ON)	190	395
Kitchener-Cambridge-Waterloo (ON)	645	3 060
London (ON)	330	1 415
Montréal (QC)	3 935	10 180
Oshawa (ON)	95	645
Ottawa-Gatineau (ON-QC)	3 245	7 275
Regina (SK)	130	505
Saskatoon (SK)	250	790
St. Catharines - Niagara (ON)	160	555
Toronto (ON)	35 315	100 045
Vancouver (BC)	47 165	90 190
Victoria (BC)	890	1 945
Windsor (ON)	425	1 085
Winnipeg (MB)	685	2 080

Source: Based on Statistics Canada, 2011 Census of Population and Statistics Canada, catalogue no. 98-314-XCB. Last modified: 2013-02-13.

a) Plot the 2011 count against the 2001 count. Would you fit a regression line for predicting 2011 counts from 2001 counts? Explain.

b) Take the log (base 10) of both columns of numbers using your computer or calculator. Re-plot the data. Now would you use regression for prediction purposes?

   Stick to this log-log model for the remaining parts of this exercise.

c) Predict the 2011 log-count in a city where the log of the 2001 count is 4.0. What would you predict for the actual 2011 count?

d) Which city's 2011 count is the most surprising, and why? Give the residual for this city. How would the fitted line change if you removed this city?

e) Which cities have the greatest leverage on the fit?

f) Take anti-logs of both sides of your fitted equation, i.e., rewrite it without any logs in the equation. If the slope of the fitted log-log model happened to be 1.00, what simple interpretation would you now get of this prediction equation?

g) What is the correlation between the two log-counts? What is one obvious reason (for example, a lurking variable) for such a high correlation? Suggest another useful way of defining the two variables that would produce a smaller correlation. (In real life studies, there are always issues about how to best define and measure the relevant variables.)

T 55. Census at school The International *CensusAtSchool* Project collects data on primary and secondary school students from various countries, including Canada. We selected a random sample of 111 Canadian secondary school students, age 14 and over, from 2006–2007. Below are the first four rows of the 111 row worksheet of data. See Chapter 6, Exercise 47, for more information. All measurements are in centimetres and were taken in the classroom. Middlefinger is distance from wrist to tip of middle finger. Wristbone is distance from wrist bone to elbow. Armspan is distance from tip of left hand to tip of right hand with arms spread wide. Foot is length of right foot.

Gender	Age	Height	Armspan	Wrist-bone	Middle-finger	Foot
Boy	14	166	169	25	19	26
Boy	14	160	170	27	19	28
Boy	14	167	163	26	20	26
Girl	16	165	165	24	17	22

Source: The Royal Statistical Society Centre for Statistical Education, CensusAtSchool project.

Below is a table of all the pairwise correlations:

	Age	Wrist-bone	Middle-finger	Foot	Armspan
Wristbone	−0.008				
Middlefinger	0.059	0.360			
Foot	0.188	0.436	0.461		
Armspan	0.219	0.449	0.565	0.548	
Height	0.272	0.397	0.560	0.542	0.781

a) Which variable appears to be the single best predictor of middlefinger? What percentage of the variation in middlefinger measurements can be explained by a linear association with this predictor variable?

b) Which variable appears to be the single best predictor of armspan? What percentage of the variation in armspan measurements can be explained by a linear association with this predictor variable?

Below is a plot of armspan versus height, including information about gender.

c) Regress armspan on height and give the prediction equation. Interpret the slope.

d) Examine a histogram and Normal probability plot of the residuals from the regression above and describe the distribution of the residuals.

e) Plot the residuals against height or the fitted (predicted) values. What do you learn from this plot?

f) Which child has the most unusually long or short arm-span considering his or her height? (Give the row number and gender.) How much longer or shorter than the (appropriate) average is it?

g) It is not clear from the plot above whether gender is an important variable in understanding the relationship between armspan and height. Plot the residuals from the regression above versus gender. Is there any pattern? For either gender, is there a tendency for the model to over-predict or under-predict armspan? If so, what would be your modelling recommendation?

T 56. Census at school, boys versus girls Separate the *CensusAtSchool* data in the previous exercise according to gender, and run the two regressions of armspan on height, once for each gender.

a) Describe the differences suggested by the two fitted models.

b) Estimate the difference in armspan between a boy and girl if each of them is 160 cm tall. What if each of them is 170 cm tall?

c) For the girls, which observation appears to have the most leverage? Do you think it is influential? Check to confirm your suspicion.

d) Which girl has the most unusually long or short arm-span, considering her height? How much longer or shorter than the (appropriate) average is it?

e) Predict the armspan of a boy who is 130 cm tall. Predict the armspan of a boy who is 185 cm tall. Is either prediction more reliable than the other? Why?

T 57. Planet distances and years 2006 At a meeting of the International Astronomical Union (IAU) in Prague in 2006, Pluto was determined not to be a planet, but rather the largest member of the Kuiper belt of icy objects (http://www.enchantedlearning.com/subjects/astronomy/planets/). Let's examine some facts.

The table below shows the nine sun-orbiting objects formerly known as planets:

Planet	Position Number	Distance from Sun (million miles)	Length of Year (Earth years)
Mercury	1	36	0.24
Venus	2	67	0.61
Earth	3	93	1.00
Mars	4	142	1.88
Jupiter	5	484	11.86
Saturn	6	887	29.46
Uranus	7	1784	84.07
Neptune	8	2796	164.82
Pluto	9	3707	247.68

a) Plot the length of the year against the distance from the sun. Describe the shape of your plot.
b) Re-express one or both variables to straighten the plot (with logs). Use the re-expressed data to create a model describing the length of a planet's year based on its distance from the sun.
c) Comment on how well your model fits the data.

T 58. Asian immigrants 2012 Below are data on immigration to Canada from countries in Asia:

Year	Total immigrants	Total Asian immigrants	Percentage from Asia
1963	93 151	3 912	4.2
1964	112 606	6 526	5.8
1965	146 758	11 684	8
1966	194 743	14 238	7.3
1967	222 876	21 247	9.5
1968	183 974	22 076	12
1969	161 531	23 706	14.7
1970	147 713	21 451	14.5
1971	121 900	22 369	18.4
1972	122 006	23 325	19.1
1973	184 200	43 193	23.4
1974	218 465	50 566	23.1
1975	187 881	47 382	25.2
1976	149 429	44 091	29.5
1977	114 914	31 380	27.3
1978	86 313	24 029	27.8
1979	112 096	46 871	41.8
1980	143 498	71 793	50
1981	128 794	48 898	38
1982	121 331	41 749	34.4
1983	89 377	36 998	41.4
1984	88 599	42 091	47.5
1985	84 339	38 614	45.8
1986	99 343	41 650	41.9
1987	152 031	67 306	44.3
1988	161 534	80 938	50.1
1989	191 516	93 026	48.6
1990	216 424	112 902	52.2
1991	232 776	121 000	52
1992	254 856	140 385	55.1
1993	256 754	147 847	57.6
1994	224 395	141 916	63.2
1995	212 875	129 435	60.8
1996	226 061	144 397	63.9
1997	216 034	138 319	64
1998	174 184	101 492	58.3
1999	189 971	113 086	59.5
2000	227 429	140 178	61.6
2001	250 571	155 682	62.1
2002	229 071	141 358	61.7
2003	221 352	132 851	60
2004	235 824	135 658	57.5
2005	262 239	159 968	61
2006	251 652	149 523	59.4
2007	236 762	133 383	56.3
2008	247 247	140 961	57.0
2009	252 177	141 196	56.0
2010	280 682	164 408	58.6
2011	248 746	146 951	59.1
2012	257 730	149 235	57.9

Source: Adapted from Statistics Canada, CANSIM database, http://cansim2.statcan.gc.ca, table 051-0006, July 2, 2013.

a) We want to predict the percentage of total immigrants who are from Asia based on the year. You may use two straight line models (which we call *piecewise linear*) depending on the time period. Draw your fitted models on the scatterplot.
b) Predict the percent of Asian immigrants in 1955 using your model. The percent was in fact 3.3. What was the error of your prediction? Are you surprised?
c) Use your model to predict Asian immigration for 2027. On your graph from part a), draw a (straight or not straight) line out from 2007 to 2027 indicating where you think things are headed. Can you see the future? (Good luck! Check www.statcan.gc.ca/ to see how well you did.)

T 59. Placental mammals In the introductory example in Chapter 6, we plotted (average) neonatal brain weight versus (average) gestation period for a number of placental mammal species. We are interested in finding a good model for predicting the neonatal brain weight of a species from its gestation period. Below are the data for a sample of 23 species.

Brain weight (g)	Gestation (days)	Brain weight (g)	Gestation (days)
8.78	135	129.00	270
4.00	90	128.00	230
4.00	135	227.00	265
30.80	139	335.00	270
29.00	180	6.80	63
64.00	140	3.82	52
58.00	175	7.73	77
75.00	180	3.34	60
53.00	180	35.30	92
33.50	195	49.50	108
43.00	168	30.00	104
65.00	210		

Source: J.A. Witmer, 1992, *Data Analysis: An Introduction.* Table 4-1 "Brain Weight and Gestation Period of Various Mammals" p. 50, © 1992 Prentice-Hall, Inc. Reproduced by permission of Pearson Education, Inc.

a) Plot neonatal brain weight versus gestation. Fit a straight line, show it on the data scatterplot, and report R^2. Plot the residuals versus the predicted values. What do you learn?

b) Compute the logarithms of the neonatal brain weights, and plot this variable versus gestation. Fit a straight line, show it on the scatterplot, and report R^2. Plot residuals versus predicted values. Compare this model with the one in part a). Which model do you prefer and why? From your plots, do you see any possible problem(s) with the log-transformed model?

c) Take the square roots of brain weights, plot them versus gestation, and fit a straight line. Compare your results with the log transformation of part b). Which transformation do you think works best and why?

d) Using the log-transformed model of part b), predict the neonatal brain weight of an individual species whose gestation period is 200 days (start by predicting the log brain weight).

e) Using the square root model of part c), predict the neonatal brain weight of an individual species whose gestation period is 200 days.

T 60. **Placental mammals again** Continuing with the previous exercise, the species listed in the last 7 rows of the table were all carnivores, while primates occupy the first 16 rows. Enter into a new column in your data worksheet—"p" or "c" for each species, to denote whether primate or carnivore.

a) Plot brain weight versus gestation, using different symbols for the primates and carnivores. Repeat using log of brain weight. What do you learn from these plots?

b) Use the logarithm of brain weight. Fit a model to predict brain weight for the primates only. Fit another model using only the carnivore data. Does fitting separate models make sense from a statistical standpoint? From a practical standpoint? Draw by hand the two fitted straight lines on the scatterplot with all the (log-transformed) data. How do carnivores differ from primates?

c) Are humans (row 16) an outlier?

Just Checking ANSWERS

1. Not high leverage, not influential, large residual
2. High leverage, not influential, small residual
3. High leverage, influential, not large residual

MathXL

MyLab Statistics

Go to MathXL at **www.mathxl.com** or MyLab™ Statistics at **www.mystatlab.com**. You can practise exercises for this chapter as often as you want. The guided solutions will help you find answers step by step. You'll find a personalized study plan available to you too!

Review

Exploring Relationships Between Variables

Quick Review

You have now survived your second major unit of Statistics. Here's a brief summary of the key concepts and skills:

- We treat data two ways: as categorical and as quantitative.

- To explore relationships in categorical data, check out Chapter 2.

- To explore relationships in quantitative data:
  - Make a picture. Use a scatterplot. Put the explanatory variable on the x-axis and the response variable on the y-axis. (If there is a categorical variable present, forming distinct groups, show this in your scatterplot using different symbols for each group.)
  - Describe the association between two quantitative variables in terms of direction, form, and strength.
  - The amount of scatter determines the strength of the association.
  - If, as one variable increases so does the other, the association is positive. If one increases as the other decreases, it's negative.
  - If the form of the association is linear, calculate a correlation to measure its strength numerically, and possibly do a regression analysis to model it.
  - Correlations closer to −1 or +1 indicate stronger linear associations. Correlations near 0 indicate weak linear relationships, but other forms of association may still be present.
  - The line of best fit is also called the least squares regression line because it minimizes the sum of the squared residuals.
  - The regression line predicts values of the response variable from values of the explanatory variable.

- A residual is the difference between the true value of the response variable and the value predicted by the regression model.

- The slope of the line is a rate of change, best described in "y-units" per "x-unit."

- R^2 gives the fraction of the variation in the response variable that is accounted for by the model.

- The standard deviation of the residuals measures the amount of scatter around the line.

- Outliers and influential points can distort any of our models.

- If you see a pattern (curve) in the plot of residuals versus explanatory variable, your chosen model is not appropriate; try a different model. You may, for example, be able to straighten the relationship by re-expressing one or both of the variables.

- To straighten bent relationships or to ameliorate trends in dispersion, re-express the data using logarithms or a power (squares, square roots, reciprocals, etc.).

- Always remember that an association is not necessarily an indication that one of the variables causes the other.

Need more help with some of this? Try rereading some sections of Chapters 6 through 8. And see below for more opportunities to review these concepts and skills.

"One must learn by doing the thing; though you think you know it, you have no certainty until you try."

— *Sophocles (495–406 B.C.E.)*

Review Exercises

1. College Every year, *US News and World Report* publishes a special issue on many U.S. colleges and universities. The scatterplots below have *Student/Faculty Ratio* (number of students per faculty member) for the colleges and universities on the *y*-axes plotted against 4 other variables. The correct correlations for these scatterplots appear in the list of six possible correlations below. Match them.

$$-0.98 \quad -0.71 \quad -0.51 \quad 0.09 \quad 0.23 \quad 0.69$$

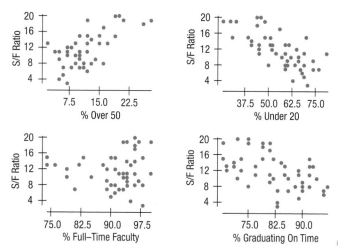

2. Togetherness Are good grades in high school associated with family togetherness? A random sample of 142 high school students was asked how many meals per week their families ate together. Their responses produced a mean of 3.78 meals per week, with a standard deviation of 2.2. Researchers then matched these responses against the students' grade point averages (GPAs). The scatterplot appeared to be reasonably linear, so they created a line of regression. No apparent pattern emerged in the residuals plot. The equation of the line was $\widehat{GPA} = 2.73 + 0.11 \, Meals$.

a) Interpret the *y*-intercept in this context.

b) Interpret the slope in this context.

c) What was the mean GPA for the students in this study?

d) If a student in this study had a negative residual, what did that mean?

e) Upon hearing of this study, a counsellor recommended that parents who want to improve their children's grades should get the family to eat together more often. Do you agree with this interpretation? Explain.

T 3. Vineyards Here are the scatterplot and regression analysis for *Case Prices* of 36 wines from vineyards in the

Finger Lakes region of New York State and the *Ages* of the vineyards.

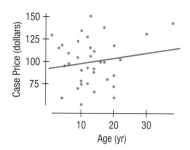

Dependent variable is: Case Price
R-squared = 2.7%

Variable	Coefficient
Constant	92.7650
Age	0.567284

a) Does it appear that vineyards in business longer get higher prices for their wines? Explain.

b) What does this analysis tell us about vineyards in the rest of the world?

c) Write the regression equation.

d) Explain why that equation is essentially useless.

T 4. Vineyards again Instead of *Age,* perhaps the *Size* of the vineyard (in acres) is associated with the price of the wines. Look at the scatterplot:

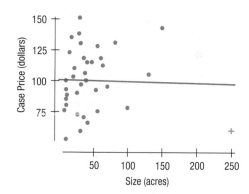

a) Do you see any evidence of an association?

b) What concern do you have about this scatterplot?

c) If the red "+" data point is removed, would you expect the correlation to become stronger or weaker? Explain.

d) If the red "+" data point is removed, would you expect the slope of the line to increase or decrease? Explain.

5. Lurking variables Each of the following scatterplots contains two distinct subgroups (+'s for males, o's for

females). In each case, estimate the correlation between
y and *x*:
a) ignoring gender
b) taking gender into account, i.e., separately by gender

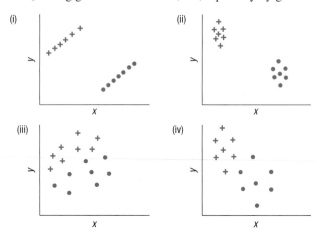

(i)

(ii)

(iii)

(iv)

6. Residuals Below is a plot of primary school enrollment
percentage of females (in the population in the correct
age group) versus gross national income per capita for a
number of countries. Also shown is the least squares
straight-line fit.
a) Show and describe by drawing on the plot and with
words what it is that makes the drawn line the "least
squares" fit.
b) If we construct the usual graph of residuals versus
predictor, what would it look like? Draw it, roughly,
but put some numbers on each axis. If we were
instead to construct the graph of residuals versus
predicted values, how would it compare with the first
residual plot?
c) What is the approximate numerical value of the biggest (in magnitude) residual?
d) Would this model predict too high or too low for
countries with incomes per capita in the \$10 000 to
\$30 000 range? Do you get positive or negative resid-
uals for these countries?

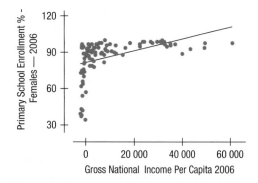

7. Acid rain Biologists studying the effects of acid rain on
wildlife collected data from 163 streams in the
Adirondack Mountains. They recorded the *pH* (acidity)
of the water and the *BCI*, a measure of biological diver-

sity, and they calculated $R^2 = 27\%$. Here's a scatterplot
of *BCI* against *pH*:

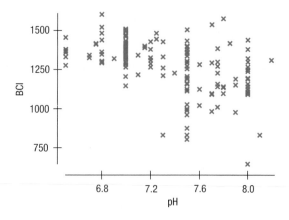

a) What is the correlation between *pH* and *BCI*?
b) Describe the association between these two variables.
c) If a stream has average *pH*, what would you predict
about the *BCI*?
d) In a stream where the *pH* is 3 standard deviations
above average, what would you predict about the *BCI*?

T 8. Manatees 2013 Marine biologists warn that the growing
number of powerboats registered in Florida threatens the
existence of manatees. The data below come from the
Florida Fish and Wildlife Conservation Commission
(myfwc.com/research/manatee/) and the National
Marine Manufacturers Association (www.nmma.org/).

Year	Manatees Killed	Powerboat Registrations (in 1000s)
1982	13	447
1983	21	460
1984	24	481
1985	16	498
1986	24	513
1987	20	512
1988	15	527
1989	34	559
1990	33	585
1992	33	614
1993	39	646
1994	43	675
1995	50	711
1996	47	719
1997	53	716
1998	38	716
1999	35	716
2000	49	735
2001	81	860
2002	95	923

2003	73	940
2004	69	946
2005	79	974
2006	92	988
2007	73	992
2008	90	932
2009	97	949
2010	83	914
2011	88	890
2012	81	870
2013	72	865

a) In this context, which is the explanatory variable?
b) Make a scatterplot of these data and describe the association you see.
c) Find the correlation between *Boat Registrations* and *Manatee Deaths*.
d) Interpret the value of R^2.
e) Does your analysis prove that powerboats are killing manatees?

T 9. A manatee model 2013 Continue your analysis of the manatee situation from the previous exercise.
a) Create a linear model of the association between *Manatee Deaths* and *Powerboat Registrations*.
b) Interpret the slope of your model.
c) Interpret the *y*-intercept of your model.
d) How accurately did your model predict the high number of manatee deaths in 2009?
e) Which is better for the manatees, positive residuals or negative residuals? Explain.
f) What does your model suggest about the future for the manatee?

10. Grades A Statistics instructor created a linear regression equation to predict students' final exam scores from their midterm exam scores. The regression equation was $\widehat{Fin} = 10 + 0.9\,Mid$.
a) If Susan scored a 70 on the midterm, what did the instructor predict for her score on the final?
b) Susan got an 80 on the final. How big is her residual?
c) Suppose that the standard deviation of the final was 12 points and the standard deviation of the midterm was 10 points. What is the correlation between the two tests?
d) How many points would someone need to score on the midterm to have a predicted final score of 100?
e) Suppose someone scored 100 on the final. Explain why you can't estimate this student's midterm score from the information given.
f) One of the students in the class scored 100 on the midterm but got overconfident, slacked off, and scored only 15 on the final exam. What is the residual for this student?
g) No other student in the class "achieved" such a dramatic turnaround. If the instructor decides not to

include this student's scores when constructing a new regression model, will the R^2 value of the regression increase, decrease, or remain the same? Explain briefly.
h) Will the slope of the new line increase or decrease?

11. Traffic Highway planners investigated the relationship between traffic density (number of automobiles per mile) and the average speed (in miles per hour) of the traffic on a moderately large city thoroughfare. The data were collected at the same location at 10 different times over a span of three months. They found a mean traffic density of 68.6 cars per mile (cpm) with a standard deviation of 27.07 cpm. Overall, the cars' average speed was 26.38 mph, with a standard deviation of 9.68 mph. The researchers found the regression line for these data to be $\widehat{Speed} = 50.55 - 0.352\,Density$.
a) What is the value of the correlation coefficient between speed and density?
b) What percent of the variation in average speed is explained by traffic density?
c) Predict the average speed of traffic on the thoroughfare when the traffic density is 50 cpm.
d) What is the value of the residual for a traffic density of 56 cpm with an observed speed of 32.5 mph?
e) The data set initially included the point (density = 125 cpm, speed = 55 mph). This point was considered an outlier and was not included in the analysis. Will the slope increase, decrease, or remain the same if we redo the analysis and include this point?
f) Will the correlation become stronger, weaker, or remain the same if we redo the analysis and include this point (125, 55)?
g) A Canadian member of the research team measured the speed of the cars in kilometres per hour (1 km ≈ 0.62 miles) and the traffic density in cars per kilometre. Find the value of his calculated correlation between speed and density.

T 12. Cramming One Thursday, researchers gave students enrolled in a section of basic Spanish a set of 50 new vocabulary words to memorize. On Friday, the students took a vocabulary test. When they returned to class the following Monday, they were retested—without advance warning. Here are the test scores for the 25 students (where each row gives us paired scores for the same student).

Fri.	Mon.	Fri.	Mon.	Fri.	Mon.
42	36	48	37	39	41
44	44	43	41	46	32
45	46	45	32	37	36
48	38	47	44	40	31
44	40	50	47	41	32
43	38	34	34	48	39
41	37	38	31	37	31
35	31	43	40	36	41
43	32				

a) What is the correlation between *Friday* and *Monday* scores?

b) What does a scatterplot show about the association between the scores?

c) What does it mean for a student to have a positive residual?

d) What would you predict about a student whose *Friday* score was one standard deviation below average?

e) Write the equation of the regression line.

f) Predict the *Monday* score of a student who earned a 40 on Friday.

T 13. Indigenous populations rising 2011 The data below show Indigenous populations in a sample of Canadian cities.

a) Plot the 2011 percents versus the 2001 percents. Guess the correlation, and confirm your guess.

b) Fit a regression model to predict the 2011 percent from the 2001 percent.

c) Interpret the slope, assuming the intercept is close enough to zero to be considered zero.

d) What distinction is held by Prince Albert? If you set it aside, are there still problems? (Check the residual scatterplot. Does the plot thicken?)

e) Including Prince Albert, take the log (base 10) of both variables and refit the model. Did taking logs help?

f) If a census area had an Indigenous percent of 10% in 2001, what would you predict for the percent in 2011?

	2001		2011	
CMA/CA	Number of Indigenous Persons	Percentage of Population	Number of Indigenous Persons	Percentage of Population
Prince Albert	11 640	29.2	15 780	38.5
Sault Ste. Marie	5 610	7.2	8 070	10.3
Prince George	7 980	9.4	9 930	12.0
Greater Sudbury	7 385	4.8	13 410	8.5
Saskatoon	20 275	9.1	23 890	9.3
Winnipeg	55 755	8.4	78 415	11.0
Regina	15 685	8.3	19 785	9.5
Kamloops	5 470	6.4	8 265	8.6
Thunder Bay	8 200	6.8	11 675	9.8
Edmonton	40 930	4.4	61 770	5.4
Victoria	8 695	2.8	14 200	4.2
Calgary	21 915	2.3	33 375	2.8
Hamilton	7 270	1.1	11 980	1.7
Vancouver	36 860	1.9	52 375	2.3
London	5 640	1.3	8 475	1.8
Ottawa-Hull	13 485	1.3	30 570	2.5
Toronto	20 300	0.4	36 990	0.7
Montreal	11 085	0.3	26 285	0.7

Source: Based on Statistics Canada: Aboriginal peoples of Canada: A demographic profile, 2003, and Statistics Canada, 2011 National Household Survey, Statistics Canada Catalogue no. 99-011-X2011028.

T 14. Quebec abortions Below are annual rates of abortion performed on females under the age of 20, per 1000 Quebec females (aged 15–49) over the period 1974–2005 (data not available for 1991).

Year	Rate	Year	Rate	Year	Rate
1974	2.4	1985	9.9	1996	18.7
1975	2.9	1986	11.1	1997	18.7
1976	3.7	1987	10.7	1998	21.0
1977	3.9	1988	12.2	1999	21.8
1978	6.1	1989	14.1	2000	21.8
1979	7.6	1990	16.3	2001	20.9
1980	8.8	1991	*	2002	20.4
1981	8.7	1992	17.3	2003	19.4
1982	9.4	1993	17.3	2004	19.7
1983	8.4	1994	19.1	2005	18.6
1984	8.8	1995	16.3		

Source: Adapted from Statistics Canada CANSIM database, table 106-9002, (accessed: 2013-07-05).

a) Plot the rate versus the year. Describe the trend.

b) Calculate the correlation.

c) Fit a straight line model, and use it to predict the abortion rate in 2007.

d) Examine the plot of residuals versus year. What does this plot tell you? Are you likely to over- or underpredict the rate in 2007?

e) Is there another approach you might take to constructing a model for the relationship between abortion rates and year in Quebec (say, fitting two different models, depending on the time period)? Try it and give your best prediction for the year 2007.

15. Cars with horses Can we predict the horsepower of the engine that manufacturers will put in a car by knowing the weight of the car? Here are the regression analysis and residual plot:

Dependent variable is: Horsepower
R-squared = 84.1%

Variable	Coefficient
Intercept	3.49834
Weight	34.3144

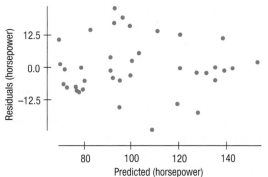

a) Write the equation of the regression line.
b) Do you think the car's weight is measured in pounds or thousands of pounds? Explain.
c) Do you think this linear model is appropriate? Explain.
d) The highest point in the residuals plot, representing a residual of 22.5 horsepower, is for a Chevy weighing 2595 pounds. How much horsepower does this car have?

16. **Colourblind** Although some women are colourblind, this condition is found primarily in men. Why is it wrong to say there's a strong correlation between sex and colourblindness?

T 17. **Old Faithful** There is evidence that eruptions of Old Faithful can best be predicted by knowing the duration of the previous eruption.
a) Describe what you see in the scatterplot of *Intervals* between eruptions vs. *Duration* of the previous eruption.

b) Write the equation of the line of best fit. Here's the regression analysis:

Dependent variable is: Interval
R-squared = 77.0%   s = 6.16 min

Variable	Coefficient
Intercept	33.9668
Duration	10.3582

c) Carefully explain what the slope of the line means in this context.
d) How accurate do you expect predictions based on this model to be? Cite statistical evidence.
e) If you just witnessed an eruption that lasted 4 minutes, how long do you predict you'll have to wait to see the next eruption?
f) So you waited, and the next eruption came in 79 minutes. Use this as an example to define a residual.

T 18. **Crocodile lengths** The ranges inhabited by the Indian gharial crocodile and the Australian saltwater crocodile overlap in Bangladesh. Suppose a very large crocodile skeleton is found there, and we wish to determine the species of the animal. Wildlife scientists have measured the lengths of the heads and the complete bodies of several crocs (in centimetres) of each species, creating the following regression analyses:

Indian Crocodile		Australian Crocodile	
Dependent variable is: IBody		Dependent variable is: ABody	
R-squared = 97.2%		R-squared = 98.0%	
Variable	**Coefficient**	**Variable**	**Coefficient**
Intercept	−69.3693	Intercept	−20.2245
IHead	7.40004	AHead	7.71726

a) Do the associations between the sizes of the heads and bodies of the two species appear to be strong? Explain.
b) In what ways are the two relationships similar? Explain.
c) What is different about the two models? What does that mean?
d) The crocodile skeleton found had a head length of 62 cm and a body length of 380 cm. Which species do you think it was? Explain why.

T 19. **How old is that tree?** One can determine how old a tree is by counting its rings, but that requires cutting the tree down. Can we estimate the tree's age simply from its diameter? A forester measured 27 trees of the same species that had been cut down, and counted the rings to determine the ages of the trees.

Diameter (inches)	Age (yr)	Diameter (inches)	Age (yr)
1.8	4	10.3	23
1.8	5	14.3	25
2.2	8	13.2	28
4.4	8	9.9	29
6.6	8	13.2	30
4.4	10	15.4	30
7.7	10	17.6	33
10.8	12	14.3	34
7.7	13	15.4	35
5.5	14	11.0	38
9.9	16	15.4	38
10.1	18	16.5	40
12.1	20	16.5	42
12.8	22		

a) Find the correlation between *Diameter* and *Age*. Does this suggest that a linear model may be appropriate? Explain.
b) Create a scatterplot and describe the association.
c) Create the linear model.
d) Check the residuals. Explain why a linear model is probably not appropriate.
e) If you used this model, would it generally overestimate or underestimate the ages of very large trees? Explain.

T **20. Improving trees** In the last exercise, you saw that the linear model had some deficiencies. Let's create a better model.
  a) Perhaps the cross-sectional area of a tree would be a better predictor of its age. Since area is measured in square units, try re-expressing the data by squaring the diameters. Does the scatterplot look better?
  b) Create a model that predicts *Age* from the square of *Diameter*.
  c) Check the residuals plot for this new model. Is this model more appropriate? Why?
  d) Estimate the age of a tree 18 inches in diameter.

21. Big screen An electronics website collects data on the size of new HD flat panel televisions (measuring the diagonal of the screen in inches) to predict the cost (in hundreds of dollars). Which of these is most likely to be the slope of the regression line: 0.03, 0.3, 3, 30? Explain.

T **22. Smoking and pregnancy 2011** The Child Trends Data Bank monitors issues related to children. The table shows a 50-state average of the percent of expectant mothers who smoked cigarettes during their pregnancies.

Year	% Smoking While Pregnant	Year	% Smoking While Pregnant
1990	19.2	2001	13.8
1991	18.7	2002	13.3
1992	17.9	2003	12.7
1993	16.8	2004	10.9
1994	16.0	2005	10.1
1995	15.4	2006	10.0
1996	15.3	2007	10.4
1997	14.9	2008	9.7
1998	14.8	2009	9.3
1999	14.1	2010	9.2
2000	14.0	2011	9.0

  a) Create a scatterplot and describe the trend you see.
  b) Find the correlation.
  c) How is the value of the correlation affected by the fact that the data are averages rather than percentages for each of the 50 states?
  d) Write a linear model and interpret the slope in this context.

T **23. Down the drain** Most water tanks have a drain plug so that the tank may be emptied when it's to be moved or repaired. How long it takes a certain size of tank to drain depends on the size of the plug, as shown in the table below. Create a model. (Hint: Replace *Drain Time* by $1/\sqrt{Drain\ Time}$)

Diameter of Plug (in.)	Drain Time (min.)	Diameter of Plug (in.)	Drain Time (min.)
$\frac{3}{8}$	140	$1\frac{1}{4}$	13
$\frac{1}{2}$	80	$1\frac{1}{2}$	10
$\frac{3}{4}$	35	2	5
1	20		

24. Tips It's commonly believed that people use tips to reward good service. A researcher for the hospitality industry examined tips and ratings of service quality from 2645 dining parties at 21 different restaurants. The correlation between ratings of service and tip percentages was 0.11. (Source: M. Lynn and M. McCall, "Gratitude and Gratuity." *Journal of Socio-Economics* 29: 203–214)
  a) Describe the relationship between *Quality of Service* and *Tip Size*.
  b) Find and interpret the value of R^2 in this context.

25. U.S. Cities Data from 50 large U.S. cities show the mean *January Temperature* (in Fahrenheit degrees) and *Latitude*. Describe what you see in the scatterplot.

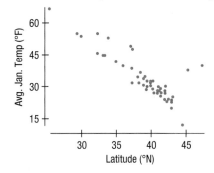

26. Correlations The study of U.S. cities in Exercise 25 found the mean *January Temperature* (degrees Fahrenheit), *Altitude* (feet above sea level), and *Latitude* (degrees north of the equator) for 55 cities. Here's the correlation matrix:

	Jan. Temp.	Latitude	Altitude
Jan. Temp.	1.000		
Latitude	−0.848	1.000	
Altitude	−0.369	0.184	1.000

  a) Which seems to be more useful in predicting *January Temperature*: *Altitude* or *Latitude*? Explain.
  b) If *Temperature* were measured in degrees Celsius, what would be the correlation between *Temperature* and *Latitude*?
  c) If *Temperature* were measured in degrees Celsius and *Altitude* in metres, what would be the correlation? Explain.
  d) What would you predict about the *January Temperature* in a city whose *Altitude* is two standard deviations higher than the average *Altitude*?

27. Winter in the city Summary statistics for the data relating the latitude and average January temperature (degrees Fahrenheit) for 55 large U.S. cities are given below.

Variable	Mean	StdDev
Latitude	39.02	5.42
JanTemp	26.44	13.49
Correlation = −0.848		

a) What percent of the variation in *January Temperature* can be explained by variation in *Latitude*?
b) What is indicated by the fact that the correlation is negative?
c) Write the equation of the line of regression for predicting *January Temperature* from *Latitude*.
d) Explain what the slope of the line means.
e) Do you think the *y*-intercept of the line is meaningful? Explain.
f) The latitude of Denver is 40° N. Predict the mean January temperature (°F) there.
g) If the residual for a city is positive, what does that mean?
h) How do your answers to parts a), c), d), and f) change if temperature is in degrees Celsius?

28. Depression The September 1998 issue of the *American Psychologist* published an article by Kraut et al. that reported on an experiment examining "the social and psychological impact of the Internet on 169 people in 73 households during their first 1 to 2 years online." In the experiment, 73 households were offered free Internet access for 1 or 2 years in return for allowing their time and activity online to be tracked. The members of the households who participated in the study were also given a battery of tests at the beginning and again at the end of the study. The conclusion of the study made news headlines: Those who spent more time online tended to be more depressed at the end of the experiment. Although the paper reports a more complex model, the basic result can be summarized in the following regression of *Depression* (at the end of the study, in "depression scale units") vs. *Internet Use* (in mean hours per week):

Dependent variable is: Depression
R-squared = 4.6%   s = 0.4563

Variable	Coefficient
Intercept	0.5655
Internet use	0.0199

The news reports about this study clearly concluded that using the Internet causes depression. Discuss whether such a conclusion can be drawn from this regression. If so, discuss the supporting evidence. If not, say why not.

T 29. Jumps 2012 How are Olympic performances in various events related? The following plot shows winning long-jump and high-jump distances, in metres, for the Summer Olympics from 1912 through 2012.

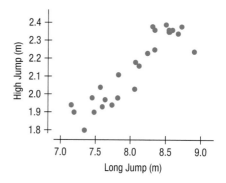

a) Describe the association.
b) Do long-jump performances somehow influence the high-jumpers? How do you account for the relationship you see?
c) The correlation for the given scatterplot is 0.920. If we converted the jump lengths to centimetres by multiplying by 100, would that make the actual correlation higher or lower?
d) What would you predict about the long jump in a year when the high-jumper jumped one standard deviation better than the average high jump?

T 30. Modelling jumps 2012 Here are the summary statistics for the Olympic long jumps and high jumps displayed in the previous exercise:

Event	Mean	StdDev
High Jump	2.148077	0.1938663
Long Jump	8.050000	0.5135913
Correlation = 0.91246		

a) Write the equation of the line of regression for estimating *High Jump* from *Long Jump*.
b) Interpret the slope of the line.
c) In a year when the long jump is 8.9 m, what high jump would you predict?
d) Why can't you use this line to estimate the Long Jump for a year when you know the High Jump was 2.25 m?
e) Write the equation of the line you need to make that prediction.

31. French Consider the association between a student's score on a French vocabulary test and the weight of the student. What direction and strength of correlation would you expect in each of the following situations? Explain.
a) The students are all in Grade 3.
b) The students are in Grades 3 through 11 in the same Chicago school district.
c) The students are in Grade 10 in Quebec.
d) The students are in Grades 3 through 11 in Quebec.

32. Dollars on education Suppose you want to investigate the relationship between standardized test scores and dollars spent for schools in your province. For each school, an average test score (for the same grade) and the average dollars spent per student in 2013 were determined. Someone suggests using the ratio of these two—dollars per test point—to compare schools for efficiency. The schools with lowest ratios will be deemed the best or most efficient. Ignoring the issue of whether standardized test scores are a good measure of education quality, comment on the appropriateness of such a metric (measure) of efficiency, and suggest a better approach, using your knowledge of regression and residuals. Draw a hypothetical data plot pointing out the best and worst schools.

T **33. Lunchtime** Does how long children remain at the lunch table help predict how much they eat? Twenty toddlers were observed over several months at a nursery school. The table and graph show the average number of minutes the kids stayed at the table and the average number of calories they consumed during lunch, calculated from careful observation of what the children ate each day. Create and interpret a model for the toddlers' lunchtime data.

Calories	Time	Calories	Time
472	21.4	450	42.4
498	30.8	410	43.1
465	37.7	504	29.2
456	33.5	437	31.3
423	32.8	489	28.6
437	39.5	436	32.9
508	22.8	480	30.6
431	34.1	439	35.1
479	33.9	444	33.0
454	43.8	408	43.7

T **34. Chips** A start-up company has developed an improved electronic chip for use in laboratory equipment. The company needs to project the manufacturing cost, so it develops a spreadsheet model that takes into account the purchase of production equipment, overhead, raw materials, depreciation, maintenance, and other business costs. The spreadsheet estimates the cost of producing 10 000 to 200 000 chips per year, as seen in the table. Develop a regression model to predict *Cost* based on the *Level* of production. (Hint: Take logs of both variables)

Chips Produced (1000s)	Cost per Chip ($)	Chips Produced (1000s)	Cost per Chip ($)
10	146.10	90	47.22
20	105.80	100	44.31
30	85.75	120	42.88
40	77.02	140	39.05
50	66.10	160	37.47
60	63.92	180	35.09
70	58.80	200	34.04
80	50.91		

T **35. Tobacco and alcohol** Are people who use tobacco products more likely to consume alcohol? Here are data on household spending (in pounds) taken by the British government on 11 regions in Great Britain. Do tobacco and alcohol spending appear to be related? What questions do you have about these data? What conclusions can you draw?

Region	Alcohol	Tobacco
North	6.47	4.03
Yorkshire	6.13	3.76
Northeast	6.19	3.77
East Midlands	4.89	3.34
West Midlands	5.63	3.47
East Anglia	4.52	2.92
Southeast	5.89	3.20
Southwest	4.79	2.71
Wales	5.27	3.53
Scotland	6.08	4.51
Northern Ireland	4.02	4.56

36. Models Find the predicted value of y, using each fitted regression model (some involving re-expressions of y) below for $x = 10$.
a) $\hat{y} = 2 + 0.8 \ln x$
b) $\widehat{\log y} = 5 - 0.23x$
c) $\widehat{1/\sqrt{y}} = 17.1 - 1.66x$

37. Vehicle weights The Minnesota Department of Transportation hoped that they could measure the weights of big trucks without actually stopping the vehicles by using a newly developed "weigh-in-motion" scale. After installation of the scale, a study was conducted to find out whether the scale's readings correspond to the true weights of the trucks being monitored. In Exercise 24 of Chapter 6, you examined the scatterplot for the data they collected, finding the association to be approximately linear with $R^2 = 93\%$. Their regression equation is $\widehat{Wt} = 10.85 + 0.64$ *Scale*, where both the scale reading and the predicted weight of the truck are measured in thousands of pounds.
a) Estimate the weight of a truck if this scale read 31 200 pounds.

b) If that truck actually weighed 32 120 pounds, what was the residual?

c) If the scale reads 35 590 pounds, and the truck has a residual of −2440 pounds, how much does it actually weigh?

d) In general, do you expect estimates made using this equation to be reasonably accurate? Explain.

e) If the police plan to use this scale to issue tickets to trucks that appear to be overloaded, will negative or positive residuals be a greater problem? Explain.

38. Profit How are a company's profits related to its sales? Let's examine data from 71 large U.S. corporations. All amounts are in millions of dollars.

a) Histograms of *Profits* and *Sales* and histograms of the logarithms of *Profits* and *Sales* are seen below. Why are the re-expressed data better for regression?

b) Here are the scatterplot and residual plot for the regression of logarithm of *Profits* vs. log of *Sales*. Do you think this model is appropriate? Explain.

c) Here's the regression analysis. Write the equation.

Dependent variable is: Log Profit
R-squared = 48.1%

Variable	Coefficient
Intercept	−0.106259
LogSales	0.647798

d) Use your equation to estimate profits earned by a company with sales of $2.5 billion (that's 2500 million).

Sample Surveys

Connections

We see surveys all the time. How can asking just a thousand people tell us much about a national election?

With this chapter, we take our first formal steps to relate our sample data to a larger population. Some of these ideas have been lurking in the background as we sought patterns and summaries for data. Even when we only worked with the data at hand, we often thought about implications for a larger population of individuals.

We often want to draw conclusions about population characteristics, called parameters, using sample statistics such as those summary measures of centre, spread, correlation, and slope that we've been calculating. To make that leap, we need to learn about proper sampling techniques and the importance of randomization.

The student remained unimpressed, saying, "You mean that just a sample of a few thousand can tell us what over 250 MILLION people are doing?" Finally, the professor, somewhat disgruntled with the skepticism, replied, "Well, the next time you go to the campus clinic and they want to do a blood test . . . tell them that's not good enough . . . tell them to TAKE IT ALL!!"

—Gary C. Ramseyer's Statistics Jokes. © Vicki S. Morrow on behalf of Dr. Gary C. Ramseyer

An Ipsos Reid poll conducted on behalf of the Dominion Institute was designed to compare Canadians' knowledge of American history with their knowledge of Canadian history. They asked a random sample of 1024 Canadian adults 20 questions—10 questions about the history of each country. The respondents had a higher average percentage of correct scores on questions about American history (47%) than they did on Canadian history (42%). Ipsos Reid went on to report that "it appears Canadians know more about the history and politics of their neighbours to the south than they do about their own country."[1] They are making a conclusion about all Canadians, not just the 1024 in the sample. That step from a small sample to the entire population is impossible without understanding Statistics. To make business decisions, to do science, to choose wise investments, or to understand how voters think they'll vote in the next election, we need to stretch beyond the data at hand to the world at large.

9.1 The Three Big Ideas of Sampling

To make that stretch, we need three ideas. You'll find the first one natural. The second may be more surprising. The third is one of the strange but true facts that often confuse those who don't know statistics.

Idea 1: Examine a Part of the Whole

The first idea is to draw a sample. We'd like to know about an entire group of individuals—a **population**—but examining all of them is usually impractical, if not impossible. So we settle for examining a smaller group of individuals—a **sample**—selected from the population.

You do this every day. For example, suppose you wonder if your friends will like the vegetable soup you're cooking for dinner tonight. To decide whether it meets your standards, you only need to try a small amount. You might taste just a spoonful or two. You certainly don't have to consume the whole pot. You trust that the taste will *represent* the flavour of the entire pot. The idea behind your tasting is that a small sample, if selected properly, can represent the entire population.

It's hard to go a day without hearing about the latest opinion poll. These polls are examples of **sample surveys**, designed to ask questions of a small group of people in the

[1] www.historicacanada.ca/content/polls

hope of learning something about the entire population. Most likely, you've never been selected to be part of one of these national opinion polls. That's true of most people. So how can the pollsters claim that a sample is representative of the entire population? The answer is that professional pollsters work quite hard to ensure that the "taste"—the sample that they take—represents the population.

Bias

Selecting a sample to represent the population fairly is more difficult than it sounds. Polls or surveys most often fail because they use a sampling method that tends to over- or under-represent parts of the population. The method may overlook subgroups that are harder to find (such as the homeless) or favour others (such as Internet users who like to respond to online surveys). Sampling methods that, by their nature, tend to over- or underemphasize some characteristics of the population are said to be **biased**. Bias is the bane of sampling—the one thing above all to avoid. Conclusions based on samples drawn with biased methods are inherently flawed. There is usually no way to fix bias after the sample is drawn and no way to salvage useful information from it.

Here's a famous example of a really dismal failure. By the beginning of the twentieth century, it was common for newspapers to ask readers to return "straw" ballots on a variety of topics. (Today's Internet surveys are the same idea, gone electronic.)

The New York Public Library/Art Resource Ny

During the period from 1916 to 1936, the magazine *Literary Digest* regularly surveyed public opinion and forecast election results correctly. During the 1936 U.S. presidential campaign between Alf Landon and Franklin Delano Roosevelt, it mailed more than 10 million ballots and got back an astonishing 2.4 million. (Polls were still a relatively novel idea, and many people thought it was important to send back their opinions.) The results were clear: Alf Landon would be the next president by a landslide, 57% to 43%. In fact, Roosevelt won, 62% to 37%, and, perhaps coincidentally, the *Digest* went bankrupt soon afterward.

What went wrong? One problem was that *Literary Digest*'s sample wasn't representative. Where would *you* find 10 million names and addresses to sample? The *Digest* used the phone book, as modern surveys do.[2] But in 1936, at the height of the Great Depression, telephones were a real luxury, so they under-sampled poor voters. The campaign of 1936 focused on the economy, and those who were less well off were more likely to vote for Roosevelt (a Democrat). So the magazine's sample was hopelessly biased.

How do modern polls get their samples to *represent* the entire population? You might think that they'd handpick individuals to sample with care and precision. But in fact, they do something quite different: They select individuals to sample *at random*.

Idea 2: Randomize

Think back to the soup sample. Suppose you add some salt to the pot. If you sample it from the top before stirring, you'll get the misleading idea that the whole pot is salty. If you sample from the bottom, you'll get an equally misleading idea that the whole pot is bland. By stirring, you *randomize* the amount of salt throughout the pot, making each taste more typical of the whole pot.

Not only does randomization protect you against factors that you know are in the data, it can also help protect against factors you don't even know are there. Suppose, while you weren't looking, a friend added a handful of peas to the soup. If they're down at the bottom of the pot, and you don't randomize the soup by stirring, your test spoonful won't have any peas. By stirring in the salt, you *also* randomize the peas throughout the pot, making your sample taste more typical of the overall pot *even though you didn't know the peas were there*. So randomizing protects us even in this case.

How do we "stir" people in a survey? We select them at random. **Randomizing** protects us from the influences of *all* the features of our population by making sure that, *on*

In 1936, a young pollster named George Gallup used a subsample of only 3000 of the 2.4 million responses that the *Literary Digest* received to reproduce the wrong prediction of Landon's victory over Roosevelt. He then used an entirely different sample of 50 000 and predicted that Roosevelt would get 56% of the vote to Landon's 44%. His sample was apparently much more *representative* of the actual voting populace. The Gallup Organization went on to become one of the leading polling companies.

Keystone Pictures USA/Alamy

---

[2]Today phone numbers are computer-generated to make sure that unlisted numbers are included. But special methods must be used for the growing part of the population who do not have "landline" access.

average, the sample looks like the rest of the population. The importance of deliberately using randomization is one of the great insights of Statistics. So, how do you actually select individuals *at random*? We'll discuss that shortly.

> **Why not match the sample to the population?** Rather than randomizing, we could try to design our sample so that the people we choose are typical in terms of every characteristic we can think of. We might want the income levels of those we sample to match the population. How about age? Political affiliation? Marital status? Having children? Living in the suburbs? We can't possibly think of all the things that might be important. Even if we could, we wouldn't be able to match our sample to the population for all these characteristics.

For Example   IS A RANDOM SAMPLE REPRESENTATIVE?

Here are summary statistics comparing two samples of 8000 drawn at random from a company's database of 3.5 million customers:

Age (yr)	White (%)	Female (%)	# of Children	Income Bracket (1–7)	Wealth Bracket (1–9)	Homeowner? (% Yes)
61.4	85.12	56.2	1.54	3.91	5.29	71.36
61.2	84.44	56.4	1.51	3.88	5.33	72.30

QUESTION: Do you think these samples are representative of the population of the company's customers? Explain.

ANSWER: The two samples look very similar with respect to these seven variables. It appears that randomizing has automatically matched them pretty closely. We can reasonably assume that since the two samples don't differ too much from each other, they don't differ much from the rest of the population either.

Idea 3: It's the Sample Size

How large a random sample do we need for the sample to be reasonably representative of the population? Most people think that we need a large percentage, or *fraction*, of the population, but it turns out that what matters is the *number of individuals in the sample*, not the size of the population. A random sample of 100 students in a university represents the student body just about as well as a random sample of 100 voters represents the entire electorate of Canada. This is the *third* idea and probably the most surprising one in designing surveys.

How can it be that only the size of the sample, and not the population, matters? Well, let's return one last time to that pot of soup. If you're cooking for a banquet rather than just for a few people, your pot will be bigger, but do you need a bigger spoon to decide how the soup tastes? Of course not. The same-size spoonful is probably enough to make a decision about the entire pot, no matter how large the pot. The *fraction* of the population that you've sampled doesn't matter.[3] It's the **sample size** - or the number of individuals in the sample - that's important.

How big a sample do you need? That depends on what you're estimating. To get an idea of what's really in the soup, you'll need a large enough taste to get a *representative* sample from the pot. For a survey that tries to find the proportion of the population falling into a category, you'll usually need several hundred respondents to say anything precise enough to be useful.[4]

---

[3]Well, that's not exactly true. If the population is small enough and the sample is more than 10% of the whole population, it *can* matter. It *doesn't* matter whenever, as usual, our sample is a very small fraction of the population.

[4]Chapter 15 gives the details behind this statement and shows how to decide on a sample size for a survey.

> **Larger samples?** A friend who knows that you are taking Statistics asks your advice on her study. What can you possibly say that will be helpful? Just say, "If you could just get a larger sample, it would probably improve your study." Even though a larger sample might not be worth the cost, it will almost always make the results more precise.

What do the pollsters do? How do professional polling agencies do their work? The most common polling method today is to contact respondents by telephone. Computers generate random telephone numbers, so pollsters can even call some people with unlisted phone numbers. The interviewer may then ask to speak with the adult who is home who had the most recent birthday or use some other essentially random way to select among those available. In phrasing questions, pollsters often list alternative responses (such as candidates' names) in different orders, to avoid biases that might favour the first name on the list.

Do these methods work? The Pew Research Center for the People and the Press, reporting on one survey, said that

Across five days of interviewing, surveys today are able to make some kind of contact with the vast majority of households (76%). But because of busy schedules, skepticism and outright refusals, interviews were completed in just 38% of households that were reached using standard polling procedures.

Nevertheless, studies indicate that those actually sampled can give a good snapshot of larger populations from which the surveyed households were drawn.

Does a Census Make Sense? Why bother determining the right sample size? Wouldn't it be better just to include everyone and "sample" the entire population? Such a special sample is called a **census**. Although a census would appear to provide the best possible information about the population, there are a number of reasons why it might not.

First, it can be difficult to complete a census. Some individuals in a population will be hard (and expensive) to locate. Or a census might just be impractical. If you were a taste-tester for the Hostess™ Company, you probably wouldn't want to census *all* the Twinkies on the production line. Not only might this be life-endangering, but you wouldn't have any left to sell.

Second, populations rarely stand still. In populations of people, babies are born and folks die or leave the country. In opinion surveys, events may cause a shift in opinion during the survey. A census takes longer to complete and the population changes while you work. A sample surveyed in just a few days may give more accurate information.

Third, taking a census can be more complex than sampling. For example, the Canadian Census may record too many university students. Students living away from home may be enumerated twice, once by their parents, and again by themselves at their student address.

Section 9.1 Exercises

1. **University of British Columbia** Administrators at University of British Columbia were interested in estimating the percentage of students who are the first in their family to go to university. The University of British Columbia student body has about 60 000 members. How might the administrators answer their question by applying the three Big Ideas of sampling?

2. **Satisfied workers** The managers of a large company wished to know the percentage of employees who feel "extremely satisfied" to work there. The company has roughly 24 000 employees. They contacted a random sample of employees and asked them about their job satisfaction, obtaining 437 completed responses. How does their study deal with the three Big Ideas of sampling?

9.2 Populations and Parameters

A study found that Canadians under 25 years old are less likely to "buckle up." In 2011, Transportation Canada reported that 7.0% of Canadians under 25 never or rarely wear seatbelts.[5] We're sure they didn't take a census, so what does the 7.0% mean? We can't know exactly what percentage of Canadians under 25 wear seatbelts. Reality is just too complex. But we can simplify the question by building a model.

---

[5]http://www.tc.gc.ca/media/documents/roadsafety/tp2436e-rs201101.pdf

■ **NOTATION ALERT**

This entire table is a notation alert.

Models use mathematics to represent reality. Parameters are the key numbers in those models. A parameter used in a model for a population is sometimes called (redundantly) a **population parameter**.

But let's not forget about the data. We use summaries of the data to estimate the population parameters. As we know, any summary found from the data is a **statistic**. Sometimes you'll see the (also redundant) term **sample statistic**.[6]

We've already met two parameters in Chapter 5: the mean, μ, and the standard deviation, σ. We'll *try* to keep denoting population model parameters with Greek letters and the corresponding statistics with Latin letters. Usually, but not always, the letter used for the statistic and the parameter correspond in a natural way. So the standard deviation of the data is s, and the corresponding parameter is σ (Greek for s). In Chapter 6, we used r to denote the sample correlation. The corresponding correlation in a model for the population would be called ρ (rho). In Chapter 7, b_1 represented the slope of a linear regression estimated from the data. But when we think about a (linear) *model* for the population, we denote the slope parameter β_1 (beta).

Get the pattern? Good. Now it breaks down. We denote the mean of a population model with μ. We denote the sample mean by putting a bar over anything when we average it, so we write $\bar{y}$ (rather than the Latin equivalent m). What about proportions? Suppose we want to talk about the proportion of Canadians under 25 years old who don't wear seatbelts. We'll use p for the population model parameter (since Greek π is already reserved for the number 3.1415926 . . .) and $\hat{p}$ for the proportion from the data (since, like $\hat{y}$ in regression, it's an estimated value).

Here's a table summarizing the notation:

Name	Statistic	Parameter
Mean	$\bar{y}$	μ (mu, pronounced "meeoo," not "moo")
Standard deviation	s	σ (sigma)
Correlation	r	ρ (rho, pronounced like "row")
Regression coefficient	b	β (beta, pronounced "baytah")
Proportion	$\hat{p}$	p (pronounced "pee"[7])

We draw samples because we can't work with the entire population. When the statistics we compute from the sample accurately reflect the corresponding parameters, then the sample is said to be **representative**, whereas a biased sampling methodology tends to over- or underestimate the parameter(s) of interest.

Just Checking

1. Various claims are often made for surveys. Why are each of the following claims not correct?

   a) It is always better to take a census than to draw a sample.

   b) Stopping students on their way out of the cafeteria is a good way to sample if we want to know about the quality of the food there.

   c) We drew a sample of 100 from the 3000 students in a school. To get the same level of precision for a town of 30 000 residents, we'll need a sample of 1000.

---

[6]Where else besides a sample *could* a statistic come from?

[7]Just in case you weren't sure.

d) A poll taken at a Statistics support website garnered 12 357 responses. The majority said they enjoy doing statistics homework. With a sample size that large, we can be pretty sure that most Statistics students feel this way too.

e) The true percentage of all Statistics students who enjoy the homework is called a "population statistic."

Section 9.2 Exercises

3. University of British Columbia again The president of the university is planning a speech for an alumni group. He plans to talk about the proportion of students who responded in the survey that they are the first in their family to attend university, but the first draft of his speech treats that proportion as the actual proportion of current University of British Columbia students who are the first in their families to attend university. Explain to the president the difference between the proportion of respondents who are first attenders and the proportion of the entire student body that are first attenders. Use appropriate statistics terminology.

4. Satisfied respondents The company's annual report states, "Our survey shows that 87.34% of our employees are 'very happy' working here." Comment on that claim. Use appropriate statistics terminology.

9.3 Simple Random Samples

How would you select a representative sample? Most people would say that every individual in the population should have an equal chance to be selected, and certainly that seems fair. But it's not sufficient. There are many ways to give everyone an equal chance that still wouldn't give a representative sample. Consider, for example, a school that has equal numbers of males and females. We could sample like this: Flip a coin. If it comes up heads, select 100 female students at random. If it comes up tails, select 100 males at random. Everyone has an equal chance of selection, but every sample is of only a single sex—hardly representative.

We need to do better. Suppose we insist that every possible *sample* of the size we plan to draw has an equal chance to be selected. This ensures that situations like the one just described are not likely to occur and still guarantees that each person has an equal chance of being selected. What's different is that with this method, each *combination* of people has an equal chance of being selected as well. A sample drawn in this way is called a **Simple Random Sample**, usually abbreviated **SRS**. Imagine writing every student's name on a small paper in a hat, shaking the hat well, then grabbing a bunch. That's not a practical way to do things, but is a good way to conceptualize an SRS. An SRS is the standard against which we measure other sampling methods, and the sampling method on which the basic theory of working with sampled data is based.

To select a sample at random, we first need to define where the sample will come from. The **sampling frame** is a list of individuals from which the sample is drawn. For example, to draw a random sample of students at a university, we might obtain a list of all registered full-time students and sample from that list. In defining the sampling frame, we must deal with the details of defining the population. Are part-time students included? How about those who are attending school elsewhere and transferring credits back to the university?

Once we have a sampling frame, we can use **random numbers** to help us select individuals in a properly randomized fashion. Even software that offers a procedure for selecting an SRS from your data will use random numbers, in the background. So just what is a random number?

Random Numbers

We all know what it means for something to be random. Or do we? Many children's games rely on chance outcomes. Rolling dice, spinning spinners, and shuffling cards all select

"The generation of random numbers is too important to be left to chance."

—Robert R. Coveyou, Oak Ridge National Laboratory

outcomes at random. Adult games use randomness as well, from card games to lotteries and bingo. What's the most important aspect of the randomness in these games? It must be fair.

What is it about random selection that makes it seem fair? It's really two things. First, nobody can guess the outcome before it happens. Second, when we want things to be fair, usually some underlying set of outcomes will be equally likely (although in many games, some combinations of outcomes are more likely than others).

Randomness is not always what we might think of as "at random." *Random outcomes* have a lot of structure, especially when viewed in the long run. You can't predict how a fair coin will land on any single toss, but you're pretty confident that if you flipped it thousands of times you'd see about 50% heads. As we will see, randomness is an essential tool of Statistics; statisticians don't think of randomness as the annoying tendency of things to be unpredictable or haphazard. In fact, without deliberately applying randomness, we couldn't do most of Statistics, and this bookwould stop right about here.

So how should we *generate random numbers*? It's surprisingly difficult to get truly random values even when they're equally likely. Computers, calculators and even smartphones have become a popular way to generate them, but they can't generate *truly* random numbers. Start a program from the same place, and it will always follow exactly the same path. So numbers generated by a program are not truly random. Technically, "random" numbers generated this way are *pseudorandom* numbers. Fortunately, pseudorandom values are virtually indistinguishable from truly random numbers even though they aren't really random. For most purposes, that's good enough.

There *are* ways to generate random numbers so that they are both equally likely and truly random. In the past, entire books of carefully generated random numbers were published. The books never made the best-seller lists and probably didn't make for great reading, but they were quite valuable to those who needed truly random values.[8] And, along with almost anything else, you can find truly random numbers on the Internet. Several sites are available that use methods like timing the decay of a radioactive element to generate truly random digits.[9] A string of random digits might look like the following (When presented in RN tables, spaces are interspersed at regular intervals to improve readability):

2217726304387410092537086270581997622725849795907032825001108963
3217535822643800292254644943760642389043766557204107354186024508
8906427308645681412198226653885873285801699027843110380420067664
8740522639824530519902027044464984322000946238678577902639002954
8887003319933147508331265192321413908608674496383528968974910533
6944182713168919406022181281304751019321546303870481407676636740

Now, let's get back to random sampling. How should we select an SRS? Random numbers are exactly what we need. One simple procedure is to number everyone in the frame (say 1 to N), and then obtain *n* distinct random numbers between 1 and N. Select those individuals possessing the selected RNs.

For Example   USING RANDOM NUMBERS TO GET AN SRS

There are 80 students enrolled in an introductory Statistics class; you are to select a sample of five.

QUESTION: How can you select an SRS of five students using these random digits found on the Internet?

05166   29305   77482

ANSWER: First I'll number the students from 00 to 79. Taking the random numbers two digits at a time gives me 05, 16, 62, 93, 05, 77, and 48. I'll ignore 93 because the students were numbered only up to 79. And so I don't pick the same person twice, I'll skip the repeated number 05. My simple random sample consists of students with the numbers 05, 16, 62, 77, and 48.

---

[8]You'll find a table of random digits of this kind in the back of this book.
[9]For example, www.random.org or www.randomnumbers.info.

- We can be more efficient when we're choosing a larger sample from a sampling frame stored in a data file. First we assign a random number with several digits (say, from 0 to 10 000) to each individual. Then we sort the random numbers in to numerical order, keeping each name with its number. Choosing the first n names from this re-arranged list will give us a random sample of that size.
- Often the sampling frame is so large that it would be too tedious to number everyone consecutively. If our intended sample size is approximately 10% of the sampling frame, we can assign each individual a single random digit 0 to 9. Then we select only those with a specific random digit, say, 5.

Samples drawn at random generally differ one from another. Each draw of random numbers selects *different* people for our sample. These differences lead to different values for the variables we measure. We call these sample-to-sample differences **sampling variability**. Surprisingly, sampling variability isn't a problem; it's an opportunity. In future chapters we'll investigate what the variation in a sample can tell us about its population.

> **ERROR OKAY, BIAS NO WAY!**
>
> Sampling variability is sometimes referred to as *sampling error*, making it sound as if it's some kind of mistake. It's not. We understand that samples will vary, so "sampling error" is to be expected. It's *bias* we must strive to avoid. Bias means our sampling method distorts our view of the population, and that will surely lead to mistakes.

Section 9.3 Exercises

5. Sampling students A professor teaching a large lecture class of 350 students samples her class by rolling a die. Then, starting with the row number on the die (1 to 6), she passes out a survey to every fourth row of the large lecture hall. She says that this is a Simple Random Sample because everyone had an equal opportunity to sit in any seat and because she randomized the choice of rows. What do you think? Be specific.

6. Sampling satisfaction A company hoping to assess employee satisfaction surveys employees by assigning computer-generated random numbers to each employee on a list of all employees and then contacting all those whose assigned random number is divisible by 7. Is this a simple random sample?

9.4 Other Sampling Designs

Simple random sampling is not the only fair or unbiased way to sample. More complicated designs may save time or money or help avoid potential bias and improve sample accuracy. All statistical sampling designs have in common the idea that chance, rather than human choice, is used to select the sample.

Stratified Sampling

Designs that are used to sample from large populations—especially populations residing across large areas—are often more complicated than simple random samples. Sometimes the population is first sliced into groups, called **strata**, before the sample is selected. We usually try to create these strata so that they are relatively homogeneous—i.e., so that individuals within the same stratum are somewhat more similar to each other than to individuals in other strata (or equivalently, so that strata differ substantially from each other), and if we succeed, significant benefits will accrue, as discussed below. However, there are also times when their formation is dictated by various practical sampling issues. Once the strata are defined, random sampling is used within each stratum, after which results from the various strata are combined to produce population estimates. This common sampling technique is called **stratified random sampling**.

Why would we want to complicate things? Here's an example. Suppose we want to learn how students feel about funding for the football team at a large university. The campus is 60% men and 40% women, and we suspect that men and women have different views on the funding. If we use simple random sampling to select 100 people for the survey, we could end up with 70 men and 30 women or 50 men and 50 women. Our resulting estimates of the level of support for the soccer funding likewise could vary widely. To help reduce this sampling variability, we can decide to force a representative balance, selecting 60 men at random and 40 women at random. This would guarantee that the proportions of men and women within our sample match the proportions in the population, and that should make such samples more accurate in representing population opinion.

You can imagine the importance of stratifying by race, income, age, and other characteristics, depending on the questions in the survey. Samples taken within a stratum vary less, so our estimates can be more precise. This reduced sampling variability is the most important benefit of effective stratifying.

There is great flexibility in this method. Since we sample independently from the various strata, we are free to use different sampling methods or sampling rates best suited to each stratum (e.g., we may choose a smaller sampling fraction in strata where the data are more difficult or expensive to collect). Another advantage is that aside from obtaining overall estimates, we will also have estimates for each stratum. For example, we might split Canada into five or six geographic regions (the strata) and have different teams run surveys separately within each region, varying sampling procedures as needed. We later combine the data from all regions to produce national estimates, and, if desired, we can easily produce regional estimates, too.

Stratified sampling can also help us notice important differences among groups. But be careful. As we saw in Chapter 2, if we unthinkingly combine group data, we risk reaching the wrong conclusion, becoming victims of Simpson's paradox.

For Example   STRATIFYING THE SAMPLE

RECAP: You're trying to find out what frosh (first year students) think of the food served on campus. Food Services believes that men and women typically have different opinions about the importance of the salad bar.

QUESTION: How should you adjust your sampling strategy to allow for this difference?

ANSWER: I will stratify my sample by drawing an SRS of men and a separate SRS of women—assuming that the data from the registrar include information about each person's sex.

Cluster and Multistage Sampling

Suppose we wanted to assess the reading level of this textbook based on the length of the sentences. Simple random sampling could be awkward; we'd have to number each sentence, then find, for example, the 576th sentence or the 2482nd sentence, and so on. Doesn't sound like much fun, does it?

It would be much easier to pick a number of *pages* at random and count the number of words for each of the sentences on those pages. Splitting the population into **clusters** (each page is a cluster of sentences) can make sampling more practical. Then we could simply select a number of clusters at random and perform a census within each of them (measure all sentences on the page). This sampling design is called simple (or one-stage) **cluster sampling**. If instead of a complete census of individuals in the cluster, we randomly select only a portion of the individuals in each selected cluster (measure only a sample of the sentences on each selected page), this introduces a second stage of sampling, and voilà, we have what is called *two-stage (cluster) sampling*. With proper random selection, cluster sampling will produce an unbiased sample.

For Example CLUSTER SAMPLING

RECAP: In trying to find out what frosh think about the food served on campus, you've considered both an SRS and a stratified sample. Now you have run into a problem: It's simply too difficult and time-consuming to track down the individuals whose names were chosen for your sample. Fortunately, frosh at this school are all housed in 12 frosh residences.

QUESTIONS: How could you use this fact to draw a cluster sample? How might that alleviate the problem? What concerns do you have?

ANSWER: To draw a cluster sample, I would select several residences at random and then try to contact everyone in each selected residence. I could save time by simply knocking on doors on a given evening and interviewing people. I hope to achieve a high response rate on that evening, or, if not, that the people I'm able to contact are representative of the residence (the usual non-response issue).

A major concern though would be the differences among residences. If students were assigned to residences pretty much at random, or if I think that students' opinions about food service will not depend much on the particular residence, this sampling approach should work well. However, if I believe that big differences among residences are possible, the sampling variation resulting from this design could be great (since if you choose different residences, the results change significantly), and it would be better to treat the residences as strata and draw a random sample from each of the 12 residences (not necessarily an SRS, maybe a systematic sample, to be discussed shortly).

Stratification vs. Clustering

It appears that we could split a population into strata or clusters. What is the difference? We create strata by dividing the entire population into subpopulations, which we try to define wisely (differing along one or more important dimensions). Then we draw a separate random sample from each and every stratum. We sample *from* or *within* strata, whereas clusters are *sampling units*—to be chosen at random from a proper sampling frame listing all the clusters. Clusters are randomly selected; strata are not. We need to collect data from each and every stratum, so we randomly select some individuals from each one, but we do not randomly select the strata themselves (instead, we *form* strata). Strata are often few and very large (e.g., geographic regions or size categories of businesses: large, medium, small), therefore we cannot skip over any of them. In contrast, clusters tend to be smaller and more numerous (such as blocks in a city or enumeration areas in a large region), so we don't need to sample all of them, just enough to ensure representativeness.

Figure 9.1 (on the next page) compares the two approaches. We have 16 city blocks with four households in each, and we wish to sample some households. Blocks may be viewed as clusters. There also may be some difference between households on the west side and households on the east side, so we can easily stratify geographically. Which scheme gives more precise estimates (with less sampling variability)? Which costs more? Is it possible to combine the two approaches?[10]

Why sample individuals in clusters, rather than directly? Generally, for practical reasons. If we want to sample households in a city, it may be too expensive or time consuming to construct a frame showing all the households, but we may be able to obtain a map showing blocks, on which we can number the blocks and then select some blocks at random. From each selected block, we can sample some or all of its households. It is easier to get at the sampled households indirectly by first sampling clusters of them, because the needed frame is readily available. As a bonus, we only need to list households in the selected blocks, and not the entire city. And then there is the improved cost efficiency, since in many

---

[10] The cluster sample would be more convenient and cheaper than the stratified sample, but less precise. We could combine methods by stratifying geographically and then selecting a cluster (block) or two at random from each stratum (geographic region).

Figure 9.1

Stratified and Cluster Sampling. Above, we stratify geographically (east, west) before drawing a sample from each geographic region. Below, we view the city blocks as clusters, then draw two of them, perhaps following up with a subsample of only half of each block (second stage of sampling).

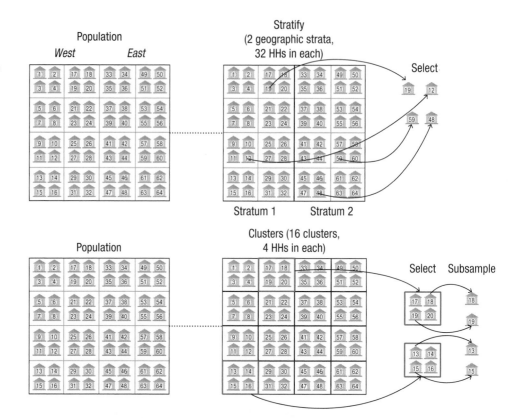

cases, continued sampling from the same cluster can add extra data very cheaply. Once you arrive at a selected city block, it is fast and easy to go from household to household on the same block. Or suppose we want to sample students. Students are clustered in schools, and it is easier to list schools than students, and more convenient to sample students from a few randomly chosen schools, than to sample students spread out all over the city. A simple random sample of students in the city would be more precise than a sample using only some selected schools, if the samples are of the same size—but what really matters is *maximizing precision given some total expenditure*, and the same expenditure may yield a much bigger cluster sample than a simple random sample.

If clusters are big, we often *subsample* the selected clusters since typically there are diminishing returns from continued observation within a cluster when the clusters are somewhat homogeneous, as is all too often the case with naturally existing clusters (for example, people in the same city block or enumeration area tend to be somewhat similar in income, ethnicity, etc.). As a general rule, the more similar the individuals in a cluster, the fewer individuals we should bother to include (and the more clusters we should select), whereas we should obtain data from many or all of the individuals in clusters that are heterogeneous, since we keep getting good new information cheap. When we randomly subsample just a portion of each selected cluster, we are introducing a *second stage* of sampling.

Often, individuals exist in a hierarchy of clusters—for example, adults residing in some areas of a city are clustered together in households, which in turn are clustered together in blocks, which are clustered in enumeration areas. To sample adults, you could start with a list of enumeration areas and randomly select some. Then, from each selected EA, get a map and select some blocks, go to each selected block and select some households, and, finally, select one adult from each household. You work your way down, in successive stages of sampling (four stages, in this case) until you finally get to the targetted population elements. This is **multistage sampling**.

Sometimes we use a variety of sampling methods together. In trying to assess the reading level of this text book, we might be concerned if it starts out easy and then gets harder as the concepts become more difficult. If so, we'd want to avoid samples that select heavily from early or late chapters. To guarantee a fair mix, we could randomly choose

> **Canadian Incidence Study of Reported Child Abuse and Neglect** This important 2008 study examined the incidence of reported child maltreatment and the characteristics of the children and families investigated by Canadian child welfare sites from all 13 provinces and territories. But how did they draw a random sample from all the investigated cases, over a three month period, across Canada? Obviously, it would be impractical to make a list of all cases and then draw a sample. They were able though to make a list of 112 child welfare sites across Canada, from which they selected a sample - after grouping them into various strata based on geography and size of organization, along with separate strata for Indigenous organizations. Sites were randomly - or occasionally 100% - sampled from each stratum (with some exceptions), and then cases were randomly - or in smaller sites 100% - sampled from the selected sites. How to describe this sampling design? - *Multi-stage sampling, after stratification of the welfare sites.* [Public Health Agency of Canada. Canadian Incidence Study of Reported Child Abuse and Neglect – 2008: Major Findings. Ottawa, 2010]

one or two chapters from each of the eight parts of the book and then randomly select a few pages from each selected chapter. If that produced too many sentences in total, we might select a few sentences at random from each of the chosen pages. In other words, we

i. stratify by the part of the book, so that each part is a stratum that we will sample from,

ii. randomly choose chapters (first-stage clusters or sampling units) from each part,

iii. then within each selected chapter, choose some pages (second-stage clusters or sampling units) at random,

iv. and perhaps finally, select an SRS (or better yet, a systematic sample, to be discussed shortly) of sentences from each selected page.

Most surveys conducted by professional polling organizations use some combination of stratified and multistage cluster sampling as well as simple random samples or, systematic samples (see next page). In the example of sampling adults in a city, it would add to our precision if we used readily available information about enumeration areas to organize them into several strata, such as high income, moderate income, and low income (or urban/suburban/rural), before sampling, and then selected, say, two enumeration areas from each of the three strata, rather than selecting an SRS of six enumeration areas from the entire list.

For Example COMBINING METHODS TO IMPROVE PRECISION

RECAP: Learning that frosh are housed in separate residences allowed you to sample their attitudes about campus food by going to residences chosen at random, but you're still concerned about possible differences in opinions between men and women. It turns out that these frosh residences house the sexes on alternate floors.

QUESTION: How can you design a sampling plan that uses this fact to your advantage?

ANSWER: Select floors at random, in two stages, and stratify where convenient. For example, we might choose four residences at random and then select two of the male residence floors and two of the female residence floors at random in each of the chosen residences. Then try to interview everyone on each selected floor, or, if that's just too many, choose an SRS of rooms on each floor. (Or, better yet, read on about systematic sampling, and consider a systematic sample along the chosen floor, taking, say, every fifth room).

In this example, we combined stratification (of floors)—based on our knowledge about possible sex differences—with two-stage cluster sampling, with residences as the first-stage sampling units (or clusters) and floors the second-stage sampling units.

Systematic Samples

Some samples select individuals systematically. For example, you might survey every tenth person on an alphabetical list of students. To make it random, you must start the systematic selection from a randomly selected individual. When the order of the list is not associated in any way with the responses sought, **systematic sampling** is similar to drawing an SRS, while it can be much less expensive. It will give even better results than an SRS when the list order is associated to some extent with the response variable, as long as the systematic sampling rule doesn't coincide with some "cycle" in the sampling frame (for example, if you decided to sample students in their dorm rooms by knocking on every other door or every fourth door, a dorm in which male and female rooms were alternated could result in a sample that was all male or all female).

Think about the textbook reading-level example again. Suppose we have chosen a chapter of the book at random, then three pages at random from that chapter, and now we want to select a sample of 10 sentences from the 73 sentences found on those pages. Instead of numbering each sentence so we can pick a simple random sample, it would be easier to sample systematically. A quick calculation shows $73/10 = 7.3$, so we can get our sample by just picking every seventh sentence on the page. But where should you start? At random, of course, so start by obtaining a random digit from 1 to 7 which will determine your first sentence. Then pick every seventh sentence after that and record all their lengths.

Just Checking

2. We need to survey a random sample of the 300 passengers on a flight from Vancouver to Tokyo. Name each sampling method described below.

   a) Pick every tenth passenger as people board the plane.

   b) From the boarding list, randomly choose 5 people flying first class and 25 of the other passengers.

   c) Randomly generate 30 seat numbers and survey the passengers who sit there.

   d) Randomly select a seat position (right window, right centre, right aisle, etc.), and survey all the passengers sitting in those seats.

Step-by-Step Example   DESIGNING A SURVEY

The assignment says, "Conduct your own sample survey to find out how many hours per week students at your school spend watching TV during the school year." Let's see how we might do this step by step. (Remember, though—actually collecting the data from your sample can be difficult and time-consuming.)

Question: How would you design this survey?

THINK ➡ **Plan** State what you want to know.

I want to design a study to find out how many hours of television students at my school watch.

Population and Parameter

Identify the *W*'s of the study. The *Why* determines the population and the associated sampling frame. The *What* identifies the parameter of interest and the variables measured. The *Who* is the sample we actually draw. The *How, When,* and *Where* are given by the sampling plan.

Often, thinking about the *Why* will help you see whether the sampling frame and plan are adequate to learn about the population.

The population studied was students at our school. I obtained a list of all students currently enrolled and used it as the sampling frame. The parameter of interest was the number of television hours watched per week during the school year, which I attempted to measure by asking students how much television they watched during the previous week.

Sampling Plan Specify the sampling method and the sample size, *n*. Specify how the sample was actually drawn. What is the sampling frame? How was the randomization performed?

The description should, be complete enough to allow someone to replicate the procedure, drawing another sample from the same population in the same manner.

I decided against stratifying by class or sex because I didn't think TV watching would differ much between males and females or across classes. I selected a simple random sample of students from the list. I obtained an alphabetical list of students, assigned each a random digit between 0 and 9, and then selected all students who were assigned a "4." This method generated a sample of 212 students from the population of 2133 students.

SHOW ➡ Sampling Practice Specify *When*, *Where*, and *How* the sampling was performed. Specify any other details of your survey, such as how respondents were contacted, what incentives were offered to encourage them to respond, how nonrespondents were treated, and so on.

The survey was taken over the period October 15 to October 25, 2016. Surveys were sent to selected students by e-mail, with the request that they respond by e-mail as well. Students who could not be reached by e-mail were handed the survey in person.

TELL ➡ Summary and Conclusion This report should include a discussion of all the elements. In addition, it's good practice to discuss any special circumstances. Professional polling organizations report the *When* of their samples but will also note, for example, any important news that might have changed respondents' opinions during the sampling process. In this survey, perhaps, a major news story or sporting event might change students' TV viewing behaviour.

The question you ask also matters. It's better to be specific ("How many hours did you watch television last week?") than to ask a general question ("How many hours of television do you usually watch in a week?").

The report should show a display of the data, provide and interpret the statistics from the sample, and state the conclusions that you reached about the population.

During the period October 15 to October 25, 212 students were randomly selected using a simple random sample from a list of all students currently enrolled. The survey they received asked the following question: "How many hours did you spend watching television last week?"

Of the 212 students surveyed, 110 responded. It's possible that the nonrespondents differ in the number of TV hours watched from those who responded, but I was unable to follow up on them due to limited time and funds. The 110 respondents reported an average 3.62 hours of TV watching per week. The median was only 2 hours per week. A histogram of the data shows that the distribution is highly right-skewed, indicating that the median might be a more appropriate measure of typical TV watching of the students.

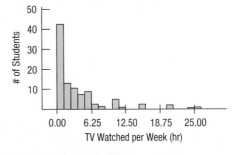

Most of the students (90%) watched between 0 and 10 hours per week, while 30% reported watching less than 1 hour per week. A few watched much more. About 3% reported watching more than 20 hours per week.

Section 9.4 Exercises

7. Sampling University of British Columbia students For each scenario, identify the kind of sample used by the university administrators from Exercise 1:

a) Select several dormitories at random and contact everyone living in the selected dorms.

b) Using a computer-based list of registered students, contact 200 first year students, 200 second year students, 200 third year students, and 200 fourth year students selected at random from each class.

c) Using a computer-based alphabetical list of registered students, select one of the first 25 on the list by random and then the student whose name is 50 names later, and then every 50 names beyond that.

8. Satisfactory satisfaction samples For each scenario, determine the sampling method used by the managers from Exercise 2.

a) Use the company e-mail directory to contact 150 employees from among those employed for less than 5 years, 150 from among those employed for 5–10 years, and 150 from among those employed for more than 10 years.

b) Use the company e-mail directory to contact every 50th employee on the list.

c) Select several divisions of the company at random. Within each division, draw an SRS of employees to contact.

9.5 From the Population to the Sample: You Can't Always Get What You Want

What's the Sample?
The population we want to study is determined by asking *why*. When we design a survey, we use the term "sample" to refer to the individuals selected, from whom we hope to obtain responses. Unfortunately, the real sample is just those we can reach to obtain responses—the *who* of the study. These are slightly different uses of the same term *sample*. The context usually makes clear which we mean, but it's important to realize that the difference between the two samples could undermine even a well-designed study.

Before you start a survey, think first about the population you want to study. You may find that it's not the well-defined group you thought it was. Who, exactly, is a student, for example? Even if the population seems well-defined, it may not be a practical group from which to draw a sample. For example, election polls want to sample from all those who will vote in the next election—a population that is impossible to identify before election day.

Next, you must specify the sampling frame—the individuals who are actually available to be sampled. (Do you have a list of students to sample from? How about a list of registered voters?) Often individuals in institutions such as prisons and military bases, and individuals in very remote regions are excluded from national surveys, for obvious practical reasons. Sometimes the sampling frame is not the group you *really* want to know about. (All those registered to vote are not equally likely to show up.) The sampling frame limits what your survey can find out.

Then there's your target sample. These are the individuals for whom you *intend* to measure responses. You're not likely to get responses from all of them. ("I know it's dinnertime, but I'm sure you wouldn't mind answering a few questions. It'll only take 20 minutes or so. Oh, you're busy?") Nonresponse is a problem in many surveys.

Finally, there's your sample—the actual respondents. These are the individuals about whom you *do* get data and can draw conclusions. Unfortunately, they might not be representative of the sampling frame or the population.

At each step, the group we can study may be constrained further. The *Who* keeps changing, and each constraint can introduce biases. Whenever you work with or learn about a sample, pay attention not only to who was sampled but to who was left out of the study. A careful study should address the question of how well each group matches the population of interest. One of the main benefits of simple random sampling is that it never loses its sense of who's *Who*. The *Who* in an SRS is the population of interest from which we've drawn a representative sample. That's not always true for other kinds of samples.

9.6 The Valid Survey

It isn't sufficient to just draw a sample and start asking questions. We'll want our survey to be *valid*. A valid survey yields the information we are seeking about the population we are interested in. Before setting out to survey, ask yourself:

- What do I want to know?
- Am I asking the right respondents?
- Am I asking the right questions?
- What would I do with the answers if I had them; would they address the things I want to know?

These questions may sound obvious, but there are a number of pitfalls to avoid.

Know what you want to know. Before considering a survey, understand what you hope to learn and about whom you hope to learn it.

Use the right sampling frame. A valid survey obtains responses from the appropriate respondents. Be sure you have a suitable *sampling frame*. Have you identified the population of interest and sampled from it appropriately? A company might survey customers who returned warranty registration cards—after all, that's a readily available sampling frame. But if the company wants to make their product more attractive, the most important population is the customers who rejected their product in favour of one from a competitor.

Tune your instrument. You may be tempted to ask questions you don't really need, but longer survey instruments yield fewer responses and thus a greater chance of nonresponse bias.

Ask specific rather than general questions. People are not very good at estimating their typical behaviour. It is better to ask "How many hours did you sleep last night?" than "How much do you usually sleep?" Sure, some responses will include some unusual events (My dog was sick; I was up all night.) but overall, you'll get better data.

Ask for quantitative results when possible. "How many magazines did you read last week?" is better than "How much do you read: A lot, A moderate amount, A little, or None at all?"

Be careful in phrasing questions. A respondent may not understand the question—or may understand the question differently than the researcher intended it. ("Does anyone in your family ride a motorcycle?" Do you mean just me, my spouse, and my children? Or does "family" include my father, my siblings, and my second cousin once removed? And does a motor scooter count?) Respondents are unlikely (or may not have the opportunity) to ask for clarification. A question like "Do you approve of the recent actions of the Canadian ambassador to the USA?" is likely not to measure what you want if many respondents don't know who the ambassador is or what he or she recently did.

Respondents may even lie or shade their responses if they feel embarrassed by the question ("Did you have too much to drink last night?"), are intimidated or insulted by the question ("Could you understand our new *Instructions for Dummies* manual, or was it too difficult for you?"), or if they want to avoid offending the interviewer ("Would you hire a

A short survey Given that the *New York Times* reports[11] that statisticians can earn $125,000 at top companies their first year on the job, do you think this course will be valuable to *you*?

Pilot testing a survey question A researcher distributed a survey to an organization before some economizing changes were made. She asked how people felt about a proposed cutback in secretarial and administrative support on a seven-point scale from Very Happy to Very Unhappy.

But virtually all respondents were very unhappy about the cutbacks, so the results weren't particularly useful. If she had pretested the question, she might have chosen a scale that ran from unhappy to outraged.

man with a tattoo?" asked by a tattooed interviewer). Also, be careful to avoid phrases that have double or regional meanings. "How often do you go to town?" might be interpreted differently by different people and cultures.

Even subtle differences in phrasing can make a difference. In January 2006, the *New York Times*[12] asked half of the 1229 U.S. adults in their sample the following question:

After 9/11, President Bush authorized government wiretaps on some phone calls in the U.S. without getting court warrants, saying this was necessary to reduce the threat of terrorism. Do you approve or disapprove of this?

They found that 53% of respondents approved. But when they asked the other half of their sample a question with only slightly different phrasing,

After 9/11, George W. Bush authorized government wiretaps on some phone calls in the U.S. without getting court warrants. Do you approve or disapprove of this?

only 46% approved.

Be careful in phrasing answers. It's often a good idea to offer choices rather than inviting a free response. Open-ended answers can be difficult to analyze. "How did you like the movie?" may start an interesting debate, but it may be better to give a range of possible responses. Be sure to phrase them in a neutral way. When asking "Do you support higher school taxes?" positive responses could be worded "Yes," "Yes, it is important for our children," or "Yes, our future depends on it." But those are not equivalent answers.

The best way to protect a survey from such unanticipated measurement errors is to perform a pilot survey. A **pilot** is a trial run of the survey you eventually plan to give to a larger group, using a draft of your survey questions administered to a small sample drawn from the same sampling frame you intend to use. By analyzing the results from this smaller survey, you can often discover ways to improve your instrument.

Section 9.6 Exercises

9. Survey students What problems do you see with asking the following question of students? "Are you the first member of your family to seek higher education?"

10. Happy employees The company plans to have the head of each corporate division hold a meeting of their employees to ask whether they are happy on their jobs. They will ask people to raise their hands to indicate whether they are happy. What problems do you see with this plan?

9.7 Common Sampling Mistakes, or How to Sample Badly

Bad sample designs yield worthless data. Many of the most convenient forms of sampling can be seriously biased. And there is no way to correct for the bias from a bad sample. So it's wise to pay attention to sample design—and to beware of reports based on poor samples.

Mistake 1: Sample Volunteers

One of the most common dangerous sampling methods is a voluntary response sample. In a **voluntary response sample**, a large group of individuals is invited to respond, and those who choose to respond are counted. The respondents, rather than the researcher, decide

---

[11]www.nytimes.com/2009/08/06/technology/06stats.html

[12] Excerpt from "The New York Times/CBS News Poll, January 20–25, 2006." Published in The New York Times, © 2006.

who will be in the sample. This method is used by call-in shows, 900 numbers, Internet polls, and letters written to Members of Parliament. Voluntary response samples are almost always biased, and so conclusions drawn from them are almost always wrong.

It's often hard to define the sampling frame of a voluntary response study. Practically, the frames are groups, such as Internet users who frequent a particular website, or those who happen to be watching a particular TV show at the moment. But those sampling frames don't correspond to the population of interest.

Even within the sampling frame, voluntary response samples are often biased toward those with strong opinions or those who are strongly motivated. People with very negative opinions tend to respond more often than those with equally strong positive opinions. The sample is not representative, even though every individual in the population may have been offered the chance to respond. The resulting **voluntary response bias** invalidates the survey.

> **If you had it to do over again, would you have children?** Ann Landers, the advice columnist, asked parents this question. The overwhelming majority—70% of the more than 10 000 people who wrote in—said no, kids weren't worth it. A more carefully designed survey later showed that about 90% of parents are actually happy with their decision to have children. What accounts for the striking difference in these two results? What kind of parents do you think are most likely to respond to the original question?

For Example VOLUNTARY RESPONSE SAMPLE

RECAP: You're trying to find out what frosh think of the food served on campus, and have thought of a variety of sampling methods, all time-consuming. A friend suggests that you set up a "Tell Us What You Think" website and invite frosh to visit the site to complete a questionnaire.

QUESTION: What's wrong with this idea?

ANSWER: Letting each frosh decide whether to participate makes this a voluntary response survey. Students who are dissatisfied might be more likely to go to the website to record their complaints, and this could give me a biased view of the opinions of all frosh.

Mistake 2: Sample Conveniently

Another sampling method that doesn't work is convenience sampling. As the name suggests, in **convenience sampling** we simply include the individuals who are convenient for us to sample. Unfortunately, this group may not be representative of the population of interest.

Many surveys conducted at shopping malls suffer from this problem. People in shopping malls are not necessarily representative of the population of interest. Mall shoppers tend to be more affluent and include a larger percentage of teenagers and retirees than the population at large. To make matters worse, survey interviewers tend to select individuals who look "safe," or easy to interview.

> **Do you use the Internet?**
> Click here ◯ for yes
> Click here ◯ for no

Internet Surveys Internet convenience surveys are worthless. As voluntary response surveys, they have no well-defined sampling frame (all those who use the Internet and visit their site?) and thus report no useful information. Do not believe them.

For Example CONVENIENCE SAMPLE

RECAP: To try to gauge frosh opinion about the food served on campus, Food Services suggests that you just stand outside a school cafeteria at lunchtime and stop people to ask them questions.

QUESTIONS: What's wrong with this sampling strategy?

ANSWER: This would be a convenience sample, and it's likely to be biased. I would miss people who use the cafeteria for dinner, but not for lunch, and I'd never hear from anyone who hates the food so much that they have stopped coming to the school cafeteria.

Mistake 3: Use a Bad Sampling Frame

An SRS from an incomplete or out-of-data sampling frame introduces bias because the individuals included may differ from the ones not in the frame. People in prison, homeless people, students, and long-term travellers are all likely to be missed. Professional polling companies now need special procedures to be sure they include in their sampling frame people who can be reached only by cell phone.

Mistake 4: Undercoverage

Many survey designs suffer from **undercoverage**, in which some portion of the population is not sampled at all or has a smaller representation in the selected sample than it has in the population. Undercoverage can arise for a number of reasons, but it's always a potential source of bias. When a national census misses many of the homeless, undocumented immigrants, etc., we have undercoverage. The Literary Digest election poll suffered from undercoverage of low-income voters because they were underrepresented in the lists that they used. Inadequate or out of date frames and convenience samples (and other non-random sampling procedures) are particularly prone to undercoverage of the target population, with some individuals or groups having little or no chance of selection. (Mistakes 1-4 all produce forms of what may be called *selection bias*.)

Nonresponse Bias

A common and serious potential source of bias for most surveys is **nonresponse bias**. No survey succeeds in getting responses from every individual selected. The problem is that those who don't respond may differ from those who do. And they may differ on just the variables we care about. The lack of response will bias the results. Rather than sending out a large number of surveys for which the response rate will be low, it is often better to design a smaller randomized survey for which you have the resources to ensure a high response rate (with multiple call-backs, etc). One of the problems with nonresponse bias is that it's usually impossible to tell what the nonrespondents might have said.

It turns out that the *Literary Digest* survey was wrong on two counts. First, their list of 10 million people was not representative. There was a selection bias in their sampling frame. There was also a nonresponse bias. We know this because the *Digest* also surveyed a *systematic* sample in Chicago, sending the same question used in the larger survey to every third registered voter. They *still* got a result in favor of Landon, even though Chicago voted overwhelmingly for Roosevelt in the election. This suggests that the Roosevelt supporters were less likely to respond to the *Digest* survey. There's a modern version of this problem: It's been suggested that those who screen their calls with caller ID or an answering machine, and so might not talk to a pollster, may differ in wealth or political views from those who just answer the phone.

Response Bias

Response bias[13] refers to anything in the survey design that influences the responses. Response biases include the tendency of respondents to tailor their responses to try to please the interviewer, the natural unwillingness of respondents to reveal personal facts or admit to illegal or unapproved behaviour, and the ways in which the wording of the questions can influence responses.

How to Think About Biases

■   **Look for biases in any survey you encounter.** If you design one of your own, ask someone else to help look for biases that may not be obvious to you. And do this

---

[13]Response bias is not the opposite of nonresponse bias. (We don't make these terms up; we just try to explain them.)

before you collect your data. There's no way to recover from a biased sampling method or a survey that asks biased questions.

Sorry, it just can't be done.

A bigger sample size for a biased study just gives you a bigger useless study. A really big sample gives you a really big useless study. (Think of the 2.4 million *Literary Digest* responses.)

- **Spend your time and resources reducing biases.** No other use of resources is as worthwhile as reducing the biases.
- **Think about the members of the population who could have been excluded from your study.** Be careful not to claim that you have learned anything about them.
- **If you can, pilot-test your survey.** Administer the survey in the exact form that you intend to use it to a small sample drawn from the population you intend to sample. Look for misunderstandings, misinterpretation, confusion, or other possible biases. Then refine your survey instrument.
- **Always report your sampling methods in detail.** Others may be able to detect biases where you did not expect to find them.

Section 9.7 Exercises

11. Student samples The university administration of Exercise 1 is considering a variety of ways to sample students for a survey. For each of these proposed survey designs, identify the problem.
a) Publish an advertisement inviting students to visit a website and answer questions.
b) Set up a table in the student union and ask students to stop and answer a survey.

12. Surveying employees The company of Exercise 2 is considering ways to survey their employees. For each of these proposed designs, identify the problem.
a) Leave a stack of surveys out in the employee cafeteria so people can pick them up and return them.
b) Stuff a questionnaire in the mailbox of each employee with the request that they fill it out and return it.

WHAT CAN GO WRONG?

- The principal thing that can go wrong in sampling is that the sample can fail to represent the population. Unfortunately, this can happen in many different ways and for many different reasons. We've considered many of them in the chapter.

- It is also a mistake to draw too small a sample for your needs regardless of how well you draw the sample. However, generally it is more worthwhile to devote effort on improving the quality of your sample than on expanding its size.

What Have We Learned?

Know the three ideas of sampling:

- Examine a part of the whole: A sample can give information about the population.
- Randomize to make the sample representative.
- The sample size is what matters. It's the size of the sample—and not its fraction of the larger population—that determines the precision of the statistics it yields.

Be able to draw a Simple Random Sample (SRS) using a table of random digits or a list of random numbers from technology or an Internet site:

- In a **simple random sample** (SRS), every possible group of *n* individuals has an equal chance of being our sample.

Understand the advantages and disadvantages of other sampling methods:

- **Stratified samples** can reduce sampling variability by identifying homogeneous subgroups (the strata) and then selecting, independently, one random sample from each stratum. This gives us the flexibility to use different sampling methods and different sampling fractions for each stratum, when advantageous.
- **Cluster samples** select a random sample of conveniently available subgroups (our clusters), making sampling tasks more manageable, since often we can more readily find a sampling frame of clusters than of individuals.
- **Systematic samples** can work in some situations and are often the least expensive method of sampling. But we still want to start them randomly.
- **Multistage samples** is an extension of simple cluster sampling, involving multiple stages of random sampling, working down through successively smaller clusters of population elements. When clusters are somewhat homogeneous, this sub-sampling of the clusters is much more efficient than sampling entire clusters.

Identify and avoid causes of bias:

- **Voluntary response** samples are almost always biased and should be avoided and distrusted.
- **Convenience samples** are likely to be flawed for similar reasons.
- **Bad sampling frames** can lead to samples that don't represent the population of interest.
- **Undercoverage** occurs when the sampling frame or procedure doesn't 'cover' or catch all individuals in the target population. Often these are individuals of some particular type or subgroup, which then will tend to be underrepresented in the sample.
- **Nonresponse bias** can arise when sampled individuals will not or cannot respond.
- **Response bias** arises when respondents' answers might be affected by external influences, such as question wording or interviewer behaviour.
- Use best practices when designing a sample survey to improve the chance that your results will be valid.

Review of Terms

Population	The entire group of individuals or instances about whom we hope to learn (p. 290).
Sample	A subset of a population, examined in hope of learning about the population (p. 290).
Sample survey	A study that asks questions of a sample drawn from some population in the hope of learning something about the entire population. Polls taken to assess voter preferences are common sample surveys (p. 290).

Bias	Any systematic failure of a sampling method to represent its population. Biased sampling methods tend to either overestimate or underestimate parameters. It is almost impossible to recover from bias, so efforts to avoid it are well spent. Common errors include
	
	  ■ relying on voluntary response
	  ■ undercoverage of the population
	  ■ nonresponse bias
	  ■ response bias (p. 291)

Randomization	The best defence against bias is randomization, in which each individual is given a fair, random chance of selection (p. 291).
Sample size	The number of individuals in a sample. The sample size determines how well the sample represents the population, not the fraction of the population sampled (p. 292).
Census	A sample that consists of the entire population (p. 293).

Population parameter	A numerically valued attribute of a model for a population. We rarely expect to know the true value of a population parameter, but we do hope to estimate it from sampled data. For example, the mean income of all employed people in the country is a population parameter (p. 294).
Statistic, sample statistic	A numerical value calculated from sampled data. Those that correspond to, and thus estimate, a population parameter are of particular interest. For example, the mean income of all employed people in a representative sample can provide a good estimate of the corresponding population parameter. The term "sample statistic" is sometimes used, usually to parallel the corresponding term "population parameter" (p. 294).
Representative	A sample is said to be representative if the statistics computed from it accurately reflect the corresponding population parameters (p. 294).
Simple random sample (SRS)	A simple random sample of sample size n is a sample in which each set of n elements in the population has an equal chance of selection (p. 295).
Sampling frame	A list of individuals (or any materials used, such as maps, to view and enumerate the individuals) from whom the sample is drawn. Individuals who may be in the population of interest, but who are not in the sampling frame, will have no chance of selection in any sample (p. 295).
Random Numbers (RNs)	Integers generated in a random or pseudorandom manner, typically to assist with selecting an SRS. (p. 295)
Sampling variability	The natural tendency of randomly drawn samples to differ, one from another. Sometimes, unfortunately, called *sampling error*, sampling variability is no error at all, but just the natural result of random sampling (p. 297).
Stratified random sample	A sampling design in which the population is divided into several subpopulations, or **strata**, and random samples are then drawn, independently, from each stratum. If the strata are homogeneous, but are different from each other, a stratified sample may yield more consistent (less variable) results than an SRS (p. 297).
Cluster sample	A sampling design in which entire groups, called **clusters**, are chosen at random. Cluster sampling is usually selected as a matter of convenience, practicality, or cost (p. 298).
Multistage sample	A sampling scheme that involves multiple stages of random sampling, where at each successive stage, we sample from lists of ever smaller clusters (hierarchical in nature). For example, a government agency concerned with gas leaks in city homes might select an SRS of some of the current enumeration areas in the city, followed by an SRS of blocks in each selected area. They might then visit an SRS, or perhaps a systematic sample, of homes in each selected block (p. 300).
Systematic sample	A sample drawn by selecting individuals systematically from a sampling frame (every k^{th} item). When there is no relationship between the order of the sampling frame and the variables of interest, a systematic sample operates essentially like a SRS, but when there a relationship, a systematic sample may be superior to a SRS. (p. 302).
Pilot (study)	A small trial run of a survey to check whether questions are clear. A pilot study can reduce errors due to ambiguous questions (p. 306).
Voluntary response sample	A sample in which individuals can choose on their own whether to participate. Samples based on voluntary response are always invalid and cannot be recovered, no matter how large the sample size (p. 306).
Voluntary response bias	Bias introduced to a sample when individuals can choose on their own whether to participate in the sample. Samples based on voluntary response are always invalid and cannot be recovered, no matter how large the sample size (p. 307).
Convenience sample	A sample of individuals who are conveniently available. Convenience samples often fail to be representative because every individual in the population is not equally convenient to sample (p. 307).
Undercoverage	A sampling scheme that biases the sample in a way that gives a part of the population less possibility of representation in the sample than it has in the population suffers from undercoverage. This results from an inadequate or incomplete frame. (p. 308).

Nonresponse bias	Bias introduced when a large fraction of those sampled fails to respond. Those who do respond are likely to not represent the entire population. For example, those who are selected but at work during the day won't respond to a telephone survey conducted only during working hours (p. 308).
Response bias	Anything in a survey design that influences responses falls under the general heading of response bias. One typical response bias arises from the wording of questions, which may suggest a favoured response. Voters, for example, are more likely to express support of "the prime minister" than support of the particular person holding that office at the moment Interviewer bias is another example (p. 308).

On the Computer GENERATING RANDOM NUMBERS

Your software may have a procedure for directly drawing an SRS from the data. If not, here's one convenient way to draw an SRS of a specified size using a computer-based sampling frame. The sampling frame can be a list of names or of identification numbers arrayed, for example, as a column in a spreadsheet, statistics program, or database:

1. Generate random numbers of enough digits so that each exceeds the size of the sampling frame list by several digits. This makes duplication unlikely.
2. Assign the random numbers arbitrarily to individuals in the sampling frame list. For example, put them in an adjacent column.
3. Sort the list of random numbers, *carrying* along the sampling frame list.
4. Now the first *n* values in the sorted sampling frame column are an SRS of *n* values from the entire sampling frame.

All technology-generated random numbers are *pseudorandom*. The random numbers available on the Internet may technically be better, but the differences won't matter for any simulation of modest size. Pseudorandom numbers generate the next random value from the previous one by a specified algorithm. But they have to start somewhere. This starting point is called the "seed." Most programs let you set the seed. There's usually little reason to do this, but if you wish to, go ahead. If you reset the seed to the same value, the programs will generate the same sequence of "random" numbers.

EXCEL

The **RAND** function generates a random value between 0 and 1. Multiply to scale it up to any range you like. Use the **INT** function to turn the result into an integer.

COMMENTS

Published tests of Excel's random-number generation have declared it to be inadequate. However, for simple simulations, it should be okay.

JMP

- In a new column, in the **Cols** menu choose **Column Info**. . .
- In the dialogue, click the **New Property** button, and choose **Formula** from the drop-down menu.
- Click the **Edit Formula** button.

- In the **Functions (grouped)** window, click on **Random**. Random Integer 10, for example, will generate a random integer between 1 and 10.

MINITAB

- In the **Calc** menu, choose **Random Data**. . .
- In the **Random Data** submenu, choose **Integer**. . .

A dialog guides you in specifying details of range and number of columns to generate.

COMMENTS

You can directly draw an SRS from your data by choosing **Calc > Random Data > Sample From Columns**

R

R has many built-in random number generators.
To generate *k* random integers between 1 and *n*,

- **sample(1:n,k,replace =T)** where *k* and *n* are numbers.
- **sample(1:n,k)** will ensure that there are no repeats. Here $k \leq n$.

COMMENTS

To generate *k* random digits, use **sample(0:9,k,replace =T)**.

SPSS

The **RV.UNIFORM(min, max)** function returns a random value that is equally likely between the min and max limits.

STATCRUNCH

To generate a list of random numbers:

- Click on **Data**.
- Choose **Simulate data » Discrete Uniform**.

Enter the number of rows and columns of random numbers you want. (Often you'll specify the desired number of random values as **Rows** and just 1 **Column**.)

- Fill in the desired **Minimum** integer and **Maximum** integer. For example, use **0** and **9** for random digits.
- Click on **Simulate**. The random numbers will appear in the data table.

TI-83/84 PLUS

To generate random numbers, use **5:RandInt** from the **Math PRB** menu. This command will produce any number of random integers in a specified range.

COMMENTS

For example:
RandInt(0,9,50) randomly returns 50 integers between 0 and 9.

Chapter Exercises

13. Roper Through their *Roper Reports Worldwide*, GfK Roper conducts a global consumer survey to help multinational companies understand different consumer attitudes throughout the world. The researchers interview 1000 people aged 13–65 from each of 30 countries. Their samples are designed so that they get 500 males and 500 females in each country.[14]
  a) Are they using a simple random sample? How do you know?
  b) What kind of design do you think they are using?

14. Student centre survey For their class project, a group of Statistics students decide to survey the student body to assess opinions about the proposed new student centre. Their sample of 200 includes 50 first-year students, 50 sophomores, 50 juniors, and 50 seniors.
  a) Do you think the group was using an SRS? Why?
  b) What kind of sampling design do you think they used?

15. Emoticons As their question of the day for website visitors, www.gamefaqs.com asked, *"Do you ever use emoticons when you type online?"* Of the 87 262 respondents, 27% said that they did not use emoticons.
  a) What kind of sample was this?
  b) How much confidence would you place in using 27% as an estimate of the fraction of people who use emoticons?

16. Drug tests Major League Baseball tests players to see whether they are using performance-enhancing drugs. Officials select a team at random, and a drug-testing crew shows up unannounced to test all 40 players on the team. Each testing day can be considered a study of drug use in Major League Baseball.
  a) What kind of sample is this?
  b) Is that choice appropriate?

17. Gallup At its website (www.gallup.com), the Gallup Poll publishes results of a new survey each day. Scroll down to the end, and you'll find a statement that includes words such as these:

Results are based on telephone interviews with 1016 national adults, aged 18 and older, conducted March 7–10, 2014. . . . In addition to sampling error, question wording and practical difficulties in conducting surveys can introduce error or bias into the findings of public opinion polls.

  a) For this survey, identify the population of interest.
  b) Gallup performs its surveys by phoning numbers generated at random by a computer program. What is the sampling frame?

  c) What problems, if any, would you be concerned about in matching the sampling frame with the population?

18. Gallup World The Gallup World Poll reports results on its website of surveys conducted in various places around the world. At the end of one of these reports, they describe their methods, including explanations, such as the following:

Results are based on face-to-face interviews with randomly selected national samples of approximately 1000 adults, aged 15 and older, who live permanently in each of the 21 sub-Saharan African nations surveyed. Those countries include Angola (areas where land mines might be expected were excluded), Benin, Botswana, Burkina Faso, Cameroon, Ethiopia, Ghana, Kenya, Madagascar (areas where interviewers had to walk more than 20 kilometres from a road were excluded), Mali, Mozambique, Niger, Nigeria, Senegal, Sierra Leone, South Africa, Tanzania, Togo, Uganda (the area of activity of the Lord's Resistance Army was excluded from the survey), Zambia, and Zimbabwe . . . . In all countries except Angola, Madagascar, and Uganda, the sample is representative of the entire population.

  a) Gallup is interested in sub-Saharan Africa. What kind of survey design are they using?
  b) Some of the countries surveyed have large populations. (Nigeria is estimated to have about 130 million people.) Some are quite small. (Togo's population is estimated at 5.4 million.) Nonetheless, Gallup sampled 1000 adults in each country. How does this affect the precision of its estimates for these countries?

19–28. What did they do? For each of the following reports about statistical studies, identify the following items (if possible). If you can't tell, then say so—this often happens when we read about a survey.
  a) The population
  b) The population parameter of interest
  c) The sampling frame
  d) The sample
  e) The sampling method, including whether randomization was employed
  f) Any potential sources of bias you can detect and any problems you see in generalizing to the population of interest

19. A business magazine mailed a questionnaire to the human resource directors of all *Fortune 500* companies and received responses from 23% of them. Those responding reported that they did not find that such surveys intruded significantly on their work day.

[14]www.gfkamerica.com

20. A question posted on the Lycos Website on June 18, 2000, asked visitors to say whether they thought that marijuana should be legally available for medicinal purposes.[15]

21. Consumers Union asked all subscribers whether they had used alternative medical treatments and, if so, whether they had benefitted from them. For almost all of the treatments, approximately 20% of those responding reported cures or substantial improvement in their condition.

22. The Gallup Poll interviewed 1007 randomly selected U.S. adults aged 18 and older from March 23 to 25, 2007. Gallup reports that when asked when (if ever) the effects of global warming will begin to happen, 60% of respondents said the effects had already begun. Only 11% thought they would never happen.

23. Researchers waited outside a bar they had randomly selected from a list of such establishments. They stopped every tenth person who came out of the bar and asked whether he or she thought drinking and driving was a serious problem.

24. Hoping to learn what issues may resonate with voters in the coming election, the campaign director for a mayoral candidate selects one block from each of the city's election districts. Staff members go there and interview all the residents they can find.

25. The Environmental Protection Agency took soil samples at 16 locations near a former industrial waste dump and checked each for evidence of toxic chemicals. They found no elevated levels of any harmful substances.

26. Provincial police set up roadblocks to estimate the percentage of trucks with up-to-date registration, insurance, and safety inspection stickers. They usually find problems with about 10% of the trucks they stop.

27. A company packaging snack foods maintains quality control by randomly selecting 10 cases from each day's production and weighing the bags. Then they open one bag from each case and inspect the contents.

28. Dairy inspectors visit farms unannounced and take samples of the milk to test for contamination. If the milk is found to contain dirt, antibiotics, or other foreign matter, the milk will be destroyed and the farm reinspected until purity is restored.

29. **Mistaken poll** A local television station conducted a "PulsePoll" to predict the winner in the upcoming mayoral election. Evening news viewers were invited to phone in their votes, with the results to be announced on the late-night news. Based on the phone calls, the station predicted that Amabo would win the election with 52%

of the vote. They were wrong: Amabo lost, getting only 46% of the vote. Do you think the station's faulty prediction is more likely to be a result of bias or sampling error? Explain.

30. **Another mistaken poll** Prior to the mayoral election discussed in Exercise 29, the newspaper also conducted a poll. The paper surveyed a random sample of registered voters stratified by political party, age, sex, and area of residence. This poll predicted that Amabo would win the election with 52% of the vote. The newspaper was wrong: Amabo lost, getting only 46% of the vote. Do you think the newspaper's faulty prediction is more likely to be a result of bias or sampling error? Explain.

31. **Parent opinion, part 1** In a large city school system with 20 elementary schools, the school board is considering the adoption of a new policy that would require elementary students to pass a test in order to be promoted to the next grade. The PTA wants to find out whether parents agree with this plan. Listed below are some of the ideas proposed for gathering data. For each, indicate what kind of sampling strategy is involved and what (if any) biases might result.
    a) Put a big ad in the newspaper asking people to log their opinions on the PTA website.
    b) Randomly select one of the elementary schools and contact every parent by phone.
    c) Send a survey home with every student, and ask parents to fill it out and return it the next day.
    d) Randomly select 20 parents from each elementary school. Send them a survey, and follow up with a phone call if they do not return the survey within a week.

32. **Parent opinion, part 2** Let's revisit the school system described in Exercise 31. Four new sampling strategies have been proposed to help the PTA determine whether parents favour requiring elementary students to pass a test in order to be promoted to the next grade. For each, indicate what kind of sampling strategy is involved and what (if any) biases might result.
    a) Run a poll on the local television news, asking people to dial one of two phone numbers to indicate whether they favour or oppose the plan.
    b) Hold a PTA meeting at each of the 20 elementary schools, and tally the opinions expressed by those who attend the meetings.
    c) Randomly select one class at each elementary school and contact each of those parents.
    d) Go through the district's enrollment records, selecting every 40th parent. PTA volunteers will go to those homes to interview the people chosen.

33. **Roller coasters** Canada's Wonderland has opened a new roller coaster. It is so popular that people are waiting for up to three hours for a two-minute ride. Concerned about how patrons (who paid a large amount to enter the park

and ride on the rides) feel about this, park staff survey every 10th person in the line for the roller coaster, starting from a randomly selected individual.
a) What kind of sample is this?
b) Is it likely to be representative?
c) What is the sampling frame?

34. **Playground** Some people have been complaining that the children's playground at a municipal park is too small and is in need of repair. Managers of the park decide to survey city residents to see if they believe the playground should be rebuilt. They hand out questionnaires to parents who bring children to the park. Describe possible biases in this sample.

35. **Social life** A question posted on the gamefaqs.com website on August 1, 2011, asked visitors to the site, "Do you have an active social life outside the Internet?" 22% of the 55 581 respondents said "No" or "Not really, most of my personal contact is online."
a) What kind of sample was this?
b) How much confidence would you place in using 22% as an estimate of the fraction of people who have no active social life outside the Internet. Explain.

36. **Churches** For your political science class, you'd like to take a survey from a sample of all the Catholic Church members in your city. A list of churches shows 17 Catholic churches within the city limits. Rather than try to obtain a list of all members of all these churches, you decide to pick three churches at random. For those churches, you'll ask to get a list of all current members and contact 100 members at random.
a) What kind of design have you used?
b) What could go wrong with your design?

37. **Wording the survey** Two members of the PTA committee in Exercises 31 and 32 have proposed different questions to ask in seeking parents' opinions.

Question 1: Should elementary school-age children have to pass high-stakes tests in order to remain with their classmates?

Question 2: Should schools and students be held accountable for meeting yearly learning goals by testing students before they advance to the next grade?
a) Do you think responses to these two questions might differ? How? What kind of bias is this?
b) Propose a question with more neutral wording that might better assess parental opinion.

38. **Banning Ephedra** An online poll at a popular website asked:

A nationwide ban of the diet supplement Ephedra went into effect recently. The herbal stimulant has been linked to 155 deaths and many more heart attacks and strokes. Ephedra manufacturer NVE Pharmaceuticals, claiming that the FDA lacked proof that Ephedra is

dangerous if used as directed, was denied a temporary restraining order on the ban yesterday by a federal judge. Do you think that Ephedra should continue to be banned nationwide? Sixty-five percent of 17 303 respondents said "yes."

Comment on each of the following statements about this poll:
a) With a sample size that large, we can be pretty certain we know the true proportion of Americans who think Ephedra should be banned.
b) The wording of the question is clearly very biased.
c) The sampling frame is all Internet users.
d) This is a voluntary response survey, so the results can't be reliably generalized to any population of interest.

39. **Another ride** The survey of patrons waiting in line for the roller coaster in Exercise 33 asks whether they think it is worthwhile to wait a long time for the ride and whether they'd like the amusement park to install more roller coasters. What biases might cause a problem for this survey?

40. **Play it again** The survey described in Exercise 34 asked,

Many people believe this playground is too small and in need of repair. Do you think the playground should be repaired and expanded even if that means raising the entrance fee to the park?

Describe two ways this question may lead to response bias.

41. **Survey questions** Examine each of the following questions for possible bias. If you think the question is biased, indicate how and propose a better question.
a) Should companies that pollute the environment be compelled to pay the costs of cleanup?
b) Given that 18-year-olds are old enough to vote and to serve in the military, is it fair to set the drinking age at 21?

42. **More survey questions** Examine each of the following questions for possible bias. If you think the question is biased, indicate how and propose a better question.
a) Do you think high school students should be required to wear uniforms?
b) Given humanity's great tradition of exploration, do you favour continued funding for space flights?

43. **Phone surveys** Anytime we conduct a survey, we must take care to avoid undercoverage. Suppose we plan to select 500 names from the city phone book, call their homes between noon and 4 p.m., and interview whoever answers, anticipating contact with at least 200 people.
a) Why is it difficult to use a simple random sample here?
b) Describe a more convenient, but still random, sampling strategy.

c) What kinds of households are likely to be included in the eventual sample of opinion? Who will be excluded?

d) Suppose, instead, that we continue calling each number, perhaps in the morning or evening, until an adult is contacted and interviewed. How does this improve the sampling design?

e) Random-digit dialling machines can generate the phone calls for us. How would this improve our design? Is anyone still excluded?

44. **Cell phone survey** What about drawing a random sample only from cell phone exchanges? Discuss the advantages and disadvantages of such a sampling method compared with surveying randomly generated telephone numbers from non-cell phone exchanges. Do you think these advantages and disadvantages have changed over time? How do you expect they'll change in the future?

45. **Arm length** How long is your arm compared with your hand size? Put your right thumb at your left shoulder bone, stretch your hand open wide, and extend your hand down your arm. Put your thumb at the place where your little finger is, and extend down the arm again. Repeat this a third time. Now your little finger will probably have reached the back of your left hand. If the fourth hand width goes past the end of your middle finger, turn your hand sideways and count finger widths to get there.[16]

a) How many hand and finger widths is your arm?

b) Suppose you repeat your measurement 10 times and average your results. What parameter would this average estimate? What is the population?

c) Suppose you now collect arm lengths measured in this way from 9 friends and average these 10 measurements. What is the population now? What parameter would this average estimate?

d) Do you think these 10 arm lengths are likely to be representative of the population of arm lengths in your community? In the country? Why or why not?

46. **Fuel economy** Occasionally, when I fill my car with gas, I figure out how many litres/100 kilometres my car got. I wrote down those results after six fill-ups in the past few months. Overall, it appears my car gets 8.7 litres per 100 kilometres.

a) What statistic have I calculated?

b) What is the parameter I'm trying to estimate?

c) How might my results be biased?

d) When the Office of Energy Efficiency (of Natural Resources Canada) checks a car like mine to predict its fuel economy, what parameter is it trying to estimate?

47. **Accounting** Between quarterly audits, a company likes to check on its accounting procedures to address any problems before they become serious. The accounting staff processes payments on about 120 orders each day. The next day, the supervisor rechecks 10 of the transactions to be sure they were processed properly.

a) Propose a sampling strategy for the supervisor.

b) How would you modify that strategy if the company makes both wholesale and retail sales, requiring different bookkeeping procedures?

48. **Happy workers?** A manufacturing company employs 14 project managers, 48 foremen, and 377 labourers. In an effort to keep informed about any possible sources of employee discontent, management wants to conduct job satisfaction interviews with a sample of employees every month.

a) Do you see any danger of bias in the company's plan? Explain.

b) Propose a sampling strategy that uses a simple random sample.

c) Why do you think a simple random sample might not provide the representative opinion the company seeks?

d) Propose a better sampling strategy.

e) Listed below are the last names of the project managers. Use random numbers to select two people to be interviewed. Be sure to explain your method carefully.

Barrett	Bowman	Chen
DeLara	DeRoos	Grigorov
Maceli	Mulvaney	Pagliarulo
Rosica	Smithson	Tadros
Williams	Yamamoto	

49. **Quality control** Sammy's Salsa, a small local company, produces 20 cases of salsa a day. Each case contains 12 jars and is imprinted with a code indicating the date and batch number. To help maintain consistency, at the end of each day, Sammy selects three bottles of salsa, weighs the contents, and tastes the product. Help Sammy select the sample jars. Today's cases are coded 07N61 through 07N80.

a) Carefully explain your sampling strategy.

b) Show how to use random numbers to pick the three jars for testing.

c) Did you use a simple random sample? Explain.

50. **A fish story** Concerned about reports of discoloured scales on fish caught downstream from a newly sited chemical plant, scientists set up a field station in a shoreline public park. For one week they asked fishers there to bring any fish they caught to the field station for a brief inspection. At the end of the week, the scientists said that 18% of the 234 fish that were submitted for inspection displayed the discolouration. From this information,

---

[16]According to Da Vinci, the distance from your elbow to the tip of your hand should be 2.5 hand lengths.

can the researchers estimate what proportion of fish in the river have discoloured scales? Explain.

51. **Sampling methods** Consider each of these situations. Do you think the proposed sampling method is appropriate? Explain.
   a) We want to know what percentage of local family doctors are accepting new patients. We call the offices of 50 doctors randomly selected from local Yellow Page listings.
   b) We want to know what percentage of local businesses anticipate hiring additional employees in the upcoming month. We randomly select a page in the Yellow Pages and call every business listed there.

52. **More sampling methods** Consider each of these situations. Do you think the proposed sampling method is appropriate? Explain.
   a) We want to know if there is neighbourhood support to turn a vacant lot into a playground. We spend a Saturday afternoon going door-to-door in the neighbourhood, asking people to sign a petition.
   b) We want to know if students at our university are satisfied with the selection of food available on campus. We go to the largest cafeteria and interview every 10th person in line.

Just Checking ANSWERS

1. a) It can be hard to reach all members of a population, and it can take so long that circumstances change, affecting the responses. A well-designed sample is often a better choice.
   b) This sample is probably biased—students who didn't like the food at the cafeteria might not choose to eat there.
   c) No, only the sample size matters, not the fraction of the overall population.
   d) Students who frequent this website might be more enthusiastic about Statistics than the overall population of Statistics students. A large sample cannot compensate for bias.
   e) It's the population "parameter." "Statistics" describe samples.

2. a) systematic
   b) stratified
   c) simple
   d) cluster

MathXL

MyLab Statistics

Go to MathXL at **www.mathxl.com** or MyLab™ Statistics at **www.mystatlab.com**. You can practise exercises for this chapter as often as you want. The guided solutions will help you find answers step by step. You'll find a personalized study plan available to you too!

CHAPTER

10

LEARNING OBJECTIVES

1 Know the difference between observational studies and randomized experiments

2 Understand and be able to apply the basic principles of experimental design:
  i. Control what you can,
  ii. Randomize the rest,
  iii. Replicate the trials,
  iv. Block when feasible

3 Know how to work with blinding and control groups

4 Know how to add more factors

5 Be alert for possible confounding

Experiments and Observational Studies

You should treat as many patients as possible with the new drugs while they still have the power to heal.

—*nineteenth-century physician Armand Trusseau*

Pavel Losevsky/Fotolia

Connections

Sample surveys are one way to gather data in order to answer important questions about populations of interest, such as what Canadians think about issues of the day. Other types of questions though can only be answered clearly by gathering data from a well designed experiment. No drug comes to market before demonstrating its safety and effectiveness in a well designed experiment.

Just as randomization was essential in producing a useful survey, randomization will again be key to obtaining experimental results of value. Randomized blinded experiments are the 'gold standard' of data collection. Much of what we know in the natural and social

Who gets good grades? And, more importantly, why? Is there something schools and parents could do to help weaker students improve their grades? Some people think they have an answer: music! No, not your iPod, but an instrument. In a study conducted at a secondary school in the province of Québec with students enrolled in the International Baccalaureat program who had just completed either their third or fourth or final year of study, researchers compared the scholastic performance of students who had elected to take music classes with that of academically similar students who had chosen plastic art (painting or sculpture) or dramatic art classes instead of music. All students had to choose one of these options after their first two years of study. Guess what? The students studying music who completed their third year of study had higher mean grades in all seven subjects (Math, Science, English, etc.) taken that year. The fourth year students who studied music had higher mean grades in seven of eight subjects, and the graduating students who studied music had higher mean grades in nine of ten final year subjects! The biggest differences were in Math and Sciences. [The two exceptions had non-music students' means higher by barely one mark.][1]

[1]Cabanac A, Perlovsky L, Bonniot-Cabanac M-C, Cabanac M. Music and Academic Performance. Behavioural Brain Research 256 (2013) 257–260.

sciences comes from carefully designed experiments.

Experiments compare groups of subjects that have been treated differently. We often *make a picture* using boxplots, the perfect tool for a visual comparison of several groups.

As a result of various studies similar to this one, many parent groups and educators have pressed for expanded music programs in our schools. They argue that the work ethic, discipline, and feeling of accomplishment fostered by learning to play an instrument also enhance a person's ability to succeed in school. They believe that involving more students in music will raise academic performance. What do you think? Does this study provide solid evidence? Or are there other possible explanations for the difference in grades? Is there any way to really prove such a conjecture?

10.1 Observational Studies

This research tried to show an association between music education and grades. But it wasn't a random sample of Canadian students. Nor did it assign students to music or plastic arts or dramatic arts classes. Instead, it simply observed a conveniently accessible group of students recording the choices they made and the outcome. Such studies are called **observational studies**. In observational studies, researchers don't *assign* choices; they simply observe them. In addition, this was a **retrospective study**, because researchers first identified subjects who studied music and then collected data on their past grades.

What's wrong with concluding that music education causes good grades? One secondary school during one academic year in one province (Québec) may not be representative of a whole country. That's true, but the real problem is that the claim that music study *caused* higher grades depends on there being *no other differences* between the groups that could account for the differences in grades, but can we really say that studying music was the *only* difference between these two groups of students?

One can think of plenty of lurking variables that might cause the groups to perform differently. Students who choose to study music may have better work habits to start with, which may make them successful in both music and course work. The home environments, parents, friends, etc. of the music students may differ in ways that can affect academic performance. Music students may have more parental encouragement and support, and that may enhance academic performance. Maybe the students came from wealthier homes and had other advantages. Or it could be that smarter kids just like to play musical instruments. Who knows, but we just can't rule out all sorts of possible differences!

> **RETROSPECTIVE STUDIES CAN GIVE VALUABLE CLUES**
>
> For rare illnesses, it's not practical to draw a large enough sample to see many ill respondents, so the only option remaining is to use retrospective data. For example, researchers can interview those who have become ill. The likely causes of both Legionnaires' disease and HIV were initially identified from retrospective studies of the small populations that were initially infected. But to confirm the causes, researchers needed laboratory-based experiments.

Observational studies are valuable for discovering trends and possible relationships. They are used widely in public health and marketing. Observational studies that try to discover variables related to rare outcomes, such as specific diseases, are often retrospective. They first identify people with the disease and then look into their history and heritage in search of things that may be related to their condition. But retrospective studies have a restricted view of the world because they are usually restricted to a small part of the entire population. And because retrospective records are based on historical data and often memories, they can have errors. (Do you recall *exactly* what you ate even yesterday? How about last Wednesday?)

A more reliable approach is to observe individuals over time, recording the variables of interest and ultimately seeing how things turn out. For example, we might start by selecting young students who have not begun to study music. We could then track their academic performance over several years, comparing those who later choose to study music with those who do not. Identifying subjects in advance and collecting data as events unfold would make this a **prospective study**.

Although an observational study may identify important variables related to the outcome we are interested in, there is no guarantee that we have found the right or the most important related variables. Students who choose to study an instrument might still differ from the others in some important way that we failed to observe. It may be this difference—whether we know what it is or not—rather than music itself that leads to better grades. It's just not possible for observational studies, whether prospective or retrospective, to demonstrate a causal relationship.

shimmo/E+/Getty Image

For Example DESIGNING AN OBSERVATIONAL STUDY

In early 2007, a larger-than-usual number of cats and dogs developed kidney failure; many died. Initially, researchers didn't know why, so they used an observational study to investigate.

QUESTION: Suppose you were called on to plan a study seeking the cause of this problem. Would your design be retrospective or prospective? Explain why.

ANSWER: I would use a retrospective observational study. Even though the incidence of disease was higher than usual, it was still rare. Surveying all pets would have been impractical. Instead, it makes sense to locate some who were sick, and ask about their diets, exposure to toxins, and other possible causes.

Section 10.1 Exercises

1. **Steroids** The 1990s and early 2000s could be considered the steroids era in Major League Baseball, as many players have admitted to using the drug to increase performance on the field. If a sports writer wanted to compare home run totals from the steroids era to an earlier decade, say the 1960s, explain why this would be an observational study. Could the writer conclude that it was the steroids that caused the increase in home runs? Why or why not?

2. **E-commerce** A business student conjectures that the Internet caused companies to become more profitable, since many transactions previously handled "face-to-face" could now be completed online. The student compares earnings from a sample of companies from the 1980s to a sample from the 2000s. Explain why this is an observational study. If indeed profitability increased, can she conclude the Internet was the cause? Why or why not?

10.2 Randomized, Comparative Experiments

Is it *ever* possible to get convincing evidence of a cause-and-effect relationship? Well, yes it is, but we would have to take a different approach. We could take a group of third graders, randomly assign half to take music lessons, and forbid the other half from doing so. Then we could compare their grades several years later. This kind of study design is an **experiment**.

An experiment requires a **random assignment** of subjects to treatments. Only an experiment can justify a claim like "Music lessons cause higher grades." Questions such as "Does taking vitamin C reduce the chance of getting a cold?" and "Does working with computers improve performance in Statistics class?" and "Is this drug a safe and effective treatment for that disease?" require a designed experiment to establish cause and effect.

Experiments study the relationship between two or more variables. An experimenter must identify at least one explanatory variable, called a **factor**, to manipulate, and at least one **response variable** to measure. What distinguishes an experiment from other types of investigation is that the experimenter actively and deliberately manipulates the factors to control the details of the possible treatments, and assigns the subjects to those treatments *at random*. The experimenter then observes the response variable and *compares* responses for different groups of subjects who have been treated differently. For example, we might design an experiment to see whether the amount of sleep and exercise you get affects your grades.

The individuals on whom or which we experiment are referred to as **experimental units**; however, humans who are experimented on are commonly called *subjects* or *participants*. When we recruit subjects for our sleep deprivation experiment, we'll probably have better luck inviting participants than adverting for experimental units!

The specific values that the experimenter chooses for a factor are called the **levels** of the factor. We might assign our participants to sleep for 4, 6, or 8 hours. Often there are several factors at a variety of levels. (Our subjects could also be assigned to a treadmill for 0 or 30 minutes.) The combination of specific levels from all the factors that an experimental unit receives is known as its **treatment**. (Our subjects could have any one of six different treatments—three sleep levels, each at two exercise levels.)

How should we assign our participants to these treatments? Some students prefer 4 hours of sleep, while others need 8. Some exercise regularly; others are couch potatoes. Should we let the students choose the treatments they'd prefer? No. That would not be a good idea. Nor should we assign them according to what we feel might be best. To have any hope of drawing a fair conclusion, we must assign our participants to their treatments at *random*.

It may be obvious to you that we shouldn't let the students choose the treatment they'd prefer, but the need for random assignment is a lesson that was once hard for some to accept. For example, physicians naturally prefer to assign patients to the therapy that they think best rather than have a random element such as a coin flip determine the treatment [If we (or the physician) knew which was the better treatment, we wouldn't be performing the experiment.] But we've known for more than a century that for the results of an experiment to be valid, we must use deliberate randomization.

> **AN EXPERIMENT**
>
> *Manipulates* the factor levels to create treatments. *Randomly assigns* subjects to these treatment levels. *Compares* the responses of the subject groups across treatment levels.

> Health Canada   No drug can be sold in Canada without first showing, in a suitably designed experiment approved by Health Canada, that it's safe and effective. The small print on the booklet that comes with many prescription drugs usually describes the outcomes of that experiment.

> **The Canadian National Breast Screening Study** shocked the world in 2014 when the researchers reported that mammography for women aged 40–59 does not save lives! Whoa... what? Really? Are screening mammograms (as opposed to diagnostic mammograms) just a waste of time and money? Many were shocked, and some radiologists in particular quickly reacted with scorn for this study, one suggesting that these researchers just don't care about saving women's lives! Breast cancer mortality has been falling, but is that due to earlier detection or better treatment?
>
> The study began in 1980, taking place in 15 different screening centres spread out across Canada, and was intended to give more reliable results than what was known from the available observational studies. Hence, it was designed as a randomized trial with nearly 90,000 women randomly divided into two groups, with one group having annual mammography screening and breast exams for five years, and another (control) group not receiving mammography screening (but receiving 'usual care' or 'breast examination alone' depending on the age group). They followed these women for 25 years reporting periodically on the results, most recently in 2014. After 25 years, they reported that the rate of death from breast cancer was the same in both groups. And many of the women who were screened underwent unnecessary treatment. ('Overdiagnosis' is a serious issue.) False positives and false negatives are additional downsides of mammography.
>
> The results of this study were controversial and hotly debated, and the protocols examined carefully. Criticisms focused mostly on design issues (such as randomization) and the quality of the mammography equipment. One radiologist even accused them of non-random preferential assignment of subjects, in effect, of scientific fraud. The criticisms though appear to have been well rebutted by the study authors.
>
> The designers of this study may well have been surprised and disappointed by the negative mammography results (Rarely do researchers spend huge amounts of time and money in expectation of negative results!). It is just one more piece of evidence in an ongoing discussion about what is best for women's health. The authors concluded that "the data suggest that the value of mammography screening should be reassessed."[2] Hopefully in the end, *evidence based medicine* will prevail. According to Cornelia J Barnes of the Dalla Lana School of Public Health, University of Toronto, "There is much to be learned from the breast screening controversy. It boils down to always being ready to question your own beliefs, to look for conflict of interest in those with the loudest voices, to be curious about the unexpected and to do more good than harm."[3]

---

[2]Miller Anthony B, Wall Claus, Baines Cornelia J, Sun Ping, To Teresa, Narod Steven A et al. Twenty five year follow-up for breast cancer incidence and mortality of the Canadian National Breast Screening Study: randomised screening trial BMJ 2014; 348 :g366

[3]Rational and Irrational Issues in Breast Cancer Screening, Cornelia J. Baines, Cancers 2011, 3, 252–266

> **For Example** DETERMINING THE TREATMENTS AND RESPONSE VARIABLE
>
> **RECAP:** In 2007, deaths of a large number of pet dogs and cats were ultimately traced to contamination of some brands of pet food. The manufacturer now claims that the food is safe, but before it can be released, it must be tested.
>
> **QUESTION:** In an experiment to test whether the food is now safe for dogs to eat,[4] what would be the treatments and what would be the response variable?
>
> **ANSWER:** The treatments would be ordinary size portions of two dog foods: the new one from the company (the test food) and one that I was certain was safe (perhaps prepared in my kitchen or laboratory). The response would be a veterinarian's assessment of the health of the test animals.

Section 10.2 Exercises

3. Tips A pizza delivery driver, always trying to increase tips, runs an experiment on his next 40 deliveries. He flips a coin to decide whether or not to call a customer from his mobile phone when he is five minutes away, hoping this slight bump in customer service will lead to a slight bump in tips. After 40 deliveries, he will compare the average tip percentage between the customers he called and those he did not. What are the experimental units and how did he randomize treatments?

4. Tomatoes You want to compare the tastiness and juiciness of tomatoes grown with three amounts of a new fertilizer: none, half the recommended amount, and the full recommended amount. You allocate 6 tomato plants to receive each amount of fertilizer, assigning them at random. What are the experimental units? What is the response variable?

5. Tips II For the experiment described in Exercise 3, list the factor, the levels, and the response variable.

6. Tomatoes II For the experiment described in Exercise 4, name the factor and its levels. How might the response be measured?

10.3 The Four Principles of Experimental Design

A methodology for designing experiments was proposed by Sir R.A. Fisher in his innovative book *The Design of Experiments* (1935). His idea of experimental design can be summarized by the following four principles.

1. **Control.** We control sources of variation other than the factors we are testing by making conditions as similar as possible for all treatment groups. For human subjects, we try to treat them alike. However, there is always a question of degree and practicality. Controlling extraneous sources of variation reduces the variability of the responses, making it easier to detect differences among the treatment groups.

   Of course, when a controlled factor has only particular levels it is hard to generalize to levels that haven't been tested. For example, if we test two laundry detergents and carefully control the water temperature at 80°C, we would reduce the variation in our results due to water temperature, but we couldn't say much about the detergents' performance in cold water.

   Although we control both experimental factors and other sources of variation, we think of them and deal with them very differently. We control a factor by assigning

---

[4]It may disturb you (as it does us) to think of deliberately putting dogs at risk in this experiment, but in fact that is what is done. The risk is borne by a small number of dogs so that the far larger population of dogs can be kept safe.

subjects to different factor levels so we can see how the response changes at those different levels. But we control other sources of variation to *prevent* them from changing and affecting the response.

2. **Randomize**. As in sample surveys, *randomization* allows us to equalize the effects of unknown or uncontrollable sources of variation. It does not eliminate the effects of these sources, but by distributing them equally (on average) across the treatment groups, it makes comparisons among the treatments fair. Assigning experimental units to treatments at random allows us to use the powerful methods of Statistics to draw conclusions from an experiment. When we assign subjects to treatments at random, we reduce bias due to uncontrolled sources of variation.

    Experimenters often control factors that are easy, or inexpensive to control. They randomize to protect against the effects of other factors, even factors they haven't thought about. How to choose between the two strategies is best summed up in the old adage that says "control what you can, and randomize the rest."

3. **Replicate**. Drawing conclusions about the world is impossible unless we repeat, or **replicate**, our results. Two kinds of replication show up in comparative experiments. First, we should apply each treatment to a number of subjects. Only with such replication we can estimate the variability of responses. If we have not assessed the variation, the experiment is not complete. The outcome of an experiment on a single subject is an anecdote, not data.

A second kind of **replication** shows up when the entire experiment is repeated on a different population of experimental units. We may believe that what is true of the students in Psych 101 who volunteered for the sleep experiment is true of all humans, but we'll feel more confident if our results for the experiment are *replicated* in another part of the country, with people of different ages, and at different times of the year. *Replication* of an entire experiment with the controlled sources of variation at different levels is an essential step in science.

4. **Block**. Some of the variation in our data may be due to differences among our participants that we can allow for in our experiment design. For example, suppose the participants available for a study of balance include two members of the varsity girls gymnastics team and 10 other students with no gymnastics experience. Randomizing may place both gymnasts in the same treatment group. In the long run, if we could perform the experiment over and over, it would all equalize. But wouldn't it be better to assign one gymnast to each group (at random) and five of the other students to each group (at random)? By doing this, we would improve fairness in the short run. This approach recognizes the variation due to practice and experience and allocates the participants at random *within* each experience level. When we do this, the variable "Experience" is a blocking variable, and the levels of "Experience" are called blocks.

Sometimes, attributes of the experimental units that we are not studying and that we can't control may nevertheless affect the outcomes of an experiment. If we group similar individuals together and then randomize within each of these *blocks*, we can account for much of the variability due to the difference among the blocks so it won't obscure our comparison of the treatment groups. Blocking is an important compromise between randomization and control. However, unlike the first three principles, blocking is not *required* in an experimental design. Blocking in experiments has much in common with stratification in surveys, but *be sure not to interchange them* and talk of stratification in an experiment or talk of blocking in a survey!

Iurii Davydov/Shutterstock

For Example CONTROL, RANDOMIZE, AND REPLICATE

RECAP: We're planning an experiment to see whether the new pet food is safe for dogs to eat. We'll feed some animals the new food and others a food known to be safe, comparing their health after a period of time.

QUESTIONS: In this experiment, how will you implement the principles of control, randomization, and replication?

ANSWER: I'd control the portion sizes eaten by the dogs. To reduce possible variability from factors other than the food, I'd standardize other aspects of their environments—housing the dogs in similar pens and ensuring that each got the same amount of water, exercise, play, and sleep time, for example. I might restrict the experiment to a single breed of dog and to adult dogs to further minimize variation.

To equalize traits, pre-existing conditions, and other unknown influences, I would assign dogs to the two feed treatments randomly.

I would replicate by assigning more than one dog to each treatment to allow us to see the variability among individual dogs' responses (the importance of which will become much clearer later on, when we discuss statistical significance). If I had the time and funding, I might replicate the entire experiment using, for example, a different breed of dog.

Diagrams

An experiment is carried out over time with specific actions occurring in a specified order. A diagram of the procedure can help in thinking about experiments.[5]

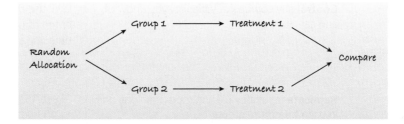

The diagram emphasizes the random allocation of subjects to treatment groups, the separate treatments applied to these groups, and the ultimate comparison of results. It's best to specify the responses that will be compared. A good way to start comparing results for the treatment groups is with boxplots.

---

[5]Diagrams of this sort were introduced by David Moore in his textbooks and are widely used.

Step-by-Step Example DESIGNING AN EXPERIMENT

Brian Chase/Shutterstock

A student, preparing for a triathlon, suspected that the 45 minutes each day she spent training in a chlorinated pool was damaging her nail polish. She wished to investigate whether the colour of the nail polish might make a difference. How could she design an experiment to investigate?

Let's work through the design, step by step. We'll design the simplest kind of experiment, a *completely randomized experiment in one factor*. The statements in the right column are the kinds of things you would need to say in proposing an experiment. You'd include them in the "methods" section of a report once the experiment is run.

> A **completely randomized design (CRD)** is the ideal simple design, just as a *simple random sample* is the ideal simple sample—and for many of the same reasons—though other more complex designs may be better choices, for practical reasons, just as was the case with sampling. And never *never* talk about using an SRS in an experiment or a CRD for a survey!

Question: How would you design an experiment to test whether two nail colours resisted chipping differently when challenged by soaks in chlorine water?

THINK ➡ **Plan** State what you want to know.	I want to know whether two colours of nail polish resist chipping differently when challenged by soaks in chlorine water.
Response Specify the response variable.	I can measure the percentage of the nail polish chipped away by examining scanned images of the nails with a computer program.
Treatments Specify the factor levels and the treatments.	The factor is nail polish *Colour*. The factor levels are *Red* and *Nude*, two common nail polish colours.
Experimental Units Specify the experimental units. **Experiment Design** Observe the principles of design:	To control the surface coloured by the polish, reducing variability that might be found with natural nails, I'll obtain 30 acrylic nails and glue each one to a chopstick.
Control any sources of variability you know of and can control.	I will use the same brand of acrylic nails and the same brand of nail polish. All nails will be painted at the same time and soaked at the same time in the same chlorine solution for the same amount of time.
Replicate results by placing more than one nail in each treatment group.	I'll use 15 nails in each group.
Randomly assign experimental units to treatments to equalize the effects of unknown or uncontrollable sources of variation.	I will randomly divide the nails into two groups for painting with each colour.
Describe how the randomization will be accomplished.	I will number the nails from 1 to 30. I will obtain 30 random numbers from an Internet site and sort them, carrying along the nail ID numbers. Then I'll assign the first 15 nails to one colour and the remaining ones to the other.

Make a Picture A diagram of your design can help you think about it clearly.

Specify any other experiment details. Give enough details so that another experimenter could exactly replicate your experiment. It's generally better to include details that might seem irrelevant than to leave out matters that could turn out to make a difference.

All nails will be soaked together for 45 minutes in a chlorine solution that matches typical swimming pool levels of chlorine. After the nails dry, I will tap each one 100 times on a computer keyboard to simulate daily stresses. I will repeat this treatment for 5 days and measure the result at the end of that time.

Specify how to measure the response.

I will scan each nail and use a digitized image of the nail to measure the fraction of the polish that has chipped off by comparing the before and after areas of nail polish using Adobe Illustrator.

SHOW ➡ Once you collect the data, you'll need to display them and compare the results for the treatment groups.

I will display the results with side-by-side boxplots to compare the treatment groups.

I will then compare the means of the groups.

TELL ➡ To answer the initial question, we ask whether the differences observed in the means of the groups are meaningful.

Because this is a randomized experiment, we can attribute significant differences to the treatments. To do this properly, we'll need methods from "statistical inference," the subject of the rest of this book.

If the difference in mean percent chipped between the groups is greater than I would expect from the variation I observe among the nails within each treatment group, I may be able to conclude that these difference can be attributed to the differences in nail polish colour.

Does the Difference Make a Difference?

If the differences among the treatment groups are big enough, we'll attribute the differences to the treatments—but how can we decide whether the differences are big enough?

We wouldn't expect the group means to be identical. Not really. Even if the treatment made no difference at all, there would still be some variation. We assigned the nails to treatments at random. But a different random assignment would have led to different results. Even a repeat of the *same* treatment on different randomly assigned sets of nails would lead to a different mean. The real question is whether the differences we observed between treatment groups are about as big as we might get just from the randomization alone, or whether they're bigger than that. If we decide that they're bigger, we'll attribute the differences to the treatments and say the differences are **statistically significant**.

How can we decide whether the treatment means differ enough to be considered statistically significant? Later chapters will offer methods to help answer that question, but to get some intuition, think about deciding whether a coin is fair and random. If we flip a fair coin 100 times, we expect, *on average*, to get 50 heads. Suppose we get 54 heads out of 100. That doesn't seem very surprising. It's well within the bounds of ordinary random fluctuations. What if we'd seen 94 heads? That's clearly outside the bounds. We'd be pretty sure that the coin flips were not random. But what about 74 heads? Is that far enough from 50% to arouse our suspicions? That's the sort of question we need to ask of our experiment results.

In Statistics terminology, 94 heads would be a statistically significant difference from 50, and 54 heads would not. Whether 74 is *statistically significant* or not would depend on the chance of getting at least 74 heads in 100 flips of a fair coin and on our tolerance for believing that rare events can happen to us.

In the nails experiment, we ask whether the differences we see among the treatment groups are the kind of differences we'd expect from randomization. A good way to get a feeling for that is to look at how much our results vary among nails that get the *same* treatment. Boxplots of our results by treatment group can give us a general idea.

For example, Figure 10.1 shows two pairs of boxplots whose centres differ by exactly the same amount. In the upper set, that difference appears to be larger than we'd expect just by chance. Why? Because the variation is quite small *within* treatment groups, so the larger difference *between* the groups is unlikely to be just from the randomization. In the bottom pair, that same difference between the centres looks less impressive. There the variation *within* each group swamps the difference *between* the two means. We'd say the difference is statistically significant in the upper pair and not statistically significant in the lower pair.

Later we'll see statistical tests that quantify this intuition. For now, the important point is that a difference is statistically significant if we don't believe that it's likely to have occurred only by chance.

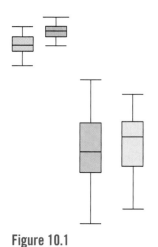

Figure 10.1

The boxplots in both pairs have centres the same distance apart, but when the spreads are large, the observed difference can be more readily explained just by random fluctuation. (And what about the sample sizes? Do you think that might matter too?)

 Just Checking

1. At one time, a method called "gastric freezing" was used to treat people with peptic ulcers. An inflatable bladder was inserted down the esophagus and into the stomach, and then a cold liquid was pumped into the bladder. Now you can find the following notice on the Internet site of a major insurance company:

   [Our company] does not cover gastric freezing (intragastric hypothermia) for chronic peptic ulcer disease . . . . Gastric freezing for chronic peptic ulcer disease is a non-surgical treatment which was popular about 20 years ago but now is seldom performed. It has been abandoned due to a high complication rate, only temporary improvement experienced by patients, and a lack of effectiveness when tested by double-blind, controlled clinical trials.

   What did that "controlled clinical trial" (experiment) probably look like? (Don't worry about "double-blind"; we'll get to that soon.)

   a) What was the factor in this experiment?

   b) What was the response variable?

   c) What were the treatments?

   d) How did researchers decide which subjects received which treatment?

   e) Were the results statistically significant?

Experiments and Samples

Both experiments and sample surveys use randomization to get unbiased data. But they do so in different ways and for different purposes. Sample surveys try to estimate population parameters, so the sample needs to be as representative of the population as possible. By

A Random Sample? Probably not Experiments are rarely performed on random samples from a population. Don't describe the subjects in an experiment as a random sample unless they really are. More likely, the only randomization was in assigning subjects to treatments.

contrast, experiments try to assess the effects of treatments. Experimental units are rarely drawn randomly from a population. For example, a medical experiment may deal only with local patients who have the disease under study. The randomization is in the assignment of their therapy. We want a sample to exhibit the diversity and variability of the population, but for an experiment the more homogeneous the subjects, the more easily we'll spot differences in the effects of the treatments.

Unless this is one of these rare studies in which the experimental units are chosen from the population of interest at random, you should be cautious about generalizing experiment results to larger populations until the experiment has been repeated under different circumstances. Results become more persuasive if they remain the same in completely different settings, such as in a different season, in a different country, or for a different species. Hence, the *Who* of an experiment always needs to be described carefully.

Nevertheless, experiments can draw stronger conclusions than surveys. By looking only at the differences across treatment groups, experiments cancel out many sources of bias. For example, the entire pool of subjects may be a biased representation of the population of interest. (University students may need more sleep, on average, than the general population.) But when we assign subjects randomly to treatment groups, all the groups are biased, *in the same way*. When we consider the differences in their responses, these biases cancel out, allowing us to see the *differences* due to treatment effects more clearly.

Section 10.3 Exercises

7. Tips again For the experiment of Exercise 3, name some variables the driver did or should have controlled. Was the experiment randomized and replicated?

8. Tomatoes again For the experiment of Exercise 4, discuss variables that could be controlled or that could not be controlled. Is the experiment randomized and replicated?

10.4 Control Treatments

Suppose you wanted to test an expensive piece of software designed to shorten download times. You could just try it on several files and record the download times, but you probably want to *compare* the speed with what would happen *without* the software installed. Such a baseline measurement is called a *control treatment*, and the experimental units to whom it is applied are called a **control group**.

This is a use of the word "control" in an entirely different context. Previously, we controlled extraneous sources of variation by keeping them constant. Here, we use a control treatment as another *level* of the factor included so we can compare the treatment results to a situation in which "nothing happens."

Without a control group in many experiments, you would not know if the (let's say) good results you got were a consequence of the treatment or just a consequence of some special attributes of your study, related to the particular study administrators, evaluators, subjects, location, weather, time of year, etc.—all of which get neutralized once you compare with a control group subjected to the same experiment characteristics. For example, the well-known Physicians Health Study established the effectiveness of aspirin for preventing heart attacks. Without a control group, good results in the aspirin group might not be due to aspirin. Subjects might want to 'help out' the treatment by exercising more, or improving their diet. A control group can bring into the study all these factors, ensuring they operate equally on all subjects in all groups—though that often requires additional measures, such as 'blinding' which we discuss next.

Blinding

Humans are notoriously susceptible to errors in judgment.[6] All of us. When we know what treatment was assigned, it's difficult not to let that knowledge influence our assessment of the response, even when we try to be careful.

Suppose you were trying to investigate which brand of cola students really prefer. You set up an experiment to see which of the three competing brands students prefer (or whether they can tell the difference at all). But people have brand loyalties. You probably prefer one brand already. If you knew which brand you were tasting, it might influence your rating. To avoid this problem, it would be better to disguise the brands as much as possible. This strategy is called **blinding** the participants to the treatment.

But it isn't just the subjects who should be blind. Experimenters themselves often subconsciously behave in ways that favour what they believe. Even technicians may treat plants or test animals differently if, for example, they expect them to die. An animal that starts doing a little better than others by showing an increased appetite may get fed a bit more than the experimental protocol specifies.

People are so good at picking up subtle cues about treatments that the best (in fact, the *only*) defense against such biases in experiments on human subjects is to keep *anyone* who could affect the outcome or the measurement of the response from knowing which subjects have been assigned to which treatments. So, not only should your cola-tasting subjects be blinded, but also *you*, as the experimenter, shouldn't know which drink is which, either—at least until you're ready to analyze the results.

There are two main classes of individuals who can affect the outcome of an experiment:

- those who could influence the results (the subjects, treatment administrators, or technicians)
- those who evaluate the results (judges, treating physicians, etc.)

When all the individuals in one of these classes are blinded, an experiment is said to be **single-blind**. When everyone in *both* classes is blinded, we call the experiment **double-blind**. Even if several individuals in one class are blinded—for example, both the patients and the technicians who administer the treatment—the study would still be just single-blind. If only some of the individuals in a class are blind—for example, if subjects are not told of their treatment but the administering technician is not blind—there is a substantial risk that subjects can discern their treatment from subtle cues in the technician's behaviour or that the technician might inadvertently treat subjects differently. Such experiments cannot be considered truly blind.

To improve the nail polish experiment design, the experimenter could have the amount of chipping measured by an independent judge who was shown only monochrome images and didn't know how the nails had been treated. That would make the experiment single-blind. To make the experiment double-blind, we could put coloured glasses on the person painting the nails, soaking them in the chlorine water, and tapping them on the keyboard so she couldn't see the colours of the nails.

> ### BLINDING BY MISLEADING
>
> Social science experiments can sometimes blind subjects by misleading them about the purpose of a study. One of the authors participated as an undergraduate volunteer in a (now infamous) psychology experiment using such a blinding method. The subjects were told that the experiment was about three-dimensional spatial perception and were assigned to draw a model of a horse. While they were busy drawing, they heard a loud noise and groaning from the next room. The *real* purpose of the experiment was to see whether the social pressure of being in groups made people react to the disaster differently. Subjects had been randomly assigned to draw either in groups or alone; that was the treatment. The experimenter was not interested in the drawings, but the subjects were blinded to the treatment because they were misled.

For Example BLINDING

RECAP: In our experiment to see if the new pet food is now safe, we're feeding one group of dogs the new food, and another group a food we know to be safe. Our response variable is the health of the animals as assessed by a veterinarian.

QUESTIONS: Should the vet be blinded? Why or why not? How would you do this? (Extra credit: Can this experiment be double-blind? Would that mean that the test animals wouldn't know what they were eating?)

---

[6]For example, here we are in Chapter 10 and you're still reading the footnotes.

ANSWER: Whenever the response variable involves judgment, it is a good idea to blind the evaluator to the treatments. The veterinarian should not be told which dogs ate which foods.

EXTRA CREDIT: There is a need for double-blinding. In this case, the workers who care for and feed the animals should not be aware of which dogs are receiving which food. We'll need to make the "safe" food look as much like the "test" food as possible.

Active placebos The placebo effect is stronger when placebo treatments are administered with authority or by a figure who appears to be an authority. "Doctors" in white coats generate a stronger effect than salespeople in polyester suits. But the placebo effect is not reduced much even when subjects know that the effect exists. People often suspect that they've gotten the placebo if nothing at all happens. So, drug manufacturers have recently gone so far in making placebos realistic that they cause the same side effects as the drug being tested! Such "active placebos" usually induce a stronger placebo effect. When those side effects include loss of appetite or hair, the practice may raise ethical questions.

Placebos

Often, simply applying *any* treatment can induce an improvement. Every parent knows the medicinal value of a kiss to make a toddler's scrape or bump stop hurting. Some of the improvement seen with a treatment—even an effective treatment—can be due simply to the act of treating. In addition, we know that regression to the mean may account for some patients' improvement. To separate the effect of the treatment from these other effects, we can use a control treatment that mimics the treatment itself.

A "fake" treatment that looks just like the treatments being tested is called a **placebo**. Placebos are the best way to blind human participants from knowing whether they are receiving the treatment or not. One common version of a placebo in drug testing is a "sugar pill." The fact is that subjects treated with a placebo sometimes improve. It's not unusual for 20% or more of subjects given a placebo to report reduction in pain, improved movement, or greater alertness, or even to demonstrate improved health or performance. This **placebo effect** highlights both the importance of effective blinding and the importance of comparing treatments with a control. Placebo controls are so effective that you should use them as an essential tool for blinding whenever possible.

In summary, the best experiments are usually

- randomized
- double-blind
- comparative
- placebo-controlled

> **Real world example: Does *ginkgo biloba* improve memory?**
> Researchers investigated the purported memory-enhancing effect of ginkgo biloba tree extract.[7] In a randomized, comparative, double-blind, placebo-controlled study, they administered treatments to 230 elderly community members. One group received Ginkoba™ according to the manufacturer's instructions. The other received a similar-looking placebo. Thirteen different tests of memory were administered before and after treatment. The placebo group showed greater improvement on seven of the tests; the treatment group on the other six. None showed any significant differences. In the margin are boxplots of one measure.

By permission of John L. Hart FLP and Creators Syndicate, Inc.

[7]P. R. Solomon, F. Adams, A. Silver, J. Zimmer, R. De Veaux, 2002, "Ginkgo for memory enhancement. A randomized controlled trial." *JAMA* 288: 835–840.

Section 10.4 Exercises

9. **More tips** Is the experiment of Exercise 3 blinded? Can it be double-blinded? Explain.

10. **More tomatoes** If the tomato taster doesn't know how the tomatoes have been treated, is the experiment single or double-blind? How might the blinding be improved further?

10.5 Blocking

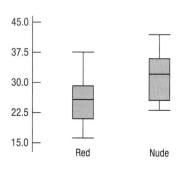

The nail polish experiment did show a difference in the durability of the two colours (see the boxplots in the margin). But the student wondered whether her artificial treatments in fact reflected the experience of swimming and dealing with the everyday stresses on her nails. So she designed a follow-up study. Her plan was to paint her own nails and record chipping during several days as she swam each day and went about her usual work schedule. In this experiment, her experimental units would be her own fingernails. One possible design would have been to paint her left hand with one colour and her right hand with the other. But she is right-handed and thought that the stresses on her nails might be different between the two hands.

How could she design the experiment so that the differences in hands don't mess up her attempts to see differences in the nail polish colours? When groups of experimental units are similar, it's often a good idea to gather them together into **blocks** and then randomly assign the experimental units within each block to the treatments. By blocking, we isolate the variability attributable to the differences between the blocks, so that we can see the differences caused by the treatments more clearly.

Here, she could define each hand to be a block. She could then randomly assign half the fingers of a hand to one colour and half to the other and randomly do the same for the other hand. Each treatment (colour) is still assigned to the same number of fingers. The randomization occurs only within the blocks. The design is a **randomized block design**.

In effect, she can conduct a separate experiment for the fingers of each hand and then combine the results. If her right hand is harder on nail polish because she is right-handed, she can still compare the two polish colours within that hand. The picture depicts the design.

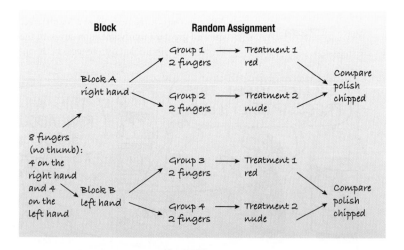

In an observational study, we sometimes pair participants that are similar in ways *not* under study. By **matching** participants in this way, we can reduce variation in much the same way as blocking. For example, a retrospective study of music education and grades might match each student who studies an instrument with someone of the same sex who is similar in family income but didn't study an instrument. We could then compare grades of music students with those of non-music students. The matching would reduce the variation due to income and sex differences (But other important differences may remain due to the lack of randomization. Don't confuse this with blocking in an experiment.)

Blocking for experiments is the same idea as stratifying is for sampling. Both methods group together subjects that are similar and randomize within those groups as a way to remove unwanted variation. (But be careful to keep the terms straight. Don't say that we "stratify" an experiment or "block" a sample.) We use blocks to reduce variability so we can see the effects of the factors; we're not usually interested in studying the effects of the blocks themselves.

Artem Kursin/Shutterstock

For Example BLOCKING

RECAP: In 2007, pet food contamination put cats at risk, as well as dogs. Our experiment should probably test the safety of the new food on both animals.

QUESTIONS: Why shouldn't we randomly assign a mix of cats and dogs to the two treatment groups? What would you recommend instead?

ANSWER: Dogs and cats might respond differently to the foods, and that variability could obscure my results. Blocking by species can remove that superfluous variation. I'd randomize cats to the two treatments (test food and safe food) separately from the dogs. I'd measure their responses separately and look at the results afterward.

Just Checking

2. Recall the experiment about gastric freezing, an old method for treating peptic ulcers that you read about in the first Just Checking. Doctors would insert an inflatable bladder down the patient's esophagus and into the stomach and then pump in a cold liquid. A major insurance company now states that it doesn't cover this treatment because "double-blind, controlled clinical trials" failed to demonstrate that gastric freezing was effective.

   a) What does it mean that the experiment was double-blind?

   b) Why would you recommend a placebo control?

   c) Suppose researchers suspected that the effectiveness of the gastric freezing treatment might depend on whether a patient had recently developed the peptic ulcer or had been suffering from the condition for a long time. How might the researchers have designed the experiment?

Adding More Factors

But why vary only one factor (colour) when there are other factors one can play with in this experiment, such as type of finish, number of coats, presence or absence of base coat (or top coat), choice of brand, etc. Say you want to also compare creme and shimmer finishes. We then introduce a second factor into our design: *Finish*, with two levels: Creme, Shimmer. Now we need to use polishes of four types, corresponding to the four treatment combinations shown in the diagram below:

	Creme	Shimmer
Nude	1	2
Red	3	4

If we use 32 acrylic nails, the experiment now assigns eight nails to each of these four treatments at random. This experiment utilizes a *completely randomized two-factor design* because any nail could end up assigned at random to any of the four treatments (and we have two factors). This also may be described as a 2 by 2 factorial arrangement, because there are two levels of one factor and two levels of the second factor.

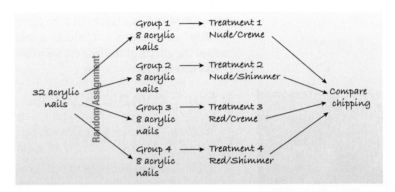

It's often important to include several factors in the same experiment in order to see what happens when the factor levels are applied in different *combinations*. A common misconception is that applying several factors at once makes it difficult to separate the effects of the individual factors. You may hear people say that experiments should always be run "one factor at a time." In fact, just the opposite is true: Experiments with more than one factor are both more efficient and provide more information than one-at-a-time experiments. There are many ways to design efficient multifactor experiments.

In our experiment above, I'm assuming we can find the four colour-finish combinations of the same brand. If not, we are mixing up or *confounding* (discussed shortly) brand differences with the factors under investigation. Of course we could also introduce *Brand* as a third factor. If using two brands, we would need to form $2 \times 2 \times 2 = 8$ treatment combinations of Colour/Finish/Brand for comparison. We also earlier discussed running this experiment using her own nails and treating each hand as a block. Blocking of experimental units is an issue quite separate from how we define the factors under investigation. For the two factor Colour/Finish experiment, we would just randomize the four treatments to four nails on each hand. For a three factor study though, we would encounter a problem—her hand does not have enough fingers for all (at least) eight treatment combinations. There are ways to deal with this, using more complicated designs, but this falls beyond the scope of this course.

Section 10.5 Exercises

11. Block that tip The driver of Exercise 3 wants to know about tipping in general. So he recruits several other drivers to participate in the experiment. Each driver randomly decides whether to phone customers before delivery and records the tip percentage. Is this experiment blocked? Is that a good idea?

12. Blocking tomatoes To obtain enough plants for the tomato experiment of Exercise 4, experimenters have to purchase plants from two different garden centres. They then randomly assign the plants from each garden centre to all three fertilizer treatments. Is the experiment blocked? Is that a good idea?

10.6 Confounding

Professor Stephen Ceci of Cornell University performed an experiment to investigate the effect of a teacher's classroom style on student evaluations. He taught a class in developmental psychology during two successive terms to a total of 472 students in two very similar classes. He kept everything about his teaching identical (same text, same syllabus, same office hours, etc.) and modified only his style in class. During the fall term, he maintained a subdued demeanour. During the spring term, he used expansive gestures and lectured with more enthusiasm, varying his vocal pitch and using more hand gestures. He administered a standard student evaluation form at the end of each term.

The students in the fall term class rated him only an average teacher. Those in the spring term class rated him an excellent teacher, praising his knowledge and accessibility, and even the quality of the textbook. On the question "How much did you learn in the course?" the average response changed from 2.93 to 4.05 on a five-point scale.[8]

How much of the difference he observed was due to his difference in manner, and how much might have been due to the season of the year? Fall term in Ithaca, New York (home of Cornell University), starts out colourful and pleasantly warm, but ends cold and bleak. Spring term starts out bitter and snowy and ends with blooming flowers and singing birds. Might students' overall happiness have been affected by the season and reflected in their evaluations?

Unfortunately, there's no way to tell. Nothing in the data enables us to tease apart these two effects, because all the students who experienced the subdued manner did so during the fall term and all who experienced the expansive manner did so during the spring. When the levels of one factor are associated with the levels of another factor, we say that these two factors are **confounded**.

In some experiments, such as this one, it's just not possible to avoid some confounding. Professor Ceci could have randomly assigned students to one of two classes during the same term, but then we might question whether mornings or afternoons were better, or whether he really delivered the same class the second time (after practising on the first class). Or he could have had another professor deliver the second class, but that would have raised more serious issues about differences in the two professors and concern over more serious confounding.

For Example CONFOUNDING

RECAP: After many dogs and cats suffered health problems caused by contaminated foods, we're trying to find out whether a newly formulated pet food is safe. Our experiment will feed some animals the new food and others a food known to be safe, and a veterinarian will check the response.

QUESTION: Why would it be a bad design to feed the test food to some dogs and the safe food to cats?

ANSWER: This would create confounding. We would not be able to tell whether any differences in animals' health were attributable to the food they had eaten or to differences in how the two species responded.

Our discussion of confounding may remind you of the problem of lurking variables we discussed in Chapters 6 and 8, where we explained how lurking variables can operate via common response or confounding to make causal connections difficult to establish. In regression and observational studies, we can only observe associations between variables. Although we can't demonstrate a causal relationship, we often imagine whether x could cause y. We can be misled by a lurking variable associated with both. In a designed experiment, we hope to show that the factor causes a response. But we can be misled by a confounding variable that's associated with the factor and may also cause or contribute to the differences we observe in the response. Effects of different variables then become difficult or impossible to disentangle.

---

[8]But the two classes performed almost identically well on the final exam.

In a designed experiment, the experimenter *assigns* treatments at random to subjects rather than just observing them, in an attempt to neutralize the effects of all possible lurking variables. But problems can arise that ruin that perfect plan. For example, because Professor Ceci used one teaching style in the fall and the other in spring, he was unable to tell how much of his students' reactions was attributable to his teaching and how much to the weather. A confounding variable in an experiment, is a variable associated with an experimental factor that may also have some effect on the response. Because of the confounding, we find that we can't tell whether any effect we see was caused by our factor or by the confounding variable—or even by both working together.

It's worth noting that the role of blinding in an experiment is to combat a possible source of confounding. There's a risk that knowledge about the treatments could lead the subjects or those interacting with them to behave differently or could influence judgments made by the people evaluating the responses. That means we won't know whether the treatments really do produce different results or if we're being fooled by these confounding influences.

A two-factor example. Confounding can also arise from a badly designed multifactor experiment. Here's a classic. A credit card bank wanted to test the sensitivity of the market to two factors: the annual fee charged for a card and the annual percentage rate charged. Not wanting to scrimp on sample size, the bank selected 100 000 people at random from a mailing list. It sent out 50 000 offers with a low rate and no fee, and 50 000 offers with a higher rate and a $50 annual fee. Guess what happened? That's right—people preferred the low-rate, no-fee card. No surprise. In fact, they signed up for that card at over twice the rate as the other offer. And because of the large sample size, the bank was able to estimate the difference precisely. But the question the bank really wanted to answer was, "How much of the change was due to the rate, and how much was due to the fee?" Unfortunately, there's simply no way to separate out the two effects. If the bank had sent out all four possible different treatments—low rate with no fee, low rate with $50 fee, high rate with no fee, and high rate with $50 fee—each to 25 000 people, it could have learned about both factors and could have also seen what happens when the two factors occur in combination. (referred to as their *interaction*, it can be the most important discovery in a two-factor study).

ETHICAL EXPERIMENTS

The ethical treatment of both human and animal subjects in experiments has been a matter of international concern. The *Declaration of Helsinki*, developed by the World Medical Association, has been the foundation for most ethical policies for experiments on human subjects since it was originally proposed in 1964. The Declaration is founded on respect for the individual, subjects' rights to self-determination, and the right to make informed decisions about participating in the research. The subject's welfare must always take precedence over the interests of science and safety, and ethical considerations always take precedence over laws and regulations, according to the Declaration. It has been revised six times, and was last revised in 2013; the 1975 revision introduced the importance of oversight by an "independent committee," which led to the development of Institutional Review Boards (IRBs), found today at all research universities and centres in Canada.

Later revisions dealt with (among other issues) the use of placebo controls in medical trials. Is it ethical to administer an inert placebo to some subjects if a proven effective alternative exists that could be used as a control treatment? Knowledge of the principles of experiment design can help you understand this question.

The 2008 revision stated:
The benefits, risks, burdens and effectiveness of a new intervention must be tested against those of the best current proven intervention, except in the following circumstances:

- *The use of placebo, no treatment, is acceptable in studies where no current proven intervention exists; or*
- *Where for compelling and scientifically sound methodological reasons the use of placebo is necessary to determine the efficacy or safety of an intervention and the patients who receive placebo or no treatment will not be subject to any risk of serious or irreversible harm. Extreme care must be taken to avoid abuse of this option.*

In April 2008, the U.S. Food and Drug Administration (FDA) declared that clinical trials conducted outside of the U.S. did not need to comply with the Declaration of Helsinki. Agency officials said that the Declaration favours available treatments over placebos, which the FDA prefers in some circumstances, and that the periodic amendments to the Declaration could create confusion among drug companies. FDA spokesperson Rachel Behrman said, "We're saying that our regulations should not depend on any documents outside our control."[9] But others (see, for example, an editorial in the *British Medical Journal*[10]) claim that the decision was a response to pressures from drug companies that wished to be free to conduct tests in poor countries using inert placebos rather than more expensive alternative therapies, and not be required to provide effective drugs to all participants in a study at its conclusion, as the Declaration demands.

The 2013 revision includes increased protection for vulnerable groups and strengthens the explicit ethical requirements for placebo use when a known effective treatment is available. In Canada, the 'Tri-Council Policy Statement: Ethical Conduct for Research Involving Humans' is a joint policy statement by Canada's three federal research agencies incorporating the Helsinki principles as well as issues of inclusion of women in research, research involving indigenous peoples, human genetic research, stem cell research, and more.[11]

Section 10.6 Exercises

13. Confounded tips For the experiment of Exercise 3, name some confounding variables that might influence the experiment's results?

14. Tomatoes finis What factors might confound the results of the experiment in Exercise 4?

WHAT CAN GO WRONG?

- **Don't give up just because you can't run an experiment.** Sometimes we can't run an experiment because we can't identify or control the factors. Sometimes it would simply be unethical to run the experiment. (Consider randomly assigning students to take—and be graded in—a Statistics course deliberately taught to be boring and difficult or one that had an unlimited budget to use multimedia, real-world examples, and field trips to make the subject more interesting.) If we can't perform an experiment, often an observational study is a good choice.

- **Beware of confounding.** Use randomization whenever possible to ensure that the factors not in your experiment are not confounded with your treatment levels. Be alert to confounding that cannot be avoided, and report it along with your results.

- **Bad things can happen even to good experiments.** Protect yourself by recording additional information. An experiment in which the air conditioning failed for two weeks, affecting the results, was saved by recording the temperature (although that was not originally one of the factors) and estimating the effect the higher temperature had on the response.[12]

(continued)

---

[9]*Science*, 9 May 2008: Vol. 320: 5877, p. 731.

[10]*British Medical Journal*, 2009: 338:b1559.

[11]World Medical Association, Declaration of Helsinki, https://www.wma.net/policies-post/wma-declaration-of-helsinki-ethical-principles-for-medical-research-involving-human-subjects/

[12]R. D. DeVeaux and M. Szelewski, 1989, "Optimizing automatic splitless injection parameters for gas chromatographic environmental analysis." *Journal of Chromatographic Science* 27:9, 513–518.

It's generally good practice to collect as much information as possible about your experimental units and the circumstances of the experiment. For example, in the nail polish experiment, it would be wise to record details (temperature, humidity) that might affect the durability of the polish on the acrylic nails. Sometimes we can use this extra information during the analysis to reduce biases.

■ **Don't spend your entire budget on the first run.** Just as it's a good idea to pretest a survey, it's always wise to try a small pilot experiment before running the full-scale experiment. You may learn, for example, how to choose factor levels more effectively, about effects you forgot to control, and about unanticipated confounding.

What Have We Learned?

Recognize observational studies:

■ A retrospective study looks at an outcome in the present and looks for facts in the past that relate to it.

■ A prospective study selects subjects and follows them as events unfold.

Know the elements of a designed randomized experiment:

■ *Experimental units* (sometimes called *subjects or participants*, when humans) are assigned at random to *treatments*.

■ The experimenter manipulates *factors*, setting them to specified *levels* to establish the treatments.

■ A quantitative *response variable* is measured or observed for each experimental unit.

■ We can attribute statistically significant differences in the response to the differences among the treatments.

State and apply the Four Principles of Experimental Design:

■ *Control* sources of variation other than the factors being tested. Make the conditions as similar as possible for all treatment groups except for differences among the treatments.

■ *Randomize* the assignment of participants to treatments.

■ *Replicate* by applying each treatment to more than one experimental unit.

■ *Block* the experiment by grouping together experimental units that are similar in important ways that you cannot control.

Work with *blinding* and *control groups*:

■ A *single-blind* study is one in which *either* all those who can affect the results *or* all those who evaluate the results are kept ignorant of which subjects receive which treatments.

■ A *double-blind* study is one in which *both* all who can affect the results *and* all who evaluate the results are ignorant of which subjects receive which treatments.

■ A *control group* is assigned to a null treatment or to the best available alternative treatment.

■ Control participants are often administered a *placebo* or null treatment that mimics the treatment being studied but is known to be inactive. This is one way to blind participants.

Understand the differences between experiments and surveys.

■ Surveys try to estimate facts (parameter) about a population, so they require a representative random sample from that population.

■ Experiments try to estimate the differences in the effects of treatments. They randomize a group of experimental units to treatments, but there is no need for the experimental units to be a representative sample from a population.

■ Be alert for possible confounding due to a variable that is not under control affecting the responses differentially.

Review of Terms

Observational study	A study based on data in which no manipulation of factors has been employed (p. 320).
Retrospective study	An observational study in which subjects are selected and then their previous conditions or behaviours are determined. Retrospective studies need not be based on random samples and they usually focus on estimating differences between groups or associations between variables (p. 320).
Prospective study	An observational study in which subjects are followed to observe future outcomes. Because no treatments are deliberately applied, a prospective study is not an experiment. Nevertheless, prospective studies typically focus on estimating differences among groups that might appear as the groups are followed during the course of the study (p. 320).
Experiment	An experiment *manipulates* factor levels to create treatments, *randomly assigns* subjects to these treatment levels, and then *compares* the responses of the subject groups across treatment levels (p. 321).
Random assignment	To be valid, an experiment must assign experimental units to treatment groups at random using some form of randomization (p. 321).
Factor	A variable whose levels are manipulated by the experimenter. Experiments attempt to discover the effects that differences in factor levels may have on the responses of the experimental units (p. 321).
Response variable	A variable whose values are compared across different treatments. In a randomized experiment, large response differences can be attributed to the effect of differences in treatment level (p. 321).
Experimental units	Individuals on whom an experiment is performed. Usually called *subjects* or *participants* when they are human (p. 321).
Level	The specific values that the experimenter chooses for a factor (p. 321).
Treatment	The process, intervention, or other controlled circumstance applied to randomly assigned experimental units. Treatments are the different levels of a single factor or are made up of combinations of levels of two or more factors (p. 322).
Principles of experimental design	■ **Control** aspects of the experiment that we know may have an effect on the response, but that are not the factors being studied (p. 323). ■ **Randomize** subjects to treatments to even out effects that we cannot control (p. 324). ■ **Replicate** over as many subjects as possible. Results for a single subject are just anecdotes. If, as often happens, the subjects of the experiment are not a representative sample from the population of interest, replicate the entire study with a different group of subjects, preferably from a different part of the population (p. 324). ■ **Block** whenever possible to reduce the effects of identifiable attributes of the subjects that cannot be controlled (p. 324).
Completely randomized design (CRD)	An experimental design in which experimental units are assigned to treatments completely at random, with all possible divisions of subjects into groups equally likely. (p. 326).
Statistically significant	When an observed difference is too large for us to believe that it is likely to have occurred just naturally or purely by chance, we consider the difference to be statistically significant. Subsequent chapters will show specific calculations and give rules, but the principle remains the same (p. 327).
Control group	The experimental units assigned to a baseline treatment level (called the "control treatment"), typically either the default treatment, which is well understood, or a null, placebo treatment. Their responses provide a basis for comparison (p. 329).
Blinding	Any individual associated with an experiment who is not aware of how subjects have been allocated to treatment groups is said to be blinded (p. 330).
Single-blind/Double-blind	There are two main classes of individuals who can affect the outcome of an experiment: ■ those who could *influence the results* (the subjects, treatment administrators, or technicians) ■ those who *evaluate the results* (judges, treating physicians, etc.)

When every individual in *either* of these classes is blinded, an experiment is said to be single-blind. When everyone in *both* classes is blinded, we call the experiment double-blind (p. 330).

Placebo	A treatment known to have no effect, administered so that all groups experience the same conditions. Many subjects respond to such a treatment (a response known as the placebo effect). Only by comparing with a placebo we can be sure that the observed effect of a treatment is not due simply to the placebo effect (p. 331).
Placebo effect	The tendency of many human subjects (often 20% or more of experiment subjects) to show a response when administered a placebo (p. 331).
Block	When groups of experimental units are similar, it is often a good idea to gather them together into blocks. By blocking, we isolate the variability attributable to the differences between the blocks so that we can see the differences caused by the treatments more clearly (p. 332).
Randomized block design	An experiment design in which participants are randomly assigned to treatments within each block (p. 332).
Matching	In a retrospective or prospective study, subjects who are similar in ways not under study may be matched and then compared with each other on the variables of interest. Matching, like blocking, reduces unwanted variation (p. 332).
Confounding	When the levels of one factor are associated with the levels of another factor in such a way that their effects cannot be separated, we say that these two factors are confounded (p. 335).

On the Computer EXPERIMENTS

Most experiments are analyzed with a Statistics package. You should almost always display the results of a comparative experiment with a quantitative response variable using side-by-side boxplots. You may also want to display the means and standard deviations of the treatment groups in a table.

The analyses offered by statistics packages for comparative randomized experiments fall under the general heading of Analysis of Variance, usually abbreviated ANOVA. These analyses are beyond the scope of this chapter, but will be discussed in later chapters of this book.

Chapter Exercises

15. Social Media usage and grades For her Statistics class experiment, researcher K. Morrison decided to study how usage of social media affected Grade 12 students' performance in school. She proposed to collect information from a random sample of Grade 12 students and examine the relationship between the average amount of time spent on social media each day and the average graduating grade.

a) Is this an experiment? If not, what kind of study is it?

b) If there is a relationship between the average amount of time spent on social media each day and the average graduating grade, why can't we conclude that the differences in average graduating grades are caused by differences in average amount of time spent on social media each day?

16. Heart attacks and height Researchers who examined the health records of thousands of males found that men who died of myocardial infarction (heart attack) tended to be shorter than men who did not.

a) Is this an experiment? If not, what kind of study is it?

b) Is it correct to conclude that shorter men are at higher risk for heart attack? Explain.

17. MS and vitamin D Multiple sclerosis (MS) is an autoimmune disease that strikes more often the farther people live from the equator. Could vitamin D—which most people get from the sun's ultraviolet rays—be a factor? Researchers compared vitamin D levels in blood samples from 150 U.S. military personnel who have developed MS with blood samples of nearly 300 who have not. The samples were taken, on average, five years before the disease was diagnosed. Those with the highest blood vitamin D levels had a 62% lower risk of MS than those with the lowest levels. (The link was only in whites, not in blacks or Hispanics.)

a) What kind of study was this?

b) Is that an appropriate choice for investigating this problem? Explain.

c) Who were the subjects?

d) What were the variables?

18. Super Bowl commercials When spending large amounts to purchase advertising time, companies want to know what audience they'll reach. In January 2013, a poll asked 1008 American adults whether they planned to watch the upcoming Super Bowl. Men and women were asked separately whether they were looking forward more to the football game or to watching the commercials. Among the men, 16% were planning to watch and were looking forward primarily to the commercials. Among women, 30% were looking forward primarily to the commercials.

a) Was this a stratified sample or a blocked experiment? Explain.

b) Was the design of the study appropriate for the advertisers' question?

19. Menopause Researchers studied the herb black cohosh as a treatment for hot flashes caused by menopause. They randomly assigned 351 women aged 45 to 55 who had reported at least two hot flashes a day to one of five groups: (1) black cohosh, (2) a multiherb supplement with black cohosh, (3) the multiherb supplement plus advice to consume more soy foods, (4) estrogen replacement therapy, or (5) a placebo. After a year, only the women given estrogen replacement therapy had symptoms different from those of the placebo group.[13]

a) What kind of study was this?

b) Is that an appropriate choice for this problem?

c) Who were the subjects?

d) Identify the treatment and response variables.

20. Honesty Coffee stations in offices often ask users to leave money in a tray to pay for their coffee, but many people cheat. Researchers at Newcastle University replaced the picture of flowers on the wall behind the coffee station with a picture of staring eyes. They found that the average contribution increased significantly above the well-established standard when people felt they were being watched, even though the eyes were patently not real.[14]

a) Was this a survey, an observational study, or an experiment? How can you tell?

b) Identify the variables.

c) What does "increased significantly" mean in a statistical sense?

21–40. What's the design? *Read each brief report of statistical research, and identify*

a) whether it was an observational study or an experiment.

If it was an observational study, identify (if possible)

b) whether it was retrospective or prospective.

c) the subjects studied and how they were selected.

d) the response variable of interest.

e) the nature and scope of the conclusion the study can reach.

If it was an experiment, identify (if possible)

b) the subjects studied.

c) the factor(s) in the experiment and the number of levels for each.

d) the number of treatments.

e) the response variable measured.

f) the design (completely randomized, blocked, or matched).

g) whether it was blind (or double-blind).

h) the nature and scope of the conclusion the experiment can reach.

21. Over a four-month period, among 30 people with bipolar disorder, patients who were given a high dose (10 g/day) of omega-3 fats from fish oil improved more than those given a placebo.[15]

22. The leg muscles of men aged 60 to 75 were 50% to 80% stronger after they had participated in a 16-week, high-intensity resistance-training program twice a week.[16]

23. In a test of roughly 200 men and women, those with moderately high blood pressure (averaging 164/89 mm Hg) did worse on tests of memory and reaction time than those with normal blood pressure.[17]

24. Among a group of disabled women aged 65 and older who were tracked for several years, those who had a

[13]*Annals of Internal Medicine* 145:12, 869–897.

[14]*New York Times*, December 10, 2006.

[15]*Archives of General Psychiatry* 56 [1999], 407.

[16]*Journal of Gerontology* 55A [2000], B336

[17]*Hypertension* 36 [2000], 1079.

vitamin B12 deficiency were twice as likely to suffer severe depression as those who did not.[18]

25. An examination of the medical records of more than 360 000 Swedish men showed that those who were overweight or who had high blood pressure had a higher risk of kidney cancer.[19]

26. To research the effects of "dietary patterns" on blood pressure in 459 subjects, subjects were randomly assigned to three groups and had their meals prepared by dieticians. Those who were fed a diet low in fat and cholesterol and high in fruits, vegetables, and low-fat dairy foods (known as the DASH diet) lowered their systolic blood pressure by an average of 6.7 points compared with subjects fed a control diet.

27. After menopause, some women take supplemental estrogen. There is some concern that if these women also drink alcohol, their estrogen levels will rise too high. Twelve volunteers who were receiving supplemental estrogen were randomly divided into two groups, as were 12 other volunteers not on estrogen. In each case, one group drank an alcoholic beverage, the other a nonalcoholic beverage. An hour later, everyone's estrogen level was checked. Only those on supplemental estrogen who drank alcohol showed a marked increase.

28. Is diet or exercise effective in combating insomnia? Some believe that cutting out desserts can help alleviate the problem, while others recommend exercise. Forty volunteers suffering from insomnia agreed to participate in a month-long test. Half were randomly assigned to a special no-desserts diet; the others continued desserts as usual. Half of the people in each of these groups were randomly assigned to an exercise program, while the others did not exercise. Those who ate no desserts and engaged in exercise showed the most improvement.

29. Some gardeners prefer to use nonchemical methods to control insect pests in their gardens. Researchers have designed two kinds of traps and want to know which design will be more effective. They randomly choose 10 locations in a large garden and place one of each kind of trap at each location. After a week, they will count the number of bugs in each trap.

30. Researchers have linked an increase in the incidence of breast cancer in Italy to dioxin released by an industrial accident in 1976. The study identified 981 women who lived near the site of the industrial explosion and were under age 40 at the time. Fifteen of the women had developed breast cancer at an unusually young average age of 45. Medical records showed that these women had heightened concentrations of dioxin in their blood and that each tenfold increase in dioxin level was associated with a doubling of the risk of breast cancer.[20]

31. In 2002, the journal *Science* reported that a study of women in Finland indicated that having sons shortened the lifespans of mothers by about 34 weeks per son, but that daughters helped to lengthen the mothers' lives. The data came from church records from the period 1640 to 1870.

32. In 2001, a report in the *Journal of the American Cancer Institute* indicated that women who work nights have a 60% greater risk of developing breast cancer. Researchers based these findings on the work histories of 763 women with breast cancer and 741 women without the disease.

33. The May 4, 2000, issue of *Science News* reported that, contrary to popular belief, depressed individuals cry no more often in response to sad situations than nondepressed people. Researchers studied 23 men and 48 women with major depression, and 9 men and 24 women with no depression. They showed the subjects a sad film about a boy whose father had died, noting whether or not the subjects cried. Women cried more often than men, but there were no significant differences between the depressed and nondepressed groups.

34. Scientists at a major pharmaceutical firm investigated the effectiveness of an herbal compound to treat the common cold. They exposed each subject to a cold virus, then gave him or her either the herbal compound or a sugar solution known to have no effect on colds. Several days later they assessed the patients' conditions, using a cold severity scale ranging from 0 to 5. They found no evidence of benefits associated with the compound.

35. Scientists examined the glycogen content of rats' brains at the rats' normal bedtimes and after they had been kept awake for an extra 6, 12, or 24 hours. The scientists found that glycogen was 38% lower among rats that had been sleep-deprived for 12 hours or more, and that the levels recovered during subsequent sleep. These researchers speculate that we may need to sleep in order to restore the brain's energy fuel.[21]

36. Some people who race greyhounds give the dogs large doses of vitamin C in the belief that the dogs will run faster. Investigators at the University of Florida tried three different diets in random order on each of five

[18]*American Journal of Psychiatry* 157 [2000], 715.
[19]*New England Journal of Medicine* 3434 [2000], 1305.
[20]*Science News*, Aug. 3, 2002.
[21]*Science News*, July 20, 2002.

racing greyhounds. They were surprised to find that when the dogs ate high amounts of vitamin C they ran more slowly.[22]

37. Some people claim they can get relief from migraine headache pain by drinking a large glass of ice water. Researchers plan to enlist several people who suffer from migraines in a test. When a participant experiences a migraine headache, he or she will take a pill that may be a standard pain reliever or a placebo. Half of each group will also drink ice water. Participants will then report the level of pain relief they experience.

38. Weight is an issue for both humans and their pets. A dog food company wants to compare a new lower-calorie food with their standard dog food to see if it's effective in helping inactive dogs maintain a healthy weight. They have found several dog owners willing to participate in the trial. The dogs have been classified as small, medium, or large breeds, and the company will supply some owners of each size of dog with one of the two foods. The owners have agreed not to feed their dogs anything else for a period of 6 months, after which the dogs' weights will be checked.

39. Athletes who had suffered hamstring injuries were randomly assigned to one of two exercise programs. Those who engaged in static stretching returned to sports activity in a mean of 37.4 days ($SD = 27.6$ days). Those assigned to a program of agility and trunk stabilization exercises returned to sports in a mean of 22.2 days ($SD = 8.3$ days).[23]

40. Pew Research compared respondents to an ordinary five-day telephone survey with respondents to a four-month-long rigorous survey designed to generate the highest possible response rate. They were especially interested in identifying any variables for which those who responded to the ordinary survey were different from those who could be reached only by the rigorous survey.

41. **Omega-3** Exercise 21 described an experiment that showed that high doses of omega-3 fats might be of benefit to people with bipolar disorder. The experiment involved a control group of subjects who received a placebo. Why didn't the experimenters just give everyone the omega-3 fats to see if they improved?

42. **Insomnia** Exercise 28 described an experiment that showed that getting exercise helped people sleep better. The experiment involved other groups of subjects who didn't exercise. Why didn't the experimenters just have everyone exercise and see if their ability to sleep improved?

43. **Omega-3 revisited** Exercises 21 and 41 described an experiment investigating a dietary approach to treating bipolar disorder. Researchers randomly assigned 30 subjects to two treatment groups, one group taking a high dose of omega-3 fats and the other a placebo.
    a) Why was it important to randomize in assigning the subjects to the two groups?
    b) What would be the advantages and disadvantages of using 100 subjects instead of 30?

44. **Insomnia again** Exercises 28 and 42 described an experiment investigating the effectiveness of exercise in combating insomnia. Researchers randomly assigned half of the 40 volunteers to an exercise program.
    a) Why was it important to randomize in deciding who would exercise?
    b) What would be the advantages and disadvantages of using 100 subjects instead of 40?

45. **Omega-3, finis** Exercises 21, 41, and 43 described an experiment investigating the effectiveness of omega-3 fats in treating bipolar disorder. Suppose some of the 30 subjects were very active people who walked a lot or got vigorous exercise several times a week, while others tended to be more sedentary, working office jobs and watching a lot of television. Why might researchers choose to block the subjects by activity level before randomly assigning them to the omega-3 and placebo groups?

46. **Insomnia, at last** Exercises 28, 42, and 44 described an experiment investigating the effectiveness of exercise in combating insomnia. Suppose some of the 40 subjects had maintained a healthy weight, but others were quite overweight. Why might researchers choose to block the subjects by weight level before randomly assigning some of each group to the exercise program?

47. **Tomatoes encore** Describe a strategy to randomly split 24 tomato plants into the three groups for the completely randomized single factor experiment of Exercise 4.

48. **Tomatoes another encore** You want to compare the tastiness and juiciness of tomatoes grown with three amounts of a new fertilizer: none, half the recommended amount, and the full recommended amount. Each tomato plant has two possible levels of irrigation: daily watering and no watering. Describe a strategy to randomly assign 24 tomato plants to the six treatments.

49. **Shoes** A running-shoe manufacturer wanted to test the effect of its new sprinting shoe on 100-metre dash times. The company sponsored five athletes who were running the 100-metre dash in the 2012 Summer Olympic Games. To test the shoe, it had all five runners run the race with a competitor's shoe and then again with its new shoe. The company used the difference in times as the response variable.

[22]*Science News*, July 20, 2002.

[23]*Journal of Orthopaedic & Sports Physical Therapy* 34 [March 2004], 3.

a) Suggest some improvements to the design.

b) Why might the shoe manufacturer not be able to generalize the results they found to all runners?

50. Swimsuits A swimsuit manufacturer wants to test the speed of its newly designed suit. The company designs an experiment by having six randomly selected Olympic swimmers swim as fast as they can with their old swimsuit first and then swim the same event again with the new, expensive swimsuit. The company will use the difference in times as the response variable. Criticize the experiment and point out some of the problems with generalizing the results.

51. Hamstrings Exercise 39 discussed an experiment to see if the time it took athletes with hamstring injuries to be able to return to sports was different depending on which of two exercise programs they engaged in.

a) Explain why it was important to *randomly* assign the athletes to the two different treatments.

b) There was no control group consisting of athletes who did not participate in a special exercise program. Explain the advantage of including such a group in this experiment.

c) How might blinding have been used in this experiment?

d) One group returned to sports activity in a mean of 37.4 days ($SD = 27.6$ days) and the other in a mean of 22.2 days ($SD = 8.3$ days). Do you think this difference is statistically significant? Explain.

52. Diet and blood pressure Exercise 26 reported on an experiment that showed that subjects fed the DASH diet were able to lower their blood pressure by an average of 6.7 points compared to a group fed a "control diet." All meals were prepared by dieticians.

a) Why were the subjects randomly assigned to the diets instead of letting people pick what they wanted to eat?

b) Why were the meals prepared by dieticians?

c) Why did the researchers need the control group? If the DASH diet group's blood pressure was lower at the end of the experiment than at the beginning, wouldn't that prove the effectiveness of that diet?

d) What additional information would you want to know to decide whether an average reduction in blood pressure of 6.7 points is statistically significant?

53. Mozart Will listening to a Mozart piano sonata make you smarter? In a 1995 study published in the journal *Neuroscience Letters*, Rauscher, Shaw, and Ky reported that when students were given a spatial reasoning section of a standard IQ test, those who listened to Mozart for 10 minutes improved their scores more than those who simply sat quietly.

Source: Rauscher FH, Shaw GL, Ky KN. Listening to Mozart enhances spatial-temporal reasoning: toward a neurophysiological basis. *Neurosci Lett* 1995;185: 44-7

a) These researchers said the differences were statistically significant. Explain what that means in this context.

b) Steele, Bass, and Crook tried to replicate the original study. In their study, also published in *Neuroscience Letters* (1999), the subjects were 125 college students who participated in the experiment for course credit. Subjects first took the test. Then they were assigned to one of three groups: listening to a Mozart piano sonata, listening to music by Philip Glass, and sitting for 10 minutes in silence. Three days after the treatments, they were retested. Draw a diagram displaying the design of this experiment.

c) These boxplots show the differences in score before and after treatment for the three groups. Did the Mozart group show improvement?

d) Do you think the results prove that listening to Mozart is beneficial? Explain.

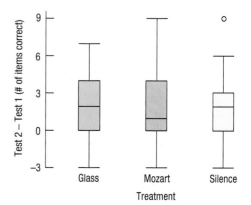

54. Mozart, 2nd movement The *New York Times* (Sept. 9, 1991) reported that "A sake brewer [near] Tokyo is marketing a Mozart brand of sake, asserting that Mozart is regularly played at the brewery." "It seems that the malt grows well especially with Mozart music in the background," the company director said. Suppose, just for the sake (as it were) of discussion, you wished to design an experiment to test whether this is true. Assume you have the full cooperation of the sake brewery. Specify how you would design the experiment. Indicate factors and response and how they would be measured, controlled, or randomized.

55. Frumpies The makers of Frumpies, "the breakfast of rug rats," want to improve their marketing, so they consult you.

a) They first want to know what fraction of children ages 10 to 13 like their celery-flavoured cereal. What kind of study should they perform?

b) They are thinking of introducing a new flavour, maple-marshmallow Frumpies, and want to know whether children will prefer the new flavour to the old one. Design a completely randomized experiment to investigate this question.

c) They suspect that children who regularly watch the Saturday morning cartoon show starring Frump, the flying teenage warrior rabbit who eats Frumpies in every episode, may respond differently to the new flavour. How would you take that into account in your design?

56. Contrast Bath treatments use the immersion of an injured limb alternately in water of two contrasting temperatures. Those who use the method claim that it can reduce swelling. Researchers compared three treatments: contrast baths and exercise (1), contrast baths alone (2), and exercise alone (3).[24] They reported the following boxplots comparing the change in hand volume after treatment:

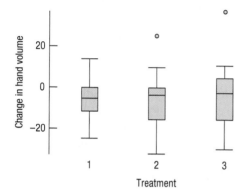

a) The researchers concluded that the differences were not statistically significant. Explain what that means in context.

b) The title says that the study was randomized and controlled. Explain what that probably means for this study.

c) The study did not use a placebo treatment. What was done instead? Do you think that was an appropriate choice? Explain.

57. Wine A 2001 Danish study published in the *Archives of Internal Medicine* casts significant doubt on suggestions that adults who drink wine have higher levels of "good" cholesterol and fewer heart attacks. These researchers followed a group of individuals born at a Copenhagen hospital between 1959 and 1961 for 40 years. Their study found that in this group the adults who drank wine were richer and better educated than those who did not.

a) What kind of study was this?

b) It is generally true that people with high levels of education and high socio-economic status are healthier than others. How does this call into question the supposed health benefits of wine?

c) Can studies such as these prove causation (that wine helps prevent heart attacks, that drinking wine makes one richer, that being rich helps prevent heart attacks, etc.)? Explain.

58. Swimming Recently, a group of adults who swim regularly for exercise were evaluated for depression. It turned out that these swimmers were less likely to be depressed than the general population. The researchers said the difference was statistically significant.

a) What does "statistically significant" mean in this context?

b) Is this an experiment or an observational study? Explain.

c) News reports claimed this study proved that swimming can prevent depression. Explain why this conclusion is not justified by the study. Include an example of a possible confounding variable.

d) But perhaps it is true. We wonder if exercise can ward off depression, and whether anaerobic exercise (like weight training) is as effective as aerobic exercise (like swimming). We find 120 volunteers not currently engaged in a regular program of exercise. Design an appropriate experiment.

59. Dowsing A water dowser claims to be able to sense the presence of water using a forked stick. Suppose we wish to set up an experiment to test his ability. We get 20 identical containers, fill some with water, and ask the dowser to tell which ones are full and which ones are empty.

a) How will we randomize this procedure?

b) The dowser correctly identifies the contents of 12 out of 20 containers. Do you think this level of success is statistically significant? Explain.

c) How many correct identifications (out of 20) would the dowser have to make to convince you that the forked-stick trick works? Explain.

60. Healing A medical researcher suspects that giving post-surgical patients large doses of vitamin E will speed their recovery times by helping their incisions heal more quickly. Design an experiment to test this conjecture. Be sure to identify the factors, levels, treatments, response variable, and the role of randomization.

61. Reading Some schools teach reading using phonics (the sounds made by letters) and others using whole language (word recognition). Suppose a school district wants to know which method works better. Suggest a design for an appropriate experiment.

62. Gas mileage Do cars get better gas mileage with premium instead of regular unleaded gasoline? While it might be possible to test some engines in a laboratory setting, we'd rather use real cars and real drivers in real day-to-day driving, so we get 20 volunteers. Design the experiment.

---

[24]R. G. Janssen, D. A. Schwartz, and P.F. Velleman, 2009, "A randomized controlled study of contrast baths on patients with carpal tunnel syndrome." *Journal of Hand Therapy.*

63. Weekend deaths A study published in the *New England Journal of Medicine* (August 2001)[25] suggests that it's dangerous to enter a hospital on a weekend. During a 10-year period, researchers tracked over 4 million emergency admissions to hospitals in Ontario, Canada. Their findings revealed that patients admitted on weekends had a much higher risk of death than those who went to the emergency room on weekdays.
a) The researchers said the difference in death rates was "statistically significant." Explain what that means in this context.
b) What kind of study was this? Explain.
c) If you think you're quite ill on a Saturday, should you wait until Monday to seek medical help? Explain.
d) Suggest some possible explanations for this troubling finding.

64. Shingles A research doctor has discovered a new ointment that she believes will be more effective than the current medication in the treatment of shingles (a painful skin rash). Eight patients have volunteered to participate in the initial trials of this ointment. You are the statistician hired as a consultant to help design a completely randomized experiment.
a) Describe how you will conduct this experiment.
b) Suppose the eight patients' last names start with the letters A to H. Using the random numbers listed below, show which patients you will assign to each treatment. Explain your randomization procedure clearly.

    41 098   18 329   78 458   31 685   55 259

c) Can you make this experiment double-blind? If so, explain how.
d) The initial experiment revealed that males and females may respond differently to the ointment. Further testing of the drug's effectiveness is now planned, and many patients have volunteered. What changes in your first design, if any, would you make for this second stage of testing?

65. Beetles Hoping to learn how to control crop damage by a certain species of beetle, a researcher plans to test two different pesticides in small plots of corn. A few days after application of the chemicals, he'll check the number of beetle larvae found on each plant. The researcher wants to know whether either pesticide works and whether there is a significant difference in effectiveness between them. Design an appropriate experiment.

66. LSAT prep Can test prep courses actually help raise LSAT scores? One organization says that the 30 students they tutored achieved an average gain of 15 points when they retook the test.
a) Explain why this does not necessarily prove that the test prep course caused the scores to go up.
b) Propose a design for an experiment that could test the effectiveness of the test prep course.
c) Suppose you suspect that the test prep course might be more helpful for students whose initial scores were particularly low. How would this affect your proposed design?

67. Safety switch An industrial machine requires an emergency shutoff switch that must be designed so that it can be easily operated with either hand. Design an experiment to find out whether workers will be able to deactivate the machine as quickly with their left hands as with their right hands. Be sure to explain the role of randomization in your design.

68. Washing clothes A consumer group wants to test the effectiveness of a new "organic" laundry detergent and make recommendations to customers about how to best use the product. They intentionally get grass stains on 30 white T-shirts to see how well the detergent will clean them. They want to try the detergent in cold water and in hot water on both the "regular" and "delicates" wash cycles. Design an appropriate experiment, indicating the number of factors, levels, and treatments. Explain the role of randomization in your experiment.

69. Skydiving, anyone? A humour piece published in the *British Medical Journal* notes that we can't tell for sure whether parachutes are safe and effective because there has never been a properly randomized, double-blind, placebo-controlled study of parachute effectiveness in skydiving.[26] (Yes, this is the sort of thing statisticians find funny . . .) Suppose you were designing such a study:
a) What is the factor in this experiment?
b) What experimental units would you propose?[27]
c) Explain what would serve as a placebo for this study.
d) What would the treatments be?
e) What would be the response variable for such a study?
f) What sources of variability would you control?
g) How would you randomize this "experiment"?
h) How would you make the experiment double-blind?

[25]*Source:* Mortality among Patients Admitted to Hospitals on Weekends as Compared with Weekdays Chaim M. Bell, M.D., and Donald A. Redelmeier, M.D. N Engl J Med 2001; 345:663-668August 30, 2001DOI: 10.1056/NEJMsa003376

[26]Gordon Smith and Jill Pell, "Parachute use to prevent death and major trauma related to gravitational challenge: Systematic review of randomized control trials," *BMJ*, 2003:327.

[27]Don't include your Statistics instructor!

Just Checking ANSWERS

1. **a)** The factor was type of treatment for peptic ulcer.

   b) The response variable could be a measure of relief from gastric ulcer pain or an evaluation by a physician of the state of the disease.

   c) Treatments would be gastric freezing and some alternative control treatment.

   d) Treatments should be assigned randomly.

   e) No. The website reports "lack of effectiveness," indicating that no significant differences in patient healing were noted.

2. **a)** Neither the patients who received the treatment nor the doctor who evaluated them to see if they had improved knew what treatment they had received.

   b) The placebo is needed to accomplish blinding. The best alternative would be using body-temperature liquid rather than the freezing liquid.

   c) The researchers should block the subjects by the length of time they had had the ulcer, then randomly assign subjects in each block to the freezing and placebo groups.

MathXL **MyLab Statistics**	Go to MathXL at **www.mathxl.com** or MyLab™ Statistics at **www.mystatlab.com**. You can practise exercises for this chapter as often as you want. The guided solutions will help you find answers step by step. You'll find a personalized study plan available to you too!

Gathering Data

Quick Review

Before you can make a boxplot, calculate a mean, describe a distribution, or fit a line, you must have meaningful data to work with. Getting good data is essential to any investigation. No amount of clever analysis can make up for badly collected data. Here's a brief summary of the key concepts and skills:

- The way you gather data depends both on what you want to discover and on what is practical.

- To answer questions about a target population, collect information from a sample with a **survey** or poll.

  - Choose the sample randomly. Random sampling designs include simple, stratified, systematic, cluster, and multistage.

  - A simple random sample draws without restriction from the entire target population.

  - When there are subgroups within the population that may respond differently, use a stratified sample.

  - Avoid bias, a systematic distortion of the results. Sample designs that allow undercoverage or response bias and designs such as voluntary response or convenience samples don't faithfully represent the population.

  - Samples will naturally vary one from another. This sample-to-sample variation is called sampling error. Each sample only approximates the target population.

- **Observational studies** collect information from a sample, sometimes drawn from a target population

(like in a sample survey), sometimes just a sample of convenience ('e.g.,' sick patients in a hospital).

  - Retrospective studies collect and examine data on events that have already taken place. Prospective studies identify subjects in advance, then follow them to collect data as the data are created, perhaps over many years.

  - Observational studies can spot associations between variables but cannot establish cause and effect. It's impossible to eliminate the possibility of lurking or confounding variables.

- To see how different treatments influence a response variable, design an **experiment**.

  - Assign subjects to treatments randomly. If you don't assign treatments randomly, your experiment is not likely to yield valid results.

  - Control known sources of variation as much as possible. Reduce variation that cannot be controlled by using blocking, if possible.

  - Replicate the experiment, assigning several subjects to each treatment level.

  - If possible, replicate the entire experiment with an entirely different collection of subjects.

  - A well-designed experiment can provide evidence that changes in the factors cause changes in the response variable.

  Now for more opportunities to review these concepts and skills . . .

Review Exercises

1–18. What design? *Analyze the design of each research example reported. Is it a sample, an observational study, or an experiment? If a sample, what is the population, the parameter of interest, and the sampling procedure? If an observational study, was it retrospective or prospective? If*

an experiment, describe the factors, treatments, randomization, response variable, and any blocking, matching, or blinding that may be present. In each, what kind of conclusions can be reached?

1. Researchers identified 242 children in the Cleveland area who had been born prematurely (at about 29 weeks). They examined these children at age 8 and again at age 20, comparing them to another group of 233 children not born prematurely. According to their report, published in the *New England Journal of Medicine*, the "preemies" engaged in significantly less risky behaviour than the others. Differences between the groups showed up in the use of alcohol and marijuana, conviction of crimes, and teenage pregnancy.

2. The journal *Circulation* reported that among 1900 people who had heart attacks, those who drank an average of 19 cups of tea a week were 44% more likely than nondrinkers to survive at least three years after the attack.

3. Researchers at the Purina Pet Institute studied Labrador retrievers for evidence of a relationship between diet and longevity. At eight weeks of age, puppies of the same sex and weight were randomly assigned to one of two groups—a total of 48 dogs in all. One group was allowed to eat all they wanted, while the other group was fed a low-calorie diet (about 75% as much as the others). The median lifespan of dogs fed the restricted diet was 22 months longer than that of other dogs. (Source: *Science News* 161, no. 19)

4. The radioactive gas radon, found in some homes, poses a health risk to residents. To assess the level of contamination in their area, a municipal health department wants to test a few homes. If the risk seems high, they will publicize the results to emphasize the need for home testing. Officials plan to use the local property tax list to randomly choose 25 homes from various areas of the municipality.

5. Data were collected over a decade from 1021 men and women with a recent history of precancerous colon polyps. Participants were randomly assigned to receive folic acid (a B vitamin) or a placebo. The study concluded that those receiving the folic acid may actually increase their risk of developing additional precancerous growths. Previous studies suggested that taking folic acid may help to prevent colorectal cancer. (Source: *JAMA*, 2007, 297)

6. In a study appearing in the journal *Science*, a research team reported that plants in southern England are flowering earlier in the spring. Records of the first flowering dates for 385 species over a period of 47 years indicate that flowering has advanced an average of 15 days per decade, an indication of climate warming, according to the authors.

7. Fireworks manufacturers face a dilemma. They must be sure that the rockets work properly, but test firing a rocket essentially destroys it. On the other hand, not testing the product leaves open the danger that they sell a bunch of duds, leading to unhappy customers and loss of future sales. The solution, of course, is to test a few of the rockets produced each day, assuming that if those tested work properly, the others are ready for sale.

8. People who read the last page of a mystery novel first generally like stories better. Researchers recruited 819 college students to read short stories. For one story, they were given a spoiler paragraph beforehand. On the second and third story, the spoiler was incorporated as the opening paragraph or not given at all. Overall, participants liked the stories best after first reading spoilers. (Source: *Psychological Science*, August 12, 2011)

9. Does keeping a child's lunch in an insulated bag, even with ice packs, protect the food from warming to temperatures where germs can proliferate? Researchers used an electric temperature gun on 235 lunches at preschools 90 minutes before they were to be eaten. Of the lunches with ice packs, over 90% of them were at unsafe temperatures. The study was of particular interest because preschoolers develop up to four times as many foodborne infections as do adults. (Source: *Science News*, August 9, 2011)

10. Some doctors have expressed concern that men who have vasectomies seem more likely to develop prostate cancer. Medical researchers used a national cancer registry to identify 923 men who had had prostate cancer and 1224 men of similar ages who had not. Roughly one quarter of the men in each group had undergone a vasectomy, many more than 25 years before the study. The study's authors concluded that there is strong evidence that having the operation presents no long-term risk for developing prostate cancer. (Source: *Science News*, July 20, 2002)

11. Widely used antidepressants may reduce ominous brain plaques associated with Alzheimer's disease. In the study, mice genetically engineered to have large amounts of brain plaque were given a class of antidepressants that boost serotonin in the brain. After a single dose, the plaque levels dropped. After four months, the mice had about half the brain plaques as the mice that didn't take the drug. (Source: *Proceedings of the National Academy of Sciences*, August 22, 2011)

12. An artisan wants to create pottery that has the appearance of age. He prepares several samples of clay with four different glazes and test fires them in a kiln at three different temperature settings.

13. Tests of gene therapy on laboratory rats have raised hopes of stopping the degeneration of tissue that

characterizes chronic heart failure. Researchers at the University of California, San Diego, used hamsters with cardiac disease, randomly assigning 30 to receive the gene therapy and leaving the other 28 untreated. Five weeks after treatment, the gene therapy group's heart muscles stabilized, while those of the untreated hamsters continued to weaken. (Source: *Science News*, July 27, 2002)

14. People aged 50 to 71 were initially contacted in the mid-1990s to participate in a study about smoking and bladder cancer. Data were collected from more than 280 000 men and 186 000 women from eight U.S. states who answered questions about their health, smoking history, alcohol intake, diet, physical activity, and other lifestyle factors. When the study ended in 2006, about half the bladder cancer cases in adults age 50 and older were traceable to smoking. (Source: *Journal of the American Medical Association*, August 17, 2011)

15. An orange-juice processing plant will accept a shipment of fruit only after several hundred oranges selected from various locations within the truck are carefully inspected. If too many of those checked show signs of unsuitability for juice (bruised, rotten, unripe, etc.) the whole truckload is rejected.

16. A soft-drink manufacturer must be sure the bottle caps on the soft drinks are fully sealed and will not come off easily. Inspectors pull a few bottles off the production line at regular intervals and test the caps. If they detect any problems, they will stop the bottling process to adjust or repair the machine that caps the bottles.

17. Physically fit people are less likely to die of cancer. A report in the May 2002 issue of *Medicine and Science in Sports and Exercise* followed 25 892 men aged 30 to 87 for 10 years. The most physically fit men had a 55% lower risk of death from cancer than the least fit group.

18. Does the use of computer software in Introductory Statistics classes lead to better understanding of the concepts? A professor teaching two sections of Statistics decides to investigate. She teaches both sections using the same lectures and assignments, but gives one class statistics software to help them with their homework. The classes take the same final exam, and graders do not know which students used computers during the semester. The professor is also concerned that students who have had calculus may perform differently from those who have not, so she plans to compare software versus no-software scores separately for these two groups of students.

19. **Tips** In restaurants, servers rely on tips as a major source of income. Does serving candy after the meal produce larger tips? To find out, two waiters determined randomly whether or not to give candy to 92 dining parties. They recorded the sizes of the tips and reported that guests getting candy tipped an average of 17.8% of the bill, compared with an average tip of only 15.1% from those who got no candy.[1]
   a) Was this an experiment or an observational study? Explain.
   b) Is it reasonable to conclude that the candy caused guests to tip more? Explain.
   c) The researchers said the difference was statistically significant. Explain what that means in this context.

20. **Tips, second course** In another experiment to see if getting candy after a meal would induce customers to leave a bigger tip, a waitress randomly decided what to do with 80 dining parties. Some parties received no candy, some just one piece, and some two pieces. Others initially got just one piece of candy, and then the waitress suggested that they take another piece. She recorded the tips received, finding that, in general, the more candy, the higher the tip, but the highest tips (23%) came from the parties who got one piece and then were offered more.[2]
   a) Diagram this experiment.
   b) How many factors are there? How many levels?
   c) How many treatments are there?
   d) What is the response variable?
   e) Did this experiment involve blinding? Double blinding?
   f) In what way might the waitress, perhaps unintentionally, have biased the results?

21. **Timing** A Léger Marketing online poll conducted in June of 2013 showed Justin Trudeau of the Liberal party with a four-point lead over current Prime Minister Stephen Harper. Considering the survey design, can you think of any reasons why Stephen Harper might still win the election, were it to be held immediately?

22. **Laundry** An experiment to test a new laundry detergent, SparkleKleen, is being conducted by a consumer advocate group. They would like to compare its performance with that of a laboratory standard detergent they have used in previous experiments. They will stain 16 swatches of cloth with 2 tsp of a common staining compound and then use a well-calibrated optical scanner to detect the amount of the stain that is left after washing with detergent. To save time in the experiment, several suggestions have been made. Comment on the possible merits and drawbacks of each one.

---

[1]"Sweetening the till: The use of candy to increase restaurant tipping." *Journal of Applied Social Psychology* 32, no. 2 [2002]: 300–309.

[2]"Sweetening the till: The use of candy to increase restaurant tipping." *Journal of Applied Social Psychology* 32, no. 2 [2002]: 300–309.

a) Since data for the laboratory standard detergent are already available from previous experiments, for this experiment wash all 16 swatches with SparkleKleen, and compare the results with the previous data.

b) Use both detergents with eight separate runs each, but to save time, use only a 10-second wash time with very hot water.

c) To ease bookkeeping, successively run all of the standard detergent washes on eight swatches, then run all of the SparkleKleen washes on the other eight swatches.

d) Rather than run the experiment, use data from the company that produced SparkleKleen, and compare them with past data from the standard detergent.

23. **Homecoming** A university Statistics class conducted a survey concerning community attitudes about the school's large homecoming celebration. That survey drew its sample in the following manner: Telephone numbers were generated at random by selecting one of the local telephone exchanges (first three digits) at random and then generating a random four-digit number to follow the exchange. If a person answered the phone and the call was to a residence, then that person was taken to be the subject for interview. (Undergraduate students and those under voting age were excluded, as was anyone who could not speak English.) Calls were placed until a sample of 200 eligible respondents had been reached.

a) Did every telephone number that could occur in that community have an equal chance of being generated?

b) Did this method of generating telephone numbers result in a simple random sample (SRS) of local residences? Explain.

c) Did this method generate an SRS of local voters? Explain.

d) Did this method generate an unbiased sample of households? Explain.

24. **Youthful appearance** *Readers' Digest* reported results of several surveys that asked graduate students to examine photographs of men and women and try to guess their ages. Researchers compared these guesses with the number of times the people in the pictures reported having sexual intercourse. It turned out that those who had been more sexually active were judged as looking younger, and that the difference was described as "statistically significant." Psychologist David Weeks, who compiled the research, speculated that lovemaking boosts hormones that "reduce fatty tissue and increase lean muscle, giving a more youthful appearance."

a) What does "statistically significant" mean in this context?

b) Explain in statistical terms why you might be skeptical about Dr. Weeks's conclusion. Propose an alternative explanation for these results.

25. **Smoking and Alzheimer's** Medical studies indicate that smokers are less likely to develop Alzheimer's disease than people who never smoked.

a) Does this prove that smoking may offer some protection against Alzheimer's? Explain.

b) Offer an alternative explanation for this association.

c) How would you conduct a study to investigate this issue?

26. **Antacids** A researcher wants to compare the performance of three types of antacid in volunteers suffering from acid reflux disease. Because men and women may react differently to this medication, the subjects are split into two groups, by sex. Subjects in each group are randomly assigned to take one of the antacids or to take a sugar pill made to look the same. The subjects will rate their level of discomfort 30 minutes after eating.

a) What kind of design is this?

b) The experiment uses volunteers rather than a random sample of all people suffering from acid reflux disease. Does this make the results invalid? Explain.

c) How may the use of the placebo confound this experiment? Explain.

27. **Sex and violence** Does the content of a television program affect viewers' memory of the products advertised in commercials? Design an experiment to compare the ability of viewers to recall brand names of items featured in commercials during programs with violent content, sexual content, or neutral content.

28. **Pubs** In England, a Leeds University researcher said that the local watering hole's welcoming atmosphere helps men get rid of the stresses of modern life and is vital for their psychological well-being. Author of the report, Dr. Colin Gill, said rather than complain, women should encourage men to "pop out for a swift half." "Pub-time allows men to bond with friends and colleagues," he said. "Men need break-out time as much as women and are mentally healthier for it." Gill added that men might feel unfulfilled or empty if they had not been to the pub for a week. The report, commissioned by alcohol-free beer brand Kaliber, surveyed 900 men on their reasons for going to the pub. More than 40% said they went for the conversation, with relaxation and a friendly atmosphere being the other most common reasons. Only 1 in 10 listed alcohol as the overriding reason.

Let's examine this news story from a statistical perspective.

a) What are the Ws: *who, what, when, where, why?*

b) What population does the researcher think the study applies to?

c) What is the most important thing about the selection process that the article does *not* tell us?

d) How do *you* think the 900 respondents were selected? (Name a method of drawing a sample that is likely to have been used.)

e) Do you think the report that only 10% of respondents listed alcohol as an important reason for going to the pub might be a biased result? Why?

29. Age and party 2008 The Pew Research Center collected data from national exits polls conducted by NBC News after the 2008 U.S. presidential election. The following table shows information regarding voter age and party preference:

Age	Republican	Democrat	Other	Total
18–29	260	390	351	1001
30–44	320	379	300	999
45–64	329	369	300	998
65 +	361	392	251	1004
Total	1270	1530	1202	4002

a) What sampling strategy do you think the pollsters used? Explain.

b) What percentage of the people surveyed were Democrats?

c) Do you think this is a good estimate of the percentage of voters in the United States who are registered Democrats? Why or why not?

d) In creating this sample design, what question do you think the pollsters were trying to answer?

30. Bias? Political analyst Michael Barone has written that "conservatives are more likely than others to refuse to respond to polls, particularly those polls taken by media outlets that conservatives consider biased" (*The Weekly Standard,* March 10, 1997). The Pew Research Foundation tested this assertion by asking the same questions in a national survey run by standard methods and in a more rigorous survey that was a true SRS with careful follow-up to encourage participation. The response rate in the "standard survey" was 42%. The response rate in the "rigorous survey" was 71%.

a) What kind of bias does Barone claim may exist in polls?

b) What is the population for these surveys?

c) On the question of political position, the Pew researchers report the following table:

	Standard Survey	Rigorous Survey
Conservative	37%	35%
Moderate	40%	41%
Liberal	19%	20%

What makes you think these results are incomplete?

d) The Pew researchers report that differences between opinions expressed on the two surveys were not statistically significant. Explain what "statistically significant" means in this context.

31. Save the grapes Vineyard owners have problems with birds that like to eat the ripening grapes. Grapes damaged by birds cannot be used for winemaking (or much of anything else). Some vineyards use scarecrows to try to keep birds away. Others use netting that covers the plants. Owners really would like to know if either method works and, if so, which one is better. One owner has offered to let you use his vineyard this year for an experiment. Propose a design. Carefully indicate how you would set up the experiment, specifying the factor(s) and response variable.

32. Bats It's generally believed that baseball players can hit the ball farther with aluminum bats than with the traditional wooden ones. Is that true? And, if so, how much farther? Players on your local high school baseball team have agreed to help you find out. Design an appropriate experiment.

33. Acupuncture. Research reported in 2008 brings to light the effectiveness of treating chronic lower back pain with different methods. One-third of nearly 1200 volunteers were administered conventional treatment (drugs, physical therapy, and exercise). The remaining patients got 30-minute acupuncture sessions. Half of these patients were punctured at sites suspected of being useful and half received needles at other spots on their bodies. Comparable shares of each acupuncture group, roughly 45%, reported decreased back pain for at least six months after their sessions ended. This was almost twice as high as those receiving the conventional therapy, leading the researchers to conclude that results were statistically significant.

a) Why did the researchers feel it was necessary to have some of the patients undergo a "fake" acupuncture?

b) Because patients had to consent to participate in this experiment, the subjects were essentially self-selected—a kind of voluntary response group. Explain why that does not invalidate the findings of the experiment.

c) What does "statistically significant" mean in the context of this experiment?

34. Cell phone risks Researchers at the Washington University School of Medicine randomly placed 480 rats into one of three chambers containing radio antennas. One group was exposed to digital cell phone radio waves, the second to analog cell phone waves, and the third group to no radio waves. Two years later the rats were examined for signs of brain tumors. In June 2002 the scientists said that differences among the three groups were not statistically significant.

a) Is this a study or an experiment? Explain.

b) Explain what "statistically significant" means in this context.

c) Comment on the fact that this research was supported by funding from Motorola, a manufacturer of cell phones.

35. Security Twenty first-class passengers and 120 coach passengers are scheduled on a flight. In addition to the usual security screening, 10% of the passengers will be subjected to a more complete search.

a) Describe a sampling strategy to randomly select those to be searched.

b) Here is the first-class passenger list and a set of random digits. Select two passengers to be searched, carefully demonstrating your process.

65 436   71 127   04 879   41 516   20 451
02 227   94 769   23 593

Bergman	Cox	Fontana	Perl
Bowman	DeLara	Forester	Rabkin
Burkhauser	Delli-Bovi	Frongillo	Roufaiel
Castillo	Dugan	Furnas	Swafford
Clancy	Febo	LePage	Testut

c) Explain how you would use a random number table to select the coach passengers to be searched.

CHAPTER

11

From Randomness to Probability

LEARNING OBJECTIVES

1. Recognize probability as long-run relative frequency for repeatable random phenomena

2. Know what is a legitimate assignment of probability

3. Understand The Law of Large Numbers and its misunderstanding

4. Perform simple probability calculations using Venn diagrams, the Complement Rule, and the Addition Rule

Connections

Flip a coin. Can you predict the outcome? It's hard to guess the outcome of just one flip because the outcome is random. If it's a fair coin, though, you can predict the *proportion* of heads you're likely to see in the long run. It's this long-term predictability of randomness that we'll use throughout the rest of the book.

We saw in the previous two chapters that randomness plays a critical role in gathering data. So it's important that we understand how random events behave, and the rules and concepts of probability give us a language to talk and think about random phenomena.

At the start of the book, we talked about "relative frequencies." They are just the same as probabilities once we talk about selecting one individual at random. And "relative frequency" is the right intuitive way to think about probability.

… Nature almost surely operates by combining chance with necessity, randomness with determinism …

—*Eric Chaisson (2007). Epic of Evolution: Seven Ages of the Cosmos, Columbia University Press*

Early humans saw a world filled with random events. To help them make sense of the chaos around them, they sought out seers, consulted oracles, and read tea leaves. As science developed, we learned to recognize some events as predictable. We can now forecast the change of seasons, tell precisely when eclipses will occur, and even make a reasonably good guess at how warm it will be tomorrow. But many other events are still essentially random. Will the coffee shop have your favourite doughnut in stock? Will you be able to get tickets for The Weeknd concert? Will the Toronto Maple Leafs make it to the playoffs this year? Will the stock market go up or down today?

But the surprising fact is that in the long run, many truly random phenomena settle down in a way that's consistent and predictable. It's this property of random phenomena that will make the future steps we'll soon be taking in Statistics possible. Randomization plays a critical role in gathering data, and if we are to make accurate inferences from random samples and randomized experiments, we need to understand how random events and random samples behave. From here on, randomness and probabilities will be fundamental to how we think about data.

Jessica Bethke/Shutterstock

11.1 Random Phenomena

Every day you drive through the intersection at Portage and Main. Even though it may seem that the light is never green when you get there, you know this can't really be true. In fact, if you try really hard, you can recall just sailing through the green light once in a while.

What's random here? The light itself is governed by a timer. Its pattern isn't haphazard. In fact, the light may even be red at precisely the same times each day. It's the time you arrive at the light that is *random*. It isn't that your driving is erratic. Even if you try to leave your house at exactly the same time every day, whether the light is red or green as *you* reach the intersection is a random phenomenon.[1] For us, a **random phenomenon** is a situation in which we know what outcomes can possibly occur, but we don't know which particular outcome will happen. Even though the colour of the light is random[2] as you approach it, some *fraction* of the time, the light will be green. How can you figure out what that fraction is?

You might record what happens at the intersection each day and graph (Figure 11.1) the *accumulated percentage* of green lights:

Figure 11.1

The accumulated percentage of times the light is green settles down as you see more outcomes.

Day	Light	% Green
1	Green	100
2	Red	50
3	Green	66.7
4	Green	75
5	Red	60
6	Red	50
⋮	⋮	⋮

The first day you recorded the light, it was green. Then for the next five days, it was red, then green again, then green, red, and red. If we plot the percentage of green lights against days, the graph would start at 100% (the first time, the light was green, so 1 out of 1 equals 100%). The next day it was red, so the accumulated percentage dropped to 50% (1 out of 2 green). The third day it was green again (2 out of 3, or 67% green), then green (3 out of 4, or 75% green), then red twice in a row (3 out of 5, for 60% green, and then 3 out of 6, for 50% green), and so on. As you collect a new observation for each day, each new outcome becomes a smaller and smaller fraction of the accumulated experience, so, in the long run, the graph settles down. As it settles down, we can see that in fact the light is green about 35% of the time.

In general, each occasion upon which we observe a random phenomenon is called a **trial**. At each trial, we note the result of the random phenomenon, and call that the trial's **outcome**.

TRIALS, OUTCOMES, AND EVENTS

A *random phenomenon* consists of *trials*. Each *trial* has an outcome. *Outcomes* combine to make *events*.

For the traffic light, there are really three possible outcomes: red, yellow, or green. Often we're more interested in a combination of outcomes rather than in the individual ones. When you see the light turn yellow, what do *you* do? If you race through the intersection, then you treat the yellow more like a green light. If you step on the brakes, you treat it more like a red light. Either way, you might want to group the yellow with one or the other. When we combine outcomes like that, the resulting combination is an **event**[3]. We'll call the collection of *all possible outcomes* the **sample space**[4]. We'll denote the sample space **S**. (Some books are even fancier and use the Greek letter Ω). For the traffic light, **S** = {red, green, yellow}. If you flip a coin once, the sample space is very simple: **S** = {H, T}. If you

THE SAMPLE SPACE

For a random phenomenon, the sample space, S, is the set of all possible outcomes of each trial.

---

[1] If you somehow managed to leave your house at *precisely* the same time every day and there was *no* variation in the time it took you to get to the light, then there wouldn't be any randomness, but that's not very realistic.

[2] Even though the randomness here comes from the uncertainty in your arrival time, we can think of the light itself as showing a colour at random.

[3] Each individual outcome is also an event.

[4] Mathematicians like to use the term "space" as a fancy name for a set. Sort of like referring to that closet they gave you to live in, in residence, as your "living space." But it's really just the set of all outcomes.

"For even the most stupid of men … is convinced that the more observations have been made, the less danger there is of wandering from one's goal."

—Jacob Bernoulli, 1713, discoverer of the LLN

Interfoto/Personalities/Alamy

Although he could have said it much more gently, Bernoulli is trying to tell us how intuitive the LLN actually is, even though it wasn't formally proved until the 18th century. Most of us would guess that the law is true from our everyday experiences.

EMPIRICAL PROBABILITY

For any event A,

$$P(A) = \frac{\text{\# times A occurs}}{\text{total \# of trials}}$$

in the long run.

---

flip two coins, it's more complicated because now there are four outcomes, so S = {HH, HT, TH, TT}. If *CBC News* takes a sample of 1000 randomly chosen Canadian adults for a poll, the sample space is incomprehensibly enormous because it would list every combination of 1000 adults you could take from the approximately 28 million adults in Canada.

The Law of Large Numbers

What's the *probability* of a green light at Portage and Main? Based on the graph, it looks like the relative frequency of green lights settles down to about 35%, so saying that the probability is about 0.35 seems like a reasonable answer. But do random phenomena always behave well enough for this to make sense? Might the relative frequency of an event bounce back and forth between two values forever, never settling on just one number?

Fortunately, a principle called the **Law of Large Numbers (LLN)** gives us the guarantee we need. The LLN says that as we repeat a random process over and over, the proportion of times that an event occurs does settle down to one number. We call this number the **probability** of the event. But the law of large numbers requires two key assumptions. First, the random phenomenon we're studying must not change—the outcomes must have the same probabilities for each trial. And, the events must be **independent**.[5] Informally, independence means that the outcome of one trial doesn't affect the outcomes of the others. (We'll see a formal definition of independent events in the next chapter.) The LLN says that as the number of independent trials increases, the long-run *relative frequency* of repeated events gets closer and closer to a single value.

Because the LLN guarantees that relative frequencies settle down in the long run, we can give a name to the value that they approach. We call it the **probability** of the event. If the relative frequency of green lights at that intersection settles down to 35% in the long run, we say that the probability of encountering a green light is 0.35, and write $P(\text{green}) = 0.35$. Because this definition is based on repeatedly observing the event's outcome, this definition of probability is often called **empirical probability**.

The *Nonexistent* Law of Averages

Even though the LLN seems natural, it is often misunderstood because the idea of the *long run* is hard to grasp. Many people believe, for example, that an outcome of a random event that hasn't occurred in many trials is "due" to occur. Many gamblers bet on numbers that haven't been seen for a while, mistakenly believing that they're likely to come up sooner. A common term for this is the "Law of Averages." After all, we know that in the long run the relative frequency will settle down to the probability of that outcome, so now we have some "catching up" to do, right?

Wrong. The Law of Large Numbers says nothing about short-run behaviour. Relative frequencies even out *only in the long run*. And, according to the LLN, the long run is *really* long (*infinitely* long, in fact).

Misconceptions in everyday life "Dear Abby: My husband and I just had our eighth child. Another girl, and I am really one disappointed woman. I suppose I should thank God she was healthy, but, Abby, this one was supposed to have been a boy. Even the doctor told me that the law of averages was in our favour 100 to one." (Excerpt from "The Law of Averages" by Abigail Van Buren from The Nature of Mathematics, by Karl J. Smith. Published by Brooks/Cole, Cengage Learning, © 1991.)

The so-called Law of Averages doesn't actually exist. But you'll hear people talk about it as if it does. A good hockey team has lost their last six games—are they *due* for a win in their next game? If you've been doing particularly well in weekly quizzes in Statistics class,

---

[5]There are stronger forms of the Law that don't require independence, but for our purposes, this form is general enough.

The Law of Averages
Don't let yourself think that there's a Law of Averages that promises short-term compensation for recent deviations from expected behaviour. A belief in such a "Law" can lead to money lost in gambling and to poor business decisions.

are you *due* for a bad grade? No. This isn't the way random phenomena work. There is *no* Law of Averages for short runs.

The lesson of the LLN is that sequences of random events don't compensate in the *short* run and don't need to do so to get back to the right long-run probability. If the probability of an outcome doesn't change and the events are independent, the probability of any outcome in another trial is *always* what it was, no matter what has happened in other trials.

For Example COINS AND THE LAW OF AVERAGES

You've just flipped a fair coin and seen six heads in a row.

QUESTION: Does the coin "owe" you some tails? Suppose you spend that coin and your friend gets it in change. When she starts flipping the coin, should she expect a run of tails?

ANSWER: Of course not. Each flip is a new event. The coin can't "remember" what it did in the past, so it can't "owe" any particular outcomes in the future.

Just to see how this works in practice, the authors ran a simulation of 100 000 flips of a fair coin. We collected 100 000 random numbers, letting the numbers 0 to 4 represent heads and the numbers 5 to 9 represent tails. In our 100 000 "flips," there were 2981 streaks of at least 5 heads. The "Law of Averages" suggests that the next flip after a run of 5 heads should be tails more often to even things out. Actually, the next flip was heads more often than tails: 1550 times to 1431 times. That's 52% heads. You can perform a similar simulation easily on a computer. Try it!

Of course, sometimes an apparent drift from what we expect means that the probabilities are, in fact, *not* what we thought. If you get 10 heads in a row, maybe the coin has heads on both sides!

 Just Checking

1. One common proposal for beating the lottery is to note which numbers have come up lately, eliminate those from consideration, and bet on numbers that have not come up for a long time. Proponents of this method argue that in the long run, every number should be selected equally often, so those that haven't come up are due. Explain why this is faulty reasoning.

Beat the casino Keno is a simple casino game in which numbers from 1 to 80 are chosen. The numbers, as in most lottery games, *are supposed to be equally likely.* Payoffs are made depending on how many of those numbers you match on your card. A group of graduate students from a Statistics department decided to take a field trip to Reno, Nevada.[6] They (*very* discreetly) wrote down the outcomes of the games for a couple of days, then drove back to test whether the numbers were, in fact, equally likely. It turned out that some numbers were *more likely* to come up than others. Rather than bet on the Law of Averages and put their money on the numbers that were "due," the students put their faith in the LLN—and all their (and their friends') money on the numbers that had come up before. After they had pocketed more than $50 000, they were escorted off the premises and invited never to show their faces in that casino again.

---

[6]Although Nevada legalized gambling in 1931, it was not until Raymond Smith, the father of Nevada's gambling industry, opened Harold's Club in Reno that Reno's gambling industry became a tourist attraction. Las Vegas was up and coming but relatively small and quiet back then.

Section 11.1 Exercises

1. **Flipping a coin** Flipping a fair coin is said to randomly generate heads and tails with equal probability. Explain what random means in this context.

2. **Dice** Rolling a fair six-sided die is supposed to randomly generate the numbers 1 through 6. Explain what random means in this context.

3. **Flipping a coin II** Your friend says: "I flipped five heads in a row! The next one has to be tails!" Explain why this thinking is incorrect.

4. **Dice II** After rolling doubles on a pair of dice three times in a row, your friend exclaims, "I can't get doubles four times in a row!" Explain why this thinking is incorrect.

11.2 Modelling Probability

Probability was first studied extensively by a group of French mathematicians who were interested in games of chance.[7] Rather than *experiment* with the games (and risk losing their money), they developed mathematical models. When the probability comes from a mathematical model and not from observation, it is called **theoretical probability**. To make things simple (as we usually do when we build models), they started by looking at games in which the different outcomes were equally likely. Fortunately, many games of chance are like that. Any of 52 cards is equally likely to be the next one dealt from a well-shuffled deck. Each face of a die is equally likely to land up (or at least it *should* be).

> **Playing with a full deck?** An ordinary deck of playing cards, like the ones used in bridge and many other card games, consists of 52 cards. There are numbered cards (2 through 10), as well as face (Jack, Queen, King) and Ace cards whose value depends on the game you are playing. Each card is also marked by one of four suits (clubs, diamonds, hearts, or spades) whose significance is also game-specific.

When outcomes are equally likely, their probability is easy to compute—it's just 1 divided by the number of possible outcomes. So the probability of rolling a three with a fair die is 1/6. The probability of picking the ace of spades from the top of a well-shuffled deck is 1/52.

It's almost as simple to find probabilities for events that are made up of several *equally likely* outcomes. We simply count all the outcomes that the event contains. The probability of the event is the number of outcomes in the event divided by the total number of possible outcomes. We can write

$$P(\mathbf{A}) = \frac{\#\text{ outcomes in A}}{\#\text{ of possible outcomes}}$$

For example, the probability of drawing a face card (Jack, Queen, King) from a deck is

$$P(\text{face card}) = \frac{\#\text{ face cards}}{\#\text{ cards}} = \frac{12}{52} = \frac{3}{13}$$

■ NOTATION ALERT

We often use capital letters—and usually from the beginning of the alphabet—to denote events. We *always* use P to denote probability. So,

$$P(\mathbf{A}) = 0.35$$

means "the probability of the event **A** is 0.35."

When being formal, use decimals (or fractions) for the probability values, but sometimes, especially when talking more informally, it's easier to use percentages.

---

[7]Ok, gambling.

> **How hard can counting be?** Finding the probability of any event when the out-
> comes are equally likely is straightforward, but not necessarily easy. It gets hard when the num-
> ber of outcomes in the event (and in the sample space) gets big. Think about flipping two
> coins. The sample space may be represented as **S** = {HH, HT, TH, TT} where the first H or T
> is the outcome of one of the coins, and the second H or T is the outcome of the other coin.
> Each outcome is equally likely. So what's the probability of getting exactly one head and one
> tail? Let's call that event **A**. Well, there are two outcomes in the event **A** = {HT, TH} out of the
> four possible equally likely ones in **S**, so $P(\mathbf{A}) = \frac{2}{4}$, or $\frac{1}{2}$.
>
> Okay, now flip 100 coins (or one coin 100 times). What's the probability of exactly 67
> heads? Well, first, how many outcomes are in the sample space? **S** = {HHHHHHH … HHH,
> HHHHHHH … HHT, …} Hmm…, a lot. In fact, there are 1 267 650 600 228 229 401 496 703
> 205 376 different outcomes possible when flipping 100 coins. And that's just the denominator
> of the probability! To answer the question, we'd still have to figure out how many ways there
> are to get 67 heads. Actually, we'll see how in Chapter 13; stay tuned!

Don't get trapped into thinking that random events are always equally likely. The chance of winning the Lotto 6/49 jackpot is close to 1 in 14 million[8]—that's a really small chance. Regardless, people continue to buy tickets. In an attempt to understand why, an interviewer asked someone who had just purchased a lottery ticket, "What do you think your chances are of winning the lottery?" The reply was, "Oh, about 50–50." The shocked interviewer asked, "How do you get that?" to which the response was, "Well, the way I figure it, either I win or I don't!" The moral of the story is that events are *not* always equally likely.

How then can we get theoretical probabilities for events when outcomes are not equally likely? Keep reading this chapter, and wait for the next couple of chapters, for some useful ways to handle this, though we'll avoid dealing with the messier probability calculations in this course.

Personal Probability

What's the probability that your grade in this Statistics course will be an A? You may be able to come up with a number that seems reasonable. Of course, no matter how confident or depressed you feel about your chances of success, your probability should be between 0 and 1. How did you come up with this probability? It can't be an empirical probability? For that, you'd have to take the course over and over (and over…) and forget everything after each time so the probability of getting an A would stay the same, then calculate the proportion of times you got an A.[9] Is it a theoretical probability? Unless you assume the outcomes (grades A, B, C, and so on) are equally likely, it will be hard to find some theoretically based probability. But people use probability in a third sense as well.

We use the language of probability in everyday speech to express a degree of uncertainty *without* basing it on long-run relative frequencies or mathematical models. Your personal assessment of your chances of getting an A expresses your uncertainty about the outcome. That uncertainty may be based on how comfortable you're feeling in the course or about your midterm grade, but it can't be based on long-run behaviour. We call this third kind of probability a **subjective** or **personal probability**.

Although personal probabilities may be based on experience, they're not based either on long-run relative frequencies or equally likely events. So they don't display the kind of consistency that we'll need probabilities to have. For that reason, we'll stick to formally defined probabilities. You should be alert to the difference.

---

[8]There are actually 13 983 816 different ways you can select your numbers. There are equally likely outcomes, but only one of them wins the top prize.

[9]And unless you *really* forget what you learned over the summer, your chances of getting an A won't stay the same, so your course grades probably won't be independent either (as required by the LLN)!

> **Which kind of probability?** The line between personal probability and the other two types of probability can be a little fuzzy. When a weather forecaster predicts a 40% probability of rain, is this a personal probability or a relative frequency probability? The claim may be that 40% of the time, when the weather radar map looks like this, it has rained (over some period of time). Or the forecaster may be stating a personal opinion based on years of experience and reflects a sense of what has happened in the past in similar situations. (And likely personal judgment was supplemented by mathematical modelling and computer simulation, taking into account various atmospheric measurements, adding a theoretical dimension to the final determination.) When you hear a probability stated, try to ascertain what kind of probability is intended.

The First Three Rules for Working with Probability

1. Make a picture.
2. Make a picture.
3. Make a picture.

We're dealing with probabilities now, not data, but the three rules don't change. The most common kind of picture to make is called a Venn diagram. We'll use Venn diagrams throughout the rest of this chapter. Even experienced statisticians make Venn diagrams to help them think about probabilities of compound and overlapping events. You should, too.

Section 11.2 Exercises

5. **Wardrobe** In your dresser are five blue shirts, three red shirts, and two black shirts.
   a) What is the probability of randomly selecting a red shirt?
   b) What is the probability that a randomly selected shirt is not black?

6. **Playlists** Your list of favourite songs contains 10 rock songs, 7 rap songs, and 3 country songs.
   a) What is the probability that a randomly played song is a rap song?
   b) What is the probability that a randomly played song is not country?

11.3 Formal Probability Rules

For some people, the phrase "50/50" means something vague like "I don't know" or "whatever." But when we discuss probabilities of outcomes, it takes on the precise meaning of *equally likely*. Speaking vaguely about probabilities will get us into trouble, so whenever we talk about probabilities, we'll need to be precise.[10] And to do that, we'll need to develop some formal rules[11] about how probability works, rules such as the following:

■ If the probability is 0, the event never occurs (but be sure to read about *surprising probabilities* below), and likewise, if the probability is 1, it *always* occurs. Even if you think an event is very unlikely, its probability can't be negative, and even if you're sure it will happen, its probability can't be greater than 1. (Think about relative frequencies.) Making this more formal gives the **Probability Assignment Rule**.

A probability is a number between 0 and 1.

For any event **A**, $0 \leq P(\mathbf{A}) \leq 1$.

---

[10]And to be precise, we will be talking only about sample spaces where we can enumerate, all the outcomes (which means we can begin to make a list, but may need to write something like 1, 2, 3 ... in cases when all integers are possible, without upper limit). Mathematicians call this a countable (even possibly infinite) number of outcomes.

[11]Actually, in mathematical terms, these are axioms—statements that we assume to be true of probability. We'll derive other rules from these in this chapter and the next.

SURPRISING PROBABILITIES

We've been careful to discuss probabilities only for situations in which the outcomes were finite, or even countably infinite. But things can get surprising if the outcomes can take on *any* numerical value referred to as *continuous*. For example, what is the probability that a randomly selected child will be *exactly* 1 metre tall? Well, if we mean 1.00000 . . . metres, the answer is zero. No randomly selected child—even one whose height would be recorded as 1 metre, will be *exactly* 1 metre tall (to an infinite number of decimal places). But, if you've grown taller than 1 metre, there must have been a time in your life when you actually *were* exactly 1 metre tall, even if only for a fraction of a second. So this is an outcome with probability 0 that not only has happened—it has happened to *you*.

Pearson Education

John Venn (1834–1923) created the Venn diagram. His book on probability, The Logic of Chance, was "strikingly original and considerably influenced the development of the theory of Statistics,"[12] *according to John Maynard Keynes, one of the luminaries of Economics.*

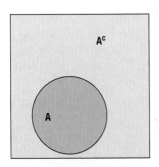

The set **A** and its complement **A**^{C}. Together, they make up the entire sample space **S**. So the complement rule is really just an application of the total probability rule, $P(\mathbf{A}) + P(\mathbf{A}^C) = P(\mathbf{S}) = 1$.

■ If a random phenomenon has only one possible outcome, it's not very interesting (or very random). So we need to distribute the probabilities among all the outcomes a trial can have. How can we do that so that it makes sense? For example, consider what you're doing as you read this book. The possible outcomes might be:

A. You read to the end of this chapter before stopping.
B. You finish this section but stop reading before the end of the chapter.
C. You bail out before the end of this section.

When we assign probabilities to these outcomes, the first thing to be sure of is that we distribute all of the available probability. Something in the sample space always occurs, so the probability of the entire sample space is 1.

Making this more formal gives the **Total Probability Rule**.

> The set of all possible outcomes of a trial must have probability 1.

$$P(\mathbf{S}) = 1$$

■ Suppose the probability that you get to class on time is 0.8. What's the probability that you don't get to class on time? Yes, it's 0.2. The set of outcomes that are *not* in the event **A** is called the **complement** of **A**, and is denoted **A**^{C}. This leads to the **Complement Rule**:

> The probability of an event occurring is 1 minus the probability that it doesn't occur.

$$P(\mathbf{A}) = 1 - P(\mathbf{A}^C)$$

For Example APPLYING THE COMPLEMENT RULE

RECAP: We opened the chapter by looking at the traffic light at the corner of Portage and Main, observing that when we arrive at that intersection, the light is green about 35% of the time.

QUESTION: If $P(\text{green}) = 0.35$, what's the probability the light isn't green when you get to Portage and Main?

ANSWER: "Not green" is the complement of "green," so $P(\text{not green}) = 1 - P(\text{green})$
$$= 1 - 0.35 = 0.65$$

There's a 65% chance I won't have a green light.

■ Suppose the probability that a randomly selected student is in second year (event **A**) is 0.20, and the probability that he or she is in third year (event **B**) is 0.30. What is the probability that the student is in *either* second or third year, written $P(\mathbf{A} \text{ or } \mathbf{B})$? If you guessed 0.50, you've deduced the *addition rule*, which says that you can

---

[12]http://mathforum.org/library/drmath/view/52427.html.

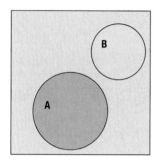

Two disjoint events, **A** and **B**.

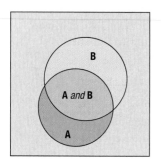

Two events **A** and **B** that are not disjoint. The event (**A** and **B**) is their intersection.

Figure 11.2
"Red light" and "green or yellow light" are complementary events. Their probabilities add to one.

add the probabilities of events that are disjoint.[13] To see whether two events are disjoint, we break them apart into their component outcomes and check whether they have any outcomes in common. **Disjoint** (or **mutually exclusive**) events have no outcomes in common. More intuitively, if the occurrence of one event excludes any possibility of the other occurring (in the same trial), we say the two events are disjoint or mutually exclusive. The **Addition Rule for disjoint events states**:

> For two disjoint events **A** and **B**, the probability that one or the other occurs is the sum of the probabilities of the two events.

$$P(\mathbf{A} \text{ or } \mathbf{B}) = P(\mathbf{A}) + P(\mathbf{B}), \text{ provided that } \mathbf{A} \text{ and } \mathbf{B} \text{ are disjoint.}$$

For Example APPLYING THE ADDITION RULE

RECAP: When you get to the light at Portage and Main, it's red, green, or yellow. We know that $P(green) = 0.35$.

QUESTION: Suppose we find out that $P(yellow)$ is about 0.04. What's the probability the light is red?

ANSWER: The light must be red, green, or yellow, so if we can figure out the probability that the light is green or yellow, we can use the Complement Rule to find the probability that it is red. To find the probability that the light is green or yellow, I can use the Addition Rule because these are disjoint events: The light can't be both green and yellow at the same time[14].

$$P(green \text{ or } yellow) = 0.35 + 0.04 = 0.39$$

Red is the only remaining alternative (Figure 11.2), which makes it the complement of 'green or yellow', and thus by the Complement Rule,

$$
\begin{aligned}
P(red) &= P(\text{not 'green or yellow'}) \\
&= 1 - P(green \text{ or } yellow) \\
&= 1 - 0.39 = 0.61.
\end{aligned}
$$

The Addition Rule can be extended to any number of disjoint events, and that's helpful for checking probability assignments. Because individual sample space outcomes are always disjoint, we have an easy way to check whether the probabilities we've assigned to the possible outcomes are legitimate. The Total Probability Rule tells us that to be a **legitimate assignment of probabilities**, the sum of the probabilities of all possible outcomes must be exactly 1. No more, no less. For example, if we were told the probabilities of selecting at random a first year, second year, third year, or fourth year student from all the undergraduates at a school were 0.25, 0.23, 0.22, and 0.20, respectively, we would know that something was wrong. These "probabilities" sum to only 0.90, so this is not a legitimate probability assignment. Either a value is wrong or we just missed some possible outcomes, like the "special student," or "more than four years" categories that soak up the remaining 0.10. Similarly, a claim that the probabilities were 0.26, 0.27, 0.29, and 0.30 would be wrong because these "probabilities" sum to more than 1.

---

[13]You may see $P(\mathbf{A} \text{ or } \mathbf{B})$ written as $P(\mathbf{A} \cup \mathbf{B})$. The symbol $\cup$ means "union," representing the outcomes in event **A** or event **B** (or both). You may see $P(\mathbf{A} \text{ and } \mathbf{B})$ written as $P(\mathbf{A} \cap \mathbf{B})$. The symbol $\cap$ means "intersection," representing outcomes that are in both event **A** and event **B**.

[14]At least that's the assumption for Portage and Main. There are some places, especially in Europe, where traffic lights are sometimes simultaneously green and yellow, but we'll ignore that here.

But be careful. The Addition Rule for disjoint events doesn't work for events that aren't disjoint. If the probability of a randomly chosen adult owning a smartphone is 0.50 and the probability of owning a computer is 0.90, the probability of owning either a smartphone or a computer may be pretty high, but it is *not* 1.40! Why can't you add probabilities like this? Because these events are not disjoint. You *can* own both. Shortly, we'll see how to find probabilities for events like these, but we'll need another, more general rule.

Section 11.3 Exercises

7. Wardrobe II In Exercise 5, you have five blue shirts, three red shirts and two black shirts in your dresser. What is the probability that a randomly selected shirt is either a blue or a black shirt?

8. Playlist II In Exercise 6, your list of favorite songs includes 10 rock songs, 7 rap songs and 3 country songs. What is the probability that a randomly played song is either a rap song or a country song?

11.4 The General Addition Rule

When you get to the light at Portage and Main, there is a 65% chance you have to stop because the light is red. But if a bus is blocking the intersection you still have to stop, even if the light is green. Suppose the empirical research has been done, and there is a 16% chance you have to stop behind a bus regardless of the colour of the light. Only 3% of the time will you have a double reason to stop: The light is red and a bus is blocking your way.

Now what is the probability you have to stop at the intersection? What is the probability of **A** = {stopping for a red light} *or* **B** = {stopping for a bus}? We know $P(\mathbf{A}) = 0.35$ and $P(\mathbf{B}) = 0.16$. But $P(\mathbf{A}\ or\ \mathbf{B})$ is not simply the sum $P(\mathbf{A}) + P(\mathbf{B})$, because the events **A** and **B** are not disjoint. Sometimes the light is red and a bus is blocking the intersection. So what can we do? We'll need a new probability rule.

As Figure 11.3 shows, we can't use the addition rule and add the two probabilities because the events are not disjoint; they overlap. There's an outcome in the *intersection* of **A** and **B**. Figure 11.3 represents the sample space. In reality, of course, there could be other reasons for stopping aside from a red light or bus, but let's assume we will only stop for these two reasons. Sitting outside the circle, then, is simply the event that we do not stop.

Figure 11.3

Red light and **Bus** are non-disjoint events. A red light and a bus obstruction can both happen. Some probabilities also correspond to specific disjoint regions. Notice that we broke up the probability of a red light, 0.65, into the two disjoint regions composing that event (and likewise for the bus).

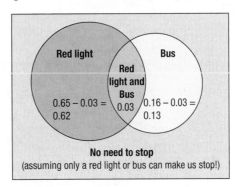

The reason we can't simply add the probabilities of **A** and **B** is that we'd count the probability of stopping for a red light *and* stopping for a bus twice. If we add the two probabilities, we have to compensate by *subtracting out* the probability of stopping for a red light *and* stopping for a bus. So,

$$P(\text{stop}) = P(\text{stopping for a red light or stopping for a bus})$$
$$= P(\text{stopping for a right light}) + P(\text{stopping for a bus}) -$$
$$P(\text{stopping for a red light } and \text{ stopping for a bus})$$
$$= 0.65 + 0.16 - 0.03$$
$$= 0.78$$

This method works in general. We add the probabilities of the two events and then subtract out the probability of their intersection. This approach gives us the **General Addition Rule**, which does not require disjoint events:

$$P(\textbf{A } or \textbf{ B}) = P(\textbf{A}) + P(\textbf{B}) - P(\textbf{A } and \textbf{ B}).$$

The addition rule for disjoint events is not unrelated to the general addition rule. In fact, the general addition rule automatically becomes the addition rule for disjoint events. When events **A** and **B** are disjoint, it's impossible for them both to happen at the same time. So when **A** and **B** are disjoint, $P(\textbf{A } and \textbf{ B}) = 0$, and the last term in the general addition rule disappears, leaving the addition rule for disjoint events.

For Example USING THE GENERAL ADDITION RULE

A survey of university students asked the questions: "Are you currently in a relationship?" and "Are you involved in intercollegiate or club sports?" The survey found that 33% were currently in a relationship, and 25% were involved in sports. Eleven percent responded "yes" to both.

QUESTION: What's the probability that a randomly selected student either is in a relationship or is involved in athletics?

ANSWER: Let **R** = {student in a relationship} and **S** = {student involved in sports}.

P(a student either is in a
   relationship or involved in sports) $= P(\textbf{R } or \textbf{ S})$

$$= P(\textbf{R}) + P(\textbf{S}) - P(\textbf{R } and \textbf{ S})$$

$$= 0.33 + 0.25 - 0.11$$

$$= 0.47$$

There's a 47% chance that a randomly selected college student either is in a relationship or is involved in sports. Notice that this use of the word "or" includes "or both." That's the usual meaning of *or* in English. If we wanted to exclude those students who were involved in sports and in a relationship we'd have to ask, "What's the probability that a selected student either is in a relationship or is involved in sports, but not both?"

Would you like dessert or coffee? Natural language can be ambiguous. In this question, what are your choices? Can you have only one of the two, or could you simply answer "yes" and have both dessert and coffee?

In Statistics, when we say "or" we always mean the inclusive version. In other words, the probability of **A** or **B** means the probability of either **A** or **B** or both. But in everyday language, "or" is usually the exclusive version meaning one or the other but not both. For example, if you are asked "Would you like the steak or the vegetarian entrée?" you can't get both. This difference can be confusing for students seeing the statistical language for the first time.

The General Addition Rule subtracts the probability of the outcomes in **A** *and* **B** because we've counted those outcomes *twice*. But they're still there. On the other hand, if we really want the probability of **A** or **B**, but NOT both, we have to get rid of the outcomes in {**A** *and* **B**}. So $P(\textbf{A } or \textbf{ B}, but \ not \ \text{both}) = P(\textbf{A } or \textbf{ B}) - P(\textbf{A } and \textbf{ B}) = P(\textbf{A}) + P(\textbf{B}) - 2 \times P(\textbf{A } and \textbf{ B})$. Now we've subtracted $P(\textbf{A } and \textbf{ B})$ twice—once because we don't want to double-count these events and a second time because we really didn't want to count them at all. Confused? *Make a picture.*

Drawing a picture using Venn diagrams often makes not just the General Additive Rule easier to see and understand, but can also make it quite easy to do some other related probability calculations.

For Example USING VENN DIAGRAMS

RECAP: We return to our survey of university students: 33% are in a relationship, 25% are involved in sports, and 11% are in both.

QUESTIONS: Based on a Venn diagram, what is the probability that a randomly selected student

a. is not in a relationship and is not involved in sports?
b. is in a relationship but not involved in sports?
c. either is in a relationship or is involved in sports, but not both?

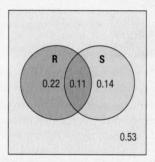

ANSWER: Let **R** = {student is in a relationship} and **S** = {student is involved in sports}. In the Venn diagram, the intersection of the circles is $P(\mathbf{R} \text{ and } \mathbf{S})$ = 0.11. Since $P(\mathbf{R})$ = 0.33, $P(\mathbf{R} \text{ and } \mathbf{S}^C)$ = 'relationship but not sports') = 0.33 − 0.11 = 0.22. Also, $P(\mathbf{R}^C \text{ and } \mathbf{S})$ = 'no relationship but sports') = 0.25 − 0.11 = 0.14. Now, 0.22 + 0.11 + 0.14 = 0.47, leaving 1 − 0.47 = 0.53 for the region outside both circles.

Now, P(not in a relationship and not involved in sports)
 = $P(\mathbf{R}^C \text{ and } \mathbf{S}^C)$ = 0.53

P(in a relationship but not involved in sports) = $P(\mathbf{R} \text{ and } \mathbf{S}^C)$ = 0.22

P(in a relationship or involved in sports, but not both)
 = $P(\mathbf{R} \text{ and } \mathbf{S}^C) + P(\mathbf{R}^C \text{ and } \mathbf{S})$ = 0.22 + 0.14 = 0.36.

Or

$P(\mathbf{R}) + P(\mathbf{S}) - 2P(\mathbf{R} \text{ and } \mathbf{S})$ = 0.33 + 0.25 − 2(0.11) = 0.36.

 # Just Checking

2. Back in Chapter 1, we suggested that you sample some pages of this book at random to see whether they held some sort of data display or an equation. We actually did just that. We drew a representative sample and found the following:

   48% of pages had some kind of data display,
   27% of pages had an equation, and
   7% of pages had both a data display and an equation.

   a) Display these results in a Venn diagram.

   b) What is the probability that a randomly selected sample page had neither a data display nor an equation?

   c) What is the probability that a randomly selected sample page had a data display but no equation?

Step-by-Step Example USING THE GENERAL ADDITION RULE

You almost certainly qualify as an "online adult"—an adult who uses the Internet. Pew Research tracks the behaviour of online adults. In 2013 they found that 71% were Facebook users, 18% used Twitter, and 15% used both. (http://www.pewinternet .org/Commentary/2012/March/Pew-Internet-Social-Networking-full-detail.aspx) (updated in September 2013)

Questions: For a randomly selected online adult, what is the probability that the person

1. uses either Facebook or Twitter?
2. uses either Facebook or Twitter, but not both?
3. doesn't use Facebook or Twitter?

Aspen Photo/Shutterstock

THINK ➡ **Plan** Define the events we're interested in. There are no conditions to check; the General Addition Rule works for any events!

Plot Make a picture, and use the given probabilities to find the probability for each region.

The blue region represents **A** but not **B**. The green intersection region represents **A** *and* **B**. Note that since $P(\textbf{A}) = 0.71$ and $P(\textbf{A} \text{ and } \textbf{B}) = 0.15$, the probability of **A** but not **B** must be $0.71 - 0.15 = 0.56$.

The yellow region is **B** but not **A**.

The gray region outside both circles represents the outcome neither **A** nor **B**. All the probabilities must total 1, so you can determine the probability of that region by subtraction.

Now, figure out what you want to know. The probabilities can come from the diagram or a formula. Sometimes translating the words to equations is the trickiest step.

Let $\textbf{A} = \{\text{respondent uses Facebook}\}$.

Let $\textbf{B} = \{\text{respondent uses Twitter}\}$.

I know that

$$P(\textbf{A}) = 0.71$$
$$P(\textbf{B}) = 0.18$$
$$P(\textbf{A} \text{ and } \textbf{B}) = 0.15$$

So

$$P(\textbf{A} \text{ and } \textbf{B}^{\textbf{C}}) = 0.71 - 0.15 = 0.56$$
$$P(\textbf{B} \text{ and } \textbf{A}^{\textbf{C}}) = 0.18 - 0.15 = 0.03$$
$$P(\textbf{A}^{\textbf{C}} \text{ and } \textbf{B}^{\textbf{C}}) = 1 - (0.56 + 0.15 + 0.03)$$
$$= 0.26$$

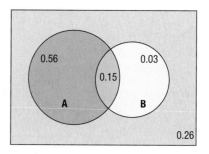

Question 1. What is the probability that the person uses either Facebook or Twitter?

SHOW ➡ **Mechanics** The probability that the respondent uses Facebook or Twitter is $P(\textbf{A} \text{ or } \textbf{B})$. We can use the General Addition Rule, or we can add the probabilities seen in the diagram.

$$P(\textbf{A} \text{ or } \textbf{B}) = P(\textbf{A}) + P(\textbf{B}) - P(\textbf{A} \text{ and } \textbf{B})$$
$$= 0.71 + 0.18 - 0.15$$
$$= 0.74$$

OR

$$P(\textbf{A} \text{ or } \textbf{B}) = 0.56 + 0.15 + 0.03 = 0.74$$

TELL ➡ **Conclusion** Don't forget to interpret your result in context.	76% of online adults use Facebook or Twitter.

Question 2. What is the probability that the person uses either Facebook or Twitter, but not both?

SHOW ➡ **Mechanics** We can use the rule, or just add the appropriate probabilities seen in the Venn diagram.	$P(\mathbf{A} \text{ or } \mathbf{B} \text{ but NOT both}) = P(\mathbf{A} \text{ or } \mathbf{B}) - P(\mathbf{A} \text{ and } \mathbf{B})$ $= 0.74 - 0.15 = 0.59$ **OR** $P(\mathbf{A} \text{ or } \mathbf{B} \text{ but NOT both})$ $= P(\mathbf{A} \text{ and } \mathbf{B^c}) + P(\mathbf{B} \text{ and } \mathbf{A^c})$ $= 0.56 + 0.03 = 0.59$
TELL ➡ **Conclusion** Interpret your result in context.	59% of online adults use Facebook or Twitter, but not both.

Question 3. What is the probability that the person doesn't use Facebook or Twitter?

SHOW ➡ **Mechanics** Not using Facebook or Twitter is the complement of using one or the other. Use the Complement Rule or just notice that "not using Facebook or Twitter" is represented by the region outside both circles.	$P(\text{not Facebook or Twitter}) =$ $1 - P(\text{either Facebook or Twitter})$ $= 1 - P(\mathbf{A} \text{ or } \mathbf{B})$ $= 1 - 0.74 = 0.26$ **OR** $P(\mathbf{A^c} \text{ and } \mathbf{B^c}) = 0.26$
TELL ➡ **Conclusion** Interpret your result in context.	Only 26% of online adults use neither Facebook nor Twitter.

Section 11.4 Exercises

9. Pet ownership Suppose that 25% of people have a dog, 29% of people have a cat, and 12% of people own both. What is the probability that someone owns a dog or a cat?

10. Cooking and shopping Forty-five percent of Americans like to cook and 59% of Americans like to shop, while 23% enjoy both activities. What is the probability that a randomly selected American either enjoys cooking or shopping or both?

WHAT CAN GO WRONG?

- **Beware of probabilities that don't add up to 1.** To be a legitimate probability assignment, the sum of the probabilities for all possible outcomes must total 1. If the sum is less than 1, you may need to add another category ("other") and assign the remaining probability to that outcome. If the sum is more than 1, check that the outcomes are disjoint. If they're not, then you can't assign probabilities by just counting relative frequencies.

- **Don't use the *addition rule for disjoint events* when the events are not disjoint, in which case you need to use the *general addition rule*.** Events must be disjoint to use the Addition Rule for disjoint events. The probability of being under 80 *or* a female is not the probability of being under 80 *plus* the probability of being a female; this sum may be more than 1. To find the probability of being under 80 *or* a female, first add the two individual probabilities and then subtract the probability of being under 80 *and* a female (if you have all these probabilities, *including* the intersection).

What Have We Learned?

Apply the facts about probability to determine whether an assignment of probabilities is legitimate.

- Probability is long-run relative frequency.
- Individual probabilities must be between 0 and 1.
- The sum of probabilities assigned to all outcomes must be 1.

Understand the Law of Large Numbers and that the common understanding of the "Law of Averages" is false.

Know the rules of probability and how to apply them.

- A probability is a number between 0 and 1.
  For any event A, $0 \le P(A) \le 1$.
- **Probability Assignment Rule:**
  $$P(S) = 1.$$
- The **Complement Rule** says that $P(not\ A) = P(A^C) = 1 - P(A)$.
- The **Addition Rule** for disjoint events says that $P(A\ or\ B) = P(A) + P(B)$ provided events A and B are disjoint.
- The **General Addition Rule** says that $P(A\ or\ B) = P(A) + P(B) - P(A\ and\ B)$ for any two events, disjoint or not.

Review of Terms

Random phenomenon   A phenomenon is random if we know what outcomes could happen, but not which particular values will happen (p. 355).

Trial   A single attempt or realization of a random phenomenon (p. 355).

Outcome   The value measured, observed, or reported for an individual instance of a trial (p. 355).

Event   A collection of outcomes. Usually, we identify events so that we can attach probabilities to them. We denote events with bold capital letters, such as A, B, or C (p. 355).

Sample Space   The collection of all possible outcome values. The collection of values in the sample space has a total probability of 1. We denote the sample space with a boldface capital S (p. 355).

Law of Large Numbers (LLN)	The *relative frequency* of an event's occurrence in repeated independent trials gets closer and closer to its *true* (long-run) relative frequency as the number of trials increases (p. 356).
Independence (informally)	Trials are *independent* if the outcome of one trial has absolutely no influence on the outcomes of other trials (p. 356).
Probability	The probability of an event is a number between 0 and 1 that reports the likelihood of an event's occurrence. We write $P(A)$ for the probability of the event A (p. 356).
Empirical probability	When the probability comes from the long-run relative frequency of an event's occurrence, it is an **empirical probability** (p. 356).
Theoretical probability	When the probability comes from a model (such as equally likely outcomes), it is a **theoretical probability** (p. 358).
Subjective or personal probability	When the probability is subjective and represents your personal degree of belief, it is a **personal probability** (p. 359).
Probability assignment rule	A probability is a number between 0 and 1. For any event A, $0 \leq P(A) \leq 1$ (p. 360).
Total probability rule	The probability of all the outcomes in the entire sample space must sum to 1. $P(S) = 1$ (p. 361).
Complement rule	The probability of an event occurring is 1 minus the probability that it doesn't occur. $P(A) = 1 - P(A^C)$ (p. 361).
Disjoint or mutually exclusive	Two events are disjoint if they share no outcomes in common. If A and B are disjoint, then knowing that A occurs tells us that B cannot occur. Disjoint events are also called *mutually exclusive* events (p. 362).
Addition rule for disjoint events	If A and B are disjoint events, then the probability of A *or* B is given by: $P(A \text{ or } B) = P(A) + P(B)$ (p. 362).
Legitimate assignment of probabilities	An assignment of probabilities to outcomes is legitimate (though not necessarily accurate) if ▪ each probability is between 0 and 1 (inclusive) ▪ the sum of the probabilities is 1 (p. 362)
General addition rule	For any two events, A and B (disjoint or not), the probability of A *or* B is given by: $P(A \text{ or } B) = P(A) + P(B) - P(A \text{ and } B)$ (p. 364).

Chapter Exercises

11. Sample spaces For each of the following, list the sample space and tell whether you think the events are equally likely.
a) Toss two coins; record the order of heads and tails.
b) A family has three children; record the number of boys.
c) Flip a coin until you get a head or three consecutive tails; record the order of head and tails.
d) Roll two dice; record the larger number.

12. Sample spaces II For each of the following, list the sample space and tell whether you think the events are equally likely.
a) Roll two dice; record the sum of the numbers.
b) A family has three children; record each child's gender in order of birth.

c) Toss four coins; record the number of tails.
d) Toss a coin 10 times; record the longest run of heads.

13. Roulette A casino claims that its roulette wheel is truly random. What should that claim mean?

14. Rain The weather reporter on television makes predictions such as a 25% chance of rain. What do you think such a phrase means?

15. Winter Comment on the following quotation:

"What I think is our best determination is it will be a colder than normal winter," said Pamela Naber Knox, a Wisconsin state climatologist. "I'm basing that on a couple of different things. First, in looking at the past few winters, there has been a lack of really cold weather. Even though we are not supposed to use the law of averages, we are due."[15]

16. Snow After an unusually dry autumn, a radio announcer is heard to say, "Watch out! We'll pay for these sunny days later on this winter." Explain what he's trying to say, and comment on the validity of his reasoning.

17. Cold streak A batter who failed to get a hit in seven consecutive times at bat hits a game-winning home run. When speaking with reporters after the game, he says he was very confident on his last time up at bat because he knew he was "due for a hit." Comment on the batter's reasoning.

18. Crash Commercial airplanes have an excellent safety record. Nevertheless, there are crashes occasionally, with the loss of many lives. In the weeks following a crash, airlines often report a drop in the number of passengers, probably because people are afraid to risk flying.
a) A travel agent suggests that, since the law of averages makes it highly unlikely to have two plane crashes within a few weeks of each other, flying soon after a crash is the safest time. What do you think?
b) If the airline industry proudly announces that it has set a new record for the longest period of safe flights, would you be reluctant to fly? Are the airlines due to have a crash?

19. Fire insurance Insurance companies collect annual payments from homeowners in exchange for paying to rebuild houses that burn down.
a) Why should you be reluctant to accept a $300 payment from your neighbour to replace his house should it burn down during the coming year?
b) Why can the insurance company make that offer?

20. Jackpot On February 11, 2009, the AP news wire released the following story:

(*LAS VEGAS, Nev.*)—A man in town to watch the NCAA basketball tournament hit a $38.7 million jackpot on Friday, the biggest slot machine payout ever. The 25-year-old software engineer from Los Angeles, whose name was not released at his request, won after putting three $1 coins in a machine at the Excalibur hotel-casino, said Rick Sorensen, a spokesman for slot machine maker International Game Technology.[16]
a) How can the Excalibur afford to give away millions of dollars on a $3 bet?
b) Why was the maker willing to make a statement? Wouldn't most businesses want to keep such a huge loss quiet?

21. Spinner The plastic arrow on a spinner for a child's game stops rotating to point at a colour that will determine what happens next. Which of the following probability assignments are possible?

	Red	Yellow	Green	Blue
a)	0.25	0.25	0.25	0.25
b)	0.10	0.20	0.30	0.40
c)	0.20	0.30	0.40	0.50
d)	0	0	1.00	0
e)	0.10	0.20	1.20	−0.50

Probabilities of ...

22. Scratch off Many stores run "secret sales": Shoppers receive cards that determine how large a discount they get, but the percentage is revealed by scratching off that black stuff (what *is* that?) only after the purchase has been totalled at the cash register. The store is required to reveal (in the fine print) the distribution of discounts available. Which of these probability assignments are plausible?

	10% off	20% off	30% off	50% off
a)	0.20	0.20	0.20	0.20
b)	0.50	0.30	0.20	0.10
c)	0.80	0.10	0.05	0.05
d)	0.75	0.25	0.25	−0.25
e)	1.00	0	0	0

Probabilities of ...

23. Car repairs A consumer organization estimates that over a one-year period, 17% of cars will need to be repaired once, 7% will need repairs twice, and 4% will require three or more repairs. What is the probability that a car chosen at random will need
a) No repairs?
b) No more than one repair?
c) Some repairs?

24. Stats projects In a large Introductory Statistics lecture hall, the professor reports that 55% of the students enrolled have never taken a Calculus course, 32% have taken only one semester of Calculus, and the rest have taken two or more semesters of Calculus. The professor randomly assigns students to groups of three to work on a project. What is the probability that the first groupmate you meet has studied
a) Two or more semesters of Calculus?
b) Some Calculus?
c) No more than one semester of Calculus?

---

[15]Associated Press, Fall 1992, quoted by Schaeffer et al.

[16]Reprinted by permission from AP news wire, February 11, 2009 © 2005 The Associated Press. All rights reserved.

25. Ethanol A Nanos Poll in August 2008 asked 500 adult Ontarians if they thought Ontario should produce more, produce less, or maintain current production of ethanol. Ethanol is made from corn, and can be used as a gasoline additive to increase fuel efficiency.

Here's how they responded:

Response	Number
More ethanol production	160
Less ethanol production	136
Maintain current production level	155
No opinion	49
Total	**500**

If we select a person at random from this sample of 500 adults,

a) What is the probability that the person responded, "More ethanol production"?

b) What is the probability that the person responded, "Maintain current production level" or had no opinion?

26. Optimism A Nanos Poll conducted in May 2012 asked 1000 Canadians whether they think the next generation will have a standard of living that is higher, the same, or lower than what Canadians have today. Here's how they responded:

Response	Number
Higher	255
Same	263
Lower	373
Unsure	110
Total	**1000**

If we select a person at random from this sample of 1000 adults,

a) What is the probability that the person responded "Higher"?

b) What is the probability that the person responded "Higher" or "Same"?

27. Census Canada conducts a census every five years. Federal, provincial, and local governments use census data to shape public policy. In 2006, Canada tried something no other country had ever tried—census questionnaires that could be completed online. Sixty-six percent of households completed their census questionnaire by the May 16 deadline. (Households that didn't respond by the deadline received follow-up letters and phone calls.) Nineteen percent of households responded online; sixteen percent of households responded on time and online.

a) What percent of households filed on time or filed online?

b) What percent of households did not file on time but used the online service when they eventually responded?

c) What percent of households filed on time but did not file online?

d) What percent of households neither filed on time nor online?

28. Census returns In 2006, 80% of households received a short census form with only eight questions. The remainder received a long census form with 53 questions. As previously mentioned in Exercise 25, census questionnaires could be completed online. Fifteen percent of households received the short form and completed their census online; 4% of households received the long form and completed their census online.

a) What percent of households filed online?

b) What percent of households received the short form and filed online?

c) What percent of households received the short form or filed online?

d) What percent of households received the long form and did not file online?

29. Homes Real estate ads suggest that 64% of homes for sale have garages, 21% have swimming pools, and 17% have both features. What is the probability that a home for sale has

a) A pool or a garage?

b) Neither a pool nor a garage?

c) A pool but no garage?

30. Immigration In 2014, 52.3% of all immigrants to Canada were females, 18.6% were under 18 years old, and 9.0% were females under 18 years old. Find the probability that a randomly selected person who immigrated to Canada in 2014 was

a) Female and at least 18 years old.

b) Either female or under 18 years old.

c) Male and at least 18 years old.

(*Source*: http://www.cic.gc.ca/english/pdf/2014-Facts-Permanent.pdf)

31. Emigration In 2007, 51% of all emigrants from Canada were males, 82% were at least 18 years old, and 42% were males of at least 18 years of age. Find the probability that a randomly selected person who emigrated from Canada in 2007 was

a) Male and under 18 years old.

b) Either male or at least 18 years of age.

c) Female and under 18 years old.

32. Workers Employment data at a large company reveal that 72% of the workers are married, 44% are university graduates, and half of the university graduates are married. What is the probability that a randomly chosen worker
a) Is neither married nor a university graduate?
b) Is married but not a university graduate?
c) Is married or a university graduate?

33. Cosmetic marketing A GfK Roper Worldwide survey asked consumers in five countries whether they agreed with the statement "I follow a skin care routine every day." Here are the responses classified by the country of the respondent.

Country	Agree	Disagree	Don't Know	Total
China	361	988	153	1502
France	695	763	81	1539
India	828	689	18	1535
UK	597	898	62	1557
US	668	841	48	1557
Total	3149	4179	362	7690

If we select a person at random from this sample,
a) What is the probability that the person agrees with the statement?
b) What is the probability that the person is from France?
c) What is the probability that the person is from France and agrees with the statement?
d) What is the probability that the person is from France or agrees with the statement?

Note: This question is modified from Business Statistics 2nd Canadian edition (by Sharpe, DeVeaux, Velleman and Wright) p. 257 #52.

34. Birth order A survey of students in a large Introductory Statistics class asked about their birth order (1 = oldest or only child) and which college of the university they were enrolled in. Here are the data:

Birth Order	1 or only	2 or more	Total
Arts & Sciences	34	23	57
Agriculture	52	41	93
Human Ecology	15	28	43
Other	12	18	30
Total	113	110	223

Suppose we select a student at random from this class.
a) What is the probability we select a Human Ecology student?

b) What is the probability that we select a first-born student?
c) What is the probability that the person is first-born *and* a Human Ecology student?
d) What is the probability that the person is first-born *or* a Human Ecology student?

35. Gambling The game of Craps starts with a player tossing a pair of dice. If the dice add up to 2 or 3 or 12, the player's turn is up and they "crap out." What is the probability of crapping out?

36. Red cards You shuffle a deck of cards, and then start turning them over one at a time. The first one is red. So is the second. And the third. In fact, you are surprised to get 10 red cards in a row. You start thinking, "The next one is due to be black!"
a) Are you correct in thinking that there's a higher probability that the next card will be black than red? Explain.
b) Is this an example of the Law of Large Numbers? Explain.

37. Poker A poker hand consists of five cards. You are dealt all four aces and the king of spades. Your buddy is dealt 10 ♥, 9 ♠, 7 ♣, 4 ♦, 2 ♥. Does your hand have a smaller probability than your buddy's? Explain.

38. Lottery Lilith never selects 1, 2, 3, 4, 5, and 6 when she gets a Lotto 6/49 ticket. After all, everyone knows this will never win. Next time she buys a ticket, however, she selects 4, 9, 14, 22, 36, and 41, believing her choices are better than 1, 2, 3, 4, 5, and 6. Is Lilith's reasoning correct? Explain.

39. Simulations Probabilities at times can be hard to determine mathematically. By building a simulation to model the trials, we can estimate probabilities using computer power instead. Use simulations to estimate the probability that at least two people have the same birthday in a random group of 40 people. Assume there are 365 equally likely birthdays for any individual in the group. Use software to draw randomly from the integers 1–365 (1 = January 1,... 365 = December 31). Repeat your simulations at least 30 times. Combine your results with classmates if possible.

40. Simulations again Suppose two hockey teams are equally matched, and play a series of 10 games. Use simulations to estimate the probability that at least one of the two teams will have a win streak of at least four games at some time in the series. Use software to draw randomly from the digits 0 (loss) and 1 (win). Repeat your simulation at least 50 times. Combine results with classmates if possible. Are you making any assumptions? (If a team wins four straight, is it a hot streak or might it just be chance?)

Just Checking ANSWERS

1. The Law of Large Numbers works only in the long run, not in the short run. The random methods for selecting lottery numbers have no memory of previous picks, so there is no change in the probability that a certain number will come up.

2. **a)**

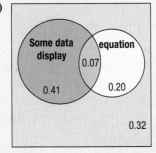

   b) 0.32

   c) 0.41

CHAPTER

12

LEARNING OBJECTIVES

1. Define and calculate conditional probability
2. Understand and use independence
3. Apply the General Multiplication Rule
4. Picture probability with tables, Venn diagrams, and tree diagrams
5. Use trees and Bayes' Rule to reverse the conditioning

Probability Rules!

Natural selection is a mechanism for generating an exceedingly high degree of improbability.

—*Ronald Fisher*

Pecold/Shutterstock

Connections

In this chapter, we develop additional probability concepts and tools, essential for dealing with complex real-world situations.

Is the probability of a car accident the same for all age groups? Insurance companies want to know so they can set rates. Is everyone at equal risk for getting the flu next winter? Medical researchers use factors like age, sex, lifestyle, and family history to estimate different flu probabilities for different individuals. These are examples of *conditional* probabilities. We'll see how probabilities can change when we learn more about the situation at hand.

The concept of independence first arose in Chapter 2 when we looked at contingency tables and asked whether the distribution of one variable was the same for each category of another variable. In

12.1 Probability on Condition

Imagine that while touring England, you spend a day exploring the ancient monument of Stonehenge. As you wander the site, wondering about its purpose, you hear a voice from home. You introduce yourself and are surprised to find out that the person you've just met is from the same province as you. Should you be surprised by this coincidence? Probably not, but it depends on your home province. If you live in Ontario, you shouldn't be at all surprised. But if you're an Inuktituk-speaking Nunavummiuq, meeting another Inuktituk-speaking Nunavummiuq, you should be totally surprised!

A lot of people live in Ontario (about 14 million). Every day, thousands of Ontarians are on holiday, many of them in England. There are probably hundreds of Ontarians at Stonehenge every single day. So if you're from Ontario, meeting another Ontarian shouldn't be at all surprising. By comparison, very few people live in Nunavut (about 38 000). Not many of these people will be exploring England. There likely will be days when there are no Inuktituk-speaking Nunavummiut in England, let alone at Stonehenge.

If you ask a travelling Canadian the name of their home province, it could be said that you are conducting a trial. The sample space would be all the possible outcomes for this trial, **S** = {Alberta, British Columbia, Manitoba, New Brunswick, Newfoundland and Labrador, Northwest Territories, Nova Scotia, Nunavut, Ontario, Prince Edward Island, Quebec, Saskatchewan, Yukon}. The probability of meeting an Ontarian is higher than the probability of meeting a Nunavummiuq, as there are simply more travelling Ontarians to be met. The 13 different outcomes in this sample space are not *equally likely.*

Chapter 9, independently drawn random numbers helped us select random samples. Now we'll see how useful an *assumption of independence* can be in computing probabilities.

Independence is a key assumption of many statistical analyses we'll perform in later chapters, but usually the one most in need of critical appraisal.

	English	French	Total
Atlantic	15	9	24
Quebec	5	50	55
Ontario	89	4	93
West	70	6	76
Total	179	69	248

Table 12.1

A contingency table giving counts of the visitors by official language most commonly used and region.

We can combine the outcomes in different ways to make different events. For example, **A** = {New Brunswick, Newfoundland and Labrador, Nova Scotia, Prince Edward Island} defines the event that the person is from the Atlantic provinces, and **B** = {New Brunswick, Northwest Territories, Nunavut, Quebec, Yukon} may be viewed as the event that French is an official language of the person's home province/territory. (Did you know there are 11 official languages in the Northwest Territories? English, French, and nine Indigenous languages.)

What about the probabilities of events **A** and **B**? The outcomes in the sample space are not equally likely, so even though event **A** contains four outcomes, it will not have probability 4/13. Similarly, event **B** will not have probability 5/13.

Conditional Probability—It Depends . . .

Suppose a keen Statistics student polled all Canadians visiting Stonehenge on Canada Day. He asked the Canadians which official language they prefer to use, French or English, and their province of residence. (Sadly, he met no Nunavummiut that day.)

Table 12.1 is a *contingency table* giving counts of the visitors by their language preference and region of residence. We looked at contingency tables and calculated *conditional distributions* back in Chapter 2. It's only a short step from counts and relative frequencies to probabilities, once we define an appropriate random phenomenon.

Let's focus on this group of people, and suppose we select one person at random from all 248 of them. The probability that we select a French speaker is just the corresponding relative frequency (since we're equally likely to select any of them). There are 69 French speakers out of a total of 248, giving a probability of

$$P(\text{French}) = 69/248 = 0.278$$

The same simple method works for more complicated events, like intersections. For example, what's the probability of selecting a French speaker from Atlantic Canada? Well, 9 French speakers of the 248 sampled live in Atlantic Canada, so the probability is

$$P(\text{French and Atlantic}) = 9/248 = 0.036$$

The probability of selecting a Quebecer is

$$P(\text{Quebecer}) = 55/248 = 0.222$$

What if we are given the information that the selected person is from Quebec? Would that change the probability that the person is a French speaker? You bet it would! The table shows that Quebecers are much more likely to speak French than people from the rest of Canada. When we restrict our focus to Quebecers, we look only at the Quebecer's row of the table, which gives the conditional distribution of language preference given "Quebec." Of the 55 Quebecers in the sample, a whopping 50 of them prefer to use French.

We write the probability that a selected person prefers French *given that we have selected a Quebecer* as

$$P(\text{French} \mid \text{Quebec}) = 50/55 = 0.909$$

For Atlantic Canada, we look at the conditional distribution of language given "Atlantic" shown in the top row of the table. There, of the 24 people from Atlantic Canada in the sample, 9 said French was their preferred language. So, $P(\text{French} \mid \text{Atlantic}) = 9/24 = 0.375$.

In general, when we want the probability of an event from a *conditional* distribution, we write $P(\mathbf{B} \mid \mathbf{A})$ and say "the probability of **B** *given* **A**." A probability that takes into account a given *condition* such as this is called a **conditional probability**. When we work with conditional probabilities, we are assuming that the event after the vertical line has occurred. Saying *given A* is really the same as saying *if A*.

Let's look at what we did. We worked with the counts, but we could work with the probabilities just as well.

To find the probability of the event **B** *given* the event **A**, we restrict our attention to the outcomes in **A**. We then find in what fraction of *those* outcomes **B** also occurred. Formally, we write:

$$P(\mathbf{B}\,|\,\mathbf{A}) = \frac{P(\mathbf{A} \text{ and } \mathbf{B})}{P(\mathbf{A})}.$$

Thinking this through, we can see that it's just what we've been doing, but with probabilities rather than with counts. Look back at the French speakers who were from Quebec. How did we calculate the probability of selecting a French speaker given that we know we've selected a Quebecer, $P(\text{French}\,|\,\text{Quebec})$? There were 55 visitors from Quebec and 50 visitors who are from Quebec *and* are also French speakers. So we calculated the probability to be $50/55$.

Our general rule uses probabilities. It says to find $P(\mathbf{A} \text{ and } \mathbf{B})/P(\mathbf{A})$. We know that we've selected a person from Quebec, so let **A** be the event that the person is from Quebec. We want to find the probability that the person (from Quebec) speaks French, so let **B** be the event that the person is a French speaker.

$$P(\mathbf{B}\,|\,\mathbf{A}) = \frac{P(\mathbf{A} \text{ and } \mathbf{B})}{P(\mathbf{A})} = \frac{P(\text{Quebec } and \text{ French})}{P(\text{Quebec})} = \frac{50/248}{55/248} = \frac{50}{55} = 0.909$$

Applying the conditional probability rule gives the same probabilities we arrived at when we used counts. It doesn't matter which we use, because the total number in the sample cancels out. [Things are fairly simple here since we have equally likely outcomes, so counting is the only tool you really need to calculate probabilities. Probability formulas and rules are essential when outcomes are not equally likely.]

Let's take our rule out for another spin. What's the probability that we have selected a Westerner *given* that the person speaks English? Applying the rule, we get

$$P(\mathbf{West}\,|\,\mathbf{English}) = \frac{P(\text{English } and \text{ West})}{P(\text{English})} = \frac{70/248}{179/248} = \frac{70}{179} = 0.391$$

To use the formula for conditional probability, we have to insist on one restriction. The formula doesn't work if $P(\mathbf{A})$ is 0. But this restriction is fine, since $P(\mathbf{A}) = 0$ means that **A** can't possibly occur. Of course, this means that thinking about the probability that **B** occurs given that **A** has occurred would be nonsense anyway. After all, we can't be "given" the fact that **A** was true if the probability of **A** is 0! (Unless we are dealing with continuous sample spaces, in which case zero probability events do occur!—but the math changes as well.)

For Example FINDING A CONDITIONAL PROBABILITY

RECAP: Recall that the survey discussed in the previous chapter found that 33% of university students are in a relationship, 25% are involved in sports, and 11% are both.

QUESTION: While dining in a campus facility open only to students, you meet a student who is on a sports team. What is the probability that your new acquaintance is in a relationship?

ANSWER: Let **R** = {student in a relationship} and **S** = {student is involved in sports}.

$P(\text{student is in a relationship given that the student is involved in sports}) = P(\mathbf{R}\,|\,\mathbf{S})$

$$= \frac{P(\mathbf{R} \text{ and } \mathbf{S})}{P(\mathbf{S})}$$

$$= \frac{0.11}{0.25}$$

$$\approx 0.44$$

There's a probability of about 0.44 that a student involved with sports is in a relationship. Notice that this is higher than the probability for all students.

Section 12.1 Exercises

1. Sports What is the probability that a person likes to watch hockey, given that she also likes to watch basketball?

	Hockey	No Hockey
Basketball	27	13
No Basketball	38	22

2. Sports again From Exercise 1, if someone doesn't like to watch basketball, what is the probability that she will be a hockey fan?

12.2 Independence and the Multiplication Rule

Statistical independence is a critically important statistical concept, often used as an assumption in probability calculations and statistical analyses, though there are other (rarer) times when independence is a verifiable condition. Informally speaking, two events are *independent* if the occurrence of one event *does not influence* or change the probability of the other event. Conditional probability is the driving force behind the concept of statistical independence, and with our new notation for conditional probabilities, we can write a formal definition:

Events **A** and **B** are **independent** if (and only if) the probability of **B** is the same when we are given that **A** has occurred. That is:

$$P(\mathbf{B}\,|\,\mathbf{A}) = P(\mathbf{B}).$$

In the definition above, you can interchange the roles of the two events **A** and **B**, which would seem to make some intuitive sense (and you can even prove it, by plugging in for $P(\mathbf{B}\,|\,\mathbf{A})$ the conditional probability formula, followed by a little doodling). For example, let's go back to the sample space of the 248 Canadians polled at Stonehenge. Are speaking French and living in the West independent? We need to check if the unconditional probability of either event is unchanged once we condition on the other event. Does $P(\textbf{West}\,|\,\textbf{French}) = P(\textbf{West})$?

$$P(\textbf{West}\,|\,\textbf{French}) = \frac{P(\text{French } and \text{ West})}{P(\text{French})} = \frac{6/248}{69/248} = \frac{6}{69} = 0.087$$

$$P(\textbf{West}) = \frac{76}{248} = 0.306$$

Since these two probabilities are unequal we conclude that speaking French and being a Westerner are dependent events. Alternatively, we could have shown that $P(\text{French|West}) \neq P(\text{French})$. The numbers may differ, but the conclusion has to be the same. Of course, it is no surprise that knowing a person's preferred language would affect the likelihood of being from some region.

Many statistics methods require an **Independence Assumption**. In those situations, it's just not possible to check the formal definition of independence (as we did above), because the required conditional probabilities cannot be mathematically calculated. So we have to argue logically for independence (that occurrence of one event cannot influence chances for the other event), based on the study design. *Assuming* independence doesn't make it true. Always *think* about whether an assumption is reasonable.

Suppose you conduct a survey asking respondents if they support the Green Party. Consider two respondents: Hector and Seema. Will their answers be independent events? Let **A** be the event "Seema answers yes" and let **B** be the event "Hector answers yes." Does $P(\mathbf{B}\,|\,\mathbf{A}) = P(\mathbf{B})$? What if they know each other or are neighbors or siblings? Wouldn't their responses be dependent or related to some degree? Well, that's the beauty of random sampling! As long as the sample is drawn using random numbers, just like the random numbers, the individual selections are independent of each other, and it doesn't matter who is selected at each independent random draw from the population. Anyone, even a neighbor, may be selected, independently of the previous selection. But... but if Hector and Seema know each other and talk to each other *during* the survey, we may lose some of the independence inherent in the selection process, since Seema's responses might be influenced by Hector's opinions, or vice versa. When assessing independence, ask if the events are akin to 'Head on first flip of coin' and 'Head on second flip of coin.' Are there possible *carry-over effects* from one event to the next? If so, the events are not independent.

Independence is the most important idea in this chapter. Spend some time getting your head around its definition and meaning. We'll be using independence in all remaining chapters of this book. And don't be fooled by the word 'dependent' into thinking that a causal connection may exist between events that are dependent. *Statistical* dependence only means association, and as we discussed in Chapter 6, there are many possible explanations for associations other than cause-and-effect.

For Example CHECKING FOR INDEPENDENCE

RECAP: Our survey told us that 33% of university students are in a relationship, 25% are involved in sports, and 11% are both.

QUESTION: Are being in a relationship and being involved in sports independent? Are they disjoint?

ANSWER: Let **R** = {student is in a relationship} and **S** = {student is involved in sports}. If these events are independent, then knowing that a student does sports doesn't affect the probability that he or she is in a relationship.

But we saw that $P(\mathbf{R}\,|\,\mathbf{S}) = P(\mathbf{R} \text{ and } \mathbf{S})/P(\mathbf{S}) = 0.44$ and we know that $P(\mathbf{R}) = 0.33$, so the events are not independent. Students who are involved in sports are more likely to be in a relationship on this campus.

These events are not disjoint either. In fact, 11% of all students on this campus are both involved in sports and in a relationship.

The General Multiplication Rule

Sometimes, we are interested in finding *joint probabilities*, i.e. probabilities for events of the form '**A** and **B**,' which says that both of two events occur. For example, we might want the probability that we will get *both* rain and hail tomorrow. Actually one rule for this is practically staring right at you. Look at the definition of a conditional probability. That equation contains the probability of **A** *and* **B**. Rearranging the equation gives

$$P(\mathbf{A} \text{ and } \mathbf{B}) = P(\mathbf{A}) \times P(\mathbf{B}\,|\,\mathbf{A}).$$

This is the **General Multiplication Rule**. It even seems to make some sense. The probability that two events, **A** and **B**, *both* occur is the probability that event **A** occurs multiplied by the probability that event **B** *also* occurs—that is, by the probability that event **B** occurs *given* that event **A** occurs.

Of course, there's nothing special about which set we call **A** and which one we call **B**. So we should be able to state this the other way around, and indeed we can. It is equally true that

$$P(\mathbf{A} \text{ and } \mathbf{B}) = P(\mathbf{B}) \times P(\mathbf{A}\,|\,\mathbf{B}).$$

Note that this formula relates a conditional probability to a joint probability ('**A** and **B**' is a *joint* event), regardless of how you write or rearrange the terms. Sometimes you know up front the joint probabilities, like the Stonehenge example, and might use this equation to find conditional probabilities (as we did earlier), while at other times, the conditional probabilities are obvious and we might wish to use them to compute joint probabilities. For example, if you draw one card, and then a second card from a standard (well shuffled) deck, it is easy to calculate joint probabilities, using some obvious conditional probabilities. The probability of two aces is just:

P('First card ace' and 'Second card ace') =
P(First card ace) $\times P$(Second card ace | First card ace) = $4/52 \times 3/51$.

The latter conditional probability $3/51$ was obvious, because you know that one card is gone, and one ace is gone, before making the second draw. It's always a matter of utilizing what is known, in the problem, to calculate what you need to know and then finding the tools or equations that may help. Yep, this can get complicated!

The Multiplication Rule for Independent Events

We can always use the general multiplication rule: $P(\mathbf{A} \ and \ \mathbf{B}) = P(\mathbf{A}) \times P(\mathbf{B} \,|\, \mathbf{A})$. But *when events **A** and **B** are independent*, we can write $P(\mathbf{B})$ for $P(\mathbf{B} \,|\, \mathbf{A})$ because the probability of **B** does not depend on **A**, and we can simplify the rule to

$$P(\mathbf{A} \ and \ \mathbf{B}) = P(\mathbf{A}) \times P(\mathbf{B}).$$

The **multiplication rule for independent events** says that for independent events, to find the probability that both events occur, we simply multiply their probabilities together. Formally:

If **A** and **B** are independent events, then $P(\mathbf{A} \ and \ \mathbf{B}) = P(\mathbf{A}) \times P(\mathbf{B})$.

Sometimes people even use this statement as the definition of independent events, but we find the other definition more intuitive.

Fortunately, the Multiplication Rule easily extends to any number of independent events. For example, if we have five independent events A, B, C, D, E, and we want the probability that all of them happen, we just multiply their respective probabilities: P(A and B and C and D and E) = P(A) $\times P$(B) $\times P$(C) $\times P$(D) $\times P$(E). But always *Think* critically about how to justify an assumption of independence. Be ready to explain why the occurrence of any one of these events cannot possibly influence or change the likelihood of any other event's occurrence.

Let's return to the traffic light example from the previous chapter.

For Example APPLYING THE MULTIPLICATION RULE

RECAP: In the previous chapter, we determined that the probability that we encounter a green light at the corner of Portage and Main is 0.35, a yellow light 0.04, and a red light 0.61. Let's think about how many times during your morning commute in the week ahead you might hit a red light there.

QUESTION: What's the probability you find the light red both Monday and Tuesday?

ANSWER: Because the colour of the light I see on Monday doesn't influence the colour I'll see on Tuesday, these are independent events; I can use the Multiplication Rule:

$$P(\text{red Monday AND red Tuesday}) = P(\text{red}) \times P(\text{red})$$
$$= (0.61)(0.61)$$
$$= 0.3721$$

There's about a 37% chance I'll hit red lights both Monday and Tuesday mornings.

QUESTION: What's the probability you don't encounter a red light until Wednesday?

ANSWER: For that to happen, I'd have to see green or yellow on Monday, green or yellow on Tuesday, and then red on Wednesday. I can simplify this by thinking of it as not red on Monday, not red on Tuesday, and then red on Wednesday. Again assuming independence of events on different days,

$$P(\text{not red}) = 1 - P(\text{red}) = 1 - 0.61 = 0.39, \text{ so}$$
$$P(\text{not red Monday AND not red Tuesday AND red Wednesday})$$
$$= P(\text{not red}) \times P(\text{not red}) \times P(\text{red})$$
$$= (0.39)(0.39)(0.61)$$
$$= 0.093$$

There's about a 9% chance that this week I'll hit my first red light there on Wednesday morning.

QUESTION: What's the probability that you'll have to stop *at least once* during the week?

ANSWER: Having to stop at least once means that I have to stop for the light 1, 2, 3, 4, or 5 times next week. It's easier to think about the complement: never having to stop at a red light. Having to stop at least once means that I didn't make it through the week with no red lights. Again assuming independence of events on different days,

P (having to stop at the light at least once in 5 days)
$$= 1 - P(\text{no red lights for 5 days in a row})$$
$$= 1 - P(\text{not red AND not red AND not red AND not red AND not red})$$
$$= 1 - (0.39)(0.39)(0.39)(0.39)(0.39)$$
$$= 1 - (0.39)^5$$
$$= 1 - 0.009$$
$$= 0.991$$

Better be prepared to stop! There's over a 99% chance I'll hit at least one red light there.

> In informal English, you may see "some" used to mean "at least one." "What is the probability that some of the eggs in that carton are broken?" means at least one.

AT LEAST

Note that the phrase "at least" is often a tip-off to think about the complement. Something that happens *at least once* <u>does</u> happen. Happening at least once is the complement of not happening at all, and that's easier to find.

 ## Just Checking

1. Opinion polling organizations contact their respondents by telephone, calling the households using randomly generated telephone numbers. According to the Pew Research Center for the People and the Press, in 2012, the contact rate was about 62%. We can reasonably assume each household's response to be independent of the others. What is the probability that

   a) the interviewer successfully contacts the next household on her list?
   b) the interviewer successfully contacts both of the next two households on her list?
   c) the interviewer's first successful contact is the third household on the list?
   d) the interviewer makes at least one successful contact among the next five households on the list?

Step-by-Step Example   MULTIPLICATION RULE

In 2002, Mars™, the parent company of M&M's®, decided to add a seventh colour to the standard colour lineup of brown, yellow, red, orange, blue, and green, for M&M's Milk Chocolate candies. To decide which colour to add, they surveyed kids in nearly every country of the world and asked them to vote for purple, pink, or teal. The global winner was purple! In Japan, the percentages of responses were 38% pink, 36% teal, and only 16% purple. Let's use Japan's percentages to ask some questions:

1. What is the probability that a Japanese M&M's survey respondent selected at random chose either pink or teal?
2. If we pick two respondents at random, what is the probability that they *both* chose purple?
3. If we pick three respondents at random, what is the probability that *at least one* chose purple?

THINK ➡ The probability of an event is its long-term relative frequency. It can be determined in several ways: by looking at many replications of an event, by deducing it from equally likely events, or by using some other information. Here, we are told the relative frequencies of the three responses.	According to the M&M's website: $P(\text{pink}) = 0.38$ $P(\text{teal}) = 0.36$ $P(\text{purple}) = 0.16$ Each is between 0 and 1, but they don't add up to 1. The remaining 10% of the voters must not have expressed a preference or must have written in another colour. I'll put them together into "other" and add: $P(\text{other}) = 0.10$.
Make sure the probabilities are legitimate. Here, they're not. Either there was a mistake or the other voters must have chosen a colour other than the three given. A check of the reports from other countries shows a similar deficit, so probably we're seeing those who had no preference or who wrote in another colour.	Now I have a legitimate assignment of probabilities.

Question 1. What is the probability that a Japanese M&M's survey respondent selected at random chose either pink or teal?

THINK ➡ Plan Decide which rules to use and check the conditions they require.	The events "pink" and "teal" are disjoint (because one respondent can't choose both colours). I can apply the addition rule.
SHOW ➡ Mechanics Show your work.	$P(\text{pink or teal}) = P(\text{pink}) + P(\text{teal})$ $= 0.38 + 0.36 = 0.74$
TELL ➡ Conclusion Interpret your results in the proper context.	The probability that the respondent chose pink or teal is 0.74.

Question 2. If we pick two respondents at random, what is the probability that they both chose purple?

THINK ➡ Plan The word "both" suggests we want $P(\text{A and B})$, which calls for the multiplication rule. Think about the assumption.	For both respondents to pick purple, the first respondent has to pick purple AND the second has to pick purple. **Independence Assumption:** The choice made by one respondent tells us nothing about the choice of the other, so the events are independent. I can use the multiplication rule.

SHOW ➡ **Mechanics** Show your work.	$P(\text{both purple})$ $= P(\text{first respondent picks purple AND second respondent picks purple})$ $= P(\text{first respondent picks purple}) \times P(\text{second respondent picks purple})$ $= 0.16 \times 0.16 = 0.0256$
TELL ➡ **Conclusion** Interpret your results in the proper context.	The probability that both respondents pick purple is 0.0256.

Question 3. If we pick three respondents at random, what is the probability that at least one chose purple?

THINK ➡ **Plan** The phrase "at least . . ." often flags a question best answered by looking at the complement, and that's the best approach here. The complement of "at least one chose purple" is "none of them chose purple." Think about the assumption.	$P(\text{at least one picked purple})$ $= P(\{\text{none picked purple}\}^{C})$ $= 1 - P(\text{none picked purple}).$ $P(\text{none picked purple}) = P(\text{not purple AND not purple AND not purple})$ **Independence Assumption**: These are independent events because they are choices by three randomly selected respondents. I can use the Multiplication Rule for independent events.
SHOW ➡ **Mechanics** We calculate $P(\text{none purple})$ using the Multiplication Rule. Then we can use the complement rule to get the probability we want.	$P(\text{none picked purple}) = P(\text{first not purple}) \times P(\text{second not purple}) \times P(\text{third not purple})$ $= [P(\text{not purple})]^{3}.$ $P(\text{not purple}) = 1 - P(\text{purple})$ $\qquad\qquad = 1 - 0.16 = 0.84.$ So $P(\text{none picked purple}) = (0.84)^{3} = 0.5927.$ $P(\text{at least one picked purple})$ $= 1 - P(\text{none picked purple})$ $= 1 - 0.5927 = 0.4073.$
TELL ➡ **Conclusion** Interpret your results in the proper context.	There is about a 40.7% chance that at least one of the respondents picked purple.

Independent ≠ Disjoint

Are disjoint events independent? Both concepts seem to imply separation and distinctness, but in fact disjoint events *cannot* be independent.[1] Let's see why.

---

[1] Well, technically, two disjoint events can be independent, but only if the probability of one of the events is 0. For practical purposes, though, we can ignore this case. After all, as statisticians we don't anticipate having data about things that don't happen.

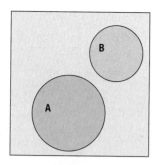

Figure 12.1

Because events **A** and **B** are mutually exclusive, learning that **A** happened tells us that **B** didn't. The probability of **B** has changed from whatever it was to zero. So disjoint events **A** and **B** are not independent.

Consider the two disjoint events {you get an A in this course} and {you get a B in this course}. They're disjoint because they have no outcomes in common. Suppose you learn that you *did* get an A in the course. Now what is the probability that you got a B? You can't get both grades, so it must be 0.

Think about what that means. Knowing that the first event (getting an A) occurred changes your probability for the second event (down to 0). So these events aren't independent.

Mutually exclusive events can't be independent. They have no outcomes in common, so if one occurs, the other doesn't (Figure 12.1). Be sure not to make the mistake of treating disjoint events as if they were independent and applying the multiplication rule for independent events.

 ## Just Checking

2. Typically only about 10% of randomly generated phone calls made by polling firms result in a completed interview. Reasons are varied, but some of the most common include no answer, refusal to cooperate, and failure to complete the call.

    Which of the following events are independent, which are disjoint, and which are neither independent nor disjoint?

    a) **A** = Your telephone number is randomly selected. **B** = You're not at home at dinner time when they call.

    b) **A** = As a selected subject, you complete the interview. **B** = As a selected subject, you refuse to cooperate.

    c) **A** = You are not at home when they call Monday at 11 a.m. **B** = You are employed full time.

Depending on Independence

It may seem easier to think about independent events than to deal with conditional probabilities, but treating all events as independent can lead to wrong, and possibly costly, decisions.

For example, insurance companies that insure against natural disasters like hurricanes or earthquakes might want to avoid insuring too many buildings in the same city. For an event like the Japan earthquake, the probability of two buildings suffering damage is not independent if they're in the same city. Even though the probability of a building being destroyed by an earthquake may be small, the probability that two neighbouring buildings are both destroyed by an earthquake is not the product of the two. In fact, it may be close to the probability that a single one is destroyed. The probabilities can't be multiplied unless the events are independent. For that reason, it's safer for a company to hold policies on buildings that are geographically spread out because then the events are closer to being independent.

Whenever you see probabilities multiplied together, stop and ask whether you think the corresponding events are really independent.

Section 12.2 Exercises

3. **Late to the bus** A student figures that he has a 30% chance of being let out of class late. If he leaves class late, there is a 45% chance that he will miss his bus. What is the probability that he is let out of class late and misses the bus?

4. **Field goals** A nervous kicker usually makes 70% of his first field goal attempts. If he makes his first attempt, his success rate rises to 90%. What is the probability that he makes his first two kicks?

5. Titanic On the *Titanic*, the probability of survival was 0.323. Among first class passengers, it was 0.625. Were *survival* and *ticket class* independent? Explain.

6. Births If the sex of a child is independent of all other births, is the probability of a woman giving birth to a girl after having four boys greater than it was on her first birth? Explain.

7. On time The Sleep Factory in Woodbridge, Ontario, claims that 99.5% of all mattress deliveries are on time. Assume that deliveries of mattresses are independent of each other. Suppose two mattress deliveries are selected at random.
   a) What is the probability that both mattresses will be delivered on time?
   b) What is the probability that both mattresses will be delivered late?

8. Smokers in Nova Scotia
(source: http://www.statcan.gc.ca/tables-tableaux/sum-som/l01/cst01/health74b-eng.htm)

   It is reported that 22.1% of adults in Nova Scotia are smokers. Suppose two adults in Nova Scotia are selected at random.
   a) What is the probability that both are smokers?
   b) What is the probability that both are non-smokers?

12.3 Picturing Probability: Tables, Venn Diagrams, and Trees

In the Facebook–Twitter Step-By-Step example in the previous chapter, a social networking survey found that 71% of online adults use Facebook, 18% use Twitter, and 15% use both. That may not look like enough information to answer all the questions you[2] might have about the problem. But often, the right picture can help. Surprisingly there are three very different ways to do this. We've seen two of them already: Venn diagrams and two-way probability tables. We'll meet a third in a minute.

Let's try to put what we know into a table. Translating percentages to probabilities, what we know looks like this:

| | | Use Facebook | | |
		Yes	No	Total
Use Twitter	Yes	0.15		0.18
	No			
	Total	0.71		1.00

Notice that the 0.71 and 0.18 are *marginal* probabilities and so they go into the *margins*. The 0.15 is the probability of using both sites—Facebook and Twitter—so that's a *joint* probability. Those belong in the interior of the table.

Because the cells of the table show disjoint events, the probabilities always add to the marginal totals going across rows or down columns. So, filling in the rest of the table is quick:

| | | Use Facebook | | |
		Yes	No	Total
Use Twitter	Yes	0.15	0.03	0.18
	No	0.56	0.26	0.82
	Total	0.71	0.29	1.00

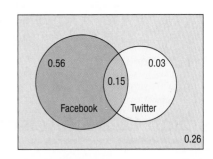

Compare this with the Venn diagram above. Notice which entries in the table match up with the sets in this diagram. Whether a Venn diagram or a table is better to use will depend on what you are given and the questions you're being asked. Try both.

---
[2] or your instructor.

Step-by-Step Example   ARE THE EVENTS DISJOINT? INDEPENDENT?

Let's take another look at the social networking survey. Researchers report that 71% of online adults use Facebook, 18% use Twitter, and 15% use both.

Questions:
1. Are using Facebook and using Twitter mutually exclusive?
2. Are Facebooking and tweeting independent?

ventdusud/Shutterstock

THINK ➡ **Plan** Define the events we're interested in. State the given probabilities.	Let A = {respondent uses Facebook}. Let B = {respondent uses Twitter}. I know that $P(A) = 0.71$ $P(B) = 0.18$ $P(A \text{ and } B) = 0.15$

Question 1. Are using Facebook and using Twitter mutually exclusive?

SHOW ➡ **Mechanics** Disjoint events cannot *both* happen at the same time, so check to see if $P(A \text{ and } B) = 0$.	$P(A \text{ and } B) = 0.15$. Since some adults use both Facebook and Twitter, $P(A \text{ and } B) \neq 0$. The events are not mutually exclusive.
TELL ➡ **Conclusion** State your conclusion in context.	15% of online adults use both Facebook and Twitter, so these are not disjoint events.

Question 2. Are Facebooking and tweeting independent?

THINK ➡ **Plan** Make a table.	<table><tr><td></td><td></td><td colspan="3">**Use Facebook**</td></tr><tr><td></td><td></td><td>Yes</td><td>No</td><td>Total</td></tr><tr><td rowspan="3">**Use Twitter**</td><td>Yes</td><td>0.15</td><td>0.03</td><td>0.18</td></tr><tr><td>No</td><td>0.56</td><td>0.26</td><td>0.82</td></tr><tr><td>Total</td><td>0.71</td><td>0.29</td><td>1.00</td></tr></table>
SHOW ➡ **Mechanics** Does using Facebook change the probability of using twitter? That is, does $P(B\mid A) = P(B)$? Because the two probabilities are *not* the same, the events are not independent.	$P(B\mid A) = \dfrac{P(A \text{ and } B)}{P(A)} = \dfrac{0.15}{0.71} \approx 0.21$ $P(B) = 0.62$ $P(B\mid A) \neq P(B)$
TELL ➡ **Conclusion** Interpret your results in context.	Overall, 18% of online adults use Twitter, but 21% of Facebook users use Twitter. Since respondents who use Facebook are more likely to tweet, the two events are not independent.

Just Checking

3. Canadian farmers grow a large quantity of genetically modified corn. According to recent estimates,

   19% of corn is genetically modified to be resistant to insects.
   70% of corn is genetically modified to be resistant to herbicides.
   13.3% of corn is genetically modified to be resistant to both.

   a) Make a contingency table for the variables *resistant to insects* and *resistant to herbicides*.
   b) If you randomly select an ear of corn, what is the probability it has been genetically modified to be resistant to either insects or herbicides?
   c) If you randomly select an ear of corn that you know has been modified to be resistant to insects, what is the probability it has also been modified to be resistant to herbicides?
   d) Are the genetic modifications of resistance to insects and resistance to herbicides independent?

Tree Diagrams

For men, binge drinking is defined as having five or more drinks in a row, and for women it is defined as having four or more drinks in a row. (The difference is because of the average difference in weight.) According to a study by the Harvard School of Public Health,[3] 44% of university students engage in binge drinking, 37% drink moderately, and 19% abstain entirely. Another study, published in the *American Journal of Health Behavior*, finds that among binge drinkers aged 21 to 34, 17% have been involved in an alcohol-related automobile accident, while among non-bingers of the same age, only 9% have been involved in such accidents.

What is the probability that a randomly selected university student will be a binge drinker who has had an alcohol-related car accident?

To start, we see that the probability of selecting a binge drinker is about 44%. To find the probability of selecting someone who is both a binge drinker and a driver with an alcohol-related accident, we would need to pull out the General Multiplication Rule and multiply the probability of one of the events by the conditional probability of the other given the first.

Or we *could* make a picture. Which would you prefer?

We thought so.

Venn diagrams can't handle conditional probabilities, and tables can get confusing. The kind of picture that helps us look at conditional probabilities is called a **tree diagram**, because it shows sequences of events as paths that look like branches of a tree. It is a good idea to make a tree diagram almost any time you plan to use the General Multiplication Rule. The number of different paths we can take can get large, so we usually draw the tree starting from the left and growing vine-like across the page, although sometimes you'll see them drawn from the bottom up or top down.

The first branch of our tree separates students according to their drinking habits. We label each branch of the tree with a possible outcome and its corresponding probability (Figure 12.2).

Notice that we cover all possible outcomes with the branches. The probabilities add up to one. But we're also interested in car accidents. The probability of having an alcohol-related

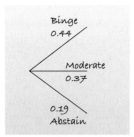

Figure 12.2

We can diagram the three outcomes of drinking and indicate their respective probabilities with a simple tree diagram.

---

[3]H. Wechsler, G. W. Dowdall, A. Davenport, & W. DeJong, "Binge drinking on campus: Results of a national study."

Figure 12.3

Extending the tree diagram, we can show both drinking and accident outcomes. The accident probabilities are conditional on the drinking outcomes, and they change depending on which branch we follow. Because we are concerned only with alcohol-related accidents, the conditional probability P(Accident | Abstinence) must be 0.

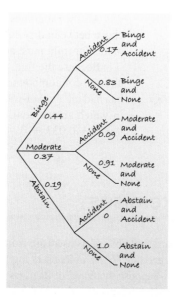

accident *depends* on one's drinking behaviour. Because the probabilities are *conditional,* we draw the alternatives separately on each branch of the tree:

On each of the second set of branches, we write the possible outcomes associated with having an alcohol-related car accident (having an accident or not) and the associated probability (Figure 12.3). These probabilities are different because they are *conditional* depending on the student's drinking behaviour. (It shouldn't be too surprising that those who binge drink have a higher probability of alcohol-related accidents.) The probabilities add up to one, because given the outcome on the first branch, these outcomes cover all the possibilities. Looking back at the General Multiplication Rule, we can see how the tree depicts the calculation. To find the probability that a randomly selected student will be a binge drinker who has had an alcohol-related car accident, we follow the top branches. The probability of selecting a binger is 0.44. The conditional probability of an accident *given* binge drinking is 0.17. The General Multiplication Rule tells us that to find the *joint* probability of being a binge drinker and having an accident, we multiply these two probabilities together:

$$P(\text{Binge } and \text{ Accident}) = P(\text{Binge}) \times P(\text{Accident} | \text{Binge})$$
$$= 0.44 \times 0.17 = 0.075$$

And we can do the same for each combination of outcomes (Figure 12.4):

Figure 12.4

We can find the probabilities of compound events by multiplying the probabilities along the branch of the tree that leads to the event, just the way the General Multiplication Rule specifies. The probability of abstaining and having an alcohol-related accident is, of course, zero.

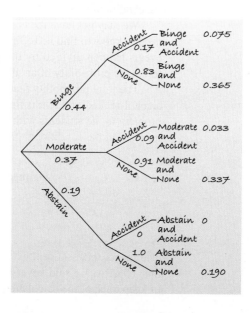

All the outcomes at the far right are disjoint because at each branch of the tree, we chose between disjoint alternatives. And they are *all* the possibilities, so the probabilities on the far right must add up to one.

Because the final outcomes are disjoint, we can add up their probabilities to get probabilities for compound events. For example, what's the probability that a selected student has had an alcohol-related car accident? We simply find *all* the outcomes on the far right in which an accident has happened. There are three and we can add their probabilities: $0.075 + 0.033 + 0 = 0.108$—almost an 11% chance.

Section 12.3 Exercises

9. **Facebook** Facebook reports that 86% of its users are from outside of North America and that 46.2% of its users log on to Facebook every day. Suppose that 9.3% of its users are North American users who log on every day. Make a probability table. Why is a table better than a tree here?

10. **Online banking** A national survey indicated that 30% of adults conduct their banking online. It also found that 40% are younger than 50, and that 25% are younger than 50 and conduct their banking online. Make a probability table. Why is a table better than a tree here?

11. **Facebook again** Suppose that the information in Exercise 9 had been presented in the following way. Facebook reports that 86% of its users are from outside of North America. Of the North American users, two-thirds log on every day. Of the non-North American users, three-sevenths log on every day. Draw a tree for this situation. Why is the tree better than the table in this case? Where are the joint probabilities found?

12. **Online banking again** Suppose that the information in Exercise 8 had been presented in the following way. A national survey of bank customers finds that 40% are younger than 50. Of those younger than 50, 5 of 8 conduct their banking online. Of those older than 50, only 1 of 12 banks online. Draw a tree for this situation. Why is the tree better than the table in this case? Where are the joint probabilities found?

12.4 Reversing the Conditioning and Bayes' Rule

If we know a student has had an alcohol-related accident, what's the probability that the student is a binge drinker? That's an interesting question, but we can't just read it from the tree. The tree gives us $P(\text{Accident} | \text{Binge})$, but we want $P(\text{Binge} | \text{Accident})$—conditioning in the other direction. The two probabilities are definitely *not* the same. We have reversed the conditioning.

We may not have the conditional probability we want, but we do know everything we need to know to find it. To find a conditional probability, we need the probability that both events happen divided by the probability that the given event occurs. We have already found the probability of an alcohol-related accident: $0.075 + 0.033 + 0 = 0.108$.

The joint probability that a student is both a binge drinker and someone who's had an alcohol-related accident is found at the top branch: 0.075. We've restricted the *Who* of the problem to the students with alcohol-related accidents, so we divide the latter by the former to find the conditional probability:

$$P(\text{Binge} | \text{Accident}) = \frac{P(\text{Binge } and \text{ Accident})}{P(\text{Accident})}$$

$$= \frac{0.075}{0.108} = 0.694$$

The chance that a student who has an alcohol-related car accident is a binge drinker is more than 69%! As we said, reversing the conditioning is rarely intuitive, but tree diagrams help us keep track of the calculation when there aren't too many alternatives to consider.

Step-by-Step Example   REVERSING THE CONDITIONING

Alexander Raths/Shutterstock

When two of the authors were in university, there were only three requirements for graduation that were the same for all students: You had to be able to tread water for 2 minutes, you had to learn a foreign language, and you had to be free of tuberculosis. For the last requirement, all freshmen had to take a TB screening test that consisted of a nurse jabbing what looked like a corncob holder into your forearm. You were then expected to report back in 48 hours to have it checked. If you were healthy and TB-free, your arm was supposed to look as though you'd never had the test.

Sometime during the 48 hours, one of us had a reaction. When he finally saw the nurse, his arm was about 50% bigger than normal and a very unhealthy red. Did he have TB? The nurse had said that the test was about 99% effective, so it seemed that the chances must be pretty high that he had TB. How high do you think the chances were? Go ahead and guess. Guess low.

We'll call **TB** the event of actually having TB and **+** the event of testing positive. To start a tree, we need to know $P(\textbf{TB})$, the probability of having TB.[4] We also need to know the conditional probabilities $P(+\mid\textbf{TB})$ and $P(+\mid\textbf{TB}^{\textbf{C}})$. Diagnostic tests can make two kinds of errors. They can give a positive result for a healthy person (a *false positive*) or a negative result for a sick person (a *false negative*). Being 99% accurate usually means a false-positive rate of 1%. That is, someone who doesn't have the disease has a 1% chance of testing positive anyway. We can write $P(+\mid\textbf{TB}^{\textbf{C}}) = 0.01$.

Since a false negative is more serious (because a sick person might not get treatment), tests are usually constructed to have a lower false-negative rate. We don't know exactly, but let's assume a 0.1% false-negative rate. So only 0.1% of sick people test negative. We can write $P(-\mid\textbf{TB}) = 0.001$.

THINK ➡ **Plan** Define the events we're interested in and their probabilities.	Let **TB** = {having **TB**} and **TB**^{C} = {no TB}
	
	   + = {testing positive} and
	
	   − = {testing negative}

Figure out what you want to know in terms of the events. Use the notation of conditional probability to write the event whose probability you want to find.	I know that $P(+\mid\textbf{TB}^{\textbf{C}}) = 0.01$ and $P(-\mid\textbf{TB}) = 0.001$. I also know that $P(\textbf{TB}) = 0.00005$. I'm interested in the probability that the author had TB given that he tested positive: $P(\textbf{TB}\mid +)$.

SHOW ➡ **Plot** Draw the tree diagram. When probabilities are very small like these are, be careful to keep all the significant digits.
 	
To finish the tree we need $P(\textbf{TB}^{\textbf{C}})$, $P(-\mid\textbf{TB}^{\textbf{C}})$, and $P(+\mid\textbf{TB})$. We can find each of these from the Complement Rule:
 	
$P(\textbf{TB}^{\textbf{C}}) = 1 - P(\textbf{TB}) = 0.99995$
 	
$P(-\mid\textbf{TB}^{\textbf{C}}) = 1 - P(+\mid\textbf{TB}^{\textbf{C}})$
    $= 1 - 0.01 = 0.99$ and
 	
$P(+\mid\textbf{TB}) = 1 - P(-\mid\textbf{TB})$
    $= 1 - 0.001 = 0.999$
	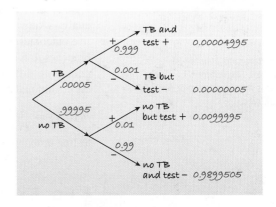

---

[4]This isn't given, so we looked it up. Although TB is a matter of serious concern to public health officials, it is a fairly uncommon disease, with an incidence of about 5 cases per 100,000 in the United States (see www.cdc .gov/tb/default.htm).

Mechanics Multiply along the branches to find the probabilities of the four possible outcomes. Check your work by seeing if they total 1.

Add up the probabilities corresponding to the condition of interest—in this case, testing positive. We can add the probabilities from the tree twigs that correspond to testing positive because the tree shows disjoint events.

Divide the probability of both events occurring (here, having TB and a positive test) by the probability of satisfying the condition (testing positive).

(Check: $0.00004995 + 0.00000005 + 0.0099995 + 0.98995050 = 1$)

$$P(+) = P(\mathbf{TB} \text{ and } +) + P(\mathbf{TB}^C \text{ and } +)$$
$$P(+) = 0.00004995 + 0.0099995$$
$$= 0.01004945$$

$$P(\mathbf{TB} \mid +) = \frac{P(\mathbf{TB} \text{ and } +)}{P(+)}$$
$$= \frac{0.00004995}{0.01004945}$$
$$= 0.00497$$

TELL ➡ **Conclusion** Interpret your result in context.

The chance of having TB after a positive test is less than 0.5%.

When we reverse the order of conditioning, we change the *Who* we are concerned with. With events of low probability, the result can be surprising. That's the reason patients who test positive for HIV, for example, are always told to seek medical counseling. They may have only a small chance of actually being infected. That's why global drug or disease testing can have unexpected consequences if people interpret *testing* positive as *being* positive.

Bayes' Rule

When we have $P(\mathbf{A} \mid \mathbf{B})$ but want the *reverse* probability $P(\mathbf{B} \mid \mathbf{A})$, we need to find $P(\mathbf{A} \text{ and } \mathbf{B})$ and $P(\mathbf{A})$. A tree is often a convenient way of finding these probabilities. It can work even when we have more than two possible events, as we saw in the binge-drinking example. Instead of using the tree, we *could* write the calculation algebraically, showing exactly how we found the quantities that we needed: $P(\mathbf{A} \text{ and } \mathbf{B})$ and $P(\mathbf{A})$. The result is a formula known as Bayes' Rule, after the Reverend Thomas Bayes (1702?–1761), who was credited with the rule after his death, when he could no longer defend himself. Bayes' Rule is quite important in Statistics and is the foundation of an approach to statistical analysis known as Bayesian Statistics. Although the simple rule deals with two alternative outcomes, the rule can be extended to the situation in which there are more than two branches to the first split of the tree. The principle remains the same (although the math gets more difficult). Bayes' Rule is just a formula for reversing the probability from the conditional probability that you're originally given. Bayes' Rule for two events says that

Pearson Education

The Reverend Thomas Bayes is credited posthumously with the rule that is the foundation of Bayesian Statistics.

$P(\mathbf{B} \mid \mathbf{A}) = \dfrac{P(\mathbf{A} \mid \mathbf{B})P(\mathbf{B})}{P(\mathbf{A} \mid \mathbf{B})P(\mathbf{B}) + P(\mathbf{A} \mid \mathbf{B}^C)P(\mathbf{B}^C)}$. Masochists may wish to try it with the TB testing probabilities. (It's easier to just draw the tree, eh?)

For Example REVERSING THE CONDITIONING

A recent Maryland highway safety study found that in 77% of all accidents, the driver was wearing a seatbelt. Accident reports indicated that 92% of those drivers escaped serious injury (defined as hospitalization or death), but only 63% of the non-belted drivers were so fortunate.

QUESTION: What's the probability that a driver who was seriously injured wasn't wearing a seatbelt?

ANSWER: Let B = the driver was wearing a seatbelt, and NB = no belt.

Let I = serious injury or death, and OK = not seriously injured.

I know $P(B) = 0.77$ so $P(NB) = 1 - 0.77 = 0.23$.

Also, $P(OK|B) = 0.92$, so $P(I|B) = 0.08$

and $P(OK|NB) = 0.63$, so $P(I|NB) = 0.37$

$$P(NB\,|\,I) = \frac{P(NB \text{ and } I)}{P(I)} = \frac{0.0851}{0.0616 + 0.0851} = 0.58$$

Even though only 23% of drivers weren't wearing seatbelts, they accounted for 58% of all the deaths and serious injuries.

Just some advice from your friends, the authors: *Please buckle up!* (We want you to finish this course.)

Photos.com/Getty Images

Nicholas Saunderson (1682–1739) was a blind English mathematician who invented a tactile board to help other blind people do mathematics. And he may have been the true originator of "Bayes' Rule."

Who discovered Bayes' Rule? Stigler's "Law of Eponymy" states that discoveries named for someone (eponyms) are usually named for the wrong person. Steven Stigler, who admits he didn't originate the law, is an expert on the history of Statistics, and he suspected that the law might apply to Bayes' Rule. He looked at the possibility that another candidate—one Nicholas Saunderson—was the real discoverer, not the Reverend Bayes. He assembled historical evidence and compared probabilities that the historical events would have happened *given* that Bayes was the discoverer of the rule, with the corresponding probabilities *given* that Saunderson was the discoverer. Of course, what he really wanted to know were the probabilities that Bayes or Saunderson was the discoverer *given* the historical events. How did he *reverse* the conditional probabilities? He used Bayes' Rule and concluded that, actually, it's more likely that Saunderson is the real originator of the rule.

But that doesn't change our tradition of naming the rule for Bayes and calling the branch of Statistics arising from this approach Bayesian Statistics.

Section 12.4 Exercises

13. Facebook final Given the probabilities in Exercise 11, what is the probability that a person is from North America given that he logs on to Facebook every day? Has the probability that he is from North America increased or decreased with the additional information?

14. Online banking last time Given the probabilities in Exercise 12, what is the probability that a person is younger than 50 given that she uses online banking? Has the probability that she is younger than 50 increased or decreased with the additional information?

WHAT CAN GO WRONG?

- **Don't use a simple (special case) probability rule where a general rule is appropriate.** Don't assume independence without reason to believe it. Don't assume that outcomes are disjoint without checking that they are. Remember that the general rules always apply, even when outcomes are in fact independent or disjoint.

- **Don't reverse conditioning naively.** As we have seen, the probability of **A** *given* **B** may not, and, in general does not, resemble the probability of **B** *given* **A**. The true probability may be counterintuitive.

- **Don't confuse "disjoint" with "independent."** Disjoint events *cannot* occur together or jointly. They exclude each other. When one happens, you know the other does not, so $P(\mathbf{B}\,|\,\mathbf{A}) = 0$. Independent events though *must* be able to occur together. When one happens, you know it has no effect on the probability of the other, so $P(\mathbf{B}\,|\,\mathbf{A}) = P(\mathbf{B})$.

- **Don't multiply (unconditional) probabilities of events if they're not independent.** The probability of selecting a student at random who is over 2 metres tall *and* on the basketball team is *not* the probability the student is over 2 metres tall *times* the probability he's on the basketball team. Knowing that the student is over 2 metres tall changes the probability of his being on the basketball team. You can't multiply these probabilities. The multiplication of probabilities of events that are not independent is one of the most common errors people make when trying to find probabilities of *joint* events (A *and* B). If the events are not independent, you need to know the appropriate conditional probabilities to use in the General Multiplication Rule.

- **Don't confuse addition and multiplication when finding probabilities.** To find the probability that events **A** and **B** *both occur*, use the general multiplication rule (or multiplication rule for independent events if appropriate). To find the probability that *either* event **A** or **B** (or both of them) occur, use the general additive rule or (additive rule for disjoint events if appropriate). When we say 'and', expect to multiply probabilities, when we say 'or' expect to add them:

  'And' → Multiply, 'Or' → Add (and maybe subtract).

  Of course, you are only able to use these rules when you know the relevant probabilities in the formulas (and if you don't, try alternative approaches such as the complementary rule!).

What Have We Learned?

Know that the conditional probability of an event **B** given the event **A** is

$$P(\mathbf{B}\,|\,\mathbf{A}) = \frac{P(\mathbf{A} \textit{ and } \mathbf{B})}{P(\mathbf{A})}$$

Know how to define and use independence.

- Events **A** and **B** are independent if $P(\mathbf{A}\,|\,\mathbf{B}) = P(\mathbf{A})$

Know the rules of probability for joint events (**A** *and* **B**) and how to apply them.

- The General Multiplication Rule says that $P(\mathbf{A} \textit{ and } \mathbf{B}) = P(\mathbf{A}) \times P(\mathbf{B}\,|\,\mathbf{A})$
- For independent events, the Multiplication Rule simplifies to $P(\mathbf{A} \textit{ and } \mathbf{B}) = P(\mathbf{A}) \times P(\mathbf{B})$

Know how to make and use tables, Venn diagrams, and tree diagrams to solve problems involving probability.

Know how to use tree diagrams and Bayes' Rule to reverse the conditioning.

Review of Terms

Conditional probability   The conditional probability of the event **B** given the event **A** has occurred is:

$$P(\mathbf{B}\,|\,\mathbf{A}) = \frac{P(\mathbf{A} \textit{ and } \mathbf{B})}{P(\mathbf{A})}$$

$P(\mathbf{B}\,|\,\mathbf{A})$ is read "the probability of **B** *given* **A**." (p. 375)

Independence	Events **A** and **B** are independent when $P(B \mid A) = P(B)$ (p. 377).
Independence Assumption	We often assume certain events to be independent. (So you should think very carefully about whether this assumption is reasonable.) (p. 377)
General Multiplication Rule	For any two events, **A** and **B**, the probability of **A** *and* **B** is $P(A \text{ and } B) = P(A) \times P(B \mid A)$ (p. 378).
Multiplication Rule for independent events	If **A** and **B** are independent events, then the probability of **A** *and* **B** is $P(A \text{ and } B) = P(A) \times P(B)$. This is easily extended to any number of independent events. (p. 379).
Tree diagram	A display of conditional events and probabilities that is helpful in thinking through conditioning (p. 386).

Chapter Exercises

15. Ethanol A Nanos Poll in August 2008 asked 500 adult Ontarians if they thought Ontario should produce more, produce less, or maintain current production of ethanol. Ethanol is made from corn, and can be used as a gasoline additive to increase fuel efficiency.
Here's how they responded:

Response	Number
More ethanol production	160
Less ethanol production	136
Maintain current production level	155
No opinion	49
Total	**500**

Suppose we select three people at random from this sample.
a) What is the probability that all three responded, "Less ethanol production"?
b) What is the probability that none responded, "Maintain current production level"?
c) What assumption did you make in computing these probabilities?
d) Explain why you think that assumption is reasonable.

16. Optimistic A Nanos Poll in November 2008 asked 1000 Canadian adults to compare Canada's international reputation in 2008 with its international reputation in 2007.
Here's how they responded:

Response	Number
Not improved	251
No change	313
Somewhat improved	288
Improved	120
No opinion	28
Total	**1000**

Let's call someone who responded "Somewhat improved" or "Improved" an "optimist." If we select two people at random from this sample,
a) What is the probability that both are optimists?
b) What is the probability that neither is an optimist?
c) What is the probability that one is an optimist and one isn't?
d) What assumption did you make in computing these probabilities?
e) Explain why you think that assumption is reasonable.

17. M&Ms The Masterfoods company says that before the introduction of purple, yellow candies made up 20% of their plain M&Ms; red another 20%; and orange, blue, and green each made up 10%. The rest were brown. If you pick three M&Ms in a row, what is the probability that
a) They are all brown?
b) The third one is the first red one?
c) None are yellow?
d) At least one is green?

18. Blood Canadian Blood Services says that about 46% of the Canadian population have Type O blood, 42% have Type A, 9% have Type B, and the rest have Type AB. Among four potential donors, what is the probability that
a) All are Type O?
b) No one is Type AB?
c) They are not all Type A?
d) At least one person is Type B?

19. Disjoint or independent? In Exercise 17, you calculated probabilities of getting various M&Ms. Some of your answers depended on the assumption that the outcomes described were *disjoint*; that is, they could not both happen at the same time. Other answers depended on the assumption that the events were *independent*; that is, the occurrence of one of them doesn't affect the probability of the other. Do you understand the difference between disjoint and independent?
a) If you draw one M&M, are the events of getting a red one and getting an orange one disjoint or independent or neither?

b) If you draw two M&Ms one after the other, are the events of getting a red on the first and a red on the second disjoint or independent or neither?

c) Can disjoint events ever be independent? Explain.

20. Disjoint or independent, again In Exercise 18, you calculated probabilities involving various blood types. Some of your answers depended on the assumption that the outcomes described were *disjoint*; that is, they could not both happen at the same time. Other answers depended on the assumption that the events were *independent*; that is, the occurrence of one of them doesn't affect the probability of the other. Do you understand the difference between disjoint and independent?

a) If you examine one person, are the events that the person is Type A and that the person is Type B disjoint or independent or neither?

b) If you examine two people, are the events that the first is Type A and the second is Type B disjoint or independent or neither?

c) Can disjoint events ever be independent? Explain.

21. Dice You roll a fair die three times. What is the probability that

a) You roll all 6s?

b) You roll all odd numbers?

c) None of your rolls gets a number divisible by 3?

d) You roll at least one 5?

e) The numbers you roll are not all 5s?

22. Slot machine A slot machine has three wheels that spin independently. Each has 10 equally likely symbols: four bars, three lemons, two cherries, and a bell. If you play, what is the probability that

a) You get three lemons?

b) You get no fruit symbols?

c) You get three bells (the jackpot)?

d) You get no bells?

e) You get at least one bar (an automatic loser)?

23. Champion bowler A certain bowler can bowl a strike 70% of the time. What is the probability that she

a) Goes three consecutive frames without a strike?

b) Makes her first strike in the third frame?

c) Has at least one strike in the first three frames?

d) Bowls a perfect game (12 consecutive strikes)?

24. The train To get to work, a commuter must cross train tracks. The time the train arrives varies slightly from day to day, but the commuter estimates he'll get stopped on about 15% of work days. During a certain five-day work week, what is the probability that he

a) Gets stopped on Monday and again on Tuesday?

b) Gets stopped for the first time on Thursday?

c) Gets stopped every day?

d) Gets stopped at least once during the week?

25. Ottawa Seventy-six percent of Ottawans speak English only when they're at home, 10% speak only French, 11% speak a non-official language, and the remainder speak more than one language. You are conducting a poll by calling Ottawans at random. In your first three calls, what is the probability you talk to

a) Only people who speak only English at home?

b) No people who speak only French at home?

c) At least one person who speaks a non-official language at home?

26. Ottawa revisited Sixty percent of Ottawans drive their own vehicles to work, 7% ride as passengers in someone else's vehicle, 22% take public transit, 10% walk or cycle, and the remainder get to work some other way. You are conducting a poll by calling Ottawans at random. In your first four calls, what is the probability you talk to

a) Only people who drive their own vehicles to work?

b) No one who takes public transit to work?

c) At least one person who walks or cycles to work?

27. Tires You bought a new set of four tires from a manufacturer who just announced a recall because 2% of those tires are defective. What is the probability that at least one of yours is defective?

28. Pepsi For a sales promotion, the manufacturer places winning symbols under the caps of 10% of all Pepsi bottles. You buy a six-pack. What is the probability that you win something?

29. Lottery tickets The Ontario Lottery Association claims that your odds of winning a prize on an Instant Win Crossword game are 1 in 3.09. This means that any ticket has probability $1/3.09 = 0.324$ of winning a prize. Every Friday you buy one Crossword game.

a) What is the probability you don't win a prize next Friday?

b) What is the probability you don't win a prize six Fridays in a row?

c) If you haven't won a prize for the past six Fridays, what is the probability you will win a prize on your next game?

d) What is the probability you will win on two of your next three games?

30. Craps The game of Craps starts with a player tossing a pair of dice. If dice add up to 2 or 3 or 12, the player's turn is up and they "crap out." You're determined to win at this game.

a) What is the probability that you crap out on your next turn?

b) What is the probability that you crap out on your next four turns?

c) If you've crapped out on your last three turns, what is the probability that you don't crap out on your next turn?

d) What is the probability that you crap out on two of your next three turns?

31. **Cards** You draw a card at random from a standard deck of 52 cards. Find each of the following conditional probabilities:
    a) The card is a heart, given that it is red.
    b) The card is red, given that it is a heart.
    c) The card is an ace, given that it is red.
    d) The card is a queen, given that it is a face card.

32. **Pets** In its monthly report, a local animal shelter states that it currently has 24 dogs and 18 cats available for adoption. Eight of the dogs and six of the cats are male. Find each of the following conditional probabilities if an animal is selected at random:
    a) The pet is male, given that it is a cat.
    b) The pet is a cat, given that it is female.
    c) The pet is female, given that it is a dog.

33. **Men's health** The table shows the approximate probabilities that an adult Canadian male has high blood pressure and/or high cholesterol.

		Blood Pressure	
		High	OK
Cholesterol	High	0.11	0.21
	OK	0.16	0.52

    a) What is the probability that a man has both conditions?
    b) What is the probability that he has high blood pressure?
    c) What is the probability that a man with high blood pressure has high cholesterol?
    d) What is the probability that a man has high blood pressure if it's known that he has high cholesterol?

34. **Death Penalty** The following table from a 2013 Angus Reid poll shows the political affiliation of Canadian voters and their positions on the death penalty (in percentage).

	In Favour of the Death Penalty (%)	Not in Favour of the Death Penalty (%)
Conservatives	31.2	8.8
Liberal	10.2	9.8
NDP	17.1	12.9
Bloc Quebecois	3.5	2.5
Green	2.0	2.0

    a) What is the probability that a randomly chosen voter favours the death penalty?
    b) What is the probability that a Liberal favours the death penalty?

    c) What is the probability that a voter who favours the death penalty is a Conservative?
    d) A candidate thinks she has a good chance of gaining the votes of anyone who is a NDP or in favour of the death penalty. What portion of the votes is that?
    Source: http://angusreidglobal.com/wp-content/uploads/2013/03/2013.03.20_Death_CAN.pdf

35. **Cosmetic marketing** A GfK Roper Worldwide survey asked consumers in five countries whether they agreed with the statement "I follow a skin care routine every day." Here are the responses classified by the country of the respondent.

Country	Agree	Disagree	Don't Know	Total
China	361	988	153	1502
France	695	763	81	1539
India	828	689	18	1535
UK	597	898	62	1557
US	668	841	48	1557
Total	3149	4179	362	7690

    a) Among the French, what is the probability that a person agreed with the statement?
    b) Among people who agreed with the statement, what is the probability that the person was French?
    c) If the person was an Asian (either Chinese or Indian), what is the probability that the person disagreed with the statement?
    d) If the person disagreed with the statement, what is the probability that the person was an Asian (either Chinese or Indian)?

36. **Birth order** A survey of students in a large Introductory Statistics class were asked about their birth order (1 = oldest or only child) and which faculty of the university they were studying under. Here are the data:

Faculty	Birth Order		Total
	1 or only	2 or more	
Arts & Sciences	34	23	57
Agriculture	52	41	93
Human Ecology	15	28	43
Other	12	18	30
Total	113	110	223

    a) If we select a student at random, what is the probability that the person is an Arts and Sciences student who is a second child (or more)?

b) Among the Arts and Sciences students, what is the probability that a student is a second child (or more)?

c) Among second children (or more), what is the probability that the student is enrolled in Arts and Sciences?

d) What is the probability that a first or only child is enrolled in the Agriculture faculty?

e) What is the probability that an Agriculture student is a first or only child?

37. Sick kids Seventy percent of kids who visit a doctor have a fever, and 30% of kids with a fever have sore throats. What is the probability that a kid who goes to the doctor has a fever and a sore throat? What did you assume to answer this question?

38. Sick cars Twenty percent of cars that are inspected have faulty pollution control systems. The cost of repairing a pollution control system exceeds $100 about 40% of the time. When a driver takes her car in for inspection, what is the probability that she will end up paying more than $100 to repair the pollution control system? What did you assume to answer this question?

39. Cards again You are dealt a hand of three cards, one at a time. Find the probability of each of the following.

a) The first heart you get is the third card dealt.

b) Your cards are all red (that is, all diamonds or hearts).

c) You get no spades.

d) You have at least one ace.

40. Another hand You pick three cards at random from a deck. Find the probability of each event described below.

a) You get no aces.

b) You get all hearts.

c) The third card is your first red card.

d) You have at least one diamond.

41. Batteries A junk box in your room contains a dozen old batteries, five of which are totally dead. You start picking batteries one at a time and testing them. Find the probability of each outcome.

a) The first two you choose are both good.

b) At least one of the first three works.

c) The first four you pick all work.

d) You have to pick five batteries to find one that works.

42. Shirts The soccer team's shirts have arrived in a big box, and people just start grabbing them, looking for the right size. The box contains 4 medium, 10 large, and 6 extra-large shirts. You want a medium for you and one for your sister. Find the probability of each event described.

a) The first two you grab are the wrong sizes.

b) The first medium shirt you find is the third one you check.

c) The first four shirts you pick are all extra-large.

d) At least one of the first four shirts you check is a medium.

43. Cards III If you draw a card at random from a well-shuffled deck, is getting an ace independent of the suit? Explain.

44. Pets again The local animal shelter in Exercise 32 reported that it currently has 24 dogs and 18 cats available for adoption; eight of the dogs and six of the cats are male. Are the species and sex of the animals independent? (For example, are the events *dog* and *male* independent events?) Explain.

45. Cosmetic marketing again In Exercise 35, we looked at the results of a survey reported by GfK Roper Worldwide about cosmetic marketing.

Country	Agree	Disagree	Don't Know	Total
China	361	988	153	1502
France	695	763	81	1539
India	828	689	18	1535
UK	597	898	62	1557
US	668	841	48	1557
Total	3149	4179	362	7690

a) Are being from India and being from US disjoint? Explain.

b) Are being from India and being from US independent? Explain.

c) Are being from the UK and disagreeing with the statement disjoint? Explain.

d) Are being from the UK and disagreeing with the statement independent? Explain.

46. Birth order, finis In Exercise 36, we looked at the birth orders and faculty choices of some Intro Stats students.

a) Are enrolling in Agriculture and Human Ecology disjoint? Explain.

b) Are enrolling in Agriculture and Human Ecology independent? Explain.

c) Are being first-born and enrolling in Human Ecology disjoint? Explain.

d) Are being first-born and enrolling in Human Ecology independent? Explain.

47. Men's health, again Given the table of probabilities from Exercise 33, are high blood pressure and high cholesterol independent? Explain.

		Blood Pressure	
		High	OK
Cholesterol	High	0.11	0.21
	OK	0.16	0.52

48. Death Penalty, reborn Given the table of percentages from Exercise 34, are being a Conservative and favouring the death penalty independent events? Explain.

	In Favour of the Death Penalty (%)	Not in Favour of the Death Penalty (%)
Conservatives	31.2	8.8
Liberal	10.2	9.8
NDP	17.1	12.9
Bloc Quebecois	3.5	2.5
Green	2.0	2.0

49. Television According to Statistics Canada, about 46% of the television we watch is Canadian programming on Canadian channels. (If this seems high to you, just think of news programs and hockey games.) Close to 15% of what we watch is foreign programming on foreign channels, and we don't watch any Canadian programming on foreign channels.
a) What proportion of our television watching is foreign programming on Canadian channels?
b) Are watching Canadian programming and watching a Canadian channel independent? Explain.

50. Snoring After surveying 995 adults, 81.5% of whom were over 30, the National Sleep Foundation reported that 36.8% of all the adults snored. Thirty-two percent of the respondents were snorers over the age of 30.
a) What percent of the respondents were under 30 and did not snore?
b) Is snoring independent of age? Explain.

51. Luggage Leah is flying from Moncton to Vancouver with a connection in Montreal. The probability that her first flight leaves on time is 0.15. If the flight is on time, the probability that her luggage will make the connecting flight in Montreal is 0.95, but if the first flight is delayed, the probability that the luggage will make it is only 0.65.
a) Are the first flight leaving on time and the luggage making the connection independent events? Explain.
b) What is the probability that Leah's luggage arrives in Vancouver with her?

52. Postsecondary A private high school report contains these statistics:
70% of incoming Grade 9 students attended public primary schools.
75% of the students who came from public primary school go on to university after completing high school.
90% of the students who did not come from public primary schools go on to university after completing high school.

a) Is there any evidence that a private-school student's chances of going to university may depend upon what kind of primary school the student attended? Explain.
b) What percent of this private high school's students go on to university?

53. Late luggage Remember Leah from Exercise 51? Suppose you pick her up at the Vancouver airport and her luggage is not there. What is the probability that Leah's first flight was delayed?

54. Further postsecondary What percent of students who go on to university from the private high school in Exercise 52 attended a public primary school?

55. Absenteeism A company's records indicate that on any given day, about 1% of their day shift employees and 2% of the night shift employees will miss work. Sixty percent of the employees work the day shift.
a) Is absenteeism independent of shift worked? Explain.
b) What percent of employees are absent on any given day?

56. Lungs and smoke Suppose that 23% of adults smoke cigarettes. It's known that 57% of smokers and 13% of nonsmokers develop a certain lung condition by age 60.
a) Explain how these statistics indicate that lung condition and smoking are not independent.
b) What is the probability that a randomly selected 60-year-old has this lung condition?

57. Absenteeism, part II At the company described in Exercise 55, what percent of the absent employees are on the night shift?

58. Lungs and smoke, second puff Based on the statistics in Exercise 56, what is the probability that someone with the lung condition was a smoker?

59. Drunks Police often set up impaired driving checkpoints—roadblocks where drivers are asked a few brief questions to allow the officer to judge whether the person may have been drinking. If the officer does not suspect a problem, drivers are released to go on their way. Otherwise, drivers are detained for a Breathalyzer test that will determine whether they will be arrested. The police say that based on the brief initial stop, trained officers can make the right decision 80% of the time. Suppose the police operate a sobriety checkpoint after 9:00 p.m. on a Saturday night, a time when national traffic safety experts suspect that about 12% of drivers have been drinking.
a) You are stopped at the checkpoint and, of course, have not been drinking. What is the probability that you are detained for further testing?
b) What is the probability that any given driver will be detained?

c) What is the probability that a driver who is detained has actually been drinking?

d) What is the probability that a driver who was released had actually been drinking?

60. Polygraphs Lie detectors are controversial instruments, barred from use as evidence in Canadian courts. Nonetheless, some employers use lie detector screening as part of their hiring process in the hope that they can avoid hiring people who might be dishonest. There has been some research, but no agreement, about the reliability of polygraph tests. Based on this research, suppose that a polygraph can detect 65% of lies, but incorrectly identifies 15% of true statements as lies.

A certain company believes that 95% of its job applicants are trustworthy. The company gives everyone a polygraph test, asking, "Have you ever stolen anything from your place of work?" Naturally, all the applicants answer, "No," but the polygraph identifies some of those answers as lies, making the person ineligible for a job. What is the probability that a job applicant rejected under suspicion of dishonesty was actually trustworthy?

61. Dishwashers Digby's Diner employs three dishwashers. Alma washes 40% of the dishes and breaks only 1% of those she handles. Filip and Kai each wash 30% of the dishes. Filip breaks only 1% of the dishes he washes, but Kai breaks 3% of the dishes he washes. (He, of course, will need a new job soon. . . .) You go to Digby's for dinner one night and hear a dish break at the sink. What is the probability that Kai is on the job?

62. Parts A company manufacturing electronic components for home entertainment systems buys electrical connectors from three suppliers. The company prefers to use supplier A because only 1% of those connectors prove to be defective, but supplier A can deliver only 70% of the connectors needed. The company also must purchase connectors from two other suppliers: 20% from supplier B and the rest from supplier C. The rates of defective connectors from B and C are 2% and 4%, respectively. You buy one of these components, and when you try to use it you find that the connector is defective. What is the probability that your component came from supplier A?

   Just Checking ANSWERS

1. a) 0.62

b) $(0.62)(0.62) = 0.3844$

c) $(1 - 0.62)^2(0.62) = 0.089528$

d) $1 - (1 - 0.62)^5 = 0.992076$

2. a) Independent

b) Disjoint

c) Neither

3. a)

		Resistant to Insects		
		Yes	**No**	
Resistant to	Yes	0.133	0.567	0.7
Herbicides	No	0.057	0.243	0.3
		0.19	0.81	1

b) P(resistant to herbicide OR resistant to insects) $= P$(resistant to herbicide) $+ P$(resistant to insects) $- P$(resistant to herbicide AND resistant to insects) $= 0.70 + 0.19 - 0.133 = 0.757$ (or $0.133 + 0.567 + 0.057$)

c) P(resistant to herbicides | resistant to insects) $= P$(resistant to herbicide AND resistant to insects)$/P$(resistant to insects) $= 0.133/0.19 = 0.70$

d) Yes they are.
P(resistant to herbicides | resistant to insects) $= P$(resistant to herbicide) $= 0.70$

MathXL

MyLab Statistics

Go to MathXL at **www.mathxl.com** or MyLab™ Statistics at **www.mystatlab.com**. You can practise exercises for this chapter as often as you want. The guided solutions will help you find answers step by step. You'll find a personalized study plan available to you too!

1 Learn to work with random variables:
- Describe the probability distribution of a random variable
- Find the expected value (mean) and standard deviation of a random variable
- Calculate the mean and variance of a linear combination of random variables

2 Recognize and work with specific models:
- Understand Bernoulli trials as building blocks for some random phenomena
- Use the Binomial Model to calculate probabilities for counts, and check its assumptions
- Use density curves, such as the Uniform or the Normal, to model distributions of continuous random variables
- Approximate the Binomial distribution with a Normal Model

Connections

How long do products last? Should you expect your computer to die just after the warranty runs out? How can you reduce your risk for developing hepatitis? Is finding 10 leukemia cases in a neighbourhood adjacent to a toxic waste site grounds for a lawsuit? Businesses, medical researchers, and other scientists answer questions like these by modelling the probability of random outcomes

Random Variables

A statistics major was completely hungover the day of his final exam. It was a true/false test, so he decided to flip a coin for the answers. The professor watched the student the entire time as he was flipping the coin . . . writing the answer . . . flipping the coin . . . writing the answer. At the end of two hours, everyone else had left except this student. The professor walked over and said, 'Listen, I see that you didn't study, you didn't even look at the questions. If you are just flipping a coin for your answers, what is taking so long?' Still flipping the coin, the student bitterly replied, 'Shhh! I am checking my answers!'

—*Friedman, Friedman, and Amoo (2002)*

Insurance companies make bets. They bet that you're going to live a long life. You bet that you're going to die sooner. Both you and the insurance company want the company to stay in business, so it's important to find a "fair price" for your bet. Of course, the right price for *you* depends on many factors, and nobody can predict exactly how long you'll live. But when the company averages over enough customers, it can make reasonably accurate estimates of the amount it can expect to collect on a policy before it has to pay its benefit.

> **WHAT IS AN ACTUARY?**
>
> Actuaries are the daring people who put a price on risk, estimating the likelihood and costs of rare events so that they can be insured. That takes financial, statistical, and business skills. It also makes them invaluable to many businesses. Actuaries are rather rare themselves; only about 20 000 work in North America. Perhaps because of this, they are well paid. If you enjoy this course and are good at math, you may want to look into a career as an actuary. Contact the Society of Actuaries or the Canadian Institute of Actuaries (who, despite what you may think, did not pay for this blurb).

Here's a simple example. An insurance company offers a "death and disability" policy that pays $10 000 when you die or $5000 if you are permanently disabled. It charges a premium of only $50 a year for this benefit. Is the company likely to make a profit selling such a plan? To answer this question, the company needs to know the *probability* that its clients will die or be disabled in any year. From actuarial information like this, the company can calculate an average or mean outcome for such policies.

13.1 Centre: The Expected Value or Mean

We'll want to build a probability model in order to answer the questions about the insurance company's risk. First we need to define a few terms. The amount the company pays out on an individual policy is called a **random variable** because its numeric value is based on the outcome of a random event. We use a capital letter, like X, to denote a random variable.

using a special kind of variable—a *random* variable. We'll learn about discrete and continuous probability models for random variables, and meet some of the most important probability models, such as the Binomial model for counts. We'll revisit the Normal model too.

We worked with real data earlier in this book, and before long, we'll get back to dealing with actual data and the link between sample statistics and model parameters, but for now we need to continue studying properties and models of random behaviour.

We'll denote a particular value that it can have by the corresponding lower-case letter, in this case *x*. For the insurance company, *x* can be $10 000 (if you die that year), $5000 (if you are disabled), or $0 (if neither occurs). Because we can list all the outcomes, we might formally call this random variable a **discrete random variable**. Otherwise, we'd call it a **continuous random variable**. (We will say more about the properties of continuous random variables later in this chapter.) The collection of all the possible values and the probabilities that they occur is called the **probability model or probability distribution** for a discrete random variable.

Suppose, for example, that the death rate in any year is 1 out of every 1000 people, and that another 2 out of 1000 suffer some kind of disability. Then we can display the probability model for the random variable X = company payout in a table like:

Policyholder Outcome	Payout x	Probability $P(X = x)$
Death	10 000	$\dfrac{1}{1000}$
Disability	5 000	$\dfrac{2}{1000}$
Neither	0	$\dfrac{997}{1000}$

Table 13.1
The probability model shows all the possible values of the random variable and their associated probabilities.

■ **NOTATION ALERT**

The most common letters for random variables are *X*, *Y*, and *Z*. But be cautious. If you see any capital letter, it just might denote a random variable.

In this chapter, we will use the corresponding lower-case letter to represent one possible outcome of the random variable in some charts and formulas.

Given these probabilities, what should the insurance company expect to pay out? They can't know exactly what will happen in any particular year, but they can calculate what to expect, at least on average. In a probability model, we call this the **expected value** and denote it in either of two ways: $E(X)$ or μ. $E(X)$ is short hand for expected value of X. The symbol μ makes sense since it's a parameter of a model, and not an average calculated from observed data. To understand the expected value calculation, it's useful to imagine some convenient number of clients. In this case, 1000 works because the probabilities all have 1000 as the denominator. If the outcomes in a year matched the probabilities exactly, then out of 1000 clients, 1 died, 2 were disabled, and 997 were safe. The average payout per client would be:

$$\mu = E(X) = \frac{10\ 000(1) + 5000(2) + 0(997)}{1000}$$

So our expected total payout comes to $20 000, or, dividing by 1000, $20 per policy. We can rewrite it as separate terms each divided by 1000.

$$\mu = E(X)$$
$$= \$10\ 000\left(\frac{1}{1000}\right) + \$5000\left(\frac{2}{1000}\right) + \$0\left(\frac{997}{1000}\right)$$
$$= \$20.$$

■ **NOTATION ALERT**

The expected value (or mean of the distribution) of a random variable is written $E(X)$ or μ_x or just μ.

If they sell policies for $50 each, that's an expected profit of $30 per customer—not bad!

Take a good look at the expression now. It's easy to calculate the **expected value** of a (discrete) random variable, given a probability model like the one in Table 13.1—just multiply each possible outcome by the probability that it occurs, and find the sum of those products

$$\mu = E(X) = \sum xP(x)$$

where $P(x)$ is short for $P(X = x)$. Be sure that every possible outcome is included in the sum. And verify that you have a valid probability model to start with—the probabilities should each be between 0 and 1 and should sum to one.

For Example LOVE AND EXPECTED VALUES

On Valentine's Day, the Quiet Nook restaurant offers a Lucky Lovers Special that could save couples money on their romantic dinners. When the waiter brings the bill, he also *brings the four aces* from a deck of cards. He'll shuffle them and lay them out face down on the table. The couple then gets to turn one card over. If it's a black ace, they'll owe the full amount; but if it's the ace of hearts, the waiter will give them a $20 Lucky Lovers discount. If they first turn over the ace of diamonds (hey—at least it's red!), they'll then get to turn over one of the remaining cards, earning a $10 discount for pulling the ace of hearts this time (instead of one of the two remaining black aces).

QUESTION: Based on a probability model for the size of the Lucky Lovers discounts the restaurant will award, what's the expected discount for a couple?

ANSWER: Let $X =$ the Lucky Lovers discount. The probabilities of the three outcomes are:

$$P(X = 20) = P(A\heartsuit) = \frac{1}{4}$$

$$P(X = 10) = P(A\diamondsuit, \text{ then } A\heartsuit) = P(A\diamondsuit) \cdot P(A\heartsuit \mid A\diamondsuit) = \frac{1}{4} \cdot \frac{1}{3} = \frac{1}{12}$$

By the General Multiplication Rule

$$P(X = 0) = P(X \neq 20 \text{ or } 10) = 1 - \left(\frac{1}{4} + \frac{1}{12} \right) = \frac{2}{3}$$

My probability model is:

Outcome	A♥	A♦, then A♥	Black Ace
x	20	10	0
$P(X = x)$	$\frac{1}{4}$	$\frac{1}{12}$	$\frac{2}{3}$

$$E(X) = 20 \cdot \frac{1}{4} + 10 \cdot \frac{1}{12} + 0 \cdot \frac{2}{3} = \frac{70}{12} \approx 5.83$$

The Quiet Nook can expect to give couples a discount of $5.83, on average.

The Law of Large Numbers for Sample Means

Can you recall the Law of Large Numbers (LLN) from Chapter 11? It tells us that the relative frequency of an event converges—gets closer and closer—to a number called its probability when we repeat the random phenomenon over and over in an identical independent fashion. Well, we have a similar result for sample means. As we take more and more independent observations on a random variable X, the sample average of all those observations also must converge in the long run to a number! Voilà—its expected value! This gives us a tremendously useful and practical interpretation of an expected value—as the long run average for X under repeated observations.

This result actually follows easily from our earlier version of the LLN. Consider the Valentine's Day example above. The expected value of $5.83 is certainly not what we expect to see from one customer—in fact, it's not even a possible outcome. But suppose we observe many customers and average their discounts—what will this sample average converge to in the long run? For example, suppose for the next 100 customers, we find that

30 get the $20 discount, 8 get the $10 discount, and 62 get no discount. For this sample of data, the sample mean $\bar{x}$ is:

$$\frac{20 + 20 + \cdots + 10 + 10 + \cdots + 0 + 0 + \cdots}{100} = \frac{20(30) + 10(8) + 0(62)}{100}$$

$$= 20 \cdot \frac{30}{100} + 10 \cdot \frac{8}{100} + 0 \cdot \frac{62}{100}$$

Where are those relative frequencies headed? Of course, by the LLN, toward the probabilities, so if we increase the sample size from 100, this sample mean calculation will look more and more like:

$$20 \cdot \frac{1}{4} + 10 \cdot \frac{1}{12} + 0 \cdot \frac{2}{3}$$

which is just the expected value of X!

 ## Just Checking

1. One of the authors took his minivan in for repair recently because the air conditioner was cutting out intermittently. The mechanic identified the problem as dirt in a control unit. He said that in about 75% of such cases, drawing down and then recharging the coolant a couple of times cleans up the problem—and costs only $60. If that fails, then the control unit must be replaced at an additional cost of $100 for parts and $40 for labour.

   a) Define the random variable and construct the probability model.

   b) What is the expected value of the cost of this repair?

   c) What does that mean in this context?

   Oh—in case you were wondering—the $60 fix worked!

Section 13.1 Exercises

1. **Oranges** A citrus farmer has observed the following distribution for the number of oranges per tree. How many oranges does he expect on average?

Oranges	25	30	35	40
Probability	0.10	0.40	0.30	0.20

2. **Caffeinated** A coffee shop tracks sales and has observed the distribution shown in the following table. What is the average daily sales that it can expect?

# of Sales	145	150	155	160	170
Probability	0.15	0.22	0.37	0.19	0.07

13.2 Spread: The Variance and Standard Deviation

■ **NOTATION ALERT**

The standard deviation of a random variable X is written as $SD(X)$ or σ.

Of course, this expected value (or mean) is not what actually happens to any *particular* insurance policyholder. No individual policy actually costs the company $20. We are dealing with random events, so some policyholders receive big payouts, others nothing. Because the insurance company must anticipate this variability, it also needs to know the *standard deviation* of the random variable. We denote it as $SD(X)$ or σ_x or just σ.

For data, we calculated the **standard deviation** by first computing each observation's deviation from the mean and squaring it. We do that with (discrete) random variables as well. First, we find the deviation of each payout from the mean (expected value):

Policyholder Outcome	Payout x	Probability $P(X = x)$	Deviation $(x - \mu)$
Death	10 000	$\dfrac{1}{1000}$	$(10\,000 - 20) = 9980$
Disability	5000	$\dfrac{2}{1000}$	$(5000 - 20) = 4980$
Neither	0	$\dfrac{997}{1000}$	$(0 - 20) = -20$

Next, we square each deviation. The **variance** of X is just the average of those squared deviations. Taking an expected value is the way we average in a probability model, so we multiply each squared deviation by the corresponding probability and sum those products:

$$Var(X) = 9980^2\left(\frac{1}{1000}\right) + 4980^2\left(\frac{2}{1000}\right) + (-20)^2\left(\frac{997}{1000}\right) = 149\,600$$

Finally, we take the square root to get the standard deviation:

$$SD(X) = \sqrt{149\,600} \approx \$386.78$$

The insurance company can expect an average payout of $20 per policy, with a standard deviation of $386.78.

Think about that. The company charges $50 for each policy and expects to pay out $20 per policy. Sounds like an easy way to make $30. In fact, most of the time (probability 997/1000) the company pockets the entire $50. But would you consider selling your roommate such a policy? The problem is that occasionally the seller of the policy loses big. With probability 1/1000, it will pay out $10 000, and with probability 2/1000, it will pay out $5000. That may be more risk than *you* would be willing to take on. The standard deviation of $386.78 gives an indication that it's no sure thing. That's a pretty big spread (and risk) for an average profit of $30. (But remember that the insurance company sells thousands of such policies—what would be its expected value and standard deviation then? See Section 13.3 and the chapter exercises.)

Here are the explicit formulas for the calculations shown above. You can also write the variance and standard deviation as σ^2 and σ respectively, reflecting their roles as parameters of a probability model

$$\sigma^2 = Var(X) = \sum (x - \mu)^2 P(x)$$
$$\sigma = SD(X) = \sqrt{Var(X)}$$

There is an alternative formula for the variance that can make computations quicker: $\sigma^2 = \sum x^2 P(x) - \mu^2$. If there are other random variables in the study, add subscripts for clarity, like $\mu_X, \mu_Y, \sigma_X, \sigma_Y$. And just as with sample statistics, the variance is in squared units, while the standard deviation is in the same units as the observations (as is the expected value).

For Example FINDING THE STANDARD DEVIATION

RECAP: The probability model for the Lucky Lovers restaurant discount is

Outcome	A♥	A♦, then A♥	Black Ace
x	20	10	0
$P(X = x)$	$\dfrac{1}{4}$	$\dfrac{1}{12}$	$\dfrac{2}{3}$

We found that the restaurant grants an average discount of $\mu = \$5.83$.

QUESTION: What's the standard deviation of the restaurant's discount?

ANSWER: First find the variance:

$$Var(X) = \sum (x - \mu)^2 \cdot P(x)$$

$$= (20 - 5.83)^2 \cdot \frac{1}{4} + (10 - 5.83)^2 \cdot \frac{1}{12} + (0 - 5.83)^2 \cdot \frac{2}{3}$$

$$= 74.306$$

So, $SD(X) = \sqrt{74.306} \approx \$8.62$

Couples can expect on average $5.83 in discounts, with a standard deviation of $8.62.

Step-by-Step Example   EXPECTED VALUES AND STANDARD DEVIATIONS FOR DISCRETE RANDOM VARIABLES

As the head of inventory for Knowway Computer Company, you were thrilled that you had managed to ship two computers to your biggest client the day the order arrived. You are horrified, though, to find out that someone had restocked refurbished computers in with the new computers in your storeroom. The shipped computers were selected randomly from the 15 computers in stock, but 4 of those were actually refurbished.

If your client gets two new computers, things are fine. If the client gets one refurbished computer, it will be sent back at your expense—$100—and you can replace it. However, if both computers are refurbished, the client will cancel the order this month and you'll lose a total of $1000.

Question: What's the expected value and the standard deviation of the company's loss?

THINK ➡ **Plan** State the problem.

Variable Define the random variable.

Plot Make a picture. This is another job for tree diagrams.

If you prefer calculation to drawing, find $P(\mathbf{NN})$ and $P(\mathbf{RR})$, then use the complement rule to find $P(\mathbf{NR}$ *or* $\mathbf{RN})$.

I want to find the company's expected loss for shipping refurbished computers and the standard deviation.

Let X = amount of loss.

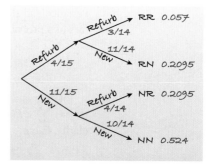

Model List the possible values of the random variable, and determine the probability model.

Outcome	x	$P(X = x)$
Two refurbs	1000	$P(\mathbf{RR}) = 0.057$
One refurb	100	$P(\mathbf{NR}$ *or* $\mathbf{RN}) = 0.2095$ $+ 0.2095 = 0.419$
New/new	0	$P(\mathbf{NN}) = 0.524$

SHOW ➡ **Mechanics** Find the expected value.	$E(X) = 0(0.524) + 100(0.419) + 1000(0.057)$ $= \$98.90$
Find the variance.	$\begin{aligned} Var(X) = &\ (0-98.90)^2(0.524) \\ &+ (100-98.90)^2(0.419) \\ &+ (1000-98.90)^2(0.057) \\ =&\ 51\ 408.79 \end{aligned}$
Find the standard deviation.	$SD(X) = \sqrt{51408.79} = \$226.735$

TELL ➡ **Conclusion** Interpret your results in context.	The cost to the firm has an expected value of \$98.90, with a large standard deviation of \$226.74 reflecting the big range of possible losses.
REALITY CHECK Both numbers seem reasonable. The expected value of \$98.90 is between the extremes of \$0 and \$1000, and there's great variability in the outcome values.	In fact, neither the expected cost nor the standard deviation are particularly informative here since this is a one-time only situation, with just three possible outcomes. The expected value is more usefully interpreted as a *long-term average outcome*. (If the distribution were closer to bell-shaped, these parameters would be much more useful as a summary.)

Section 13.2 Exercises

3. Oranges again What is the standard deviation for Exercise 1?

4. Caffeinated again What is the standard deviation for Exercise 2?

13.3 Combining Random Variables

Our insurance company expects to pay out an average of \$20 per policy, with a standard deviation of about \$387. If we take the \$50 premium into account, we see the company makes a profit of $\$50 - \$20 = \$30$ per policy on average. Suppose the company lowers the premium by \$5 to \$45. It's pretty clear that the expected profit also drops an average of \$5 per policy, to $\$45 - \$20 = \$25$.

What about the standard deviation? We know that adding or subtracting a constant from data shifts the mean but doesn't change the variance or standard deviation. The same is true of random variables.[1]

$$E(X \pm c) = E(X) \pm c \qquad Var(X \pm c) = Var(X) \qquad SD(X \pm c) = SD(X)$$

---

[1] The rules in this section are true for both discrete *and* continuous random variables.

For Example   ADDING A CONSTANT

RECAP: We've determined that couples dining at the Quiet Nook can expect Lucky Lovers discounts averaging $5.83 with a standard deviation of $8.62. Suppose that for several weeks the restaurant has also been distributing coupons worth $5 off any one meal (one discount per table).

QUESTION: If a couple dining there on Valentine's Day brings a coupon, what will be the mean and standard deviation of the total discounts they'll receive?

ANSWER: Let D = total discount (Lucky Lovers plus the coupon); then $D = X + 5$.

$$E(D) = E(X + 5) = E(X) + 5 = 5.83 + 5 = \$10.83$$
$$Var(D) = Var(X + 5) = Var(X) = 8.62^2$$
$$SD(D) = \sqrt{Var(X)} = \$8.62$$

Couples with the coupon can expect on average $10.83 in discounts. The standard deviation is still $8.62.

Back to insurance. . . . What if the company decides to double all the payouts—that is, pay $20 000 for death and $10 000 for disability? This would double the average payout per policy and also increase the variability in payouts. We have seen that multiplying or dividing all data values by a constant changes both the mean and the standard deviation by the same factor. Variance, being the square of standard deviation, changes by the square of the constant. The same is true of random variables. In general, multiplying each value of a random variable by a constant multiplies the mean by that constant and the variance by the *square* of the constant.

$$E(aX) = aE(X) \qquad Var(aX) = a^2 Var(X) \qquad SD(aX) = |a| SD(X)$$

For Example   DOUBLE THE LOVE

RECAP: On Valentine's Day at the Quiet Nook, couples may get a Lucky Lovers discount averaging $5.83 with a standard deviation of $8.62. When two couples dine together on a single bill, the restaurant doubles the discount offer—$40 for the ace of hearts on the first card and $20 on the second.

QUESTION: What are the mean and standard deviation of discounts for such foursomes?

$$E(2X) = 2E(X) = 2(5.83) = \$11.66$$
$$Var(2X) = 2^2 Var(X) = 2^2 \cdot 8.62^2 = 297.2176, \text{ so}$$
$$SD(2X) = \sqrt{297.2176} = \$17.24$$

ANSWER: If the restaurant doubles the discount offer, two couples dining together can expect on average $11.66 in savings with a standard deviation of $17.24.

An insurance company sells policies to more than just one person. How can the company find the total expected payout (and standard deviation) for policies taken over all policyholders? Consider a simple case: just two customers, Mr. Ecks and Ms. Wye. With an expected payout of $20 on each policy, we might predict a total of expected payout $20 + $20 = $40 on the two policies. Nothing surprising there. The expected value of the sum is the sum of the expected values.

$$E(X + Y) = E(X) + E(Y)$$

What about the variability? Is the risk of insuring two people the same as the risk of insuring one person for twice as much? We wouldn't expect both clients to die or become disabled in the same year. Because we've spread the risk, the standard deviation should be smaller than when insuring one person for twice as much. Indeed, this is the fundamental principle behind insurance. By spreading the risk among many policies, a company can keep the standard deviation quite small and predict costs more accurately.

But how much smaller is the standard deviation of the sum? It turns out that, if the random variables are independent (so that the outcome of one variable tells us absolutely nothing about the likely outcome of the other variable), there is a simple Addition Rule for variances: *The variance of the sum of two independent random variables is the sum of their individual variances.*

For Mr. Ecks and Ms. Wye, the insurance company can expect their outcomes to be independent, so (using X for Mr. Ecks's payout and Y for Ms. Wye's)

$$\begin{aligned} Var(X + Y) &= Var(X) + Var(Y) \\ &= 149\,600 + 149\,600 \\ &= 299\,200 \end{aligned}$$

If they had insured only Mr. Ecks for twice as much, there would only be one outcome rather than two *independent* outcomes, so the variance would have been

$$Var(2X) = 2^2 Var(X) = 4 \times 149\,600 = 598\,400, \text{ or}$$

twice as big as with two independent policies.

Of course, variances are in squared units. The company would prefer to know standard deviations, which are in dollars. The standard deviation of the payout for two independent policies is $\sqrt{299\,200} = \$546.99$. But the standard deviation of the payout for a single policy of twice the size is $\sqrt{598\,400} = \$773.56$, or about 40% more.

So, if the company has only two customers, it will have an expected annual total payout of $40 with a standard deviation of about $547.

For Example ADDING THE DISCOUNTS

RECAP: The Valentine's Day Lucky Lovers discount for couples averages $5.83 with a standard deviation of $8.62. We've seen that if the restaurant doubles the discount offer for two couples dining together on a single bill, the expected savings is $11.66 with a standard deviation of $17.24. Some couples may decide instead to get separate bills and pool their two discounts.

QUESTION: You and your amour go to this restaurant with another couple and agree to share any benefit from this promotion. Does it matter whether you pay separately or together?

ANSWER: Let X_1 and X_2 represent the two separate discounts, and T the total; then $T = X_1 + X_2$.

$$E(T) = E(X_1 + X_2) = E(X_1) + E(X_2) = 5.83 + 5.83 = \$11.66,$$

so the expected saving is the same either way.

The cards are reshuffled for each couple's turn, so the discounts couples receive are *independent*. It's okay to add the variances:

$$Var(T) = Var(X_1 + X_2) = Var(X_1) + Var(X_2) = 8.62^2 + 8.62^2 = 148.6088$$

$$SD(T) = \sqrt{148.6088} = \$12.19$$

When two couples get separate bills, there's less variation in their total discount. The standard deviation is $12.19, compared to $17.24 for couples who play for the double discount on a single bill.

In general,

- *The mean of the sum of two random variables is the sum of the means.*
- *The mean of the difference of two random variables is the difference of the means.*
- *If the random variables are independent, the variance of their sum or difference is the sum of the variances.*

$$E(X \pm Y) = E(X) \pm E(Y)$$

$$Var(X \pm Y) = Var(X) + Var(Y) \text{ when the variables are independent}$$

Wait a minute! Is that third part correct? Do we always *add* variances? Yes. Think about the two insurance policies. Suppose we want to know the mean and standard deviation of the *difference* in payouts to the two clients. Since each policy has an expected payout of $20, the expected difference is $20 − $20 = $0. If we also subtract variances, we get $0 too, and that surely doesn't make sense. Note that if the outcomes for the two clients are independent, the difference in payouts could range from $10 000 − $0 = $10 000 to $0 − $10 000 = −$10 000, a spread of $20 000 (same as for the sum). The variability in differences increases as much as the variability in sums. If the company has two customers, the difference in payouts has a mean of $0 and a standard deviation of about $547 (again).

For Example WORKING WITH DIFFERENCES

RECAP: The Lucky Lovers discount at the Quiet Nook averages $5.83 with a standard deviation of $8.62. Just up the street, the Wise Fool restaurant has a competing Lottery of Love promotion. There a couple can select a specially prepared chocolate from a large bowl and unwrap it to learn the size of their discount. The restaurant's manager says the discounts vary with an average of $10.00 and a standard deviation of $15.00.

QUESTION: How much more can you expect to save at the Wise Fool? With what standard deviation?

ANSWER: Let W = discount at the Wise Fool, X = the discount at the Quiet Nook, and D = the difference: $D = W - X$. These are different promotions at separate restaurants, so the outcomes may be assumed to be independent.

$$E(W - X) = E(W) - E(X) = 10.00 - 5.83 = \$4.17$$

$$\begin{aligned} SD(W - X) &= \sqrt{Var(W - X)} \\ &= \sqrt{Var(W) + Var(X)} \\ &= \sqrt{15^2 + 8.62^2} \\ &\approx \$17.30 \end{aligned}$$

Discounts at the Wise Fool will average $4.17 more than at the Quiet Nook, with a standard deviation for the difference of $17.30.

For random variables, does *X* + *X* + *X* = 3*X*? Be careful. Insuring one person for $30 000 is not the same risk as insuring three people for $10 000 each (as we saw earlier in the case of two people). When each instance represents a different outcome for the same random variable, it's easy to fall into the trap of writing all of them with the same symbol. Don't make this common mistake. Make sure you write each instance as a *different* random variable. Just because each random variable describes a similar situation (drawing from the same probability distribution) doesn't mean that each random outcome will be the same.

These are *random* variables, not the variables you saw in Algebra. Being random, they take on different values each time they're evaluated. So what you really mean is $X_1 + X_2 + X_3$ where the different subscripts indicate that there are separate first, second, and third observations (on the same variable X). Written this way, it's clear that the sum shouldn't necessarily equal three times *anything*. [For another example, think about the difference between tossing three dice and summing three outcomes versus tossing one die and tripling the outcome.]

For Example SUMMING A SERIES OF RANDOM OUTCOMES

RECAP: The Quiet Nook's Lucky Lovers promotion offers couples discounts averaging $5.83 with a standard deviation of $8.62. The restaurant owner is planning on serving 40 couples on Valentine's Day.

QUESTION: What's the expected total of the discounts the owner will give? With what standard deviation?

ANSWER: Let $X_1, X_2, X_3, \ldots X_{40}$ represent the discounts to the 40 couples, and T the total of all the discounts. Then:

$$T = X_1 + X_2 + X_3 + \ldots + X_{40}$$
$$E(T) = E(X_1 + X_2 + X_3 + \ldots + X_{40})$$
$$= E(X_1) + E(X_2) + E(X_3) + \ldots + E(X_{40})$$
$$= 5.83 + 5.83 + 5.83 + \ldots + 5.83$$
$$= \$233.20$$

Reshuffling cards between couples makes the discounts independent (Don't forget to think about this assumption before adding variances!), so:

$$SD(T) = \sqrt{Var(X_1 + X_2 + X_3 + \ldots + X_{40})}$$
$$= \sqrt{Var(X_1) + Var(X_2) + Var(X_3) + \ldots + Var(X_{40})}$$
$$= \sqrt{8.62^2 + 8.62^2 + 8.62^2 + \ldots + 8.62^2}$$
$$\approx 54.52$$

The expected total discount to the 40 couples is $233.20 on average, with a standard deviation of $54.52.

Just Checking

2. Suppose the time it takes a customer to get and pay for seats at the ticket window of a baseball park is a random variable with a mean of 100 seconds and a standard deviation of 50 seconds. When you get there, you find only two people in line in front of you.

    a) How long is your expected wait for your turn to get tickets?

    b) What's the standard deviation of your wait time?

    c) What assumption did you make about the two customers in finding the standard deviation?

What is the Actual Probability Model?

Although we know how to find the expected value and the variance for the sum of two independent random variables, that may not be all we'd like to know. It would be great if we could go directly from models of each random variable to a model for their sum. But the probability model for the sum of two random variables is *not* necessarily the same as the model we started with *even when the variables are independent.*

That's easy to see for discrete random variables because we can just think about their possible values. When we looked at insuring one person, either we paid out $0, $5000, or $10 000. Suppose we insure three people. The possibilities are not just $0, $15 000, or $30 000. We might have to pay out in total $5000, or $20 000, or other possible combinations of the three outcomes. Even though the expected values simply add, the probability model is different and keeps changing depending on how many random variables we sum. Nor is the (simpler) probability model for $3X$ the same as the model for $X_1 + X_2 + X_3$. That's another reason to use the proper subscripted form rather than write $X + X + X$.

Fortunately, there are some special cases where we readily know the model for a sum or difference, such as for Normal random variables (Section 13.6). As well as some shocking (and very useful) news about what happens to the model when you add a large number of random variables together (Central Limit Theorem in Chapter 14).

*Correlation and Covariance

In Chapter 6, we saw that the association of two variables could be measured with their correlation. What about random variables? We can talk about the correlation between random variables, too. But it's easier to start with the related concept of *covariance*.

If X is a random variable with expected value $E(X) = \mu$ and Y is a random variable with expected value $E(Y) = v$, then the **covariance** of X and Y is defined as

$$Cov(X, Y) = E[(X - \mu)(Y - v)]$$

The covariance, like the correlation, measures how X and Y vary together ($co = together$). Think about a scatterplot of two highly positively correlated variables. When X is above its mean, Y tends to be above its mean. And when X is below its mean, Y tends to be below its mean. So, on average, the product in the covariance formula is positive.

Covariance has a few properties worth remembering:

- $Cov(X, Y) = Cov(Y, X)$
- $Cov(X, X) = Var(X)$
- $Cov(cX, dY) = cdCov(X, Y)$, for any constants c and d
- If X and Y are independent, then $Cov(X, Y) = 0$, but be careful not to assume that the converse is true; it is possible for two variables to have zero covariance yet not be independent.

The covariance gives us the extra information we need to find the variance of the sum or difference of two random variables when they are not independent:

$$Var(X \pm Y) = Var(X) + Var(Y) \pm 2Cov(X, Y)$$

When X and Y are independent, their covariance is zero, so we have the Pythagorean Theorem of statistics, as we saw earlier.[2]

Covariance can take any number between $-\infty$ and $+\infty$. If X and Y have large values, the covariance will be large as well. And that's its problem. If we know the covariance between X and Y is -64.5, it doesn't give us a real sense of how negatively related they are. To fix the "problem," we can divide the covariance by each of the standard deviations to get the **correlation**:

$$Corr(X, Y) = \frac{Cov(X, Y)}{\sigma_x \, \sigma_y}$$

Correlation of random variables is usually denoted as ρ. Correlation of random variables has many of the properties we saw in Chapter 6 for the sample correlation coefficient, r.

- The sign of a correlation coefficient gives the direction of the association between random variables.
- Correlation is always between -1 and $+1$.
- Correlation treats X and Y symmetrically. The correlation of X with Y is the same as the correlation of Y with X.
- Unlike covariance, correlation has no units.
- Correlation is not affected by changes in the centre or scale of either variable.

---

[2]If you want to follow the geometry one step further, the more general form for correlated variables is related to the Law of Cosines.

Section 13.3 Exercises

5. Salary An employer's total salary expenses for a 5-day workweek have a mean of $1250 with a standard deviation of $129. On the weekends, his total salary expenses have a mean of $450 with a standard deviation of $57. What is the mean and standard deviation of his total weekly salaries?

6. Golf scores A golfer keeps track of his score for playing nine holes of golf (half a normal golf round). His mean score is 85 with a standard deviation of 11. Assuming that the second nine has the same mean and standard deviation, what is the mean and standard deviation of his total score if he plays a full 18 holes?

13.4 The Binomial Model

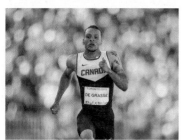

Christopher Morris/ZUMA Wire/Alamy

Suppose a cereal manufacturer puts pictures of Canadian medal winners from the 2016 Summer Olympic Games on cards in boxes of cereal, in the hope of increasing sales. The manufacturer announces that 20% of the boxes contain a picture of Andre De Grasse, 30% a picture of Christine Sinclair, and the rest a picture of Penny Oleksiak.

You're a huge Andre De Grasse fan so you really hope to find some De Grasse pictures. If you open, say, five boxes, how likely is it that you will find one De Grasse card? Two De Grasse cards? No De Grasse cards? We'll see that the Binomial Model is just what we need to answer these questions. But first we need to define Bernoulli trials (named after Jacob Bernoulli, whom you can recall from Chapter 11), which are the building blocks for this model.

We'll keep the assumption that pictures are distributed at random and we'll trust the manufacturer's claim that 20% of the cards are De Grasse. So when you open a box, the probability that you succeed in finding De Grasse is 0.20. Now we'll call the act of opening *each* box a trial, and note that:

- There are only two possible outcomes (called *success* and *failure*) on each trial. Either you get De Grasse's picture (success), or you don't (failure).
- The probability of success, denoted p, is the same on every trial. Here $p = 0.20$.
- The trials are independent. Finding De Grasse in the first box does not change what might happen when you reach for the next box.

Situations like this occur often, and are called **Bernoulli trials**. Common examples of Bernoulli trials include tossing a coin, looking for defective products rolling off an assembly line, or even shooting free throws in a basketball game. There may well be more than only two distinct outcomes possible in a trial—in our example there are three possible outcomes when opening a box—but if we are only interested in *whether or not* one particular outcome occurs (De Grasse picture), we have effectively *dichotomized* the trials into Success and Failure. We can use Bernoulli trials to build a wide variety of useful probability models.

The Binomial Model

Suppose you buy five boxes of cereal. What is the probability you get *exactly* two pictures of Andre De Grasse? We want to find the probability of getting two successes among five Bernoulli trials.

We're interested in the *number of successes* in the five trials, so we'll call it $X =$ number of successes. We want to find $P(X = 2)$. This is an example of a **Binomial probability**. It takes two parameters to define this **Binomial model**: the number of trials, n, and the probability of success, p. We denote this model Binom(n, p). Here, $n = 5$ trials, and $p = 0.2$, the probability of finding a Andre De Grasse card in any single trial.

Exactly two successes in five trials means two successes and three failures. It seems logical that the probability should be $(0.2)^2(0.8)^3$. Too bad! It's not that easy. That calculation would give you the probability of finding De Grasse in the first two boxes and not in the next three—*in that order*. But you could find De Grasse in the third and fifth boxes and still have two successes. The probability of those outcomes in that particular order is $(0.8)(0.8)(0.2)$

(0.8)(0.2). That's also $(0.2)^2(0.8)^3$. In fact, the probability will always be the same, no matter what order the successes and failures occur in. Anytime we get two successes in five trials, no matter what the order, the probability will be $(0.2)^2(0.8)^3$. We just need to take into account all of the possible orders in which the outcomes can occur.

Fortunately, these possible orders are *disjoint*. (For example, if your two successes came on the first two trials, they couldn't come on the last two.) So we can use the Addition Rule and add up the probabilities for all the possible orderings. Since the probabilities are all the same, we only need to know how many orders are possible. For small numbers, we can just make a tree diagram and count the branches. For larger numbers this isn't practical, so we let the computer or calculator do the work.

Each different arrangement of k successes in n trials may be called a "combination." The total number of ways that can happen is written $\binom{n}{k}$ or $_nC_k$ and pronounced "n choose k." We evaluate this with the following formula:

$$_nC_k = \frac{n!}{k!(n-k)!} \text{ where } n! \text{ (pronounced "}n\text{ factorial") } = n \times (n-1) \times \cdots \times 1$$

For two successes in five trials,

$$_5C_2 = \frac{5!}{2!(5-2)!} = \frac{5 \times 4 \times 3 \times 2 \times 1}{2 \times 1 \times 3 \times 2 \times 1} = \frac{5 \times 4}{2 \times 1} = 10.$$

So there are 10 ways to get two De Grasse pictures in five boxes, and the probability of each is $(0.2)^2(0.8)^3$. Now we can find what we wanted. If $X = \#$ successes, then:

$$P(X = 2) = 10(0.2)^2(0.8)^3 = 0.2048$$

In general, the probability of exactly k successes in n trials is $_nC_kp^kq^{n-k}$ since p^kq^{n-k} is the probability of any one sequence of k successes (and hence $n-k$ failures), while $_nC_k$ is the number of such sequences.

Using this formula, we could find the expected value by adding up $kP(X = k)$ for all values, but it would be a long, hard way to get an answer that you already know intuitively. What's the expected value? If we have five boxes and De Grasse's picture is in 20% of boxes, then we would expect on average to have $5(0.2) = 1$ success. If we had 100 trials with probability of success 0.2, how many successes would you expect? Can you think of any reason not to say 20? It seems so simple that most people wouldn't even stop to think about it. Multiply the probability of success by n. In other words, $E(X) = np$.[3]

The standard deviation is less obvious; you can't just rely on your intuition (but you can see a proof in the upcoming Math Box). Fortunately, the formula for the standard deviation also boils down to something simple: $SD(X) = \sqrt{npq}$. In 100 boxes of cereal, we expect to find 20 Andre De Grasse cards on average, with a standard deviation of $\sqrt{100 \times 0.8 \times 0.2} = 4$ pictures.

Time to summarize. A Binomial probability model describes the probability distribution for the number of successes in a specified number of trials. It takes two parameters to specify this model: the number of trials n and the probability of success p.

BINOMIAL PROBABILITY MODEL FOR BERNOULLI TRIALS: BINOM(n, p)

n = number of trials
p = probability of success (and $q = 1 - p$ = probability of failure)
X = number of successes in n trials

$$P(X = x) = {}_nC_xp^xq^{n-x}, \text{ where } {}_nC_x = \frac{n!}{x!(n-x)!} \text{ for } x = 0, 1, \ldots n$$

Mean or Expected number of successes in n trials: $\mu = np$
Standard Deviation: $\sigma = \sqrt{npq}$

---

[3] It's an amazing fact that this does work out the long way. That is, $\sum_{x=0}^{n} x \cdot {}_nC_xp^xq^{n-x} = np$, something that is not obvious. In the Math Box, we work it out another way.

Binom(n, p) will be our shorthand for saying "Binomial Model with parameters *n* and *p*." But be sure to plug in the values for *n* and *p* that match your actual study. The assumptions you need to check are those for Bernoulli trials. Often, the most critical assumption *and* the one most likely to be violated is the assumption that the trials are independent. We'll say more about this shortly, and also introduce a useful *condition* which will help us check the reasonableness of this assumption in certain situations.

MATH BOX

To derive the formulas for the mean and standard deviation of a Binomial model, we start with the most basic situation.

Consider a single Bernoulli trial with probability of success *p*. Let's find the mean and variance of the number of successes.

Here's the probability model for the number of successes:

x	0	1
$P(X = x)$	q	p

Find the expected value: $E(X) = 0q + 1p$
$$E(X) = p$$

And now the variance: $Var(X) = (0 - p)^2 q + (1 - p)^2 p$
$$= p^2 q + q^2 p$$
$$= pq(p + q)$$
$$= pq(1)$$
$$Var(X) = pq$$

What happens when there is more than one trial, though? A Binomial model simply counts the number of successes in a series of *n* independent Bernoulli trials. That makes it easy to find the mean and standard deviation of a Binomial random variable, *Y*.

Let $Y = X_1 + X_2 + X_3 + \cdots X_n$ where $X_i = 1$ if we have a success in the ith trial, and $= 0$ if we have a failure. Thus *Y* counts the number of successes. Then

$$E(Y) = E(X_1 + X_2 + X_3 + \cdots + X_n)$$
$$= E(X_1) + E(X_2) + E(X_3) + \cdots + E(X_n)$$
$$= p + p + p + \cdots + p \text{ (There are } n \text{ terms.)}$$

So, as we thought, the mean is $E(Y) = np$.

And since the trials are by assumption independent, the variances add:

$$Var(Y) = Var(X_1 + X_2 + X_3 + \cdots + X_n)$$
$$= Var(X_1) + Var(X_2) + Var(X_3) + \cdots + Var(X_n)$$
$$= pq + pq + pq + \cdots + pq \text{ (Again, } n \text{ terms.)}$$
$$Var(Y) = npq$$

Voilà! The standard deviation is $SD(Y) = \sqrt{npq}$

The Critical Assumption of Independence

One of the important requirements for Bernoulli trials is that the trials be independent. Sometimes that's a reasonable assumption. Is it true for our example? We said that whether we find a Andre De Grasse card in one box has no effect on the probabilities in other boxes. This is *almost* true. Technically, if exactly 20% of the boxes have Andre De Grasse cards, then when you find one, you've reduced the number of remaining Andre De Grasse cards. With a few million boxes of cereal, though, the difference is hardly worth mentioning.

■ **NOTATION ALERT**

Now we have two more reserved letters. Whenever we deal with Bernoulli trials, *p* represents the probability of success, and *q* the probability of failure. (Of course, $q = 1 - p$.)

But if you knew there were two Andre De Grasse cards hiding in the 10 boxes of cereal on the grocery store shelf, then finding one in the first box would clearly change your chances of finding De Grasse in the next box.

If we had an infinite number of boxes, there wouldn't be a problem. It's selecting from a finite population that causes the probabilities to change, making the trials not independent. Obviously, taking two out of 10 boxes changes the success probability from trial to trial. Taking even a few hundred out of millions, though, makes very little difference. Fortunately, we have a rule of thumb for the in-between cases. It turns out that if we randomly sample less than 10% of the population, we can pretend that the trials are independent and still calculate probabilities that are sufficiently accurate.

The 10% Condition: Bernoulli trials must be independent. If that assumption is violated, it is still okay to proceed as long as the sample is smaller than 10% of the population.[4]

Dependence among trials can arise in other ways as well, such as if there is a learning or practice effect from trial to trial (in say a memory training experiment). So, also check that there is little "carry-over" effect from one trial to the next.

For Example SPAM AND THE BINOMIAL MODEL

RECAP: The communications monitoring company Postini has reported that 91% of email messages are spam. Suppose your inbox contains 25 messages.

QUESTIONS: What are the expected value and standard deviation of the number of real messages you'll find in your inbox? What is the probability that you'll find only one or two real messages?

ANSWER: I assume that messages arrive independently and at random, with the probability of success (a real message) $p = 1 - 0.91 = 0.09$. Let $X =$ the number of real messages among 25. I can use the model Binom(25, 0.09).

$$E(X) = np = 25(0.09) = 2.25$$
$$SD(X) = \sqrt{npq} = \sqrt{25(0.09)(0.91)} = 1.43$$
$$P(X = 1 \text{ or } 2) = P(X = 1) + P(X = 2)$$
$$= {}_{25}C_1 (0.09)^1(0.91)^{24} + {}_{25}C_2(0.09)^2(0.91)^{23}$$
$$= 0.2340 + 0.2777$$
$$= 0.5117$$

Among 25 email messages, I expect on average 2.25 that aren't spam, with a standard deviation of 1.43 messages. There's just over a 50% chance that one or two of my 25 emails will be real messages.

Step-by-Step Example WORKING WITH A BINOMIAL MODEL

Canadian Blood Services is a not-for-profit organization that collects nearly one million units of blood per year. About 6% of people have O-negative blood. They are called *universal donors* because O-negative blood can be given to anyone, regardless of the recipient's blood type. Suppose that 20 blood donors arrive at one of this organization's donation centres.

Questions:

1. What are the mean and standard deviation of the number of universal donors among them?
2. What is the probability that there are two or three universal donors?

---

[4]If the sample is bigger than 10% of the population, we would need to use the Hypergeometric model, which is even more complicated than the Binomial.

THINK ➡ **Plan** State the questions.	I want to know the mean and standard deviation of the number of universal donors among 20 people, and the probability that there are two or three of them.
Check to see that these are Bernoulli trials.	✓ There are two outcomes: success = O-negative failure = other blood types ✓ $p = 0.06$ because volunteers tend to show up more or less at random (at least with respect to blood type). ✓ **10% Condition**: Trials are not independent because the population is finite, but fewer than 10% of all possible donors are lined up.
Variable Define the random variable. **Model** Specify the model.	Let X = number of O-negative donors among $n = 20$ people. I can model the probability distribution of X with the Binom(20, 0.06).
SHOW ➡ **Mechanics** Find the expected value and standard deviation.	$E(X) = np = 20(0.06) = 1.2$ $SD(X) = \sqrt{npq} = \sqrt{20(0.06)(0.94)} \approx 1.06$ $P(X = 2 \text{ or } 3) = P(X = 2) + P(X = 3)$ $\qquad = {}_{20}C_2(0.06)^2(0.94)^{18}$ $\qquad\qquad + {}_{20}C_3(0.06)^3(0.94)^{17}$ $\qquad \approx 0.2246 + 0.0860$ $\qquad = 0.3106$
TELL ➡ **Conclusion** Interpret your results in context.	In groups of 20 randomly selected blood donors, I expect to find on average 1.2 universal donors, with a standard deviation of 1.06. About 31% of the time, I'd find two or three universal donors among the 20 people.

Section 13.4 Exercises

7. Toasters A manufacturer ships toasters in cartons of 20. In each carton, they estimate a 5% chance that one of the toasters will need to be sent back for minor repairs. What is the probability that in a carton, there will be exactly 3 toasters that need repair?

8. Soccer A soccer team estimates that they will score on 8% of the corner kicks. In next week's game, the team hopes to kick 15 corner kicks. What are the chances that they will score on 2 of those opportunities?

*13.5 The Poisson Model

In the early 1990s, a leukemia cluster was identified in the Massachusetts town of Woburn. Many more cases of leukemia, a cancer that originates in a cell in the marrow of bone, appeared in this small town than would be predicted. Was it evidence of a problem in the town, or was it chance? That question led to a famous trial in which the families of eight leukemia victims sued, and became grist for the book and movie *A Civil Action*. Following an 80-day trial, the judge called for a retrial after dismissing the jury's contradictory and confusing findings. Shortly thereafter, the chemical companies and the families settled.

When rare events occur together or in clusters, people often want to know if that happened just by chance or whether something else is going on. If we assume that the events occur independently, we can use a Binomial model to find the probability that a cluster of events like this might occur. For rare events, p will be quite small, and when n is large, it may be difficult to compute the exact probability that a certain size cluster might occur.

To see why, let's try to compute the probability that a cluster of cases of size x occurs in Woburn. We'll use the U.S. *national average* of leukemia incidence to get a value for p, and the population of Woburn as our value for n. In the United States in the early 1990s, there were about 30 800 new cases of leukemia each year and about 280 000 000 people, giving a value for p of about 0.00011. The population of Woburn was about $n = 35\,000$. We'd expect $np = 3.85$ new cases of leukemia in Woburn. How unlikely would eight or more new cases be? To answer that, we'll need to calculate the complement, adding the probabilities of no cases, exactly one case, etc., up to seven cases. Each of those probabilities would have the form $_{35000}C_x\, p^x q^{35000-x}$. To find $_{35000}C_x$, we'll need 35 000!—a daunting number even for computers. In the 18th century, when people first were interested in computing such probabilities, it was impossible. We can easily go beyond the capabilities of today's computers by making n large enough and p small enough.

Simeon Denis Poisson[5] was a French mathematician interested in events with very small probability. He originally derived his model to approximate the Binomial model when the probability of a success, p, is very small and the number of trials, n, is very large. Poisson's contribution was providing a simple approximation to find that probability. When you see the formula, however, you won't necessarily see the connection to the Binomial. The Poisson's parameter is often denoted by λ, the mean of the distribution. To use the **Poisson model** to approximate the Binomial, we'll make their means match, so we set $\lambda = np$.

Trinity Mirror/Mirrorpix/Alamy

Poisson in History In Thomas Pynchon's book *Gravity's Rainbow*, Roger Mexico is a statistician using the Poisson model for the bombings over London during World War II. "The rockets *are* distributing about London just as Poisson's equation in the textbook predicts. As the data keep coming in, Roger looks more and more like a prophet . . . all I'm doing is plugging numbers into a well-known equation, you can look it up in the book and do it yourself. . . ."

POISSON PROBABILITY MODEL FOR SUCCESSES: POISSON (λ)

λ = mean number of successes
X = number of successes

$$P(X = x) = \frac{e^{-\lambda}\lambda^x}{x!} \text{ for } x = 0, 1, \ldots \text{(recall that } e = 2.71828 \ldots)$$

Expected value: $\qquad E(X) = \lambda$

Standard deviation: $\qquad SD(X) = \sqrt{\lambda}$

The Poisson model is a reasonably good approximation of the Binomial when $n \geq 20$ and $p \leq 0.05$.

Using Poisson's model, we can easily find the probabilities of a given size cluster of leukemia cases. Using $\lambda = np = 35\,000 \times 0.00011 = 3.85$ new cases a year, we find that the probability of seeing exactly x cases in a year is $P(X = x) = \dfrac{e^{-3.85}3.85^x}{x!}$. By adding up these probabilities for $x = 0, 1, \ldots 7$, we'd find that the probability of eight or more cases with $\lambda = 3.85$ is about 0.043. That's small but not terribly unusual.

---

[5]*Poisson* is a French name, properly pronounced "pwa sohn" and not with an "oy" sound as in poison.

In spite of its origins as an approximation to the Binomial, the Poisson model is also used directly to model the probability of the occurrence of events for a variety of phenomena. It's a good model to consider whenever your data consist of counts of occurrences (over time or space). It requires only that the events be independent and that the mean number of occurrences stays constant for the duration of the data collection (and beyond, if we hope to make predictions).

One nice feature of the Poisson model is that it scales according to the sample size.[6] For example, if we know that the average number of occurrences in a town the size of Woburn, with 35 000 people, is 3.85, we know that the average number of occurrences in a town of only 3500 residents is 0.385. We can use that new value of λ to calculate the probabilities for the smaller town. Using the Poisson formula, we'd find that the probability of 0 occurrences for a town of 3500 is about 0.68.

The Poisson distribution, with some help from Professor Jeff Rosenthal of the University of Toronto (or maybe it was the other way around), determined that the 200 major Lotto 6/49 prizes won by retail lottery sellers during the period of 1999–2006 had a probability of 0.00000000013 of occurring, if chance alone were producing winners (as reported on CBC's *The Fifth Estate*). Clearly, so many prizes were not won purely by luck, and fraud investigations—and reforms—followed. Probability or statistical analysis can often uncover serious problems, though the underlying explanation of a rare phenomenon may be difficult or impossible to discover (for another example, see the butterfly ballot example in Chapter 8).

One of the features of the Poisson model is that, as long as the mean rate of occurrences stays constant, the occurrence of past events doesn't change the probability of future events. That is, even though events that occur according to the Poisson model may *appear* to cluster, the probability of *another* event occurring is still the same (just as with the Binomial model). This is counterintuitive and may be one reason why so many people believe in the form of the "Law of Averages" that we argued against in Chapter 11. When the Poisson model was used to model bombs dropping over London in World War II, the model said that even though several bombs had hit a particular sector, the probability of *another* hit was still the same. So there was no point in moving people from one sector to another, even after several hits. You can imagine how difficult it must have been trying to convince people of that! This same phenomenon leads many people to think they see patterns in which an athlete, a slot machine, or a financial market gets "hot." But the occurrence of several successes in a row is not unusual and is rarely evidence that the mean has changed. Careful studies have been made of "hot streaks," and the evidence suggests that "hot streaks" are a vastly overrated notion.

 Just Checking

The Pew Research Center (www.pewresearch.org) reports that they are actually able to contact 76% of the randomly selected households drawn for a telephone survey.

3. Explain why these phone calls can be considered Bernoulli trials.

4. How would you model the number of successful contacts from a list of 1000 sampled households? Explain.

5. Suppose that in the course of a year, legitimate survey organizations (not folks pretending to take a survey but actually trying to sell you something or change your vote) sample 70 000 of the approximately 107 000 000 households in the United States. You'll be living in the States for the next year, in a city with 10 000 households, and you wonder if people from that city are represented in these polls. How would you model the number of households in your new city that will be contacted next year by a legitimate poll?

---

[6]Because *Poisson* means "fish" in French, you can remember this fact by the bilingual pun "*Poisson* scales!"

Section 13.5 Exercises

9. **Sell!** A car dealership sells an average of 5 cars in a day. Using the Poisson model, what is the probability that the dealer sells 3 cars tomorrow?

10. **Passing on** A large hospital has an average of 7 fatalities in a week. Using the Poisson model, what is the probability that this week has 10 fatalities?

13.6 Continuous Models

Up to now, we have dealt with discrete random variables, variables whose outcomes can be listed. We then assign probabilities one at a time to each outcome. But suppose instead the random variable is to represent the survival time of a mouse given some treatment for a deadly disease. We might measure it in weeks, or more accurately in days, or perhaps in hours, or minutes, or seconds. Unless we measure it very coarsely, there are a huge number of possible outcomes. We can even imagine the actual outcome as any real number from zero on up, making the sample space a portion of the real line. Now, what is the probability the mouse survives exactly 40 hours? Exactly 40 hours means 40.0000000 . . . hours—with zeros out to infinity. How about assigning a zero probability to such an outcome? Seems reasonable enough, since you can also think of this outcome as an infinitesimally small point on the real line. Indeed, with *continuous* models, it is necessary to assign zero probability to any particular value, so $P(X = x) = 0$ for any single numerical outome x. The probabilities that we typically need though are the probabilities of intervals, such as the probability of surviving at most 40 hours (0 to 40). The probability of surviving for 40 hours, when rounded to the nearest hour, is also just an interval: from 39.5 to 40.5 (or if measured to the nearest tenth of an hour, 39.95 to 40.05). To assign probabilities to intervals rather than to individual outcomes, we need to use a *density curve*. Does that sound familiar?

Recall that most famous density curve, the Normal curve, discussed in Chapter 5. There, we used it to model relative frequencies (or proportions) of numerical outcomes. Well, probabilities are just relative frequencies, in the long run. So, no surprise, we can also use density curves to model probabilities. An area under the curve will now represent probability instead of relative frequency (see Figure 13.1). A density curve used as a probability model for a continuous random variable is often referred to as a *probability density function (or curve)*. Finding areas under a curve in general requires the use of a tool from calculus (called integration), but for important models, tables or software should be readily available.

Figure 13.1

Hypothetical density curve model for *X* where $P(0 < X < 40)$ = area under curve between 0 and 40 = 0.7452.

The probability distributions of some continuous random variables may be well modelled with Normal density curves; others may need to be modelled with skewed or, uniform, or other-shaped curves. Regardless of shape, all continuous random variables have means (which we also call *expected values*) and variances. In this book we don't need to know how to calculate them (again, involving integration), but we can still work with models for continuous random variables when we know the parameters of the model. Let's start by examining one simple example of a continuous model, the Uniform model, and then we'll revisit the Normal model.

The Uniform Model

One of the authors often takes the Bloor subway line westbound from the Bay Street station in Toronto when going to the airport. His wait time for the next train is a random variable. Subway trains run at regular intervals; let's say it is very off-peak hours and the trains are only running every 8 minutes. We might model his wait time for the next subway with a uniform model on the interval 0 to 8 minutes, since any wait time is as likely as any other

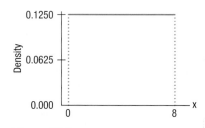

Figure 13.2

Uniform model over the interval (0, 8).

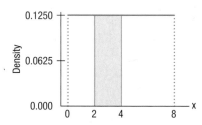

Figure 13.3

Find $P(2 < X < 4)$ by using the correct area under the curve.

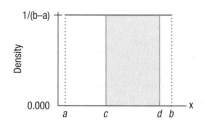

Figure 13.4

Uniform model on (*a*, *b*) showing $P(c < X < d)$.

wait time, given his rather random arrival at the station. The density curve should not clump up anywhere, and should look like Figure 13.2.

What is the height of the curve? Recall that *total probability must equal 1*. So the total area under the curve must equal one. Since the area of a rectangle is the height multiplied by the width, the height of this flat curve must equal 1/8 or 0.1250. The probability that the wait will be between 2 and 4 minutes is the area under the curve between these two numbers (Figure 13.3), which is easily seen to be $(4 - 2) \times 1/8$ or 0.25. The probability of waiting exactly 4 minutes $P(X = 4) = 0$, since the area under the density curve between 4 and 4 is zero. And I doubt anyone will ever wait exactly 4 minutes—to say one googol decimal places! [One googol is 10^{100}].

In general, we say that *X* has a uniform distribution or follows a **Uniform model** on the interval (*a*, *b*) if the density curve is flat above the interval (a, b) as shown in Figure 13.4.

Then $P(c < X < d) = \dfrac{d - c}{b - a}$ by working out the area of the proper rectangle. Even without calculus, you should be able to guess that the mean of this distribution, or $E(X)$, must equal $(a + b)/2$. The variance is $(b - a)^2/12$, but this requires calculus to confirm.[7]

If only all continuous models were so easy to work with! Flat density curve models are rare. Most are not flat, but clump up somewhere, with either a symmetrical spread around the centre or some skewness. There are some important skewed density curves that we will discuss much later in this text (like the chi-squared and F curves), but for now, the key model we need to understand and work with is the Normal model.

The Normal Probability Model

In Chapter 5, we introduced the Normal curve and learned how to calculate areas under the curve, as well as how to find percentile points. Also recall Figure 13.5. Now, we can also consider the Normal as a potential probability model for continuous random variables. There are two additional properties of the Normal probability model that will be very useful:

■ Shifting or scaling a Normal random variable—that is, adding a constant to *X* or multiplying *X* by a constant—produces another Normal random variable, but with a new mean and standard deviation.

■ We know that when we add two random variables together, the sum generally has a different probability model than the one we started with. Amazingly, independent Normal random variables behave better:

The probability model for the sum of any number of independent Normal random variables is also Normal.

And we *already* know what the mean and variance of that sum will be because those both add.

Figure 13.5

The Normal Model

[7] For those of you who have studied some calculus, the formulas for the mean and variance are:

μ_x or $E(X) = \displaystyle\int xf(x)dx$, and σ_x^2 or $Var(X) = \displaystyle\int (x - \mu_x)^2 f(x)dx$, where $f(x)$ is the density curve for X.

Step-by-Step Example PACKAGING STEREOS

N-Photos/Fotolia

A company manufactures small stereo systems. At the end of the production line, the stereos are packaged and prepared for shipping. Stage 1 of this process is called "packing." Workers must collect all the system components (a main unit, two speakers, a power cord, an antenna, and some wires), put each in plastic bags, and then place everything inside a protective styrofoam form. The packed form then moves on to Stage 2, called "boxing." There, workers place the form and a packet of instructions in a cardboard box, close it, then seal and label the box for shipping.

The company says that times required for the packing stage can be described by a Normal model with a mean of 9 minutes and standard deviation of 1.5 minutes. The times for the boxing stage can also be modelled as Normal, with a mean of 6 minutes and standard deviation of 1 minute.

Questions:

1. What is the probability that packing two consecutive systems takes over 20 minutes?
2. What percentage of the stereo systems take longer to pack than to box?

Question 1: What is the probability that packing two consecutive systems takes over 20 minutes?

THINK ➡ **Plan** State the problem.	I want to estimate the probability that packing two consecutive systems takes over 20 minutes.
Variables Define your random variables.	Let P_1 = time for packing the first system P_2 = time for packing the second T = total time to pack two systems
Write an appropriate equation.	$T = P_1 + P_2$
Think about the assumptions. Sums of independent Normal random variables follow a Normal model. Such simplicity isn't true in general.	✓ **Normal Model Assumption:** We are told that both random variables follow Normal models. ✓ **Independence Assumption:** We can reasonably assume that the two packing times are independent.

SHOW ➡ **Mechanics** Find the expected value.	$E(T) = E(P_1 + P_2)$ $\quad = E(P_1) + E(P_2)$ $\quad = 9 + 9 = 18 \ minutes$
For sums of independent random variables, variances add. (We don't need the variables to be Normal for this to be true—just independent.)	Since the times are independent, $Var(T) = Var(P_1 + P_2)$ $\quad = Var(P_1) + Var(P_2)$ $\quad = 1.5^2 + 1.5^2$
Find the standard deviation.	$Var(T) = 4.50$ $SD(T) = \sqrt{4.50} \approx 2.12 \ minutes$
Now we use the fact that both random variables follow Normal models to say that their sum is also Normal. Sketch a picture of the Normal model for the total time, shading the region representing over 20 minutes. Find the z-score for 20 minutes. Use technology or Table Z to find the probability.	I'll model T with $N(18, 2.12)$. 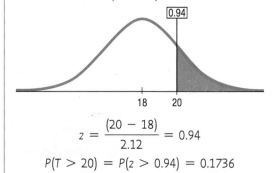 $z = \dfrac{(20 - 18)}{2.12} = 0.94$ $P(T > 20) = P(z > 0.94) = 0.1736$

TELL ➡ Conclusion Interpret your results in context.

There's a little more than a 17% chance that it will take a total of over 20 minutes to pack two consecutive stereo systems.

Question 2: What percentage of the stereo systems take longer to pack than to box?

THINK ➡ Plan State the question.

I want to estimate the percentage of the stereo systems that take longer to pack than to box.

Variables Define your random variables.

Let P = time for packing a system
B = time for boxing a system
D = difference in times to pack and box a system

Write an appropriate equation.

$D = P - B$

What are we trying to find? Notice that we can tell which of two quantities is greater by subtracting and asking whether the difference is positive or negative.
Don't forget to think about the assumptions.
Is D Normal? A difference is also a sum; D is the sum of the independent Normals P and −B, hence Normal.

The probability that it takes longer to pack than to box a system is the probability that the difference $P - B$ is greater than zero.

✓ **Normal Model Assumption:** We are told that both random variables follow Normal models.

✓ **Independence Assumption:** We can assume that the times it takes to pack and to box a system are independent.

SHOW ➡ Mechanics Find the expected value.

$$E(D) = E(P - B)$$
$$= E(P) - E(B)$$
$$= 9 - 6 = 3 \text{ minutes}$$

For the difference of independent random variables, variances add.

Since the times are independent,

$$Var(D) = Var(P - B)$$
$$= Var(P) + Var(B)$$
$$= 1.5^2 + 1^2$$
$$Var(D) = 3.25$$

Find the standard deviation.

$$SD(D) = \sqrt{3.25} \approx 1.80 \text{ minutes}$$

State what model you will use.

I'll model D with $N(3, 1.80)$

Sketch a picture of the Normal model for the difference in times, and shade the region representing a difference greater than zero.

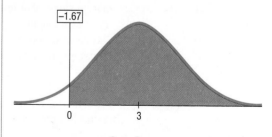

Find the z-score for 0 minutes, then use Table Z or technology to find the probability.

$$z = \frac{0 - 3}{1.80} = -1.67$$
$$P(D > 0) = P(z > -1.67) = 0.9525$$

TELL ➡ Conclusion Interpret your result in context.

About 95% of all the stereo systems will require more time for packing than for boxing.

Section 13.6 Exercises

11. **Battery** The life span of a calculator battery is normally distributed with a mean of 45 hours and a standard deviation of 5 hours. What is the probability that a battery lasts more than 53 hours?

12. **Ketchup** An automatic filling machine in a factory fills bottles of ketchup with a mean of 16.1 oz and a standard deviation of 0.05 oz with a distribution that can be well modelled by a Normal model. What is the probability that your bottle of ketchup contains less than 16 oz?

13.7 Approximating the Binomial with a Normal Model

Suppose Canadian Blood Services anticipates the need for at least 1850 units of O-negative blood this year. It estimates that it will collect blood from 32 000 donors. How great is the risk that they will fall short of meeting their need? We've just learned how to calculate such probabilities. We can use the Binomial model with $n = 32\ 000$ and $p = 0.06$. The probability of falling short with *exactly* 1849 units of O-negative blood from 32 000 donors is $_{32000}C_{1849} \times 0.06^{1849} \times 0.94^{30151}$. No calculator on earth can calculate that first term (it has more than 100 000 digits).[8] And that's just the beginning. Falling short means any number of units *less than* 1850, so we have to do it again for 1848, for 1847, and all the way down to 0. No thanks.

When we're dealing with a large number of trials like this, making direct calculations of the probabilities becomes tedious (or outright impossible). Here our old friend—the Normal model—comes to the rescue.

The Binomial model has mean $np = 1920$ and standard deviation $\sqrt{npq} \approx 42.48$. We could try approximating its distribution with a Normal model, using the same mean and standard deviation. Remarkably enough, that turns out to be a very good approximation. (We'll see why in the next chapter.) With that approximation, we find the *probability*:

$$P(X < 1850) = P\left(z < \frac{1850 - 1920}{42.28}\right) \approx P(z < -1.65) \approx 0.05$$

There seems to be about a 5% chance that Canadian Blood Services will run short of O-negative blood.

Can we always use a Normal model to make estimates of Binomial probabilities? No. Consider the Andre De Grasse situation—pictures in 20% of the cereal boxes. If we buy five boxes, the actual Binomial probabilities that we get for 0, 1, 2, 3, 4, or 5 pictures of De Grasse are 33%, 41%, 20%, 5%, 1%, and 0.03%, respectively. The first histogram in the margin shows that this probability model is skewed. That makes it clear that we should not try to estimate these probabilities by using a Normal model.

Now suppose we open 50 boxes of this cereal and count the number of Andre De Grasse pictures we find. The second histogram shows this probability model. It is centred at $np = 50(0.2) = 10$ pictures, as expected, and it appears to be fairly symmetric around that centre. Let's have a closer look.

The third histogram again shows Binom(50, 0.2), this time magnified somewhat and centred at the expected value of 10 pictures of De Grasse. It looks close to Normal, for sure. With this larger sample size, it appears that a Normal model might be a useful approximation.

A Normal model, then, is a close enough approximation only for a large enough number of trials. And what we mean by "large enough" depends on the probability of success. We'd need a larger sample if the probability of success were very low (or very high). It

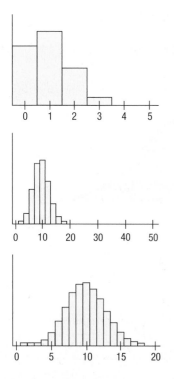

─────────

[8] If your calculator *can* find Binom(32 000,0.06), then it's smart enough to use an approximation. Read on to see how you can, too.

turns out that a Normal model works pretty well if we expect to see at least 10 successes and 10 failures. That is, we check the **Success/Failure Condition**.

The **Success/Failure Condition:** A Binomial model is approximately Normal if we expect at least 10 successes and 10 failures:

$$np \geq 10 \text{ and } nq \geq 10.$$

MATH BOX

It's easy to see where the magic number 10 comes from. You just need to remember how Normal models work. The problem is that a Normal model extends infinitely in both directions. But a Binomial model must have between 0 and n successes, so if we use a Normal to approximate a Binomial, we have to cut off its tails. That's not very important if the centre of the Normal model is so far from 0 and n that the lost tails have only a negligible area. More than three standard deviations should do it, because a Normal model has little probability past that.

So the mean needs to be at least three standard deviations from 0 and at least three standard deviations from n. Let's look at the 0 end.

We require:	$\mu - 3\sigma > 0$
Or, in other words:	$\mu > 3\sigma$
For a Binomial, that's:	$np > 3\sqrt{npq}$
Squaring yields:	$n^2p^2 > 9npq$
Now simplify:	$np > 9q$
Since $q \leq 1$, we can require:	$np > 9$

For simplicity, we usually require that np (and nq for the other tail) be at least 10 to use the Normal approximation, the Success/Failure Condition.[9]

For Example SPAM AND THE NORMAL APPROXIMATION TO THE BINOMIAL

RECAP: The communications monitoring company Postini has reported that 91% of email messages are spam. Recently, you installed a spam filter. You observe that over the past week it okayed only 151 of 1422 emails you received, classifying the rest as junk. Should you worry that the filtering is too aggressive?

QUESTION: What is the probability that no more than 151 of 1422 emails are real messages?

ANSWER: As before, I assume that messages arrive randomly and independently, with a probability of success (a real message) $p = 0.09$. So the model Binom(1422, 0.09) applies, but will be hard to work with. Checking conditions for the Normal approximation, I see that:

✓ I expect $np = (1422)(0.09) = 127.98$ real messages and $nq = (1422)(0.91) = 1294.02$ spam messages, both far greater than 10.

[9]Looking at the final step, we see that we need $np > 9$ in the worst case, when q (or p) is near 1, making the Binomial model quite skewed. When q and p are near 0.5—say between 0.4 and 0.6—the Binomial model is nearly symmetric and $np > 5$ ought to be safe enough. Although we'll always check for 10 expected successes and failures, keep in mind that for values of p near 0.5, we can be somewhat more forgiving.

It's okay to approximate this Binomial probability by using a Normal model.

$$\mu = np = 1422(0.09) = 127.98$$

$$\sigma = \sqrt{npq} = \sqrt{1422(0.09)(0.91)} \approx 10.79$$

$$P(x \le 151) = P\left(z \le \frac{151 - 127.98}{10.79}\right)$$

$$= P(z \le 2.13)$$

$$= 0.9834$$

Among my 1422 emails, there's over a 98% chance that no more than 151 of them were real messages, so the filter may be working properly.

Section 13.7 Exercises

13. **Toasters again** In a batch of 10,000 toasters, what are the chances that fewer than 450 need to be returned?

14. **Soccer again** If this team has 200 corner kicks over the season, what are the chances that they score more than 22 times?

*13.8 The Continuity Correction

When we used the Normal model to approximate the Binomial probabilites in the last section, we actually pulled a fast one on you. The Binomial is discrete, giving probabilities for specific counts, but the Normal is a **continuous random variable** that can take on *any value*. As we've seen, the probability that a Normal takes on any *specific* value is 0. So, for a Binomial with $n = 50$ and $p = 0.2$ the probability of exactly 10 successes is $P(X = 10) = 0.1398$. But for the Normal, $P(X = 10) = 0$. So there's a problem. The box below describes a solution, called the **continuity correction**.

THE CONTINUITY CORRECTION

When we use a continuous model to model a set of discrete events, we may need to make an adjustment called the **continuity correction**. We approximated the Binomial distribution (50, 0.2) with a Normal distribution. But what does the Normal distribution say about the probability that $X = 50$? Every specific value in the Normal probability model has probability 0. That's not the answer we want.

Because X is really discrete, it takes on the exact values 0, 1, 2, . . ., 50, each with positive probability. The histogram holds the secret to the correction. Look at the bin corresponding to $X = 10$ in the histogram. It goes from 9.5 to 10.5. What we really want is to find the area under the Normal curve *between* 9.5 and 10.5. So when we use the Normal distribution to approximate discrete events, we go halfway to the next value on the left and/or the right. We approximate $P(X = 10)$ for the Binomial model by finding $P(9.5 \leq X \leq 10.5)$ for the Normal model. For a Binomial (50, 0.2), $\mu = 10$ and $\sigma = 2.83$.

$$\text{So } P(9.5 \leq X \leq 10.5) \approx P\left(\frac{9.5 - 10}{2.83} \leq z \leq \frac{10.5 - 10}{2.83}\right)$$
$$= P(-0.177 \leq z \leq 0.177)$$
$$= 0.1405$$

By comparison, the *exact* Binomial probability is 0.1398.

If computing the probability for a range of Binomial outcomes, for example $P(7 \leq X \leq 10)$, we can improve the accuracy of the Normal approximation by *correcting* for continuity both limits of this interval before standardizing, so as to include the complete (and not half) bins at 7 and 10. So we would change this statement to $P(6.5 \leq X \leq 10.5)$, view X as Normal, and proceed to standardize, etc. Also see the next For Example.

However, when n is large and we calculate the probability for a large range of Binomial outcomes, each with a very small probability, the continuity correction may have very little effect, so we often just omit it.

We have only discussed the continuity correction when using a Normal curve to approximate Binomial probabilities, but there may be other situations where a Normal curve is used to calculate probabilities for a genuinely discrete random variable, and in those cases, a continuity correction may also be helpful.

For Example USING THE NORMAL APPROXIMATION WITH CONTINUITY CORRECTION

Some LCD panels have stuck or "dead" pixels that have defective transistors and are permanently unlit. If a panel has too many dead pixels, it must be rejected. A manufacturer knows that, when the production line is working correctly, the probability of rejecting a panel is 0.07.

QUESTIONS:

1. How many screens do they expect to reject in a day's production run of 500 screens? What is the standard deviation?
2. If they reject 40 screens today, is that a large enough number that they should be concerned that something may have gone wrong with the production line?
3. In the past week of 5 days of production (2500 screens), they've rejected 200 screens—an average of 40 per day. Should that raise concerns?

ANSWERS:

1. The number of screens rejected should be Binomially distributed, so using Binomial formulas, we get:

   $\mu = 0.07 \times 500 = 35$ is the expected number of rejects

   $\sigma = \sqrt{npq} = \sqrt{500 \times 0.07 \times 0.93} = 5.7$

2. Using the Normal approximation for the Binomial (since np and nq are bigger than 10) with continuity correction, we get:

   $$P(X \geq 40) = P\left(z \geq \frac{39.5 - 35}{5.7}\right) = P(z \geq 0.789) \approx 1 - 0.7852 \text{ (using Normal Tables)}$$
   $= 0.215$, not an extraordinarily large number of rejects. [The continuity correction may have been worthwhile here; without it, we get only 0.19]

3. Again using the Normal approximation with continuity correction:

   $\mu = 0.07 \times 2500 = 175$

   $\sigma = \sqrt{2500 \times 0.07 \times 0.93} = 12.757$

   $$P(X \geq 200) = P\left(z \geq \frac{199.5 - 175}{12.757}\right) = P(z \geq 1.92) \approx 1 - 0.9726 \text{ (from Tables)} = 0.0274$$

Yes, this seems to be a number of rejects that would occur by chance rarely if nothing were wrong [0.025 without continuity correction, not a big difference].

WHAT CAN GO WRONG?

■ **Probability models are still just models.** Models can be useful, but they are not reality. Think about the assumptions behind your models. Are your dice really perfectly fair? (They are probably pretty close.) But when you hear that the probability of a nuclear accident is 1/10 000 000 per year, is that likely to be a precise value? Question probabilities as you would data.

■ **If the model is wrong, so is everything else.** Before you try to find the mean or standard deviation of a random variable, check to make sure the probability model is reasonable. As a start, the probabilities in your model should add up to 1. If not, you may have calculated a probability incorrectly or left out a value of the random variable. For instance, in the insurance example, the description mentions only death and disability. Good health is by far the most likely outcome, not to mention the best for both you and the insurance company (who gets to keep your money). Don't overlook that.

To find the expected value of the sum or difference of random variables, we simply add or subtract means. Centre is easy; spread is trickier. Watch out for some common traps such as the first four items below:

■ **Watch out for variables that aren't independent.** You can add expected values of any two random variables, but you can only add variances of independent random variables. Suppose a survey includes questions about the number of hours of sleep people get each night and also the number of hours they are awake each day. From their answers, we find the mean and standard deviation of hours asleep and hours awake. The expected total must be 24 hours; after all, people are either asleep or awake.[10] The means still add just fine. Since all the totals are exactly 24 hours, however, the standard deviation of the total will be 0. We can't add variances here because the number of hours you're awake depends on the number of hours you're asleep. Be sure to check for independence before adding variances.

■ **Don't forget: Variances of independent random variables add. Standard deviations don't.**

■ **Don't forget: Variances of independent random variables add, even when you're looking at the difference between them.**

■ **Don't write independent instances of a random variable with notation that looks like they are the same variables.** Make sure you write each instance as a different random variable. Just because each random variable describes a similar situation doesn't mean that each random outcome will be the same. These are *random* variables, not the variables you saw in Algebra. Write $X_1 + X_2 + X_3$ rather than $X + X + X$.

■ **Don't assume you have the Bernoulli trials required for the Binomial model without checking.** Having two possible outcomes isn't enough. Be sure to check the requirements: two possible outcomes per trial ("success" and "failure," though a failure can simply be all the outcomes other than the one deemed a success), a constant probability of success, and independence. Remember to check the 10% Condition when sampling without replacement.

■ **Don't use the Normal approximation to the Binomial with small *n*.** To use a Normal approximation in place of a Binomial model, there must be at least 10 expected successes and 10 failures.

■ **Don't assume everything's Normal.** Just because a random variable is continuous or you happen to know a mean and standard deviation doesn't mean that a Normal model will be useful. You must *Think* about whether the **Normality Assumption** is justified. Using a Normal model when it is a poor representation of the random behaviour will lead to wrong answers and misleading conclusions.

---

[10]Although some students do manage to attain a state of consciousness somewhere between sleeping and wakefulness during Statistics class.

What Have We Learned?

Understand how probability models relate probabilities to outcomes.

- For discrete random variables, probability models assign a probability to each possible outcome.

- For continuous random variables, areas under density curves assign probabilities to intervals of outcomes. And we've learned that Normal models are once again special. Sums or differences of Normally distributed random variables also follow Normal models.

Know how to find the mean (expected value) of a discrete probability model from $\mu = \Sigma x P(x)$ and the standard deviation from $\sigma = \sqrt{\Sigma (x - \mu)^2 P(x)}$.

Foresee the consequences of shifting and scaling random variables, specifically

$$E(X \pm c) = E(X) \pm c \qquad E(aX) = aE(X)$$
$$Var(X \pm c) = Var(X) \qquad Var(aX) = a^2 Var(X)$$
$$SD(X \pm c) = SD(X) \qquad SD(aX) = |a| SD(X)$$

Understand that when adding or subtracting random variables, the expected values add or subtract as well: $E(X \pm Y) = E(X) \pm E(Y)$. However, when adding or subtracting independent random variables, the variances *add:*

$$Var(X \pm Y) = Var(X) + Var(Y)$$

We use a Binomial model when we're interested in the number of successes in a certain number of Bernoulli trials.

- If the number of trials n is large and we expect at least 10 successes and 10 failures, we can use a Normal model to approximate the Binomial model.

- If n is large and the probability of success p is very small, we can use a Poisson model to approximate the Binomial model.

Review of Terms

Random variable
A random variable assumes any of several different values as a result of some random event. Random variables are often denoted by a capital letter, such as X (p. 399).

Discrete random variable
A random variable that can take one of a finite (or countably infinite) number[11] of distinct outcomes (p. 400).

Continuous random variable
A random variable that can take on any of an uncountably infinite number of outcomes, typically, an interval (bounded or infinite) of values on the real line (p. 400).

Probability model or probability distribution
A function that associates a probability P with each value of a discrete random variable X, denoted $P(X = x)$ or $P(x)$, or with any interval of values of a continuous random variable, using a density curve (p. 400).

Expected value
The expected value of a random variable is its theoretical long-run average value, the mean of its probability distribution model. Denoted μ or $E(X)$, it is found (if the random variable is discrete) by summing the products of variable values and their respective probabilities (p. 400):

$$\mu = E(X) = \Sigma x P(x)$$

Variance
The variance of a random variable is the expected value of the squared deviation from the mean. For discrete random variables, it can be calculated as (p. 403):

$$\sigma^2 = Var(X) = \Sigma (x - \mu)^2 P(x)$$

Standard deviation
The standard deviation of a random variable describes the spread in the model and is the square root of the variance, denoted $SD(X)$ or σ (p. 403).

---

[11] There can be an infinite number of outcomes as long as they're *countable*. Essentially, that means we can imagine listing them all in order, like the counting numbers 1, 2, 3, 4, 5, ...

Changing a random variable by a constant	$E(X \pm c) = E(X) \pm c$ $Var(X \pm c) = Var(X)$ $SD(X \pm c) = SD(X)$ (p. 405) $E(aX) = aE(X)$ $Var(aX) = a^2 Var(X)$ $SD(aX) = \lvert a \rvert SD(X)$ (p. 406).
Adding or subtracting random variables	$E(X \pm Y) = E(X) \pm E(Y)$ (p. 408). If X and Y are independent: $Var(X \pm Y) = Var(X) + Var(Y)$, and $SD(X \pm Y) = \sqrt{Var(X) + Var(Y)}$. <div align="center">(Pythagorean Theorem of Statistics) (p. 408).</div>
***Covariance**	The covariance of random variables X and Y is $Cov(X, Y) = E[(X - \mu)(Y - \nu)]$ where $\mu = E(X)$ and $\nu = E(Y)$. Dividing by the two standard deviations yields a *correlation*, with similar properties to the sample correlation defined in Chapter 6. And in general (no need to assume independence) we also know that <div align="center">$Var(X \pm Y) = Var(X) + Var(Y) \pm 2Cov(X, Y)$ (p. 410).</div>
Bernoulli trials	A sequence of trials where (p. 411) 1. There are exactly two possible outcomes (usually denoted *success* and *failure*). 2. The probability of success is constant. 3. The trials are independent.
Binomial probability model	A Binomial distribution is appropriate for a random variable that counts the number of successes in a fixed number n of Bernoulli trials. The probability formula is below: $P(X = x) = {}_nC_x p^x q^{n-x}$, where ${}_nC_x = \dfrac{n!}{x!(n-x)!}$ (p. 411).
***Poisson probability model**	A discrete model used to model counts for a wide variety of phenomena, where we are counting events occurring over time or space, such as the number of arrivals of customers arriving in a queue or calls arriving into a call centre. Also sometimes used to approximate Binomial probabilites for rare events (p. 416).
Uniform probability model	A Uniform model gives equal probability to intervals of the same size (p. 419).
Success/Failure Condition	For a Normal model to be a good approximation of a Binomial model, we must expect at least 10 successes and 10 failures. That is, $np \ge 10$ and $nq \ge 10$ (p. 423).

On the Computer RANDOM VARIABLES

Statistics packages deal mostly with data, less with random variables. Nevertheless, the calculations needed to find means and standard deviations of random variables are little more than weighted means. Most packages can manage that, but then they are just being overblown calculators. For technological assistance with these calculations, we recommend you pull out your calculator.

Most statistics packages offer functions that compute probabilities for various probability models. Some technology solutions automatically use the Normal approximation for the Binomial when the exact calculations become unmanageable. The only important differences among these functions are how they are named and the order of their arguments. In these functions, pdf stands for "probability density function"—what we've been calling a probability model. For a discrete random variable (e.g., Binomial or Poisson) the pdf(x) gives the probability $P(X = x)$. For a continuous random variable (e.g., Normal), pdf(x) gives the value of the curve under which the probability lies (generally not useful). To find the probability $P(X \le x)$, use the cdf(x), the "cumulative distribution function," that accumulates the probability and gives $P(X \le x)$ for discrete or continuous variables.

For example, in Excel, **Binomdist(x, n, prob, cumulative)** computes Binomial probabilities. If cumulative is set to false, the calculation is only for one value of x. That is **Binomdist(2,5,0.2,false)** would be the probability of finding Andre De Grasse's picture twice in five boxes with probability of success = 0.2. **Binomdist(2,5,.2,true)** would be the probability of finding 0, 1, or 2 (up to and including 2) De Grasse pictures out of 5.

EXCEL

Binomdist(*x, n, prob, cumulative*)
Norm.dist(*x, mean, standard_dev, cumulative*)

COMMENTS

Set cumulative = *true* for cdf, *false* for pdf.
Excel's function fails when *x* or *n* is large.
It may not use the Normal approximation.

JMP

Binomial Probability (*prob, n, x*) (pdf)
Binomial Distribution (*prob, n, x*) (cdf)
Poisson Probability (*mean, k*) (pdf)
Poisson Distribution (*mean, k*) (cdf)
Normal Distribution(*x, mean = 0, sd = 1*) (cdf)

MINITAB

- Choose **Probability Distributions** from the Calc menu.
- Choose **Binomial** from the Probability Distributions submenu.
- To calculate the probability of getting *x* successes in *n* trials, choose **Probability**.

- To calculate the probability of getting *x* or fewer successes among *n* trials, choose **Cumulative Probability**.
- For Poisson, choose **Poisson** from the Probability Distribution submenu.

R

The standard library stats contains the pdfs and cdfs of several common distributions

- library(stats)
- **dbinom(x,n,p)** # Gives $P(X = x)$ for binomial with n and p.
- **pbinom(x,n,p)** # Gives the cumulative distribution function $P(X \leq x)$ for the binomial with *n* and *p*.

- **dpois(x, λ)** # Gives $P(X = x)$ for the Poisson with mean lambda.
- **ppois(x, λ)** # Gives the cumulative distribution function $P(X \leq x)$ for the Poisson with mean lambda.
- **dnorm(x,mu,sig)** # Gives the pdf for the Normal model
- **pnorm(x,mu,sig)** # Gives the cdf for the Normal model

SPSS

PDF.BINOM(*x, n, prob*)
CDF.BINOM(*x, n, prob*)
PDF.Poisson(*x, mean*)
CDF.Poisson(*x, mean*)
CDF.NORMAL(*quant, mean, stddev*)
PDF.NORMAL(*quant, mean, stddev*)

STATCRUNCH

To calculate binomial probabilities:

- Click on **Stat**.
- Choose **Calculators » Binomial**.
- Enter the parameters, **n** and **p**.
- Choose a specific outcome (=) or a lower tail (≤ or <), or upper tail (≥ or >) sum.
- Enter the number of successes *x*.
- Click on **Compute**.

To calculate Poisson probabilities:

- Click on **Stat**.
- Choose **Calculators » Poisson**.
- Enter the **mean**.
- Choose a specific outcome (=) or a lower tail (≤ or <) or upper tail (≥ or >) sum.
- Enter the number of successes **x**.
- Click on **Compute**.

TI-83/84 PLUS

To calculate the mean and standard deviation of a discrete random variable, enter the probability model in two lists:

- In one list (say, L1), enter the *x*-values of the variable.
- In a second list (say, L2), enter the associated probabilities $P(X = x)$.
- From the **STAT CALC** menu, select **1-VarStats** and specify the two lists. You'll see the mean and standard deviation.

For calculations with Binomial or Poisson distributions, use:

binompdf(*n, prob, x*)
binomcdf(*n, prob, x*)
poissonpdf(*mean, x*)
poissoncdf(*mean, x*)

COMMENTS

You can enter the probabilities as fractions; the calculator will change them to decimals for you.
Notice that the calculator knows enough to call the standard deviation σ, but mistakenly says $\bar{x}$ when it should say μ. Make sure you don't make that mistake!

Find the distribution commands in the 2nd DISTR menu (the calculator refers to models as "distributions").

Chapter Exercises

15. Expected value Find the expected value of each random variable:

a)

x	10	20	30
$P(X = x)$	0.3	0.5	0.2

b)

x	2	4	6	8
$P(X = x)$	0.3	0.4	0.2	0.1

16. Expected value II Find the expected value of each random variable:

a)

x	0	1	2
$P(X = x)$	0.2	0.4	0.4

b)

x	100	200	300	400
$P(X = x)$	0.1	0.2	0.5	0.2

17. Spinning the wheel A wheel comes up green 50% of the time and red 50% of the time. If it comes up green, you win $100. If it comes up red, you win nothing.
a) Intuitively, how much do you expect to win per spin of the wheel, on average?
b) Calculate the expected value of the game.

18. Stock market A day trader buys an option on a stock that will return $100 profit if the stock goes up today and lose $200 if it goes down. If the trader thinks there is a 75% chance that the stock will go up, what is his expected value of the option?

19. Pick a card, any card You draw a card from a deck. If you get a red card, you win nothing. If you get a spade, you win $5. For any club, you win $10 plus an extra $20 for the ace of clubs.
a) Create a probability model for the amount you win at this game.
b) Find the expected amount you'll win.
c) How much would you be willing to pay to play this game?

20. You bet! You roll a die. If it comes up a six, you win $100. If not, you get to roll again. If you get a six the second time, you win $50. If not, you lose.
a) Create a probability model for the amount you win at this game.
b) Find the expected amount you'll win.
c) How much would you be willing to pay to play this game?

21. Kids A couple plans to have children until they get a girl, but they agree that they will not have more than three children even if all are boys. (Assume boys and girls are equally likely.)
a) Create a probability model for the number of children they'll have.
b) Find the expected number of children.
c) Find the expected number of boys they'll have.

22. Carnival A carnival game offers a $100 cash prize for anyone who can break a balloon by throwing a dart at it. It costs $5 to play, and you're willing to spend up to $20 trying to win the cash prize. You estimate that you have about a 10% chance of hitting the balloon on any throw.
a) Create a probability model for the possible outcomes when you play this carnival game.
b) Find the expected number of darts you'll throw.
c) Find your expected winnings.

23. Software A small software company bids on two contracts. It anticipates a profit of $50 000 if it gets the larger contract and a profit of $20 000 on the smaller contract. The company estimates there is a 30% chance it will get the larger contract and a 60% chance it will get the smaller contract. Assuming the contracts will be awarded independently, what is the expected profit?

24. Racehorse A man buys a racehorse for $20 000 and enters it in two races. He plans to sell the horse afterward, hoping to make a profit. If the horse wins both races, its value will jump to $100 000. If it wins one of the races, it will be worth $50 000. If it loses both races, it will be worth only $10 000. The man believes there's a 20% chance that the horse will win the first race and a 30% chance it will win the second one. Assuming that the two races are independent events, find the man's expected profit.

25. Variation 1 Find the standard deviations of the random variables in Exercise 15.

26. Variation 2 Find the standard deviations of the random variables in Exercise 16.

27. Pick another card Find the standard deviation of the amount you might win drawing a card in Exercise 19.

28. The die Find the standard deviation of the amount you might win rolling a die in Exercise 20.

29. Kids again Find the standard deviation of the number of children the couple in Exercise 21 may have.

30. Darts Find the standard deviation of your winnings throwing darts in Exercise 22.

31. Repairs The probability model below describes the number of repair calls that an appliance repair shop may receive during an hour.

Repair Calls	0	1	2	3
Probability	0.1	0.3	0.4	0.2

a) How many calls should the shop expect per hour, on average?
b) What is the standard deviation?

32. Red lights A commuter must pass through five traffic lights on her way to work and will have to stop at each one that is red. She estimates the probability model for the number of red lights she hits, as shown below.

X = # of red	0	1	2	3	4	5
$P(X = x)$	0.05	0.25	0.35	0.15	0.15	0.05

a) How many red lights should she expect to hit each day, on average?
b) What is the standard deviation?

33. Defects A consumer organization inspecting new cars found that many had appearance defects (dents, scratches, paint chips, etc.). While none had more than three of these defects, 7% had three, 11% had two, and 21% had one defect. Find the expected number of appearance defects in a new car, and the standard deviation.

34. Insurance An insurance policy costs $100 and will pay policyholders $10 000 if they suffer a major injury (resulting in hospitalization) or $3000 if they suffer a minor injury (resulting in lost time from work). The company estimates that each year, 1 in every 2000 policyholders may have a major injury, and 1 in 500 may have a minor injury.
a) Create a probability model for the profit on a policy.
b) What is the company's expected profit on this policy?
c) What is the standard deviation?

35. Cancelled flights Mary is deciding whether to book the cheaper flight home from university after her final

exams, but she's unsure when her last exam will be. She thinks there is only a 20% chance that the exam will be scheduled after the last day she can get a seat on the cheaper flight. If it is and she has to cancel the flight, she will lose $150. If she can take the cheaper flight, she will save $100.
a) If she books the cheaper flight, what can she expect to gain, on average?
b) What is the standard deviation?

36. Day trading again An option to buy a stock is priced at $200. If the stock closes above 30 on May 15, the option will be worth $1000. If it closes below 20, the option will be worth nothing. If it closes between 20 and 30 (inclusively), the option will be worth $200. A trader thinks there is a 50% chance that the stock will close in the 20–30 range, a 20% chance that it will close above 30, and a 30% chance that it will fall below 20 on May 15.
a) How much is her expected gain?
b) What is the standard deviation of her gain?
c) Should she buy the stock option?

37. Contest You play two games against the same opponent. The probability that you win the first game is 0.4. If you win the first game, the probability that you also win the second is 0.2. If you lose the first game, the probability that you win the second is 0.3.
a) Are the two games independent? Explain your answer.
b) What is the probability that you lose both games?
c) What is the probability that you win both games?
d) Let random variable X be the number of games you win. Find the probability model for X.
e) What are the expected value and standard deviation of X?

38. Contracts Your company bids for two contracts. You believe the probability that you will get contract 1 is 0.8. If you get contract 1, the probability that you will also get contract 2 will be 0.2. If you do not get contract 1, the probability that you will get contract 2 will be 0.3.
a) Are the two contracts independent? Explain.
b) Find the probability that you get both contracts.
c) Find the probability that you get no contract.
d) Let X be the number of contracts you get. Find the probability model for X.
e) Find the expected value and standard deviation of X.

39. Batteries In a group of 10 batteries, 3 are dead. You choose two batteries at random.
a) Create a probability model for the number of good batteries you get.
b) What is your expected number of good ones?
c) What is the standard deviation?

40. Kittens In a litter of seven kittens, three are female. You pick two kittens at random.

a) Create a probability model for the number of male kittens you get.
b) What is the expected number of males?
c) What is the standard deviation?

41. Random variables Given independent random variables X and Y, with means and standard deviations as shown, find the mean and standard deviation of each of the random variables below. Also, X_1 and X_2 are independent variables with the same distribution as X.
a) $3X$
b) $Y + 6$
c) $X + Y$
d) $X - Y$
e) $X_1 + X_2$

	Mean	SD
X	10	2
Y	20	5

42. Random variables II Given independent random variables X and Y, with means and standard deviations as shown, find the mean and standard deviation of each of the variables below. Also, Y_1 and Y_2 are independent variables with the same distribution as Y.
a) $X - 20$
b) $0.5Y$
c) $X + Y$
d) $X - Y$
e) $Y_1 + Y_2$

	Mean	SD
X	80	12
Y	12	3

43. Random variables III Given independent random variables X and Y, with means and standard deviations as shown, find the mean and standard deviation of each of the variables below. Also, Y_1, Y_2, Y_3, and Y_4 are independent variables with the same distribution as Y.
a) $0.8Y$
b) $2X - 100$
c) $X + 2Y$
d) $3X - Y$
e) $Y_1 + Y_2 + Y_3 + Y_4$

	Mean	SD
X	120	12
Y	300	16

44. Random variables fini Given independent random variables X and Y, with means and standard deviations as shown, find the mean and standard deviation of each of the variables below. X_1, X_2, and X_3 are independent variables with the same distribution as X.
a) $2Y + 20$
b) $3X$
c) $0.25X + Y$
d) $X - 5Y$
e) $X_1 + X_2 + X_3$

	Mean	SD
X	80	12
Y	12	3

45. Eggs A grocery supplier believes that in a dozen eggs, the mean number of broken ones is 0.6 with a standard deviation of 0.5 eggs. You buy three dozen eggs without checking them.
a) What is your expected number of broken eggs?
b) What is the standard deviation?
c) What assumptions did you have to make about the eggs to answer this question? Do you think that assumption is warranted? Explain.

46. **Garden** A company selling vegetable seeds in packets of 20 estimates that the mean number of seeds that will actually grow is 18, with a standard deviation of 1.2 seeds. You buy five different seed packets.
   a) What is your expected number of bad (non-growing) seeds?
   b) What is the standard deviation of the number of bad seeds?
   c) What assumption did you make about the seeds? Do you think that assumption is warranted? Explain.

47. **Repair calls** Find the mean and standard deviation of the number of repair calls the appliance shop in Exercise 31 will receive during an 8-hour day. State any assumptions you are making.

48. **Stop!** Find the mean and standard deviation of the number of red lights the commuter in Exercise 32 will hit on her way to work during a five-day work week. State any assumptions you are making.

49. **Tickets** A delivery company's trucks occasionally get parking tickets. Based on past experience, the company expects that the trucks will average 1.3 tickets a month, with a standard deviation of 0.7 tickets.
   a) If they have 18 trucks, what are the mean and standard deviation of the total number of parking tickets the company will have to pay this month?
   b) What assumption did you make in answering this question?
   c) What would you consider to be an unusually bad month for the company?

50. **Donations** Organizers of a televised fundraiser know from past experience that most people donate small amounts ($10–$25), some donate larger amounts ($50–$100), and a few people make very generous donations of $250, $500, or more. Historically, pledges average about $32 with a standard deviation of $54.
   a) If 120 people call in pledges, what are the mean and standard deviation of the total amount raised?
   b) What assumption did you make in answering this question?
   c) The organizers are worried that people are pledging less than normal. How low would the 120-person total need to be to convince you that this is the case?

51. **Fire!** An insurance company estimates that it should make an expected annual profit of $150 on each homeowner's policy written, with a standard deviation of $6000.
   a) Why is the standard deviation so large?
   b) If it writes only two of these policies, what are the mean and standard deviation of the annual profit?
   c) If it writes 10 000 of these policies, what are the mean and standard deviation of the annual profit?
   d) Do you think the company is likely to be profitable? Explain.

   e) What assumptions underlie your analysis? Can you think of circumstances under which those assumptions might be violated? Explain.

52. **Casino** A casino knows that people play the slot machines in hopes of hitting the jackpot, but that most of them lose their dollar. Suppose a certain machine pays out an average of $0.92, with a standard deviation of $120.
   a) Why is the standard deviation so large?
   b) If you play five times, what are the mean and standard deviation of the casino's profit?
   c) If gamblers play this machine 1000 times in a day, what are the mean and standard deviation of the casino's profit?
   d) Do you think the casino is likely to be profitable? Explain.

53. **Bernoulli** Can we use probability models based on Bernoulli trials to investigate the following situations? Explain.
   a) We roll 50 dice to find the distribution of the number of spots on the faces.
   b) How likely is it that in a group of 120 people the majority will have Type A blood, given that Type A is found in 43% of the population?
   c) We deal five cards from a deck and get all hearts. How likely is that?
   d) To predict the outcome of a vote on the school budget, we poll 500 of the 3000 likely voters to see how many favour the proposed budget.
   e) A company realizes that about 10% of its packages are not being sealed properly. In a case of 24, is it likely that more than 3 are unsealed?

54. **Bernoulli 2** Can we use probability models based on Bernoulli trials to investigate the following situations? Explain.
   a) You are rolling five dice and need to get at least two 6s to win the game.
   b) We record the eye colours found in a group of 500 people.
   c) A manufacturer recalls a doll because about 3% have buttons that are not properly attached. Customers return 37 of these dolls to the local toy store. Is the manufacturer likely to find any dangerous buttons?
   d) A city council of 11 Conservatives and 8 Liberals picks a committee of 4 at random. What's the probability they choose all Liberals?
   e) A 2002 Rutgers University study found that 74% of high-school students have cheated on a test at least once. Your local high-school principal conducts a survey in homerooms and 322 of the 481 students admit to cheating.

55. **De Grasse again** Let's take one last look at the De Grasse picture search. You know his picture is in 20% of the cereal boxes. You buy five boxes to see how many pictures of De Grasse you might get.

a) Describe how you would simulate the number of pictures of De Grasse you might find in five boxes of cereal (for example, using random numbers, where some represent a success, others a failure, or using software to sample from a Bernoulli distribution).
b) Run at least 30 simulations.
c) Based on your simulation, estimate the probabilities that you get no pictures of De Grasse, one picture, two pictures, etc.
d) Calculate the actual probability model.
e) Compare the distribution of outcomes in your simulation to the probability model.

56. **Seatbelts** Suppose 75% of all drivers always wear their seatbelts. Let's investigate how many of the drivers might be belted among five cars waiting at a traffic light.
a) Describe how you would simulate the number of seatbelt-wearing drivers among the five cars (for example, using random numbers, where some represent a success, others a failure, or using software to sample from a Bernoulli distribution).
b) Run at least 30 simulations.
c) Based on your simulation, estimate the probabilities there are no belted drivers, exactly one, two, etc.
d) Calculate the actual probability model.
e) Compare the distribution of outcomes in your simulation to the probability model.

57. **On time** A Transport Canada report about air travel found that, nationwide, 76% of all flights are on time. Suppose you are at the airport and your flight is one of 50 scheduled to take off in the next two hours. Can you consider these departures to be Bernoulli trials? Explain.

58. **Lost luggage** A Transport Canada report about air travel found that airlines misplace about 5 bags per 1000 passengers. Suppose you are travelling with a group of people who have checked 22 pieces of luggage on your flight. Can you consider the fate of these bags to be Bernoulli trials? Explain.

59. **Coins and intuition** You flip a fair coin 100 times.
a) Intuitively, how many heads do you expect?
b) Use the formula for expected value to verify your intuition.

60. **Roulette and intuition** An American roulette wheel has 38 slots, of which 18 are red, 18 are black, and 2 are green (0 and 00). You spin the wheel 38 times.
a) Intuitively, how many times would you expect the ball to wind up in a green slot?
b) Use the formula for expected value to verify your intuition.

61. **Lefties** Assume that 13% of people are left-handed. We select 12 people at random.
a) Find the mean and standard deviation of the number of right-handers in the group.
b) What is the probability that they're not all right-handed?

c) What is the probability that there are no more than 10 righties?
d) What is the probability that there is exactly six of each?
e) What is the probability that the majority is right-handed?

62. **Arrows** Kateri Vrakking, an Olympic archer, is able to hit the bull's-eye 80% of the time. Assume each shot is independent of the others. She shoots 10 arrows.
a) Find the mean and standard deviation of the number of bull's-eyes she may get.
b) What is the probability that she never misses?
c) What is the probability that there are no more than eight bull's-eyes?
d) What is the probability that there are exactly eight bull's-eyes?
e) What's the probability that she hits the bull's-eye more often than she misses?

63. **Vision** It is generally believed that nearsightedness affects about 12% of all children. A school district tests the vision of 169 incoming kindergarten children. How many would you expect to be nearsighted, on average? With what standard deviation?

64. **International students** At a certain university, 6% of all students come from outside Canada. Incoming students are assigned at random to freshman dorms, where they live in residential clusters of 40 freshmen sharing a common lounge area. How many international students would you expect to find in a typical cluster, on average? With what standard deviation?

65. **Tennis, anyone?** A particular tennis player makes a successful first serve 70% of the time. Assume that each serve is independent of the others. If she serves six times, what is the probability that she gets
a) all six serves in?
b) exactly four serves in?
c) at least four serves in?
d) no more than four serves in?

66. **Frogs** A wildlife biologist examines frogs for a genetic trait he suspects may be linked to sensitivity to industrial toxins in the environment. Previous research had established that this trait is usually found in one of every eight frogs. He collects and examines a dozen frogs. If the frequency of the trait has not changed, what is the probability that he finds the trait in
a) none of the 12 frogs?
b) at least two frogs?
c) three or four frogs?
d) no more than four frogs?

67. **And more tennis** Suppose the tennis player in Exercise 65 serves 80 times in a match.
a) What are the mean and standard deviation of the number of good first serves expected?
b) Verify that you can use a Normal model to approximate the distribution of the number of good first serves.
c) Use the 68–95–99.7 Rule to describe this distribution.

d) What is the probability that she makes at least 65 first serves?

68. More arrows Kateri Vrakking, the archer in Exercise 62, will be shooting 200 arrows in a large competition.

a) What are the mean and standard deviation of the number of bull's-eyes she might get?

b) Is a Normal model an appropriate approximation here? Explain.

c) Use the 68–95–99.7 Rule to describe the distribution of the number of bull's-eyes she may get.

d) Would you be surprised if she made only 140 bull's-eyes? Explain.

69. Apples An orchard owner knows that he'll have to use about 6% of the apples he harvests for cider because they will have bruises or blemishes. He expects a tree to produce about 300 apples.

a) Describe an appropriate model for the number of cider apples that may come from that tree. Justify your model.

b) Find the probability there will be no more than a dozen cider apples.

c) Is it likely there will be more than 50 cider apples? Explain.

70. Frogs, part II Based on concerns raised by his preliminary research, the biologist in Exercise 66 decides to collect and examine 150 frogs.

a) Assuming the frequency of the trait is still one in eight, determine the mean and standard deviation of the number of frogs with the trait that he will find in his sample.

b) Verify that he can use a Normal model to approximate the distribution of the number of frogs with the trait.

c) He found the trait in 22 of his frogs. Do you think this proves that the trait has become more common? Explain.

71. Lefties again A lecture hall has 200 seats with folding arm tablets, 30 of which are designed for left-handers. The average size of classes that meet there is 188, and we can assume that about 13% of students are left-handed. What is the probability that a right-handed student in one of these classes is forced to use a lefty arm tablet?

72. No-shows An airline, believing that 5% of passengers fail to show up for flights, overbooks (sells more tickets than there are seats). Suppose a plane will hold 265 passengers, and the airline sells 275 tickets. What is the probability that the airline will not have enough seats, so someone gets bumped?

73. Annoying phone calls A newly hired telemarketer is told he will probably make a sale on about 12% of his phone calls. The first week he called 200 people, but only made 10 sales. Should he suspect he was misled about the true success rate? Explain.

74. The euro Shortly after the introduction of the euro coin in Belgium, newspapers around the world published articles claiming the coin is biased. The stories were based on reports that someone had spun the coin 250 times and gotten 140 heads—that's 56% heads. Do you think this is evidence that spinning a euro is unfair? Explain.

*75. Hurricanes 2015, redux We first looked at the occurrences of hurricanes in Chapter 3 exercises and found that they arrive with a mean of 2.65 per year. Suppose the number of hurricanes can be modelled by a Poisson distribution with this mean.

a) What is the probability of no hurricanes next year?

b) What is the probability that during the next two years, there will be exactly one hurricane?

*76. Bank tellers You are the only bank teller on duty at your local bank. You need to run out for 10 minutes, but you don't want to miss any customers. Suppose the arrival of customers can be modelled by a Poisson distribution with a mean of two customers per hour.

a) What is the probability that no one will arrive in the next 10 minutes?

b) What is the probability that two or more people arrive in the next 10 minutes?

c) You've just served two customers who came in one after the other. Is this a better time to run out?

*77. HIV It is given that the probability of contracting HIV is small, with p about 0.0005 for a new case in a given year. In a town of 8000 people,

a) What is the expected number of new cases?

b) Use the Poisson model to approximate the probability that there will be at least one new case of HIV next year.

*78. Lotto 6/49 scandal Earlier in the chapter, we mentioned Professor Jeff Rosenthal's work on the Lotto 6/49 scandal. He estimated—conservatively—that for the 5713 major-prize lottery wins over the period 1996–2006, the expected number of wins by retailers was 123.2. In fact, retailers won 200 times. Using the Poisson distribution, Rosenthal calculated the chances of their winning this many times or more. The calculation is a bit messy, so let's do a similar but simpler calculation. Assume that conditions (purchasing patterns, etc.) have not changed since Rosenthal's study was conducted.

a) What is the expected number of retailer wins among the next 40 major-prize wins in Ontario if just chance is at work? (Adjust the expected wins in 5713 prizes to get the expected wins in 40 prizes.)

b) Use the expected value from part a) in the Poisson model to approximate the probability that retailers will win two or more prizes among the next 40 major wins.

79. Seatbelts II Police estimate that 80% of drivers now wear their seatbelts. They set up a safety roadblock, stopping cars to check for seatbelt use.

a) What is the probability that the first 10 drivers are all wearing their seatbelts?

b) If police stop 30 cars during the first hour, find the mean and standard deviation of the number of drivers who will be wearing seatbelts.

c) If they stop 120 cars during this safety check, what's the probability they find at least 20 drivers not wearing their seatbelts?

80. **Rickets** Vitamin D is essential for strong, healthy bones. Our bodies produce vitamin D naturally when the skin is exposed to sunlight, or it can be taken as a dietary supplement. Although the bone disease rickets was largely eliminated in England during the 1950s, some people there are concerned that this generation of children is at increased risk because they are more likely to watch television or play computer games than spend time outdoors. Recent research indicated that about 20% of British children are deficient in vitamin D. Suppose doctors test a group of elementary school children.

a) What is the probability that the first 10 children tested are all okay?

b) They test 50 Grade 3 students. Find the mean and standard deviation of the number who are deficient in vitamin D.

c) If they test 320 children at this school, what is the probability that no more than 50 of them have the vitamin deficiency?

81. **ESP** Scientists wish to test the mind-reading ability of a person who claims to "have ESP." They use five cards with different and distinctive symbols (square, circle, triangle, line, squiggle). Someone picks a card at random and thinks about the symbol. The "mind reader" must correctly identify which symbol was on the card. If the test consists of 100 trials, how many would this person need to get right to convince you that ESP may actually exist? Explain.

82. **True–False** A true–false test consists of 50 questions. How many does a student have to get right to convince you that he is not merely guessing? Explain.

83. **Hot hand** A basketball player who ordinarily makes about 55% of his free throw shots has made four in a row. Is this evidence that he has a "hot hand" tonight? That is, is this streak so unusual that it means the probability he makes a shot must have changed? Explain.

84. **New bow** Kateri Vrakking, the archer in Exercise 62, purchases a new bow, hoping that it will improve her success rate to more than 80% bull's-eyes. She is delighted when she first tests her new bow and hits six consecutive bull's-eyes. Do you think this is compelling evidence that the new bow is better? In other words, is a streak like this unusual for her? Explain.

85. **Hotter hand** Our basketball player in Exercise 83 has new sneakers, which he thinks improve his game. Over his past 40 shots, he's made 32—much better than the 55% he usually shoots. Do you think his chances of

making a shot really increased? In other words, is making at least 32 of 40 shots really unusual for him? (Do you think it's his sneakers?)

86. **New bow, again** Kateri Vrakking in Exercise 84 continues shooting arrows, ending up with 45 bull's-eyes in 50 shots. Now are you convinced that the new bow is better? Explain.

87. **Continuous uniform model** Suppose that X is continuous and uniformly distributed over the outcomes 0 to 360. (For example, you might be spinning a bottle and observing the final angle.)

a) Draw a picture of the flat density curve to model this random variable. What is the correct height for the curve (recall that total probability is always equal to 1)?

b) Find $P(20 < X < 110)$ using the correct area under the density curve. Shade in this area in your picture of the density curve.

88. **Continuous models** Suppose that X is a continuous random variable with the following density curve model:

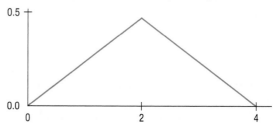

a) Find $P(0 < X < 1)$ and shade in the relevant area.

b) Find $P(1 \leq X \leq 3)$ and shade in the relevant area.

89. **Cereal** The amount of cereal that can be poured into a small bowl varies with a mean of 1.5 ounces and a standard deviation of 0.3 ounces. A large bowl holds a mean of 2.5 ounces with a standard deviation of 0.4 ounces. You open a new box of cereal and pour one large and one small bowl.

a) How much more cereal do you expect there to be in the large bowl than in the small bowl?

b) What's the standard deviation of this difference?

c) If the difference follows a Normal model, what is the probability that the small bowl contains more cereal than the large one?

d) What are the mean and standard deviation of the total amount of cereal in the two bowls?

e) If the total follows a Normal model, what is the probability you poured out more than 4.5 ounces of cereal in the two bowls together?

f) The amount of cereal the manufacturer puts in the boxes is a random variable with a mean of 16.3 ounces and a standard deviation of 0.2 ounces. Find the expected amount of cereal left in the box, and the standard deviation.

90. **Pets** The Canadian Veterinary Medical Association claims that the annual cost of medical care for dogs

averages $100 with a standard deviation of $30, and for cats averages $120 with a standard deviation of $35.

a) What is the expected difference in the cost of medical care for dogs and cats?

b) What is the standard deviation of that difference?

c) If the difference in costs can be described by a Normal model, what is the probability that medical expenses are higher for someone's dog than for her cat?

d) Do you have any concerns regarding your answers above?

91. More cereal In Exercise 89, we poured a large and a small bowl of cereal from a box. Suppose the amount of cereal that the manufacturer puts in the boxes is a random variable with mean 16.2 ounces, and standard deviation 0.1 ounces.

a) Find the expected amount of cereal left in the box.

b) What is the standard deviation?

c) If the weight of the remaining cereal can be described by a Normal model, what is the probability that the box still contains more than 13 ounces?

92. More pets You're thinking about getting two dogs and a cat. Assume that annual veterinary expenses are independent and have a Normal model with the means and standard deviations described in Exercise 90.

a) Define appropriate variables and express the total annual veterinary costs you might accrue.

b) Describe the model for this total cost. Be sure to specify the model's name, expected value, and standard deviation.

c) What is the probability that your total expenses will exceed $400?

d) About how much would a typical year cost you in expenses (give a range)? How much money should you set aside if you wanted to be fairly sure that your annual bill will be covered?

93. Medley In the 4×100 medley relay event, four swimmers swim 100 yards, each using a different stroke. A college team preparing for the national championship looks at the times their swimmers have posted and creates a model based on the following assumptions:

■ The swimmers' performances are independent.

■ Each swimmer's times follow a Normal model.

■ The means and standard deviations of the times (in seconds) are as shown:

Swimmer	Mean	SD
1 (backstroke)	50.72	0.24
2 (breaststroke)	55.51	0.22
3 (butterfly)	49.43	0.25
4 (freestyle)	44.91	0.21

a) What are the mean and standard deviation for the relay team's total time in this event?

b) The team's best time so far this season was 3:19.48. (That's 199.48 seconds.) Do you think the team is likely to swim faster than this at the championship? Explain.

94. Bikes Bicycles arrive at a bike shop in boxes. Before they can be sold, they must be unpacked, assembled, and tuned (lubricated, adjusted, etc.). Based on past experience, the shop manager makes the following assumptions about how long this may take:

■ The times for each setup phase are independent.

■ The times for each phase follow a Normal model.

■ The means and standard deviations of the times (in minutes) are as shown:

Phase	Mean	SD
Unpacking	3.5	0.7
Assembly	21.8	2.4
Tuning	12.3	2.7

a) What are the mean and standard deviation for the total bicycle set-up time?

b) A customer decides to buy a bike like one of the display models but wants a different colour. The shop has one, still in the box. The manager says they can have it ready in half an hour. Do you think the bike will be set up and ready to go as promised? Explain.

c) Most days, the bike shop completes this process on three bikes. The manager schedules 90 minutes per day in total for store employees to complete this task. Are the workers going to be able to complete this task in 90 minutes (of their combined work time)?

95. Farmers' market A farmer has 100 lb of apples and 50 lb of potatoes for sale. The market price for apples (per pound) each day is a random variable with a mean of 0.5 dollars and a standard deviation of 0.2 dollars. Similarly, for a pound of potatoes, the mean price is 0.3 dollars and the standard deviation is 0.1 dollars. It also costs him $2 to bring all the apples and potatoes to the market. The market is busy with eager shoppers, so we can assume that he'll be able to sell all of each type of produce at that day's price.

a) Define your random variables, and use them to express the farmer's net income.

b) Find the mean.

c) Find the standard deviation of the net income.

d) Do you need to make any assumptions in calculating the mean? How about the standard deviation?

96. Bike sale The bicycle shop in Exercise 94 will be offering two specially priced children's models at a sidewalk sale. The basic model will sell for $120 and the deluxe model for $150. Past experience indicates that sales of the basic model will have a mean of 5.4 bikes with a standard deviation of 1.2, and sales of the deluxe model will have a mean of 3.2 bikes with a standard deviation of 0.8 bikes. The cost of setting up for the sidewalk sale is $200.

a) Define random variables and use them to express the bicycle shop's net income.

b) What is the mean of the net income?

c) What is the standard deviation of the net income?

d) Do you need to make any assumptions in calculating the mean? How about the standard deviation?

97. Geometric model Suppose we keep flipping a loonie until we see a head. Let $X =$ the flip on which we observe that first head. For example, if we get a tail, then a head, $X = 2$. Assume this is a typical loonie, flipped high and randomly.

a) Are these Bernoulli trials?

b) Can the Binomial model be applied here? Explain.

c) Find the probability that the first head occurs on the fifth flip.

d) The actual probability model for X is called the Geometric model, with parameter p, where here $p = 0.5$, the probability of success. The general formula is $p(x) = p\, q^{x-1}$ for $x = 1, 2, \ldots$ Use this formula to find the probability that the first head occurs on the third flip. (It is not so easy to prove, but $E(X) = 1/p$.)

e) Do you think this is a fairly symmetric distribution or a very skewed distribution? Why?

98. Geometric model again Suppose we inspect objects produced by a process in which 10% on average are defective. We will keep sampling and testing objects one at a time until we find a defective one.

a) Are these Bernoulli trials?

b) Can the Binomial model be applied here? Explain.

c) Find the probability that we find the first defective object upon inspecting the fourth article.

d) The actual probability model for X is called the Geometric model, with parameter p, where here $p = 0.1$, the probability of success. The general formula is $p(x) = p\, q^{x-1}$ for $x = 1, 2, \ldots$ Use this formula to find the probability that the first defective is found on the

tenth inspection. (Note that even though the number of possible outcomes for X is infinite, X is still a discrete *RV*, since we can list and eventually reach any outcome via that list. We say that such a set of outcomes is *countably infinite*, whereas values on the real line are *uncountably infinite*.)

e) It can be shown that the mean of this distribution is $1/p$. What is the mode?

*99. Correlated RVs Investors often purchase negatively correlated investment vehicles to minimize risk. Suppose you purchase two investments. Based on past experience, you believe that each has a mean gain of 5% with standard deviation of 2% (over some time frame). Let $X =$ the percentage return from investment A, and let $Y =$ the percentage return from investment B. If you invest equally, your percentage return will be $W = (X + Y)/2$. Find the variance of W if:

a) The returns are independent.

b) The returns have a correlation of -0.5 (covariance of -2).

c) The returns have a correlation of 0.5 (covariance of 2).

d) The returns have a correlation of 1.0 (covariance of 4).

e) Which of the above situations is least risky (the mean return is the same in each case)?

*100. Correlated RVs again Suppose you purchase two investments. Based on past experience, you believe that each has a mean gain of 10% with standard deviation of 4% (over some time frame). Let $X =$ the percentage return from investment A, and let $Y =$ the percentage return from investment B. If you invest 70% in A and 30% in B, your percentage return will be $W = .7X + .3Y$. Find the variance of W if:

a) The returns are independent.

b) The returns have a correlation of -0.8 (covariance of -12.8).

c) The returns have a correlation of 0.2 (covariance of 3.2).

d) The returns have a correlation of -1.0 (covariance of -16).

e) Which of the above situations is least risky (the mean return is the same in each case)?

Just Checking ANSWERS

1. a)

Outcome	$X =$ cost of repair	Probability
Recharging works	$ 60	0.75
Replace control unit	$200	0.25

b) $60(0.75) + 200(0.25) = \$95$

c) Car owners with this problem will spend $95 on average to get it fixed.

2. a) $100 + 100 = 200$ seconds

   b) $\sqrt{50^2 + 50^2} = 70.7$ seconds

   c) The times for the two customers are independent.

3. There are two outcomes (contact, no contact). The probability of contact is 0.76, and random calls should be independent.

4. Binomial, with $n = 1000$ and $p = 0.76$. [For actual calculations—see Section 13.7—we could approximate using a Normal model with $\mu = np = 1000(0.76) = 760$ and $\sigma = \sqrt{npq} = \sqrt{1000(0.76)(0.24)} \approx 13.5$.]

5. Poisson, with $\lambda = np = 10\,000\left(\dfrac{70\,000}{107\,000\,000}\right) = 6.54$.

MathXL	Go to MathXL at **www.mathxl.com** or MyLab™ Statistics at **www.mystatlab.com**. You can practise exercises for this chapter as often as you want. The guided solutions will help you find answers step by step. You'll find a
MyLab Statistics	personalized study plan available to you too!

Randomness and Probability

Quick Review

Here's a brief summary of the key concepts and skills in probability and probability modelling:

- The Law of Large Numbers says that the more times we try something, the closer the results will come to theoretical prediction.
  - Don't mistakenly misinterpret the Law of Large Numbers as the "Law of Averages." There's no such thing.

- Rules of probability can handle most situations.
  - To find the probability that an event OR another event happens, add their probabilities and subtract the probabilities that both happen.
  - To find the probability that one event AND another event both happen, multiply the probabilities (the second probability being conditional on the first event).
  - Conditional probabilities tell you how likely one event is to happen, knowing that another event has happened.
  - Mutually exclusive events (also called "disjoint") cannot both happen at the same time.
  - Two events are independent if the occurrence of one doesn't change the probability that the other happens.

- A probability model for a random variable describes the theoretical distribution of outcomes.
  - The mean of the theoretical distribution of a random variable is also called the expected value of the random variable.
  - For sums or differences of independent random variables, variances add.
  - To estimate probabilities involving quantitative variables, you may be able to use a Normal model—but only if the distribution of the variable is unimodal and symmetric.
  - To estimate the probability that you'll get a certain number of successes in a specified number of independent (Bernoulli) trials, use a Binomial model.
  - (Optional) To estimate the probability that you'll get a certain number of occurrences of a relatively rare phenomenon or a certain number of occurrences of a phenomenon occurring over time or space, consider using a Poisson model.

Ready? Here are some opportunities to check your understanding of these ideas.

Review Exercises

1. **Quality control** A consumer organization estimates that 29% of new cars have a cosmetic defect, such as a scratch or a dent, when they are delivered to car dealers. This same organization believes that 7% have a functional defect—something that does not work properly—and that 2% of new cars have both kinds of problems.
   a) If you buy a new car, what's the probability that it has some kind of defect?
   b) What's the probability it has a cosmetic defect but no functional defect?
   c) If you notice a dent on a new car, what's the probability it has a functional defect?
   d) Are the two kinds of defects disjoint events? Explain.
   e) Do you think the two kinds of defects are independent events? Explain.

2. **Workers** A company's human resources officer reports a breakdown of employees by job type and sex as shown in the table.

		Sex	
		Male	Female
Job Type	Management	7	6
	Supervision	8	12
	Production	45	72

   a) What's the probability that a worker selected at random is
      i. female?
      ii. female or a production worker?

   iii. female, if the person works in production?

   iv. a production worker, if the person is female?

  b) Do these data suggest that job type is independent of being male or female? Explain.

3. Airfares Each year a company must send three officials to a meeting in China and five officials to a meeting in France. Airline ticket prices vary from time to time, but the company purchases all tickets for a country at the same price. Past experience has shown that tickets to China have a mean price of $1200, with a standard deviation of $150, while the mean airfare to France is $900, with a standard deviation of $100.

  a) Define random variables and use them to express the total amount the company will have to spend to send these officials to the two meetings.

  b) Find the mean and standard deviation of this total cost.

  c) Find the mean and standard deviation of the difference in price of a ticket to China and a ticket to France.

  d) Do you need to make any assumptions in calculating these means? How about the standard deviations?

4. Autism Psychiatrists estimate that about 1 in 100 adults has autism. What's the probability that in a city of 20 000, there are more than 300 people with this condition? Be sure to verify that a Normal model can be used here.

5. A game To play a game, you must pay $5 for each play. There is a 10% chance you will win $5, a 40% chance you will win $7, and a 50% chance you will win only $3.

  a) What are the mean and standard deviation of your net winnings?

  b) You play twice. Assuming the plays are independent events, what are the mean and standard deviation of your total winnings?

6. Emergency switch Safety engineers must determine whether industrial workers can operate a machine's emergency shutoff device. Among a group of test subjects, 66% were successful with their left hands, 82% with their right hands, and 51% with both hands.

  a) What percent of these workers could not operate the switch with either hand?

  b) Are success with right and left hands independent events? Explain.

  c) Are success with right and left hands mutually exclusive? Explain.

7. Facebook According to Pew Research, 50% of adults and 75% of teenagers were using a social networking site in early 2012. Most of that activity was on Facebook, so let's assume these probabilities apply strictly to Facebook. Among a group of 10 people, what's the probability that

  a) at least one of the people was not on Facebook if they were all adults?

  b) at least one of the people was not on Facebook if they were all teenagers?

  c) at least one of the people was not on Facebook if half were teenagers?

8. Deductible A car owner may buy insurance that will pay the full price of repairing the car after an at-fault accident, or save $12 a year by getting a policy with a $500 deductible. Her insurance company says that about 0.5% of drivers in her area have an at-fault auto accident during any given year. Based on this information, should she buy the policy with the deductible or not? How does the value of her car influence this decision?

9. More Facebook Using the percentages from Exercise 7, suppose there is a group of five teenagers. What's the probability that

  a) all will be on Facebook?

  b) exactly one will be on Facebook?

  c) at least three will be on Facebook?

10. At fault The car insurance company in Exercise 8 believes that about 0.5% of drivers have an at-fault accident during a given year. Suppose the company insures 1355 drivers in that city.

  a) What are the mean and standard deviation of the number who may have at-fault accidents?

  b) Can you describe the distribution of these accidents with a Normal model? Explain.

11. Friend me? One hundred fifty-eight teenagers are standing in line for a big movie premier night. Refer to Exercise 7.

  a) What are the mean and standard deviation of the number of Facebook users we might expect?

  b) Can we use a Normal model in this situation?

  c) What's the probability that no more than 110 of the teenagers are Facebook users?

12. Child's play In a board game, you determine the number of spaces you may move by spinning a spinner and rolling a die. The spinner has three regions: Half of the spinner is marked "5," and the other half is equally divided between "10"and "20." The six faces of the die show 0, 0, 1, 2, 3, and 4 spots. When it's your turn, you spin and roll, adding the numbers together to determine how far you may move.

  a) Create a probability model for the outcome on the spinner.

  b) Find the mean and standard deviation of the spinner results.

  c) Create a probability model for the outcome on the die.

  d) Find the mean and standard deviation of the die results.

  e) Find the mean and standard deviation of the number of spaces you get to move.

13. Language Neurological research has shown that in about 80% of people, language abilities reside in the brain's left side. Another 10% display right-brain language centres, and the remaining 10% have two-sided language control. (The latter two groups are mainly left-handers.)[1]

  a) Assume that a first-year composition class contains 25 randomly selected people. What's the probability

---

[1] *Science News*, 161 no. 24 [2002].

that no more than 15 of them have left-brain language control?

b) In a randomly assigned group of five of these students, what's the probability that no one has two-sided language control?

c) In the entire first-year class of 1200 students, how many would you expect to find of each type?

d) What are the mean and standard deviation of the number of these students who will be right-brained in language abilities?

e) If an assumption of Normality is justified, use the 68–95–99.7 Rule to describe how many students in the first-year class will have right-brain language control.

14. **Play again** If you land in a "penalty zone" on the game board described in Exercise 12, your move will be determined by subtracting the roll of the die from the result on the spinner. Now what are the mean and standard deviation of the number of spots you may move?

15. **Beanstalks** In some cities, tall people who want to meet and socialize with other tall people can join Beanstalk Clubs. To qualify, a man must be over 6'2" tall and a woman over 5'10." According to the National Health Survey, heights of adults may have a Normal model with mean heights of 69.1" for men and 64.0" for women. The respective standard deviations are 2.8" and 2.5".

a) You're probably not surprised to learn that men are generally taller than women, but what does the greater standard deviation for men's heights indicate?

b) Are men or women more likely to qualify for Beanstalk membership?

c) Beanstalk members believe that height is an important factor when people select their spouses. To investigate, we select at random a married man and, independently, a married woman. Define two random variables, and use them to express how many inches taller the man is than the woman.

d) What's the mean of this difference?

e) What's the standard deviation of this difference?

f) What's the probability that the man is taller than the woman (that the difference in heights is greater than 0)?

g) Suppose a survey of married couples reveals that 92% of the husbands were taller than their wives. Based on your answer to part f), do you believe that people choose spouses independent of height? Explain.

16. **Stocks** Since the stock market began in 1872, stock prices have risen in about 73% of the years. Assuming that market performance is independent from year to year, what's the probability that

a) the market will rise for three consecutive years?

b) the market will rise three out of the next five years?

c) the market will fall during at least one of the next five years?

d) the market will rise during a majority of years over the next decade?

17. **Multiple choice** A multiple choice test has 50 questions, with 4 answer choices each. You must get at least 30 correct to pass the test, and the questions are very difficult.

a) Are you likely to be able to pass by guessing on every question? Explain.

b) Suppose, after studying for a while, you believe you have raised your chances of getting each question right to 70%. How likely are you to pass now?

18. **Stock strategy** Many investment advisors argue that after stocks have declined in value for two consecutive years, people should invest heavily because the market rarely declines three years in a row.

a) Since the stock market began in 1872, there have been two consecutive losing years eight times. In six of those cases, the market rose during the following year. Does this confirm the advice?

b) Overall, stocks have risen in value during 95 of the 130 years since the market began in 1872. How is this fact relevant in assessing the statistical reasoning of the advisors?

19. **Insurance** A 65-year-old woman takes out a $100 000 term life insurance policy. The company charges an annual premium of $520. Estimate the company's expected profit on such policies if mortality tables indicate that only 2.6% of women age 65 die within a year.

20. **Teen smoking** The Centers for Disease Control say that about 23% of high-school students smoke tobacco (down from a high of 38% in 1997). Suppose you randomly select high school students to survey them on their attitudes toward scenes of smoking in the movies. What's the probability that

a) none of the first four students you interview is a smoker?

b) there are no more than two smokers among 10 people you choose?

21. **Passing stats** Molly's university offers two sections of Statistics 101. From what she has heard about the two professors listed, Molly estimates that her chances of passing the course are 0.80 if she gets Professor Scedastic and 0.60 if she gets Professor Kurtosis. The registrar uses a lottery to randomly assign the 120 enrolled students based on the number of available seats in each class. There are 70 seats in Professor Scedastic's class and 50 in Professor Kurtosis's class.

a) What's the probability that Molly will pass Statistics?

b) At the end of the semester, we find out that Molly failed. What's the probability that she had Professor Kurtosis?

22. **Teen smoking II** Suppose that, as reported by the Centers for Disease Control, about 23% of high school students smoke tobacco. You randomly select 120 high school students to survey them on their attitudes toward scenes of smoking in the movies.

a) What's the expected number of smokers?

b) What's the standard deviation of the number of smokers?

c) The number of smokers among 120 randomly selected students will vary from group to group. Explain why that number can be described with a Normal model.

d) Using the 68–95–99.7 Rule, create and interpret a model for the number of smokers among your group of 120 students.

23. **Random variables** Given independent random variables with means and standard deviations as shown, find the mean and standard deviation of each of the variables below. The variables X_1 and X_2 are independent with the same distribution as X.
   a) $X + 50$
   b) $10Y$
   c) $X + 0.5Y$
   d) $X - Y$
   e) $X_1 + X_2$

	Mean	SD
X	50	8
Y	100	6

24. **Merger** Explain why the facts you know about variances of independent random variables might encourage two small insurance companies to merge. (*Hint:* Think about the expected amount and potential variability in payouts for the separate and the merged companies.)

25. **Youth survey** According to a recent Gallup survey, 93% of teens use the Internet, but there are differences in how teen boys and girls say they use computers. The telephone poll found that 77% of boys had played computer games in the past week, compared with 65% of girls. On the other hand, 76% of girls said they had emailed friends in the past week, compared with only 65% of boys.
   a) For boys, the cited percentages are 77% playing computer games and 65% using email. That total is 142%, so there is obviously a mistake in the report. No? Explain.
   b) Based on these results, do you think playing games and using email are mutually exclusive? Explain.
   c) Do you think whether a teen emails friends is independent of being a boy or a girl? Explain.

26. **Meals** A university student on a seven-day meal plan reports that the amount of money he spends daily on food varies with a mean of $13.50 and a standard deviation of $7.
   a) What are the mean and standard deviation of the amount he might spend in two consecutive days?
   b) What assumption did you make in order to find that standard deviation? Are there any reasons you might question that assumption?
   c) Estimate his average weekly food costs and the standard deviation.
   d) Do you think it likely he might spend less than $50 in a week? Explain, including any assumptions you make in your analysis.

27. **Travel to Kyrgyzstan** Your pocket copy of *Kyrgyzstan on 4237 ± 360 Soms a Day* claims that you can expect to spend about 4237 soms each day with a standard deviation of 360 soms. How well can you estimate your expenses for the trip?
   a) Your budget allows you to spend 90 000 soms. To the nearest day, how long can you afford to stay in Kyrgyzstan, on average?
   b) What's the standard deviation of your expenses for a trip of that duration?
   c) You doubt that your total expenses will exceed your expectations by more than two standard deviations. How much extra money should you bring? On average, how much of a "cushion" will you have per day?

28. **Picking melons** Two stores sell watermelons. At the first store, the melons weigh an average of 22 pounds, with a standard deviation of 2.5 pounds. At the second store, the melons are smaller, with a mean of 18 pounds and a standard deviation of 2 pounds. You select a melon at random at each store.
   a) What's the mean difference in weights of the melons?
   b) What's the standard deviation of the difference in weights?
   c) If a Normal model can be used to describe the difference in weights, what's the probability that the melon you got at the first store is heavier?

29. **Home sweet home 2011** According to Statistics Canada, 67% of Canadian households own the home they live in. A mayoral candidate conducts a survey of 820 randomly selected homes in your city and finds only 523 owned by the current residents. The candidate then attacks the incumbent mayor, saying that there is an unusually low level of home ownership in the city. Do you agree? Explain.

30. **Buying melons** The first store in Exercise 28 sells watermelons for 32 cents a pound. The second store is having a sale on watermelons—only 25 cents a pound. Find the mean and standard deviation of the difference in the price you may pay for melons randomly selected at each store.

31. **Who's the boss 2011?** According to Industry Canada, 17% of all small businesses were owned by women at the end of 2011. You call some small business firms doing business locally, assuming that the national percentage is true in your area.
   a) What's the probability that the first three you call are all owned by women?
   b) What's the probability that none of your first four calls finds a firm that is owned by a woman?
   c) Suppose none of your first five calls found a firm owned by a woman. What's the probability that your next call does?

32. **Jerseys** A Statistics professor comes home to find that all four of his children got white team shirts from soc-

cer camp this year. He concludes that this year, unlike other years, the camp must not be using a variety of colours. But then he finds out that in each child's age group there are four teams, only one of which wears white shirts. Each child just happened to get on the white team at random.

a) Why was he so surprised? If each age group uses the same four colours, what's the probability that all four kids would get the same colour shirt?

b) What's the probability that all four would get white shirts?

c) We lied. Actually, in the oldest child's group there are six teams instead of the four teams in each of the other three groups. How does this change the probability you calculated in part b)?

Courtesy of Richard De Veaux

33. When to stop? You play a game that involves rolling a die. You can roll as many times as you want, and your score is the total for all the rolls. But . . . if you roll a 6, your score is 0 and your turn is over. What might be a good strategy for a game like this?

You attempted to devise a good strategy by simulating several plays to see what might happen. Let's try calculating a strategy.

a) It can be shown that on average, a 6 would appear for the first time on the 6th role. So, roll *one time less* than that. Assuming all those rolls were not 6s, what's your expected score?

b) What's the probability that you can roll that many times without getting a 6?

34. Plan B Here's another attempt at developing a good strategy for the dice game in Exercise 33. Instead of stopping after a certain number of rolls, you could decide to stop when your score reaches a certain number of points.

a) How many points would you expect a roll to *add* to your score?

b) In terms of your current score, how many points would you expect a roll to *subtract* from your score?

c) Based on your answers in parts a) and b), at what score will another roll "break even"?

d) Describe the strategy this result suggests.

35. Technology on campus 2005 Every five years, the Conference Board of the Mathematical Sciences surveys college math departments. In 2005, the board reported that 51% of all undergraduates taking Calculus I were in classes that used graphing calculators and 21% were in classes that used computer assignments. Suppose that 10% used both calculators and computers.

a) What percent used neither kind of technology?

b) What percent used calculators but not computers?

c) What percent of the calculator users had computer assignments?

d) Based on this survey, do calculator and computer use appear to be independent events? Explain.

36. Dogs A census by the municipal dog control officer found that 18% of homes kept one dog as a pet, 4% had two dogs, and 1% had three or more. If a salesperson visits two homes selected at random, what's the probability he encounters

a) no dogs?

b) some dogs?

c) dogs in each home?

d) more than one dog in each home?

*37. O-rings Failures of O-rings on the space shuttle are fairly rare, but often disastrous, events. If we are testing O-rings, suppose that the probability of a failure of any one O-ring is 0.01. Let X be the number of failures in the next 10 O-rings tested.

a) Clearly the Binomial model can be applied here. What other model might you use to model X?

b) What is the mean number of failures in the next 10 O-rings?

c) What is the probability that there is exactly one failure in the next 10 O-rings? Use the model from your answer to part a).

d) What is the probability that there is at least one failure in the next 10 O-rings? Use the model from your answer to part a).

*38. Volcanoes Almost every year, there is some incidence of volcanic activity on the island of Japan. In 2005, there were five volcanic episodes, defined as either eruptions or sizable seismic activity. Suppose the mean number of episodes is 2.4 per year. Let X be the number of episodes in the two-year period 2013–2014.

a) What model might you use to model X?

b) What is the mean number of episodes in this period?

c) What is the probability that there will be no episodes in this period?

d) What is the probability that there are more than three episodes in this period?

e) How much activity would convince you that there is a significant increase in volcanic activity in the region?

39. Socks In your sock drawer you have four blue socks, five grey socks, and three black ones. Half asleep one morning, you grab two socks at random and put them on. Find the probability you end up wearing

a) two blue socks.

b) no grey socks.

c) at least one black sock.

d) a green sock.

e) matching socks.

40. Coins A coin is tossed 36 times.

a) What are the mean and standard deviation of the number of heads?

b) Suppose the resulting number of heads is unusual, two standard deviations above the mean. How many "extra" heads were observed?

c) If the coin were tossed 100 times, would you still consider the same number of extra heads unusual? Explain.

d) In the 100 tosses, how many extra heads would you need to observe in order to say the results were unusual?

e) Explain how these results refute the Law of Averages but confirm the Law of Large Numbers.

41. The Drake equation Not the latest hit by the Toronto rapper, but an equation developed in 1961 by astronomer Frank Drake. He was trying to estimate the number of extraterrestrial civilizations in our galaxy that might be able to communicate with us via radio transmissions. Now largely accepted by the scientific community, the Drake equation has helped spur efforts by radio astronomers to search for extraterrestrial intelligence.

Here is the equation:

$$N_c = N \cdot f_p \cdot n_e \cdot f_l \cdot f_i \cdot f_c \cdot f_L$$

Okay, it looks a little messy, but here's what it means:

Factor	What It Represents	Possible Value
N	Number of stars in the Milky Way Galaxy	200–400 billion
f_p	Probability that a star has planets	20%–50%
n_e	Number of planets in a solar system capable of sustaining earth-type life	1? 2?
f_l	Probability that life develops on a planet with a suitable environment	1%–100%
f_i	Probability that life evolves intelligence	50%?
f_c	Probability that intelligent life develops radio communication	10%–20%
f_L	Fraction of the planet's life for which the civilization survives	$\frac{1}{1\,000\,000}$?
N_c	Number of extraterrestrial civilizations in our galaxy with which we could communicate	?

So, how many ETs are out there? That depends; values chosen for the many factors in the equation depend on ever-evolving scientific knowledge and one's personal guesses. But now, some questions.

a) What quantity is calculated by the first product $N \cdot f_p$?

b) What quantity is calculated by the product $N \cdot f_p \cdot n_e \cdot f_l \cdot f_i$?

c) What probability is calculated by the product $f_l \cdot f_i$?

d) Which of the factors in the formula are conditional probabilities? Restate each in a way that makes the condition clear.

42. Recalls In a car rental company's fleet, 70% of the cars are American brands, 20% are Japanese, and the rest are German. The company notes that manufacturers' recalls seem to affect 2% of the American cars, but only 1% of the others.

a) What's the probability that a randomly chosen car is recalled?

b) What's the probability that a recalled car is American?

43. Pregnant? Suppose that 70% of the women who suspect they may be pregnant and purchase an in-home pregnancy test are actually pregnant. Further suppose that the test is 98% accurate. What's the probability that a woman whose test indicates that she is pregnant actually is?

44. Door prize You are among 100 people attending a charity fundraiser at which a large-screen television will be given away as a door prize. To determine who wins, 99 white balls and 1 red ball have been placed in a box and thoroughly mixed. The guests will line up and, one at a time, pick a ball from the box. Whoever gets the red ball wins the television, but if the ball is white, it is returned to the box. If none of the 100 guests gets the red ball, the television will be auctioned off for additional benefit to the charity.

a) What's the probability that the first person in line wins the television?

b) You are the third person in line. What's the probability that you win the television?

c) What's the probability that the charity gets to sell the television because no one wins?

d) Suppose you get to pick your spot in line. Where would you want to be to maximize your chances of winning?

e) After hearing some protest about the plan, the organizers decide to award the prize by not returning the white balls to the box, thus ensuring that 1 of the 100 people will draw the red ball and win the television. Now what position in line would you choose to maximize your chances?

Sampling Distribution Models

LEARNING OBJECTIVES

1 Understand how a sampling distribution models sample to sample variation in a statistic

2 Recognize that different statistics have different sampling distributions

3 Know the sampling distribution of a sample proportion and related assumptions and conditions

4 Know the sampling distribution of a sample mean and related assumptions and conditions

5 Understand the effect of sample size

6 Appreciate the surprising Central Limit Theorem and its consequences

The n's justify the means.

—Apocryphal statistical saying

"First-past-the-post" sounds like the hoped for outcome in a horse race, but it is also describes the current electoral system in Canada. In the 2015 federal election, the Liberals, NDP, and Green parties all came out in favour of changes to this system, to either some type of Proportional Representation system or a Ranked/Preferential Ballot system. The Liberals even pledged that 2015 would be the last federal election conducted under the first-past-the-post voting system[2]. But making such changes are no simple task. Should there be a national referendum on any proposed changes? A poll of 1429 randomly selected Canadian adults by Forum Research Inc in July 2016 asked if a national referendum is necessary before changing the way MPs are chosen. 65% said 'Yes.' Another similar size poll by Global News Ipsos, several weeks earlier had 73% wanting a referendum before any changes are made. Should we be surprised to find that we could get proportions this different from properly selected random samples drawn from the same population? Since samples are composed of different people, we expect some variation in results, but just how much variability among polls should we expect to see?

14.1 Sampling Distribution of a Proportion

If we (or Forum) had surveyed every possible sample of 1429 Canadian adults, we could find the proportion $\hat{p}$ of each sample who believe a referendum is needed. Then we could see how much those sample proportions varied. Of course, we can't do that, but we can *imagine* drawing those samples. In fact, we can do better than that. We can simulate what the distribution will look like. For a simulation, we first have to pick a particular value to be the "true" population proportion. Recall that we denote the true population proportion with the letter p. We can pretend for now that Forum got it exactly right and the true proportion is 65%. Then we can draw 1429 values that have a 35% chance of saying "No, a referendum is not needed" and a 65% chance of saying "Yes, it is."

The first three times we drew the 1429 values, we got 946, 899, and 911 simulated respondents (for $\hat{p} = 0.662, 0.629, 0.638$) supporting a referendum. When we ran the simulation 2000 times, a histogram of all the sample proportions looked liked Figure 14.1.

It should be no surprise that we don't get the same proportion for each sample we draw, even though the underlying true value is the same for the population. Each $\hat{p}$ comes from a different simulated sample. The histogram is a simulation of what we'd get if we could see *all the proportions from all possible samples*. That distribution has a special name. It is called the **sampling distribution** of the proportion.

---

[1]In the same study, 1000 Americans were polled and 45% of them thought Canada's healthcare system was superior.

[2]On Feb 1, 2017, the Liberals abandoned this pledge. Trudeau stated "I'm not going to do something that is wrong for Canadians just to tick off a box on an electoral platform."

Figure 14.1

A histogram of sample proportions for 2000 simulated samples of 1429 adults drawn from a population with $p = 0.65$. The sample proportions vary, but their distribution is centred at the true proportion, p.

random variable—and hence may be studied using methods from the previous chapter. We'll need to develop models for their random behaviour—called sampling distribution models—which will (in later chapters) enable us to start generalizing from the sample we have at hand to the population at large.

And shockingly often, the Normal Model provides exactly the model that we need for the random behaviour of these statistics! (A consequence of The Central Limit Theorem.)

The histogram is unimodal, symmetric, and centred at p. That's probably not surprising. But it's an amazing and fortunate fact that a Normal model is just the right one for the histogram of sample proportions. And in fact, this was proved mathematically over 300 years ago by the French mathematician Abraham de Moivre. There, is no reason you should guess that the Normal model would be the one we need here,[3] and indeed the importance of De Moivre's result was not immediately understood by his contemporaries. But (unlike De Moivre's contemporaries in 1718) we know how useful the Normal model can be.

Modelling how sample statistics, such as proportions or means vary from sample to sample is one of the most powerful ideas we'll see in this course. A **sampling distribution model** for how a statistic from a sample varies from sample to sample allows us to quantify that variation and to make statements about where we think the corresponding population parameter is.

Who	Canadian adults
What	Whether to have a referendum on federal electoral reform
When	July 2016
Where	Canada
Why	Public attitudes

North Wind Picture Archives/Alamy

Abraham de Moivre (1667–1754)

For Example USING THE SAMPLING DISTRIBUTION MODEL FOR A PROPORTION

According to Statistics Canada, about 18% of Canadians aged 12 and over still smoke.

QUESTION: How much would we expect the proportion of smokers in a sample of size 1000 to vary from sample to sample?

ANSWER: We'll simulate by taking random integers between 1 and 100. We'll assign values between 1 and 18 to represent a smoker, and the rest to represent a non-smoker. Then we'll repeat the process, drawing samples of size 1000 over and over and recording the sample proportion of smokers for each sample. If we did this 10 000 times, we might get a histogram of sample proportions that looks like this:

---

[3]Well, the fact that we spent most of Chapter 5 on the Normal model might have been a hint.

IMAGINE

We see only the sample that we actually drew, but by simulating or modelling, we can *Imagine* what we might have seen had we drawn other possible random samples.

The mean of this distribution is located at the population proportion, 0.18. The standard deviation of these 10 000 sample proportions is 0.0122 or 1.22%. Because the shape is symmetric and roughly Normal, the 68–95–99.7% Rule should work quite well. For example, we would expect about 95% of these samples to have proportions within 2 SDs (2.44%) of 18%. In fact, 9541 or 95.41% of them did.

Looking at lots of sample proportions is a big shift in our point of view. Up to now, you've thought of the sample proportion as a fixed value that comes from one particular sample. Given the number of successes and the size of your sample, the proportion was a number, a value that could be right or wrong on your homework. But now we're trying to understand how proportions from different random samples might behave. Each random sample has its own proportion and each of those proportions may be different, so we really have to think of proportions as random quantities. That's why simulating helps; it lets us see how sample proportions vary from sample to sample. We know (since 1718) that the Normal distribution provides a good model for the sampling distribution, but simulating different situations and seeing it can help you to believe it.[5]

Which Normal?

To use a Normal model, we need to specify two parameters: its mean and standard deviation. The centre of the histogram is naturally at p, so that's what we'll use for the mean of the Normal model.

What about the standard deviation? For quantitative data, the mean gives us no information about the standard deviation. If we told you that a batch of bike helmets had a mean diameter of 26 centimetres and asked what the standard deviation was, you'd probably have no idea. And there's no reason why you should. There's no information about σ from knowing the value of μ for quantitative data.

But there's a special fact about proportions that makes the standard deviation easy to find. Once we know the mean, p, we automatically also know the standard deviation of $\hat{p}$. We saw in Chapter 13 that for a Binomial model, the standard deviation of the *number* of successes is $\sqrt{npq}$ (where $q = 1 - p$). Now we want the standard deviation of the *proportion* of successes $\hat{p}$. The proportion is just the number of successes divided by the number of trials, n, so the standard deviation is also divided by n[6]:

$$\sigma(\hat{p}) = SD(\hat{p}) = \frac{\sqrt{npq}}{n} = \sqrt{\frac{pq}{n}}$$

NOTATION ALERT

We use p as our symbol for the true (generally unknown) proportion of successes in the population.[4] This violates our "Greek letters for parameters" rule, but if we stuck to that, our natural choice would be π, which we prefer to reserve for the universal constant 3.1415926 ... Recall that in Chapter 7 we introduced $\hat{y}$ as the predicted value for y. The "hat" over p here plays a similar role. It indicates an *estimate* of the parameter p. Likewise, the sample proportion of failures estimates the true failure proportion and is written $\hat{q}$. The "hat" signifies an estimate of what is underneath.

Figure 14.2

A Normal model centred at p with a standard deviation of $\sqrt{\dfrac{pq}{n}}$ is a good model for a collection of proportions found for many random samples of size n from a population with success probability p.

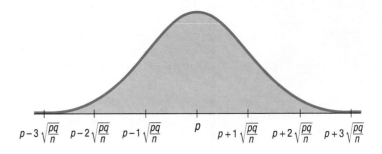

$p-3\sqrt{\dfrac{pq}{n}}$ $p-2\sqrt{\dfrac{pq}{n}}$ $p-1\sqrt{\dfrac{pq}{n}}$ p $p+1\sqrt{\dfrac{pq}{n}}$ $p+2\sqrt{\dfrac{pq}{n}}$ $p+3\sqrt{\dfrac{pq}{n}}$

---

[4]But be careful. We've already used capital P for a general probability. And we'll soon see another use of P in Chapter 16. There are a lot of p's in this course; you'll need to think clearly about the context to keep them straight.

[5]The saying isn't "being told is believing," but "seeing is believing" and there's a good reason for that.

[6]And likewise we can confirm mathematically that the centre of the distribution of $\hat{p}$ values is p since E(#Successes/n) = E(#Successes)/n = np/n = p. When the expected value of a sample statistic equals a population parameter, we say that the statistic is an *unbiased estimator* of the parameter.

When we draw simple random samples of n individuals, the proportions we find will vary from sample to sample. As long as n is reasonably large,[7] we can model the distribution of these sample proportions with a probability model (Figure 14.2) that is

$$N\left(p, \sqrt{\frac{pq}{n}}\right)$$

Although we'll never know the true proportion of adults who favour a referendum, we're supposing it for the moment to be 65%. Once we put the centre at $p = 0.65$, the standard deviation for the Forum poll is

$$SD(\hat{p}) = \sqrt{\frac{pq}{n}} = \sqrt{\frac{(0.65)(0.35)}{1429}} = 0.0126, \text{ or } 1.26\%$$

Here (Figure 14.3) is a picture of the Normal model for our simulation histogram:

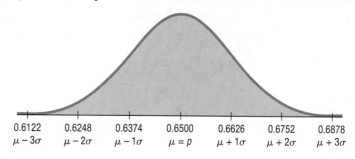

Figure 14.3

Using 0.65 for p gives this Normal model for Figure 14.1's histogram of the sample proportions of adults favouring early withdrawal ($n = 1000$).

| 0.6122 | 0.6248 | 0.6374 | 0.6500 | 0.6626 | 0.6752 | 0.6878 |
| $\mu - 3\sigma$ | $\mu - 2\sigma$ | $\mu - 1\sigma$ | $\mu = p$ | $\mu + 1\sigma$ | $\mu + 2\sigma$ | $\mu + 3\sigma$ |

Let's see how well the theory works. In our simulation of smokers (For Example, page 448), the SD of $\hat{p}$ was 0.0122. The formula says it actually should be[8]

$$SD(\hat{p}) = \sqrt{\frac{pq}{n}} = \sqrt{\frac{(0.18)(0.82)}{1000}} = 0.0121.$$

That works.

Because we have a Normal model, we can use the 68–95–99.7 Rule or look up other probabilities using a table or technology. For example, we know that 95% of Normally distributed values are within two standard deviations of the mean, so we should expect about 19 out of 20 similarly randomized polls at the same time as the Forum poll, of the same size[9], and asking the exact same question to give results within $2 \times 1.26\% = 2.52\%$[10] of the true percentage, or to vary by as much as 5.04%—the 4σ width of 95% of the histogram values. We are using a *hypothetical* value for p, but small changes to the value of p have very little effect on the calculated *SD* value. It certainly looks like the difference between the Forum and Ipson polls reflects more than just sampling variation[11].

---

[7]For smaller n, we could use a Binomial model, which we saw in Chapter 13.

[8]The standard deviation is 0.012149 or 1.21%. Remember that the standard deviation always has the same units as the data. Here our units are %, or "percentage points." The standard deviation isn't 1.21% of anything, it is just 1.21 percentage points. If that's confusing, try writing the units out as "percentage points" instead of using the symbol %. Many polling agencies now do that too.

[9]The Ipsos poll was actually a bit smaller in size, making the standard deviation slightly greater, but not by enough to matter.

[10]The standard deviation always has the same units as the data. Here, our units are percentage points. But that can be confusing, because the standard deviation is not 1.26% of anything; it is 1.26 percentage points. If that's confusing, write the units as "percentage points" instead of using the symbol %.

[11]Actually, these polls were about one month apart, so intervening events can change some opinions. There may have been fine differences in the wording of the questions, small differences in the populations sampled due to different sampling methods. And what about the 'not sure' numbers? How would throwing those out affect the percentages? Other polls around the same time indicated in the question that the alternative to a referendum would be a high degree of public consultation and Parliamentary hearings. This simple change in wording reduced the percentage for a referendum to about 50% in several polls! The lesson here is that in addition to *sampling error*, there is always *nonsampling error* affecting results. In this chapter, though, we focus on understanding sampling errors—which can be modelled precisely (what a relief!).

Such sample-to-sample variation is sometimes called **sampling error**. It's not really an *error* at all, but just *variability* you'd expect to see from one sample to another. A better term would be **sampling variability**.

Section 14.1 Exercises

1. **Website** An investment website can tell what devices are used to access the site. The site managers wonder whether they should enhance the facilities for trading via smartphones so they want to estimate the proportion of users who access the site that way (even if they also use their computers sometimes). They draw a random sample of 200 investors from their customers. Suppose that the true proportion of smartphone users is 36%.
   a) What would you expect the shape of the sampling distribution for the sample proportion to be?
   b) What would be the mean of this sampling distribution?
   c) What would be the standard deviation of the sampling distribution?

2. **Marketing** The proportion of adult women in Canada is approximately 51%. A marketing survey company telephones 400 people at random.
   a) What proportion of the sample of 400 would you expect to be women?
   b) What would the standard deviation of the sampling distribution be?
   c) How many women, on average, would you expect to find in a sample of that size?

3. **Send money** When it sends out its fundraising letter, a philanthropic organization typically gets a return from about 5% of the people on their mailing list. To see what the response rate might be for future appeals, it did a simulation using samples of size 20, 50, 100, and 200. For each sample size, it simulated 1000 mailings with success rate $p = 0.05$, and constructed the histogram of the 1000 sample proportions, shown below. Explain how these histograms change as the sample size increases. Be sure to talk about shape, centre, and spread.

4. **Character recognition** An automatic character recognition device can successfully read about 85% of handwritten credit card applications. To estimate what might happen when this device reads a stack of applications, the company did a simulation using samples of size 20, 50, 75, and 100. For each sample size, they simulated 1000 samples with success rate $p = 0.85$, and constructed the histogram of the 1000 sample proportions, shown here. Explain how these histograms change as the sample size increases. Be sure to talk about shape, centre, and spread.

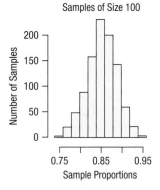

14.2 When Does the Normal Model Work? Assumptions and Conditions

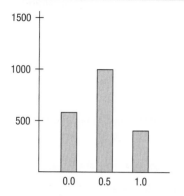

Figure 14.4

Proportions from samples of size two can take on only three possible values. A Normal model does not work well.

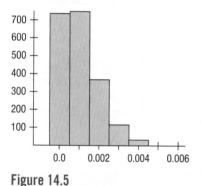

Figure 14.5

The distribution of sample proportions for 2000 samples of size 1000 with $p = 0.001$. Because the true proportion is so small, the sampling distribution is skewed to the right and the Normal model won't work well.

> **Successes and Failures** The terms "success" and "failure" for the outcomes that have probability p and q are common in statistics. But they are completely arbitrary labels. When we say that a disease occurs with probability p, we certainly don't mean that getting sick is a "success" in the ordinary sense of the word.

De Moivre claimed that the sampling distribution can be modelled well by a Normal model. But, does it always work? Well, no. For example, if we drew samples of size 2, the only possible proportion values would be 0, 0.5, and 1. There's no way the histogram (Figure 14.4) consisting only of those values could look like a Normal model.

In fact, De Moivre's claim is only approximately true. (But that's okay. After all, models are only supposed to be approximately true.) The Normal model becomes a better and better representation of the sampling distribution as the size of the samples gets bigger.[12] Samples of size 1 or 2 won't work at all, but the distributions of sample proportions of larger samples are modelled well by the Normal model as long as p isn't too close to 0 or 1.

Populations with a true proportion, p, close to 0 or 1 can be a problem. Suppose a basketball coach surveys students to see how many male high school seniors are over 2.0 metres. What will the proportions in samples of size 1000 look like? If the true proportion of students that tall is 0.001, then the coach is likely to get only a few seniors over 2.0 m in any random sample of 1000. Most samples will have proportions of 0/1000, 1/1000, 2/1000, 3/1000, 4/1000 with only a very few samples having a higher proportion. Figure 14.5 shows results from a simulation of 2000 surveys of size 1000 with $p = 0.001$.

The problem is that the distribution is skewed to the right because p is so close to 0. (Had p been very close to 1, it would have been skewed to the left). So, even though n is large, p is too small, and so the Normal model still won't work well.

Assumptions and Conditions

When does the Normal model with mean p and standard deviation $\sqrt{\dfrac{pq}{n}}$ work well as a model for the sampling distribution of a sample proportion? We can summarize those cases by checking the following assumptions and conditions:

The Independence Assumption: The individuals in the samples (whose proportion we are finding) must be independent of each other. If not, the observations will resemble each other too much and distort the standard deviation. You can't know if an assumption is true or not, but you *can* check to see whether the data were collected in a way that makes this assumption plausible.

If your data come from an experiment in which subjects were randomly assigned to treatments or from a survey based on a simple random sample, then the data satisfy the **Randomization Condition**. In practice, other more practical sampling designs are often used (e.g., multistage samples), and while the sampling distribution will still tend to be Normal for large enough n, our formula for the SD of the proportion falls apart and often gives an underestimate of the true SD. Statisticians then have to derive the correct formula mathematically, or they may decide to "bootstrap" it (a computer-intensive approach, discussed in Chapter 29). Statistics Canada bootstraps nearly all their SDs due to complex sample designs. And if not a proper random sample, you'll need to think carefully about whether it is honest to consider your cases independent with respect to the variables of interest and whether the cases you have are representative of the population you care about.

We've seen that n has to be large enough and that p (and $q = 1 - p$ as well) can't be too close to 0 or 1. We put those two concerns together and check for at least 10 "successes" and 10 "failures." That is, make sure that $n\hat{p}$ and $n\hat{q}$ are both at least 10.

Success/Failure Condition: You should have at least 10 successes and 10 failures in your data.

---

[12]Formally, we say the claim is true in the limit as n grows.

We have now answered the question raised at the start of the chapter. To know how variable a sample proportion (from an SRS) is, we need to know the proportion and the size of the sample. That's all.

10% (and least important) Condition: It can also be a problem, of sorts, to sample too much of the population. Once you've sampled more than about 10% of the population, the remaining individuals are no longer really independent of each other, which messes up our SD formula. There are special formulas that you can use to adjust for sampling a large part of a population[13], but they are rarely needed, and this 10% condition is usually checked more casually than other assumptions and conditions.

The Success/Failure condition encourages a large sample size, but the 10% condition cautions about sampling a big proportion of a population. This apparent conflict is not really a problem. Most populations are safely large enough, and it always makes sense to devote your effort and expense to gathering a larger sample or more experimental units.

In summary, if the assumptions and conditions are met, then you can use the Normal model to model the distribution of the sample proportion. In other words:

> **THE SAMPLING DISTRIBUTION MODEL FOR A PROPORTION**
>
> Provided that the sampled values are independent and the sample size is large enough, the sampling distribution of $\hat{p}$ is modelled by a Normal model with mean p and standard deviation $SD(\hat{p}) = \sqrt{\dfrac{pq}{n}}$.

Without the sampling distribution model, the rest of Statistics just wouldn't exist. Sampling models are what make Statistics work. They inform us about the amount of variation we should expect when we sample. With a sampling distribution model, we can make informed decisions about how precise our estimate of the true value of a parameter might be. That's exactly what we'll be doing for the rest of this book.

Sampling distrbution models enable us to say something about the population when all we have is data from a sample. By imagining what *might* happen if we were to draw many, many samples from the same population, we can learn a lot about how close the statistics computed from our one particular sample may be to the corresponding population parameters they estimate. That's the path to the *margin of error* you hear about in polls and surveys. We'll see how to determine that in the next chapter.

For Example USING THE SAMPLING DISTRIBUTION MODEL FOR PROPORTIONS

The Centers for Disease Control and Prevention report that 22% of 18-year-old women in the United States have a body mass index (BMI)[14] of 30 or more—a value associated with increased health risk.

As part of a routine health check at a large university, the physical education department usually requires students to come in to be measured and weighed. This year, the department decided to try out a self-report system. It asked 200 randomly selected female students to report their heights and weights (from which their BMIs could be calculated). Only 31 of these students had BMIs greater than 30.

QUESTION: Is this proportion of high-BMI students unusually small?

ANSWER: First, check:

✓ **Randomization Condition:** The department drew a random sample, so the respondents should be independent and randomly selected from the target population.

✓ **Success/Failure Condition:** There were 31 "successes"—students with BMIs greater than 30 and 169 "failures." Both are at least 10.

✓ **10% Condition:** 200 respondents is less than 10% of all the female students at a "large university." [For a small university, we might need some adjustment to our *SD* formula.]

---

[13]See optional Section 14.5: Finite Population Correction.

[14]BMI = mass in kg/(height in m)^2

It's okay to use a Normal model to describe the sampling distribution of the proportion of respondents with BMIs above 30.

The phys ed department observed $\hat{p} = \dfrac{31}{200} = 0.155$.

The department expected $p = 0.22$, with $SD(\hat{p}) = \sqrt{\dfrac{pq}{n}} = \sqrt{\dfrac{(0.22)(0.78)}{200}} = 0.029$,

so $z = \dfrac{\hat{p} - p}{SD(\hat{p})} = \dfrac{0.155 - 0.22}{0.029} = -2.24$.

By the 68–95–99.7 Rule, I know that values more than 2 standard deviations below the mean of a Normal model show up less than 2.5% of the time. Perhaps women at this university differ from the general population, or self-reporting may not provide accurate heights and weights.

Just Checking

1. You want to poll a random sample of 100 students on campus to see if they are in favour of the proposed location for the new student centre. Of course, you'll get just one number, your sample proportion, $\hat{p}$. But if you imagined all the possible samples of 100 students you could draw and imagined the histogram of all the sample proportions from these samples, what shape would it have?

2. Where would the centre of that histogram be?

3. If you think that about half the students are in favour of the plan, what would the standard deviation of the sample proportions be?

Step-by-Step Example WORKING WITH SAMPLING DISTRIBUTION MODELS FOR PROPORTIONS

Suppose that about 13% of the population is left-handed.[15] A 200-seat school auditorium has been built with 15 "lefty seats," seats that have the built-in desk on the left rather than the right arm of the chair. (For the right-handed readers among you, have you ever tried to take notes in a chair with the desk on the left side?)

Question: In a class of 90 students, what's the probability that there will not be enough seats for the left-handed students?

THINK ➡ **Plan** State what we want to know.	I want to find the probability that in a group of 90 students, more than 15 will be left-handed. Since 15 out of 90 is 16.7%, I need the probability of finding more than 16.7% left-handed students out of a sample of 90 if the proportion of lefties is 13%.
Model Think about the assumptions and check the conditions. You might be able to think of cases where the **Independence Assumption** is not plausible—for example, if the students are all related, or if they were selected for being left- or	✓ **Independence Assumption:** The 90 students in the class are not a random sample, but it is reasonable to assume that the probability that one student is left-handed is not changed by the fact that another student is right- or left-handed. It is also reasonable to

[15]Actually, there is little agreement among experts about the proportion of lefties in the population—or even how to define "left-handedness." Estimates range from 8% to 15%.

right-handed. But for students in a class, the assumption of independence with respect to handedness seems reasonable.

assume that the students in the class are representative of students in general with respect to handedness.

✓ **Success/Failure Condition:**
$$np = 90(0.13) = 11.7 \geq 10$$
$$nq = 90(0.87) = 78.3 \geq 10$$

✓ **10% Condition:** 90 is surely less than 10% of all students so there is no need for special formulas. (Even if the school itself is small, I'm thinking of the population of all possible students who could have gone to the school.)

State the parameters and the sampling distribution model.

[We could also work with the count instead of the proportion and use the Binomial model with Normal approximation, as in Chapter 13]

The population proportion is $p = 0.13$. The conditions are satisfied, so I'll model the sampling distribution of $\hat{p}$ with a Normal model with mean 0.13 and a standard deviation of

$$SD(\hat{p}) = \sqrt{\frac{pq}{n}} = \sqrt{\frac{(0.13)(0.87)}{90}} \approx 0.035$$

My model for $\hat{p}$ is N(0.13, 0.035).

SHOW ➡ **Plot** Make a picture. Sketch the model and shade the area we're interested in, in this case the area to the right of 16.7%.

Mechanics Use the standard deviation as a ruler to find the z-score of the cutoff proportion. We see that 16.7% lefties would be just over one standard deviation above the mean.

Find the resulting probability from a table of Normal probabilities, a computer program, or a calculator.

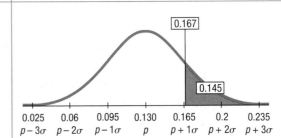

$$z = \frac{\hat{p} - p}{SD(\hat{p})} = \frac{0.167 - 0.13}{0.035} = 1.06$$

$$P(\hat{p} > 0.167) = P(z > 1.06) = 0.1446$$

TELL ➡ **Conclusion** Interpret the probability in the context of the question.

There is about a 14.5% chance that there will not be enough seats for the left-handed students in the class.

Section 14.2 Exercises

5. **Marriage** According to a Pew Research survey, 27% of American adults are pessimistic about the future of marriage and the family. That is based on a random sample of about 1500 people. Is it reasonable for Pew Research to use a Normal model for the sampling distribution of the sample proportion? Why or why not?

6. **Campus sample** For her final project, Stacy plans on surveying a random sample of 50 students on whether they

plan to go to Florida for Spring Break. From past years, she guesses that about 10% of the class goes. Is it reasonable for her to use a Normal model for the sampling distribution of the sample proportion? Why or why not?

7. **Send more money** The philanthropic organization in Exercise 3 expects about a 5% success rate when they send fundraising letters to the people on their mailing list. In Exercise 3, you looked at the histograms showing

distribution of sample proportions from 1000 simulated mailings for samples of size 20, 50, 100, and 200. Here are both the theoretical and observed statistics for these simulated samples:

n	Observed Mean	Theoretical Mean	Observed St. Dev.	Theoretical St. Dev.
20	0.0497	0.05	0.0479	0.0487
50	0.0516	0.05	0.0309	0.0308
100	0.0497	0.05	0.0215	0.0218
200	0.0501	0.05	0.0152	0.0154

a) Looking at the histograms of Exercise 3, at what sample size would you be comfortable using the Normal model as an approximation for the sampling distribution?

b) What does the Success/Failure Condition say about the choice you made in part a?

8. Character recognition, again The automatic character recognition device discussed in Exercise 4 successfully reads about 85% of handwritten credit card applications. In Exercise 4, you looked at the histograms showing distributions of sample proportions from 1000 simulated samples of size 20, 50, 75, and 100. The sample statistics and theoretical values for each simulation are as follows:

n	Observed Mean	Theoretical Mean	Observed St. Dev.	Theoretical St. Dev.
20	0.8481	0.85	0.0803	0.0799
50	0.8507	0.85	0.0509	0.0505
100	0.8481	0.85	0.0406	0.0412
200	0.8488	0.85	0.0354	0.0357

a) Looking at the histograms in Exercise 4, at what sample size would you be comfortable using the Normal model as an approximation for the sampling distribution?

b) What does the Success/Failure Condition say about the choice you made in part a?

14.3 The Sampling Distribution of Other Statistics

The sampling distribution for the proportion is especially useful because the Normal model provides such a good approximation. But, it might be useful to know the sampling distribution for *any* statistic that we can calculate, not just the sample proportion. Is the Normal model a good model for all statistics? Would you expect the sampling distribution of the minimum or the maximum or the variance of a sample to be Normally distributed? What about the median? Here's an example to help us find out.

For Example SIMULATING THE SAMPLING DISTRIBUTIONS OF OTHER STATISTICS

Earlier in this book, we looked at a study of body measurements of 250 men (full data are in **Body fat complete**). The median of all 250 men's weights is 176.125 lbs and the variance is 730.9 lbs^2. We can treat these 250 men as a population, draw repeated random samples of 10, and compute the median, the variance, and the minimum for each sample.

Each of these histograms depicts the estimated sampling distribution of its respective statistic. And it is easy to see that they aren't all the same. The sampling

Figure 14.6

Figure 14.7

distribution of the medians is unimodal and symmetric. The sampling distribution of the variances is skewed to the right. And the sampling distribution of the minimums is, well, messy.

We can simulate to find the sampling distribution of *any* statistic we like: the maximum, the IQR, the 37th percentile, anything. But, both the proportion and the mean (as we'll see in the next section) have sampling distributions that are well approximated by a Normal model. And that observation hints at a fundamental fact about the sampling distribution models for the two summary statistics that we use most often.

Simulating the Sampling Distribution of a Mean

Here's a simple simulation. Let's start with one fair die. If we toss this die 10 000 times, what should the histogram of the numbers on the face of the die look like? Figure 14.6 shows the results of a simulated 10 000 tosses.

Now let's toss a *pair* of dice and record the average of the two. If we repeat this (or at least simulate repeating it) 10 000 times, recording the average of each pair, what will the histogram of these 10 000 averages look like? Before you look, think a minute. Is getting an average of 1 on *two* dice as likely as getting an average of 3 or 3.5?

Let's see (Figure 14.7). We're much more likely to get an average near 3.5 than we are to get one near 1 or 6. Without calculating those probabilities exactly, it's fairly easy to see that the *only* way to get an average of 1 is to get two 1's. To get a total of 7 (for an average of 3.5), though, there are many more possibilities. This distribution even has a name: the *triangular* distribution.

What if we average three dice? We'll simulate 10 000 tosses of three dice and take their average. Figure 14.8 shows the results. What's happening? First notice that it's getting harder to have averages near the ends. Getting an average of 1 or 6 with three dice requires all three to come up 1 or 6, respectively. That's less likely than for two dice to come up both 1 or both 6. The distribution is being pushed toward the middle. But what's happening to the shape? (This distribution doesn't have a name, as far as we know.)

Figure 14.8

Figure 14.9

Let's continue this simulation to see what happens with larger samples. Figure 14.9 shows a histogram of the averages for 10 000 tosses of five dice. The pattern is becoming clearer. Two things continue to happen. The first fact we knew already from the Law of Large Numbers. It says that as the sample size (number of dice) gets larger, each sample average is more likely to be closer to the population mean. So we see the means continuing to tighten around 3.5. More important, the shape of the distribution is approaching the Normal model.

Let's skip ahead and try 20 dice. The histogram of averages for 10 000 throws of 20 dice is in Figure 14.10. Now we see the Normal shape clearly (and notice how much smaller the spread is). But can we count on this happening for situations other than dice throws?

Figure 14.10

Section 14.3 Exercises

9. Sample maximum The distribution of scores on a Statistics test for a particular class is skewed to the left. The professor wants to predict the maximum score and so wants to understand the distribution of the sample maximum. She simulates the distribution of the maximum of the test for 30 different tests (with $n = 5$). The histogram below shows a simulated sampling distribution of the sample maximum from these tests.

a) Would a Normal model be a useful model for this sampling distribution? Explain.

b) The mean of this distribution is 46.3 and the SD is 3.5. Would you expect about 95% of the samples to have their maximums within 7 of 46.3? Why or why not?

10. **Soup** A machine is supposed to fill cans with 16 oz of soup. Of course there, will be some variation in the amount actually dispensed, and measurement errors are often approximately normally distributed. The manager would like to understand the variability of the variances of the samples, so he collects information from the last 250 batches of size 10 and plots a histogram of the variances:

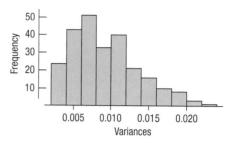

a) Would a Normal model be a useful model for this sampling distribution? Explain.

b) The mean of this distribution is 0.009 and the SD is 0.004. Would you expect about 95% of the samples to have their variances within 0.008 of 0.009? Why or why not?

14.4 The Central Limit Theorem: The Fundamental Theorem of Statistics

Georgios Kollidas/Shutterstock

"The theory of probabilities is at bottom nothing but common sense reduced to calculus."

—**Laplace,** in *Théorie analytique des probabilités,* **1812**

What we saw with dice is true for means of repeated samples for almost every situation. When we looked at the sampling distribution of a proportion, we had to check only a few conditions. For means, the result is even more remarkable. *There are almost no conditions at all.*

Let's say that again: The sampling distribution of *any* mean becomes more nearly Normal as the sample size grows. All we need is for the observations to be independent and collected with randomization. We don't even care about the shape of the population distribution![16] This surprising fact was proved by the famous French mathematician Pierre-Simon Laplace in a fairly general form in 1810. At the time, Laplace's theorem caused quite a stir (at least in mathematics circles) because the fact that it works no matter how the data are distributed seems remarkable. Laplace's result is called the **Central Limit Theorem**[17] (CLT).

Why should the Normal model show up again for the sampling distribution of means as well as proportions? We're not going to try to persuade you that it is obvious, clear, simple, or straightforward. In fact, the CLT is surprising and a bit weird. Not only does the distribution of means of repeated random samples get closer and closer to a Normal model as the sample size grows, *this is true regardless of the shape of the population distribution!* Even if we sample from a skewed or bimodal population, the CLT tells us that means of repeated random samples will tend to follow a Normal model for a sufficiently large sample size. If the sample size is small, it will work better the closer the population distribution is to a Normal model. If the data come from a population that's exactly Normal to start with, then the observations themselves are Normal. If we take samples of size 1, their "means" are just the observations—so, of course, they have a Normal sampling distribution.

> **Pierre-Simon Laplace** Laplace was one of the greatest scientists and mathematicians of his time. In addition to his contributions to probability and statistics, he published many new results in mathematics, physics, and astronomy (where his nebular theory was one of the first to describe the formation of the solar system in much the way it is understood today). He also played a leading role in establishing the metric system of measurement.
>
> His brilliance, though, sometimes got him into trouble. A visitor to the Academie des Sciences in Paris reported that Laplace let it be widely known that he considered himself the best mathematician in France. The effect of this on his colleagues was not eased by the fact that Laplace was right.

---

[16]Okay, one technical condition. The data must come from a population with a finite variance. You probably can't imagine a population with an infinite variance, but statisticians can construct such things, so we have to discuss them in footnotes like this. It really makes no difference in how you think about the important stuff, so you can just forget we mentioned it.

[17]The word "central" in the name of the theorem means "fundamental." It doesn't refer to the centre of a distribution.

But what if we start with a distribution that's strongly skewed (Figure 14.11), like the CEO compensations we saw back in Chapter 4?

Figure 14.11

The distribution of CEO compensations is highly right skewed.

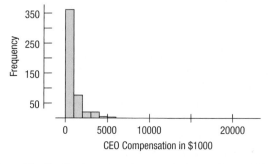

If we sample from this distribution, there's a good chance that we'll get an exceptionally large value. So some samples will have sample means much larger than others. Figure 14.12 shows the distribution we get for the means from 1000 repeated samples of size 10 from this population:

Figure 14.12

Samples of size 10 from the CEOs have means whose distribution is still right skewed, but less so than the distribution of the individual values shown in Figure 14.11.

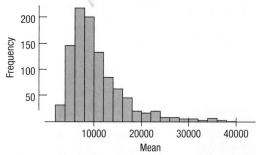

This distribution is not *as* skewed as the original distribution, but still strongly right skewed. We wouldn't ordinarily find a mean of such a skewed distribution, but bear with us for a few steps. The mean of these values is around 10 000 (in $1000s), which actually matches the mean of all 500 CEOs in our population. What happens if we take a larger sample? Figure 14.13 shows the sampling distribution of means from samples of size 50:

Figure 14.13

Means of samples of size 50 have a distribution that is only moderately right skewed.

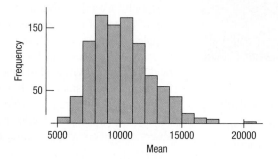

This distribution is less skewed than the corresponding distribution from smaller samples and its mean is again near 10 000. Will this continue as we increase the sample size? Let's take many samples of size 100 and examine their means (Figure 14.14):

Figure 14.14

The means of samples of size 100 are nearly symmetric.

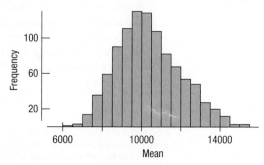

As we take larger samples, the distribution of means becomes more and more symmetric. By the time we get to samples of size 200, the distribution (Figure 14.15) is quite symmetrical and, as in all the previous simulations, has a mean quite close to 10 000.

Figure 14.15

When the sample size is 200, even though we started with a highly skewed distribution, the sampling distribution of means is now almost perfectly symmetrical.

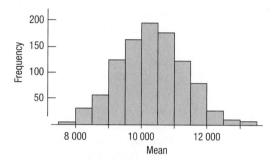

What about a really bimodal population, one that consists of only 0's and 1's. The CLT says that even means of (big) samples from this population will follow a Normal sampling distribution model. But wait. Suppose we have a categorical variable and we assign a 1 to each individual in the category and a 0 to each individual not in the category. And then we find the mean of these 0's and 1's. That's the same as counting the number of individuals who are in the category and dividing by n. That mean will be . . . the *sample proportion, p,* of individuals who are in the category (a "success"). So maybe it wasn't so surprising after all that proportions, like means, have Normal sampling distribution models; they are actually just a special case of Laplace's remarkable theorem. Of course, for such an extremely bimodal population, we'll need a reasonably large sample size—and that's where the special conditions for proportions come in.

THE CENTRAL LIMIT THEOREM (CLT)

The mean[18] of a random sample is a random variable whose sampling distribution can be approximated by a Normal model if the sample size is big enough. The larger the sample, the better the approximation will be.

Assumptions and Conditions

The CLT requires essentially the same assumptions as we saw for modelling proportions:

Independence Assumption: The sampled values must be independent of each other. If they satisfy the **Randomization Condition** you may safely assume independence. Otherwise you must think carefully about whether they can be considered independent with respect to the variable of interest and whether the cases are representative of the population you care about. (Sampling more than 10% of the population would compromise independence, but we'll omit the 10% condition here since it is too rarely relevant when estimating means.)

Sample Size Condition: Although the CLT tells us that a Normal model is useful in thinking about the behaviour of sample means when the sample size is large enough, it doesn't tell us how large a sample we need. The truth is, it depends; there's no one-size-fits-all rule. If the population is unimodal and symmetric, even a fairly small sample (like five or ten) is OK. If the population is strongly skewed, like the compensation for CEOs, we saw that it can take a pretty large sample (like 100, 200, or even more) for a Normal model to adequately describe the distribution of sample means. For proportions, we have the success/failure conditions to guide us. For now, you'll just need to think about your sample size in the context of what you know about the population, and then tell whether you believe the **Sample Size Condition** has been met.

---

[18]And hence the sample total, too, since the mean and the total only differ by the multiplicative constant n.

But Which Normal?

The CLT says that the sampling distribution of any mean or proportion is approximately Normal for a sufficiently large sample size. But which Normal model? We know that any Normal is specified by its mean and standard deviation. For proportions, the sampling distribution is centred at the population proportion. For means, it's centred at the population mean.

What about the standard deviations? We noticed in our dice simulation that the histograms got narrower as we averaged more and more dice together. This shouldn't be surprising. Means vary less than the individual observations. Think about it for a minute. Which would be more surprising, having *one* person in your Statistics class who is over 205 cm tall or having the *mean* of 100 students taking the course be over 205 cm? The first event is fairly rare.[19] You may have seen somebody this tall in one of your classes sometime. But finding a class of 100 whose mean height is over 205 cm tall just won't happen. Why? Because *means have smaller standard deviations than individuals.*

How much smaller? Well, we have good news and bad news. The good news is that the standard deviation of the sample mean $\bar{y}$ falls as the sample size grows. The bad news is that it doesn't drop as fast as we might like. It only goes down by the *square root* of the sample size. Why? The Math Box will show you that the Normal model for the sampling distribution of the mean has a standard deviation equal to

$$SD(\bar{y}) = \frac{\sigma}{\sqrt{n}}$$

where σ is the standard deviation of the population. To emphasize that this is a standard deviation *parameter* of the sampling distribution model for the sample mean, $\bar{y}$, we write $SD(\bar{y})$ or $\sigma(\bar{y})$.

On very rare occasions, we may sample more than 10% of the population. In this case, the SD formula above overestimates the true SD (but the formula can be corrected—if you're interested, see the Finite Population Correction discussion at the end of this chapter). Since these

---

THE SAMPLING DISTRIBUTION MODEL FOR A MEAN

When a simple random sample is drawn from any (much bigger) population with mean μ and standard deviation σ, its sample mean, $\bar{y}$, has a sampling distribution with the same *mean* μ but whose *standard deviation* is $\frac{\sigma}{\sqrt{n}}$ (and we write $\sigma(\bar{y}) = SD(\bar{y}) = \frac{\sigma}{\sqrt{n}}$). No matter what population the random sample comes from, the *shape* of the sampling distribution is approximately Normal as long as the sample size is large enough. The larger the sample used, the more closely the Normal model approximates the sampling distribution for the mean.

MATH BOX

Why is $SD(\bar{y}) = \frac{\sigma}{\sqrt{n}}$? We know that $\bar{y}$ is a sum divided by n:

$$\bar{y} = \frac{y_1 + y_2 + y_3 + \ldots + y_n}{n}$$

As we saw in Chapter 13, when, a random variable is divided by a constant its variance is divided by the *square* of the constant:

$$Var(\bar{y}) = \frac{Var(y_1 + y_2 + y_3 + \ldots + y_n)}{n^2}$$

---

[19] If students are a random sample of adults, fewer than 1 out of 10 000 should be taller than 205 cm. Why might university students not really be a random sample with respect to height? Even if they're not a perfectly random sample, a university student over 205 cm tall is still rare.

We draw the y's randomly from a big population, ensuring they are at least approximately independent. For independent random variables, variances add:

$$Var(\bar{y}) = \frac{Var(y_1) + Var(y_2) + Var(y_3) + \ldots + Var(y_n)}{n^2}$$

All n of the y's were drawn from our sampled population, so they all have the same variance, σ^2:

$$Var(\bar{y}) = \frac{\sigma^2 + \sigma^2 + \sigma^2 + \ldots + \sigma^2}{n^2} = \frac{n\sigma^2}{n^2} = \frac{\sigma^2}{n}$$

The standard deviation of $\bar{y}$ is the square root of this variance:

$$SD(\bar{y}) = \sqrt{\frac{\sigma^2}{n}} = \frac{\sigma}{\sqrt{n}}$$

In a very similar manner, we can also prove that the expected value of the sample mean, $E(\bar{y})$ is equal to μ, which shows that the mean of the measurements, $\bar{y}$, in a simple random sample is an *unbiased estimator* of the population mean, μ (in the absence of non-sampling sources of bias, naturally).

We now have two closely related sampling distribution models that we can use *when the appropriate assumptions and conditions are met*. Which one we use depends on which kind of data we have:

■ When we have categorical data, we calculate a sample proportion, $\hat{p}$; the sampling (or probability) distribution of this random variable is approximately Normal with a mean at the true proportion p, and a standard deviation of $SD(\hat{p}) = \sqrt{\dfrac{pq}{n}} = \dfrac{\sqrt{pq}}{\sqrt{n}}$.

■ When we have quantitative data, we calculate a sample mean, $\bar{y}$; the sampling (or probability) distribution of this random variable is approximately Normal with a mean at the true mean, μ, and a standard deviation of $SD(\bar{y}) = \dfrac{\sigma}{\sqrt{n}}$.

The means of these models are easy to remember, so all you need to be careful about are the standard deviations. Remember that these are standard deviations of the *statistics* $\hat{p}$ and $\bar{y}$. They both have a square root of n in the denominator. That tells us that the larger the sample, the less either statistic will vary. The only difference is in the numerator. If you just start by writing $SD(\bar{y})$ for quantitative data and $SD(\hat{p})$ for categorical data, you'll be able to remember which formula to use.

For Example USING THE CLT FOR MEANS

RECAP: A university physical education department asked a random sample of 200 female students to self-report their heights and weights, but the percentage of students with body mass indexes over 30 seemed suspiciously low. One possible explanation may be that the respondents "shaded" their weights down a bit. The CDC reports that the mean weight of 18-year-old women is 65.4 kg, with a standard deviation of 23.45 kg, but these 200 randomly selected women reported a mean weight of only 63.7 kg.

QUESTION: Based on the CLT and the 68–95–99.7 Rule, does the mean weight in this sample seem exceptionally low, or might this just be random sample-to-sample variation?

ANSWER: The conditions check out okay:

✓ Randomization Condition: The women were a random sample and their weights can be assumed to be independent.
✓ Sample Size Condition: The distribution of university women's weights is likely to be unimodal and reasonably symmetric, so the CLT applies to means of even small samples; 200 values is plenty.

The sampling model for sample means is approximately Normal with $E(\bar{y}) = 65.4$ and

$$SD(\bar{y}) = \frac{\sigma}{\sqrt{n}} = \frac{23.45}{\sqrt{200}} = 1.66.$$ The expected distribution of sample means is

The 68–95–99.7 Rule suggests that although the reported mean weight of 63.7 kg is somewhat lower than expected, it does not appear to be unusual. Such variability is not all that extraordinary for samples of this size.

Step-by-Step Example WORKING WITH THE SAMPLING DISTRIBUTION MODEL FOR THE MEAN

The mean weight of adult men in Canada is 82 kg with a standard deviation of 16 kg.[20]

Question: An elevator in our building has a weight limit of 1000 kg. What's the probability that if 10 men get on the elevator, they will overload its weight limit?

THINK ➡ Plan State what we want to know.	Asking the probability that the total weight of a sample of 10 men exceeds 1000 kg is equivalent to asking the probability that their mean weight is greater than 100 kg.
Model Think about the assumptions and check the conditions.	✓ **Independence Assumption:** We don't have a random sample, but it's reasonable to think that the weights of 10 men entering the elevator will be independent of each other and representative of the elevator-user population. (But there could be exceptions—for example, if they were all from the same family or if the elevator were in a building with a diet clinic!)
If the sample were larger, we'd be less concerned about the shape of the distribution of all weights.	✓ **Sample Size Condition:** I don't have data, but I believe the distribution of population weights is unimodal and only mildly skewed, so my sample of 10 men seems large enough to use the CLT.
State the parameters and the sampling model.	The mean for all weights is $\mu = 82$ kg and the standard deviation is $\sigma = 16$ kg. Since the conditions are satisfied, the CLT says that the sampling distribution of $\bar{y}$ has a Normal model with mean 82 and standard deviation $$SD(\bar{y}) = \frac{\sigma}{\sqrt{n}} = \frac{16}{\sqrt{10}} \approx 5.06$$

---

[20]Author's estimate, using Canadian Community Health Survey data and other sources.

SHOW➡ Plot Make a picture. Sketch the model and shade the area we're interested in. Here the mean weight of 100 kg appears to be far out on the right tail of the curve.

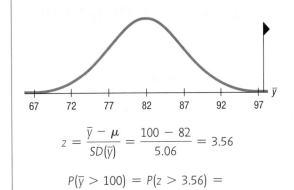

67 72 77 82 87 92 97 $\bar{y}$

Mechanics Use the standard deviation as a ruler to find the z-score of the cutoff mean weight. We see that an average of 100 kg is more than 3 standard deviations above the mean.

Find the resulting probability from a table of Normal probabilities such as Table Z, a computer program, or a calculator.

$$z = \frac{\bar{y} - \mu}{SD(\bar{y})} = \frac{100 - 82}{5.06} = 3.56$$

$$P(\bar{y} > 100) = P(z > 3.56) = 1 - 0.9998 = 0.0002$$

TELL➡ Conclusion Interpret your result in the proper context, being careful to relate it to the original question.

The chance that a random collection of 10 men will exceed the elevator's weight limit is only 0.0002. So, if they are akin to a random sample, it is quite unlikely that 10 people will exceed the total weight allowed on the elevator.

Section 14.4 Exercises

11. Tips A waiter believes the distribution of his tips has a model that is slightly skewed to the right, with a mean of $9.60 and a standard deviation of $5.40.
   a) Explain why you cannot determine the probability that a given party will tip him at most $20.
   b) Can you estimate the probability that the next 4 parties will tip an average of at most $20? Explain.
   c) Is it likely that his 10 parties today will tip an average of at most $15? Explain.

12. Groceries A grocery store's receipts show that Sunday customer purchases have a skewed distribution with a mean of $30 and a standard deviation of $20.
   a) Explain why you cannot determine the probability that the next Sunday customer will spend at least $40.
   b) Can you estimate the probability that the next 10 Sunday customers will spend an average of at least $40? Explain.
   c) Is it likely that the next 50 Sunday customers will spend an average of at least $40? Explain.

13. More tips The waiter in Exercise 11 usually waits on about 40 parties over a weekend of work.
   a) Estimate the probability that he will earn at most $500 in tips.
   b) How much does he earn on the best 10% of such weekends?

14. More groceries Suppose the store in Exercise 12 had 312 customers this Sunday.
   a) Estimate the probability that the store's revenues were at least $10,000.

   b) If, on a typical Sunday, the store serves 312 customers, how much does the store take in on the worst 10% of such days?

15. Coin tosses In a large class of introductory Statistics students, the professor has each student toss a coin 16 times and calculate the proportion of his or her tosses that were heads. The students then report their results, and the professor plots a histogram of these several proportions.
   a) What shape would you expect this histogram to be? Why?
   b) Where do you expect the histogram to be centred?
   c) How much variability would you expect among these proportions?
   d) Explain why a Normal model should not be used here.

16. M&M's The candy company claims that 10% of the M&M's it produces are green. Suppose that the candies are packaged at random in small bags containing about 50 M&M's. A class of elementary school students learning about percents opens several bags, counts the various colours of the candies, and calculates the proportion that are green.
   a) If we plot a histogram showing the proportions of green candies in the various bags, what shape would you expect it to have?
   b) Can that histogram be approximated by a Normal model? Explain.
   c) Where should the centre of the histogram be?
   d) What should the standard deviation of the proportion be?

14.5 Sampling Distributions: A Summary

Means vary less than individual data values. That makes sense. If the same test is given to many sections of a large course and the class average is, say, 80%, some students may score 95% because individual scores vary a lot. But we'd be shocked (and pleased!) if the *average* score of the students in any section was 95%. Averages are much less variable.

Not only do group averages vary less than individual values, but common sense suggests that averages should be more consistent for larger groups. The fact that $SD(\bar{y}) = \dfrac{\sigma}{\sqrt{n}}$ and $SD(\hat{p}) = \sqrt{pq/n}$ confirms this hunch because n is in the denominator in both cases. This shows that the variability of sample means decreases as the sample size increases. But the standard deviation of the sampling distribution declines only with the square root of the sample size and not, for example, with $1/n$.

The mean of a random sample of four has half $\left(\dfrac{1}{\sqrt{4}} = \dfrac{1}{2} \right)$ the standard deviation of an individual data value. To cut the standard deviation in half again, we'd need a sample of 16, and a sample of 64 to halve it once more.

If only we had a much larger sample, we could get the standard deviation of the sampling distribution *really* under control so that the sample mean could tell us still more about the unknown population mean, but larger samples cost more and take longer to survey. And there are often practical limits to sample sizes. While we're gathering all that extra data, the population itself may change, or a news story may alter opinions. As we shall see, that nasty square root limits how much we can make a sample tell about the population. This is an example of the Law of Diminishing Returns.

A Billion Dollar Misunderstanding? In the late 1990s, the Bill and Melinda Gates Foundation began funding an effort to encourage the breakup of large schools into smaller schools. Why? It had been noticed that smaller schools were more common among the best-performing schools than one would expect. In time, the Annenberg Foundation, the Carnegie Corporation, the Center for Collaborative Education, the Center for School Change, Harvard's Change Leadership Group, the Open Society Institute, Pew Charitable Trusts, and the U.S. Department of Education's Smaller Learning Communities Program all supported the effort. Well over a billion dollars was spent to make schools smaller.

But was it all based on a misunderstanding of sampling distributions? Statisticians Howard Wainer and Harris Zwerling[21] looked at the mean test scores of schools in Pennsylvania. They found that indeed 12% of the top-scoring 50 schools were from the smallest 3% of Pennsylvania schools—substantially more than the 3% we'd naively expect. But then they looked at the *bottom* 50. There they found that 18% were small schools! The explanation? Mean test scores, behave like well, means. We are looking at a rough real-world simulation in which each school is a trial. Even if all Pennsylvania schools were equivalent, we'd expect their mean scores to vary. How much? The CLT tells us that means of test scores vary according to $\dfrac{\sigma}{\sqrt{n}}$. Smaller schools have (by definition) smaller n's, so the sampling distributions of their mean scores naturally have larger standard deviations. It's natural, then, that small schools have both higher and lower mean scores.

On October 26, 2005, the *Seattle Times* reported:[22]

The Gates Foundation announced last week it is moving away from its emphasis on converting large high schools into smaller ones and instead giving grants to specially selected school districts with a track record of academic improvement and effective leadership. Education leaders at the Foundation said they concluded that improving classroom instruction and mobilizing the resources of an entire district were more important first steps to improving high schools than breaking down the size.

---

[21]Wainer, H. and Zwerling, H., 2006, "Legal and empirical evidence that smaller schools do not improve student achievement." *The Phi Delta Kappan* 87:300–303. Discussed in Howard Wainer, 2007, "The most dangerous equation." *American Scientist* pp. 249–256; also at www.americanscientist.org.

[22]Copyright 2005, Seattle Times Company. Used with permission.

The Real World and the Model World

Be careful. We have been slipping smoothly between the real world, in which we draw random samples of data, and a magical mathematical model world, in which we describe how the sample means and proportions we observe in the real world behave as random variables in all the random samples that we *might* have drawn. Now we have *two* distributions to deal with. The first is the real world distribution of the sample, which we might display with a histogram (for quantitative data) or with a bar chart or table (for categorical data). The second is the math world *sampling distribution model* of the statistic, often a Normal model based on the CLT. Don't confuse the two.

For example, don't mistakenly think the CLT says that the *data* are Normally distributed as long as the sample is large enough. In fact, as samples get larger, we expect the distribution of the data to look more and more like the population from which they are drawn—skewed, bimodal, whatever—but not necessarily Normal. You can collect a sample of CEO salaries for the next 1000 years,[23] but the histogram will never look Normal. It will be skewed to the right. The CLT doesn't talk about the distribution of the data from the sample. It talks about the distribution of sample *means* and sample *proportions* of many different random samples drawn from the same population. Of course, the CLT does require that the sample be big enough when the population shape is not unimodal and symmetric, but the fact that, even then, a Normal model is useful is still a very surprising and powerful result.

The CLT in Nature

The CLT we have given here is adequate for our purposes, assuming independent observations (as is common in statistical studies), but actually there have been many generalizations of this theorem over the years. It turns out that even if the individual random variables exhibit some dependence (though not too great), or do not all come from the same distribution (as long as each has but a small effect on the total), Normality will still arise if enough variables are averaged (or summed). Perhaps this more general CLT effect helps explain all those variables in nature that possess an approximately Normal frequency distribution. Are heights in a genetically homogenous population nearly Normal because the final result comes from summing (or averaging) many small genetic and environmental influences? Possibly.[24] Was your height a sum of such random influences, somewhat like summing the random values in a sample?

Many other variables are formed by multiplying many small influences together rather than adding them. But guess what? Take the log (converting multiplication to addition), and you may just get Normality!

Sampling Distribution Models—The Big Picture

Let's summarize what we've learned about sampling distributions. At the heart is the idea that *the statistic itself is a random quantity*. We can't know what our statistic will be because it comes from a random sample. It's just one instance of something that happened for our particular random sample. A different random sample would have given a different result. This sample-to-sample variability is what generates the sampling distribution. The sampling distribution shows us the distribution of possible values that the statistic could have had.

We could simulate that distribution by pretending to take lots of samples. While a sampling distribution can take on pretty much any form depending on the particular statistic and study design, fortunately, for the mean or the proportion, the CLT tells us that we can usually model the sampling distribution with a Normal model.

---

[23]Don't forget to adjust for inflation.

[24]But add gender to the mix of influences on height, and if the average height difference between males and females is very big, we can lose the Normality, possibly ending up with bi-modality. What went wrong? One variable had a big effect on the final result.

The two basic truths about sampling distributions are:

1. Sampling distributions arise because samples vary. Each random sample will contain different cases, and so a different value of the statistic.

2. Although we can simulate a sampling distribution, the CLT saves us the trouble for means and proportions.

Figure 14.16 is a picture showing the process going into the sampling distribution model:

Figure 14.16

We start with a population model, which can have any shape. It can even be bimodal or skewed (as is this one). We label the mean of this model μ and its standard deviation, σ.

We draw one real sample (solid line) of size n and show its histogram and summary statistics. We imagine the other possible samples (dotted lines), with their own histograms and summary statistics. [These sample histograms should have spreads very similar to the population.]

We (imagine) gathering the means of all the samples into a histogram.

The CLT tells us we can model the shape of this histogram with a Normal model. The mean of this Normal is μ, and the standard deviation is

$$SD(\bar{y}) = \frac{\sigma}{\sqrt{n}}.$$

Just Checking

4. Human gestation times have a mean of about 266 days, with a standard deviation of about 16 days. If we record the gestation times of a sample of 100 women, do we know that a histogram of the times will be well modelled by a Normal model?

5. Suppose we look at the *average* gestation times for a sample of 100 women. If we imagined all the possible random samples of 100 women we could take and looked at the histogram of all the sample means, what shape would it have?

6. Where would the centre of that histogram be?

7. What would be the standard deviation of that histogram?

*Finite Population Correction

When drawing a simple random sample from a finite population of size N, the actual variance of the sample proportion and sample mean are given by the formulas:

$$SD(\hat{p}) = \sqrt{\frac{pq}{n}} \times \sqrt{\frac{N-n}{N-1}} \quad and \quad SD(\bar{y}) = \frac{\sigma}{\sqrt{n}} \times \sqrt{\frac{N-n}{N-1}}$$

The latter multiplicative part, $\sqrt{\dfrac{N-n}{N-1}}$, appended to our earlier formulas, is called the *finite population correction*. This factor is very close to one if N is much bigger than n, in which case these formulas reduce to our earlier formulas for SD's. Also note that they give you the correct result of $SD = 0$ if you sample the entire population, since then the sample mean (or proportion) must equal the true mean (or proportion), and hence has no variability at all.

The limitation of our earlier formulas and derivations is that we assumed independent observations. But when you sample (without replacement) from the population, you change it. Imagine sampling 5 cards from a deck of 52. Every time you draw, say, a face card, you reduce the number of face cards left and the chance that the next card will be a face card. Suppose instead that you draw 5 cards from 10 decks shuffled together. Every time you draw a card now, the remaining deck composition changes little, and the draws are nearly independent.

Our earlier formulas are good enough most of the time, but what if we fail to correct for finiteness when we should (if we violate that *10% Condition*)? The result is that we overestimate the variability of our sample, and our results will actually be more precise or accurate than suggested by our calculations.[25] Not a bad way to err; we call this a *conservative error* (sorry, but there is no place for *liberal* thinking in Statistics). But what is this worth if we may lose the Normality? Fear not—more general versions of the CLT (relaxing the Independence Assumption) assure us of Normality even if we take a very big bite out of the population.

Section 14.5 Exercises

17. More coins Suppose the class in Exercise 15 repeats the coin-tossing experiment.
   a) The students toss the coins 25 times each. Use the 68–95–99.7 Rule to describe the sampling distribution model.
   b) Confirm that you can use a Normal model here.
   c) They increase the number of tosses to 64 each. Draw and label the appropriate sampling distribution model. Check the appropriate conditions to justify your model.
   d) Explain how the sampling distribution model changes as the number of tosses increases.

18. Bigger bag Suppose the class in Exercise 16 buys bigger bags of candy, with 200 M&M's each. Again the students calculate the proportion of green candies they find.

   a) Explain why it's appropriate to use a Normal model to describe the distribution of the proportion of green M&M's they might expect.
   b) Use the 68–95–99.7 Rule to describe how this proportion might vary from bag to bag.
   c) How would this model change if the bags contained even more candies?

19. Just (un)lucky? One of the students in the introductory Statistics class in Exercise 17 claims to have tossed her coin 200 times and found only 42% heads. What do you think of this claim? Explain.

20. Too many green ones? In a really large bag of M&Ms, the students in Exercise 18 found 500 candies, and 12% of them were green. Is this an unusually large proportion of green M&Ms? Explain.

WHAT CAN GO WRONG?

■ **Don't confuse the sampling distribution with the distribution of the sample.** When you take a sample, you always look at the distribution of the values, usually with a histogram, and you may calculate summary statistics. Examining the distribution of the sample like this is wise. But that's not the sampling distribution. The sampling distribution is an imaginary collection of the values that a statistic *might* have taken for all the random samples—the one you got and the ones that you didn't get. We use the sampling distribution model to make statements about how the statistic varies.

■ **Beware of observations that are not independent.** The CLT depends crucially on the assumption of independence. If our elevator riders are related, are all from the same school (for example, an elementary school), or in some other way aren't similar enough to a random sample, then the statements we try to make about the mean are going to be wrong. Unfortunately, this isn't something you can check in your data. You have to think about how the data were gathered. Good sampling practice and well-designed randomized experiments ensure independence.

(continued)

---

[25]And the same goes for two sample procedures to be discussed in later chapters.

■ **Watch out for small samples from skewed or bimodal populations.** The CLT assures us that the sampling distribution model is Normal if n is large enough. If the population is nearly Normal, even small samples (like our 10 elevator riders) work. If the population is very skewed or bimodal then n will have to be large before the Normal model will work well. If we sampled 15 or even 20 CEOs and used a Normal model to calculate some probabilities involving $\bar{y}$ we'd likely get into trouble because the underlying data distribution is so skewed. Unfortunately, there's no good rule of thumb.[26] It just depends on how skewed the data distribution is. Always plot the data to check.

What Have We Learned?

Model the variation in statistics from sample to sample with a sampling distribution.

■ The CLT tells us that the sampling distribution of both the sample proportion and the sample mean are Normal, for sufficiently large samples.

Understand that, usually, the mean of a sampling distribution is the value of the parameter estimated.

■ For the sampling distribution of $\hat{p}$ the mean is p.
■ For the sampling distribution of $\bar{y}$, the mean is μ

Interpret the standard deviation of a sampling distribution.

■ The standard deviation of a sampling model is the most important information about it.
■ The standard deviation of the sampling distribution of a proportion is $\sqrt{\dfrac{pq}{n}}$, where $q = 1 - p$.
■ The standard deviation of the sampling distribution of a mean is $\dfrac{\sigma}{\sqrt{n}}$, where σ is the population standard deviation.

Understand that the CLT is a *limit* theorem.

■ The sampling distribution of the mean is Normal, *no matter what the underlying distribution of the data is, however . . .*
■ The CLT says that this only happens in the limit, as the sample size grows. The Normal model applies sooner when sampling from a unimodal, symmetric population and more gradually when the population is very non-Normal.

Review of Terms

Sampling distribution Different random samples give different values for a statistic. The distribution of all those values over all possible samples (for the same sample size n) is called the sampling distribution of the statistic. When we obtain a random sample and calculate a statistic, the value we get is a random selection from its sampling distribution. (p. 446).

Sampling distribution model Because in practice we only have one sample to work with and can never actually see all possible samples, we use math or simulations to find a theoretical model that can adequately approximate the true sampling distribution (p. 447).

Sampling variability (sampling error) The variability we expect to see from one random sample to another. It is sometimes called sampling error, but sampling variability is the better term (p. 450).

Sampling distribution model for a proportion If assumptions of independence and random sampling are met, and we observe or expect at least 10 successes and 10 failures, then the sampling distribution of a proportion is modelled by a Normal model with a mean equal to the true proportion value, p, and a standard deviation equal to $\sqrt{\dfrac{pq}{n}}$, (p. 452).

---

[26]For proportions, of course, there is a rule: the Success/Failure Condition. That works for proportions because the standard deviation of a proportion is linked to its mean.

Central Limit Theorem (CLT)	States that the sampling distribution model of the sample mean (and proportion) from a random sample is approximately Normal for large n, *regardless of the distribution of the population*, as long as the observations are independent (p. 459).
Sampling distribution model for a mean	If assumptions of independence and random sampling are met, and the sample size is large enough, the sampling distribution of the sample mean is modelled by a Normal model with a mean equal to the population mean, μ, and a standard deviation equal to $\dfrac{\sigma}{\sqrt{n}}$ (p. 460).

Chapter Exercises

21. **Speeding** Provincial police believe that 70% of the drivers travelling on a particular section of the Trans-Canada Highway exceed the speed limit. They plan to set up a radar trap and check the speeds of 80 cars.
    a) Using the 68–95–99.7 Rule, draw and label the distribution of the proportion of these cars the police will observe speeding.
    b) Do you think the appropriate conditions necessary for your analysis are met? Explain.

22. **Smoking pot** The Canadian Drug Use and Monitoring Survey estimates that 33% of 15- to 24-year-olds used cannabis in 2008. Using the 68–95–99.7 Rule, describe the sampling distribution model for the proportion of cannabis users among a randomly selected group of fifty 15- to 24-year-olds. Be sure to discuss your assumptions and conditions.

23. **Vision** It is generally believed that nearsightedness affects about 12% of all children. A school district has registered 170 incoming kindergarten children.
    a) Can you apply the Central Limit Theorem to describe the sampling distribution model for the sample proportion of children who are nearsighted? Check the conditions and discuss any assumptions you need to make.
    b) Sketch and clearly label the sampling model, based on the 68–95–99.7 Rule.
    c) How many of the incoming students might the school expect to be nearsighted? Explain.

24. **Mortgages** In early 2009, rising unemployment and the house-price slump in Alberta led to a mortgage delinquency rate of 0.54% of all big bank mortgages (rising for the nineteenth straight month), the highest delinquency rate in Canada. Suppose a large bank holds 1731 mortgages.
    a) Can you apply the Central Limit Theorem to describe the sampling distribution model for the sample pro-

portion of delinquencies? Check the conditions and discuss any assumptions you need to make.
    b) Sketch and clearly label the sampling model, based on the 68–95–99.7 Rule.
    c) How many of these homeowners might the bank expect to be delinquent on their mortgages? Explain.

25. **Loans** Based on past experience, a bank believes that 7% of the people who receive loans will not make payments on time. The bank has recently approved 200 loans.
    a) What are the mean and standard deviation of the proportion of clients in this group who may not make timely payments?
    b) What assumptions underlie your model? Are the conditions met? Explain.
    c) What is the probability that over 10% of these clients will not make timely payments?

26. **Contacts** Assume that 30% of students at a university wear contact lenses.
    a) We randomly pick 100 students. Let $\hat{p}$ represent the proportion of students in this sample who wear contacts. What is the appropriate model for the distribution of $\hat{p}$? Specify the name of the distribution, the mean, and the standard deviation. Be sure to verify that the conditions are met.
    b) What is the approximate probability that more than one third of this sample wear contacts?

27. **Big families** According to Statistics Canada, 8.3% of Canadian families have five or more members. If we take a random sample of 400 families, use the 68–95–99.7 Rule to describe the sampling distribution model for the percentage of those 400 families that will be of size five or bigger. Are the appropriate conditions met?

28. **Binge drinking** A national study found that 44% of university students engage in binge drinking (five drinks at

a sitting for men, four for women). Use the 68–95–99.7 Rule to describe the sampling distribution model for the proportion of students in a randomly selected group of 200 who engage in binge drinking. Do you think the appropriate conditions are met?

29. **Big families, again** Exercise 27 mentioned that 8.3% of Canadian families have five or more members. Suppose we take a random sample of 625 rural farm families and find that 16% of those families have five or more members. Is this a very unusual result? Would you conclude that farm families differ from other Canadian families? Why?

30. **Binge sample** After hearing of the national result that 44% of students engage in binge drinking, a professor surveyed a random sample of 244 students at his university and found that 96 of them admitted to binge drinking in the past week. Should he be surprised at this result? Explain.

31. **Polling** Just before a Quebec referendum on sovereignty, a local newspaper polls 400 voters in an attempt to predict whether it would succeed. Suppose that sovereignty has the support of 48% of the voters. What is the probability that the newspaper's sample will lead them to predict victory? Be sure to verify that the assumptions and conditions necessary for your analysis are met.

32. **Seeds** Information on a packet of seeds claims that the germination rate is 92%. What is the probability that more than 95% of the 160 seeds in the packet will germinate? Be sure to discuss your assumptions and check the conditions that support your model.

33. **Apples** When a truckload of apples arrives at a packing plant, a random sample of 150 is selected and examined for bruises, discolouration, and other defects. The whole truckload will be rejected if more than 5% of the sample is unsatisfactory. Suppose that 8% of the apples on the truck do not meet the desired standard. What is the probability that the shipment will be accepted anyway?

34. **Airline no-shows** About 12% of ticketed passengers do not show up for their flights, so airlines like to overbook for flights. If an airplane has sold tickets to 400 people for the next flight, how many seats, at the least, would you hope to be available on the plane to be very sure of having enough seats for all who show up? Comment on the assumptions and conditions that support your model, and explain what "very sure" means to you.

35. **Nonsmokers** You are planning to open a new "Peking Duck" specialty restaurant in Beijing, and need to plan the seating arrangement. Unlike in Canada, you'll need to plan for smokers as well as nonsmokers. While some nonsmokers do not mind being seated in the smoking section of a restaurant, you estimate that about 60% of the customers will demand a smoke-free area. Your restaurant will have 120 seats. How many seats should be in the nonsmoking area to be very sure of having enough seating there? Comment on the assumptions and conditions that support your model, and explain what "very sure" means to you.

36. **Meals** A restauranteur anticipates serving about 180 people on a Friday evening, and believes that about 20% of the patrons will order the chef's steak special. How many of those meals should he plan on serving to be pretty sure of having enough steaks on hand to meet customer demand? Justify your answer, including an explanation of what "pretty sure" means to you.

37. **Sampling** A sample is chosen randomly from a population that can be described by a Normal model.
   a) What is the sampling distribution model for the sample mean? Describe shape, centre, and spread.
   b) If we choose a larger sample, what is the effect on this sampling distribution model?

38. **Sampling, part II** A sample is chosen randomly from a population that was strongly skewed to the left.
   a) Describe the sampling distribution model for the sample mean if the sample size is small.
   b) If we make the sample larger, what happens to the sampling distribution model's shape, centre, and spread?
   c) As we make the sample larger, what happens to the expected distribution of the data in the sample?

39. **Waist size** A study measured the waist size of 250 men, finding a mean of 36.33 inches and a standard deviation of 4.02 inches. Here is a histogram of these measurements

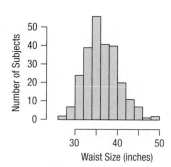

a) Describe the histogram of waist size.
b) To explore how the mean might vary from sample to sample, the researchers simulated by drawing many samples of size 2, 5, 10, and 20, with replacement, from the 250 measurements. Here are histograms of the sample means for each simulation. Explain how these histograms demonstrate what the Central Limit Theorem says about the sampling distribution model for sample means.

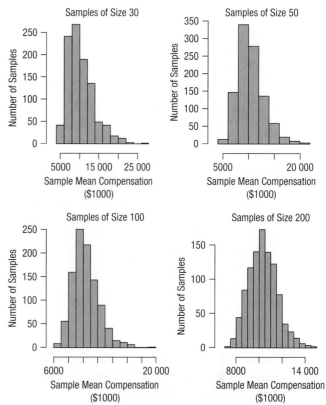

40. CEO compensation In a recent year, the total annual compensations of the chief executive officers (CEOs) of the 800 largest U.S. companies (the *Fortune* 800) averaged (in thousands of dollars) 10 307.31 with a standard deviation of 17 964.62 (also in $1000). Here is a histogram of their annual compensations:

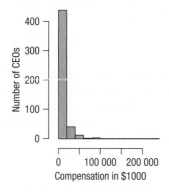

a) Describe the histogram of *Compensation*.
   A research organization simulated sample means by drawing samples of 30, 50, 100, and 200, with replacement, from the 800 CEOs. The histograms further below show the distributions of means for many samples of each size.
b) Explain how these histograms demonstrate what the Central Limit Theorem says about the sampling distribution model for sample means. Be sure to talk about shape, centre, and spread.

c) Comment on the "rule of thumb" that "With a sample size of at least 30, the *sampling* distribution of the mean is nearly Normal"[27]?

41. Waist size revisited Researchers measured the waist sizes of 250 men in a study on body fat. The true mean and standard deviation of the waist sizes for the 250 men are 36.33 inches and 4.019 inches, respectively. In Exercise 41, you looked at the histograms of simulations that drew samples of sizes 2, 5, 10, and 20 (with replacement). The summary statistics for these simulations were as follows:

n	mean	st. dev.
2	36.314	2.855
5	36.314	1.805
10	36.341	1.276
20	36.339	0.895

a) According to the Central Limit Theorem, what should the theoretical mean and standard deviation be for each of these sample sizes?
b) How close are the theoretical values to what was observed in the simulation?

---

[27]Actually, this rule of thumb is generally qualified with the condition that the population not be severely skewed or have extreme outliers.

c) Looking at the histograms in Exercise 39, at what sample size would you be comfortable using the Normal model as an approximation for the sampling distribution?

d) What about the shape of the distribution of waist size explains your choice of sample size in part c)?

42. CEOs revisited In Exercise 40, you looked at the annual compensation for 800 CEOs, for which the true mean and standard deviation were (in thousands of dollars) 10 307.31 and 17 964.62, respectively. A simulation drew samples of sizes 30, 50, 100, and 200 (with replacement) from the total annual compensations of the *Fortune* 800 CEOs. The summary statistics for these simulations were as follows:

n	mean	st. dev.
30	10 251.73	3359.64
50	10 343.93	2483.84
100	10 329.94	1779.18
200	10 340.37	1230.79

a) According to the Central Limit Theorem, what should the theoretical mean and standard deviation be for each of these sample sizes?

b) How close are the theoretical values to what was observed from the simulation?

c) Looking at the histograms in Exercise 40, at what sample size would you be comfortable using the Normal model as an approximation for the sampling distribution?

d) What about the shape of the distribution of total compensation explains your answer in part c)?

43. GPAs A university's data about incoming high school graduates indicates that the mean of their high school averages was 73, with a standard deviation of 7; the distribution was roughly mound-shaped and only slightly skewed. The students are randomly assigned to first-year writing seminars in groups of 25. What might the mean high school average of one of these seminar groups be? Describe the appropriate sampling distribution model—shape, centre, and spread—with attention to assumptions and conditions. Make a sketch using the 68–95–99.7 Rule.

44. Home values Assessment records indicate that the value of homes in a small city is skewed right, with a mean of $280 000 and standard deviation of $90 000. To check the accuracy of the assessment data, officials plan to conduct a detailed appraisal of 100 homes selected at random. Using the 68–95–99.7 Rule, draw and label an appropriate sampling model for the mean value of the homes selected.

45. The trial of the pyx In 1150, it was recognized in England that coins should have a standard weight of precious metal as the basis for their value. A guinea, for example, was supposed to contain 128 grains of gold. (There are 12.7 grains in a gram.) In the "trial of the pyx," coins minted under contract to the Crown were weighed and compared to standard coins (which were kept in a wooden box called the pyx). Coins were allowed to deviate by no more than 0.28 grains—roughly equivalent to specifying that the standard deviation should be no greater than 0.09 grains (although they didn't know what a standard deviation was in 1150). In fact, the trial was performed by weighing 100 coins at a time and requiring the *sum* to deviate by no more than 100×0.28 or 28 grains—equivalent to the sum having a standard deviation of about 9 grains.

a) In effect, the trial of the pyx required that the mean weight of the sample of 100 coins have a standard deviation of 0.09 grains. Explain what was wrong with performing the trial in this manner.

b) What should the limit have been on the standard deviation of the mean?

Note: Because of this error, the Crown was exposed to being cheated by private mints that could mint coins with greater variation and then, after their coins passed the trial, select out the heaviest ones and recast them at the proper weight, retaining the excess gold for themselves. The error persisted for over 600 years, until sampling distributions became better understood.

46. Safe cities Allstate Insurance Company identified the 10 safest and 10 least-safe cities from among the 200 largest cities in the United States, based on the mean number of years that drivers went between automobile accidents. The cities on both lists were all smaller than the 10 largest cities. Using facts about the sampling distribution model of the mean, explain why this is not surprising.

Exercises 47–64 require the use of Normal tables or technology.

47. Pregnancy Assume that the duration of human pregnancies can be described by a Normal model with mean 266 days and standard deviation 16 days.

a) What percentage of pregnancies should last between 270 and 280 days?

b) At least how many days should the longest 25% of all pregnancies last?

c) Suppose a certain obstetrician is currently providing prenatal care to 60 pregnant women. Let $\bar{y}$ represent the mean length of their pregnancies. According to the Central Limit Theorem, what is the distribution of this sample mean, $\bar{y}$? Specify the model, mean, and standard deviation.

d) What is the probability that the mean duration of these patients' pregnancies will be less than 260 days?

48. Calgary in January The average January temperature in Calgary varies from year to year. These average January temperatures themselves have an average, over many

past years, of –7.9 degrees Celsius, with a standard deviation of 5.0 degrees. Assume a Normal model applies.
a) During what percentage of years does Calgary have a January average above 0 degrees?
b) In the 20% coldest Januaries, the average January temperature reaches no higher than how many degrees?
c) Suppose you will be in Calgary for the next four years. Describe the sampling distribution model for the average of the next four January average temperatures. Assume climatic conditions are not changing.
d) What is the probability that the average of the next four January average temperatures will exceed –4 degrees? If this happens, would you consider this evidence of climatic warming (at least locally, in Calgary), or just Normal variation?

49. **Pregnant again** The duration of human pregnancies may not actually follow a Normal model, as described in Exercise 47.
a) Explain why it may be somewhat skewed to the left.
b) If the correct model is in fact skewed, does that change your answers to parts a), b), and c) of Exercise 47? Explain why or why not for each.

50. **At work** Some business analysts estimate that the length of time people work at a job has a mean of 6.2 years and a standard deviation of 4.5 years.
a) Explain why you suspect this distribution may be skewed to the right.
b) Explain why you could estimate the probability that 100 people selected at random had worked for their employers an average of 10 years or more, but you could not estimate the probability that an individual had done so.

51. **Dice and dollars** You roll a die. You win nothing if the number of spots is odd. You win $1 for a two or a four, and $10 for a six.
a) Find the expected value and standard deviation of your prospective winnings.
b) You play twice. Find the mean and standard deviation of your total winnings.
c) You play 40 times. What is the probability that you win at least $100?

52. **New game** You pay $10 and roll a die. If you get a six, you win $50. If not, you get to roll again. If you get a six this time, you get your $10 back.
a) Create a probability model for this game.
b) Find the expected value and standard deviation of your prospective winnings.
c) You play this game five times. Find the expected value and standard deviation of your average winnings.
d) 100 people play this game. What is the probability that the person running the game makes a profit?

53. **Canadian families** In the 2011 census, Statistics Canada reported the following distribution for family size in Canada.

Size	Percent of Families
6 or more	2.9
5	5.4
4	14.3
3	15.6
2	34.1
1	27.6

Source: Extracted from: http://www.statcan. gc.ca/tables-tableaux/sum-som/l01/cst01/famil53a-eng.htm

a) Find the mean and standard deviation of family size (treat "6 or more" as "6").
b) If we select a random sample of 40 families, would you expect their sizes to follow a Normal model? Explain.
c) Consider the mean sizes of random samples of 40 families. Describe the sampling model for these means (shape, centre, and spread).

54. **Museum membership** A museum offers several levels of membership, as shown in the table.

Member Category	Amount of Donation ($)	Percent of Members
Individual	50	41
Family	100	37
Sponsor	250	14
Patron	500	7
Benefactor	1000	1

a) Find the mean and standard deviation of the donations.
b) During the museum's annual membership drive, it hopes to sign up 50 new members each day. Would you expect the distribution of the donations for a day to follow a Normal model? Explain.
c) Consider the mean donation of the 50 new members each day. Describe the sampling model for these means (shape, centre, and spread).

55. **Canadian families, again** In a certain rural area, there are 36 farm families whose average size is 3.0. What is the probability that the average family size will be at least 3.0 if we were to draw a random sample of 36 families from the entire Canadian population, as described in Exercise 53? Does this calculation tell you anything about farm families?

56. **Joining the museum** One of the museum's phone volunteers sets a personal goal of getting an average donation of at least $100 from the new members she enrols during the membership drive from Exercise 54. If she

gets 80 new members and they can be considered a random sample of all the museum's members, what is the probability that she can achieve her goal?

57. **Pollution** Carbon monoxide (CO) emissions for a certain kind of truck vary with mean 2.9 L/km and standard deviation 0.4 L/km. A company has 80 of these trucks in its fleet. Let $\bar{y}$ represent the mean CO level for the company's fleet.
   a) What is the approximate model for the distribution of $\bar{y}$? Explain.
   b) Estimate the probability that $\bar{y}$ is between 3.0 and 3.1 L/km.
   c) There is only a 5% chance that the fleet's mean CO level is greater than what value?

58. **Potato chips** The weight of potato chips in a medium-size bag is stated to be 300 g. The amount that the packaging machine puts in these bags is believed to have a Normal model with mean 306 g and standard deviation 3.6 g.
   a) What fraction of all bags sold are underweight?
   b) Some of the chips are sold in "bargain packs" of three bags. What is the probability that none of the three is underweight?
   c) What is the probability that the mean weight of the three bags is below the stated amount?
   d) What is the probability that the mean weight per bag of a 24-bag case of potato chips is below 300 g?

59. **Big stats course** STA220H is a large introductory Statistics course at the University of Toronto, with 720 students enrolled. On the term test, the average score was 72, with a standard deviation of 16. There are 24 tutorials, with 30 students in each. Xiu Xiu teaches one, and Sabri teaches another. Assume that tutorial groups are formed quite randomly.
   a) We select a student at random from Sabri's tutorial. Find the probability that this student's test score is at least 80. Assume that scores are Normally distributed.
   b) We select one student at random from each tutorial. Find the probability that Sabri's student outperformed Xiu Xiu's student by at least seven marks on the test. (Hint: Work with the difference between the two scores.) Assume that scores are Normally distributed.
   c) What is the probability that the average for Sabri's tutorial is at least 80? Do you need to assume that scores are Normally distributed?
   d) What is the probability that the average test score in Sabri's tutorial is at least seven marks higher than the average test score for Xiu Xiu's tutorial? (Hint: Work with the difference between the two tutorial averages.) Do you need to assume that scores are Normally distributed?

60. **Milk** Although most of us buy milk by the litre, farmers measure daily production in kilograms or pounds. Ayrshire cows average 47 pounds of milk a day, with a standard deviation of six pounds. For Jersey cows, the mean daily production is 43 pounds, with a standard deviation of 5 pounds. Assume that Normal models describe milk production for these breeds.
   a) We select an Ayrshire at random. What is the probability that she produces more than 50 pounds of milk in one day?
   b) What is the probability that a randomly selected Ayrshire gives more milk than a randomly selected Jersey on one day? (Hint: Work with the difference in milk production between the two cows.)
   c) A farmer has 20 Jerseys. What is the probability that the average production for this small herd exceeds 45 pounds of milk on one day?
   d) A neighbouring farmer has 10 Ayrshires. What is the probability that his herd average is at least five pounds higher than the average for the Jersey herd on one day? (Hint: Work with the difference of the two means.)

*For the exercises below, go to www.onlinestatbook.com/ stat_sim/sampling_dist/index.html where there is a Java applet illustrating sampling distributions. Click on the **Begin** button. According to the questions below, select first the desired population shape, at the top. Next, where you see **Mean** and N = 5, change these according to the question (click on the small arrows). Below you can also change **None** and N = 5 if the question asks you to compare two different statistics or two different sample sizes. After these are set, you are ready to sample and learn. Always start by choosing **Sample: Animated**, which produces a single sample and calculates the chosen statistic. Then click on **Sample: 1000** several times to add more samples, until the picture of the sampling distribution stabilizes nicely. Note that measures of centre and spread appear to the side of all distributions.*

61. **Mean or median?** Choose the (default) Normal population at the top. What is the centre of this population? Choose **Mean** and N = 25, and below choose **Median** and N = 25. Draw samples and examine the two sampling distributions. Which of these sampling distributions are centred at the population centre (unbiased)? Which statistic would you prefer to use as an estimator of the true population centre—if unknown to you—for this type of population? Explain.

62. **Effect of sample size** Choose the (default) Normal population. Choose **Mean** at both lower levels, but choose N = 5 at one level and N = 25 at the other. Predict the mean and standard deviation of each of the two sampling distributions you generate. Now generate the sampling distributions and confirm your predictions.

63. **Central Limit Theorem** Choose the skewed population at the top, and N = 5 and N = 25 at the two levels below, along with the **Mean** in both cases. Predict the shapes, means, and standard deviations of the two sampling distributions you generate. Generate the sampling distributions and compare with your predictions.

64. Central Limit Theorem again Use your computer mouse to create an odd-looking population at the top. Choose **Mean** and **N = 25** at the first level below. Generate the sampling distribution. What phenomenon does its shape reflect? Try again to create another odd population that you think might lead to a very different sampling distribution below. Did you succeed?

65. Central Limit Theorem, finis Compare the sampling distribution of the sample mean when sampling from the uniform population versus sampling from the skewed population (both provided at the top). Start with a small sample size and gradually increase it, first for the uniform population, then for the skewed population. About how big does the sample size have to be to produce a nearly Normal sampling distribution for the mean, for each type of population? Why the difference?

66. Another sampling distribution Use the Normal population at the top. Every statistic has its own sampling distribution. Let's examine the sampling distribution of the sample variance. Choose **N = 5** and **Variance** at one of the levels below. The sample variance calculated uses the denominator n rather than the $n - 1$ we are accustomed to using. Is the centre (mean) of the sampling distribution of this sample variance equal to the actual population variance (printed out at the top next to the population)? If so, we say that this statistic is an unbiased estimator of the population variance. Try changing the statistic to **Var (U)**. This statistic uses $n - 1$ as its denominator. Which version of the sample variance do you prefer? Explain. (We'll learn more about this denominator, which we call the *degrees of freedom*, in Chapter 18.)

Just Checking ANSWERS

1. A Normal model (approximately).
2. At the actual proportion of all students who are in favour.
3. $SD(\hat{p}) = \sqrt{\dfrac{(0.5)(0.5)}{100}} = 0.05$
4. No, this is a histogram of individuals. It may or may not be Normal, but we can't tell from the information provided.
5. A Normal model (approximately).
6. 266 days.
7. $\dfrac{16}{\sqrt{100}} = 1.6$ days.

MathXL

MyLab Statistics

Go to MathXL at **www.mathxl.com** or MyLab™ Statistics at **www.mystatlab.com**. You can practise exercises for this chapter as often as you want. The guided solutions will help you find answers step by step. You'll find a personalized study plan available to you too!

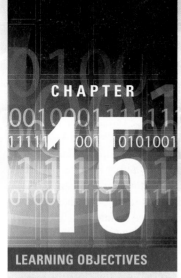

Confidence Intervals
for Proportions

Far better an approximate answer to the right question . . . than an exact answer to the wrong question.

—John Tukey

The city of Aleppo in Syria was all too often in the news during 2016, with frequent stories of war atrocities. The Syrian civil war, which began in 2011 as an offshoot of the Arab Spring uprisings in the region, produced one of the worst humanitarian crises of our time, as increasingly large numbers of refugees fled Syria. During the 2015 federal election, the Liberals promised to take in 25 000 Syrian refugees by the end of 2015. They won a majority and proceeded to kept this promise (though a couple of months late).

Not all Canadians though were enthusiastic about taking in so many refugees so quickly. Many other nations such as the USA took the opposite stance about accepting large numbers of refugees, believing that terrorists might slip in among the refugees or that many were economic migrants and not genuine refugees. Nanos Research (joint with the Globe and Mail) conducted a poll of 1000 randomly selected adult Canadians in late August 2016 and reported that 66% of adult Canadians (strongly or somewhat) support the federal government's overall response to the Syrian refugee crisis.

The authors of the report discuss this finding as if these 1000 respondents can tell us what adult Canadians *in general* are thinking. Of course, that would be far more interesting than knowing the opinions of only about 1000 adults, however carefully they may have been sampled. So can a study like this actually tell us what millions of Canadians are thinking?

We have a sample proportion, which we write as $\hat{p}$, of 0.66, or 66%. Our first guess might be that this observed proportion is close to the population proportion, p. But we also know that because of natural sampling variability, if the researchers had drawn a second sample of 1000 adult Canadians at roughly the same time, the proportion of Canadians in that sample supporting the government probably wouldn't have been 66% again.

What *can* we say about the population proportion, p? To start to answer this question, think about how different the sample proportion might have been if we'd taken another random sample from the same population. But wait. Remember—we aren't actually going to take more samples. We just want to *imagine* how the sample proportions might vary from sample to sample. In other words, we want to know about the *sampling distribution* of the sample proportion who support the government's approach.

---

[1]The top sources of national pride were the flag (86%), the Armed Forces (84%), and hockey (72%).

those sampling distributions studied in the previous chapter. We'll use the standard deviation as a ruler again, but now the standard deviation we'll need is the standard deviation of the sampling distribution—which tells how much a proportion can vary from sample to sample.

We are now starting to discuss formal *statistical inference*—drawing precise conclusions about a population based on a properly selected random sample.

15.1 A Confidence Interval

Let's look at our model for the sampling distribution. We know it's approximately Normal (under certain assumptions, which we should be careful to check) and that its mean is the proportion of all Canadian adults who support the government's response. Is that proportion 66%? No, that's not the proportion of the entire population, that's just $\hat{p}$, our estimate. We don't know the proportion, p, of all Canadians who are supportive; that's what we're trying to find out. We do know, though, that for a simple random sample, the sampling distribution model of $\hat{p}$ is centred at p, and that the standard deviation of the sampling distribution is $\sqrt{\dfrac{pq}{n}}$.

Now we have a problem: Since we don't know p, we can't find the true standard deviation of the sampling distribution model. We do know the observed proportion, $\hat{p}$, so of course we just use what we know, and we estimate. That may not seem like a big deal, but it gets a special name. Whenever we estimate the standard deviation of a sampling distribution, we will call it a **standard error**.[2] For a sample proportion, $\hat{p}$, the standard error is

$$SE(\hat{p}) = \sqrt{\dfrac{\hat{p}\hat{q}}{n}}.$$

For the Nanos poll on support for government policy, then:

$$SE(\hat{p}) = \sqrt{\dfrac{\hat{p}\hat{q}}{n}} = \sqrt{\dfrac{(0.66)(0.34)}{1000}} = 0.015 = 1.5\%$$

Now we know that the sampling model for $\hat{p}$ should look like Figure 15.1:

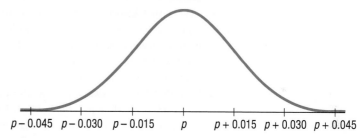

Figure 15.1

The sampling distribution model for $\hat{p}$ is Normal (since n is big) with a mean of p and a standard deviation we estimate to be 0.015.

Who	Canadian adults
What	Opinion about the federal government's handling of the Syrian refugee crisis
When	August 2016
Where	Canada
Why	To assess public support for government policy

■ NOTATION ALERT

Remember that $\hat{p}$ is our sample-based estimate of the true proportion p. Recall also that q is just shorthand for $1 - p$ and $\hat{q} = 1 - \hat{p}$.

When we use $\hat{p}$ to estimate the standard deviation of the sampling distribution model, we call that the **standard error**, and write $SE(\hat{p}) = \sqrt{\dfrac{\hat{p}\hat{q}}{n}}$.

What does that tell us? Well, because it's Normal, it says that about 68% of all samples of 1000 Canadians will have $\hat{p}$'s within 1 SE, 0.015, of p. And about 95% of all these samples will be within $p \pm 2$ SEs. But where is *our* sample proportion in this picture? And what value does p have? We still don't know!

We do know that for 95% of random samples, the sample proportion $\hat{p}$ will be no more than 2 SEs away from the true population proportion p. So let's look at this from $\hat{p}$'s point of view (Figure 15.2). If I'm $\hat{p}$ there's a 95% chance that p is no more than 2 SEs away from me. If I reach out 2 SEs, or 2×0.015 away from me on both sides, I'm 95% sure that p will be within my grasp. Now, I've got him! Probably. Of course, even if my interval does catch p, I still don't know its true value. The best I can do is an interval, and even then I can't be certain it contains p.

---

[2]Some statisticians (and books) refer to the standard deviation of the sampling distribution as the (theoretical) standard error, and refer to our estimate as the (estimated) standard error. But your statistics software (whatever software you are using) just calls the estimated version the "standard error." We think that's easier, so we will too.

Using the term 'standard error' may also help avoid confusion with population and sample standard deviations (which measure variation among individuals, not the possible values of a statistic). And when the statistic is estimating a parameter, a *deviation* from the parameter is just an *error* in the estimate.

Figure 15.2

Reaching out two SEs on either side of $\hat{p}$ makes us 95% confident that we'll trap the true proportion, p.

ACME *p*-trap: Guaranteed*
to capture *p*.

*with 95% confidence

$\hat{p} - 2\,SE$ $\hat{p}$ $\hat{p} + 2\,SE$

So what can we really say about p? Here's a list of things we'd like to be able to say, in order of strongest to weakest, and the reasons we can't say most of them:

1. **"66% of *all* adult Canadians support the government's response to the Syrian refugee crisis."** No. It would be nice to be able to make absolute statements about population values with certainty, but we just don't have enough information to do that. There's no way to be sure that the population proportion is the same as the sample proportion; in fact, it almost certainly isn't (particularly if we calculated the sample percentage to more decimal places[3]). Observations vary. Another sample would yield a different sample proportion.

2. **"It is *probably* true that 66% of all Canadian adults are supporters of the government's response."** No again. In fact, we can be pretty sure that whatever the true proportion is, it's not exactly equal to the sample percentage. So the statement is not true.

3. **"We don't know exactly what proportion of Canadians support the government's response but we *know* that it's within the interval 66% $\pm$ 2 $\times$ 1.5%. That is, it's between 63% and 69%."** This is getting closer, but it's still wrong. We just can't be certain. We can't know *for sure* that the true proportion is in this interval—or in any particular interval.

4. **"We don't know exactly what the proportion of supporters is, but the interval from 63% to 69% *probably* contains the true proportion."** Now we're getting closer. We've admitted that we're unsure about two things. First, we need an interval, not just a value, to try to capture the true proportion. Second, we aren't even certain that the true proportion is in that interval, but we're "pretty sure" that it is.

That last statement may be true, but we can tighten it up by quantifying what we mean by "probably." We saw that 95% of the time when we reach out 2 *SEs* from $\hat{p}$ we capture p, *so we can be 95% confident that this is one of those times*. After putting a number on the probability that this interval covers the true proportion, we've given our best guess of where the parameter is and how certain we are that it's within some range.

5. **"We are 95% confident that between 63% and 69% of all adult Canadians support the government's response to the refugee crisis."** Statements like these are called **confidence intervals**. They're the best we can do.

Each confidence interval discussed in the book has a name. You'll see many different kinds of confidence intervals in the following chapters. Some will be about more than one sample, some will be about statistics other than proportions, and some will use models other than the Normal. The interval calculated and interpreted here is sometimes called a **one-proportion z-interval.**[4]

---

[3]The report only gives percentages rounded to the nearest 1%, so the actual percentage could have been anywhere between 65.5% and 66.4%.

[4]In fact, this confidence interval is so standard for a single proportion that you may see it simply called a "confidence interval for the proportion."

Just Checking

The Angus Reid Strategies Canada Day poll asked (adult) Canadians about various sources of national pride. Fifty-eight percent of respondents said they were ("very or moderately") proud of the Canadian healthcare system. We can estimate a 95% confidence interval to be 0.58 $\pm$ 0.03, or between 55% and 61%. Are the following statements correct interpretations of the confidence interval? Explain.

1. In the Angus Reid sample, somewhere between 55% and 61% of respondents reported that they were proud of Canadian healthcare.

2. We can be 95% confident that 58% of Canadians are proud of their health-care system.

3. We are 95% confident that between 55% and 61% of all Canadians are proud of their healthcare system.

4. We know that between 55% and 61% of all Canadians are proud of their healthcare system.

5. Ninety-five percent of all Canadians are proud of their healthcare system.

Section 15.1 Exercises

1. **Lying about age** Pew Research, in November 2011, polled a random sample of 799 teens about Internet use. 49% of those teens reported that they had misrepresented their age online to gain access to websites and online services.
   a) Explain the meaning of $\hat{p} = 0.49$ in the context of this situation.
   b) Calculate the standard error of $\hat{p}$.
   c) Explain what this standard error means in the context of this situation.

2. **Economic conditions** Nano Research periodically asks a random sample of adults whether they think the Canadian economic conditions are getting better. When it polled 1,000 Canadians in November 2016, 194 thought the economic conditions in Canada were getting better.
   a) Explain the meaning of $\hat{p} = 0.194$" in the context in this situation.
   b) Calculate the standard error of $\hat{p}$.
   c) Explain what this standard error means in the context of this situation.

15.2 Interpreting Confidence Intervals: What Does "95% Confidence" Really Mean?

What do we mean when we say we have 95% confidence that our interval contains the true proportion? Formally, what we mean is that "95% of samples of this size will produce confidence intervals that capture the true proportion." This is correct, but a little long winded, so we sometimes say, "we are 95% confident that the true proportion lies in our interval." Our uncertainty is about whether the particular sample we have at hand is one of the successful ones or one of the 5% that fail to produce an interval that captures the true value.

We know that proportions vary from sample to sample. If other researchers select their own samples of Canadians, each person's sample proportion will almost certainly differ. When they each try to estimate the true proportion of supporters in the entire population, they'll centre *their* confidence intervals at the proportions they observed in their own samples. Each will end up with a different interval.

Our interval guessed the true proportion of supporters to be between about 63% and 69%. Another survey of 1000 Canadians might find 68% of the individuals polled to be supportive of the government, with a confidence interval from 65% to 71%. Still another survey of 1000 Canadians might find only 630 supporters of the government, and get a confidence interval of 60% to 66%. And so on. We don't expect all of these intervals to contain the true proportion of government supporters in the Canadian adult population. Some will be duds, missing the population proportion entirely. But we expect, on average, that 95% of them will be winners and capture the true proportion.

The figure below shows confidence intervals produced by simulating 20 different random samples. The red dots are the proportions of government supporters in each sample, and the blue segments show the confidence intervals found for each. The green line represents the true proportion of government supporters in the entire population, so you can see that most of the intervals caught it—but a few missed. (And notice again that it is the *intervals* that vary from sample to sample; the green line doesn't move.)

The horizontal green line shows the true percentage of Canadians who support the government's response to the refugee crisis. Most of the 20 simulated samples produced confidence intervals that captured the true value, but a few missed.

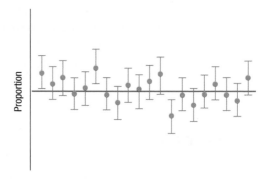

Of course, there's a huge number of possible samples that *could* be drawn, each with its own sample proportion. These are just some of them. Each sample proportion can be used to make a confidence interval. That's a large pile of possible confidence intervals, and ours is just one of those in the pile. Did *our* confidence interval "work"? We can never be sure, because we'll never know the true proportion. However, the Central Limit Theorem assures us that (in the long run) 95% of the intervals are winners, covering the true value, and only 5% are duds. *That's* why we're 95% confident that our interval is a winner!

For Example POLLS AND CONFIDENCE INTERVALS

Four in five Canadians believe in aliens!— according to a February 2016 survey of 1515 Canadians conducted by the Angus Reid Institute. More precisely, 79% of the 1515 Canadians randomly selected for this survey[5] agreed that it is definitely or probably true that intelligent life exists elsewhere in the universe.

QUESTION: It is standard among pollsters to use a 95% confidence level unless otherwise stated. If the pollsters also reported a 95% confidence interval, how would you interpret the confidence interval in this context?

ANSWER: If this polling were done repeatedly, 95% of all random samples would yield confidence intervals that contain the true proportion of all Canadians who believe in intelligent extraterrestrial life.

---

[5] Actually, this was a random sample from members of the Angus Reid Forum, consisting of 130 000 households, but they claim good representativeness of the Canadian population, further ensured by some special statistical techniques (differential weighting to correspond to current demographic and other data) as well as much higher response rates than typical online or phone surveys.

So, What *Can* I Say?

Confidence intervals are based on random samples, so the interval is random, too. The Central Limit Theorem tells us that 95% of random samples will yield intervals that capture the true value. That's what we mean by being 95% confident.

Technically, we should say, "I am 95% confident that the interval from 63% to 69% captures the true proportion of adult Canadians who support the government's response to the refugee crisis." That formal phrasing emphasizes that *our confidence (and our uncertainty) is about the interval, not the true proportion.* But you may choose a more casual phrasing like "I am 95% confident that between 63% and 69% of adult Canadians support the government's response to the refugee crisis." Because you've made it clear that the uncertainty is yours and you didn't suggest that the randomness is in the true proportion, this is OK. Keep in mind that it's the interval that's random and is the focus of both our confidence and doubt.

Section 15.2 Exercises

3. **Lying about age again** The 95% confidence interval for the number of teens in Exercise 1 who reported that they had misrepresented their age online is from 45.6% to 52.5%.
   a) Interpret the interval in this context.
   b) Explain the meaning of "95% confident" in this context.

4. **Single moms** In a 2010 Pew Research study on trends in marriage and family, 4% of randomly selected respondents said that the trend of more single women having children is "a good thing." The 95% confidence interval is from 2.9% to 5.1% ($n = 1229$).
   a) Interpret this interval in this context.
   b) Explain the meaning of "95% confident" in this context.

15.3 Margin of Error: Certainty versus Precision

We've just claimed that with a certain confidence we've captured the true proportion of all Canadians who support the government's response to the refugee crisis. Our confidence interval had the form

$$\hat{p} \pm 2\,SE(\hat{p})$$

The extent of the interval on either side of $\hat{p}$ is called the **margin of error** (*ME*). We'll want to use the same approach for many other situations besides estimating proportions. In fact, almost any population parameter—a proportion, a mean, or a regression slope, for example—can be estimated with some margin of error. The margin of error is a way to describe our uncertainty in estimating the population value. We'll see how to find a margin of error for each of these values and for others.

For all of those statistics, regardless of how we calculate the margin of error, we'll be able to construct a confidence interval that looks like this:

$$Estimate \pm ME$$

The margin of error for our 95% confidence interval was 2 *SE*. What if we wanted to be more confident? To be more confident, we'll need to capture *p* more often, and to do that we'll need to make the interval wider (Figure 15.3). For example, if we want to be 99.7% confident, the margin of error will have to be 3 *SE*.

The more confident we want to be, the larger the margin of error must be. We can be 100% confident that the proportion of Canadians who support the government's Syrian crisis response is between 0% and 100%, but this isn't very useful. On the other hand, we could give a very narrow confidence interval like 65.9% to 66.1%, but we can't be very confident about a precise statement like this. Every confidence interval is a balance between certainty and precision.

Figure 15.3

Reaching out three SEs on either side of $\hat{p}$ makes us 99.7% confident we'll trap the true proportion, p. Compare with Figure 15.2.

The tension between certainty and precision is always there. Fortunately, in most cases we can be both sufficiently certain and sufficiently precise to make useful statements. There is no simple answer to the conflict. You must choose a confidence level yourself. The data can't do it for you. The choice of confidence level is somewhat arbitrary. The most commonly chosen confidence levels are 90%, 95%, and 99%, but, in theory, any percentage can be used. (In practice, though, using something like 92.9% or 97.2% is likely to make people think you're up to something.)

Paws, Inc./Universal Uclick

For Example   FINDING THE MARGIN OF ERROR (TAKE 1)

RECAP: A February 2016 Angus Reid poll of 1515 Canadians reported a margin of error of ± 2.5 percentage points. It is a convention among many pollsters to use a 95% confidence level and to report the "worst case" margin of error, based on $p = 0.5$.

QUESTION: With a 'worst case' approach, what do you get for the margin of error?

ANSWERS: Assuming $p = 0.5$, for simple random samples of size $n = 1515$,

$$SD(\hat{p}) = \sqrt{\frac{pq}{n}} = \sqrt{\frac{(0.5)(0.5)}{1515}} = 0.0128$$

For a 95% confidence level, $ME = 2(0.0128) = 0.026$.[6] So we get 2.6% as the margin of error.

Critical Values

In the survey about belief in aliens, we used 2 *SE* to give us a 95% confidence interval. To change the confidence level, we'd need to change the *number* of SEs so that the size of the margin of error corresponds to the new level. This number of SEs is called the **critical value**.

---

[6]Using 1.96 *SE* instead of 2 *SE* would produce their reported *ME* of ± 2.5%. See section "Critical Values" for an explanation of this small difference.

■ NOTATION ALERT

We'll put an asterisk on a letter to indicate a critical value, so z^* is always a critical value from a standard Normal model. Critical values are just percentile points of the model, chosen to match the probability requirements.

Figure 15.4

For a 90% confidence interval, the critical value is 1.645 because, for a Normal model, 90% of the values are within 1.645 standard deviations from the mean.

Here, it's based on the standard Normal model, so we denote it z^*. For any confidence level, we can find the corresponding critical value from a computer, a calculator, or a Normal probability table, such as Table Z.

For a 95% confidence interval, you'll find the precise critical value is $z^* = 1.96$. That is, 95% of a Normal model is found within ± 1.96 standard deviations of the mean. We've been using $z^* = 2$ from the 68–95–99.7 Rule because it's easy to remember. But what if we decide on 90% confidence? See Figure 15.4.

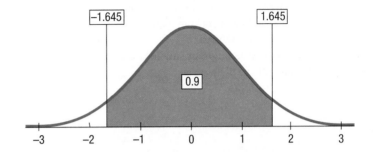

For Example FINDING THE MARGIN OF ERROR (TAKE 2)

RECAP: The Angus Reid poll of 1515 Canadians estimated that 79% of Canadians believe in aliens. They reported a 95% level of confidence with a margin of error of 2.5%.

QUESTIONS: Using the critical value of z and the standard error based on the observed proportion, what would be the margin of error for a 90% level of confidence? What's good and bad about this change?

ANSWER: With $n = 1515$ and $\hat{p} = 0.79$,

$$SE(\hat{p}) = \sqrt{\frac{\hat{p}\hat{q}}{n}} = \sqrt{\frac{(0.79)(0.21)}{1515}} = 0.0105$$

For a 90% confidence level, $z^* = 1.645$, so

$$ME = 1.645(0.0105) = 0.017$$

Now the margin of error is only about 1.7%, producing a narrower interval. What's good about the change is that we now have a smaller interval, but what's bad is that we are less certain that the interval actually contains the true proportion of Canadians who believe in aliens.

 ## Just Checking

Think some more about the 95% confidence interval originally created for the proportion of Canadian adults who believe in aliens.

6. If we wanted to be 98% confident, would our confidence interval need to be wider or narrower?

7. The study's reported margin of error was 2.5%. If we wanted to reduce it to 1% by varying the confidence level (something that would be unacceptable in practice—one should increase the sample size instead), would our level of confidence be higher or lower?

8. If Angus Reid had polled more people, would the interval's margin of error have been larger or smaller?

Section 15.3 Exercises

5. Consulted? In November, 2016 a York University Faculty Association survey asked a random sample of 1552 eligible voters to what extent they had been consulted about the search for the new President of York University. Only 6.5% said they had been consulted.
a) Calculate the margin of error for the proportion of all eligible voters who had been consulted about the search for a 90% confidence interval.
b) Would the margin of error be larger or smaller for a 95% confidence interval? Explain.

6. Economic conditions again In the Nano poll described in Exercise 2, 25.2% of those polled believed economic conditions in Canada were getting weaker.
a) Calculate the margin of error for the proportion of Canadians who expressed concern about a weakening economy for a 99% confidence interval.
b) Would the margin of error be larger or smaller for a 95% confidence interval? Explain.

15.4 Assumptions and Conditions

We've just made some pretty sweeping statements about Canadians' opinions of government policy on Syrian refugees. Those statements were possible because we used a Normal model for the sampling distribution. But is that model appropriate?

We've also said that the same basic ideas will work for other statistics. One difference for those statistics is that some will have sampling distribution models that are different than the Normal. But the background theory (which we won't bother with) is so similar for all of those models that they share the same basic assumptions and conditions about independence and sample size. Then for each statistic, we usually tack on a special-case assumption or two. We'll deal with those as we get to them in later chapters.

We saw the assumptions and conditions for using the Normal to model the sampling distribution for a proportion in the last chapter. Because they are so crucial to making sure our confidence interval is useful, we'll repeat them here.

Independence Assumption

Independence Assumption: If the data were sampled at random or generated from a properly randomized experiment, they satisfy the **Randomization Condition** (below). Otherwise, we need to *think* about whether the independence assumption is plausible. We often look for reasons to suspect that it fails. We wonder whether there is any reason to believe that the observations somehow affect each other. (For example, in a survey, is there time and opportunity for some respondents to communicate with each other before responding?) Whether you decide that the **Independence Assumption** is plausible depends on your knowledge of the situation. It's not one you can check by looking at the data.

However, now that we have data, there are two conditions that we can check:

Randomization Condition: Were the data sampled at random or generated from a properly randomized experiment? Proper randomization can help ensure independence. However, if a multi-stage or other complex sampling design was used, observations may be dependent and the simple SE formula presented in this chapter will no longer be correct, often understating the true SE (so get some statistical advice about how to calculate the correct SE).

10% Condition:[7] If you sample more than 10% of the population, our SE formula falls apart (due to lack of independence among the observations), producing an overestimate

---

[7]To be clear, sampling more than 10% of the population is definitely a good thing. But you need to recognize that the SE formula presented here overestimates the true SE. We will gladly sacrifice independence for precision any day!

of the true SE. In that case, if you report the Confidence Interval formula given in this chapter, be sure to note that the nominal confidence level is in fact *conservative*, that is, lower than the true level of confidence. (Or apply the "finite population correction" as discussed in the optional section in Chapter 14.) This issue only arises when your target population is small—less than 10 times bigger than the sample—and is not a design defect capable of invalidating a study, but rather a small nuisance to deal with.

Sample Size Assumption

The model we use for inference for proportions is based on the Central Limit Theorem. We need to know whether the sample is large enough to make the sampling model for the sample proportions approximately Normal (and to allow us to estimate its SD with the SE). It turns out that we need more data as the proportion gets closer and closer to either extreme (0 or 1). We can check the:

Success/Failure Condition: We require at least 10 "successes" and at least 10 "failures" in our sample. Recall that, by tradition, we arbitrarily label one alternative (usually the outcome being counted) as a "success" even if it's something bad (like getting a disease). The other alternative is, of course, then a "failure." Recall that the Condition required for the Normal model to apply to the sampling distribution of a proportion (Section 14.2) was that both $np \geq 10$ and $nq \geq 10$. Here p and q are unknown parameters, so we use $\hat{p}$ and $\hat{q}$ in place of p and q, and check that $n\hat{p} \geq 10$ and $n\hat{q} \geq 10$. But $n\hat{p}$ and $n\hat{q}$ are just the number of successes and failures, respectively.

> **ONE-PROPORTION Z-INTERVAL**
>
> When the conditions are met, we are ready to find a level C confidence interval for the population proportion, p. The confidence interval is $\hat{p} \pm z^* \times SE(\hat{p})$, where the standard deviation of the proportion is estimated by $SE(\hat{p}) = \sqrt{\dfrac{\hat{p}\hat{q}}{n}}$ and the critical value, z^*, specifies the number of SEs needed for C% of random samples to yield confidence intervals that capture the true parameter value.

Step-by-Step Example A CONFIDENCE INTERVAL FOR A PROPORTION

Who	Adults in the United States
What	Response to a question about the death penalty
When	October 2010
Where	United States
How	510 adults were randomly sampled and asked by the Gallup Poll
Why	Public opinion research

The last execution in Canada for the crime of murder took place in 1962. There is little support in Canada for bringing back capital punishment, unlike in the United States where capital punishment has considerable public support (there were 3125 inmates on death row as of January 1, 2013).

In October 2010, the Gallup Poll[8] asked 510 randomly sampled U.S. adults the question, "Generally speaking, do you believe the death penalty is applied fairly or unfairly in this country today?" Of these, 58% answered "fairly," 36% said "unfairly," and 7% said they didn't know (percentages add up to 101% due to rounding).

Question: From this survey, what can we conclude about the opinions of *all* adults?

To answer this question, we'll build a confidence interval for the proportion of all U.S. adults who believe the death penalty is applied fairly. There are four steps to building a confidence interval for proportions: Plan, Model, Mechanics, and Conclusion.

---

[8]www.gallup.com/poll/1606/death-penalty.aspx

THINK ➡ **Plan** State the problem and the *W*'s.

Identify the *parameter* you wish to estimate.

Identify the *population* about which you wish to make statements.

Choose and state a confidence level.

Model Think about the assumptions and check the conditions.

I want to find an interval that is likely with 95% confidence to contain the true proportion, *p*, of U.S. adults who think the death penalty is applied fairly. I have a random sample of 510 U.S. adults.

✓ **Randomization Condition**: Gallup drew a random sample (which we'll assume was similar to an SRS) from all U.S. adults. I can be confident that the respondents are independent.

✓ **10% Condition**: The sample is certainly less than 10% of the population.

✓ **Success/Failure Condition**

$n\hat{p} = 510(58\%) = 296 \geq 10$ *and*

$n\hat{q} = 510(42\%) = 214 \geq 10,$

State the sampling distribution model for the statistic.

Choose your method.

so the sample appears to be large enough to use the Normal model.

The conditions are satisfied, so I can use a Normal model to find a **one-proportion z-interval**.

SHOW ➡ **Mechanics** Construct the confidence interval.

First find the standard error. (Remember: It's called the "standard error" because we don't know *p* and have to use $\hat{p}$ instead.)

Next find the margin of error. We could informally use 2 for our critical value, but 1.96 is more accurate.

Write the confidence interval (CI).

REALITY CHECK   The CI is centred at the sample proportion and about as wide as we might expect for a sample of 510.

$n = 510, \hat{p} = 0.58$

$$SE(\hat{p}) = \sqrt{\frac{\hat{p}\hat{q}}{n}} = \sqrt{\frac{(0.58)(0.42)}{510}} = 0.022$$

Because the sampling model is Normal, for a 95% confidence interval, the critical value $z^* = 1.96$.

The margin of error is

$$ME = z^* \times SE(\hat{p}) = 1.96(0.022) = 0.043$$

So the 95% confidence interval is

$$0.58 \pm 0.043 \text{ or } (0.537, 0.623)$$

TELL ➡ **Conclusion** Interpret the confidence interval in the proper context. We're 95% confident that our interval captured the true proportion.

I am 95% confident that between 53.7% and 62.3% of all U.S. adults think that the death penalty is applied fairly.

Choosing Your Sample Size

The question of how large a sample to take is an important step in planning any study. We weren't ready to make that calculation when we first looked at study design in Chapter 9, but now we can—*and we always should.*

Suppose a candidate is planning a poll and wants to estimate voter support within 3 percentage points with 95% confidence. How large a sample does she need?

Let's look at the margin of error:

$$ME = z^*\sqrt{\frac{\hat{p}\hat{q}}{n}}$$

And now plug in to get: $0.03 = 1.96\sqrt{\frac{\hat{p}\hat{q}}{n}}$

We want to find n, the sample size. To find n we need a value for $\hat{p}$. We don't know $\hat{p}$ because we don't have a sample yet, but we can probably guess a value. To be safe, we can take—the value that makes $\hat{p}\hat{q}$ (and therefore n) largest. That value is $\hat{p} = \hat{q} = 0.50$, so if we use that value for $\hat{p}$, we'll certainly be safe. Our candidate probably expects to be near 50% voter support anyway.

Our equation, then, is

$$0.03 = 1.96\sqrt{\frac{(0.5)(0.5)}{n}}$$

To solve for n, we first multiply both sides of the equation by $\sqrt{n}$ and then divide by 0.03:

$$0.03\sqrt{n} = 1.96\sqrt{(0.5)(0.5)}$$

$$\sqrt{n} = \frac{1.96\sqrt{(0.5)(0.5)}}{0.03} \approx 32.667$$

Notice that evaluating this expression tells us the *square root* of the sample size. We need to square that result to find n:

$$n \approx (32.667)^2 \approx 1067.1$$

When computing sample sizes, we *always round up* (to ensure the confidence level we wanted) and conclude that we need at least 1068 respondents to keep the margin of error as small as 3 percentage points with a confidence level of 95%.

> ### WHAT DO I USE INSTEAD OF $\hat{P}$?
>
> Often we have an estimate of the population proportion based on experience or perhaps a previous study. If so, use that value as $\hat{p}$ in calculating what size sample you need. If not, the cautious approach is to use $\hat{p} = 0.5$ in the sample size calculation; that will determine the largest sample necessary regardless of the true proportion.

For Example CHOOSING A SAMPLE SIZE

RECAP: The Angus Reid poll that estimated that 79% of all Canadians believe in aliens had a margin of error of $\pm 2.5\%$. Suppose another organization was planning a follow-up survey of opinions on alien existence and wanted to determine a 95% confidence interval with a margin of error of no more than 1.0%.

QUESTION: How large a sample would they need? Use the Angus Reid estimate as the basis for your calculation.

ANSWER:

$$ME = z^*\sqrt{\frac{\hat{p}\hat{q}}{n}}$$

$$0.01 = 1.96\sqrt{\frac{(0.79)(0.21)}{n}}$$

$$\sqrt{n} = \frac{1.96\sqrt{(0.79)(0.21)}}{0.01} \approx 79.8324$$

$$n = 79.8324^2 = 6373.2$$

This organization's survey will need about 6374 respondents. Note that to reduce the ME by 60% (from 2.5% to 1%), we would need to increase the sample size a whopping 6-fold! ($n = 1020$ if you redo this calculation with ME = 0.025.)

Unfortunately, bigger samples cost more money and require more effort. Because the standard error declines only with the *square root* of the sample size, to cut the standard error (and thus the ME) in half, we must *quadruple* the sample size.

Often a margin of error of 5 percentage points or less is acceptable, but different circumstances call for different standards. For a pilot study, a margin of error of 10% may be fine, so a sample of 100 will do quite well. In a close election, a polling organization might want to get the margin of error down to 2%. Drawing a large sample to get a smaller ME, however, can run into trouble. It takes time to survey 2400 people, and a survey that extends over a week or more may be trying to hit a target that moves during the time of the survey. An important event can change public opinion in the middle of the survey process. Professional polling organizations often report both an instantaneous estimate and a rolling average of several recent polls in order to increase the sample size.

The sample size for a survey is the number of respondents, not the number of people to whom questionnaires were sent or whose phone numbers were dialled. And keep in mind that *a low response rate turns any study essentially into a voluntary response study*, which is of little worth for inferring population values. It's almost always better to spend resources on increasing the response rate than on surveying a larger group. A full or nearly full response by a modest-size sample can yield more useful results.

Surveys are not the only place where proportions pop up. Banks sample huge mailing lists to estimate what proportion of people will accept a credit card offer. Even pilot studies may mail offers to over 50 000 customers. Most don't respond; but, in this case, that doesn't make the sample smaller—they simply said, "No, thanks." Those who do respond want the card. To the bank, the response rate[9] is $\hat{p}$. With a typical success rate around 0.5%, the bank needs a very small margin of error—often as low as 0.1%—to make a sound business decision. That calls for a large sample, and the bank must take care in estimating the size needed. For our election poll calculation we used $\hat{p} = 0.5$, both because it's safe and because we honestly believed p to be near 0.5. If the bank used that value, they'd get an absurd answer. Instead, they base their calculation on a proportion closer to the one they expect to find.

> **How big should a margin of error be?**
> Public opinion polls often sample 1000 people, which gives an ME of 3 percentage points when $\hat{p} = 0.5$. But businesses and nonprofit organizations typically use much larger samples to estimate the proportion of people who will accept a direct mail offer. Why? Because that proportion is very low—often far below 5%. An ME of 3% wouldn't be precise enough. An ME like 0.1% would be more useful, and that requires a very large sample size.

For Example SAMPLE SIZE REVISITED

A credit card company is about to send out a mailing to test the market for a new credit card. From that sample, they want to estimate the true proportion of people who will sign up for the card nationwide. A pilot study suggests that about 0.5% of the people receiving the offer will accept it.

QUESTION: To be within a tenth of a percentage point (0.001) of the true rate with 95% confidencee, how big does the test mailing have to be?

ANSWER: Using the estimate $\hat{p} = 0.5\%$:

$$ME = 0.001 = z^* \sqrt{\frac{\hat{p}\hat{q}}{n}} = 1.96 \sqrt{\frac{(0.005)(0.995)}{n}}$$

$$(0.001)^2 = 1.96^2 \frac{(0.005)(0.995)}{n} \Rightarrow n = \frac{1.96^2 (0.005)(0.995)}{(0.001)^2}$$

$$= 19\,111.96 \text{ or } 19\,112$$

That's a lot, but it's actually a reasonable size for a trial mailing such as this. But if they had assumed a conservative 0.50 for the value of p, they would have found

$$ME = 0.001 = z^* \sqrt{\frac{pq}{n}} = 1.96 \sqrt{\frac{(0.5)(0.5)}{n}}$$

$$(0.001)^2 = 1.96^2 \frac{(0.5)(0.5)}{n} \Rightarrow n = \frac{1.96^2 (0.5)(0.5)}{(0.001)^2} = 960\,400$$

Quite a different (and expensive) result.

---

[9]In marketing studies, every mailing yields a response—"yes" or "no"—and "response rate" means the proportion of customers who accept an offer. That's not the way we use the term for survey response.

Try to be conservative with your guesses when planning the sample size. Figure 15.5 graphs pq versus p and shows that pq peaks at $(0.5)(0.5) = 0.25$, so the closer p is to 0.5, the bigger pq is. If, for example, you think p should be somewhere between 0.7 and 0.9, let $pq = (0.7)(0.3) = 0.21$ in your sample size calculation. In other words, choose the value for p from your assumed range that lies closest to 0.5. Even if you guess badly, your confidence interval will still be valid, just wider than hoped.

Figure 15.5

Graph of pq versus p. Choose the p from its plausible range that maximizes pq.

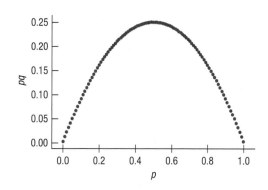

Section 15.4 Exercises

7. Consulted again? Consider the York University Faculty Association survey of Exercise 5.
a) Are the necessary assumptions and conditions met?
b) How many people would need to be surveyed for a 90% confidence interval to ensure the margin of error would be less than 2%?

8. More single moms In Exercise 4, we saw that 4% of people in the sample think the trend of more single women having children is a "good thing."

a) Are the conditions for constructing a confidence interval met?
b) How many people would need to be surveyed for a 95% confidence interval to ensure the margin of error would be less than 3%? (Hint: Don't use $p = 0.5$. Be more realistic and use the data we already have.)

*15.5 The Plus Four Confidence Interval for Small Samples

When the **Success/Failure Condition** fails, all is not lost. The Normal sampling model fails, but we know the exact sampling distribution model for the sample count—the Binomial distribution. We can use this underlying distribution to derive what are called *exact Binomial confidence intervals* for the parameter p. Many statistics software programs will compute these exact confidence intervals, but there is another approach, involving a simple adjustment to the Normal-based confidence interval, which seems to work even better (though that may seem strange—how something *approximate* can be better than something said to be *exact*!). It's called the *Plus Four CI* for p.

All we do is add four *phony* observations to our sample—two successes and two failures. So instead of the proportion $\hat{p} = \dfrac{y}{n}$, we use the adjusted proportion $\tilde{p} = \dfrac{y + 2}{n + 4}$ and, for convenience, we write the phony sample size as $\tilde{n} = n + 4$. We modify the interval by using these adjusted values for both the centre of the interval and the margin of error. Now the adjusted interval is

$$\tilde{p} \pm z^* \sqrt{\dfrac{\tilde{p}(1 - \tilde{p})}{\tilde{n}}}$$

This method gives better performance overall,[10] and especially for proportions near 0 or 1. It has the additional advantage that we no longer need to check the Success/Failure Condition that the two counts are both greater than 10.

For Example A PLUS FOUR CONFIDENCE INTERVAL

Surgeons examined their results to compare two methods for a surgical procedure used to alleviate pain on the outside of the wrist. A new method was compared with the traditional "freehand" method for the procedure. Of 45 operations using the "freehand" method, 3 were unsuccessful, for a failure rate of 6.7%. With only three failures, the data don't satisfy the Success/Failure Condition, so we can't use the regular Normal-based confidence interval.

QUESTION: What is the confidence interval using the Plus Four method?

ANSWER: There were 42 successes and 3 failures. Adding two "pseudo-successes" and two "pseudo-failures," we find

$$\tilde{p} = \frac{3 + 2}{45 + 4} = 0.102$$

A 95% confidence interval for the failure rate is then

$$0.102 \pm 1.96\sqrt{\frac{0.102(1 - 0.102)}{49}} = 0.102 \pm 0.085$$

or (0.017, 0.187).

Notice that although the observed failure rate of 0.067 is contained in the interval, it is not at the centre of the interval—something we haven't seen with any of the other confidence intervals we've considered (reflecting the asymmetry of the underlying Binomial distribution in this case).

---

*15.6 Large Sample Confidence Intervals

Looking back at the derivation of the confidence interval for a proportion suggests that this method may be broadly applicable and easily adapted for other parameters. The key features are as follows:

I. We can construct an estimator of the parameter that has an approximately *Normal sampling distribution* centred at the parameter of interest (so Figure 15.1 applies); and

II. We can derive a formula for the *standard deviation of this sampling distribution* and evaluate it with sufficient accuracy, generally using the sample data, to give us the SE, from which we can then calculate the $ME = z^* \times SE$

What do large samples have to do with this? Well, many estimators (such as means or proportions or differences thereof) will be approximately Normal if the sample is big, regardless of the population distribution, due to the Central Limit Theorem. And for a big enough sample, we can also use the sample information to estimate the standard deviation of the sampling distribution with sufficient accuracy. A non-Normal sampling distribution

---

[10]By "better performance," we mean that a 95% confidence interval has more nearly a 95% chance of covering the true population proportion. Simulation studies have shown that our original unadjusted confidence interval is in fact less likely than 95% to cover the true population proportion when the sample size is small or the proportion is very close to 0 or 1 while 'exact Binomial' confidence intervals have slightly more than 95% confidence. The original idea for this method can be attributed to E. B. Wilson. The approach discussed here was proposed by Agresti and Coull (A. Agresti and B. A. Coull 1998, "Approximate is better than 'exact' for interval estimation of binomial proportions." *The American Statistician*, 52: 119–129).

or unreliable SE estimate will make this approach fall apart (in which case, you may need small sample methods, like the Plus Four method discussed in the previous section or the methods for means in Chapters 18–20).

If we are estimating a parameter θ, using an estimator $\hat{\theta}$ whose sampling distribution is Normal with mean equal to θ and standard deviation denoted $SD(\hat{\theta})$, the *confidence interval for* θ takes the form:

$$Estimate \pm ME = \hat{\theta} \pm z^* SD(\hat{\theta})$$

However, this equation can only be used if you can evaluate the $SD(\hat{\theta})$. Usually you will not know the true value of $SD(\hat{\theta})$, but if you can estimate it accurately enough with an appropriate formula and data from a big enough sample, you can then plug in its estimate, denoted $SE(\hat{\theta})$, to get the *large sample confidence interval for* θ:

$$\hat{\theta} \pm z^* SE(\hat{\theta})$$

Now let's apply the above approach to the estimation of the mean of a quantitative population.

Confidence Interval for the Mean of a Quantitative Population When the Sample Is Big

To estimate the mean μ of a quantitative population, assume we draw an SRS. From the sampled items, we take measurements on the variable of interest: $y_1, y_2, \ldots, y_n$, and compute their mean, $\bar{y}$, which will be used to estimate the population mean. By the Central Limit Theorem, for size n, the sampling distribution of $\bar{y}$ is approximately Normal. In Chapter 14, we derived a formula for the standard deviation of this sampling distribution:

$$SD(\bar{y}) = \frac{\sigma}{\sqrt{n}}, \text{ which we can estimate from the data by } \frac{s}{\sqrt{n}}.$$

Now plugging into the earlier generic confidence interval formula, we can derive our

Confidence Interval for μ: $Estimate \pm ME = \hat{\theta} \pm z^* SD(\hat{\theta})$

$$= \hat{\mu} \pm z^* SD(\hat{\mu}) = \bar{y} \pm z^* SD(\bar{y}) = \bar{y} \pm z^* \frac{\sigma}{\sqrt{n}}$$

And the latter, we can estimate by $\bar{y} \pm z^* \dfrac{s}{\sqrt{n}}$ for *big enough n*.[11]

How big should n be in order to achieve approximate Normality and a sufficiently accurate estimate of the SD? We'll say that n *should be at least 60* (other texts may differ in their advice, ranging from $n = 30$ to $n = 100+$, depending on how accurate we need to be in our level of confidence). For smaller n, use methods described in Chapter 18. And even if $n > 60$, *inspect the data* for severe outliers or extreme skewness. If extremely skewed, a bigger n might be needed, or think about a transformation of the data—like with a logarithm—before doing any analysis.

And—of course,—just as with proportions, make sure that the observations are independent (of each other) and that you are poised to generalize beyond your sample to a larger population.

---

[11] Why not use the true population standard deviation, σ, instead of its estimate, s? Since σ is a parameter of the population, you need the entire population in order to calculate it. The best we usually can do is a sample estimate—the sample standard deviation, s. Like the sample mean, it will not exactly equal the corresponding population characteristic due to sampling variation. However, if you find yourself in the rare situation where you actually believe you know the true population σ (perhaps from prior experience with the process), then go ahead and use it. The real thing is always better than an estimate.

Consider an example. A team of researchers tested 150 farm-raised salmon (collected from various sources) for organic contaminants, in particular the level of the carcinogenic insecticide mirex.[12] We expect some variation in the individual salmon measurements, but we may be primarily interested in the mean concentration for salmon raised in this type of environment. The researchers found the mean mirex concentration in this sample to be 0.0913 ppm with a standard deviation of 0.0495 ppm. The histogram of the data showed no extreme skewness or outliers. Assuming this to be a simple random sample of farm-raised salmon, a 95% confidence interval for the true mean level of mirex in farm-raised salmon would be (for convenience, we'll use $z^* = 2$ instead of $z^* = 1.96$):

$$\bar{y} \pm z^* \frac{s}{\sqrt{n}} = 0.0913 \pm 2 \times \frac{0.0495}{\sqrt{150}}$$

$$= 0.0913 \pm 2 \times 0.00404 = 0.0913 \pm 0.00808$$

$$= (0.0832, 0.0994)$$

So we can say with 95% confidence that the true mean concentration for salmon raised in this type of environment is between 0.0832 ppm and 0.0994 ppm.

Choosing the Sample Size

Just as with proportions, we should think carefully about the sample size before running a study. To fix the margin of error, proceed just as you did with proportions, setting the formula for the margin of error equal to the requirement and then solving for n. So set:

$$z^* \frac{\sigma}{\sqrt{n}} = required \; margin \; or \; error$$

and solve for n. But what should we do about that σ? Make an educated guess, perhaps based on some past experience, a pilot study, or any similar studies that are known. Or guess the likely range of the data, call it R, and then use R/4 as a rough ballpark estimate of the standard deviation of the measurements. Guessing badly will not affect the validity of your conclusions, just the actual margin of error attained.

As a final comment, we should note that not all confidence intervals are formed in this way, nor are they all symmetric about the estimate, but, fortunately, they may all be interpreted exactly the same way—If the procedure were repeated over and over, 95 times in 100, the true value of the parameter being estimated would be caught between the limits of a 95% confidence interval!

WHAT CAN GO WRONG?

Confidence intervals are powerful tools. Not only do they tell what we know about the parameter value, but—more important—they also tell what we *don't* know. To use confidence intervals effectively, you must be clear about what you say about them.

Don't Misstate What the Interval Means

■ **Don't suggest that the parameter varies.** A statement like, "There is a 95% chance that the true proportion is between 63% and 69%" sounds as though you think the population

---

[12]Ronald A. Hites, Jeffrey A. Foran, David O. Carpenter, M. Coreen Hamilton, Barbara A. Knuth, and Steven J. Schwager, 2004, "Global assessment of organic contaminants in farmed salmon," *Science* Vol. 303, no. 5655, pp. 226–229.

proportion wanders around and sometimes happens to fall between 63% and 69%. When you interpret a confidence interval, make it clear that *you* know that the population parameter is fixed and that it is the interval that varies from sample to sample.

■ **Don't claim that other samples will agree with yours.** Keep in mind that the confidence interval makes a statement about the true population proportion. An interpretation such as, "In 95% of samples of Canadian adults, the proportion who support the government's response will be between 63% and 69%" is just wrong. The interval isn't about sample proportions but about the population proportion.

■ **Don't be certain about the parameter.** Saying, "Between 63% and 69% of Canadian adults support the government's response asserts that the population proportion cannot be outside that interval. Of course, we can't be absolutely certain of that. (Just reasonably sure.)

■ **Don't forget: It's about the parameter.** Don't say, "I'm 95% confident that $\hat{p}$ is between 63% and 69%." Of course you are—in fact, we calculated that $\hat{p} = 66\%$ so we know the sample proportion is in the interval. The confidence interval is about the (unknown) population parameter, p.

■ **Don't claim to know too much.** Be sure to specify the *Who, Where, and When* as clearly as possible. Our Nanos poll example was only about adults (age 18 and above), in all of Canada, in August 2016 (and opinions may have changed by the time you read this). Some surveys report results just for one or several age groups, or only sample conveniently accessible groups, so be sure to specify the *Who* clearly.

■ **Do take responsibility.** Confidence intervals are about *un*certainty. *You* are the one who is uncertain, not the parameter. You have to accept the responsibility and consequences of the fact that not all the intervals you compute will capture the true value. In fact, about 5% of the 95% confidence intervals you find will fail to capture the true value of the parameter. You *can* say, "I am 95% confident that between 63% and 69% of Canadian adults support the government's response to the refugee crisis."[13]

■ **Do treat the whole interval equally.** Although a confidence interval is a set of plausible values for the parameter, don't think that the values in the middle of a confidence interval are somehow "much more plausible" than values nearer the edges.

Plan for an Appropriate Sized Margin of Error

■ **Beware of margins of error that are too large to be useful**. A confidence interval that says that the percentage of Canadians who support the government's response is between 10% and 90% wouldn't be of much use. Most likely, you have some sense of how large a margin of error you can tolerate. What can you do?

One way to make the margin of error smaller is to reduce your level of confidence. But that is usually a very bad idea. It's a rare study that reports confidence levels lower than 80%. Levels of 95% or 99% are more common.

The time to think about whether your margin of error is small enough to be useful is when you design your study. Don't wait until you compute your confidence interval. To get a narrower interval without giving up confidence, you need to have less variability in your sample proportion. How can you do that? Choose a larger sample.

Violations of Assumptions

Confidence intervals and margins of error are often reported along with poll results and other analyses. But it's easy to misuse them and wise to be aware of the ways things can go wrong.

■ **Watch out for biased sampling.** Don't forget about the potential sources of bias in surveys we discussed in Chapter 9. Just because we have more statistical machinery now doesn't mean we can forget what we've already learned. A questionnaire that finds that most people enjoy filling out surveys suffers from nonresponse bias if only 10% of the surveys are returned. Even though now we're able to put confidence intervals around this (biased) estimate, the nonresponse has made it useless.

(continued)

---

[13]When we are being very careful, we say, "95% of samples of this size will produce confidence intervals that capture the true proportion of Canadian adults who support the government's approach to the refugee crisis."

■ **Think about independence.** The assumption that the values in our sample are mutually independent is one that we usually cannot check. It always pays to think about it, though. For example, suppose we wanted to poll Facebook users to estimate the percent who update their profiles daily. If we sample some Facebook users and then include all of the friends in their network, that would violate the Independence Assumption, because friends in the same network may behave similarly. This could severely affect our sample proportion. It could be that the proportion of daily updaters is actually quite small, but the researchers just happened to find a network of friends who spend all their time on Facebook. To avoid this, the researchers should be careful in their sample design to ensure proper randomization. This is always an issue with any sort of convenience sample.

What Have We Learned?

Construct a confidence interval for a proportion, p, as the statistic, $\hat{p}$, plus or minus a **margin of error**.

- The margin of error consists of a **critical value** based on the sampling model times a **standard error** based on the sample.
- The critical value is found from the standard Normal model.
- The standard error of a sample proportion is calculated as $\sqrt{\dfrac{\hat{p}\hat{q}}{n}}$.

Interpret a confidence interval correctly.

- You can claim to have the specified level of confidence that the interval you have computed actually covers the true value.

Understand the relationship of the sample size, n, to both the certainty (confidence level) and precision (margin of error).

- For the same sample size and true population proportion, more certainty means less precision (wider interval) and more precision (narrower interval) implies less certainty.

Know and check the assumptions and conditions for finding and interpreting confidence intervals.

- Independence Assumption and Randomization Condition
- Success/Failure Condition
- 10% Condition (least important)

Be able to invert the calculation of the margin of error to find the sample size required, given a guess at the proportion, a confidence level, and a desired margin of error.

Review of Terms

Standard error   When we estimate the standard deviation of a sampling distribution using statistics found from the data, the estimate is called a standard error (p. 477).

$$SE(\hat{p}) = \sqrt{\frac{\hat{p}\hat{q}}{n}}$$

Confidence interval   A level C confidence interval for a model parameter is an interval of values often of the form

$$estimate \pm margin\ of\ error$$

found from data in such a way that C% of all random samples will yield intervals that capture the true parameter value (p. 478).

One-proportion *z*-interval   A confidence interval for the true value of a proportion based on the Normal approximation. The confidence interval is

$$\hat{p} \pm z^{*}SE(\hat{p}),$$

where z^{*} is a critical value from the standard Normal model corresponding to the specified confidence level (p. 478).

Margin of error   In a confidence interval, the extent of the interval on either side of the observed statistic value is called the margin of error. A margin of error is typically the product of a critical value from the sampling distribution and a standard error from the data. A small margin of error corresponds to a confidence interval that pins down the parameter precisely. A large margin of error corresponds to a confidence interval that gives relatively little information about the estimated parameter. For a proportion, (p. 481)

$$ME = z^{*}\sqrt{\frac{\hat{p}\hat{q}}{n}}$$

Critical value   The number of standard errors to move away from the mean of the sampling distribution to correspond to the specified level of confidence. The critical value, denoted z^{*}, is usually found from a table or with technology (p. 482).

On the Computer CONFIDENCE INTERVALS FOR PROPORTIONS

Confidence intervals for proportions are so easy and natural that many statistics packages don't offer special commands for them. Statistics programs can use the "raw data" for computations. For proportions, the raw data are the "success" and "failure" status for each case. Usually, these are given as 1 or 0, but they might be category names like "yes" and "no." Often we just know the proportion of successes, $\hat{p}$, and the total count, n. Computer packages usually deal easily with summary data like this, and the statistics routines found on many graphing calculators also allow you to create confidence intervals from summaries of the data—usually all you need to enter are the number of successes and the sample size.

In some programs, you can reconstruct variables of 0s and 1s with the given proportions. But even when you have (or can reconstruct) the raw data values, you may not get exactly the same margin of error from a computer package as you would find working by hand. The reason is that some packages make approximations or use other methods. The result is very close but not exactly the same. Fortunately, Statistics means never having to say you're certain, so the approximate result is good enough.

EXCEL

Inference methods for proportions are not part of the standard Excel tool set.

COMMENTS

For summarized data, you can type the calculation into any cell and evaluate it.

JMP

For a **categorical** variable that holds category labels, the **Distribution** platform includes tests and intervals for proportions.

For summarized data, put the category names in one variable and the frequencies in an adjacent variable. Designate the frequency column to have the **role of frequency**. Then use the **Distribution** platform.

COMMENTS

JMP uses slightly different methods for proportion inferences than those discussed in this text . Your answers are likely to be slightly different, especially for small samples.

MINITAB

Choose **Basic Statistics** from the **Stat** menu.

- Choose **1Proportion** from the Basic Statistics submenu.
- If the data are category names in a variable, assign the variable from the variable list box to the **Samples in columns** box. If you have summarized data, click the **Summarized Data** button and fill in the number of trials and the number of successes.
- Click the **Options** button and specify the remaining details.
- If you have a large sample, check **Use test and interval based on Normal distribution.**

Click the **OK** button.

COMMENTS

When working from a variable that names categories, MINITAB treats the last category as the "success" category. You can also specify how the categories should be ordered.

R

To create a confidence interval for *k* successes in *n* trials, use prop.test(k,n).

SPSS

SPSS does not find confidence intervals for proportions.

STATCRUNCH

To create a confidence interval for a proportion using summaries:

- Click on **Stat**.
- Choose **Proportions » One sample » with summary**.
- Enter the **Number of successes** (*x*) and **Number of observations** (*n*).
- Click on **Next**.
- Indicate **Confidence Interval** (Standard-Wald), and then enter the Level of confidence.
- Click on **Calculate**.

To create a confidence interval for a proportion using data:

- Click on **Stat**.
- Choose **Proportions » One sample » with data**.
- Choose the variable **Column** listing the **Outcomes**.
- Enter the outcome to be considered a **Success**.
- Click on **Next**.
- Indicate **Confidence Interval**, and then enter the **Level** of confidence.
- Click on **Calculate**.

TI-83/84 PLUS

To calculate a confidence interval for a population proportion:

- Go to the **STATS TESTS** menu and select **A:1-PropZInt.**
- Enter the number of successes observed and the sample size.
- Specify a confidence level.
- Calculate the interval.

COMMENTS

Beware: When you enter the value of *x*, you need the count, not the percentage. The count must be a whole number. If the number of successes is given as a percentage, you must first form the product *np* and then round the result.

Chapter Exercises

9. Margin of error A television newscaster reports the results of a poll of voters, and then says, "The margin of error is plus or minus 4%." Carefully explain what that means.

10. Margin of error II A medical researcher estimates the percentage of children who are exposed to lead-based paint, adding that he believes his estimate has a margin of error of about 3%. Explain what the margin of error means.

11. Conditions Consider each situation described below. Identify the population and the sample, explain what p and $\hat{p}$ represent, and tell whether the methods of this chapter can be used to create a confidence interval.
   a) Police set up an auto checkpoint at which drivers are stopped and their cars inspected for safety problems. They find that 14 of the 134 cars stopped have at least one safety violation. They want to estimate the percentage of all cars that may be unsafe.
   b) A television talk show asks viewers to register their opinions on prayer in schools by logging on to a website. Of the 602 people who voted, 488 favoured prayer in schools. They want to estimate the level of support among the general public.
   c) A school is considering requiring students to wear uniforms. It surveys parent opinion by sending a questionnaire home with all 1245 students. Three hundred and eighty surveys are returned, with 228 families in favour of the change.
   d) A college admits 1632 new first-year students, and four years later 1388 of them graduate on time. The college wants to estimate the percentage of all their first year enrollees who graduate on time.

12. More conditions Consider each situation described. Identify the population and the sample, explain what p and $\hat{p}$ represent, and tell whether the methods of this chapter can be used to create a confidence interval.
   a) A consumer group hoping to assess customer experiences with auto dealers surveys 167 people who recently bought new cars. Three percent of them expressed dissatisfaction with the salesperson.
   b) What percent of university students have cell phones? Researchers questioned 2883 students as they entered a football stadium, and 243 indicated that they had cell phones with them.
   c) Researchers randomly check 240 potato plants in a field in P.E.I. and only seven show signs of blight. How severe is the blight problem for the Canadian potato industry?
   d) Twelve of the 309 employees of a small company suffered an injury on the job last year. What can the company expect in future years?

13. Conclusions A catalogue sales company promises to deliver orders placed on the Internet within three days. Follow-up calls to a few randomly selected customers show that a 95% confidence interval for the proportion of all orders that arrive on time is 88% ± 6%. What does this mean? Are these conclusions correct? Explain.
   a) Between 82% and 94% of all orders arrive on time.
   b) 95% of all random samples of customers will show that 88% of orders arrive on time.
   c) 95% of all random samples of customers will show that 82% to 94% of orders arrive on time.
   d) We are 95% sure that between 82% and 94% of the orders placed by the customers in this sample arrived on time.
   e) On 95% of the days, between 82% and 94% of the orders will arrive on time.

14. More conclusions In January 2002, two students made worldwide headlines by spinning a Belgian euro 250 times and getting 140 heads—that's 56%. That makes the 90% confidence interval (51%, 61%). What does this mean? Are these conclusions correct? Explain.
   a) Between 51% and 61% of all Belgian euros are unfair.
   b) We are 90% sure that in this experiment this euro landed heads on between 51% and 61% of the spins.
   c) We are 90% sure that spun Belgian euros will land heads between 51% and 61% of the time.
   d) If you spin a Belgian euro many times, you can be 90% sure of getting between 51% and 61% heads.
   e) 90% of all spun Belgian euros will land heads between 51% and 61% of the time.

15. Confidence intervals Several factors are involved in the creation of a confidence interval. Among them are the sample size, the level of confidence, and the margin of error. Which statements are true?
   a) For a given sample size, higher confidence means a smaller margin of error.
   b) For a specified confidence level, larger samples provide smaller margins of error.
   c) For a fixed margin of error, larger samples provide greater confidence.
   d) For a given confidence level, halving the margin of error requires a sample twice as large.

16. Confidence intervals, again Several factors are involved in the creation of a confidence interval. Among them are the sample size, the level of confidence, and the margin of error. Which statements are true?
   a) For a given sample size, reducing the margin of error will mean lower confidence.
   b) For a certain confidence level, you can get a smaller margin of error by selecting a bigger sample.

c) For a fixed margin of error, smaller samples will mean lower confidence.

d) For a given confidence level, a sample nine times as large will make a margin of error one third as big.

17. Cars What fraction of cars have a made in Japan classification? The computer output below summarizes the results of a random sample of 50 autos. Explain carefully what it tells you.

z-Interval for proportion

With 90.00% confidence,

$0.29938661 < p(\text{Japan}) < 0.46984416$

18. Parking in Toronto In a June 2009 *Toronto Star* study of 100 downtown Toronto parkers, researchers counted how many parkers, out of 100 observed, bumped or mounted the curb while parking.[14] Below is the computer output for these 100 observations. Assuming these parkers are representative of all those who might be attempting to park in this area at the same time of day, what can you conclude?

z-Interval for proportion

With 95.00% confidence,

$0.237840 < p(\text{bump or mount}) < 0.422160$

19. Mislabelled seafood In December 2011, *Consumer Reports* published their study of labelling of seafood sold in New York, New Jersey, and Connecticut. They purchased 190 pieces of seafood from various kinds of food stores and restaurants in the three states and genetically compared the pieces to standard gene fragments that can identify the species. Laboratory results indicated that 22% of these packages of seafood were mislabelled, incompletely labelled, or misidentified by store or restaurant employees.

a) Construct a 95% confidence interval for the proportion of all seafood packages in those three states that are mislabelled or misidentified.

b) Explain what your confidence interval says about seafood sold in these three states.

c) A 2009 report by the Government Accountability Board says that the Food and Drug Administration has spent very little time recently looking for seafood fraud. Suppose an official said, "That's only 190 packages out of the billions of pieces of seafood sold in a year. With the small number tested, I don't know that one would want to change one's buying habits." (An official was quoted similarly in a different but similar context). Is this argument valid? Explain.

20. Mislabelled seafood, second course The December 2011 *Consumer Reports* study described in Exercise 19 also found that 12 of the 22 "red snapper" packages tested were a different kind of fish.

a) Are the conditions for creating a confidence interval satisfied? Explain.

b) Construct a 95% confidence interval.

c) Explain what your confidence interval says about "red snapper" sold in these three states.

21. Canadian pride In the 2009 Canada Day poll of 1000 Canadians by Angus Reid Strategies, 58% said they were proud of Canadian health care. A year earlier, 50% of a similar-size sample reported being proud of Canadian health care.

a) Find the margin of error for the 2009 poll if we want 90% confidence in our estimate of the percentage of national adults who are proud of Canadian heath care.

b) Explain what that margin of error means.

c) If we wanted to be 99% confident, would the margin of error be larger or smaller? Explain.

d) Find that margin of error.

e) In general, if all other aspects of the situation remain the same, will smaller margins of error produce greater or less confidence in the interval?

f) Do you think there's been a change from 2008 to 2009 in the real proportion of Canadian adults who are proud of the Canadian health care system? Explain.

22. Canada's birthday 2009 A June 2009 Harris-Decima poll found that only 21% of a random sample of 1000 adults knew that Canada was turning 142 years old.

a) Find the margin of error for this poll if we want 95% confidence in our estimate of the percent of Canadian adults who know Canada's age.

b) Explain what that margin of error means.

c) If we only need to be 90% confident, will the margin of error be larger or smaller? Explain.

d) Find that margin of error.

e) In general, if all other aspects of the situation remain the same, would smaller samples produce smaller or larger margins of error?
Source: http://www.pressreader.com/canada/ the-amherst-news/20090630/282059092985295

23. Contributions, please Philanthropic organizations, like the MS Society, rely on contributions. They send free mailing labels and greeting cards to potential donors on their list and ask for a voluntary contribution. To test a new campaign, one philanthropic organization sent letters to a random sample of 100 000 potential donors and received 4781 donations.

a) Give a 95% confidence interval for the true proportion of those from their entire mailing list who may donate.

b) A staff member thinks that the true rate is 5%. Given the confidence interval you found, do you find that percentage plausible?

24. Take the offer A major credit card company is planning a new offer for their current cardholders. The offer will give double airline miles on purchases for the next six

---

[14]Diana Zlomislic, *Toronto Star*, June 6, 2009.

months if the cardholder goes online and registers for the offer. To test the effectiveness of the campaign, the company sent out offers to a random sample of 50 000 cardholders. Of those, 1184 registered.

a) Give a 95% confidence interval for the true proportion of cardholders who will register for the offer.

b) If the acceptance rate is only 2% or less, the campaign won't be worth the expense. Given the confidence interval you found, what would you suggest?

25. **Teenage drivers** An insurance company checks police records on 582 accidents selected at random and notes that teenagers were at the wheel in 91 of them.

a) Create a 95% confidence interval for the percentage of all auto accidents that involve teenage drivers.

b) Explain what your interval means.

c) Explain what "95% confidence" means.

d) A politician urging tighter restrictions on drivers' licenses issued to teens says, "In one of every five auto accidents, a teenager is behind the wheel." Does your confidence interval support or contradict this statement? Explain.

26. **Junk mail** Direct mail advertisers send solicitations (a.k.a. "junk mail") to thousands of potential customers in the hope that some will buy the company's product. The acceptance rate is usually quite low. Suppose a company wants to test the response to a new flyer, and sends it to 1000 people randomly selected from their mailing list of over 200 000 people. They get orders from 123 of the recipients.

a) Create a 90% confidence interval for the percentage of people the company contacts who may buy something.

b) Explain what this interval means.

c) Explain what "90% confidence" means.

d) The company must decide whether or not to do a mass mailing. The mailing won't be cost-effective unless it produces at least a 5% return. What does your confidence interval suggest? Explain.

27. **Safe food** Some food retailers propose subjecting food to a low level of radiation to improve safety, but sale of such "irradiated" food is opposed by many people. Suppose a grocer wants to find out what his customers think. He has cashiers distribute surveys at checkouts and ask customers to fill them out and drop them in a box near the front door. He gets responses from 122 customers, of whom 78 oppose the radiation treatments. What can the grocer conclude about the opinions of all his customers?

28. **Local news** The mayor of a small city suggests that the federal government locate a new prison there, arguing that the construction project and resulting jobs will be good for the local economy. A total of 183 residents show up for a public hearing on the proposal, and a show of hands finds only 31 in favour of the prison project.

What can the city council conclude about public support for the mayor's initiative?

29. **Death penalty, again** In the survey on the death penalty you read about in the chapter, the Gallup Poll actually split the sample at random, asking 510 respondents the question quoted earlier: "Generally speaking, do you believe the death penalty is applied fairly or unfairly in this country today?" The other 510 were asked, "Generally speaking, do you believe the death penalty is applied unfairly or fairly in this country today?" Seems like the same question, but sometimes the order of the choices matters. Suppose that for the second way of phrasing it, only 48% said they thought the death penalty was fairly applied.

a) What kind of bias may be present here?

b) If we combine them, considering the overall group to be one larger random sample, what is a 95% confidence interval for the proportion of the general public that thinks the death penalty is being fairly applied?

c) How does the margin of error based on this pooled sample compare with the margins of error from the separate groups? Why?

30. **Gambling** A city ballot includes an initiative that would permit casino gambling. The issue is hotly contested, and two groups decide to conduct polls to predict the outcome. The local newspaper finds that 53% of 1200 randomly selected voters plan to vote "yes," while a university Statistics class finds 54% of 450 randomly selected voters in support. Both groups will create 95% confidence intervals.

a) Without finding the confidence intervals, explain which one will have the larger margin of error.

b) Find both confidence intervals.

c) Which group concludes that the outcome is too close to call? Why?

31. **Rickets** Vitamin D, whether ingested as a dietary supplement or produced naturally when the skin is exposed to sunlight, is essential for strong, healthy bones. The bone disease rickets was largely eliminated in England during the 1950s, but now there is concern that a generation of children more likely to watch television or play computer games than spend time outdoors is at increased risk. A recent study of 2700 children randomly selected from all parts of England found 20% of them deficient in vitamin D.

a) Find a 98% confidence interval.

b) Explain carefully what your interval means.

c) Explain what "98% confidence" means.

32. **Pregnancy** In 1998, a San Diego reproductive clinic reported 49 live births to 207 women under the age of 40 who had previously been unable to conceive.

a) Find a 90% confidence interval for the success rate at this clinic.

b) Interpret your interval in this context.

c) Explain what "90% confidence" means.

d) Do these data refute the clinic's claim of a 25% success rate? Explain.

33. Studying poverty To study poverty in the city of Halifax, researchers will randomly select 30 blocks from the city, then randomly select 10 households from each of those 30 blocks to be interviewed, producing a total of 300 interviews. Based on the interviews, the families will be classified as living below the poverty line or not. Suppose that we find that 45 families out of the 300 meet the criterion for living in poverty. Give a 95% confidence interval for the true proportion of families living in poverty in Halifax, with the usual approach discussed in this chapter. What is wrong with this approach here? Do you think you have over or underestimated the true SE of the sample proportion? Explain.

34. Legal music A random sample of 168 students were asked how many songs were in their digital music library and what fraction of them were legally purchased. Overall, they reported having a total of 117 079 songs, of which 23.1% were legal. The music industry would like a good estimate of the fraction of songs in students' digital music libraries that are legal.

a) Think carefully. What is the parameter being estimated? What is the population? What is the sample size?

b) Construct a 95% confidence interval for the fraction of legal digital music, if you can. If not, explain the problem.

35. Deer ticks Wildlife biologists inspect 153 deer taken by hunters and find 32 of them carrying ticks that test positive for Lyme disease.

a) Create a 90% confidence interval for the percentage of deer that may carry such ticks.

b) If the scientists want to cut the margin of error in half, how many deer must they inspect?

c) What concerns do you have about this sample?

36. Pregnancy II The San Diego reproductive clinic in Exercise 32 wants to publish updated information on its success rate.

a) The clinic wants to cut the stated margin of error in half. How many patients' results must be used?

b) Do you have any concerns about this sample? Explain.

37. Graduation It's believed that as many as 25% of adults over 50 never graduated from high school. We wish to see if this percentage is the same among the 25 to 30 age group.

a) How many of this younger age group must we survey to estimate the proportion of nongraduates to within 6 percentage points, with 90% confidence?

b) Suppose we want to cut the margin of error to 4 percentage points. What's the necessary sample size?

c) What sample size would produce a margin of error of 3 percentage points?

38. Hiring In preparing a report on the economy, we need to estimate the percentage of businesses that plan to hire additional employees in the next 60 days.

a) How many randomly selected employers must we contact to create an estimate in which we are 98% confident with a margin of error of 5%?

b) Suppose we want to reduce the margin of error to 3%. What sample size will suffice?

c) Why might it not be worth the effort to try to get an interval with a margin of error of only 1%?

39. Graduation, again As in Exercise 37, we hope to estimate the percentage of adults aged 25 to 30 who never graduated from high school. What sample size would allow us to increase our confidence level to 95% while reducing the margin of error to only 2 percentage points?

40. Better hiring info Editors of the business report in Exercise 38 are willing to accept a margin of error of 4% but want 99% confidence. How many randomly selected employers will they need to contact?

41. Pilot study A province's environmental ministry worries that many cars may be violating clean air emissions standards. The agency hopes to check a sample of vehicles in order to estimate that percentage with a margin of error of 3% and 90% confidence. To gauge the size of the problem, the agency first picks 60 cars and finds 9 with faulty emissions systems. How many should be sampled for a full investigation?

42. Another pilot study During routine screening, a doctor notices that 22% of her adult patients show higher-than-normal levels of glucose in their blood—a possible warning signal for diabetes. Hearing this, some medical researchers decide to conduct a large-scale study, hoping to estimate the proportion to within 4 percentage points with 98% confidence. How many randomly selected adults must they test?

43. Approval rating A newspaper reports that the Premier's approval rating stands at 65%. The article adds that the poll is based on a random sample of 972 adults and has a margin of error of 2.5%. What level of confidence did the pollsters use?

44. Amendment A television news reporter says that a proposed constitutional amendment is likely to win approval in an upcoming referendum because a poll of 1505 likely voters indicated that 52% would vote in favour. The reporter goes on to say that the margin of error for this poll was 3.0%.

a) Explain why the poll is actually inconclusive.

b) What confidence level did the pollsters use?

45. Bad countries In the Canada's World poll, conducted by Environics in January 2008 (pre-Obama), 52% of respondents pointed to the United States as a negative force in the world (4% named Canada as a negative force). Among young people (aged 15–24), the percentage rose to 63%.

A total of 2001 Canadians were randomly sampled for the poll. Assume it was a simple random sample. Answer the following questions at the 99% confidence level.

a) What is the margin of error for the estimate of the proportion pointing to the United States as a negative force? Give a 99% confidence interval for the true population proportion.

b) The figure for British Columbia was the highest, at 58%. Can we safely take the margin of error to be the same as in part a)? Explain.

c) If young people constituted, say, 25% of the sample, what would be the margin of error for their reported 63% figure?

d) How big would the sample have to be to reduce by half the margin of error for the overall 52% estimate?

46. More Canadian pride In the Angus Reid 2009 Canada Day poll of 1000 adult Canadians, hockey was a source of pride for 78% of males and 66% of females. Recall that the reported margin of error for the total sample is 3.1% (you may round this to 3%) at the 95% confidence level.

a) Can we safely say with 95% confidence that between 75% and 81% of all adult Canadian males feel pride in the sport of hockey? Explain.

b) Determine a 99% confidence interval for the true population proportion of Canadian females who feel pride in our national game of hockey. You may make any reasonable assumptions, but state them.

c) To reduce the margin of error in part b) by three-quarters, or 75%, how big should the entire sample of Canadians have been?

47. Simulations Use your computer software to generate a sample of size 100 from a Bernoulli distribution. This distribution has only two outcomes, 0 and 1, with complementary probabilities. Set $p(1) = 0.8$. Consider "1" to be a success and "0" to be a failure.

a) From the sample, calculate a 90% confidence interval for the true population proportion of successes. Does it contain the true population proportion of 0.8?

b) Repeat part a) 99 more times, obtaining 99 additional samples and confidence intervals. How many confidence intervals contained the true proportion? What percentage of confidence intervals would you expect to contain the true proportion if you repeated these simulations many times? For 100 simulated samples, if $X = $ number of confidence intervals that contain 0.8, what is the probability distribution of X?

48. More simulations Use your computer software to generate a sample of size 50 from a Bernoulli distribution. This distribution has only two outcomes, 0 and 1, with complementary probabilities. Set $p(1) = 0.3$. Consider "1" to be a success and "0" to be a failure.

a) From the sample, calculate an 80% confidence interval for the true population proportion of successes. Does it contain the true population proportion of 0.3?

b) Repeat part a) 49 more times, obtaining 49 additional samples and confidence intervals. How many confidence intervals contained the true proportion? What percentage of confidence intervals would you expect to contain the true proportion if you repeated these simulations many times? For 50 simulated samples, if $X = $ number of confidence intervals that contain 0.3, what is the probability distribution of X?

49. Dogs Canine hip dysplasia is a degenerative disease that causes pain in many dogs. Sometimes advanced warning signs appear in puppies as young as six months. A veterinarian checked 42 puppies and found 5 with early hip dysplasia. She considers this group to be a random sample of all puppies.

a) Explain why we cannot construct the customary Normal approximation based confidence interval for the rate of occurrence of early hip dysplasia among all puppies.

b) Construct a "Plus Four" confidence interval and interpret it in context.

50. Fans A survey of 81 randomly selected people standing in line to enter a football game found that 73 of them were home-team fans.

a) Explain why we cannot construct the customary Normal approximation based confidence interval for the proportion of all people at the game who are fans of the home team.

b) Construct a "Plus Four" confidence interval and interpret it in context.

*51. Departures 2011 What are the chances your flight will leave on time? The U.S. Bureau of Transportation Statistics of the Department of Transportation publishes information about airline performance. Here are a histogram and summary statistics for the percentage of flights departing on time each month from 1995 through 2011.[15]

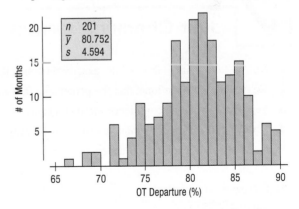

There is no evidence of a trend over time.

a) Check the assumptions and conditions for inference.

b) Find a 90% confidence interval for the true mean monthly on-time departure rate.

c) Interpret this interval for a traveller planning to fly.

[15]www.transtats.bts.gov/HomeDrillChart.asp

*52. Late arrivals 2011 Will your flight get you to your destination on time? The U.S. Bureau of Transportation Statistics reported the percentage of flights that were late each month from 1995 through 2011. Here is a histogram, along with some summary statistics:

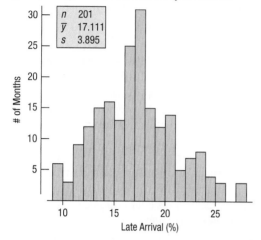

n	201
$\bar{y}$	17.111
s	3.895

We can consider these data to be a representative sample of all months. There is no evidence of a time trend ($r = 0.07$).
a) Check the assumptions and conditions for inference about the mean.
b) Find a 99% confidence interval for the true mean monthly late-arrival rate.
c) Interpret this interval for a traveller planning to fly.

*53. Speed of light In 1897, Michelson conducted an experiment to measure the speed of light (which is known to be a physical constant throughout the universe). He reported the results of 100 independent trials. The measurements are in kilometres per second and were coded by subtracting

299 000 from each. The average of the 100 coded measurements was 852.4, with a standard deviation of 79.0.
a) Estimate the standard deviation of the mean speed of light measurement.
b) Report a 99% confidence interval for the true speed of light.
c) State in words what this interval means.
d) In order to trust your reported level of confidence, is there anything else you might like to see or check?
e) The true speed of light is now known to be 299 710.5 km/sec, corresponding to a coded value of 710.5. What does this indicate about Michelson's experiment?

*54. Hot dogs A nutrition lab tested 60 "reduced sodium" hot dogs and found the mean sodium content to be 318 mg, with a standard deviation of 32 mg.
a) Estimate the standard deviation of the mean sodium content of the sample.
b) Find and interpret a 90% confidence interval for the true mean sodium content for this type of hot dog.
c) In order to trust your reported level of confidence, is there anything more you need to do or know?

*55. Hot dog margin of error To reduce the margin of error by 50% in the example of "reduced sodium" hot dogs from Exercise 54, how much bigger would the sample need to be? How big would it need to be to produce a margin of error equal to 4.0 mg?

*56. Speed of light margin of error To reduce the margin of error in Exercise 53 by 2/3 (to 1/3 of what was actually obtained), how much larger would the sample need to be? How much larger would it have to be to produce a margin of error equal to 10.0 km/sec?

 Just Checking ANSWERS

1. No. We know that in the sample, exactly 58% said they were proud; there's no need for a margin of error.
2. No, we are 95% confident that the percentage falls in some interval, not exactly on a particular value.
3. Yes. That's what the confidence interval means.
4. No. We don't know for sure that's true; we are only 95% confident.
5. No. That's our level of confidence, not the proportion who are proud.
6. Wider.
7. Lower.
8. Smaller.

MathXL   Go to MathXL at **www.mathxl.com** or MyLab™ Statistics at **www.mystatlab.com**. You can practise exercises for this chapter as often as you want. The guided solutions will help you find answers step by step. You'll find a **MyLab Statistics**   personalized study plan available to you too!

"Half the money I spend on advertising is wasted; the trouble is I don't know which half."

—*John Wanamaker*

Ingots are huge pieces of metal, sometimes weighing more than 9000 kilograms, made in a giant mould. They must be cast in one large piece for use in fabricating large structural parts for cars and planes. As the liquid metal cools, cracks can develop on the surface of the ingot, which can propagate into the zone required for the part, compromising its integrity. Airplane manufacturers insist that metal for their planes be defect-free, so the ingot must be remade if any cracking is detected.

Even though the metal from the cracked ingot is recycled, the scrap cost runs into the tens of thousands of dollars to recast an ingot, not to mention the energy waste. About 2/3 of all aluminum ingots produced use a process called the "direct chill" method designed to reduce recasting. Metal manufacturers would like to avoid cracking if at all possible. But the casting process is complicated and not everything is completely under control. It's estimated that about 5% of aluminum ingots need to be recast because of cracking. That rate depends on the size of the ingot cast. In one plant that specializes in very large (over 14 000 kg) ingots designed for the airplane industry, about 20% of the ingots have had some kind of surface crack. In an attempt to reduce the cracking proportion, the plant engineers and chemists recently tried out some changes in the casting process. Since then, 400 ingots have been cast and only 17% of them have cracked. Has the new method worked? Has the cracking rate really decreased, or was 17% just due to luck?

We can treat the 400 ingots cast with the new method as a random sample. We know that each random sample will have a somewhat different proportion of cracked ingots. Is the 17% we observe merely a result of natural sampling variability, or is this lower cracking rate strong enough evidence to assure management that the true cracking rate now is really below 20%?

People are constantly trying to find suitable answers to these types of questions. Has the Prime Minister's approval rating changed since last month? Has teenage smoking decreased in the past five years? Is the global temperature increasing? Did the Super Bowl ad we bought actually increase sales? To answer such questions, we test *hypotheses* about models.

16.1 Hypotheses

We want to know if the changes made by the engineers have lowered the cracking rate from 20%. Humans are natural skeptics, so to test whether the changes have worked, we'll assume that they haven't. We'll make the **hypothesis** that the cracking rate is still 20% and see if the data can convince us otherwise. Hypotheses are models that we adopt temporarily—until we can test them once we have data. This starting hypothesis to be tested is called the **null hypothesis**—null because it assumes that nothing has changed.

Like the confidence interval, our hypothesis test for p will use the standard deviation of the sampling distribution as a ruler, and the Normal model once again will provide the needed sampling model.

We denote it H_0. It specifies a parameter value—here that the cracking rate is 20%—which we usually write in the form H_0: *parameter = hypothesized value*. So, for the ingots, we would write H_0: $p = 0.20$.

The **alternative hypothesis**, which we denote H_A, contains the values of the parameter that we consider plausible if we reject the null hypothesis. Our null hypothesis is that $p = 0.20$. What's the alternative? Management is interested in *reducing* the cracking rate, so their alternative is H_A: $p < 0.20$.

What would convince you that the cracking rate had actually gone down? If only 4 out of the next 400 ingots crack (for a rate of 1%), most folks would conclude that the changes helped. But if the sample cracking rate is 19.8%, you should be skeptical. After all, observed proportions do vary, so we wouldn't be surprised to see some difference. How much smaller must the sample cracking rate be before we *are* convinced that things have changed? That's the crucial question in a hypothesis test. As usual in Statistics, when we think about how big the change has to be, we think of using the standard deviation as the ruler to measure that change. So let's start by finding the standard deviation of the sample cracking rate.

Since the company changed the process, 400 new ingots have been cast of which 68 have visible surface cracks, for a sample proportion of 17%. The sample size of 400 is big enough to satisfy the **Success/Failure Condition**. (We expect $0.20 \times 400 = 80$ ingots to crack under H_0.) Although not a random sample, the engineers think that whether an ingot cracks should be independent from one ingot to the next, so the Normal sampling distribution model should work well. If the null hypothesis is true, the standard deviation of the sampling model is

$$SD(\hat{p}) = \sqrt{\frac{pq}{n}} = \sqrt{\frac{(0.20)(0.80)}{400}} = 0.02$$

■ NOTATION ALERT

Capital H is the standard letter for hypotheses. H_0 always labels the null hypothesis and H_A (sometimes written H_1) labels the alternative hypothesis.

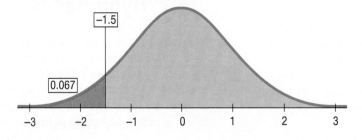

I CAN'T BELIEVE SCHOOLS ARE STILL TEACHING KIDS ABOUT THE NULL HYPOTHESIS.

I REMEMBER READING A BIG STUDY THAT CONCLUSIVELY DISPROVED IT *YEARS* AGO.

> ### WHY DID WE WRITE SD($\hat{p}$) AND NOT SE($\hat{p}$)?
>
> Remember that we reserve the term standard error for the *estimate* of the standard deviation of the sampling distribution. But we're not estimating here—we have a value of p from our null hypothesis model. To remind us that the parameter value comes from the null hypothesis, it is sometimes written as p_0 and the standard deviation as $SD(\hat{p}) = \sqrt{p_0 q_0/n}$. That's different than when we found a confidence interval for p. In that case we couldn't assume that we knew its value, so we estimated the standard deviation from the sample value $\hat{p}$.

Now we know both parameters of the Normal sampling distribution model: $p = 0.20$ and $SD(\hat{p}) = 0.02$, so we can find out how likely it would be to see the observed rate of 17%. Since we are using a Normal model, we find the z-score:

$$z = \frac{0.17 - 0.20}{0.02} = -1.5$$

Then we ask, "How likely is it to observe a value at least 1.5 standard deviations below the mean of a Normal model?" The answer (from a calculator, computer program, or the Normal table) is about 0.067, as shown in Figure 16.1. This is the probability of observing a cracking rate of 17% or less in a sample of 400 if the null hypothesis is true and the true cracking rate is still 20%. Management now must decide whether an event that would happen 6.7% of the time by chance is strong enough evidence to conclude that the true cracking proportion has decreased.

Figure 16.1

How likely is a z-score of -1.5 (or lower)? This is what it looks like. The red area is 0.067, or 6.7%, of the total area under the curve.

A Trial as a Hypothesis Test

Does the reasoning of hypothesis tests seem backwards? That could be because we usually prefer to think about getting things right rather than getting them wrong. You have seen this reasoning before because it's the logic of jury trials.

Let's suppose a defendant has been accused of robbery. In British common law and systems derived from it, the null hypothesis is that the defendant is innocent. Instructions to juries are quite explicit about this.

How is the null hypothesis tested? The prosecution first collects evidence, in the form of facts that seem to contradict the presumption of innocence. ("If the defendant were innocent, wouldn't it be remarkable that the police found him at the scene of the crime with a bag full of money in his hand, a mask on his face, and a getaway car parked outside?") For us, the data is the evidence.

The next step is to judge the evidence. Evaluating the evidence is the responsibility of the jury in a trial, but it falls on your shoulders in hypothesis testing. The jury considers the evidence in light of the *presumption* of innocence and judges whether the evidence against the defendant would be plausible *if the defendant were in fact innocent*.

Like the jury, you ask, "Could these data plausibly have happened by chance if the null hypothesis were true?" If they are very unlikely to have occurred, then the evidence raises serious doubt about the null hypothesis.

> **Beyond a Reasonable Doubt** We ask whether the data were unlikely beyond a reasonable doubt. We've just calculated that probability. The probability that the observed statistic value (or an even more extreme value) could occur if the null model were true—in this case, 0.067—is the P-value.

Ultimately, you must make a decision. The standard of "beyond a reasonable doubt" is wonderfully ambiguous because it leaves the jury to decide the degree to which the evidence contradicts the hypothesis of innocence. Juries don't explicitly use probability to help them decide whether to reject that hypothesis. But when you ask the same question of your null hypothesis, you have the advantage of being able to quantify exactly how surprising the evidence would be if the null hypothesis were true.

How unlikely is unlikely? Some people set rigid standards, like 1 time out of 20 (0.05) or 1 time out of 100 (0.01). But if *you* have to make the decision, you must judge for yourself in each situation whether the probability of observing your data is small enough to leave you with little "reasonable doubt."

Section 16.1 Exercises

1. **Better than aspirin?** A very large study showed that aspirin reduced the rate of first heart attacks by 44%. A pharmaceutical company thinks they have a drug that will be more effective than aspirin, and plans to do a randomized clinical trial to test the new drug.
a) What is the null hypothesis the company will use?
b) What is their alternative hypothesis?

2. **Psychic** A friend of yours claims to be psychic. You are skeptical. To test this you take a stack of 100 playing cards and have your friend try to identify the suit (hearts, diamonds, clubs, or spades), without looking, of course!
a) State the null hypothesis for your experiment.
b) State the alternative hypothesis.

16.2 P-Values

The fundamental step in our reasoning is the question: "Are the data surprising, given the null hypothesis?" And the key calculation is to determine exactly how likely the data we observed would be given that the null hypothesis were a true model of the world. Specifically, we want to find the *probability* of seeing data like these (or something even less likely) *given* that the null hypothesis is true. This probability tells us how surprised we'd be to see the data we collected if the null hypothesis is true. It's so important that it gets a special name. It's called the **P-value**.

When the P-value *is* small enough, it says that we are very surprised. It means that it's very unlikely we'd observe data like these if our null hypothesis were true. The model we started with (the null hypothesis) and data we collected are at odds with each other, so we have to make a choice. Either the null hypothesis is correct and we've just seen something remarkable, or the null hypothesis is wrong, and we were wrong to use it as the basis for computing our P-value. If you believe in data more than in assumptions, then, given that choice, you should reject the null hypothesis.

When the P-value is high, we haven't seen anything unlikely or surprising at all. Events that have a high probability of happening happen often. The data are consistent with the model from the null hypothesis, and we have no reason to reject the null hypothesis. But many other similar hypotheses could also account for the data we've seen, so *we haven't proven that the null hypothesis is true*. The most we can say is that it doesn't appear to be false. Formally, we "fail to reject" the null hypothesis. That's a pretty weak conclusion, but it's all we can do with a high P-value.

What to Do with an "Innocent" Defendant

If the evidence is not strong enough to reject the defendant's presumption of innocence, what verdict does the jury return? They say "not guilty." Notice that they do not say that the defendant is innocent. All they say is that they have not seen sufficient evidence to convict, to reject innocence. The defendant may, in fact, be innocent, but the jury has no way to be sure.

Said statistically, the jury's null hypothesis is H_0: innocent defendant. If the evidence is too unlikely given this assumption - if the P-value is too small - the jury rejects the null hypothesis and finds the defendant guilty. But—and this is an important distinction—if there is *insufficient evidence* to convict the defendant, the jury does not decide that H_0 is true and declare the defendant innocent. Juries can only *fail to reject* the null hypothesis and declare the defendant "not guilty."

In the same way, if the data are not particularly unlikely under the assumption that the null hypothesis is true, then the most we can do is to "fail to reject" our null hypothesis. We never declare the null hypothesis to be true (or "accept the null"), because we simply do not know whether it's true or not. (After all, more evidence may come along later.)

In the trial, the burden of proof is on the prosecution. In a hypothesis test, the burden of proof is on the unusual claim. The null hypothesis is the ordinary state of affairs, so it's the alternative to the null hypothesis that we consider unusual and for which we must marshal evidence.

Imagine a clinical trial testing the effectiveness of a new headache remedy. In Chapter 10 we saw the value of comparing such treatments to a placebo. We hope to prove that the new treatment is more effective than the placebo. To do this, we set up its negation as our null hypothesis—"the new treatment is no more effective than the placebo"—and then do our best to destroy its credibility. How we choose the null is important, because results are variable and some patients will improve even when administered the placebo treatment. If we use only six people to test the drug, the results are likely *not to be clear* and we'll be unable to discredit the null hypothesis. Does this mean the drug doesn't work? Of course not. It simply means that we don't have enough evidence to reject our (null) assumption. That's why we don't start by assuming that the drug *is more effective*. If we were to do that, then we could test just a few people, find that the results aren't clear, and claim that since we've been unable to reject our assumption of drug effectiveness, it must be so! Health Canada is unlikely to be impressed by that argument.

"You must find the accused not guilty of the offence unless the Crown proves beyond a reasonable doubt that he/she is guilty of it."

—*Canadian Judicial Council*

DON'T "ACCEPT" THE NULL HYPOTHESIS

Think about the null hypothesis that H_0: All swans are white. Does collecting a sample of 100 white swans prove the null hypothesis? The data are *consistent* with this hypothesis and seem to lend support to it, but they don't *prove* it. In fact, all we can do is disprove the null hypothesis—for example, by finding just one non-white swan.

Just Checking

1. A research team wants to know if aspirin helps to thin blood. The null hypothesis says that it doesn't. They test 12 patients, observe the proportion with thinner blood, and get a P-value of 0.32. They proclaim that aspirin doesn't work. What would you say?

2. An allergy drug has been tested and found to give relief to 75% of the patients in a large clinical trial. Now the scientists want to see if the new, improved version works even better. What would the null hypothesis be?

3. The new drug is tested and the P-value is 0.0001. What would you conclude about the new drug?

Section 16.2 Exercises

3. Better than aspirin II? A clinical trial compares the new drug described in Exercise 1 to aspirin. The group using the new drug had somewhat fewer heart attacks than those in the aspirin group.
a) The P-value from the hypothesis test was 0.28. What do you conclude?
b) What would you have concluded if the P-value had been 0.004?

4. Psychic after all? The "psychic" friend from Exercise 2 correctly identified more than 25% of the cards.
a) A hypothesis test gave a P-value of 0.014. What do you conclude?
b) What would you conclude if the P-value had been 0.245?

16.3 The Reasoning of Hypothesis Testing

Hypothesis tests follow a carefully structured path. To avoid getting lost as we navigate down it, we divide that path into four distinct sections.

1. Hypotheses

"The null hypothesis is never proved or established, but is possibly disproved, in the course of experimentation. Every experiment may be said to exist only in order to give the facts a chance of disproving the null hypothesis."

—*Sir Ronald Fisher,*
The Design of Experiments

First, we state the null hypothesis. That's usually the skeptical claim that nothing's different. Are we considering a (New! Improved!) possibly better method? The null hypothesis says, "Oh yeah? Convince me!" To convert a skeptic, we must pile up enough evidence against the null hypothesis that we can reasonably reject it.

In statistical hypothesis testing, hypotheses are almost always about model parameters. To assess how unlikely our data may be, we need a null model. The null hypothesis specifies a particular parameter value to use in our model. In the usual shorthand, we write H_0: *parameter = hypothesized value*. The alternative hypothesis, H_A, contains the values of the parameter we consider plausible when we reject the null.

For Example WRITING HYPOTHESES

The provincial Ministry of Transportation claims that 80% of candidates pass driving tests, but a newspaper reporter's survey of 90 randomly selected local teens who have taken the test found only 61 who passed.

QUESTION: Does this finding suggest that the passing rate for teenagers is lower than the Ministry's reported overall rate? Write appropriate hypotheses.

ANSWER: I'll assume that the passing rate for teenagers is the same as the Ministry's overall rate of 80%, unless there's strong evidence that it's lower.

$$H_0: p = 0.80$$

$$H_A: p < 0.80$$

2. Model

WHEN THE CONDITIONS FAIL . . .

You might proceed with caution, explicitly stating your concerns. Or you may need to do the analysis with and without an outlier, or on different subgroups, or after re-expressing the response variable. Or you may not be able to proceed at all. (In which case, try searching for a test that makes different or fewer assumptions. You may find one.)

To plan a statistical hypothesis test, specify the *model* you will use to test the null hypothesis and the parameter of interest. Of course, all models require assumptions, so you will need to state them and check any corresponding conditions.

Your Model step should end with a statement such as

Because the conditions are satisfied, I can model the sampling distribution of the proportion with a Normal model.

Watch out, though. Your Model step could end with

Because the conditions are not satisfied, I can't proceed with the test.

If that's the case, stop and reconsider.

Each test in the book has a name that you should include in your report. We'll see many tests in the chapters that follow. Some will be about more than one sample, some will involve statistics other than proportions, and some will use models other than the Normal (and so will not use *z*-scores). The test about proportions that we discuss here is called a **one-proportion *z*-test**.[1]

ONE-PROPORTION *Z*-TEST

The conditions for the one-proportion *z*-test are the same[2] as for the one-proportion *z*-interval. We test the hypothesis $H_0: p = p_0$ using the statistic $z = \dfrac{(\hat{p} - p_0)}{SD(\hat{p})}$. We use the hypothesized proportion to find the standard deviation, $SD(\hat{p}) = \sqrt{\dfrac{p_0 q_0}{n}}$.

When the conditions are met and the null hypothesis is true, this statistic follows the standard Normal model, so we can use that model to obtain a P-value.

For Example   CHECKING THE CONDITIONS

RECAP: The provincial Ministry of Transportation claims that 80% of candidates pass driving tests. A reporter has results from a survey of 90 randomly selected local teens who had taken the test.

QUESTION: Are the conditions for inference satisfied?

✓ **Randomization Condition:** The 90 teens surveyed were a random sample of local teenage driving candidates.

✓ **Success/Failure Condition:** We expect $np_0 = 90(0.80) = 72$ successes and $nq_0 = 90(0.20) = 18$ failures. Both are at least 10.

✓ **10% Condition:** 90 is fewer than 10% of the teenagers who take driving tests in a province.

ANSWER: The conditions are satisfied, so it's okay to use a Normal model and perform a one-proportion *z*-test.

3. Mechanics

Under "Mechanics," we place the actual calculation of our test statistic from the data. Different tests we encounter will have different formulas and different test statistics. Usually,

---

[1]It's also called the "one-sample test for a proportion."

[2]For a test, we check the Success/Failure Condition using expected counts assuming the null hypothesis, whereas for a confidence interval, we have no choice but to use the observed counts.

the mechanics are handled by a statistics program or calculator, but it's good to have the formulas recorded for reference and to know what's being computed. The ultimate goal of the calculation is to find out how surprising our data would be if the null hypothesis were true. We measure this by the P-value—the probability that the observed statistic value (or an even more extreme value) occurs if the null model is correct. If the P-value is small enough, we'll reject the null hypothesis. And draw a picture. You can usually catch a silly mistake in the calculations simply by sketching the Normal distribution and shading the desired area on one side of your z-score.

For Example FINDING A P-VALUE

RECAP: The Ministry of Transportation claims that 80% of candidates pass driving tests, but a survey of 90 randomly selected local teens who have taken the test found only 61 who passed.

QUESTION: What's the P-value for the one-proportion z-test?

ANSWER: I have $n = 90$, $x = 61$, and a hypothesized $p = 0.80$.

$$\hat{p} = \frac{61}{90} \approx 0.678$$

$$SD(\hat{p}) = \sqrt{\frac{p_0 q_0}{n}} = \sqrt{\frac{(0.8)(0.2)}{90}} \approx 0.042$$

$$z = \frac{\hat{p} - p_0}{SD(\hat{p})} = \frac{0.678 - 0.800}{0.042} \approx -2.90$$

$$\text{P-value} = P(z < -2.90) = 0.002$$

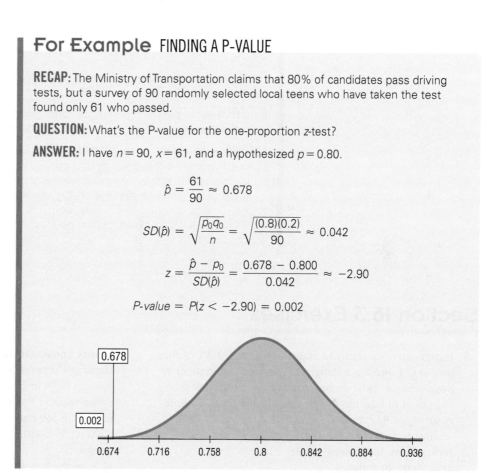

4. Conclusion

The conclusion in a hypothesis test is always a statement about the null hypothesis. The conclusion must state how strong the evidence is for rejection of the null hypothesis, as well as any decision we make (to reject or fail to reject the null hypothesis) based on the evidence - in those circumstances where a decision is required. And, as always, the conclusion should be stated in context.

Your conclusion about the null hypothesis should never be the end of a testing procedure. Often, there are actions to take or policies to change. In our ingot example, management must decide whether to continue the changes proposed by the engineers. The decision always includes the practical consideration of whether the new method is worth the cost. Suppose management concludes that the cracking percentage has been reduced. They must still evaluate how much the cracking rate will be reduced and how much it will cost to accomplish the reduction. The *size of the effect* is always a concern when we test hypotheses. A good way to look at the **effect size** is to examine a confidence interval.

HOW MUCH DOES IT COST?

Formal tests of a null hypothesis base the decision of whether to reject the null hypothesis solely on the size of the P-value. But in real life, we want to evaluate the costs of our decisions as well. How much would you be willing to pay for a faster computer? Shouldn't your decision depend on how much faster? And on how much more it costs? Costs are not just monetary, either. Would you use the same standard of proof for testing the safety of an airplane as for the speed of your new computer?

For Example STATING THE CONCLUSION

RECAP: The Ministry of Transportation claims that 80% of candidates pass driving tests. Data from a reporter's survey of randomly selected local teens who had taken the test produced a P-value of 0.002.

QUESTION: What can the reporter conclude? And how might the reporter explain what the P-value means for the newspaper story?

ANSWER: Because the P-value of 0.002 is very low, there is good evidence to reject the null hypothesis. These survey data provide strong evidence that the passing rate for teenagers taking the driving test is lower than 80%.

If the passing rate for teenage driving candidates were actually 80%, we'd expect to see success rates this low in only about 1 in 500 samples (0.2%).

This seems quite unlikely, casting doubt that the Ministry's stated success rate applies to teens.

Section 16.3 Exercises

5. Indigenous According to Statistics Canada, 4.3% of the people in Canada are Indigenous people. The Premier of Saskatchewan believes his province has a different proportion of Indigenous people than the nation as a whole. A survey of 947 residents, selected at random from the (entire) province, reveals that 96 of the 947 residents surveyed are Indigenous people.
a) State the hypotheses.
b) Name the model and check the appropriate conditions for a hypothesis test.
c) Draw and label a sketch, and then calculate the test statistic and P-value.
d) State your conclusion.

6. Empty houses According to a Canada Mortgage and Housing Corporation 2015 report, 3.5% of all rental housing units in Canada were vacant in 2015. A real estate agent in Toronto believes Toronto has a lower rental housing unit vacancy rate. She randomly selects 1240 rental housing units, and finds 17 of them are vacant.
a) State the hypotheses.
b) Name the model and check the appropriate conditions for a hypothesis test.
c) Draw and label a sketch, and then calculate the test statistic and P-value.
d) State your conclusion.

16.4 Alternative Alternatives

Tests on the ingot data can be viewed in two different ways. We know the old cracking rate is 20%, so the null hypothesis is

$$H_0: p = 0.20$$

But we have a choice of alternative hypotheses. A metallurgist working for the company might be interested in *any* change in the cracking rate due to the new process.

Even if the rate got worse, she might learn something useful from it. She's interested in possible changes on both sides of the null hypothesis. So she would write her alternative hypothesis as

$$H_A: p \neq 0.20$$

An alternative hypothesis such as this is known as a **two-sided alternative**,[3] because we are equally interested in deviations on either side of the null hypothesis value. For two-sided alternatives, the P-value is the probability of deviating at least as much as our actual sample in *either* direction from the null hypothesis value.

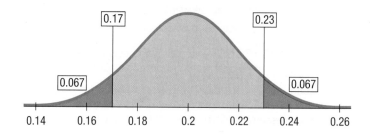

But management is really interested only in *lowering* the cracking rate below 20%. The scientific value of knowing how to *increase* the cracking rate may not appeal to them. The only alternative of interest to them is that the cracking rate *decreases*. They would write their alternative hypothesis as

$$H_A: p < 0.20$$

An alternative hypothesis that focuses on deviations from the null hypothesis value in only one direction is called a **one-sided alternative**.[4]

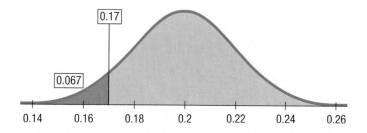

For a hypothesis test with a one-sided alternative, the P-value is the probability of deviating *only in the direction of the alternative* away from the null hypothesis value. For the same data, the one-sided P-value is half the two-sided P-value (when the data point in the same direction as the one-sided alternative). So, a one-sided test will reject the null hypothesis more often. *If you aren't sure which to use, a two-sided test is always more conservative.* Be sure you can justify the choice of a one-sided test from the *Why* of the situation (and not from the data).

Some statisticians believe that you should pretty much never do one-sided tests, as they consider them an artificial way to "buy" yourself more power, or more significant results. We'll leave that discussion to a more philosophical course, but one thing you should certainly never do is decide on the alternative hypothesis *after* looking at your data or the observed test statistic.[5] If a one-sided alternative is of interest to you, it should be of interest before you see the direction of the data!

---

[3]It is also called a *non-directional alternative*.

[4]Also called a directional alternative.

[5]Besides, you'll never see the test statistic go in both directions!

Step-by-Step Example TESTING A HYPOTHESIS

Advances in medical care, such as prenatal ultrasound examination, now make it possible to determine a child's sex early in a pregnancy. There is a fear that in some cultures some parents may use this technology to select the sex of their children. A study from Punjab, India,[6] reported that, in one hospital, 56.9% of the 550 live births that year were boys. It's a medical fact that male babies are slightly more common than female babies. The study's authors report a baseline for this region of 51.7% male live births.

Question: Is there evidence that the proportion of male births is different for this hospital?

THINK ➡ Plan State what we want to know. Define the variables and discuss the W's.

Hypotheses The null hypothesis makes the claim of no difference from the baseline.

Before seeing the data, we were interested in any difference from the baseline in male births, so the alternative hypothesis is two-sided.[7]

Model Think about the assumptions and check the appropriate conditions.

For testing proportions, the conditions are the same ones we had for making confidence intervals, except that we check the **Success/Failure Condition** with the *hypothesized* proportions rather than with the *observed* proportions.

I want to know whether the proportion of male births at a hospital is different from the established baseline of 51.7%. The data are the recorded sexes of the 550 live births from a hospital in Punjab, India, in one year, collected for a study on fetal sex determination. The parameter of interest, p, is the proportion of male births:

$$H_0: p = 0.517$$

$$H_A: p \neq 0.517$$

✓ **Independence Assumption**: There is no reason to think that the sex of one baby can affect the sex of other babies, so births can reasonably be assumed to be independent with regard to the sex of the child.

✓ **Randomization Condition**: The 550 live births are not a random sample, so I must be cautious about any general conclusions, but I believe that this is a reasonably representative period of time for this hospital.

✓ **Success/Failure Condition**: Both
$np_0 = 550(0.517) = 284.35$ and
$nq_0 = 550(0.483) = 265.65$ are
greater than 10; I expect the births of at least 10 boys and at least 10 girls, so the sample is large enough.

✓ **10% Condition**: These 550 births may be viewed as fewer than 10% of all possible births at this hospital around this time period.

Specify the sampling distribution model.

Tell what test you plan to use.

The conditions are satisfied, so I can use a Normal model and perform a *one-proportion z-test*.

---

[6]E. Booth, M. Verma, and R. S. Beri, "Fetal sex determination in infants in Punjab, India: Correlations and implications," *BMJ* 309: 1259–1261.

[7]If we had a clear theory or concern *a priori* (before seeing the data) about a possible excess of male births in this region, a one-sided alternative could be justified. The P-value, calculated at the end of this example, would then be only 0.0073.

SHOW ➡ Mechanics The null model gives us the mean, and (because we are working with proportions) the mean gives us the standard deviation.

The null model is a Normal distribution with a mean of 0.517 and a standard deviation of

$$SD(\hat{p}) = \sqrt{\frac{p_0 q_0}{n}} = \sqrt{\frac{(0.517)(1 - 0.517)}{550}}$$

$$= 0.0213$$

We find the z-score for the observed proportion to find out how many standard deviations it is from the hypothesized proportion.

The observed proportion, $\hat{p}$, is 0.569, so

$$z = \frac{\hat{p} - p_0}{SD(\hat{p})} = \frac{0.569 - 0.517}{0.0213} = 2.44$$

The sample proportion lies 2.44 standard deviations above the mean.

From the z-score, we can find the P-value, which tells us the probability of observing a value that extreme (or more). Use technology or a table (see further below).

Make a picture. Sketch a Normal model centred at $p_0 = 0.517$. Shade the region to the right of the observed proportion, and because this is a two-tail test, also shade the corresponding region in the other tail. (Or sketch a standard Normal curve, and shade the regions more extreme than z = 2.44.)

$$P = 2P(z > 2.44) = 2(0.0073) = 0.0146$$

The P-value is the probability of observing an outcome more than 2.44 standard deviations from the mean of a Normal model *in either direction*. We must therefore *double* the probability we find in the upper tail.

TELL ➡ Conclusion Report and interpret the P-value. Discuss the strength of the evidence against the null hypothesis. State your conclusion, in context, with any qualifications needed.

The P-value of 0.0146 says that if the true proportion of male babies at this hospital were 51.7%, then an observed proportion as different as 56.9% male babies would occur at random only about 15 times in 1000. With a P-value this small, we have strong evidence that the birth ratio of boys to girls is not equal to the baseline level for this region. It appears that the proportion of boys may be larger[8], at least around the time period of the study.

Here's a portion of a Normal table (Figure 16.2) that gives the probability we needed for the hypothesis test. At $z = 2.44$, the table gives the percentile (shaded blue) as 0.9927. The upper-tail probability (shaded red) is, therefore, $1 - 0.9927 = 0.0073$; so, for our two-sided test, the P-value is $2(0.0073) = 0.0146$.

---

[8]So there was an increase. Just how big was it? Use a confidence interval.

Figure 16.2

A portion of a Normal table with the look-up for $z = 2.44$ indicated. Since $1 - 0.9927 = 0.0073$, the P-value we want is $2(0.0073) = 0.0146$.

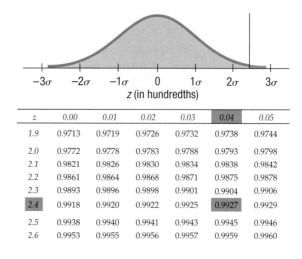

z (in hundredths)

z	0.00	0.01	0.02	0.03	0.04	0.05
1.9	0.9713	0.9719	0.9726	0.9732	0.9738	0.9744
2.0	0.9772	0.9778	0.9783	0.9788	0.9793	0.9798
2.1	0.9821	0.9826	0.9830	0.9834	0.9838	0.9842
2.2	0.9861	0.9864	0.9868	0.9871	0.9875	0.9878
2.3	0.9893	0.9896	0.9898	0.9901	0.9904	0.9906
2.4	0.9918	0.9920	0.9922	0.9925	0.9927	0.9929
2.5	0.9938	0.9940	0.9941	0.9943	0.9945	0.9946
2.6	0.9953	0.9955	0.9956	0.9957	0.9959	0.9960

Section 16.4 Exercises

7. Psychic again (you should have seen this coming) If you were to do a hypothesis test on your experiment with your "psychic" friend from Exercise 2, would your alternative hypothesis be one- or two-sided? Explain why.

8. Indigenous II The Premier of Saskatchewan in Exercise 5 is going to perform a hypothesis test to see if he has evidence that the proportion of people in Saskatchewan that are Indigenous is different from the entire nation. Should his alternative hypothesis be one-sided or two-sided? Explain.

16.5 P-Values and Decisions: What to Tell About a Hypothesis Test

TELL ➡

MORE

Hypothesis tests are particularly useful when there is a decision of consequence that must be made based on the available evidence. Is the defendant guilty or not? Should we choose print advertising or television? The absolute nature of the hypothesis test decision - to reject or fail to reject the null hypothesis, however, makes some people (including the authors) uneasy. Whenever possible, it's a good idea to report a confidence interval for the parameter of interest as well.

How small should the P-value be for you to reject the null hypothesis? A jury needs enough evidence to show the defendant as guilty "beyond a reasonable doubt." How does that translate to P-values? The answer is that there is no good, universal answer. How small the P-value has to be to reject the null hypothesis is highly context-dependent. When we're screening for a disease and want to be sure we treat all those who are sick, we may be willing to reject the null hypothesis of no disease with a P-value as large as 0.10. That would mean that 10% of the healthy people would be treated as sick and subjected to further testing. We might rather treat (or recommend further testing for) the occasional healthy person than fail to treat someone who was really sick. But a long-standing hypothesis, believed by many to be true, needs stronger evidence (and a correspondingly small P-value) to reject it.

See if you require the same P-value to reject each of the following null hypotheses:

■ A renowned musicologist claims that she can distinguish between the works of Mozart and Haydn simply by hearing a randomly selected 20 seconds of music from any work by either composer. What's the null hypothesis? If she's just guessing, she'll get 50% of the pieces correct, on average. So our null hypothesis is that p equals 50%. If she's for real, she'll get more than 50% correct. We present her with 10 pieces of Mozart or Haydn chosen at random. She gets 9 out of 10 correct. It turns out that the P-value associated with that result is 0.011. (In other words, if you tried to just guess, you'd get at least 9 out of 10 correct only about 1% of the time[9].) What

---

[9]To calculate this P-value, use the Binomial distribution, and find P(X = 9 or 10) when n = 10, p = 0.5.

would *you* conclude? Most people would probably reject the null hypothesis and be convinced that she has some ability to do as she claims. Why? Because the P-value is small and we don't have any particular reason to doubt the alternative.

- On the other hand, imagine a student who bets that he can make a flipped coin land the way he wants just by thinking hard. To test him, we flip a fair coin 10 times. Suppose he gets 9 out of 10 right. This also has a P-value of 0.011. Are you willing now to reject this null hypothesis? Are you convinced that he's not just lucky? What amount of evidence *would* convince you? We require more evidence if rejecting the null hypothesis would contradict long-standing beliefs or other scientific results. Of course, with sufficient evidence we should revise our opinions (and scientific theories). That's how science makes progress.

Another factor in choosing a P-value criterion is the importance of the issue being tested. Consider the following two tests:

- A researcher claims that the proportion of university students who hold part-time jobs now is higher than the proportion known to hold such jobs a decade ago. You might be willing to believe the claim (and reject the null hypothesis of no change) with a P-value of 0.10.
- An engineer claims that even though there were several previous problems with the rivets holding the wing on an airplane in their fleet, they've retested the proportion of faulty rivets and now the P-value is small enough to conclude that the proportion of rivets that fail is below the proportion at which the wing would fall off. What P-value would be small enough to get you to fly on that plane?

Your conclusion about any null hypothesis should always be accompanied by the P-value of the test. Don't just declare the null hypothesis rejected or not rejected. Report the P-value to show the strength of evidence against the null hypothesis. This will let each reader use their own criteria for making a decision.

Below is a general guideline for reporting the strength of evidence against H_0 based on the P-value, though as our examples above illustrate, this may not help you very much when it comes to making decisions:

P-Value	Strength of evidence against the null hypothesis
P-value < 0.001	Very strong
0.001 < P-value < 0.01	Strong
0.01 < P-value < 0.05	Moderate
0.05 < P-value < 0.1	Weak
P-value < 0.1	None

To complete your analysis (particularly if the null is rejected), follow up your test with a confidence interval to help provide a good idea of the size of the effect, and hence whether or not the result is truly of practical importance.

 Just Checking

4. A bank is testing a new method for getting delinquent customers to pay their past-due credit card bills. The standard way was to send a letter (costing about $0.40) asking the customer to pay. That worked 30% of the time. They want to test a new method that involves sending a DVD to customers encouraging them to contact the bank and set up a payment plan. Developing and sending the video costs about $10.00 per customer. What is the parameter of interest? What are the null and alternative hypotheses?

5. The bank sets up an experiment to test the effectiveness of the DVD. They mail it out to several randomly selected delinquent customers and keep track of how many actually do contact the bank to arrange payments. The bank's statistician calculates a P-value of 0.003. What does this P-value suggest about the DVD?

6. The statistician tells the bank's management that the results are clear and that they should switch to the DVD method. Do you agree? What else might you want to know?

Step-by-Step Example AFTER A SIGNIFICANT TEST, REPORT THE CONFIDENCE INTERVAL

In the previous Step-by-Step Example, we discovered evidence of an increased proportion of male babies in a hospital in Punjab, India. This should not be the end of our analysis. The next logical question is:

Question: How big an increase are we talking about? Let's find a confidence interval for the proportion of male births.

THINK AGAIN ➡ **Model** Check the conditions.

The conditions are identical to those for the hypothesis test, with one difference. Now we are not given a hypothesized proportion, p_0, so we must instead work with the observed proportion $\hat{p}$.

Specify the sampling distribution model.

Tell what method you plan to use.

✓ **Success/Failure Condition:** Both $n\hat{p} = 550(0.569) = 313$ male births and $n\hat{q} = 237$ female births are both at least 10.

The conditions are satisfied, so I can model the sampling distribution of the proportion with a Normal model and find a *one-proportion z-interval*.

SHOW MORE ➡ **Mechanics** We can't find the sampling model standard deviation from the null model proportion. (In fact, we've just rejected it.) Instead, we find the standard error of $\hat{p}$ from the *observed* proportion.

$$SE(\hat{p}) = \sqrt{\frac{\hat{p}\hat{q}}{n}} = \sqrt{\frac{(0.569)(1 - 0.569)}{550}}$$
$$= 0.0211$$

The sampling model is Normal, so for a 95% confidence interval, the critical value $z^* = 1.96$.

The margin of error is

$$ME = z^* \times SE(\hat{p}) = 1.96(0.0211) = 0.041$$

So the 95% confidence interval is

$$0.569 \pm 0.041 \text{ or } (0.528, 0.610)$$

TELL ALL ➡ **Conclusion** Confidence intervals help us think about the size of the effect. Comment on the possible magnitude of the change from the baseline.

We are 95% confident that the true proportion of male births is between 52.8% and 61.0%. The change from the baseline of 51.7% male births might be quite substantial.

Follow Your Plan A study published in the *Canadian Medical Association Journal* (CMAJ) revealed that many published studies reported hypothesis tests that were not part of the original research plan. The researchers compared the original protocols for randomized clinical trials approved by the Canadian Institutes of Health Research (CIHR) between 1990 and 1998 with the subsequent reports published in journals. They found that 40% of the publications focused on research outcomes that were different from the variable the researchers originally set out to analyze. The authors suggested that this was done when the data failed to fit the original hypothesis, declaring, "This is an ethical and scientific issue." CMAJ editor Dr. John Hoey says, "Once you change the objective of the trial so it fits with the data, it's no longer a randomized trial, it's a fishing expedition."[10]

Ideally, hypotheses should be stated even before the data are collected, but in reality, you often have the data before you formulate your hypotheses. What you may *not* do is analyze the data first, and *then* state your hypotheses to fit what you find.

That's "drawing the target around the arrow." And it's wrong.

Section 16.5 Exercises

9. Bad medicine Occasionally, a report comes out that a drug that cures some disease turns out to have a nasty side effect. For example, some antidepressant drugs may cause suicidal thoughts in younger patients. A researcher wants to study such a drug and look for evidence that such side effects exist.

a) If the test yields a low P-value and the researcher rejects the null hypothesis, but there is actually no ill side effect of the drug, what are the consequences of such an error?

b) If the test yields a high P-value and the researcher fails to reject the null hypothesis, but there *is* a bad side effect of the drug, what are the consequences of such an error?

10. Expensive medicine Developing a new drug can be an expensive process, resulting in high costs to patients. A pharmaceutical company has developed a new drug to reduce cholesterol, and it will conduct a clinical trial to compare the effectiveness to the most widely used current treatment. The results will be analyzed using a hypothesis test.

a) If the test yields a low P-value and the researcher rejects the null hypothesis that the new drug is not more effective, but it actually is not better, what are the consequences of such an error?

b) If the test yields a high P-value and the researcher fails to reject the null hypothesis, but the new drug *is* more effective, what are the consequences of such an error?

*16.6 Large Sample Tests of Hypothesis

Just as with confidence intervals, the procedure discussed in this chapter may be easily extended to test hypotheses about other parameters, such as means, as long as we have a sufficiently large sample (for small sample tests of means, use the methods described in Chapters 18–20).

Suppose we wish to test a hypothesis about a parameter θ, and have an estimator $\hat{\theta}$ whose distribution is approximately Normal, centred at θ, with a standard deviation denoted $SD(\hat{\theta})$. The test statistic is just the estimate of the parameter, $\hat{\theta}$, standardized, using the value for θ specified by our null hypothesis, denoted θ_0:

$$z = \frac{(\hat{\theta} - \theta_0)}{SD(\hat{\theta})} \quad \text{or in words:} \quad \frac{\text{Estimate} - \text{Hypothesized Value}}{\text{Standard Deviation of the Estimator}}$$

But as with confidence intervals, we can only use this method if we can accurately evaluate all parts of it, and the denominator presents a challenge. First, we need a formula

---

[10]http://news.utoronto.ca/_bulletin/2004-10-12.pdf

for this standard deviation, and then we must be able to plug in the values for all of its parts. However, if the sample is big enough, we just plug in sample estimates for any unknowns. Calculate the numerical value of this z-statistic, and then, as illustrated in Figure 16.1, find the P-value to base our test on.

Testing for the Mean of a Quantitative Population With a Big Sample

Suppose we wish to test a hypothesis about the mean of a quantitative population, denoted μ, where the null hypothesized value is μ_0. We may use the sample mean $\bar{y}$ to estimate μ without bias, and, for a large sample, we know that $\bar{y}$ will be nearly Normal due to the Central Limit Theorem. The standard deviation of the sampling distribution of $\bar{y}$ has the formula $SD(\bar{y}) = \dfrac{\sigma}{\sqrt{n}}$, which may be estimated from the sample data by $\dfrac{s}{\sqrt{n}}$, which we call the standard error of $\bar{y}$, or $SE(\bar{y})$.

Our test statistic then becomes $z = \dfrac{\bar{y} - \mu_0}{SE(\bar{y})}$, where the SE is $\dfrac{s}{\sqrt{n}}$. Substituting the SE for the true standard deviation in the denominator has some effect on the distribution of this statistic, but the effect is negligible if the sample is sufficiently large—say, greater than 60. If we introduce too much variability into the denominator, the distribution changes into something other than a standard Normal. In fact, it becomes the "t" distribution, a sampling model that will be discussed in great detail in Chapter 18, where we will deal with the small sample problem. Be sure to inspect the data for extreme outliers or severe skewness - which might force you to transform the data or focus inference on the median rather than the mean (in which case, the methods in Chapters 28 and 29 might help).

And just as with proportions, make sure that the observations are mutually independent and that your sampling procedure will allow you to generalize beyond your sample to a bigger population—do you have a valid random sample or a reasonably representative sample?

Continuing the example from Chapter 15, a team of researchers had tested 150 farm-raised salmon for organic contaminants. They found a mean concentration of the carcinogenic insecticide mirex to be 0.0913 ppm with a standard deviation of 0.0495 ppm. A histogram of the data showed no extreme skewness or outliers. A level higher than 0.08 ppm is considered dangerous. Is there *clear evidence* of excessively high levels of mirex? This requires a test of hypothesis. After all, even if the true mean level were 0.08 ppm, half of the time we would get a sample mean bigger than 0.08 purely by chance. Is the observed mean level too high to just be chance variation at work?

$$\bar{y} = 0.0913 \text{ ppm, with an estimated standard deviation} = \frac{s}{\sqrt{n}} = \frac{0.0495}{\sqrt{150}} = 0.00404,$$

Our test statistic is $z = \dfrac{\bar{y} - \mu_0}{SE(\bar{y})} = \dfrac{0.0913 - 0.08}{0.00404} = 2.80$, from which it follows that the P-value $= P(Z > 2.80) = (1 - 0.9974) = 0.0026$. This is small enough to provide very convincing evidence of an excessively high mean level of the mirex contaminant.

WHAT CAN GO WRONG?

Hypothesis tests are so widely used—and so widely misused—that we've devoted all of the next chapter to discussing the pitfalls involved, but there are a few issues that we can talk about already.

- **Don't base your null hypotheses on what you see in the data.** You are not allowed to look at the data first and then adjust your null hypothesis so that it will be rejected. When your sample value turns out to be $\hat{p} = 51.8\%$, with a standard deviation of 1%, don't form a null hypothesis like H_0: $p = 49.8\%$, knowing that you can reject it. You should always

Think about the situation you are investigating and make your null hypothesis describe the "nothing interesting" or "nothing has changed" scenario. No peeking at the data!

- **Don't base your alternative hypothesis on the data, either.** Again, you need to *Think* about the situation. Are you interested only in knowing whether something has *increased*? Then write a one-sided (upper-tail) alternative. Or would you be equally interested in a change in either direction? Then you want a two-sided alternative. You should decide whether to do a one- or two-sided test based on what results would be of interest to you, not what you see in the data.

- **Don't make your null hypothesis what you want to show to be true.** Remember, the null hypothesis is the status quo, the nothing-is-strange-here position a skeptic would take. You wonder whether the data cast doubt on that. You can reject the null hypothesis, but you can never "accept" or "prove" the null.

- **Don't forget to check the conditions.** The reasoning of inference depends on randomization. No amount of care in calculating a test result can recover from biased sampling. The probabilities we compute depend on the Independence Assumption. And the sample must be large enough to justify our use of a Normal model.

- **Don't accept the null hypothesis.** You may not have found enough evidence to reject it, but you surely have *not* proven it's true!

What Have We Learned?

Know how to formulate a null and an alternative hypothesis for a question of interest.

- The null hypothesis specifies a parameter and a (null) value for that parameter.
- The alternative hypothesis specifies a range of plausible values should we reject the null.

Be able to perform a hypothesis test for a proportion.

- The null hypothesis has the form $H_0: p = p_0$.
- We find the standard deviation of the sampling distribution of the sample proportion by assuming that the null hypothesis is true:

$$SD(\hat{p}) = \sqrt{\frac{p_0 q_0}{n}}$$

- We refer the statistic $z = \dfrac{\hat{p} - p_0}{SD(\hat{p})}$ to the standard Normal model.

Understand P-values.

- A P-value is the estimated probability of observing a statistic value at least as far from the (null) hypothesized parameter as the one we have actually observed, assuming the null hypothesis to actually be true.
- A small P-value indicates that the statistic we have observed would be unlikely were the null hypothesis true. That leads us to doubt the null.
- A large P-value just tells us that we have insufficient evidence to doubt the null hypothesis. In particular, it does not prove the null to be true.

Know the reasoning of hypothesis testing.

- State the **hypotheses**.
- Determine (and check assumptions for) the sampling distribution **model**.
- Calculate the test statistic—the **mechanics**.
- State your **conclusions and decisions**.

Be able to decide on a one-sided or two-sided alternative hypothesis, and justify your decision. Know that confidence intervals and hypothesis tests go hand in hand in helping us think about models.

- A hypothesis test assesses the plausibility of the value of a parameter value, by determining the strength of the contradictory evidence present in the sample with a *P*-value.
- A confidence interval shows the range of plausible values for the parameter.

Review of Terms

Null hypothesis
The claim being assessed in a hypothesis test. Usually, the null hypothesis is a statement of "no change from the traditional value," "no effect," "no difference," or "no relationship." For a claim to be a testable null hypothesis, it must specify a value for some population parameter that can form the basis for assuming a sampling distribution for a test statistic (p. 503).

Alternative hypothesis
Proposes what we should conclude if we find the null hypothesis to not be plausible (p. 504).

P-value
The probability of observing a value for a test statistic at least as far from the hypothesized parameter as the statistic value actually observed if the null hypothesis is true. A small P-value indicates either that the observation is improbable or that the probability calculation was based on incorrect assumptions. The assumed truth of the null hypothesis is the assumption under suspicion (p. 505).

One-proportion z-test
A test of the null hypothesis that the proportion parameter for a single population equals a specified value (H_0: $p = p_0$) by referring the statistic $z = \dfrac{\hat{p} - p_0}{SD(\hat{p})}$ to a standard Normal model (p. 508).

Effect size
The difference between the null hypothesis value and the true value of a model parameter (p. 509).

Two-sided alternative (Non-directional alternative)
An alternative hypothesis is two-sided (H_A: $p \neq p_0$) when we are interested in deviations in *either* direction away from the null hypothesized parameter value (p. 511).

One-sided alternative (Directional alternative)
An alternative hypothesis is one-sided (H_A: $p > p_0$ or H_A: $p < p_0$) when we are interested in deviations in *only* one direction away from the hypothesized parameter value (p. 511).

On the Computer   HYPOTHESIS TESTS FOR PROPORTIONS

Hypothesis tests for proportions are so easy and natural that many statistics packages don't offer special commands for them. Most statistics programs want to know the "success" and "failure" status for each case. Usually these are given as 1 or 0, but they might be category names like "yes" and "no." Often, we just know the proportion of successes, $\hat{p}$ and the total count, n. Computer packages don't usually deal naturally with summary data like this, but the statistics routines found on many graphing calculators do. These calculators allow you to test hypotheses from summaries of the data—usually, all you need to enter are the number of successes and the sample size.

In some programs, you can reconstruct the original values. But even when you have reconstructed (or can reconstruct) the raw data values, often you won't get exactly the same test statistic from a computer package as you would find working by hand. The reason is that when the packages treat the proportion as a mean, they make some approximations. The result is very close, but not exactly the same.

EXCEL

Inference methods for proportions are not part of the standard Excel tool set.

COMMENTS

For summarized data, type the calculation into any cell and evaluate it.

JMP

For a **categorical** variable that holds category labels, the **Distribution** platform includes tests and intervals of proportions. For summarized data,

- Put the category names in one variable and the frequencies in an adjacent variable.
- Designate the frequency column to have the role of **frequency**. Then use the **Distribution** platform.

COMMENTS

JMP uses slightly different methods for proportion inferences than those discussed in this text. Your answers are likely to be slightly different.

MINITAB

- Choose **Basic Statistics** from the **Stat** menu.
- Choose **1Proportion** from the Basic Statistics submenu.
- If the data are category names in a variable, assign the variable from the variable list box to the **Samples in columns** box.
- If you have summarized data, click the **Summarized Data** button and fill in the number of trials and the number of successes.
- Click the **Options** button and specify the remaining details.
- If you have a large sample, check **Use test and interval based on Normal distribution**.
- Click the **OK** button.

COMMENTS

When working from a variable that names categories, MINITAB treats the last category as the "success" category. You can specify how the categories should be ordered.

R

In library(stats):

- prop.test(X, n, p = NULL, alternative = c("two.sided," "less," "greater"), conf.level = 0.95, correct = FALSE) will test the hypothesis that $p = p_0$ against

various alternatives with $\alpha = 0.05$. For example, with 260 successes out of 500, at $\alpha = 0.05$, to test that $p = 0.5$ vs. $p \neq 0.5$, use:

prop.test(260,500,0.5,"two.sided," correct = FALSE)

SPSS

SPSS does not find hypothesis tests for proportions.

STATCRUNCH

To test a hypothesis for a proportion using summaries:
- Click on **Stat**.
- Choose **Proportions » One sample » with summary**.
- Enter the **Number of successes** (x) and **Number of observations** (n).
- Click on **Next**.
- Indicate **Hypothesis Test**, then enter the hypothesized **Null** proportion, and choose the **Alternative** hypothesis.
- Click on **Calculate**.

To test a hypothesis for a proportion using data:
- Click on **Stat**.
- Choose **Proportions » One sample » with data**.
- Choose the variable **Column** listing the **Outcomes**.
- Enter the outcome to be considered a **Success**.
- Click on **Next**.
- Indicate **Hypothesis Test**, then enter the hypothesized **Null** proportion, and choose the **Alternative** hypothesis.
- Click on **Calculate**.

TI-83/84 PLUS

To do the mechanics of a hypothesis test for a proportion:
- Select **5:1-PropZTest** from the **STAT TESTS** menu.
- Specify the hypothesized proportion.
- Enter the observed value of x.
- Specify the sample size.
- Indicate what kind of test you want: one-tail lower tail, two-tail, or one-tail upper tail.
- Calculate the result.

COMMENTS

Beware: When you enter the value of x, you need the *count*, not the percentage. The count must be a whole number. If the number of successes is given as a percent, you must first multiply np and round the result to obtain x.

Chapter Exercises

11. **Hypotheses** Write the null and alternative hypotheses you would use to test each of the following situations:
    a) The Premier of the province is concerned about his "negatives"—the percentage of provincial residents who express disapproval of his job performance. His political committee pays for a series of television ads, hoping that they can keep the negatives below 30%. They will use follow-up polling to assess the ads' effectiveness.
    b) Is a coin fair?
    c) Only about 20% of people who try to quit smoking succeed. Sellers of a motivational tape claim that listening to the recorded messages can help people quit.

12. **More hypotheses** Write the null and alternative hypotheses you would use to test each of the following situations.
    a) In the 1950s, only about 40% of high school graduates went on to college or university. Has the percentage changed?
    b) Twenty percent of cars of a certain model needed costly transmission work after being driven between 50 000 and 100 000 miles. The manufacturer hopes

that a redesign of a transmission component will solve this problem.
    c) We field-test a new flavour of soft drink, planning to market it only if we are sure that over 60% of the people tested like the flavour.

13. **Negatives** After the political ad campaign described in Exercise 11a), pollsters check the Premier's negatives. They test the hypothesis that the ads produced no change against the alternative that the negatives are now below 30% and find a P-value of 0.22. Which conclusion is appropriate? Explain.
    a) There's a 22% chance that the ads worked.
    b) There's a 78% chance that the ads worked.
    c) There's a 22% chance that the poll they conducted is correct.
    d) There's a 22% chance that natural sampling variation could produce poll results like these if there's really no change in public opinion.

14. **Dice** The seller of a loaded die claims that it will favour the outcome 6. We don't believe that claim, and roll the

die 200 times to test an appropriate hypothesis. Our P-value turns out to be 0.03. Which conclusion is appropriate? Explain.

a) There's a 3% chance that the die is fair.

b) There's a 97% chance that the die is fair.

c) There's a 3% chance that a loaded die could randomly produce the results we observed, so it's reasonable to conclude that the die is fair.

d) There's a 3% chance that a fair die could randomly produce the results we observed, so it's reasonable to conclude that the die is loaded.

15. **Relief** A company's old antacid formula provided relief for 70% of the people who used it. The company tests a new formula to see if it is better and gets a P-value of 0.27. Is it reasonable to conclude that the new formula and the old one are equally effective? Explain.

16. **Cars** A survey investigating whether the proportion of today's Grade 12 students who own their own cars is higher than it was a decade ago finds a P-value of 0.017. Is it reasonable to conclude that more Grade 12 students have cars? Explain.

17. **He cheats!** A friend of yours claims that when he tosses a coin he can control the outcome. You are skeptical and want him to prove it. He tosses the coin, and you call heads; it's tails. You try again and lose again.

a) Do two losses in a row convince you that he really can control the toss? Explain.

b) You try a third time, and again you lose. What's the probability of losing three tosses in a row if the process is fair?

c) Would three losses in a row convince you that your friend cheats? Explain.

d) How many times in a row would you have to lose to be pretty sure that this friend really can control the toss? Justify your answer by calculating a probability and explaining what it means.

18. **Candy** Someone hands you a box of a dozen chocolate-covered candies, telling you that half are vanilla creams and the other half are peanut butter. You pick candies at random and discover that the first three you eat are all vanilla.

a) If there really were six vanilla and six peanut butter candies in the box, what is the probability that you would have picked three vanillas in a row?

b) Do you think there really might have been six of each? Explain.

c) Would you continue to believe that half are vanilla if the fourth one you try is also vanilla? Explain.

19. **Smartphones** Many people have trouble setting up all the features of their smartphones, so a company has developed what it hopes will be easier instructions. The goal is to have at least 96% of customers succeed. The company tests the new system on 200 people, of whom 188 were successful. Is this strong evidence that the new system

fails to meet the company's goal? A student's test of this hypothesis is shown. How many mistakes can you find?

$H_0: \hat{p} = 0.96$

$H_A: \hat{p} \neq 0.96$

SRS, $0.96(200) > 10$

$\dfrac{188}{200} = 0.94; SD(\hat{p}) = \sqrt{\dfrac{(0.94)(0.06)}{200}} = 0.017$

$z = \dfrac{0.96 - 0.94}{0.017} = 1.18$

$P = P(z > 1.18) = 0.12$

There is strong evidence that the new instructions do not work.

20. **Got milk?** In November 2001, the *Ag Globe Trotter* newsletter reported that 90% of adults drink milk. A regional farmers' organization planning a new marketing campaign across its multicounty area polls a random sample of 750 adults living there. In this sample, 657 people said that they drink milk. Do these responses provide strong evidence that the 90% figure is not accurate for this region? Correct the mistakes you find in a student's attempt to test an appropriate hypothesis.

$H_0: \hat{p} = 0.9$

$H_A: \hat{p} < 0.9$

SRS, $750 > 10$

$\dfrac{657}{750} = 0.876; SD(\hat{p}) = \sqrt{\dfrac{(0.88)(0.12)}{750}} = 0.012$

$z = \dfrac{0.876 - 0.90}{0.012} = -2$

$P = P(z > -2) = 0.977$

There is more than a 97% chance that the stated percentage is correct for this region.

21. **Dowsing** In a rural area, only about 30% of the wells that are drilled find adequate water at 100 metres deep or less. A local man claims to be able to find water by "dowsing"—using a forked stick to indicate where the well should be drilled. You check with 80 of his customers and find that 27 have wells less than 100 metres deep. What do you conclude about his claim?

a) Write appropriate hypotheses.

b) Check the necessary assumptions.

c) Perform the mechanics of the test. What is the P-value?

d) Explain carefully what the P-value means in this context.

e) What is your conclusion?

22. **Abnormalities** In the 1980s, it was generally believed that congenital abnormalities affect about 5% of the nation's children. Some people believe that the increase in the number of chemicals in the environment in recent

years has led to an increase in the incidence of abnormalities. A recent study examined 384 children and found that 46 of them showed signs of an abnormality. Is this strong evidence that the risk has increased?
a) Write appropriate hypotheses.
b) Check the necessary assumptions.
c) Perform the mechanics of the test. What is the P-value?
d) Explain carefully what the P-value means in this context.
e) What is your conclusion?
f) Do environmental chemicals cause congenital abnormalities?

23. **Educated** Statistics Canada monitors many aspects of secondary and post-secondary education in the regular census. The 2011 National Household Survey (NHS) released in June 2013 showed that 64.1% of the total population aged 25–64 had a post-secondary degree. In comparison, the 2006 census showed that 60.7% held a similar qualification.

   Is it appropriate to conduct a hypothesis test to determine if there is a change in the proportion of degrees awarded between 2006 and 2011? Discuss why or why not.

24. **More educated** To test the figures quoted in the previous question, you run an SRS of 4000 people aged 25–64 to see if the proportion with a post-secondary degree in 2013 has changed since 2011. You find that 2630 have a degree.
a) Write appropriate hypotheses.
b) Check the assumptions and conditions.
c) Perform the test and find the P-value.
d) State your conclusion.
e) Do you think this difference is meaningful? Explain.

25. **Contributions, please, part II** In Exercise 23 of Chapter 15, you learned that charitable organizations like the MS Society send free mailing labels and greeting cards to potential donors on their list and ask for a voluntary contribution. Suppose they test a new campaign by sending letters to a random sample of 100 000 potential donors and receive 4781 donations. A staff member worries that the contribution rate will be lower than their past average rate of 5% if they run this campaign as currently designed.
a) What are the hypotheses?
b) Are the assumptions and conditions for inference met?
c) Do you think the rate would drop? Explain.

26. **Take the offer, part II** In Exercise 24 of Chapter 15, you learned that a major credit card company is planning a new offer for their current cardholders. They will give double airline miles on purchases for the next six months if the cardholder goes online and registers for this offer. To test the effectiveness of this campaign, the company recently sent out offers to a random sample of 50 000 cardholders. Of those, 1184 registered. A staff member suspects that the success rate for the full campaign will be comparable to the standard 2% rate that they are used to seeing in similar campaigns. What do you predict?

a) What are the hypotheses?
b) Are the assumptions and conditions for inference met?
c) Do you think the rate would change if they use this fundraising campaign? Explain.

27. **Law school** According to the Ontario Universities' Application Centre, in the year 2012, 29.6% of law school applicants were accepted to an Ontario school.[11] A training program claims that 83 of the 240 students trained in 2012 were admitted to law school. You can safely consider these trainees to be representative of the population of law school applicants. Has this program demonstrated a real improvement over the provincial average?
a) What are the hypotheses?
b) Check the conditions and find the P-value.
c) Would you recommend this program based on what you see here? Explain.

28. **Rural med school applicants** In 2002, 2246 students from Ontario applied to the province's medical schools; in 2003, 2702 applied. Of the two-year total (4948), there were 4588 applicants from urban areas and 360 from rural areas.[12] GPAs and MCAT scores were similar for the urban and rural applicants, who had similar acceptance rates (about 20% of applicants). Since Ontario's population is 13% rural, is there reason for concern about a lack of rural applicants for medical school? Are rural students underrepresented among the applicants, or are we just seeing some natural random variation in the results for this two-year period?
a) What are the hypotheses?
b) Verify that the conditions are satisfied for a test, and find the P-value.
c) Should we be concerned about a lack of rural applicants to medical school?

29. **Pollution** A company with a fleet of 150 cars found that the emissions systems of seven out of the 22 they tested failed to meet pollution control guidelines. Is this strong evidence that more than 20% of the fleet might be out of compliance? Test an appropriate hypothesis and state your conclusion. Be sure the appropriate assumptions and conditions are satisfied before you proceed.

30. **Scratch and dent** An appliance manufacturer stockpiles washers and dryers in a large warehouse for shipment to retail stores. Sometimes, in handling, the appliances get damaged. Even though the damage may be minor, the company must sell those machines at drastically reduced prices. The company goal is to keep the level of damaged machines below 2%. One day an inspector randomly checks 60 washers and finds that five of them have scratches or dents. Is this strong evidence that the warehouse is failing to meet the company goal? Test an appropriate hypothesis and state your conclusion. Be

---

[11]www.ouac.on.ca/statistics/law-school-application-statistics/
[12]www.pubmedcentral.nih.gov/articlerender.fcgi?artid=1479467

sure the appropriate assumptions and conditions are satisfied before you proceed.

31. Twins In 2001, a U.S. national vital statistics report indicated that about 3% of all births produced twins. Is the rate of twin births the same among very young mothers? Data from a large city hospital found that only seven sets of twins were born to 469 teenage girls. Test an appropriate hypothesis and state your conclusion. Be sure the appropriate assumptions and conditions are satisfied before you proceed.

32. Hockey 2013 During the 2013 National Hockey League playoffs, the home team won 59 of the 86 games. Is this strong evidence of a home-ice advantage in professional hockey? Test an appropriate hypothesis and state your conclusion. Be sure the appropriate assumptions and conditions are satisfied before you proceed.

33. WebZine A magazine is considering the launch of an online edition. It plans to go ahead only if it's convinced that more than 25% of current readers would subscribe. The magazine contacted a simple random sample of 500 current subscribers, and 137 of those surveyed expressed interest. What should the company do? Test an appropriate hypothesis and state your conclusion. Be sure the appropriate assumptions and conditions are satisfied before you proceed.

34. Seeds Garden centre staff want to store leftover packets of vegetable seeds for sale the following spring, but are concerned that the seeds may not germinate at the same rate a year later. The manager finds a packet of last year's green bean seeds and plants them as a test. Although the packet claims a germination rate of 92%, only 171 of 200 test seeds sprout. Is this evidence that the seeds have lost viability during a year in storage? Test an appropriate hypothesis and state your conclusion. Be sure the appropriate assumptions and conditions are satisfied before you proceed.

35. Female executives A company is criticized because only 13 of 43 people in executive-level positions are women. The company explains that although this proportion is lower than it might wish, it's not surprising given that only 40% of all its employees are women. What do you think? Test an appropriate hypothesis and state your conclusion. Be sure the appropriate assumptions and conditions are satisfied before you proceed.

36. Jury Census data for a certain municipality show that 19% of the adult residents are South Asian. Suppose 72 people are called for jury duty and only nine of them are South Asian. Does this apparent underrepresentation of South Asians call into question the fairness of the jury selection system? Explain.

37. Hockey injuries and relative age Is there a relationship between (relative) age and injury prevalence in Canadian youth hockey? Below are summary data from the Canadian Hospitals Injury Reporting and Prevention Program (for boys 10–15 years of age, 1995–2002). Quartile 1 included those with January, February, and March birth dates and quartile 4 included those with October, November, and December birth dates; hence those born in quartile 1 were relatively older than their peers born in quartile 4. The other two quartiles do not affect any of the conclusions that may be drawn from this study, and hence are omitted to simplify things.

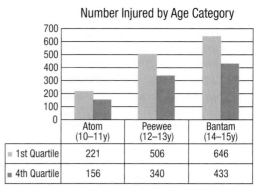

	Atom (10–11y)	Peewee (12–13y)	Bantam (14–15y)
■ 1st Quartile	221	506	646
■ 4th Quartile	156	340	433

a) Test, at the Atom level, whether an injured youth is equally likely to be from either quartile. If not, then there may be a "relative age" effect on likelihood of injury. "Equally likely" means that half of injured players in the two quartiles (377 in total) should be found in quartile 4, on average.

b) Repeat part a) for the Bantam level injured players.

c) The question of interest here is whether relative age influences chance of injury, but researchers have also found that there is some selection bias for relatively older players. How does this affect conclusions drawn from these data about the question of interest? What additional data would you try to obtain to more accurately assess the increased risk for older players?

38. Acid rain A study of the effects of acid rain on trees in the Hopkins Forest shows that 25 of 100 trees sampled exhibit some sort of damage from acid rain. This rate seemed to be higher than the 15% quoted in a recent *Environmetrics* article on the average proportion of damaged trees in the Northeast U.S. Does the sample suggest that trees in the Hopkins Forest are more susceptible than trees from the rest of the region? Comment, and write up your own conclusions based on an appropriate confidence interval as well as a hypothesis test. Include any assumptions you made about the data.

39. Lost luggage An airline's public relations department says that the airline rarely loses passengers' luggage. It further claims that on those occasions when luggage is lost, 90% is recovered and delivered to its owner within 24 hours. A consumer group that surveyed a large number of air travellers found that only 103 of 122 people who lost luggage on that airline were reunited with the missing items by the next day. Does this cast doubt on the airline's claim? Explain.

40. TV ads A start-up company is about to market a new computer printer. It decides to gamble by running commercials during the Super Bowl. The company hopes that name recognition will be worth the high cost of the ads. The goal of the company is that over 40% of the public recognize its brand name and associate it with computer equipment. The day after the game, a pollster contacts 420 randomly chosen adults and finds that 181 of them know that this company manufactures printers. Would you recommend that the company continue to advertise during Super Bowls? Explain.

41. John Wayne Like many others, the famous actor John Wayne died of cancer. But is there more to this story? In 1955, Wayne was in Utah shooting the film *The Conqueror*. Across the state line, in Nevada, the United States military was testing atomic bombs. Radioactive fallout from those tests drifted across the filming location. A total of 46 of the 220 people working on the film eventually died of cancer. Cancer experts estimate that one would expect only about 30 cancer deaths in a group this size.
a) Is the death rate observed in the movie crew unusually high?
b) Does this prove that exposure to radiation increases the risk of cancer?
Source: https://en.wikipedia.org/wiki/The_Conqueror_(film)

42. TOEFL scores Of those undergraduate-level students who took the 2008 paper-based TOEFL exam, 47% earned scores of 540 or higher. One teacher wondered if the performance of her students was different. Thirty of her 54 students achieved scores of 540 or better. Does she have cause to brag? Explain.

43. Obama or Harper? A poll by the Strategic Counsel, conducted June 12–22, 2008 (before Obama was elected President and while Stephen Harper was Prime Minister) surveyed random samples of 1000 Canadians and 1000 Americans.[13] When asked about which political figure they admired most, the top two responses from Canadians were:

> Barack Obama: 26%
>
> Stephen Harper: 21%

Would it be fair to say that more Canadians would choose Obama as their most admired political figure than would choose Harper (at the time of this survey)? Do a (two-sided) test, after throwing out the 53% (round to 530) who chose someone else.

44. Your health care system or mine? In the same poll mentioned in Exercise 43, it was reported that 45% of Americans felt Canada had the superior healthcare system, while 42% thought the United States should stick with its own. Would it be fair to say that more Americans preferred the Canadian healthcare system over their own (at the time of this survey)? Do a (two-sided) test, after throwing out the 13% (round to 130) who were undecided. (Considering the misinformation in some U.S. media about Canadian health care, the 45% figure is surprisingly high!)

[13]www.ctv.ca/servlet/ArticleNews/story/CTVNews/20080629/poll_us_canada_080629/20080629?hub=Politics

 Just Checking ANSWERS

1. You can't conclude that the null hypothesis is true. You can conclude only that the experiment was unable to disprove the null hypothesis. They were unable, on the basis of 12 patients, to show that aspirin was effective.

2. The null hypothesis is H_0: $p = 0.75$.

3. A P-value of 0.0001 strongly contradicts the assumed null hypothesis, that is, there is strong evidence that the improved version of the drug gives relief to a higher proportion of patients.

4. The parameter of interest is the proportion, p, of all delinquent customers who will pay their bills.

$$H_0: p = 0.30 \text{ and } H_A: p > 0.30.$$

5. The very low P-value leads us to reject the null hypothesis. There is strong evidence that the DVD is more effective in getting people to start paying their debts than just sending a letter had been.

6. All we know is that there is strong evidence to suggest that $p > 0.30$. We don't know how much higher than 30% the new proportion is. We'd like to see a confidence interval to see if the new method is worth the cost.

More About Tests

1 Know:

- The reasoning and mechanics of hypothesis testing
- How to formulate null and alternative hypotheses
- How to interpret P-values

2 Know how to make decisions based on a predetermined *alpha level*

3 Be able to test hypotheses using critical values

4 Be aware of the two types of decision errors: Type I and Type II errors

5 Understand the concept of power and the relationship between power and sample size

6 Understand the relationship between a confidence interval and a test

Connections

A news headline reports a "statistically significant" increase in global temperatures. Another says that studies have found no "statistically significant" benefits of taking vitamin C to prevent colds. What does *significance* really mean?

In the two previous chapters, we started discussing statistical inference—which enables us to generalize beyond the sample at hand. In this chapter, we'll look at hypothesis testing in more depth, and in future chapters, we'll expand

… They make things admirably plain, But one hard question will remain: If one hypothesis you lose, Another in its place you choose …

—*James Russell Lowell, Credidimus Jovem Regnare*

Are you planning on becoming a medical doctor (and reading this book because of the Statistics requirement)? Most Canadian doctors choose to work in the large urban areas, while rural areas often have a shortage of doctors. Ready access to health care for rural populations depends on the availability of family physicians, so increasing interest in rural medicine is a worthy challenge for policy makers. Choosing family practice and a rural practice location are complex decisions, but one factor associated with these choices is whether a person comes from a rural background.

In 2002–2003, the Ontario Medical School Application Service (OMSAS) requested the postal codes of the residences where applicants lived when they attended secondary school. These postal codes were linked to community locations that could be categorized as either rural or urban, using Statistics Canada's definition of rural areas and small towns. Applicants of rural origin were defined by having attended secondary school while residing in communities with core populations of less than 10 000.

In Ontario, 13% of the population is rural, yet in 2002–2003, of 4948 applicants to Ontario medical schools (who attended high school in Ontario), only 360 applicants were of rural origin, or 7.3%[1]. Is this evidence of a shortage of rural applicants, or are we just observing the natural fluctuation of such statistics?

17.1 Choosing the Hypotheses

How do we choose the null hypothesis? The appropriate null arises directly from the context of the problem. It is dictated not by the data, but by the situation. One good way to identify both the null and alternative hypotheses is to think about why the study is being done and what we hope to learn from the test. In this case we are interested in seeing whether the proportion of rural applicants is below the rural population proportion, but our null hypothesis is what we want to disprove, namely that the proportion equals the population proportion of 13%. The alternative is what we hope to prove—that the proportion is too low.

To write a null hypothesis, identify a parameter and choose a null value that relates to the question at hand. Even though the null usually means no difference or no change, you can't automatically interpret "null" to mean zero. A claim that there are no rural applicants would be absurd. The null hypothesis for the Ontario study is that the true rural application

---

[1]www.ncbi.nlm.nih.gov/pmc/articles/PMC1479467/

the discussion of inference to more complicated studies.

All the hypothesis tests will boil down to the same question: "Is an observed difference or pattern too large to be attributed to chance?" We measure 'how large' by putting our sample results in the context of a sampling distribution model—often the Normal model—but we'll see other sampling models (such as the 't') very soon, like next chapter!

Who	Ontario medical school applicants, who attended Ontario secondary schools
What	Percentage of medical school applicants who are of rural origin
When	2002–2003
Where	Ontario
Why	Assessment of the shortage of rural applicants to Ontario medical schools

rate properly reflect the rural population proportion at $p = 0.13$. You need to find the value for the parameter in the null hypothesis from the context of the problem.

There is a temptation to state your *claim* as the null hypothesis. As we have seen, however you cannot prove a null hypothesis true any more than you can prove a defendant innocent. So, it makes more sense to use what you want to show as the *alternative*. This way, if the data support your claim, the P-value will be small. When you then reject the null, you will be left with what you want to show.

For Example WRITING HYPOTHESES

The diabetes drug Avandia® was approved to treat Type 2 diabetes in 1999. But in 2007 an article in the *New England Journal of Medicine* (NEJM)[2] raised concerns that the drug might carry an increased risk of heart attack. This study combined results from a number of other separate studies to obtain an overall sample of 4485 diabetes patients taking Avandia®. People with Type 2 diabetes are known to have about a 20.2% chance of suffering a heart attack within a seven-year period. According to the article's author, Dr. Steven E. Nissen,[3] the risk found in the NEJM study was equivalent to a 28.9% chance of heart attack over seven years. The FDA is the U.S. government agency responsible for relabelling Avandia® to warn of any increased risks. Although the statistical methods they use are more sophisticated, we can get an idea of their reasoning with the tools we have learned.

QUESTION: What null hypothesis and alternative hypothesis about seven-year heart attack risk would you test? Explain.

$$H_0: p = 0.202$$

$$H_A: p > 0.202$$

ANSWER: The parameter of interest is the proportion of diabetes patients suffering a heart attack in seven years. The FDA is concerned only with whether Avandia® *increases* the seven-year risk of heart attacks above the baseline value of 20.2%, so a one-sided upper-tail test is appropriate.

One-sided or two? In the 1930s, a series of experiments was performed at Duke University in an attempt to see whether humans were capable of extrasensory perception, or ESP. Psychologist Karl Zener designed a set of cards with five symbols, later made infamous in the movie *Ghostbusters*:

In the experiment, the "sender" selects one of the five cards at random from a deck and then concentrates on it. The "receiver" tries to determine which card it is. If we let p be the proportion of correct responses, what's the null hypothesis? The null hypothesis is that ESP makes no difference. Without ESP, the receiver would just be guessing, and since there are five possible responses, there would be a 20% chance of guessing each card correctly. So, H_0 is $p = 0.20$. What's the alternative? It seems that it should be $p > 0.20$, a one-sided alternative. But some ESP researchers have expressed the claim that if the proportion guessed were much lower than expected, that would show an "interference" and should be considered evidence for ESP as well. So they argue for a two-sided alternative.

---

[2]Steven E. Nissen and Kathy Wolski, 2007, "Effect of rosiglitazone on the risk of myocardial infarction and death from cardiovascular causes," *NEJM*: 356.

[3]Interview reported in the *New York Times* (May 26, 2007).

Section 17.1 Exercises

1. **Parameters and hypotheses** For each of the following situations, define the parameter (proportion or mean) and write the null and alternative hypotheses in terms of parameter values. Example: We want to know if the proportion of up days in the stock market is 50%. Answer: Let $p = $ the proportion of up days. H_0: $p = 0.5$ vs. H_A: $p \neq 0.5$.

   a) A casino wants to know if their slot machine really delivers the 1 in 100 win rate that it claims.

   b) Last year, customers spent an average of $35.32 per visit to the company's website. Based on a random sample of purchases this year, the company wants to know if the mean this year has changed.

   c) A pharmaceutical company wonders if their new drug has a cure rate different from the 30% reported by the placebo.

   d) A bank wants to know if the percentage of customers using their website has changed from the 40% that used it before their system crashed last week.

2. **Hypotheses and parameters** As in Exercise 1, for each of the following situations, define the parameter and write the null and alternative hypotheses in terms of parameter values.

   a) Seat-belt compliance in Massachusetts was 65% in 2008. The state wants to know if it has changed.

   b) Last year, a survey found that 45% of the employees were willing to pay for on-site day care. The company wants to know if that has changed.

   c) Regular card customers have a default rate of 6.7%. A credit card bank wants to know if that rate is different for their Gold card customers.

17.2 How to Think About P-Values

WHICH CONDITIONAL?

Suppose that as a political science major you are offered the chance to be an intern at the Office of the Prime Minister. There would be a very high probability that next summer you'd be in Ottawa. That is, $P($Ottawa | Intern$)$ would be high. But if we find a student in Ottawa, is it likely that he's an intern at the Office of the Prime Minister? Almost surely not; $P($Intern | Ottawa$)$ is low. You can't switch around conditional probabilities. The P-value is $P($data | $H_0)$. We might wish we could report $P(H_0 |$ data$)$, but these two quantities are NOT the same.

"The wise man proportions his belief to the evidence."

—*David Hume,*
Enquiry Concerning Human
Understanding, 1748

A P-value is actually a conditional probability. It tells us the probability of getting results at least as unusual as the observed statistic, *given* that the null hypothesis is true. We can write P-value $= P($observed statistic value [or even more extreme] | $H_0)$.

Writing the P-value this way helps to make clear that the P-value is *not* the probability that the null hypothesis is true. It is a probability about the data. Let's say that again:

The P-value is not the probability that the null hypothesis is true.

The P-value is not even the conditional probability that the null hypothesis is true given the data. We would write that probability as $P(H_0 |$ observed statistic value$)$. This is a conditional probability but in reverse. It would be nice to know this, but it's impossible to calculate without making additional assumptions. As we saw in Chapter 12, reversing the order in a conditional probability is difficult, and the results can be counterintuitive.

We can find the P-value, $P($observed statistic value | $H_0)$, because H_0 gives the parameter values that we need to find the required probability. But there's no direct way to find $P(H_0 |$ observed statistic value$)$.[4] As tempting as it may be to say that a P-value of 0.03 means there's a 3% chance that the null hypothesis is true, that just isn't right. All we can say is that, given the null hypothesis, there's a 3% chance of observing the statistic value that we have actually observed (or one even more inconsistent with the null value).

What to Do with a Small P-Value

We know that a small P-value means that the result we just observed is unlikely to occur if the null hypothesis is true. So we have evidence against the null hypothesis. An even smaller P-value implies stronger evidence against the null hypothesis, but it doesn't mean that the null hypothesis is "less true" (see "How Guilty Is the Suspect" on page 532).

---

[4]The approach to statistical inference known as Bayesian Statistics addresses the question in just this way, but it requires more advanced mathematics and more assumptions. See p. 390 for more about the founding father of this approach.

How small the P-value has to be for you to reject the null hypothesis depends on a lot of things, not all of which can be precisely quantified. Your belief in the null hypothesis will influence your decision. Your trust in the data, in the experimental method if the data come from a planned experiment, or in the survey protocol if the data come from a designed survey, all influence your decision. The P-value should serve as a measure of the strength of the evidence against the null hypothesis, but should never serve as a hard and fast rule for decisions. You have to take that responsibility on yourself.

As a review, let's look at the medical school example from the chapter opener. Do students of rural origin in Ontario apply to medical schools less frequently than students of urban origin? What is the evidence?

Step-by-Step Example ANOTHER ONE-PROPORTION Z-TEST

Question: Are rural students under-represented in Ontario medical school applications?

THINK ➡ Plan State the problem and discuss the variables and the *W*'s.	I want to know if Ontario students of rural origin are under-represented in medical school applications. I have data from OMSAS showing 360 of 4948 applicants were of rural origin (using postal codes provided by applicants).
Hypotheses The null hypothesis is established by the proportion of the population that is rural. The study was concerned with possible under-representation of rural applicants, so they'll want to know if too few applicants are of rural origin. Hence a one-sided alternative is appropriate.	The percentage of the population that is rural is 13% so I'll use that as my null hypothesis value. The alternative is one-sided because I'm only interested in seeing if the proportion of rural applicants is less than the rural population percentage of 13%. $$H_0: p = 0.13$$ $$H_A: p < 0.13$$
SHOW ➡ Model Check the conditions.	✓ **Independence Assumption:** The data are for applicants during a two-year period, who should pretty much be independent of one another.
	✓ **Randomization Condition:** No randomization was applied, but we may consider these applicants to be reasonably representative of Ontario medical school applicants around this general time.
	✓ **Success/Failure Condition:** We'd expect $np = 4948(0.13) = 643$ rural applicants and $nq = 4948(0.87) = 4305$ urban applicants. Both are at least 10.
Specify the sampling distribution model and name the test.	The conditions are reasonably satisfied, so I will use the Normal model and perform **a one-proportion z-test.**

SHOW ➡ **Mechanics** Find the standard deviation of the sampling model using the hypothesized proportion. (Given the large sample size here, we should expect the SD of $\hat{p}$ to be quite small)	There were 360 rural applicants among the 4948 med school applicants. $$\hat{p} = \frac{360}{4948} = 0.0728$$ $$SD(\hat{p}) = \sqrt{\frac{p_0 q_0}{n}}$$
Find the z-score for the observed proportion.	$$= \sqrt{\frac{(0.13)(0.87)}{4948}} = 0.0048$$ $$z = \frac{\hat{p} - p_0}{SD(\hat{p})}$$ $$= \frac{0.0728 - 0.13}{0.0048} \approx -11.92$$
Make a picture. Sketch a Normal model centred at the hypothesized percentage of rural applicants rate of 13%. This is a lower-tail test, so shade the region to the left of the observed rate. Given this z-score, the P-value is obviously very low.	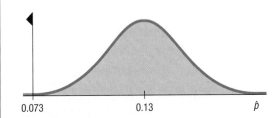 0.073 0.13 $\hat{p}$ The observed percentage of rural applicants rate is almost 12 standard deviations below the rural percentage of the Ontario population. The corresponding P-value is much less than 0.0000001.
TELL ➡ **Conclusion** Link the P-value to your decision about the null hypothesis, and then state your conclusion in context.	The P-value says that if the true overall percentage of rural applicants were 13%, the probability of observing a rural application percentage no higher than 7.3% in a sample like this is less than 1 chance in 1000, so I reject the null hypothesis. There is strong evidence of under-representation of rural applicants to Ontario med schools.

The P-value in the example above is extremely small—less than 0.0000001. That's strong evidence to suggest under-representation of rural applicants. But it doesn't say that the percentage of rural applicants was "a lot lower than the rural population percentage." After all, the sample size was very large here, so we can detect even small differences from the null value. To assess the difference in practical terms, we should also construct a confidence interval:

$$\hat{p} \pm z^* \sqrt{\frac{\hat{p}\hat{q}}{n}} = 0.0728 \pm 1.96(0.0048) = (0.063, 0.082)$$

(using 95% confidence).

There is strong evidence that the rural application rate is not 13%, but the small P-value by itself says nothing about how much lower the rate might be. The confidence interval provides that information; the rural application rate seems to be a bit below 10% of total applications. Whether policy makers consider this an important difference in the rate of applicants is a judgment that depends on the situation, but not on the P-value. Canadian literature on choosing rural practice is limited, but we know more rural family doctors are needed. Have you (future doctors) considered the benefits of rural practice?

HOW GUILTY IS THE SUSPECT?

We might like to know $P(H_0 \mid data)$, but when you think about it, we can't talk about the probability that the null hypothesis is true. The null is not a random event, so either it is true or it isn't. The data, however, are random in the sense that if we were to repeat a randomized experiment or draw another random sample, we'd get different data and expect to find a different statistic value. So we can talk about the probability of the data given the null hypothesis, and that's the P-value.

But it does make sense that the smaller the P-value, the more confident we can be in declaring that we doubt the null hypothesis. Think again about the jury trial. Our null hypothesis is that the defendant is innocent. Then the evidence starts rolling in. A car the same colour as his was parked in front of the bank. Well, there are lots of cars that colour. The probability of that happening (given his innocence) is pretty high, so we're not persuaded that he's guilty. The bank's security camera showed the robber was male and about the defendant's height and weight. Hmm. Could that be a coincidence? If he's innocent, then it's a little less likely that the car and description would *both* match, so our P-value goes down. We're starting to question his innocence a little. Witnesses said the robber wore a blue jacket just like the one the police found in a garbage can behind the defendant's house. Well, if he's innocent, then that doesn't seem very likely, does it? If he's really innocent, the probability that all of these could have happened is getting pretty low. Now our P-value may be small enough to be called "beyond a reasonable doubt" and lead to a conviction. Each new piece of evidence strains our skepticism a bit more. The more compelling the evidence—the more *unlikely* it would be were he innocent—the more convinced we become that he's guilty.

But even though it may make us more confident in declaring him guilty, additional evidence does not make him any guiltier. Either he robbed the bank or he didn't. Additional evidence (like the teller picking him out of a police lineup) just makes us more confident that we did the right thing when we convicted him. The lower the P-value, the more comfortable we feel about our decision to reject the null hypothesis, but the null hypothesis doesn't get any more false.

For Example THINKING ABOUT THE P-VALUE

RECAP: A *New England Journal of Medicine* paper reported that the seven-year risk of heart attack in diabetes patients taking the drug Avandia® was increased from the baseline of 20.2% to an estimated risk of 28.9% and said the P-value was 0.03.

QUESTION: How should the P-value be interpreted in this context?

ANSWER: The P-value $= P(\hat{p} \geq 28.9\% \mid p = 20.2\%)$. That is, it's the probability of seeing such a high heart attack rate among the people studied if, in fact, taking Avandia® really didn't increase the risk at all [Assuming a one-sided alternative].

What to Do with a High P-Value

Therapeutic touch (TT), taught in some schools of nursing, is a therapy in which the practitioner moves her hands near, but does not touch, a patient in an attempt to manipulate a "human energy field." Therapeutic touch practitioners believe that by adjusting this field they can promote healing. However, no instrument has ever detected a human energy field, and no experiment has ever shown that TT practitioners can detect such a field.

In 1998, the *Journal of the American Medical Association* published a paper reporting work by a then nine-year-old girl.[5] She had performed a simple experiment in which she

---

[5] L. Rosa, E. Rosa, L. Sarner, and S. Barrett, "A Close Look at Therapeutic Touch," *JAMA* 279(13) [1 April 1998]: 1005–1010.

Paul Velleman

challenged 15 TT practitioners to detect whether her unseen hand was hovering over their left or right hand (selected by the flip of a coin).

The practitioners "warmed up" with a period during which they could see the experimenter's hand, and each said that they could detect the girl's human energy field. Then a screen was placed so that the practitioners could not see the girl's hand, and they attempted 10 trials each. Overall, of 150 trials, the TT practitioners were successful only 70 times—a success proportion of 46.7%.

The null hypothesis here is that the TT practitioners were just guessing. If that were the case, since the hand was chosen using a coin flip, the practitioners would guess correctly 50% of the time. So the null hypothesis is that $p = 0.5$ and the alternative that they could actually detect a human energy field is (one-sided) $p > 0.5$.

What would constitute evidence that they weren't guessing? Certainly, a very high proportion of correct guesses out of 150 would convince most people. Exactly how high the proportion of correct guesses has to be for you to reject the null hypothesis depends on how small a P-value you need to be convinced (which, in turn, depends on how often you're willing to make mistakes—a topic we'll discuss later in the chapter).

But let's look again at the TT practitioners' proportion. Does it provide any evidence that they weren't guessing? The proportion of correct guesses is 46.7%—that's *less* than the hypothesized value, not greater! When we find $SD(\hat{p}) = 0.041$ (or 4.1%), we can see that 46.7% is almost 1 SD *below* the hypothesized proportion:

$$SD(\hat{p}) = \sqrt{\frac{p_0 q_0}{n}} = \sqrt{\frac{(0.5)(0.5)}{150}} \approx 0.041$$

The observed proportion, $\hat{p}$ is 0.467.

$$z = \frac{\hat{p} - p_0}{SD(\hat{p})} = \frac{0.467 - 0.5}{0.041} = -0.805$$

The observed success rate is 0.805 standard deviations below the hypothesized mean.

$$P = P(z > -0.805) = 0.790$$

If the practitioners had been highly successful, we would have seen a low P-value. In that case, we would then have concluded that they could actually detect a human energy field.

But that's not what happened. What we observed was a $\hat{p} = 0.467$ success rate. The P-value for this proportion is greater than 0.5 because the observed value is on the "wrong" side of the null hypothesis value. Wrong, that is, if you wanted to find evidence against the null hypothesis. To convince us, the practitioners should be doing better than guessing, not worse!

Obviously, we won't be rejecting the null hypothesis; for us to reject it, the P-value would have to be quite small. But a P-value of 0.790 seems so big it is almost strange. With a success rate even lower than chance, we could have concluded right away that we have no evidence for rejecting H_0. In fact, for a one-sided test, any time our test statistic is on the "wrong" side of the hypothesis, we know immediately that the P-value is at least 0.50 and we know that we have no evidence to reject the null hypothesis.

Big P-values just mean that what we've observed isn't surprising. That is, the results are in line with our assumption that the null hypothesis models the world, so we have no reason to reject it. A big P-value doesn't prove that the null hypothesis is true, but it certainly offers no evidence that it's *not* true. When we see a large P-value, all we can say is that we "don't reject the null hypothesis."

For Example   MORE ABOUT P-VALUES

RECAP: The question of whether the diabetes drug Avandia® increased the risk of heart attack was raised by a study in the *New England Journal of Medicine*. This study estimated the seven-year risk of heart attack to be 28.9% and reported a P-value of 0.03 for a test of whether this risk was higher than the baseline seven-year risk of 20.2%. An earlier study (the ADOPT study) had estimated the seven-year risk to be 26.9% and reported a P-value of 0.27.

QUESTION: Why did the researchers in the ADOPT study not express alarm about the increased risk they had seen?

ANSWER: A P-value of 0.27 means that a heart attack rate at least as high as the one they observed could be expected in 27% of similar experiments even if, in fact, there were no increased risk from taking Avandia®. That's not remarkable enough to reject the null hypothesis. In other words, the ADOPT study wasn't convincing.

Section 17.2 Exercises

3. P-values Which of the following are true? If false, explain briefly.
   a) A very high P-value is strong evidence that the null hypothesis is false.
   b) A very low P-value proves that the null hypothesis is false.
   c) A high P-value shows that the null hypothesis is true.
   d) A P-value below 0.05 is always considered sufficient evidence to reject a null hypothesis.

4. More P-values Which of the following are true? If false, explain briefly.
   a) A very low P-value provides evidence against the null hypothesis.
   b) A high P-value is strong evidence in favour of the null hypothesis.
   c) A P-value above 0.10 shows that the null hypothesis is true.
   d) If the null hypothesis is true, you can't get a P-value below 0.01.

5. Hypotheses For each of the following, write out the null and alternative hypotheses, being sure to state whether the alternative is one-sided or two-sided.
   a) A company reports that last year, 40% of their reports in accounting were on time. From a random sample

this year, they want to know if that proportion has changed.
   b) Last year, 42% of the employees enrolled in at least one wellness class at the company's site. Using a survey, they want to know if a greater percentage is planning to take a wellness class this year.
   c) A political candidate wants to know from recent polls if she's going to garner a majority of votes in next week's election.

6. More hypotheses For each of the following, write out the alternative hypothesis, being sure to indicate whether it is one-sided or two-sided.
   a) *Consumer Reports* discovered that 20% of a certain computer model had warranty problems over the first three months. From a random sample, the manufacturer wants to know if a new model has improved that rate.
   b) The last time a philanthropic agency requested donations, 4.75% of people responded. From a recent pilot mailing, they wonder if that rate has increased.
   c) A student wants to know if other students on her campus prefer Coke or Pepsi.

17.3 Alpha Levels and Significance

A. Barrington Brown/Science Source

Sir Ronald Fisher (1890–1962) was one of the founders of modern Statistics.

■ NOTATION ALERT

The first Greek letter, α, is used in statistics for the threshold value of a hypothesis test. You'll hear it referred to as the alpha level. Common values are 0.10, 0.05, 0.01, and 0.001.

> **It could happen to you!** Of course, if the null hypothesis *is* true, no matter what alpha level you choose, you still have a probability α of rejecting a true null hypothesis by mistake. This is the rare event we want to protect ourselves against. When we do reject the null hypothesis, no one ever thinks that *this* is one of those rare times. As statistician Stu Hunter notes, "*The statistician says 'rare events do happen—but not to me!'*"

Sometimes we need to make a firm decision about whether to reject the null hypothesis. A jury must *decide* whether the evidence reaches the level of "beyond a reasonable doubt." A business must *select* a Web design. You need to decide which section of Statistics to enrol in.

When the P-value is small, it tells us that our data are rare, *given the null hypothesis*. As humans, we are suspicious of rare events. If the data are "rare enough," we just don't think that could have happened due to chance. Since the data *did* happen, something must be wrong. All we can do now is to reject the null hypothesis.

But how rare is "rare"?

We can define "rare event" arbitrarily by setting a threshold for our P-value. If our P-value falls below that point, we'll reject the null hypothesis. We call such results **statistically significant**. The threshold is called an **alpha level**. Not surprisingly, it's labelled with the Greek letter α. Common α levels are 0.10, 0.05, and 0.01. You have the option—almost the *obligation*—to consider your alpha level carefully and choose an appropriate one for the situation. If you're assessing the safety of air bags, you'll want a low alpha level; even 0.01 might not be low enough. If you're just wondering whether folks prefer their pizza with or without pepperoni, you might be happy with $\alpha = 0.10$. It can be hard to justify your choice of α, though, so often we arbitrarily choose 0.05. Note, however: You must select the alpha level *before* you look at the data. Otherwise, you can be accused of cheating by tuning your alpha level to suit the data.

> **Where did the value 0.05 come from?** In 1931, in a famous book called *The Design of Experiments*, Sir Ronald Fisher discussed the amount of evidence needed to reject a null hypothesis. He said that it was *situation dependent*, but remarked, somewhat casually, that for many scientific applications, 1 out of 20 *might be* a reasonable value. Since then, some people—indeed some entire disciplines—have treated the number 0.05 as sacrosanct.

The alpha level of the test is also called its **significance level**. When we reject the null hypothesis, we say that the test is "significant at that level." For example, we might say that we reject the null hypothesis "at the 5% level of significance."

What can you say if the P-value does not fall below α?

When you have not found sufficient evidence to reject the null according to the standard you have established, you should say that "the data have failed to provide sufficient evidence to reject the null hypothesis." Don't say that you "accept the null hypothesis." You certainly haven't proven or established it; it was merely assumed to begin with. Say that you've failed to reject it.

Think again about the therapeutic touch example. The P-value was 0.79. This is so much larger than any reasonable alpha level that we can't reject H_0. For this test, we'd conclude, "We fail to reject the null hypothesis. There is insufficient evidence to conclude that the practitioners are performing better than they would if they were just guessing."

The automatic nature of the reject/fail-to-reject decision when we use an alpha level may make you uncomfortable. If your P-value falls just slightly above your alpha level, you're not allowed to reject the null. Yet a P-value just barely below the alpha level leads to rejection. If this bothers you, you're in good company. Many statisticians think it better to report the P-value than to base a decision on an arbitrary alpha level. When you do decide to declare a verdict, it's always a good idea to report the P-value as an indication of the strength of the evidence. (The P-value is also commonly referred to as the *observed level of significance*.)

Statistically significant? Yes, but is it important? A large insurance company mined its data and found a statistically significant ($P = 0.04$) difference between the mean value of policies sold in 2015 and 2016. The difference in the mean values was $9.83. Even though it was statistically significant, management did not see this as an important difference when a typical policy sold for more than $1000. On the other hand, even a clinically important improvement of 10% in the cure rate with a new treatment is not likely to produce statistically significant results in a study of fewer than 225 patients. A small clinical trial would probably not be conclusive.

> **IT'S IN THE STARS**
>
> Some disciplines carry the idea further and code P-values by their size. In this scheme, a P-value between 0.05 and 0.01 gets highlighted by *. A P-value between 0.01 and 0.001 gets **, and a P-value less than 0.001 gets ***. This can be a convenient summary of the weight of evidence against the null hypothesis if it's not taken too literally. But we warn you against making a black-and-white decision near the boundaries. The boundaries are a matter of tradition, not science; there is nothing special about 0.05. A P-value of 0.051 should be looked at very seriously and not casually thrown away just because it's larger than 0.05, and one that's 0.009 is not very different from one that's 0.011.

Sometimes it's best to report that the conclusion is not yet clear and to suggest that more data be gathered. (In a trial, a jury may "hang" and be unable to return a verdict.) In these cases, the P-value is the best summary we have of what the data say or fail to say about the null hypothesis.

Practical vs. Statistical Significance

What do we mean when we say that a test is statistically significant? All we mean is that the test statistic had a P-value lower than our alpha level. Don't be lulled into thinking that statistical significance carries with it any sense of practical importance or impact.

For large samples, even small, unimportant ("insignificant") deviations from the null hypothesis can be *statistically* significant. On the other hand, if the sample is not large enough, even large financially or scientifically "significant" differences may not produce *statistically* significant test results.

In addition to the P-value on which we base statistical significance, it's good practice to report a confidence interval, to indicate the range of plausible values for the parameter.

Section 17.3 Exercises

7. Alpha true and false Which of the following statements are true? If false, explain briefly.
   a) Using an alpha level of 0.05, a P-value of 0.04 results in rejecting the null hypothesis.
   b) The alpha level depends on the sample size.
   c) With an alpha level of 0.01, a P-value of 0.10 results in rejecting the null hypothesis.
   d) Using an alpha level of 0.05, a P-value of 0.06 means the null hypothesis is true.

8. Alpha false and true Which of the following statements are true? If false, explain briefly.
   a) It is better to use an alpha level of 0.05 than an alpha level of 0.01.
   b) If we use an alpha level of 0.01, then a P-value of 0.001 is statistically significant.
   c) If we use an alpha level of 0.01, then we reject the null hypothesis if the P-value is 0.001.
   d) If the P-value is 0.01, we reject the null hypothesis for any alpha level greater than 0.01.

17.4 Critical Values for Hypothesis Tests

When we make a confidence interval, we find a **critical value**, z^*, to correspond to our selected confidence level. Critical values can also be used as a shortcut for hypothesis tests. Before computers and calculators were common, P-values were hard to find. It was easier to select a few common alpha levels (0.05, 0.01, 0.001, for example) and learn the corresponding critical values for the Normal model (z-values that produce a P-value exactly equal to alpha). Rather than looking up the probability corresponding to your calculated z-score in the table, you'd just check your z-score directly against these z^* values. Any z-score larger in magnitude (that is, more extreme) than a particular critical value has to be less likely, so it will have a P-value smaller than the corresponding alpha. If we are willing to settle for a flat reject/fail-to-reject decision, comparing an observed z-score with the critical value for a specified alpha level gives a shortcut path to that decision.

For the medical school example, if we choose $\alpha = 0.05$, we then need to find the critical z-value that cuts off an area of 0.05 to the left under a standard Normal curve. Using the Normal table, we get $z^* = -1.645$. To reject H_0, our z-score has to be less than the critical value. Our calculated z-score is -11.92, so clearly we can reject the null hypothesis at the 5% level. This is perfectly correct and does give us a yes/no decision, but it gives us less information about the hypothesis because we don't have the P-value to think about. With technology, P-values are easy to find. And since they give more information about the strength of the evidence, you should always report them. In the medical school example, the very tiny P-value that we calculated tells us that the evidence of a low percentage of rural applicants is not just adequate at the 5% level, but is in fact overwhelming!

Below are the traditional critical values from the standard Normal model[6] and Figure 17.1 and 17.2 show the corresponding critical regions.

α	1-sided	2-sided
0.05	1.645	1.96
0.01	2.33	2.576
0.001	3.09	3.29

Figure 17.1

When the alternative is one-sided, the critical value puts all of α on one properly chosen side.

Figure 17.2

When the alternative is two-sided, the critical value splits α equally into two tails.

Confidence Intervals and Hypothesis Tests

For the medical school example, a 95% confidence interval would give $0.0728 \pm 1.96 \times 0.0048 = (0.063, 0.082)$, or 6.3% to 8.2%. If the (null) rural population percentage had been, say, 8.0%, we would not have been able to reject the null hypothesis because 8.0% is in the interval, so it's a plausible value. Indeed, *any* hypothesized value for the true proportion of rural applicants in this interval is consistent with the data. Any value outside the confidence interval would make a null hypothesis that we would reject, but we'd feel more strongly about values far outside the interval.

Confidence intervals and hypothesis tests are built from the same calculations.[7] They have the same assumptions and conditions. As we have just seen, you can approximate a hypothesis test by examining the confidence interval. Just ask whether the null hypothesis value is consistent with a confidence interval for the parameter at the corresponding confidence level. Because confidence intervals are naturally two-sided, they correspond to two-sided tests. For example, a 95% confidence interval corresponds to a two-sided hypothesis

---

[6]In a sense, these are the flip side of the 68–95–99.7 Rule. When we studied that rule, we chose simple statistical distances from the mean and recorded the areas between the tails. Now we select convenient tail areas (0.05, 0.01, and 0.001, either on one side or adding both together) and record the corresponding statistical distances.

[7]As we saw in Chapter 16, this is not *exactly* true for proportions. For a confidence interval, we estimate the standard deviation of $\hat{p}$ from $\hat{p}$ itself. For the corresponding hypothesis test, we use the model's standard deviation for $\hat{p}$, based on the null hypothesis value p_0. These calculations usually give similar results. When the SE and the null model's SD differ considerably, you're likely to reject H_0 (because the observed proportion is far from your hypothesized value) and in that case we recommend not using the confidence interval to perform a test.

test at $\alpha = 5\%$. In general, a confidence interval with a confidence level of $C\%$ corresponds to a two-sided hypothesis test with an α level of $(100 - C)\%$.

The relationship between confidence intervals and one-sided hypothesis tests is a little more complicated. One-sided tests are often accompanied by *one-sided confidence intervals*. These intervals take the form of $(-\infty, a)$ or (b, ∞). The former says that the parameter is no bigger than a, while the latter interval says that the parameter is at least as big as b. If you accept the alternative hypothesis that $p < 0.6$, for example, you know p is smaller than 0.6, but how much smaller? A one-sided confidence interval from the data for example might be $(-\infty, 0.53)$, which tells you that p is *no bigger than 0.53*. Just as with two-sided tests and confidence intervals, a precise relationship exists, such that values lying in, for example, the 95% one-sided confidence interval are precisely those values for the null hypothesis that we could not reject in a one-sided 5% level test. We will not explore this topic any further here. In practice, for one-sided tests, you may see either type of confidence interval reported. However, in this text we will stick to the more traditional two-sided confidence intervals regardless of the nature of the test alternative.

If you want to use a two-sided confidence interval to perform a one-sided test, how might you proceed? If we have a 95% confidence interval (more generally, a $C\%$ confidence interval), and the entire interval lies within the one-sided alternative, you may then reject the null at not just the 5% level, but at the more impressive 2.5% level (more generally, at the $\frac{1}{2}(100 - C)\%$ level). Because you specified a direction from the null *a priori*, you earn some benefit.

MATH BOX

Relating the test to the confidence interval

Suppose we are testing a null hypothesis against a two-sided alternative hypothesis:

$$H_0: p = p_0 \text{ vs. } H_A: p \neq p_0$$

and we use $\alpha = 0.05$.

Our test statistic is $z = \dfrac{\hat{p} - p_0}{SD(\hat{p})}$. We would fail to reject the null hypothesis if

$$-1.96 < \frac{\hat{p} - p_0}{SD(\hat{p})} < 1.96.$$

We can perform some simple algebraic manipulations (cross multiply by the SD, etc.) on the above, until we catch p_0 in the middle:

$$\hat{p} - 1.96 SD(\hat{p}) < p_0 < \hat{p} + 1.96 SD(\hat{p})$$

So, we cannot reject H_0 if the null value p_0 falls inside $\hat{p} \pm 1.96 SD(\hat{p})$.

But that looks remarkably like a 95% confidence interval for p:

$$\hat{p} \pm 1.96 SE(\hat{p})$$

Except for the fact that in the confidence interval we use $SE(\hat{p})$ instead of $SD(\hat{p})$, they are the same statement. In most cases, the difference between the SE and the SD is very small (and for many other tests that we will discuss in future chapters, the same SE is used for both the test and confidence interval).

So we cannot *reject* H_0 at significance level α *when the $100(1-\alpha)\%$ confidence interval covers the hypothesized value p_0*.

We said that a confidence interval holds the *plausible* values of p, so we shouldn't be surprised to find that we would fail to reject a hypothesized p_0 if it is a plausible value (found in the confidence interval), and reject it if it is not a plausible value (outside of the confidence interval).

For Example   MAKING A DECISION BASED ON A CONFIDENCE INTERVAL

RECAP: The baseline seven-year risk of heart attacks for diabetics is 20.2%. In 2007, a *NEJM* study reported a 95% confidence interval equivalent to 20.8%–40.0% for the risk among patients taking the diabetes drug Avandia®.

QUESTION: What did this confidence interval suggest to the FDA about the safety of the drug?

ANSWER: The *FDA* could be 95% confident that the interval from 20.8% to 40.0% included the true risk of heart attack for diabetes patients taking Avandia®. Because the lower limit of this interval was higher than the baseline risk of 20.2%, there was evidence of an increased risk (at the 5% level, for a two-sided alternative, or at the 2.5% level, for a one-sided alternative).

 Just Checking

1. An experiment to test the fairness of a roulette wheel gives a *z*-score of 0.62. What would you conclude?

2. In the last chapter, we encountered a bank that wondered if it could get more customers to make payments on delinquent balances by sending them a DVD urging them to set up a payment plan. Well, the bank just got back the results of their test of this strategy and reported that a 90% confidence interval for the success rate is (0.29, 0.45). Their old send-a-letter method had worked 30% of the time. Can you reject the null hypothesis that the proportion is still 30%? Explain.

3. Given the confidence interval the bank found in their trial of DVDs, what would you recommend that they do? Should they scrap the DVD strategy?

Section 17.4 Exercises

9. Critical values For each of the following situations, find the critical value(s) for *z*
   a) $H_0: p = 0.5$ vs. $H_A: p \neq 0.5$ at $\alpha = 0.05$.
   b) $H_0: p = 0.4$ vs. $H_A: p > 0.4$ at $\alpha = 0.05$.
   c) $H_0: p = 0.5$ vs. $H_A: p > 0.5$ at $\alpha = 0.01$.

10. More critical values For each of the following situations, find the critical value for *z*
   a) $H_0: p = 0.05$ vs. $H_A: p > 0.05$ at $\alpha = 0.05$.
   b) $H_0: p = 0.6$ vs. $H_A: p \neq 0.6$ at $\alpha = 0.01$.
   c) $H_0: p = 0.5$ vs. $H_A: p < 0.5$ at $\alpha = 0.01$.
   d) $H_0: p = 0.2$ vs. $H_A: p < 0.2$ at $\alpha = 0.01$.

17.5 Decision Errors

Nobody's perfect. Even with lots of evidence, we can still make the wrong decision. In fact, when we perform a hypothesis test, we can make mistakes in *two* ways:

  I. The null hypothesis is true, but we mistakenly reject it.
  II. The null hypothesis is false, but we fail to reject it.

These two types of errors are known as **Type I** and **Type II errors**. One way to keep the names straight is to remember that we start by assuming the null hypothesis is true, so a Type I error is the first kind of error we could make.

In medical disease testing, the null hypothesis is usually the assumption that a person is healthy. The alternative is that he or she has the disease we're testing for. So a Type I error is a *false positive*: A healthy person is diagnosed with the disease. A Type II error, in which an infected person is diagnosed as disease-free, is a *false negative*. These errors have other names, depending on the particular discipline and context.

Which type of error is more serious depends on the situation. In the jury trial, a Type I error occurs if the jury convicts an innocent person. A Type II error occurs if the jury fails to convict a guilty person. Which seems more serious? In medical diagnosis, a false negative could mean that a sick patient goes untreated. A false positive might mean that the person must undergo further tests. But for a drug test or a disease like AIDS, a false-positive result that is not kept confidential could have serious consequences. If you are testing a new treatment for epilepsy, a Type I error will lead to future patients getting a useless treatment; a Type II error means a useful treatment will remain undiscovered. Which of these errors seems more serious? It depends on the situation, the cost, and your point of view.

Here's an illustration of the situations:

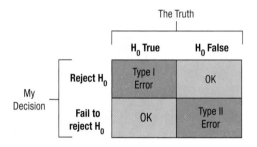

How often will a Type I error occur? It happens when the null hypothesis is true but we've had the bad luck to draw an unusual sample. To reject H_0, the P-value must fall below α. When H_0 is true, that happens *exactly* with probability α. So when you choose level α, you're setting the probability of a Type I error to α.

What if H_0 is not true? Then we can't possibly make a Type I error. You can't get a false positive from a sick person. A Type I error can happen only when H_0 is true.

When H_0 is false and we *fail* to reject it, we have made a Type II error. We assign the letter $\boldsymbol{\beta}$ to the probability of this mistake. What's the value of β? That's harder to assess than α because we don't know the true value of the parameter. When H_0 is true, it specifies a single parameter value. But when H_0 is false, we don't know the parameter value and there are many possible values. We can compute the probability β for any parameter value in H_A.

So β is not one single number but is actually a function of the parameter values contained in H_A. Not surprisingly, the value of β gets smaller as the true parameter value moves away from the null hypothesized value—if a coin produces heads 90% of the time on average, it is less likely that any data produced will support a null hypothesis of 50–50 balance, than if the coin produces heads 51% of the time. β when $p = 0.9$ is smaller than β when $p = 0.51$. The further the truth (true value of p) lies from the null, the easier it is to make the right decision, the less likely to make the wrong decision. A small β is desirable, but we can't make β small for all values in H_A, and fortunately we never need to. Some values in H_A will be of more practical importance than others—we'll say more about this shortly.

We could reduce β for *all* alternative parameter values by increasing α. By making it easier to reject the null, we'd be more likely to reject it whether it's true or not. So we'd reduce β—the chance that we fail to reject a false null—but we'd make more Type I errors. This tension between Type I and Type II errors is inevitable. In the political arena, consider the ongoing debate between those who favour provisions to reduce Type I errors in the courts (supporting rights of accused persons, providing legal representation for those who can't afford it, etc.) and those who advocate changes to reduce Type II errors (admitting into evidence confessions made when no lawyer is present, eavesdropping on conferences with lawyers, restricting paths of appeal, etc.).

The only way to reduce *both* types of error is to collect more evidence or, in statistical terms, to collect more data. Too often, studies fail because their sample sizes are too small to detect the changes or effects the researchers are looking for.

Of course, what we really want is to detect false null hypotheses. When H_0 is false and we reject it, we have done the right thing. A test's ability to detect a false null hypothesis is called the *power* of the test. In a jury trial, power is the ability of the criminal justice system to convict people who are guilty—a good thing!

For Example THINKING ABOUT ERRORS

RECAP: A published study found the risk of heart attack to be increased in patients taking the diabetes drug Avandia®. The issue of the *New England Journal of Medicine* (NEJM) in which that study appeared also included an editorial that said, in part, "A few events either way might have changed the findings for myocardial infarction (medical jargon for "heart attack") or for death from cardiovascular causes. In this setting, the possibility that the findings were due to chance cannot be excluded."

QUESTION: What kind of error would the researchers have made if, in fact, their findings were due to chance? What could be the consequences of this error?

ANSWER: The null hypothesis said the risk didn't change, but the researchers rejected that model and claimed evidence of a higher risk. If these findings were just due to chance, they rejected a true null hypothesis—a Type I error.

If, in fact, Avandia® carried no extra risk, then patients might be deprived of its benefits for no good reason.

Fisher and $\alpha = 0.05$ Why did Sir Ronald Fisher suggest 0.05 as a criterion for testing hypotheses? It turns out that he had in mind small initial studies. Small studies have relatively little power. Fisher was concerned that they might make too many Type II errors—failing to discover an important effect—if too strict a criterion were used.

The increased risk of Type I errors arising from a generous criterion didn't concern him as much because he expected a significant result to be followed with a replication or a larger study. The probability that two independent studies tested at $\alpha = 0.05$ would both make Type I errors is $0.05 \times 0.05 = 0.0025$, so Fisher didn't think that Type I errors in initial studies were a major concern.

The widespread use of the relatively generous 0.05 criterion even in large studies is most likely not what Fisher had in mind.

Section 17.5 Exercises

11. Errors For each of the following situations, state whether a Type I, a Type II, or neither error has been made. Explain briefly.

a) A bank wants to know if the enrollment on their website is above 30% based on a small sample of customers. They test $H_0: p = 0.3$ vs. $H_A: p > 0.3$ and reject the null hypothesis. Later they find out that actually 28% of all customers enrolled.

b) A student tests 100 students to determine whether other students on her campus prefer Coke or Pepsi and finds no evidence that preference for Coke is not 0.5. Later, a marketing company tests all students on campus and finds no difference.

c) A pharmaceutical company tests whether a drug lifts the headache relief rate from the 25% achieved by the placebo. They fail to reject the null hypothesis because the P-value is 0.465. Further testing shows that the drug actually relieves headaches in 38% of people.

12. More errors For each of the following situations, state whether a Type I, a Type II, or neither error has been made.

a) A test of $H_0: p = 0.8$ vs. $H_A: p < 0.8$ fails to reject the null hypothesis. Later it is discovered that $p = 0.9$.

b) A test of $H_0: p = 0.5$ vs. $H_A: p \neq 0.5$ rejects the null hypothesis. Later is it discovered that $p = 0.65$.

c) A test of $H_0: p = 0.7$ vs. $H_A: p < 0.7$ fails to reject the null hypothesis. Later it is discovered that $p = 0.6$.

17.6 Power and Sample Size

When we failed to reject the null hypothesis about TT practitioners, did we prove that they were just guessing? No, it could be that they actually *can* discern a human energy field, but we just couldn't tell. For example, suppose they really have the ability to get 53% of the trials right but just happened to get only 47% in our particular experiment (well within the bounds of chance variation here). With these data the *P*-value would be too large for us to reject the null. And if we retained the null even though the true proportion was actually greater than 50%, we would have made a Type II error because we failed to detect their ability.

Remember, we can never prove a null hypothesis true. We can only fail to reject it. But when we fail to reject a null hypothesis, it's natural to wonder whether we looked hard enough. Might the null hypothesis actually be false and our test too weak to tell?

When the null hypothesis actually *is* false, we hope our test is strong enough to reject it. We'd like to know how likely we are to succeed. The power of the test gives us a way to think about that. The **power** of a test is the probability that it correctly rejects a false null hypothesis. When the power is high, we can be confident that we've looked hard enough. We know that β is the probability that a test *fails* to reject a false null hypothesis, so the power of the test is the probability that it *succeeds* in rejecting the false null hypothesis: $1 - \beta$. We might have just written $1 - \beta$, but power is such an important concept that it gets its own name.

Whenever a study fails to reject its null hypothesis, the test's power comes into question. Was the sample size big enough to detect an effect had there been one? Might we have missed an effect large enough to be interesting just because we failed to gather sufficient data or because there was too much variability in the data we could gather? The therapeutic touch experiment failed to reject the null hypothesis that the TT practitioners were just guessing. Might the problem be that the experiment simply lacked adequate power to detect their ability?

For Example ERRORS AND POWER

RECAP: The study of Avandia® published in the NEJM combined results from 47 different trials—a method called meta-analysis.[8] The drug's manufacturer, GlaxoSmithKline (GSK), issued a statement that pointed out, "Each study is designed differently and looks at unique questions: For example, individual studies vary in size and length, in the type of patients who participated, and in the outcomes they investigate." Nevertheless, by combining data from many studies, meta-analyses can effectively achieve a much larger effective sample size.

QUESTION: How could this larger sample size help?

ANSWER: If Avandia® really did increase the seven-year heart attack rate, doctors needed to know. To overlook that would be a Type II error (failing to detect a false null hypothesis), resulting in patients being put at greater risk. Increasing the sample size could increase the power of the analysis, making it more likely that researchers will detect the danger if there is one.

Sensitivity and specificity The terms *sensitivity* and *specificity* often appear in medical studies. In these studies, the null hypothesis is that the person is healthy and the alternative is that the person is sick. The *specificity* of a test measures its ability to correctly identify only the healthy (the ones who should test negative). It's defined as

$$\frac{\text{number of true negatives}}{(\text{number of true negatives } + \text{ number of false positives})}$$

---

[8]One of the challenges of a good meta-analysis is digging out the studies that may have been *filed away* and not published, possibly due to lack of statistical significance (the "file drawer" problem). Yet, without those studies, conclusions of a meta-analysis can be seriously flawed and biased.

So, *specificity* = $1 - \alpha$.

Specificity gives you the probability that the test will correctly identify you as healthy when you are healthy.

The *sensitivity* is the ability to detect the disease. It's defined as:

$$\frac{\text{number of true positives}}{(\text{number of true positives } + \text{ number of false negatives})}$$

So *sensitivity* = $1 - \beta$ = power of the test. Sensitivity gives you the probability that the test will correctly identify you as sick when you are sick.

Effect Size

When we think about power, we imagine that the null hypothesis is false. The value of the power depends on how far the truth lies from the value we hypothesize. We call the distance between the null hypothesis value (for example), p_0, and the truth, p, the **effect size**. The effect size is unknown, of course, since it involves the true p, but we can *estimate* the effect size as the difference between the null value and the observed estimate.

The effect size is central to how we think about the power of a hypothesis test. A larger effect is easier to see and results in larger power. Small effects are naturally more difficult to detect. They'll result in more Type II errors and therefore lower power. When we design a study we won't know the effect size, but we can imagine possible effect sizes and look at their consequences. How can we decide what effect sizes to look at?

One way to think about the effect size is to ask, "*How big a difference would matter?*" The answer depends on who is asking it and why. For example, if therapeutic touch practitioners could detect a human energy field but only 53% of the time, that might not be a sufficient improvement over chance for health insurers to conclude it was worth covering. But *any* real ability to detect a previously unknown human energy field would be of great interest to scientists.[9]

The power of the test depends on both the size of the effect and the amount of variability in the sampling model. For proportions, we use a Normal sampling model (for $\hat{p}$) with standard deviation inversely proportional to the square root of the sample size, n. When researchers design a study, they often know what effect size would matter to their conclusions, based on which they can estimate, using a bit of algebra, the sample size needed. This is a common calculation to perform when designing a study so you can be sure to gather a large enough sample (and, speaking practically, to gather enough funds to pay for that large a sample). In Chapter 15 we made a similar calculation of a sample size, based on how large a margin of error we could accept. This is essentially the same idea.

Effect size and power are also an important part of our conclusions whenever we fail to reject the null hypothesis. The natural question to wonder about in that event is whether we tried hard enough—whether we had sufficient power to discern a difference. After all, if we based our test on only three observations, we might easily have missed even a large effect.

Whenever a hypothesis test fails to reject the null, the question of power can arise. In the therapeutic touch experiment, if the researchers considered 75% a reasonably interesting effect size (keeping in mind that 50% is the level of guessing), they could determine that the TT experiment would have been able to detect such an ability with a power of 99.99%[10] (assuming an α-level of 0.01). So there is only a very small chance that the study would have failed to detect a practitioner's ability at that level, had it existed. If, on the other hand, they thought that even an increase to 51% was interesting, then they would need a sample size of over 20 000 trials to have a 90% chance of detecting it!

---

[9]And probably lead to a Nobel Prize.

[10]See upcoming Math Box for exact calculation.

Just Checking

4. Remember the bank that is sending out DVDs to try to get customers to make payments on delinquent loans? It is looking for evidence that the costlier DVD strategy produces a higher success rate than the letters it has been sending. Explain what a Type I error is in this context and what the consequences would be to the bank.

5. What would be a Type II error in the bank experiment context, and what would the consequences be?

6. For the bank, which situation would have higher power: a strategy that works really well, actually getting 60% of people to pay off their balances, or a strategy that barely increases the payoff rate to 32%? Explain briefly.

A Picture Worth $\dfrac{1}{P(z > 3.09)}$ Words

It makes intuitive sense that the larger the effect size, the easier it should be to see it. Obtaining a larger sample size decreases the probability of a Type II error, so it increases the power. It also makes sense that the more we're willing to accept a Type I error, the less likely we will be to make a Type II error.

Figure 17.3

The power of a test is the probability that it rejects a false null hypothesis. The upper figure shows the null hypothesis model for a test about a proportion. We'll reject the null in a one-sided test if an observed proportion $\hat{p}$ falls to the right of the vertical line placed at a critical value, p^*. The lower figure shows an assumed true model, in which case, observing a proportion that exceeds the critical value would lead us to the correct decision to reject the null hypothesis. So the power of the test is the purple region on the right in the lower figure. Of course, when drawing samples whose observed proportions are distributed around p, we'll sometimes get a value in the red region on the left and make a Type II error of failing to reject the null.

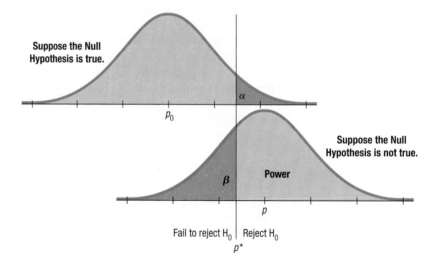

■ NOTATION ALERT

We've attached symbols to many of the p's. Let's keep them straight. p is a true proportion parameter. p_0 is a hypothesized value of p. $\hat{p}$ is an observed proportion. p^* is the test's critical value for an observed proportion corresponding to a specified α.

Figure 17.3 above shows a good way to visualize the relationships among these concepts. Suppose we are testing $H_0: p = p_0$ against the alternative $H_A: p > p_0$. We'll reject the null if the observed proportion, $\hat{p}$, is big enough. By big enough, we mean $\hat{p} > p^*$ for some critical value p^*, which can be determined from our α-level (shown as the red region in the right tail of the upper curve). For example, suppose that p^* were 0.65, that is, we would be willing to believe the ability of therapeutic touch practitioners if they were successful in at least 65% of our trials. This is what the upper model in Figure 17.3 shows. It's a picture of the sampling distribution model for the proportion if the null hypothesis were true. We'd make a Type I error whenever the sample gives us $\hat{p} > p^*$, because we would reject the (true) null hypothesis. And unusual samples like that would happen only with probability α.

In reality, though, the null hypothesis is rarely *exactly* true. The lower probability model in Figure 17.3 supposes that H_0 is not true. In particular, it supposes that the true value is p, not p_0. (Perhaps the TT practitioner really can detect the human energy field 75% of the time.) It shows a distribution of possible observed $\hat{p}$ values around this true value. Because of sampling variability, sometimes $\hat{p} < p^*$ and we fail to reject the (false) null hypothesis. Suppose a TT practitioner with a true ability level of 75% is actually successful on fewer than 65% of our tests. Then we'd make a Type II error. The area under the curve to the left of p^* in the bottom model represents how often this happens. The probability is β. In this picture, β is less than half, so most of the time we *do* make the right decision. The *power* of the test—the probability that we make the right decision—is shown as the region to the right of p^*. It's $1 - \beta$.

MATH BOX

Calculating the Power of the Test

Power calculations are always relevant, but they can be quite difficult to perform for many tests. In this case, however, it's not so hard. Let's give it a try, using the TT example on page 544, where the null hypothesis is H_0: $p = 0.5$, with $\alpha = 0.01$, and we are interested in the power of the test when the true $p = 0.75$, for $n = 150$ trials.

First, restate the criterion for rejection of the null hypothesis, in terms of a critical value p^*:

We reject H_0 at the 1% level if

$$z = \frac{\hat{p} - p_0}{SD(\hat{p})} = \frac{\hat{p} - 0.5}{0.0408} > z^* = 2.33 \text{ (from tables), since}$$

$$SD(\hat{p}) = \sqrt{\frac{(0.5)(0.5)}{150}} = 0.0408, \text{ under the null.}$$

Or, rewriting the above, we reject H_0 if $\hat{p} > 0.595$ (the critical value for $\hat{p}$, denoted p^* in our diagram).

Power = Probability of rejecting H_0 when H_A is true, and in particular, when p equals 0.75

$$= P(\hat{p} > 0.595 \text{ when } p = 0.75)$$

or standardized:

$$= P\left(\frac{\hat{p} - p}{SD(\hat{p})} > \frac{0.595 - p}{SD(\hat{p})} \right)$$

Since we are assuming that $p = 0.75$, plug this in for p in the top and also in the SD formula in the bottom, which gives $SD(\hat{p}) = \sqrt{\frac{(0.75)(0.25)}{150}} = 0.0354$.

$$\text{Power} = P\left(\frac{\hat{p} - p}{SD(\hat{p})} > \frac{0.595 - 0.75}{0.0354} \right) = P(z > -4.38) = 0.99999$$

(to 5 decimal places, using software).

So, if we have 150 trials, and the true success rate is 75%, we are essentially certain of making the correct decision!

We calculate $p*$ based on the upper model because $p*$ depends only on the null model and the alpha level. No matter what the true proportion, no matter whether the practitioners can detect a human energy field 90%, 53%, or 2% of the time, $p*$ doesn't change. After all, we don't *know* the truth, so we can't use it to determine the critical value. But we always reject H_0 when $\hat{p} > p*$.

How often we correctly reject H_0 when it's *false* depends on the effect size. We can see from the picture that if the effect size were larger (the true proportion were farther from the hypothesized value), the bottom curve would shift to the right, making the power greater.

We can see several important relationships from this figure:

- Power $= 1 - \beta$.
- Reducing α to lower the chance of committing a Type I error will move the critical value, $p*$, to the right (in this example). This will have the effect of increasing β, the probability of a Type II error, and correspondingly reducing the power.
- The larger the real difference between the hypothesized value p_0 and the true population value p, the smaller the chance of making a Type II error and the greater the power of the test. If the two proportions are very far apart, the two models will barely overlap, and we will not be likely to make any Type II errors at all—but then, we are unlikely to really need a formal hypothesis-testing procedure to see such an obvious difference. If the TT practitioners were successful almost all the time, we'd be able to see that with even a small experiment.

Determining the Sample Size

Figure 17.3 shows that we can reduce the chances of a Type II error (β), by increasing our risk of a Type I error (α). But this would not be a practical way to reduce Type II errors, since Type I errors are usually too important in their own right. But in fact there is a way to reduce β, while maintaining our desired α. Can you think of it?

If we can make both curves shown in Figure 17.4 narrower, with the size of the red α area fixed, the red β area will decrease and the power of the test will increase.

Figure 17.4

Making the standard deviations of these curves smaller will increase the power without changing the alpha level. The bar at $p*$ will slide to the left as the curves get narrower, increasing the power. The two proportions are just as far apart as in Figure 17.3, but the Type II error rate is reduced.

Suppose the Null Hypothesis is true.

α

p_0

Suppose the Null Hypothesis is not true.

Power

β

p

Fail to Reject H_0 | Reject H_0

$p*$

How can we accomplish that? The only way is to reduce the standard deviations of these sampling distributions by *increasing the sample size*. Compare the distributions in Figure 17.4 to those in Figure 17.3. You'll see that both sampling distributions are much narrower. The means haven't moved, but if we keep α the same size, the critical value, $p*$, moves closer to p_0 and farther from p. That means that a larger sample size increases the power of the test as well, as you can see by the smaller β.

So sample size is at issue once again! Just as with confidence intervals, at the design stage of a study we give careful thought to choosing an appropriate sample size. But instead of controlling the margin of error, now we want to control the Type II error rate or, equivalently, the power. How should we proceed?

First, we set α appropriately depending on the consequences of a Type I error. Then we determine a sample size big enough to keep the Type II error rate (β) low, or the power $(1 - \beta)$ high, for alternatives to the null (*effect sizes*) of practical importance. Before beginning the TT study, we should ask what departures from $p = 0.5$ are so big and important that we cannot allow them to slip by undiscovered. Suppose we decide that a success rate of 75%[11] is so remarkable that we (or health insurers) need to know about it. We should choose n so that the power is high when $p = 0.75$. While it is possible to develop a precise formula for this simple test, we will emphasize the use of software here (as well as for later tests). One hundred and fifty trials were conducted in this particular study, which produced a power of essentially 1.0 (see Math Box), but if researchers decided that a power of, say, 0.9 was adequate, then from the output below, we can see that 31 trials would have been sufficient.

POWER AND SAMPLE SIZE

Test for One Proportion

Testing proportion = 0.5 (versus > 0.5), Alpha = 0.05

Alternative Proportion	Sample Size	Target Power	Actual Power
0.75	31	0.9	0.905785

On the other hand, if a scientist suggests that even a value as small as $p = 0.51$ is hugely important, constituting the discovery of a previously unknown energy field, how many trials might now be needed? Likely many more in order to discover such a small effect. Below is the answer—over 21 400 trials!

POWER AND SAMPLE SIZE

Test for One Proportion

Testing proportion = 0.5 (versus > 0.5), Alpha = 0.05

Alternative Proportion	Sample Size	Target Power	Actual Power
0.51	21 406	0.9	0.900002

For Example SAMPLE SIZE, ERRORS, AND POWER

RECAP: The meta-analysis of the risks of heart attacks in patients taking the diabetes drug Avandia® combined results from 47 smaller studies. As GlaxoSmith-Kline (GSK), the drug's manufacturer, pointed out in their rebuttal, "Data from the ADOPT clinical trial did show a small increase in reports of myocardial infarction among the Avandia®-treated group . . . however, the number of events is too small to reach a reliable conclusion about the role any of the medicines may have played in this finding."

QUESTION: Why would this smaller study have been less likely to detect the difference in risk? What are the appropriate statistical concepts that underlie combining the smaller studies? What are the possible drawbacks of a very large sample?

---

[11]Technically, our power calculation is accurate only if we use one practitioner for all trials or a different practitioner for each trial. If we have multiple trials for each of multiple practioners, there is a clustering effect or dependence for data from any one practitioner, invalidating our independence assumption, and complicating any calculations.

ANSWER: Smaller studies are subject to greater sampling variability; that is, the sampling distributions they estimate have a larger standard deviation for the sample proportion. That gives small studies less power: They'd be less able to discern whether an apparently higher risk was merely the result of chance variation or evidence of real danger. Health Canada doesn't want to restrict the use of a drug that's safe and effective (Type I error), nor do they want patients to continue taking a medication that puts them at risk (Type II error). Larger sample sizes can reduce the risk of both kinds of error. Greater power (the probability of rejecting a false null hypothesis) means a better chance of spotting a genuinely higher risk of heart attacks. Of course, larger studies are more expensive. Their larger power also makes them able to detect differences that may not be clinically or financially significant. This tension among the power, cost, and sample size lies at the heart of most clinical trial designs.

On September 23, 2010, "The U.S. Food and Drug Administration announced that it will significantly restrict the use of the diabetes drug Avandia (rosiglitazone) to patients with Type 2 diabetes who cannot control their diabetes on other medications. These new restrictions are in response to data that suggest an elevated risk of cardiovascular events, such as heart attack and stroke, in patients treated with Avandia." (www.fda.gov/drugs/DrugSafety/PostmarketDrugSafetyInformationfor PatientsandProviders/ucm226956.htm)

Section 17.6 Exercises

13. (Power) Consider testing $H_0 : p = 0.3$ vs. $H_A : p > 0.3$. Assume n = 500 and $\alpha = 0.01$. Find the power of the test when the true $p = 0.35$.

14. (More power) Consider testing $H_0 : p = 0.7$ vs. $H_A : p < 0.7$. Assume n = 100 and $\alpha = 0.05$. Find the power of the test when the true $p = 0.6$.

WHAT CAN GO WRONG?

- **Don't interpret the P-value as the probability that H_0 is true.** The P-value is about the data, not the hypothesis. It's the probability of observing data this unusual, *given* that H_0 is true, not the other way around.

- **Don't believe too strongly in arbitrary alpha levels.** There's not really much difference between a P-value of 0.051 and a P-value of 0.049, but sometimes it's regarded as the difference between night (having to retain H_0) and day (being able to shout to the world that your results are "statistically significant"). It may just be better to report the P-value and a confidence interval and let the world decide along with you.

- **Don't confuse practical and statistical significance.** A large sample size can make it easy to discern even a trivial change from the null hypothesis value. On the other hand, an important difference can be missed if your test lacks sufficient power.

- **Don't forget that in spite of all your care, you might make a wrong decision.** We can never reduce the probability of a Type I error (α) or of a Type II error (β) to zero (but increasing the sample size helps).

- **If you fail to reject the null hypothesis, don't think that a bigger sample would be more likely to lead to rejection.** If the results you looked at were "almost" significant, it's enticing to think that because you would have rejected the null had the sample size n been larger, that a larger sample would likely lead to rejection. Don't be misled. Remember, each sample is different, and a larger sample won't duplicate your current observations. Maybe they'd be statistically significant, maybe not. On the other hand, if a small study (small n) yields weak evidence ($P < 0.10$, for example) for the hoped-for effect, proceeding to a bigger future study can make sense. Examining the confidence interval can also help. If none of the plausible parameter values in the interval would matter to you (because none would be *practically* significant), then even a larger study is unlikely to be worthwhile.

What Have We Learned?

Review: Know the reasoning of hypothesis testing.

- State the **hypotheses**.
- Determine (and check assumptions for) the sampling distribution **model**.
- Calculate the test statistic—the **mechanics**.
- State your **conclusions and decisions**.

Know how to formulate a null and alternative hypothesis for a question of interest.

- The null hypothesis specifies a parameter and a (null) value for that parameter.
- The alternative hypothesis specifies a range of plausible values should we fail to reject the null.
- Be able to decide on a two-sided or one-sided alternative hypothesis, and justify your decision.

Understand P-values.

- A P-value is the estimated probability of observing a statistic value at least as far from the (null) hypothesized value as the one we have actually observed.
- A small P-value indicates that the statistic we have observed would be unlikely were the null hypothesis true. That leads us to doubt the null.
- A large P-value just tells us that we have insufficient evidence to doubt the null hypothesis. In particular, it does not prove the null to be true.

Know that sometimes P-values are compared to a pre-determined α-level to decide whether to reject the null hypothesis.

Know that α is the probability of making a Type I error—rejecting the null hypothesis when it is, in fact, true.

Know that statistical significance simply implies strong evidence that the null hypothesis is false, not that the difference is important.

Know the value of reporting the estimated effect size.

- A test may be statistically significant, but practically meaningless if the estimated effect is of trivial importance.

Be aware of the risks of making errors when testing hypotheses.

- A Type I error can occur when rejecting a null hypothesis if that hypothesis is, in fact, true. The probability of this is α.
- A Type II error can occur when failing to reject a null hypothesis if that hypothesis is, in fact, false. The probability of this is β (not one number, but a function of the alternatives to the null!).

Understand the concept of the power of a test.

- We are particularly concerned with power when we fail to reject a null hypothesis.
- The power of a test report, for some specified effect size, the probability that the test would correctly reject a false null hypothesis.
- Remember that increasing the sample size will generally improve the power of any test.

Review of Terms

Statistically significant When the P-value falls below the alpha level, we say that the test is "statistically significant" at that alpha level (p. 535).

Alpha level (α) The threshold P-value that determines when we reject a null hypothesis. If we observe a statistic whose P-value based on the null hypothesis is less than α, we reject that null hypothesis (p. 535).

Significance level The alpha level is also called the significance level of the test, most often in a phrase such as a conclusion that a particular test is "significant at the 5% significance level" (p. 535).

Critical value	The value in the sampling distribution model of the statistic whose P-value is equal to the alpha level. The critical value is often denoted with an asterisk, as z^*, for example (p. 536).
Type I error	The error of rejecting a null hypothesis when in fact it is true (called a "false positive" in diagnostic tests). The probability of a Type I error is α (p. 539).
Type II error	The error of failing to reject a null hypothesis when in fact it is false (called a "false negative" in diagnostic tests) (p. 539).
β	The probability of a Type II error. β depends on the specified alternative value for the parameter (or effect size) (p. 540).
Power	The probability that a hypothesis test will correctly reject a false null hypothesis is the power of the test. To find power, we must specify a particular alternative parameter value as the "true" value. For any specific value in the alternative, the power is $1 - \beta$ (p. 542).
Effect size	The difference between the null hypothesis value and true value of a model parameter (p. 543).

On the Computer   HYPOTHESIS TESTS

Reports about hypothesis tests generated by technologies don't follow a standard form. Most will name the test and provide the test statistic value, its standard deviation, and the P-value. But these elements may not be labelled clearly. For example, the expression "Prob > |z|" means the probability (the "Prob") of observing a test statistic whose magnitude (the absolute value tells us this) is larger than that of the one (the "z") found in the data (which, because it is written as "z," we know follows a Normal model). That is a fancy (and not very clear) way of saying P-value. In some packages, you can specify that the test be one-sided. Others might report three P-values, covering the ground for both one-sided test and the two-sided test.

Sometimes a confidence interval and hypothesis test are automatically given together. The confidence interval ought to be for the corresponding confidence level: $1 - \alpha$.

Often, the standard deviation of the statistic is called the "standard error," and usually that's appropriate because we've had to estimate its value from the data. That's not the case for proportions, however: We get the standard deviation for a proportion from the null hypothesis value. Nevertheless, you may see the standard deviation called a "standard error," even for tests with proportions.

It's common for statistics packages and calculators to report more digits of "precision" than could possibly have been found from the data. You can safely ignore them. Round values such as the standard deviation to one digit more than the number of digits reported in your data.

Here are the kind of results you might see. This is not from any program or calculator we know of, but it shows some of the things you might see in typical computer output.

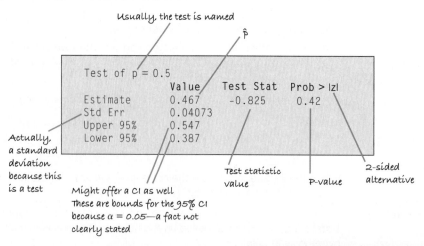

For information on hypothesis testing with particular statistics packages, see the On the Computer section in the previous chapter.

CALCULATING POWER AND SAMPLE SIZE

Good statistical packages will readily compute the power of a test for a given sample size, or the sample size needed to achieve a certain power, for any specified effect size. Common test procedures should be listed in the *Power and Sample Size* menu for you to choose from.

EXCEL

Power calculations are not in Excel's standard data analysis toolkit, but you can try to imitate the individual steps shown in the Math Box titled Calculating the Power of the Test.

JMP

- Choose **Power and Sample Size** from the **DOE** menu.
- Choose **One Sample Proportion** from the submenu.
- Indicate the **Null Proportion** and alternative **Proportion** that you are testing, select the **Alpha** value, and choose a one- or two-sided alternative.
- Fill in your desired **Sample size** or **Power**. Whichever you leave blank will be calculated.
- Click **Continue**.

COMMENTS

Specify the **alternative** Proportion that represents a departure from the null hypothesis that is large enough to be of practical importance. If the sample size is fixed and you need to know the power, leave **Power** blank, but if you are planning a study and need to know the sample size required to achieve desired power, fill in the **Power** sought and leave **Sample size** blank. If you leave both **Sample size** and **Power** blank, you will get a power curve, showing power as a function of sample size.

MINITAB

- Choose **Power and Sample Size** from the **Stat** menu.
- Choose **1 Proportion** from the submenu.
- Fill in the **Hypothesized** *p* that you are testing.
- Fill in any two of: **Sample sizes, Alternative values of** *p*, **Power values**.
- Use **Options** if you want a one-sided alternative or a significance level different from 0.05.
- Use **Graphs** if you want to produce an entire power curve of power values as a function of *p*.
- Click **OK**.

COMMENTS

Specify the **alternative value of** *p* that represents a departure from the null big enough to be of practical importance. If the sample size is fixed and you need to know the power, leave **Power values** blank, but if you are planning a study and need to know the sample size required to achieve desired power, fill in the **Power values** sought (multiple entries are fine) and leave **Sample sizes** blank.

R

R has a package {pwr} (not installed by default) that does various power calculations.

SPSS

The IBM SPSS SamplePower add-on makes power calculations simple. Otherwise, you can try to imitate the individual steps shown in the Math Box titled Calculating the Power of the Test.

STATCRUNCH

Select **Stat > Proportion Statistics > One Sample > Power/Sample size**. At the top of the *one sample p calculator*, click on **Hypothesis Test Power**. This allows you to find the effect size, power, or sample size given any two of the values as inputs. Proceed as follows:

- Enter **Alpha**, the **Target** (null) proportion, and the type of **Alternative** hypothesis.
- Enter two of the following three values: **True proportion**, **Power**, and **Sample size**.

- Click **Compute** to calculate the value left blank while updating the power curve.
- Click and drag the red dot in the power curve to interactively investigate the power/difference relationship. Note how the values below for difference and power change accordingly.

T1-83/84 PLUS

Power calculations are not automatic, but you can imitate the individual calculation steps shown in the Math Box titled Calculating the Power of the Test.

Chapter Exercises

15. One-sided or two? In each of the following situations, is the alternative hypothesis one-sided or two-sided? What are the hypotheses?
  a) A business student conducts a taste test to see whether students prefer Diet Coke or Diet Pepsi.
  b) PepsiCo recently reformulated Diet Pepsi in an attempt to appeal to teenagers. They run a taste test to see if the new formula appeals to teenagers more than the standard formula.
  c) A budget override in a small town requires a two-thirds majority to pass. A local newspaper conducts a poll to see if there's evidence it will pass.
  d) One financial theory states that the stock market will go up or down with equal probability. A student collects data over several years to test the theory.

16. Which alternative? In each of the following situations, is the alternative hypothesis one-sided or two-sided? What are the hypotheses?
  a) A university dining service wants to know if students prefer plastic or metal cutlery. They conduct a survey to find out.

  b) In recent years, 10% of third-year students have applied for study abroad. The dean's office conducts a survey to see if the percentage changed this year.
  c) A pharmaceutical company conducts a clinical trial to see if more patients who take a new drug experience headache relief than the (previously established rate of) 22% who claim relief after taking a placebo.
  d) At a small computer peripherals company, only 60% of the hard drives produced passed all their performance tests the first time. Management recently invested a lot of resources into the production system and now conducts a test to see if it helped.

17. P-value A medical researcher has tested a new treatment for poison ivy against the traditional ointment. He concludes that the new treatment, with a P-value of 0.047, is more effective. Explain what the P-value means in this context.

18. Another P-value Have harsher penalties and ad campaigns increased seatbelt use among drivers and passengers? Observations of commuter traffic failed to find evidence of a significant change compared with three years ago. Explain what the study's P-value of 0.17 means in this context.

19. Alpha A researcher developing scanners to search for hidden weapons at airports has concluded that a new device is significantly better than the current scanner. He made this decision based on a test using $\alpha = 0.05$. Would he have made the same decision at $\alpha = 0.10$? How about $\alpha = 0.01$? Explain.

20. Alpha again Environmentalists concerned about the impact of high-frequency radio transmissions on birds found that there was no evidence of a higher mortality rate among hatchlings in nests near cell towers. They based this conclusion on a test using $\alpha = 0.05$. Would they have made the same decision at $\alpha = 0.10$? How about $\alpha = 0.01$? Explain.

21. Significant? Public health officials believe that 90% of children have been vaccinated against measles. A random survey of medical records at many schools across the country found that, among more than 13 000 children, only 89.4% had been vaccinated. A statistician would reject the 90% hypothesis with a P-value of $P = 0.011$.
a) Explain what the P-value means in this context.
b) The result is statistically significant, but is it important? Comment.

22. Significant again? A new reading program may reduce the number of elementary school students who read below grade level. The company that developed this program supplied materials and teacher training for a large-scale test involving nearly 8500 children in several school districts. Statistical analysis of the results showed that the percentage of students who did not attain the grade-level standard was reduced from 15.9% to 15.1%. The hypothesis that the new reading program produced no improvement was rejected with a P-value of 0.023.
a) Explain what the P-value means in this context.
b) Even though this reading method has been shown to be significantly better, why might you not recommend that your local school adopt it?

23. Groceries In January 2011, Yahoo surveyed 2400 U.S. men. A total of 1224 of the men identified themselves as the primary grocery shopper in their household.
a) Estimate the percentage of all U.S. males who identify themselves as the primary grocery shopper. Use a 98% confidence interval. Check the conditions first.
b) A grocery store owner had been told that 45% of men are the primary grocery shopper for their family, and targets his advertising accordingly. He wishes to conduct a hypothesis test to see if the fraction is in fact different from 45%. What does your confidence interval indicate? Explain.
c) What is the level of significance of this test? Explain.

Source: https://www.naturalproductsinsider.com/news/2011/01/51-of-men-are-primary-grocery-shoppers.aspx

24. Is the euro fair? Soon after the euro was introduced as currency in Europe, it was widely reported that someone had spun a euro coin 250 times and gotten heads 140 times. We wish to test a hypothesis about the fairness of spinning the coin.
a) Estimate the true proportion of heads. Use a 95% confidence interval. Don't forget to check the conditions first.
b) Does your confidence interval provide evidence that the coin is unfair when spun? Explain.
c) What is the significance level of this test? Explain.

25. Rural med school applicants, again In Exercise 28 in Chapter 16, we examined medical school applications from the rural Ontario population. In 2002, 2246 students from Ontario applied to the province's medical schools; in 2003, 2702 applied. Of the two-year total (4948), there were 4588 applicants from urban areas and 360 from rural areas.[12] Of course, there are many sources of random variation in the results from any one-year or two-year period.
a) Give a 95% confidence interval for the percentage of applications generally coming from rural students around this time period.
b) Since the population of Ontario is 13% rural, is there reason for concern about a lack of rural applicants for medical school? Use your 95% confidence interval to test the null hypothesis that the general percentage of applications coming from rural students is no different than the rural population percentage of 13%.

26. Hard times In June 2010, a random poll of 800 working men found that 9% had taken on a second job to help pay the bills. (www.careerbuilder.com)
a) Estimate the true percentage of men that are taking on second jobs by constructing a 95% confidence interval.
b) A pundit on a TV news show claimed that only 6% of working men had a second job. Use your confidence interval to test whether his claim is plausible given the poll data.

27. Loans Before lending someone money, banks must decide whether they believe the applicant will repay the loan. One strategy used is a point system. Loan officers assess information about the applicant, totalling points they award for the person's income level, credit history, current debt burden, and so on. The higher the point total, the more convinced the bank is that it's safe to make the loan. Any applicant with a lower point total than a certain cutoff score is denied a loan.

We can think of this decision as a hypothesis test. Since the bank makes its profit from the interest collected on repaid loans, their null hypothesis is that the applicant will repay the loan and therefore should get the money. Only if the person's score falls below the minimum cutoff will the bank reject the null and deny the loan. This system is reasonably reliable, but, of course, sometimes there are mistakes.

a) When a person defaults on a loan, which type of error did the bank make?

b) Which kind of error is it when the bank misses an opportunity to make a loan to someone who would have repaid it?

c) Suppose the bank decides to lower the cutoff score from 250 points to 200. Is that analogous to choosing a higher or lower value of α for a hypothesis test? Explain.

d) What impact does this change in the cutoff value have on the chance of each type of error?

28. **Spam** Spam filters try to sort your e-mails, deciding which are real messages and which are unwanted. One method used is a point system. The filter reads each incoming e-mail and assigns points to the sender, the subject, key words in the message, and so on. The higher the point total, the more likely it is that the message is unwanted. The filter has a cutoff value for the point total; any message rated lower than that cutoff passes through to your inbox, and the rest, suspected to be spam, are diverted to the junk mailbox.

We can think of the filter's decision as a hypothesis test. The null hypothesis is that the e-mail is a real message and should go to your inbox. A higher point total provides evidence that the message may be spam; when there's sufficient evidence, the filter rejects the null, classifying the message as junk. This usually works pretty well, but, of course, sometimes the filter makes a mistake.

a) When the filter allows spam to slip through into your inbox, which kind of error is that?

b) Which kind of error is it when a real message gets classified as junk?

c) Some filters allow the user (that's you) to adjust the cutoff. Suppose your filter has a default cutoff of 50 points, but you reset it to 60. Is that analogous to choosing a higher or lower value of α for a hypothesis test? Explain.

d) What impact does this change in the cutoff value have on the chance of each type of error?

29. **Second loan** Exercise 27 described the loan score method a bank uses to decide which applicants to lend money to. Only if the total points awarded for various aspects of an applicant's financial condition fail to add up to a minimum cutoff score set by the bank will the loan be denied.

a) In this context, explain what is meant by the power of the test.

b) What could the bank do to increase the power?

c) What is the disadvantage of doing that?

30. **More spam** Consider again the points-based spam filter described in Exercise 28. When the points assigned to various components of an e-mail exceed the cutoff value you've set, the filter rejects its null hypothesis (that the message is real) and diverts that e-mail to a junk mailbox.

a) In this context, explain what is meant by the power of the test.

b) What could you do to increase the filter's power?

c) What is the disadvantage of doing that?

31. **Obesity in First Nations youth** According to the March 2007 "Healthy Weights for Healthy Kids" report of the Standing Committee on Health, 55% of First Nations children living on reserves are either overweight or obese, which is far above the (already too high) national average.[13] Suppose the federal and provincial governments decide to allocate funds for a new program targeted specifically at First Nations youth to help reduce obesity in this community. They decide to make it a trial program, to be evaluated after four years when data will be available. They will continue with the program only if there is strong evidence that the obesity rate in First Nations youth has decreased.

a) In words, what will their hypotheses be?

b) What would a Type I error be?

c) What would a Type II error be?

d) For each type of error, tell who would be harmed.

e) What would the power of the test represent in this context?

32. **Alzheimer's** Testing for Alzheimer's disease can be a long process, consisting of lengthy tests and medical diagnosis. Recently, a group of researchers (Solomon et al., 1998) devised a seven-minute test to serve as a quick screen for the disease for use in the general population of senior citizens. A patient who tested positive would then go through the more expensive battery of tests and medical diagnosis. The authors reported a false positive rate of 4% and a false negative rate of 8%.

a) Put this in the context of a hypothesis test. What are the null and alternative hypotheses?

b) What would a Type I error mean?

c) What would a Type II error mean?

d) Which is worse here, a Type I or Type II error? Explain.

e) What is the power of this test?

33. **Testing cars** A clean air standard requires that vehicle exhaust emissions not exceed specified limits for various pollutants. Suppose government regulators double-check a random sample of cars that a suspect repair shop has certified as okay. They will revoke the shop's license if

---

[13]www.parl.gc.ca/HousePublications/Publication.aspx?DocId=2795145&Language=E&Mode=1&Parl=39&Ses=1

they find significant evidence that the shop is certifying vehicles that do not meet standards.
a) In this context, what is a Type I error?
b) In this context, what is a Type II error?
c) Which type of error would the shop's owner consider more serious?
d) Which type of error might environmentalists consider more serious?

34. **Quality control** Production managers on an assembly line must monitor the output to be sure that the level of defective products remains small. They periodically inspect a random sample of the items produced. If they find a significant increase in the proportion of items that must be rejected, they will halt the assembly process until the problem can be identified and repaired.
a) In this context, what is a Type I error?
b) In this context, what is a Type II error?
c) Which type of error would the factory owner consider more serious?
d) Which type of error might customers consider more serious?

35. **Cars again** As in Exercise 33, government regulators are checking up on repair shops to see if they are certifying vehicles that do not meet pollution standards.
a) In this context, what is meant by the power of the test the regulators are conducting?
b) Will the power be greater if they test 20 or 40 cars? Why?
c) Will the power be greater if they use a 5% or a 10% level of significance? Why?
d) Will the power be greater if the repair shop's inspectors are only a little out of compliance or a lot? Why?

36. **Production** Consider again the task of the quality control inspectors in Exercise 34.
a) In this context, what is meant by the power of the test the inspectors conduct?
b) They are currently testing five items each hour. Someone has proposed that they test 10 each hour instead. What are the advantages and disadvantages of such a change?
c) Their test currently uses a 5% level of significance. What are the advantages and disadvantages of changing to an alpha level of 1%?
d) Suppose that, as a day passes, one of the machines on the assembly line produces more and more items that are defective. How will this affect the power of the test?

37. **Equal opportunity?** A company is sued for job discrimination because only 19% of the newly hired candidates were minorities when 27% of all applicants were minorities. Is this strong evidence that the company's hiring practices are (prima facie) discriminatory?
a) Is this a one-tailed or a two-tailed test? Why?
b) In this context, what would a Type I error be?

c) In this context, what would a Type II error be?
d) In this context, describe what is meant by the power of the test.
e) If the hypothesis is tested at the 5% level of significance instead of 1%, how will this affect the power of the test?
f) The lawsuit is based on the hiring of 37 employees. Is the power of the test higher than, lower than, or the same as it would be if it were based on 87 hires?

38. **Stop signs** Highway safety engineers test new road signs, hoping that increased reflectivity will make them more visible to drivers. Volunteers drive through a test course with several of the new- and old-style signs and rate which kind shows up the best.
a) Is this a one-tailed or a two-tailed test? Why?
b) In this context, what would a Type I error be?
c) In this context, what would a Type II error be?
d) In this context, describe what is meant by the power of the test.
e) If the hypothesis is tested at the 1% level of significance instead of 5%, how will this affect the power of the test?
f) The engineers hoped to base their decision on the reactions of 50 drivers, but time and budget constraints may force them to cut back to 20. How would this affect the power of the test? Explain.

39. **Dropouts** A Statistics professor has observed that for several years about 13% of the students who initially enrol in his Introductory Statistics course withdraw before the end of the semester. A salesperson suggests that he try a statistics software package that gets students more involved with computers, predicting that it will cut the dropout rate. The software is expensive, and the salesperson offers to let the professor use it for a semester to see if the dropout rate goes down significantly. The professor will have to pay for the software only if he chooses to continue using it.
a) Is this a one-tailed or two-tailed test? Explain.
b) Write the null and alternative hypotheses.
c) In this context, explain what would happen if the professor makes a Type I error.
d) In this context, explain what would happen if the professor makes a Type II error.
e) What is meant by the power of this test?

40. **Ads** A company is willing to renew its advertising contract with a local radio station only if the station can prove that more than 20% of the residents of the city have heard the ad and recognize the company's product. The radio station conducts a random phone survey of 400 people.
a) What are the hypotheses?
b) The station plans to conduct this test using a 10% level of significance, but the company wants the significance level lowered to 5%. Why?

c) What is meant by the power of this test?

d) For which level of significance will the power of this test be higher? Why?

e) They finally agree to use $\alpha = 0.05$, but the company proposes that the station call 600 people instead of the 400 initially proposed. Will that make the risk of Type II error higher or lower? Explain.

41. Dropouts, part II Initially 203 students signed up for the Stats course in Exercise 39. They used the software suggested by the salesperson, and only 11 dropped out of the course.

a) Should the professor spend the money for this software? Support your recommendation with an appropriate test.

b) Explain carefully what your P-value means in this context.

42. Testing the ads The company in Exercise 40 contacts 600 people selected at random, and only 133 remember the ad.

a) Should the company renew the contract? Support your recommendation with an appropriate test.

b) Explain carefully what your P-value means in this context.

43. Two coins Two coins are in a drawer. They look the same, but one coin produces heads 90% of the time when spun, while the other one produces heads only 30% of the time. You select one of the coins. You are allowed to spin it *once* and then must decide whether the coin is the 90%- or the 30%-head coin. Your null hypothesis is that your coin produces 90% heads.

a) What is the alternative hypothesis?

b) Given that the outcome of your spin is tails, what would you decide? What if it were heads?

c) How large is α in this case?

d) How large is the power of this test? (*Hint:* How many possibilities are in the alternative hypothesis?)

e) How could you lower the probability of a Type I error and increase the power of the test at the same time?

44. Faulty or not? You are in charge of shipping computers to customers. You learn that a faulty disk drive was put into some of the machines. There's a simple test you can perform, but it's not perfect. All but 4% of the time, a good disk drive passes the test, but unfortunately, 35% of the bad disk drives pass the test, too. You have to decide on the basis of one test whether the disk drive is good or bad. Make this a hypothesis test.

a) What are the null and alternative hypotheses?

b) Given that a computer fails the test, what would you decide? What if it passes the test?

c) How large is α for this test?

d) What is the power of this test? (*Hint:* How many possibilities are in the alternative hypothesis?)

45. Hoops A basketball player with a poor foul-shot record practises intensively during the off-season. He tells the coach that he has raised his proficiency from 60% to 80%. Dubious, the coach asks him to take 10 shots, and is surprised when the player hits 9 out of 10. Did the player prove that he has improved?

a) Suppose the player really is no better than before—still a 60% shooter. What is the probability that he can hit at least 9 of 10 shots anyway? (*Hint:* Use a Binomial model.)

b) If that is what happened, now the coach thinks the player has improved when he has not. Which type of error is that?

c) If the player really can hit 80% now, and it takes at least 9 out of 10 successful shots to convince the coach, what's the power of the test?

d) List two ways the coach and player could increase the power to detect any improvement.

46. Pottery An artist experimenting with clay to create pottery with a special texture has been experiencing difficulty with these pieces. About 40% break in the kiln during firing. Hoping to solve this problem, she buys some more expensive clay from another supplier. She plans to make and fire 10 pieces and will decide to use the new clay if at most one of them breaks.

a) Suppose the new, expensive clay really is no better than her usual clay. What is the probability that this test convinces her to use it anyway? (*Hint:* Use a Binomial model.)

b) If she decides to switch to the new clay and it is no better, what kind of error did she commit?

c) If the new clay really can reduce breakage to only 20%, what is the probability that her test will not detect the improvement?

d) How can she improve the power of her test? Offer at least two suggestions.

47. Hoops II In Exercise 45, the basketball player was hoping to improve his foul-shot percentage. Suppose the coach asks him to take 80 shots, and we test the null hypothesis that his percentage is still 60% against the alternative that he has improved, at the 5% significance level.

a) What percentage of shots does he have to hit for us to reject the null hypothesis?

b) Calculate the power of the test if he now can hit on 75% of shots. Imitate the calculation in this chapter's second Math Box (page 545).

c) If your software includes power calculations, confirm your calculation from part b). Also find the number of shots necessary to raise the power of the test to 0.99 when his true foul-shot percentage has risen to 75%.

48. Pottery II In Exercise 46, the artist was hoping to reduce the breakage by switching to a new type of clay.

Suppose she fires up 60 pieces, and tests the null hypothesis that the breakage rate is still 40% against the alternative of reduced breakage using the new clay, at the 5% significance level.

a) What percentage of the 60 tested pieces have to break for her to reject the null hypothesis?

b) Calculate the power of the test if breakage in fact is only 30% for the new clay. Imitate the calculation in this chapter's second Math Box (pages 545).

c) If your software includes power calculations, confirm your calculation in part b). Also find the number of pieces needed in the study to raise the power to 0.80 when the new clay's true breakage rate is 30%.

49. Simulations Use your computer software to generate a sample of size 75 from a Bernoulli distribution. This distribution has only two outcomes, 0 and 1, with complementary probabilities. Set the probability of the outcome "1" equal to 0.8. Consider "1" to be a success and "0" to be a failure.

a) For your sample, test the null hypothesis that $p = 0.8$, at the 5% significance level, with a two-sided alternative. What do you conclude? Did you make the right decision?

b) Draw 99 additional samples and repeat part a) for each, performing 99 additional 5% level tests. How often did you make the correct decision? The wrong decision? What percent of tests would you expect to produce a correct decision if you repeated these simulations many times? For 100 simulated samples, if $X =$ number wrong decisions, what is the probability distribution (model) of X?

50. More simulations Use your computer software to generate a sample of size 60 from a Bernoulli distribution. This distribution has only two outcomes, 0 and 1, with complementary probabilities. Set the probability of "1" equal to 0.3. Consider "1" to be a success and "0" to be a failure.

a) For your sample, test the null hypothesis that $p = 0.3$, at the 10% significance level, with a two-sided alternative. What do you conclude? Did you make the right decision?

b) Draw 49 additional samples and repeat part a) for each, performing 49 additional 10% level tests. How often did you make the correct decision? The wrong decision? What percent of tests would you expect to produce a correct decision if you repeated these simulations many times? For 50 simulated samples, if $X =$ number wrong decisions, what is the probability distribution (model) of X?

51. Still more simulations Use your computer software to generate a sample of size 75 from a Bernoulli distribution. This distribution has only two outcomes, 0 and 1, with complementary probabilities. Set the probability of

"1" equal to 0.8. Consider "1" to be a success and "0" to be a failure.

a) For your sample, test the null hypothesis that $p = 0.7$, at the 5% significance level, for a two-sided alternative. What do you conclude? Did you make the correct decision?

b) Draw 99 additional samples and repeat part a) for each, performing 99 additional 5% level tests. How often did you make the correct decision? The wrong decision? Based on your simulations, estimate the Type II error probability and the power of the test when the true $p = 0.8$. Combine with results of other students in your class to get a better estimate of the power.

c) Optional: Use software or follow the Math Box calculation on page 545 to find the exact power of the test that you estimated by simulations in part b).

52. Yet more simulations Use your computer software to generate a sample of size 60 from a Bernoulli distribution. This distribution has only two outcomes, 0 and 1, with complementary probabilities. Set the probability of "1" equal to 0.3. Consider "1" to be a success and "0" to be a failure.

a) For your sample, test the null hypothesis that $p = 0.5$, at the 10% significance level, for a two-sided alternative. What do you conclude? Did you make the correct decision?

b) Draw 49 additional samples and repeat part a) for each, performing 49 additional 10% level tests. How often did you make the correct decision? The wrong decision? Based on your simulations, estimate the Type II error probability and the power of the test when the true $p = 0.3$. Combine with results of other students in your class to get a better estimate of the power.

c) Optional: Use software or follow the Math Box calculation on page 545 to find the exact power of the test that you estimated by simulations in part b).

53. Multiple tests Suppose that 10 different research centres are investigating the same null hypothesis, but with different groups of subjects. Each tests the same null hypothesis at the 5% level. If the null is true, what is the chance of one or more centres producing a significant study (at 5% level)? What is the distribution of the number of centres that produce a (5% level) significant study? If (exactly) one centre gets a P-value < 0.05, what should you conclude? (What we really need is a *meta-analysis*, which is a methodology for combining the evidence from many studies properly.)

54. Lots of tests Suppose you run a study to compare a new drug with a placebo, and run a new test at the 5% level every day on the evolving data. The null hypothesis says there is no difference. On the 20th day, you find a significance difference at the 5% level. Should you celebrate the drug's success? Explain.

Just Checking ANSWERS

1. With a z-score of 0.62, you can't reject the null hypothesis. The experiment shows no evidence that the wheel is not fair.

2. At $\alpha = 0.10$ for a two-sided alternative, or at $\alpha = 0.05$ for a one-sided alternative, you can't reject the null hypothesis because 0.30 is contained in the 90% confidence interval—it's plausible that sending the DVDs is no more effective than just sending letters.

3. The confidence interval is from 29% to 45%. The DVD strategy is more expensive and may not be worth it. We can't distinguish the success rate from 30% given the results of this experiment, but 45% would represent a large improvement. The bank should consider another trial, increasing their sample size to get a narrower confidence interval.

4. A Type I error would mean deciding that the DVD success rate is higher than 30% when it really isn't. They would adopt a more expensive method for collecting payments that would be no better than the less expensive strategy.

5. A Type II error would mean deciding that there's not enough evidence to say that the DVD strategy works when in fact it does. The bank would fail to discover an effective method for increasing their revenue from delinquent accounts.

6. 60%; the larger the effect size, the greater the power. It's easier to detect an improvement to a 60% success rate than to a 32% rate.

MathXL	Go to MathXL at **www.mathxl.com** or MyLab™ Statistics at **www.mystatlab.com**. You can practise exercises for this chapter as often as you want. The guided solutions will help you find answers step by step. You'll find a
MyLab Statistics	personalized study plan available to you too!

Inferences About Means

LEARNING OBJECTIVES

1. Know the sampling distribution of the mean

2. Understand the need for Gosset's *t* model

3. Construct a *t*-Interval for the mean

4. Test hypotheses about the mean using
   - *P*-values
   - Critical values

5. Check the assumptions and conditions

6. Determine the sample size for specified
   - Margin of error
   - Power of test

A curve has been found representing the frequency distribution of values of the means of such samples (from a normal population), when these values are measured from the mean of the population in terms of the standard deviation of the sample . . .

—William Gosset

Connections

We've learned how to generalize from the data at hand to the world at large for proportions. But not all data are as simple as "yes" or "no." In this chapter, we'll learn how to make confidence intervals and test hypotheses for the mean of a quantitative variable.

The steps for finding a confidence interval or hypothesis test for means are just like the corresponding steps for proportions. But we'll need a new sampling model, called the '*t*' distribution. As the *z*-statistic did for proportions, the *t*-statistic will tell us how many standard errors our sample mean is from the hypothesized mean.

As with all of our inference methods, the randomization applied in drawing a

Travelling back and forth to work or school can be a real pain (though a good seat on the bus or subway can provide a chance to read, study, maybe snooze . . .). Since 2000, the International Census at School Project has surveyed over a million school students from Canada, the U.K., Australia, New Zealand, and South Africa using educational activities conducted in class. School participation is on a voluntary basis. Tens of thousands of Canadian students participate each year. One question commonly asked in the survey is, "How long does it usually take you to travel to school?"

So just how long does it take Ontario students to get to school? Times vary from student to student, but knowing the average would be helpful. As we've learned, a single number or estimate that is almost surely wrong is not as useful as a range of values (or confidence interval) that is almost surely correct. Using the random data selector from the Census at School project, the responses (in minutes) were obtained for a random sample of 40 participating Ontario secondary school students from 2007 to 2008.[1] Figure 18.1 shows their commute times and a histogram (Data in **Ontario_Students**)

Figure 18.1

The travel times (to school) of Ontario secondary students appear to be unimodal and perhaps slightly right-skewed.

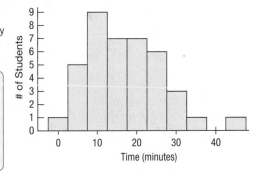

Time						
30	20	18	25	30	20	10
8	13	5	10	10	15	25
15	10	15	25	22	5	20
25	7	20	2	25	35	12
15	10	8	47	15	20	
25	30	15	5	5	8	

These data differ from data on proportions in one important way. Proportions are summaries of individual responses, which have two possible values, such as "yes" and "no," "male" and "female," or "1" and "0." Quantitative data, however, usually report a quantitative value for each individual, in which case, you should recall the three rules of data analysis and *plot the data,* as we have done here.

With quantitative data, we summarize with measures of centre and spread, such as the mean and standard deviation. Because we want to make inferences, we'll need to think about sampling distributions, which will lead us to a new sampling model. But first, let's review what we learned earlier about quantitative variables and their sample means.

---

[1]www.censusatschool.ca. We thought about updating to a more recent school year, but in October 2016, we read at the site: "This summer, the funding for the international Census at School Project ended abruptly, and with it the random data sampler." The good news is that they hope to fix things.

random sample or in randomizing a comparative experiment is what generates the sampling distribution and makes inference possible.

There's just no escaping histograms and Normal probability plots - since in order to use the *t* sampling model, we will need to check that the data are roughly Normal.

Who	Secondary school students
What	Time to travel to school
Units	Minutes
When	2007–2008
Where	Ontario
Why	To learn about the time spent by Ontario students travelling to school
How	Taking a random sample from the Census At School data base

18.1 The Sampling Model for the Sample Mean

You've learned how to create confidence intervals and test hypotheses about proportions. We always centre confidence intervals at our best guess of the unknown parameter. Then, we add and subtract a margin of error. For proportions, we write $\hat{p} \pm ME$.

We found the margin of error as the product of the standard error, $SE(\hat{p})$, and a critical value, z^*, from the Normal table. So we had $\hat{p} \pm z^*SE(\hat{p})$.

We knew we could use a z^* value from the Normal distribution because the Central Limit Theorem (CLT) told us (in Chapter 14) that the sampling distribution model for proportions is Normal.

Now we want to do exactly the same thing for means, and the CLT (still in Chapter 14) told us that the same Normal model works as the sampling distribution for means of sufficiently large samples. Here again is what we know about the sample mean:

THE SAMPLING DISTRIBUTION OF THE SAMPLE MEAN

When a random sample is drawn from any population with mean μ and standard deviation σ, its sample mean, $\bar{y}$, has a sampling distribution with the same *mean μ* but whose *standard deviation* is $\frac{\sigma}{\sqrt{n}}$ (and we write $\sigma(\bar{y})$ or $SD(\bar{y}) = \frac{\sigma}{\sqrt{n}}$).

The CLT also tells us that no matter what population the random sample comes from, the *shape* of the sampling distribution is approximately Normal as long as the sample size is large enough. The larger the sample used, the more closely the Normal approximates the sampling distribution of the sample mean.

For Example WORKING WITH THE SAMPLE MEAN (WHEN WE KNOW σ)

Based on weighing thousands of animals, the American Angus Association reports that mature Angus cows have a mean weight of 1309 pounds (1 pound = 0.4536 kg) with a standard deviation of 157 pounds. This result was based on a very large sample of animals from many herds over a period of 15 years, so let's assume that these summaries are the population parameters and that the distribution of the weights was unimodal and not severely skewed.

QUESTION: What can we predict about the mean weight seen in random samples of 100 mature Angus cows?

ANSWER: It's given that weights of all mature Angus cows have $\mu = 1309$ and $\sigma = 157$ pounds. I expect the resulting sample means $\bar{y}$ will average 1309 pounds and have a standard deviation of $SD(\bar{y}) = \frac{\sigma}{\sqrt{n}} = \frac{157}{\sqrt{100}} = 15.7$ pounds.

Because $n = 100$ animals is a fairly large sample, I can apply the CLT, which says that the distribution of sample means approximately follows a Normal model, so the 68–95–99.7 Rule applies. I'd expect that

- in 68% of random samples of 100 mature Angus cows, the mean weight will be between $1309 - 15.7 = 1293.3$ and $1309 + 15.7 = 1324.7$ pounds;
- in 95% of such samples, $1277.6 \le \bar{y} \le 1340.4$ pounds;
- in 99.7% of such samples, $1261.9 \le \bar{y} \le 1356.1$ pounds.

So, in order to model the sampling distribution of $\bar{y}$ with a Normal curve, we need (by the CLT) a sufficiently large random sample of quantitative data.

And the true population standard deviation, σ.

Uh oh. That could be a problem. Unlike the example above, in an inferential situation, we will only have a sample and will not know the true population mean or standard deviation. With proportions, we had a link between the proportion value and the standard deviation of the sample proportion: $SD(\hat{p}) = \sqrt{\dfrac{pq}{n}}$. And there was an obvious way to estimate the standard deviation from the data: $SE(\hat{p}) = \sqrt{\dfrac{\hat{p}\hat{q}}{n}}$. For means, $SD(\bar{y}) = \dfrac{\sigma}{\sqrt{n}}$. We know n, the sample size, but the population standard deviation, σ, could be *anything*. So what should we do? We do what any sensible person would do: We estimate the population parameter σ with s, the sample standard deviation based on the data. The resulting standard error is $SE(\bar{y}) = \dfrac{s}{\sqrt{n}}$.

> **STANDARD ERROR**
>
> Because we estimate the standard deviation of the sampling distribution model from the data, we'll call it a *standard error*. So we use the $SE(\bar{y})$ notation. Remember, though, that it's just the estimated standard deviation of the sampling distribution model for means.

A century ago, people used this standard error with the Normal model, assuming it would work. And for large sample sizes it *did* work pretty well (as mentioned in optional Section 15.6). But they began to notice problems with smaller samples. The sample standard deviation, s, like any other statistic, varies from sample to sample. And this extra variation in the standard error was messing up the P-values and margins of error.

William S. Gosset is the man who first investigated this problem. He realized that not only do we need to allow for the extra variation with larger margins of error and P-values, but we even need a new sampling distribution model. In fact, we need a whole *family* of models, depending on the sample size, n. These models are unimodal, symmetric, bell-shaped models, but the smaller our sample, the more we must stretch out the tails. Gosset's work transformed Statistics, but most people who use his work don't even know his name.

Section 18.1 Exercises

1. Salmon A specialty food company sells whole King Salmon to various customers. The mean weight of these salmon is 35 pounds with a standard deviation of 2 pounds. The company ships them to restaurants in boxes of 4 salmon, to grocery stores in cartons of 16 salmon, and to discount outlet stores in pallets of 100 salmon. To forecast costs, the shipping department needs to estimate the standard deviation of the mean weight of the salmon in each type of shipment
   a) Find the standard deviations of the mean weight of the salmon in each type of shipment.
   b) The distribution of the salmon weights turns out to be skewed to the high end. Would the distribution of shipping weights be better characterized by a Normal model for the boxes or pallets? Explain.

2. LSAT The LSAT (a test taken for law school admission) has a mean score of 151 with a standard deviation of 9 and a unimodal, symmetric distribution of scores. A test preparation organization teaches small classes of 9 students at a time. A larger organization teaches classes of 25 students at a time. Both organizations publish the mean scores of all their classes.
   a) What would you expect the distribution of mean class scores to be for each organization?
   b) If either organization has a graduating class with a mean score of 160, they'll take out a full-page ad in the local school paper to advertise. Which organization is more likely to have that success? Explain.
   c) Both organizations advertise that if any class has an average score below 145, they'll pay for everyone to retake the LSAT. Which organization is at greater risk of having to pay?

18.2 Gosset's *t*

Gosset had a job that made him the envy of many. He was the chief Experimental Brewer for the Guinness Brewery in Dublin, Ireland. The brewery was a pioneer in scientific brewing and Gosset's job was to meet the demands of the brewery's many discerning customers by developing the best stout (a thick, dark beer) possible.

Gosset's experiments often required as much as a day to make the necessary chemical measurements or a full year to grow a new crop of hops. For these reasons, not to mention his health, his sample sizes were small—often as small as three or four.

International Statistical Institute (ISI)

To find the sampling distribution of $\dfrac{\bar{y} - \mu}{s/\sqrt{n}}$, Gosset simulated it *by hand*.

He drew 750 samples of size 4 by shuffling 3000 cards on which he'd written the heights of some prisoners and computed the means and standard deviations with a mechanically cranked calculator. (He knew μ because he was simulating and knew the population from which his samples were drawn.) Today, you could repeat in seconds on a computer the experiment that took him over a year. Gosset's work was so meticulous that not only did he get the shape of the new histogram approximately right, but he even figured out the exact *formula* for it from his sample. The formula was not confirmed mathematically until years later by Sir R. A. Fisher.

When he calculated means of these small samples, Gosset wanted to compare them to a target mean to judge the quality of the batch. To do so, he followed common statistical practice of the day, which was to calculate *z*-scores and compare them to the Normal model. But Gosset noticed that with samples of this size, his tests weren't quite right. He knew this because when the batches that he rejected were sent back to the laboratory for more extensive testing, too often they turned out to be OK. In fact, about three times more often than he expected. Gosset knew something was wrong, and it bugged him.

Guinness granted Gosset time off to earn a graduate degree in the emerging field of Statistics, and naturally he chose this problem to work on. He figured out that when he used the standard error, $\dfrac{s}{\sqrt{n}}$, as an estimate of the standard deviation of the mean, the shape of the sampling model changed. He even figured out what the new model should be.

The Guinness Company may have been ahead of its time in using statistical methods to manage quality, but they also had a policy that forbade their employees to publish. Gosset pleaded that his results were of no specific value to brewers and was allowed to publish under the pseudonym "Student," chosen for him by Guinness's managing director. Accounts differ about whether the use of a pseudonym was to avoid ill feelings within the company or to hide from competing brewers the fact that Guinness was using statistical methods. In fact, Guinness was alone in applying Gosset's results in their quality assurance operations.

It was a number of years before the true value of "Student's" results was recognized. By then, statisticians knew Gosset well, as he continued to contribute to the young field of Statistics. But this important result is still widely known as **Student's *t***.

Gosset's sampling distribution model is always bell-shaped, but the details change with different sample sizes. When the sample size is very large, the model is nearly Normal, but when it's small, the tails of the distribution are much heavier than the Normal. That means that values far from the mean are more common and that can be important for small samples (see Figure 18.2). So the Student's *t*-models form a whole *family* of related distributions that depend on a parameter known as **degrees of freedom**. The degrees of freedom of a distribution represent the number of independent quantities that are left after we've estimated the parameters. Here it's simply the number of data values, n, minus the number of estimated parameters. When we estimate one mean, that's just $n - 1$. We often denote degrees of freedom as *df* and the model as t_{df} with the degrees of freedom as a subscript.

Figure 18.2

The *t*-model (solid curve) on 2 degrees of freedom has fatter tails than the Normal model (dashed curve). So the 68–95–99.7 Rule doesn't work for *t*-models with only a few degrees of freedom. It may not look like a big difference, but a *t* with 2 df is more than 4 times as likely to have a value greater than 2 compared to a standard Normal.

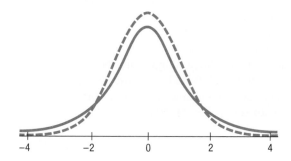

What Did Gosset See?

We can reproduce the simulation experiment that Gosset performed to get an idea of what he saw and how he reached some of his conclusions. Gosset drew 750 samples of size 4 from data on the heights of 3000 convicts. That population looks like this:[2]

---

[2]You might have noticed some gaps in the histogram. Gosset's height data were rounded to the nearest 1/8 inch, which made for some gaps. Gosset noted that flaw in his paper.

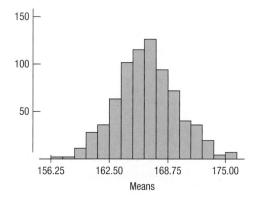

Following Gosset's example, we drew 750 independent random samples of size 4 and found their means and standard deviations.[3] As we (and Gosset) expected, the distribution of the means was even more Normal.[4]

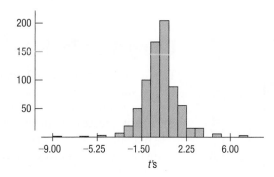

Gosset's concern was for the distribution of $\dfrac{\bar{y} - \mu}{s/\sqrt{n}}$. We know $\mu = 166.301$ cm because we know the population, and $n = 4$. The values of $\bar{y}$ and s we find for each sample. Here's what the distribution looks like:

It's easy to see that this distribution is much thinner in the middle and longer in the tails than a Normal model we saw for the means themselves. This was Gosset's principal result.

---

[3] In fact, Gosset shuffled 3000 cards with the numbers on them and then dealt them into 750 piles of four each. That's not quite the same thing as drawing independent samples, but it was quite good enough for his purpose. We've drawn these samples in the same way.

[4] Of course, we don't know the means that Gosset actually got because we randomized using a computer and he shuffled 3000 cards, but this is one of the distributions he might have gotten, and we're pretty sure most of the others look like this as well.

18.3　A *t*-Interval for the Mean

To make confidence intervals or test hypotheses for means, we need to use Gosset's model. Which one? Well, for means, it turns out the right value for degrees of freedom is $df = n - 1$.

■ NOTATION ALERT

Ever since Gosset, *t* has been reserved in Statistics for his distribution.

> **A PRACTICAL SAMPLING DISTRIBUTION MODEL FOR MEANS**
>
> When certain assumptions and conditions[5] are met, the standardized sample mean,
>
> $$t = \frac{\bar{y} - \mu}{SE(\bar{y})}$$
>
> follows a Student's *t*-model with $n - 1$ degrees of freedom. We estimate the standard deviation of $\bar{y}$ with
>
> $$SE(\bar{y}) = \frac{s}{\sqrt{n}}$$

When Gosset corrected the model for the extra uncertainty, the margin of error got bigger, as you might have guessed. When you use Gosset's model instead of the Normal model, your confidence intervals will be just a bit wider and your P-values just a bit larger. That's the correction you need. By using the *t*-model, you've compensated for the extra variability in precisely the right way.[6]

■ NOTATION ALERT

When we found critical values from a standard Normal model, we called them $z*$. When we use a Student's *t*-model, we'll denote the critical values $t*$.

> **ONE-SAMPLE *t*-INTERVAL FOR THE MEAN**
>
> When the assumptions and conditions[7] are met, we are ready to find the confidence interval for the population mean, μ. The confidence interval is
>
> $$\bar{y} \pm t_{n-1}^{*} \times SE(\bar{y})$$
>
> where the standard error of the mean $SE(\bar{y}) = \dfrac{s}{\sqrt{n}}$.
>
> The critical value t_{n-1}^{*} depends on the particular confidence level, C, that you specify and on the number of degrees of freedom, $n - 1$, which we get from the sample size.

Two tail probability	0.20	0.10	0.05
One tail probability	0.10	0.05	0.025
Table T　df			
Values of t_α　1	3.078	6.314	12.706
2	1.886	2.920	4.303
3	1.638	2.353	3.182
4	1.533	2.132	2.776
5	1.476	2.015	2.571
6	1.440	1.943	2.447
7	1.415	1.895	2.365
8	1.397	1.860	2.306
9	1.383	1.833	2.262
10	1.372	1.812	2.228
11	1.363	1.796	2.201
12	1.356	1.782	2.179
13	1.350	1.771	2.160
14	1.345	1.761	2.145
15	1.341	1.753	2.131
16	1.337	1.746	2.120
17	1.333	1.740	2.110
18	1.330	1.734	2.101
19	1.328	1.729	2.093

Part of Table T

Using the *t*-Table to Find Critical Values

The Student's *t*-model is different for each value of degrees of freedom. Usually we find critical values and margins of error for Student's *t*-based intervals with technology. Calculators or statistics programs can give critical values for a *t*-model for any number of degrees of freedom and for any confidence level you choose.

But you can also use tables, such as Table T at the back of this book. The tables run down the page for as many degrees of freedom as can fit. For *enough* degrees of freedom, the *t*-model gets closer and closer to the Normal, so the tables give a final row with the critical values from the Normal model and label it "∞ *df*."

These tables are only a portion of the full tables, such as the one we used for the Normal model. We could have printed a book with a table like Table Z for every df, but that's

---

[5]You can probably guess what they are. We'll see them in the next section.

[6]Gosset, as the first to recognize the consequence of using *s* rather than σ, was also the first to give the sample standard deviation, *s*, a different letter than the population standard deviation σ.

[7]Yes, the same ones, and they're still coming in the next section.

a lot of pages and not likely to be a bestseller. To shorten the book we could limit ourselves to only a few confidence levels. Although it might be nice to be able to get a critical value for a 93.4% confidence interval with 179 df, in practice we usually limit ourselves to 90%, 95%, 99%, and 99.9% and selected degrees of freedom. So, Table T fits on a single page with columns for selected confidence levels and rows for selected df's.[8]

For confidence intervals, the values in the table are usually enough to cover most cases of interest. If you can't find a row for the df you need, just be conservative and use the next smaller df in the table.

> If your df's aren't listed, take the conservative approach by using the *next lower df value*, or use technology.

For Example A ONE-SAMPLE *t*-INTERVAL FOR THE MEAN

In 2004, a team of researchers published a study of contaminants in farmed salmon.[9] Fish from many sources were analyzed for 14 organic contaminants. The study expressed concerns about the level of contaminants found. One of those was the insecticide mirex, which has been shown to be carcinogenic and is suspected to be toxic to the liver, kidneys, and endocrine system. One farm in particular produced salmon with very high levels of mirex. After those outliers are removed, summaries for the mirex concentrations (in parts per million) in the rest of the farmed salmon are (Data in **Farmed_Salmon**):

$$n = 150 \quad \bar{y} = 0.0913 \text{ ppm} \quad s = 0.0495 \text{ ppm}.$$

QUESTION: What does a 95% confidence interval say about mirex?

$$df = 150 - 1 = 149$$

$$SE(\bar{y}) = \frac{s}{\sqrt{n}} = \frac{0.0495}{\sqrt{150}} = 0.0040$$

$$t^*_{149} \approx 1.977 \text{ (from table } T, \text{ using 140 df)}$$

(actually, $t^*_{149} \approx 1.976$ from technology)

ANSWER: So the confidence interval for μ is

$$\bar{y} \pm t^*_{149} \times SE(\bar{y})$$

$$0.0913 \pm 1.977 (0.0040)$$

$$0.0913 \pm 0.0079$$

$$(0.0834, 0.0992)$$

I'm 95% confident that the mean level of mirex concentration in farm-raised salmon is between 0.0834 and 0.0992 parts per million.

Student's *t*-models are all unimodal, symmetric, and bell-shaped, like the Normal. But *t*-models with only a few degrees of freedom have noticeably longer tails and larger standard deviation than the Normal. (That's what makes the margin of error bigger.) As the degrees of freedom increase, the *t*-models look more and more like the standard Normal. In fact, the *t*-model with infinite degrees of freedom is exactly the standard Normal.[10] This is great news if you happen to have an infinite number of data values, but that's not likely. *However, above about 60 degrees of freedom, it's very hard to tell the difference.* Of course, in the very rare situation that we *know* σ, it would be foolish not to use that information, and if we don't have to estimate σ, we can use the Normal model.

---

[8]You can also find tables and interactive tools on the Internet. Search for terms like "statistical tables t."

[9]Ronald A. Hites, Jeffery A. Foran, David O. Carpenter, M. Coreen Hamilton, Barbara A. Knuth, and Steven J. Schwager, 2004, "Global assessment of organic contaminants in farmed salmon," *Science* 303: 5655, pp. 226–229.

[10]Formally, in the limit as *n* goes to infinity.

z OR t?

If you know σ, use the Normal curve as your sampling model. (That's rare!) Whenever you use s to estimate σ, use t (though for large df, the t is well approximated by the standard Normal)

When σ is known. Administrators of a hospital were concerned about the prenatal care given to mothers in their part of the city. To study this, they examined the gestation times of babies born there. They drew a sample of 25 babies born in their hospital in the previous six months. Human gestation times for healthy pregnancies are thought to be well modelled by a Normal curve with a mean of 280 days and a standard deviation of 10 days. The hospital administrators wanted to test the mean gestation time of their sample of babies against the known standard. For this test, they used the established value for the standard deviation, 10 days, rather than estimating the standard deviation from their sample. Because they used the model parameter value for σ, they based their test on the standard Normal model rather than Student's t.

Assumptions and Conditions

Gosset found the t-model by simulation. Years later, Sir Ronald A. Fisher showed mathematically that Gosset was right and confirmed the assumptions needed by Gosset in discovering the t curve—that we are making *repeated independent draws from a Normally distributed population*. Now for our practical advice:

Independence Assumption   The data values should be mutually independent. There's really no way to check independence of the data by looking at the sample, but you should think about whether the assumption is reasonable.

   Randomization Condition: This condition is satisfied if the data arise from a random sample or suitably randomized experiment. Randomly sampled data, especially data from a simple random sample, are ideal—almost surely independent, with well-defined target population. If the data don't satisfy the **Randomization Condition**, then you should think about whether the values are likely to be independent for the variables you are concerned with and whether the sample you have is likely to be representative of the population you wish to learn about. Cluster and multistage samples, though, may have bigger SEs than suggested by our formula.

   In the rare case that you have a sample that is a more than 10% of the population, you may want to consider using special formulas that adjust for that. But that's not a common concern for means. Without the corrected formula, your SE will just err on the conservative side (be too high). This is technically a violation of the independence assumption, but a good one, since the effects are known and beneficial.

Normal Population Assumption   Student's t-models won't work for data that are badly skewed. How skewed is too skewed? Formally, we assume that the data are from a population that follows a Normal model. Practically speaking, there's no way to be sure this is true.

   And it's almost certainly *not* true. Models are idealized; real data are, well, real. The good news, however, is that even for small samples, it's sufficient to check the . . .

   Nearly Normal Condition: The data come from a distribution that is unimodal and reasonably symmetric.

   Check this condition by making a histogram or Normal probability plot. Normality of the individual observations is less important, to Student's t sampling model, for larger sample sizes. Just our luck: It matters most when it's hardest to check.[11]

   For very small samples ($n < 15$ or so), the data should follow a Normal model fairly closely. Of course, with so little data, it's rather hard to tell. But if you do find outliers or clear skewness, don't trust these methods.

   For moderate sample sizes (n between 15 and 40 or so), the t methods will work reasonably well for mildly to moderately skewed unimodal data, but would perform badly in the presence of strong skewness or outliers. Make a histogram.

WHEN THE ASSUMPTIONS FAIL

When you check conditions, you usually hope to make a meaningful analysis of your data. The conditions serve as *disqualifiers*—keep going unless there's a serious problem. If you find minor issues, note them and express caution about your results. If the sample is not an SRS, but you believe it's representative of some population, limit your conclusions accordingly. If there are outliers, perform the analysis both with and without them. If the sample looks bimodal, try to analyze subgroups separately. Only when there's major trouble—like a strongly skewed small sample or an obviously nonrepresentative sample—are you unable to proceed at all.

[11]There are formal tests of Normality, but they don't really help. When we have a small sample—just when we really care about checking Normality—these tests are not very effective (low power). So it doesn't make much sense to use them in deciding whether to perform a t-test. We don't recommend that you use them.

When the sample size is larger than 40 or 50, the *t* methods are generally quite safe to use, though very severe skewness can require much larger sample sizes (in which case it might be better to apply a non-linear transformation, like a logarithm, to reduce skewness; or you might want to focus inference on the median rather than the mean). Be sure to make a histogram. If you find outliers in the data, it's always a good idea to perform the analysis twice, once with and once without the outliers, even for large samples. Outliers may well hold additional information about the data, but you may decide to give them individual attention and then summarize the rest of the data. If you find multiple modes, you may have different groups that should be analyzed and understood separately.

Guinness stout may be hearty, but the *t*-procedure is robust!
The one-sample *t*-test is an example of a *robust statistical test*. We say that it is robust with respect to the assumption of Normality, or against violations of Normality. This means that although the procedure is derived mathematically from an assumption of Normality, it can still often produce accurate results even when this assumption is violated. How well does the procedure tolerate violations of assumptions? How greatly do violations perturb the accuracy of P-values and confidence levels? These are questions about the robustness of the procedure.

The bigger the violations that can be tolerated, the greater is the robustness of the procedure. Robustness for most procedures will increase with the size of the sample. The robustness of the one-sample *t*-procedure is described above, where we see moderate robustness for sample sizes as small as 15 and remarkable robustness for samples over size 40. Pretty impressive! And the two-sample *t*-procedure to be discussed in the next chapter is even more robust. The usefulness of these *t*-procedures would be greatly compromised were it not for their high level of robustness. Similarly, many other common statistical procedures have good robustness, increasing their utility and value.

For Example CHECKING ASSUMPTIONS AND CONDITIONS FOR STUDENT'S *t*

RECAP: Researchers purchased whole farmed salmon from 51 farms in eight regions in six countries. The following histogram shows the concentrations of the insecticide mirex in 150 farmed salmon (after removing some outliers, as mentioned earlier).

QUESTION: Are the assumptions and conditions for inference about the mean satisfied?

ANSWER:

- *Independence/Randomization:* The fish were not a random sample because no simple population existed to sample from. But they were raised in many different places, and samples were purchased independently from several sources, so they were likely to be nearly independent and to reasonably represent the population of farmed salmon worldwide.

- *Nearly Normal Condition:* The histogram of the data is unimodal. Although it may be somewhat skewed to the right, this is not a concern with a sample size of 150.

It's okay to use these data for inference about farm-raised salmon. Whew, now we know that we can actually trust that mechanical confidence interval calculation done earlier! Anyone can plug into a formula; the hard part is determining whether your procedure gives trustworthy results and answers.

Just Checking

Every five years, Statistics Canada conducts a census in order to compile a statistical portrait of Canada and its people. There are two forms: the short questionnaire, distributed to 80% of households, and the long questionnaire[12] (short-form questions plus additional questions), slogged through by the remaining one in five households, chosen at random. For estimates resulting from the additional questions appearing only on the long form, Statistics Canada calculates a standard error.

1. Why does Statistics Canada need to calculate a standard error for long-form information, but not for the questions that appear on both the long and short forms?

2. The standard errors are calculated after re-weighting the individual results, to correct for differences between the sample proportion who are male, aged 15–24, etc., and the known (from the long form) population proportions who are male, aged 15–24, etc., so that the resulting estimates will be more precise (for example, if we know that 50% of residents in a region are male, and 52% of the 20% sample are male, each male is given a slightly lower weight or multiplier than each female to "correct" for the overrepresentation of males). Hence, a simple average (for quantitative variables) or simple proportion is not used. Can Statistics Canada use the t-model for standard errors and associated confidence intervals (for quantitative variables)? If simple (unweighted) averages were used instead, could we employ the t-model?

3. The standard errors calculated by Statistics Canada are bigger for geographic areas with smaller populations and for characteristics of small sub-groups in the area examined (such as people living in poverty in a middle-income neighbourhood). Why is this so? For example, why should a confidence interval for mean family income be wider for a sparsely populated area of farms in the Prairies than for a densely populated area in an urban centre? How does the t-model formula show that this will happen?

   [To deal with this problem, Statistics Canada classifies estimates based on "data quality" (the size of the associated standard error relative to the estimate), warns of low-quality (high standard error) estimates, and omits low-quality estimates with excessively high standard errors, since the latter are essentially noninformative and also have the potential to compromise confidentiality, due to the small number of cases.]

4. Suppose that in one census tract, there were 200 First Nations individuals in the long-form sample, and we calculate their mean annual earnings, along with a standard error and 95% confidence interval, using the simple t-model. In another census tract, we would like to calculate a similar confidence interval, but there were only 50 First Nations people in the sample. What effect would the smaller number of First Nations people in the second tract have on the 95% confidence interval? Specifically, which values used in the formula for the margin of error would change a lot and which would change only slightly? Approximately how much wider would the confidence interval based on 50 individuals be than the one based on 200 individuals?

---

[12]In June 2010, the minority Conservative government decided to do away with the mandatory long form and to replace it with the voluntary National Household Survey, in spite of significant opposition, citing privacy concerns. In November 2015, the newly elected Liberal government restored the mandatory long form census.

Step-by-Step Example A ONE-SAMPLE *t*-INTERVAL FOR THE MEAN

Let's build a 90% confidence interval for the mean travel time to school for Ontario secondary school students.

Question: What can we say about the mean travel time to school for secondary school students in Ontario?

THINK ➡ **Plan** State what we want to know. Identify the parameter of interest.

Identify the variables and review the *W*'s.

Make a picture. Check the distribution shape and look for skewness, multiple modes, and outliers.

REALITY CHECK The histogram centres around 15–20 minutes, and the data lie between 0 and 50 minutes. We'd expect a confidence interval to place the population mean close to 15 or 20 minutes.

I want to find a 90% confidence interval for the mean travel time to school for Ontario secondary school students. I have data on the travel time of 40 students in 2007–2008.

Here's a histogram of the 40 travel times.

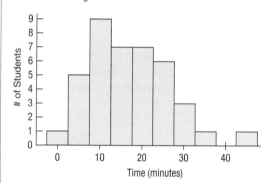

Model Think about the assumptions and check the conditions.

✓ **Independence Assumption**: These are independent selections from the stored data.

✓ **Randomization Condition**: Participation was voluntary but very broad-based across the province. I believe the students we randomly selected from the database should be reasonably representative of Ontario students.

✓ **Nearly Normal Condition**: The histogram of the speeds is unimodal and slightly right-skewed, but not enough to be a concern.

State the sampling distribution model for the statistic.

Choose your method.

The conditions are satisfied, so I will use a Student's *t*-model with

$$n - 1 = 39 \text{ degrees of freedom}$$

and find a **one-sample *t*-interval for the mean**.

SHOW ➡ **Mechanics** Construct the confidence interval.

Be sure to include the units along with the statistics.

Calculating from the data given at the beginning of this chapter:

$$n = 40 \text{ students}$$
$$\bar{y} = 17.00 \text{ minutes}$$
$$s = 9.66 \text{ minutes}$$

The standard error of $\bar{y}$ is

$$SE(\bar{y}) = \frac{s}{\sqrt{n}} = \frac{9.66}{\sqrt{40}} = 1.53 \text{ minutes}$$

The critical value we need to make a 90% interval comes from a Student's *t* table, a computer program, or a calculator. We have $40 - 1 = 39$ degrees of freedom. The selected confidence level says that we want 90% of the probability caught in the middle, or 5% in each tail, which is all we need to know to find the critical value.

REALITY CHECK   The result looks plausible and in line with what we thought.

The 90% critical value is $t^*_{39} = 1.685$ (using software), so the margin of error is

$$ME = t^*_{39} \times SE(\bar{y})$$
$$= 1.685(1.53)$$
$$= 2.58 \; minutes$$

The 90% confidence interval for the mean travel time is 17.0 ± 2.6 minutes $= (14.4, 19.6)$ minutes.

TELL ➡ **Conclusion** Interpret the confidence interval in the proper context.

When we construct confidence intervals in this way, we expect 90% of them to cover the true mean and 10% to miss the true value. That's what "90% confident" means.

I am 90% confident that the interval from 14.4 to 19.6 minutes contains the true mean travel time to school for Ontario secondary school students.

Here's the part of the Student's *t* table, in the back of the book, that gives the critical value we needed. Locate the row of the table corresponding to the degrees of freedom and the column corresponding to the probability you want. Our 90% confidence interval leaves 5% of the values on either side, so look for a one-tail probability of 0.05 at the top of the column or 90% confidence level at the bottom (not shown in excerpt below). The value in the table at that intersection is the critical value we need, but unfortunately, this concise table omits 39 *df*. The correct value lies between 1.684 and 1.690. Either be conservative and go with the bigger value, 1.690 (smaller df), or use software (as we did above).

Using Table T to look up the critical value t* for a 90% confidence level with 39 degrees of freedom.

| Probability | **1.684–1.690** |

| Two-tail | 0.20 | 0.10 | 0.05 | 0.02 | 0.01 |
One-tail	0.10	0.05	0.025	0.01	0.005
28	1.313	1.701	2.048	2.467	2.763
29	1.311	1.699	2.045	2.462	2.756
30	1.310	1.697	2.042	2.457	2.750
32	1.309	1.694	2.037	2.449	2.738
35	1.306	1.690	2.030	2.438	2.725
40	1.303	1.684	2.021	2.423	2.704
45	1.301	1.679	2.014	2.412	2.690
50	1.299	1.676	2.009	2.403	2.678
60	1.296	1.671	2.000	2.390	2.660

Of course, you can also create the entire confidence interval with the right computer software or calculator.

Make a Picture, Make a Picture, Make a Picture

The only reasonable way to check the Nearly Normal Condition is with graphs of the data. Make a histogram and Normal probability plot of the data to catch possible problems, like severe skewness, bimodality, and outliers. You can spot deviations from the Normal model more easily with a Normal probability plot, but it's easier to understand the particular nature of the deviations from a histogram. If you have a computer or graphing calculator doing the work, there's no excuse not to look at *both* displays. Figure 18.3 shows an NPP for the student travel times.

Figure 18.3

A Normal probability plot of travel times looks a bit curved but close enough to straight (considering the sample size).

Interpreting Confidence Intervals

Confidence intervals for means offer new, tempting, and wrong interpretations. Here are some things you *shouldn't* say:

- ■ *Don't say,* "*90% of all Ontario secondary school students* take between 14.4 and 19.6 minutes to get to school." The confidence interval is about the *mean* travel time, not about the times of *individual* students.

- ■ *Don't say,* "We are 90% confident that *a randomly selected student* will take between 14.4 and 19.6 minutes to get to school." This false interpretation is also about individual students rather than about the *mean* of their times. We are 90% confident that the *mean* travel time of all secondary school students is between 14.4 and 19.6 minutes.

- ■ *Don't say,* "The mean student travel time is 17.0 minutes, 90% *of the time.*" That's about means, but still wrong. It implies that the true mean varies, when in fact it is the confidence interval that would have been different had we gotten a different sample.

- ■ Finally, *don't say,* "90% *of all samples* will have mean travel times between 14.4 and 19.6 minutes." That statement suggests that this interval somehow sets a standard for every other interval. In fact, this interval is no more (or less) likely to be correct than any other. You could say that 90% of all possible samples will produce intervals that actually do contain the true mean time. (The problem is that, because we'll never know where the true mean time really is, we can't know if our sample was one of those 90%.)

- ■ *Do say,* "90% of intervals found in this way cover the true value." Or make it more personal and say, "I am 90% confident that the true mean student travel time is between 14.4 and 19.6 minutes."

SO WHAT *SHOULD* WE SAY?

Since 90% of random samples yield an interval that captures the true mean, we *should* say, "I am 90% confident that the interval from 14.4 to 19.6 minutes contains the mean travel time for Ontario secondary school students." It's also okay to say something less formal: "I am 90% confident that the average travel time for Ontario secondary school students is between 14.4 and 19.6 minutes." Remember: *Our uncertainty is about the interval, not the true mean.* The interval varies randomly. The true mean travel time is neither variable nor random—just unknown.

Just Checking

One disadvantage of using both long and short census forms is that estimates of characteristics that are reported on the short form will not exactly match the long-form estimates.

Short form summary measures are computed from a complete census, so they are the "true" values—something we don't usually have when we do inference.

5. Suppose we use long-form data to make 95% confidence intervals for the mean age of residents for each of 100 census tracts. How many of these 100 intervals should we expect will fail to include the true mean age (as determined from the complete census data)?

Section 18.3 Exercises

3. *t*-models, part I Using the *t* tables, software, or a calculator, estimate
   a) the critical value of *t* for a 90% confidence interval with df = 17.
   b) the critical value of *t* for a 98% confidence interval with df = 88.

4. *t*-models, part II Using the *t* tables, software, or a calculator, estimate
   a) the critical value of *t* for a 95% confidence interval with df = 7.
   b) the critical value of *t* for a 99% confidence interval with df = 102.

5. *t*-models, part III Describe how the shape, centre, and spread of *t*-models change as the number of degrees of freedom increases.

6. *t*-models, part IV Describe how the critical value of *t* for a 95% confidence interval changes as the number of degrees of freedom increases.

7. **Home sales** The housing market has recovered slowly from the economic crisis of 2008. Recently, in one large community, realtors randomly sampled 36 bids from potential buyers to estimate the average loss in home value. The sample showed the average loss from the peak in 2008 was $9560 with a standard deviation of $1500.
   a) What assumptions and conditions must be checked before finding a confidence interval? How would you check them?
   b) Find a 95% confidence interval for the mean loss in value per home.

c) Interpret this interval and explain what 95% confidence means in this context.

8. **Home sales again** In the previous exercise, you found a 95% confidence interval to estimate the average loss in home value.
   a) Suppose the standard deviation of the losses had been $3000 instead of $1500. What would the larger standard deviation do to the width of the confidence interval (assuming the same level of confidence)?
   b) Your classmate suggests that the margin of error in the interval could be reduced if the confidence level were changed to 90% instead of 95%. Do you agree with this statement? Why or why not?
   c) Instead of changing the level of confidence, would it be more statistically appropriate to draw a bigger sample?

18.4 Hypothesis Test for the Mean

Students and their parents are naturally concerned about how long the commute to school takes. Suppose the Ministry of Education claims that the average commute time for secondary school students is no greater than 15 minutes. But you're not so sure, particularly after collecting some data and finding a sample mean higher than 15 minutes. Maybe this is just chance variation or maybe you've found real evidence to dispute the Ministry's claim about the mean commute time. How can we tell the difference? A question like this about the mean of a population calls for a hypothesis test, and the one we need here is the **one-sample *t*-test for the mean**.

You already know enough to construct this test. The test statistic looks just like the others we've seen. It compares the difference between the observed statistic and a hypothesized value to the standard error of the observed statistic. We've seen that, for means, the appropriate probability model to use for P-values is Student's *t* with $n - 1$ degrees of freedom.

ONE-SAMPLE *t*-TEST FOR THE MEAN

The assumptions and conditions for the one-sample *t*-test for the mean are the same as for the one-sample *t*-interval. We test the hypothesis $H_0: \mu = \mu_0$ using the statistic

$$t = \frac{\bar{y} - \mu_0}{SE(\bar{y})}$$

The standard error of $\bar{y}$ is $SE(\bar{y}) = \dfrac{s}{\sqrt{n}}$

When the conditions are met and the null hypothesis is true, this statistic follows a Student's *t*-model with $n - 1$ degrees of freedom. We use that model to obtain a P-value.

If you have to make a decision, set your α-level *a priori*, and reject H_0 if $P < \alpha$.

For Example   A ONE-SAMPLE *t*-TEST FOR THE MEAN

RECAP: Researchers tested 150 farm-raised salmon for organic contaminants. They found the mean concentration of the carcinogenic insecticide mirex to be 0.0913 parts per million, with standard deviation 0.0495 ppm. As a safety recommendation to recreational fishers, the Environmental Protection Agency's (EPA) recommended "screening value" for mirex is 0.08 ppm.

QUESTION: Are farmed salmon contaminated beyond the level permitted by the EPA?

ANSWER: (We've already checked the conditions; see p. 567.)

$$H_0: \mu = 0.08^{13}$$
$$H_A: \mu > 0.08$$

---

[13]The *true* null hypothesis is $H_0: \mu \leq 0.08$, but we can only test one null value for μ. $\mu = 0.08$ is the *conservative* choice, since if we can reject $\mu = 0.08$ in favour of a larger μ, we can even more convincingly reject any μ smaller than 0.08. Just plug in something smaller than 0.08 for μ_0 and you can see the *t*-statistic get bigger and more statistically significant.

These data satisfy the conditions for inference; I'll do a one-sample t-test for the mean:

$$n = 150, \; df = 149$$
$$\bar{y} = 0.0913, \; s = 0.0495$$
$$SE(\bar{y}) = \frac{0.0495}{\sqrt{150}} = 0.0040$$
$$t = \frac{0.0913 - 0.08}{0.0040} = 2.825$$

$$P(t_{149} > 2.825) = 0.0027 \text{ (from technology)}$$

Such a low P-value provides overwhelming evidence that, in farm-raised salmon, the mirex contamination level does exceed the EPA screening value.

What if, in the example above about farm-raised salmon, you had used the standard Normal distribution instead of the t distribution? You would get essentially the same P-value. This is sometimes referred to as the *large sample z-test*, since the Normal distribution will work just fine as the sampling model when you plug $SE(\bar{y}) = \dfrac{s}{\sqrt{n}}$ in place of $SD(\bar{y}) = \dfrac{\sigma}{\sqrt{n}}$ in the denominator of the standardized statistic—*when n is large. Only, only, only* when n is large.

Step-by-Step Example A ONE-SAMPLE t-TEST FOR THE MEAN

The Ministry of Transportation claims that secondary school students can get to their schools in 15 minutes or less, on average (we confess, we made up this claim).

Question: Do the data convincingly refute this claim?

THINK ➡ Plan State what we want to know. Make clear what the population and parameter are. Identify the variables and review the *W*'s.	I want to know whether the mean travel time for students exceeds the Ministry's claim. I have a sample of 40 travel times from 2007–2008.
Hypotheses The null hypothesis is that the true mean travel time is as claimed (but we use the borderline value of 15 minutes for our null). Because we're interested in whether travel times are excessive, the alternative is one-sided. Make a picture. Check the distribution for major skewness, multiple modes, and outliers. **REALITY CHECK** The histogram is centred in the general vicinity of 15 minutes, so the null hypothesis looks plausible (and the fact that 15 is within the confidence interval we've just found also suggests its plausibility).	H_0: Mean travel time $\mu = 15$ minutes H_A: Mean travel time $\mu > 15$ minutes

Model Think about the assumptions and check the conditions.

State the sampling distribution model. (Be sure to include the degrees of freedom.)

Choose your method.

✓ **Independence Assumption**: Discussed earlier.

✓ **Randomization Condition**: Discussed earlier.

✓ **Nearly Normal Condition**: Discussed earlier.

The conditions are satisfied, so I'll use a Student's t-model with $(n - 1) = 39$ degrees of freedom to do a **one-sample t-test for the mean**.

SHOW ➡ Mechanics Be sure to include the units when you write down what you know from the data.

From the data,

$$n = 40 \text{ students}$$
$$\bar{y} = 17.0 \text{ minutes}$$
$$s = 9.66 \text{ minutes}$$
$$SE(\bar{y}) = \frac{s}{\sqrt{n}} = \frac{9.66}{\sqrt{40}} = 1.53 \text{ minutes.}$$
$$t = \frac{\bar{y} - \mu_0}{SE(\bar{y})} = \frac{17.0 - 15.0}{1.53} = 1.31$$

The t-statistic calculation is just a standardized value, like z. We subtract the hypothesized mean and divide by the standard error.

The observed mean is 1.31 standard errors above the hypothesized value.

We use the null model to find the P-value. Make a picture of the t-model centred at zero. Since this is an upper-tail test, shade the region to the right of the observed t-value.

The P-value is the probability of observing a t-value as large as 1.31 (or larger). We can find this P-value from a table, calculator, or computer program.

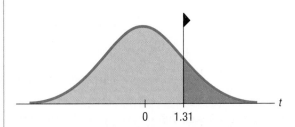

REALITY CHECK We're not surprised that the difference isn't very statistically significant.

P-value $= P(t_{39} > 1.31) = 0.099$ (using software)

TELL ➡ Conclusion Link the P-value to your decision about H_0, and state your conclusion in context.

The P-value of 0.099 says that if the true mean student travel time were 15 minutes, samples of 40 students can be expected to produce a t-statistic of 1.31 or bigger 9.9% of the time. This P-value is not very small, so I won't reject the hypothesis of a mean travel time of 15 minutes. These data do not provide enough evidence to convince me to reject the Ministry's claim with any real conviction.

For hypothesis tests, the computed t-statistic can take on any value, so the value you get is not likely to be one found in the table. The best we can do is to trap a calculated t-value between two columns. Just look across the row with the appropriate degrees of freedom to find where the t-statistic falls. The P-value will be between the two values at the heads of the columns. Report that the P-value falls between these two values. Usually that's good enough.

For Example ESTIMATING P-VALUES FROM TABLE T

RECAP: We've computed a one-sample t-test for the mean mirex contamination in farmed salmon, finding $t = 2.825$ with 149 df. In the earlier example, we found the P-value with technology.

QUESTION: How can we estimate the P-value for this upper-tail test using Table T?

ANSWER: I seek $P(t_{149} > 2.825)$. Table T has neither a row for 149 df nor an entry that is exactly 2.825. Here's the part of Table T where I'll need to work; roughly the right degrees of freedom and t-values:

Two-tail probability One-tail probability		0.20 0.10	0.10 0.05	0.05 0.025	0.02 0.01	0.01 0.005
Values of t_α	df					
	140	1.288	1.656	1.977	2.353	2.611
	180	1.286	1.653	1.973	2.347	2.603

Since 149 df doesn't appear in the table, I'll be conservative and use the next lower df value that does appear (since the t sampling model is more dispersed with bigger critical values in this row). In this table, that's 140 df. Looking across the row for 140 df, I see that the largest t-value in the table is 2.611. According to the column heading, a t-value this large or larger will occur with probability 0.005. My t-value of 2.825 is larger than this, even further out in the tail, so I know that the probability of a t-value that large must be even smaller. I can report P < 0.005.[14]

Statistical Significance and Importance

Recall that "statistically significant" does not mean "actually important" or "meaningful," even though it sort of sounds that way. In the student commute time example, it does seem possible that travel times may average to a bit above 15 minutes. If so, perhaps a larger sample would show statistical significance, due to an increase in power.

So, should we try to obtain a bigger sample? The difference between 17 minutes and 15 minutes doesn't seem very *meaningful*, and even if statistically significant, it would be hard to consequently convince the government of a need to build more schools or the public to spend more money on improving transportation modes. Looking back at the confidence interval, we can say with 90% confidence that the mean travel time is somewhere between 14.4 and 19.6 minutes. Even in the worst case, if the mean travel time is around 19-20 minutes, would this be a bad enough situation to convince anyone to spend more money? Probably not. It's always a good idea when we test a hypothesis to also check the confidence interval and think about the possible values for the mean.

While examining the confidence interval after a non-significant test may be useful, particularly with small sample sizes where you suspect low power, *reporting a confidence interval after a significant test is essential*! Rejecting your null hypothesis only tells you that the parameter falls in the range specified by the alternative hypothesis, which is *infinite* and too broad to be very informative, but a confidence interval narrows it down to a useful range of plausible parameter values, which will help you judge the practical importance of your findings.

The Relationship Between Intervals and Tests

Confidence intervals and hypothesis tests look at the data from different perspectives. A hypothesis test starts with a *proposed parameter value* and asks if the *data* are consistent with that value. If the data are inconsistent with the proposed parameter value, the

---

[14]If the alternative was instead $H_A: \mu \neq 0.08$, we would report $p < 2(0.005) = 0.01$, since values in both tails would now support H_A.

proposed value no longer seems plausible, so we reject the null hypothesis. By contrast, a confidence interval starts with the *data* and finds an interval of plausible values for where the parameter may lie.

The 90% confidence interval for the mean school travel time was 17.0 ± 2.6 minutes, or (14.4 minutes, 19.6 minutes). If someone hypothesized that the mean time was really 18 minutes, how would you feel about it? How about 25 minutes? Because the confidence interval included the time of 18.0 minutes, it certainly looks like 18 minutes might be a plausible value for the true mean school travel time. "Plausible" sounds rather like "acceptable" as a null hypothesis, and indeed this is the case. If we wanted to test the null hypothesis that the true mean is 18 minutes, and we find that 18 lies within some confidence interval, it follows that 18 minutes is a 'plausible' null hypothesis, i.e. a null value we would not reject in a test—at some alpha level—but what alpha level? This depends on the confidence level of the confidence interval.

Confidence intervals and significance tests are built from the same calculations. Here's the connection: The confidence interval contains all possible values for the parameter that would not be rejected, as null hypotheses, in a test (after matching up the test alpha level and confidence level, as discussed below).

To be precise, a level C confidence interval contains *all* of the null hypothesis values that would *not* be rejected by a two-sided hypothesis test at alpha level $1 - C$. So a 95% confidence interval contains all the null hypothesis values we can't reject when performing a test at an alpha level of $1 - 0.95 = 0.05$, or 5%.

Confidence intervals are naturally two-sided, so they match up exactly with two-sided hypothesis tests. When the hypothesis is one-sided, as in our example, it matches up exactly with a one-sided confidence interval (which we are not covering in this text).

To relate a one-sided hypothesis test to a two-sided confidence interval, proceed as follows: Check to see if the level C confidence interval misses the null value and supports the alternative hypothesis—that is, lies entirely within the range of values of the alternative hypothesis. If so, you can reject the null hypothesis at the $(1 - C)/2$ level of significance. If not, the test will fail to reject the null hypothesis at the $(1 - C)/2$ level.

So if we were to use our 90% confidence interval of (14.4, 19.6) to test $H_0: \mu = \mu_0$ vs $H_A: \mu > \mu_0$, then any value for μ_0 smaller than 14.4 would have to be rejected as a null hypothesis, not just at the 10% level, but also at the 5% level of significance ($P < 0.05$), since $(1 - 0.90)/2 = 0.05$.

Degrees of Freedom

> **Don't divide by n.**
> Some calculators offer an alternative button for standard deviation that divides by n instead of $n - 1$. Try sticking a wad of gum over the "n" button so you won't be tempted to use it. Use $n - 1$.

The parameter of the t curve, its $df = n - 1$, might have reminded you of the value we divide by to find the standard deviation of the data (since, in fact, it's the same number). When we introduced that formula, we promised to later say more about why we divide by $n - 1$ rather than by n.

If only we knew the true population mean, μ, we would use it to calculate the sample standard deviation, giving us:[15]

$$s = \sqrt{\frac{\Sigma(y - \mu)^2}{n}} \qquad \text{(Equation 18.1)}$$

But we don't know μ, so we naturally use $\bar{y}$ instead, and that causes a problem. For any sample, the data values will generally be closer to their own sample mean than to the true population mean, μ. Why is that? Imagine that we take a simple random sample of 10 students who just wrote the final exam in your very large introductory Statistics course. Suppose that the mean test score (for all students) was 70. The sample mean, $\bar{y}$, for *these* 10 students won't be exactly 70. Are the 10 students' scores closer to 70 or $\bar{y}$? They'll tend to be closer to their own average $\bar{y}$. So, when we calculate s using $\Sigma(y - \bar{y})^2$ instead of $\Sigma(y - \mu)^2$ in Equation 18.1, our standard deviation estimate is too small. The amazing

---

[15]Statistics textbooks often use equation numbers so they can talk about equations by name. We haven't needed equation numbers yet, but we admit it's useful here, so this is our first.

mathematical fact is that we can fix it by dividing by $n - 1$ instead of by n. This difference is much more important when n is small than when it's big. The t-distribution inherits this same number and we call $n - 1$ the degrees of freedom.[16]

Section 18.4 Exercises

9. *t*-models, again Using the *t* tables, software, or a calculator, estimate the
   a) $P(t \geq 2.09)$ with 4 degrees of freedom.
   b) $P(|t| > 1.78)$ with 22 degrees of freedom.

10. *t*-models, last time Using the *t* tables, software, or a calculator, estimate the
    a) $P(t \leq 2.19)$ with 41 degrees of freedom.
    b) $P(|t| > 2.33)$ with 12 degrees of freedom.

11. **Home prices** The last census of the entire community from which the sample in Exercise 7 was taken was 2010 when the mean loss was $10 200. From the sample in Exercise 7, use a *t*-test to test whether the mean loss has decreased.

12. **Home prices II** Suppose the standard deviation of home price losses had been $3000, as in Exercise 8? What would your conclusion be then?

18.5 Determining the Sample Size

How large a sample do we need? The simple answer is "more." But more data cost money, effort, and time, so how much is enough? Suppose your computer just took half an hour to download a movie you want to watch. You're not happy. You hear about a program that claims to download movies in less than 15 minutes. You're interested enough to spend $29.95 for it, but only if it really delivers. So you get the free evaluation copy and test it by downloading that movie 10 different times. Of course, the mean download time is not exactly 15 minutes as claimed. Observations vary. But if the margin of error were not too large, you'd probably be able to decide whether the software is worth the money. Would a margin of error of 4 minutes suffice? Perhaps. Doubling the sample size would double the hours spent testing and would only reduce your margin of error from 4 minutes to a bit under 3 minutes. You'll need to decide whether that's worth the effort.

As we make plans to collect data, we usually have some idea of how small a margin of error we need to be able to draw a useful conclusion. Armed with the target *ME* and confidence level, we can find the sample size we'll need. Approximately.

For a mean, $ME = t^*_{n-1} \times SE(\bar{y})$ and $SE(\bar{y}) = \dfrac{s}{\sqrt{n}}$, so to find the sample size, solve this equation for n (see the margin note):

$$ME = t^*_{n-1}\frac{s}{\sqrt{n}}$$

SOLVING FOR *n*

$$ME = t^*_{n-1}\frac{s}{\sqrt{n}}$$

$$\sqrt{n} = t^*_{n-1}\frac{s}{ME}$$

$$n = \left(t^*_{n-1}\frac{s}{ME}\right)^2$$

But the equation needs s and before we collect the data, we don't know s. A guess is often good enough, but if you have no idea what the standard deviation might be, or if the sample size really matters (for example, because each additional individual is very expensive to sample or experiment on), a small *pilot study* can provide you with a rough estimate of the standard deviation.

That's not all. Without knowing n, we don't know the degrees of freedom so we can't find the critical value, t^*_{n-1}. One common approach is to use the corresponding z^* value from the Normal model. If you've chosen a 95% confidence level, then just use 2, following

---

[16]Here is another way to think about *df*. If the data are say: 4, 5, 9, the mean is 6, and the deviations are -2, -1, $+3$. The sum of deviations from the sample mean must equal zero, so since the first two deviations sum to -3, the last one *must* be $+3$. Only $n - 1$ deviations are truly *free to vary* (unlike the deviations about μ, all n of which are free to vary). Dividing a sum of squared deviations by its *df* is generally the best way to convert such a sum to an average.

the 68–95–99.7 Rule. If your estimated sample size is, say, 60 or more, it's probably okay—z^* was a good guess. If it's smaller than that, you may want to add a step, using z^* at first, finding n, and then replacing z^* with the corresponding t^*_{n-1} and calculating the sample size once more.

For Example FINDING SAMPLE SIZE

A company claims its program will allow your computer to download movies quickly. We'll test the free evaluation copy by downloading a movie several times, hoping to estimate the mean download time with a margin of error of only 4 minutes. We think the standard deviation of download times is about 5 minutes.

QUESTION: How many trial downloads must we run if we want 95% confidence in our estimate with a margin of error of only 4 minutes?

ANSWER: Using $z^* = 1.96$, solve

$$4 = 1.96 \frac{5}{\sqrt{n}}$$

$$\sqrt{n} = \frac{1.96 \times 5}{4} = 2.45$$

$$n = (2.45)^2 = 6.0025$$

That's a small sample size, so we'll use $(6 - 1) = 5$ degrees of freedom[17] to substitute an appropriate t^* value. At 95%, $t^*_5 = 2.571$. Solving the equation one more time:

$$4 = 2.571 \frac{5}{\sqrt{n}}$$

$$\sqrt{n} = \frac{2.571 \times 5}{4} \approx 3.214$$

$$n = (3.214)^2 \approx 10.33$$

To make sure the *ME* is no larger, we'll round up, which gives $n = 11$ runs. So, to get an *ME* of 4 minutes, we'll find the downloading times for 11 movies.

Sample size calculations are *never* exact. But, it's always a good idea to know whether the sample size is large enough to give you a good chance of being able to tell you what you want to know, before you collect any data.

On the other hand, when we are testing a null hypothesis, we will be concerned with our ability to detect departures from the null that might be of considerable practical importance, so our focus shifts from the margin of error to *the power of the test*. Power calculations for the *t*-test are more complicated than those for the single proportion test (as illustrated in Chapter 16), but the basic idea is the same. Specify the difference from the null that you believe is big enough to be of practical importance, then determine the sample size (using software) that achieves the desired power (such as 0.8 or 0.9).

Make sure your test is adequately powered for important alternatives, or you risk letting those big important effects (departures from the null)—that you were looking for—slip by undiscovered. Turning up the power of a test is like turning up the power of a microscope, allowing you to see and discern even small things more clearly. In the case of tests, it allows you to see genuine effects or patterns more clearly. With low power, you may end up seeing nothing clearly, so you fall back on the status quo of the null.

Let's return to the chapter example where we tested the null hypothesis of a mean mirex content in farm-raised salmon of 0.08 ppm, the recommended screening level for this contaminant. True levels slightly above 0.08 ppm might not matter all that much, but suppose that a level as high as 0.10 ppm was considered dangerously high, meriting major

---

[17]Ordinarily we'd round the sample size up. But at this stage of the calculation, rounding *down* is the safer choice. Can you see why?

remedial action. We would want to ensure that our test will lead us to correctly reject the null hypothesis when such a high contamination level is actually present. Tell your software the following:

- *Statistical test to be used.* The one-sample *t*-test.
- *Alpha level of your test.* Let's choose the very common $\alpha = 0.05$.
- *The null hypothesis.* In this example, it is the screening level of $\mu = 0.08$ ppm.
- *Directionality of alternative.* Make it one-sided: $\mu > 0.08$, since we only care to detect high levels of mirex.
- *Particular alternative (effect size) considered to be of practical importance.* This is the alternative (to the null) that we want to have a good chance to detect, should it be true. We decided that $\mu = 0.10$ ppm was a dangerously high level, so that is the alternative that we enter. For purposes of comparison, we'll also consider $\mu = 0.09$ ppm.
- *Your guess at the standard deviation of the measurements.* Let's guess that approximately $\sigma = 0.05$, perhaps from some available data or pilot study, or by making an educated guess. Why can't we use the *s* from our study? Well, remember that this is usually a planning exercise, so the study hasn't been run yet!
- *Desired power.* Let's aim for a high power of 0.95. This means that 95 times in 100 when we have a situation as bad as 0.10 ppm, we will correctly reject the null hypothesis and conclude that mirex levels are too high.

Okay! Ready . . . aim . . . fire (up your software). Below is typical output:

Power and Sample Size

One-Sample *t* Test
Testing mean = null (versus > null)
Calculating power for mean = null + difference
Alpha = 0.05 Assumed standard deviation = 0.05

Difference	Sample Size	Target Power	Actual Power
0.01	272	0.95	0.950054
0.02	70	0.95	0.952411

The sample size is for each group.

It appears that the researchers didn't need to test 150 salmon; 70 would have sufficed, if 0.02 ppm above the screening value is where serious problems occur. But if they felt they needed to detect a lower mirex level, like 0.09 ppm (0.01 above the screening level), 272 salmon would be required for testing. Note we have demanded a rather high power of 0.95. If you reduce this target power, smaller sample sizes will result.

Actually, this power calculation is quite doable without using the computer if the sample size is not too small—at least 40 or 50 (since for big samples, we can use the N(0,1) model rather than the *t* distribution). Check the two exercises at the end of this chapter if you'd like to go through these calculations.

Section 18.5 Exercises

13. Jelly A consumer advocate wants to collect a sample of jelly jars and measure the actual weight of the product in the container. He needs to collect enough data to construct a confidence interval with a margin of error of no more than 2 grams with 99% confidence. The standard deviation of these jars is usually 4 grams. What do you recommend for his sample size?

14. A good book An English professor is attempting to estimate the mean number of novels that the student body reads during their time in college. He is conducting an exit survey with seniors. He hopes to have a margin of error of 3 books with 95% confidence. From reading previous studies, he expects a large standard deviation and is going assume it is 10. How many students should he survey?

WHAT CAN GO WRONG?

The most fundamental issue you face is knowing when to use Student's t methods.

■ **Don't confuse proportions and means.** When you treat your data as categorical, counting successes and summarizing with a sample proportion, make inferences using the (usually Normal model based) methods you learned about in Chapters 15 through 17. When you treat your data as quantitative, summarizing with a sample mean, make your inferences using Student's t methods.

Student's t methods work well when the Normality Assumption is roughly true. Naturally, many of the ways things can go wrong turn out to be different ways that the Normality Assumption can fail. It's always a good idea look for the most common kinds of failure. It turns out that you can even fix some of them.

■ **Beware of multimodality.** The Nearly Normal Condition clearly fails if a histogram of the data has two or more modes. When you see this, look for the possibility that your data come from two groups. If so, your best bet is to try to separate the data into different groups. (Use the variables to help distinguish the modes, if possible. For example, if the modes seem to be composed mostly of men in one and women in the other, split the data according to sex.) Then you could analyze each group separately.

■ **Beware of severely skewed data.** Make a Normal probability plot and a histogram of the data. If the data are very skewed, you might try re-expressing the variable. Re-expressing may yield a distribution that is more nearly unimodal and symmetric, more appropriate for Student's t inference methods for means. Re-expression cannot help if the sample distribution is not unimodal. Unless your sample is very large, you just can't use the methods of this chapter on data that are severely skewed (in which case, you might well be more interested in the median than the mean).

> **DON'T IGNORE OUTLIERS**
>
> As tempting as it is to get rid of annoying values, you can't just throw away outliers and not discuss them. It isn't appropriate to lop off the highest or lowest values just to improve your results.

■ **Set outliers aside—respectfully.** Student's t methods are built on the mean and standard deviation, so we should beware of outliers when using them. When you make a histogram to check the Nearly Normal Condition, be sure to check for outliers as well. If you find some, consider doing the analysis twice, with the outliers excluded and with them included in the data, to get a sense of how much they affect the results.

The suggestion that you can perform an analysis with outliers removed may be controversial in some disciplines. Setting aside outliers is seen by some as "cheating." But an analysis of data with outliers left in place is *always* wrong. The outliers violate the Nearly Normal Condition, so they invalidate inference procedures. An analysis of the nonoutlying points, along with a separate discussion of the outliers, is often much more informative and can reveal important aspects of the data.

How can you tell whether there are outliers in your data? The "outlier nomination rule" of boxplots can offer some guidance, but it's just a very rough rule of thumb and not an absolute definition. The best practical definition is that a value is an outlier if removing it substantially changes your conclusions about the data. You won't want a single value to determine your understanding of the world unless you are very, very sure that it is absolutely correct and truly "belongs" to your target population. Of course, when the outliers affect your conclusion, this can lead to the uncomfortable state of not really knowing what to conclude. Such situations call for you to use your knowledge of the real world and your understanding of the data you are working with.[18]

Of course, Normality issues aren't the only risks you face when doing inferences about means. Remember to *Think* about the usual suspects.

■ **Watch out for bias.** Measurements of all kinds can be biased. If your observations differ from the true mean in a systematic way, your confidence interval may not capture the true mean. And there is no sample size that will save you. A bathroom scale that's 5 pounds off will be 5 pounds off even if you weigh yourself 100 times and take the average. We've seen several sources of bias in surveys, and measurements can be biased, too. Be sure to think about possible sources of bias in your measurements.

---

[18]An important reason for you to know Statistics rather than let someone else analyze your data.

- **Make sure cases are independent.** Student's *t* methods also require the sampled values to be mutually independent. Think hard about whether there are likely violations of independence in the data collection method. If there are, be very cautious about using these methods.

- **Check whether the data are from an appropriately randomized sample.** Ideally, all data that we analyze are drawn from a simple random sample or are generated by a completely randomized experimental design. When they're not, be careful about making inferences from them (and think carefully about the *Who* of your study). You may still compute a confidence interval or get the mechanics of the P-value right, but this might not save you from making a serious mistake in inference. For other types of random samples, more complicated SE formulas apply. Cluster sampling in particular may have a much bigger SE than given by our formula.

- **Interpret your confidence interval correctly.** Many statements that sound tempting are, in fact, misinterpretations of a confidence interval for a mean. You might want to have another look at some of the common mistakes (as explained on p. 571). Keep in mind that a confidence interval is about the mean of the population, not about the means of samples, individuals in samples, or individuals in the population.

- **Choose your alternative hypothesis based only on what you are trying to prove.** Never choose a one-sided alternative after seeing which way the data are pointing, or you will incorrectly report a P-value half its true size. If you have any doubt about the nature of the alternative, go with the conservative choice of a two-sided alternative.

What Have We Learned?

Know the sampling distribution of the mean.

- To make inferences using the sample mean, we need to estimate its standard deviation. This *standard error* is given by:

$$SE(\bar{y}) = \frac{s}{\sqrt{n}}.$$

- When we use the *SE* instead of the *SD* (which is unknown when σ is unknown), the sampling distribution model that allows for the additional uncertainty is Student's *t*.

Construct confidence intervals for the true mean, μ.

- A confidence interval for the mean has the form $\bar{y} \pm ME$.
- The Margin of Error is $ME = t^*_{df}SE(\bar{y})$.
- Find *t** values by technology or from tables (like our table T).
- When constructing confidence intervals for means, the correct degrees of freedom is $n - 1$.
- Check the Assumptions and Conditions before using any sampling distribution such as Student's *t* for inference.

Perform hypothesis tests for the mean using the standard error of $\bar{y}$ as a ruler and then finding the *P*-value from Student's *t* model on $n - 1$ degrees of freedom. Obtain the *P*-value from technology or estimate it from tables.

Write clear summaries to interpret a confidence interval or state a hypothesis test's conclusion. Find the sample size needed to produce a given margin of error or to produce desired power in a test of hypothesis.

Review of Terms

Student's *t* A family of distributions indexed by its degrees of freedom. The *t*-models are unimodal, symmetric, and bell-shaped, but generally have fatter tails and a narrower centre than the Normal model. As the degrees of freedom increase, *t*-distributions approach the standard Normal model (p. 562).

Degrees of freedom for Student's *t*-distribution

For the application of the *t*-distribution in this chapter, the degrees of freedom are equal to $n - 1$, where n is the sample size (p. 562).

One-sample *t*-interval for the mean

A one-sample *t*-interval for the population mean is

$$\bar{y} \pm t_{n-1}^{*} \times SE(\bar{y}), \text{ where } SE(\bar{y}) = \frac{s}{\sqrt{n}}$$

The critical value t_{n-1}^{*} depends on the particular confidence level, *C*, that you specify and on the number of degrees of freedom, $n - 1$ (p. 564).

One-sample *t*-test for the mean

The one-sample *t*-test for the mean tests the hypothesis $H_0: \mu = \mu_0$ using the statistic

$$t = \frac{\bar{y} - \mu_0}{SE(\bar{y})} \text{ where the standard error of } \bar{y} \text{ is } SE(\bar{y}) = \frac{s}{\sqrt{n}} \text{ (p. 572).}$$

On the Computer INFERENCE FOR MEANS

Statistics packages offer convenient ways to make histograms of the data. Even better for assessing near-Normality is a Normal probability plot (or the similar Normal Q-Q plot). When you work on a computer, there is simply no excuse for skipping the step of plotting the data to check that it is nearly Normal. Beware: Statistics packages don't agree on whether to place the Normal scores on the *x*-axis (as we have done) or the *y*-axis. Read the axis labels.

Any standard statistics package can compute a hypothesis test. Here's what the package output might look like in general (although no package we know gives the results in exactly this form):[19]

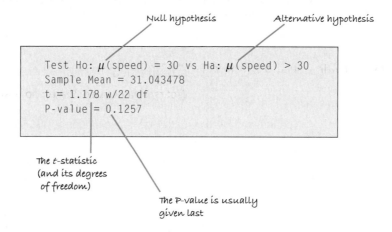

The package computes the sample mean and sample standard deviation of the variable and finds the P-value from the *t*-distribution based on the appropriate number of degrees of freedom. All modern statistics packages report P-values. The package may also provide additional information, such as the sample mean, sample standard deviation, *t*-statistic value, and degrees of freedom. These are useful for interpreting the resulting P-value and telling the difference between a meaningful result and one that is merely statistically significant. Statistics packages that report the estimated standard deviation of the sampling distribution usually label it "standard error" or "SE."

---

[19]Many statistics packages keep as many as 16 digits for all intermediate calculations. If we had kept as many, our results in the Step-By-Step section would have been closer to these.

Inference results are also sometimes reported in a table. You may have to read carefully to find the values you need. Often, test results and the corresponding confidence interval bounds are given together. And often you must read carefully to find the alternative hypotheses. Here's an example of that kind of output:

μ_0

Calculated mean, $\bar{y}$

Hypothesized value	30	
Estimated mean	31.043478	
DF	22	
Std Error	0.886	
Alpha	0.05	

The alpha level often defaults to 0.05. Some packages let you choose a different alpha level

t-statistic

	tTest		tinterval
Statistic	1.178		
Prob > |t|	0.2513	Upper 95%	32.880348
Prob > t	0.1257	Lower 95%	29.206608
Prob < t	0.8743		

1-sided $H_A: \mu > 30$

1-sided $H_A: \mu < 30$

2-sided alternative (note the $|t|$)

P-values for each alternative

Corresponding confidence interval

EXCEL

Specify formulas. Find t^* with the TINV(alpha, df) function.

COMMENTS

Not really automatic. There's no easy way to find P-values or to perform power and sample size calculations in Excel.

JMP

From the **Analyze** menu, select **Distribution**.
For a confidence interval, scroll down to the "Moments" section to find the interval limits.
For a hypothesis test, click the red triangle next to the variable's name, and choose **Test Mean** from the menu. Then, fill in the resulting dialogue.
For power and sample size calculations, proceed as follows:
- Choose **Power and Sample Size** from the **DOE** menu.
- Choose **One Sample Mean** from the submenu.
- Indicate the **Difference** in the means that you are hoping to detect, the **Alpha value**, and choose a one-or two-sided alternative.

- Guess at the **Std Dev**.
- Fill in either your desired **Sample size** or **Power**. The one you leave blank will be calculated.
- Click **Continue**.

COMMENTS

"Moment" is a fancy statistical term for means, standard deviations, and other related statistics.

MINITAB

- From the **Stat** menu, choose the **Basic Statistics** submenu.

- From that menu, choose **1-sample t. . . .**
- Then, fill in the dialogue.

For power and sample size calculations:

- From the **Stat** menu, choose the **Basic Statistics**, then **Power and Sample Size**, then **1-Sample t. . . .**
- In the dialogue box, fill in any two of **Sample Sizes, Differences, Power values**.
- Make your best guess at the value for the **Standard deviation**.
- Be sure to check the **Options** for the correct alternative hypothesis and significance level.

- For **Difference**, fill in the difference between the null value for the mean and the alternative value of the mean at which you are doing the calculation. No need to indicate the null value anywhere, as only the difference matters.

COMMENTS

The dialogue offers a clear choice between confidence interval and test.

R

To test the hypothesis that $\mu = $ mu (default is mu = 0) against an alternative (default is two-sided) and to produce a confidence interval (default is 95%), create a vector of data in x and then:

- **t.test**(x, alternative = c("two.sided," "less," "greater"), mu = 0, conf.level = 0.95)

provides the *t*-statistic, P-value, degrees of freedom, and the confidence interval for a specified alternative.

In package {pwr} (not installed by default), to perform a sample size or power calculation for a one-sample *t*-test,

- **pwr.t.test**(n= , d= , sig.level= , power= , type= "one. sample") returns **one** argument that is not specified in the function.

For example, for fixed $\alpha = 5\%$, 80% power, and effect size, *d*, of 0.5,

- **pwr.t.test**(d=0.5, sig.level=0.05, power=0.8, type="one.sample")

will return a sample size of 33.36713 (i.e., 34). Use the **alternative= "two.sided," "less,"** or **"greater"** attribute to perform one-sided tests. In the case of using "less," your effect size should be negative.

COMMENTS

The effect size, *d*, required by *R* is equal to the difference between the alternative and null means *divided by the population standard deviation*. Since you often won't have data when doing this calculation, the SD needs to be guessed.

SPSS

From the **Analyze** menu, choose the **Compare Means** submenu.
From that, choose the **One-Sample *t*-test** command.
You need the IBM SPSS SamplePower add-on for power and sample size calculations.

COMMENTS

The commands suggest neither a single mean nor an interval. But the results provide both a test and an interval.

STATCRUNCH

To do inference for a mean using summaries:

- Click on **Stat**.
- Choose **T Statistics » One sample » with summary**.
- Enter the **Sample mean, Sample std dev**, and **Sample size**.
- Click on **Next**.
- Indicate **Hypothesis Test**, then enter the hypothesized **Null mean**, and choose the **Alternative** hypothesis.
                    OR
- Indicate **Confidence Interval**, and then enter the **Level** of confidence.
- Click on **Calculate**.

To do inference for a mean using data:

- Click on **Stat**.
- Choose **T Statistics » One sample » with data**.
- Choose the variable **Column**.
- Click on **Next**.
- Indicate **Hypothesis Test**, then enter the hypothesized **Null mean**, and choose the **Alternative** hypothesis.
                    OR
- Indicate **Confidence Interval**, then entre the **Level** of confidence.
- Click on **Calculate**.

Power & Sample size calculations are readily available, using **Stat » T Statistics » One Sample » Power/Sample size**.

■ Click on **Hypothesis Test Power**.

■ Fill in all the boxes except for the one that you want to determine, either **Power** or **Sample Size**.
■ Make a guess at the Standard deviation.

TI-83/84 PLUS

Finding a confidence interval:
In the **STAT TESTS** menu, choose **8:TInterval**. You may specify that you are using data stored in a list, or you may enter the mean, standard deviation, and sample size. You must also specify the desired level of confidence.

Power and sample size calculations are not provided.

Testing a hypothesis:
In the **STAT TESTS** menu, choose **2:T-Test**. You may specify that you are using data stored in a list, or you may enter the mean, standard deviation, and size of your sample. You must also specify the hypothesized model mean and whether the test is to be two-tail, lower-tail, or upper-tail.

Chapter Exercises

Ⓣ indicates data is available in a saved file at the Companion Website at www.pearsoncanada.ca/deveaux

15. **Cattle** Researchers give livestock a special feed supplement to see if it will promote weight gain. They report that the 77 cows studied gained an average of 56 pounds, and that a 95% confidence interval for the mean weight gain this supplement produces has a margin of error of ± 11 pounds. Some students wrote the following conclusions. Did anyone interpret the interval correctly? Explain any misinterpretations.
   a) 95% of the cows studied gained between 45 and 67 pounds.
   b) We're 95% sure that a cow fed with this supplement will gain between 45 and 67 pounds.
   c) We're 95% sure that the average weight gain among the cows in this study was between 45 and 67 pounds.
   d) The average weight gain of cows fed with this supplement will be between 45 and 67 pounds 95% of the time.
   e) If this supplement is tested on another sample of cows, there is a 95% chance that their average weight gain will be between 45 and 67 pounds.

16. **Viewing hours** Software analysis of the weekly hours spent by Canadian secondary school students viewing television, videos, or movies from a random sample of 200 students produced the t-interval shown below. Which conclusion, from the choices below, is correct? What's wrong with the others?

   With 90% Confidence,
   $8.6 < \mu$ (weekly viewing hours) < 10.8

   a) If we took many random samples of Canadian secondary students, about 9 out of 10 of them would produce this confidence interval.

   b) If we took many random samples of Canadian secondary students, about 9 out of 10 of them would produce a confidence interval that contained the mean weekly television, video, or movie viewing time of all Canadian secondary students.
   c) About 9 out of 10 Canadian secondary students spend between 8.6 and 10.8 hours per week on television, videos, or movies.
   d) About 9 out of 10 of the students surveyed spend between 8.6 and 10.8 hours per week on television, video, or movie viewing.
   e) We are 90% confident that the average time spent viewing television, videos, or movies by secondary students in Canada is between 8.6 and 10.8 hours per week.

17. **Meal plan** After surveying students at Dartmouth College, a campus organization calculated that a 95% confidence interval for the mean cost of food for one term (of three in the Dartmouth trimester calendar) is ($1372, $1562). Now the organization is trying to write its report and is considering the following interpretations. Comment on each.
   a) 95% of all students pay between $1372 and $1562 for food.
   b) 95% of the sampled students paid between $1372 and $1562.
   c) We're 95% sure that students in this sample averaged between $1372 and $1562 for food.
   d) 95% of all samples of students will have average food costs between $1372 and $1562.
   e) We're 95% sure that the average amount all students pay is between $1372 and $1562.

18. Snow Based on meteorological data for the past century, a local television weather forecaster estimates that the region's average winter snowfall is 58 cm, with a margin of error of 5 cm. Assuming he used a 95% confidence interval, how should viewers interpret this news? Comment on each of these statements (assuming a lack of systematic climate change):
a) During 95 of the past 100 winters, the region got between 53 cm and 63 cm of snow.
b) There's a 95% chance that the region will get between 53 cm and 63 cm of snow this winter.
c) There will be between 53 cm and 63 cm of snow on the ground for 95% of winter days.
d) Residents can be 95% sure that the area's average snowfall is between 53 cm and 63 cm.
e) Residents can be 95% confident that the average snowfall during the past century was between 53 cm and 63 cm per winter.

T 19. Pulse rates A medical researcher measured the pulse rates (beats per minute) of a sample of randomly selected adults and found the following Student's *t*-based confidence interval:

With 95.00% Confidence,
$70.887604 < \mu(\text{Pulse}) < 74.497011$

a) Explain carefully what the software output means.
b) What is the margin of error for this interval?
c) If the researcher had calculated a 99% confidence interval, would the margin of error be larger or smaller? Explain.

20. Crawling Data collected by child development scientists produced this confidence interval for the average age (in weeks) at which babies begin to crawl:

t-Interval for μ
(95.00% Confidence): $29.202 < \mu(\text{age}) < 31.844$

a) Explain carefully what the software output means.
b) What is the margin of error for this interval?
c) If the researcher had calculated a 90% confidence interval, would the margin of error be larger or smaller? Explain.

21. CEO compensation A sample of 20 CEOs from the *Forbes* 500 shows total annual compensations ranging from a minimum of $0.1 million to $62.24 million. The average for these 20 CEOs is $7.946 million. Here's a histogram:

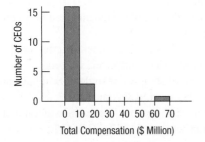

Total Compensation ($ Million)

Based on these data, a computer program found that a 95% confidence interval for the mean annual compensation of all *Forbes* 500 CEOs is (1.69, 14.20) $ million. Why should you be hesitant to trust this confidence interval?

22. Credit card charges A credit card company takes a random sample of 100 cardholders to see how much they charged on their card last month. Here's a histogram:

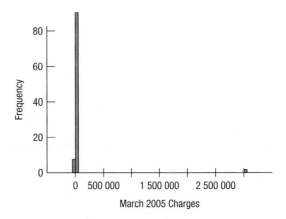

March 2005 Charges

A computer program found that the resulting 95% confidence interval for the mean amount spent in March 2013 is (–$28 366.84, $90 691.49). Explain why the analysts didn't find the confidence interval useful, and explain what went wrong.

23. Normal temperature The researcher described in Exercise 19 also measured the body temperatures of that randomly selected group of adults. The data he collected are summarized below. We wish to estimate the average (or "normal") temperature among the adult population.

Summary	
Count	52
Mean	36.83°C
Median	36.78°C
MidRange	37.00°C
StdDev	0.38
Range	1.55
IntQRange	0.58

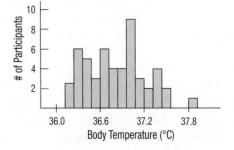

Body Temperature (°C)

a) Are the necessary conditions for a *t*-interval satisfied? Explain.

b) Find a 98% confidence interval for mean body temperature.

c) Explain the meaning of that interval.

d) Explain what "98% confidence" means in this context.

e) 37°C is commonly assumed to be "normal." Do these data suggest otherwise? Explain.

24. **Parking** Hoping to lure more shoppers downtown, a city builds a new public parking garage in the central business district. The city plans to pay for the structure through parking fees. During a two-month period (44 weekdays), daily fees collected averaged $126, with a standard deviation of $15.

a) What assumptions must you make in order to use these statistics for inference?

b) Write a 90% confidence interval for the mean daily income this parking garage will generate.

c) Explain in context what this confidence interval means.

d) Explain what "90% confidence" means in this context.

e) The consultant who advised the city on this project predicted that parking revenues would average $130 per day. Based on your confidence interval, do you think the consultant was correct? Why?

25. **Normal temperatures, part II** Consider again the statistics about human body temperature in Exercise 23.

a) Would a 90% confidence interval be wider or narrower than the 98% confidence interval you calculated before? Explain. (You should not need to compute the new interval.)

b) What are the advantages and disadvantages of the 98% confidence interval?

c) If we conduct further research, this time using a sample of 500 adults, how would you expect the 98% confidence interval to change? Explain.

d) How large a sample would you need to estimate the mean body temperature to within 0.05 degrees with 98% confidence?

26. **Parking II** Suppose that, for budget planning purposes, the city in Exercise 24 needs a better estimate of the mean daily income from parking fees.

a) Someone suggests that the city uses its data to create a 95% confidence interval instead of the 90% interval first created. How would this interval be better for the city? (You need not actually create the new interval.)

b) How would the 95% interval be worse for the planners?

c) How could they achieve an interval estimate that would better serve their planning needs?

d) How many days' worth of data must they collect to have 95% confidence of estimating the true mean to within $3?

27. **Speed of light** In 1882, Michelson measured the speed of light (usually denoted *c* as in Einstein's famous equation $E = mc^2$). His values are in km/sec and have 299 000 subtracted from them. He reported the results of 23 trials with a mean of 756.22 and a standard deviation of 107.12.

a) Find a 95% confidence interval for the true speed of light from these statistics.

b) State in words what this interval means. Keep in mind that the speed of light is a physical constant that, as far as we know, has a value that is true throughout the universe.

c) What assumptions must you make in order to use your method?

T 28. **Better light** After his first attempt to determine the speed of light (described in Exercise 27), Michelson conducted an "improved" experiment. In 1897, he reported results of 100 trials with a mean of 852.4 km/sec and a standard deviation of 79.0.

a) What is the standard error of the mean for these data?

b) Without computing it, how would you expect a 95% confidence interval for the second experiment to differ from the confidence interval for the first? Note at least three specific reasons why they might differ, and indicate the ways in which these differences would change the interval.

c) According to Stigler (who reports these values), the true speed of light is 299 710.5 km/sec, corresponding to a value of 710.5 for Michelson's 1897 measurements. What does this indicate about Michelson's two experiments? Explain, using your confidence interval.

T 29. **On time 2013** What are the chances your flight will leave on time? The U.S. Bureau of Transportation Statistics of the Department of Transportation publishes information about airline performance. Here are a histogram and summary statistics for the percentage of flights departing on time each month from 1994 through October 2013. (www.transtats.bts.gov/HomeDrillChart.asp)

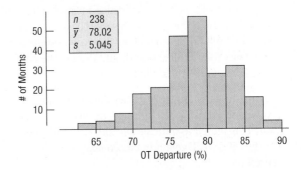

There is no evidence of a trend over time.

a) Check the assumptions and conditions for inference.

b) Find a 90% confidence interval for the true percentage of flights that depart on time.

c) Interpret this interval for a traveller planning to fly.

T **30. Delayed 2013** Will your flight get you to your destination on time? The U.S. Bureau of Transportation Statistics reported the percentage of flights that were delayed each month from 1994 through October of 2013. Here's a histogram, along with some summary statistics:

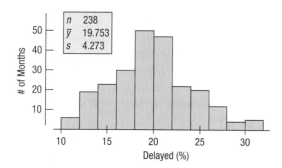

We can consider these data to be a representative sample of all months. There is no evidence of a time trend ($r = 0.07$).
a) Check the assumptions and conditions for inference about the mean.
b) Find a 99% confidence interval for the true percentage of flights that arrive late.
c) Interpret this interval for a traveller planning to fly.

T **31. Farmed salmon, second look** This chapter's For Examples looked at mirex contamination in farmed salmon. We first found a 95% confidence interval for the mean concentration to be 0.0834 to 0.0992 parts per million. Later, we rejected the null hypothesis that the mean did not exceed the EPA's recommended safe level of 0.08 ppm based on a P-value of 0.0027. Explain how these two results are consistent. Your explanation should discuss the confidence level, the P-value, and the decision.

32. Hot dogs A nutrition lab tested 40 hot dogs to see if their mean sodium content was less than the 325 mg upper limit set by regulations for "reduced sodium" franks. The lab failed to reject the null hypothesis that the hot dogs did not meet this requirement, with a P-value of 0.142. A 90% confidence interval estimated the mean sodium content for this kind of hot dog at 317.2 to 326.8 mg. Explain how these two results are consistent. Your explanation should discuss the confidence level, the P-value, and the decision.

33. Pizza A researcher tests whether the mean cholesterol level among those who eat frozen pizza exceeds the value considered to indicate a health risk. She gets a P-value of 0.07. Explain in this context what the "7%" represents.

34. Golf balls The United States Golf Association (USGA) sets performance standards for golf balls. For example, the initial velocity of the ball may not exceed 250 feet per second when measured by an apparatus approved by the USGA. Suppose a manufacturer introduces a new kind of

ball and provides a sample for testing. Based on the mean speed in the test, the USGA comes up with a P-value of 0.34. Explain in this context what the "34%" represents.

35. TV safety The manufacturer of a metal stand for home television sets must be sure that its product will not fail under the weight of the television. Since some larger sets weigh nearly 300 pounds (about 136 kg), the company's safety inspectors have set a standard of ensuring that the stands can support an average of over 500 pounds. Their inspectors regularly subject a random sample of the stands to increasing weight until they fail. They test the hypothesis $H_0: \mu = 500$ against $H_A: \mu > 500$, using the level of significance $\alpha = 0.01$. If the stands in the sample fail to pass this safety test, the inspectors will not certify the product for sale to the general public.
a) Is this an upper-tail or lower-tail test? In the context of the problem, why do you think this is important?
b) Explain what will happen if the inspectors commit a Type I error.
c) Explain what will happen if the inspectors commit a Type II error.

36. Catheters During an angiogram, heart problems can be examined via a small tube (a catheter) threaded into the heart from a vein in the patient's leg. It's important that the company that manufactures the catheter maintain a diameter of 2.00 mm. (The standard deviation is quite small.) Each day, quality control personnel make several measurements to test $H_0: \mu = 2.00$ against $H_A: \mu \neq 2.00$ at a significance level of $\alpha = 0.05$. If they discover a problem, they will stop the manufacturing process until it is corrected.
a) Is this a one-sided or two-sided test? In the context of the problem, why do you think this is important?
b) Explain in this context what happens if the quality control people commit a Type I error.
c) Explain in this context what happens if the quality control people commit a Type II error.

37. TV safety revisited The manufacturer of the metal television stands in Exercise 35 is thinking of revising its safety test.
a) If the company's lawyers are worried about being sued for selling an unsafe product, should they increase or decrease the value of α? Explain.
b) In this context, what is meant by the power of the test?
c) If the company wants to increase the power of the test, what options does it have? Explain the advantages and disadvantages of each option.

38. Catheters again The catheter company in Exercise 36 is reviewing its testing procedure.
a) Suppose the significance level is changed to $\alpha = 0.01$. Will the probability of a Type II error increase, decrease, or remain the same?
b) What is meant by the power of the test the company conducts?

c) Suppose the manufacturing process is slipping out of proper adjustment. As the actual mean diameter of the catheters produced gets farther and farther above the desired 2.00 mm, will the power of the quality control test increase, decrease, or remain the same?

d) What could they do to improve the power of the test?

39. **Marriage** In 1960, census results indicated that the age at which Canadian women first married had a mean of 22.6 years. It is widely suspected that young people today are waiting longer to get married. We want to find out if the mean age of first marriage has increased during the past several decades.

a) Write appropriate hypotheses.

b) We plan to test our hypotheses by selecting a random sample of 40 women who married for the first time last year. Do you think the necessary assumptions for inference are satisfied? Explain.

c) Describe the approximate sampling distribution model for the mean age in such samples.

d) The women in our sample married at an average age of 27.2 years, with a standard deviation of 5.3 years. What is the P-value for this result?

e) Explain (in context) what this P-value means.

f) What is your conclusion?

40. **Fuel economy** A company with a large fleet of cars hopes to keep gasoline costs down and sets a goal of attaining a fleet average of at most 9 litres per 100 km. To see if the goal is being met, they check the gasoline usage for 50 company trips chosen at random, finding a mean of 9.40 L/100 km and a standard deviation of 1.81 L/100 km. Is this strong evidence that they have failed to attain their fuel economy goal?

a) Write appropriate hypotheses.

b) Are the necessary assumptions to make inferences satisfied?

c) Describe the sampling distribution model of mean fuel economy for samples like this.

d) Find the P-value.

e) Explain what the P-value means in this context.

f) State an appropriate conclusion.

T 41. **Ruffles** Students investigating the packaging of potato chips purchased six bags of Lay's Ruffles marked with a net weight of 28.3 grams. They carefully weighed the contents of each bag, recording the following weights (in grams): 29.3, 28.2, 29.1, 28.7, 28.9, 28.5.

a) Do these data satisfy the assumptions for inference? Explain.

b) Find the mean and standard deviation of the observed weights.

c) Create a 95% confidence interval for the mean weight of such bags of chips.

d) Explain in context what your interval means.

e) Comment on the company's stated net weight of 28.3 grams.

T 42. **Doritos** Some students checked six bags of Doritos marked with a net weight of 28.3 grams. They carefully weighed the contents of each bag, recording the following weights (in grams): 29.2, 28.5, 28.7, 28.9, 29.1, 29.5.

a) Do these data satisfy the assumptions for inference? Explain.

b) Find the mean and standard deviation of the observed weights.

c) Create a 95% confidence interval for the mean weight of such bags of chips.

d) Explain in context what your interval means.

e) Comment on the company's stated net weight of 28.3 grams.

T 43. **Popcorn** Yvon Hopps ran an experiment to test optimum power and time settings for microwave popcorn. His goal was to find a combination of power and time that would deliver high-quality popcorn with only 10% of the kernels left unpopped, on average. After experimenting with several bags, he determined that power 9 at four minutes was the best combination.

a) He concluded that this popping method achieved the 10% goal. If it really does not work that well, what kind of error did Hopps make?

b) To be sure that the method was successful, he popped eight more bags of popcorn (selected at random) at this setting. All were of high quality, with the following percentages of unpopped popcorn: 7, 13.2, 10, 6, 7.8, 2.8, 2.2, 5.2. Does this provide evidence that he met his goal of an average of no more than 10% unpopped kernels? Explain.

T 44. **Ski wax** Bjork Larsen was trying to decide whether to use a new racing wax for cross-country skis. He decided that the wax would be worth the price if he could average less than 55 seconds on a course he knew well, so he planned to test the wax by racing on the course eight times.

a) Suppose that he eventually decides not to buy the wax, but it really would lower his average time to below 55 seconds. What kind of error would he have made?

b) His eight race times were 56.3, 65.9, 50.5, 52.4, 46.5, 57.8, 52.2, and 43.2 seconds. Should he buy the wax? Explain.

T 45. **Chips Ahoy** In 1998, as an advertising campaign, the Nabisco Company announced a "1000 Chips Challenge," claiming that every 18-ounce (about 625 grams) bag of their Chips Ahoy cookies contained at least 1000 chocolate chips. Dedicated Statistics students at the Air Force Academy (no kidding) purchased some randomly selected bags of cookies and counted the chocolate chips. Some of their data are given below. (*Chance*, 12, no. 1[1999])

1219	1214	1087	1200	1419	1121	1325	1345
1244	1258	1356	1132	1191	1270	1295	1135

a) Check the assumptions and conditions for inference. Comment on any concerns you have.
b) Create a 95% confidence interval for the average number of chips in bags of Chips Ahoy cookies.
c) What does this evidence say about Nabisco's claim? Use your confidence interval to test an appropriate hypothesis and state your conclusion.

T 46. Yogourt *Consumer Reports* tested 14 brands of vanilla yogourt and found the following numbers of calories per serving:

160	200	220	230	120	180	140
130	170	190	80	120	100	170

a) Check the assumptions and conditions for inference.
b) Create a 95% confidence interval for the average calorie content of vanilla yogourt.
c) A diet guide claims that you will get an average of 120 calories from a serving of vanilla yogourt. What does this evidence indicate? Use your confidence interval to test an appropriate hypothesis and state your conclusion.

T 47. Maze Psychology experiments sometimes involve testing the ability of rats to navigate mazes. The mazes are classified according to difficulty, as measured by the mean length of time it takes rats to find the food at the end. One researcher needs a maze that will take rats an average of about one minute to solve. He tests one maze on several rats, collecting the data shown.

Time (sec)	
38.4	57.6
46.2	55.5
62.5	49.5
38.0	40.9
62.8	44.3
33.9	93.8
50.4	47.9
35.0	69.2
52.8	46.2
60.1	56.3
55.1	

a) Plot the data. Do you think the conditions for inference are satisfied? Explain.
b) Test the hypothesis that the mean completion time for this maze is 60 seconds. What is your conclusion?
c) Eliminate the outlier, and test the hypothesis again. What is your conclusion?
d) Do you think this maze meets the "one-minute average" requirement? Explain.

48. Braking A tire manufacturer is considering a newly designed tread pattern for its all-weather tires. Tests have indicated that these tires will provide better gas mileage and longer tread life. The last remaining test is for braking effectiveness. The company hopes the tire will allow a car travelling at 100 km/h to come to a complete stop within an average of 38 metres after the brakes are applied. They will adopt the new tread pattern unless there is strong evidence that the tires do not meet this objective. The distances (in metres) for 10 stops on a test track were 39.3, 39.0, 39.6, 40.2, 41.1, 37.5, 31.1, 38.1, 39.0, and 39.6. Should the company adopt the new tread pattern? Test an appropriate hypothesis and state your conclusion. Explain how you dealt with the outlier and why you made the recommendation you did.

T 49. Driving distance 2011 How far do professional golfers drive a ball? (For non-golfers, the drive is the shot hit from a tee at the start of a hole and is typically the longest shot.) Here's a histogram of the average driving distances of the 186 leading professional golfers by end of November 2011 along with summary statistics (www.pgatour.com).

a) Find a 95% confidence interval for the mean drive distance.
b) Interpreting this interval raises some problems. Discuss.
c) The data are the mean driving distance for each golfer. Is that a concern in interpreting the interval? (*Hint:* Review the What Can Go Wrong warnings of Chapter 8. Chapter 8?! Yes, Chapter 8.)
d) If instead we used these golfers' individual drive distances, what problem would this create for our inferential procedures?

T 50. Wind power Should you generate electricity with your own personal wind turbine? That depends on whether you have enough wind on your site. To produce enough energy, your site should have an annual average wind speed of at least eight miles per hour (mph), according to the Wind Energy Association. One candidate site was monitored for a year, with wind speeds recorded every six hours. A total of 1114 readings of wind speed averaged 8.019 mph with a standard deviation of 3.813 mph.

You've been asked to make a statistical report to help the landowner decide whether to place a wind turbine at this site.

a) Discuss the assumptions and conditions for using Student's *t* inference methods with these data. Here are some plots that may help you decide whether the methods can be used:

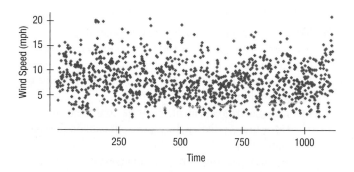

b) What would you tell the landowner about whether this site is suitable for a small wind turbine? Explain.

c) Why could we easily analyze data like this even before Gosset's discovery of the *t* distribution?

51. Worst of times Below is a sample randomly selected from all the Census Metropolitan Areas (CMAs) and Census agglomerations (CAs) in Canada showing the percentage change in the number of persons unemployed between May 2008 and May 2009 (during the deep 2008–2009 recession) for each area.

Area	% Change
Victoria	182.8
Alma	10.4
Salaberry-de-Valleyfield	64.0
Penticton	180.0
Campbell River	139.6
Woodstock	127.1
Baie-Comeau	6.6
Whitehorse	62.2
Hawkesbury, Ont. part	60.0
London	96.5
Prince Albert	62.2
Red Deer	304.3
Swift Current	200.0
Port Hope	122.2
Port Alberni	102.7
Ottawa-Gatineau, Gatineau part	53.7
Norfolk	129.3
Trois-Rivières	29.3
Labrador City	110.3
Nanaimo	136.6

Source: Adapted from Statistics Canada, Employment Insurance Statistics Maps, 73-002-XWE2009002 June 2009, Released August 25, 2009.

a) Estimate with a 95% confidence interval the true mean percentage increase in the number of unemployed persons per CMA/CA in Canada over this period.

b) Is your 95% confidence level quoted in part a) trustworthy? Check what you should.

c) The overall Canadian change in unemployment numbers was an increase of 71.5%. If we took similar repeated random samples and calculated such 95% confidence intervals over and over, would you expect them to catch this 71.5% figure 95% of the time? Why or why not?

52. Mercury sushi Torontonians (including one of your authors) seem to love their sushi, but is it always safe? The *New York Times* bought pieces of tuna sushi from a number of restaurants and stores in New York City in October 2007 and tested them for mercury levels. The results were not good. At most, consuming just six pieces per week would put you beyond an acceptable consumption level of mercury (49 micrograms of mercury per week for a person of average weight of 70 kg). Let's hope Toronto would fare better—but then again, the article states that experts believe similar results would be observed elsewhere, particularly for bluefin tuna sushi (the most common type in the survey).

Analysts examined at least two pieces from each place and calculated the methylmercury level in parts per million. Results below are for the piece of sushi with the highest mercury level for the restaurants surveyed. The pieces vary in size, so also shown is how many pieces per week it would take to exceed the acceptable mercury intake of 49 micrograms per week (*Rfd* = reference dose).

Restaurants	Methylmercury (parts per million)	Number of pieces to reach Rfd
Bar Masa	0.49	8.6
Blue Ribbon Sushi	1.40	2.6
Japonica	0.86	1.6
Jewel Bako	0.83	5.2
Megu	0.87	7.7
Monster Sushi (22 West 46th Street)	0.56	4.7
New York Times cafeteria	0.50	6.4
Nobu Next Door	1.00	6.2
Sushi of Gari	1.04	3.6
Sushi Seki	1.04	4.9
Sushi Yasuda	0.79	9.9
Yuka	0.61	3.3
Yuki Sushi	0.86	4.1

a) Give a 95% confidence interval for the mean mercury concentration level (per *worst* piece) if we can consider this to be a representative sample of New York City restaurants. Now check to see if that figure of 95% confidence is really trustworthy (that is, check and comment on any necessary conditions).

b) Give a 95% confidence interval for the mean number of (worst) pieces required to exceed health guidelines if we can assume this to be a representative sample of New York City restaurants. Now check to see if that 95% confidence level figure is really trustworthy (that is, check any necessary conditions).

53. **Simulations** Use your computer software to generate a sample of size 20 from a Normal distribution with a mean of 50 and a standard deviation of 10.
   a) From the sample, calculate a 90% confidence interval for the population mean. Does it contain the number 50?
   b) Repeat part a) for 99 fresh samples. How many confidence intervals out of 100 contained the number 50?

What percent of confidence intervals would you expect to contain the number 50 if you repeated these simulations many times? If X = the number of confidence intervals out of 100 that contain 50, what is the distribution of X?

54. **More simulations** Use your computer software to generate a sample of size 30 from a (continuous) uniform distribution on the interval 0 to 1.
   a) From the sample, calculate an 80% confidence interval for the population mean. Does it contain the true mean?
   b) Repeat part a) for 49 fresh samples. How many confidence intervals out of 50 contained the true mean? What percent of confidence intervals would you expect to contain the true mean if you repeated these simulations many times? If X = the number of confidence intervals out of 50 that contain the true mean of this uniform distribution, what is the distribution of X?

55. **Still more simulations** Use your computer software to generate a sample of size 15 from an exponential distribution with a mean of 1 (if a parameter is requested, set it equal to 1.0). Plot the data and describe the shape of this distribution.
   a) From the sample, calculate a 90% confidence interval for the population mean. Does it contain the true mean?
   b) Repeat part a) for 99 fresh samples. How many confidence intervals out of 100 contained the true mean of 1.0? What percent of confidence intervals would you expect to contain 1.0 if you repeated these simulations many times? If there are difficulties answering this question, explain. If we changed the sample size to 100, would that affect your answer?

56. **Yet more simulation.** Use your computer software to generate a sample of size 100 from an exponential distribution with a mean of 1 (if requested, set scale = 1.0 and threshold = 0.0). Plot the data and describe the shape of this distribution.
   a) From the sample, calculate a 90% confidence interval for the population mean. Does it contain the true mean?
   b) Repeat part a) for 99 fresh samples. How many confidence intervals out of 100 contained the true mean of 1.0? What percent of confidence intervals would you expect to contain 1.0 if you repeated these simulations many times? Justify your answer.

57. **Even more simulations** Use your computer software to generate a sample of size 5 from a Normal distribution with mean of 50 and a standard deviation of 5. (For example, these might be Guinness stout measurements for a batch that you are checking for adequate quality.)
   a) From the sample, test the null hypothesis that the population mean is 50 (our requirement for passing the batch) at the 10% significance level versus a

two-sided alternative. Did you reject the null hypothesis? Did you make an incorrect decision? Did you pass or fail a good or bad batch of stout? An incorrect decision here would constitute what type of error?

b) Repeat part a) for 99 fresh samples. In how many tests did you reject the null hypothesis? In what percent of tests would you expect to reject this null hypothesis at the 10% level if you repeated these simulations many times? If $X =$ the number of tests out of 100 in which you reject the null hypothesis at the 10% significance level, what is the distribution of X?

58. Final simulations Use your computer software to generate a sample of size 5 from a Normal distribution with a mean of 50 and a standard deviation of 5. (For example, these might be Guinness stout measurements for a batch that you are checking for adequate quality.)

a) From the sample, test the null hypothesis that the population mean is 60 (our requirement for passing the batch) at the 10% significance level versus a two-sided alternative. Did you reject the null hypothesis? Was this a correct decision or an error? Did you pass or fail a good or bad batch of stout? What type of error would a wrong decision here constitute?

b) Repeat part a) for 99 fresh samples. In how many tests did you reject the null hypothesis? This number is an estimate of something—what, exactly? If you have the appropriate software, use it to determine what would be the long-run percentage of such rejections.

59. Calculating power For the chapter example about salmon mirex levels, let's do an approximate calculation of the power of the test for an alternative of 0.09 ppm. For samples of size n $\geq$ 30, we can approximate the t distribution by the standard Normal distribution in these rough calculations. We also need to make a guess at the value of the unknown parameter σ, but here a study has already been run, so let's just take that $\sigma = 0.0495$ as the guess for σ and round it to 0.05. Since we are guessing at σ, this calculation is just an approximation, but usually that's all we need.

a) Setting alpha at 0.05, find the critical value for a standard Normal z-statistic. Write the criterion for rejection of the null in terms of the t-statistic. You should have a criterion like $\dfrac{\bar{y} - 0.08}{s/\sqrt{150}} > z^*$, with a specific number for the critical value.

b) Now another approximation. Pretend that the sample standard deviation s will equal the true population standard deviation σ. Funny thing to do, since s is random, not a constant, but this works well enough as an approximation when n is not too small (and otherwise we'd be stuck!). Rewrite the criterion above with just the sample mean on the left side, that is, find

out just how big the sample mean must be for you to reject the null hypothesis (after setting $s = \sigma$). You should have a criterion like $\bar{y} > \bar{y}^*$ with a specific number for the $\bar{y}^*$ critical value.

c) Calculate the probability that $\bar{y}$ is bigger than $\bar{y}^*$, assuming the true mean is equal to 0.09 (standardize properly and use the standard Normal table). What you've now got is the power—the probability of making the right decision (to reject $\mu = 0.08$ ppm) should the true mean $\mu = 0.09$ ppm.

d) For a small n, though, this does not work well, since you have to take into account more properly the random variation present in the sample standard deviation s, in which case using statistical software is recommended. If your software does power calculations for the one-sample t-test, use it to confirm your calculation above. Also using your software, determine how low the power drops:

i. if you halve your sample size ($n = 75$).
ii. if you halve the sample size yet again (to $n = 38$).

60. More power to you For the chapter example about school travel times, let's do an approximate calculation of the power of the test for an alternative of 20 minutes. For samples of size $n \geq 30$, we can approximate the t distribution by the standard Normal distribution in these rough calculations. We also need to make a guess at the value of the unknown parameter σ, but here, a study has already been run so let's just make that $s = 9.66$ minutes as the guess for σ and round it to 10 minutes. Since we are guessing at σ, this calculation is just an approximation, but usually that's all we need.

a) Setting alpha at 0.05, find the critical value for a standard Normal z-statistic. Write the criterion for rejection of the null in terms of the t-statistic. You should have a criterion like $\dfrac{\bar{y} - 15}{s/\sqrt{40}} > z^*$, with a specific number for the critical value.

b) Now another approximation. Pretend that the sample standard deviation s will equal the true population standard deviation σ. Funny thing to do, since s is random, not a constant, but this works well enough as an approximation when n is not too small (and otherwise we'd be stuck!). Rewrite the criterion above with just the sample mean on the left side; that is, find out just how big the sample mean must be for you to reject the null hypothesis (after setting $s = \sigma$). You should have a criterion like $\bar{y} > \bar{y}^*$ with a specific number for the $\bar{y}^*$ critical value.

c) Calculate the probability that $\bar{y}$ is bigger than $\bar{y}^*$, assuming the true mean is equal to 20 minutes (standardize properly and use the standard Normal table). What you've now got is the power—the probability of making the right decision (to reject $\mu = 15$ minutes) should the true mean $\mu = 20$ minutes.

d) For small n, this does not work well, since you have to take into account more properly the random variation present in the sample standard deviation s, in which case using statistical software is recommended. If your software does power calculations for the one-sample t-test, use it to confirm your calculation above. Also using your software, determine how low the power drops:

i. if you halve the sample size ($n = 20$).

ii. if you double the sample size (to $n = 80$).

Just Checking ANSWERS

1. Questions on the short form are answered by everyone in the population. This is a census, so means or proportions *are* the true population values. The long forms are given to just a sample of the population. When we estimate parameters from a sample, we use a confidence interval to take sample-to-sample variability into account.

2. They don't know the population standard deviation, so they must use the sample standard deviation as an estimate. The additional uncertainty is taken into account by t-models if we are using an unweighted average.[20] We don't know what model to use for a weighted average (perhaps a t model but with a different SE formula).

3. The margin of error for a confidence interval for a mean depends, in part, on the standard error, $SE(\bar{y}) = \dfrac{s}{\sqrt{n}}$.

   Since n is in the denominator, smaller sample sizes lead to larger SEs and correspondingly wider intervals. Long forms returned by one in every five households in a less populous area will produce a smaller sample.

4. The t^* value would change a little, while n and its square root change a lot, making the interval much narrower for the larger sample. The smaller sample is one fourth as large, so the confidence interval would be roughly twice as wide.

5. We expect 95% of such intervals to cover the true value on average, so about 5 of the 100 intervals might be expected to miss.

| **MathXL** | Go to MathXL at **www.mathxl.com** or MyLab™ Statistics at **www.mystatlab.com**. You can practise exercises for this chapter as often as you want. The guided solutions will help you find answers step by step. You'll find a personalized study plan available to you too! |
| **MyLab Statistics** | |

---

[20]Though ideally a finite population correction factor should be applied to the SE formula, as discussed in Chapter 14, since the sample is more than 10% of the population.

Review

From the Data at Hand to the World at Large

Quick Review

What do samples really tell us about the populations from which they are drawn? Are the results of an experiment meaningful, or are they just sampling error? Statistical inference based on our understanding of sampling models can help answer these questions. Here's a brief summary of the key concepts and skills:

- Sampling models describe the variability of sample statistics using a remarkable result called the Central Limit Theorem.

  - When the number of trials is sufficiently large, proportions found in different samples vary according to an approximately Normal model.

  - When samples are sufficiently large, the means of different samples vary, with an approximately Normal model.

  - The variability of sample statistics decreases as sample size increases.

  - Statistical inference procedures are often based on the Central Limit Theorem.

  - No inference procedure is valid unless the underlying assumptions are true. Always check the conditions before proceeding.

- Statistical inference procedures for means are also based in part on the Central Limit Theorem, but we don't usually know the population standard deviation. Student's *t*-models take the additional uncertainty of independently estimating the standard deviation into account.

- A confidence interval uses a sample statistic (such as a proportion or mean) to estimate a range of plausible values for the parameter of a population model.

  - All confidence intervals involve an estimate of the parameter, a margin of error, and a level of confidence.

  - For confidence intervals based on a given sample, the higher the confidence level, the bigger the margin of error.

  - At a given level of confidence, the larger the sample, the smaller the margin of error.

- A hypothesis test proposes a model for the population, then examines the observed statistics to see if that model is plausible.

- A null hypothesis suggests a parameter value for the population model. Assuming the null hypothesis to be true, there should be nothing unusual, or surprisingly different about the sample results.

- The alternative hypothesis states what we will believe if the sample results turn out to be inconsistent with our null model.

- We compare the difference between the statistic and the hypothesized value with the standard deviation of the statistic. It's the sampling distribution of this ratio that gives us a P-value.

- The P-value of the test is the probability of obtaining results at least as extreme as those observed in the sample or the experiment, assuming the null hypothesis to be in fact true.

- A low P-value indicates evidence against the null model. If it is sufficiently low, we have evidence to reject the null model.

- A high P-value indicates that the sample results are not inconsistent with the null model, so we cannot reject it. However, this does not prove the null model is true.

- Sometimes we will mistakenly reject the null hypothesis even though it's actually true—that's called a Type I error. If we fail to reject a false null hypothesis, we commit a Type II error.

- The power of a test measures its ability to detect a false null hypothesis.

- You can lower the risk of a Type I error by requiring a higher standard of proof (lower P-value) before rejecting the null hypothesis. But this will raise the risk of a Type II error and decrease the power of the test.

- The simplest way to increase the power of a test is to design a study based on a larger sample (though in more complicated studies, 'noise-reducing' techniques, such as blocking or stratification, can also increase power and precision).

And now for some opportunities to review these concepts and skills . . .

Review Exercises

1. **Whole wheat bread?** A researcher wants to examine the possibility that children are equally likely to choose between whole wheat bread and white bread for breakfast. 86 children were presented with a choice between a whole wheat bread and a white bread for breakfast. 48 of them chose to have white bread. State the null and alternative hypotheses the researcher used in her study.

2. **Colour-blind** Medical literature says that about 8% of males are colour-blind. A university's introductory Psychology course is taught in a large lecture hall. Among the students, there are 325 males. Each semester when the professor discusses visual perception, he shows the class a test for colour-blindness. The percentage of males who are colour-blind varies from semester to semester.
   a) Is the sampling distribution model for the sample proportion likely to be Normal? Explain.
   b) What are the mean and standard deviation of this sampling distribution model?
   c) Sketch the sampling model, using the 68–95–99.7 Rule.
   d) Write a few sentences explaining what the model says about this professor's class.

3. **Birth days** During a two-month period in 2002, 72 babies were born at the Tompkins Community Hospital in upstate New York. The table shows how many babies were born on each day of the week.

Day	Births
Mon.	7
Tues.	17
Wed.	8
Thurs.	12
Fri.	9
Sat.	10
Sun.	9

   a) If births are uniformly distributed across all days of the week, how many would you expect on each day?
   b) Only seven births occurred on a Monday. Does this indicate that women might be less likely to give birth on a Monday? Explain.
   c) Are the 17 births on Tuesdays unusually high? Explain.
   d) Can you think of any reasons why births may not occur completely at random?

4. **Third parties** Support for the New Democratic Party[1] has come in at around 15% in federal polls and elections

   for decades. Assuming the party's support remains at 15.0% in the future, and that 1000 individuals are randomly sampled in each poll, answer the following:
   a) How low will NDP support drop in 2.5% of future polls?
   b) How high will NDP support rise in about 4 of the next 25 polls?
   c) The Liberals tend to get 35% or so of popular support. In future polls, would the Liberal percent tend to vary more, less, or about the same as the NDP percent? Explain.

5. **Leaky gas tanks** It is estimated that 40% of service stations have gas tanks that leak to some extent. A new program in California is designed to lessen the prevalence of these leaks. We want to assess the effectiveness of the program by seeing if the percentage of service stations whose tanks leak has decreased. To do this, we randomly sample 27 service stations in California and determine whether there is any evidence of leakage. In our sample, only seven of the stations exhibit any leakage. Is there evidence that the new program is effective?
   a) What are the null and alternative hypotheses?
   b) Check the assumptions necessary for inference.
   c) Test the null hypothesis.
   d) What do you conclude (in plain English)?
   e) If the program actually works, have you made an error? What kind?
   f) What two things could you do to decrease the probability of making this kind of error?
   g) What are the advantages and disadvantages of taking those two courses of action?

6. **Hamsters** How large are hamster litters? Among 47 golden hamster litters recorded, there were an average of 7.72 baby hamsters, with a standard deviation of 2.5 hamsters per litter.
   a) Create and interpret a 90% confidence interval.
   b) Would a 98% confidence interval have a larger or smaller margin of error? Explain.
   c) How many litters must be used to estimate the average litter size to within 1 baby hamster with 95% confidence?

7. **Scrabble** Using a computer to play many simulated games of Scrabble, researcher Charles Robinove found that the letter "A" occurred in 54% of the hands. This study had a margin of error of $\pm 10\%$.[2]
   a) Explain what the margin of error means in this context.
   b) Why might the margin of error be so large?
   c) Probability theory predicts that the letter "A" should appear in 63% of the hands. Does this make you concerned that the simulation might be faulty? Explain.

8. **Dice** When one die is rolled, the number of spots showing has a mean of 3.5 and a standard deviation of 1.7. Suppose

---

[1]Third parties can make important contributions. The NDP's predecessor, the CCF, fought through a doctor's strike in Saskatchewan in July 1962 to implement North America's first universal health insurance system, leading soon thereafter to similar plans across Canada.

[2]*Chance*, 15, no. 1 [2002].

you roll 10 dice. What's the approximate probability that your total is between 30 and 40 (that is, the average for the 10 dice is between 3 and 4)? Specify the model you use and the assumptions and conditions that justify your approach.

9. **Net-newsers** In June 2008, the Pew Research Foundation sampled 3615 U.S. adults and asked about their choice of news sources. They identified 13% as "Net-newsers" who regularly get their news from online sources rather than TV or newspapers.
   a) Pew reports a margin of error of $\pm 2\%$ for this result. Explain what the margin of error means.
   b) Pew's survey included 2802 respondents contacted by landline and 813 contacted by cell phone. If the percentage of Net-newsers is the same in both groups and Pew estimated those percentages separately, which group would have the larger margin of error? Explain.
   c) Pew reports that 82% of the 470 Net-newsers in their survey get news during the course of the day, far more than other respondents. Find a 95% confidence interval for this statistic.
   d) How does the margin of error for your confidence interval compare with the values in parts a) and b)? Explain why.

10. **Rally tally** "Rally tally was really 52 800, experts' averaged figures show" reported a Toronto newspaper one day about the size of a rally at Queen's Park in Toronto. Reporters used the average of six experts' assessments, all of whom were given the same photos of the crowd that day. With reference to appropriate statistical concepts (like bias and variability), answer the following:
    a) Why do you think they averaged the six estimates?
    b) If they could average not just six, but hundreds of experts' assessments, could they expect this average to stabilize around some value as they average more and more assessments? Explain.
    c) Could they expect to hit on the true rally size (at the time the photos were taken) if enough assessments are averaged? Explain.

11. **Bimodal** We are sampling randomly from a distribution known to be bimodal.
    a) As our sample size increases, what's the expected shape of the sample's distribution?
    b) What's the expected value of our sample's mean? Does the size of the sample matter?
    c) How is the variability of this sample's mean related to the standard deviation of the population? Does the size of the sample matter?
    d) How is the shape of the sampling distribution model affected by the sample size?

12. **Vitamin D 2012** In 2012, the *American Journal of Clinical Nutrition* reported that 31% of Australian adults over age 25 have a vitamin D deficiency. The data came from the AusDiab study of 11 218 Australians.
    a) Do these data meet the assumptions necessary for inference? What would you like to know that you don't?
    b) Create a 95% confidence interval.

    c) Interpret the interval in this context.
    d) Explain what "95% confidence" means in this context.

13. **Fritos®** As a project for an Introductory Statistics course, students checked 6 bags of Fritos marked with a net weight of 35.4 grams. They carefully weighed the contents of each bag, recording the following weights (in grams): 35.5, 35.3, 35.1, 36.4, 35.4, 35.5. Is there evidence that the mean weight of bags of Fritos is less than advertised?
    a) Write appropriate hypotheses.
    b) Do these data satisfy the assumptions for inference? Explain.
    c) Test your hypothesis using all 6 weights.
    d) Retest your hypothesis with the one unusually high weight removed.
    e) What would you conclude about the stated net weight?

14. **And it means?** Every statement about a confidence interval contains two parts—the level of confidence and the interval. Suppose that an insurance agent estimating the mean loss claimed by clients after home burglaries created the 95% confidence interval ($1644, $2391).
    a) What's the margin of error for this estimate?
    b) Carefully explain what the interval means.
    c) Carefully explain what the 95% confidence level means.

15. **Batteries** We work for the "Watchdog for the Consumer" consumer advocacy group. We've been asked to look at a battery company that claims its batteries last an average of 100 hours under normal use. There have been several complaints that the batteries don't last that long, so we decide to test them. To do this we select 16 batteries and run them until they die. They lasted a mean of 97 hours, with a standard deviation of 12 hours.
    a) One of the editors of our newsletter (who does not know statistics) says that 97 hours is a lot less than the advertised 100 hours, so we should reject the company's claim. Explain to him the problem with doing that.
    b) What are the null and alternative hypotheses?
    c) What assumptions must we make in order to proceed with inference?
    d) At a 5% level of significance, what do you conclude?
    e) Suppose that, in fact, the average life of the company's batteries is only 98 hours. Has an error been made in part d? If so, what kind?

16. **Eggs** The ISA Babcock Company supplies poultry farmers with hens, advertising that a mature B300 Layer produces eggs with a mean weight of 60.7 grams. Suppose that egg weights follow a Normal model with standard deviation 3.1 grams.
    a) What fraction of the eggs produced by these hens weigh more than 62 grams?
    b) What's the probability that a dozen randomly selected eggs average more than 62 grams?
    c) Using the 68–95–99.7 Rule, sketch a model of the total weights of a dozen eggs.

17. No appetite for privatization In a June 2009 Harris-Decima poll,[3] 82% of Canadians indicated that they felt the Canadian health care system is superior to the American health care system, while 8% felt that the American system is superior. Twelve percent felt that there should be more privatization (fee-based services) of the Canadian health care system, while 55% wanted more public elements. However, in Quebec, 19% felt the American system is superior, and 16% wanted more fee-based services. It was reported that the poll was based on a random national sample of 1000 Canadians, and that a sample of the same size has a margin of error of 3.1%, 19 times in 20.
a) Is the actual margin of error for the 12% figure (for more privatization) equal to 3.1%? If not, what is it?
b) Can you determine the margin of error for the Quebec figure of 16%? If not, what do you need to know? Would it help if we told you that the population of Quebec is 23% of the Canadian population?
c) To reduce the margin of error for Quebec by 75%, how large would the total sample have to be? (Of course, it would be cheaper to over-sample from just Quebec.)

18. Enough eggs? One of the important issues for poultry farmers is the production rate—the percentage of days on which a given hen actually lays an egg. Ideally, that would be 100% (an egg every day), but realistically, hens tend to lay eggs on about three of every four days. ISA Babcock wants to advertise the production rate for the B300 Layer (see Exercise 16) as a 95% confidence interval with a margin of error of $\pm 2\%$. How many hens must they collect data on?

19. Teen deaths 2005 Traffic accidents are the leading cause of death among people aged 16 to 20. In May 2005, the National Highway Traffic Safety Administration reported that even though only 6.3% of licensed drivers are between 16 and 20 years old, they were involved in 12.6% of all fatal crashes. Insurance companies have long known that teenage boys were high risks, but what about teenage girls? One insurance company found that the driver was a teenage girl in 44 of the 388 fatal accidents they investigated. Is this strong evidence that the accident rate is lower for girls than for teens in general?
a) Test an appropriate hypothesis and state your conclusion.
b) Explain what your P-value means in this context.

20. Hours spent on the Internet The Seneca Student Federation surveyed 25 Seneca College students to try to understand current trends among college students. They reported that the average amount of time spent per week on the Internet was 19.0 hours with a standard deviation of 12.5 hours.
a) Create a 95% confidence interval for the mean amount of time spent per week on the Internet by Seneca College students.

b) Interpret your interval.
c) Explain what "95% confidence" means.
d) Would a 90% confidence interval be wider or narrower? Explain without actually calculating the interval.

21. Largemouth bass Organizers of a fishing tournament believe that the lake holds a sizable population of largemouth bass. They assume that the weights of these fish have a model that is skewed to the right with a mean of 3.5 pounds and a standard deviation of 2.2 pounds.
a) Explain why a skewed model makes sense here.
b) Explain why you cannot determine the probability that a largemouth bass randomly selected ("caught") from the lake weighs over three pounds.
c) Each fisherman in the contest catches five fish each day. Can you determine the probability that someone's catch averages over three pounds? Explain.
d) The 12 fishermen competing each caught the limit of five fish. What's the probability that the total catch of 60 fish averaged more than three pounds per fish?

22. Cheating A Rutgers University study found that many high school students cheat on tests. The researchers surveyed a random sample of 4500 high school students across the U.S.; 74% of them said they had cheated at least once.
a) Create a 90% confidence interval for the level of cheating among high school students. Don't forget to check the appropriate conditions.
b) Interpret your interval.
c) Explain what "90% confidence" means.
d) Would a 95% confidence interval be wider or narrower? Explain without actually calculating the interval.

23. Language Neurological research has shown that in about 80% of people, language abilities reside in the brain's left side. Another 10% display right-brain language centres, and the remaining 10% have two-sided language control. (The latter two groups are mainly left-handers.)[4]
a) We select 60 people at random. Is it reasonable to use a Normal model to describe the possible distribution of the proportion of the group that has left-brain language control? Explain.
b) What's the probability that our group has at least 75% left-brainers?
c) If the group had consisted of 100 people, would that probability be higher, lower, or about the same? Explain why, without actually calculating the probability.
d) How large a group would almost certainly guarantee at least 75% left-brainers? Explain.

24. Following RIM It seems hard to believe now, but back in 2009, the price of Research In Motion (now renamed Blackberry) shares on the TSE was about $80 in late April, and was at $80 again in late July. Sometimes it appears that there may be *autocorrelation* in stock price daily movements, which means that what happens on one day may affect or correlate with what happens on the next day. Good days may have follow-through or carry-over effects to the next day or two (or, maybe after a good day, we tend to "correct" to the downside the next day, which would be negative autocorrelation). Tracking RIM's up and down days, there were 16 times it followed an up day with an up day, and 13 times it followed an up day with a down day. There were 15 times it followed a down day by a down day and 13 times it followed a down day by an up day. If you want to play the short-term trends, should you, say, buy on an up day, hoping for positive follow-through the next day? Test the null hypothesis that there is no relationship between up days and price direction on the next day using these data.

25. **Crohn's disease** In 2002, the medical journal *The Lancet* reported that 335 of 573 patients suffering from Crohn's disease responded positively to injections of the arthritis-fighting drug infliximab.
    a) Create a 95% confidence interval for the effectiveness of this drug.
    b) Interpret your interval in context.
    c) Explain carefully what "95% confidence" means in this context.

    Source: The Lancet Volume 359, Issue 9300, 5 January 2002, Pages 62–69

26. **Narrower or wider?** A marketing researcher is running a study to estimate the proportion of adults in Canada who spent more than $1000 on electronics in the past six months. He obtained a simple random sample of 4000 Canadian adults and calculated a 95% confidence interval estimate of this proportion. Estimates were also obtained, from the same sample, for the proportion of men who spent more than $1000 on electronics, the proportion of women who spent more than $1000 on electronics, and the proportion of Ontarians who spent more than $1000 on electronics. A similar study was conducted in the U.S. that selected 16 000 adults at random. Assume that there are equal numbers of males and females in each population, that Ontario's population is 25% of the Canadian population, and that the U.S. population is nine times bigger than the Canadian population. Answer the following questions, assuming that the true proportion of the population who spent more than $1000 is actually fairly similar for each population studied.
    a) Since we selected a bigger percentage of the Canadian population than the U.S. population, the precision should be greater, resulting in a narrower confidence interval for the Canadian estimate. Do you agree? Explain why or why not.
    b) The margin of error for the U.S. estimate will be about one-fourth of the margin of error for the Ontario estimate. Do you agree? Explain why or why not.
    c) The margin of error for the estimate of the proportion of Canadian women who spent more than $1000 will be about half as big as the margin of error for the proportion of Ontarians who spent more than $1000. Do you agree? Explain why or why not.

27. **Alcohol abuse** Growing concern about binge drinking among university students has prompted one large school to conduct a survey to assess the size of the problem on its campus. The university plans to randomly select students and ask how many have been drunk during the past week. If the school hopes to estimate the true proportion among all its students with 90% confidence and a margin of error of ±4%, how many students must be surveyed?

28. **Waiting for homicide** In London, England, the number of days between homicides was recorded for the period April 2004–March 2007, and is displayed below.[5] Days is on the *x*-axis, with counts on the *y*-axis.
    a) Estimate with a 95% confidence interval the average time between homicides in London (estimate as best you can what you need from the graph).

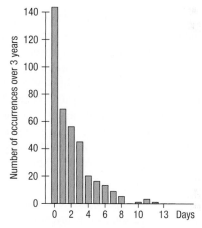

    b) Can you trust the reported confidence level of 95% for your interval?

29. **Bread** Tobermory Bakery is trying to predict how many loaves to bake. In the past 100 days, the bakery has sold between 95 and 140 loaves per day. Here are a histogram and the summary statistics for the number of loaves sold for the past 100 days.[5]

Summary of sales	
Mean	103
Median	100
StdDev	9.000
Min	95
Max	140
Lower 25th %tile	97
Upper 25th %tile	105.5

[5]D. Spiegelhalter & A. Barnett, 2009, "London murders: A predictable pattern?" *Significance*. Reprinted with permission of John Wiley & Sons Inc.

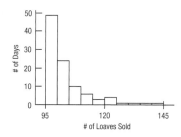

a) Can you use these data to estimate the number of loaves sold on the busiest 10% of all days? Explain.
b) Explain why you can use these data to construct a 95% confidence interval for the mean number of loaves sold per day.
c) Calculate a 95% confidence interval and carefully interpret what that confidence interval means.
d) If the bakery would have been satisfied with a confidence interval whose margin of error was twice as wide, how many days' data could they have used?
e) When the bakery opened, the owners estimated that they would sell an average of 100 loaves per day. Does your confidence interval provide strong evidence that this estimate was incorrect? Explain.

30. Safety Observers in Texas watched children at play in eight communities. Of the 814 children seen biking, roller skating, or skateboarding, only 14% wore a helmet.
a) Create and interpret a 95% confidence interval.
b) What concerns do you have about this study that might make your confidence interval unreliable?
c) Suppose we want to do this study again, picking various communities and locations at random, and hope to end up with a 98% confidence interval having a margin of error of ±4%. How many children must we observe?

31. Fried PCs A computer company recently experienced a disastrous fire that ruined some of its inventory. Unfortunately, during the panic of the fire, some of the damaged computers were sent to another warehouse, where they were mixed with undamaged computers. The engineer responsible for quality control would like to check out each computer to decide whether it's undamaged or damaged. Each computer undergoes a series of 100 tests. The number of tests it fails will be used to make the decision. If it fails more than a certain number, it will be classified as damaged and then scrapped. From past history, the distribution of the number of tests failed is known for both undamaged and damaged computers. The probabilities associated with each outcome are listed in the table below:

Number of tests failed	0	1	2	3	4	5	>5
Undamaged (%)	80	13	2	4	1	0	0
Damaged (%)	0	10	70	5	4	1	10

The table indicates, for example, that 80% of the undamaged computers have no failures, while 70% of the damaged computers have two failures.
a) To the engineers, this is a hypothesis-testing situation. State the null and alternative hypotheses.
b) Someone suggests classifying a computer as damaged if it fails any of the tests. Discuss the advantages and disadvantages of this test plan.
c) What number of tests would a computer have to fail to be classified as damaged if the engineers want to have the probability of a Type I error equal to 5%?
d) What's the power of the test plan in part c)?
e) A colleague points out that increasing α by just 2% would increase the power substantially. Explain.

32. Power We are replicating an experiment. How will each of the following changes affect the power of our test? Indicate whether it will increase, decrease, or remain the same, assuming that all other aspects of the situation remain unchanged.
a) We increase the number of subjects from 40 to 100.
b) We require a higher standard of proof, changing from $\alpha = 0.05$ to $\alpha = 0.01$.

33. Name recognition An advertising agency won't sign an athlete to do product endorsements unless it is sure the person is known to more than 25% of its target audience. The agency always conducts a poll of 500 people to investigate the athlete's name recognition before offering a contract. Then it tests H_0: $p = 0.25$ against H_A: $p > 0.25$ at a 5% level of significance.
a) Why does the company use upper-tail tests in this situation?
b) Explain what Type I and Type II errors would represent in this context, and describe the risk that each error poses to the company.
c) The company is thinking of changing its test to use a 10% level of significance. How would this change the company's exposure to each type of risk?

34. Name recognition, part II The advertising company described in Exercise 33 is thinking about signing an Olympic athlete to an endorsement deal. In its poll, 27% of the respondents could identify her.
a) Fans who never took Statistics can't understand why the company did not offer this Olympic athlete an endorsement contract even though the 27% recognition rate in the poll is above the 25% threshold. Explain it to them.
b) Suppose that further polling reveals that this Olympic athlete really is known to about 30% of the target audience. Did the company initially commit a Type I or Type II error in not signing her?
c) Would the power of the company's test have been higher or lower if the player were more famous? Explain.

T 35. Forest fires Below is a random sample of fire sizes in hectares (ha) from the Large Fire Database at Natural

Resources Canada for fires that covered over 200 ha that occurred in Canada in the late 1990s.[6]

900.0	2 000.0	804.2	1 682.4	559.0	5 000.0	309.0
3 790.0	2 450.0	1 380.0	12 630.0	4 243.6	557.9	1 560.0
485.6	3 215.0	12 492.0	2 300.0	389.9	1 366.0	2 612.0
1 200.0	208.0	55 000.0	40 759.7	250.0	6 607.2	774.1
6 284.0	349.8	4 381.0	7 400.0	255.3	304.0	1 200.0
545.9	627.0	1 777.0	400.0	1 657.2	2 443.4	547.8
14 200.0	631.0	1 280.0	221.0	450.0	1 550.7	17 400.0
459.0	2 513.0	34 209.0	4 274.1	356.0	4 4261.0	1 400.0
347.9	2 800.0	5 650.0	12 557.0	19 320.8	30 070.0	3 261.2
321.3	19 735.6	435.0	6 465.5	1 817.0	13 433.0	339.0
3 580.7	739.0	799.5	4 170.0	3 000.0	414.9	14 072.5
500.0	27 500.0	225.4	750.0	611.0	300.0	1 642.4
10 240.0	447.6	48 451.1	1 500.0	297.0	825.0	850.0
333.2	17 597.0	431.1	300.0	3267.1	604.6	10 600.0
17 531.0	520.0					

a) Estimate with a 90% confidence interval the true mean fire size for all fires in Canada over this period.

b) The actual mean of all fires over 200 ha during this period was 6398. Did your interval in a) catch this value? If you were to repeat the sampling procedure over and over, obtaining independent random samples from the population of fires, and then calculate confidence intervals from each sample, how often would your confidence interval contain the value 6398? Suppose you obtained 200 independent random samples from the entire population and counted the number of samples that contained the value 6398; call it X. What would be the distribution of the random variable X?

c) Examine the shape of the distribution for the sample. Do you have any qualms about the validity of your analysis in part a)?

36. Dropouts One study comparing various treatments for the eating disorder anorexia nervosa initially enlisted 198 subjects, but found overall that 105 failed to complete their assigned treatment programs. Construct and interpret an appropriate confidence interval. Discuss any reservations you have about this inference.

37. Sodium Chloride Measurements of water intake, obtained from a sample of 17 rats that had been injected with a sodium chloride solution, produced a mean and standard deviation of 31.0 and 6.2 cubic centimetres, respectively. It is given that the average water intake for non-injected rats observed over a comparable period of time is 22.0 cubic centimetres.

a) Do injected rats drink more water than non-injected rats based on the information provided?

b) Create a 99% confidence interval for the mean water intake for injected rats.

38. Two-stage sample To study the problem of bullying, we sample 20 schools at random from Toronto, and then randomly select 60 students from each school to interview. We want to estimate the proportion of students who have been bullied in the past month. If we find that 300 out of the 1200 selected students have been bullied, can you determine the margin of error for this 25% estimate? If so, do it. If not, explain why not.

39. CADUMS survey The Canadian Alcohol and Drug Use Survey (CADUMS) from 2008 included random samples of about 1000 respondents from each Canadian province, except for Alberta and B.C., where larger samples of about 4600 and 4000, respectively, were taken, producing a total sample size of about 16 600. All were Canadian residents aged 15 years and over. The researchers collected information about alcohol and drug related behaviours. The survey included 1443 individuals aged 15–24. Researchers estimated that 52.9% percent of 15–24 year olds had used cannabis in their lifetime. The reported margin of error was 4.7% with 95% confidence.

a) If you try to confirm the margin of error, do you run into any problem? Explain. Why is the reported margin of error bigger or smaller than one might expect for a random sample of this size?

b) Using the given margin of error, calculate a 99% confidence interval for the true proportion and interpret your interval in context.

40. Splitting hairs Samples of five hairs are taken from each of four cancer patients' heads, and the zinc content is measured for each of the 20 specimens. We want to see if the zinc levels are lower in the hair of cancer patients than in the general population (as is the case for blood zinc levels). From the data, would you be able to construct a 95% confidence interval for the mean hair zinc concentration in the patient population sampled, by computing the mean and standard deviation of the 20 numbers, plugging into the usual one-sample t-statistic, and then obtaining the P-value from the t-distribution with 19 degrees of freedom? Explain. If you did proceed in this way, would your margin of error be predictably off the mark in some way? (Too large? Too small?)

[6]Reproduced with permission from Natural Resources Canada, Canadian Forest Service, 2013. http://cwfis.cfs.nrcan.gc.ca/home.

Comparing Means

Nothing is good or bad, but by comparison.

—Thomas Fuller

LEARNING OBJECTIVES

1 For two independent samples of quantitative data,

- Estimate the difference in means using a confidence interval
- Test for a significant difference in mean

2 Know the assumptions and conditions and how to check them

3 Be familiar with the 'pooled variance' approach

4 Determine the sample size needed to attain a required:

- Margin of error, or
- Power for a test

Connections

Does taking echinacea help you get over a cold faster? Does playing Mozart to babies in the womb make them smarter? Do students who use technology learn Statistics better than those who do not? It's much more common to compare the means of two groups than to test whether a single mean might equal some number.

We compared groups with side-by-side boxplots in Chapter 4, where we asked whether their centres looked different—compared to how spread out the distributions were. Here we'll make that kind of thinking precise, using confidence intervals or tests for the difference in mean response. Standard errors

hould you buy generic rather than brand-name batteries? A Statistics student designed a study to test battery life. He wanted to know whether there was any real difference between brand-name batteries and a generic brand. To estimate the difference in mean lifetimes, he kept a battery-powered CD player[1] continuously playing the same CD, with the volume control fixed at level five, and measured the time until no more music was heard through the headphones. For his trials, he used six sets of AA alkaline batteries from two major battery manufacturers: a well-known brand name and a generic brand. He measured the time in minutes until the sound stopped. To account for possible changes in the CD player's performance over time, he randomized the run order by choosing sets of batteries at random.

Here are his data (times in minutes):

Brand Name	Generic
194.0	190.7
205.5	203.5
199.2	203.5
172.4	206.5
184.0	222.5
169.5	209.4

Studies that compare two groups are common throughout science and industry. We might want to compare the effects of a new drug with the traditional therapy, the fuel efficiency of two car engine designs, or the sales of new products in two different test cities. In fact, battery manufacturers do research like this on their products and competitors' products.

19.1 Comparing Means of Independent Samples

Plot the Data

The natural display for comparing two groups is boxplots of the data for the two groups, placed side by side. Although we can't make a confidence interval or test a hypothesis from the boxplots themselves, you should always start with boxplots when comparing groups. Let's look at the boxplots of the battery test data, in Figure 19.1 on the next page.

---

[1]Once upon a time, not so very long ago, there were no iPods. At the turn of the century, people actually carried CDs around, and devices to play them. We bet you can find one on eBay.

again provide the ruler needed for determining significance or margins of error, and Student's *t* provides the requisite sampling model.

Comparing groups may also be viewed as assessing the association between a categorical grouping variable and the within-group response variable. Part VI of this text focuses on inference for associations, but the type of analysis will depend on whether the variables are quantitative or categorical.

Who	AA alkaline batteries
What	Length of battery life while playing a CD continuously
Units	Minutes
Why	Class project
When	1998

Figure 19.1

Boxplots comparing the brand-name and generic batteries suggest a difference in duration.

It sure looks like the generic batteries lasted longer. But is the difference large enough to change our battery-buying behaviour? Can we be confident that the observed difference is more than just random fluctuation? That's why we need statistical inference.

The boxplot for the generic data identifies two possible outliers. That's interesting, but with only six measurements in each group, the outlier nomination rule is not very reliable. Both of the extreme values are plausible results (barely exceeding the 1.5 IQR outlier criterion), and the range of the generic values is smaller than the range of the brand-name values, even with those boxplot outliers. So if they are not very suspicious or unusual, we're better off leaving these values in the data.

Standard Error and Sampling Model

The population model parameter of interest is the difference between the *mean* battery lifetimes of the two brands, $\mu_1 - \mu_2$. The statistic of interest is the difference in the two observed means, $\bar{y}_1 - \bar{y}_2$. We'll start with this statistic to build our confidence interval, but we'll need to know its standard deviation and its sampling model. Then we can build confidence intervals and find P-values for hypothesis tests.

We know how to estimate the standard deviation of the mean in each group, but how about their difference? The answer comes from realizing that we have to start with the variances and the Pythagorean Theorem of Statistics from Chapter 13

The variance of the sum or difference of two independent random variables is the sum of their variances.

The diagram above shows SD's. If you square them, you get the variances. The hypotenuse represents the SD of the sum or difference.

This is such an important (and powerful) idea in Statistics that it's worth pausing a moment to review the reasoning. Here's some intuition about why variation increases even when we subtract two independent random quantities:

> **Variances Add for Sums and Differences** Here's an example. Let X = the time you'll wait for the subway today, and let Y = the time you'll wait for the subway tomorrow. Assume the subway runs precisely every 4 minutes, in which case your wait time is uniformly distributed between 0 and 4 minutes, if your arrival at the station is rather random. X + Y is your total time waited. Its mean is 2 + 2 = 4 minutes, but it can be anywhere from 0 + 0 = 0 to 4 + 4 = 8 minutes. What about the difference between your wait times X − Y? Its mean is 2 − 2 = 0 minutes, and this difference can be as small as 0 − 4 = −4 or as large as 4 − 0 = +4. Both distributions have identical symmetric (triangular) shapes, with X + Y ranging from 0 to 8, centred at 4, while X − Y ranges from −4 to +4, centred at 0. Different centres but the exact same range of 8 minutes in both cases, so it should come as no surprise that other measures of spread such as the standard deviation will also be the same in both cases. X − Y is just X + (−Y), so both X + Y and X − Y involve adding two independent sources of variation.

To find the standard deviation of the difference between two independent sample means, we add their variances and then take a square root:

$$SD(\bar{y}_1 - \bar{y}_2) = \sqrt{Var(\bar{y}_1) + Var(\bar{y}_2)}$$

$$= \sqrt{\frac{\sigma_1^2}{n_1} + \frac{\sigma_2^2}{n_2}}$$

We don't know the true standard deviations of the two groups, σ_1 and σ_2, so as usual, we'll use the estimates, s_1 and s_2 in the formula above, which gives us the *standard error* of the difference:

$$SE(\bar{y}_1 - \bar{y}_2) = \sqrt{\frac{s_1^2}{n_1} + \frac{s_2^2}{n_2}}$$

We'll use the standard error to assess how big the difference really is. Because we are working with means and estimating the standard error of their difference using the data, we shouldn't be surprised that the sampling model is a Student's t.

For Example FINDING THE STANDARD ERROR OF THE DIFFERENCE BETWEEN INDEPENDENT SAMPLE MEANS

Can you tell how much you are eating from how full you are? Or do you need visual cues? Researchers[2] constructed a table with two ordinary 18 fluid oz.[3] soup bowls and two identical-looking bowls that had been modified to slowly, imperceptibly, refill as they were emptied. They randomly assigned experiment participants to the bowls, and served them tomato soup. Those eating from the ordinary bowls had their bowls refilled by ladle whenever they were one-quarter full. If people judge their portions by internal cues, they should eat about the same amount. How big a difference was there in the amount of soup consumed? The table summarizes their results.

	Ordinary bowl	Refilling bowl
n	27	27
$\bar{y}$	8.5 oz	14.7 oz
s	6.1 oz	8.4 oz

QUESTION: How much variability do we expect in the difference between the two means? Find the standard error.

ANSWER: Participants were randomly assigned to bowls, so the two groups should be independent. It's okay to add variances.

$$SE(\bar{y}_{refill} - \bar{y}_{ordinary}) = \sqrt{\frac{s_r^2}{n_r} + \frac{s_o^2}{n_o}} = \sqrt{\frac{8.4^2}{27} + \frac{6.1^2}{27}} = 2.0 \ oz.$$

The confidence interval we build is called a **two-sample t-interval** (for the difference in means). The corresponding hypothesis test is called a **two-sample t-test**. The interval looks just like all the others we've seen—the statistic plus or minus an estimated margin of error:

$$(\bar{y}_1 - \bar{y}_2) \pm ME$$

$$\text{where } ME = t^* \times SE(\bar{y}_1 - \bar{y}_2)$$

The ME is just a critical value times the standard error. Here we use a Student's t-model to find the appropriate critical t^* value corresponding to our chosen confidence level, just as with the one-sample t procedure. It is essential to use the t model here, and not the standard normal (unless the df are quite large, in which case z^* and t^* values are essentially the same).

What are we missing? Only the degrees of freedom for the Student's t-model. Unfortunately, *that* formula is strange.

---

[2]Brian Wansink, James E. Painter, and Jill North, 2005, "Bottomless bowls: Why visual cues of portion size may influence intake," *Obesity Research*, Vol. 13, No. 1.

[3]1 fluid oz. = 29.6 mL and 18 fluid oz. = 532 mL.

The deep, dark secret is that the sampling model isn't *really* Student's *t*, but only something close. The trick is that by using a special, adjusted degrees-of-freedom value, we can make it so close to a Student's *t*-model that nobody can tell the difference. The adjustment formula is straightforward but doesn't help our understanding much, so we leave it to the computer or calculator. (If you are curious and really want to see the formula, look in the footnote.[4])

The key sampling result that we need for inference is given below. This describes the sampling model used both for deriving a confidence interval and for testing hypotheses.

A SAMPLING DISTRIBUTION FOR THE DIFFERENCE BETWEEN TWO MEANS

When the assumptions and conditions (described below) are met, the sampling distribution of the difference between the sample means of two independent groups, divided by its standard error

$$t = \frac{(\bar{y}_1 - \bar{y}_2) - (\mu_1 - \mu_2)}{SE(\bar{y}_1 - \bar{y}_2)}$$

can be modelled by a Student's *t*-model with a number of degrees of freedom found with a special formula. We estimate the standard error with

$$SE(\bar{y}_1 - \bar{y}_2) = \sqrt{\frac{s_1^2}{n_1} + \frac{s_2^2}{n_2}}$$

Assumptions and Conditions

Before we can make a two-sample *t*-interval or perform a two-sample *t*-test, we have to check some assumptions and conditions.

Independence Assumptions

Independent Responses Assumption: Within each group, we need independent responses from the individuals. We can't check that for certain, but randomization provides evidence of independence, so we *can* check the following:

Randomization Condition: Ideally, the data in each group should be drawn independently and at random from a population[5] or generated by a completely randomized comparative experiment.

If the samples are bigger than 10% of the target populations in a survey, random draws are no longer approximately independent. No problem, just note that your results are conservative (and tend to overestimate the SE).

Independent Groups Assumption: To use the **two-sample *t* methods**, the two groups we are comparing must be independent of each other. In fact, this test is sometimes called the *independent samples t-test*. Usually, the independence of the groups from each other is evident from the way the data were collected. The assumption would be violated, for example, if one group consisted of husbands and the other group of their wives. Whatever we measure on couples might naturally be related. Similarly, if we compared subjects' performances before some treatment with their performances afterward, we'd expect a

---

[4]
$$df = \frac{\left(\frac{s_1^2}{n_1} + \frac{s_2^2}{n_2}\right)^2}{\frac{1}{n_1 - 1}\left(\frac{s_1^2}{n_1}\right)^2 + \frac{1}{n_2 - 1}\left(\frac{s_2^2}{n_2}\right)^2}$$

Are you sorry you looked? This formula usually doesn't even give a whole number. If you are using a table, you'll need a whole number, so round down to be safe. If you are using technology, it's even easier. The approximation formulas that computers and calculators use for the Student's *t*-distribution deal with fractional degrees of freedom automatically.

[5]Unfortunately, for cluster or multistage random sampling, the SE formula given above will *underestimate* the true variability present when the clusters are somewhat homogeneous (as is often the case). A more complicated formula for the SE should be used.

relationship of each "before" measurement with its corresponding "after" measurement. *In cases such as these, where the observational units in the two groups are related or matched, the two-sample methods of this chapter can't be applied*, since the SE formula above falls apart, (as it assumes independence in order to add two variances together). When this happens, we need to use a different (actually quite simple) procedure, which we'll see in the next chapter.

Normal Population Assumption

As we did before with Student's *t*-models, we should check the assumption that the underlying populations of individual responses are *each* Normally distributed. We check the:

Nearly Normal Condition: We must check this for *both* groups; a violation by either one violates the condition. As we saw for single sample means, the Normality Assumption matters most when sample sizes are very small. For samples of $n < 10$ in either group, you should not use these methods if the histogram or Normal probability plot shows clear skewness[6]. For *n*'s of 10 or so, a moderately skewed histogram is okay, but you should not work with strongly skewed data or data containing clear outliers. For larger samples (*n*'s ≥ 20), data skewness is even less of an issue, but you should still be on the lookout for outliers, extreme skewness, and multiple modes.

> **Robustness.** The two-sample *t* procedure is even more *robust* against violations of Normality than the one-sample *t*, particularly for equal *n*'s and similarly shaped distributions, in which case even $n_1 = n_2 = 5$ may be sufficient to give reliable conclusions.

For Example CHECKING ASSUMPTIONS AND CONDITIONS

RECAP: Researchers randomly assigned people to eat soup from one of two bowls: 27 got ordinary bowls that were refilled by ladle, and 27 others ate from bowls that secretly refilled slowly.

QUESTION: Can the researchers use their data to make inferences about the role of visual cues in determining how much people eat?

ANSWER:

- **Independence:** Since subjects were randomly assigned to treatments, independence between subjects is plausible. The researchers should also conduct the experiment in such a way that individuals do not influence each other's consumption.
- **Independent Groups Assumption:** Randomization to treatment groups guarantees this.
- **Nearly Normal Condition:** The histograms for both groups look unimodal but somewhat skewed to the right. However, both groups are large enough (27) to allow use of *t*-methods.

It's okay to construct a two-sample *t*-interval for the difference in means.

---

[6]Consider a re-expression (like logs) if both groups exhibit similar skewness.

Confidence Interval

When the above assumptions hold, we can use the *t* sampling distribution model stated earlier to derive the following confidence interval.

TWO-SAMPLE *t*-INTERVAL FOR THE DIFFERENCE BETWEEN MEANS

When the conditions are met, we are ready to find the confidence interval for the difference between means of two independent groups, $\mu_1 - \mu_2$. The confidence interval is

$$(\bar{y}_1 - \bar{y}_2) \pm t^*_{df} \times SE(\bar{y}_1 - \bar{y}_2)$$

where the standard error of the difference of the means is

$$SE(\bar{y}_1 - \bar{y}_2) = \sqrt{\frac{s_1^2}{n_1} + \frac{s_2^2}{n_2}}$$

The critical value t^*_{df} depends on the particular confidence level, C, that you specify and on the number of degrees of freedom, which we get from the sample sizes and a special formula.

AN EASIER RULE?

The formula for the degrees of freedom of the sampling distribution of the difference between two means is long, but the number of degrees of freedom is always at least the smaller of the two *n*'s minus 1. Wouldn't it be easier to just use that value? You could, but *that* approximation can be a poor choice because it can give fewer than *half* the degrees of freedom you're entitled to from the correct formula. Underestimating the *df* gives you a *conservative* result (wider confidence interval than necessary).

For Example   FINDING A CONFIDENCE INTERVAL FOR THE DIFFERENCE IN SAMPLE MEANS

RECAP: Researchers studying the role of internal and visual cues in determining how much people eat conducted an experiment in which some people ate soup from bowls that secretly refilled. The results are summarized in the table.

	Ordinary bowl	Refilling bowl
n	27	27
$\bar{y}$	8.5 oz.	14.7 oz.
s	6.1 oz.	8.4 oz.

We've already checked the assumptions and conditions, and have found the standard error for the difference in means to be $SE(\bar{y}_{refill} - \bar{y}_{ordinary}) = 2.0$ oz.

QUESTION: What does a 95% confidence interval say about the difference in mean amounts eaten?

ANSWER: The observed difference in means is

$$\bar{y}_{refill} - \bar{y}_{ordinary} = (14.7 - 8.5) = 6.2 \text{ oz.}$$

$$df = 47.46 \quad t^*_{47.46} = 2.011 \text{ from software (Table T gives } t^*_{45} = 2.014)$$

$$ME = t^* \times SE(\bar{y}_{refill} - \bar{y}_{ordinary}) = 2.011(2.0) = 4.02 \text{ oz.}$$

The 95% confidence interval for $\mu_{refill} - \mu_{ordinary}$ is

$$6.2 \pm 4.02, \text{ or } (2.18, 10.22) \text{ oz.}$$

I am 95% confident that people (well, people with similar characteristics—age, etc.—to those in this study) eating from a subtly refilling bowl will eat an average of between 2.18 and 10.22 more ounces of soup than those eating from an ordinary bowl.

Step-by-Step Example A TWO-SAMPLE *t*-INTERVAL

Let's return to the chapter opening example, where we were comparing lifetimes of generic and brand-name batteries. Judging from the boxplot, the generic batteries seem to have lasted about 20 minutes longer than the brand-name batteries. Before we change our buying habits, what should we expect to happen more generally?

Question: How much longer might the generic batteries last, on average?

THINK ➡ Plan State what we want to know.

Identify the *population(s)* about which you wish to make statements. We hope to make decisions about purchasing batteries, so we're interested in all the AA batteries of these two brands.

Identify the variables and review the *W*'s.

Identify the *parameter* you wish to estimate. Here our parameter is the *difference* in the means, not the two individual group means.

Make a picture. Boxplots are the display of choice for comparing groups. We'll also want to check the distribution of each group. Histograms or Normal probability plots do a better job there.

REALITY CHECK From the boxplots, it appears our confidence interval should be centred near a difference of 20 minutes. We don't have a lot of intuition about how far the interval should extend on either side of 20.

Model Think about the appropriate assumptions and check the conditions to be sure that a Student's *t*-model for the sampling distribution is appropriate.

I have measurements of the lifetimes (in minutes) of six sets of generic and six sets of brand-name AA batteries from a randomized experiment. I want to find an interval that is likely, with 95% confidence, to contain the true difference $\mu_G - \mu_B$ between the mean lifetime of the generic AA batteries and the mean lifetime of the brand-name batteries.

✓ **Randomization Condition:** The sets of batteries were selected at random from those available for sale. Not exactly an SRS, but a reasonably representative random sample. And randomizing the run order helps ensure independence of the observations.

✓ **Independent Groups Assumption:** Batteries purchased in separate packages should be two independent groups.

✓ **Nearly Normal Condition:** Considering how small the sample sizes, the histograms below look reasonably unimodal and symmetric (and Normal probability plots also show no clear curving pattern):

Under these conditions, it's okay to use a Student's *t*-model.

State the sampling distribution model for the statistic.

Specify your method.

I'll use a two-sample *t*-interval.

SHOW ➡ **Mechanics** Construct the confidence interval.

Be sure to include the units along with the statistics. Use meaningful subscripts to identify the groups.

Use the sample standard deviations to find the standard error of the sampling distribution.

I know

$$n_G = 6 \qquad n_B = 6$$

$$\bar{y}_G = 206.0 \text{ min} \quad \bar{y}_B = 187.4 \text{ min}$$

$$s_G = 10.3 \text{ min} \quad s_B = 14.6 \text{ min}$$

The groups are independent, so

$$SE(\bar{y}_G - \bar{y}_B) = \sqrt{SE^2(\bar{y}_G) + SE^2(\bar{y}_B)}$$

$$= \sqrt{\frac{s_G^2}{n_G} + \frac{s_B^2}{n_B}}$$

$$= \sqrt{\frac{10.3^2}{6} + \frac{14.6^2}{6}}$$

$$= \sqrt{\frac{106.09}{6} + \frac{213.16}{6}}$$

$$= \sqrt{53.208}$$

$$= 7.29 \text{ min}$$

We have three choices for degrees of freedom. The best alternative is to let the computer or calculator use the aforementioned special formula for *df*. This gives a fractional degree of freedom (here *df* = 8.98), and technology can find a corresponding critical value. In this case, it is $t^* = 2.263$.

Or we could round the formula's *df* value down to an integer so we can use a *t* table. That gives 8 *df* and a critical value $t^* = 2.306$.

The very conservative *easy rule* says to use only 6 − 1 = 5 *df*. That gives a critical value $t^* = 2.571$. The corresponding confidence interval is about 14% wider—a high price to pay for a small savings in effort.

Use the SE and t^* to find the ME. Wrap the ME around the sample difference to form the confidence interval.

The *df* formula calls for 8.98 degrees of freedom.[7]

The corresponding critical value for a 95% confidence level from a Student's *t*-model with 8.98 *df* is $t^* = 2.263$.

So the margin of error is

$$ME = t^* \times SE(\bar{y}_G - \bar{y}_B)$$

$$= 2.263(7.29)$$

$$= 16.50 \text{ min}$$

The 95% confidence interval is

$$(206.0 - 187.4) \pm 16.5 \text{ min}$$

$$\text{or } 18.6 \pm 16.5 \text{ min}$$

$$= (2.1, 35.1) \text{ min}.$$

TELL ➡ **Conclusion** Interpret the confidence interval in the proper context.

I am 95% confident that generic batteries last an average of 2.1 to 35.1 minutes longer than brand-name batteries for this task (a rather wide interval, due to the small sample sizes, but note that it is completely above zero, giving good evidence that the generic do last longer, on average). If generic batteries are cheaper, there seems little reason not to use them.

---

[7]If you try to find the degrees of freedom with that messy special formula (We dare you! It's in the footnote on page 605) using the values above, you'll get 8.99. The minor discrepancy is because we rounded the standard deviations to make the exposition clearer.

Just Checking

Carpal tunnel syndrome (CTS) causes pain and tingling in the hand. It can be bad enough to keep sufferers awake at night and restrict their daily activities. Researchers studied the effectiveness of two alternative surgical treatments for CTS (Mackenzie, Hainer, and Wheatley, *Annals of Plastic Surgery*, 2000). Patients were randomly assigned to have endoscopic or open-incision surgery. Four weeks later the endoscopic surgery patients demonstrated a mean pinch strength of 9.1 kg compared to 7.6 kg for the open-incision patients.

1. Why is the randomization of the patients into the two treatments important?

2. A 95% confidence interval for the difference in mean strength is about (0.04 kg, 2.96 kg). Explain what this interval means.

3. Why might we want to examine such a confidence interval in deciding between these two surgical procedures?

4. Why might you want to see the data before trusting the confidence interval?

Section 19.1 Exercises

1. **Cost of shopping** Do consumers spend more on a trip to Walmart or Target? Suppose researchers interested in this question collected a systematic sample from 85 Walmart customers and 80 Target customers by asking customers for their purchase amount as they left the stores. The data collected are summarized in the table below.

	Walmart	Target
n	85	80
$\bar{y}$	$45	$53
s	$21	$19

To perform inference on these two samples, what conditions must be met? Are they? Explain.

2. **Athlete ages** A sports reporter suggests that professional baseball players must be, on average, older than professional football players, since football is a contact sport and players are more susceptible to concussions and serious injuries (www.sports.yahoo.com). One player was selected at random from each team in both professional baseball (MLB) and professional football (NFL). The data are summarized below.

	MLB	NFL
n	30	32
$\bar{y}$	27.5	26.16
s	3.94	2.78

To perform inference on these two samples, what conditions must be met? Are they? Explain.

3. **Cost of shopping, again** Using the summary statistics provided in Exercise 1, researchers calculated a 95% confidence interval for the mean difference between Walmart and Target purchase amounts. The interval was (−$14.15, −$1.85). Explain in context what this interval means.

4. **Athlete ages, again** Using the summary statistics provided in Exercise 2, the sports reporter calculated the following 95% confidence interval for the mean difference between major league baseball players and professional football players. The 95% interval for $\mu_{MLB} - \mu_{NFL}$ was (−0.41, 3.09). Summarize in context what the interval means.

19.2 The Two-Sample *t*-Test

If you bought a used camera in good condition from a friend, would you pay the same as you would if you bought the same camera from a stranger? A researcher at Cornell University[8] wanted to know how friendship might affect simple sales such as this.

[8]J. J. Halpern, "The Transaction Index: A method for standardizing comparisons of transaction characteristics across different contexts," *Group Decision and Negotiation*, 6: 557–572.

She randomly divided subjects into two groups and gave each group descriptions of items they might want to buy. One group was told to imagine buying from a friend they expected to see again. The other group was told to imagine buying from a stranger.

Here are the prices they offered for a used camera in good condition:

Who	University students
What	Prices offered for a used camera
Units	$
Why	Study of the effects of friendship on transactions
When	1990s
Where	Cornell University

Price Offered for a Used Camera ($)	
Buying from a Friend	**Buying from a Stranger**
275	260
300	250
260	175
300	130
255	200
275	225
290	240
300	

The researcher who designed this study had a specific concern. Previous theories had doubted that friendship had a measurable effect on pricing. She hoped to find an effect of friendship. This calls for a hypothesis test—in this case, a **two-sample *t*-test for the difference between means**.[9]

A Test for the Difference between Two Means

When we write the null hypothesis for comparing two means, we could write: $H_0: \mu_1 = \mu_2$. But sometimes we'll want to see if the difference is something other than 0, so we usually write it slightly differently. We'll focus on the hypothesized *difference* between the two means and we'll call that hypothesized difference Δ_0 ("delta naught"). It's so common for that hypothesized difference to be zero that we often just assume $\Delta_0 = 0$. In that case, we'll write: $H_0: \mu_1 - \mu_2 = 0$, but in general we'll write: $H_0: \mu_1 - \mu_2 = \Delta_0$.

You already know enough to construct this test. The test statistic looks just like the others we've seen. First, we find the difference between the observed group means and compare this with a hypothesized value for that difference (usually zero). Then, as usual, we compare the observed difference with its standard error (the same standard error that we used for the confidence interval) by forming their ratio. We find P-values from a Student's *t*-model on that same special number of degrees of freedom.

NOTATION ALERT

Δ_0—delta naught—isn't so standard that you can assume everyone will understand it. We use it because it's the Greek letter (good for a parameter) "D" for "difference."

TWO-SAMPLE *t*-TEST FOR THE DIFFERENCE BETWEEN MEANS

The conditions for the two-sample *t*-test for the difference between the means of two independent groups are the same as for the two-sample *t*-interval. We test the hypothesis

$$H_0: \mu_1 - \mu_2 = \Delta_0$$

where the hypothesized difference is almost always 0, using the statistic

$$t = \frac{(\bar{y}_1 - \bar{y}_2) - \Delta_0}{SE(\bar{y}_1 - \bar{y}_2)}$$

The standard error of $\bar{y}_1 - \bar{y}_2$ is

$$SE(\bar{y}_1 - \bar{y}_2) = \sqrt{\frac{s_1^2}{n_1} + \frac{s_2^2}{n_2}}$$

When the conditions are met and the null hypothesis is true, this statistic can be closely modelled by a Student's *t*-model with a number of degrees of freedom given by the special formula. We use that model to obtain a P-value.

---

[9]Because it is performed so often, this test is usually just called a "two-sample *t*-test."

Step-by-Step Example   A TWO-SAMPLE *t*-TEST FOR THE DIFFERENCE BETWEEN TWO MEANS

Recall: We want to know if people will pay the same amount, on average, when buying from a friend or a stranger. The usual null hypothesis is that there's no difference in means. That's just the right null hypothesis for the camera purchase prices.

Question: Is there a difference, on average, in the price people would offer a friend rather than a stranger?

THINK ➡ Plan State what we want to know.

Identify the *parameter* you wish to estimate. Here our parameter is the difference in the means, not the individual group means.

Identify the variables and check the *W*'s.

Hypotheses State the null and alternative hypotheses.

The research claim is that friendship changes what people are willing to pay.[10] The natural null hypothesis is that friendship makes no difference.

We didn't start with any knowledge of whether friendship might increase or decrease the price, so we choose a two-sided alternative.

Make a picture. Boxplots are the display of choice for comparing groups. We'll also want to check the distribution of each group. Histograms or Normal probability plots do a better job there.

REALITY CHECK   Looks like the prices are higher if you buy from a friend, but we can't be sure. You can't easily tell from looking at the boxplots whether the difference is statistically significant—the plot shows spreads on the same scale as the data, whereas we'll need to add the *variances* of the means (square-rooted) to get a suitable ruler for comparing the difference.

Model Think about the assumptions and check the conditions.

I want to know whether people are likely to offer a different amount for a used camera when buying from a friend than when buying from a stranger. I wonder whether the difference between mean amounts is zero. I have bid prices from eight subjects buying from a friend and seven buying from a stranger, found in a completely randomized experiment.

H_0: The difference in mean price offered to friends and mean price offered to strangers is zero:

$$\mu_F - \mu_S = 0.$$

H_A: The difference in mean prices is not zero:

$$\mu_F - \mu_S \neq 0.$$

✓ **Randomization Condition**: The experiment was randomized. Subjects were assigned to treatment groups at random and so the prices paid should be independent.

✓ **Independent Groups Assumption**: Randomizing the subjects gives two independent groups.

✓ **Nearly Normal Condition**: Histograms of the two sets of prices are reasonably

---

[10]This claim is a good example of what is called a "research hypothesis" in many social sciences. The only way to prove it is to deny that it's true and see whether we can reject that resulting null hypothesis.

unimodal and close to symmetric, given the small samples sizes:

State the sampling distribution model. Specify your method.

The assumptions are reasonable and the conditions are okay, so I'll use a Student's *t*-model to perform a *two-sample t-test*.

SHOW ➡ **Mechanics** List the summary statistics. Be sure to use proper notation.

Use the null model to find the P-value. First determine the standard error of the difference between sample means.

From the data:

$$n_F = 8 \qquad n_S = 7$$
$$\bar{y}_F = \$281.88 \quad \bar{y}_S = \$211.43$$
$$s_F = \$18.31 \quad s_S = \$46.43$$

For independent groups,

$$SE(\bar{y}_F - \bar{y}_S) = \sqrt{SE^2(\bar{y}_F) + SE^2(\bar{y}_S)}$$

$$= \sqrt{\frac{s_F^2}{n_F} + \frac{s_S^2}{n_S}}$$

$$= \sqrt{\frac{18.31^2}{8} + \frac{46.43^2}{7}}$$

$$= 18.70$$

The observed difference is

$$(\bar{y}_F - \bar{y}_S) = 281.88 - 211.43 = \$70.45$$

Find the *t*-value.

Make a picture. Sketch the *t*-model centred at the hypothesized difference of zero. Because this is a two-tailed test, shade the region to the right of the observed *t* value and the corresponding region in the other tail.

A statistics program or graphing calculator can find the P-value using the fractional degrees of freedom from the special formula. If you are doing a test like this without technology, you could use the smaller sample size to quickly determine conservative degrees of freedom, which would equal $n_2 - 1 = 6$.[11]

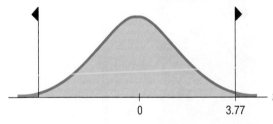

$$t = \frac{(\bar{y}_F - \bar{y}_S) - (0)}{SE(\bar{y}_F - \bar{y}_S)} = \frac{70.45}{18.70} = 3.77$$

The approximation formula gives 7.62 degrees of freedom.[12]

$$\text{P-value} = 2P(t_{7.62} > 3.77) = 0.006$$

---

[11]Keep in mind that you err conservatively with this formula; your reported P-value would be inflated. With 6 degrees of freedom, *P* would equal 0.009 here. If that is small enough for you to reject the null hypothesis, there is little need for a more accurate calculation!

[12]If you were daring enough to calculate that messy degrees-of-freedom formula by hand with the values given here, you'd get about 7.74. Without a computer, you would likely round this down to 7 df and use the tables. Computers also work with more precision for the standard deviations than we gave in our example. All give about the same result for the P-value, so it doesn't really matter—the conclusion would be the same.

TELL ➡ **Conclusion** Link the P-value to your decision about the null hypothesis, and state the conclusion in context.

Be cautious about generalizing to items whose prices are outside the range of those in this study.

If there were no difference in the mean prices, a difference this large would occur only 6 times in 1000. That's too rare to believe, so I reject the null hypothesis and conclude that people are likely to offer a friend more than they'd offer a stranger for a used camera (and possibly for other, similar items).

✔ Just Checking

Recall the experiment comparing patients four weeks after surgery for carpal tunnel syndrome. The patients who had endoscopic surgery demonstrated a mean pinch strength of 9.1 kg compared to 7.6 kg for the open-incision patients.

5. What hypotheses would you test?

6. The P-value of the test was less than 0.05. State a brief conclusion.

7. The study reports work on 36 "hands," but there were only 26 patients. In fact, 7 of the endoscopic surgery patients had both hands operated on, as did 3 of the open-incision group. Does this alter your thinking about any of the assumptions? Explain.

Pearson Education

For Example A TWO-SAMPLE *t*-TEST

Many office "coffee stations" collect voluntary payments for the food consumed. Researchers at the University of Newcastle upon Tyne performed an experiment to see whether the image of eyes watching would change employee behaviour.[13] They alternated pictures (seen here) of eyes looking at the viewer with pictures of flowers each week on the cupboard behind the "honesty box." They measured the consumption of milk to approximate the amount of food consumed and recorded the contributions (in £) each week per litre of milk. The table summarizes their results.

	Eyes	Flowers
n (# weeks)	5	5
$\bar{y}$	0.417 £/L	0.151 £/L
s	0.1811 £/L	0.067 £/L

QUESTION: Do these results provide evidence that there really is a difference in honesty even when it's only photographs of eyes that are "watching"?

ANSWER:

$$H_0: \mu_{eyes} - \mu_{flowers} = 0$$
$$H_A: \mu_{eyes} - \mu_{flowers} \neq 0$$

■ **Independence Assumption:** Average contributions from week to week should be fairly unrelated or independent.

■ **Nearly Normal Condition:** I don't have the data to check, but it seems unlikely there would be outliers in either group. I could be more certain if I could see histograms for both groups.

---

[13]Melissa Bateson, Daniel Nettle, and Gilbert Roberts, "Cues of being watched enhance cooperation in a real-world setting," *Biol. Lett. doi*:10.1098/rsbl.2006.0509.

- **Independent Groups Assumption:** Though there is pairing of an 'eyes' week with a 'flowers' week (due to alternation) the results from weeks with eyes should be pretty much independent of the results from weeks with flowers (but if we believed things were trending over time, we could apply a more sophisticated analysis to the data).

It's okay to do a two-sample t-test for the difference in means:

$$SE(\bar{y}_{eyes} - \bar{y}_{flowers}) = \sqrt{\frac{s^2_{eyes}}{n_{eyes}} + \frac{s^2_{flowers}}{n_{flowers}}} = \sqrt{\frac{0.1811^2}{5} + \frac{0.067^2}{5}} = 0.0864$$

$$df = 5.07$$

$$t = \frac{\bar{y}_{eyes} - \bar{y}_{flowers} - 0}{SE(\bar{y}_{eyes} - \bar{y}_{flowers})} = \frac{0.417 - 0.151}{0.0864} = 3.08$$

$$P\text{-value} = 2 \times P(t_5 > 3.08) = 2 \times 0.0137 = 0.027$$

Assuming the data were free of outliers, the low P-value leads me to reject the null hypothesis. This study provides evidence that people will leave higher average voluntary payments for food if pictures of eyes are "watching."

(Note: In Table T we can see that at 5 df, $t = 3.08$ lies between the critical values for $P = 0.02$ and $P = 0.05$, so if we were just using that table, we could report $P < 0.05$.)

Section 19.2 Exercises

5. Cost of shopping, three The researchers in Exercise 1 decide to test the hypothesis that the means are equal. The degrees of freedom formula gives 162.75 df. Test the null hypothesis at $\alpha = 0.05$.

6. Athlete ages, three The researchers in Exercise 2 decide to test the hypothesis. The degrees of freedom formula gives 51.83 df. Test the null hypothesis at $\alpha = 0.05$. Is the alternative one or two-sided?

19.3 The Pooled t-Test

You may encounter another way to test whether the means of two groups are different called the **pooled t-test**. We present it both for historical reasons and because it's important in the analysis of designed experiments (Chapter 24). Before 1949, it wasn't known that the statistic we saw on page 605 could be modelled (approximately) by a Student's t distribution—even one with that crazy number of degrees of freedom. So, another, simpler test was used. Although simpler, this test has a big assumption—it assumes that the variances of the two groups are the same.

That might not seem like a big deal, but it's a very restrictive assumption. Just because the means of two groups are assumed equal in a test doesn't mean that their variances are as well. Generally, knowing the mean doesn't tell you *anything* about the variance. But there are cases where the assumption makes some sense—especially in the analysis of designed experiments. If you're trying to decide whether the new computer code you designed actually makes downloading files faster, the null hypothesis is that the mean of your method is the same as the standard code mean. But, if it has no effect on the mean, it might be safe to assume that it has no effect on the variance either. In such a case, the pooled t-test may be useful.

Pooled t-tests have a couple of advantages. They often have a few more degrees of freedom than the corresponding two-sample test and a much simpler degrees of freedom formula. But these advantages come at a price: You have to think about another assumption.

The assumption of equal variances is a strong one, is often not true, and is difficult to check. For these reasons, we recommend that you generally use the previously discussed (unpooled) two-sample *t*-test instead.

The pooled *t*-test is the theoretically correct method only when we have good reason to believe that the variances are equal. And (as we will see shortly) there are times when this makes sense. Keep in mind, however, that *it's never wrong not to pool.*

Details of the Pooled *t*-Test

Termites cause billions of dollars of damage each year to homes and other buildings, but some tropical trees seem to be able to resist termite attack. A researcher extracted a compound from the sap of one such tree and tested it by feeding it at two different concentrations to randomly assigned groups of 25 termites.[14] After five days, eight groups fed the lower dose had an average of 20.875 termites alive, with a standard deviation of 2.23. But six groups fed the higher dose had an average of only 6.667 termites alive, with a standard deviation of 3.14. Is this a large enough difference to declare the sap compound effective in killing termites? To use the pooled *t*-test, we must make the **Equal Variance Assumption** that the variances of the two populations from which the samples have been drawn are equal. That is, $\sigma_1^2 = \sigma_2^2$. (Of course, we could think about the standard deviations being equal instead.) The corresponding **Similar Spreads Condition** really just consists of looking at the boxplots to check that the spreads are not wildly different. We were going to make boxplots anyway, so there's really nothing new here.

Once we decide to pool, we estimate the common variance by combining numbers we already have:

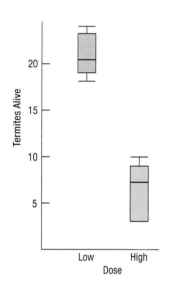

$$s_{\text{pooled}}^2 = \frac{(8-1)2.23^2 + (6-1)3.14^2}{(8-1) + (6-1)}$$
$$= 7.01$$

$$s_{\text{pooled}}^2 = \frac{(n_1 - 1)s_1^2 + (n_2 - 1)s_2^2}{(n_1 - 1) + (n_2 - 1)}$$

(If the two sample sizes are equal, this is just the average of the two variances.)

Now, we just substitute this pooled variance in place of each of the variances in the standard error formula.

$$SE_{\text{pooled}}(\bar{y}_1 - \bar{y}_2) = \sqrt{\frac{7.01}{8} + \frac{7.01}{6}} = 1.43$$

$$SE_{\text{pooled}}(\bar{y}_1 - \bar{y}_2) = \sqrt{\frac{s_{\text{pooled}}^2}{n_1} + \frac{s_{\text{pooled}}^2}{n_2}} = s_{\text{pooled}}\sqrt{\frac{1}{n_1} + \frac{1}{n_2}}$$

The formula for degrees of freedom for the Student's *t*-model is simpler, too. It was so complicated for the unpooled two-sample *t* that we stuck it in a footnote.[15] Now it's just df = $n_1 + n_2 - 2$.

Substitute the pooled-*t* estimate of the standard error and its degrees of freedom into the steps of the confidence interval or hypothesis test, and you'll be using the pooled-*t* method for two independent samples. For the termites, $\bar{y}_1 - \bar{y}_2 = 14.208$, giving a *t*-value = 9.936 with 12 df and a P-value < 0.0001. Of course, if you decide to use a pooled-*t* method, you must defend your assumption that the variances of the two groups are equal.

$$t = \frac{20.875 - 6.667}{1.43} = 9.936$$

POOLED *t*-TEST AND CONFIDENCE INTERVAL FOR MEANS

The conditions for the pooled *t*-test for the difference between the means of two independent groups (commonly called a "pooled *t*-test") are the same as for the two-sample *t*-test discussed earlier with the additional assumption that the variances of the two groups are the same. We test the hypothesis

$$H_0: \mu_1 - \mu_2 = \Delta_0$$

---

[14]Adam Messer, Kevin McCormick, H. H. Sunjaya, Ferny Tumbel Hagedorm, and J. Meinwald, "Defensive role of tropical tree resins: antitermitic sesquiterpenes from Southeast Asian Dipterocarpaceae," *J Chem Ecology*, 16:122, pp. 3333–3352.

[15]But not this one. See page 605.

where the hypothesized difference, Δ_0, is almost always 0, using the statistic

$$t = \frac{(\bar{y}_1 - \bar{y}_2) - \Delta_0}{SE_{pooled}(\bar{y}_1 - \bar{y}_2)}$$

The standard error of $\bar{y}_1 - \bar{y}_2$ is

$$SE_{pooled}(\bar{y}_1 - \bar{y}_2) = \sqrt{\frac{s^2_{pooled}}{n_1} + \frac{s^2_{pooled}}{n_2}} = s_{pooled}\sqrt{\frac{1}{n_1} + \frac{1}{n_2}}$$

where the pooled variance is

$$s^2_{pooled} = \frac{(n_1 - 1)s_1^2 + (n_2 - 1)s_2^2}{(n_1 - 1) + (n_2 - 1)}$$

When the conditions are met and the null hypothesis is true, we can model this statistic's sampling distribution with a Student's t-model with $(n_1 - 1) + (n_2 - 1)$ degrees of freedom. We use that model to obtain a P-value for a test or a margin of error for a confidence interval.

The corresponding confidence interval is

$$(\bar{y}_1 - \bar{y}_2) \pm t^*_{df} \times SE_{pooled}(\bar{y}_1 - \bar{y}_2)$$

where the critical value t^* depends on the confidence level and is found with $(n_1 - 1) + (n_2 - 1)$ degrees of freedom.

Everybody Out of the Pool!

We're testing whether the means are equal, so we admit that we don't *know* whether they are equal. Doesn't it seem a bit much to just *assume* that the variances are equal? Well, yes—but there are some special cases to consider. So when *should* you use the pooled method rather than the unpooled method?

Never.

What, never?

Well, hardly ever.

You see, when the variances of the two groups are in fact equal, the two methods give pretty much the same result. (For the termites, the two-sample t statistic is not much different—9.436 with 8 df—and the P-value is still < 0.0001.) Pooled methods have a small advantage (slightly narrower confidence intervals, slightly more powerful tests) mostly because they usually have a few more degrees of freedom, but the advantage is slight.

When the variances are *not* equal, the pooled method is just not valid and can give poor results. You have to use the unpooled method instead.

As the sample sizes get bigger, the advantages that come from a few more degrees of freedom make less and less difference. So the advantage (such as it is) of the pooled method is greatest when the samples are small—just when it's hardest to check the conditions. And the difference in the degrees of freedom is greatest when the variances are not equal—just when you can't use the pooled method anyway. Our advice is to use the unpooled methods to compare means.

Why did we devote a whole section to a method that we don't recommend using? The answer is that pooled methods are actually very important in Statistics. It's just that the simplest of the pooled methods—those for comparing two means—have good alternatives in the unpooled methods that don't require the extra assumption. Lacking the burden of the Equal Variances Assumption, the unpooled methods apply to more situations and are safer to use.

Pooling

Pooled methods show up in several places in Statistics, and often in more complex situations than a two-sample test. As you've seen, the most sensitive part of most inference procedures is how we estimate the standard deviation of the test statistic so we can use it as a ruler to judge significance or construct a confidence interval. Anything we can do to improve our standard

---

WHEN SHOULD I USE THE POOLED-t TEST?

Because the advantages of pooling are small, and you are allowed to pool only rarely (when the Equal Variances Assumption is met), *don't*. **It's never wrong *not* to pool.**

> **Why not test whether the variances are equal?** There are tests of whether two variances are equal, but they are very sensitive to their assumptions and work poorly for small sample sizes. So they don't help in choosing between the pooled and the unpooled methods.

error estimate pays us back with improved power, shorter confidence intervals, and sometimes, simply with a valid method. So if you think the groups you are comparing have the same standard deviation, you should look for a pooled method that takes advantage of that.

Here's a common situation. In a randomized experiment, we start by assigning our experimental units to treatments at random, as the experimenter did with the termites. Because of this random assignment, each treatment group is a random sample from the same population,[16] so before they were treated, each group had not just the same mean, but also the same variance. The pooled *t*-test assumes that the treatment doesn't change the variance. Our null hypothesis says that the treatment didn't change the means, so it's not much of a stretch to think that the treatments made no difference *at all*. For example, we might suspect that the treatment is no different from the placebo. So, it might make sense to assume that the variances have remained equal. It's still an assumption, and there are conditions that need to be checked (make the boxplots, make the boxplots, make the boxplots), but at least it's a plausible assumption.

This line of reasoning is important. The methods used to analyze comparative experiments *do* pool variances in exactly this way and defend the pooling with a version of this argument. Chapter 24 (Analysis of Variance) introduces these methods.

For Example A POOLED *t*-TEST

RECAP: Remember the experiment showing that people would consume more soup if their bowls were secretly being refilled as they ate? Could it be that those subjects with the refilling bowl just enjoyed eating soup more? The experimenters also asked the participants to estimate how many ounces of soup they had eaten. Their responses are summarized in the table.

	Ordinary bowl	Refilling bowl
n	27	27
$\bar{y}$	8.2	9.8
s	6.9	9.2

QUESTION: Is there evidence that those who ate more knew what they had done?

ANSWER:

$$H_0: \mu_{refill} - \mu_{ordinary} = 0$$

$$H_A: \mu_{refill} - \mu_{ordinary} \neq 0$$

We'll play it *safe* here with a two-sided alternative.

The Independence Assumptions checked out earlier. Also,

Nearly Normal Condition: The histograms for both groups look unimodal and skewed to the right. The similar shapes and lack of clear outliers (the value around 30 in the left histogram appears to be just part of the long tail extending to the right) in groups *as large as these* suggest that *t*-methods should be very safe to use.

---

[16]That is, the population of experimental subjects. Remember that to be valid, experiments do not need a representative sample drawn from a population because we are not trying to estimate a population model parameter.

■ **Equal Variances Assumption:** Subjects were assigned randomly to the two treatments, so I can assume that before they ate any soup, both groups' volume-estimating abilities had equal variances.

■ **Similar Spreads Condition:** The boxplots show comparable spreads for the two groups.

A pooled *t*-test should be quite reliable here.

$$s^2_{pooled} = \frac{(n_r - 1)s^2_r + (n_0 - 1)s^2_0}{(n_r - 1) + (n_0 - 1)} = \frac{26 \times 9.2^2 + 26 \times 6.9^2}{(26) + (26)} = 66.125$$

$$SE_{pooled}(\bar{y}_o - \bar{y}_r) = \sqrt{\frac{s^2_{pooled}}{n_r} + \frac{s^2_{pooled}}{n_o}} = s_{pooled}\sqrt{\frac{1}{n_r} + \frac{1}{n_o}}$$

$$= 8.13\sqrt{\frac{1}{27} + \frac{1}{27}} = 2.213 \text{ (where 8.13 is } \sqrt{66.125})$$

$$df = (27 - 1) + (27 - 1) = 52$$

$$t = \frac{(9.8 - 8.2) - 0}{2.213} = 0.723$$

$$P\text{-value} = P(|t_{52}| > 0.723) = 0.473$$

An event that happens nearly half of the time by chance isn't very extraordinary, so I would not reject the null hypothesis. There's no evidence that the subjects who ate more soup from the refilling bowl could tell that they did so.

> **Nonparametrics.** Tired of making and checking so many assumptions and conditions? What if there are severe violations? If a re-expression (like log, e.g.) doesn't fix things, or you just don't feel right about changing the measurement scale, you can track down and try an alternative *nonparametric* (distribution-free) procedure. The proper one here is called the Wilcoxon rank sum test, which is discussed in Chapter 28. Nonparametric procedures though do have some disadvantages.

Section 19.3 Exercises

7. Cost of shopping, once more Repeat the test you did in Exercise 5, but assume that the variances of purchase amounts is the same at Target and Walmart. Did your conclusion change? Why do you think that is?

8. Athlete ages, one more time Repeat the test you did in Exercise 6, but assume that the variances of ages is the same in the two leagues. Did your conclusion change? Why do you think that is?

19.4 Determining the Sample Size

When planning a study, it's important to try to determine an optimal sample size, though this usually requires making some guesses (for example, at standard deviations). Too small a sample, and your margin of error is too big or the power of the test is inadequate to discover important effects that may be present. Too big a sample, and time, money, and resources are needlessly wasted.

To control the margin of error, set the appropriate formula equal to the desired margin of error and solve for sample size. Guess at what you do not know, perhaps using some prior knowledge or studies. This is quite similar to the discussion of *sample size* for the one-sample case in Chapter 18 (see page 577), but now there are two sample sizes to determine, so let's make the simplifying assumption that the two sample sizes are equal. And there are good reasons for aiming for equal sample sizes; for example, it improves the robustness of the procedure. You may work with unequal sample sizes if you wish: just set $n_1 = 2n_2$ if you want sample 1 to be, say, twice as big as sample 2. But henceforth we will assume $n_1 = n_2$, and let $n =$ the common sample size (so that $2n$ will be the total number of subjects or responses actually required).

Consider the earlier example comparing brand-name and generic batteries. The margin of error computed was 16.5 minutes—rather large, considering that the difference in means was nearly as large at 18.6 minutes. We know that we could decrease the margin of error roughly by half by simply multiplying both sample sizes by 2^2 or 4, but let's learn to proceed in a more precise manner. Suppose we want the margin of error to equal 5.0 minutes. Let the common sample size equal n. Choose 95% confidence. We'll view the earlier data as coming from a small pilot study, and take the two sample standard deviations (rounded up, to be conservative) as estimates of the true standard deviations needed in the margin of error formula. Now set up and solve the following:

$$\text{Set } 5.0 = 1.96 \sqrt{\frac{10.3^2}{n} + \frac{14.6^2}{n}} \text{ and then square both sides and solve.}$$

We get $n = 50$. So we need 50 batteries of each type (of course, you'd better weigh your desire for precision against the costs incurred and time involved in draining so many batteries). Note that we use a z-value of 1.96 to start with and only need to change this if our sample size turns out to be fairly small (as in the Chapter 18 example). $n = 50$ is big enough to justify use of the standard Normal distribution instead of the t-distribution, particularly since we know that the degrees of freedom cannot be less than $n - 1 = 49$, and likely should be considerably more.

On the other hand, when performing a *test of significance*, our concern will be about having *adequate power* to detect important differences. Consider the example of killing termites with a sap compound, where we wanted to see if a higher concentration would kill more termites. The investigators might feel that a difference of less than 2 or 3 termites killed (out of the 25 termites at the start), on average, would not be a big enough effect to consider this compound a success, but that a difference of, say, 5 termites (20% of the original population size) would be a large enough effect to make this compound a great success. We should then choose a sample size that will give us a high power when the (unknown) true difference in the means is 5.0.

Power calculations are not possible unless we assume equal variances (another reason to at least be aware of the pooled test), but even if we believe that the variances are not equal, we can err conservatively by guessing at both, then using the higher one (or less conservatively, using the average of the two standard deviations). The standard deviations from our (assumed) pilot study were 2.23 and 3.14, so let's take 3.0 as a rough guess at a common or average standard deviation. Now *fire up your statistics software*, look for *power and sample size* calculations for the *two-sample t-test*, and input the following:

- *Alpha level of the test.* Perhaps 5%.
- *Alternative hypothesis* (the null is a difference of zero). Let's choose the safe two-sided option (though one could justify a one-sided alternative here).

- *Size of the difference you want to detect, if present.* Set this equal to 5.0 killed termites.
- *Your guess at the common sigma.* We'll go with a guess of 3.0 killed termites.
- *Desired power.* Let's aim for 0.90.

Here is some typical output (from Minitab):

Power and Sample Size

Two-Sample *t*-Test
Testing mean 1 = mean 2 (versus not =)
Calculating power for mean 1 = mean 2 + difference
Alpha = 0.05 Assumed standard deviation = 3

Difference	Sample Size	Target Power	Actual Power
5	9	0.9	0.912548

The sample size is for each group.

So, we only need nine groups fed the lower dose and nine groups fed the higher dose to achieve the desired power for a difference of 5 dead termites, on average. In other words, with this sample size, if the true mean difference in dead termites between the two dosages is 5 dead termites (out of 25), then 91 times out of 100 we will make the correct decision and correctly detect a dosage effect.

Section 19.4 Exercises

9. Cost of shopping, last look The researchers in Exercise 1 want the margin of error for the estimate of the mean difference between Walmart and Target purchase amounts to be $4.50, with 95% confidence. How big should the sample size be (assuming two equal sample sizes)? Use the standard deviations given in Exercise 1 as your estimates.

10. Athletes, final inning The researchers in Exercise 2 want the margin of error for the estimate of the mean difference between major league baseball players and professional football players age to be 0.63, with 90% confidence. How big should the sample size be (assuming two equal sample sizes)? Use the standard deviations given in Exercise 2 as your estimates.

WHAT CAN GO WRONG?

- **Don't use two-sample methods when the samples aren't independent of each other.** These methods give wrong answers when this assumption of independence is violated. Good randomization is usually the best assurance of independent groups. Make sure there is no relationship between the two groups. Matched-pairs designs in which observational units are deliberately linked between the two samples arise often and are important, but can't be analyzed by the methods of this chapter. The next chapter deals with them.

- **Look at the plots.** The usual (by now) cautions about checking for outliers and non-Normal distributions apply, of course. The simple defense is to make and examine boxplots. You may be surprised how often this step saves you from the wrong or even absurd conclusions that can be generated by a single undetected outlier. You don't want to conclude that two methods have very different means just because one observation is atypical.

- **Be cautious in drawing conclusions where there was no randomization.** If the data do not come from representative random samples or a properly randomized experiment, conclusions about differences between the groups may be completely unjustified. In a prospective or retrospective study, there is always the danger that some lurking variables are the real reason for an observed difference. Conclusions about causality require well-designed controlled randomized experiments.

> **Do what we say, not what we do . . .** Precision machines used in industry often have a bewildering number of parameters that have to be set, so experiments are performed in an attempt to try to find the best settings. Such was the case for a hole-punching machine used by a well-known computer manufacturer to make printed circuit boards. The data were analyzed by one of the authors, but because he was in a hurry, he didn't look at the box-plots first and just performed *t*-tests on the experimental factors. When he found extremely small P-values even for factors that made no sense, he plotted the data. Sure enough, there was one observation 1 000 000 times bigger than the others. It turns out that it had been recorded in microns (millionths of an inch), while all the rest were in inches.

What Have We Learned?

Know how to construct and interpret the two-sample *t*-interval for the difference between the means of two populations.

- Know the requisite Assumptions and Conditions: (i) Independence and Normality of the individual responses, (ii) Independence of the samples.

Be able to perform and interpret a two-sample *t*-test of the difference between the means of two populations.

- Understand the relationship between testing and providing a confidence interval.
- The most common null hypothesis is that the means are equal.

Recognize that in special cases in which it is reasonable to assume equal variances between the groups, we can also use a pooled *t*-test for testing the difference between means.

- This may make sense particularly for randomized experiments in which the randomization has produced groups with equal variance to start with and the null hypothesis is that a treatment under study has had no effect.

Keep in mind that these two sample methods require both independence of observations within each group and independence between the groups themselves.

- These methods are not appropriate for paired or matched data (discussed in next chapter).

Review of Terms

Two-sample *t* methods   Two-sample *t* methods use the Student's *t* sampling model to draw conclusions about the difference between the means of two quantitative populations, based on *independent samples*. These methods also require independent Normally distributed responses in each sample, but possess considerable robustness against violations of Normality. (p. 605).

Two-sample *t*-interval for the difference between means   A confidence interval for the difference between the means of two populations based on independent samples found as

$$(\bar{y}_1 - \bar{y}_2) \pm t^*_{df} \times SE(\bar{y}_1 - \bar{y}_2)$$

where

$$SE(\bar{y}_1 - \bar{y}_2) = \sqrt{\frac{s_1^2}{n_1} + \frac{s_2^2}{n_2}}$$

and the number of degrees of freedom is given by a special formula (see footnote 4 on page 605) (p. 604).

Two-sample *t*-test for the difference between means   A hypothesis test for the difference between the means of two populations based on independent samples. It tests the null hypothesis

$$H_0: \mu_1 - \mu_2 = \Delta_0$$

where the hypothesized difference, Δ_0, is almost always 0, using the statistic

$$t = \frac{(\bar{y}_1 - \bar{y}_2) - \Delta_0}{SE(\bar{y}_1 - \bar{y}_2)}$$

with the number of degrees of freedom given by the special formula (p. 611).

Pooled *t*-test
A hypothesis test for the difference in the means of two populations, based on independent samples when we are willing and able to assume that the variances of the populations are equal. It tests the null hypothesis

$$H_0: \mu_1 - \mu_2 = \Delta_0$$

where the hypothesized difference, Δ_0, is almost always 0, using the statistic

$$t = \frac{(\bar{y}_1 - \bar{y}_2) - \Delta_0}{SE_{\text{pooled}}(\bar{y}_1 - \bar{y}_2)}$$

where the pooled standard error is defined as for the pooled interval (see below) and the degrees of freedom is $(n_1 - 1) + (n_2 - 1)$ (p. 615).

Pooled *t*-interval
A confidence interval for the difference between the means of two populations, based on independent samples used when we are willing and able to assume that the variances of the groups are equal. It is found as

$$(\bar{y}_1 - \bar{y}_2) \pm t^*_{df} \times SE_{\text{pooled}}(\bar{y}_1 - \bar{y}_2)$$

where

$$SE_{\text{pooled}}(\bar{y}_1 - \bar{y}_2) = \sqrt{\frac{s^2_{\text{pooled}}}{n_1} + \frac{s^2_{\text{pooled}}}{n_2}} = s_{\text{pooled}}\sqrt{\frac{1}{n_1} + \frac{1}{n_2}}$$

and the pooled variance is

$$s^2_{\text{pooled}} = \frac{(n_1 - 1)s^2_1 + (n_2 - 1)s^2_2}{(n_1 - 1) + (n_2 - 1)}$$

The number of degrees of freedom is $(n_1 - 1) + (n_2 - 1)$ (p. 616).

Pooling
We may sometimes combine, or *pool*, data from two or more populations to estimate a parameter (such as a common variance) when we are willing to assume that the parameter is the same in both populations. Using the additional data may lead to a more reliable estimate. However, pooled estimates are appropriate only when the required assumptions are believed to be (nearly) true (p. 617).

On the Computer TWO-SAMPLE METHODS FOR MEANS

Here's some typical computer package output with comments:

May just say "difference of means"

Test Statistic

```
2-Sample t-Test of μ1-μ2 = 0 vs ≠ 0

Difference Between Means = 0.99145299  t-Statistic = 1.540
w/196 df
Fail to reject Ho at Alpha = 0.05
P = 0.1251
```

Some programs will draw a conclusion about the test. Others just give the P-value and let you decide for yourself.

df found from approximation formula and rounded down. The unrounded value may be given, or may be used to find the P-value.

Many programs give far too many digits. Ignore the excess digits.

Most statistics packages compute the test statistic for you and report a P-value corresponding to that statistic. And, of course, statistics packages make it easy to examine the boxplots and histograms of the two groups, so you have no excuse for skipping this important check.

Some statistics software automatically tries to test whether the variances of the two groups are equal. Some automatically offer both the unpooled two-sample-t and the pooled-t results. Ignore any test for the variances. If the pooled and unpooled two-sample methods differ in any important way, you should stick with the unpooled two-sample method. Most likely, the equal variance assumption needed for the pooled method has failed.

The degrees of freedom approximation usually gives a fractional value. Some packages may round the approximate value down to the next smallest integer (though most will actually compute the P-value with the fractional value).

There are two ways to organize data when we want to compare two independent groups. The data can be in two columns, as in the table at the start of this chapter. Each column can be thought of as a variable. In this method, the variables in the batteries example would be *Brand-name* and *Generic*. Graphing calculators usually prefer this form, and some computer programs can use it as well.

There's another way to think about the data. What is the response variable for the battery life experiment? It's the *Time* until the music stopped. But the values of this variable are in both columns, and actually there's an experiment factor here, too—namely, the *Brand* of the battery. So, we could put the data into two different columns, one with the *Time* in it and one with the *Brand*. Then the data would look like this:

Time	Brand
194.0	Brand name
205.5	Brand name
199.2	Brand name
172.4	Brand name
184.0	Brand name
169.5	Brand name
190.7	Generic
203.5	Generic
203.5	Generic
206.5	Generic
222.5	Generic
209.4	Generic

This way of organizing the data makes sense as well. Now the factor and the response variables are clearly visible. You'll have to see which method your program requires. Some packages allow you to structure the data either way.

EXCEL

- From the **Data** Tab, Analysis Group, choose **Data Analysis**.
- Alternatively (if the Data Analysis Tool Pack is not installed), in the Formulas tab, choose **More functions.** > **Statistical** > **TTEST**, and specify Type = 3 in the resulting dialogue.
- Fill in the cell ranges for the two groups, the hypothesized difference, and the alpha level.

Power and sample size calculations are not included in the standard data analysis toolkit.

COMMENTS

Excel expects the two groups to be in separate cell ranges. Notice that, contrary to Excel's wording, we do not need to assume that the variances are *not* equal; we simply choose not to assume that they *are* equal.

JMP

- From the **Analyze** menu, select **Fit y by x**.
- Select variables: a **Y, Response** variable that holds the data and an **X, Factor** variable that holds the group names. JMP will make a dotplot.
- Click the **red triangle** in the dotplot title, and choose **Unequal variances**. The *t*-test is at the bottom of the resulting table.
- Find the P-value from the Prob > F section of the table (they are the same).

For power and sample size calculations, select **Power and Sample Size** from the **DOE** menu. Choose **Two Sample Means** from the submenu. Indicate the **Difference** in the means that you are hoping to detect, the **Alpha** value, and choose a one- or two-sided alternative. Guess at the common (or larger) **Std Dev**. Fill in either your desired **Sample size** or **Power**. The one you leave blank will be calculated. Click **Continue**.

COMMENTS

JMP expects data in one variable and category names in the other. Don't be misled: There is no need for the variances to be unequal to use two-sample *t* methods.

MINITAB

- From the **Stat** menu, choose the **Basic Statistics** submenu.
- From that menu, choose **2-sample t . . .** Then, fill in the dialogue.

For power and sample size calculations, choose **Basic Statistics** from the **Stat** menu, then **Power and Sample Size**, then **2-Sample t**. In the dialogue box, fill in two of **Sample sizes**, **Differences**, **Power values**, and Minitab will calculate the remaining one.

COMMENTS

The dialogue offers a choice of data in two variables, or data in one variable and category names in the other.

The power and sample size calculation is based on the pooled test, so you have to give a guess for only one standard deviation. If you believe that the two standard deviations differ substantially, guess at the average of the two, or be conservative and guess at the bigger one.

R

To test the hypothesis that $\mu_1 = \mu_2$ against an alternative (default is two-sided):

- Create a vector for the data of each group, say X and Y, and produce the confidence interval (default is 95%).
- t.test(X, Y, alternative = c("two.sided", "less", "greater"), conf.level = 0.95)

will produce the *t*-statistic, degrees of freedom, P-value, and confidence interval for a specified alternative.

In package {pwr}, to perform a sample size or power calculation for a two-sample t-test:

- pwr.t.test(n =, d =, sig.level =, power =, type = "two.sample") returns **one** argument that is not specified in the function.

For unequal groups sizes, use:

- pwr.t2n.test(n1 =, n2 =, d =, sig.level =, power =) instead.

COMMENTS

This is the same function as for the one-sample *t*-test, but with two vectors. In R, the default *t*-test does not assume equal variances in the two groups (default is **var.equal = FALSE**). To get the "pooled" *t*-test, add **var.equal = TRUE** to the function call.

The effect size (d) is equal to the difference between the two group means, *divided by the pooled standard deviation*—which usually needs to be guessed (perhaps from a pilot study). The calculation is based on the pooled *t*-test, hence you can only specify one SD. A pilot study can help.

SPSS

- From the **Analyze** menu, choose the **Compare Means** submenu.
- From that, choose the **Independent-Samples t-test** command. Specify the data variable and "group variable."
- Then, type in the labels used in the group variable. SPSS offers both the two-sample and pooled-*t* results in the same table.

For power and sample size calculations, you will need the IBM SPSS SamplePower add-on.

COMMENTS

SPSS expects the data in one variable and group names in the other. If there are more than two group names in the group variable, only the two that are named in the dialogue box will be compared.

STATCRUNCH

To do inference for the difference between two means *using summaries*:
- Click on **Stat**.
- Choose **T Statistics** » **Two sample** » **with summary**.
- Enter the **Sample mean, Standard deviation**, and sample **Size** for each group.
- De-select **Pool variances**.
- Click on **Next**.
- Indicate **Hypothesis Test**, then enter the hypothesized **Null mean** difference (usually 0), and choose the **Alternative** hypothesis.

OR

   Indicate **Confidence Interval**, and then enter the **Level** of confidence.
- Click on **Calculate**.

To do inference for the difference between two means *using data*:
- Click on **Stat**.

- Choose **T Statistics** » **Two sample** » **with data**.
- Choose the variable Column for each group.
- De-select **Pool variances**.
- Click on **Next**.
- Indicate **Hypothesis Test**, then enter the hypothesized **Null mean** difference (usually 0), and choose the **Alternative** hypothesis.

OR

   Indicate **Confidence Interval**, and then enter the **Level** of confidence.
- Click on **Calculate**

Power and Sample size calculations are readily available using Stat > T Statistics > Two Sample > Power/Sample size.
- Click on **Hypothesis Test Power**.
- Fill in all the boxes except for the one that you want to determine, either **Power** or **Sample Size**.
- Make a guess at the **Pooled std. dev.**

TI-83/84 PLUS

For a confidence interval:
- In the **STAT TESTS** menu, choose **0:2-SampTInt**.
- Specify if you are using data stored in two lists, or enter the means, standard deviations, and sizes of both samples.
- Indicate whether to pool the variances (when in doubt, say no), and specify the desired level of confidence.

To test a hypothesis:
- In the **STAT TESTS** menu, choose **4:2-SampTTest**.

- Specify if you are using data stored in two lists, or enter the means, standard deviations, and sizes of both samples.
- Indicate whether to pool the variances (when in doubt, say no) and specify whether the test is to be two-tail, lower-tail, or upper-tail.

Power and sample size calculations are not available.

Chapter Exercises

T indicates data is available in a saved file at the Companion Website at www.pearsoncanada.ca/deveaux

11. **Hot dogs** In its July 2007 issue, *Consumer Reports* examined the calorie content of two kinds of hot dogs: meat (usually a mixture of pork, turkey, and chicken) and all-beef. The researchers purchased samples of several different brands. The meat hot dogs averaged 111.7 calories, compared to 135.4 for the beef hot dogs. A test of the null hypothesis that there's no difference in mean calorie content yields a P-value of 0.124. Would a 95% confidence interval for $\mu_{meat} - \mu_{beef}$ include 0? Explain.

12. **Hot dogs and sodium** The *Consumer Reports* article described in Exercise 11 also listed the sodium content (in mg) for the various hot dogs tested. A test of the null hypothesis that beef hot dogs and meat hot dogs don't differ in the mean amounts of sodium yields a P-value of 0.11. Would a 95% confidence interval for $\mu_{meat} - \mu_{beef}$ include 0? Explain.

13. **Hot dogs and fat** The *Consumer Reports* article described in Exercise 11 also listed the fat content (in grams) for samples of beef and meat hot dogs. The resulting 90% confidence interval for $\mu_{meat} - \mu_{beef}$ is $(-6.5, -1.4)$.
    a) The endpoints of this confidence interval are negative numbers. What does that indicate?
    b) What does the fact that the confidence interval does not contain 0 indicate?
    c) If we use this confidence interval to test the hypothesis that $\mu_{meat} - \mu_{beef} = 0$, what is the corresponding alpha level?

14. **Washers** In its June 2007 issue, *Consumer Reports* examined the relative merits of top-loading and front-loading washing machines, testing samples of several different brands of each type. One of the variables the article reported was "cycle time,"—the number of minutes it took each machine to wash a load of clothes. Among the machines rated good to excellent, the 98% confidence interval for the difference in mean cycle time $(\mu_{top} - \mu_{front})$ is $(-40, -22)$.
    a) The endpoints of this confidence interval are negative numbers. What does that indicate?
    b) What does the fact that the confidence interval does not contain 0 indicate?
    c) If we use this confidence interval to test the hypothesis that $\mu_{top} - \mu_{front} = 0$ what is the corresponding alpha level?

15. **Hot dogs, 2nd helping** In Exercise 13, we saw a 90% confidence interval of $(-6.5, -1.4)$ grams for $\mu_{meat} - \mu_{beef}$, the difference in mean fat content for meat versus all-beef hot dogs. Explain why you think each of the following statements is true or false:

    a) If I eat a meat hot dog instead of a beef hot dog, there's a 90% chance I'll consume less fat.
    b) 90% of meat hot dogs have between 1.4 and 6.5 grams less fat than a beef hot dog.
    c) I'm 90% confident that meat hot dogs average 1.4–6.5 grams less fat than the beef hot dogs.
    d) If I were to get more samples of both kinds of hot dogs, 90% of the time the meat hot dogs would average 1.4–6.5 grams less fat than the beef hot dogs.
    e) If I tested more samples, I'd expect about 90% of the resulting confidence intervals to include the true difference in mean fat content between the two kinds of hot dogs.

16. **Second load of wash** In Exercise 14, we saw a 98% confidence interval of $(-40, -22)$ minutes for $\mu_{top} - \mu_{front}$, the difference in time it takes top-loading and front-loading washers to do a load of clothes. Explain why you think each of the following statements is true or false:
    a) 98% of top loaders are 22 to 40 minutes faster than front loaders.
    b) If I choose the laundromat's top loader, there's a 98% chance that my clothes will be done faster than if I had chosen the front loader.
    c) If I tried more samples of both kinds of washing machines, in about 98% of these samples I'd expect the top loaders to be an average of 22 to 40 minutes faster.
    d) If I tried more samples, I'd expect about 98% of the resulting confidence intervals to include the true difference in mean cycle time for the two types of washing machines.
    e) I'm 98% confident that top loaders wash clothes an average of 22 to 40 minutes faster than front-loading machines.

17. **Learning math** The Core Plus Mathematics Project (CPMP) is an innovative approach to teaching Mathematics that engages students in group investigations and mathematical modelling. After field tests in 36 high schools over a three-year period, researchers compared the performances of CPMP students with those taught using a traditional curriculum. In one test, students had to solve applied algebra problems using calculators. Scores for 320 CPMP students were compared to those of a control group of 273 students in a traditional Math program. Computer software was used to create a confidence interval for the difference in mean scores.[17]

    Conf level: 95% Parameter: μ(CPMP) $- \mu$(Ctrl)
    Interval: (5.573, 11.427)

---

[17] *Journal for Research in Mathematics Education*, 31, no. 3 [2000].

a) What is the margin of error for this confidence interval?
b) If we had created a 98% CI, would the margin of error be larger or smaller?
c) Explain what the calculated interval means in this context.
d) Does this result suggest that students who learn Mathematics with CPMP will have significantly higher mean scores in Algebra than those in traditional programs? Explain.

T 18. Stereograms Stereograms appear to be composed entirely of random dots. However, they contain separate images that a viewer can "fuse" into a three-dimensional (3D) image by staring at the dots while defocusing the eyes. An experiment was performed to determine whether knowledge of the form of the embedded image affected the time required for subjects to fuse the images. One group of subjects (group NV) received no information or just verbal information about the shape of the embedded object. A second group (group VV) received both verbal information and visual information (specifically, a drawing of the object). The experimenters measured how many seconds it took for the subject to report that he or she saw the 3D image.

> 2-Sample t-Interval
>
> Conf level $= 90\%$ df $= 70$
>
> $\mu(\text{NV}) - \mu(\text{VV})$ interval: $(0.55, 5.47)$

a) Interpret your interval in context.
b) Does it appear that viewing a picture of the image helps people "see" the 3D image in a stereogram?
c) What is the margin of error for this interval?
d) Explain carefully what the 90% confidence level means.
e) Would you expect a 99% confidence level to be wider or narrower? Explain.
f) Might that change your conclusion in part b)? Explain.

19. CPMP, again During the study described in Exercise 17, students in both CPMP and traditional classes took another Algebra test that did not allow them to use calculators. The table below shows the results. Are the mean scores of the two groups significantly different?

Performance on Algebraic Symbolic
Manipulation Without Use of Calculators

Math Program	n	Mean	SD
CPMP	312	29.0	18.8
Traditional	265	38.4	16.2

a) Write an appropriate hypothesis.
b) Do you think the assumptions for inference are satisfied? Explain.

c) Here is computer output for this hypothesis test. Explain what the P-value means in this context.

> 2-Sample t-Test of $\mu_1 - \mu_2 \neq 0$
>
> t-Statistic $= -6.451$ w/574.8761 df
>
> $P < 0.0001$

d) State a conclusion about the CPMP program.

T 20. Memory Does ginkgo biloba enhance memory? In an experiment to find out, subjects were assigned randomly to take ginkgo biloba supplements or a placebo. Their memory was tested to see whether it improved. Here are boxplots comparing the two groups and some computer output from a two-sample t-test computed for the data.

> 2-Sample t-Test of $\mu_\text{G} - \mu_\text{P} > 0$
>
> Difference Between Means $= -0.9914$
>
> t-Statistic $= -1.540$ w/196 df
>
> $P = 0.9374$

a) Explain what the P-value means in this context.
b) State your conclusion about the effectiveness of ginkgo biloba.
c) Proponents of ginkgo biloba continue to insist that it works. What type of error do they claim your conclusion makes? Explain.

21. Commuting A man who moves to a new city sees that there are two routes he could take to work. A neighbour who has lived there a long time tells him Route A will average five minutes faster than Route B. The man decides to experiment. Each day, he flips a coin to determine which way to go, driving each route for 20 days. He finds that Route A takes an average of 40 minutes, with standard deviation 3 minutes, and Route B takes an average of 43 minutes, with standard deviation 2 minutes. Histograms of travel times for the routes are roughly symmetric and show no outliers.

a) Find a 95% confidence interval for the difference in average commuting time for the two routes. (From technology, df $= 33.1$.)
b) Should the man believe the neighbour's claim that he can save an average of five minutes a day by always driving Route A? Explain.

22. Pulse rates A researcher wanted to see whether there is a significant difference in resting pulse rates for men and women. The data she collected are displayed in the boxplots and summarized below.

Sex		
	Male	**Female**
Count	28	24
Mean	72.75	72.625
Median	73	73
StdDev	5.37225	7.69987
Range	20	29
IQR	9	12.5

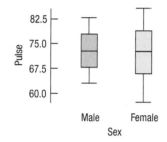

a) What do the boxplots suggest about differences between male and female pulse rates?
b) Is it appropriate to analyze these data using the methods of inference discussed in this chapter? Explain.
c) Create a 90% confidence interval for the difference in mean pulse rates.
d) Does the confidence interval confirm your answer to part a)? Explain.

T 23. Cereal The data below show the sugar content (as a percentage of weight) of several national brands of children's and adults' cereals. Create and interpret a 95% confidence interval for the difference in mean sugar content. Be sure to check the necessary assumptions and conditions.

Children's cereals: 40.3, 55, 45.7, 43.3, 50.3, 45.9, 53.5, 43, 44.2, 44, 47.4, 44, 33.6, 55.1, 48.8, 50.4, 37.8, 60.3, 46.6

Adults' cereals: 20, 30.2, 2.2, 7.5, 4.4, 22.2, 16.6, 14.5, 21.4, 3.3, 6.6, 7.8, 10.6, 16.2, 14.5, 4.1, 15.8, 4.1, 2.4, 3.5, 8.5, 10, 1, 4.4, 1.3, 8.1, 4.7, 18.4

T 24. Egyptians Some archaeologists theorize that ancient Egyptians interbred with several different immigrant populations over thousands of years. To see if there is any indication of changes in body structure that might have resulted, they measured 30 skulls of male Egyptians dated from 4000 B.C.E. and 30 others dated from 200 B.C.E. (*Source*: A. Thomson and R. Randall-Maciver, *Ancient Races of the Thebaid*, Oxford: Oxford University Press, 1905.)

Maximum Skull Breadth (mm)			
4000 B.C.E.	**4000 B.C.E.**	**200 B.C.E.**	**200 B.C.E.**
131	131	141	131
125	135	141	129
131	132	135	136
119	139	133	131
136	132	131	139
138	126	140	144
139	135	139	141
125	134	140	130
131	128	138	133
134	130	132	138
129	138	134	131
134	128	135	136
126	127	133	132
132	131	136	135
141	124	134	141

a) Are these data appropriate for inference? Explain.
b) Create a 95% confidence interval for the difference in mean skull breadth between these two eras.
c) Do these data provide evidence that the mean breadth of males' skulls changed over this period? Explain.
d) The standard deviations are very similar, so pooled methods should work well here. Calculate a 95% pooled *t* confidence interval for the difference and compare with the interval in part b.

T 25. Hurricanes to 2013 The data below show the number of hurricanes (category 3 or higher) recorded annually before and after 1970. Create an appropriate visual display, and determine whether these data are appropriate for testing whether there has been a change in the frequency of hurricanes.

1944–1969	1970–2013
3, 2, 1, 2, 4, 3, 7, 2, 3, 3, 2, 5, 2, 2, 4, 2, 2, 6, 0, 2, 5, 1, 3, 1, 0, 3	2, 1, 0, 1, 2, 3, 2, 1, 2, 2, 2, 3, 1, 1, 1, 3, 0, 1, 3, 2, 1, 2, 1, 1, 0, 5, 6, 1, 3, 5, 3, 3, 2, 3, 6, 7, 2, 2, 5, 2, 5, 4, 2, 0

T 26. Streams Researchers collected samples of water from streams in the Adirondack Mountains to investigate the effects of acid rain. They measured the pH (acidity) of the water and classified the streams with respect to the kind of substrate (type of rock over which they flow). A lower pH means the water is more acidic. Here is a plot of the pH of the streams by substrate (limestone, mixed, or shale):

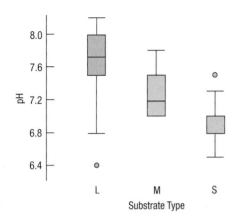

Here are selected parts of a software analysis comparing the pH of streams with limestone and shale substrates:

2-Sample t-Test of $\mu_1 - \mu_2$

Difference Between Means $= 0.735$

t-Statistic $= 16.30$ w/133 df

$P < 0.0001$

a) State the null and alternative hypotheses for this test.
b) From the information you have, do the assumptions and conditions appear to be met?
c) What conclusion would you draw?

T **27.** **Reading** An educator believes that new reading activities for elementary schoolchildren will improve reading comprehension scores. She randomly assigns Grade 3 students to an eight-week program in which some will use these activities and others will experience traditional teaching methods. At the end of the experiment, both groups take a reading comprehension exam. The back-to-back stem-and-leaf display shows their scores. Do these results suggest that the new activities are better? Test an appropriate hypothesis and state your conclusion.

New Activities		Control
	1	07
4	2	068
3	3	377
96333	4	12222238
9876432	5	355
721	6	02
1	7	
	8	5

28. **CPMP and word problems** The study of the new CPMP Mathematics methodology described in Exercise 17 also tested students' abilities to solve word problems. This table shows how the CPMP and traditional groups performed. What do you conclude?

Math Program	n	Mean	SD
CPMP	320	57.4	32.1
Traditional	273	53.9	28.5

T **29.** **Home runs 2013** American League baseball teams play their games with the designated hitter rule, meaning that pitchers do not bat. The league believes that replacing the pitcher, traditionally a weak hitter, with another player in the batting order produces more runs and generates more interest among fans. Below are the average numbers of home runs hit per game in American League and National League stadiums for the 2013 season (www.hittrackeronline.com/stadiums.php).

American		National	
2.86	1.94	2.30	1.80
2.64	1.94	2.28	1.67
2.31	1.93	2.17	1.60
2.21	1.89	2.16	1.35
2.10	1.89	1.96	1.33
2.06	1.75	1.84	1.31
2.04	1.54	1.84	1.04
1.94		1.83	

a) Create an appropriate display of these data. What do you see?
b) With a 95% confidence interval, estimate the mean number of home runs hit in American League games.
c) Coors Field, in Denver, stands a mile above sea level, an altitude far greater than that of any other major league ball park. Some believe that the thinner air makes it harder for pitchers to throw curve balls and easier for batters to hit the ball a long way. Do you think the 1.96 home runs hit per game at Coors is unusual? (Denver is a National League team.) Explain.
d) Explain why you should not use two separate confidence intervals to decide whether the two leagues differ in average number of runs scored.
e) Using a 95% confidence interval, estimate the difference between the mean number of home runs hit in American and National League games.
f) Interpret your interval.
g) Does this interval suggest that the two leagues may differ in average number of home runs hit per game?

T **30.** **Hard water** In an investigation of environmental causes of disease, data were collected on the annual mortality rate (deaths per 100 000) for males in 61 large towns in England and Wales. In addition, the water hardness was recorded as the calcium concentration (parts per million, or ppm) in the drinking water. The data set also notes, for each town, whether it was south or north of Derby. Is there a significant difference in mortality rates in the two regions? Here are the summary statistics.

Summary of: For categories in:	**Mortality** **Derby**		
Group	**Count**	**Mean**	**Median**	**StdDev**
North	34	1631.59	1631	138.470
South	27	1388.85	1369	151.114

a) Test appropriate hypotheses, and state your conclusion.
b) The boxplots of the two distributions show an outlier among the data north of Derby. What effect might that have had on your test?

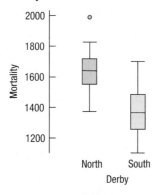

c) Repeat part a) using the pooled *t*-test, and compare results. Since the standard deviations are so similar, we would expect the pooled test to work well here.

T 31. Job satisfaction A company institutes an exercise break for its workers to see if it will improve job satisfaction, as measured by a questionnaire that assesses workers' satisfaction. Scores for 10 randomly selected workers before and after implementation of the exercise program are shown below. The company wants to assess the effectiveness of the exercise program. Explain why you can't use the methods discussed in this chapter to do that. (Don't worry, we'll give you another chance to do this the right way.)

Worker Number	Job Satisfaction Index	
	Before	**After**
1	34	33
2	28	36
3	29	50
4	45	41
5	26	37
6	27	41
7	24	39
8	15	21
9	15	20
10	27	37

32. Summer school Having done poorly on their Math final exams in June, six students repeat the course in summer school, then take another exam in August. Both results are shown below for each student. If we consider these students as representative of all students who might attend this summer school in other years, do these results provide evidence that the program is worthwhile? Can you analyze the data with methods discussed in this chapter?

June exam	54	49	68	66	62	62
Aug. exam	50	65	74	64	68	72

33. Sex and violence In June 2002, the *Journal of Applied Psychology* reported on a study that examined whether the content of television shows influenced the ability of viewers to recall brand names of items featured in the commercials. The researchers randomly assigned volunteers to watch one of three programs, each containing the same nine commercials. One of the programs had violent content, another sexual content, and the third neutral content. After the shows ended, the subjects were asked to recall the brands of products that were advertised. The table below summarizes the results.

	Program Type		
	Violent	**Sexual**	**Neutral**
No. of subjects	108	108	108
Brands recalled			
Mean	2.08	1.71	3.17
SD	1.87	1.76	1.77

a) Do these results indicate that viewer memory for ads may differ depending on program content? A test of the hypothesis that there is no difference in ad memory between programs with sexual content and those with violent content has a P-value of 0.136. State your conclusion.
b) Is there evidence that viewer memory for ads may differ between programs with sexual content and those with neutral content? Test an appropriate hypothesis, and state your conclusion.
c) Repeat part a) using the pooled *t*-test, and compare the P-values. Why might the pooled *t*-test be justifiable here?

34. Ad campaign You are a consultant to the marketing department of a business preparing to launch an ad campaign for a new product. The company can afford to run ads during one television show, and has decided not to sponsor a show with sexual content. You read the study described in Exercise 33, then use a computer to create a confidence interval for the difference in mean number of brand names remembered between the groups watching violent shows and those watching neutral shows.

TWO-SAMPLE t

95% CI FOR $\mu_{\text{viol}} - \mu_{\text{neut}}$: $(-1.578, -0.602)$

a) At the meeting of the marketing staff, you have to explain what this output means. What will you say?

b) What advice would you give the company about the upcoming ad campaign?

35. Sex and violence II In the study described in Exercise 33, the researchers also contacted the subjects again, 24 hours later, and asked them to recall the brands advertised. Results are summarized below.

	Program Type		
	Violent	**Sexual**	**Neutral**
No. of subjects	101	106	103
Brands recalled			
Mean	3.02	2.72	4.65
SD	1.61	1.85	1.62

a) Is there a significant difference in viewers' abilities to remember brands advertised in shows with violent versus neutral content?

b) Find a 95% confidence interval for the difference in mean number of brand names remembered between the groups watching shows with sexual content and those watching neutral shows. Interpret your interval in this context.

36. Ad recall In Exercises 33 and 35, we see the number of advertised brand names people recalled immediately after watching television shows and 24 hours later. Strangely enough, it appears that they remembered more about the ads the next day. Should we conclude this is true in general about people's memory of television ads?

a) Suppose one analyst conducts a two-sample hypothesis test, using the two means displayed earlier for violent show viewers, to see if memory of brands advertised during violent television shows is higher 24 hours later. The P-value of his test is 0.00013. What might he conclude?

b) Explain why his procedure was inappropriate. Which of the assumptions for inference was violated?

c) Suggest a design that could compare immediate brand-name recall with recall one day later, if we are restricted to using methods discussed in this chapter.

37. Hungry? Researchers investigated how the size of a bowl affects how much ice cream people tend to scoop when serving themselves.[18] At an "ice cream social,"

people were randomly given either a 17 fluid oz (503 mL) or a 34 fluid oz (1006 mL) bowl (both large enough that they would not be filled to capacity). They were then invited to scoop as much ice cream as they liked. Did the bowl size change the selected portion size? Here are the summaries:

Small Bowl		Large Bowl	
n	26	n	22
$\bar{y}$	5.07 oz	$\bar{y}$	6.58 oz
s	1.84 oz	s	2.91 oz

Test an appropriate hypothesis and state your conclusions. For assumptions and conditions that you cannot test, you may assume that they are sufficiently satisfied to proceed.

38. Thirsty? Researchers randomly assigned participants either a tall, thin "highball" glass or a short, wide "tumbler," each of which held 355 mL. Participants were asked to pour a shot (44.3 mL) into their glass. Did the glass shape make a difference in how much liquid they poured?[19] Here are the summaries:

highball		tumbler	
n	99	n	99
$\bar{y}$	42.2 mL	$\bar{y}$	60.9 mL
s	16.2 mL	s	17.9 mL

Test an appropriate hypothesis and state your conclusions. For assumptions and conditions that you cannot test, you may assume that they are sufficiently satisfied to proceed.

T 39. Battle of the sexes A June 2009 *Toronto Star* study of 100 downtown Toronto parkers attempted "to resolve scientifically the question of who is most competent at pulling into a parking space . . . men or women?"[20] It reported that women are better at parallel parking, though they take longer. Researchers counted the distance from the curb, in inches, and reported an average distance of 9.3 inches for 47 women and 11.1 inches for 46 men. Below are the data:

Distances for males:

0.5 1.0 1.5 1.5 1.5 3.0 3.5 5.0 6.0 6.0 7.0
7.0 7.0 7.0 7.0 8.0 8.0 8.5 8.5 9.0 9.0 9.5
10.0 10.0 11.0 12.0 12.0 12.0 13.0 14.0 14.0 14.0
14.0 14.0 14.0 14.5 14.5 14.5 17.0 18.0 18.0 19.0
19.0 20.0 21.0 48.0

---

[18]Brian Wansink, Koert van Ittersum, and James E. Painter, 2006, "Ice cream illusions: Bowls, spoons, and self-served portion sizes," *Am J Prev Med.*

[19]Brian Wansink and Koert van Ittersum, 2005, "Shape of glass and amount of alcohol poured: Comparative study of effect of practice and concentration," *BMJ*:331; 1512–1514.

[20]Diana Zlomislic, "A new angle on the battle of the sexes." *Toronto Star*, June 6, 2009: A23. Data supplied by Andrew Bailey.

Distances for females:

2.0 2.5 3.0 3.0 3.0 3.0 3.0 4.0 4.5 4.5 4.5
5.0 5.0 5.0 6.0 6.0 7.0 7.0 7.0 7.0 7.5 8.0
8.5 8.5 9.0 9.0 9.0 9.0 10.0 10.0 12.0 12.0 12.0
12.0 12.5 13.0 13.0 13.0 14.0 14.0 14.0 15.5 16.0
18.0 20.0 21.0 25.0

Respond to the *Star*'s claim. Do you accept it? Make your case carefully and properly.

T 40. Ontario students, U.K. students Below are usual travel times to school, in minutes, for 40 British secondary school students selected using the CensusAtSchool random data selector (www.censusatschool.com). Participation was voluntary but there was broad participation from schools, teachers, and students in the U.K in this exercise in statistical literacy.

45 5 4 15 50 20 20 20 20 20 25 35 15 30 20
10 45 10 3 60 25 20 5 15 5 15 17 30 40 20
10 30 10 15 20 10 15 17 10 25

In Chapter 18, we selected 40 Ontario students in a similar fashion (www.censusatschool.ca), with the following travel times:

30 10 8 30 5 8 7 15 10 35 15 10 25 22 20
25 30 10 25 8 15 18 25 15 10 25 5 2 5 25
20 15 47 20 20 13 20 5 15 12

a) Test for a difference in mean travel time between Ontario students and British students. Also give a 95% confidence interval for the difference.
b) Are your P-value and confidence level, in part a) trustworthy?

T 41. Swim the Lake 2013 revisited As we saw in chapter 8, Exercise 40, between 1954 and 2013, swimmers have crossed Lake Ontario 58 times. Both women and men have made the crossing. Here are some plots (we've omitted a crossing by Vikki Keith, who swam a round trip—North to South to North—in 3390 minutes):

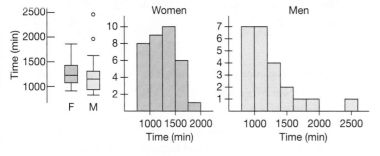

Summary statistics are as follows:

	Summary of Time (min)		
Group	Count	Mean	StdDev
F	34	1257.09	261.10
M	23	1226.04	382.85

Is there a difference between the mean amount of time (in minutes) it takes female and male swimmers to swim the lake?
a) Construct and interpret a 95% confidence interval for the difference between female and male crossing times (technology gives 35.7 df).
b) Comment on the assumptions and conditions and your conclusions.

T 42. Still swimming Here's some additional information about the Ontario crossing times presented in the previous exercise. It is generally thought to be harder to swim across the lake from north to south. Indeed, this has been done only 5 times. Every one of those crossings was by a woman. If we omit those 5 crossings, the boxplots look like this:

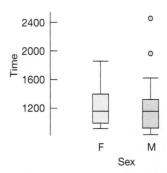

Although not designated as an outlier, the slowest female time belongs to Vikki Keith, who crossed the lake swimming only butterfly stroke. Omitting that extraordinary swim gives the following summary statistics:

	Summary of Time (min)		
Sex	Count	Mean	StdDev
F	29	1195.34	216.326
M	23	1226.04	382.85

a) Do women or men appear to be faster at swimming across the lake? Support your answer by interpreting a confidence interval (Use 32 df).
b) Some might argue that the men are being unfairly slowed down by two extraordinarily slow swims. Here are the summaries without those two swims. Repeat a) based on these data. Does your conclusion change? (Use 42 df).

	Summary of Time (min)		
Sex	Count	Mean	StdDev
F	29	1195.34	216.326
M	21	1232.14	220.36

c) Vikki Keith was responsible for two of the more remarkable crossings, but she also swam Lake Ontario two other times. In fact, of the 50 crossings in this analysis, 7 were repeat crossings by a swimmer who'd crossed the lake before. How does this fact affect your thoughts about the confidence interval?

T **43. Running heats London** In Olympic running events, preliminary heats are determined by random draw, so we should expect that the abilities of runners in the various heats to be about the same, on average. Here are the times (in seconds) for the 400-m women's run in the 2012 Olympics in London for preliminary heats 2 and 5. Is there any evidence that the mean time to finish is different for randomized heats? Explain. Be sure to include a discussion of assumptions and conditions for your analysis. (Note: one runner in heat 2 did not finish and one runner in heat 5 did not start.)

Country	Name	Time	Heat
BOT	Amantle Montsho	50.4	2
JAM	Christine Day	51.05	2
GBR	Shana Cox	52.01	2
GUY	Aliann Pompey	52.1	2
ANT	Afia Charles	54.25	2
NCA	Ingrid Narvaez	59.55	2
BAH	Shaunae Miller	DNF	2
RUS	Antonina Krivoshapka	50.75	5
UKR	Alina Lohvynenko	52.08	5
GBR	Lee McConnell	52.23	5
SWE	Moa Hjelmer	52.86	5
MAW	Ambwene Simukonda	54.2	5
FIJ	Danielle Alakija	56.77	5
GRN	Kanika Beckles	DNS	5

T **44. Swimming heats London** In the previous exercise, we looked at the times in two different heats for the 400-m women's run from the 2012 Olympics. Unlike track events, swimming heats are *not* determined at random. Instead, swimmers are seeded so that better swimmers are placed in later heats. Here are the times (in seconds) for the women's 400-m freestyle from heats 2 and 5. Do these results suggest that the mean times of seeded heats are not equal? Explain. Include a discussion of assumptions and conditions for your analysis.

Country	Name	Time	Heat
BUL	Nina Rangelova	251.7	2
CHI	Kristel Köbrich	252.0	2
JPN	Aya Takano	252.3	2

LIE	Julia Hassler	253.0	2
MEX	Susana Escobar	254.8	2
THA	Nathanan Junkrajang	256.5	2
SIN	Lynette Lim	258.6	2
KOR	Kim Ga-Eul	283.5	2
FRA	Camille Muffat	243.3	5
USA	Allison Schmitt	243.3	5
NZL	Lauren Boyle	243.6	5
DEN	Lotte Friis	244.2	5
ESP	Melanie Costa	246.8	5
AUS	Bronte Barratt	248.0	5
HUN	Boglarka Kapas	250.0	5
RUS	Elena Sokolova	252.2	5

45. Tees Does it matter what kind of tee a golfer places the ball on? The company that manufactures Stinger tees claims that the thinner shaft and smaller head will lessen resistance and drag, reducing spin and allowing the ball to travel farther. In August 2003, Golf Laboratories, Inc. compared the distance travelled by golf balls hit off regular wooden tees to those hit off Stinger tees. All the balls were struck by the same golf club using a robotic device set to swing the club head at approximately 95 miles per hour. Below are some summary statistics from the test. Assume that six balls were hit off each tee and that the data were suitable for inference.

		Total Distance (yards)	Ball Velocity (mph)	Club Velocity (mph)
Regular	Avg.	227.17	127.00	96.17
tee	SD	2.14	0.89	0.41
Stinger	Avg.	241.00	128.83	96.17
tee	SD	2.76	0.41	0.52

Is there evidence that balls hit off the Stinger tees would on average have a higher initial velocity?

46. Golf again Given the test results on golf tees described in the previous exercise, is there evidence that balls hit off Stinger tees would travel farther? Again, assume that six balls were hit off each tee and that the data were suitable for inference.

47. Statistics journals When a professional statistician has information to share with colleagues, he or she will submit an article to one of several Statistics journals for publication. This can be a lengthy process; typically, the article must be circulated for peer review and perhaps edited before being accepted for publication. Then the article must wait in line with other articles before actually appearing in print. In *Chance* magazine, Eric Bradlow and Howard Wainer reported on this delay for several journals. For 288 articles published in *The American Statistician*, the mean length of time between initial submission and publication was 21 months, with a standard deviation of 8 months. For 209 *Applied Statistics* articles, the mean time to publication was 31 months, with a standard deviation of 12 months. Create and interpret a 90% confidence interval for the difference in mean delay, and comment on the assumptions that underlie your analysis.

48. Music and memory Is it a good idea to listen to music when studying for a big test? In a study conducted by some Statistics students, 62 people were randomly assigned to listen to rap music, music by Mozart, or no music while attempting to memorize objects pictured on a page. They were then asked to list all the objects they could remember. Here are summary statistics for each group:

	Rap	Mozart	No Music
Count	29	20	13
Mean	10.72	10.00	12.77
SD	3.99	3.19	4.73

a) Does it appear that it is better to study while listening to Mozart than to rap music? Test an appropriate hypothesis, and state your conclusion.

b) Create a 90% confidence interval for the mean difference in memory score between students who study to Mozart and those who listen to no music at all. Interpret your interval.

49. Rap Using the results of the experiment described in the previous exercise, does it matter whether one listens to rap music while studying, or is it better to study without music at all?

a) Test an appropriate hypothesis and state your conclusion.

b) If you concluded that there is a difference, estimate the size of that difference with a confidence interval and explain what your interval means.

T 50. Cuckoos Cuckoos lay their eggs in the nests of other (host) birds. The eggs are then adopted and hatched by the host birds. But the potential host birds may lay eggs of different sizes. Does the cuckoo change the size of her eggs for different foster species? The numbers in the table are lengths (in mm) of cuckoo eggs found in nests of three different species of other birds. The data are drawn from the work of O.M. Latter in 1902 and were used in a fundamental textbook on statistical quality

control by L.H.C. Tippett (1902–1985), one of the pioneers in that field.

Cuckoo Egg Length (mm)		
Foster Parent Species		
Sparrow	**Robin**	**Wagtail**
20.85	21.05	21.05
21.65	21.85	21.85
22.05	22.05	21.85
22.85	22.05	21.85
23.05	22.05	22.05
23.05	22.25	22.45
23.05	22.45	22.65
23.05	22.45	23.05
23.45	22.65	23.05
23.85	23.05	23.25
23.85	23.05	23.45
23.85	23.05	24.05
24.05	23.05	24.05
25.05	23.05	24.05
	23.25	24.85
	23.85	

Investigate the question of whether the mean length of cuckoo eggs is the same for different species, and state your conclusion.

51. Tee it up again Consider again Exercise 45 about the Stinger golf tees. Six balls were used on each type of tee. If we wanted the margin of error in the estimate of the mean difference to be just 1.0 yard, with 95% confidence, how many balls should have been struck off each type of tee? Be conservative and set each standard deviation to 3.0 yards.

T 52. Harder water Consider again Exercise 30 about hard water in England and Wales. If we want the margin of error for our estimate of the difference in mortality rates to be only 50 (per 100 000), with 95% confidence, how big should the sample sizes be (assume two equal sample sizes)? You can use the standard deviations given previously as your estimates.

53. Fore! Consider again Exercise 45 about the Stinger golf tees (by the way, "fore" is shouted in golf to warn of a ball flying your way). Perhaps a difference of as little as one yard, on average, is important enough to justify using a new tee type. Using software, determine how many balls should be driven off each type of tee so that the test will have a power of 0.95 when the true mean difference is three yards. Assume, conservatively, that both standard deviations equal about 3.0 yards.

T 54. Hard water again Consider again Exercise 30 about hard water in England and Wales. If we want to successfully detect a difference of 100 (per 100 000), in mortality rates, 9 times in 10, how many towns should actually be sampled (assume two equal sample sizes)? You can use 150.0 as your estimate of both standard deviations.

Just Checking ANSWERS

1. Randomization should balance unknown sources of variability in the two groups of patients and helps us believe the two groups are independent.

2. We can be 95% confident that after four weeks, endoscopic surgery patients will have a mean pinch strength between 0.04 kg and 2.96 kg higher than open-incision patients.

3. The lower bound of this interval is close to 0, so the difference may not be great enough that patients could actually notice the difference. We may want to consider other issues such as cost or risk in making a recommendation about the two surgical procedures.

4. Without data, we can't check the Nearly Normal Condition.

5. H_0: Mean pinch strength is the same after both surgeries ($\mu_E - \mu_0 = 0$).

   H_A: Mean pinch strength is different after the two surgeries ($\mu_E - \mu_0 \neq 0$).

6. With a P-value this low, we reject the null hypothesis. We can conclude that mean pinch strength differs after four weeks in patients who undergo endoscopic surgery versus patients who have open-incision surgery. Results suggest that the endoscopic surgery patients may be stronger, on average.

7. If some patients contributed two hands to the study, then the observations within the groups may not be mutually independent. It is reasonable to assume that two hands from the same patient might respond in similar ways to similar treatments.

MathXL

MyLab Statistics

Go to MathXL at **www.mathxl.com** or MyLab™ Statistics at **www.mystatlab.com**. You can practise exercises for this chapter as often as you want. The guided solutions will help you find answers step by step. You'll find a personalized study plan available to you too!

Paired Samples and Blocks

LEARNING OBJECTIVES

1 Recognize paired data designs

2 Construct a confidence interval for the mean difference

3 Perform a hypothesis test about the mean difference

4 Know and check the assumptions and conditions for inference

5 Plan the sample size

6 Know the advantages of pairing/blocking

7 Perform a non-parametric Sign Test (optional)

To call in the statistician after the experiment is done may be no more than asking him to perform a post-mortem examination: he may be able to say what the experiment died of.

—Ronald Fisher

peed-skating races are run in pairs. Two skaters start at the same time, one on the inner lane and one on the outer lane. Halfway through the race, they cross over, switching lanes so that each will skate the same distance in each lane. Even though this seems fair, at the 2006 Olympics some fans thought there might have been an advantage to starting on the outside. After all, the winner, Canadian Cindy Klassen,[1] started on the outside and skated a remarkable 1.47 seconds faster than the silver medalist.

Here are the data for the women's 1500-m race:

Inner Lane		Outer Lane	
Name	**Time**	**Name**	**Time**
OLTEAN Daniela	129.24	(no competitor)	
ZHANG Xiaolei	125.75	NEMOTO Nami	122.34
ABRAMOVA Yekaterina	121.63	LAMB Maria	122.12
REMPEL Shannon	122.24	NOH Seon Yeong	123.35
LEE Ju-Youn	120.85	TIMMER Marianne	120.45
ROKITA Anna Natalia	122.19	MARRA Adelia	123.07
YAKSHINA Valentina	122.15	OPITZ Lucille	122.75
BJELKEVIK Hedvig	122.16	HAUGLI Maren	121.22
ISHINO Eriko	121.85	WOJCICKA Katarzyna	119.96
RANEY Catherine	121.17	BJELKEVIK Annette	121.03
OTSU Hiromi	124.77	LOBYSHEVA Yekaterina	118.87
SIMIONATO Chiara	118.76	JI Jia	121.85
ANSCHUETZ THOMS Daniela	119.74	WANG Fei	120.13
BARYSHEVA Varvara	121.60	VAN DEUTEKOM Paulien	120.15
GROENEWOLD Renate	119.33	GROVES Kristina	116.74
RODRIGUEZ Jennifer	119.30	NESBITT Christine	119.15
FRIESINGER Anni	117.31	KLASSEN Cindy	115.27
WUST Ireen	116.90	TABATA Maki	120.77

Connections

How much will an LSAT prep course raise students' scores? Do married women do more housework than their husbands? (But seriously, just how much more do they do per week?) Questions like these look at paired variables. When pairs of quantitative observations go together naturally, they can't be independent, so the methods we used in the last chapter won't work. In this chapter, you'll see what to do with paired data.

Pairing is just a basic form of *blocking* - an experimental design concept first discussed in Chapter 10. The one-sample *t* methods learned previously are really all we need to analyze paired data, once we compute the paired differences.

---

[1]Winnipeg native and tied with Clara Hughes as Canada's all-time most decorated Olympian.

Who	Olympic speed-skaters
What	Time for women's 1500 m
Units	Seconds
When	2006
Where	Torino, Italy
Why	To see whether one lane is faster than the other

Figure 20.1

Using boxplots to compare times in the inner and outer lanes shows little because it ignores the fact that the skaters raced in pairs (with one exception, Daniela Oltean, who raced alone and whose time of 129.24 is not shown above).

We can view this skating event as an experiment testing whether the lanes were equally fast. Skaters were randomly assigned to lanes. The boxplots of times recorded in the inner and outer lanes (Figure 20.1 in the margin) don't show much difference. But that's not the right way to compare these times. Conditions can change during the day. The data are recorded for races run *in pairs* (with the exception of Daniela Oltean, who raced alone), so the two groups of race times are not independent, and the methods of Chapter 19 should not be applied here.

20.1 Paired *t*-Test

Paired Data

Data such as these are called **paired**. We have the times for skaters in each lane for each race. The races are run in pairs, so results can't be independent, so instead we focus on the *differences* in times for each racing pair.

Paired data arise in a number of ways. Perhaps the most common way is to compare subjects with themselves before and after a treatment. When pairs arise from an experiment, the pairing is a type of *blocking*. When they arise from an observational study, it is a form of *matching*. Paired data may be quantitative or categorical, but we'll only discuss quantitative data in this text (for categorical data, look up McNemar's test).

For Example IDENTIFYING PAIRED DATA

Do flexible schedules reduce the demand for resources? The Lake County, Illinois, Health Department experimented with a flexible four-day work week. For a year, the department recorded the mileage driven by 11 field workers on an ordinary five-day work week. Then, it changed to a flexible four-day work week and recorded mileage for another year.[2] Here are the data in miles (1 mile = 1.6 km):

Name	5-Day Mileage	4-Day Mileage
Jeff	2798	2914
Betty	7724	6112
Roger	7505	6177
Tom	838	1102
Aimee	4592	3281
Greg	8107	4997
Larry G.	1228	1695
Tad	8718	6606
Larry M.	1097	1063
Leslie	8089	6392
Lee	3807	3362

QUESTION: Why are these data paired?

ANSWER: The mileage data are paired because each driver's mileage is measured before and after the change in schedule. I'd expect drivers who drove more than others before the schedule change to continue to drive more afterward, so the two sets of mileages can't be considered independent.

---

[2]Charles S. Catlin, "Four-day work week improves environment," *Journal of Environmental Health*, Denver, 59:7.

Jerry Lampen/Reuters/Alamy

Pairing isn't a problem; it's an opportunity. If you know the data are paired, you can take advantage of that fact—in fact, you *must* take advantage of it. You *may not* use the two-sample methods of the previous chapter when the data are paired. Remember: Those methods rely on the Pythagorean Theorem of Statistics, and that *requires the two samples be independent.* Paired data aren't. You must decide whether the data are paired from understanding how they were collected and what they mean (check the *W*'s). There is no test to determine whether the data are paired.

Once we recognize that the speed-skating data are matched pairs, it makes sense to consider the difference in times for each two-skater race. So we look at the *pairwise* differences:

Skating Pair	Inner Time	Outer Time	Inner – Outer
1	129.24		
2	125.75	122.34	3.41
3	121.63	122.12	−0.49
4	122.24	123.35	−1.11
5	120.85	120.45	0.40
6	122.19	123.07	−0.88
7	122.15	122.75	−0.60
8	122.16	121.22	0.94
9	121.85	119.96	1.89
10	121.17	121.03	0.14
11	124.77	118.87	5.90
12	118.76	121.85	−3.09
13	119.74	120.13	−0.39
14	121.60	120.15	1.45
15	119.33	116.74	2.59
16	119.30	119.15	0.15
17	117.31	115.27	2.04
18	116.90	120.77	−3.87

One skater raced alone, so we'll omit that race. Because it is the *differences* we care about, we'll treat them as if *they* were the data, ignoring the original two columns. Now that we have only one column of values to consider, we can use a simple one-sample *t*-test. Mechanically, **a paired *t*-test** is just a one-sample *t*-test for the means of these pairwise differences. The sample size is the number of pairs.

Section 20.1 Exercises

1. **Which method?** Which of the following scenarios should be analyzed as paired data?
   a) Students take a MCAT prep course. Their before and after scores are compared.
   b) 20 male and 20 female students in class take a midterm. We compare their scores.
   c) A group of college freshmen are asked about the quality of the university cafeteria. A year later, the same students are asked about the cafeteria again. Do student's opinions change during their time at school?

2. **Which method II?** Which of the following scenarios should be analyzed as paired data?
   a) Spouses are asked about the number of hours of sleep they get each night. We want to see if husbands get more sleep than wives.
   b) 50 insomnia patients are given a placebo and 50 are given a mild sedative. Which subjects sleep more hours?
   c) A group of college freshmen and a group of sophomores are asked about the quality of the university cafeteria. Do students' opinions change during their time at school?

20.2 Assumptions and Conditions

Paired Data Condition

Paired Data Condition: The data must be quantitative and paired. You can't just decide to pair data when in fact the samples are independent. When you have two groups with the same number of observations, it may be tempting to match them up. Don't, unless you are prepared to justify your claim that the data are paired.

On the other hand, be sure to recognize paired data when you have them. Remember, two-sample *t* methods of Chapter 19 aren't valid without independent groups, and groups with paired observations aren't independent. Although this is a strict requirement, it is one that can be easy to check if you understand how the data were collected.

Independence Assumption

Independence Assumption: If the data are paired, the *groups* are not independent. For these methods, it's the *differences* that must be independent of each other. In our example, there's no reason to believe that the difference in speeds of one pair of races could affect the difference in speeds for another pair.

Suitable randomization can both prevent bias *and* produce independence where needed in a design. With paired data, randomness can arise in many ways. The pairs may be a random sample. In an experiment, the order of the two treatments may be randomly assigned, or the treatments may be randomly assigned to one member of each pair. In a before-and-after study, we may believe that the observed differences are a representative sample from a population of interest. If we have any doubts, we'll need to include a control group to be able to draw conclusions. What we want to know usually focuses our attention on where the randomness should be. In our example, skaters were assigned to the lanes at random.

If the study involves sampling, and your sample is bigger than 10% of the population, which is rare, be sure to acknowledge that your standard error and P-value are in fact conservative (inflated) estimates.

Normal Population Assumption

We need to assume that the population of *differences* follows a Normal model. We don't need to check the individual groups.

Nearly Normal Condition: This condition can be checked with a histogram or Normal probability plot of the *differences*—but not of the individual groups. As with the one-sample *t*-methods, *robustness* against departures from Normality increases with the sample size; in other words, the Normality assumption matters less the more pairs we have to consider. You may be pleasantly surprised when you check this condition. Even if your original measurements are skewed or bimodal, the *differences* may be nearly Normal. After all, the individual who was way out in the tail on an initial measurement is likely to still be out there on the second one, giving a perfectly ordinary difference.

For Example CHECKING ASSUMPTIONS AND CONDITIONS

RECAP: Field workers for a health department compared driving mileage on a five-day work schedule with mileage on a new four-day schedule. To see if the new schedule changes the amount of driving they do, we'll look at paired differences in mileages before and after.

Name	5-Day Mileage	4-Day Mileage	Difference
Jeff	2798	2914	−116
Betty	7724	6112	1612
Roger	7505	6177	1328
Tom	838	1102	−264
Aimee	4592	3281	1311
Greg	8107	4997	3110
Larry G.	1228	1695	−467
Tad	8718	6606	2112
Larry M.	1097	1063	34
Leslie	8089	6392	1697
Lee	3807	3362	445

QUESTION: Is it okay to use these data to test whether the new schedule changes the amount of driving?

ANSWER:

✓ **Paired Data Condition:** The data are paired because each value is the mileage driven by the same person before and after a change in work schedule.

✓ **Independence/Randomization:** The driving behaviour of any individual worker is independent of the others, so the differences are mutually independent. And the results may be reasonably representative of what might happen for other field workers.

✓ **Nearly Normal Condition:** The histogram of the mileage differences is reasonably unimodal and symmetric:

Since the assumptions and conditions are satisfied, it's okay to use paired-t methods for these data.

The steps of testing a hypothesis for paired differences are very much like the steps for a one-sample t-test for a mean. After all, it's just the one-sample t-test applied to a single sample of *differences*.

THE PAIRED t-TEST

When the assumptions and conditions are met, we are ready to test whether the mean of paired differences is significantly different from Δ_0. We test the hypothesis

$$H_0: \mu_d = \Delta_0$$

where the d's are the pairwise differences and Δ_0 is almost always 0.

We use the statistic

$$t = \frac{\bar{d} - \Delta_0}{SE(\bar{d})}$$

where $\bar{d}$ is the mean of the pairwise differences and

$$SE(\bar{d}) = \frac{s_d}{\sqrt{n}}$$

is the ordinary standard error of the mean, of the differences, with $s_d =$ the standard deviation of the differences, and $n =$ the number of pairs.

When the conditions are met and the null hypothesis is true, we can model the sampling distribution of this statistic with a Student's t-model with $n - 1$ degrees of freedom, and use that model to obtain a P-value.

Step-by-Step Example A PAIRED *t*-TEST

Question: Was there a difference in speeds between the inner and outer speed-skating lanes at the 2006 Winter Olympics?

THINK ➡ **Plan** State what we want to know.

Identify the *parameter* we wish to estimate. Here our parameter is the mean difference in race times.

Identify the variables and check the *W*'s.

Hypotheses State the null and alternative hypotheses.

Although fans suspected one lane was faster, we can't use the data we have to specify the direction of a test. We (and Olympic officials) would be interested in a difference in either direction, so we'd better test a two-sided alternative.

Model Think about the assumptions and check the conditions.

State why you think the data are paired. Simply having the same number of individuals in each group and displaying them in side-by-side columns doesn't make them paired.

Think about what we hope to learn and where the randomization comes from.

Make a picture—just one. There is no need to plot separate distributions of the two groups[3]— that entirely misses the pairing. For paired data, it's the Normality of the *differences* that we care about. Treat those paired differences as you would a single variable, and check the Nearly Normal Condition with a histogram or a Normal probability plot.

Specify the sampling distribution model.
Choose the method.

I want to know whether there really was a difference in the speeds of the two lanes for speed-skating at the 2006 Olympics. I have data for 17 pairs of racers at the women's 1500-m race.

H_0: Neither lane offered an advantage:

$$\mu_d = 0$$

H_A: The mean difference is different from zero:

$$\mu_d \neq 0$$

✓ **Paired Data Condition**: The data are paired because racers compete in pairs.

✓ **Independence/Randomization**: Lane assignments are random, and each race should be independent of the others, so the differences are mutually independent and free of bias.

✓ **Nearly Normal Condition**: The histogram of the differences is unimodal and fairly symmetric:

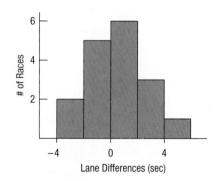

The conditions are met, so I'll use a Student's *t*-model with $(n - 1)$ degrees of freedom, and perform a **paired t-test**.

SHOW ➡ **Mechanics** *n* is the number of *pairs*—in this case, the number of races.

$\overline{d}$ is the mean difference.

s_d is the standard deviation of the differences.

The data give

$$n = 17 \ pairs$$

$$\overline{d} = 0.499 \ seconds$$

$$s_d = 2.333 \ seconds$$

---

[3]Though often done to provide a more complete visual impression of both the absolute and *relative* effect size.

Find the standard error and the *t*-score of the observed mean difference. There is nothing new in the mechanics of the paired-*t* methods. These are the mechanics of the *t*-test for a mean applied to the differences.	I estimate the standard deviation of $\bar{d}$ using $$SE(\bar{d}) = \frac{s_d}{\sqrt{n}} = \frac{2.333}{\sqrt{17}} = 0.5658$$ So $\quad t = \dfrac{\bar{d} - 0}{SE(\bar{d})} = \dfrac{0.499}{0.5658} = 0.882$
Make a picture. Sketch a *t*-model for the null hypothesis. Because this is a two-tail test, shade both the region to the right of the calculated *t* value and the corresponding region in the lower tail.	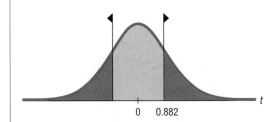
Find the P-value, using technology, or estimate it from a table.	$P\text{-value} = 2P(t_{16} > 0.882) = 0.39$
REALITY CHECK The mean difference is 0.499 seconds. That may not seem like much, but a smaller difference determined the silver and bronze medals. The standard error is about this big, so a *t*-value less than 1.0 isn't surprising. Nor is a large P-value.	

 Conclusion Link the P-value to your decision about H_0, and state your conclusion in context.

The P-value is large. Events that happen more than a third of the time are not remarkable. So, even though there is an observed difference between the lanes, I can't conclude that it isn't due simply to random chance. The fans may have interpreted a random fluctuation in the data as one lane's advantage. There's insufficient evidence to declare any lack of fairness.

For Example DOING A PAIRED *t*-TEST

RECAP: We want to test whether a change from a five-day work week to a four-day work week could change the amount driven by field workers of a health department. We've already confirmed that the assumptions and conditions for a paired *t*-test are met.

QUESTION: Is there evidence that a four-day work week would change how many miles workers drive?

ANSWERS: H_0: The change in the health department workers' schedules didn't change the mean mileage driven; the mean difference is zero:

$$\mu_d = 0$$

H_A: The mean difference is different from zero:

$$\mu_d \neq 0$$

The conditions are met, so I'll use a Student's *t*-model with $(n - 1) = 10$ degrees of freedom and perform a paired *t*-test.

The data give

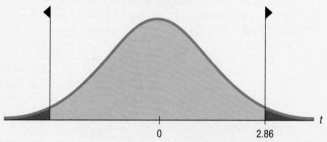

$$n = 11 \text{ pairs}$$
$$\bar{d} = 982 \text{ miles}$$
$$s_d = 1139.6 \text{ miles}$$
$$SE(\bar{d}) = \frac{s_d}{\sqrt{n}} = \frac{1139.6}{\sqrt{11}} = 343.6$$
$$t = \frac{\bar{d} - 0}{SE(\bar{d})} = \frac{982.0}{343.6} = 2.86$$

P-value $= 2P(t_{10} > 2.86) = 0.017$

The P-value is small, so I reject the null hypothesis and conclude that the change in work week did lead to a change in average driving mileage.

It appears that changing the work schedule may reduce the mileage driven by workers, but it's hard to tell from the hypothesis test whether the reduction matters. Is the difference in mileage important in the sense of reducing air pollution or costs, or is it merely statistically significant? To help make that decision, we should look at a confidence interval. If the difference in mileage proves to be large in a practical sense, then we might recommend a change in schedule for the rest of the department.

Section 20.2 Exercises

T **3. Cars and trucks** We have data on the city and highway fuel efficiency of 581 cars and 581 trucks.

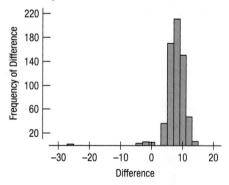

a) Would it be appropriate to use paired t methods to compare the cars and the trucks?
b) Would it be appropriate to use paired t methods to compare the city and highway fuel efficiencies of these vehicles?
c) A histogram of the differences (highway − city) is shown. Are the conditions for inference satisfied?

T **4. Weighing trucks** One kind of scale for weighing trucks can measure their weight as they drive across a plate. Is this method consistent with the traditional method of static weighing? Are the conditions for matched pairs inference satisfied? Weights are in 1000s of pounds.

Weight-in-Motion	Static Weight	Diff (Static − Motion)
26.0	27.9	−1.9
29.9	29.1	0.8
39.5	38.0	1.5
25.1	27.0	−1.9
31.6	30.3	1.3
36.2	34.5	1.7
25.1	27.8	−2.7
31.0	29.6	1.4
35.6	33.1	2.5
40.2	35.5	4.7

20.3 Paired *t* Confidence Interval

Who	170 randomly sampled couples
What	Ages
Units	Years
When	Recently
Where	Britain

In developed countries, the average age of women is generally higher than that of men. After all, women tend to live longer. But if we look at *married couples*, husbands tend to be slightly older than wives. How much older, on average, are husbands? We have data from a random sample of 200 British couples, the first 7 of which are shown below. Only 170 couples provided ages for both husband and wife, so we can work only with that many pairs. Let's form a confidence interval for the mean difference of husband's and wife's age for these 170 couples. Here are the first seven pairs:

Wife's Age	Husband's Age	Difference (husband − wife)
43	49	6
28	25	−3
30	40	10
57	52	−05
52	58	6
27	32	5
52	43	−9
⋮	⋮	⋮

Clearly, these data are paired. The survey selected *couples* at random, not individuals. We're interested in the mean age difference within couples. How would we construct a confidence interval for the true mean difference in ages?

> **PAIRED *t*-INTERVAL**
>
> When the assumptions and conditions are met, we are ready to find the confidence interval for the mean of the paired differences. The confidence interval is
>
> $$\bar{d} \pm t^*_{n-1} \times SE(\bar{d})$$
>
> where the standard error of the mean difference is $SE(\bar{d}) = \dfrac{s_d}{\sqrt{n}}$.
>
> The critical value t^* from the Student's *t*-model depends on the particular confidence level, *C*, that you specify and on the degrees of freedom, $n-1$, which is based on the number of pairs, *n*.

Making confidence intervals for matched pairs follows exactly the steps for a one-sample *t*-interval.

Step-by-Step Example A PAIRED *t*-INTERVAL

Question: How big a difference is there, on average, between the ages of husbands and wives?

THINK ➡ **Plan** State what we want to know. Identify the variables and check the *W*'s. Identify the parameter you wish to estimate. For a paired analysis, the parameter of interest is the mean of the differences in the entire population.	I want to estimate the mean difference in age between husbands and wives. I have a random sample of 200 British husbands and wives, 170 of whom provided both ages.

Model Think about the assumptions and check the conditions.

✓ **Paired Data Assumption**: The data are paired because they are on husbands and wives.

✓ **Independence Assumption**: The data are from a randomized survey, so couples should be independent of each other.

Make a picture. We focus on the differences, so a histogram or Normal probability plot is best here.

✓ **Nearly Normal Condition**: The histogram of the husband–wife differences is fairly unimodal and symmetric:

REALITY CHECK The histogram shows husbands are often older than wives (because most of the differences are greater than 0). The mean difference seen here of around 2 years is reasonable.

State the sampling distribution model. Choose your method.

The conditions are met, so I can use a Student's t-model with $(n-1) = 169$ degrees of freedom and find a **paired t-interval**.

SHOW ➡ **Mechanics** n is the number of *pairs*, here the number of couples.

$\overline{d}$ is the mean difference.

s_d is the standard deviation of the differences.

Be sure to include the units along with the statistics.

The critical value we need to make a 95% interval comes from a Student's t table, a computer program, or a calculator.

$n = 170$ *couples*

$\overline{d} = 2.2$ *years*

$s_d = 4.1$ *years*

I calculate the standard error of $\overline{d}$ as

$$SE(\overline{d}) = \frac{s_d}{\sqrt{n}} = \frac{4.1}{\sqrt{170}} = 0.31 \text{ years}.$$

The df for the t-model is $n - 1 = 169$.

The 95% critical value for t_{169} (from the table) is conservatively estimated to be 1.977.

The margin of error is

$$ME = t^*_{169} \times SE(\overline{d}) = 1.977(0.31) = 0.61$$

REALITY CHECK This result makes sense. Our everyday experience confirms that an average age difference of about 2 years is reasonable.

So the 95% confidence interval is

$$2.2 \pm 0.6 \text{ years,}$$

or an interval of (1.6, 2.8) years.

TELL ➡ **Conclusion** Interpret the confidence interval in context.

I am 95% confident that British husbands are, on average, 1.6 to 2.8 years older than their wives.

Section 20.3 Exercises

T **5. Cars and trucks again** In Exercise 3, after deleting an outlying value of -27, the mean difference in fuel efficiencies for the 1162 vehicles was 7.55 mpg with a standard deviation of 2.53 mpg. Find a 95% confidence interval for this difference and interpret it in context.

T **6. Weighing trucks II** Find a 98% confidence interval of the weight differences in Exercise 4. Interpret this interval in context.

20.4 Effect Size and Sample Size

When we examined the speed-skating times, we failed to reject the null hypothesis, so we couldn't be certain whether there really was a difference between the lanes. Maybe there wasn't any difference, or maybe whatever difference there might have been was just too small to be detected without a bigger sample. Were the fans right to be concerned?

We can't tell from the hypothesis test, but using the same summary statistics, we can find that the corresponding 95% confidence interval for the mean difference is $(-0.70 < \mu_d < 1.70)$ seconds.

A confidence interval is a good way to get a sense for the size of the effect we're trying to understand. That gives us a plausible range of values for the true mean difference in lane times. If differences of 1.7 seconds were too small to matter in 1500-m Olympic speed skating, we'd be pretty sure there was no need for concern.

But in fact, except for the gold–silver gap, the successive gaps between each skater and the next-faster one were *all* less than the high end of this interval, and most were right around the middle of the interval:

So even though we were unable to discern a real difference, the confidence interval shows that the effects we're considering *just may be* big enough to be important. We may want to continue this investigation by checking out other races on this ice and being alert for possible differences at other venues (effectively increasing the sample size).

Effect size issues should make you think about *sample size* issues. Here, the sample size may have been in some sense fixed, but since it is possible to continue this investigation by monitoring other races on this ice, we can think about the usual questions:

■ How small a margin of error or confidence interval should we plan for in order to have sufficiently precise and useful information for decision-making purposes?

Or thinking in terms of hypothesis testing, we might ask:

■ How big an effect size (difference) is sufficiently important so that its discovery should be assured with high probability by our test?

We can then determine the sample size accordingly. And since this is just a one-sample *t*-procedure (based on differences), the issues are exactly the same as when we discussed sample size determination in Chapter 18. And we attack it the same way.

If we wanted the margin of error to equal, say, 0.50 seconds, with 95% confidence, we would solve the following equation for *n* (using the sample standard deviation of 2.333 from this study as our estimated standard deviation; otherwise, make an educated guess):

$$0.50 = \frac{1.96(2.333)}{\sqrt{n}}$$

Solving for *n* gives us $n = 83.6$, or 84 paired skater races that should be observed to attain the desired margin of error of one-half second.

On the other hand, suppose we decided that a difference of one second was large enough to make an important practical difference to the racers. In that case, we might decide to set the power of a test of hypothesis equal to some high probability, such as 0.90, when the true difference equals one second. This would give us a good chance of detecting a problem of this magnitude (and correctly rejecting the null hypothesis). Using Minitab software, we see below that 60 races would need to be observed to attain the desired power when the *true difference* is 1.0 seconds. But if a half-second difference was considered too important to miss, the required sample size, shown below, jumps to 231.

Power and Sample Size

1-Sample *t*-Test
Testing mean = null (versus not = null)
Calculating power for mean = null + difference
Alpha = 0.05 Assumed standard deviation = 2.333

Difference	Sample Size	Target Power	Actual Power
1.0	60	0.9	0.904173
0.5	231	0.9	0.900382

For Example LOOKING AT EFFECT SIZE WITH A PAIRED-*t* CONFIDENCE INTERVAL

RECAP: We know that, on average, the switch from a five-day work week to a four-day work week reduced the amount driven by field workers in that health department. However, finding that there is a significant difference doesn't necessarily mean that difference is meaningful or worthwhile. To assess the size of the effect, we need a confidence interval. We already know the assumptions and conditions are met.

QUESTION: By how much, on average, might a change in work week schedule reduce the amount driven by workers?

ANSWER: $\bar{d} = 982 \, mi$ $SE(\bar{d}) = 343.6$ $t_{10}^* = 2.228 \, (for \, 95\%)$

$$ME = t_{10}^* \times SE(\bar{d}) = 2.228(343.6) = 765.54$$

So the 95% confidence interval for μ_d is

$$982 \pm 765.54 \text{ or } (216.46, 1747.54) \text{ fewer miles.}$$

With 95% confidence, I estimate that by switching to a four-day work week, employees would drive on average 216 to 1748 fewer miles per year. With high gas prices, this could save a lot of money.

This is a very wide confidence interval, and we were fortunate to find a significant difference with such a small sample. Planning the sample size (number of workers) more carefully could have narrowed the interval, making it more informative and useful.

Section 20.4 Exercises

7. Bench press A researcher asked subjects to perform several tasks before and after 24 hours of sleep deprivation. One task measured the number of bench presses performed by the subject until muscle failure occurred. She used 1.8 as her estimated standard deviation, and wanted the margin of error in the difference in the number of bench presses before and after 24 hours of sleep deprivation to be 0.3, with 95% confidence. What is the required sample size?

8. Age of couples Section 20.3 examined the difference in the mean age of married couples. The researcher wanted the margin of error in the difference in age to be 0.5 years, with 99% confidence. What sample size is required?

20.5 Blocking

Figure 20.2

This display is interesting but lacks important information. It doesn't do much good to compare all the wives as a group with all the husbands. We care about the paired differences.

Because the sample of British husbands and wives includes both older and younger couples, there's a lot of variation in the ages of the men and in the ages of the women. In fact, that variation is so great that a boxplot of the two groups (Figure 20.2) would show little difference. But that would not be the most appropriate plot. It's the *difference* we care about. Pairing isolates the extra variation and allows us to focus on the individual differences. In Chapter 10, we saw how to design an experiment with blocking to filter out the variability between identifiable groups of subjects, allowing us to better see variability among treatment groups that is attributable to their responses to the treatment. A paired design is an example of blocking.

When we pair, we have roughly half the degrees of freedom of a two-sample test. You may see discussions that suggest that in "choosing" a paired analysis we "give up" these degrees of freedom. This isn't really true, though. If the data are inherently paired, then there never were additional degrees of freedom, and we have no "choice."

Matching pairs generally removes so much extra variation that it more than compensates for having only half the degrees of freedom, so it is usually a good design choice when you plan a study. Of course, matching, on some poorly thought out basis, likely to be unrelated to the response variable, would just cost degrees of freedom without the benefit of reducing the variance.

What's Independent?

The methods of this chapter require the pairs to be independent of each other. They make no assumptions about the individuals—only about the pairs. By contrast, many other inference methods require independence among all the individuals.

You will sometimes see the paired *t* methods of this chapter referred to as methods for "dependent" samples to differentiate them from the independence we required in the previous chapter. But in fact, the only "dependence" in paired data is the pairing itself. That's not a failure of statistical independence; it's a feature (and a beneficial one) of the design for the data collection.

 Just Checking

Think about each of the situations described below.

■   Would you use a two-sample *t* or paired-*t* method (or neither)? Why?
■   Would you perform a hypothesis test or find a confidence interval?

1. Random samples of 50 men and 50 women are asked to imagine buying a birthday present for their best friend. We want to estimate the difference in how much they are willing to spend.

2. Mothers of twins were surveyed and asked how often in the past month strangers had asked whether the twins were identical.

3. Are parents equally strict with boys and girls? In a random sample of families, researchers asked a brother and sister from each family to rate how strict their parents were.

4. Forty-eight overweight subjects are randomly assigned to either aerobic or stretching exercise programs. They are weighed at the beginning and at the end of the experiment to see how much weight they lost.

   a. We want to estimate the mean amount of weight lost by those doing aerobic exercise.

   b. We want to know which program is more effective at reducing weight.

5. Couples at a dance club were separated and each person was asked to rate the band. Do men or women like this band more?

Section 20.5 Exercises

9. **Blocking cars and trucks** Thinking about the data on fuel efficiency in Exercise 3, why is the blocking accomplished by a matched pairs analysis particularly important for a sample that has both cars and trucks?

Ⓣ 10. **Weighing trucks III** Consider the weights from Exercise 4. The side-by-side boxplots below show little difference between the two groups. Should this be sufficient to draw a conclusion about the accuracy of the weigh-in-motion scale?

*20.6 A Non-Parametric Alternative: The Sign Test

Another and perhaps more simple way to test the speed skating null hypothesis would be to ignore the actual numerical values of the differences and for each just ask, "Is the difference positive?" So rather than record the numerical differences in seconds, we could just record a "yes" for a positive difference and a "no" for a negative difference (and we'll ignore any zero differences).

But what is the actual null hypothesis that could be tested from such Yes/No data? Well, 0 seconds would be some sort of centre if roughly equal numbers of differences were positive and negative. Aha! That would then make 0 seconds not the mean, but rather the median difference, and so our null hypothesis would say that the median of the differences is 0 seconds. If this null hypothesis were true, we'd expect the proportion of positive differences to be 50%. On the other hand, if the true median difference were different from 0, we'd expect to have positive differences more than 50% of the time, or negative differences more than 50% of the time.

What we've done is turn the quantitative data about race time differences into a set of yes-or-no values (Bernoulli trials from Chapter 13). And we've turned a question about the median difference into a test of a proportion: 'Is the proportion of positive differences equal to 0.50?'. We already know how to conduct a test of proportions, so this isn't a new situation. (Can you see why we had to throw out the values exactly equal to 0?)

When we test a hypothesized median by counting the number of values above (or below) that value, it's called a **sign test**. The sign test is a *distribution-free method* (or *nonparametric* method), so called because there are no *distributional* assumptions or conditions on the data. Specifically, because we are no longer working with the original quantitative data, we aren't requiring the Nearly Normal Condition.

We already know all we need for the sign test Step-by-Step:

Step-by-Step Example   *A SIGN TEST

THINK ➡ Plan State what we want to know.

I want to know whether there is evidence that the difference in speeds between the inner and outer lanes has a median that differs from 0 seconds.

Identify the parameter of interest. Here, it is the population median of the differences.

Identify the variables and review the W's.

I have differences for 17 skating pairs and have recorded whether or not each one has a positive *sign*.

Hypotheses Write the null and alternative hypotheses.

There is not a great need to plot the data. Medians are resistant to the effects of skewness or outliers.

H_0: The median difference in speeds is 0 seconds. Equivalently, the proportion p, of positive differences is 50%:

$$H_0: p = 0.50.$$

H_A: The median difference is not equal to 0 seconds, or equivalently, the true proportion of positive differences does not equal 0.50, or $p \neq 0.50$.

Model Think about the assumptions and check the conditions. The sign test doesn't require the Nearly Normal Condition.

If the **Success/Failure Condition** fails, we can still calculate a P-value using the Binomial model for the observed count of Successes.

✓ **Independence/Randomization**: Previously checked.

✗ **Success/Failure Condition**: Both $np_0 = 17(0.5) = 8.5$ and $nq_0 = 17(0.5) = 8.5$ are less than 10, showing that I expect fewer than 10 successes or 10 failures. Hence the Normal model for proportions may not be very accurate and should not be used.

Choose your method.

I'll do a **sign test**. This is just a test that the proportion of positive signs $p = 0.5$. I'll use the Binomial model rather than a Normal model to calculate the P-value, to ensure accuracy.

SHOW ➡ Mechanics We use the null model to find the P-value—the probability of observing a proportion as far from the hypothesized proportion as the one we observed, or even farther. To use the Binomial distribution, we'll work with the count instead of the proportion.

Below is a picture of the null distribution for $X =$ number of positive differences. It centres at $E(X) = 17 \times 0.5 = 8.5$

Of the 17 differences, 10 are positive, so the observed count $X = 10$.

Hence, the P-value is the probability of observing a sample count as extreme as 10 (or even more extreme) when the null hypothesis is true. Let $X =$ number of positive differences. 10 is bigger than $E(X) = 17 \times 0.5 = 8.5$ under the null, so we ask how big this outcome is, and double that upper tail probability

$$P = 2 \times P(X \geq 10 \mid p = 0.5)$$

Under the null, X has a Binomial ($n = 17$, $p = 0.5$) distribution, so the *P*-value can be found by computer, calculator, or table. [If the sample had been slightly bigger, we could have used the Normal model to evaluate significance, by standardizing the sample proportion $= 10/17$, and placing its z value on a N(0,1) curve.]

The P-value is $2 \times P(X \geq 10) = 2 \times 0.315 = 0.63.$

TELL ➡	**Conclusion** Link the P-value to your decision, then state your conclusion in the proper context.	The P-value of 0.63 is not small, so I fail to reject the null hypothesis. There is insufficient evidence of a non-zero median for the differences, so, just as with the t test, we have no evidence of an 'inner vs outer' lane effect.

The sign test is simpler than the *t*-test, and it requires fewer assumptions. We need only yes/no data. We still should check for Independence and the Randomization Condition, but we no longer need the Nearly Normal Condition. When the data satisfy all the assumptions and conditions for a *t*-test on the mean, we usually prefer the *t*-test because it is more powerful than the sign test; for the same data, the P-value from the *t*-test would be smaller than the P-value from the sign test. (Recall that the P-value for the *t*-test here was 0.39.) That's because the *t*-test uses the actual quantitative data values, which contain much more information than just knowing the signs of those same values. The more information we use, the more potential for statistical significance.

On the other hand, the sign test works even when the data have outliers or a skewed distribution—problems that can distort the results of the *t*-test and reduce its power. When we have doubts whether the conditions for the *t*-test are satisfied, it's a good idea to perform a sign test.[4]

Let's also try it on the married couples data. Of the 170 couples, there are 119 with the husband older, 32 with the wife older, and 19 in which the husband and wife are the same age. So, we test $p = 0.5$, with a sample proportion of $119/151 = 0.788$. The Success/Failure Condition is easily satisfied. Applying the one-proportion *z*-test gives a *z*-score of 7.08, with a P-value < 0.00001. We can be pretty confident that the median age difference is not 0. (Just for comparison, the *t*-statistic for testing $\mu_d = 0$ is 7.097 with 169 df—almost an identical result.)

The sign test can in fact be used on any data set to test for any hypothesized median, not just for a zero median. Instead of counting the number of positive values, just count the number of data values above the hypothesized median (ignore values exactly equal to it). If each value above the hypothesized median is labelled a *Success*, then the null hypothesis says that $P(\text{Success}) = 0.5$, by definition of the median, so we test the null hypothesis by testing $H_0: p = 0.5$.

Section 20.6 Exercises

11. **Cars and trucks III** Of the 1162 cars and trucks in Exercise 3, 9 of them have better fuel efficiency when driving on city streets, 4 of them have no difference in fuel efficiency between city and highway driving, and the rest have better fuel efficiency when driving on the highway. Use the sign test to test whether the median of the difference in fuel efficiency is 0.

T 12. **Weighing trucks IV** Consider the weights from Exercise 4, and use the sign test to test whether the median of the difference in weight is 0.

WHAT CAN GO WRONG?

■ **Don't use a two-sample *t*-test when you have paired data.** See the What Can Go Wrong? discussion in Chapter 19.

■ **Don't use a paired-*t* method when the samples aren't paired.** Just because two groups have the same number of observations doesn't mean they can be paired, even if they

---

[4]It's probably a good idea to routinely compute both. If they agree, then the inference is clear. If they differ, it may be interesting and important to see why.

are shown side-by-side in a table. We might have 25 men and 25 women in our study, with both groups completely independent of one another. But if they were paired siblings or spouses, the data would likewise be paired. Remember that you cannot *choose* which method to use based on your preferences. If the data are from two independent samples, use two-sample *t* methods. If the data are from an experiment in which observations were paired, you must use a paired method. If the data are from an observational study, you must be able to defend your decision to use matched pairs or independent groups.

■ **Don't forget outliers.** The outliers we care about now are in the differences. A subject who is extraordinary both before and after a treatment may still have a perfectly typical difference. But one outlying difference can completely distort your conclusions. Be sure to plot the differences (even if you also plot all the data).

■ **Don't look for the difference between the means of paired groups with side-by-side boxplots.** The point of the paired analysis is to remove extra variation. The boxplots of each group still contain that variation. Comparing them is often done, but such a graph gives a very incomplete and possibly misleading picture.

What Have We Learned?

Recognize when data are paired or matched.

Know how to construct a confidence interval for the mean difference in paired data.

Know how to perform a hypothesis test about the mean difference (usually with a null of zero representing no difference between the groups).

Know how to plan the sample size to control the margin of error or the power of a test.

Review of Terms

Paired data Data are paired when the observations are collected in pairs or the observations in one group are naturally related to observations in the other. The simplest form of pairing is when we measure each subject twice—often before and after a treatment is applied. More sophisticated forms of pairing in experiments are a form of blocking and arise in other contexts. Pairing in observational and survey data is a form of matching (p. 638).

Paired *t*-test A hypothesis test for the mean of the pairwise differences of two groups. It tests the null hypothesis

$$H_0: \mu_d = \Delta_0$$

where the hypothesized difference is almost always 0, using the statistic

$$t = \frac{\bar{d} - \Delta_0}{SE(\bar{d})}$$

which has a t_{n-1} sampling model under the null hypothesis, where $SE(\bar{d}) = \frac{s_d}{\sqrt{n}}$, and n is the number of pairs (p. 639).

Paired-*t* confidence interval A confidence interval for the mean of the pairwise differences between paired groups found as

$$\bar{d} \pm t_{n-1}^* \times SE(\bar{d}), \text{ where } SE(\bar{d}) = \frac{s_d}{\sqrt{n}} \text{ and } n \text{ is the number of pairs (p. 645).}$$

Sign test A distribution-free (non-parametric) test of a hypothesized median, often used as an alternative to the one-sample or paired *t*-test for the mean (p. 650).

On the Computer PAIRED *t*

Most statistics programs can compute paired-*t* analyses. Some may want you to find the differences yourself and use the one-sample *t* methods. Those that perform the entire procedure will need to know the two variables to compare. The computer, of course, cannot verify that the variables are naturally paired. Most programs will check whether the two variables have the same number of observations, but some stop there, and that can cause trouble. Most programs will automatically omit any pair that is missing a value for either variable (as we did with the British couples). You must look carefully to see whether that has happened.

As we've seen with other inference results, some packages pack a lot of information into a simple table, but you must locate what you want for yourself. Here's a generic example with comments:

Other packages try to be more descriptive. It may be easier to find the results, but you may get less information from the output table.

Computers make it easy to examine the boxplots of the two groups and the histogram of the differences. Some programs offer a scatterplot of the two variables. That can be helpful. In terms of the scatterplot, a paired *t*-test is about whether the points tend to be above or below the 45° line $y = x$. (Note, though, that pairing says nothing about whether the scatterplot should be straight. That doesn't matter for our *t*-methods.)

EXCEL

- In Excel 2003 and earlier, select **Data Analysis** from the **Tools** menu.
- In Excel 2007, select **Data Analysis** from the **Analysis** group on the **Data** tab.
- From the **Data Analysis** menu, choose *t*-test: **paired two-sample for Means**. Fill in the cell ranges for the two groups, the hypothesized difference, and the alpha level.

COMMENTS

Excel expects the two groups to be in separate cell ranges.

Warning: Do not compute this test in Excel without checking for missing values. If there are any missing values (empty cells), Excel will usually give a wrong answer. Excel compacts each list, pushing values up to cover the missing cells, and then checks only that it has the same number of values in each list. The result is mismatched pairs and an entirely wrong analysis.

JMP

- From the **Analyze** menu, select **Matched Pairs**.
- Specify the columns holding the two groups in the **Y Paired Response** dialogue.
- Click **OK**.

For power and sample size calculations, see Chapter 18.

MINITAB

- From the **Stat** menu, choose the **Basic Statistics** submenu. From that menu, choose **Paired t. . .**
- Then, fill in the dialogue.

For power and sample size calculations, see Chapter 18.

COMMENTS

Minitab takes "First sample" minus "Second sample."

R

To test the hypothesis that $\mu_1 = \mu_2$ for paired data against an alternative (default is two-sided), create a vector for the data of each group, such as x and y, and produce the confidence interval (default is 95%):

- t.test(x,y, alternative = c("two sided," "less," "greater," paired =TRUE, conf.level = 0.095)

will produce the *t*-statistic, degrees of freedom, P-value, and confidence interval for a specified alternative.

In package {pwr}, to perform a sample size or power calculation for a matched pairs *t*-test, either recognize this as just a one-sample *t* test (on differences), or use:

- pwr.t.test(n= , d= , sig.level= , power= , type= "paired") returns **one** argument that is not specified in the function.

COMMENTS

t-test is the same function for both the one- and two-sample *t*-test, but with paired=TRUE.

SPSS

- From the **Analyze** menu, choose the **Compare Means** submenu.
- From that, choose the **Paired-Samples *t*-test** command.
- Select pairs of variables to compare, and click the arrow to add them to the selection box.

For power and sample size calculations, you will need the IBM SPSS SamplePower add-on.

COMMENTS

You can compare several pairs of variables at once. Options include the choice to exclude cases missing in any pair from all tests.

STATCRUNCH

To do inference for the mean of paired differences:

- Click on **Stat**.
- Choose **T Stats » Paired**.
- Choose the **Column** for each variable.
- Check **Save differences** so you can look at a histogram to be sure the Nearly Normal condition is satisfied.
- Click on **Next**.
- Indicate **Hypothesis Test**, then enter the hypothesized **Null mean** difference (usually 0), and choose the **Alternative** hypothesis.

OR

- Indicate **Confidence Interval**, and then enter the **Level** of confidence.
- Click on **Calculate**.

For power and sample size calculations, use **Stat>T Stats>One Sample> Power/Sample Size** after storing the differences in one column.

TI-83/84 PLUS

If the data are stored in two lists, say, L1 and L2, create a list of the differences:

L1 − L2 → L3. (The arrow is the STO button.)

- Since inference for paired differences uses one-sample *t*-procedures, select **2:T-Test** or **8:TInterval** from the **STAT TESTS** menu.

- Specify as your data the list of differences you just created in L3, and apply the procedure.

Chapter Exercises

Be alert! Both paired and independent samples designs will appear in these exercises.

T indicates data is available in a saved file at the Companion Website at www.pearsoncanada.ca/deveaux

13. **More eggs?** Can a food additive increase egg production? Agricultural researchers want to design an experiment to find out. They have 100 hens available. They have two kinds of feed: the regular feed and the new feed with the additive. They plan to run their experiment for a month, recording the number of eggs each hen produces.
    a) Design an experiment that will require a two-sample *t*-procedure to analyze the results.
    b) Design an experiment that will require a matched-pairs *t*-procedure to analyze the results.
    c) Which experiment would you consider the stronger design? Why?

14. **MTV** Some students do homework with the television on. (Anyone come to mind?) Some researchers want to see if people can work as effectively with distraction as they can without it. The researchers will time some volunteers to see how long it takes them to complete some relatively easy crossword puzzles. During some of the trials, the room will be quiet; during other trials in the same room, a television will be on, tuned to MTV.
    a) Design an experiment that will require a two-sample *t*-procedure to analyze the results.

    b) Design an experiment that will require a matched-pairs *t*-procedure to analyze the results.
    c) Which experiment would you consider the stronger design? Why?

15. **Sex sells?** Ads for many products use sexual images to try to attract attention to the product. But do these ads bring people's attention to the item that was being advertised? We want to design an experiment to see if the presence of sexual images in an advertisement affects people's ability to remember the product.
    a) Describe an experimental design that would require a matched-pairs *t*-procedure to analyze the results.
    b) Describe an experimental design that would require an independent sample procedure to analyze the results.

16. **Freshman 15** Many people believe that students tend to gain weight during first year at university (allegedly 15 pounds, on average). Suppose we plan to conduct a study to see if this is true.
    a) Describe a study design that would require a matched-pairs *t*-procedure to analyze the results.
    b) Describe a study design that would require a two-sample *t*-procedure to analyze the results.

17. Diabetes and weight loss Researchers at the University Obesity Research Centre in Australia were interested in the effect on Type II diabetes of surgically induced weight loss. One group of 30 obese patients received conventional therapy with a focus on weight loss, while laparoscopic adjustable gastric banding (a surgical intervention) with conventional diabetes care was used for another group of 30 obese patients. One concern was weight loss, as researchers did find a strong association between diabetes remission and weight loss. Weights of all subjects were recorded at the start (baseline) and after two years.

a) What procedure would you use to test if subjects given the surgical treatment weigh less on average after two years as compared with baseline (at entry)? State your null hypothesis. What would be the distribution of the test statistic? The P-value was < 0.001. What can you conclude?

b) What procedure would you use to test if surgical treatment produces greater weight loss, on average, than conventional treatment? State your null hypothesis. What would be the distribution of the test statistic? The P-value was < 0.001. What can you conclude?

T 18. Rain Simpson, Alsen, and Eden (*Technometrics* 1975) reported the results of trials in which clouds were seeded and the amount of rainfall recorded. The authors reported on 26 seeded and 26 unseeded clouds in order of the amount of rainfall, largest amount first. Here are two possible tests to study the question of whether cloud seeding works. Which test is appropriate for these data? Explain your choice. Using the test you select, state your conclusion.

Paired t-Test of $\mu(1-2)$

Mean of Paired Differences $= -277.39615$

t-Statistic $= -3.641$ w/25 df

$P = 0.0012$

2-Sample t-Test of $\mu 1 - \mu 2$

Difference Between Means $= -277.4$

t-Statistic $= -1.998$ w/33 df

$P = 0.0538$

a) Which of these tests is appropriate for these data? Explain.

b) Using the test you selected, state your conclusion.

T 19. Friday the 13th, I The *British Medical Journal* published an article titled, "Is Friday the 13th Bad for Your Health?" Researchers in Britain examined how Friday the 13th affects human behaviour. One question was whether people tend to stay at home more on that day. The data below are the number of cars passing Junctions 9 and 10 on the M25 motorway for consecutive Fridays (the 6th and 13th) for five different periods.

Year	Month	6th	13th
1990	July	134 012	132 908
1991	September	133 732	131 843
1991	December	121 139	118 723
1992	March	124 631	120 249
1992	November	117 584	117 263

Here are summaries of two possible analyses:

Paired t-Test of $\mu(1-2) = 0$ vs. $\mu(1-2) > 0$

Mean of Paired Differences: 2022.4

t-Statistic $= 2.9377$ w/4 df

$P = 0.0212$

2-Sample t-Test of $\mu 1 = \mu 2$ vs. $\mu 1 > \mu 2$

Difference Between Means: 2022.4

t-Statistic $= 0.4273$ w/7.998 df

$P = 0.3402$

a) Which of the tests is appropriate for these data? Explain.

b) Using the test you selected, state your conclusion.

c) Are the assumptions and conditions for inference met?

Source: Adapted from T J Scanlon, R N Luben, F L Scanlon, N Singleton (1993). Is Friday the 13th Bad for Your Health?, *British Medical Journal*, vol 307, 18–25.

T 20. Friday the 13th, II The researchers in Exercise 19 also examined the number of people admitted to emergency rooms for vehicular accidents on 12 Friday evenings (six each on the 6th and 13th).

Year	Month	6th	13th
1989	October	9	13
1990	July	6	12
1991	September	11	14
1991	December	11	10
1992	March	3	4
1992	November	5	12

Based on these data, is there evidence that more people are admitted, on average, on Friday the 13th? Here are two possible analyses of the data:

Paired t-Test of $\mu(1-2) = 0$ vs. $\mu(1-2) < 0$

Mean of Paired Differences $= 3.333$

t-Statistic $= 2.7116$ w/5 df

$P = 0.0211$

2-Sample t-Test of $\mu 1 = \mu 2$ vs. $\mu 1 < \mu 2$

Difference Between Means $= 3.333$

t-Statistic $= 1.6644$ w/9.940 df

$P = 0.0636$

a) Which of these tests is appropriate for these data? Explain.

b) Using the test you selected, state your conclusion.

c) Are the assumptions and conditions for inference met?

Source: Adapted from T J Scanlon, R N Luben, F L Scanlon, N Singleton (1993). Is Friday the 13th Bad for Your Health?, *British Medical Journal*, vol 307, 18-25.

T 21. Online insurance I After seeing countless commercials claiming people can get cheaper car insurance from an online company, a local insurance agent was concerned that he might lose some customers. To investigate, he randomly selected profiles (type of car, coverage, driving record, etc.) for 10 of his clients and checked online price quotes for their policies. The comparisons are shown in the table. His statistical software produced the following summaries (where *PriceDiff = Local – Online*):

Variable	Count	Mean	StdDev
Local	10	799.200	229.281
Online	10	753.300	256.267
PriceDiff	10	45.9000	175.663

Local	Online	PriceDiff
568	391	177
872	602	270
451	488	−37
1229	903	326
605	677	−72
1021	1270	−249
783	703	80
844	789	55
907	1008	−101
712	702	10

At first, the insurance agent wondered whether there was some kind of mistake in this output. He thought the Pythagorean Theorem of Statistics should work for finding the standard deviation of the price differences—in other words, that

$$SD(Local - Online) = \sqrt{SD^2(Local) + SD^2(Online)}.$$

But when he checked, he found that

$\sqrt{(229.281)^2 + (256.267)^2} = 343.864$, not 175.663 as given by the software. Tell him where his mistake is.

T 22. Windy, part I To select the site for a wind turbine, wind speeds were recorded at several potential sites every six hours for a year. Two sites not far from each other looked good. Each had a mean wind speed high enough to qualify, but we should choose the site with a higher average daily wind speed. Because the sites are near each other and the wind speeds were recorded at the same times, we should view the speeds as paired. Here are the summaries of the speeds (in kilometres per hour):

Variable	Count	Mean	StdDev
site2	1114	11.993	5.771
site4	1114	11.664	5.505
site2 − site4	1114	0.328	4.105

Is there a mistake in this output? Why doesn't the Pythagorean Theorem of Statistics work here? In other words, shouldn't

$$SD(site2 - site4) = \sqrt{SD^2(site2) + SD^2(site4)}?$$

But $\sqrt{(5.771)^2 + (5.505)^2} = 7.976$, not 4.105 as given by the software. Explain why this happened.

T 23. Online insurance II In Exercise 21, we saw summary statistics for 10 drivers' car insurance premiums quoted by a local agent and an online company. Here are displays for each company's quotes and for the difference (*Local – Online*):

a) Which of the summaries would help you decide whether the online company offers cheaper insurance? Why?

b) The standard deviation of *PriceDiff* is quite a bit smaller than the standard deviation of prices quoted by either the local or online companies. Discuss why.

c) Using the information you have, discuss the assumptions and conditions for inference with these data.

T 24. Windy, part II In Exercise 22, we saw summary statistics for wind speeds at two sites near each other, both being considered as locations for a wind turbine. The data, recorded every six hours for a year, showed each of the sites had a mean wind speed high enough to qualify, but how can we tell which site is best? Here are some displays:

a) The boxplots show outliers for each site, yet the histogram shows none. Discuss why.
b) Which of the summaries would you use to select between these sites? Why?
c) Using the information you have, discuss the assumptions and conditions for paired *t* inference for these data. (*Hint:* Think hard about the independence assumption in particular.)

T 25. Online insurance III Exercises 21 and 23 give summaries and displays for car insurance premiums quoted by a local agent and an online company. Test an appropriate hypothesis to see if there is evidence that drivers might save money by switching to the online company.

T 26. Windy, part III Exercises 22 and 24 give summaries and displays for two potential sites for a wind turbine. Test an appropriate hypothesis to see if there is evidence that either of these sites has a higher average wind speed.

T 27. Temperatures The table below gives the average high temperatures in January and July for several European cities. Write a 90% confidence interval for the mean temperature difference between summer and winter in Europe. Be sure to check conditions for inference, and clearly explain what your interval means.

| City | Mean High Temperatures (°C) | |
	Jan.	July
Vienna	1	24
Copenhagen	2	22
Paris	6	24
Berlin	2	23
Athens	12	32
Rome	12	31
Amsterdam	4	21
Madrid	8	31
London	7	23
Edinburgh	6	18
Moscow	−6	24
Belgrade	3	29

T 28. NY Marathon 2013 The table below shows the winning times (in minutes) for men and women in the New York City Marathon between 1978 and 2013. (The race was not run in 2012 because of Superstorm Sandy.) Assuming that performances in the Big Apple resemble performances elsewhere, we can think of these data as a sample of performance in marathon competitions. Create a 90% confidence interval for the mean difference in winning times for male and female marathon competitors (www.nycmarathon.org).

Year	Men	Women	Year	Men	Women
1978	132.2	152.5	1996	129.9	148.3
1979	131.7	147.6	1997	128.2	148.7
1980	129.7	145.7	1998	128.8	145.3
1981	128.2	145.5	1999	129.2	145.1
1982	129.5	147.2	2000	130.2	145.8
1983	129.0	147.0	2001	127.7	144.4
1984	134.9	149.5	2002	128.1	145.9
1985	131.6	148.6	2003	130.5	142.5
1986	131.1	148.1	2004	129.5	143.2
1987	131.0	150.3	2005	129.5	144.7
1988	128.3	148.1	2006	130.0	145.1
1989	128.0	145.5	2007	129.1	143.2
1990	132.7	150.8	2008	128.7	143.9
1991	129.5	147.5	2009	129.3	148.9
1992	129.5	144.7	2010	128.3	148.3
1993	130.1	146.4	2011	125.1	143.3
1994	131.4	147.6	2012	<cancelled>	
1995	131.1	148.1	2013	128.4	140.1

T 29. Push-ups Every year the students at Central Secondary School take a physical fitness test during their gym classes. One component of the test asks them to do as many push-ups as they can. Results for one class are shown below, separately for boys and girls. Assuming that students at Central are assigned to gym classes at random, create a 90% confidence interval for how many more push-ups boys can do than girls, on average, at that school.

Boys	17	27	31	17	25	32	28	23	25	16	11	34
Girls	24	7	14	16	2	15	19	25	10	27	31	8

T 30. Brain waves An experiment was performed to see whether sensory deprivation over an extended period of time has any effect on the alpha-wave patterns produced by the brain. To determine this, 20 subjects, inmates in a Canadian prison, were randomly split into two groups. Members of one group were placed in solitary confinement. Those in the other group were allowed to remain in their own cells. Seven days later, alpha-wave frequencies were measured for all subjects, as shown in the following table.[5]

Nonconfined	Confined
10.7	9.6
10.7	10.4
10.4	9.7
10.9	10.3
10.5	9.2
10.3	9.3
9.6	9.9
11.1	9.5
11.2	9.0
10.4	10.9

a) What are the null and alternative hypotheses? Be sure to define all the terms and symbols you use.
b) Are the assumptions necessary for inference met?
c) Perform the appropriate test, indicating the formula you used, the calculated value of the test statistic, and the P-value.
d) State your conclusion.

T 31. Job satisfaction (When you first read about this exercise break plan in Chapter 19, you did not have an

inference method that would work. Try again now.) A company institutes an exercise break for its workers to see if it will improve job satisfaction, as measured by a questionnaire that assesses workers' satisfaction. Scores for 10 randomly selected workers before and after the implementation of the exercise program are shown in the table.

a) Identify the procedure you would use to assess the effectiveness of the exercise program, and check to see if the conditions allow the use of that procedure.
b) Test an appropriate hypothesis and state your conclusion.
c) If your conclusion turns out to be incorrect, what kind of error would you be making?
*d) Use the sign test to test the appropriate hypothesis. Do your conclusions change from those in part b)?

Worker Number	Job Satisfaction Index	
	Before	After
1	34	33
2	28	36
3	29	50
4	45	41
5	26	37
6	27	41
7	24	39
8	15	21
9	15	20
10	27	37

T 32. Summer school (When you first read about the summer school issue in Chapter 19, you did not have an inference method that would work. Try again now.) Having done poorly on their Math final exams in June, six students repeat the course in summer school and take another exam in August.

June	54	49	68	66	62	62
Aug.	50	65	74	64	68	72

a) If we consider these students to be representative of all students who might attend this summer school in other years, do these results provide evidence that the program is worthwhile?
b) Your conclusion, of course, may be incorrect. If so, which type of error would you be making?

T 33. Yogourt Do these data suggest that there is a significant difference in calories between servings of strawberry and vanilla yogourt? Test an appropriate hypothesis and state your conclusion. Don't forget to check assumptions and conditions!

	Calories per Serving	
Brand	**Strawberry**	**Vanilla**
Astro	210	200
Breyer's Lowfat	220	220
Columbo	220	180
Dannon Light'n Fit	120	120
Dannon Lowfat	210	230
Dannon la Crème	140	140
Great Value	180	80
La Yogurt	170	160
Mountain High	200	170
Stonyfield Farm	100	120
Yoplait Custard	190	190
Yoplait Light	100	100

T 34. Gasoline Many drivers of cars that can run on regular gas actually buy premium in the belief that they will get better gas mileage. To test that belief, we use 10 cars from a company fleet in which all the cars run on regular gas. Each car is filled first with either regular or premium gasoline, decided by a coin toss, and the mileage for that tank is recorded. Then, the mileage is recorded again for the same cars for a tank of the other kind of gasoline. We don't let the drivers know about this experiment.

Here are the results (litres per 100 km):

Car #	1	2	3	4	5	6	7	8	9	10
Regular	14.7	11.8	11.5	10.7	10.2	10.7	8.7	9.4	8.7	8.4
Premium	12.4	10.7	9.8	9.8	9.4	9.4	9.0	9.0	8.4	7.4

a) Is there evidence that cars get significantly better fuel economy with premium gasoline?
b) How big might that difference be? Check a 90% confidence interval.
c) Even if the difference is significant, why might the company choose to stick with regular gasoline?
d) Suppose you had done a "bad thing." (We're sure you didn't.) Suppose you had mistakenly treated these data as two independent samples instead of matched pairs. What would the significance test have found? Carefully explain why the results are so different.
e) *Use the sign test to test the appropriate hypothesis. Do your conclusions change from those in part a)?

T 35. Braking In a test of braking performance, a tire manufacturer measured the stopping distance for one of its tire models. On a test track, a car made repeated stops from a speed of 100 km per hour. The test was run on both dry and wet pavement, with results as shown in the table. (Note that actual *braking distance*, which takes into account the driver's reaction time, is much longer, typically nearly 300 feet, or 91.5 metres, at 100 km/h!)
a) Write a 95% confidence interval for the mean dry pavement stopping distance. Be sure to check the appropriate assumptions and conditions, and explain what your interval means.
b) Write a 95% confidence interval for the mean increase in stopping distance due to wet pavement. Be sure to check the appropriate assumptions and conditions, and explain what your interval means.

Stopping Distance (ft)	
Dry Pavement	**Wet Pavement**
145	211
152	191
141	220
143	207
131	198
148	208
126	206
140	177
135	183
133	223

T 36. Braking, test 2 For another test of the tires in Exercise 35, the company tried them on 10 different cars, recording the stopping distance for each car on both wet and dry pavement. The following table shows the results.

	Stopping Distance (ft)	
Car	**Dry Pavement**	**Wet Pavement**
1	150	201
2	147	220
3	136	192
4	134	146
5	130	182
6	134	173
7	134	202
8	128	180
9	136	192
10	158	206

a) Write a 95% confidence interval for the mean dry pavement stopping distance. Be sure to check the appropriate assumptions and conditions, and explain what your interval means.

b) Write a 95% confidence interval for the mean increase in stopping distance due to wet pavement. Be sure to check the appropriate assumptions and conditions, and explain what your interval means.

T 37. Math and gender 2009 Below are mean PISA (Programme for International Student Assessment) math scores for very large national samples of 15-year-old male and female students in a random sample of the Organisation for Economic Co-operation and Development (OECD) and various other industrialized nations.[6]

Country	Male Average	Female Average
Singapore	565	559
Norway	500	495
Korea	548	544
Switzerland	544	524
Sweden	493	495
Australia	519	509
Brazil	394	379
Argentina	394	383
Liechtenstein	547	523
Trinidad and Tobago	410	418
Italy	490	475
Tunisia	378	366
Dubai (UAE)	454	451
Hungary	496	484
Colombia	398	366
Romania	429	425

a) Create a 90% confidence interval for the mean difference in average math scores between genders. Be sure to justify the procedure that you use.

b) What does your confidence interval say about the claim that there is no difference between genders? About the claim that the mean difference is at least 10 points, with males scoring higher? (Actually, the biggest gender gap was in reading scores, with females scoring higher on average in all OECD countries, and more than 50 points higher in some countries.)

---

[6]Based on data from OECD PISA 2009 online database, http://pisa2009. acer.edu.au, accessed on Feb 22, 2013.

c) Is your confidence interval an estimate of the average difference between male and female scores for students aged 15 in all the participating countries? Or is it an estimate of something else? Explain.

T 38. Sex sells, part II In Exercise 15, you considered the question of whether sexual images in ads affected people's abilities to remember the item being advertised. To investigate, a group of Statistics students cut ads out of magazines. They were careful to find two ads for each of 10 similar items, one with a sexual image and one without. They arranged the ads in random order and had 39 subjects look at them for one minute. Then, they asked the subjects to list as many of the products as they could remember. The following table shows their data. Is there evidence that the sexual images mattered?

Subject Number	Ads Remembered		Subject Number	Ads Remembered	
	Sexual Image	No Sex		Sexual Image	No Sex
1	2	2	21	2	3
2	6	7	22	4	2
3	3	1	23	3	3
4	6	5	24	5	3
5	1	0	25	4	5
6	3	3	26	2	4
7	3	5	27	2	2
8	7	4	28	2	4
9	3	7	29	7	6
10	5	4	30	6	7
11	1	3	31	4	3
12	3	2	32	4	5
13	6	3	33	3	0
14	7	4	34	4	3
15	3	2	35	2	3
16	7	4	36	3	3
17	4	4	37	5	5
18	1	3	38	3	4
19	5	5	39	4	3
20	2	2			

T 39. Strike three Advertisements for an instructional video claim that the techniques will improve the ability of young amateur baseball pitchers to throw strikes and that, after undergoing the training, players will be able to throw strikes on at least 60% of their pitches. To test this claim, we have 20 players throw 50 pitches each, and we record the number of strikes. After the players participate in the training program, we repeat the test. The table below shows the number of strikes each player threw before and after the training.

a) Is there evidence that after training players can throw strikes more than 60% of the time?

b) Is there evidence that the training is effective in improving a player's ability to throw strikes?

*c) Use the sign test to test the appropriate hypothesis. Do your conclusions change from those in part b)?

Number of Strikes (out of 50)		Number of Strikes (out of 50)	
Before	After	Before	After
28	35	33	33
29	36	33	35
30	32	34	32
32	28	34	30
32	30	34	33
32	31	35	34
32	32	36	37
32	34	36	33
32	35	37	35
33	36	37	32

T 40. Freshman 15, revisited In Exercise 16, you thought about how to design a study to see if it's true that students tend to gain weight during their first year at university. Well, Cornell Professor of Nutrition David Levitsky did just that. He recruited students from two large sections of an introductory health course. Although they were volunteers, they appeared to match the other first-year students in terms of demographic variables, such as sex and ethnicity. The students were weighed during the first week of the semester, then again 12 weeks later. Based on Professor Levitsky's data, estimate the mean weight gain in first-year students with a confidence interval. What can you conclude? Is there a serious problem? (Weights are in pounds.)

Subject Number	Initial Weight	Terminal Weight	Subject Number	Initial Weight	Terminal Weight
1	171	168	35	148	150
2	110	111	36	164	165
3	134	136	37	137	138
4	115	119	38	198	201
5	150	155	39	122	124
6	104	106	40	146	146
7	142	148	41	150	151
8	120	124	42	187	192
9	144	148	43	94	96
10	156	154	44	105	105
11	114	114	45	127	130
12	121	123	46	142	144
13	122	126	47	140	143
14	120	115	48	107	107
15	115	118	49	104	105
16	110	113	50	111	112
17	142	146	51	160	162
18	127	127	52	134	134
19	102	105	53	151	151
20	125	125	54	127	130
21	157	158	55	106	108
22	119	126	56	185	188
23	113	114	57	125	128
24	120	128	58	125	126
25	135	139	59	155	158
26	148	150	60	118	120
27	110	112	61	149	150
28	160	163	62	149	149
29	220	224	63	122	121
30	132	133	64	155	158
31	145	147	65	160	161
32	141	141	66	115	119
33	158	160	67	167	170
34	135	134	68	131	131

T 41. Wheelchair marathon 2013 The Boston Marathon has had a wheelchair division since 1977. Who do you think is typically faster, the men's marathon winner on foot or the women's wheelchair marathon winner? Because the conditions differ from year to year, and speeds have improved over the years, it seems best to treat these as paired measurements. Here are summary statistics for the pairwise differences in finishing time (in minutes):

Andy Crawford/DK Images

Summary of wheelchrF — runM
$n = 37$
Mean $= -4.88$
SD $= 35.227$

a) Comment on the assumptions and conditions.
b) Assuming that these times are representative of such races and the differences appeared acceptable for inference, construct and interpret a 95% confidence interval for the mean difference in finishing times.
c) Would a hypothesis test at $\alpha = 0.05$ reject the null hypothesis of no difference? What conclusion would you draw?
(www.boston.com/sports/marathon/history/champions)

T **42. Marathon start-up years 2013** When we considered the Boston Marathon in Exercise 41, we were unable to check the Nearly Normal Condition. Here's a histogram of the differences:

F Wheelchair – M Runners (minutes)

The three largest differences are the first three years of wheelchair competition: 1977, 1978, and 1979. Often the start-up years of new events are different; later on, more athletes train and compete. If we omit those three years, the summary statistics change as follows:

Summary of wheelchrF — runM
$n = 34$
Mean $= -13.40$
SD $= 20.57$

a) Comment on the assumptions and conditions.
b) Assuming that these times are representative of such races, construct and interpret a 95% confidence interval for the mean difference in finishing time.
c) Would a hypothesis test at $\alpha = 0.05$ reject the null hypothesis of no difference? What conclusion would you draw?

43. BST In the U.S., some dairy cows receive injections of BST, a hormone intended to spur greater milk production. After the first injection, a test herd of 60 Ayrshire cows increased their mean daily production from 47 pounds to 61 pounds of milk. The standard deviation of the increases was 5.2 pounds. We want to estimate the mean increase a farmer could expect in his own cows.
a) Check the assumptions and conditions for inference.
b) Write a 95% confidence interval.
c) Explain what your interval means in this context.
d) Given the cost of BST, a farmer believes he cannot afford to use it unless he is sure of attaining at least a 25% increase in milk production. Based on your confidence interval, what advice would you give him?

44. BST II In the experiment about hormone injections in cows described in Exercise 43, a group of 52 Jersey cows increased average milk production from 43 pounds to 52 pounds per day, with a standard deviation of 4.8 pounds. Is this evidence that the hormone may be more effective in one breed than the other? Test an appropriate hypothesis and state your conclusion. Be sure to discuss any assumptions you make.

45. Sample size & margin of error For the chapter example about switching from a five-day to a four-day work week, how big should the sample have been to narrow the width of the reported confidence interval to 500 miles? Use the sample standard deviation given in the example as your standard deviation guess.

46. BST III Review Exercise 43 about BST injections for Ayrshire cows. Approximately how big would the sample have to be in order to narrow the width of the reported confidence interval to about 1.00 pounds?

47. Sample size & power For the chapter example about switching from a five-day to a four-day work week, *use software* to determine how big the sample size should have been in order to attain a power of 0.90 for the test of hypothesis (at the 5% level) when there is an actual annual mean mileage reduction of 500 miles. Use the sample standard deviation given in the example as your estimate of the true standard deviation. Explain in words what the probability of 0.90 represents here.

48. **BST IV** Review Exercise 43 about BST injections for Ayrshire cows. *Use software* to estimate how big the sample should have been if we wanted to be 90% certain of correctly detecting an increase in average milk production when in fact the true mean increase is 2.0 pounds. Assume this to be a one-sided hypothesis test, at the 5% level, and use the sample standard deviation calculated in Exercise 43 as an estimate of the true standard deviation.

Just Checking ANSWERS

1. These are independent groups sampled at random, so use a two-sample t confidence interval to estimate the size of the difference.

2. There is only one sample. Use a one-sample t-interval.

3. A brother and sister from the same family represent a matched pair. The question calls for a paired t-test.

4. **a.** A before-and-after study calls for paired t-methods. To estimate the loss, find a confidence interval for the before–after differences.

   b. The two treatment groups were assigned randomly, so they are independent. Use a two-sample t-test to assess whether the mean weight losses differ.

5. Most likely, couples would discuss the band or even decide to go to the club because they both like a particular band. While we may not know to what degree their responses may be related, the sampling design has us selecting the individuals in linked pairs, so use paired-t.

MathXL **MyLab Statistics**	Go to MathXL at **www.mathxl.com** or MyLab™ Statistics at **www.mystatlab.com**. You can practise exercises for this chapter as often as you want. The guided solutions will help you find answers step by step. You'll find a personalized study plan available to you too!

Comparing Two Proportions

Statistics is not a discipline like physics, chemistry, or biology where we study a subject to solve problems in the same subject. We study statistics with the main aim of solving problems in other disciplines.

—C. R. Rao

Connections

Is the proportion of men who like our new webpage the same as the proportion of women who like it? Do people really feel better about the economy this month compared to last month, or was that increase just the result of sampling variation? In the two previous chapters, we compared means of quantitative measurements. Now, we'll discuss comparing proportions calculated from categorical data, which is very much like inference for one proportion. The standard error is different, but the concepts are essentially the same.

In Chapter 2, we looked at contingency tables and compared various proportions. Now we'll make those comparisons more precise using inferential methods. In the next chapter, we'll extend this discussion of associations between categorical variables even further.

D o men take more risks than women? Psychologists have documented that, in many situations, men choose riskier behaviour than women do. But what is the effect of having a woman by their side? A recent seatbelt observation study in Massachusetts[1] found that, not surprisingly, male drivers wear seatbelts less often than women do. The study also noted that men's belt-wearing jumped more than 16 percentage points when they had a female passenger. Seatbelt use was recorded at 161 locations in Massachusetts, using random-sampling methods developed by the National Highway Traffic Safety Administration (NHTSA). Female drivers wore belts more than 70% of the time, regardless of the sex of their passengers. Of 4208 male drivers with female passengers, 2777 (66%) were belted. But among 2763 male drivers with male passengers only, 1363 (49.3%) wore seatbelts. This was only a random sample, but it suggests there may be a shift in men's risk-taking behaviour when women are present. What would we estimate the true size of that gap to be?

Comparisons between two percentages are very common and can be quite important. We often want to know how two groups differ, whether a treatment is better than a placebo control, or whether this year's results are better than last year's.

21.1 The Standard Deviation of the Difference Between Two Proportions

We know the difference between the two proportions of men wearing seatbelts seen in the *sample*. It's 16.7%. But what's the *true* difference for all men (in Massachusetts anyway)? We know that our estimate probably isn't exactly the same as the overall or true value. To say more, we'll need to estimate its standard deviation.

We know how to estimate the standard deviation of the proportion in each group, but how about their difference? Just as was the case when comparing two sample means, the answer comes from realizing that we have to start with the variances and use the Pythagorean Theorem of Statistics from Chapter 13:

> *The variance of the sum or difference of two independent random variables is the sum of their variances.*

---

[1]Massachusetts Traffic Safety Research Program (June 2007).

Fortunately, proportions observed in independent random samples *are* independent, so we can find the variance of the difference by adding their individual variances.

We will represent the two sample proportions as $\hat{p}_1$ and $\hat{p}_2$, and the two sample sizes as n_1 and n_2.

The standard deviations of the sample proportions are $SD(\hat{p}_1) = \sqrt{\dfrac{p_1 q_1}{n_1}}$ and $SD(\hat{p}_2) = \sqrt{\dfrac{p_2 q_2}{n_2}}$, so the variance of the difference in the proportions is

$$Var(\hat{p}_1 - \hat{p}_2) = \left(\sqrt{\frac{p_1 q_1}{n_1}}\right)^2 + \left(\sqrt{\frac{p_2 q_2}{n_2}}\right)^2 = \frac{p_1 q_1}{n_1} + \frac{p_2 q_2}{n_2}$$

The standard deviation is the square root of that variance:

$$SD(\hat{p}_1 - \hat{p}_2) = \sqrt{\frac{p_1 q_1}{n_1} + \frac{p_2 q_2}{n_2}}$$

But we don't know the true values of p_1 and p_2. When we have the sample proportions in hand from the data, we can use them to estimate the true proportions in the SD formula, giving us the standard error of the difference:

$$SE(\hat{p}_1 - \hat{p}_2) = \sqrt{\frac{\hat{p}_1 \hat{q}_1}{n_1} + \frac{\hat{p}_2 \hat{q}_2}{n_2}}$$

For Example FINDING THE STANDARD ERROR OF A DIFFERENCE IN PROPORTIONS

A recent survey of 886 randomly selected teenagers (aged 12–17) found that more than half of them had online profiles.[2] Some researchers and privacy advocates are concerned about the possible access to personal information about teens in public places on the Internet. There appear to be differences between boys and girls in their online behaviour. Among teens aged 15–17, 57% of the 248 boys had posted profiles, compared to 70% of the 256 girls. Let's start the process of estimating how large the true gender gap might be.

QUESTION: What is the standard error of the difference in sample proportions?

ANSWER: Because the boys and girls were selected at random, it's reasonable to assume their behaviours are independent, so it's okay to use the Pythagorean Theorem of Statistics and add variances:

$$SE(\hat{p}_{boys}) = \sqrt{\frac{0.57 \times 0.43}{248}} = 0.0314$$

$$SE(\hat{p}_{girls}) = \sqrt{\frac{0.70 \times 0.30}{256}} = 0.0286$$

$$SE(\hat{p}_{girls} - \hat{p}_{boys}) = \sqrt{0.0314^2 + 0.0286^2} = 0.0425$$

---

[2]Princeton Survey Research Associates International for the Pew Internet & American Life Project.

Section 21.1 Exercises

1. **Canada** An advocacy organization surveyed 960 Canadians and 1250 Americans and reported that 192 of the Canadians were born outside Canada, whereas 170 Americans were born outside the USA. Find the standard error of the difference in sample proportions.

2. **Non-profits** Do people who work for non-profit organizations differ from those who work at for-profit companies when it comes to personal job satisfaction? Separate random samples were collected by a polling agency to investigate the difference. Data collected from 422 employees at non-profit organizations revealed that 377 of them were "highly satisfied." From the for-profit companies, 431 out 518 employees reported the same level of satisfaction. Find the standard error of the difference in sample proportions.

21.2 Assumptions and Conditions When Comparing Proportions

Before we complete our example, we need to check assumptions and conditions.

Independence Assumptions

Independent Responses Assumption: Within each group, the data should be based on results for independently responding individuals. We can't check that for certain, but we *can* check the following:

Randomization Condition: Ideally, the data in each group should be drawn independently and at random[3] from the target population or generated by a randomized comparative experiment. If not, you need to argue that the cases may be considered reasonably representative of some larger populations to which your conclusions may apply.

The (least important) 10% Condition: If the sample is bigger than 10% of the population, the individual observations will not be approximately independent and our SE formula requires an adjustment—but you may still proceed using our SE formula as long as you note that your P-value and confidence level will be conservative (since the calculated SE will be an overestimate).

Our analysis makes use of an additional and critical Independence Assumption. If it is violated, these methods just won't work.

Independent Groups Assumption: The two groups we're comparing must also be independent of each other. Usually, the independence of the groups from each other is evident from the way the data were collected. (Note this is identical to the independent groups assumption for quantitative data in Chapter 19.)

Why is the independent groups assumption so important? If we compare husbands with their wives, or a group of subjects before and after some treatment, we can't just add the variances (to find the correct SE). Subjects' performance before a treatment might very well be related to their performance after the treatment. So the proportions are not independent and the Pythagorean-style variance formula does not hold. A different method of analysis would be required if the subjects or cases are linked between the two samples in some systematic fashion (a method we won't be discussing in this book, except for the case of quantitative data, back in Chapter 20).

Sample Size Condition

Each of the groups must be big enough. As with individual proportions, we need larger groups to estimate proportions that are near 0% or 100%. We usually check the Success/Failure Condition for each group.

---

[3]Our SE formula assumes simple random samples. It should be a good approximation or a bit conservative for stratified or systematic samples, but could be a severe underestimate for cluster or multistage samples.

Success/Failure Condition: Both groups are big enough that at least 10 successes and at least 10 failures have been observed in each, or will be expected in each (when testing a hypothesis).

For Example CHECKING ASSUMPTIONS AND CONDITIONS

RECAP: Among randomly sampled teens aged 15–17, 57% of the 248 boys had posted online profiles, compared with 70% of the 256 girls.

QUESTION: Can we use these results to make inferences about all 15–17-year-olds?

ANSWER:

- **Randomization Condition:** The sample of boys and the sample of girls were both chosen randomly.
- **10% Condition:** 248 boys and 256 girls are each less than 10% of all teenage boys and girls, so no adjustment to the SE is needed.
- **Independent Groups Assumption:** Because the samples were selected at random, it's reasonable to believe the boys' online behaviours are independent of the girls' online behaviours.
- **Success/Failure Condition:** Among the boys, 248(0.57) = 141 had online profiles and the other 248(0.43) = 107 did not. For the girls, 256(0.70) = 179 successes and 256(0.30) = 77 failures. All counts are at least 10.

Because all the assumptions and conditions are satisfied, it's okay to proceed with inference for the difference in proportions.

(Note that when we find the observed counts of successes and failures, we round off to whole numbers. We're using the reported percentages to recover the actual counts.)

Section 21.2 Exercises

3. **Canada, deux** If the information in Exercise 1 is to be used to make inferences about the proportion of all Canadians and all Americans born in other countries, what conditions must be met before proceeding? Are they met? Explain.

4. **Non-profits, part 2** If the information in Exercise 2 is to be used to make inferences about all people who work at non-profits and for-profit companies, what conditions must be met before proceeding? List them and explain if they are met.

21.3 A Confidence Interval for the Difference Between Two Proportions

WHY NORMAL?

In Chapter 13, we learned that sums and differences of independent Normal random variables also follow a Normal model. That's the reason we use a Normal model for the difference of two independent proportions.

We're almost there. We just need one more fact about proportions. We already know that for large enough samples, each of our proportions has an approximately Normal sampling distribution. The same is true of their difference.

> **THE SAMPLING DISTRIBUTION MODEL FOR A DIFFERENCE BETWEEN TWO INDEPENDENT PROPORTIONS**
>
> Provided that the sampled responses are independent, the samples are independent, and the sample sizes are large enough, the sampling distribution of $\hat{p}_1 - \hat{p}_2$ can be modelled by a Normal model with mean $\mu = p_1 - p_2$ and standard deviation
>
> $$SD(\hat{p}_1 - \hat{p}_2) = \sqrt{\frac{p_1 q_1}{n_1} + \frac{p_2 q_2}{n_2}}.$$

The sampling distribution model and the standard deviation give us all we need to find a margin of error for the difference in proportions—or at least they would if we knew the true proportions, p_1 and p_2. However, we don't know the true values, so we'll work with the observed proportions, $\hat{p}_1$ and $\hat{p}_2$, and use $SE(\hat{p}_1 - \hat{p}_2)$ to estimate the standard deviation. The rest is just like a one-proportion z-interval.

A TWO-PROPORTION Z-INTERVAL

When the conditions are met,[4] we are ready to find the confidence interval for the difference of two proportions, $p_1 - p_2$. The confidence interval is

$$(\hat{p}_1 - \hat{p}_2) \pm z^* \times SE(\hat{p}_1 - \hat{p}_2)$$

where we find the standard error of the difference,

$$SE(\hat{p}_1 - \hat{p}_2) = \sqrt{\frac{\hat{p}_1 \hat{q}_1}{n_1} + \frac{\hat{p}_2 \hat{q}_2}{n_2}}.$$

from the observed proportions.
The critical value z^* depends on the particular confidence level, C, that we specify.

For Example FINDING A TWO-PROPORTION Z-INTERVAL

RECAP: Among randomly sampled teens aged 15–17, 57% of the 248 boys had posted online profiles, compared to 70% of the 256 girls. We calculated the standard error for the difference in sample proportions to be $SE(\hat{p}_{girls} - \hat{p}_{boys}) = 0.0425$ and found that the assumptions and conditions required for inference checked out okay.

QUESTION: What does a confidence interval say about the difference in online behaviour?

ANSWER: A 95% confidence interval for $p_{girls} - p_{boys}$ is

$$(\hat{p}_{girls} - \hat{p}_{boys}) \pm z^* SE(\hat{p}_{girls} - \hat{p}_{boys})$$
$$(0.70 - 0.57) \pm 1.96(0.0425)$$
$$0.13 \pm 0.083$$
$$(4.7\%, 21.3\%)$$

We can be 95% confident that among teens aged 15–17, the proportion of girls who post online profiles is between 4.7 and 21.3 percentage points higher than the proportion of boys who do. It seems clear that teen girls are more likely to post profiles than are boys of the same age (since the confidence interval lies entirely above zero).

Step-by-Step Example A TWO-PROPORTION Z-INTERVAL

Who	6971 male drivers
What	Seatbelt use
Why	Highway safety
When	2007
Where	Massachusetts

Now we are ready to be more precise about the passenger-based gap in male drivers' seatbelt use. We'll estimate the difference with a confidence interval using a **two-proportion z-interval**.

Question: How much difference is there in the proportion of male drivers who wear seatbelts when sitting next to a male passenger and the proportion who wear seatbelts when sitting next to a female passenger?

THINK ➡ **Plan** State what you want to know. Discuss the variables and the *W*'s.

Identify the parameter you wish to estimate. (It really doesn't matter in which direction

I want to estimate the true difference in the population proportion, p_M, of male drivers who wear seatbelts when sitting next to a man, and p_F, the proportion who wear seatbelts when

---

[4]What if the sample size condition is not met? Just as in the case of one proportion, there is a "plus four" adjustment that works surprisingly well. Add one phony success and one phony failure to each of the samples, and apply the confidence interval formula given in this box to your artificially enhanced samples (altering the $\hat{p}_i, \hat{q}_i,$ and n_i values).

we subtract, but it is much easier for people to understand the meaning of the interval when we choose the direction with a positive difference.)

sitting next to a woman. The data are from a random sample of drivers in Massachusetts in 2007, observed according to procedures developed by the NHTSA. The parameter of interest is the difference $p_F - p_M$.

Choose and state a confidence level.

I will find a 95% confidence interval for this parameter.

Model Think about the assumptions and check the conditions.

✓ **Randomization Condition**: The NHTSA methods are more complex than an SRS but result in a suitable random sample allowing us to generalize beyond the sample, and to assume that driver behaviour is independent from car to car. Lacking more knowledge of their actual sampling design, we'll assume the resultant precision is similar to that of an SRS, justifying our SE formula. (However, if cluster sampling was used, our SE estimate may be too small.)

✓ **10% Condition**: The samples include far fewer than 10% of all male drivers accompanied by male or by female passengers. Our SE formula needs no special adjustment.

✓ **Independent Groups Assumption**: It's reasonable to believe that seatbelt use among drivers with male passengers and those with female passengers are independent.

The Success/Failure Condition must hold for each group.

✓ **Success/Failure Condition**: Among male drivers with female passengers, 2777 wore seatbelts and 1431 did not; of those driving with male passengers, 1363 wore seatbelts and 1400 did not. Each group contained far more than 10 successes and 10 failures.

State the sampling distribution model for the statistic.

Choose your method.

Under these conditions, the sampling distribution of the difference between the sample proportions is approximately Normal, so I'll find a **two-proportion z-interval**.

SHOW ➡ **Mechanics** Construct the confidence interval.

I know

$n_F = 4208$, $n_M = 2763$

The observed sample proportions are

As often happens, the key step in finding the confidence interval is estimating the standard

$$\hat{p}_F = \frac{2777}{4208} = 0.660, \hat{p}_M = \frac{1363}{2763} = 0.493$$

deviation of the sampling distribution model of the statistic. Here the statistic is the difference in the proportions of men who wear seatbelts when they had a female passenger and those who had a male passenger. Substitute the data values into the formula.

I'll estimate the SD of the difference with

$$SE(\hat{p}_F - \hat{p}_M) = \sqrt{\frac{\hat{p}_F \hat{q}_F}{n_F} + \frac{\hat{p}_M \hat{q}_M}{n_M}}$$

$$= \sqrt{\frac{(0.660)(0.340)}{4208} + \frac{(0.493)(0.507)}{2763}}$$

$$= 0.012$$

$$ME = z^* \times SE(\hat{p}_F - \hat{p}_M)$$

$$= 1.96(0.012) = 0.024$$

The sampling distribution is Normal, so the critical value for a 95% confidence interval, z^*, is 1.96. The margin of error is the critical value times the SE.

The confidence interval is the *statistic* $\pm$ *ME*.

The observed difference in proportions is $\hat{p}_F - \hat{p}_M = 0.660 - 0.493 = 0.167$, so the 95% confidence interval is

$$0.167 \pm 0.024$$

or 14.3% to 19.1%

TELL ➡ **Conclusion** Interpret your confidence interval in the proper context. (Remember: We're 95% confident that our interval captured the true difference.)

I am 95% confident that in Massachusetts the proportion of male drivers who wear seatbelts when driving next to a female passenger is between 14.3 and 19.1 percentage points higher than the proportion who wear seatbelts when driving next to a male passenger.[5]

This is an interesting result—but be careful not to try to say too much! In Massachusetts, overall seatbelt use is relatively low, so we can't be certain that these results generalize to other regions. And these were two different groups of men, so we can't say that, individually, men are more likely to buckle up when they have a woman passenger. You can probably think of several alternative explanations; we'll suggest just a couple. Perhaps age is a lurking variable: Maybe older men are more likely to wear seatbelts and also more likely to be driving with their wives. Or maybe men who don't wear seatbelts have trouble attracting women!

 Just Checking

A listener-sponsored radio station plans to launch a special appeal for additional contributions from current members. Unsure of the most effective way to contact people, they run an experiment. They randomly select two groups of current members. They send the same request for donations to everyone, but it goes to one group by e-mail and to the other group by regular mail. The station was successful in getting contributions from 26% of the members they e-mailed but only from 15% of

---

[5]And since the interval lies *entirely above zero*, we have some evidence of a passenger gender effect on male driver seatbelt use.

those who received the request by regular mail. A 90% confidence interval estimated the difference in donation rates to be $11\% \pm 7\%$.

1. Interpret the confidence interval in this context.

2. Based on this confidence interval, what conclusion would we reach if we tested the hypothesis that there's no difference in the response rates to the two methods of fundraising? Explain.

Section 21.3 Exercises

5. Canada, trois The information in Exercise 1 was used to create a 95% two-proportion confidence interval for the difference between Canadians and Americans who were born in foreign countries. Interpret this interval with a sentence in context.

95% confidence interval for

$$p_{\text{Canadians}} - p_{\text{Americans}} \text{ is } (3.24\%, 9.56\%)$$

6. Non-profits, part 3 The researchers from Exercise 2 created a 95% two-proportion confidence interval for the difference in those who are "highly satisfied" when comparing people who work at non-profits to people who work at for-profit companies. Interpret the interval with a sentence in context.

95% confidence interval for

$$p_{\text{non-profits}} - p_{\text{for-profits}} = (1.77\%, 10.50\%)$$

7. Canada, encore For the interval given in Exercise 5, explain what "95% confidence" means.

8. Non-profits, again For the interval given in Exercise 6, explain what "95% confidence" means.

21.4 The *z*-Test for a Difference Between Proportions

Prejudice in the Penitentiary?

Who	Women serving time in federal penitentiaries
What	Security risk classification
When	2003
Where	Federal penitentiaries in Canada
Why	To compare the treatment of Indigenous and non-Indigenous women inmates

Correctional Services Canada uses a rating system to classify prison inmates according to security risk level. A higher security risk level classification has consequences such as withdrawal of certain privileges and increased likelihood of serving out a longer sentence.

In a study published in the *Canadian Journal of Criminology and Criminal Justice*, University of Toronto criminologists Anthony Doob and Cheryl Webster contend that Indigenous women are often classified at a higher security risk level compared with other female prisoners. In 2003, they found that 60% (41) of 68 Indigenous women inmates were classified as medium security risk, while 42% (112) of 266 non-Indigenous women inmates were similarly classified, though fewer infractions had actually been committed by the Indigenous women while incarcerated. (31% of the Indigenous women inmates had committed infractions, compared with 53% of the non-Indigenous women inmates.)[6]

How should we interpret the observed 18% difference? Is it indicative of a real difference in general medium risk classification rates between Indigenous and non-Indigenous women—or might this difference be the result of some random variation that naturally occurs from sample to sample (due to numerous chance factors at work in any given year)? After all, if you toss two loonies 100 times each, would you expect to see equal counts of heads in the two samples? I hope not. So we have to rule out random sampling variation before we can conclude that a *real* difference exists.

The question calls for a hypothesis test. The parameter of interest is the true *difference* between the medium security risk classification rates of the two groups.

---

[6]www.magazine.utoronto.ca/leading-edge/prejudice-in-the-penitentiary

What is the appropriate null hypothesis? That's easy: We hypothesize that there is no difference in the proportions. This is such a natural null hypothesis that we rarely consider any other. But instead of writing $H_0: p_1 = p_2$, we hypothesize that the difference in proportions is zero:

$$H_0: p_1 - p_2 = 0$$

We know that the standard error of the difference in proportions is

$$SE(\hat{p}_1 - \hat{p}_2) = \sqrt{\frac{\hat{p}_1\hat{q}_1}{n_1} + \frac{\hat{p}_2\hat{q}_2}{n_2}}$$

and we could just plug in the numbers, but we can do slightly better. There are two proportions in the standard error formula—but look at the null hypothesis. It says that these proportions are equal. To do a hypothesis test, we *assume* that the null hypothesis is true. So there should be just a single value of $\hat{p}$ in the SE formula (and, of course, $\hat{q}$ is just $1 - \hat{p}$).

How would we find this estimate of the allegedly common p for this risk classification example? If the null hypothesis is true, then among all female inmates, the two groups have the same medium-risk proportion. Overall, we saw $41 + 112 = 153$ medium risk inmates out of a total of $68 + 266 = 334$ inmates. The overall proportion of medium-risk inmates was $153/334 = 0.4581$.

Combining the counts like this to get an overall proportion is called **pooling**. Whenever we have data from different sources or different groups that we believe (by some hypothesis or assumption) came from the same underlying population, we can pool them to get better estimates.

When we have counts for each group, we can find the pooled proportion as

$$\hat{p}_{\text{pooled}} = \frac{Success_1 + Success_2}{n_1 + n_2}$$

where $Success_1$ is the number of successes in group 1 and $Success_2$ is the number of successes in group 2. That's the overall proportion of success.

When we have only proportions and not the counts, as in the inmate risk classification example, we have to reconstruct the number of successes by multiplying the sample sizes by the proportions:[7]

$$Success_1 = n_1\hat{p}_1 \text{ and } Success_2 = n_2\hat{p}_2$$

If these calculations don't come out to whole numbers, round them first. There must have been a whole number of successes, after all. (This is the *only* time you should round values in the middle of a calculation.)

We then put this pooled value into the formula, substituting it for *both* sample success proportions in the standard error formula (and subtract it from 1 for the pooled failure proportion):

$$SE_{\text{pooled}}(\hat{p}_1 - \hat{p}_2) = \sqrt{\frac{\hat{p}_{\text{pooled}}\hat{q}_{\text{pooled}}}{n_1} + \frac{\hat{p}_{\text{pooled}}\hat{q}_{\text{pooled}}}{n_2}}$$

$$= \sqrt{\frac{0.4581 \times (1 - 0.4581)}{68} + \frac{0.4581 \times (1 - 0.4581)}{266}}$$

This works out to 0.068. (In the formula above, you may factor out the product of the two proportions if you wish.)

Naturally, we'll reject our null hypothesis if we see a large enough difference in the two proportions. How can we decide whether the observed difference $\hat{p}_1 - \hat{p}_2$ is large? The answer is the same as always: We compare it to its standard deviation.

---

[7]Hence we may also calculate the pooled p using $\hat{p}_{\text{pooled}} = \dfrac{n_1\hat{p}_1 + n_2\hat{p}_2}{n_1 + n_2}$, so our pooled estimate is just a weighted average of the two sample proportions.

Unlike the one-sample test, the null hypothesis here doesn't provide a standard deviation, so we'll use a standard error (here, pooled). Since the sampling distribution is Normal, divide the observed difference by its standard error to get a z-score. Then find the corresponding P-value so you can assess the evidence. We call this procedure a **two-proportion z-test**.

> We didn't pool proportions for the confidence interval, because we weren't assuming the two proportions were equal. Combining them doesn't usually make a big practical difference, but we should take advantage of it for the hypothesis test.

TWO-PROPORTION Z-TEST

The conditions for the two-proportion z-test are the same as for the two-proportion z-interval. We are testing the hypothesis

$$H_0: p_1 - p_2 = 0$$

Because we hypothesize that the proportions are equal, we pool the groups to find

$$\hat{p}_{pooled} = \frac{Success_1 + Success_2}{n_1 + n_2}$$

and use that pooled value to estimate the standard error:

$$SE_{pooled}(\hat{p}_1 - \hat{p}_2) = \sqrt{\frac{\hat{p}_{pooled}\hat{q}_{pooled}}{n_1} + \frac{\hat{p}_{pooled}\hat{q}_{pooled}}{n_2}}$$

which is often simplified, by factoring out the square root of the product of the two proportions. Now we find the test statistic,

$$z = \frac{(\hat{p}_1 - \hat{p}_2) - 0}{SE_{pooled}(\hat{p}_1 - \hat{p}_2)}$$

When the conditions are met and the null hypothesis is true, this statistic follows the standard Normal model, so we can use that model to obtain a P-value.

Step-by-Step Example A TWO-PROPORTION Z-TEST

Question: Are the medium security classification rates of the Indigenous and non-Indigenous women inmates really different?

THINK ➡ Plan State what you want to know. Discuss the variables and the W's.

I want to know whether medium risk classification rates differ in general for Indigenous and non-Indigenous women inmates. The data are for female federal penitentiary inmates in Canada in 2003.

Hypotheses Did the researchers have a clear *a priori* hypothesis of unfavourable treatment of Indigenous women and set out to prove this? If so, a one-sided alternative could be justified. But the details are not so clear, so we will take the conservative approach of a two-sided alternative, reflecting our uncertainty about what to expect and an interest in detecting any difference in classification practices.

H_0: There is no difference in medium risk classification rates in the two groups:

$$p_{Indigenous} - p_{Other} = 0$$

H_A: The rates are different: $p_{Indigenous} - p_{Other} \neq 0$

Model Think about the assumptions and check the conditions.

✓ **Independence Assumption**: Classifications of different inmates should be pretty much independently determined.

✓ **Randomization Condition**: The inmates were not randomly selected, but we can view them as representative of a larger collection of past, present, and future inmates incarcerated under similar conditions.

✓ **10% Condition**: We can imagine many more incarcerated under similar conditions.

✓ **Independent Groups Assumption**: We have two unrelated groups of women (no linking or pairing of women in one group with women in the other group).

✓ **Success/Failure Condition**: For Indigenous women, the counts were 41 and $68 - 41 = 27$. For non-Indigenous women, the counts were 112 and $266 - 112 = 154$. The observed numbers of both successes and failures are much more than 10 for both groups.[8]

State the null model.

Choose your method.

Because the conditions are satisfied, I'll use a Normal model and perform a **two-proportion z-test**.

SHOW ➡ Mechanics The hypothesis is that the proportions are equal, so pool the sample data.

Use the pooled SE to estimate

$$SD(\hat{p}_{Indigenous} - \hat{p}_{Other})$$

$$n_{Indigenous} = 68, y_{Indigenous} = 41, \hat{p}_{Indigenous} = 0.603$$

$$n_{Other} = 266, y_{Other} = 112, \hat{p}_{Other} = 0.421$$

$$\hat{p}_{pooled} = \frac{y_{Indigenous} + y_{Other}}{n_{Indigenous} + n_{Other}} = \frac{41 + 112}{68 + 266} = 0.4581$$

$$SE_{pooled}(\hat{p}_{Indigenous} - \hat{p}_{Other})$$

$$= \sqrt{\frac{\hat{p}_{pooled}\hat{q}_{pooled}}{n_{Indigenous}} + \frac{\hat{p}_{pooled}\hat{q}_{pooled}}{n_{Other}}}$$

$$= \sqrt{\frac{(0.4581)(0.5419)}{68} + \frac{(0.4581)(0.5419)}{266}}$$

$$\approx 0.0677$$

The observed difference in sample proportions is

$$\hat{p}_{Indigenous} - \hat{p}_{Other} = 0.603 - 0.421 = 0.182$$

Make a picture. Sketch a Normal model centred at the hypothesized difference of 0. Shade the region to the right of the observed difference, and because this is a two-tailed test, also shade the corresponding region in the other tail.

The observed difference in proportions is 0.182. Find its SE and standardize to get a z-score.

Find the P-value using Table Z or technology. Because this is a two-tailed test, we must *double* the probability we find in the upper tail.

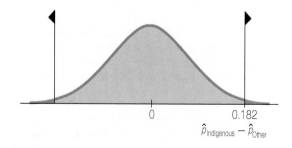

$$z = \frac{(\hat{p}_{Indigenous} - \hat{p}_{Other}) - 0}{SE_{pooled}(\hat{p}_{Indigenous} - \hat{p}_{Other})} = \frac{0.182 - 0}{0.0677} = 2.69$$

$$P = 2P(z \geq 2.69) = 0.0072$$

---

[8]This is one of those situations in which the traditional term "success" seems a bit weird. A success here is an inmate deemed a security risk. "Success" and "Failure" are arbitrary labels left over from studies of gambling games.

TELL ➡ **Conclusion** Link the P-value to your decision about the null hypothesis, and state your conclusion in context.

The P-value of 0.0072 says that if there really were no genuine difference in security risk between the two groups, then the difference observed in this study would happen only 7 times in 1000 by chance. This is so small that I reject the null hypothesis of no difference and conclude that there is a difference in the rate of medium-security risk classification, with Indigenous women more likely to be classified as a medium security risk. It appears that a *prima facie* case exists for the possibility of systemic bias, though other lurking variables may explain this statistically significant difference.

Improving the Success/Failure Condition

The vaccine Gardasil® was introduced to prevent the strains of human papillomavirus (HPV) that are responsible for almost all cases of cervical cancer. In randomized placebo-controlled clinical trials,[9] only one case of HPV was diagnosed among 7897 women who had received the vaccine, compared with 91 cases diagnosed among 7899 women who had received a placebo. The one observed HPV case ("success") doesn't appear to meet the at-least-10-successes criterion mentioned earlier for confidence intervals.

However, for the two-proportion z-test, the *proper* success/failure test stems from the null hypothesis and hence uses the *expected* frequencies, which we can find from the pooled proportion. In this case,

$$\hat{p}_{\text{pooled}} = \frac{91 + 1}{7899 + 7897} = 0.0058$$

$$n_1\hat{p}_{\text{pooled}} = 7899(0.0058) = 46$$

$$n_2\hat{p}_{\text{pooled}} = 7897(0.0058) = 46$$

so we can proceed with the hypothesis test.

Often it is easier just to check the observed numbers of successes and failures. If they are both greater than 10, you don't need to look further. But keep in mind that the correct condition actually uses the *expected* frequencies rather than the *observed* ones.

 Just Checking

3. The Canadian Drug Use and Monitoring Survey (CADMUS) 2011 survey estimated that 2.6% of Canadians aged 15–24 had used the drug ecstasy in the previous year, based on its sample of 671 Canadians in this age group. A similar study in 2004 estimated that 4.4% of Canadians in this age group had used ecstasy in the previous year, based on its sample of 2085 Canadians in this age group. To see if the percentage using ecstasy has decreased over time, why can't we just use a one-proportion z-test of H_0: $p = 0.044$ and see what the P-value for $\hat{p} = 0.026$ is?

4. For surveys like this, which has more variability: the percentage of respondents who say they have used ecstasy in either year or the difference in the percentages between the two years?

[9]*Quadrivalent Human Papillomavirus Vaccine: Recommendations of the Advisory Committee on Immunization Practices (ACIP)*, National Center for HIV/AIDS, Viral Hepatitis, STD and TB Prevention [May 2007].

For Example A TWO-PROPORTION Z-TEST

RECAP: Another concern of the study on teens' online profiles was safety and privacy. In the random sample, girls were less likely than boys to say that they are easy to find online from their profiles. Only 19% (62 girls) of 325 teen girls with profiles say that they are easy to find, while 28% (75 boys) of the 268 boys with profiles say the same.

QUESTION: Are these results evidence of a real difference between boys and girls? Perform a two-proportion z-test and discuss what you find.

ANSWER:

$$H_0: p_{boys} - p_{girls} = 0$$

$$H_A: p_{boys} - p_{girls} \neq 0$$

✓ **Randomization Condition:** The sample of boys and the sample of girls were both chosen randomly, ensuring independence of the individual observations and generalizability.

✓ **10% Condition:** 268 boys and 325 girls are each less than 10% of all teenage boys and girls with online profiles (so our simple SE formula will work well).

✓ **Independent Groups Assumption:** Because the samples were selected at random, it's reasonable to believe the boys' perceptions are independent of the girls'.

✓ **Success/Failure Condition:** Among the girls, there were 62 "successes" and 263 failures, and among boys, 75 successes and 193 failures. These counts are at least 10 for each group.

Because all the assumptions and conditions are satisfied, it's okay to do a **two-proportion z-test:**

$$\hat{p}_{pooled} = \frac{75 + 62}{268 + 325} = 0.231$$

$$SE_{pooled}(\hat{p}_{boys} - \hat{p}_{girls}) = \sqrt{\frac{0.231 \times 0.769}{268} + \frac{0.231 \times 0.769}{325}} = 0.0348$$

$$z = \frac{(0.28 - 0.19) - 0}{0.0348} = 2.59$$

$$P(z > 2.59) = 0.0048$$

This is a two-tailed test, so the P-value = 2(0.0048) = 0.0096. Because this P-value is very small, I reject the null hypothesis. This study provides strong evidence that there really is a difference in the proportions of teen girls and boys who say they are easy to find online.[10]

Sample Size Determination

When designing a study, we need to carefully plan the sample size, or we may end up with not very useful results, lacking either in precision (with a big margin of error) or in power (with little ability to detect important differences).

To control the margin of error, follow the basic approach discussed in earlier chapters: Set the formula for the margin of error equal to the desired margin of error and solve for the sample size after guessing at unknowns in the formula. Here you will have two proportions to guess, or you can go with the conservative approach and plug in 0.5 for any unknown proportion.

Suppose that in the example of teen online profile privacy, we want the width of the confidence interval to be no more than 6 percentage points, or, equivalently, we want a margin of error equal to 0.03 or less, with 95% confidence. How large would the samples

---

[10]How big is the difference? A confidence interval would be a good way to wrap things up.

have to be? Assume equal sample sizes and denote the common sample size by n. Using the worst case scenario with all proportions equal to 0.5, we set:

$$\textit{\textbf{Desired margin of error}} = 0.03 = 2\sqrt{\frac{p_1 q_1}{n_1} + \frac{p_2 q_2}{n_2}} = 2\sqrt{\frac{(.5)(.5)}{n} + \frac{(.5)(.5)}{n}}$$

Then solve for n and we get $n = 2222$ (2222 girls and 2222 boys).

To control the power of the test, we recommend the use of software. Unfortunately, this calculation depends not just on the difference between the two proportions, but also on the specific values of the two proportions, so we have to do a bit of guesswork. You have to make a guess at one proportion,[11] then think about how big a difference from that proportion would represent a clinically or practically important change or difference in order to get the second proportion.

Consider the teen online profile privacy example again. How big should the samples be if we consider a difference of 5 percentage points to be important and we want a 90% chance of discovering a difference of this magnitude when it actually exists?

Using your software, specify all of the following:

■ *The nature of the alternative (one-sided or two-sided).* Choose a two-sided alternative here.
■ *The alpha level of the test.* Let's go with $\alpha = 0.05$.
■ *The values of both proportions.* This is the hard part. Let's guess that about a quarter of teens have easy-to-find profiles (by looking at the actual study data here, but if no data is available yet, make a guess or use 50%, to be conservative). Set one proportion at 0.25 and the other 5 percentage points lower, at 0.20.
■ *The desired power of the test.* Let's target the power at 0.9.

We see from the (Minitab) computer output below that a sample of 1464 boys and 1464 girls would be required to achieve the desired power for a 5% difference in the true proportions.

POWER AND SAMPLE SIZE

Test for Two Proportions

Testing proportion 1 = proportion 2 (versus not =)
Calculating power for proportion 2 = 0.25
Alpha = 0.05

Proportion 1	Sample Size	Target Power	Actual Power
0.2	1464	0.9	0.900057

The sample size is for each group.

Section 21.4 Exercises

9. Canada, la fin The researchers from Exercise 1 want to test if the proportions of foreign born are the same in the United States as in Canada.
a) What is the difference in the proportions of foreign born residents from both countries?
b) What is the pooled proportion of foreign born in both countries combined?
c) What is the standard error of the difference in part a)?
d) What is the value of the z-statistic?
e) What do you conclude at $\alpha = 0.05$?

---

[11]If one treatment is a control group, a standard treatment, or a "before" treatment, making a reasonable guess may not be so hard. Using a value of 0.5 would produce the most conservative calculation, just as with margin of error calculations.

10. Non-profits, last one The researchers from Exercise 2 want to test if the proportions of satisfied employees are the same at for-profit companies as at not-for-profit companies.
a) What is the difference in the proportions of the two types of companies?
b) What is the pooled proportion of satisfied employees in both types of companies combined?
c) What is the standard error of the difference in part a)?
d) What is the value of the z-statistic?
e) What do you conclude at $\alpha = 0.05$?

WHAT CAN GO WRONG?

- **Don't use two-sample proportion methods when the samples aren't independent.** These methods give wrong answers when this assumption of independence is violated. Make sure there is no relationship between the two groups. For example, you can't compare the proportion in your sample who favour the Liberals with the proportion *in the same sample* who favour the Conservatives! The responses are not independent because you've asked the same people!

  Randomized block designs, where we pair subjects before administering treatments (Chapter 10), arise often and are important, but they can't be analyzed with the methods of this chapter (Look up McNemar's test). The pairing of subjects creates a dependence between observations in the two treatment groups.

- **Interpret inference methods with caution where there was no randomization.** If the data do not come from random samples or from a properly randomized experiment, then inference about the differences in proportions must be interpreted cautiously. Statistical significance means that something more than chance is at work. A genuine difference has been demonstrated, but in just what populations?

- **Don't interpret a significant difference in proportions causally unless the study is a randomized experiment.** In an example in this chapter, we discovered that Indigenous women in prison are more likely to be classified as high risk than non-Indigenous women. The data, though, are based on an observational study, so the observed difference here is only suggestive of possible systemic bias based on race; there may be other differences between Indigenous women inmates and other female inmates affecting risk classification. In an observational study, there is always the danger that other lurking variables not accounted for are the real reason for an observed difference. They may be difficult to guess, but that doesn't mean they don't exist, or can't be found with further careful investigation. Be careful not to jump to conclusions about causality based on a significant association.

What Have We Learned?

Know how to construct and interpret confidence intervals for the difference between proportions of two independent groups.

- Check the Assumptions and Conditions before making inferences.

Be able to perform and interpret a two-sample z-test of whether proportions in two independent groups are equal.

- Because we are hypothesizing that the proportions are equal, we pool the counts to find a *pooled standard error* estimate of the standard deviation of the difference.

Keep in mind that we require both independence of responses within each group and independence between the groups themselves.

- The methods discussed in this chapter are not appropriate for paired or matched data.

Review of Terms

Sampling distribution of the difference between two proportions

The sampling distribution $\hat{p}_1 - \hat{p}_2$ is, under appropriate assumptions, modelled by a Normal model with mean $\mu = p_1 - p_2$ and standard deviation $SD(\hat{p}_1 - \hat{p}_2) = \sqrt{\dfrac{p_1 q_1}{n_1} + \dfrac{p_2 q_2}{n_2}}$ (p. 669).

Two-proportion z-interval

Gives a confidence interval for the true difference in proportions, $p_1 - p_2$, based on two independent groups.

The confidence interval is $(\hat{p}_1 - \hat{p}_2) \pm z^* \times SE(\hat{p}_1 - \hat{p}_2)$, where z^* is a critical value from the standard Normal model corresponding to the specified confidence level (p. 670).

Pooling

Data from two populations may sometimes be combined, or *pooled*, to estimate a parameter when the parameter is assumed to be the same in both populations. If we test a null hypothesis that states that two proportions are equal, we can pool the data to provide an estimate of the common proportion, and then use that pooled value in the SE calculation to make it more precise for the test (p. 674).

Two-proportion z-test

Test the null hypothesis $H_0\colon p_1 - p_2 = 0$ by referring the statistic

$$z = \frac{\hat{p}_1 - \hat{p}_2}{SE_{\text{pooled}}(\hat{p}_1 - \hat{p}_2)}$$

to a standard Normal model in order to get the *P*-value (p. 675).

On the Computer TWO SAMPLE METHODS FOR PROPORTIONS

It is so common to test against the null hypothesis of no difference between the two true proportions that most statistics programs simply assume this null hypothesis. And most will automatically use the pooled SE. If you wish to test a different null (say, that the true difference is 0.3), you may have to search for a way to do it.

As with inference for single proportions, most statistics programs want the raw data with the "success" and "failure" status for each case, usually specified by category names like "yes" and "no." Often we just know the proportions of successes, $\hat{p}_1$ and $\hat{p}_2$, and the sample sizes, n_1 and n_2. Computer packages may or may not deal with summary statistics like these.

EXCEL

Inference methods for proportions are not part of the standard Excel tool set, nor are power & sample size calculations.

COMMENTS

For summarized data, type the calculation into any cell and evaluate it.

JMP

For a **categorical variable** that holds category labels, the **Distribution** platform includes tests and intervals of proportions.

- For summarized data, put the category names in one variable and the frequencies in an adjacent variable.

- Designate the frequency column to have the **role** of **frequency**. Then use the **Distribution** platform.

For power & sample size calculations, proceed as follows:
- Choose **Power and Sample Size** from the **DOE** menu.
- Choose **Two Sample Proportions** from the submenu.

- Choose your **Alpha** value and select a one- or two-sided alternative.
- Indicate the null difference between the **Proportions** (likely zero).
- Next, choose two reasonable values for **Proportion 1** and **Proportion 2**, which are separated enough to be of practical importance.
- Fill in either your desired **Sample sizes** or **Power**. What you leave blank will be calculated.
- Click **Continue**.

COMMENTS

JMP uses slightly different methods for proportion inferences than those discussed in this text. Your answers are likely to be slightly different.

MINITAB

To find hypothesis tests and confidence intervals for proportions:

- Choose **Basic Statistics** from the **Stat** menu.
- Choose 2 **Proportions** . . . from the Basic Statistics submenu. If the data are organized with the categorical response variable in one column and group ID in another, assign the variables from the variable list box to the **Samples in one column** box. "Subscripts" are the category names.
- If the data are organized as two separate columns of categorical responses, click on **Samples in different columns:** and assign the variables from the variable list box. If you have summarized data, click the **Summarized Data** button and fill in the number of trials and the number of successes for each group.
- Click the **Options** button and specify the remaining details. Remember to click the **Use pooled estimate of *p* for test** box when testing the null hypothesis of no difference between proportions.
- Click the **OK** button.

To compute the power of the two independent proportions test for a given sample size, or to find the sample size needed to achieve the target power:

- Choose **Power and Sample Size** from the **Stat** menu.
- Choose 2 **Proportions** from the submenu.
- The null hypothesis is zero difference between the proportions. Fill in your guess or estimate for one of the proportions labelled **Proportion 2**.

- Fill in a value for **Proportion 1 value** reflecting a difference from **Proportion 2** that matters.
- Fill in either **Sample size(s)** or **Power value(s)**. The sample size is assumed to be the same for each group.
- Use **Options** . . . if you want a one-sided alternative, or significance level different from 0.05.
- Use **Graph.** . . if you want to produce an entire power curve of power values as a function of **Proportion 1** (Release 16 or later).
- Click **OK**.

COMMENTS

When working from a variable that names categories, MINITAB treats the last category as the "success" category. You can specify how the categories should be ordered.

For power calculations, fill **Proportion 1** and **Proportion 2** to reflect the type of difference you want to detect. Make reasonable guesses. If one group is a control group or standard treatment, estimate **Proportion 2** for that treatment based on prior knowledge, and then think how big a difference would be important to get **Proportion 1**. If the sample size is fixed and you need to know the power, leave **Power value** blank. If you are planning a study and need to know the sample size required to achieve desired power, fill in the **Power value(s)** sought (multiple entries are fine), and leave **Sample size(s)** blank.

R

- Create a **vector** of **successes** = c(**X1,X2**) where X1 is the number of successes in group 1 and X2 is the number of successes in group 2.
- Create a vector of **sample.sizes** = c(**n1,n2**) where n1 and n2 are the sample sizes for the two groups, respectively.

- To test the equality of the two proportions:
  - prop.test(successes,sample.sizes,alternative = "two.sided",conf.level = 0.95,correct = FALSE)

will test the null hypothesis of equal proportions against the two-sided alternative and give a 95% confidence level.

The package {pwr} does various power calculations, but they are not straightforward in the case of proportions, due to a definition of effect size that uses the arcsine transformation (which stabilizes variances of proportions).

COMMENTS

R uses the χ^2 statistic (discussed in the next chapter) instead of the z-statistic for testing. To find the corresponding z-statistic, take the square root of the X-squared (χ^2) statistic in the output. The P-values are the same. Other possibilities for alternative are "greater" and "less."

SPSS

SPSS does not provide hypothesis tests for proportions. However, power and sample size calculations can be performed with the IBM SPSS SamplePower add-on.

STATCRUNCH

To do inference for the difference between two proportions *using summaries*:

- Click on **Stat**.
- Choose **Proportions » Two sample » with summary**.
- Enter the **Number of successes** (x) and **Number of observations** (n) in each group.
- Click on **Next**.
- Indicate **Hypothesis Test**, then enter the hypothesized **Null proportion** difference (usually 0), and choose the **Alternative** hypothesis.

OR

Indicate **Confidence Interval**, and then enter the **Level** of confidence.
- Click on **Calculate**.

To do inference for the difference between two proportions *using raw data*:

- Click on **Stat**.

- Choose **Proportions » Two sample » with data**.
- For each group, choose the variable **Column** listing the **Outcomes**, and enter the **outcome** to be considered a **Success**.
- Click on **Next**.
- Indicate **Hypothesis Test**, then enter the hypothesized **Null** proportion difference (usually 0), and choose the **Alternative** hypothesis.

OR

Indicate **Confidence Interval**, and then enter the **Level** of confidence.
- Click on **Calculate**.

Power calculations may be done using **Stat>Proportion Statistics>Two sample>Power/Sample size**. Click on **Hypothesis Test Power**. Fill in all the boxes except for the one that you want to determine, either **Power** or **Sample Size**

TI-83/84 PLUS

To calculate a confidence interval for the difference between two population proportions:

- Select **B:2-PropZInt** from the **STAT TESTS** menu.
- Enter the observed counts and the sample sizes for both samples.
- Specify a confidence level.
- Calculate the interval.

To do the mechanics of a hypothesis test for equality of population proportions:

- Select **6:2-PropZTest** from the **STAT TESTS** menu.
- Enter the observed counts and sample sizes.

- Indicate what kind of test you want: one-tail upper tail, lower tail, or two-tail.
- Calculate the result.

Power and sample size calculations are not available.

COMMENTS

Beware: When you enter the value of x, you need the count, not the percentage. The count must be a whole number.

Chapter Exercises

11. **Online social networking** In April 2011, the Pew Internet and American Life Project surveyed 799 teenagers aged 12 to 17. Among the study's findings, 93% of teen social media users have a Facebook account, while only 24% have a MySpace account. What does it mean to say that the difference in proportions is "significant"?

12. **Science news** At the end of 2010, the Pew Project for Excellence in Journalism investigated where people are getting their news. In the study, 34% of responders said they read news online as opposed to 31% who favoured newspapers. What does it mean to say that the difference is not significant?

13. **Name recognition** A political candidate runs a week-long series of television ads designed to attract public attention to his campaign. Polls taken before and after the ad campaign show some increase in the proportion of voters who now recognize this candidate's name, with a P-value of 0.033. Is it reasonable to believe the ads may be effective?

14. **Origins** In a 1993 Gallup poll, 47% of U.S. respondents agreed with the statement: "*God created human beings pretty much in their present form at one time within the last 10 000 years or so.*" When Gallup asked the same question in 2001, only 45% of respondents agreed. Is it reasonable to conclude that there was a change in public opinion given that the P-value for the difference is 0.37? Explain.

15. **Revealing information** A random sample of 886 teens were asked which of several personal items of information they thought were okay to share with someone they had just met. Forty-four percent said it was okay to share their e-mail addresses, but only 29% said they would give out their cell phone numbers. A researcher claims that a two-proportion z-test could tell whether there was a real difference among all teens. Explain why that test would not be appropriate for these data.

16. **Regulating access** When a random sample of 935 parents were asked about rules in their homes, 77% said they had rules about the kinds of television shows their children could watch. Among the 790 parents whose teenage children had Internet access, 85% had rules about the kinds of Internet sites their teens could visit. That looks like a difference, but can we tell? Explain why a two-sample z-test would not be appropriate here.

17. **Gender gap** A local member of parliament fears he has a problem with women voters. His campaign staff plans to run a poll to assess the situation. They'll randomly sample 300 men and 300 women, asking if they have a favourable impression of the candidate. Obviously the staff can't know this, but suppose the candidate has a positive image with 59% of males but only with 53% of females.
    a) What kind of sampling design is his staff planning to use?
    b) What difference would you expect the poll to show?
    c) Sampling error means the poll won't reflect the difference perfectly. What is the standard deviation for the difference in the proportions?
    d) Sketch a sampling model for the size difference in proportions of men and women with favourable impressions of the candidate that might appear in a poll like this.
    e) Could the campaign staff be misled by the poll, concluding in error that there really is no gender gap? Explain.

18. **Buy it again?** A consumer magazine plans to poll car owners to see if they are happy enough with their vehicles that they would purchase the same model again. They'll randomly select 450 owners of North American–made cars and 450 owners of Japanese models. Obviously the actual opinions of the entire population can't be known, but suppose 76% of owners of North American cars and 78% of owners of Japanese cars would purchase another.
    a) What kind of sampling design is the magazine planning to use?
    b) What difference would you expect their poll to show?
    c) Sampling error means the poll won't reflect the difference perfectly. What is the standard deviation for the difference in the proportions?
    d) Sketch a sampling model for the difference in proportions that might appear in a poll like this.
    e) Could the magazine be misled by the poll, concluding that owners of North American cars are much happier with their vehicles than are owners of Japanese cars? Explain.

19. **Arthritis** The U.S. Centers for Disease Control and Prevention reported the results of a survey of randomly selected Americans aged 65 and older, which found that 411 of 1012 men and 535 of 1062 women suffered from some form of arthritis.
    a) Are the assumptions and conditions necessary for inference satisfied? Explain.
    b) Create a 95% confidence interval for the difference in the proportions of senior men and women who have this disease.
    c) Interpret your interval in this context.
    d) Does this confidence interval suggest that arthritis is more likely to afflict women than men? Explain.

20. Prostate cancer There has been debate among doctors over whether surgery can prolong life among men suffering from prostate cancer, a type of cancer that typically develops and spreads very slowly. Recently, the *New England Journal of Medicine* published results of some Scandinavian research. Men diagnosed with prostate cancer were randomly assigned to either undergo surgery or not. Among the 347 men who had surgery, 16 eventually died of prostate cancer, compared with 31 of the 348 men who did not have surgery.
a) Was this an experiment or an observational study? Explain.
b) Are the assumptions and conditions necessary for inference satisfied? Explain.
c) Create a 95% confidence interval for the difference in rates of death for the two groups of men.
d) Based on your confidence interval, is there evidence that surgery may be effective in preventing death from prostate cancer? Explain.

21. Pets Researchers at the U.S. National Cancer Institute released the results of a study that investigated the effect of weed-killing herbicides on house pets. They examined 827 dogs from homes where an herbicide was used on a regular basis, diagnosing malignant lymphoma in 473 of them. Of the 130 dogs from homes where no herbicides were used, only 19 were found to have lymphoma.
a) What is the standard error of the difference in the two proportions?
b) Construct a 95% confidence interval for this difference.
c) State an appropriate conclusion.

22. Carpal tunnel The painful wrist condition called carpal tunnel syndrome can be treated with surgery or less invasive wrist splints. Recently, *Time* magazine reported on a study of 176 patients. Among the half that had surgery, 80% showed improvement after three months, but only 48% of those who used the wrist splints improved.
a) What is the standard error of the difference in the two proportions?
b) Construct a 95% confidence interval for this difference.
c) State an appropriate conclusion.

23. Ear infections A new vaccine was recently tested to see if it could prevent the painful and recurrent ear infections that many infants suffer from. *The Lancet*, a medical journal, reported a study in which babies of about a year old were randomly divided into two groups. One group received vaccinations; the other did not. During the following year, only 333 of 2455 vaccinated children had ear infections, compared to 499 of 2452 unvaccinated children in the control group.
a) Are the conditions for inference satisfied?
b) Find a 95% confidence interval for the difference in rates of ear infection.

c) Use your confidence interval to explain whether you think the vaccine is effective.

24. Anorexia The *Journal of the American Medical Association* reported on an experiment intended to see if the drug Prozac® could be used as a treatment for the eating disorder anorexia nervosa. The subjects, women being treated for anorexia, were randomly divided into two groups. Of the 49 who received Prozac®, 35 were deemed healthy a year later, compared to 32 of the 44 who got the placebo.
a) Are the conditions for inference satisfied?
b) Find a 95% confidence interval for the difference in outcomes.
c) Use your confidence interval to explain whether you think Prozac® is effective.

25. Another ear infection In Exercise 23 you used a confidence interval to examine the effectiveness of a vaccine against ear infections in babies. Suppose that instead you had conducted a hypothesis test. (Answer these questions *without* actually doing the test.)
a) What hypotheses would you test?
b) State a conclusion based on your confidence interval.
c) If that conclusion is wrong, which type of error did you make?
d) What would be the consequences of such an error?

26. Anorexia again In Exercise 24 you used a confidence interval to examine the effectiveness of Prozac® in treating anorexia nervosa. Suppose that instead you had conducted a hypothesis test. (Answer these questions *without* actually doing the test.)
a) What hypotheses would you test?
b) State a conclusion based on your confidence interval.
c) If that conclusion is wrong, which type of error did you make?
d) What would be the consequences of such an error?

27. Teen smoking, part I A Vermont study published by the American Academy of Pediatrics examined parental influence on teenagers' decisions to smoke. A group of students who had never smoked were questioned about their parents' attitudes toward smoking. These students were questioned again two years later to see if they had started smoking. The researchers found that, among the 284 students who had indicated that their parents disapproved of kids smoking, 54 had become established smokers. Among the 41 students who had initially said their parents were lenient about smoking, 11 became smokers. Do these data provide strong evidence that parental attitude influences teenagers' decisions about smoking?
a) What kind of design did the researchers use?
b) Write appropriate hypotheses.
c) Are the assumptions and conditions necessary for inference satisfied?
d) Test the hypothesis and state your conclusion.

e) Explain what your P-value means in this context.

f) If it is later found that parental attitudes actually do influence teens decisions to smoke, which type of error did you commit?

28. Depression A study published in the *Archives of General Psychiatry* examined the impact of depression on a patient's ability to survive cardiac disease. Researchers identified 450 people with cardiac disease, evaluated them for depression, and followed the group for four years. Of the 361 patients with no depression, 67 died. Of the 89 patients with minor or major depression, 26 died. Among people who suffer from cardiac disease, are depressed patients more likely to die than nondepressed ones?

a) What kind of design was used to collect these data?

b) Write appropriate hypotheses.

c) Are the assumptions and conditions necessary for inference satisfied?

d) Test the hypothesis and state your conclusion.

e) Explain what your P-value means in this context.

f) If your conclusion is actually incorrect, which type of error did you commit?

29. Teen smoking, part II Consider again the Vermont study discussed in Exercise 27.

a) Create a 95% confidence interval for the difference in the proportion of teens who may smoke and have approving parents and those who may smoke and have disapproving parents.

b) Interpret your interval in this context.

c) Carefully explain what "95% confidence" means.

30. Depression revisited Consider again the study of the association between depression and cardiac disease survivability in Exercise 28.

a) Create a 95% confidence interval for the difference in survival rates.

b) Interpret your interval in this context.

c) Carefully explain what "95% confidence" means.

31. Pregnancy A reproductive clinic reported 42 live births to 157 women under the age of 38, but only 7 live births for 89 clients aged 38 and older. Is this strong evidence of a difference in the effectiveness of the clinic's methods between younger and older women?

a) Was this an experiment? Explain.

b) Test an appropriate hypothesis and state your conclusion in context.

c) If you concluded that there was a difference, estimate that difference with a confidence interval and interpret your interval in context.

32. Birthweight The *Journal of the American Medical Association* reported a study examining the possible impact of air pollution caused by the 9/11 attack on New York's World Trade Center on the weight of babies. Researchers found that 8% of 182 babies born to mothers who were exposed to heavy doses of soot and ash on

September 11 were classified as having low birth weight. Only 4% of 2300 babies born in another New York City hospital whose mothers had not been near the site of the disaster were similarly classified. Does this indicate a possibility that air pollution might be linked to a significantly higher proportion of low-weight babies?

a) Was this an experiment? Explain.

b) Test an appropriate hypothesis, and state your conclusion in context.

c) If you concluded there is a difference, estimate that difference with a confidence interval and interpret that interval in context.

33. Politics and sex One month before an election, a poll of 630 randomly selected voters showed 54% planning to vote for a certain candidate. A week later it became known that the candidate was involved in a scandal, and a new poll showed only 51% of 1010 voters supporting him. Do these results indicate a decrease in voter support for his candidacy?

a) Test an appropriate hypothesis, and state your conclusion.

b) If your conclusion turns out to be wrong, would you be making a Type I or a Type II error?

c) If you concluded there was a difference, estimate that difference with a confidence interval and interpret your interval in context.

34. Mammograms Regular mammogram screening may detect breast cancer early, resulting in fewer deaths from that disease. One study that investigated this issue over a period of 18 years was published during the 1970s. Among 30 565 women who had never had mammograms, 196 died of breast cancer, while only 153 of 30 131 who had undergone screening died of breast cancer.

a) Do these results support that mammograms may be an effective screening tool to reduce breast cancer deaths?

b) If your conclusion is incorrect, what type of error have you committed?

c) If you found a difference, estimate it with a confidence interval and interpret your interval in context.

35. Pain Researchers comparing the effectiveness of two pain medications randomly selected a group of patients who had been complaining of a certain kind of joint pain. They randomly divided these people into two groups, then administered the pain killers. Of the 112 people in the group who received medication A, 84 said this pain reliever was effective. Of the 108 people in the other group, 66 reported that pain reliever B was effective.

a) Write a 95% confidence interval for the percentage of people who may get relief from this kind of joint pain by using medication A. Interpret your interval.

b) Write a 95% confidence interval for the percentage of people who may get relief by using medication B. Interpret your interval.

c) Do the intervals for A and B overlap? What do you think this means about the comparative effectiveness of these medications?

d) Find a 95% confidence interval for the difference in the proportions of people who may find these medications effective. Interpret your interval.

e) Does this interval contain zero? What does that mean?

f) Why do the results in parts c) and e) seem contradictory? If we want to compare the effectiveness of these two pain relievers, which is the correct approach? Why?

36. **Gender gap** Candidates for political office realize that different levels of support among men and women may be a crucial factor in determining the outcome of an election. One candidate found that 52% of 473 men polled said they will vote for him, but only 45% of the 522 women in the poll expressed support.

a) Write a 95% confidence interval for the percentage of male voters who may vote for this candidate. Interpret your interval.

b) Write a 95% confidence interval for the percentage of female voters who may vote for him. Interpret your interval.

c) Do the intervals for males and females overlap? What do you think this means about the gender gap?

d) Find a 95% confidence interval for the difference in the proportions of males and females who will vote for this candidate. Interpret your interval.

e) Does this interval contain zero? What does that mean?

f) Why do the results in parts c) and e) seem contradictory? If we want to see if there is a gender gap among voters with respect to this candidate, which is the correct approach? Why?

37. **Sensitive men** A *Time* magazine article about a survey of men's attitudes noted that "Young men are more comfortable than older men talking about their problems." The survey reported that 80 of 129 men between 18 and 24 years of age and 98 of 184 men between 25 and 34 years of age said they were comfortable talking about their problems. What do you think? Is *Time's* interpretation justified by these numbers?

38. **Mammograms redux** In 2001, the conclusion of the study outlined in Exercise 34 was questioned. A new nine-year study was conducted in Sweden, comparing 21 088 women who had mammograms with 21 195 who did not. Of the women who underwent screening, 63 died of breast cancer, compared to 66 deaths among the control group. (Source: *The New York Times*, Dec. 9, 2001.) Do these results support the effectiveness of regular mammograms in preventing deaths from breast cancer?

39. **Online activity checks** Are more parents checking up on their teens' online activities? A Pew survey in 2004 found that 33% of 868 randomly sampled teens said that their parents checked to see what websites they visited. In 2006, the same question posed to 811 teens found 41% reporting such checks. Do these results provide evidence that more parents are checking?

40. **Computer gaming** Who plays online or electronic games? A survey in 2006 found that 69% of 223 boys aged 12–14 said they "played computer or console games like Xbox or PlayStation . . . or games online." Of 248 boys aged 15–17, only 62% played these games. Is this evidence of a real age-based difference?

41. **Food preference** GfK Roper Consulting gathers information on consumer preferences around the world to help companies monitor attitudes about health, food, and healthcare products. They asked people in many different cultures how they felt about the following statement:

I have a strong preference for regional or traditional products and dishes from where I come from.

In a random sample of 800 respondents, 417 of 646 people who live in urban environments agreed (either completely or somewhat) with that statement, compared to 78 out of 154 people who live in rural areas. Based on this sample, is there evidence that the percentage of people agreeing with the statement about regional preferences differs between all urban and rural dwellers?

42. **Fast food** The global survey we learned about in Exercise 41 also asked respondents how they felt about the statement, "I try to avoid eating fast foods." The random sample of 800 included 411 people 35 years old or younger, and of those, 197 agreed (completely or somewhat) with the statement. Of the 389 people over 35 years old, 246 people agreed with the statement. Is there evidence that the percentage of people avoiding fast food is different in the two age groups?

43. **Lake Ontario fish** According to Environmental Defense's "Up to the Gills" 2009 update on pollution in Great Lakes fish species, mercury contamination was the cause of four fish advisories issued for 44 various fish-size combinations in Lake Ontario in 2009, whereas there were 11 mercury-caused fish advisories for the same 44 species-size combinations in 2003.[12] Test for a significant difference between years, if you can. If you cannot, explain precisely why not.

44. **First openly gay Premier** In a February 2013 Harris-Decima poll, it was reported that 31% of a random sample of Canadians felt that Kathleen Wynne's swearing in as the first female Ontario Premier and the first openly gay Premier in Canada was a very significant breakthrough for women in Canadian politics, while 11% felt

[12]www.environmentaldefence.ca/reports/pdf/UpToTheGills_2009%20Final.pdf

that it was not at all significant (a majority felt it was somewhat or not that significant). At the end of the report, it was stated that 1000 randomly selected Canadians were interviewed and that a sample of this size has a margin of error of 3.1%, 19 times in 20. (www.harris-decima.ca/news/releases)

a) Give a 95% confidence interval for the difference between the percentage of Canadians who felt this event was a very significant breakthrough and the percentage who felt it was not significant at all, if you can. If you cannot, explain clearly why not.

b) In the same poll, it was reported that the 40% of those aged 18–24 felt it was a very significant breakthrough, while 28% of those over the age of 50 felt it was very significant. Give a 95% confidence interval for the true population difference in this opinion between 18 and 24 year olds and those over 50, if you can. If you cannot, explain clearly why not.

45. Sex and reading preference For a class project, a University of Toronto student studied the difference in reading preferences between males and females observed in the Toronto subway system.

Reading Material	Male	Female
Novels	23	30
School Notes	8	4
Newspaper	18	10
Magazine	3	5
Other	4	1

a) Test for a difference in the proportion of males and females who read newspapers on the subway.

b) Test for a difference in the proportion of males and females who read novels on the subway.

c) Is there a danger in comparing one category at a time via a series of 5% level tests? Explain. What would happen if there were not five but 40 categories of reading material to compare in this way?

46. Stabbed in London, shot in Toronto In London, England, during 2006–07, there were 167 homicides: 68 by stabbing and 29 by shooting. In the same period in Toronto, there were 154 homicides: 47 by stabbing and 72 by shooting.[13]

a) Test for a significant difference in the proportions of murders committed by shooting between the two cities. If significant, report and interpret the appropriate 95% confidence interval.

b) Test for a significant difference in the proportions of murders committed by stabbing between the two cities.

If significant, report and interpret the appropriate 95% confidence interval.

c) We could also test for significant differences in the proportions of murders committed by several other methods. If we did this for several methods, say four different ones, each at the 5% level, what would happen, roughly, to the chance of making one or more Type I errors while performing the four tests? In other words, if there are no differences in any of these average rates between the two cities, would there be a 5% chance of a falsely significant test result? Explain.

47. Homicide victims Cases of female victims of male-perpetrated homicide are well publicized, but the majority of victims of male violence—at the homicide level—are in fact male. In Canada in 2007, 90% of those accused of homicide were male and 73% of victims were male. Gender clearly matters. What about age? Below are some data on the victims in 2007 by gender and age:

	Male victim	Female victim
0–11 years	14	11
12–29 years	202	45
30 or more years	216	106
Total	432	162

Source: Adapted from Statistics Canada CANSIM database, http://cansim2.statcan.gc.ca, table 253-0003, October 23, 2008.

a) Compute and compare the (conditional) victim age distributions for males and females.

b) Test for a significant difference in the proportions of victims aged 30 or more.

c) Test for a significant difference in the proportions of victims aged 11 or less.

d) Test for a significant difference in the proportions of victims aged 12–29.

e) Describe your findings from the tests above—in other words, describe the nature of the association between gender and age for the victims of homicide. Instead of doing all these tests, an alternative would be the chi-square test of association, to be discussed in Chapter 22. What is one key issue of concern when examining a question via multiple tests of hypothesis (hunting for significance)?

f) Considering the young children murdered (age 0–11), is there a significant gender effect? Are such victims equally divided between the genders? Do a test.

48. Hockey injuries and relative age Is there a relationship between (relative) age and injury prevalence in Canadian youth hockey? Below are summary data from the Canadian Hospitals Injury Reporting and Prevention Program (boys 10–15 years of age; 1995–2002). Quartile 1 included those with January, February, and March birth

---

[13]D. Spiegelhalter and A. Barnett, "London murders: A predictable pattern?" *Significance*, March 2009.

dates and quartile 4 included those with October, November, and December birth dates; hence those born in quartile 1 were relatively older than their peers born in quartile 4. The other two quartiles do not affect any of the conclusions that may be drawn from this study and are omitted to simplify things.

Word Number of Injuries by Age Category

	Atom (10–11y)	Peewee (12–13y)	Bantam (14–15y)
1st Quartile	221	506	646
4th Quartile	156	340	433

a) Test whether the relative age effect is the same at the Atom level and the Bantam level; that is, whether the likelihood that the injured player (given he is from one of the two age groups shown) is relatively older is the same at both levels.

b) Repeat part a), but compare Peewee and Bantam.

c) Repeat part a), but compare Atom and Peewee.

d) We are investigating the relationship between level of play and the relative age effect on injury. Comment on any possible problems inherent in examining a single question via multiple tests.

49. Carpal tunnel again In Exercise 22, two methods for treating carpal tunnel syndrome (wrist splints and surgery) were compared.

a) Instead of 176 patients, about how many patients in total would have been needed to reduce the margin of error to 5 percentage points (with 95% confidence)?

b) We think that about half, or 50%, of patients improve using wrist splints. An improvement of 10% (from say 50% to 60%) from use of surgery would be big enough to matter clinically. For a test of hypothesis (two-sided), how many patients should be included if we wanted an 80% chance of detecting an improvement of this magnitude (power = 0.80) when using surgery? (You need software for this.)

50. More pain In Exercise 35, we compared the effectiveness of two pain medications.

a) About how many patients in each group (assuming equal sizes) would have been needed to bring down the margin of error to 10 percentage points (with 95% confidence)?

b) We think that a difference of about 20% in effectiveness would be a clinically important difference between the medications. If we think one will be effective in about 50% of patients, how many subjects would be needed to have a 90% probability of detecting a genuine difference of 20% in effectiveness (power = 0.90) in a two-sided test of hypothesis? (You need software for this.)

Just Checking ANSWERS

1. We're 90% confident that if members are contacted by e-mail, the donation rate will be between 4 and 18 percentage points higher than if they received regular mail.

2. Since a difference of 0 is not in the confidence interval, we'd reject the null hypothesis. There is evidence that more members will donate if contacted by e-mail.

3. The proportion from the sample in 2004 has variability, too. If we do a one-proportion z-test, we won't take that variability into account and our P-value will be incorrect.

4. The difference in the proportions between the two years has more variability than either individual proportion. The variance of the difference is the sum of the two variances.

MathXL

MyLab Statistics

Go to MathXL at **www.mathxl.com** or MyLab™ Statistics at **www.mystatlab.com**. You can practise exercises for this chapter as often as you want. The guided solutions will help you find answers step by step. You'll find a personalized study plan available to you too!

CHAPTER

22

LEARNING OBJECTIVES

1 Recognize when it is appropriate to use a chi-square test of
  • Goodness of fit
  • Homogeneity
  • Independence

2 Know the assumptions and conditions

3 Know how to:
  • Calculate expected counts and the chi-square statistic
  • Find the *P*-value using the chi-square sampling model

4 Examine and interpret the standardized residuals

Comparing Counts

Life is the art of drawing sufficient conclusions from insufficient premises.

—*Samuel Butler*

Does your zodiac sign predict how successful you will be later in life? *Fortune* magazine collected the zodiac signs of 256 heads of the largest 400 companies. Here are the number of births for each sign:

Births	Sign	Births	Sign
23	Aries	18	Libra
20	Taurus	21	Scorpio
18	Gemini	19	Sagittarius
23	Cancer	22	Capricorn
20	Leo	24	Aquarius
19	Virgo	29	Pisces

Birth totals by sign for 256 *Fortune* 400 executives.

We can see some variation in the number of births per sign, and there *are* more Pisces, but is that enough to claim that successful people are more likely to be born under some signs than others?

22.1 Goodness-of-Fit

If births of *Fortune* 400 CEOs were distributed uniformly across the year, we would expect about 1/12 of them to occur under each sign of the zodiac. That suggests 256/12, or about 21.3 births per sign. How closely do the observed numbers of births per sign fit this simple "null" model?

A hypothesis test to address this question is called a test of "**goodness-of-fit**." The name suggests a certain badness-of-grammar, but it is quite standard. After all, we are asking whether the model that births are uniformly distributed over the signs fits the data good . . . er, well. Goodness-of-fit involves testing a hypothesis. We have specified a model for the distribution and want to know whether it fits. There is no single parameter to estimate, so a confidence interval wouldn't make much sense.

If the question were about only one astrological sign (for example, "Are executives more likely to be Pisces?"[1]), we could use a one-proportion *z*-test and ask if the true proportion of executives with that sign is equal to 1/12. However, here we have

Connections

Is your favourite colour related to how much education you've had? A survey found a higher percentage of adults with only a high school education named blue and a lower percentage said red compared to adults with more education. Could this be just chance, or are colour preferences genuinely different across different education levels?

We first asked such questions in Chapter 2, where we examined the association between *categorical variables* using tables and charts. In this chapter, we'll learn how to test hypotheses about independence or association between two categorical

---

[1]A question actually posed to us by someone who was undoubtedly a Pisces.

variables, as well as hypotheses about individual variables.

Chi-square methods generalize the inference methods for proportions discussed earlier, allowing us to step up to studies involving more than two proportions or more than two categories of interest.

12 hypothesized proportions, one for each sign. We need a test that considers all of them together and gives an overall idea of whether the observed distribution differs from the hypothesized one.

For Example FINDING EXPECTED COUNTS

Birth month may not be related to success as a CEO, but what about on the hockey rink? It has been proposed by some researchers that children who are the older ones in their class at school naturally perform better in sports, and that these children then get more coaching and encouragement. Could that make a difference in who makes it to the professional level in sports?

Canada contributed 71.7% of the NHL players from the start of the league until the end of the 2009 season (though that percentage has fallen to around 50% in recent years). We also have demographic statistics for the population birth percentages for the years 1970–2005, excepting '72, '91 and '98.[2] We'll consider the birth months of the players born in those years, and test whether the observed distribution of hockey players' birth months shows just random fluctuations or whether it represents a real deviation from the national pattern.

Month	Hockey Player Count	National Birth %	Month	Hockey Player Count	National Birth %
1	127	8.0	7	88	8.7
2	116	7.6	8	74	8.5
3	117	8.7	9	84	8.5
4	127	8.6	10	79	8.2
5	118	8.9	11	75	7.8
6	110	8.5	12	69	7.9
			Total	1184	

Month	Expected	Month	Expected
1	94.72	7	103.01
2	89.98	8	100.64
3	103.01	9	100.64
4	101.82	10	97.09
5	105.38	11	92.35
6	100.64	12	93.54

QUESTION: We will need expected counts. How can we find them?

ANSWER: There are 1184 players in this set of data. I'd expect 8% of them to have been born in January, or 1184(0.08) = 94.72. I won't round off, because expected "counts" needn't be integers. Multiplying 1184 by each of the birth percentages gives the expected counts shown in the *table in the margin.*

Assumptions and Conditions

The raw data are categorical in nature, but rather than displaying each individual's response, we often work with just the summary counts for the various categories, as we did in Chapter 2. In our example, we didn't display the raw data showing the birth signs of each of the 256 executives, only the totals for each sign. Our assumptions and conditions will reflect this organization of the raw data into tables of summary counts.

Counted Data Condition The numbers displayed in the table must be *counts* of individuals that were classified into the categories corresponding to a categorical variable. This might seem a simplistic—even silly—condition. But many kinds of values can be assigned to categories, and it is unfortunately common to find the methods of this chapter applied incorrectly to proportions, percentages, or measurements just because they happen

[2]Addona, Vittorio and Yates, Philip A. (2010) "A Closer Look at the Relative Age Effect in the National Hockey League." *Journal of Quantitative Analysis in Sports*: Vol. 6: Iss. 4, Article 9. Available at: www.bepress.com/jqas/vol6/iss4/9

to be organized in a similar looking table. So, check that the values in each **cell** of the table really are counts.

Independence Assumption   The responses of the *individuals* who make up the *counts* in the cells should be independent of each other. The easiest case is when the individuals who are counted in the cells are sampled independently from some population. That's what we'd like to have if we want to draw conclusions about that population. Randomness can arise in other ways, though. For example, these *Fortune* 400 executives are not a random sample of company executives, but we have no reason to think that their birth dates should not be randomly distributed throughout the year. If we want to generalize to a large population, we should check the usual **Randomization Condition**.

We can use chi-square tests to assess patterns in a table provided the individuals counted are independent. But if we want to generalize our conclusions to a larger population, the individuals who have been classified and counted should be representative of that population. The easiest way to be sure of this is if they are a random sample from the population of interest (though once again, be aware that our *P*-values for all tests in this chapter will err on the conservative side when sampling more than 10% of a population).

Sample Size Assumption   We must have enough data for the methods to work. We usually check the

Expected Cell Frequency Condition: The expected frequency (count) for every cell in the table should be at least 5.0.

The Expected Cell Frequency Condition sounds like—and is, in fact, quite similar to—the condition that *np* and *nq* be at least 10 when we tested proportions. Just as that condition allowed us to approximate the sampling model with a Normal curve, the (chi-square) sampling model that we will use here is also a large sample approximation, that falls apart when the sample is too small. In our astrology example, assuming equal births under each zodiac sign leads us to expect 21.3 births per sign, so the condition is easily met here.

> ## For Example   CHECKING ASSUMPTIONS AND CONDITIONS
>
> **RECAP:** Are Canadian professional hockey players more likely to be born in some months than in others? We have observed and expected counts for 1184 players.
>
> **QUESTION:** Are the assumptions and conditions met for performing a goodness-of-fit test?
>
> **ANSWER:**
>
> ✓ Counted Data Condition: I have month-by-month counts of hockey player births.
>
> ✓ Independence/Randomization: These births were independent. Even though these players are not from a random sample, we can think of them as representative of many more players past and future. I'll proceed with caution
>
> ✓ Expected Cell Frequency Condition: The expected counts range from 89.98 to 105.38, all much greater than 5.
>
> It's okay to use these data for a goodness-of-fit test.

Calculations

We have observed a count in each category from the data, and have an expected count for each category from the hypothesized proportions. Are the differences just natural sampling variability, or are they so large that they indicate something important? It's natural to look at the *differences* between these observed and expected counts, denoted (*Obs − Exp*). We'd like to think about the total of the differences, but just adding them won't work because some differences are positive and others negative. We've been in this predicament before—once when we looked at deviations from the mean, and again when we dealt with residuals. In fact, these *are* residuals. They're just the differences between the observed data and the expected counts given by the (null) model. We handle these residuals in essentially the same way we did in regression: We square them. That gives us positive values and focuses attention on any cells with large differences from what we expected. Because the differences between observed and expected counts generally get larger the more data we have, we also need to get an idea of the *relative* sizes of the differences. To do that, we divide each squared difference by the expected count for that cell.

The test statistic, called the **chi-square (or chi-squared) statistic**, is found by adding up the sum of the squares of the deviations between the observed and expected counts divided by the expected counts:

$$\chi^2 = \sum_{all\ cells} \frac{(Obs - Exp)^2}{Exp}$$

The chi-square statistic is denoted χ^2, where χ is the Greek letter chi (pronounced "ky" as in "sky"). It refers to a family of sampling distribution models we have not seen before called (remarkably enough) the **chi-square models**.

This is a one-parameter family of models, like the Student's *t*-models, differing only in the number of *degrees of freedom*. The number of degrees of freedom for a goodness-of-fit test is *the number of categories minus 1*. For the zodiac example, we have 12 signs, so our χ^2 statistic has 11 degrees of freedom. You can also reason that we have 12 *Obs-Exp* terms in our statistic, and all but the last difference are free to vary (since these differences must sum to zero).

NOTATION ALERT

The only use of the Greek letter χ in Statistics is to represent this statistic and the associated sampling distribution. This is another violation of our "rule" that Greek letters represent population parameters. Here we are using a Greek letter simply to name a family of distribution models and a statistic.

Chi-Square P-Values

The chi-square statistic is used only for testing hypotheses, not for constructing confidence intervals. If the observed counts don't closely match the expected, the statistic will be large. It can't be "too small." That would just mean that our model *really* fits the data well. So the chi-square test is always one-sided. If the calculated statistic value is large enough, we'll reject the null hypothesis. What could be simpler?

If you don't have technology handy, it's easy to read the χ^2 table (Table X in Appendix C).

A portion of Table X.

Right-Tail Probability		0.10	0.05	0.025	0.01	0.005
	df					
Values of χ^2_α	1	2.706	3.841	5.024	6.635	7.879
	2	4.605	5.991	7.378	9.210	10.597
	3	6.251	7.815	9.348	11.345	12.838
	4	7.779	9.488	11.143	13.277	14.860
	5	9.236	11.070	12.833	15.086	16.750
	6	10.645	12.592	14.449	16.812	18.548
	7	12.017	14.067	16.013	18.475	20.278
	8	13.362	15.507	17.535	20.090	21.955
	9	14.684	16.919	19.023	21.666	23.589
	10	15.987	18.307	20.483	23.209	25.188
	11	17.275	19.675	21.920	24.725	26.757
	12	18.549	21.026	23.337	26.217	28.300
	13	19.812	22.362	24.736	27.688	29.819
	14	21.064	23.685	26.119	29.141	31.319

The usual selected upper tail probability values are at the top of the columns. As with the t-tables, we have only selected probabilities, so the best we can do is to trap a P-value between two of the values in the table. Just find the row for the correct number of degrees of freedom, and read across to find where your calculated χ^2 value falls. Of course, technology can find an exact P-value, and that's usually what we'll see.

Even though its mechanics work like a one-sided hypothesis test, the interpretation of a chi-square test is in some sense *many*-sided. With more than two proportions, there are many ways the null hypothesis can be wrong. By squaring the differences, we made all the deviations positive, whether our observed counts were higher or lower than expected. There's no direction to the rejection of the null model. All we know is that it doesn't fit.

For Example   DOING A GOODNESS-OF-FIT TEST

RECAP: We're looking at data on the birth months of Canadian NHL players. We've checked the assumptions and conditions for performing a χ^2 test.

QUESTIONS: What are the hypotheses, and what does the test show?

ANSWER: H_0: The distribution of birth months for Canadian NHL players is the same as that for the general population.

H_A: The distribution of birth months for Canadian NHL players differs from that of the rest of the country.

$$\chi^2 = \sum \frac{(Obs - Exp)^2}{Exp} = \frac{(127 - 94.72)^2}{94.72} + \frac{(116 - 89.98)^2}{89.98} + \cdots = 54.09$$

with $df = 12 - 1 = 11$

$P(\chi^2_{11} \geq 54.09) < 0.005$ (from our table).

Because of the small P-value, I reject H_0; there's evidence that birth months of Canadian NHL players have a different distribution from the rest of us.

Step-by-Step Example   A CHI-SQUARE TEST FOR GOODNESS-OF-FIT

Who	Executives of Fortune 400 companies
What	Zodiac birth sign
Why	Maybe the researcher was a Gemini and naturally curious?

We have counts of 256 executives in 12 zodiac sign categories. The natural null hypothesis is that birth dates of executives are divided equally among all the zodiac signs. The test statistic looks at how closely the observed data match this idealized situation.

Question: Are zodiac signs of CEOs distributed uniformly over the 12 signs?

THINK ➡ Plan State what you want to know. Identify the variables and check the *W*'s.

Hypotheses State the null and alternative hypotheses. For χ^2 tests, it's usually easier to do that in words than in symbols.

I want to know whether births of successful people are uniformly distributed across the signs of the zodiac. I have counts of 256 *Fortune* 400 executives, categorized by their birth sign.

H_0: Births are uniformly distributed over zodiac signs.[3]

---

[3]It may seem that we have broken our rule of thumb that null hypotheses should specify parameter values. If you want to get formal about it, the null hypothesis is that

$$p_{Aries} = p_{Taurus} = \cdots = p_{Pisces} = 1/12.$$

That is, we hypothesize that the true proportions of births of CEOs under each sign are equal. The role of the null hypothesis is to specify the model so that we can compute the test statistic. That's what this one does.

H_A: Births are not uniformly distributed over zodiac signs.

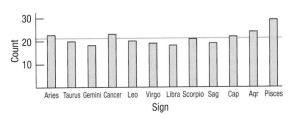

The bar chart shows some variation from sign to sign, and Pisces is the most frequent. But it is hard to tell whether the variation is more than I'd expect from purely random variation.

Model Make a picture. The null hypothesis is that the frequencies are equal, so a bar chart (with a line at the hypothesized "equal" value) is a good display.

Think about the assumptions and check the conditions.

We don't need random sampling to assert that birth dates are independent. But we do want our sample to be representative of CEOs so we can draw inferences about the population of CEOs. That's a reasonable assumption about this sample.

✓ **Counted Data Condition**: I have counts of the number of executives in 12 categories.

✓ **Independence/Randomization**: The birth dates of executives should be independent of each other. This is a convenience sample of CEOs, not an SRS, but there's no reason to suspect bias.

✓ **Expected Cell Frequency Condition**: The null hypothesis expects that 1/12 of the 256 births, or 21.333, should occur under each sign. These expected values are all at least five, so the condition is satisfied.

Specify the sampling distribution model.

Name the test you will use.

The conditions are satisfied, so I'll use a χ^2 model with $12 - 1 = 11$ degrees of freedom and do a chi-square goodness-of-fit test.

SHOW ➡ Mechanics Each cell contributes an $\dfrac{(Obs - Exp)^2}{Exp}$ value to the chi-square sum.

We add up these components for each zodiac sign. If you do it by hand, it can be helpful to arrange the calculation in a table. We show that after this Step-By-Step.

The P-value is the area in the upper tail of the χ^2 model above the computed χ^2 value.

The χ^2 models are skewed to the high end, and change shape depending on the degrees of freedom. The P-value considers only the right tail. Large χ^2 statistic values correspond to small P-values, which lead us to reject the null hypothesis.

The expected value for each zodiac sign is 21.333.

$$\chi = \sum \frac{(Obs - Exp)^2}{Exp} = \frac{(23 - 21.333)^2}{21.333}$$

$$+ \frac{(20 - 21.333)^2}{21.333} + \cdots$$

$$= 5.094 \text{ for all 12 signs.}$$

$$P\text{-}value = P(\chi^2 > 5.094) = 0.926$$

TELL ➡ **Conclusion** Link the P-value to your decision. Remember to state your conclusion in terms of what the data mean, rather than just making a statement about the distribution of counts.

> The P-value of 0.926 says that if the zodiac signs of executives were in fact distributed uniformly, an observed chi-square value of 5.09 or higher would occur about 93% of the time. This certainly isn't unusual, so I fail to reject the null hypothesis, and conclude that these data show virtually no evidence of nonuniform distribution of zodiac signs among executives.

The Chi-Square Calculation

Let's make the chi-square procedure very clear. After clearly stating H_0 and H_A, the steps for calculations are:

1. **Find the expected values.** These come from the null hypothesis model. Every model gives a hypothesized proportion for each cell. The expected value is the product of the *total number of observations times this proportion.*

   For our example, the null model hypothesizes *equal* proportions. With 12 signs, 1/12 of the 256 executives should be in each category. The expected number for each sign is 21.333.

2. **Compute the residuals.** Once you have expected values for each cell, find the residuals, *Observed – Expected.*

3. **Square the residuals.**

4. **Compute the components.** Now find the **component**, $\dfrac{(Observed - Expected)^2}{Expected}$, for each cell.

5. **Find the sum of the components.** That's the chi-square statistic.

6. **Find the degrees of freedom.** It's equal to *the number of cells minus one.* For the zodiac signs, that's $12 - 1 = 11$ degrees of freedom.

7. **Test the hypothesis.** Large chi-square values mean lots of deviation from the hypothesized model, so they give small P-values. Look up the critical value from a table of chi-square values, or use technology to find the P-value directly.

The steps of the chi-square calculations are often laid out in tables. If you don't have a statistics program handy, these calculations are particularly well-suited to being performed on a spreadsheet. Use one row for each category, and columns for observed counts, expected counts, residuals, squared residuals, and the contributions to the chi-square total, like this:

Sign	Observed	Expected	Residual = $(Obs - Exp)$	$(Obs - Exp)^2$	Component = $\dfrac{(Obs - Exp)^2}{Exp}$
Aries	23	21.333	1.667	2.778889	0.130262
Taurus	20	21.333	−1.333	1.776889	0.083293
Gemini	18	21.333	−3.333	11.108889	0.520737
Cancer	23	21.333	1.667	2.778889	0.130262
Leo	20	21.333	−1.333	1.776889	0.083293
Virgo	19	21.333	−2.333	5.442889	0.255139
Libra	18	21.333	−3.333	11.108889	0.520737
Scorpio	21	21.333	−0.333	0.110889	0.005198
Sagittarius	19	21.333	−2.333	5.442889	0.255139
Capricorn	22	21.333	0.667	0.444889	0.020854
Aquarius	24	21.333	2.667	7.112889	0.333422
Pisces	29	21.333	7.667	58.782889	2.755491
					$\Sigma = 5.094$

How big is big? When we calculated χ^2 for the zodiac sign example, we got 5.094. That value would have been big for z or t, leading us to reject the null hypothesis. Not here. Were you surprised that $\chi^2 = 5.094$ had a huge P-value of 0.926? What *is* big for a χ^2 statistic, anyway?

Think about how χ^2 is calculated. In every cell, any deviation from the expected count contributes to the sum. Large deviations generally contribute more, but if there are a lot of cells, even small deviations can add up, making the χ^2 value larger. So the more cells there are, the higher the value of χ^2 has to get before it becomes noteworthy. For χ^2, then, the decision about how big is big depends on the number of degrees of freedom.

Unlike the Normal and t families, χ^2 models are skewed. Curves in the χ^2 family change both shape and centre as the number of degrees of freedom grows. Here, for example, are the χ^2 curves for 5 and 9 degrees of freedom.

Notice that the value $\chi^2 = 10$ might seem somewhat extreme when there are 5 degrees of freedom, but appears to be rather ordinary for 9 degrees of freedom. Here are two simple facts to help you think about χ^2 models:

- The mode is at $\chi^2 = df - 2$. (Look at the curves; their peaks are at 3 and 7, see?)
- The expected value (mean) of a χ^2 model is its number of degrees of freedom. That's a bit to the right of the mode—as we would expect for a skewed distribution.

Our test for zodiac birthdays had 11 df, so the relevant χ^2 curve peaks at 9 and has a mean of 11. Knowing that, we might have easily guessed that the calculated χ^2 value of 5.094 wasn't going to be significant.

But I Believe the Null Model . . .

Goodness-of-fit tests are likely to be performed by people who have a theory of what the proportions *should* be in each category and who believe their theory to be true. Unfortunately, the only *null* hypothesis available for a goodness-of-fit test is that the theory is true. And as we know, the hypothesis-testing procedure allows us only to *reject* the null or *fail to reject* it. We can never confirm that a theory is, in fact, true, which is often what people want to do.

Unfortunately, they're stuck. At best, we can point out that the data are consistent with the proposed theory. But this doesn't *prove* the theory. The data *could* be consistent with the model even if the theory were wrong. In that case, we fail to reject the null hypothesis, but can't conclude anything for sure about whether the theory is true.

And we can't fix the problem by turning things around. Suppose we try to make our favoured hypothesis the alternative. Then, it is impossible to pick a single null. For example, suppose, as a doubter of astrology, you want to prove that the distribution of executive births is uniform. If you choose uniform as the null hypothesis, you can only *fail* to reject it. So you'd like uniformity to be your alternative hypothesis. Which particular violation of equally distributed births would you choose as your null? The problem is that the model can be wrong in many, many ways. There's no way to frame a null hypothesis the other way around. There's just no way to prove that a favoured model is true.

WHY CAN'T WE PROVE THE NULL?

A biologist wanted to show that her inheritance theory about fruit flies was valid. It says that 10% of the flies should be type 1, 70% type 2, and 20% type 3. After her students collected data on 100 flies, she did a goodness-of-fit test and found a P-value of 0.07. She started celebrating, since her null hypothesis wasn't rejected—that is, until her students collected data on 100 more flies. With 200 flies, the P-value dropped to 0.02. Although she knew the answer was probably no, she asked the statistician, somewhat hopefully, if she could just ignore half the data and stick with the original 100. By this reasoning we could always "prove the null" just by not collecting much data. With only a little data, the chances are good that they'll be consistent with almost anything. But they also have little chance of disproving anything, either. In this case, the test has almost no power. Don't let yourself be lured into this scientist's reasoning. With data, more is always better (if you can afford it). But you can't ever prove that your null hypothesis is true.

Section 22.1 Exercises

1. **Human births** If there is no seasonal effect on human births, we would expect equal numbers of children to be born in each season (winter, spring, summer, and fall). A student takes a census of her statistics class and finds that of the 120 students in the class, 25 were born in winter, 35 in spring, 32 in summer, and 28 in fall. She wonders if the excess in the spring is an indication that births are not uniform throughout the year.
   a) What is the expected number of births in each season if there is no "seasonal effect" on births?
   b) Compute the χ^2 statistic.
   c) How many degrees of freedom does the χ^2 statistic have?

2. **Bank cards** At a major credit card bank, the percentages of people who historically apply for the Silver, Gold, and Platinum cards are 60%, 30%, and 10%, respectively. In a recent sample of customers responding to a promotion, of 200 customers, 110 applied for Silver, 55 for Gold, and 35 for Platinum. Is there evidence to suggest that the percentages for this promotion may be different from the historical proportions?
   a) What is the expected number of customers applying for each type of card in this sample if the historical proportions are still true?
   b) Compute the χ^2 statistic.
   c) How many degrees of freedom does the χ^2 statistic have?

3. **Human births, again** For the births in Exercise 1,
   a) If there is no seasonal effect, about how big, on average, would you expect the χ^2 statistic to be (what is the mean of the χ^2 distribution)?
   b) Does the statistic you computed in Exercise 1 seem large in comparison to this mean? Explain briefly.
   c) What does that say about the null hypothesis?
   d) Find the $\alpha = 0.05$ critical value for the χ^2 distribution with the appropriate number of df.
   e) Using the critical value, what do you conclude about the null hypothesis at $\alpha = 0.05$?

4. **Bank cards, again** For the customers in Exercise 2,
   a) If the customers apply for the three cards according to the historical proportions, about how big, on average, would you expect the χ^2 statistic to be (what is the mean of the χ^2 distribution)?
   b) Does the statistic you computed in Exercise 2 seem large in comparison to this mean? Explain briefly.
   c) What does that say about the null hypothesis?
   d) Find the $\alpha = 0.05$ critical value for the χ^2 distribution with the appropriate number of df.
   e) Using the critical value, what do you conclude about the null hypothesis at $\alpha = 0.05$?

22.2 Chi-Square Test of Homogeneity

Who	Graduates from 4 colleges at an upstate New York university
What	Postgraduation activities
When	2011
Why	Survey for general information

Many universities survey graduating classes to determine the plans of the graduates. We might wonder whether the plans of students are the same at different colleges within a university. Table 22.1 is a **two-way table** for the Class of 2011 graduates from several colleges at one university. Each cell of the table shows how many students from a particular college made a certain choice.

	Agriculture	Arts & Sciences	Engineering	ILR	Total
Employed	209	198	177	101	685
Grad School	104	171	158	33	466
Other	135	115	39	16	305
Total	448	484	374	150	1456

Table 22.1

Postgraduation activities of the class of 2011 for several colleges of a large university. (Note: ILR is the School of Industrial and Labor Relations.)

Because class sizes are so different, we see differences better by examining the proportions for each class rather than the counts (Table 22.2):

Table 22.2

Activities of graduates as a percentage of respondents from each college.

	Agriculture	Arts & Sciences	Engineering	ILR	Total
Employed	46.7%	40.9%	47.3%	67.3%	**47.0%**
Grad School	23.2%	35.3%	42.2%	22.0%	**32.0%**
Other	30.1%	23.8%	10.4%	10.7%	**20.9%**
Total	**100.0%**	**100.0%**	**100.0%**	**100.0%**	**100.0%**

We already know how to test whether *two* proportions are the same. For example, we could use a two-proportion *z*-test to see whether the proportion of students choosing graduate school is the same for Agriculture students as for Engineering students. But now we have more than two groups. We want to test whether the students' choices are the same across all four colleges. The *z*-test for two proportions generalizes to a **chi-square test of homogeneity**.

Chi-square again? It turns out that the basic mechanics of this test are *identical* to the chi-square test for goodness-of-fit that we just saw. (How similar can you get?) Why a different name, then? The tests are really quite different. The goodness-of-fit test compared counts with a theoretical model. But here we're asking whether the distribution of choices is the same among different groups, so we find the expected counts for each category directly from the data. As a result, we count the degrees of freedom slightly differently as well.

The term "homogeneity" means that things are the same. Here, we ask whether the post-graduation choices made by students are the *same* for these four colleges. The homogeneity test comes with a built-in null hypothesis: We hypothesize that the distribution does not change from group to group. The test looks for differences large enough to step beyond what we might expect from random sample-to-sample variation. It can reveal a large deviation in a single category or small, but persistent, differences over all the categories—or anything in between.

Assumptions and Conditions

The assumptions and conditions are the same as for the chi-square test for goodness-of-fit. The **Counted Data Condition** says that these data must be counts (which summarize categorical observations). You can't do a test of homogeneity on proportions, so you have to work with the counts of graduates given in the first table. Also, you can't do a chi-square test on measurements. For example, if we had recorded GPAs for these same groups, we wouldn't be able to determine whether the mean GPAs were different using this test.[4]

Ideally, when we compare the proportions across several groups, we would like the cases within each group to be selected randomly. We need to know whether the **Independence Assumption** *both within and across groups* is reasonable. As usual, randomization is a good condition to check to make the assumption plausible.

We still must be sure we have enough data for this method to work. The **Expected Cell Frequency Condition** says that the expected count in each cell must be at least 5. We'll confirm that as we do the calculations.

Calculations

The null hypothesis says that the distribution of the proportions of graduates choosing each alternative is the same for all four colleges, so we can estimate those overall proportions by pooling our data from the four colleges together. Within each college, the expected proportion for each choice is just the overall proportion of all students making that choice.

---

[4]To do that, you'd use a method called Analysis of Variance (Chapter 24).

The expected counts are those proportions applied to the number of students in each graduating class.

For example, overall, 685, or about 47.0%, of the 1456 students who responded to the survey were employed. If the distributions are homogeneous (as the null hypothesis asserts), then 47% of the 448 Agriculture school graduates (or about 210.77 students) should be employed. Similarly, 47% of the 374 Engineering grads (or about 175.95) should be employed. (Equivalently: Expected cell count = row total × column total ÷ grand total)

Working in this way, we (or, more likely, the computer) can fill in expected values for each cell (Table 22.3). Because these are theoretical values, they don't have to be integers. The expected values look like this:

Table 22.3

Expected values for the '11 graduates.

	Agriculture	Arts & Sciences	Engineering	ILR	Total
Employed	210.769	227.706	175.955	70.570	**685**
Grad School	143.385	154.907	119.701	48.008	**466**
Other	93.846	101.387	78.345	31.422	**305**
Total	**448**	**484**	**374**	**150**	**1456**

Now check the **Expected Cell Frequency Condition**. Indeed, there are at least 5 individuals expected in each cell.

Following the pattern of the goodness-of-fit test, we compute the component for each cell of the table. For the highlighted cell, employed students graduating from the Agriculture school, that's

$$\frac{(Obs - Exp)^2}{Exp} = \frac{(209 - 210.769)^2}{210.769} = 0.0148.$$

Summing these components across all cells gives

$$\chi^2 = \sum_{all\,cells} \frac{(Obs - Exp)^2}{Exp} = 93.66.$$

How about the degrees of freedom? We don't really need to calculate all the expected values in the table. We know there is a total of 685 employed students, so once we find the expected values for three of the colleges, we can determine the expected number for the fourth by just subtracting. Similarly, we know how many students graduated from each college, so after filling in two rows, we can find the expected values for the remaining row by subtracting. To fill out the table, we need to know the counts in only $R - 1$ rows and $C - 1$ columns (see the Notation Alert). So the table has $(R - 1)(C - 1)$ degrees of freedom.

In our example, we need to calculate only 2 choices in each column and counts for 3 of the 4 colleges, for a total of $2 \times 3 = 6$ degrees of freedom. We'll need the degrees of freedom to find a P-value for the chi-square statistic.

■ NOTATION ALERT

For a contingency table, R represents the number of rows and C the number of columns.

Step-by-Step Example A CHI-SQUARE TEST FOR HOMOGENEITY

We have reports from four colleges on the post-graduation activities of their 2011 graduating classes.

Question: Are the distributions of students' choices of post-graduation activities the same across all the colleges?

THINK ➡ Plan State what you want to know.

Identify the variables and check the W's.

I want to test whether post-graduation choices are the same for students from each of four colleges. I have a table of counts classifying each college's Class of 2011 respondents according to their activities.

Hypotheses State the null and alternative hypotheses.

H_0: Students' post-graduation activities are distributed in the same way for all four colleges.

H_A: Students' plans do not have the same distribution.

Model Make a picture: A side-by-side bar chart shows the four distributions of post-graduation activities. Plot column percents to remove the effect of class size differences. A split bar chart would also be an appropriate choice.

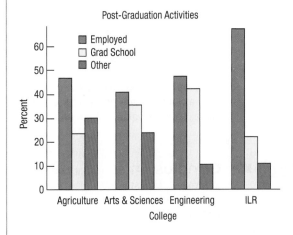

Post-Graduation Activities

A side-by-side bar chart shows how the distributions of choices differ across the four colleges.

Think about the assumptions and check the conditions.

✓ **Counted Data Condition**: I have counts of the number of students in categories.

✓ **Independence Assumption**: Even though this isn't a random sample, student plans should be largely independent of each other. The occasional friends or couples who make post-graduation decisions together are too rare to affect this analysis.

✓ **Expected Cell Frequency Condition**: The expected values (shown below) are all at least 5.

State the sampling distribution model and name the test you will use.

The conditions seem to be met, so I can use a χ^2 model with $(3 - 1) \times (4 - 1) = 6$ degrees of freedom and do a **chi-square test of homogeneity**.

SHOW ➡ Mechanics Find the expected counts for each cell of the data table (row total × column total ÷ grand total). You could make separate tables for the observed and expected counts, or put both counts in each cell as shown here. While observed counts must be whole numbers, expected counts rarely are—don't be tempted to round those off.

	Ag	A&S	Eng	ILR
Empl.	209	198	177	101
	210.769	227.706	175.955	70.570
Grad School	104	171	158	33
	143.385	154.907	119.701	48.008
Other	135	115	39	16
	93.846	101.387	78.345	31.422

Calculate χ^2.	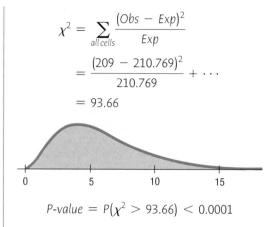 $$\chi^2 = \sum_{all\,cells} \frac{(Obs - Exp)^2}{Exp}$$ $$= \frac{(209 - 210.769)^2}{210.769} + \cdots$$ $$= 93.66$$
The shape of a χ^2 model depends on the degrees of freedom. A χ^2 model with 6 df is skewed to the high end. The P-value considers only the right tail. Here, the calculated value of the χ^2 statistic is off the scale, so the P-value is quite small.	P-value $= P(\chi^2 > 93.66) < 0.0001$
TELL ➡ **Conclusion** State your conclusion in the context of the data. You should specifically talk about whether the distributions for the groups appear to be different.	The P-value is very small, so I reject the null hypothesis and conclude that there's evidence that the post-graduation activities of students from these four colleges don't have the same distribution.

If you find that simply rejecting the hypothesis of homogeneity is a bit unsatisfying, you're in good company. OK, so the post-graduation plans are different. What we'd really like to know is what the differences are, where they're the greatest, and where they're smallest. The test for homogeneity doesn't answer these interesting questions, but it does provide evidence that *differences do indeed exist*. It now would make sense to learn more about them by constructing some confidence intervals for proportions or differences between proportions (see Chapters 15 and 21).

Section 22.2 Exercises

5. Customer ages An analyst at a local bank wonders if the age distribution of customers coming for service at his branch in town is the same as at the branch located near the mall. He selects 100 transactions at random from each branch and researches the age information for the associated customer. Here are the data:

	Age			
	Less Than 30	**30–55**	**56 or Older**	**Total**
In-Town Branch	20	40	40	**100**
Mall Branch	30	50	20	**100**
Total	**50**	**90**	**60**	**200**

a) What is the null hypothesis?
b) What type of test is this?
c) What are the expected numbers for each cell if the null hypothesis is true?
d) Find the χ^2 statistic.
e) How many degrees of freedom does it have?
f) Find the P-value.
g) What do you conclude?

6. Bank cards, once more A market researcher working for the bank in Exercise 2 wants to know if the distribution of applications by card is the same for the past three mailings. She takes a random sample of 200 from each mailing and counts the number applying for Silver, Gold, and Platinum. The data follow:

	Type of Card			
	Silver	**Gold**	**Platinum**	**Total**
Mailing 1	120	50	30	**200**
Mailing 2	115	50	35	**200**
Mailing 3	105	55	40	**200**
Total	**340**	**155**	**105**	**600**

a) What is the null hypothesis?
b) What type of test is this?
c) What are the expected numbers for each cell if the null hypothesis is true?
d) Find the χ^2 statistic.
e) How many degrees of freedom does it have?
f) Find the P-value.
g) What do you conclude?

22.3 Examining the Residuals

Whenever we reject the null hypothesis, it's a good idea to examine residuals. (We don't need to do that when we fail to reject because when the χ^2 value is small, all of its components must have been small.) For chi-square tests, we want to compare residuals for cells that may have very different counts. So we're better off standardizing the residuals. We know the mean residual is zero,[5] but we need to know each residual's standard deviation. When we tested proportions, we saw a link between the expected proportion and its standard deviation. For counts, there's a similar link. To standardize a cell's residual, we just divide by the square root of its expected value:

$$c = \frac{(Obs - Exp)}{\sqrt{Exp}}$$

Notice that these **standardized residuals** are just the square roots of the **components** we calculated for each cell, and their sign indicates whether we observed more cases than we expected or fewer.

The standardized residuals give us a chance to think about the underlying patterns and to consider the ways in which the distribution of post-graduation plans may differ from college to college. Now that we've subtracted the mean (the residual, $Obs - Exp$, has mean 0, so it's already subtracted) and divided by their standard deviations, these are z-scores. If the null hypothesis were true, we could even appeal to the Central Limit Theorem, think of the Normal model, and use the 68–95–99.7 Rule to judge how extraordinary the large ones are.

Here (Table 22.4) are the standardized residuals for the Class of 2011 data:

Table 22.4

Standardized residuals can help show how the table differs from the null hypothesis pattern.

	Ag	A&S	Eng	ILR
Employed	−0.121866	−1.96860	0.078805	3.62235
Grad School	−3.28909	1.29304	3.50062	−2.16607
Other	4.24817	1.35192	−4.44511	−2.75117

The column for Engineering students immediately attracts our attention. It holds both the largest positive and the largest negative standardized residuals. It looks like Engineering graduates are more likely to go on to graduate work and are very unlikely to take time off for "volunteering and travel, among other activities" (as the "Other" category is explained). By contrast, Agriculture graduates seem to be readily employed and less likely to pursue graduate work immediately after graduation.

For Example LOOKING AT χ^2 RESIDUALS

RECAP: Some people suggest that school children who are the older ones in their class naturally perform better in sports and therefore get more coaching and encouragement. To see if there's any evidence for this, we looked at Canadian NHL players born since 1970. A goodness-of-fit test found their birth months to have a distribution that's significantly different from the rest of us. The table shows the standardized residuals.

Month	Residual	Month	Residual
1	3.32	7	−1.48
2	2.74	8	−2.66
3	1.38	9	−1.66
4	2.49	10	−1.84
5	1.23	11	−1.81
6	0.93	12	−2.54

---

[5]Residual = observed − expected. Because the total of the expected values is set to be the same as the observed total, the residuals must sum to zero.

QUESTION: What is different about the distribution of birth months among Canadian NHL players?

ANSWER: It appears that, compared to the general population, significantly fewer hockey players than expected were born in August and December, and more than expected were born in January, February, and April. The clear pattern of positive residuals in the first half of the year and negative residuals in the second half supports the conjecture that being older is an advantage in terms of a career as a pro athlete.

 # Just Checking

Tiny black potato flea beetles can damage potato plants in a vegetable garden. These pests chew holes in the leaves, causing the plants to wither or die. They can be killed with an insecticide, but a canola oil spray has been suggested as a non-chemical "natural" method of controlling the beetles. To conduct an experiment to test the effectiveness of the natural spray, we gather 500 beetles and place them in three Plexiglas® containers. Two hundred beetles go in the first container, where we spray them with the canola oil mixture. Another 200 beetles go in the second container, and we spray them with the insecticide. The remaining 100 beetles in the last container serve as a control group; we simply spray them with water. Then, we wait six hours and count the number of surviving beetles in each container.

1. Why do we need the control group?
2. What would our null hypothesis be?
3. After the experiment is over, we could summarize the results in a table like this. How many degrees of freedom does our χ^2 test have?

	Canola oil	Insecticide	Water	Total
Survived				
Died				
Total	200	200	100	500

4. Suppose that, all together, 125 beetles survived. (That's the first-row total.) What's the expected count in the first cell—survivors among those sprayed with the canola oil?
5. If it turns out that only 40 of the beetles in the first container survived, what's the calculated component of χ^2 for that cell?
6. If the total calculated value of χ^2 for this table turns out to be around 10, would you expect the P-value of our test to be large or small? Explain.

Section 22.3 Exercises

7. **Human births, last time** For the data in Exercise 1,
   a) Compute the standardized residual for each season.
   b) Are any of these particularly large? (Compared to what?)
   c) Why should you have anticipated the answer to part b)?

8. **Bank cards, last time** For the data in Exercise 2,
   a) Compute the standardized residual for each type of card.
   b) Are any of these particularly large? (Compared to what?)
   c) What does the answer to part b) say about this new group of customers?

22.4 Chi-Square Test of Independence

A study from the University of Texas Southwestern Medical Center examined 626 people being treated for non–blood-related diseases to see whether the risk of hepatitis C was related to whether people had tattoos and to where they got their tattoos. There are over 10 000 reported cases of hepatitis C each year in Canada, but often the disease lies undetected for years after infection.

The data from this study can be summarized in a two-way table (Table 22.5), as follows:

Table 22.5

Counts of patients classified by their hepatitis C test status according to whether they had a tattoo from a tattoo parlour or from another source, or had no tattoo.

	Hepatitis C	No Hepatitis C	Total
Tattoo, parlour	17	35	52
Tattoo, elsewhere	8	53	61
None	22	491	513
Total	47	579	626

Who	Patients being treated for non-blood-related disorders
What	Tattoo status and hepatitis C status
When	1991, 1992
Where	Texas

These data differ from the kinds of data we've considered before in this chapter because they categorize subjects from a single group on two categorical random variables rather than on only one. The categorical variables here are *Hepatitis C Status* ("Hepatitis C" or "No Hepatitis C") and *Tattoo Status* ("Parlour," "Elsewhere," "None"). We've seen counts classified by two categorical variables displayed like this in Chapter 2, so we know such tables are called contingency tables. **Contingency tables** categorize counts on two (or more) variables so that we can see whether the distribution of one variable is contingent on the other.

The natural question to ask of these data is whether the chance of having hepatitis C is *independent* of one's tattoo status. Recall that for events **A** and **B** to be independent, $P(\mathbf{A})$ must equal $P(\mathbf{A}|\mathbf{B})$. Here, this means the probability that a randomly selected patient has hepatitis C should be the same regardless of the patient's tattoo status. If *Hepatitis Status* is independent of *Tattoo Status*, we'd expect the proportion of people testing positive for hepatitis to be the same for the three levels of *Tattoo Status*, except for some chance differences due to random sampling. We examined the question of independence in just this way back in Chapter 2, but we had no way then to take into account those chance differences among sampled proportions. That's just what a test will do. This sounds a lot like the test of homogeneity. In fact, the mechanics of the calculation are identical.

The difference is that now we have two categorical variables measured on a single population. For the homogeneity test, we had a single categorical variable measured separately in two or more populations. But now we ask a different question: "Are the variables independent?" rather than "Are the groups homogeneous?" These are subtle differences, but they are important when we draw conclusions. When we ask whether two variables measured on the same population are independent we're performing a **chi-square test of independence.**

The only difference between the test for homogeneity and the test for independence is in what you . . .

THINK ➡

For Example WHICH χ^2 TEST?

A York University research team examined over 81,000 Ottawa traffic stops from 2013 to 2015 where officers recorded the driver's race, gender, age, the reason for the stop and whether the stop resulted in charges (www.ottawapolice.ca/en/about-us/resources/.TSRDCP_York_Research_Report.pdf). 97% of stops were due to 'provincial and municipal offences,' and no group was stopped in disproportionate numbers. The other stops were generally for 'criminal offences' or 'suspicious activities,' and a disproportionate number of those stopped were racialized minorities. This author was curious about the outcomes of stops for suspicious activities, and tallied the summary counts below. (Using data only from year one of the study, in **OttawaTrafficStops**)

		Outcome		
		Charged	No Action	Warning
Driver Race	Black	16	95	31
	Middle East	15	126	45
	Other	9	65	33
	White/Caucasian	41	304	133

QUESTIONS: Which test would be appropriate to examine whether race is a factor in the traffic stop outcome (for those stops based on suspicious activities)? What are the hypotheses?

ANSWER: These data are categorized on two variables, *Driver Race* and *Outcome*. I'll do a chi-square test of independence.

H_0: The traffic stop outcome is independent of the race of the driver.

H_A: Traffic stop outcomes are not independent of the driver's race.

Assumptions and Conditions

Of course, we still need summary counts of independent responses from the individuals, and enough data so that the expected values are at least 5 in each cell.

If we're interested in the independence of variables, we usually want to generalize from the data to some population. In that case, we'll need to check that the data are a representative random sample from that population.

Step-by-Step Example A CHI-SQUARE TEST FOR INDEPENDENCE

We have counts of 626 individuals categorized according to their "tattoo status" and their "hepatitis status."

Question: Are tattoo status and hepatitis status independent?

THINK ➡ Plan State what you want to know. Identify the variables and check the *W*'s.

I want to know whether the categorical variables Tattoo Status and Hepatitis C Status are statistically independent. I have a contingency table of 626 patients with an unrelated disease.

H_0: Tattoo Status and Hepatitis C Status are independent.[6]

H_A: Tattoo Status and Hepatitis C Status are not independent.

Hypotheses State the null and alternative hypotheses.

We perform a test of independence when we suspect the variables may not be independent. We are on the familiar ground of making a claim (in this case, that knowing *Tattoo Status* will change probabilities for *Hepatitis C Status*) and testing the null hypothesis that it is *not* true.

---

[6]Once again, parameters are hard to express. The hypothesis of independence itself tells us how to find expected values for each cell of the contingency table. That's all we need.

Model Make a picture. Because these are only two categories—Hepatitis C and No Hepatitis C—a simple bar chart of the distribution of tattoo sources for hepatitis C patients shows all the information.

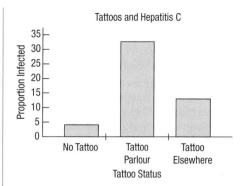

The bar chart suggests strong differences in hepatitis C risk based on tattoo status.

Think about the assumptions and check the χ^2 conditions.

✓ **Counted Data Condition**: I have counts of individuals classified jointly on two categorical variables.

✓ **Independence Assumption**: These data are from a retrospective study of patients being treated for something unrelated to hepatitis. Although they are not an SRS, their responses are likely to be independent of each other.

✗ **Expected Cell Frequency Condition**: The expected values do not meet the condition that all are at least 5.

This table shows both the observed and expected counts for each cell. The expected counts are calculated exactly as they were for a test of homogeneity; in the first cell, for example, we expect $\frac{52}{626}$ (that's 8.3%) of 47.

Warning: Be wary of proceeding when there are small expected counts. If we see expected counts that fall far short of 5 or if many cells violate the condition, we should not use χ^2. (We will soon discuss ways you can fix the problem.) If you do continue, always check the residuals to be sure those cells did not have a major influence on your result.

Specify the model.

Name the test you will use.

	Hepatitis C	No Hepatitis C	Total
Tattoo, Parlour	17 — 3.904	35 — 48.096	52
Tattoo, Elsewhere	8 — 4.580	53 — 56.420	61
None	22 — 38.516	491 — 474.484	513
Total	47	579	626

Although the Expected Cell Frequency Condition is not satisfied, the values falling short are close to 5. I'll go ahead, but I'll check the residuals carefully. I'll use a χ^2 model with $(3-1) \times (2-1) = 2$ df and do a **chi-square test of independence**.

SHOW ➡ **Mechanics** Calculate χ^2.

The shape of a chi-square model depends on its degrees of freedom. With 2 df, the model looks quite different, as you can see here. We still care only about the right tail.

$$\chi^2 = \sum_{all\,cells} \frac{(Obs - Exp)^2}{Exp}$$

$$= \frac{(17 - 3.094)^2}{3.094} + \cdots = 57.91$$

$$P\text{-}Value = P(\chi_2^2 > 57.91) < 0.0001$$

TELL ➡ **Conclusion** Link the P-value to your decision. State your conclusion about the independence of the two variables.

(We should be wary of this conclusion because of the small expected counts. A complete solution must include the additional analysis, recalculation, and final conclusion discussed in the following section.)

The P-value is very small, so I reject the null hypothesis and conclude that *Hepatitis C Status* is not independent of *Tattoo Status*. Because the Expected Cell Frequency Condition was violated, I need to check that the two cells with small expected counts did not influence this result too greatly.

For Example CHI-SQUARE MECHANICS

RECAP: We have data that allow us to investigate whether the outcomes of traffic stops for suspicious activities are independent of the race of driver.

		Outcome			
		Charged	No Action	Warning	Total
Driver Race	Black	16	95	31	142
	Middle East	15	126	45	186
	Other	9	65	33	107
	White/Caucasian	41	304	133	478
	Total	81	590	242	913

QUESTIONS: What are the degrees of freedom for this test? What is the expected frequency of warnings to Middle Eastern drivers? What is that cell's component in the χ^2 computation? And how is the standardized residual for that cell computed?

ANSWER: This is a 4 × 3 contingency table, so $df = (4 - 1)(3 - 1) = 6$. Overall, 242 of 913 stops resulted in warnings. If outcomes are independent of race, then I'd expect 242/913 of the 186 Middle Eastern drivers or $\dfrac{(186 \times 242)}{913} = 49.30$ of them to receive a warning.

That cell's term in the χ^2 calculation is

$$\frac{(Obs - Exp)^2}{Exp} = \frac{(45 - 49.30)^2}{49.30} = 0.375$$

The standardized residual for that cell is

$$-\sqrt{0.375} = -0.61$$

Examine the Residuals

Each cell of the contingency table contributes a term to the chi-square sum. As we did earlier, we should examine the residuals when we have rejected the null hypothesis. In this example, we have an additional concern that the cells with small expected frequencies might be the ones responsible for making the chi-square statistic large.

Our interest in the data arises from the potential for improving public health. If patients with tattoos are more likely to test positive for hepatitis C, perhaps physicians should be advised to suggest blood tests for such patients.

The standardized residuals are displayed below (Table 22.6):

Table 22.6

Standardized residuals for the hepatitis and tattoos data. Are any of them particularly large in magnitude?

	Hepatitis C	No Hepatitis C
Tattoo, Parlour	6.628	−1.888
Tattoo, Elsewhere	1.598	−0.455
None	−2.661	0.758

THINK ➡

AGAIN

The chi-square value of 57.91 is the sum of the squares of these six values. The cell for people with tattoos obtained in a tattoo parlour who have hepatitis C is very large and positive, indicating there are many more people in that cell than the null hypothesis of independence would predict. Maybe tattoo parlours are a source of infection or maybe those who go to tattoo parlours also engage in risky behaviour.

The second-largest component is a negative value for those with no tattoos who test positive for hepatitis C. A negative value says that there are fewer people in this cell than independence would expect. That is, those who have no tattoos are less likely to be infected with hepatitis C than we might expect if the two variables were independent.

What about the cells with small expected counts? The formula for the chi-square standardized residuals divides each residual by the square root of the expected frequency. Too small an expected frequency can arbitrarily inflate the residual and lead to an inflated chi-square statistic. Any expected count close to the arbitrary minimum of 5 calls for checking that cell's standardized residual to be sure it is not particularly large. In this case, the standardized residual for the "Hepatitis C and Tattoo, Elsewhere" cell is not particularly large, but the standardized residual for the "Hepatitis C and Tattoo, Parlour" cell is large.

SHOW ➡

MORE

We might choose not to report the results because of concern with the small expected frequency. Alternatively, we could include a warning along with our report of the results (and point out that the chi-square value would still be highly significant even without the contributions of cells with low expected counts). Yet another approach is to combine categories to get a larger count and correspondingly larger expected counts, if there are some categories that can be appropriately combined. Here, we might naturally combine the two rows for tattoos, obtaining the 2 × 2 table below (Table 22.7):

Table 22.7

Combining the two tattoo categories gives a table with all expected counts greater than 5.

	Hepatitis C	No Hepatitis C	Total
Tattoo	25	88	113
None	22	491	513
Total	47	579	626

TELL ➡

ALL

This table has expected values of at least 5 in every cell, and a chi-square value of 42.42 on 1 degree of freedom. The corresponding P-value is < 0.0001.

We conclude that *Tattoo Status* and *Hepatitis C Status* are not independent, and there is a higher hepatitis C incidence among those with tattoos. To check if tattoo parlours are a particular problem, we could now do an additional test. Remove the individuals without tattoos, to get a 2 × 2 table, showing tattoo source by hepatitis status for the 113 tattooed folks. Then do another chi-square test (or two proportions *z*-test) on that 2 × 2 table, to assess the relationship between hepatitis and tattoo source.

For Example   WRITING CONCLUSIONS FOR χ^2 TESTS

RECAP: We're looking at Ottawa traffic stop data to see if they show evidence of a relationship between race of driver and stop outcome, for the stops due to suspicious activities. With 6 df, technology calculates $\chi^2 = 4.466$, a P-value of approximately 0.614, and these standardized residuals:

		Outcome	
	Charged	**No Action**	**Warning**
Black	0.9585	0.3379	−1.0821
Middle East	−0.3697	0.5293	−0.6126
Other	−0.1600	−0.4986	0.8710
White/Caucasian	−0.2161	−0.2784	0.5598

(Driver Race)

QUESTIONS: What is your conclusion?

ANSWER: The high P-value leads me to conclude that there is no evidence that outcomes of stops for suspicious activities are associated with driver race.

It should be no surprise that none of the standardized residuals are large, since the chi-square test pooled the evidence from all the cells, and found non-significance.

Just Checking

Which of the three chi-square tests—goodness-of-fit, homogeneity, or independence—would you use in each of the following situations?

7. A restaurant manager wonders whether customers who dine on Friday nights have the same preferences among the four "chef's special" entrées as those who dine on Saturday nights. One weekend he has the wait staff record which entrées were ordered each night. Assuming these customers to be typical of all weekend diners, he'll compare the distributions of meals chosen Friday and Saturday.

8. Company policy calls for parking spaces to be assigned to everyone at random, but you suspect that may not be so. There are three lots of equal size: lot A, next to the building; lot B, a bit farther away; and lot C, on the other side of the highway. You gather data about employees at middle management level and above to see how many were assigned parking in each lot.

9. Is a student's social life affected by where the student lives? A campus survey asked a random sample of students whether they lived in a dormitory, in off-campus housing, or at home, and whether they had been out on a date 0, 1–2, 3–4, or 5 or more times in the past two weeks.

Chi-Square and Causation

Chi-square tests are common. Tests for independence are especially widespread. Unfortunately, many people interpret a small P-value as proof of causation. We know better. Just as correlation between quantitative variables does not demonstrate causation, a failure of independence between two categorical variables does not show a cause-and-effect relationship between them, nor should we say that one variable *depends* on the other.

The chi-square test for independence treats the two variables symmetrically. There is no way to differentiate the direction of any possible causation from one variable to the other. In our example, it is unlikely that having hepatitis causes one to crave a tattoo, but other examples are not so clear.

In this case, it's easy to imagine that lurking variables are responsible for the observed lack of independence. Perhaps the lifestyles of some people include both tattoos and behaviours that put them at increased risk of hepatitis C, such as body piercings or even drug use. Even a small subpopulation of people with such a lifestyle among those with tattoos might be enough to create the observed result. After all, we observed only 25 patients with both tattoos and hepatitis.

In some sense, a failure of independence between two categorical variables is less impressive than a strong, consistent, linear association between quantitative variables. Two categorical variables can fail the test of independence in many ways, including ways that show no consistent pattern of failure. Examination of the chi-square standardized residuals can help you think about the underlying patterns.

Section 22.4 Exercises

9. Internet use poll A Pew Research poll in April 2009 from a random sample of U.S. adults asked the questions "Did you use the Internet yesterday?" and "Are you White, Black, or Hispanic/Other?" Is the response to the question about the Internet independent of race? The data follow:

	Did You Use the Internet Yesterday?	
	Yes	No
White	2546	856
Black	314	146
Hispanic/Other	431	174

a) Under the null hypothesis, what are the expected values?
b) Compute the χ^2 statistic.
c) How many degrees of freedom does it have?
d) Find the P-value.
e) What do you conclude?

10. Internet use poll, II The same poll as in Exercise 9 also asked the questions "Did you use the Internet yesterday?" and "What is your educational level?" Is the response to the question about the Internet independent of educational level? The data follow:

	Did You Use the Internet Yesterday?	
	Yes	No
Less Than High School	209	131
High School	932	550
Some College	958	346
College Grad	1447	247

a) Under the null hypothesis, what are the expected values?
b) Compute the χ^2 statistic.
c) How many degrees of freedom does it have?
d) Find the P-value.
e) What do you conclude?

WHAT CAN GO WRONG?

- **Don't use chi-square methods unless you have counts.** All three of the chi-square tests apply only to counts (which are summaries of categorical data). Other kinds of data can be arrayed in two-way tables. Just because numbers are in a two-way table doesn't make them suitable for chi-square analysis. Data reported as proportions or percentages can be suitable for chi-square procedures, *but only after they are converted to counts.* If you try to do the calculations without first finding the counts, your results will be wrong.

- **Beware large samples.** Be wary of chi-square tests performed on very large samples. No hypothesized distribution fits perfectly, no two groups are exactly homogeneous, and two variables are rarely perfectly independent. The degrees of freedom for chi-square tests don't grow with the sample size. With a sufficiently large sample size, a chi-square test can

(continued)

almost always reject the null hypothesis. Confidence intervals for proportions and differences between proportions, along with bar charts or pie charts, will help you assess the practical importance of your findings.

■ **Don't say that one variable "depends" on the other just because they're not independent.** Dependence suggests a model or a pattern and implies causation, but variables can fail to be independent in many different ways. When variables fail the test for independence, you might just say they are "associated."

What Have We Learned?

Recognize hypotheses about categorical variables. The tests are based on counts of data in categories, and all rely on chi-square models, a new family indexed by its degrees of freedom.

■ Goodness-of-fit tests compare the observed distribution of a single categorical variable to an expected distribution based on a theory or model.

■ Tests of homogeneity compare the distribution of a categorical variable in several groups.

■ Tests of independence examine counts from a single group for evidence of an association between two categorical variables.

For each test, find the expected cell frequencies. Evaluate the chi-square statistic, and know the correct *df*. Find the *P*-value using the upper tail only.

For each test, check the assumptions and corresponding conditions:

■ Counted data condition.

■ Independence assumption; randomization makes independence more plausible.

■ Sample size assumption with the expected cell frequency condition: expect a count of at least 5 in each cell.

Interpret a chi-square test.

■ Even though we might believe the model, we cannot prove that the data fit the model with a chi-square test because that would mean confirming the null hypothesis.

Examine the standardized residuals to understand what cells were responsible for rejecting a null hypothesis.

Understand how the null hypothesis for a test of independence differs from the null hypothesis for a test of homogeneity.

■ Both are computed the same way. You may not find both offered by your technology.

You can use either one as long as you interpret your result correctly.

Review of Terms

Chi-square goodness-of-fit test A test of whether the distribution of counts in one categorical variable matches the distribution predicted by a model is called a test of goodness-of-fit. In a chi-square goodness-of-fit test, the expected counts come from the predicting model. The test finds a P-value from a chi-square model with k −1 degrees of freedom, where k is the number of categories of the categorical variable (and not the sample size) (p. 690).

Cell One element of a two-way table corresponding to a specific row and a specific column. Table cells can hold counts, percentages, or measurements on other variables, or they can hold several values (p. 692).

Chi-square statistic Can be used to test whether the observed counts in a frequency distribution or contingency table match the counts we would expect according to some model. It is calculated as

$$\chi^2 = \sum_{all\ cells} \frac{(Obs - Exp)^2}{Exp}$$

Chi-square statistics differ in how expected counts and degrees of freedom are found, depending on the question asked (p. 693).

Chi-square model
Chi-square models are skewed to the right. They are *parameterized by their degrees of freedom* and become less skewed with increasing degrees of freedom (p. 693).

Chi-square component
The components of a chi-square calculation are

$$\frac{(Observed - Expected)^2}{Expected}$$

found for each cell of the table (p. 696).

Two-way table
Each *cell* of a two-way table shows counts of individuals. One way classifies a sample according to a categorical variable. The other way can classify different groups of individuals on the same variable or classify the same individuals according to a different categorical variable (p. 698).

Chi-square test of homogeneity
A test comparing the distribution of *one categorical variable* in *two or more groups*. A chi-square test of homogeneity finds expected counts based on the overall frequencies, adjusted for the totals in each group under the (null hypothesis) assumption that the distributions are the same for each group. We find a P-value from a chi-square distribution with $(\#Rows - 1) \times (\#Cols - 1)$ degrees of freedom, where $\#Rows$ gives the number of categories for the categorical response variable and $\#Cols$ gives the number of independent groups (or vice versa) (p. 699).

Standardized residual
In each cell of a two-way table, a standardized residual is the square root of the chi-square component for that cell with the sign of the *Observed – Expected* difference:

$$\frac{(Obs - Exp)}{\sqrt{Exp}}$$

When we reject a chi-square test, an examination of the standardized residuals can sometimes reveal more about how the data deviate from the null model (p. 703).

Chi-square test of independence
A test of whether *two categorical random variables* are independent examines the distribution of counts for *one group of individuals* classified according to both variables. A chi-square test of independence finds expected counts by assuming that knowing the marginal totals tells us the cell frequencies, assuming that there is no association between the variables. This turns out to be the same calculation as a test of homogeneity. We find a P-value from a chi-square distribution with $(\#Rows - 1) \times (\#Cols - 1)$ degrees of freedom, where $\#Rows$ gives the number of categories in one variable and $\#Cols$ gives the number of categories in the other (p. 705).

Contingency table
A two-way table that classifies individuals according to two categorical variables (p. 705).

On the Computer CHI-SQUARE

Most statistics packages associate chi-square tests with contingency tables. Often chi-square is available as an option only when you make a contingency table. This organization can make it hard to locate the chi-square test and may confuse the three different roles that the chi-square test can take. In particular, chi-square tests for goodness-of-fit may be hard to find or missing entirely. Chi-square tests for homogeneity are computationally the same as chi-square tests for independence, so you may have to perform the mechanics as if they were tests of independence and interpret them afterward as tests of homogeneity.

Most statistics packages work well with the raw categorical data on individuals. Many packages will let you enter the summary counts, or let you choose either type of input. Some packages offer a way to reconstruct the raw data from the summary counts so they can then be passed back through the chi-square calculation, finding the cell counts again. Many packages offer chi-square standardized residuals (although they may be called something else).

EXCEL

Excel offers the function:

CHITEST (actual_ range, expected_range), which computes a chi-square value for homogeneity. Both ranges are of the form UpperleftCell:LowerRightCell, specifying two rectangular tables that must hold counts (although Excel will not check for integer values). The two tables must be of the same size and shape.

COMMENTS

Excel's documentation claims this is a test for independence and labels the input ranges accordingly, but Excel offers no way to find expected counts, so the function is not particularly useful for testing independence. You can use this function only if you already know both tables of counts or are willing to program additional calculations.

JMP

From the **Analyze** menu, select **Fit Y by X**.

- Select one variable as the Y or response variable, and the other as the X or factor variable. Both selected variables must be Nominal or Ordinal.

JMP will make a plot and a contingency table. Below the contingency table, JMP offers a **Tests** panel. In that panel, the chi-square for independence is called a **Pearson ChiSquare**. The table also offers the P-value.

- Click on the **Contingency Table** title bar to drop down a menu that offers to include a **Deviation** and **Cell Chi square** in each cell of the table.

COMMENTS

JMP will choose a chi-square analysis for a **Fit Y by X** if both variables are nominal or ordinal (marked with an N or O), but not otherwise. Be sure the variables have the right type. Deviations are the 'observed – expected' differences in counts. Cell chi-squares are the squares of the standardized residuals. Refer to the deviations for the sign of the difference.

Look under **Distributions** in the **Analyze** menu to find a chi-square test for goodness-of-fit.

MINITAB

- From the **Stat** menu, choose the **Tables** submenu.
- From that menu, choose **Chi Square Test . . .**
- In the dialogue, choose **Raw data . . .** or **Summarized data . . .** at the top. Identify the columns with the raw data or the columns that make up the table. Click on **Statistics . . .** in the dialog if you wish to request the standardized residuals.

Minitab will display the table and print the chi-square value and its P-value.

COMMENTS

Alternatively, select the **Cross Tabulation . . .** command to see even more options for the table.

R

Goodness-of-fit test:
To test the probabilities in a vector prob with observed values in a vector x,

- **chisq.test**(x,p=prob)

Test of Independence or homogeneity:

With counts in a contingency table (a matrix called say, con. table), the test of independence (or homogeneity) is found by

- **chisq.test**(con.table)

COMMENTS

Using the function xtabs you can create a contingency table from two variables x and y in a data frame called mydata by

- con.table = xtabs(x+y,data=mydata)
  then
- chisq.test(con.table)

SPSS

- From the **Analyze** menu, choose the **Descriptive Statistics** submenu.
- From that submenu, choose **Crosstabs . . . .**
- In the Crosstabs dialogue, assign the row and column variables from the variable list. Both variables must be categorical.
- Click the **Cells** button to specify that standardized residuals should be displayed.
- Click the **Statistics** button to specify a chi-square test.

COMMENTS

SPSS offers only variables that it knows to be categorical in the variable list for the Crosstabs dialogue. If the variables you want are missing, check that they have the right type.

STATCRUNCH

To perform a goodness-of-fit test:
- Enter the observed counts in one column of a data table, and the expected counts in another.
- Click on **Stat**.
- Choose **Goodness-of-fit » Chi-Square test**.
- Choose the **Observed Column** and the **Expected Column** and click on **Calculate**.

To perform a Test of Homogeneity or Independence:
- Create a table (without totals):
- Name the first column as one variable, enter the categories underneath.
- Name the adjacent columns as the categories of the other variable, entering the observed counts underneath.
- Click on **Stat**.
- Choose **Tables » Contingency » with summary**.
- Choose the **Columns** holding counts.
- Choose the **Row labels column**.
- Enter the **Column variable** name.
- Click on **Next** to ask for **expected counts**.
- Click on **Calculate**.

COMMENTS

These Chi-square tests may also be performed using the actual data table instead of summary counts. See the StatCrunch **Help page** for details.

TI-83/84 PLUS

To test goodness of fit:
- Enter the observed counts in one list and the expected counts in another list. Expected counts can be entered as n*p and the calculator will compute them for you.
- From the **STAT TESTS** menu, select χ^2 **GOF-Test**. Specify the lists where you stored the observed and expected counts.
- Enter the number of degrees of freedom, and then hit **Calculate**.

COMMENTS

The cell-by-cell components will be calculated automatically and stored in a list named **CNTRB**.

Chapter Exercises

T indicates data is available in a saved file at the Companion Website at www.pearsoncanada.ca/deveaux

11. **Which test? For each of the following situations, state whether you'd use a chi-square goodness-of-fit test, a chi-square test of homogeneity, a chi-square test of independence, or some other statistical test:**
    a) A brokerage firm wants to see whether the type of account a customer has (Silver, Gold, or Platinum) affects the type of trades that customer makes (in person, by phone, or on the Internet). It collects a random sample of trades made for its customers over the past year and performs a test.
    b) That brokerage firm also wants to know if the type of account affects the size of the account (in dollars). It performs a test to see if the mean size of the account is the same for the three account types.
    c) The academic research office at a large university wants to see whether the distribution of courses chosen (Humanities, Social Science, or Science) is different for its residential and nonresidential students. It assembles last semester's data and performs a test.

12. **Which test again?** For each of the following situations, state whether you'd use a chi-square goodness-of-fit test, a chi-square test of homogeneity, a chi-square test of independence, or some other statistical test:
    a) Is the quality of a car affected by what day it was built? A car manufacturer examines a random sample of the warranty claims filed over the past two years to test whether defects are randomly distributed across days of the work week.
    b) A medical researcher wants to know if blood cholesterol level is related to heart disease. She examines a database of 10 000 patients, testing whether the cholesterol level (in milligrams) is related to whether or not a person has heart disease.
    c) A student wants to find out whether political leaning (liberal, moderate, or conservative) is related to choice of major. He surveys 500 randomly chosen students and performs a test.

13. **Dice** After getting trounced by your little brother in a children's game, you suspect the die he gave you to roll may be unfair. To check, you roll it 60 times, recording the number of times each face appears. Do these results cast doubt on the die's fairness?

Face	Count
1	11
2	7
3	9
4	15
5	12
6	6

a) If the die is fair, how many times would you expect each face to show?
b) To see if these results are unusual, will you test goodness-of-fit, homogeneity, or independence?
c) State your hypotheses.
d) Check the conditions.
e) How many degrees of freedom are there?
f) Find χ^2 and the P-value.
g) State your conclusion.

14. **M&M's** As noted in an earlier chapter, the Masterfoods Company says that until very recently yellow candies made up 20% of its milk chocolate M&M's, red another 20%, and orange, blue, and green 10% each. The rest are brown. On his way home from work the day he was writing these exercises, one of the authors bought a bag of plain M&M's. He got 29 yellow ones, 23 red, 12 orange, 14 blue, 8 green, and 20 brown. Is this sample consistent with the company's stated proportions? Test an appropriate hypothesis and state your conclusion.
    a) If the M&M's are packaged in the stated proportions, how many of each colour should the author have expected to get in his bag?
    b) To see if his bag was unusual, should he test goodness-of-fit, homogeneity, or independence?
    c) State the hypotheses.
    d) Check the conditions.
    e) How many degrees of freedom are there?
    f) Find χ^2 and the P-value.
    g) State a conclusion.

15. **Nuts** A company says its premium mixture of nuts contains 10% Brazil nuts, 20% cashews, 20% almonds, and 10% hazelnuts, and the rest are peanuts. You buy a large can and separate the various kinds of nuts. Upon weighing them, you find there are 112 grams of Brazil nuts, 183 grams of cashews, 207 grams of almonds, 71 grams of hazelnuts, and 446 grams of peanuts. You wonder whether your mix is significantly different from what the company advertises.
    a) Explain why the chi-square goodness-of-fit test is not an appropriate way to find out.
    b) What might you do instead of weighing the nuts in order to use a χ^2 test?

16. **Mileage** A salesperson who is on the road visiting clients thinks that, on average, he drives the same distance each day of the week. He keeps track of his mileage for several weeks and discovers that he averages 122 kilometres on Mondays, 203 kilometres on Tuesdays, 176 kilometres on Wednesdays, 181 kilometres on Thursdays, and 108 kilometres on Fridays. He wonders if this evidence contradicts his belief in a uniform distribution of

kilometres across the days of the week. Explain why it is not appropriate to test his hypothesis using the chi-square goodness-of-fit test.

17. Blood type According to geneticists, in a general population, 42% of people are type A, 10% are type B, 3% are type AB, and 45% are type O. In 1919, a study was conducted for variance populations. In particular, a sample of 116 residents on Bongainville Island was tested. It was observed that 74 were type A, 12 were type B, 11 were type AB, and 19 were type O. Do the blood types of residents on Bongainville Island reflect the general population? Test an appropriate hypothesis and state your conclusion.

18. Violence against women 2009 In its study *When Men Murder Women: An Analysis of 2009 Homicide Data*, 2011, the Violence Policy Center (www.vpc.org) reported that 1818 women were murdered by men in 2009. Of these victims, a weapon could be identified for 1654 of them. Of those for whom a weapon could be identified, 861 were killed by guns, 364 by knives or other cutting instruments, 214 by other weapons, and 215 by personal attack (battery, strangulation, etc.). The FBI's Uniform Crime Report says that, among all murders nationwide, the weapon use rates were as follows: guns 63.4%, knives 13.1%, other weapons 16.8%, personal attack 6.7%. Is there evidence that violence against women involves different weapons than other violent attacks in the United States?

19. Fruit flies Offspring of certain fruit flies may have yellow or ebony bodies and normal wings or short wings. Genetic theory predicts that these traits will appear in the ratio 9:3:3:1 (9 yellow, normal: 3 yellow, short: 3 ebony, normal: 1 ebony, short). A researcher checks 100 such flies and finds the distribution of the traits to be 59, 20, 11, and 10, respectively.
a) Are the results this researcher observed consistent with the theoretical distribution predicted by the genetic model?
b) If the researcher had examined 200 flies and counted exactly twice as many in each category—118, 40, 22, 20—what conclusion would he have reached?
c) Why is there a discrepancy between the two conclusions?

T 20. Pi Many people know the mathematical constant π is approximately 3.14. But that's not exact. To be more precise, here are 20 decimal places: 3.14159265358979323846. Still not exact, though. In fact, the actual value is irrational, a decimal that goes on forever without any repeating pattern. But notice that there are no 0s and only one 7 in the 20 decimal places above. Does that pattern persist, or do all the digits show up with equal frequency? The table shows the number of times each digit appears in the first million digits. Test the hypothesis that the digits 0 through 9 are uniformly distributed in the decimal representation of π.

The first million digits of π	
Digit	Count
0	99 959
1	99 758
2	100 026
3	100 229
4	100 230
5	100 359
6	99 548
7	99 800
8	99 985
9	100 106

T 21. Hurricane frequencies The U.S. National Hurricane Center provides data that list the numbers of large (category 3, 4, or 5) hurricanes that have struck the United States by decade since 1851.[7] Here are the data:

Decade	Count	Decade	Count
1851–1860	6	1931–1940	8
1861–1870	1	1941–1950	10
1871–1880	7	1951–1960	9
1881–1890	5	1961–1970	6
1891–1900	8	1971–1980	4
1901–1910	4	1981–1990	4
1911–1920	7	1991–2000	5
1921–1930	5	2001–2010	7

Recently, there's been some concern that perhaps the number of large hurricanes has been increasing. The natural null hypothesis would be that the frequency of such hurricanes has remained constant.
a) With 96 large hurricanes observed over the 16 periods, what are the expected value(s) for each cell?
b) What kind of chi-square test would be appropriate?
c) State the null and alternative hypotheses.
d) How many degrees of freedom are there?
e) The value of χ^2 is 12.67. What is the P-value of the test?
f) State your conclusion.
g) Look again at the last decade. Suppose we only had data up to 2006 (where in fact the hurricane count was already at 7). Would this alter our conclusion at all?

T 22. Lottery numbers The fairness of Lotto 6/49 was recently called into question. The lottery publishes historical statistics on its website.[8] The table below shows the

---

[7]www.nhc.noaa.gov
[8]www.olg.ca/lotteries/viewFrequentWinningNumbers.do

number of times each of the 49 numbers was drawn in the main lottery and as the "bonus" number as of July 2009.

We wonder if all the numbers are equally likely to be the bonus number.

Number	Count	Bonus	Number	Count	Bonus
1	343	57	26	344	59
2	338	39	27	367	57
3	339	61	28	314	51
4	342	48	29	350	37
5	333	60	30	354	53
6	333	39	31	386	56
7	362	55	32	360	52
8	337	59	33	338	54
9	338	56	34	391	63
10	319	51	35	334	47
11	322	74	36	343	64
12	334	54	37	348	59
13	327	58	38	348	55
14	317	49	39	349	66
15	309	30	40	376	46
16	338	52	41	341	65
17	340	50	42	347	60
18	329	62	43	376	58
19	352	51	44	339	41
20	374	47	45	369	55
21	345	59	46	371	55
22	329	56	47	374	59
23	339	59	48	328	45
24	317	59	49	338	51
25	325	69			

a) What kind of test should we perform?
b) What are the appropriate expected value(s) for the test?
c) State the null and alternative hypotheses.
d) How many degrees of freedom are there?
e) What is the P-value of the test?
f) State your conclusion.
g) Can you see any possible problem with doing this same analysis on the non-bonus number selections?

23. Childbirth, part 1 There is some concern that if a woman has an epidural to reduce pain during childbirth, the drug can get into the baby's bloodstream, making the baby sleepier and less willing to breastfeed. In December 2006, the *International Breastfeeding Journal* published results of a study conducted at Sydney University. Researchers followed up on 1178 births, noting whether the mother had an epidural and whether the baby was still nursing after six months. The results are summarized in the following table.

		Epidural?		
		Yes	No	Total
Breastfeeding @ 6 months?	Yes	206	498	704
	No	190	284	474
	Total	396	782	1178

a) What kind of test would be appropriate?
b) State the null and alternative hypotheses.

24. Does your doctor know? A survey[9] of articles from the *New England Journal of Medicine* (NEJM) classified them according to the principal statistics methods used. The articles recorded were all noneditorial pieces appearing during the indicated years. Let's look at whether these articles used statistics or not.

	Publication Year			
	1978–79	1989	2004–05	Total
No stats	90	14	40	144
Stats	242	101	271	614
Total	332	115	311	758

Has there been a change in the use of statistics?
a) What kind of test would be appropriate?
b) State the null and alternative hypotheses.

25. Childbirth, part 2 In Exercise 23, the table shows results of a study investigating whether the after-effects of epidurals administered during childbirth might interfere with successful breastfeeding. We're planning to do a chi-square test.
a) How many degrees of freedom are there?
b) The smallest expected count will be in the epidural/no breastfeeding cell. What is it?
c) Check the assumptions and conditions for inference.

26. Does your doctor know?, part 2 The table in Exercise 24 shows whether NEJM medical articles during various time periods included statistics or not. We're planning to do a chi-square test.
a) How many degrees of freedom are there?
b) The smallest expected count will be in the 1989/no stats cell. What is it?
c) Check the assumptions and conditions for inference.

27. Childbirth, part 3 In Exercises 23 and 25, we began examining the possible impact of epidurals on successful breastfeeding.
a) Calculate the component of chi-square for the epidural/no breastfeeding cell.
b) For this test, $\chi^2 = 14.87$. What is the P-value?
c) State your conclusion.

---

[9]Suzanne S. Switzer and Nicholas J. Horton, 2007, "What your doctor should know about statistics (but perhaps doesn't)." *Chance*, 20:1.

28. Does your doctor know?, part 3 In Exercises 24 and 26, we began examining whether the use of statistics in NEJM medical articles has changed over time.
   a) Calculate the component of chi-square for the 1989/no stats cell.
   b) For this test, $\chi^2 = 25.28$. What is the P-value?
   c) State your conclusion.

29. Childbirth, part 4 In Exercises 23, 25, and 27, we tested a hypothesis about the impact of epidurals on successful breastfeeding. The table shows the test's standardized residuals.

		Epidural?	
		Yes	No
Breastfeeding at 6 months?	Yes	−1.99	1.42
	No	2.43	−1.73

   a) Show how the standardized residual for the epidural/no breastfeeding cell was calculated.
   b) What can you conclude from the standardized residuals?

30. Does your doctor know?, part 4 In Exercises 24, 26, and 28, we tested a hypothesis about whether the use of statistics in NEJM medical articles has changed over time. The table shows the test's standardized residuals.

	1978–79	1989	2004–05
No stats	3.39	−1.68	−2.48
Stats	−1.64	0.81	1.20

   a) Show how the standardized residual for the 1989/no stats cell was calculated.
   b) What can you conclude from the patterns in the standardized residuals?

31. Childbirth, part 5 In Exercises 23, 25, 27, and 29, we looked at a study examining epidurals as one factor that might inhibit successful breastfeeding of newborn babies. Suppose a broader study included several additional issues, including whether the mother drank alcohol, whether this was a first child, and whether the parents occasionally supplemented breastfeeding with bottled formula. Why would it not be appropriate to use chi-square methods on the 2 × 8 table with yes/no columns for each potential factor?

32. Does your doctor know?, part 5 In Exercises 24, 26, 28, and 30, we considered data in articles in the NEJM. The original study listed 23 different statistics methods. (The list read: t-tests, contingency tables, linear regression. . . .) Why would it not be appropriate to use chi-square methods on the 23 × 3 table with a row for each method?

T 33. Titanic Here is a table we first saw in Chapter 2 showing who survived the sinking of the *Titanic* based on whether they were crew members or passengers booked in first-, second-, or third-class staterooms:

	Crew	First	Second	Third	Total
Alive	212	202	118	178	710
Dead	673	123	167	528	1491
Total	885	325	285	706	2201

   a) If we draw an individual at random from this table, what is the probability that we will draw a member of the crew?
   b) What is the probability of randomly selecting a third-class passenger who survived?
   c) What is the probability of a randomly selected passenger surviving, given that the passenger was a first-class passenger?
   d) If someone's chances of surviving were the same regardless of their status on the ship, how many members of the crew would you expect to have lived?
   e) State the null and alternative hypotheses we would test here.
   f) Give the degrees of freedom for the test.
   g) The chi-square value for the table is 187.8, and the corresponding P-value is barely greater than 0. State your conclusions about the hypotheses.

T 34. NYPD and sex discrimination The table shows the rank attained by male and female officers in the New York City Police Department (NYPD). Do these data indicate that men and women are equitably represented at all levels of the department?

Rank	Male	Female
Officer	21 900	4 281
Detective	4 058	806
Sergeant	3 898	415
Lieutenant	1 333	89
Captain	359	12
Higher ranks	218	10

   a) What is the probability that a person selected at random from the NYPD is a female?
   b) What is the probability that a person selected at random from the NYPD is a detective?
   c) Assuming no bias in promotions, how many female detectives would you expect the NYPD to have?
   d) To see if there is evidence of differences in ranks attained by males and females, will you test goodness-of-fit, homogeneity, or independence?
   e) State the hypotheses.
   f) Test the conditions.
   g) How many degrees of freedom are there?

h) Find χ^2 and the P-value.

i) State your conclusion.

j) If you concluded that the distributions are not the same, analyze the differences using the standardized residuals of your calculations.

35. *Titanic,* **again** Examine and comment on this table of the standardized residuals for the chi-square test you looked at in Exercise 33.

	Crew	First	Second	Third
Alive	−4.35	9.49	2.72	−3.30
Dead	3.00	−6.55	−1.88	2.27

36. **NYPD again** Examine and comment on this table of the standardized residuals for the chi-square test you looked at in Exercise 34.

	Male	Female
Officer	−2.34	5.57
Detective	−1.18	2.80
Sergeant	3.84	−9.14
Lieutenant	3.58	−8.52
Captain	2.46	−5.86
Higher ranks	1.74	−4.14

37. **Birth order and study choice** Students in an Introductory Statistics class at a large university were classified by birth order and by the discipline they study.

Birth Order
(1 = oldest or only child)

Discipline	1	2	3	4 or more	Total
Arts and Sciences	34	14	6	3	57
Agriculture	52	27	5	9	93
Social Science	15	17	8	3	43
Professional	13	11	1	6	31
Total	114	69	20	21	224

Expected Values
Birth Order
(1 = oldest or only child)

Discipline	1	2	3	4 or more
Arts and Sciences	29.0089	17.5580	5.0893	5.3438
Agriculture	47.3304	28.6473	8.3036	8.7188
Social Science	21.8839	13.2455	3.8393	4.0313
Professional	15.7768	9.5491	2.7679	2.9063

a) What kind of chi-square test is appropriate—goodness-of-fit, homogeneity, or independence?

b) State your hypotheses.

c) State and check the conditions.

d) How many degrees of freedom are there?

e) The calculation yields $\chi^2 = 17.78$ with P = 0.0378. State your conclusion.

f) Examine and comment on the standardized residuals below. Do they challenge your conclusion? Explain.

Standardized Residuals
Birth Order
(1 = oldest or only child)

Discipline	1	2	3	4 or more
Arts and Sciences	0.92667	20.84913	0.40370	21.01388
Agriculture	0.67876	20.30778	21.14640	0.09525
Social Science	−1.47155	1.03160	2.12350	20.51362
Professional	−0.69909	0.46952	21.06261	1.81476

38. **Bingeing and politics** Students in a large Statistics class at a university were asked to describe their political position and whether they had engaged in binge drinking (five drinks at a sitting for a man; four for a woman).

Political Position

Binge Drinking	Conservative	Moderate	Liberal	Total
In the past 3 days	7	19	14	40
In the past week	9	19	28	56
In the past month	3	8	23	34
Some other time	8	15	22	45
Never	9	22	20	51
Total	36	83	107	226

Expected Values

	Conservative	Moderate	Liberal
In the past 3 days	6.37168	14.6903	18.9381
In the past week	8.92035	20.5664	26.5133
In the past month	5.41593	12.4867	16.0973
Some other time	7.16814	16.5265	21.3053
Never	8.12389	18.7301	24.1460

a) What kind of chi-square test is appropriate—goodness-of-fit, homogeneity, or independence?

b) State your hypotheses.

c) State and check the conditions.

d) How many degrees of freedom are there?

e) The calculation yields $\chi^2 = 10.10$, with P $= 0.2578$. State your conclusion.

f) Would you expect that a larger sample might find statistical significance? Explain.

T 39. Cranberry juice It's common folk wisdom that drinking cranberry juice can help prevent urinary tract infections in women. In 2001, the *British Medical Journal* reported the results of a Finnish study in which three groups of 50 women were monitored for these infections over six months. One group drank cranberry juice daily, another group drank a lactobacillus drink, and the third drank neither of those beverages, serving as a control group. In the control group, 18 women developed at least one infection, compared to 20 of those who consumed the lactobacillus drink and only 8 of those who drank cranberry juice. Does this study provide supporting evidence for the value of cranberry juice in warding off urinary tract infections?

a) Is this a survey, a retrospective study, a prospective study, or an experiment? Explain.

b) Will you test goodness-of-fit, homogeneity, or independence?

c) State the hypotheses.

d) Test the conditions.

e) How many degrees of freedom are there?

f) Find χ^2 and the P-value.

g) State your conclusion.

h) If you concluded that the groups are not the same, analyze the differences using the standardized residuals of your calculations.

T 40. Cars A random survey of autos parked in the student lot and in the staff lot at a large university classified the brands by country of origin, as seen in the table. Are there differences in the national origins of cars driven by students and staff?

		Driver	
		Student	**Staff**
Origin	American	107	105
	European	33	12
	Asian	55	47

a) Is this a test of independence or homogeneity?

b) Write appropriate hypotheses.

c) Check the necessary assumptions and conditions.

d) Find the P-value of your test.

e) State your conclusion and analysis.

41. Computer-assisted ordering In the book *Essentials of Marketing Research*, Dillon, Madden, and Firtle discuss the relationship between delivery time and computer-assisted ordering. The table shows data from a sample of 40 firms. Is there any evidence of an association between computer-assisted ordering and delivery time? Do a test.

	Below Industry Average	Equal to Industry Average	Above Industry Average
Not computer-assisted	4	12	8
Computer-assisted	10	4	2

a) Is this a test of homogeneity or independence?

b) Write an appropriate hypothesis.

c) Are the conditions for inference satisfied?

d) Find the P-value for your test.

e) State a complete conclusion.

T 42. Fish diet Medical researchers followed 6272 Swedish men for 30 years to see if there was any association between the amount of fish in their diet and prostate cancer.[10]

Fish Consumption		Total Subjects	Prostate Cancers
	Never/seldom	124	14
	Small part of diet	2621	201
	Moderate part	2978	209
	Large part	549	42

a) Is this a survey, a retrospective study, a prospective study, or an experiment? Explain.

b) Is this a test of homogeneity or independence?

c) Do you see evidence of an association between the amount of fish in a man's diet and his risk of developing prostate cancer?

d) Does this study prove that eating fish does not prevent prostate cancer? Explain.

43. Grades Two professors teach an introductory Statistics course. The table shows the distribution of final grades they awarded. Is one of these professors an "easier" grader?

	Prof. Alpha	Prof. Beta
A	3	9
B	11	12
C	14	8
D	9	2
F	3	1

a) Will you test goodness-of-fit, homogeneity, or independence?

b) Write appropriate null hypotheses.

c) Find the expected counts for each cell, and explain why the chi-square procedures are not appropriate for this table.

---

[10]"Fatty fish consumption and risk of prostate cancer," *Lancet*, June 2001.

T 44. **Full moon** Some people believe that a full moon elicits unusual behaviour in people. The table shows the number of arrests made in a small town during weeks of six full moons and six other randomly selected weeks in the same year. We wonder if there is evidence of a difference in the types of illegal activity that take place.

	Full Moon	Not Full
Violent (murder, assault, rape, etc.)	2	3
Property (burglary, vandalism, etc.)	17	21
Drugs/Alcohol	27	19
Domestic abuse	11	14
Other offenses	9	6

a) Will you test goodness-of-fit, homogeneity, or independence?
b) Write appropriate null hypotheses.
c) Find the expected counts for each cell, and explain why the chi-square procedures are not appropriate for this table.

45. **Grades, again** In some situations where the expected cell counts are too small, as in the case of the grades given by Professors Alpha and Beta in Exercise 43, we can complete an analysis anyway. We can often proceed after combining cells in some way that makes sense and also produces a table in which the conditions are satisfied. Here we create a new table displaying the same data, but calling D and F grades "Below C":

	Prof. Alpha	Prof. Beta
A	3	9
B	11	12
C	14	8
Below C	12	3

a) Find the expected counts for each cell in this new table, and explain why a chi-square procedure is now appropriate.
b) With this change in the table, what has happened to the number of degrees of freedom?
c) Test your hypothesis about the two professors, and state an appropriate conclusion.

T 46. **Full moon, next phase** In Exercise 44, you found that the expected cell counts failed to satisfy the conditions for inference.
a) Find a sensible way to combine some cells that will make the expected counts acceptable.
b) Test a hypothesis about the full moon and state your conclusion.

47. **Eating in front of the TV** Roper Reports asked a random sample of people in 30 countries whether they agreed with the statement "I like to nibble while reading or watching TV." Allowable responses were "Agree completely," "Agree somewhat," "Neither disagree nor agree," "Disagree somewhat," "Disagree completely," and "I Don't Know/No Response." Does a person's age influence their response? Here are data from 3792 respondents in the 2006 sample of five countries (China, India, France, United Kingdom, and United States) for three age groups (teens, 30s (30–39), and over 60):

	Agree Completely	Agree Somewhat	Neither Disagree nor Agree	Disagree Somewhat	Disagree Completely
Teen	369	540	299	175	106
30s	272	522	325	229	170
60+	93	207	153	154	178

a) Make an appropriate display of these data.
b) Does a person's age seem to affect their response to the question about nibbling?

48. **Eating in front of the TV, II** In Exercise 47, we saw a survey of people who were asked if they agreed with the statement "I like to nibble while reading or watching TV." Does a person's culture tend to influence the response? In the table are random samples from each of the five countries that responded:

	Agree Completely	Agree Somewhat	Neither Disagree nor Agree	Disagree Somewhat	Disagree Completely
China	253	479	382	244	139
France	188	407	328	296	308
India	315	395	246	227	332
U.K.	307	578	330	218	114
U.S.	308	632	331	201	73

a) Make an appropriate display of the data.
b) Does what country a person lives in seem to affect their response to the question about nibbling? Explain.

T 49. **Racial steering** A subtle form of racial discrimination in housing is "racial steering." Racial steering occurs when real estate agents show prospective buyers only homes in neighbourhoods already dominated by those buyers' race. This violates the *Canadian Human Rights Act*. According to an article in *Chance* magazine,[11] tenants at a large apartment complex recently filed a lawsuit alleging racial steering. The complex is divided into

---
[11]Vol. 14, no. 2 (2001).

two parts: Section A and Section B. The plaintiffs claimed that white potential renters were steered to Section A, while African-Americans were steered to Section B. The table displays the data that were presented in court to show the locations of recently rented apartments. Do you think there is evidence of racial steering?

New Renters

	White	Black	Total
Section A	87	8	95
Section B	83	34	117
Total	170	42	212

T **50.** *Titanic*, **redux** Newspaper headlines at the time, and traditional wisdom in the succeeding decades, have held that women and children escaped the *Titanic* in greater proportions than men. Here's a table with the relevant data. Do you think that survival was independent of whether the person was male or female? Defend your conclusion.

	Female	Male	Total
Alive	343	367	710
Dead	127	1364	1491
Total	470	1731	2201

51. **Steering, revisited** You could have checked the data in Exercise 49 for evidence of racial steering using two-proportion z procedures.
a) Find the z-value for this approach, and show that when you square your z-value, you get the value of χ^2 you calculated in Exercise 49.
b) Show that the resulting P-values are the same.

T **52.** **Survival on the *Titanic*, one more time** In Exercise 50, you could have checked for a difference in the chances of survival for men and women using two-proportion z procedures.
a) Find the z-value for this approach.
b) Show that the square of your calculated value of z is the value of χ^2 you calculated in Exercise 50.
c) Show that the resulting P-values are the same.

T **53.** **Race and education** Data from the U.S. Census Bureau show levels of education attained by age 30 for a sample of U.S. residents.

	Not HS Grad	HS Diploma	College Grad	Adv. Degree
White	810	6429	4725	1127
Black	263	1598	549	117
Hispanic	1031	1269	412	99
Other	66	341	305	197

Do these data highlight significant differences in education levels attained by these groups?

T **54.** **Pregnancies** Most pregnancies result in live births, but some end in miscarriages or stillbirths. A 2005 Statistics Canada report examined those outcomes, broken down by the age of the mother. The table shows counts consistent with that report. Is there evidence that the distribution of outcomes is not the same for these age groups? Do you have any concerns about your analysis?

		Live Births	Fetal Losses
Age of Mother	Under 20	14 013	586
	20–24	55 318	1422
	25–29	105 566	2121
	30–34	107 524	2161
	35 or over	59 672	2158

T **55.** **Race and education, part 2** Consider only the people who have graduated from high school from Exercise 53. Do these data suggest there are significant differences in opportunities for Black and Hispanic Americans who have completed high school to pursue university or advanced degrees?

	HS Diploma	College Grad	Adv. Degree
Black	1598	549	117
Hispanic	1269	412	99

T **56.** **Education by age** Use the survey results in the table to investigate differences in education level attained among different age groups in Canada.

		Age Group				
		25–34	35–44	45–54	55–64	≥ 65
Education Level	Not HS grad	27	50	52	71	101
	HS	82	19	88	83	59
	1–3 years college	43	56	26	20	20
	≥ 4 years college	48	75	34	26	20

57. **Ranking universities** In 2004, the Institute of Higher Education at Shanghai's Jiao Tong University evaluated the world's universities. Among their criteria were the size of the institution, the number of Nobel Prizes and Fields Medals won by faculty and alumni, and the faculty's research output. This ranking of the top 502 universities included 200 in North or Latin America, 209 in Europe, and 93 in the rest of the world (Asia, Pacific, and Africa). A closer examination of the top 100 showed 55 in the Americas, 37 in Europe, and 8 elsewhere.

Is there anything unusual about the geographical distribution of the world's top 100 universities?

58. **Ranking universities, redux** In Exercise 57, you read about the world's top universities, as ranked by the Institute of Higher Education. The table shows the geographical distribution of these universities by groups of 100. (Not all groups have exactly 100 because of ties in the rankings.) Do these institutions appear to be randomly distributed around the world, or does there appear to be an association between ranking and location?

		Ranking			
	1–100	101–200	201–300	301–400	401–500
North/Latin America	55	46	37	26	36
Europe	37	42	46	46	38
Asia/Africa/Pacific	8	13	17	30	25

(Location — row label for the three location rows)

Just Checking ANSWERS

1. We need to know how well beetles can survive six hours in a Plexiglas® box so that we have a baseline to compare the treatments.

2. There's no difference in survival rate in the three groups.

3. $(2-1)(3-1) = 2 \, df$

4. 50

5. 2

6. The mean value for a χ^2 with 2 df is 2, so 10 seems pretty large. The P-value is probably small.

7. This is a test of homogeneity. The clue is that the question asks whether the distributions are alike.

8. This is a test of goodness-of-fit. We want to test the model of equal assignment to all lots against what actually happened.

9. This is a test of independence. We have responses on two variables for the same individuals.

MathXL

MyLab Statistics

Go to MathXL at **www.mathxl.com** or MyLab™ Statistics at **www.mystatlab.com**. You can practise exercises for this chapter as often as you want. The guided solutions will help you find answers step by step. You'll find a personalized study plan available to you too!

CHAPTER

23

LEARNING OBJECTIVES

1. Understand the linear regression model and its assumptions

2. Assess model adequacy using scatterplots and residual plots

3. Construct a confidence interval and test hypotheses about the slope

4. Construct confidence intervals for the mean response or a single future response at some x-value

5. Perform a *t*-test for the correlation

6. Understand the Regression ANOVA table (optional)

7. Understand the difference between ordinary least-squares and logistic regression (optional)

Connections

A scatterplot of IQ versus brain size shows a mild positive association. Could this just be due to chance? A hypothesis test is needed. We know how to estimate the slope, but how reliable is our estimate? In this chapter, we'll apply what we've learned about tests and confidence intervals to making inferences about regression models.

When we have a quantitative response, the assumption of *constant variance* is very common. Here we assume a constant spread about

Inferences for Regression

Statisticians, like artists, have the bad habit of falling in love with their models.

—*George Box, famous statistician*

Three percent of a man's body is essential fat. (For a woman, the percentage is closer to 12.5%.) As the name implies, essential fat is necessary for a normal, healthy body. Fat is stored in small amounts throughout your body. Too much body fat, however, can be dangerous to your health. For men between 18 and 39 years old, a healthy percent body fat ranges from 8% to 19%. (For women of the same age, it's 21% to 32%.)

Measuring body fat can be tedious and expensive. The "standard reference" measurement is by dual-energy X-ray absorptiometry (DEXA), which involves two low-dose X-ray generators and takes from 10 to 20 minutes.

How close can we get to a useable prediction of body fat from easily measurable variables such as *Height*, *Weight*, or *Waist* size? Figure 23.1 shows a scatterplot of *%Body Fat* plotted against *Waist* size for a sample of 250 males of various ages. (Data in **Body fat complete**).

Figure 23.1

Percent *Body Fat* versus *Waist* size for 250 men of various ages. The scatterplot shows a strong, positive, linear relationship.

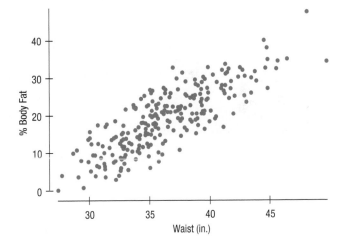

In Chapter 7, we modelled relationships like this by fitting a least squares line. The plot is clearly straight, so we can find that line. The equation of the least squares line for these data is

$$\widehat{\%Body\ Fat} = -42.7 + 1.7\ Waist$$

The slope says that, on average, *%Body Fat* is greater by 1.7% in men who have an additional inch around the waist.

How useful is this model? When we fit linear models before, we used them to describe the relationship between the variables and we interpreted the slope and intercept as descriptions of the data. Now, we'd like to know what the regression model can tell us beyond the 250 men in this study. To do that, we'll want to make confidence intervals and test hypotheses about the slope and (less frequently) intercept of the true regression line.

the line, which is essentially the same as the assumption of equal variances required for the pooled-*t* methods of Chapter 19 (and for the ANOVA methods of the next two chapters).

As usual with quantitative data, there is a Normality assumption, so histograms and normal probability plots come into play again. And as we saw in Chapter 7, scatterplots and residual plots are essential for assessing model adequacy.

23.1 A Regression Model

The Population and the Sample

When we found a confidence interval for a mean response, we could imagine a single, true underlying value for the mean. When we tested whether two means or two proportions were equal, we imagined a true underlying difference. But what does it mean to do inference for regression? We know better than to think that the data would line up perfectly on a straight line even if we knew every population value. After all, even in our sample, not all men who have 38-inch waists have the same *%Body Fat*. In fact, there's a whole distribution of *%Body Fat* for these men shown in Figure 23.2.

This is true at each *Waist* size. In fact, we could depict the distribution of *%Body Fat* at different *Waist* sizes as shown in Figure 23.3.

Who	250 male subjects
What	Body fat and waist size
Units	%Body fat and inches
When	1990s
Where	United States
Why	Scientific research

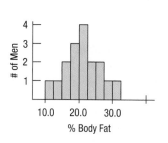

Figure 23.2

The distribution of *%Body Fat* for men with a *Waist* size of 38 inches is unimodal and symmetric.

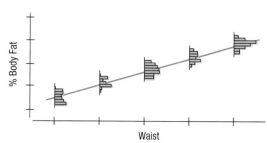

Figure 23.3

There's a distribution of *%Body Fat* for each value of *Waist* size. We'd like the means of these distributions to line up in a simple pattern.

Ivica Drusany/Shutterstock

■ **TERMINOLOGY ALERT**

The variable we are trying to predict is called the *y-variable* or *response* or *dependent variable*, while the variable whose observed value will be used to enable the prediction is called the *x-variable* or *explanatory variable* or *predictor*.

■ **NOTATION ALERT**

This time we used up only one Greek letter for two things. Lower-case Greek β (beta) is the natural choice to correspond to the *b*'s in the regression equation. We used β before for the probability of a Type II error, but there's little chance of confusion here.

But we want to *model* the relationship between *%Body Fat* and *Waist* size for all men. To do that, we imagine an idealized regression line. The model assumes that the *means* of the distributions of *%Body Fat* for each *Waist* size fall along the line, even though the individuals are scattered around it. We know that this model is not a perfect description of how the variables are associated, but it may be useful for predicting *%Body Fat* and for understanding how it's related to *Waist* size.

If only we had all the values in the population, we could find the slope and intercept of this *idealized regression line* explicitly by using least squares. Following our usual conventions, we write the idealized line with Greek letters and consider the coefficients (the slope and intercept) to be *parameters*: β_0 is the intercept and β_1 is the slope. Corresponding to our fitted line of $\hat{y} = b_0 + b_1x$, we write

$$\mu_y = \beta_0 + \beta_1x$$

Why μ_y instead of $\hat{y}$? Because this is a model. There is a distribution of *%Body Fat* for each *Waist* size. The model places the *means* of the distributions of *%Body Fat* for each *Waist* size on the same straight line.

Of course, not all the individual *y*'s are at these means. (In fact, the line will miss most—and quite possibly all—of the plotted points.) Some individuals lie above and some below the line, so, like all models, this one makes **errors**. Lots of them. In fact, one at each point. These errors are random and, of course, can be positive or negative. They are model errors, so we use a Greek letter and denote them by ε.

When we put the errors into the equation, we can account for each individual *y*:

$$y = \beta_0 + \beta_1x + \varepsilon$$

This equation is now true for each data point (since there is an ε to soak up the deviation), so the model gives a value of *y* for any value of *x*.

For the body fat data, an idealized model like this provides a summary of the relationship between *%Body Fat* and *Waist* size. Like all models, it simplifies the real situation.

We know there is more to predicting body fat than *Waist* size alone. But the advantage of a model is that the simplification might help us to think about the situation and assess how well *%Body Fat* can be predicted from simpler measurements.

We estimate the β's by finding a regression line, $\hat{y} = b_0 + b_1 x$, as we did in Chapter 7. The residuals, $e = y - \hat{y}$, are the sample-based versions of the errors, ε. We'll use them to help us assess the regression model. [We could also include a subscript 'i' to more explicitly label each response y_i and error ε_i corresponding to a particular case, i]

We know that least squares regression will give reasonable estimates of the parameters of this model from a random sample of data. Our challenge is to account for our uncertainty in how well they do. For that, we need to make some assumptions about the model and the errors.

Assumptions and Conditions

In Chapter 7, when we fit lines to data, we needed to check only the Straight Enough Condition. Now, when we want to make inferences about the coefficients of the line, we'll have to make more assumptions. Fortunately, we can check conditions to help us judge whether these assumptions are reasonable for our data. And as we've done before, we'll make some checks *after* we find the regression equation.

Also, we need to be careful about the order in which we check conditions. If our initial assumptions are not true, it makes no sense to check the later ones. So now we number the assumptions to keep them in order.

1. Linearity Assumption

If the true relationship is far from linear and we use a straight line to fit the data, our entire analysis will be useless, so we always check this first.

The **Straight Enough Condition** is satisfied if a scatterplot looks straight. It's generally not a good idea to draw a line through the scatterplot when checking. That can fool your eye into seeing the plot as more straight. Sometimes it's easier to see violations of the Straight Enough Condition by looking at a scatterplot of the residuals against x or against the predicted values, $\hat{y}$. That plot will have a horizontal direction and should have no pattern if the condition is satisfied. (For a model with only one explanatory variable x, a plotting the residuals against x will give the same information as plotting the residuals against $\hat{y}$. However, for multiple linear regression models, discussed in Chapter 26, scatterplots of residuals against individual explanatory variables and against $\hat{y}$ give different information.)

If the scatterplot is straight enough, we can go on to some assumptions about the errors. If not, stop here, or consider re-expressing the data to make the scatterplot more nearly linear. For the *%Body Fat* data, the scatterplot (Figure 23.4) is beautifully linear. Of course, the data must be quantitative for this to make sense. Check the **Quantitative Data Condition**.

2. Independence Assumption

The errors in the true underlying regression model (the ε's) must be mutually independent. As usual, there's no way to be sure that the Independence Assumption is true.

Usually, when we care about inference for the regression parameters, it's because we think our regression model might apply to a larger population. In such cases, we can check that the individuals are a representative (ideally, random) sample from that population.

We can also check displays of the regression residuals for evidence of patterns, trends, or clumping, any of which would suggest a failure of independence. In the special case when the observations are taken at different points in time, a common violation of the Independence Assumption is for the errors to be correlated. (The error our model makes this year may be similar to the one it made last year.) This violation can be checked by plotting the residuals against the order of occurrence, in other words, a time plot of the residuals, and looking for patterns. A further check can be made by plotting the residuals

CHECK THE SCATTERPLOT

The shape must be linear, or we can't use linear regression at all.

Figure 23.4

The relationship between *%Body Fat* and *Waist* size is straight enough.

CHECK THE RESIDUAL PLOT (1)

The residuals should appear to be randomly scattered.

Figure 23.5

The residuals show only random scatter when plotted against *Waist* size.

CHECK THE RESIDUAL PLOT (2)

The vertical spread of the residuals should be roughly the same everywhere.

Figure 23.6

A scatterplot of residuals against predicted values can reveal plot thickening. In this plot of the residuals from a regression of diamond prices on carat weight, we see that larger diamonds have more price variation than smaller diamonds. When the Equal Spread Assumption is violated, we can't summarize how the residuals vary with a single number.

CHECK A HISTOGRAM OF THE RESIDUALS

The distribution of the residuals should be unimodal and symmetric.

Figure 23.7

A histogram of the residuals is one way to check whether they are nearly Normal. Alternatively, we can look at a Normal probability plot.

Figure 23.8

The regression model has a distribution of *y*-values for each *x*-value. These distributions follow a Normal model with means lined up along the line and with the same standard deviations.

against those same residuals offset by one time position (also called "lagged"). Neither plot should show patterns. When residuals at time *t* are correlated with residuals at time $t - 1$, we describe this problem as *auto-correlation*, and a more advanced type of model, called a time-series model may be needed.

The *%Body Fat* data were collected on a sample of men taken to be representative. The subjects were not related in any way, so we can be pretty sure that their measurements are independent. The residual plot (Figure 23.5) shows no pattern.

3. Equal Variance Assumption

The variability of *y* should be about the same for all values of *x*. In Chapter 7, we looked at the standard deviation of the residuals (s_e) to measure the size of the scatter. Now, we'll need this standard deviation to build confidence intervals and test hypotheses. The standard deviation of the residuals is the building block for the standard errors of all the regression parameters. But it makes sense only if the scatter of the residuals is the same everywhere. In effect, the standard deviation of the residuals "pools" information across all of the individual distributions at each *x*-value, and pooled estimates are appropriate only when they combine information for groups with the same variance.

Practically, what we can check is the **Does the Plot Thicken? Condition**. A scatterplot of *y* against *x* offers a visual check. Fortunately, we've already made one. Make sure the spread around the line is nearly constant. Be alert for a "fan" shape or some other tendency for the variation to grow or shrink in one part of the scatterplot. Often, it is better to look at the residuals plotted against the predicted values, $\hat{y}$, as shown in Figure 23.6 (note a common problem—responses become more variable as they gets bigger). With the slope of the line removed, it's easier to see patterns left behind. For the body fat data, the spread of *%Body Fat* around the line is remarkably constant across Waist sizes from 30 inches to about 45 inches.

If the plot is straight enough, the data are independent, and the plot doesn't thicken, you can move on to the final assumption.

4. Normal Population Assumption

We assume the errors around the idealized regression line at each value of *x* follow a Normal model. This is the same as saying that the response *y* is Normally distributed at any *x* value, since errors are just responses shifted by a constant. We need this assumption so that we can use a Student's *t*-model for inference. The residuals estimate the errors, so we examine them.

As we have at other times when we've used Student's *t*, we'll settle for the residuals satisfying the **Nearly Normal Condition** and the **Outlier Condition**. Look at a histogram or Normal probability plot of the residuals.[1]

The histogram of residuals in the *%Body Fat* regression certainly looks nearly Normal (Figure 23.7). As we have noted before, the Normality Assumption becomes less important as the sample size grows, because the model is about means and the Central Limit Theorem takes over.

If all four assumptions were true, the idealized regression model would look like this:

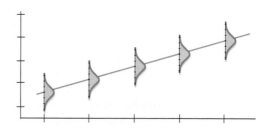

---

[1] *This* is why we have to check the conditions in order. We have to check that the residuals are independent and that the variation is the same for all *x*'s so that we can lump all the residuals together for a single check of the Nearly Normal Condition.

At each value of *x*, there is a distribution of *y*-values that follows a Normal model, and each of these Normal models is centred on the line and has the same standard deviation. Of course, we don't expect the assumptions to be exactly true, and we know that all models are wrong, but the linear model is often close enough to be very useful.

One final comment is in order. Our model only models random variation in the *y*-variable. In our example, the waist measurements were obtained from individuals in a random sample, so the *x*-variable was itself a random variable. In other regression studies, the *x*-variable is a controlled variable whose values are pre-set or controlled as in an experiment. Fortunately, it does not matter whether *x* is random or not, for regression inference—*as long as the x-values are measured accurately*. Just imagine if there were big measurement errors in the *x*-values. Move some points horizontally to the left, some to the right, and the fitted line could change quite a bit. Modelling such measurement errors in *x* would complicate things greatly!

Stocktrek/Photodisc/Getty Images

Who	39 impact craters
What	Diameter and Age
Units	km and millions of years ago
When	Past 35 million years
Where	Worldwide
Why	Scientific research

For Example CHECKING ASSUMPTIONS AND CONDITIONS

Look at the moon with binoculars or a telescope, and you'll see craters formed by thousands of impacts. Earth, being larger, has been hit even more often. Meteor Crater in Arizona was the first recognized impact crater and was identified as such only in the 1920s. With the help of satellite images, more and more craters have been identified; now more than 180 are known. These are only a small sample of all the impacts Earth has experienced: Only 29% of Earth's surface is land, and many craters have been covered or eroded away. Astronomers have recognized a roughly 35-million-year cycle in the frequency of cratering, although the cause of this cycle is not fully understood. Below is a scatterplot of the known impact craters from the most recent 35 million years.[2] (Data in **Craters**). We've taken logs of both age (in millions of years ago) and diameter (km) to make the relationship simpler and more linear. (Chapter 6 discusses re-expressing data to straighten scatterplots.)

QUESTION: Are the assumptions and conditions satisfied for fitting a linear regression model to these data?

ANSWER:

✓ **Linearity Assumption:** The scatterplot satisfies the Straight Enough Condition.

✓ **Independence Assumption:** Impacts are likely to be generally independent. However, these are the only known craters, and may differ from others that have disappeared and not yet been found. We'll need to be careful not to generalize our conclusions too broadly.

---

[2]Data, pictures, and much more information at the Earth Impact Database found at www.passc.net/EarthImpact-Database/index.html

✓ **Does the Plot Thicken? Condition:** After fitting a linear model, we find these residuals:

Two points seem to give the impression that the residuals may be more variable for higher predicted values than for lower ones, but this doesn't seem to be a serious violation of the Equal Variance Assumption.

✓ **Nearly Normal Condition:** A Normal probability plot suggests a bit of skewness in the distribution of residuals, and the histogram confirms that.

There are no violations severe enough to stop our regression analysis, but we'll be cautious about our conclusions.

> Truth will emerge more readily from error than from confusion.
>
> —*Francis Bacon*

Which Comes First: The Conditions or the Residuals?

In regression, there's a little catch. The best way to check many of the conditions is with the residuals, but we get the residuals only *after* we compute the regression. Before we compute the regression, however, we should check at least one of the conditions.

So we work in this order:

1. Make a scatterplot of the data and check the Straight Enough Condition. If the relationship looks curved, you can try re-expressing the data. If that doesn't work, you may need polynomial terms or a more complex type of model.[3]

2. If the data appear straight enough, fit a regression line and find the predicted values, $\hat{y}$, and the residuals, e. (Well, *your software will find them*, and plot them as needed or requested.)

3. Make a scatterplot of the residuals vs. x or vs. the predicted values. Most software programs will do this for you. The plot should have no obvious pattern. Check in

---

[3]Which means you'll have to continue studying Statistics!

particular for any bend (which would suggest that the data weren't all that straight after all), for any thickening (or thinning), and, of course, for any outliers. (If there are outliers, and you can correct them or justify removing them, do so and go back to step 1, or consider performing two regressions—one with and one without the outliers.)

4. If the data are measured over time, plot the residuals vs. time to check for evidence of patterns that might suggest the observations are not independent.

5. If the scatterplots look okay, then make a histogram and a Normal probability plot of the residuals to check the Nearly Normal Condition.

6. If you've made it this far, you are ready to go ahead with inference.

Step-by-Step Example REGRESSION INFERENCE

If our data can jump through all these hoops, we're ready to do regression inference. Let's see how much more we can learn about body fat and waist size from a regression model.

Questions: What is the relationship between *%Body Fat* and *Waist* size in men? What model best predicts body fat from waist size, and how well does it do the job?

THINK ➡ **Plan** Specify the question of interest.

Name the variables and report the *W*'s.

Identify the parameters you want to estimate.

Model Think about the assumptions and check the conditions.

Make pictures. For regression inference, you'll need a data scatterplot, a residual scatterplot, and either a histogram or a Normal probability plot of the residuals.

(We've seen plots of the residuals already. See Figures 23.6 and 23.7.)

I have quantitative body measurements on 250 adult males from the BYU Human Performance Research Center. I want to understand the relationship between *%Body Fat* and *Waist* size.

✓ **Straight Enough Condition**: There's no obvious bend in the original scatterplot of the data or in the plot of residuals against predicted values.

✓ **Independence Assumption**: These data are not collected over time, and there's no reason to think that the *%Body Fat* of one man influences the *%Body Fat* of another.

✓ **Does the Plot Thicken? Condition**: Neither the original scatterplot nor the residual scatterplot shows any changes in the spread about the line.

✓ **Nearly Normal Condition, Outlier Condition**: A histogram of the residuals is unimodal and symmetric. The Normal probability plot of the residuals is quite straight, indicating that the Normal model is reasonable for the errors.

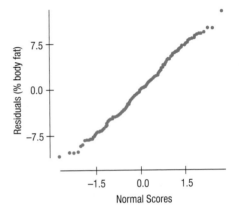

Choose your method.

Under these conditions a **regression model** is appropriate.

SHOW **Mechanics** Let's just "push the button" and see what the regression looks like.

The formula for the regression equation can be found in Chapter 7, and the standard error formulas will be shown a bit later, but regressions are almost always computed with a computer program or calculator.

Write the regression equation.

Here's the computer output for this regression:

Dependent variable is: %BF

R-squared = 67.8%

$s = 4.713$ with $250 - 2 = 248$ degrees of freedom

Variable	Coeff	SE(Coeff)	t-ratio	P-value
Intercept	−42.734	2.717	−15.7	<0.0001
Waist	1.70	0.0743	22.9	<0.0001

The estimated regression equation is

$$\widehat{\%\text{Body Fat}} = -42.73 + 1.70\ \text{Waist}$$

TELL ➡ **Conclusion** Interpret your results in context.

The R^2 for the regression is 67.8%. *Waist* size seems to account for about 2/3 of the *%Body Fat* variation in men. The slope of the regression says that *%Body Fat* increases by about 1.7 percentage points per additional inch of *Waist* size, on average.

More Interpretation We haven't worked it out in detail yet, but the output gives us numbers labelled as *t*-statistics and corresponding P-values, and we have a general idea of what those mean.

(Now it's time to learn more about regression inference so we can figure out what the rest of the output means.)

The standard error of 0.07 for the slope is much smaller than the slope itself, so it looks like the slope estimate is reasonably precise. And there are a couple of t-ratios and P-values given. Because the P-values are small, it appears that some null hypotheses can be rejected (to be discussed shortly).

Section 23.1 Exercises

1. Graduation rates 2013 A prestigious college is interested in factors that might be associated with better graduation rates. The administrators wonder whether there is a relationship between acceptance rates and graduation rates. Before proceeding with their regression inference, what conditions and assumptions must be satisfied? Are they? (Source: *US News and World Report* National University Rankings October 2013. http://colleges.usnews.rankingsandreviews.com/best-colleges/rankings/national-universities)

Response variable is Grad%
R-squared = 75.2%
s = 0.0254 with 50 − 2 = 48 degrees of freedom

Variable	Coefficient	SE(Coeff)	t-Ratio	P-Value
Intercept	0.975175	0.0074	131	≤0.0001
Acceptance %	−0.284895	0.0236	−12.1	≤0.0001

taken has an affect on the number of goals scored. The coach performed a preliminary analysis, using the scoring statistics from 293 offensive players who play professional hockey. He predicts *Goals* from number of *Shots* (taken for the season). Discuss each of the conditions and assumptions required for him to proceed with the regression analysis. (www.nhl.com)

Dependent variable is Goals
R-squared = 63.4%
s = 5.13 with 293 − 2 = 291 degrees of freedom

Variable	Coefficient	SE(Coeff)	t-Ratio	P-Value
Intercept	−1.77095	0.9087	−1.9488	0.052
Shots	0.11938	0.0053	22.460	<0.0001

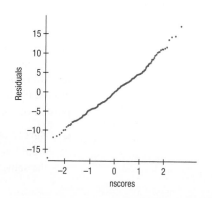

2. Shoot to score A college hockey coach collected data from the 2010–2011 National Hockey League season. He hopes to convince his players that the number of shots

23.2 Standard Errors of Parameter Estimates

Wait a minute! We've just pulled a fast one. We've pushed the "regression button" on our computer or calculator but haven't discussed where the standard errors for the slope or intercept come from. We know that if we had collected similar data on a different random sample of men, the slope and intercept would be different. Each sample would have produced its own regression line, with slightly different b_0's and b_1's. This sample-to-sample variation is what generates the sampling distributionsw for these coefficients.

There's only one regression model; each sample regression is trying to estimate the same parameters, β_0 and β_1. We expect any sample to produce a b_1 whose expected value is the true slope, β_1 (because unbiasedness is a consequence of the least-squares procedure). What about its standard deviation? What aspects of the data do you think will affect how much the slope (and intercept) vary from sample to sample?

- **Spread around the line.** Figure 23.9 below shows two situations in which we might do regression. Which situation would yield the more consistent (less variable) slope? That is, if we were to sample over and over from the two underlying populations that these samples come from and compute all the slopes, which group of slopes would vary less?

Figure 23.9

Which of these scatterplots shows a situation that would give the more consistent regression slope estimate if we were to sample repeatedly from its underlying population?

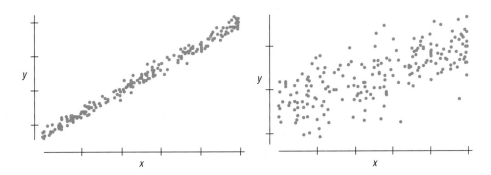

Clearly, data like those in the left plot will give more consistent slope estimates. The less variable the data, the less variable will be any estimates of parameters (all else equal).

Less scatter around the line means the slope will be more consistent from sample to sample. The spread around the line is measured with the **residual standard deviation**, s_e. You can always find s_e in the regression output, often just labelled s. Sometimes it is called the "standard error" although, as we know, that's not quite right—probably a misinterpretation of the s_e notation. You're not likely to calculate the residual standard deviation by hand. When we first saw this formula in Chapter 7, we said that it looks a lot like the standard deviation of y, only now subtracting the predicted values rather than the mean and dividing by $n - 2$ instead of $n - 1$:

$$s_e = \sqrt{\frac{\sum (y - \hat{y})^2}{n - 2}}$$

The less scatter around the line, the smaller the residual standard deviation and the stronger the relationship between x and y.

Some people prefer to assess the strength of a regression by looking at s_e rather than R^2. After all, s_e has the same units as y, and because it's the standard deviation of the errors around the line, it tells you how close the data are to our model. By contrast, R^2 is the proportion of the variation of y accounted for by x. Why not look at both?

- **Spread of the x values:** Figure 23.10 shows two more situations. Which of these would yield more consistent slope estimates (assuming the numbers on the x-axis are the same)?

$n - 2$?

For standard deviation (in Chapter 3), we divided by $n - 1$ because we didn't know the true mean and had to estimate it. Now, it's later in the course and there's even more we don't know. Here we don't know *two* things: the slope and the intercept. If we knew them both, we'd divide by n and have n degrees of freedom. When we *estimate* both, however, we adjust by subtracting 2 so we divide by $n - 2$ and (as we will soon see) have 2 fewer degrees of freedom.

Figure 23.10
Which of these scatterplots shows a situation that would give the more consistent regression slope estimates if we were to sample repeatedly from the underlying population?

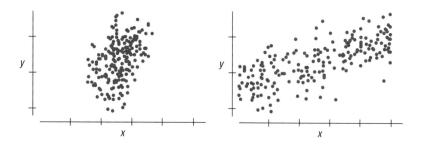

A plot like the one on the right has a broader range of x-values, so it gives a more stable base for the slope. We'd expect the slopes of samples from situations like that to vary less from sample to sample. A larger standard deviation of x, s_x, provides a more stable regression.

- **Sample size.** What about these two?

Figure 23.11
Which of these scatterplots shows a situation that would give the more consistent regression slope estimate if we were to sample repeatedly from the underlying population?

It shouldn't be a surprise that having a larger sample size, n, gives more consistent estimates from sample to sample.

First, and Most Importantly, *the Slope*

Three aspects of the scatterplot, then, affect the standard error of the regression slope:

- Spread around the line: s_e
- Spread of x values: s_x
- Sample size: n

These are in fact the *only* things that affect the standard error of the slope. Although you'll probably never have to calculate it by hand, one formula for the standard error of the slope is:

$$SE(b_1) = \frac{s_e}{\sqrt{n-1}\, s_x}$$

The error standard deviation, s_e, is in the *numerator*, since spread around the line *increases* the slope's standard error. The denominator has both a sample size term $\sqrt{n-1}$ and s_x, because increasing either of these *decreases* the slope's standard error.

We know the b_1's would vary from possible sample to possible sample. As you'd expect, their sampling distribution model is centred at β_1, the slope of the idealized regression line. Now we can estimate its standard deviation with $SE(b_1)$. What about its shape? Here William Gosset (Chapter 18) comes to the rescue again. When we standardize the slope by subtracting the model mean and dividing by its standard error, we get a Student's t-model, this time with $n - 2$ degrees of freedom:

$$\frac{b_1 - \beta_1}{SE(b_1)} \sim t_{n-2}$$

■ **NOTATION ALERT**

Don't confuse the standard deviation of the residuals, s_e, with the standard error of the slope, $SE(b_1)$. The first measures the scatter around the line, and the second tells us how reliably we can estimate the slope.

■ **NOTATION ALERT**

In Statistics, the '~' symbol is a shortcut for saying "Is distributed as." For the case on this page, it means the standardized slope is distributed as a t random variable with $n - 2$ degrees of freedom.

A SAMPLING DISTRIBUTION FOR REGRESSION SLOPES

When the assumptions and conditions are met, the standardized sample regression slope,

$$t = \frac{b_1 - \beta_1}{SE(b_1)}$$

follows a Student's t-model with $n - 2$ degrees of freedom. We estimate the standard error with

$$SE(b_1) = \frac{s_e}{\sqrt{n-1}\, s_x}, \text{ where } s_e = \sqrt{\frac{\sum(y - \hat{y})^2}{n - 2}}$$

n is the number of data values, and s_x is the ordinary standard deviation of the x-values.

For Example   FINDING STANDARD ERRORS

RECAP: Recent terrestrial impact craters seem to show a relationship between age and size that is linear when re-expressed using logarithms.

Here are summary statistics and regression output from a statistics package:

Variable	Count	Mean	StdDev
logAge	39	−0.656310	1.57682
logDiam	39	0.012600	1.04104

Dependent variable is: logDiam
R squared = 63.6%
s = 0.6362 with 39 − 2 = 37 degrees of freedom

Variable	Coefficient	SE(coeff)	t-ratio	P-value
Intercept	0.358262	0.1106	3.24	0.0025
logAge	0.526674	0.0655	8.05	<0.0001

QUESTIONS: How are the standard error of the slope and the t-ratio for the slope calculated? (And aren't you glad the software does this for you?)

ANSWER: $SE(b_1) = \dfrac{s_e}{\sqrt{n-1} \times s_x} = \dfrac{0.6362}{\sqrt{39-1} \times 1.57682} = 0.0655$

To test the hypothesis of no linear association ($\beta_1 = 0$),

$$t = \frac{b_1 - \beta_1}{SE(b_1)} = \frac{0.526674 - 0}{0.0655} = 8.05$$

What About the Intercept?

The same reasoning applies for the intercept. We could write

$$\frac{b_0 - \beta_0}{SE(b_0)} \sim t_{n-2}$$

and use it to construct confidence intervals and test hypotheses, but often the value of the intercept isn't something we care about—or have a natural hypothesis for. The intercept usually isn't interesting. Most hypothesis tests and confidence intervals for regression are about the slope. We won't even bother to display the formula for the $SE(b_0)$, since there is nothing useful to be learned from inspecting it.

Section 23.2 Exercises

3. Graduation rates 2013, part II Using the regression output in Exercise 1, identify the error standard deviation and explain what it means in the context of the problem.

4. Shoot to score another one Using the regression output from Exercise 2, identify the error standard deviation and explain its meaning with a sentence in context.

5. Graduation rates 2013, part III Continuing with the regression of Exercise 1, write a sentence that explains the meaning of the standard error of the slope of the regression line, $SE(b_1) = 0.0240$.

6. Shoot to score, hat trick Returning to the results of Exercise 2, write a sentence to explain the meaning of the standard error of the slope of the regression line, $SE(b_1) = 0.0053$.

23.3 Regression Inference

> **WHAT IF THE SLOPE WERE 0?**
>
> If $\beta_1 = 0$, our model becomes $\mu_y = \beta_0 + 0x$. The equation collapses to just $\mu_y = \beta_0$. Now, x is nowhere in sight, so y doesn't depend on x at all.
>
> And the least squares estimate of β_0 would turn out to be $\bar{y}$. So when the slope is 0, the fitted equation is just $\hat{y} = \bar{y}$; at every value of x, we always predict the mean value for y.

Now that we have the standard error of the slope and its sampling distribution, we can test a hypothesis about it and make confidence intervals. The usual null hypothesis about the slope is that it's equal to 0. Why? Well, a slope of zero would say that the response y doesn't tend to change linearly when x changes—in other words, there is no linear association between the two variables. If the slope were zero, there wouldn't be much left of our regression equation.

So a null hypothesis of a zero slope questions the entire claim of a linear relationship between the two variables—and often that's just what we want to know. In fact, every software package or calculator that does regression simply assumes that you want to test the null hypothesis that the slope is really zero.

To test $H_0: \beta_1 = 0$, we find

$$t = \frac{b_1 - 0}{SE(b_1)}$$

which has a t_{n-2} distribution under the null hypothesis. This is just like every t-test we've seen: a difference between the statistic and its hypothesized value, divided by its standard error.

For our body fat data, the computer found the slope (1.7), its standard error (0.0743), and the ratio of the two: $(1.7 - 0)/0.0743 = 22.9$ (see p. 732). Nearly 23 standard errors from the hypothesized value certainly seems big. The P-value (< 0.0001) confirms that a t-ratio this large would be very unlikely to occur if the true slope were indeed zero.

Maybe the standard null hypothesis isn't all that interesting here. Did you have any doubts that %*Body Fat* is related to *Waist* size? A more useful procedure might be to make a confidence interval for the slope, instead.

We can build a confidence interval in the usual way, as an estimate plus or minus a margin of error. As always, the margin of error is just the product of the standard error and a critical value. Here the critical value comes from the t-distribution with $n - 2$ degrees of freedom, so a 95% **confidence interval for β_1** is

$$b_1 \pm t^*_{n-2} \times SE(b_1)$$

For the body fat data, $t^*_{248} = 1.970$, so that comes to $1.7 \pm 1.97 \times 0.074$, or an interval from 1.55 to 1.85 %*Body Fat* per additional inch of *Waist* size.

For Example INTERPRETING A REGRESSION MODEL

RECAP: On a log scale, there seems to be a linear relationship between the diameter and the age of recent terrestrial impact craters. We have regression output from statistics software:

Dependent variable is: logDiam
R squared = 63.6%
s = 0.6362 with 39 − 2 = 37 degrees of freedom

Variable	Coefficient	SE(coeff)	t-ratio	P-value
Intercept	0.358262	0.1106	3.24	0.0025
logAge	0.526674	0.0655	8.05	≤ 0.0001

QUESTIONS: What is the regression model, and what can it tell us?

ANSWER: For terrestrial impact craters younger than 35 million years, the logarithm of diameter grows linearly with the logarithm of *Age*:

$$\widehat{log\ Diam} = 0.358 + 0.527\ logAge.$$

The P-value for the *logAge* coefficient's *t*-statistic is very small, so we are quite confident that the true slope is not zero. Based on our model, we can conclude that, on average, the older a crater is, the larger it tends to be. This model accounts for 63.6% of the variation in *logDiam*.

Although it is possible that impacts (and their craters) are getting smaller, it is also quite possible that we are seeing the effects of age on craters. Small craters are probably more likely to erode or become buried, or otherwise be difficult to find as they age. Larger craters may survive the huge expanses of geologic time more successfully.

 # Just Checking

Researchers in food science studied how big people's mouths tend to be. They measured mouth volume by pouring water into the mouths of subjects who lay on their backs. Unless this is your idea of a good time, it would be helpful to have a model to estimate mouth volume more simply. Fortunately, mouth volume is related to height. (Mouth volume is measured in cubic centimetres and height in metres.)

The data were checked and deemed suitable for regression. Here is some computer output:

Summary of	Mouth Volume
Mean	60.2704
StdDev	16.8777

Dependent variable is: Mouth Volume
R squared = 15.3%
s = 15.66 with 61 − 2 = 59 degrees of freedom

Variable	Coefficient	SE(coeff)	t-ratio	P-value
Intercept	−44.7113	32.16	−1.39	0.1697
Height	61.3787	18.77	3.27	0.0018

1. What does the *t*-ratio of 3.27 tell us about this relationship? How does the P-value help our understanding?

2. Would you say that measuring a person's height could reliably be used as a substitute for the wetter method of determining how big a person's mouth is? What numbers in the output helped you reach that conclusion?

3. What does the value of s_e add to this discussion?

Another Example: Breaking Up Is Hard to Predict

Every spring, Nenana, Alaska, hosts a contest in which participants try to guess the exact minute that a wooden tripod placed on the frozen Tanana River will fall through the breaking ice. The contest started in 1917 as a diversion for railroad engineers, with a jackpot of $800 for the closest guess. It has grown into an event in which hundreds of thousands of entrants enter their guesses on the Internet[4] and vie for as much as $300 000.

Because so much money and interest depends on the time of breakup, it has been recorded to the nearest minute with great accuracy ever since 1917. And because a standard measure of breakup has been used throughout this time, the data are consistent. An

---
[4]www.nenanaakiceclassic.com

article in *Science*[5] used the data to investigate global warming—whether greenhouse gasses and other human actions have been making the planet warmer. Others might just want to make a good prediction of next year's breakup time.

Of course, we can't use regression to tell the *causes* of any change. But we can estimate the *rate* of change (if any) and use it to make better predictions.

Here are some of the data; a scatterplot follows in Figure 23.12.

Who	Years
What	Year, day, and hour of ice breakup
Units	*x* is in years since 1900.
	y is in days after midnight Dec. 31.
When	1917–present
Where	Nenana, Alaska
Why	Wagering, but proposed to look at global warming (Data in **Nenana 2014**)

Year (since 1900)	Breakup Date (days since midnight Dec. 31)	Year (since 1900)	Breakup Date (days since midnight Dec. 31)
17	119.4792	30	127.7938
18	130.3979	31	129.3910
19	122.6063	32	121.4271
20	131.4479	33	127.8125
21	130.2792	34	119.5882
22	131.5556	35	134.5639
23	128.0833	36	120.5403
24	131.6319	37	131.8361
25	126.7722	38	125.8431
26	115.6688	39	118.5597
27	131.2375	40	110.6437
28	126.6840	41	122.0764
29	124.6535	⋮	⋮

Figure 23.12

Has the date of ice breakup on the Tanana River changed since 1917, when record keeping began? Earlier breakup dates might be a sign of global warming.

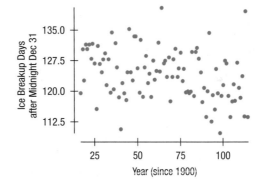

Step-by-Step Example A REGRESSION SLOPE *t*-TEST

The slope of the regression gives the change in Nenana ice breakup date per year.

Questions: Is there sufficient evidence to claim that ice breakup times are changing? If so, how rapid is the change?

THINK ➡ Plan State what you want to know.

Identify the *parameter* you wish to estimate. Here our parameter is the slope.

Identify the variables and review the *W*'s.

I wonder whether the date of ice breakup in Nenana has changed over time. The slope of that change might indicate climate change. I have the date of ice breakup annually for 98 years starting in 1917, recorded as the number of days and fractions of a day until the ice breakup.

---

[5]"Climate change in nontraditional data sets." *Science* 294 [26 October 2001]: 811

Hypotheses Write your null and alternative hypotheses.

H_0: There is no change in the date of ice breakup: $\beta_1 = 0$

H_A: Yes, there is: $\beta_1 \neq 0$

Model Think about the assumptions and check the conditions.

✓ **Straight Enough Condition:** I have quantitative data with no obvious bend in the scatterplot (in Figure 23.12 above), nor in the scatterplot of residuals vs. Year, shown below.

✓ **Independence Assumption:** These data are a time series, which raises my suspicions that adjacent annual observations may be correlated rather than independent. To check, here's a plot of the residuals against time, which here also happens to be the same as the x-variable of the regression:

Make pictures. If the scatterplot seems straight enough, we can find and plot the residuals, for a closer examination.

When the data are measured over time, it is always a good idea to plot residuals against time to look for possible 'auto-correlation,' that is, a correlation or similarity between residuals adjacent in time (a kind of wiggly or oscillating pattern, up for a while, down for a while, ...).

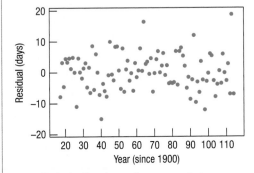

I see little indication of *auto-correlation* or similarity of time-adjacent residuals. They look pretty independent of each other. These data are not a random sample, so I'm reluctant to extend my conclusions beyond this river and these years.

✓ **Does the Plot Thicken? Condition:** Neither the scatterplot nor the plot of residuals vs *Year* shows any obvious trends in the spread.

✓ **Nearly Normal Condition, Outlier Condition:** A histogram of the residuals is unimodal and symmetric.

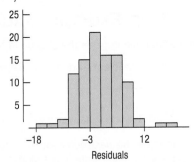

State the sampling distribution model.

Under these conditions, the sampling distribution of the regression slope can be modelled by a Student's *t*-model with $(n - 2) = 96$ degrees of freedom.

Choose your method.

I'll do a **regression slope *t*-test**.

SHOW ➡ Mechanics The regression equation can be found from the formulas in Chapter 7, but regressions are almost always found from a computer program or calculator.

The P-values given in the regression output table are from the Student's t-distribution on $(n - 2) = 96$ degrees of freedom. They are appropriate for two-sided alternatives.

Here's the computer output for this regression:

Response variable is: Breakup Date

R-squared = 9.8%

$s = 5.925$ with $98 - 2 = 96$ degrees of freedom

Variable	Coeff	SE(Coeff)	t-ratio	P-value
Intercept	128.581	1.510	85.2	<0.0001
Years Since 1900	−0.068340	0.0212	−3.23	0.0017

The estimated regression equation is

$$\widehat{\text{Date}} = 128.581 - 0.068\, \text{Years Since 1900}.$$

TELL ➡ Conclusion Link the P-value to your decision and state your conclusion in the proper context.

The P-value of 0.0017 means that the association we see in the data is unlikely to have occurred by chance *even though the R^2 is not particularly strong*. I reject the null hypothesis, and conclude that there is strong evidence that, on average, the ice breakup is occurring earlier each year.

SHOW ➡ **Create a confidence interval for the true slope**
MORE

A 95% confidence interval for β_1 is

$$b_1 \pm t^*_{96} \times SE(b_1)$$
$$-0.068 \pm (1.985)(0.0212)$$
$$\text{or } (-0.11, -0.03) \text{ days per year.}$$

TELL ➡ Interpret the interval Simply reject-
MORE ing the standard null hypothesis doesn't guarantee that the size of the effect is large enough to be important. Whether we want to know the breakup time to the nearest minute or are interested in global warming, a change measured in hours each year is big enough to be interesting.

I am 95% confident that the ice has been breaking up, on average, between 0.03 days (about 40 minutes) and 0.11 days (about 3 hours) earlier each year since 1917.

But is it global warming? So the ice is breaking up earlier. Temperatures are higher. Must be global warming, right?

Maybe.

An article challenging the original analysis of the Nenana data proposed a possible confounding variable. It noted that the city of Fairbanks is upstream from Nenana and suggested that the growth of Fairbanks could have warmed the river. So maybe it's not global warming.

Or maybe global warming is a lurking variable, leading more people to move to a now balmier Fairbanks and also leading to generally earlier ice breakup in Nenana.

Or maybe there's some other variable or combination of variables at work. We can't set up an experiment, so we may never really know.

Only one thing is for sure. When you try to explain an association by claiming cause and effect, you're bound to be on thin ice.[6]

---

[6]How *do* scientists sort out such messy situations? Even though they can't conduct an experiment, they *can* look for replications elsewhere. A number of studies of ice on other bodies of water have also shown earlier ice breakup times in recent years. That suggests they need an explanation that's more comprehensive than just Fairbanks and Nenana.

Section 23.3 Exercises

7. Graduation 2013, part IV The college administrators in Exercise 1 tested the hypotheses $H_0: \beta_1 = 0$ vs. $H_A: \beta_1 \neq 0$ and rejected the null hypothesis because the P-value was less than 0.0001. What can they conclude about the relationship between admission rates and graduation rates?

8. Shoot to score, number four What can the hockey coach in Exercise 2 conclude about shooting and scoring goals from the fact that the P-value < 0.0001 for the slope of the regression line? Write a sentence in context.

9. Graduation 2013, part V The college administrators in Exercise 1 constructed a 95% confidence interval for the slope of their regression line. Interpret the meaning of their interval $(-0.332, -0.238)$ within the context of the problem.

10. Shoot to score, overtime The coach in Exercise 2 found a 95% confidence interval for the slope of his regression line. Recall that he is trying to predict total goals scored based on shots taken. Interpret with a sentence the meaning of the interval 0.12 ± 0.01.

23.4 Confidence and Prediction Intervals for the Response Variable

Once we have a useful regression, how can we indulge our natural desire to predict, without being irresponsible? We know how to compute predicted values of the response y for any value of x. We first did that in Chapter 7. This predicted value would be our best estimate of the response, but it's still just an informed guess.

Now, however, we can find standard errors, and use those to construct confidence intervals and report our uncertainty honestly.

Two Different Standard Errors

From our model of *%Body Fat* and *Waist* size, we might want to use *Waist* size to get a reasonable estimate of *%Body Fat*. A confidence interval can tell us how precise that prediction will be. The precision depends on the question we ask, however, and there are two questions: Do we want to know the mean *%Body Fat* for *all* men with a *Waist* size of, say, 38 inches? Or do we want to predict the *%Body Fat* for a particular man with a 38-inch waist without making him climb onto the X-ray table?

What's the difference between the two questions? The predicted *%Body Fat* from our fitted model is the same, but one question leads to an answer much more precise than the other. We can estimate the *mean %Body Fat* for *all* men whose *Waist* size is 38 inches with a lot more precision than we can predict the *%Body Fat* of a *particular individual* whose *Waist* size happens to be 38 inches. Both can be interesting questions.

We start with the same prediction in both cases. We are predicting the response value for a new individual, one that was not part of the original data set. To emphasize this, we'll call his x-value "x sub new" and write it x_ν.[7] Here, x_ν is 38 inches. The regression equation predicts *%Body Fat* as $\hat{y}_\nu = b_0 + b_1 x_\nu$.

Now that we have the predicted value, we construct both intervals around this same number. Both intervals take the form

$$\hat{y}_\nu \pm t^*_{n-2} \times SE$$

Even the t^* value is the same for both. It's the critical value (from Table T or technology) for $n - 2$ degrees of freedom and the specified confidence level. The intervals differ because they have different standard errors. Our choice of ruler depends on which interval we want.

---

[7] Yes, this is a bilingual pun. The Greek letter ν is called "nu."

Each of these factors contributes uncertainty—that is, variability—to the estimate. When there's more spread around the line, the standard error will be larger. If we're less certain of the slope, we'll be less certain of our predictions. If we have more data (n larger), our predictions will be more precise. Finally, predictions farther from the mean of x will have more variability than predictions closer to it. This last factor is new, but makes intuitive sense: It's a lot easier to predict a data point near the middle of the data set than far from the centre. Because the factors are independent of each other, we can add their variances to find the total variability. The resulting formula for the standard error of the estimated *mean* response explicitly takes into account each of the factors:

$$SE(\hat{\mu}_v) = \sqrt{SE^2(b_1) \times (x_v - \bar{x})^2 + \frac{s_e^2}{n}}$$

Individual values vary more than means, so the standard error for a single predicted response has to be larger than the standard error for the estimated mean. In fact, the standard error of a single predicted response has an *extra* source of variability: the variation of individuals around their own mean. That appears as the extra variance term, s_e^2, at the end under the square root:

$$SE(\hat{y}_v) = \sqrt{SE^2(b_1) \times (x_v - \bar{x})^2 + \frac{s_e^2}{n} + s_e^2}$$

Remember to keep this distinction between the two kinds of standard errors when looking at computer output. The smaller one is for the estimated *mean* value, and the larger one is for a predicted *individual* value.[8]

Confidence Intervals for Mean or Individual Response

Now that we have standard errors, we can ask how well our analysis can estimate the mean *%Body Fat* for men with 38-inch waists. The (earlier) regression output table provides most of the numbers we need:

$$s_e = 4.713$$
$$n = 250$$
$$SE(b_1) = 0.074,$$

and from the data we need to know that $\bar{x} = 36.3$

The regression model gives a predicted value at $x_v = 38$ of

$$\hat{y}_v = -42.7 + 1.7(38) = 21.9\%$$

Let's find the 95% **confidence interval for the mean** *%Body Fat* for all men with 38-inch waists. We find the standard error from the formula:

$$SE(\hat{\mu}_v) = \sqrt{0.074^2 \times (38 - 36.3)^2 + \frac{4.713^2}{250}} = 0.32\%$$

The t^* value that excludes 2.5% in either tail with $250 - 2 = 248$ df is (according to the tables) 1.97.

Putting it all together, we find the margin of error as

$$ME = 1.97(0.32) = 0.63\%$$

So, we are 95% confident that the mean *%Body Fat* for men with 38-inch waist is

$$21.9\% \pm 0.63\%$$

---

[8]You may see the standard error expressions written in other, equivalent ways. The most common alternatives are $SE(\hat{\mu}_v) = s_e\sqrt{\dfrac{1}{n} + \dfrac{(x_v - \bar{x})^2}{\sum(x - \bar{x})^2}}$ and $SE(\hat{y}_v) = s_e\sqrt{1 + \dfrac{1}{n} + \dfrac{(x_v - \bar{x})^2}{\sum(x - \bar{x})^2}}$

MEAN VS. INDIVIDUAL PREDICTIONS For the Nenana Ice Classic, someone who planned to place a bet would want to predict this year's breakup time. By contrast, scientists studying global warming are likely to be more interested in the mean breakup time. If you want to gamble, be sure to take into account that the variability is greater when predicting for a single year.

Suppose, instead, we want to predict the *%Body Fat* for an individual man with a 38-inch waist. We need the larger standard error:

$$SE(\hat{y}_v) = \sqrt{0.074^2 \times (38 - 36.3)^2 + \frac{4.713^2}{250} + 4.713^2} = 4.72\%$$

The corresponding margin of error is

$$ME = 1.97(4.72) = 9.30\%$$

so the **prediction interval** is

$$21.9\% \pm 9.30\%$$

We can think of this interval as having a 95% chance of capturing the true *%Body Fat* of a randomly selected man whose waist is 38 inches.[9] (That R^2 of 67.8% looked promising, but is a prediction ME of more than 9% good enough? Health professionals might not be satisfied.) Notice how much wider this interval is than the previous one. As we've known since Chapter 14, the sample mean is much less variable than a randomly selected individual value, and our regression estimate of a population mean is a very similar creature.

Keep in mind this distinction between the two kinds of confidence intervals: The narrower interval is a confidence interval for the mean response at x_v, and the wider interval is a prediction interval for an individual response with that x-value.

Figure 23.13 below shows many confidence and prediction intervals, connected together into bands. Note the difference in width between the two types of intervals, and the increasing width as you move away from the centre of the data. Most of the points are contained within the prediction intervals (as one should expect). With bigger n, the confidence intervals for the means approach zero width, but not so for the prediction intervals, which approach $\mu_y \pm 1.96\sigma_\varepsilon$.

Figure 23.13

A scatterplot of *%Body Fat* versus *Waist* size with a least squares regression line. The solid green lines near the regression line show the extent of the 95% confidence intervals for mean *%Body Fat* at each *Waist* size. The dashed red lines show the prediction intervals. Notice that there is more uncertainty (wider intervals) the farther we are from the mean of the *x* values.

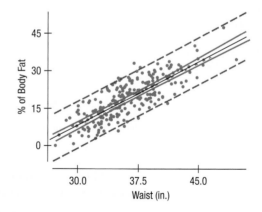

For Example INTERVALS FOR PREDICTIONS

RECAP: We've found a linear model for the relationship between the *logDiam* and *logAge* (with *Age* in millions of years) for terrestrial impact craters:

$$\widehat{logDiam} = 0.358 + 0.527logAge.$$

Let's look at some confidence intervals, keeping in mind that the model is in terms of the logs of these variables, so we must work with logs and then transform back to original units when we are done. (Many of the values substituted into the formulas are from the software outputs shown in the For Example boxes on pages 736 and 737.)

---

[9]Technically, it's a little more complicated, but it's very close to this.

QUESTION #1: If we wish to hunt for new craters that are 5 million years old, what's a 95% confidence interval for the mean size we'd expect them to be?

ANSWER #1:

$$\widehat{logDiam} = 0.358 + 0.527logAge$$

$$= 0.358 + 0.527log(5)$$

$$= 0.726$$

$$SE(\hat{\mu}) = \sqrt{SE^2(b_1) \cdot (x - \bar{x})^2 + \frac{s_e^2}{n}}$$

$$= \sqrt{0.0655^2(log(5) - (-0.65631))^2 + \frac{0.6362^2}{39}}$$

$$= 0.1351$$

From Table T, $t^* = 2.030$, so a 95% CI is $0.726 \pm 2.030(0.1351)$, or $(0.452, 1.00)$.

Since these represent logs of the diameters, we'll find the original units:

$$10^{0.452} \text{ km and } 10^{1.00} \text{ km.}$$

We are 95% confident that terrestrial craters 5 million years old average between 2.8 and 10 kilometres in diameter. But, oops, there is one little problem here. The back transformation just applied (from logs to original units) does not really work mathematically, because it is based on means, rather than individual observations. Consider the confidence interval obtained this way to be a rough approximation only. It might be best to stick to the new units and more accurately just say we are 95% confident that the mean log diameter is between 0.452 and 1.00.

QUESTION #2: There's just been a news report announcing the discovery of a new crater that is 10 million years old. How large do you expect it to be? What's a 95% interval for your prediction?

ANSWER #1: For a crater 10 million years old, $\widehat{logDiam} = 0.358 + 0.527log(10) = 0.885$ and $10^{0.885} = 7.674$. I predict the new crater to be about 7.7 km in diameter.

$$SE(\hat{y}) = \sqrt{SE^2(b_1) \cdot (x - \bar{x})^2 + \frac{s_e^2}{n} + s_e^2}$$

$$= \sqrt{0.0655^2(log(10) - (-0.65631))^2 + \frac{0.6362^2}{39} + 0.6362^2}$$

$$= 0.653$$

A 95% prediction interval is $0.885 \pm 2.030(0.653) = (-0.441, 2.211)$; in original units, that's $(0.36, 162.55)$.

I'm 95% confident that the newly discovered 10 million-year-old crater is between 0.36 and 162.55 kilometres in diameter. (Without a lot of data and a very strong association, prediction intervals often aren't very precise!)

Section 23.4 Exercises

11. Graduation 2013, part VI The college described in Exercise 1 admits about 33% of applicants each year. Which will be larger, the standard error of the predicted 6-Year Graduation Rate of this college, or the standard error of the mean 6-Year Graduation Rate of all colleges admitting 33% of their applicants? Explain briefly.

12. Shoot to score, double overtime The coach in Exercise 2 has a problem. One of his best players has been in a slump, and the coach is trying to convince him to take more shots. Last year, the player took 104 shots. Would the standard error of the predicted number of goals he shot be larger or smaller than the standard error of the mean predicted number of all players who took 104 shots? Explain briefly.

13. Graduation 2013, part VII As in Exercise 11, the college described in Exercise 1 admits about 33% of applicants each year. Using their regression model, the administrators calculated a 95% confidence interval for the mean *6-Year Graduation Rate* for all top colleges who admit 33% of students. The interval was (87.35%, 88.88%). Interpret the meaning of this interval with a sentence.

14. Shoot to score, triple overtime As in Exercise 12, one of the coach's best players has been in a slump, and the

coach is trying to convince him to take more shots. Last year, the player took 104 shots. The coach calculated a 95% confidence interval for the mean number of goals a player would score if 104 shots were taken. The interval turned out to be 9.8 to 11.5 goals. Summarize the interval with a sentence in context.

15. Graduation 2013, part VIII The administrators from Exercise 1 also determined a 95% prediction interval for the graduation rate of a school admitting 33% of applicants to be (82.96%, 93.28%). Carefully explain what the 95% indicates for this interval.

16. Shoot to score, again The coach, still working with his superstar from Exercise 12, went on to suggest the player shoot more. Hopeful for 150 shots for the season, he calculated a 95% prediction interval for the number of goals this player might score. If the prediction interval for goals scored was (6.03, 24.25), interpret the meaning of this interval with a sentence in context.

23.5 Correlation Test

In Chapter 7, we learned that the estimated regression line slope can be calculated by

$$b_1 = r\, s_y / s_x$$

with r being the sample correlation coefficient of x and y.

Sir Ronald Fisher (remember him?) showed that when the *population correlation coefficient* between two random variables denoted as ρ is 0, then, under the usual regression assumptions, the statistic

$$\frac{r\sqrt{n-2}}{\sqrt{1-r^2}}$$ has a t distribution with $n-2$ df.

When $\rho = 0$ (correlation), the regression line slope also must be zero ($\beta_1 = 0$). So the above statistic can be used for testing $H_0: \beta_1 = 0$ in a regression model.

For our body fat data, the correlation coefficient is 0.8234. Hence,

$$t = \frac{0.8234\sqrt{250-2}}{\sqrt{1-(0.8234)^2}} = 22.9$$

which is the same t value as before. So we get the same P-value (less than 0.0001), and same conclusion to reject the null hypothesis since a t-ratio this large (with $250 - 2 = 248$ df) would be very unlikely to occur if the true slope (or correlation) were zero.

For Example CORRELATION TEST

RECAP: On a log scale, the correlation coefficient of crater age and crater diameter is 0.7975, based on a sample size of 39.

QUESTION: Are the two variables linearly related?

ANSWER: Rather than testing the slope from the regression, we can test whether the true population correlation $\rho = 0$:

$$t = \frac{r\sqrt{n-2}}{\sqrt{1-r^2}} = \frac{0.7975\sqrt{39-2}}{\sqrt{1-(0.7975)^2}} = 8.05$$

> The P-value is less than 0.0001 (using the t_{37} distribution), so we are quite confident that a linear relationship exists between log age and log diameter.
>
> (Note that this t value is the same as the t-ratio for the slope when we regressed one of these variables on the other on page 736.)

Equivalently, when x and y are both *random* variables, we can test whether they are correlated, with null H_0: $\rho = 0$, simply by testing H_0: $\beta_1 = 0$, after regression of one variable on the other. We might also wish to construct a confidence interval for ρ. However, the strong conditions required and the details involved make it a topic we elect to omit in this text.

Section 23.5 Exercises

17. Graduation 2013, part IX The administrator from Exercise 1 also reported the correlation coefficient of graduation rates and acceptance rates to be -0.8672. Are the two variables linearly related?

18. Shoot to score, revisited The coach in Exercise 2 calculated the correlation coefficient of goals and shots to be 0.7962. Are the two variables linearly related?

*23.6 The Analysis of Variance (ANOVA) for Regression

When running regressions using software, you will usually see an ANOVA table in the output, so let's see what we can learn from it. We'll use the chapter opening example again, but to simplify some pictures, we took a sample of just 10 of the men to work with here. Here is a scatterplot including the least-squares fitted line for this smaller sample:

We may think of the ith individual's body fat percentage measurement, y_i, as the overall sample mean body fat percentage ($\bar{y} = 21\%$ here) plus a (blue) deviation reflecting the effect of waist size on body fat ($\hat{y}_i - \bar{y}$) plus another (green) deviation reflecting all the other influences on a man's body fat ($y_i - \hat{y}_i$), i.e.

$$y_i = \bar{y} + (\hat{y}_i - \bar{y}) + (y_i - \hat{y}_i)$$

or generally speaking:

Response = Mean Response + Regression Deviation + Residual Deviation

Now, subtract $\bar{y}$ from each side of the equation above, square each side, and sum ($i = 1$ to n). It may be shown that the cross product term, after summing, equals zero and drops out, leaving us with the following *key decomposition*:

$$\sum (y_i - \bar{y})^2 \quad = \quad \sum (\hat{y}_i - \bar{y})^2 \quad + \quad \sum (y_i - \hat{y}_i)^2$$

$$\|\qquad\qquad\qquad \|\qquad\qquad\qquad\qquad \|$$

Total SS $\qquad$ RegressionSS (or ModelSS) $\qquad$ ResidualSS (or ErrorSS) where 'SS' stands for 'Sum of Squares.' Let's look at each of these sums of squared deviations more closely, by showing them in the data plot:

Interpreting the SS's

The TotalSS is a measure of total variability or variation in the data. It is just the sample variance of the responses (*y*-values), but missing the usual $n - 1$ denominator; hence it is just a *sum* of squared deviations, rather than an average of them. We always break it down into more useful components.

The Regression (Model) SS reflects how much the predicted values vary. If there is little relationship between the variables, the predicted values will all be close to $\bar{y}$ and vary little. As the relationship gets stronger, the predicted values more closely track the observed values and their SS likewise approaches the TotalSS. This SS measures the portion of the total variation *explained* by our regression model.

The Residual (Error) SS shows variation about the predicted values, hence it reflects variation *unexplained* by the model, variation attributable to all the other influences on the response variable other than the *x*-variable. For a weak relationship, this SS is relatively high, whereas for a strong relationship, this residual or left-over variation is relatively small.

It is a fortunate mathematical fact that we get this neat breakdown of the total variation, as measured by the TotalSS, into two distinct and very useful parts—as we will see. (Indeed, there is a Pythagorean or right-triangle aspect to all this, but the underlying geometry is beyond the level of this textbook.)

Degrees of freedom

Every SS has some associated *degrees of freedom*, which is utilized when making inference. Rougly speaking, this indicates how many independent pieces of information are wrapped up in that SS. Here is some intuition about their respective df:

- TotalSS has $n - 1$ df since it is like the sample variance (except for denominator). There are *n* deviations but only $n - 1$ are free to vary, and the last deviation is completely determined by the other $n - 1$ deviations. We also say that we lose 1 df since we had to estimate one parameter—the mean—in order to form these deviations.

- ResidualSS has n-2 df, since we have n deviations but lose 2 df due to estimating two parameters (β_0, β_1) in order to form the deviations.
- RegressionSS has 1 df since all these deviations follow a simple linear pattern, completely predictable from the slope; in fact this SS is a simple function of the slope and it can be shown that the RegressionSS $= b_1^2 \sum (x_i - \bar{x})^2$. Think of estimation of a parameter as *using up* 1 df.

  Note that just as component SS's add up to the TotalSS, their respective df must also sum to the Total df, so $n - 2 + 1 = n - 1$ checks out. (When we study models with multiple predictors in Chapter 26, the Regression df will go up by +1 for every additional parameter estimated.)

The ANOVA table and F-test

Using software, here is the ANOVA Table produced for our data above:

Source	DF	SS	MS	F-Value	P-Value
Regression	1	247.7	247.71	11.72	0.009
Residual	8	169.1	21.14		
Total	9	416.8			

The first three columns are nothing new. 'Source' is short for Source of Variation. Under SS, we see each Sum of Squares, and in the next column we see its DF = degrees of freedom.

Whenever we fit a model, the *first usual question is whether the model is useful at all*—does it actually help explain or predict the response variable? In this chapter we only discussed the simple linear model in one predictor, so testing the usefulness of the model is equivalent to testing $H_0: \beta_1 = 0$. If the model is useful $(\beta_1 \neq 0)$, that means the model actually explains some of the variation in the response. In other words, we want to test if the variation explained by the model, the RegressionSS, is significantly large (it is always bigger than zero, because b_1 will not be zero, even when $\beta_1 = 0$). For this we need some benchmark or ruler—which turns out to be the ResidualSS—well, almost. To make these two SS's comparable, we first convert each Sum of Squares to an average or *Mean Square*, by dividing each SS by its degrees of freedom. We see the MS's in the 4th column. The Residual MS in fact is just the sample variance of the residuals, or square of s_e, so it measures random sampling variation or 'noise' in the data. On the other hand, the RegressionMS reflects both noise as well as regression model (slope) effects.

Finally, we divide the Regression MS by the ResidualMS to assess whether the former is larger than what we might expect simply from random sampling variation. That ratio appears in the next column under *F-Value* and here equals $\frac{247.71}{21.14} = 11.72$. If the model is useless, or $\beta_1 = 0$, we expect these two MS's to be roughly similar in magnitude, and their ratio to be in the vicinity of 1.0. If the RegressionMS is much larger than the ResidualMS, we have evidence that the Regression model does indeed explain some of the variation in the response $(\beta_1 \neq 0)$. But how big is big enough? As usual, we need a precise measure of statistical significance.

Math comes to our rescue again. It turns out that the ratio of our two MS's has a known sampling distribution, called an F distribution (named after Sir Ronald Fisher). We will study this distribution in more detail in the next chapter, but all you need to know now is that a large F ratio is evidence that $\beta_1 \neq 0$, and that the 'P-value' reported in the output assesses the likelihood of getting such a big F value (or bigger) when the null hypothesis is true. In fact, this F ratio is just the square of the t-value you would get from the t test of $\beta_1 = 0$ discussed earlier in this chapter. (So, why do we need an ANOVA F test? Mostly for testing more complicated models which use more predictors and df for fitting the model.)

Here the P-value = 0.009, hence we have good evidence that the model is useful for explaining or predicting the response variable, i.e. we have strong evidence that body fat percentage is (linearly) related to waist size.

Since we have evidence that the model is useful, we should also report a measure of its predictive utility, namely the Coefficient of Determination or R^2. We calculated it before by squaring the correlation, and indicated it may be interpreted as the proportion of the variation in the response explained by our linear model. If you just translate those words into SS's, you get a new way to calculate R^2—take the ratio of the corresponding SS's and we get $R^2 = \dfrac{RegressionSS}{TotalSS}$, which from the ANOVA table equals $247.7 \div 416.8 = 59.4\%$, so we can estimate that waist size explains (linearly) about 59% of the variation in men's body fat percentages.

Here is the ANOVA table for the entire data set of 250 men:

Source	DF	SS	MS	F-Value	P-Value
Regression	1	116 21	116 21.2	523.28	0.000
Error	248	550 8	22.2		
Total	249	171 29			

R^2 is similar at 67.8% (116 21 ÷ 171 29), but is more reliably estimated now, being based on much more data, so it's no surprise that the P-value is much more impressive now (0.000), due to greater test power.

Expect to see many more ANOVA tables in future chapters. They are useful whenever we are modelling a quantitative response variable as a function of one or more quantitative or categorical variables. The corresponding *analysis of variance* breaks down the variability in the response in order to assess the contributions from various sources or influences on the response.

Section 23.6 Exercises

19. Graduation 2013, part X The administrator from Exercise 1 obtained the following ANOVA table. What is the proportion of the total variation explained by the regression model?

Analysis of Variance

Source	DF	SS	MS	F-value	p-value
Regression	1	0.0944	0.0944	146.41	0.0000
Residual	48	0.0310	0.0006		
Total	49	0.1254			

20. Shoot to score, once more The coach in Exercise 2 obtained the following ANOVA table. What is the proportion of the total variation unexplained by the regression model?

Analysis of Variance

Source	DF	SS	MS	F-value	p-value
Regression	1	132 75.60	132 75.60	504.45	0.0000
Residual	291	765 8.22	26.32		
Total	292	209 33.82			

*23.7 Logistic Regression

The Pima Indians of southern Arizona are a unique community. Their ancestors were among the first people to cross over into the Americas some 30 000 years ago. For at least two millennia, they have lived in the Sonoran Desert near the Gila River. Known throughout history as a generous people, they have given of themselves for the past 30 years helping researchers at the National Institutes of Health study certain diseases like diabetes and obesity. Young Pima Indians often marry other Pimas, making them an ideal group for genetic researchers to study. Pimas also have an extremely high incidence of diabetes.

Researchers investigating factors for increased risk of diabetes examined data on 768 adult women of Pima Indian heritage[10]. (Data in **Pima_Indians**). One possible predictor is the body mass index, BMI, calculated as weight/height^2, where weight is measured in kilograms and height in metres. We are interested in the relationship between *BMI* and the incidence of diabetes. We might start by looking at boxplots of *BMI* for each group:

Figure 23.14

Side-by-side boxplots for the two *Diabetes* groups (1 = has diabetes; 0 = doesn't have diabetes) show significantly elevated body mass index (BMI) for the women who have (BMI) diabetes.

From the boxplots, we see that the group with diabetes has a higher mean *BMI*. (A *t*-test would show the difference to be more than 9 SEs from 0 with a P-value < 0.0001.) There is clearly a relationship. Here we've displayed *BMI* as the response and *Diabetes* as the predictor. But the researchers are interested in predicting the increased risk of *Diabetes* due to increased *BMI*, not the other way around.

Diabetes is usually treated as a categorical variable. It might seem unwise at first, but coding having Diabetes as 1 and No Diabetes as 0 gives rise to a useful model as we'll soon see. Using this coding, we can reverse roles, treating Diabetes as the response, and make this scatterplot:

Figure 23.15

A scatterplot of *Diabetes* by *BMI*, using the 0–1 coding for *Diabetes*, shows a shift in *BMI* for the two groups.

We now have the problem of how to model this relationship. Fitting a linear regression to these data has some obvious problems:

Figure 23.16

A regression of *Diabetes* on *BMI* using the 0–1 coding for *Diabetes*. The linear regression is not appropriate because even if the predicted values are interpreted as the probability of having the disease, this line will predict probabilities greater than 1 and less than 0.

---

[10]Data can be found at the UCI Machine Learning Repository http://archive.ics.uci.edu/ml/datasets/ Pima+Indians+Diabetes.

The equation says: Predicted *Diabetes* = –0.351 + 0.022 *BMI*. Does this make *any* sense? Suppose someone had a *BMI* of 44. The equation predicts 0.62. What would you guess is the chance that she will have diabetes? If you said about 62%, then you're using the line to model the *probability* of having *Diabetes*.

There are some obvious problems with this model, though. What's the probability that someone with a *BMI* of 10 has diabetes? It's low, but the equation predicts –0.13, obviously an impossible probability. And if we imagined someone with a *BMI* of 70, we might suspect that the probability of her having diabetes is pretty high, but not the predicted value of 1.19.

One simple fix to this problem might be to set all negative probabilities to 0 and all probabilities greater than 1 to 1. That would give a model that looked like this:

Figure 23.17

This model eliminates the problem of probability values that are negative or that exceed 1, but the corners are not aesthetically pleasing, nor do they make scientific sense.

This avoids one problem, but it can't really be correct. The occurrence of *Diabetes* can't be either certain ($p = 1$) or impossible ($p = 0$) based only on *BMI*. And it's not aesthetically pleasing. Real-world changes are likely to be smooth, so we prefer models with smooth transitions rather than corners. That makes good predictive sense, too. There's no reason to expect sharp changes at certain *BMI* values. So instead, we can use a smooth curve to model the probability of having *Diabetes*, like this:

Figure 23.18

The smooth curve models the probability of having *Diabetes* as a function of *BMI* in a sensible way. The logistic curve shown here is just one of a number of choices for the form of the curve, but all are fairly similar in shape and in the resulting predicted probabilities.

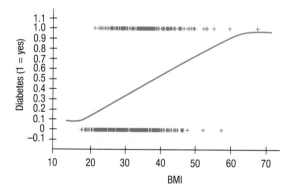

This smoother version is a sensible way to model the probability of having *Diabetes* as a function of *BMI*. There are many curves in mathematics with shapes like this that we might use for our model. One of the most common is the *logistic curve*. The regression based on this curve is called **logistic regression**.

The equation for logistic regression can be written like an ordinary regression on a transformed *y*-variable:

$$ln\left(\frac{\hat{p}}{1 - \hat{p}}\right) = b_0 + b_1 x.$$

The expression on the left-hand side gives the *logistic* curve. The logarithm used is the natural, or base *e*, log, although that doesn't really matter to the shape of the model.

It turns out that the logistic curve has a particularly useful interpretation. Racetrack enthusiasts know that when p is a probability, $\dfrac{p}{1-p}$ is the *odds* in favour of a success. For example, when the probability of success, p, $= 1/3$, we'd get the ratio $\dfrac{1/3}{2/3} = 1/2$. We'd say that the odds in favour of success are 1:2 (or we'd probably say the odds *against* it are 2:1). Logistic regression models the *logarithm* of the odds as a linear function of x. In fact, nobody really thinks in terms of the log of the odds ratio. But it's the combination of that ratio and the logarithm that gets us the nice S-curved shape. What is important is that we can work backward from a log odds ratio to get the probability—which is often easier to think about.

Because we're not fitting a straight line to the data directly, we can't use ordinary least squares to estimate the parameters b_0 and b_1.[11] Instead, we use special *nonlinear* methods that require a good deal more computation—but the computers take care of that for us.

For the Pima Indians data set, the equation is

$$\ln\left(\frac{\hat{p}}{1-\hat{p}}\right) = -4.00 + 0.1025\ BMI$$

A computer output will typically provide a table like the following:

Term	Estimate	Std Error	ChiSquare	P-value
Intercept	−3.9967	0.4288	86.8576	<0.00001
BMI	0.1025	0.0126	66.0872	<0.00001

We usually don't interpret the slope itself (unless you happen to think naturally in log odds), but we can perform a test on whether the slope is 0, similar to the t-test we did for linear regression. Unlike linear regression, the ratio of the estimate to its standard error does not have a t distribution, but the square of that ratio has a χ^2 distribution.[12] The P-values for both the slope and the intercept clearly indicate that neither is 0.

Once we have decided that the slope is not 0, we can use the model to predict the probabilities. Solving the equation $\ln\left(\dfrac{\hat{p}}{1-\hat{p}}\right) = b_0 + b_1 x$ for $\hat{p}$ gives

$$\hat{p} = \frac{1}{1 + e^{-(b_0 + b_1 x)}}$$

and the logistic equation guarantees that the estimate of $\hat{p}$ will be between 0 and 1. Fortunately, technology can provide these probability estimates, producing the curve shown previously and giving an estimate for the probability at any *BMI* value.

Response variables that are dichotomous (having only two possible values) like the variable *Diabetes* are common, so there's a widespread need to model data like these. Phone companies want to predict who will switch carriers, credit card companies want to know which transactions are likely to be fraudulent, and loan companies want to know who is most likely to declare bankruptcy. All of these are potential application areas for logistic regression. It's not surprising that logistic regression has become an important modelling tool in the toolbox of analysts in science and industry in the past decade. By understanding the basics of what logistic regression can do, you can expand your ability to apply regression to many other important applications.

---

[11]This is a tricky point. We've fit regressions with transformed y's before without special methods. If our data consisted of observed *proportions* of people, we could transform the data using $\log(p/(1-p))$ and use linear regression to fit the equation. But all we have are individual 0's and 1's. To be able to fit this equation with our raw data, we need special 'nonlinear' methods.

[12]Here's that nonlinear fitting popping up again. Don't let it bother you. You've seen χ^2 models before, and all you really need to look at for a test is the P-value.

WHAT CAN GO WRONG?

In this chapter, we've added inference to the regression explorations we looked at in Chapters 7 and 8. Everything covered in those chapters that could go wrong with regression can still go wrong. It's probably a good time to review Chapter 8. Take your time; we'll wait.

With inference, we've put numbers on our estimates and predictions, but these numbers are only as good as the model. Here are the main things to watch out for:

- **Don't fit a linear regression to data that aren't straight.** This is the most fundamental assumption. If the relationship between x and y isn't approximately linear, there's no sense in fitting a straight line to it.

- **Watch out for the plot thickening.** The common part of confidence and prediction intervals is the estimate of the error standard deviation, the spread around the line. If it changes with x, the estimate won't make sense. Imagine making a prediction interval for these data.

  When x is small, we can predict y precisely, but as x gets larger, it's much harder to pin y down. Unfortunately, if the spread changes, the single value of s_e won't pick that up. The prediction interval will use the average spread around the line, with the result that we'll be too pessimistic about our precision for low x-values and too optimistic for high x-values. A re-expression of y is often a good fix for changing spread.

- **Make sure the errors are Normal.** When we make a prediction interval for an individual, the Central Limit Theorem can't come to our rescue. For us to believe the prediction interval, the errors must be from the Normal model. Check the histogram and Normal probability plot of the residuals to see if this assumption looks reasonable.

- **Watch out for extrapolation.** It's tempting to think that because we have prediction *intervals*, they'll take care of all of our uncertainty so we don't have to worry about extrapolating. Wrong. The interval is only as good as the model. The uncertainty our intervals predict is correct only if our model is true. There's no way to adjust for wrong models. That's why it's always dangerous to predict for x-values that lie far from the centre of the data.

- **Watch out for high-influence points and outliers.** We always have to be on the lookout for a few points that have undue influence on our estimated model—and regression is certainly no exception.

- **Watch out for one-tailed tests.** Because tests of hypotheses about regression coefficients are usually two-tailed, software packages report two-tailed P-values. If you are using software to conduct a one-tailed test about slope, you may need to divide the reported P-value in half.

What Have We Learned?

Understand that the "true" population regression line does not fit the population perfectly, but is an idealized summary or model of the relationship.

Apply your understanding of inference for means using Student's t to inference about regression coefficients.

Know the Assumptions and Conditions for inference about regression coefficients and how to check them, in this order:

- Linearity Assumption, checked with the Straight Enough Condition by examining a scatterplot of y vs. x or of the residuals plotted against the predicted values.

- Independence Assumption, which can't be checked, but is more plausible if the data were collected with appropriate randomization—the Randomization Condition. Be especially careful when working with data recorded over time-plot residuals versus time and look for patterns.

- Equal Variance Assumption, which requires that the spread around the regression model be the same everywhere. We check it with the Does the Plot Thicken? Condition, assessed with a scatterplot of the residuals versus the predicted values.

- Normal Population Assumption, which is required to use Student's t-models unless the sample size is large. Check it with the Nearly Normal Condition by making a histogram or normal probability plot of the residuals.

Know the components of the standard error of the slope coefficient:

- The standard deviation of the residuals, $s_e = \sqrt{\dfrac{\sum(y - \hat{y})^2}{n - 2}}$

- The standard deviation of x, $s_x = \sqrt{\dfrac{(x - \overline{x})^2}{n - 1}}$

- The sample size, n

Be able to find and interpret the standard error of the slope.

- $SE(b_1) = \dfrac{s_e}{s_x\sqrt{n - 1}}$

- The standard error of the slope is the estimated standard deviation of the sampling distribution of the slope.

State and test the standard null hypothesis on the slope.

- $H_0: \beta_1 = 0$. This would mean that x and y are not linearly related.

- We test this null hypothesis using the t-statistic $t = \dfrac{b_1 - 0}{SE(b_1)}$

- Know where to find the relevant numbers in standard computer regression output.

Construct and interpret a confidence interval for the slope of a regression.

- $b_1 \pm t^* \times SE(b_1)$

Know how to use a t-test to test whether the true correlation is zero.

- $t = r\sqrt{\dfrac{n - 2}{1 - r^2}}$

Construct and interpret a confidence interval for the mean response corresponding to a specified value, xv.

- $\hat{y} \pm t^*_{n-2} \times SE(\hat{\mu}_v)$, where $SE(\hat{\mu}_v) = \sqrt{SE^2(b_1) \times (x_v - \overline{x})^2 + \dfrac{s_e^2}{n}}$

Construct and interpret a confidence interval (often called a prediction interval) for one individual's response corresponding to a specified value, xv.

- $\hat{y}_v \pm t^*_{n-2} \times SE(\hat{y}_v)$, where $SE(\hat{y}_v) = \sqrt{SE^2(b_1) \times (x_v - \overline{x})^2 + \dfrac{s_e^2}{n} + s_e^2}$

Review of Terms

Residual standard deviation
The spread of the data around the regression line is measured with the residual standard deviation, s_e (p. 734):
$$s_e = \sqrt{\frac{\sum(y - \hat{y})^2}{n - 2}} = \sqrt{\frac{\sum e^2}{n - 2}}$$

t-test for the regression slope
When the assumptions are satisfied, we can perform a test for the slope coefficient. We usually test the null hypothesis that the true value of the slope is zero against the alternative that it is not. A zero slope would indicate a complete absence of linear relationship between y and x.

To test $H_0: \beta_1 = 0$ we find
$$t = \frac{b_1 - 0}{SE(b_1)}$$
where
$$SE(b_1) = \frac{s_e}{\sqrt{n - 1}\,s_x}, \quad s_e = \sqrt{\frac{\sum(y - \hat{y})^2}{n - 2}}$$

n is the number of cases, and s_x is the standard deviation of the x-values. We find the P-value from the Student's t-model with $n - 2$ degrees of freedom.

Alternatively, to test $H_0: \beta_1 = 0$ (or $\rho = 0$), we find
$$t = \frac{r\sqrt{n - 2}}{\sqrt{1 - r^2}}$$

where r is the correlation coefficient. Same t-value and distribution as above. (p. 737).

Confidence interval for the regression slope

When the assumptions are satisfied, we can find a confidence interval for the slope parameter from $b_1 \pm t^*_{n-2} \times SE(b_1)$. The critical value, t^*_{n-2}, depends on the confidence level specified and on a Student's t-model with $n - 2$ degrees of freedom (p. 737).

Confidence interval for a mean response

At any x-value, the predicted y-value estimates the mean of the entire population of y-values or responses, corresponding to that particular value of x. We can form a confidence interval for the true mean response at a specified x-value, x_v, as $\hat{y}_v \pm t^*_{n-2} \times SE(\hat{\mu}_v)$, where

$$SE(\hat{\mu}_v) = \sqrt{SE^2(b_1) \cdot (x_v - \overline{x})^2 \pm \frac{s_e^2}{n}}$$

The critical value, t^*_{n-2}, depends on the specified confidence level and the Student's t-model with $n - 2$ degrees of freedom (p. 743).

Prediction interval for an individual response

We can also form a (wider) confidence interval to capture a single future observation on the response y at some particular x-value. The interval takes the form $\hat{y}_v \pm t^*_{n-2} \times SE(\hat{y}_v)$, where

$$SE(\hat{y}_v) = \sqrt{SE^2(b_1) \cdot (x_v - \overline{x})^2 + \frac{s_e^2}{n} + s_e^2}$$

The critical value, t^*_{n-2}, depends on the specified confidence level and the Student's t-model with $n - 2$ degrees of freedom (p. 744).

*Logistic regression

A regression model suitable for a categorical response variable with two categories (p. 752).

On the Computer REGRESSION ANALYSIS

All statistics packages make a table of results for a regression. These tables differ slightly from one package to another, but all are essentially the same. We've seen two examples of such tables already.

All packages offer analyses of the residuals. With some, you must request plots of the residuals as you request the regression. Others let you find the regression first and then analyze the residuals afterward. Either way, your analysis is not complete if you don't check the residuals with a histogram or Normal probability plot and a scatterplot of the residuals against x or the predicted values.

You should, of course, always look at the scatterplot of your two variables before computing a regression.

Regressions are almost always found with a computer or calculator. The calculations are too long to do conveniently by hand for data sets of any reasonable size. No matter how the regression is computed, the results are usually presented in a table that has a standard form. Here's a portion of a typical regression results table, along with annotations showing where the numbers come from:

The regression table gives the coefficients (once you find them in the middle of all this other information), so we can see that the regression equation is

$$\widehat{\%BF} = -42.73 + 1.7 \, Waist$$

and that the R^2 for the regression is 67.8%.

The column of t-ratios gives the test statistics for the respective null hypotheses that the true values of the coefficients are zero. The corresponding P-values are also usually reported.

EXCEL

- In Excel 2003 and earlier, select Data Analysis from the **Tools** menu. In Excel 2007, select Data Analysis from the **Analysis Group** on the Data tab.
- Select Regression from the **Analysis Tools** list.
- Click the **OK** button.
- Enter the data range holding the y-variable in the box labelled "Y-range."
- Enter the range of cells holding the x-variable in the box labelled "X-range."
- Select the **New Worksheet Ply** option.
- Select **Residuals** options. Click the **OK** button.

COMMENTS

The Y and X ranges do not need to be in the same rows of the spreadsheet, although they must cover the same number of cells. But it is a good idea to arrange your data in parallel columns as in a data table.

Although the dialogue offers a Normal probability plot of the residuals, the data analysis add-in does not make a correct probability plot, so don't use this option.

Excel calls the standard deviation of the residuals the "Standard Error." This is a common error. Don't be confused; it is not $SE(y)$, but rather s_e.

JMP

- From the **Analyze** menu, select **Fit Y by X**.
- Select variables: a Y, Response variable, and an X, Factor variable. Both must be continuous (quantitative).
- JMP makes a scatterplot.
- Click on the red triangle beside the heading labelled **Bivariate Fit . . .** and choose **Fit Line**. JMP draws the least squares regression line on the scatterplot and displays the results of the regression in tables below the plot.
- The portion of the table labelled "Parameter Estimates" gives the coefficients and their standard errors, t-ratios, and P-values.

COMMENTS

JMP chooses a regression analysis when both variables are "Continuous." If you get a different analysis, check the variable types.

The Parameter table does not include the residual standard deviation s_e. You can find that as Root Mean Square Error in the Summary of Fit panel of the output.

MINITAB

- Choose **Regression** from the **Stat** menu.
- Choose **Regression** and then **Fit Regression Model** from the **Regression** submenu.
- In the Regression dialogue, assign the y-variable to the Response box and assign the x-variable to the Continuous Predictors box.

- Click the **Graphs** button.
- In the Regression-Graphs dialogue, select **Standardized residuals**, and check **Normal plot of residuals** and **Residuals versus fits** or choose **Four in one**, for a useful batch of four residual plots.

- Click the **OK** button to return to the Regression dialogue.
- Click the **OK** button to compute the regression.

COMMENTS

You can also start by choosing a **Fitted Line plot** from the **Regression** submenu to see the scatterplot first—usually good practice.

R

From a data frame called mydata with variables x and y, to find the linear model:

- mylm=lm(y~x,mydata)
- summary(mylm) # shows a summary of the model including estimates, SEs, and the F-statistic

To get confidence or prediction intervals use:

- predict(mylm, interval="confidence")

or

- predict(mylm, interval="prediction")

COMMENTS

Predictions on points not found in the original data frame can be found from predict as well. In the new data frame (called, say, mynewdata), there must be a predictor variable with the same name as the original "x" variable. Then predict(mylm, newdata = mynewdata) will produce predictions at all the "x" values of mynewdata.

SPSS

- Choose **Regression** from the **Analyze** menu.
- Choose **Linear** from the **Regression** submenu.
- In the Linear Regression dialogue that appears, select the y-variable and move it to the dependent target. Then move the x-variable to the independent target.
- Click the **Plots** button.

- In the Linear Regression Plots dialogue, choose to plot the *SRESIDs against the *ZPRED values.
- Click the **Continue** button to return to the Linear Regression dialogue.
- Click the **OK** button to compute the regression.

STATCRUNCH

To compute a regression with confidence and prediction intervals:

- Click on **Stat**.
- Choose **Regression » Simple Linear**.
- Choose **X and Y variables** from the list of columns.
- Under **Perform**, choose **Hypothesis Tests** and/or **Confidence Intervals** (for both parameters).

- Choose **Prediction of Y** for prediction and confidence intervals at your choice of X value.
- Under **Graphs**, you can select various data and residual plots. You can also **Save** residuals to use them for additional plots.
- Click on Compute.

TI-83/84 PLUS

Under **STAT TESTS** choose **LinRegTTest**. Specify the two lists where the data are stored, and (usually) choose the two-tail option. In addition to reporting the calculated value of t and the P-value, the calculator will tell you the

coefficients of the regression equation (a and b), the values of r^2 and r, and the value of s you need for confidence or prediction intervals.

Chapter Exercises

T indicates data is available in a saved file at the Companion Website at www.pearsoncanada.ca/deveaux

T 21. Hurricane predictions 2012 Let's look at some data from the National Oceanic and Atmospheric Administration about their success in predicting hurricane tracks. Here is a scatterplot of the error (in nautical miles) for predicting hurricane locations 72 hours in the future versus the year in which the prediction (and the hurricane) occurred:

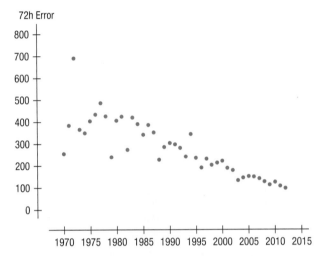

Here is the regression analysis:

Dependent variable is: 72Error
R squared = 71.1%
s = 68.7 with 43 − 2 = 41 degrees of freedom

Variable	Coefficient	SE(Coeff)	t-ratio	P-value
Intercept	454.558	20.5942	22.07	< 0.0001
Year since 1970	−8.4790	0.8443	−10.04	< 0.0001

a) Explain in words and numbers what the regression says.
b) State the hypothesis about the slope (both numerically and in words) that describes how hurricane prediction quality has changed.
c) Assuming that the assumptions for inference are satisfied, perform the hypothesis test and state your conclusion. Be sure to state it in terms of prediction errors and years.
d) Explain what the R-squared means in terms of this regression.

T 22. Drug use 2013 The *2013 World Drug Report* investigated the prevalence of drug use as a percentage of the population aged 15 to 64. Data from 32 European countries are shown in the following scatterplot and regression analysis. (*Source: World Drug Report*, 2013. www.unodc.org/unodc/en/data-and-analysis/WDR-2013.html)

Response variable is Cocaine
R-squared = 25.8%
s = 0.7076 with 32 − 2 = 30 degrees of freedom

Variable	Coefficient	SE(Coeff)	t-Ratio	P-Value
Intercept	0.296899	0.2627	1.13	0.2673
Canabis	0.123638	0.0382	3.23	0.0030

a) Explain in context what the regression says.
b) State the hypothesis about the slope (both numerically and in words) that describes how use of marijuana is associated with other drugs.
c) Assuming that the assumptions for inference are satisfied, perform the hypothesis test and state your conclusion in context.
d) Explain what R-squared means in context.
e) Do these results indicate that marijuana use leads to the use of harder drugs? Explain.

T 23. Movie budgets [13] How does the cost of a movie depend on its length? Data on the cost (millions of dollars) and the running time (minutes) for major release films of 2005 are summarized in these plots and computer output:

Dependent variable is Budget($million)
R-squared = 27.3%
s = 32.95 with 120 − 2 = 118 degrees of freedom

[13]Data have been corrected since previous editions.

Variable	Coefficient	SE(Coeff)	t-Ratio	P-Value
Intercept	−63.9981	17.12	−3.74	0.0003
Run Time	1.02648	0.1540	6.66	≤0.0001

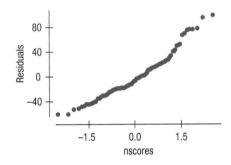

a) Explain in context what the regression says.
b) The intercept is negative. Discuss its value.
c) The output reports $s = 32.95$. Explain what that means in this context.
d) What's the value of the standard error of the slope of the regression line?
e) Explain what that means in this context.

T 24. House prices How does the price of a house depend on its size? Data from Saratoga, New York, on 1064 randomly selected sold houses include data on price ($1000s) and size (1000 ft²), producing the following graphs and computer output:

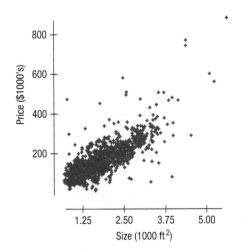

Dependent variable is: Price
R squared $= 59.5\%$
$s = 53.79$ with $1064 - 2 = 1062$ degrees of freedom

Variable	Coefficient	SE(Coeff)	t-ratio	P-value
Intercept	−3.11686	4.688	−0.665	0.5063
Size	94.4539	2.393	39.5	≤0.0001

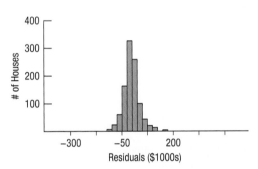

a) Explain in words and numbers what the regression says.
b) The intercept is negative. Discuss its value, taking note of its P-value.
c) The output reports $s = 53.79$. Explain what that means in this context.
d) What is the value of the standard error of the slope of the regression line?
e) Explain what that means in this context.

T 25. Movie budgets: the sequel Exercise 23 shows computer output examining the association between the length of a movie and its cost.
a) Check the assumptions and conditions for inference.
b) Find a 95% confidence interval for the slope and interpret it.

T 26. Second home Exercise 24 shows computer output examining the association between the size of houses and their sale prices.
a) Check the assumptions and conditions for inference.
b) Find a 95% confidence interval for the slope and interpret it.

T 27. Hot dogs Healthy eating probably doesn't include hot dogs, but if you are going to have one, you'd probably hope it's low in both calories and sodium. In its July 2007 issue, *Consumer Reports* listed the number of calories and sodium content (in milligrams) for 13 brands of all-beef hot dogs it tested. Examine the association, assuming that the data satisfy the conditions for inference.

Dependent variable is: Sodium
R squared = 60.5%
s = 59.66 with 13 − 2 = 11 degrees of freedom

Variable	Coefficient	SE(Coeff)	*t*-ratio	P-value
Constant	90.9783	77.69	1.17	0.2663
Calories	2.29959	0.5607	4.10	0.0018

a) State the appropriate hypotheses about the slope.
b) Test your hypotheses and state your conclusion in the proper context.

T 28. Cholesterol 2007 Does a person's cholesterol level tend to change with age? Data collected from 1406 adults aged 45 to 62 produced the regression analysis shown. Assuming that the data satisfy the conditions for inference, examine the association between age and cholesterol level.

Dependent variable is: Chol
s = 46.16

Variable	Coefficient	SE(Coeff)	*t*-ratio	P-value
Intercept	194.232	13.55	14.3	≤0.0001
Age	0.771639	0.2574	3.00	0.0028

a) State the appropriate hypothesis for the slope.
b) Test your hypothesis and state your conclusion in the proper context.

T 29. Second frank Look again at Exercise 27's regression output for the calorie and sodium content of hot dogs.
a) The output reports s = 59.66. Explain what that means in this context.
b) What is the value of the standard error of the slope of the regression line?
c) Explain what that means in this context.

T 30. More cholesterol Look again at Exercise 28's regression output for age and cholesterol level.
a) The output reports s = 46.16. Explain what that means in this context.
b) What is the value of the standard error of the slope of the regression line?
c) Explain what that means in this context.

T 31. Last dog Based on the regression output seen in Exercise 27, create a 95% confidence interval for the slope of the regression line and interpret your interval in the proper context.

T 32. Cholesterol, finis Based on the regression output seen in Exercise 28, create a 95% confidence interval for the slope of the regression line and interpret your interval in the proper context.

T 33. Marriage age 2008 The scatterplot suggests a decrease in the difference in ages at first marriage for Canadian men and women since 1921. We want to examine the regression to see if this decrease is significant.

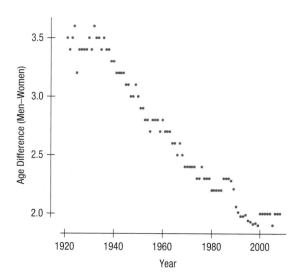

Dependent variable is: Men−Women
R squared = 95.75%
s = 0.1139 with 88 − 2 = 86 degrees of freedom

Variable	Coefficient	SE(Coeff)	*t*-ratio	P-value
Intercept	43.9879	0.9387888	46.86	<0.0001
Year	−0.021036	0.0004778	−44.02	<0.0001

a) Write appropriate hypotheses.
b) Here are the residuals plot and a histogram of the residuals. Do you think the conditions for inference are satisfied? Explain.

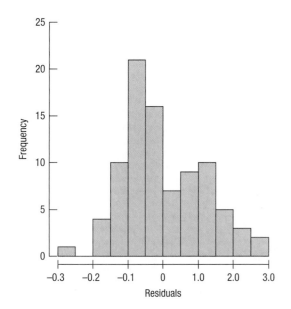

c) Test the hypothesis and state your conclusion about the trend in age at first marriage.

(Data: www4.hrsdc.gc.ca/.3ndic.1t.4r@-eng.jsp?iid=78)

T 34. Used cars 2013 Classified ads in a newspaper offered several used Toyota Corollas for sale. The following lists the ages of the cars and the advertised prices.

Age	Price
1	13 900
1	14 610
2	13 998
2	16 887
3	10 900
3	12 995
3	14 995
4	10 999
4	11 988
4	11 995
4	12 495
5	8 988
5	9 488
5	10 488
5	10 989
7	11 400
8	6 995
9	6 988
10	10 000
12	2 700

a) Make a scatterplot for these data.
b) Do you think a linear model is appropriate? Explain.
c) Find the equation of the regression line.

d) Check the residuals to see if the conditions for inference are met.

T 35. Marriage age 2008, again Based on the analysis of differences in marriage ages given in Exercise 33, give a 95% confidence interval for the rate at which the age gap is closing. Clearly explain what your confidence interval means.

T 36. Used cars 2013, again Based on the analysis of used car prices you did for Exercise 34, create a 95% confidence interval for the slope of the regression line and explain what your interval means in context.

T 37. Fuel economy A consumer organization has reported test data for 50 car models. We will examine the association between the weight of the car (in thousands of pounds) and the fuel efficiency (in miles per gallon). Shown are the scatterplot, summary statistics, and regression analysis:

Variable	Count	Mean	StdDev
MPG	50	25.0200	4.83394
wt/1000	50	2.88780	0.511656

Dependent variable is: MPG
R-squared = 75.6%
s = 2.413 with 50 − 2 = 48 df

Variable	Coefficient	SE(Coeff)	t-ratio	P-value
Intercept	48.7393	1.976	24.7	≤0.0001
Weight	−8.21362	0.6738	−12.2	≤0.0001

a) Is there strong evidence of an association between the weight of a car and its gas mileage? Write an appropriate hypothesis.
b) Are the assumptions for regression satisfied? Check the following plots.
c) Test your hypothesis and state your conclusion.

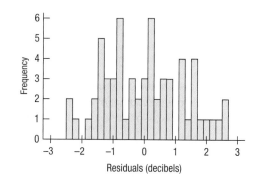

38. Cockpit noise How strong is the association between cockpit noise level (in decibels) and airspeed (in knots indicated air speed, KIAS) for a Boeing 727, an older type of aircraft lacking the design improvements of newer planes? Here are summaries and plots of the data:

Variable	Count	Mean	Median	StdDev	Range
Speed	61	332.5082	335	48.7957	190
Noise	61	89.6803	90	3.9487	15

Dependent variable is: Noise
R-squared = 89.47%
s = 1.2923 with 61 − 2 = 59 df

Variable	Coefficient	SE(Coeff)	t-ratio	P-value
Intercept	64.2294	1.1484	55.9073	≤0.0001
Speed	0.0765	0.0034	22.3867	≤0.0001

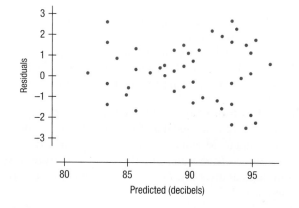

a) Is there evidence of an association between cockpit noise level and airspeed? Write an appropriate hypothesis.
b) Discuss the assumptions for inference.
c) Test your hypothesis and state an appropriate conclusion.

T 39. Fuel economy, part II Consider again the data in Exercise 37 about the gas mileage and weights of cars.
a) Create a 95% confidence interval for the slope of the regression line.
b) Explain in this context what your confidence interval means.

40. Cockpit noise, part II Consider the cockpit noise data from Exercise 38.
a) Find a 90% confidence interval for the slope of the true line describing the association between cockpit noise level and airspeed.
b) Explain in this context what your confidence interval means.

T 41. Fuel economy, part III Consider again the data in Exercise 37 about the gas mileage and weights of cars.
a) Create a 95% confidence interval for the average fuel efficiency among cars weighing 2500 pounds, and explain what your interval means.
b) Create a 95% prediction interval for the gas mileage you might get driving your new 3450-pound SUV, and explain what that interval means.

42. Cockpit noise again Consider the cockpit noise data from Exercise 38 (Challenging exercise).
a) Find a 90% confidence interval for the mean speed of all Boeing 727s with a noise level of 90 decibels.
b) Find a 90% prediction interval for the speed of a Boeing 727 if you know that its noise level is 95 decibels.

T 43. Cereal People are often concerned about the amount of calories and sodium in their cereal. Data for 77 cereals were examined and judged acceptable for inference. The 77 cereals had between 50 and 160 calories per serving and between 0 and 320 mg of sodium per serving. The regression analysis is shown:

Dependent variable is: Sodium
R-squared = 9.0%
s = 80.49 with 77 − 2 = 75 degrees of freedom

Variable	Coefficient	SE(Coeff)	t-ratio	P-value
Intercept	21.4143	51.47	0.416	0.6786
Calories	1.29357	0.4738	2.73	0.0079

a) Is there an association between the number of calories and the sodium content of cereals? Explain.

b) Do you think this association is strong enough to be useful? Explain.

T 44. Brain size Does your IQ depend on the size of your brain? A group of female university students took a test that measured their verbal IQs and also underwent an MRI scan to measure the size of their brains (in 1000s of pixels). The scatterplot and regression analysis are shown, and the assumptions for inference were satisfied.

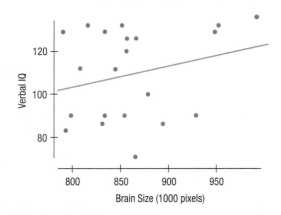

Dependent variable is: IQ_Verbal
R-squared = 6.5%

Variable	Coefficient	SE(Coeff)
Intercept	24.1835	76.38
Size	0.098842	0.0884

a) Test an appropriate hypothesis about the association between brain size and IQ.

b) State your conclusion about the strength of this association.

T 45. Another bowl Further analysis of the data for the breakfast cereals in Exercise 43 looked for an association between fibre content and calories by attempting to construct a linear model. Several graphs are shown. Which of the assumptions for inference are violated? Explain.

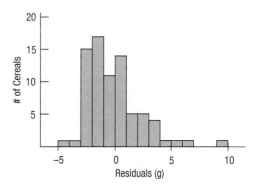

T 46. Heptathlon, round 2 Consider again the 2012 Women's heptathlon data from the Chapter 7 exercises. If we regress the High Jump result on the Long Jump (avg 6.06 m) for the top 29 women, we get the following results:

Dependent variable is: High Jump
R-squared: 38.58%

Variable	Coefficient	SE(Coeff)	t-ratio	P-value
Intercept	0.64931	0.27610	2.352	0.026237*
LongJump	0.18759	0.04555	4.119	0.000323***

Residual standard error: 0.06024 on 27 degrees of freedom

Assuming the conditions for inference are met,

a) Estimate the average high jump result for a long jump of 5.90m, and give a 95% confidence interval for this estimate.

b) Predict the high jump result for a heptathlete who jumps 6.25m in the long jump, and give a 95% confidence interval for this prediction.

c) Test the hypothesis that the correlation coefficient between high jump and long jump is zero.

T 47. Acid rain Biologists studying the effects of acid rain on wildlife collected data from 163 streams in the Adirondack Mountains. They recorded the pH (acidity) of the water and the BCI, a measure of biological diversity. Following is a scatterplot of BCI against pH:

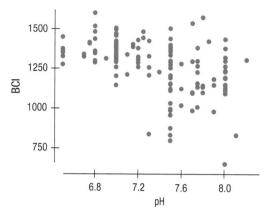

And here is part of the regression analysis:

Dependent variable is: BCI
R-squared = 27.1%
s = 140.4 with 163 − 2 = 161 degrees of freedom

Variable	Coefficient	SE(Coeff)
Intercept	2733.37	187.9
pH	−197.694	25.57

a) State the null and alternative hypotheses under investigation.
b) Assuming that the assumptions for regression inference are reasonable, find the *t*- and P-values for the test.
c) State your conclusion.

T 48. Climate change 2013 Data collected from around the globe show that the earth is getting warmer. The most common theory relates climate change to an increase in atmospheric levels of carbon dioxide (CO_2), a greenhouse gas. The mean annual CO_2 concentration in the atmosphere (parts per million) is measured at the top of Mauna Loa in Hawaii (away from any local contaminants) and available at ftp://aftp.cmdl.noaa.gov/products/trends/co2/co2_annmean_mlo.txt

The mean surface air temperature is recorded as the change in °C relative to a base period of 1951 to 1980. It is available at data.giss.nasa.gov/gistemp/graphs_v3/

Here are a scatterplot and regression for the years from 1970 to 2013:

Response variable is Global Avg Temp
R-squared = 73.6%   R-squared (adjusted) = 72.9%
s = 0.1316 with 44 − 2 = 42 degrees of freedom

Variable	Coefficient	SE(Coeff)	t-Ratio	P-Value
Intercept	10.6442	0.3418	31.1	≤0.0001
CO_2	0.010325	0.0010	10.8	≤0.0001

a) Write the equation of the regression line.
b) Is there evidence of an association between CO_2 level and global temperature?
c) Do you think predictions made by this regression will be very accurate? Explain.
d) Does this regression prove that increasing CO_2 levels are causing global warming? Discuss.

49. Ozone The Environmental Protection Agency is examining the relationship between the ozone level (in parts per million) and the population (in millions) of U.S. cities. Part of the regression analysis is shown.

Dependent variable is: Ozone
R-squared = 84.4%
s = 5.454 with 16 − 2 = 14 df

Variable	Coefficient	SE(Coeff)
Intercept	18.892	2.395
Pop	6.650	1.910

a) We suspect that the greater the population of a city, the higher its ozone level. Is the relationship significant? Assuming the conditions for inference are satisfied, test an appropriate hypothesis and state your conclusion in context.
b) Do you think that the population of a city is a useful predictor of ozone level? Use the values of both R^2 and *s* in your explanation.

T 50. Sales and profits A business analyst was interested in the relationship between a company's sales and its profits. She collected data (in millions of dollars) from a random sample of *Fortune* 500 companies, and created the regression analysis and summary statistics shown. The assumptions for regression inference appeared to be satisfied.

	Profits	Sales
Count	79	79
Mean	209.839	4178.29
Variance	635 172	49 163 000
Std Dev	796.977	7011.63

Dependent variable is: Profits
R-squared = 66.2%   s = 466.2

Variable	Coefficient	SE(Coeff)
Intercept	−176.644	61.16
Sales	0.092498	0.0075

a) Is there a significant association between sales and profits? Test an appropriate hypothesis and state your conclusion in context.
b) Do you think that a company's sales serve as a useful predictor of its profits? Use the values of both R^2 and *s* in your explanation.

51. Ozone, again Consider again the relationship between the population and ozone level of U.S. cities that you analyzed in Exercise 49.
   a) Give a 90% confidence interval for the approximate increase in ozone level associated with each additional million city inhabitants.
   b) For the cities studied, the mean population was 1.7 million people. The population of Boston is approximately 0.6 million people. Predict the mean ozone level for cities of that size with an interval in which you have 90% confidence.

T 52. More sales and profits Consider again the relationship between the sales and profits of *Fortune* 500 companies that you analyzed in Exercise 50.
   a) Find a 95% confidence interval for the slope of the regression line. Interpret your interval in context.
   b) Last year the drug manufacturer Eli Lilly, Inc., reported gross sales of $9 billion (that's $9000 million). Create a 95% prediction interval for the company's profits and interpret your interval in context.

53. Start the car! In October 2002, *Consumer Reports* listed the price (in dollars) and power (in cold cranking amps) of auto batteries. We want to know if more expensive batteries are generally better in terms of starting power. Here are several software displays:

Dependent variable is: Power
R-squared = 25.2%
s = 116.0 with 33 − 2 = 31 degrees of freedom

Variable	Coefficient	SE(Coeff)	t-ratio	P-value
Intercept	384.594	93.55	4.11	0.0003
Cost	4.14649	1.282	3.23	0.0029

a) How many batteries were tested?
b) Are the conditions for inference satisfied? Explain.
c) Is there evidence of an association between the cost and cranking power of auto batteries? Test an appropriate hypothesis and state your conclusion.
d) Is the association strong? Explain.
e) What is the equation of the regression line?
f) Create a 90% confidence interval for the slope of the true line.
g) Interpret your interval in this context.

T 54. Crawling Researchers at the University of Denver Infant Study Center wondered whether temperature might influence the age at which babies learn to crawl. Perhaps the extra clothing that babies wear in cold weather would restrict movement and delay the age at which they started crawling. Data were collected on 208 boys and 206 girls. Parents reported the month of the baby's birth and the age (in weeks) at which their child first crawled. The table gives the average outside temperature (°F) when the babies were six months old and average crawling age (in weeks) for each month of the year. Make the plots and compute the analyses necessary to answer the following questions.

Birth Month	6-Month Temperature	Average Crawling Age
Jan.	66	29.84
Feb.	73	30.52
Mar.	72	29.70
April	63	31.84
May	52	28.58
June	39	31.44
July	33	33.64
Aug.	30	32.82
Sept.	33	33.83
Oct.	37	33.35
Nov.	48	33.38
Dec.	57	32.32

a) Would this association appear to be weaker, stronger, or the same if data had been plotted for individual babies instead of using monthly averages? Explain.

b) Is there evidence of an association between temperature and crawling age? Test an appropriate hypothesis and state your conclusion. Don't forget to check the assumptions.

c) Create and interpret a 95% confidence interval for the slope of the true relationship.

T 55. Tablet computers 2014 cnet.com tests tablet computers and continuously updates its list. As of January 2014, the list included the battery life (in hours) and luminous intensity (i.e., screen brightness, in cd/m^2). We want to know if brighter screens are related to the maximum *Screen Brightness*. http://cnet.com/news/cnet-tablet-battery-life-results

Response variable is   Battery life (hrs)
R-squared = 11.3%
s = 2.128 with 34 − 2 = 32 degrees of freedom

Variable	Coefficient	SE(Coeff)	*t*-Ratio	P-Value
Intercept	5.38719	1.727	3.12	0.0038
Max Brightness	0.00904	0.0045	2.02	0.0522

a) How many tablet computers were tested?

b) Are the conditions for inference satisfied? Explain.

c) Is there evidence of an association between maximum brightness of the screen and battery life? Test an appropriate hypothesis and state your conclusion.

d) Is the association strong? Explain.

e) What is the equation of the regression line?

f) Create a 90% confidence interval for the slope of the true line.

g) Interpret your interval in this context.

T 56. Strike? Ads from a Little League instructional video claim that the techniques proposed will improve the performances of Little League pitchers. To test this claim, 20 Little Leaguers threw 50 pitches each, and we recorded the number of strikes. After the players participated in the training program, we repeated the test. The table shows the number of strikes each player threw before and after the training. A test of paired differences failed to show that this training was effective in improving a player's ability to throw strikes. Is there any evidence that the effectiveness of the video depends on the player's initial ability to throw strikes? Test an appropriate hypothesis and state your conclusion.

Number of Strikes (out of 50)			
Before	**After**	**Before**	**After**
28	35	33	33
29	36	33	35
30	32	34	32
32	28	34	30
32	30	34	33
32	31	35	34
32	32	36	37
32	34	36	33
32	35	37	35
33	36	37	32

T 57. Body fat Do the data shown in the table indicate an association between *Waist* size and *%Body Fat*?

a) Test an appropriate hypothesis and state your conclusion.

b) Give a 95% confidence interval for the mean *%Body Fat* found in people with 40-inch waists.

Waist (in.)	Weight (lb)	Body Fat (%)	Waist (in.)	Weight (lb)	Body Fat (%)
32	175	6	33	188	10
36	181	21	40	240	20
38	200	15	36	175	22
33	159	6	32	168	9
39	196	22	44	246	38
40	192	31	33	160	10
41	205	32	41	215	27
35	173	21	34	159	12
38	187	25	34	146	10
38	188	30	44	219	28

T 58. Body fat, again Use the data from Exercise 57 to examine the association between weight and *%Body Fat*.
  a) Find a 90% confidence interval for the slope of the regression line of *%Body Fat* on weight.
  b) Interpret your interval in context.
  c) Give a 95% prediction interval for the *%Body Fat* of an individual who weighs 165 pounds.

T 59. Grades The data set below shows midterm scores from an Introductory Statistics course.

First Name	Midterm 1	Midterm 2	Homework
Timothy	82	30	61
Karen	96	68	72
Verena	57	82	69
Shuang	89	92	84
Rodrigo	88	86	84
Alok	93	81	71
Zygmunt	90	83	79
Lele	83	21	51
Ashish	59	62	58
Justin	89	57	79
Zhang	83	86	78
Wang	95	75	77
Annie	81	66	66
Zheng	86	63	74
Brian	81	86	76
Gregory	81	87	75
Kristina	98	96	84
Timothy	50	27	20
Jason	91	83	71
Whitney	87	89	85
Alexis	90	91	68
Nicholas	95	82	68
Amandeep	91	37	54
Irena	93	81	82
Yvon	88	66	82
Sara	99	90	77
Yang	89	92	68
Benjamin	87	62	72
David	92	66	78
Josef	62	43	56
Rebecca	93	87	80
Joshua	95	93	87
Ian	93	65	66
Katharine	92	98	77
Emily	91	95	83
Brian	92	80	82
Li	61	58	65
Yang	55	65	51
John	76	88	67
Zhu	63	62	67
Peng	89	66	72
Peter	91	42	66
Catherine	90	85	78
Christina	75	62	72
Enrique	75	46	72
Sarah	91	65	77
Thomas	84	70	70
Sonya	94	92	81
Michael	93	78	72
Wesley	91	58	66
Xiaoling	91	61	79
Adam	89	86	62
Jared	98	92	83
Michael	96	51	83
Kathryn	95	95	87
Nicole	98	89	77
Wayne	89	79	44
Elizabeth	93	89	73
Wang	74	64	72
Valentin	97	96	80
David	94	90	88
Marc	81	89	62
Samuel	94	85	76
Brooke	92	90	86

  a) Fit a model predicting the second midterm score from the first.
  b) Comment on the model you found, including a discussion of the assumptions and conditions for regression. Is the coefficient for the slope statistically significant?
  c) A student comments that because the P-value for the slope is very small, Midterm 2 is very well predicted from Midterm 1. So, he reasons, next term the professor can give just one midterm. What do you think?

T 60. Grades? The professor teaching the Introductory Statistics class discussed in Exercise 59 wonders whether performance on homework can accurately predict midterm scores.
  a) To investigate it, she fits a regression of the sum of the two midterms scores on homework scores. Fit the regression model.
  b) Comment on the model, including a discussion of the assumptions and conditions for regression. Is the coefficient for the slope "statistically significant"?
  c) Do you think she can accurately judge a student's performance without giving the midterms? Explain.

T 61. Swimming Ontario Since Marilyn Bell first swam across Lake Ontario in 1954, swimmers have crossed the lake 58 times (as of August, 2013). Have swimmers been getting faster or slower, or has there been no change in swimming speed? A regression predicting time for the swim (in minutes) against year of the swim (for the years 1974–2010) looks like this:

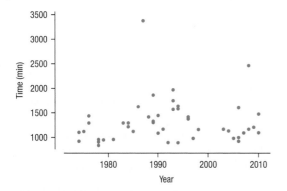

Regression Analysis: Time (min) versus Year
The regression equation is
Time (min) = −9931 + 5.64 Year

Predictor	Coef	SE Coef	T	P
Constant	−99.31	116 94	−0.85	0.400
Year	5.636	5.873	0.96	0.342

S = 449.925 R-Sq = 2.0% R-Sq(adj) = 0.0%

a) What do you hope to learn from these data? Define the variables, and state the null and alternative hypotheses about the slope.
b) Check assumptions and conditions.
c) If they are satisfied, complete the analysis.
d) The longest time is for a swim in which Vicki Keith swam round-trip, crossing the lake twice. If we remove it from the data, the new analysis is shown. Check the conditions again.

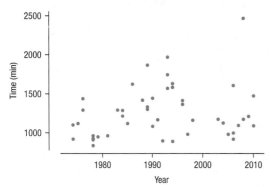

The regression equation is
Time (min) = −131 11 + 7.21 Year

Predictor	Coef	SE Coef	T	P
Constant	−131 11	849 0	−1.54	0.130
Year	7.211	4.263	1.69	0.098

S = 326.099 R-Sq = 6.0% R-Sq(adj) = 3.9%

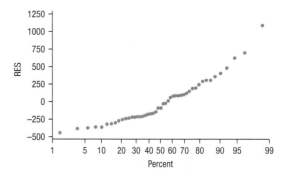

e) If the conditions are satisfied, complete the analysis.
f) The slope itself didn't change very much when we removed the outlier, but the P-value did. Consider the formula for the *t*-ratio and explain what changed.

T 62. All the efficiency money can buy 2011 A sample of 84 model-2011 cars from an online information service was examined to see how fuel efficiency (as highway mpg) relates to the cost (Manufacturer's Suggested Retail Price in dollars) of cars. Here are displays and computer output:

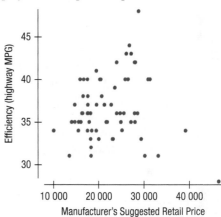

Dependent variable is MPG
R-squared = 0.0216%
s = 3.54

Variable	Coefficient	SE(Coeff)	*t*-Ratio	P-Value
Intercept	36.514	1.496	24.406	<0.0001
Slope	−8.089E−6	6.439E−5	−0.1256	0.900

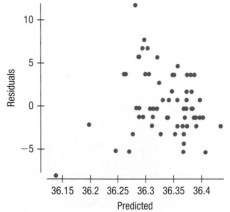

a) State what you want to know, identify the variables, and give the appropriate hypotheses.
b) Check the assumptions and conditions.
c) If the conditions are met, complete the analysis.

T 63. Education and mortality The software output below is based on the mortality rate (deaths per 100 000 people) and the education level (average number of years in school) for 58 U.S. cities.

Variable	Count	Mean	StdDev
Mortality	58	942.501	61.8490
Education	58	11.0328	0.793480

Dependent variable is: Mortality
R-squared = 41.0%
s = 47.92 with 58 − 2 = 56 degrees of freedom

Variable	Coefficient	SE(Coeff)
Intercept	1493.26	88.48
Education	249.9202	8.000

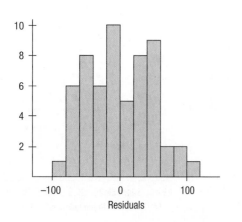

a) Comment on the assumptions for inference.
b) Is there evidence of a strong association between the level of education in a city and the mortality rate? Test an appropriate hypothesis and state your conclusion.
c) Can we conclude that getting more education is likely (on average) to prolong your life? Why or why not?
d) Find a 95% confidence interval for the slope of the true relationship.
e) Explain what your interval means.
f) Obtain the correlation coefficient of mortality rates and education levels. Are the two variables linearly related?
*g) Report the ANOVA table. What is the total variation explained by the regression model?
h) Find a 95% confidence interval for the average mortality rate in cities where the adult population completed an average of 12 years of school.

T 64. Property assessments The software outputs below provide information about the size (in square feet) of 18 homes in Ithaca, New York, and the city's assessed value of those homes.

Variable	Count	Mean	StdDev	Range
Size	18	200 3.39	264.727	890
Value	18	609 46.7	552 7.62	197 10

Dependent variable is: Value
R-squared = 32.5%
s = 4682 with 18 − 2 = 16 degrees of freedom

Variable	Coefficient	SE(Coeff)
Intercept	37 108.8	8 664
Size	11.8987	4.290

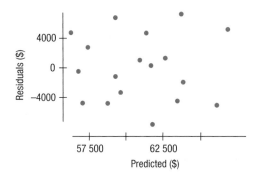

a) Explain why inference for linear regression is appropriate with these data.

b) Is there a significant association between the size of a home and its assessed value? Test an appropriate hypothesis and state your conclusion.

c) What percentage of the variability in assessed value is explained by this regression?

d) Give a 90% confidence interval for the slope of the true regression line, and explain its meaning in the proper context.

e) From this analysis, can we conclude that adding a room to your house will increase its assessed value? Why or why not?

f) The owner of a home measuring 2100 square feet files an appeal, claiming that the $70 200 assessed value is too high. Do you agree? Explain your reasoning.

g) Obtain the correlation coefficient of sizes and assessed values. Are the two variables linearly related?

*h) Report the ANOVA table. What is the total variation explained by the regression model?

T 65. Seasonal spending The authors have noticed that spending on credit cards decreases after the Christmas spending season (as measured by amount charged on a credit card in December). The data set online at www.pearsoncanada.ca/deveaux contains the monthly credit card charges of a random sample of 99 cardholders.

a) Build a regression model to predict January spending from December's spending.

b) How much, on average, will cardholders who charged $2000 in December charge in January?

c) Give a 95% confidence interval for the average January charges of cardholders who charged $2000 in December.

d) From part c), give a 95% confidence interval for the average decrease in the charges of cardholders who charged $2000 in December.

e) What reservations, if any, do you have about the confidence intervals you made in parts c) and d)?

T 66. Seasonal spending II. As we saw in Exercise 65, financial analysts know that January credit card charges will generally be much lower than those of the month before. What about the difference between January and the next month? Does the trend continue? The data set online at www.pearsoncanada.ca/deveaux contains the monthly credit card charges of a random sample of 99 cardholders.

a) Build a regression model to predict February charges from January's charges.

b) How much, on average, will cardholders who charged $2000 in January charge in February?

c) Give a 95% confidence interval for the average February charges of cardholders who charged $2000 in January.

d) From part c), give a 95% confidence interval for the average decrease in the charges of cardholders who charged $2000 in January.

e) What reservations, if any, do you have about the confidence intervals you made in parts c) and d)?

T 67. CPU Performance Below is a sample of computer processor performance and price data, obtained from CPU-Benchmark.net in July 2013. The full dataset is online at www.pearsoncanada.ca/deveaux.

CPU	Performance	Price ($)
AMD FX-6100 Six-Core	5380	104.99
Intel Celeron G530 @ 2.40GHz	2274	48.99
AMD FX-6200 Six-Core	6231	134.99
AMD FX-6300 Six-Core	6459	139.99
Intel Celeron G540 @ 2.50GHz	2297	49.99
AMD FX-8320 Eight-Core	8229	179.99
AMD FX-8350 Eight-Core	9239	209.99

a) Make a scatterplot of the data. Do you see any points of high leverage, high influence, or with large residuals? If so, remove the data points that are causing the problem (and do not draw any conclusions in this domain).

b) Fit a linear regression model to the remaining data. Check the residuals plot and the Normal probability plot. Do you see any problems?

c) AMD has just come out with a new processor that has scored 8000 on the performance test. Predict the price of this processor and give a 95% interval around this prediction.

d) Intel has just come out with a new processor that has scored 2000 on the performance test. Predict the price of this processor and give a 95% interval around this prediction.

e) You might have expected the interval widths in the previous two parts to be quite different, yet they're not. Comment on why they are so close.

*68. Right-to-work laws Are so-called "right-to-work" laws related to the percent of public sector employees in unions and the percent of private sector employees in unions? This data set looks at these percentages for the states in the United States in 1982. The dependent variable is whether the state had a right-to-work law or not. The computer output for the logistic regression is given here. (Source: N. M. Meltz, "Interstate and Interprovincial Differences in Union Density," *Industrial Relations*, 28:2 [Spring 1989], 142–158 by way of DASL.)

Logistic Regression Table

Predictor	Coeff	SE(Coeff)	Chisq	P
Intercept	6.1995	1.78724	12.04	0.001
publ	−0.1062	0.0475	5.02	0.025
pvt	−0.2230	0.0811	7.56	0.006

a) Write out the estimated regression equation.
b) The following are scatterplots of the response variable against each of the explanatory variables. Does logistic regression seem appropriate here? Explain.

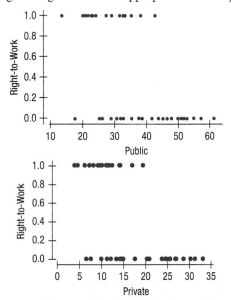

*69. **Cost of higher education** Are there fundamental differences between liberal arts colleges and universities? In this case, we have information on the top 25 liberal arts colleges and the top 25 universities in the Unites States. We will consider the type of school as our response variable and will use the percent of students who were in the top 10% of their high school class and the amount of money spent per student by the college or university as our explanatory variables. The output from this logistic regression is given on the following page.

Logistic Regression Table

Predictor	Coeff	SE(Coeff)	Chisq	P
Intercept	−13.1461	3.98629	10.89	0.001
Top 10%	0.0845	0.0396345	4.54	0.033
$/Student	0.0003	0.0000860	9.12	0.003

a) Write out the estimated regression equation.
b) Is percent of students in the top 10% of their high school class statistically significant in predicting whether or not the school is a university? Explain.
c) Is the amount of money spent per student statistically significant in predicting whether or not the school is a university? Explain.

Just Checking ANSWERS

1. A high t-ratio of 3.27 indicates that the slope is different from zero—that is, that there is a linear relationship between height and mouth size. The small P-value says that a slope this large would be very unlikely to occur by chance if, in fact, there was no linear relationship between the variables.

2. Not really. The R^2 for this regression is only 15.3%, so height doesn't account for very much of the variability in mouth size.

3. The value of s tells the standard deviation of the residuals. Mouth sizes have a mean of 60.3 cubic centimetres. A standard deviation of 15.7 in the residuals indicates that the errors made by this regression model can be quite large relative to what we are estimating. Errors of 15 to 30 cubic centimetres would be common.

MathXL

MyLab Statistics

Go to MathXL at **www.mathxl.com** or MyLab™ Statistics at **www.mystatlab.com**. You can practise exercises for this chapter as often as you want. The guided solutions will help you find answers step by step. You'll find a personalized study plan available to you too!

Review

Assessing Associations Between Variables

Quick Review

We now turn our focus to relationships between variables. This is where the effectiveness of Statistics becomes clear. Statistical inference about how variables are related is essential to science, social science, health, and business. In this part we have considered comparing groups with means and proportions, comparing groups that have a natural pairing, comparing proportions in several categories and between two categorical variables, and inferential decisions about linear regression. That may seem like a wide variety of ideas, but they are all held together by the central concept that by understanding the ways in which variables and groups relate to each other, we can understand how the world works. Here's a brief summary of the key concepts and skills:

- A confidence interval uses a sample statistic to estimate a range of possible values for the parameter of a population model.

- A hypothesis test proposes a model for the population, then examines the observed statistics to see if the model is plausible.

- Statistical inference procedures for comparing proportions are based on the Central Limit Theorem. We can make inferences about the difference of two proportions using Normal models.

- Statistical inference procedures for comparing means and for estimating regression coefficients require a new sampling model because we usually don't know the population standard deviation (hence the Central Limit Theorem's predicted Normal model cannot be specified precisely). Student's t-models take the additional uncertainty of independently estimating the standard deviation into account.

  - We can make inferences about the difference of two independent means, the mean of paired differences, or the intercept or slope of a linear regression using t-models.

  - No inference procedure is valid unless the underlying assumptions are true. Always think about the assumptions and check the conditions before proceeding. For regression, take care to check them in the right order.

  - Because t-models assume that samples are drawn from roughly Normal populations, data in the

sample (or, for regression, the residuals) should appear to be nearly Normal. Skewness and outliers are particularly problematic. Fortunately, robustness against violations of the Normality assumption increases with bigger sample sizes.

  - To identify the appropriate statistic for comparing the means of groups, you must think carefully about how the data were collected. You may use two-sample t procedures only if the groups are independent.

  - Unless there is some obvious reason to suspect that two independent populations have the same standard deviation, you should not pool the variances. It is never wrong to use unpooled t procedures.

  - If two groups are paired, the data are *not* from independent groups. You must use matched-pairs t procedures and test the mean of the paired differences (with a one-sample t-test).

  - Linear regression inference is only valid if the relationship between the two variables is straight. Examine a scatterplot of the data and also of the residuals.

- Not all sampling distributions are unimodal, symmetric, or bell-shaped. Inferences about distributions of counts use chi-square models, which are unimodal but skewed to the high end. Nevertheless, the sampling distribution plays the same role in inference, helping us to model the random behaviour of samples and translate values (statistics) calculated from data into probabilities.

  - To see if an observed distribution is consistent with a proposed model, use a chi-square goodness-of-fit test.

  - To see if two or more observed distributions could have arisen from identical populations, use a test of homogeneity.

  - To see if two categorical variables are independent, perform a chi-square test of independence.

- You can now use statistical inference to answer questions about means, proportions, distributions, associations, and linear regression models.

  - No inference procedure is valid unless the underlying assumptions are true. Always check the conditions before proceeding.

  - You can make inferences about the difference between two proportions using Normal models.

- You can make inferences about the difference between two independent means, or about the mean of paired differences using *t*-models.
- You can make inferences about distributions of categorical variables using chi-square models.
- You can make inferences about association between categorical variables using chi-square models.
- You can make inferences about the coefficients in a linear regression model using *t*-models.

Now for some opportunities to review these concepts. Be careful. You have a lot of thinking to do. These review exercises mix questions about proportions, means, chi square, and regression. You have to determine which of our inference procedures is appropriate in each situation. Then you have to check the proper assumptions and conditions. Keeping track of those can be difficult, so first we summarize the many procedures with their corresponding assumptions and conditions on the next page. Look them over carefully . . . then, on to the Exercises!

Assumptions for Inference	**And the Conditions that Support or Override them**
Proportions (*z*)	
■ **One sample**	1. SRS.
1. Individuals respond independently.	2. Successes ≥ 10 and failures ≥ 10.
2. Sample is sufficiently large.	
■ **Two sample**	1. (Think about how the data were collected.)
1. Samples are independent of each other.	2. Both are SRSs OR random allocation.
2. Individual responses in each sample are independent.	3. Successes ≥ 10 and failures ≥ 10 for both samples.
3. Both samples are sufficiently large.	
Means (*t*)	
■ **One sample (df $= n - 1$)**	1. SRS.
1. Individuals respond independently.	2. Histogram is unimodal and symmetric.*
2. Population has a Normal model.	
■ **Two independent Samples (df from technology)**	1. (Think about the design.)
1. Samples are independent of each other.	2. SRSs OR random allocation.
2. Individual responses in each sample are independent.	3. Both histograms are unimodal and symmetric.*
3. Both populations are Normal.	
■ **Matched pairs (df $= n - 1$)**	1. (Think about the design.)
1. Each individual is paired with an individual in the other sample; *n* pairs.	2. SRSs OR random allocation.
2. Individual differences are independent.	3. Histogram of differences is unimodal and symmetric.*
3. Population of differences is Normal.	
Distributions/Association (χ^2)	
■ **Goodness of fit** [df $=$ # of cells -1; one categorical variable, one sample compared with population model]	
1. Data are counts of individuals classified into categories.	1. (Are they?)
2. Individuals' responses are independent.	2. SRS.
3. Sample is sufficiently large.	3. All expected counts ≥ 5.
■ **Homogeneity** [$df = (r - 1)(c - 1)$; samples from many populations compared on one cateogorical variable]	
1. Data are counts of individuals classified into categories.	1. (Are they?)
2. Individuals' responses are independent.	2. SRSs OR random allocation.
3. Groups are sufficiently large.	3. All expected counts ≥ 5.
■ **Independence** [$df = (r - 1)(c - 1)$; sample from one population classified on two categorical variables]	
1. Data are counts of observations classified into categories.	1. (Are they?)
2. Individuals' responses are independent.	2. SRSs.
3. Group is sufficiently large.	3. All expected counts ≥ 5.
Regression with One Predictor and One Response Variable - Both Quantitative (*t*, $df = n - 2$)	
1. Form of relationship is linear.	1. Scatterplot of *y* against *x* is straight enough. Scatterplot of residuals against predicted values shows no special structure (e.g. bends).
2. Errors are independent.	2. No apparent pattern in plot of residuals against predicted values or against time (if data collected in time sequence).
3. Variability of errors is constant.	3. Plot of residuals against predicted values has constant spread, doesn't "thicken."
4. Errors follow a Normal model.	4. Histogram of residuals is approximately unimodal and symmetric, or Normal probability plot is reasonably straight.*

**Less critical as *n* increases

Note: For all of these procedures, sampling more than 10% of the population compromises independence—but in a good way! Be aware that *P*-values and confidence coefficients calculated in the manner we have discussed become overly conservative (so if your ME or *P*-value is just a bit lacking, get a statistician to help make the appropriate adjustments).

Quick Guide to Inference							
Think				**Show**			**Tell?**
Inference about?	One sample or two?	Procedure	Sampling Model	Parameter	Estimate	SE	Chapter
Proportions	One sample	1-Proportion z-Interval	z	p	$\hat{p}$	$\sqrt{\dfrac{\hat{p}\hat{q}}{n}}$	15
		1-Proportion z-Test				$\sqrt{\dfrac{p_0 q_0}{n}}$	16, 17
	Two independent groups	2-Proportion z-Interval	z	$p_1 - p_2$	$\hat{p}_1 - \hat{p}_2$	$\sqrt{\dfrac{\hat{p}_1 \hat{q}_1}{n_1} + \dfrac{\hat{p}_2 \hat{q}_2}{n_2}}$	21
		2-Proportion z-Test				$\sqrt{\dfrac{\hat{p}\hat{q}}{n_1} + \dfrac{\hat{p}\hat{q}}{n_2}}, \hat{p} = \dfrac{y_1 + y_2}{n_1 + n_2}$	21
Means	One sample	t-Interval t-Test	t df $= n - 1$	μ	$\bar{y}$	$\dfrac{s}{\sqrt{n}}$	18
	Two independent groups	2-Sample t-Test 2-Sample t-Interval	t df from technology	$\mu_1 - \mu_2$	$\bar{y}_1 - \bar{y}_2$	$\sqrt{\dfrac{s_1^2}{n_1} + \dfrac{s_2^2}{n_2}}$	19
	n Matched pairs	Paired t-Test Paired t-Interval	t df $= n - 1$	μ_d	$\bar{d}$	$\dfrac{s_d}{\sqrt{n}}$	20
Distributions (one categorical variable)	One sample	Goodness of fit	χ^2 df $= cells - 1$			Test statistic $= \sum \dfrac{(\text{obs} - \text{exp})^2}{\text{exp}}$	22
	Many independent samples	Homogeneity χ^2 Test	χ^2 df $= (r-1)(c-1)$				
Independence (two categorical variables)	One sample	Independence χ^2 Test					
Association (two quantitative variables)	One sample	Linear Regression t-Test or Confidence Interval for β	t df $= n - 2$	β_1	b_1	$\dfrac{s_e}{s_x\sqrt{n-1}}$ (compute with technology)	23
		Confidence Interval for μ_v		μ_v	$\hat{y}_v$	$\sqrt{SE^2(b_1) \cdot (x_v - \bar{x})^2 + \dfrac{s_e^2}{n}}$	
		Prediction Interval for y_v		y_v	$\hat{y}_v$	$\sqrt{SE^2(b_1) \cdot (x_n - \bar{x})^2 + \dfrac{s_e^2}{n} + s_e^2}$	
Inference about?	One sample or two?	Procedure	Sampling Model	Parameter	Estimate	SE	Chapter

Review Exercises

1. **Crawling** A study published in 1993 found that babies born at different times of the year may develop the ability to crawl at different ages! The author of the study suggested that these differences may be related to the temperature at the time the infant is six months old.[1]
   a) The study found that 32 babies born in January crawled at an average age of 29.84 weeks, with a standard deviation of 7.08 weeks. Among 21 July babies, crawling ages averaged 33.64 weeks with a standard deviation of 6.91 weeks. Is this difference significant?
   b) For 26 babies born in April, the mean and standard deviation were 31.84 and 6.21 weeks, while for 44 October babies, the mean and standard deviation of crawling ages were 33.35 and 7.29 weeks. Is this difference significant?
   c) Are these results consistent with the researchers' claim? (We'll examine these data in more detail in a later chapter.)

T 2. **Mazes and smells** Can pleasant smells improve learning? Researchers timed 21 subjects as they tried to complete paper-and-pencil mazes. Each subject attempted a maze both with and without the presence of a floral aroma. Subjects were randomized with respect to whether they did the scented trial first or second. The table shows some of the data collected. Is there any evidence that the floral scent improved the subjects' ability to complete the mazes?[2]

Time to Complete the Maze (sec)			
Unscented	Scented	Unscented	Scented
25.7	30.2	61.5	48.4
41.9	56.7	44.6	32.0
51.9	42.4	35.3	48.1
32.2	34.4	37.2	33.7
64.7	44.8	39.4	42.6
31.4	42.9	77.4	54.9
40.1	42.7	52.8	64.5
43.2	24.8	63.6	43.1
33.9	25.1	56.6	52.8
40.4	59.2	58.9	44.3
58.0	42.2		

3. **BC birth weights** Below are the means and standard deviations of birth weights of newborns born in BC during 1981–2000 to some different groups, for births following a 40-week gestation period (the modal gestation period).[3] For the entire BC population, the mean birth weight was 3558 grams.

	# Births	Mean Weight	Standard Deviation
First Nations people	20 578	3645 g	466 g
Canadians of Chinese origin	14 586	3393 g	391 g

(Group)

   a) Find a 95% confidence interval for the average birth weight for each group above.
   b) Does either group differ significantly from the general BC population?
   c) Do the means for these two groups differ significantly from each other? Include the P-value. If so, find a confidence interval for the mean difference.

T 4. **Drugs** In a full-page ad that ran in many U.S. newspapers in August 2002, a Canadian discount pharmacy listed costs of drugs that could be ordered from a website in Canada. The table compares prices (in US$) for commonly prescribed drugs.

	Cost per 100 Pills		
	United States	Canada	Percent Savings
Cardizem	131	83	37
Celebrex	136	72	47
Cipro	374	219	41
Pravachol	370	166	55
Premarin	61	17	72
Prevacid	252	214	15
Prozac	263	112	57
Tamoxifen	349	50	86
Vioxx	243	134	45
Zantac	166	42	75
Zocor	365	200	45
Zoloft	216	105	51

(Drug Name)

   a) Give a 95% confidence interval for the average savings in dollars.
   b) Give a 95% confidence interval for the average savings in percent.
   c) Which analysis do you think is more appropriate? Why?
   d) In small print the newspaper ad says, "Complete list of all 1500 drugs available on request." How does this comment affect your conclusions above?

[1] Benson and Janette, *Infant Behavior and Development*, 1993.

[2] A. R. Hirsch and L. H. Johnston, "Odors and Learning." Chicago: Smell and Taste Treatment and Research Foundation.

[3] British Columbia Vital Statistics Agency

5. Pottery Archaeologists can use the chemical composition of clay found in pottery artifacts to determine whether different sites were populated by the same ancient people. They collected five samples of Romano-British pottery from each of two sites in Great Britain—the Ashley Rails site and the New Forest site—and measured the percentage of aluminum oxide in each. Based on these data, do you think the same people used these two kiln sites? Base your conclusion on a 95% confidence interval for the difference in aluminum oxide content of pottery made at the sites.[4]

Ashley Rails	19.1	14.8	16.7	18.3	17.7
New Forest	20.8	18.0	18.0	15.8	18.3

6. Diet Thirteen overweight women volunteered for a study to determine whether eating specially prepared crackers before a meal could help them lose weight. The subjects were randomly assigned to eat crackers with different types of fibre (bran fibre, gum fibre, both, and a control cracker). Unfortunately, some of the women developed uncomfortable bloating and upset stomachs. Researchers suspected that some of the crackers might be at fault. The contingency table of cracker versus bloat shows the relationship between the four different types of crackers and the reported bloating. The study was paid for by the manufacturers of the gum fibre. What would you recommend to them about the prospects for marketing their new diet cracker?

		Bloat	
		Little/None	Moderate/Severe
Cracker	Bran	11	2
	Gum	4	9
	Combo	7	6
	Control	8	4

7. Gehrig Ever since baseball legend Lou Gehrig developed amyotrophic lateral sclerosis (ALS), this deadly condition has been commonly known as Lou Gehrig's disease. Some believe that ALS is more likely to strike athletes or the very fit. Columbia University neurologist Lewis P. Rowland recorded personal histories of 431 patients he examined between 1992 and 2002. He diagnosed 280 as having ALS: 38% of them had been varsity athletes; the other 151 had other neurological disorders, and only 26% of them had been varsity athletes. (*Science News*, Sept. 28, 2002)
a) Is there evidence that ALS is more common among athletes? Attack this two ways: using a *t*-test and a chi-square test: How are these two tests related?

b) What kind of study is this? How does that affect the inference you made in part a)?

8. Teen drinking A study of the health behaviour of school-aged children asked a sample of 15-year-olds in several different countries if they had been drunk at least twice. The table shows the results. Give a 95% confidence interval for the difference in the rates for males and females. Be sure to check the assumptions that support your chosen procedure, and explain what your interval means.[5]

	Percent of 15-Year-Olds Drunk at Least Twice	
Country	Female	Male
Denmark	63	71
Wales	63	72
Greenland	59	58
England	62	51
Finland	58	52
Scotland	56	53
No. Ireland	44	53
Slovakia	31	49
Austria	36	49
Canada	42	42
Sweden	40	40
Norway	41	37
Ireland	29	42
Germany	31	36
Latvia	23	47
Estonia	23	44
Hungary	22	43
Poland	21	39
USA	29	34
Czech Rep.	22	36
Belgium	22	36
Russia	25	32
Lithuania	20	32
France	20	29
Greece	21	24
Switzerland	16	25
Israel	10	18

9. Genetics Two human traits controlled by a single gene are the ability to roll one's tongue and whether one's ear lobes are free or attached to the neck. Genetic theory says that people will have neither, one, nor both of these traits in the ratio 1:3:3:9 (1—attached, noncurling; 3—attached, curling; 3—free, noncurling; 9—free, curling). An Introductory Biology class of 122 students collected the data shown. Are they consistent with the genetic

[4]A. Tubb, A. J. Parker, and G. Nickless, 1980, "The analysis of Romano-British pottery by atomic absorption spectrophotometry." *Archaeometry*, 22: 153–171.

[5]*Health and Health Behavior Among Young People*. Copenhagen: World Health Organization, 2000.

theory? Test an appropriate hypothesis, and state your conclusion.

	Trait			
	Attached, noncurling	Attached, curling	Free, noncurling	Free, curling
Count	10	22	31	59

10. **Twins** There is some indication in medical literature that doctors may have become more aggressive in inducing labour or doing preterm Caesarean sections when a woman is carrying twins. Records at a large hospital show that, of the 43 sets of twins born in 1990, 20 were delivered before the 37th week of pregnancy. In 2000, 26 of 48 sets of twins were born preterm. Does this indicate an increase in the incidence of early births of twins? Test an appropriate hypothesis and state your conclusion.

11. **Perfect pitch** A recent study of perfect pitch tested 2700 students in American music conservatories. It found that 7% of non-Asian and 32% of Asian students have perfect pitch. A test of the difference in proportions resulted in a P-value of <0.0001.
    a) What are the researchers' null and alternative hypotheses?
    b) State your conclusion.
    c) Explain in this context what the P-value means.
    d) The researchers claimed that the data prove that genetic differences between the two populations cause a difference in the frequency of occurrence of perfect pitch. Do you agree? Why or why not?

12. **Renters weigh less** On June 16, 2009, the *Windsor Star* described a study that found female homeowners tend to weigh more than women who are leasing a residence.[6] In this survey of 600 women, homeowners outweighed renters by 12.5 pounds (5.7 kg), on average.
    a) Are you impressed by the difference? What else might you like to know?
    b) If 300 women were in each group, what would be the P-value for the observed difference of 12.5 pounds if the pooled standard deviation was 30 pounds?
    c) If 300 women were in each group, what would the pooled standard deviation have to be in order to make a difference of 12.5 pounds statistically significant at the 1% level?

13. **Feeding fish** Large mouth bass is one type of fish raised by the aquaculture industry. Researchers wanted to know whether the fish would grow better if fed a natural diet of fathead minnows or an artificial diet of food pellets. They stocked six ponds with bass fingerlings weighing about 8 grams. For one year, the fish in three of the ponds were fed minnows, and the others were fed the commercially

prepared pellets. The fish were then weighed and measured. The bass fed a natural food source had a higher average length (19.6 cm) and weight (95.9 g) than those fed the commercial fish food (17.3 cm and 72.0 g, respectively). The researchers reported P-values for both measurements to be less than 0.001.
    a) Explain to someone who has not studied Statistics what the P-values mean here.
    b) What advice should the researchers give the people who raise largemouth bass?
    c) If that advice turns out to be incorrect, what type of error occurred?

T 14. **Risk** A study of auto safety determined the number of driver deaths per million vehicle sales for the model years 2001–2004, classified by type of vehicle. The data below are for six midsize models and six SUVs. We wonder if there is evidence that drivers of SUVs are safer, hoping to create a 95% confidence interval for the difference in driver death rates for the two types of vehicles. Are these data appropriate for this inference? Explain.[7]

Midsize	11	14	30	55	85	118
SUV	13	21	44	51	115	127

T 15. **Twins again** In 2000, *The Journal of the American Medical Association* published a study that examined a sample of pregnancies that resulted in the birth of twins. Births were classified as preterm with intervention (induced labour or Caesarean), preterm without such procedures, or term or post-term. Researchers also classified the pregnancies by the level of prenatal medical care the mother received (inadequate, adequate, or intensive). The table below summarizes the data from the years 1995–1997. Figures are in thousands of births.[8]

		Twin Births 1995–1997 (in thousands)			
		Preterm (induced or Caesarean)	Preterm (without procedures)	Term or postterm	Total
Level of Prenatal Care	Intensive	18	15	28	61
	Adequate	46	43	65	154
	Inadequate	12	13	38	63
	Total	76	71	131	278

Is there evidence of an association between the duration of the pregnancy and the level of care received by the mother?

T 16. **Twins, yet again** After reading of the *JAMA* study in Exercise 15, a large city hospital examined their records of twin births for several years, and found the data summarized

[6]M. Harris, 2009, "Women who rent weigh less, survey says." *Windsor Star*, June 16, 2009, p. B8.

[7]*Insurance Institute for Highway Safety*, Status Report, Vol. 42, No. 4, April 19, 2007.

[8]*JAMA*, 284[2002]:335–341.

in the table below. Is there evidence that the way the hospital deals with pregnancies involving twins may have changed?

		1990	1995	2000
Outcome of Pregnancy	**Preterm** (induced or Caesarean)	11	13	19
	Preterm (without procedures)	13	14	18
	Term or postterm	27	26	32

17. Age In a study of how depression may impact one's ability to survive a heart attack, researchers reported the ages of the two groups they examined. The mean age of 2397 patients without cardiac disease was 69.8 years (SD = 8.7 years), while for the 450 patients with cardiac disease the mean and standard deviation of the ages were 74.0 and 7.9, respectively.
a) Create a 95% confidence interval for the difference in mean ages of the two groups.
b) How might an age difference confound these research findings about the relationship between depression and ability to survive a heart attack?

18. Smoking In the depression and heart attack research described in Exercise 17, 32% of the diseased group were smokers, compared with only 23.7% of those free of heart disease.
a) Create a 95% confidence interval for the difference in the proportions of smokers in the two groups.
b) Is this evidence that the two groups in the study were different? Explain.
c) Could this be a problem in analyzing the results of the study? Explain.

19. Computer use A Gallup telephone poll of 1240 teens conducted in 2001 found that boys were more likely than girls to play computer games, by a margin of 77% to 65%. An equal number of boys and girls were surveyed.
a) What kind of sampling design was used?
b) Give a 95% confidence interval for the difference in game playing between boys and girls.
c) Does your confidence interval suggest that among all teens a higher percentage of boys than girls play computer games?

T 20. Hearing Fitting someone for a hearing aid requires assessing the patient's hearing ability. In one method of assessment, the patient listens to a tape of 50 English words. The tape is played at low volume, and the patient is asked to repeat the words. The patient's hearing ability score is the number of words perceived correctly. Four tapes of equivalent difficulty are available so that each ear can be tested with more than one hearing aid. These lists were created to be equally difficult to perceive in silence, but hearing aids must work in the presence of background noise. Researchers had 24 subjects with normal hearing compare two of the tapes when a background noise was present,

with the order of the tapes randomized. Is it reasonable to assume that the two lists are still equivalent for purposes of the hearing test when there is background noise? Base your decision on a confidence interval for the mean difference in the number of words people might misunderstand.[9]

	List				List	
Subject	**A**	**B**	**Subject**	**A**	**B**
1	24	26	13	36	32
2	32	24	14	32	34
3	20	22	15	38	32
4	14	18	16	14	18
5	32	24	17	26	20
6	22	30	18	14	20
7	20	22	19	38	40
8	26	28	20	20	26
9	26	30	21	14	14
10	38	16	22	18	14
11	30	18	23	22	30
12	16	34	24	34	42

T 21. Old Faithful Old Faithful eruptions do not occur at uniform intervals and may vary greatly. Can we improve our chances of predicting the time of the next eruption if we know how long the previous eruption lasted?
a) Describe what you see in this scatterplot.

b) Write an appropriate hypothesis.
c) Here are a histogram of the residuals and the residuals plot. Do you think the assumptions for inference are met? Explain.

---

[9]Faith Loven, *A Study of the Interlist Equivalency of the CID W-22 Word List Presented in Quiet and in Noise*. University of Iowa, 1981.

d) State a conclusion based on this regression analysis:

Dependent variable is Interval
R-squared = 77.0%
s = 6.159 with 222 − 2 = 220 degrees of freedom

Variable	Coefficient	SE(Coeff)	t-Ratio	P-Value
Intercept	33.9668	1.428	23.8	≤0.0001
Pop	10.3582	0.3822	27.1	≤0.0001

Variable	Mean	StdDev
Duration	3.57613	1.08395
Interval	71.0090	12.7992

e) The second table shows the summary statistics for the two variables. Create a 95% confidence interval for the mean length of time that will elapse following a 2-minute eruption.

f) You arrive at Old Faithful just as an eruption ends. Witnesses say it lasted 4 minutes. Create a 95% prediction interval for the length of time you will wait to see the next eruption.

T 22. Newspapers Who reads the newspaper more, men or women? Eurostat, an agency of the European Union (EU), conducts surveys on several aspects of daily life in EU countries. Recently, the agency asked samples of 1000 respondents in each of 14 European countries whether they read the newspaper on a daily basis. Below are the data by country and sex.

	% Reading a Newspaper Daily	
Country	Men	Women
Belgium	56.3	45.5
Denmark	76.8	70.3
Germany	79.9	76.8
Greece	22.5	17.2
Spain	46.2	24.8
Ireland	58.0	54.0
Italy	50.2	29.8
Luxembourg	71.0	67.0
Netherlands	71.3	63.0
Austria	78.2	74.1
Portugal	58.3	24.1
Finland	93.0	90.0
Sweden	89.0	88.0
U.K.	32.6	30.4

a) Examine the differences in the percentages for each country. Which of these countries seem to be outliers? What do they have in common?

b) After eliminating the outliers, is there evidence that in Europe men are more likely than women to read the newspaper?

23. Preemies Among 242 Cleveland-area children born prematurely at low birth weights between 1977 and 1979, only 74% graduated from high school. Among a comparison group of 233 children of normal birth weight, 83% were high school graduates.[10]

a) Create a 95% confidence interval for the difference in graduation rates between children of normal birth weights and children of very low birth weights. Be sure to check the appropriate assumptions and conditions.

b) Does this provide evidence that premature birth may be a risk factor for not finishing high school? Use your confidence interval to test an appropriate hypothesis.

c) Suppose your conclusion is incorrect. Which type of error did you make?

T 24. Gender and school Below are primary school enrolment percentages (enrolment of the official age group for primary education expressed as a percentage of the corresponding population) by gender for a sample of countries. Compute a 95% confidence interval for the difference between genders, interpret it, and be sure to examine any important assumptions. Is there evidence of a systemic gender effect on primary education accessibility?

Location	Net Primary School Enrolment Ratio Female (%), 2006	Net Primary School Enrolment Ratio Male (%), 2006
Algeria	94	96
Benin	73	87
Burkina Faso	42	52
Burundi	73	76
Cape Verde	87	88
Central African Republic	38	53
Congo	52	58
Eritrea	43	50
Ethiopia	62	68
Gambia	64	59
Ghana	64	63
Guinea	66	77
Kenya	76	75
Lesotho	74	71
Liberia	39	40
Madagascar	96	96
Malawi	94	88
Mali	54	67

[10]"Outcomes in young adulthood for very-low-birth-weight infants," *New England Journal of Medicine*, 346, no. 3 (2002).

Location	Net Primary School Enrolment Ratio Female (%), 2006	Net Primary School Enrolment Ratio Male (%), 2006
Mauritania	82	78
Mauritius	96	94
Mozambique	73	79
Namibia	79	74
Niger	37	50
Sao Tome and Principe	95	97
Senegal	70	71
Togo	75	86
United Republic of Tanzania	97	98
Zambia	94	90
Zimbabwe	88	87
Bahamas	89	87
Barbados	96	97
Belize	97	97
Bolivia	95	94
Colombia	88	89
Cuba	97	96
Dominica	79	75
Dominican Republic	78	76
Ecuador	97	96
El Salvador	94	94
Guatemala	92	96
Honduras	97	96
Mexico	97	98
Nicaragua	90	90
Panama	98	99
Peru	97	96
Saint Lucia	97	99
Suriname	98	95
United States	93	91
Uruguay	100	100
Venezuela	91	91
Djibouti	34	42
Egypt	92	96
Jordan	90	89
Kuwait	83	84
Lebanon	82	82
Morocco	85	91
Oman	75	73
Pakistan	57	73
Qatar	94	93
Tunisia	97	96
United Arab Emirates	88	88
Andorra	83	83
Armenia	84	80
Austria	98	97
Azerbaijan	83	86
Belarus	88	90
Belgium	97	97
Bulgaria	92	93
Croatia	90	91

Location	Net Primary School Enrolment Ratio Female (%), 2006	Net Primary School Enrolment Ratio Male (%), 2006
Cyprus	99	99
Denmark	96	95
Estonia	94	95
Finland	97	97
France	99	98
Georgia	91	88
Germany	98	98
Greece	99	100
Hungary	88	89
Iceland	97	98
Ireland	95	94
Israel	97	96
Italy	98	99
Kazakhstan	90	90
Kyrgyzstan	85	86
Lithuania	89	90
Luxembourg	98	96
Netherlands	97	99
Norway	98	98
Poland	96	96
Portugal	98	98
Republic of Moldova	88	88
Romania	93	93
Russian Federation	91	91
Slovenia	95	96
Spain	99	100
Sweden	95	95
Switzerland	89	89
Tajikistan	95	99
Turkey	89	93
Ukraine	90	90
United Kingdom	99	98
Bhutan	79	79
India	87	90
Indonesia	94	97
Maldives	97	97
Myanmar	100	99
Thailand	94	94
Australia	97	96
Brunei Darussalam	94	94
Cambodia	89	91
Fiji	91	91
Japan	100	100
Lao People's Democratic Republic	81	86
Mongolia	93	90
New Zealand	99	99
Philippines	92	90
Vanuatu	87	88

Source: Reproduced, with the permission of the publisher, from World Health Statistics-socioeconomics and demographics-net primary school enrollment for 2006.

25. Teach for America Several programs attempt to address the shortage of qualified teachers in the U.S. by placing uncertified instructors in schools with acute needs—often in inner cities. A 1999–2000 study compared students taught by certified teachers with others taught by under-certified teachers in the same schools. Reading scores of the students of certified teachers averaged 35.62 points with standard deviation 9.31. The scores of students instructed by undercertified teachers had mean 32.48 points with standard deviation 9.43 points on the same test. There were 44 students in each group. The appropriate t procedure has 86 degrees of freedom. Is there evidence of lower scores with uncertified teachers? Discuss.[11]

26. Labour force The Labour Force Survey conducted by Statistics Canada provided employment figures for men and women in 2006. The following table shows the values reported for men and women who were employed full time and part time separately. Is there evidence that employment and sex are dependent?

	Male	Female
Full time	6972	5157
Part time	483	1351

a) Test an appropriate hypothesis.
b) State your conclusion, including an analysis of differences you find (if any).

27. Streams Researchers in the Adirondack Mountains collect data on a random sample of streams each year. One of the variables recorded is the substrate of the stream—the type of soil and rock over which they flow. The researchers found that 69 of the 172 sampled streams had a substrate of shale. Construct a 95% confidence interval for the proportion of Adirondack streams with a shale substrate. Clearly interpret your interval in this context.

T 28. Legionnaires' disease In 1974, the Bellevue-Stratford Hotel in Philadelphia was the scene of an outbreak of what later became known as Legionnaires' disease. The cause of the disease was finally discovered to be bacteria that thrived in the air-conditioning units of the hotel. Owners of the Rip Van Winkle Motel, hearing about the Bellevue-Stratford, replaced their air-conditioning system. The following data are the bacteria counts in the air of eight rooms, before and after a new air-conditioning system was installed (measured in colonies per cubic foot of air). The objective is to find out whether the new system has succeeded in lowering the bacterial count. You are the statistician assigned to report to the hotel whether the strategy has worked. Base your analysis on a confidence interval. Be sure to list all your assumptions, methods, and conclusions.

Room Number	Before	After
121	11.8	10.1
163	8.2	7.2
125	7.1	3.8
264	14	12
233	10.8	8.3
218	10.1	10.5
324	14.6	12.1
325	14	13.7

29. PISA reading scale We looked at PISA math scores earlier in the text. What about reading scores? The means and SEs of PISA reading scale scores for Canadian 15-year-old students are shown below, by gender.[12]

Gender	Mean	SE
All	527	2.4
Males	511	2.8
Females	543	2.5

a) The study involved about 22 000 15-year-old students from nearly 1000 schools across Canada. The standard deviation for all the scores was reported as 97. Why do you think we can't divide the standard deviation by the square root of the sample size to get the reported SE of 2.4? Give a confidence interval for the Canadian population mean, stating any assumptions you are making.
b) The difference between average scores for females and males in the study is 32. It was also reported that the SE of the difference in means is 2.3. Why do you think this does not agree with what you would get using the Pythagorean rule? Give a confidence interval for the population difference between males and females, and state any assumptions you are making.

T 30. Dairy sales Peninsula Creameries sells both cottage cheese and ice cream. The CEO recently noticed that in months when the company sells more cottage cheese, it seems to sell more ice cream as well. Two of his aides were assigned to test whether this is true or not. The first aide's plot and analysis of sales data for the past 12 months (in millions of pounds for cottage cheese and for ice cream) appear below.

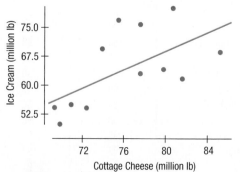

[11]*The Effectiveness of "Teach for America" and Other Under-certified Teachers on Student Academic Achievement: A Case of Harmful Public Policy.* Education Policy Analysis Archives, 2002.

[12]www.pisa.oecd.org

Dependent variable is Ice cream
R-squared = 36.9%
s = 8.320 with 12 − 2 = 10 degrees of freedom

Variable	Coefficient	SE(Coeff)	t-Ratio	P-Value
Intercept	−26.5306	37.68	−0.704	0.4975
Cottage C …	1.19334	0.4936	2.42	0.0362

The other aide looked at the differences in sales of ice cream and cottage cheese for each month and created the following output:

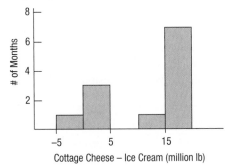

Cottage Cheese – Ice Cream (million lb)

Cottage Cheese - Ice Cream

Count	12
Mean	11.8000
Median	15.3500
StdDev	7.99386
IntQRange	14.3000
25th %tile	3.20000
75th %tile	17.5000

Test H_0: $\mu(CC − IC) = 0$ vs H_a: $\mu(CC − IC) \neq 0$

Sample Mean = 11.800000 t-Statistic = 5.113 w/11 df

Prob = 0.0003

Lower 95% bound = 6.7209429

Upper 95% bound = 16.879057

a) Which analysis would you use to answer the CEO's question? Why?
b) What would you tell the CEO?
c) Which analysis would you use to test whether the company sells more cottage cheese or ice cream in a typical year? Why?
d) What would you tell the CEO about this other result?
e) What assumptions are you making in the analysis you chose in part a? What assumptions are you making in the analysis in part c?
f) Next month's cottage cheese sales are 82 million pounds. Ice cream sales are not yet available. How much ice cream do you predict Peninsula Creameries will sell?
g) Give a 95% confidence interval for the true slope of the regression equation of ice cream sales by cottage cheese sales.
h) Explain what your interval means.

T 31. Online testing The Educational Testing Service is now administering several of its standardized tests online, such as the TOEFL exam. Since taking a test on a computer is different from taking a test with pencil and paper,

researchers wondered if the scores will be the same. To investigate this question, they created two versions of a TOEFL-type test and got 20 volunteers to participate in an experiment. Each volunteer took both versions of the test, one with pencil and paper and the other online. Subjects were randomized with respect to the order in which they sat for the tests (online/paper) and which form they took (Test A, Test B) in which environment. The table summarizes the scores (out of a possible 20).

Subject	Paper	Online
	Test A	Test B
1	14	13
2	10	13
3	16	8
4	15	14
5	17	16
6	14	11
7	9	12
8	12	12
9	16	16
10	7	14
11	8	13
12	11	13
13	15	17
14	11	13
15	13	14
16	9	9
17	15	9
18	14	15
19	16	12
20	8	10

a) Were the two forms (A/B) of the test equivalent in terms of difficulty? Test an appropriate hypothesis and state your conclusion.
b) Is there evidence that testing environment (paper/online) matters? Test an appropriate hypothesis and state your conclusion.

32. Home ice There may be some home ice advantage in hockey, and it may depend to some degree on the particular team. In 2008–09, the Vancouver Canucks won 24 of 41 home games and 21 of 41 away games. The Calgary Flames won 27 of 41 home games and 19 of 41 away games.
a) Is the difference in home-winning percentages between the two teams statistically significant?
b) Is the difference between home-and-away winning percentages statistically significant for the Flames? For the Canucks?
c) We looked through the teams' records[13] to find the team with the biggest relative home ice advantage, and discovered that this distinction belonged to the last

[13]www.nhl.com/ice/standings.htm?season=20082009&type=LEA

place Islanders who won 17 home games and 9 away, the biggest home-to-away ratio of nearly 2 to 1. How impressive is this home ice advantage, considering that we looked through all the teams' records to find the best case? Do the Islanders really have a great home ice advantage or could this just have been by chance? If you toss a coin 26 times, can you get 17 (or more) heads just by chance? What *are* the chances?

d) Researchers perform thousands of studies. Considering the above, if researchers only publish the best, most statistically significant results, might we just be reading about Type I errors? If researchers alter their hypotheses and response variables in the course of a study (not uncommon), how does this affect any reported statistical significance?

33. Irises Can measurements of the petal length of flowers be of value when you need to determine the species of a certain flower? Here are the summary statistics from measurements of the petals of two species of irises.[14]

| | Species | |
	Versicolor	Virginica
Count	50	50
Mean	55.52	43.22
Median	55.50	44
StdDev	5.519	5.362
Min	45	30
Max	69	56
Lower quartile	51	40
Upper quartile	59	47

a) Make parallel boxplots of petal lengths for the two species.
b) Describe the differences seen in the boxplots.
c) Write a 95% confidence interval for this difference.
d) Explain what your interval means.
e) Based on your confidence interval, is there evidence of a difference in petal length? Explain.

34. Insulin and diet A study published in the *Journal of the American Medical Association* examined people to see if they showed any signs of IRS (insulin resistance syndrome) involving major risk factors for Type 2 diabetes and heart disease. Among 102 subjects who consumed dairy products more than 35 times per week, 24 were identified with IRS. In comparison, IRS was identified in 85 of 190 individuals with the lowest dairy consumption, fewer than 10 times per week.

a) Is this strong evidence that IRS risk is different in people who frequently consume dairy products than in those who do not? Attack this two ways: use a *z*-test and a chi-square test. How are these tests related?

b) Does this prove that dairy consumption influences the development of IRS? Explain.

35. World Series If the two teams playing in the World Series are evenly matched, the probability that each team wins any game is 0.5. Then, the probability that the Series ends with one of the teams sweeping four straight games would be $2(0.5)^4 = 0.125$. Further probability calculations indicate that 25% of all World Series should last five games, 31.25% should last six games, and the other 31.25% should last the full seven games. The table shows the number of games it took to decide all the World Series from 1922 (when the seven-game format was set) through 2012. Do these results indicate that the teams are usually equally matched? Give statistical evidence to support your conclusion.

Length of series	4 games	5 games	6 games	7 games
Number of times	19	18	19	33

T 36. Rainmakers? In an experiment to determine whether seeding clouds with silver iodide increases rainfall, 52 clouds were randomly assigned to be seeded or not. The amount of rain they generated was then measured (in acre-feet). Create a 95% confidence interval for the average amount of additional rain created by seeding clouds. Explain what your interval means.

	Unseeded Clouds	Seeded Clouds
Count	26	26
Mean	164.588	441.985
Median	44.200	221.600
SD	278.426	650.787
IntQRange	138.600	337.600
25 %ile	24.400	92.400
75 %ile	163	430

T 37. Tableware Nambe Mills manufactures plates, bowls, and other tableware made from an alloy of several metals. Each item must go through several steps, including polishing. To better understand the production process and its impact on pricing, the company checked the polishing *Time* (in minutes) and the retail *Price* (in $) of these items. The regression analysis is shown below. The scatterplot showed a linear pattern, and residuals were deemed suitable for inference.

Dependent variable is: Price
R-squared = 84.5%
s = 20.50 with 59 − 2 = 57 degrees of freedom

Variable	Coefficient	SE(Coeff)
Intercept	−2.89054	5.730
Time	2.49244	0.1416

---

[14]R. A. Fisher, 1936, "The use of multiple measurements in axonomic problems." *Annals of Eugenics*, 7: 179–188.

a) How many different products were included in this analysis?

b) What percentage of the variation in retail price is explained by the polishing time?

c) Create a 95% confidence interval for the slope of this relationship.

d) Interpret your interval in this context.

T 38. Colour or text? In an experiment, 32 volunteer subjects are briefly shown seven cards, each displaying the name of a colour printed in a different colour (example: red, blue, and so on). The subject is asked to perform one of two tasks: memorize the order of the words or memorize the order of the colours. Researchers record the number of cards remembered correctly. Then, the cards are shuffled and the subject is asked to perform the other task. The tables display the results for each subject. Is there any evidence that either the colour or the written word dominates perception?

a) What role does randomization play in this experiment?

b) State appropriate hypotheses.

c) Are the assumptions necessary for inference reasonable here?

d) Perform the test.

e) State your conclusion.

Subject	Colour	Word	Subject	Colour	Word
1	4	7	17	4	3
2	1	4	18	7	4
3	5	6	19	4	3
4	1	6	20	0	6
5	6	4	21	3	3
6	4	5	22	3	5
7	7	3	23	7	3
8	2	5	24	3	7
9	7	5	25	5	6
10	4	3	26	3	4
11	2	0	27	3	5
12	5	4	28	1	4
13	6	7	29	2	3
14	3	6	30	5	3
15	4	6	31	3	4
16	4	7	32	6	7

39. And it means? Every statement about a confidence interval contains two parts—the level of confidence and the interval. Suppose that an insurance agent estimating the mean loss claimed by clients after home burglaries created the 95% confidence interval ($1644, $2391).

a) What's the margin of error for this estimate?

b) Carefully explain what the interval means.

c) Carefully explain what the 95% confidence level means.

T 40. Hard water In an investigation of environmental causes of disease, data were collected on the annual *Mortality* rate (deaths per 100 000) for males in 61 large towns in England and Wales. In addition, the water hardness was recorded as the *Calcium* concentration (parts per million, or ppm) in the drinking water. Here are the scatterplot and regression analysis of the relationship between mortality and calcium concentration.

Dependent variable is: Mortality
R-squared = 43%
s = 143.0 with 61 − 2 = 59 degrees of freedom

Variable	Coefficient	SE(Coeff)
Intercept	1676	29.30
Calcium	−3.23	0.48

a) Is there an association between the hardness of the water and the mortality rate? Write the appropriate hypothesis.

b) Assuming the assumptions for regression inference are met, what do you conclude?

c) Create a 95% confidence interval for the slope of the true line relating calcium concentration and mortality.

d) Interpret your interval in context.

T 41. Togetherness Are good grades in high school associated with family togetherness? A simple random sample of 142 high school students was asked how many meals per week their families ate together. Their responses produced a mean of 3.78 meals per week, with a standard deviation of 2.2. Researchers then matched these responses against the students' grade-point averages. The scatterplot appeared to be reasonably linear, so they went ahead with the regression analysis, seen below. No apparent pattern emerged in the residuals plot.

Dependent variable: GPA
R-squared = 11.0%
s = 0.6682 with 142 − 2 = 140 df

Variable	Coefficient	SE(Coeff)
Intercept	2.7288	0.1148
Meals/wk	0.1093	0.0263

a) Is there evidence of an association? Test an appropriate hypothesis and state your conclusion.

b) Do you think this association would be useful in predicting a student's grade-point average? Explain.

c) Are your answers to parts a) and b) contradictory? Explain.

42. Family planning Before the introduction of birth control pills, many young women experienced unplanned pregnancies. A 1954 study of 1438 pregnant women examined the association with the women's education levels, producing these data:

	Education Level		
	< 3 yr HS	3 + yr HS	Some college
Number of pregnancies	591	608	239
% unplanned	66.2%	55.4%	42.7%

Do these data provide evidence of an association between family planning and education level?[15]

43. Recruiting In September 2002, CNN reported on a method of grad student recruiting by the Haas School of Business at U.C.-Berkeley. The school notifies applicants by formal letter that they have been admitted, and also emails the accepted students a link to a website that greets them with personalized balloons, cheering, and applause. The director of admissions says this extra effort at recruiting has really worked well. The school accepts 500 applicants each year, and the percentage who actually choose to enrol at Berkeley has increased from 52% the year before the Web greeting to 54% in 2002.

a) Create a 95% confidence interval for the change in enrolment rates.

b) Based on your confidence interval, are you convinced that this new form of recruiting has been effective? Explain.

T 44. Education vs. income 2009 The information below examines the median income and education level (percent of population with at least a Bachelor's degree) for several large U.S. cities in 2009.

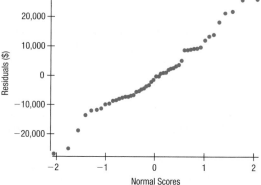

Summary Statistics:

Column	n	Mean	Std. Dev.
Median Household Income	50	46521.22	15455.175
Percentage Bachelor's Degrees	50	28.402	11.756

Dependent variable is Median Household Income
R-squared = 42.1%
s = 11,881.87 with 50 − 2 = 48 degrees of freedom

Variable	Coefficient	SE(Coeff)	t-Ratio	P-Value
Intercept	22294.24	4431.67	5.03	<0.0001
Percentage Bachelor's Degrees	853.00	144.38	5.91	<0.0001

a) Do you think the assumptions for inference are met? Explain.

b) Does there appear to be an association between education and income levels in these cities?

c) Would this association appear to be weaker, stronger, or the same if data were plotted for individual people rather than for cities in aggregate? Explain.

d) Create and interpret a 95% confidence interval for the slope of the true line that describes the association between income and education.

e) Predict the median household income for cities where an average of 25% of residents have at least a Bachelor's degree. Describe your estimate with a 90% confidence interval, and interpret that result.

---

[15]*Fertility Planning and Fertility Rates by Socio-Economic Status*, Social and Psychological Factors Affecting Fertility, 1954.

CHAPTER

24

Analysis of Variance

If you torture the data long enough, it will confess.

—Ronald Coase

LEARNING OBJECTIVES

1. Recognize when to use an Analysis of Variance to compare the means of several groups

2. Interpret and draw conclusions from the corresponding ANOVA table

3. Know how to check the required conditions

4. Construct confidence intervals for the differences between pairs of group means

5. Use a *multiple comparisons* procedure to adjust confidence intervals or tests for the number of comparisons made

Connections

In Chapter 4, we examined data on the performance of four brands of cups to see which was the most effective at keeping coffee hot. We could run a *t*-test for each of the six head-to-head comparisons (as in Chapter 19), but we'll learn a better way to *compare more than two groups* in this chapter (with a generalization of the two sample pooled *t*-test).

In the previous chapter, we used regression to model the relationship between a quantitative response variable and *quantitative* explanatory variable. In this chapter, the explanatory variable will be *categorical*, but the methods are in fact closely related to

Did you wash your hands with soap before eating? You've undoubtedly been asked that question a few times in your life. Mom knows that washing with soap eliminates most of the germs you've managed to collect on your hands. Or does it? A student decided to investigate just how effective washing with soap is in eliminating bacteria. To do this, she tested four different methods—washing with water only, washing with regular soap, washing with antibacterial soap (ABS), and spraying hands with antibacterial spray (AS) (containing 65% ethanol as an active ingredient). Her experiment consisted of one experimental factor, the washing *Method*, at four levels.

She suspected that the number of bacteria on her hands before washing might vary considerably from day to day. To help even out the effects of those changes, she generated random numbers to determine the order of the four treatments. Each morning, she washed her hands according to the treatment randomly chosen. Then, she placed her right hand on a sterile media plate designed to encourage bacteria growth. She incubated each plate for two days at 36°C, after which she counted the bacteria colonies. She replicated this procedure 8 times for each of the four treatments (Data in **Hand_washing**).

A side-by-side boxplot (Figure 24.1) of the numbers of colonies seems to show some differences among the treatments:

Figure 24.1

Boxplots of the bacteria colony counts for the four different washing methods suggest some differences between treatments.

When we first looked at a quantitative variable measured for each of several groups in Chapter 4, we displayed the data this way with side-by-side boxplots. And, when we compared the boxes, we asked whether the centres seemed to differ, using the spreads of the boxes to judge the size of the differences. Now, we want to make this more formal by testing a hypothesis. We'll make the same kind of comparison, comparing the variability among the means with the spreads of the boxes. It looks like the alcohol spray has lower bacteria counts, but as always, we're skeptical. Could it be that the four methods really have the same mean counts and we just *happened* to get a difference like this because of natural sampling variability?

the analysis in the previous two chapters.

787

regression, with similar assumptions about the random errors (including constant variance—a remarkably common assumption, which serves to make the underlying math tractable).

Chapter 10 discussed a variety of experimental designs. Each of them can be analyzed with a variant or extension of the ANOVA methods discussed in this chapter.

What is the null hypothesis here? It seems natural to start with the hypothesis that *all the group means are equal*. That would say that it doesn't matter what method you use to wash your hands because the mean bacteria count will be the same. We know that even if there were no differences at all in the *means* (for example, if someone replaced all the solutions with water), there would still be sample-to-sample differences. We want to see, statistically, whether differences as large as those observed in the experiment could naturally occur by chance in groups that have equal means. If we find that the differences in washing *Methods* are so large that they would occur only very infrequently in groups that actually have the same mean, then, as we've done with other hypothesis tests, we'll reject the null hypothesis and conclude that the washing *Methods* really have different means.[1]

For Example A ONE-FACTOR EXPERIMENT

Hand clinics commonly use contrast baths to reduce swelling and stiffness after surgery. Patients' hands are immersed alternately in warm and cool water. (That's the *contrast* in the name.) Sometimes, the treatment is combined with mild exercise. Although the treatment is widely used, it had never been verified that it accomplishes the stated outcome goal of reducing swelling.

Researchers[2] randomly assigned 59 patients who had received carpal tunnel release surgery to one of three treatments: contrast bath, contrast bath with exercise, and (as a control) exercise alone. Hand therapists, who did not know how the subjects had been treated, measured hand volumes in millilitres before and after treatments by measuring how much water the hand displaced when submerged. The change in hand volume (after treatment minus before) was reported as the outcome (Data in **Contrast_baths**).

QUESTION: Specify the details of the experiment's design. Identify the subjects, the sample size, the experimental factor, the treatment levels, and the response. What is the null hypothesis? Was randomization employed? Was the experiment blinded? Was it double-blinded?

ANSWER: Subjects were patients who received carpal tunnel release surgery. Sample size is 59 patients. The factor was contrast bath treatment with three levels: contrast baths alone, contrast baths with exercise, and exercise alone. The response variable is the change in hand volume. The null hypothesis is that the mean changes in hand volume will be the same for the three treatment levels. Patients were randomly assigned to treatments. The study was single-blind because the evaluators were blind to the treatments. It was not (and could not be) double-blind because the patients had to be aware of their treatments.

24.1 Testing Whether the Means of Several Groups Are Equal

We saw in Chapter 19 how to use a *t*-test to see whether two groups have equal means. We compared the difference in the means to a standard error estimated from all the data. And when we were willing to assume that the underlying group variances were equal, we pooled data from the two groups to find the standard error.

Now, we have more groups, so we can't just look at differences in the means.[3] But all is not lost. Even if the null hypothesis were true, and the means of the populations

---

[1]The alternative hypothesis is that "the means are not *all* equal." Be careful not to confuse that with "all the means are different." With 11 groups, we could have 10 means equal to each other and 1 different. The null hypothesis would still be false.

[2]Janssen, Robert G., Schwartz, Deborah A., and Velleman, Paul F., "A Randomized Controlled Study of Contrast Baths on Patients with Carpal Tunnel Syndrome," *Journal of Hand Therapy*, **22**:3, pp. 200–207. The data reported here differ slightly from those in the original paper because they include some additional subjects and exclude some outliers.

[3]You might think of testing all pairs, but that method generates too many Type I errors. We'll see more about this later in the chapter.

underlying the groups were equal, we'd still expect the sample means to vary a bit. We could measure that variation by finding the variance of the means. How much should they vary? Well, if we look at how much the data themselves vary, we can get a good idea of how much the means should vary. And if the underlying means are actually different, we'd expect that variation to be larger.

It turns out that we can build a hypothesis test to check whether the variation in the means is bigger than we'd expect it to be just from random fluctuations. We'll need a new sampling distribution model, called the *F*-model, but that's just a different table to look at (Table F can be found at the end of this chapter).

To get an idea of how it works, let's start by looking at the two sets of boxplots below:

Figure 24.2

It's hard to see the difference in the means in these boxplots because the spreads are large relative to the differences in the means.

Figure 24.3

In contrast with Figure 24.2, the smaller variation makes it much easier to see the differences among the group means. (Note also that the scale of the *y*-axis is considerably different from the plot on the left.)

We're trying to decide if the means are different enough for us to reject the null hypothesis. If they're close, we'll attribute the differences to natural sampling variability. What do you think? It's easy to see that the means in the Figure 24.3 differ. It's hard to imagine that the means could be that far apart just from natural sampling variability alone. How about the set in Figure 24.2? It looks as though these observations *could* have occurred from treatments with the same means.[4] This much variation among groups does seem consistent with equal group means.

Believe it or not, the two sets of treatment means in both figures are the same. (They are 31, 36, 38, and 31, respectively.) So why do the figures look so different? In Figure 24.3, the variation *within* each group is so small that the differences *between* the means stand out. This is what we looked for when we compared boxplots by eye back in Chapter 4. And it's the central idea of the *F*-test. We compare the differences *between* the means of the groups with the variation *within* the groups. When the differences between means are large compared with the variation within the groups, we reject the null hypothesis and conclude that the means are not equal. In the first figure, the differences among the means look as though they could have arisen just from natural sampling variability from groups with equal true means, so there appears to be little evidence to reject H_0.

How can we make this comparison more precise statistically? All the tests we've seen have compared differences of some kind with a ruler based on an estimate of variation. And we've always done that by looking at the ratio of the statistic to that variation estimate. Here, the differences among the means will show up in the numerator, and the ruler we compare them with will be based on the underlying standard deviation—that is, on the variability *within* the treatment groups.

---

[4]Of course, with a large enough sample, we can detect any small differences. For experiments with the same sample size, it's easier to detect the differences when the variation *within* each box is smaller.

For Example   BOXPLOTS FOR COMPARING TREATMENTS

RECAP: Fifty-nine post-surgery patients were randomly assigned to one of three treatment levels. Changes in hand volume were measured. Here are boxplots of the data. The recorded values are volume after treatment—volume before treatment, so positive values indicate swelling. Some swelling is to be expected. (Data in **Contrast_baths**)

QUESTION: What do the boxplots say about the results?

ANSWER: There doesn't seem to be much difference between the two contrast bath treatments. The exercise-only treatment may result in less swelling.

How Different Are They?

<table>
<tr><td>Who</td><td>Hand washings by four different methods, assigned randomly and replicated 8 times each</td></tr>
<tr><td>What</td><td>Number of bacteria colonies</td></tr>
<tr><td>How</td><td>Sterile media plates incubated at 36°C for two days</td></tr>
</table>

The challenge here is that we can't take a simple difference as we did when comparing two groups. In the hand-washing experiment, we have differences in mean bacteria counts across *four* treatments. How should we measure how different the four group means are? With only two groups, we naturally took the difference between their means as the numerator for the *t*-test. It's hard to imagine what else we could have done. How can we generalize that to more than two groups? When we've wanted to know how different or spread apart many individual observations were, we measured how much they vary from each other, with a sample variance, and that's what we do here.

How much natural variation should we expect among the means if the null hypothesis were true? If the null hypothesis were *true*, then each of the treatment means would estimate the *same* underlying population mean. If the washing methods are all the same, it's as if we're just estimating the mean bacteria count on hands that have been washed with plain water. And we have several (in our experiment, four) different, independent estimates of this mean. Here comes the clever part. We can treat these estimated means as if they were observations, and simply calculate their (sample) variance. This variance is the measure we'll use to assess how different the group means are from each other. It's the generalization of the difference between means for only two groups.

The more the group means resemble each other, the smaller this variance will be. The more they differ (perhaps because the treatments actually have an effect), the larger this variance will be.

For the bacteria counts, the four means are listed in the table in the margin. If you took those four values, treated them as observations, and found their sample variance, you'd get 1245.08. That's fine, but how can we tell whether it is a big value? Now, we need a model, and the model is based on our null hypothesis that all the group means are equal. Here, the null hypothesis is that it doesn't matter what washing method you use, the mean bacteria count will be about the same:

$$H_0: \mu_1 = \mu_2 = \mu_3 = \mu_4$$

Level	n	Mean
Alcohol spray	8	37.5
Antibacterial soap	8	92.5
Soap	8	106.0
Water	8	117.0

As always when testing a null hypothesis, we'll start by assuming that it is true. And if the group means are equal, then there's an overall mean, μ—the bacteria count you'd expect all the time after washing your hands in the morning. And each of the observed group means is just a sample-based estimate of that underlying mean.

We know how sample means vary. The variance of a sample mean is σ^2/n. With eight observations in a group, that would be $\sigma^2/8$. The estimate that we've just calculated, 1245.08, should estimate this quantity. If we want to get back to the variance of the *observations*, σ^2, we need to multiply it by 8. So $8 \times 1245.08 = 9960.64$ should estimate σ^2.

Is 9960.64 large for this variance? How can we tell? We'll need a hypothesis test. You won't be surprised to learn that there is just such a test. The details of the test, due to Sir Ronald Fisher in the early 20th century, are truly ingenious, and one of the most amazing statistical results of that century.

The Ruler Within

We need a suitable ruler for comparison—one based on the underlying variability in our measurements. That variability is due to the day-to-day differences in the bacteria count even when the same soap is used. Why would those counts be different? Maybe the experimenter's hands were not equally dirty, or she washed less well some days, or the plate incubation conditions varied. We randomized just so we could see past such things.

We need an independent estimate of σ^2, one that doesn't depend on the null hypothesis being true, and one that won't change if the groups have different means. As in many quests, the secret is to look "within." We could look in *any* of the treatment groups and find its variance. But which one should we use? The answer is *all* of them!

At the start of the experiment (when we randomly assigned experimental units to treatment groups), the units were drawn randomly from the same pool, so each treatment group had a sample variance that estimated the same σ^2. If the null hypothesis is true, then not much has happened to the experimental units—or at least, their means have not moved apart. It's not much of a stretch to believe that their variances haven't moved apart much either. (If the washing methods are equivalent, then the choice of method would not affect the mean *or* the variability.) So each group variance still estimates a common σ^2.

We're assuming that the null hypothesis is true. If the group variances are equal, then the common variance they all estimate is just what we've been looking for. Since all the group variances estimate the same σ^2, we can pool them to get an overall estimate of σ^2. Recall that we pooled data to estimate variances when we tested the null hypothesis that two proportions were equal—and for the same reason. It's also exactly what we did in a pooled t-test. The variance estimate we get by pooling we'll denote, as before, by s_p^2.

Level	n	Mean	Std Dev	Variance
Alcohol spray	8	37.5	26.56	705.43
Antibacterial soap	8	92.5	41.96	1760.64
Soap	8	106.0	46.96	2205.24
Water	8	117.0	31.13	969.08

For the bacteria counts, the standard deviations and variances are listed above. If we pool the four variances (here we can just average them because all the sample sizes are equal), we'd get $s_p^2 = 1410.10$. In the pooled variance, each variance is taken around its *own* treatment mean, so the pooled estimate doesn't depend on the treatment means being equal. But the estimate in which we took the four means as observations and took their variance does. That estimate gave 9960.64. That seems a lot bigger than 1410.10. Might this be evidence that the four means are not equal?

Let's see what we've got. We have an estimate of σ^2 from the variation *within* groups of 1410.10. That's just the variance of the *residuals* pooled across all groups. Because it's a pooled variance, we could write it as s_p^2. Traditionally, this quantity is also called the

Error Mean Square or **Mean Square Error** (or sometimes the **Within Mean Square**) and denoted by MS_E. These names date back to the early 20th century when the methods were developed. If you think about it, the names do make sense—variances are means of squared differences.[5]

But we also have a *separate* estimate of σ^2 from the variation *between* the groups because we know how much means ought to vary. For the hand-washing data, when we took the variance of the four means and multiplied it by n, we got 9960.64. We expect this to estimate σ^2, too, *as long as we assume the null hypothesis is true.* We call this quantity the **Treatment Mean Square** (or sometimes the **Between Mean Square**[6]) and denote it by MS_T.

The *F*-statistic

Now, we have two different estimates of the underlying variance. The first one, the MS_T, is based on the differences *between* the group means. If the group means are equal, as the null hypothesis asserts, it will estimate σ^2. But, if they are not, it will give some bigger value. The other estimate, the MS_E, is based only on the variation *within* the groups around each of their own means, and doesn't depend at all on the null hypothesis being true.

So how do we test the null hypothesis? When the null hypothesis is true, the treatment means are equal, and both MS_E and MS_T estimate σ^2. Their ratio, then, should be close to 1.0. But, when the null hypothesis is false, the MS_T will be *larger* because the treatment means are not equal. The MS_E is a pooled estimate in which the variation within each group is found around its own group mean, so differing means won't inflate it. That makes the ratio MS_T/MS_E perfect for testing the null hypothesis. When the null hypothesis is true, the ratio should be near 1. If the treatment means really are different, the numerator will tend to be larger than the denominator, and the ratio will tend to be bigger than 1.

Of course, even when the null hypothesis *is* true, the ratio will vary around 1 just due to natural sampling variability. How can we tell when it's big enough to reject the null hypothesis? To be able to tell, we need a sampling distribution model for the ratio. Sir Ronald Fisher found the sampling distribution model of the ratio in the early 20th century. In his honor, we call the distribution of MS_T/MS_E the **F-distribution**. And we call the ratio MS_T/MS_E the **F-statistic**. By comparing this statistic with the appropriate F-distribution, we (or the computer) can get a P-value.

The resulting **F-test** is one-tailed because any differences in the means make the F-statistic larger. Larger differences in the treatments' effects lead to the means being more variable, making the MS_T bigger. That makes the F-ratio grow. So the test is significant if the F-ratio is big enough. In practice, we find a P-value, and big F-statistic values go with small P-values.

The entire analysis is called the **Analysis of Variance,** commonly abbreviated **ANOVA**. You might think that it should be called the analysis of means, since it's the equality of the means we're testing. But we use the *variances* within and between the groups for the test.

Like Student's *t*-models, the *F*-models are a family. *F*-models depend on not one, but two, degrees of freedom parameters. The degrees of freedom come from the two variance estimates and are sometimes called the *numerator df* and the *denominator df*. The Treatment Mean Square, MS_T, is the sample variance of the observed treatment means. If you think of them as observations, then since there are k groups, this variance has $k - 1$ degrees

■ **NOTATION ALERT**

Capital *F* is used only for this distribution model and statistic in honour of Sir Ronald Fisher.

---

[5]Well, actually, they're sums of squared differences divided by their degrees of freedom: $(n - 1)$ for the first variance we saw back in Chapter 3, and other degrees of freedom for each of the others we've seen. But even back in Chapter 3, we said this was a "kind of" mean, and indeed, it still is.

[6]Grammarians would probably insist on calling it the *Among Mean Square*, since the variation is among *all* the group means. Traditionally, though, it's called the *Between Mean Square* and we have to talk about the variation between all the groups (as bad as that sounds).

of freedom. The Error Mean Square, MS_E, is the pooled estimate of the variance within the groups. If there are n observations in each group, then we get $n - 1$ degrees of freedom from each, for a total of $k(n - 1)$ degrees of freedom.

A simpler way of tracking the degrees of freedom is to start with all the cases. Let's denote the total number of cases as N. Each group has its own mean, costing us a degree of freedom, or k in total. So we have $N - k$ degrees of freedom for the error. When the groups all have equal sample size, that's the same as $k(n - 1)$, but this way works even if the group sizes differ.

We say that the F-statistic, MS_T/MS_E, has an F-distribution with $k - 1$ and $N - k$ degrees of freedom. We usually write these parameters as subscripts, numerator df first, then denominator df. On page 795, you can see what the density curve actually looks like.

This procedure could even be used to compare just two groups. In this case, we would get an F-statistic with 1 df in the numerator. It turns out that the F-value would be exactly the *square* of the pooled t-statistic! So, the F-test can be viewed as a generalization of the pooled t-test, which tested equality of only two means, to a test for equality of any number of means. And not surprisingly, they make exactly the same assumptions.

Back to Bacteria!

For the hand-washing experiment, the $MS_T = 9960.64$. The $MS_E = 1410.14$. If the treatment means were equal, the Treatment Mean Square should be about the same size as the Error Mean Square, about 1410. But it's 9960.64, which is 7.06 times bigger. In other words, $F = 7.06$. This F-statistic has $4 - 1 = 3$ and $32 - 4 = 28$ degrees of freedom.

An F-value of 7.06 is bigger than 1, but we can't tell for sure whether it's big enough to reject the null hypothesis until we check the $F_{3,28}$ model to find its P-value. (Usually, that's most easily done with technology, but we can use printed tables.) It turns out the P-value is 0.0011. In other words, if the treatment means were actually equal, we would expect the ratio MS_T/MS_E to be 7.06 or larger about 11 times out of 10 000, just from natural sampling variability. That's not very likely, so we reject the null hypothesis and conclude that the means are different. We have strong evidence that the four different methods of hand washing are not equally effective at eliminating germs.

Section 24.1 Exercises

1. Popcorn A student runs an experiment to test four different popcorn brands, recording the number of kernels left unpopped. She pops measured batches of each brand four times, using the same popcorn popper and randomizing the order of the brands. After collecting her data and analyzing the results, she reports that the F-ratio is 13.56.
a) What are the null and alternative hypotheses?
b) How many degrees of freedom does the treatment sum of squares have? How about the error sum of squares?
c) The P-value is 0.00037. Assuming that the conditions required for ANOVA are satisfied, what would you conclude?

2. Skating A figure skater tried various approaches to her Salchow jump in a designed experiment using five different places for her focus (arms, free leg, midsection, take-off leg, and free). She tried each jump six times in random order, using two of her skating partners to

judge the jumps on a scale from 0 to 6. After collecting the data and analyzing the results, she reports that the F-ratio is 7.43.
a) What are the null and alternative hypotheses?
b) How many degrees of freedom does the treatment sum of squares have? How about the error sum of squares?
c) The P-value is 0.00044. Assuming that the conditions are satisfied, what would you conclude?

3. Gas mileage A student runs an experiment to study the effect of three different mufflers on gas mileage. He devises a system so that his Jeep Wagoneer uses gasoline from a one-litre container. He tests each muffler eight times, carefully recording the number of kilometres he can go in his Jeep Wagoneer on one litre of gas. After analyzing his data, he reports that the F-ratio is 2.35 with a P-value of 0.1199.
a) What are the null and alternative hypotheses?

b) How many degrees of freedom does the treatment sum of squares have? How about the error sum of squares?

c) What would you conclude?

d) If your conclusion in part c) is wrong, what type of error have you made?

4. Darts A student interested in improving her dart-throwing technique designs an experiment to test four different stances to see whether they affect her accuracy. After warming up for several minutes, she randomizes the order of the four stances, throws a dart at a target using each stance, and measures the distance of the dart in centimetres from the centre of the bull's-eye. She replicates this procedure 10 times. After analyzing the data, she reports that the F-ratio is 1.41 with a P-value of 0.2557.

a) What are the null and alternative hypotheses?

b) How many degrees of freedom does the treatment sum of squares have? How about the error sum of squares?

c) What would you conclude?

d) If your conclusion in part c) is wrong, what type of error have you made?

24.2 The Analysis of Variance (ANOVA)

You'll often see the mean squares and other information put into a table called the **ANOVA table**. Here's the table for the hand-washing experiment:

Calculating the ANOVA table This table has a long tradition stretching back to when ANOVA calculations were done by hand. Major research labs had rooms full of mechanical calculators operated by women. (Yes, always women; women were thought—by the men in charge, at least—to be more careful at such an exacting task.) Three women would perform each calculation, and if any two of them agreed on the answer, it was taken as the correct value.

Analysis of Variance Table					
Source	DF	Sum of Squares	Mean Square	F-ratio	P-value
Method	3	29 882	9960.64	7.0636	0.0011
Error	28	39 484	1410.14		
Total	31	69 366			

The ANOVA table was originally designed to organize the calculations. With technology, we have much less use for that. We'll show how to calculate the sums of squares later in the chapter, but the most important quantities in the table are the F-statistic and its associated P-value. When the F-statistic is large, the Treatment (here *Method*) Mean Square is large compared to the Error Mean Square, and provides evidence that in fact the means of the groups are not all equal.

You'll almost always see ANOVA results presented in a table like this. After nearly a century of writing the table this way, statisticians (and their technology) aren't going to change. Even though the table was designed to facilitate hand calculation, computer programs that compute ANOVAs still present the results in this form. Usually the P-value is found next to the F-ratio. The P-value column may be labelled with a title such as "Prob > F," "sig," or "Prob." Don't let that confuse you; it's just the P-value corresponding to the F-statistic.

You'll sometimes see the two mean squares referred to as the *Mean Square Between* and the *Mean Square Within*—especially when we test data from observational studies rather than experiments. ANOVA is often used for such observational data, and as long as certain conditions are satisfied, there's no problem with using it in that context.

For Example F-TEST

RECAP: An experiment to determine the effect of contrast bath treatments on swelling in post-surgical patients recorded hand volume changes for patients who had been randomly assigned to one of three treatments. Here is the analysis of variance for these data:

Analysis of variance for hand volume change					
Source	DF	Sum of Squares	Mean Square	F-ratio	P-value
Treatment	2	716.159	358.080	7.4148	0.0014
Error	56	2704.38	48.2926		
Total	58	3420.54			

QUESTION: What does the ANOVA say about the results of the experiment? Specifically, what does it say about the null hypothesis?

ANSWER: The *F*-ratio of 7.4148 has a P-value that is quite small. We can reject the null hypothesis that the mean change in hand volume is the same for all three treatments.

The *F*-table

Usually, you'll get the P-value for the *F*-statistic from technology. Any software program performing an ANOVA will automatically "look up" the appropriate one-sided P-value for the *F*-statistic. If you want to do it yourself, you'll need an *F*-table. (There's one in Appendix C called **Table F**.) *F*-tables are usually printed only for a few values of α, often 0.05, 0.01, and 0.001. They give the critical value of the *F*-statistic with the appropriate number of degrees of freedom determined by your data, for the α level that you select. If your *F*-statistic is greater than that value, you know that its P-value is less than that α level. So, you'll be able to tell whether the P-value is greater or less than 0.05, 0.01, or 0.001, but to be more precise, you'll need technology (or an interactive table like the one in MyStatLab).

Figure 24.4 is an excerpt from an *F*-table for $\alpha = 0.05$:

Figure 24.4

Part of an *F*-table showing critical values for $\alpha = 0.05$, and highlighting the critical value, 2.947, for 3 and 28 degrees of freedom. We can see that only 5% of the values in the *F* distribution with 3 and 28 df will be greater than 2.947.

Notice that the critical value for 3 and 28 degrees of freedom at $\alpha = 0.05$ is 2.947. Since our *F*-statistic of 7.06 is larger than this critical value, we know that the P-value is less than 0.05. We could also look up the critical value for $\alpha = 0.01$ and find that it's 4.568 and the critical value for $\alpha = 0.001$ is 7.193. So, our *F*-statistic sits between the two critical values 0.01 and 0.001, and our P-value is slightly *greater* than 0.001. Technology can find the value precisely. It turns out to be 0.0011.

Just Checking

A student conducted an experiment to see which, if any, of four different paper air-plane designs results in the longest flights (measured in inches). The boxplots look like this (with the overall mean shown in red): (Data in **Paper_planes**)

The ANOVA table shows:

Analysis of Variance					
Source	DF	Sum of Squares	Mean Square	F Ratio	Prob > F
Design	3	51 991.778	17 330.6	37.4255	< .0001
Error	32	14 818.222	463.1		
Total	35	66 810.000			

1. What is the null hypothesis?
2. From the boxplots, do you think that there is evidence that the mean flight distances of the four designs differ?
3. Does the *F*-test in the ANOVA table support your preliminary conclusion in (2)?
4. The researcher concluded that "there is substantial evidence that all four of the designs result in different mean flight distances." Do you agree?

The ANOVA Model

There are many types of ANOVA tables, depending on the particular study design. Here, we have only one factor of interest by which to classify the responses, so we call the table a *one-way* ANOVA (*two-way* ANOVAs are in the next chapter).

To understand the ANOVA table, let's start by writing a model for what we observe. We start with the simplest interesting model: one that says that the only differences of interest among the groups are the differences in their means. This **one-way ANOVA model** characterizes each observation in terms of its mean and assumes that any variation around that mean is just random error:

$$y_{ij} = \mu_j + \varepsilon_{ij}$$

That is, each observation is the sum of the mean for the treatment it received plus a random error. Our null hypothesis is that the treatments made no difference—that is, that the means are all equal:

$$H_0: \mu_1 = \mu_2 = \cdots = \mu_k$$

It will help our discussion if we think of an overall mean of the experiment and consider the treatments as adding or subtracting an effect to this overall mean. Thinking in this way, we could write μ for the overall mean and τ_j for the deviation from this mean to get to the jth treatment mean—the *effect* of the treatment (if any) in moving that group away from the overall mean:

$$y_{ij} = \mu + \tau_j + \varepsilon_{ij}$$

Thinking in terms of the effects, we could also write the null hypothesis in terms of these treatment *effects* instead of the means:

$$H_0 : \tau_1 = \tau_2 = \cdots = \tau_k = 0$$

We now have three different kinds of components in our model: the overall mean, the treatment effects, and the errors. We'll want to estimate them from the data. Fortunately, we can do that in a straightforward way.

To estimate the overall mean, μ, we use the mean of all the observations: $\bar{\bar{y}}$ (called the "grand mean."[7]) To estimate each treatment effect, we find the difference between the mean of that particular treatment and the grand mean:

$$\hat{\tau}_j = \bar{y}_j - \bar{\bar{y}}$$

There's an error, ε_{ij}, for each observation. We estimate those with the residuals from the treatment means: $e_{ij} = y_{ij} - \bar{y}_j$.

We can write each observation as the sum of three quantities that correspond to our model:

$$y_{ij} = \bar{\bar{y}} + (\bar{y}_j - \bar{\bar{y}}) + (y_{ij} - \bar{y}_j)$$

What this says is simply that we can write each observation as the sum of

- the grand mean,
- the effect of the treatment it received, and
- the residual.

Or:

$$Observation = Grand\ mean + Treatment\ effect + Residual$$

If we look at the equivalent equation

$$y_{ij} = \bar{\bar{y}} + (\bar{y}_j - \bar{\bar{y}}) + (y_{ij} - \bar{y}_j)$$

closely, it doesn't really seem like we've done anything. In fact, collecting terms on the right-hand side will give back just the observation, y_{ij} again. But this decomposition is actually the secret of the Analysis of Variance. We've split each observation into "sources"—the grand mean, the treatment effect, and error.

Where does the residual term come from? Think of the annual report from any *Fortune* 500 company. The company spends billions of dollars each year and at the end of the year, the accountants show where each penny goes. How do they do it? After accounting for salaries, bonuses, supplies, taxes, etc., etc., etc., what's the last line? It's always labelled "other" or "miscellaneous." Using "other" as the difference between all the sources they know and the total they start with, they can always make it add up perfectly. The residual is just the statisticians' "other." It takes care of all the other sources we didn't think of or don't want to consider, and makes the decomposition work by adding (or subtracting) back in just what we need.

---

[7]The father of your father is your grandfather. The mean of the group means should probably be the grandmean, but we usually spell it as two words.

Let's explore this decomposition in some real concrete detail, using our hand-washing data. Here are the data again, displayed a little differently:

	Alcohol	AB Soap	Soap	Water
	51	70	84	74
	5	164	51	135
	19	88	110	102
	18	111	67	124
	58	73	119	105
	50	119	108	139
	82	20	207	170
	17	95	102	87
Treatment Means	37.5	92.5	106	117

The grand mean of all observations is 88.25. Let's put that into a similar table:

Alcohol	AB Soap	Soap	Water
88.25	88.25	88.25	88.25
88.25	88.25	88.25	88.25
88.25	88.25	88.25	88.25
88.25	88.25	88.25	88.25
88.25	88.25	88.25	88.25
88.25	88.25	88.25	88.25
88.25	88.25	88.25	88.25
88.25	88.25	88.25	88.25

The treatment *means* are 37.5, 92.5, 106, and 117, respectively, so the treatment *effects* are those minus the grand mean (88.25). Let's put the treatment effects into their table:

Alcohol	AB Soap	Soap	Water
−50.75	4.25	17.75	28.75
−50.75	4.25	17.75	28.75
−50.75	4.25	17.75	28.75
−50.75	4.25	17.75	28.75
−50.75	4.25	17.75	28.75
−50.75	4.25	17.75	28.75
−50.75	4.25	17.75	28.75
−50.75	4.25	17.75	28.75

Finally, we compute the residuals as the differences between each observation and its treatment mean:

Alcohol	AB Soap	Soap	Water
13.5	−22.5	−22	−43
−32.5	71.5	−55	18
−18.5	−4.5	4	−15
−19.5	18.5	−39	7
20.5	−19.5	13	−12
12.5	26.5	2	22
44.5	−72.5	101	53
−20.5	2.5	−4	−30

Now, we have four tables for which each

$$Observation = Grand\ Mean + Treatment\ Effect + Residual$$

You can verify, for example, that the first observation, $51 = 88.25 + (−50.75) + 13.5$.

Why do we want to think in this way? Think back to the boxplots in Figures 24.2 and 24.3. To *test* the hypothesis that the treatment effects are zero, we want to see whether the treatment effects are large *compared* to the errors. Our eye looks at the variation *between* the treatment means and compares it to the variation *within* each group.

The ANOVA separates those two quantities into the *treatment effects* and the *residuals*. Sir Ronald Fisher's insight was how to turn those quantities into a statistical test. We want to see if the treatment effects are large compared with the residuals. To do that, we first compute the Sums of Squares of each table. Fisher's insight was that dividing these sums of squares by their respective degrees of freedom lets us test their ratio by a distribution that he found (which was later named the *F* in his honour). When we divide a sum of squares by its degrees of freedom we get the associated *mean square.*

When the Treatment Mean Square is *large* compared to the Error Mean Square, this provides evidence that the treatment means are different. And we can use the *F*-distribution to see how large "large" is.

The sums of squares for each table are easy to calculate. Just take every value in the table, square it, and add them all up. For the *Methods,* the Treatment Sum of Squares, $SS_T = (−50.75)^2 + (−50.75)^2 + \cdots + (28.75)^2 = 29\,882$. There are four treatments (methods), and so there are 3 degrees of freedom. So,

$$MS_T = SS_T/3 = 29\,882/3 = 9960.64$$

In general, we could write the Treatment Sum of Squares as

$$SS_T = \sum \sum (\bar{y}_j - \bar{\bar{y}})^2$$

Be careful to note that the summation is over the *whole* table, rows and columns. That's why there are two summation signs.

And,

$$MS_T = SS_T/(k - 1)$$

The table of residuals shows the variation that remains after we remove the overall mean and the treatment effects. These are what's left over after we account for what we're interested in—in this case, the treatments. Their variance is the variance *within* each group that we see in the boxplots of the four groups. To find its value, we first compute the Error Sum of Squares, SS_E, by summing up the squares of every element in the residuals table. To get the Mean Square (the variance), we have to divide it by $N - k$ rather than by $N - 1$ because we found them by subtracting each of the k treatment means.

So,

$$SS_E = (13.5)^2 + (-32.5)^2 + \ldots + (-30)^2 = 39\,484$$

and

$$MS_E = SS_E/(32 - 4) = 1410.14$$

As equations:

$$SS_E = \sum\sum (y_{ij} - \bar{y}_i)^2$$

and

$$MS_E = SS_E/(N - k)$$

Now where are we? To test the null hypothesis that the treatment means are all equal, we find the F-statistic:

$$F = MS_T/MS_E$$

and compare that value to the F-distribution with $k - 1$ and $N - k$ degrees of freedom. When the F-statistic is large enough (and its associated P-value small), we reject the null hypothesis and conclude that at least one mean is different.

There's another amazing result hiding in these tables. If we take each of these tables, square every observation, and add them up, the sums add as well!

$$SS_{Observations} = SS_{Grand\,Mean} + SS_T + SS_E$$

The $SS_{Observations}$ is usually very large compared to SS_T and SS_E, so when ANOVA was originally done by hand, or even by calculator, it was hard to check the calculations using this fact. The first sum of squares was just too big. So, usually you'll see the ANOVA table use the "Corrected Total" sum of squares. If we write

Observation = Grand Mean + Treatment Effect + Residual

we can naturally write

Observation − Grand Mean = Treatment Effect + Residual

Mathematically, this is the same statement, but numerically it is more stable. What's amazing is that the sums of the squares still add up. That is, if you make the first table of observations with the grand mean subtracted from each, square those, and add them up, what you get is called the Total SS or SS_{Total} and we have the breakdown:

$$SS_{Total} = SS_T + SS_E$$

That's what the ANOVA table shows. If you find this surprising, you must be following along. The tables add up, so sums of their elements must add up. But it is not at all obvious that the sums of the *squares* of their elements should add up, and this is another great insight of the Analysis of Variance.

MATH BOX

In a general set up (with possibly unequal sample sizes $n_1, n_2, \ldots n_k$), the data can be displayed as

Group	1	2	$\ldots$	k
Observation	y_{11}	y_{12}	$\cdots$	y_{1k}
	y_{21}	y_{22}	$\cdots$	y_{2k}
	$\vdots$	$\vdots$	$\vdots$	$\vdots$

	$y_{n_1 1}$	$y_{n_2 2}$	$\cdots$	$y_{n_k k}$
Total	T_1	T_2	$\cdots$	T_k
Sample size	n_1	n_2	$\cdots$	n_k
Average	$\bar{y}_1$	$\bar{y}_2$	$\cdots$	$\bar{y}_k$

where $n_j =$ size of the jth group, $y_{ij} = i^{th}$ observation in the jth group,

$$T_j = y_{1j} + y_{2j} + \cdots + y_{n_j j}, \text{ and } \bar{y}_j = \frac{T_j}{n_j}.$$

Then the total # of observations and the grand mean are

$$N = n_1 + n_2 + \cdots + n_k \text{ and } \bar{\bar{y}} = \frac{\Sigma\Sigma y_{ij}}{N}.$$

Since we have $y_{ij} - \bar{\bar{y}} = (\bar{y}_j - \bar{\bar{y}}) + (y_{ij} - \bar{y}_j)$

we can square both sides of the equation:

$$(y_{ij} - \bar{\bar{y}})^2 = [(\bar{y}_j - \bar{\bar{y}}) + (y_{ij} - \bar{y}_j)]^2$$

By summing over the observations within each group and then over the groups, we get

$$\Sigma\Sigma(y_{ij} - \bar{\bar{y}})^2 = \Sigma\Sigma[(\bar{y}_j - \bar{\bar{y}}) + (y_{ij} - \bar{y}_j)]^2$$

Notice that the sum of squares equation can be simplified to

$$\Sigma\Sigma(y_{ij} - \bar{\bar{y}})^2 = \Sigma\Sigma(\bar{y}_j - \bar{\bar{y}})^2 + \Sigma\Sigma(y_{ij} - \bar{y}_j)^2$$

because the cross product term $\Sigma\Sigma(\bar{y}_j - \bar{\bar{y}})(y_{ij} - \bar{y}_j) = 0$. So we get

$$SS_{Total} = SS_T + SS_E$$

where $SS_T = \Sigma\Sigma(\bar{y}_j - \bar{\bar{y}})^2 = \Sigma n_j(\bar{y}_j - \bar{\bar{y}})^2$ with degrees of freedom $= k - 1$, $SS_E = \Sigma\Sigma(y_{ij} - \bar{y}_j)^2$ with degrees of freedom $= N - k$, and $SS_{Total} = \Sigma\Sigma(y_{ij} - \bar{\bar{y}})^2$ with degrees of freedom $= N - 1$.

From these formulas, the ANOVA table can be obtained (though using software is recommended).

Back to Standard Deviations

We've been using the variances because they're easier to work with. But when it's time to think about the data, we'd really rather have a standard deviation because it's in the units of the response variable. The natural standard deviation to think about is the standard deviation of the residuals.

The variance of the residuals is staring us in the face. It's the MS_E. All we have to do to get the **residual standard deviation** is take the square root of MS_E:

$$s_p = \sqrt{MS_E} = \sqrt{\frac{\Sigma e^2}{N - k}}$$

The p subscript is to remind us that this is a *pooled* standard deviation, combining residuals across all k groups. The denominator in the fraction shows that finding a mean for each of the k groups cost us one degree of freedom for each.

This standard deviation should "feel" right. That is, it should reflect the kind of variation you expect to find in any of the experimental groups. For the hand-washing data, $s_p = \sqrt{1410.14} = 37.6$ bacteria colonies. Looking back at the boxplots of the groups, we see that 37.6 seems to be a reasonable compromise standard deviation for all four groups.

Section 24.2 Exercises

T 5. **Activating baking yeast** To shorten the time it takes him to make his favourite pizza, a student designed an experiment to test the effect of sugar and milk on the activation times for baking yeast. Specifically, he tested four different recipes and measured how many seconds it took for the same amount of dough to rise to the top of a bowl. He randomized the order of the recipes and replicated each treatment four times.

Here are the boxplots of activation times from the four recipes:

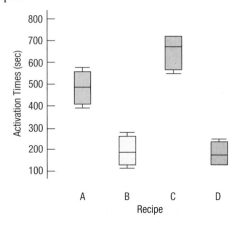

The ANOVA table is shown below.

Analysis of Variance

Source	DF	Sum of Squares	Mean Square	F-ratio	P-value
Recipe	3	638 967.69	212 989	44.7392	<0.0001
Error	12	57 128.25	4761		
Total	15	696 095.94			

a) State the hypotheses about the recipes (both numerically and in words).
b) Assuming that the assumptions for inference are satisfied, perform the hypothesis test and state your conclusion. Be sure to state it in terms of activation times and recipes.

T 6. **Frisbee throws** A student performed an experiment with three different grips to see what effect it might have on the distance of a backhanded Frisbee throw. She tried it with her normal grip, with one finger out, and with the Frisbee inverted. She measured in paces how far her throw went. The boxplots and the ANOVA table for the three grips are shown below:

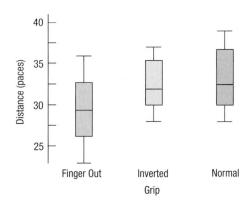

Analysis of Variance

Source	DF	Sum of Squares	Mean Square	F-ratio	P-value
Grip	2	58.58333	29.2917	2.0453	0.1543
Error	21	300.75000	14.3214		
Total	23	359.33333			

a) State the hypotheses about the grips.
b) Assuming that the assumptions for inference are satisfied, perform the hypothesis test and state your conclusion. Be sure to state it in terms of Frisbee grips and distance thrown.

24.3 Assumptions and Conditions

Just as you would never perform a linear regression without looking at the scatterplot of y versus x, you should never embark on an ANOVA without first examining side-by-side boxplots of the data comparing the responses for all of the groups. You already know what to look for—we talked about that in Chapter 4. Check for outliers within any of the groups and correct them if there are errors in the data. Get an idea of whether the groups have similar spreads (as we'll need) and whether the centres seem to be alike (as the null hypothesis claims) or different. If the spreads of the groups are very different—and especially if they seem to grow consistently as the means grow—consider re-expressing the response variable to make the spreads more nearly equal. Doing so is likely to make the analysis more powerful and more correct. Likewise, if the boxplots are skewed in the same direction, you may be able to make the distributions more symmetric with a re-expression.

Don't ever carry out an analysis of variance without looking at the side-by-side boxplots first. The chance of missing an important pattern or violation is just too great.

When we checked assumptions and conditions for regression, we had to take care to perform our checks in order. Here we have a similar concern. For regression we found that displays of the residuals were often a good way to check the corresponding conditions. That's true now as well.

Independence Assumptions

The groups must be independent of each other. No test can verify this assumption. You have to think about how the data were collected. The assumption would be violated, for example, if we measured subjects' performance before some treatment, again in the middle of the treatment period, and then again at the end.[8] Or, if we compared five diets by randomizing them to five siblings in one litter, then randomizing the diets again to five siblings in a second litter, etc., this would be a randomized block design (see Chapter 10), and the treatment samples would not be independent (but you can use methods discussed in the next chapter, to take into account the additional source of variation - the litters).

The individual responses *within* each treatment group must be independent as well. The individuals must be drawn independently and at random from the target population, or generated by a completely randomized experiment.

We check the **Randomization Condition:** Were the data collected with suitable randomization? For surveys, are the data drawn from each group a representative random sample of that group? For experiments, were the treatments assigned to the experimental units at random (that is, using a completely randomized design - see Chapter 10)?

We were told that the hand-washing experiment was randomized[9].

Equal Variance Assumption

The ANOVA requires that the variances of the treatment groups be equal. After all, we need to find a pooled variance for the MS_E. To check this assumption, we can check that the groups have similar spreads:

Similar Spread Condition: There are some ways to see whether the variation in the treatment groups seems roughly equal:

■ Look at side-by-side boxplots of the groups to see whether they have roughly the same spread. It can be easier to compare spreads across groups when they have the same centre, so consider making side-by-side boxplots of the residuals. If the groups differ substantially in sample size, unequal spread can be an especially bad problem because the test may be more conservative (with fewer Type I errors) or less (with too many Type I errors), depending on which groups have the larger variances. However, with balanced (equal size) groups, the effect of unequal sample sizes is considerably mitigated, as the F test is quite robust against small to moderate differences in spread in this case, causing only a very minor inflation of the Type I error rate. *Aim for balance in your design whenever possible*, as increased robustness is one (and not the only) advantage of such designs.

■ Look at the original boxplots of the response values again. In general, do the spreads seem to change *systematically* with the centres? One common pattern is for the boxes with bigger centres to have bigger spreads. This kind of systematic trend in the

---

[8]There is a modification of ANOVA, called *repeated measures* ANOVA, that deals with such data. (If the design reminds you of a paired-*t* situation, you're on the right track, and the lack of independence is the same kind of issue we discussed in Chapter 20.)

[9]But if the order of the 4 treatments was randomized for the first 4 days, then randomized again for the next 4 days, etc., until the 8th 4-day period - rather than having the order of all 32 runs randomized - it would be best to run a two-way ANOVA, as discussed in the next chapter, with the eight 'replications' of the experiment treated as blocks.

variances is more of a problem than random differences in spread among the groups and should not be ignored. Fortunately, such systematic violations are often helped by re-expressing the data. (If, in addition to spreads that grow with the centres, the boxplots are skewed with the longer tail stretching off to the high end, then the data are pleading for a re-expression. Try taking logs of the response variable for a start. You'll likely end up with a much cleaner analysis.)

■ Look at the residuals plotted against the predicted values. Often, larger predicted values lead to larger magnitude residuals. This is another sign that the condition is violated. (This may remind you of the **Does the Plot Thicken? Condition** of regression. And it should.) When the plot thickens (to one side or the other), it's usually a good idea to consider re-expressing the response variable. Such a systematic change in the spread is a more serious violation of the Equal Variance Assumption than slight variations of the spreads across groups.

Let's check the conditions for the hand-washing data. Figure 24.5 shows a boxplot of residuals by group and residuals against predicted value:

Figure 24.5

Boxplots of residuals for the four hand-washing methods and a plot of residuals versus predicted values. There's no evidence of a systematic change in variance from one group to the other or by predicted value.

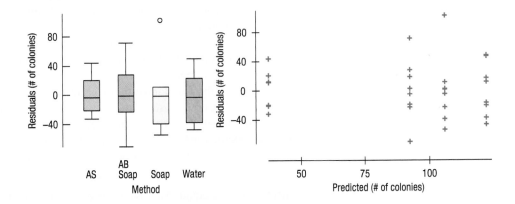

Neither plot shows a violation of the condition. The IQRs (the box heights) are quite similar and the plot of residuals versus predicted values does not show a pronounced widening to one end. The pooled estimate of 37.6 colonies for the error standard deviation seems reasonable for all four groups.

Normal Population Assumption   Like Student's t-tests, the mathematical derivation of the F-test requires the underlying errors to follow a Normal model. As before when we've faced this assumption, we'll check a corresponding **Nearly Normal Condition**.

Technically, we need to assume that the Normal model is reasonable for the populations underlying *each* treatment group. We can (and should) look at the side-by-side boxplots for indications of skewness. Certainly, if they are all (or mostly) skewed in the same direction, the Nearly Normal Condition fails (and re-expression is likely to help).

In experiments, we often work with fairly small groups for each treatment, and it's nearly impossible to assess whether the distribution of only six or eight numbers might be roughly Normal (though sometimes it's so skewed or has such an extreme outlier that we can see that it's not). Here we are saved by the Equal Variance Assumption, which we've already checked. The residuals have their group means subtracted, so the mean residual for each group is 0. If their variances are equal, we can group all the residuals together for the purpose of checking the Nearly Normal Condition.

Check Normality with a histogram or a Normal probability plot of all the residuals together. The hand-washing residuals look nearly Normal in the Normal probability plot (Figure 24.6), although, as the boxplots showed, there's a possible outlier in the Soap group (actually, just barely above the upper fence).

Because we really care about the Normal model within each group, the Normal population assumption is violated if there are outliers in any of the groups. Check for outliers in

Figure 24.6

The hand-washing residuals look nearly Normal in this Normal probability plot.

the boxplots of the values for each treatment group. The Soap group of the hand-washing data shows a possible outlier, so we might want to compute the analysis again without that observation. (For these data, it turns out to make little difference.)

ONE-WAY ANOVA *F*-TEST

We test the null hypothesis $H_0: \mu_1 = \mu_2 = \ldots = \mu_k$ against the alternative that the true group means are not all equal. We test the hypothesis with the *F*-statistic, $F = \dfrac{MS_T}{MS_E}$, where MS_T is the Treatment Mean Square, measuring variability among the means of the treatment groups, and MS_E is the Error Mean Square, found by pooling the variances within each of the treatment groups. If the *F*-statistic is large enough, we reject the null hypothesis.

Step-by-Step Example ANALYSIS OF VARIANCE

In Chapter 4, we looked at side-by-side boxplots of four different containers for holding hot beverages. The experimenter wanted to know which type of container would keep his beverages hot longest. To test it, he heated water to a temperature of 180°F, placed it in the container, and then measured the temperature of the water again 30 minutes later. He randomized the order of the trials and tested each container eight times. His response variable was the difference in temperature (in °F) between the initial water temperature and the temperature after 30 minutes. (Data in **Cups**)

Question: Do the four containers maintain temperature equally well?

THINK ➡ Plot Plot the side-by-side boxplots of the data.	

Plan State what you want to know and the null hypothesis you wish to test. For ANOVA, the null hypothesis is that all the treatment groups have the same mean. The alternative is that at least one mean is different.

I want to know whether there is any difference among the four containers in their ability to maintain the temperature of a hot liquid for 30 minutes. I'll write μ_k for the mean temperature difference for container k, so the null hypothesis is that these means are all the same:

$$H_0: \mu_1 = \mu_2 = \mu_3 = \mu_4.$$

The alternative is that the group means are not all equal.

Think about the assumptions and check the conditions.

✓ **Randomization Condition:** The experimenter performed the trials in random order, so it's reasonable to assume that the performance of one tested cup is independent of other cups.

✓ **Similar Spread Condition:** The Nissan mug variation seems to be a bit smaller than the others. I'll investigate plot thickening shortly.

SHOW ➡ **Mechanics** Fit the ANOVA model, and continue checking the conditions, using the residuals.

Analysis of Variance

Source	DF	Sum of Squares	Mean Square	F-ratio	P-value
Container	3	714.1875	238.063	10.713	<0.0001
Error	28	622.1875	22.221		
Total	31	1336.3750			

✓ **Nearly Normal Condition, Outlier Condition**: The Normal probability plot is not very straight, but there are no outliers.

The histogram shows that the distribution of the residuals is moderately skewed to the right:

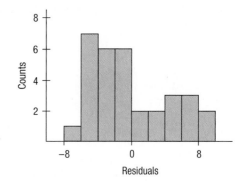

The table of means and SDs further below shows that the standard deviations grow along with the means (also evident from a plot of residuals versus predicted values). Possibly a re-expression of the data would improve matters.

Under these circumstances, I *cautiously* will find the P-value for the F-statistic from the F-model with 3 and 28 degrees of freedom (already displayed in ANOVA table above).

REALITY CHECK   Does the pooled standard deviation look reasonable?

From the ANOVA table, the Error Mean Square, MS_E, is 22.22, which means that the standard deviation of all the errors is estimated to be $\sqrt{22.22} = 4.71$ degrees F.

This seems like a reasonable value for the error standard deviation in the four treatments (with the possible exception of the Nissan mug).

SHOW ➡ Show the table of means. Report the *F*-statistic and P-value.	<table><tr><th>Level</th><th>*n*</th><th>Mean</th><th>Std Dev</th></tr><tr><td>CUPPS</td><td>8</td><td>10.1875</td><td>5.20259</td></tr><tr><td>Nissan</td><td>8</td><td>2.7500</td><td>2.50713</td></tr><tr><td>SIGG</td><td>8</td><td>16.0625</td><td>5.90059</td></tr><tr><td>Starbucks</td><td>8</td><td>10.2500</td><td>4.55129</td></tr></table> The ratio of the mean squares gives an *F*-ratio of 10.7134 with a P-value of < 0.0001.
TELL ➡ **Interpretation** Tell what the *F*-test means.	An *F*-ratio this large would be very unlikely if the containers all had the same mean temperature difference.
TELL ➡ State your conclusions. More precise conclusions might require a re-expression of the data.	**Conclusions:** Even though some of the conditions are mildly violated, the P-value is so extremely small that even if the sampling model is a bit inaccurate, the P-value should still be quite small. I conclude that the means are not all equal and that the four cups do not maintain temperature equally well.

The Balancing Act

The two examples we've looked at so far share a special feature. Each treatment group has the same number of experimental units. For the hand-washing experiment, each washing method was tested eight times. For the cups, there were also eight trials for each cup. This feature (the equal numbers of cases in each group, not the number eight) is called **balance**, and experiments that have equal numbers of experimental units in each treatment are said to be balanced or to have balanced designs.

Balanced designs are a bit easier to analyze because the calculations are simpler, so we usually try for balance. But in the real world we often encounter unbalanced data. Participants drop out or become unsuitable, plants die, or maybe we just can't find enough experimental units to fit a particular criterion.

Everything we've done so far works just fine for unbalanced designs except that the calculations get a bit more complicated. Where once we could write *n* for the number of experimental units in a treatment, now we have to write n_k and sum more carefully. Where once we could pool variances with a simple average, now we have to adjust for the different *n*'s. Technology clears these hurdles easily, so you're safe thinking about the analysis in terms of the simpler balanced formulas and trusting that the technology will make the necessary adjustments.

For Example ANALYSIS OF VARIANCE FOR HAND VOLUME CHANGE

RECAP: An ANOVA for the contrast baths experiment had a *statistically significant F-value*. Here are summary statistics for the three treatment groups:

Group	Count	Mean	StdDev
Bath	22	4.54545	7.76271
Bath + Exercise	23	8	7.03885
Exercise	14	−1.07143	5.18080

> **QUESTION:** What can you conclude about these results?
>
> **ANSWER:** We can be confident that there are some differences among the three treatments. However, it is the exercise treatment that appears to reduce swelling the most. We might conclude (as the researchers did) that contrast bath treatments are of limited value. (If you require more precise comparisons between *pairs* of treatments, read the next section.)

Section 24.3 Exercises

T **7. Fuel economy by cylinders** Here are boxplots that show the relationship between the number of cylinders a car's engine has and the car's fuel economy.

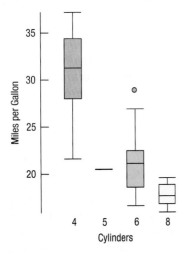

a) State the null and alternative hypotheses that you might consider for these data.
b) Do the conditions for an ANOVA seem to be met here? Why or why not?

T **8. Finger Lakes wines 2014** Here are bottle prices (in dollars) of wines produced by wineries along three of the Finger Lakes.

a) What null and alternative hypotheses would you test for these data? Talk about prices and location, not symbols.
b) Do the conditions for an ANOVA seem to be met here? Why or why not?

24.4 Comparing Means

When we reject H_0, it's natural to ask which means are different. No one would be happy with an experiment to test 10 cancer treatments that concluded only with "We can reject H_0—the treatments are different!" We'd like to know more, but the F-statistic doesn't offer that information.

What can we do? If we can't reject the null, we've got to stop. There's no point in further testing. If we've rejected the simple null hypothesis, however, we *can* do more. In particular, we can test whether any pairs or combinations of group means differ. For example, we might want to compare treatments against a control or a placebo, or against the current standard treatment.

In the hand-washing experiment, we could consider plain water to be a control. Nobody would be impressed with (or want to pay for) a soap that did no better than water alone. A test of whether the antibacterial soap (for example) was different from plain water would be a simple test of the difference between two group means. To be able to perform an ANOVA, we first check the Similar Spread Condition. If things look okay, we assume that the variances are equal. If the variances *are* equal, then a pooled t-test is appropriate. Even better (this is the special part), we already have a pooled estimate of the standard deviation based on *all* of the tested washing methods. That's s_p, which, for the hand-washing experiment, was equal to 37.55 bacteria colonies.

The null hypothesis is that there is no difference between water and the antibacterial soap. As we did in Chapter 19, we'll write that as a hypothesis about the difference in the means:

$$H_0: \mu_W - \mu_{ABS} = 0.$$

The alternative is

$$H_A: \mu_W - \mu_{ABS} \neq 0.$$

The natural test statistic is $\bar{y}_W - \bar{y}_{ABS}$ and the (pooled) standard error is

$$SE(\bar{y}_W - \bar{y}_{ABS}) = s_p\sqrt{\frac{1}{n_W} + \frac{1}{n_{ABS}}}$$

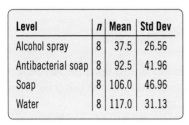

Level	n	Mean	Std Dev
Alcohol spray	8	37.5	26.56
Antibacterial soap	8	92.5	41.96
Soap	8	106.0	46.96
Water	8	117.0	31.13

The difference in the observed means is $117.0 - 92.5 = 24.5$ colonies. The standard error comes out to 18.775. The *t*-statistic, then, is $t = \dfrac{24.5}{18.775} = 1.31$. To find the P-value, we consult the Student's *t*-distribution on $N - k = 32 - 4 = 28$ degrees of freedom. The P-value is about 0.2—not small enough to impress us. So, we can't discern a significant difference between washing with the antibacterial soap and just using water.

Our *t*-test asks about a simple difference. We could also ask a more complicated question about groups of differences. Does the average of the two soaps differ from the alcohol spray, for example? Complex combinations like these are called *contrasts*. Finding the standard errors for contrasts is straightforward but beyond the scope of this book. We'll restrict our attention to the common question of comparing simple pairs of treatments after H_0 has been rejected. And even this type of question involves some important issues we will now discuss.

Bonferroni Multiple Comparisons

Our hand-washing experimenter *was* pretty sure that alcohol would kill the germs even before she started the experiment. But alcohol dries the skin and leaves an unpleasant smell. She was hoping that one of the antibacterial soaps would work as well as alcohol so she could use that instead. That means she really wanted to compare *each* of the other treatments against the alcohol spray. We know how to compare two of the means with a *t*-test. But now we want to do several tests, and each test poses the risk of a Type I error. As we do more and more tests, the risk that we might make a Type I error grows bigger than the α level of each individual test. With each additional test, the risk of making an error grows. If we do enough tests, we're almost sure to reject one of the null hypotheses by mistake—and we'll never know which one.

There is a defense against this problem. In fact, there are several defenses. As a class, they are called **methods for multiple comparisons**. All multiple comparison methods require that we first be able to reject the overall (global) null hypothesis with the ANOVA's *F*-test. Once we've rejected the overall null, then we can think about comparing several—or even all—pairs of group means.

Let's look again at our test of the water treatment against the antibacterial soap treatment. This time we'll look at a confidence interval instead of the pooled *t*-test. We did a test at significance level $\alpha = 0.05$. The corresponding confidence level is $1 - \alpha = 95\%$. For *any* pair of means, a confidence interval for their difference is $(\bar{y}_1 - \bar{y}_2) \pm ME$, where the margin of error is

$$ME = t^* \times s_p\sqrt{\frac{1}{n_1} + \frac{1}{n_2}}$$

As we did in the previous section, we get s_p as the pooled standard deviation found from *all* the groups in our analysis. Because s_p uses the information about the standard deviation from *all* the groups, it's a better estimate than we would get by combining the standard deviations of just two of the groups. This uses the **Equal Variance Assumption** and "borrows strength" in estimating the common standard deviation all the groups. We find the critical value t^*—from the Student's t-model corresponding to the specified confidence level found with $N - k$ degrees of freedom, and the n_k's are the number of experimental units in each of the treatments.

To reject the null hypothesis that the two group means are equal, the difference between them must be larger than the *ME*. That way, 0 won't be in the confidence interval for the difference. When we use it in this way, we call the margin of error the **least significant difference (LSD** for short). If two group means differ by more than this amount, then they are *significantly different at level α for each individual test*.

For our hand-washing experiment, each group has $n = 8$, $s_p = 37.55$, and df $= 32 - 4 = 28$. From technology or Table T, we can find that t^* with 28 df (for a 95% confidence interval) is 2.048. So

$$\text{LSD} = 2.048 \times 37.55 \times \sqrt{\frac{1}{8} + \frac{1}{8}} = 38.45 \text{ colonies},$$

and we could use this margin of error to make a 95% confidence interval for any difference between group means. And any two washing methods whose means differ by more than 38.45 colonies could be said to differ at $\alpha = 0.05$ by this method.

Of course, we're still just examining individual pairs. If we want to examine *many pairs simultaneously*, there are several methods that adjust the ME or LSD so that, in spite of making *many* such tests, the overall Type I error rate stays at (or below) α.

To be precise, we can define the *overall* Type I error rate for the batch of tests as the probability of one or more (falsely) significant test results when there are no actual differences between any of the true means. If we think in terms of the confidence intervals, we can define the *overall* or *simultaneous* confidence level as the probability that each and every one of our confidence intervals is correct (captures the true difference). Controlling the overall α-level or Type I error rate at 0.05 for the batch of tests is equivalent to controlling the simultaneous confidence level at 0.95 for the corresponding batch of confidence intervals.

One such method is called the **Bonferroni method**. This method adjusts the LSD to allow for making many comparisons. The result is a wider margin of error we may call the **minimum significant difference, or MSD**. The MSD is found by replacing t^* with a slightly larger number. That makes the confidence intervals wider for each contrast and the corresponding Type I error rates lower for *each* test. But it keeps the *overall* Type I error rate at or below α.

The Bonferroni method distributes the error rate equally among all the tests. It divides the error rate among J tests by calculating each confidence interval at confidence level $1 - \dfrac{\alpha}{J}$ instead of the original $1 - \alpha$. To signal this adjustment, we label the critical value t^{**} rather than t^*. For example, to make the three confidence intervals comparing the alcohol spray with the other three washing methods, and preserve our overall α risk below 5%, we'd construct each with a confidence level of

$$1 - \frac{0.05}{3} = 1 - 0.01667 = 0.98333$$

The only problem with this is that t-tables don't have a column for 98.33% confidence (or, correspondingly, for $\alpha = 0.01667$). Fortunately, technology has no such constraints.[10] For the hand-washing data, if we want to examine the three confidence intervals comparing each of the other methods with the alcohol spray, the t^{**}-value (on 28 degrees of freedom) turns out to be 2.546. That's somewhat larger than the individual t^*-value of 2.048 that we would

---

[10]The electronic t-tables provided in MyStatLab let you add new columns to the t-table at any alpha level, so you can do the Bonferroni calculation easily.

have used for a single confidence interval. And the corresponding ME is 47.69 colonies (rather than 38.45 for a single comparison). The larger critical value along with correspondingly wider intervals is the price we pay for making *multiple* comparisons.

Many statistics packages assume that you'd like to compare all pairs of means. Some will display the result of these comparisons in a table like this:

Level	n	Mean	Groups
Alcohol spray	8	37.5	A
Antibacterial soap	8	92.5	B
Soap	8	106.0	B
Water	8	117.0	B

This table shows that the alcohol spray is in a class by itself and that the other three hand-washing methods are statistically indistinguishable from one another. Other software may give you all the pair-wise confidence intervals. Then you can check which intervals miss '0', indicating a significant difference, and also use the confidence interval to say just how big the difference actually is.

With software, you may be asked either to specify the overall α-level (e.g. 0.05) or the overall confidence level (e.g. 0.95). If your software offers Tukey's method for multiple comparisons rather than the Bonferroni method, by all means, choose it! The interpretations are identical, but Tukey's method is specifically designed just for comparing means, hence will give shorter confidence intervals and more powerful tests than the Bonferroni method. The latter can be quite conservative and always results in a true overall α-level substantially less than what is requested. (So why not discuss Tukey's method? Well, for one thing, it uses yet another new sampling model.)

Section 24.4 Exercises

9. Tellers A bank is studying the time that it takes 6 of its tellers to serve an average customer. Customers line up in the queue and then go to the next available teller. Here is a boxplot of the last 200 customers and the times it took each teller:

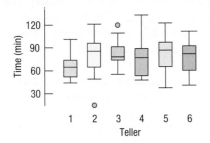

ANOVA Table

Source	DF	Sum of Squares	Mean Square	F-Ratio	P-Value
Teller	5	3 315.32	663.064	1.508	0.1914
Error	134	58 919.1	439.695		
Total	139	62 234.4			

a) What are the null and alternative hypotheses?
b) What do you conclude?

c) Would it be appropriate to run a multiple comparisons test (for example, a Bonferroni test) to see which tellers differ from each other? Explain.

10. Hearing A researcher investigated four different word lists for use in hearing assessment. She wanted to know whether the lists were equally difficult to understand in the presence of a noisy background. To find out, she tested 96 subjects with normal hearing randomly assigning 24 to each of the four word lists and measured the number of words perceived correctly in the presence of background noise. Here are the boxplots of the four lists:

ANOVA Table

Source	DF	Sum of Squares	Mean Square	F-Ratio	P-Value
List	3	920.4583	306.819	4.9192	0.0033
Error	92	5738.1667	62.371		
Total	95	6658.6250			

a) What are the null and alternative hypotheses?

b) What do you conclude?

c) Would it be appropriate to run a multiple comparisons test (for example, a Bonferroni test) to see which lists differ from each other in terms of mean percent correct? Explain.

24.5 ANOVA on Observational Data

So far we've applied ANOVA only to data from designed experiments. That's natural for several reasons. The primary one is that, as we saw in Chapter 10, randomized comparative experiments are specifically designed to compare the results for different treatments. The overall null hypothesis, and the subsequent tests on pairs of treatments in ANOVA, address such comparisons directly. In addition, as we discussed earlier, the **Equal Variance Assumption** (which we need for all of the ANOVA analyses) is often plausible in a randomized experiment because the treatment groups start out with sample variances that all estimate the same underlying variance of the collection of experimental units.

Sometimes, though, we just can't perform an experiment. When ANOVA is used to test equality of group means from observational data, there's no *a priori* reason to think the group variances might be equal at all. Even if the null hypothesis of equal means were true, the groups might easily have different variances. But if the side-by-side boxplots of responses for each group show roughly equal spreads and symmetric, outlier-free distributions, you can use ANOVA on observational data.

Observational data tend to be messier than experimental data. They are much more likely to be unbalanced. If you aren't assigning subjects to treatment groups, it's harder to guarantee the same number of subjects in each group. And because you are not controlling conditions as you would in an experiment, things tend to be, well, less controlled. The only way we know to avoid the effects of possible lurking variables is with control and randomized assignment to treatment groups, and for observational data, we have neither.

ANOVA is often applied to observational data when an experiment would be impossible or unethical. (We can't randomly break some subjects' legs, but we *can* compare pain perception among those with broken legs, those with sprained ankles, and those with stubbed toes by collecting data on subjects who have already suffered those injuries.) In such data, subjects are already in groups, but not by random assignment.

Be careful; if you have not assigned subjects to treatments randomly, you can't draw *causal* conclusions even when the *F*-test is significant. You have no way to control for lurking variables or confounding, so you can't be sure whether any differences you see among groups are due to the grouping variable or to some other unobserved variable that may be related to the grouping variable.

Because observational studies often are intended to estimate parameters, there is a temptation to construct confidence intervals for the group means for this purpose. Although these confidence intervals are statistically correct, be sure to think carefully about the population that the inference is about. The relatively few subjects that happen to be in a group may not be a simple random sample of any interesting population, so their "true" mean may have only limited meaning.

Step-by-Step Example ONE MORE EXAMPLE

Here's an example that exhibits many of the features we've been discussing. It gives a fair idea of the kinds of challenges often raised by real data.

A study in a liberal arts department attempted to find out who watches more TV in university: men or women? Varsity athletes or nonathletes? Student researchers asked 200 randomly selected students questions about their backgrounds and about their television-viewing habits and received 197 legitimate responses. The researchers found that men watch, on average, about 2.5 hours per week more TV than women, and that varsity athletes watch about 3.5 hours per week more than those who are not varsity athletes. But is this the whole story? To investigate further, they divided the students into four groups: male athletes (MA), male nonathletes (MNA), female athletes (FA), and female non-athletes (FNA). (Data in **TV watching**).

Question: Do these four groups of students spend about the same amount of time watching TV?

THINK ➡ Variables Name the variables, report the *W*'s, and specify the questions of interest.

I have the number of hours spent watching TV in a week for 197 randomly selected students. We know their sex and whether they are varsity athletes or not. I wonder whether TV watching differs according to sex and athletic status.

Plot Always start an ANOVA with side-by-side boxplots of the responses in each of the groups. Always.

These data offer a good example why.

Here are the side-by-side boxplots of the data:

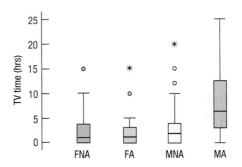

This plot suggests problems with the data. Each box shows a distribution skewed to the high end, and outliers pepper the display, including some extreme outliers. The box with the highest centre (MA) also has the largest spread. These data just don't pass our first screening for suitability. This sort of pattern calls for a re-expression.

Here are the boxplots for the square root of TV hours.

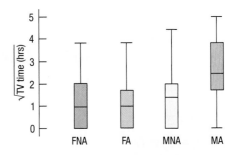

The spreads in the four groups are now very similar and the individual distributions closer to symmetric. And now there are no outliers.

Think about the assumptions and check the conditions.

✓ **Randomization Condition**: The data come from a random sample of students. Hence, the responses should be independent.

✓ **Similar Spread Condition**: The boxplots show similar spreads. I may want to check the residuals later.

SHOW ➡ Fit the ANOVA model.

The ANOVA table looks like this:

Source	DF	Sum of Squares	Mean Square	F-ratio	P-value
Group	3	47.24733	15.7491	12.8111	<0.0001
Error	193	237.26114	1.2293		
Total	196	284.50847			

✓ **Nearly Normal Condition, Outlier Condition**: A histogram of the residuals is below:

There is some granularity in the data, including a large number of respondents who reported 0 hours of TV viewing (producing a modal peak around −1.2 to −1.4). However, the histogram and NPP are still close enough to Normal. Under these conditions, it's appropriate to use Analysis of Variance.

TELL ➡ Interpretation

The F-statistic is large and the corresponding P-value small. I conclude that the TV watching behaviour is not the same among these groups.

So Do Male Athletes Watch More TV?

Here's a Bonferroni comparison of all pairs of groups:

DIFFERING STANDARD ERRORS?

In case you were wondering . . . The standard errors are different because this isn't a balanced design. Differing numbers of experimental units in the groups generate differing standard errors.

	Difference	Std. Err.	P-Value
FA–FNA	0.049	0.270	0.9999
MNA–FNA	0.205	0.182	0.8383
MNA–FA	0.156	0.268	0.9929
MA–FNA	1.497	0.250	<0.0001
MA–FA	1.449	0.318	<0.0001
MA–MNA	1.292	0.248	<0.0001

Three of the differences are very significant. It seems that among women, there's little difference in TV watching between varsity athletes and others. Among men, though, the corresponding difference is large. And among varsity athletes, men watch significantly more TV than women.

But wait. How far can we extend the inference that male athletes watch more TV than other groups? The data came from a random sample of students made during the week of March 21. If the students carried out the survey correctly using a simple random sample, we should be able to make the inference that the generalization is true for the entire student body during that week.

Is it true for other universities? Is it true throughout the year? The students conducting the survey followed it up by collecting anecdotal information about TV watching of male athletes. It turned out that during the week of the survey, the NCAA men's basketball tournament was televised. This could explain the increase in TV watching for the male athletes. It could be that the increase extends to other students at other times, but we don't know that. Always be cautious in drawing conclusions too broadly. Don't generalize from one population to another.

Section 24.5 Exercises

11. Eye and hair colour In a survey of 1021 school-age children conducted by randomly selecting children from several large urban elementary schools, two of the questions concerned eye and hair colour. In the survey, the following codes were used:

Hair Colour	Eye Colour
1 = Blond	1 = Blue
2 = Brown	2 = Green
3 = Black	3 = Brown
4 = Red	4 = Grey
5 = Other	5 = Other

The students analyzing the data were asked to study the relationship between eye and hair colour. They produced this plot:

They then ran an Analysis of Variance with *Eye Colour* as the response and *Hair Colour* as the factor. The ANOVA table they produced follows:

Source	DF	Sum of Squares	Mean Square	F-Ratio	P-Value
Hair Colour	4	1.46946	0.367365	0.4024	0.8070
Error	1016	927.45317	0.912848		
Total	1020	928.92263			

What suggestions do you have for the Statistics students? What alternative analysis might you suggest?

12. U.S. ZIP codes An intern from the marketing department at the Holes R Us online piercing salon has recently finished a study of the company's 500 customers. He wanted to know whether people's ZIP codes vary by the last product they bought. They have 16 different products, and the ANOVA table of ZIP code by product showed the following:

Source	DF	Sum of Squares	Mean Square	F-Ratio	P-Value
Product	15	3.836e10	2.55734e9	4.9422	<0.0001
Error	475	2.45787e11	517445573		
Total	490	2.84147e11			

(Nine customers were not included because of missing ZIP code or product information.)

What criticisms of the analysis might you make? What alternative analysis might you suggest?

WHAT CAN GO WRONG?

- **Watch out for outliers.** One outlier in a group can change both the mean and the spread of that group. It will also inflate the Error Mean Square, which can influence the F-test. The good news is that ANOVA fails on the safe side by losing power when there are outliers. That is, you are less likely to reject the overall null hypothesis if you have (and leave) outliers in your data. But they are not likely to cause you to make a Type I error.

- **Watch out for changing variances.** The conclusions of the ANOVA depend crucially on the assumptions of independence and constant variance, and (somewhat less seriously as n increases) on Normality. If the conditions on the residuals are violated, it may be necessary to re-express the response variable to approximate these conditions more closely. ANOVA benefits so greatly from a judiciously chosen re-expression that the choice of a re-expression might be considered a standard part of the analysis.

- **Be wary of drawing conclusions about causality from observational studies.** ANOVA is often applied to data from randomized experiments for which causal conclusions are appropriate. If the data are not from a designed experiment, however, the Analysis of Variance provides no more evidence for causality than any other method we have studied. Don't get into the habit of assuming that ANOVA results have causal interpretations.

- **Be wary of generalizing** to situations other than the one at hand. Think hard about how the data were generated to understand the breadth of conclusions you are entitled to draw.

- **Watch for multiple comparisons.** When rejecting the null hypothesis, you can conclude that the means are not *all* equal. But you can't start comparing every pair of treatments in your study with a standard t-test. You'll run the risk of inflating your Type I error rate. Use a multiple comparisons method when you want to test many pairs.

- **Watch out for more complicated designs, like blocked designs.** The groups must be independent, as in a *completely randomized design*. If not, you will need to use more advanced methods, such as those discussed in the next chapter, to take into account additional sources of variation in the data.

What Have We Learned?

We learned in Chapter 4 that a good first step in looking at the relationship between a quantitative response and a categorical grouping variable is to look at side-by-side boxplots. We've seen that it's still certainly a good first step before formally testing the null hypothesis.

Recognize when to use an Analysis of Variance (ANOVA) to test for differences among the means of several groups.

Know how to read an ANOVA table to locate the degrees of freedom, the Mean Squares, and the resulting F-statistic (usually with corresponding P-value).

Know how to check the conditions required for an ANOVA:

- Independence of the groups from each other and of the individual cases within each group.
- Equal variance of the groups.
- Normal error distribution.

Know how to create and interpret confidence intervals or tests for the differences between each pair of group means, recognizing the need to adjust the Margin of Error (or LSD) for the number of comparisons made—using a multiple comparisons procedure.

Review of Terms

Treatment (or Between) Mean Square (MS$_T$)	Another estimate of the error variance—under the assumption that the treatment means are all equal. If the (null) assumption is not true, the MS$_T$ will tend to be larger than the error variance or MS$_E$ (p. 792).
F-distribution	The sampling distribution of the F-statistic when the null hypothesis that the treatment means are equal is true. It has two degrees of freedom parameters, one for the numerator, $k - 1$, and one for the denominator, $N - k$, where N is the total number of observations and k is the number of groups (p. 792).
ANOVA	An analysis method for testing equality of means across treatment groups. ANOVA is an abbreviation for *Analysis of Variance* (p. 792).
Error (or Within) Mean Square (MS$_E$)	The estimate of the error variance obtained by *pooling* the variances of each treatment group. The square root of the MS$_E$ is the estimate of the error standard deviation, s_p (p. 792).
ANOVA table	A table that shows the degrees of freedom, the Treatment Mean Square, the Error Mean Square, their ratio, the F-statistic, and its P-value. Other quantities of lesser interest are usually included as well (p. 794).
F-statistic	The ratio MS$_T$/MS$_E$. When the F-statistic is sufficiently large, we reject the null hypothesis that the group means are equal (p. 792).
One-way ANOVA model	The model for a one-way (one response, one factor) ANOVA is $$y_{ij} = \mu_j + \varepsilon_{ij}$$ Estimating above with $y_{ij} = \bar{y}_j + e_{ij}$ gives predicted values $\hat{y}_{ij} = \bar{y}_j$ (just the group means) and residuals $e_{ij} = y_{ij} - \bar{y}_j$ (p. 796).
Residual standard deviation	The residual standard deviation, $$s_p = \sqrt{MS_E} = \sqrt{\frac{\sum e^2}{N - k}}$$ gives an idea of the underlying variability of the response values (p. 801).
F-test	Tests the null hypothesis that all the group means are equal against the alternative that they are not all equal. We reject the hypothesis of equal means if the F-statistic exceeds the critical value from the F-distribution corresponding to the specified significance level and degrees of freedom (p. 792).
Balance	An experiment's design is balanced if each treatment has the same number of experimental units. Balanced designs make calculations simpler and are generally more powerful (p. 807).
Multiple comparisons	If we reject the null hypothesis of equal means, we often then want to investigate further and compare pairs of treatment group means to see if they differ. If we want to test several such pairs, we must adjust for performing several tests to keep the overall risk of a Type I error from growing too large. Such adjustments are called methods for multiple comparisons (p. 809).
Least significant difference (LSD)	The standard margin of error in the confidence interval for the difference of two means. It has the correct Type I error rate when used to execute a single test, but not when performing more than one comparison (p. 810).
Bonferroni method	One of many methods for adjusting the size of the margin of error when testing many or all of the pairwise differences among several group means (p. 810).
Minimum significant difference (MSD)	The margin of error for the confidence interval for the difference of two group means *when* using the Bonferroni method or some other multiple comparisons method (and the MSD will vary depending on the choice of method). The MSD can be used to test (or form confidence intervals for) differences between any or all pairs of group means. If their difference exceeds the MSD, they are different at the overall α rate (p. 810).

On the Computer ONE-WAY ANOVA

Most analyses of variance are found with computers, and all statistics packages present the results in an ANOVA table much like the one we discussed. Technology also makes it easy to examine the side-by-side boxplots and check the residuals for violations of the assumptions and conditions.

Statistics packages offer different choices among possible multiple comparisons methods (although Bonferroni is quite common). This is a specialized area. Get advice or read further if you need to choose a multiple comparisons method.

As we saw in Chapter 4, there are two ways to organize data recorded for several groups. We can place the data for each group in its own column ('unstacked' data), or we can put all the response values in a single variable and use a second, "factor," variable to hold the group identities ('stacked' data). Most statistics packages expect the data to be in stacked format because this form also works for more complicated experimental designs. Some packages can work with either form, and some use one form for some things and the other for others. (Be careful, for example, when you make side-by-side boxplots; be sure to give the appropriate version of the command to correspond to the structure of your data.)

Most packages offer to save residuals and predicted values and make them available for further tests of conditions. In some packages you may have to request them specifically.

EXCEL

- In Excel 2003 and earlier, select **Data Analysis** from the Tools menu.
- In Excel 2007, select **Data Analysis** from the Analysis Group on the Data Tab.
- Select **Anova Single Factor** from the list of analysis tools.
- Click the **OK** button.
- Enter the data range in the box provided.
- Check the **Labels in First Row** box, if applicable.
- Enter an alpha level for the *F*-test in the box provided.
- Click the **OK** button.

COMMENTS

The data range should include two or more columns of data to compare. Unlike all other statistics packages, Excel expects each column of the data to represent a different level of the factor. However, it offers no way to label these levels. The columns need not have the same number of data values, but the selected cells must make up a rectangle large enough to hold the column with the most data values.

JMP

- From the **Analyze** menu select **Fit Y by X**.
- Select variables: a quantitative Y, Response variable, and a categorical X, Factor variable.
- JMP opens the **Oneway** window.
- Click on the red triangle beside the heading, select **Display Options**, and choose **Boxplots**.
- From the same menu, choose the **Means/ANOVA.t-test** command.
- JMP opens the one-way ANOVA output.

COMMENTS

JMP expects data in "stacked" format with one response and one factor variable.

MINITAB

- Choose **ANOVA** from the Stat menu.
- Choose **One-way . . .** from the **ANOVA** submenu.

- In the One-way ANOVA dialogue, assign a quantitative Y variable to the Response box and assign a categorical X variable to the Factor box.

- Click the **Graphs** button.
- In the ANOVA-Graphs dialogue, select **Data plots: Boxplot of data**, and also select **Residual plots: Four in one** for a useful batch of four residual plots.
- Click the **OK** button to return to the ANOVA dialogue. Click OK again.

COMMENTS

If your data are in unstacked format, with separate columns for each treatment level, choose **One-way (unstacked)** from the **ANOVA** submenu.

R

To perform an analysis of variance of a variable y on a categorical variable (factor) x:

- myaov = aov($y \sim x$)
- summary(myaov) # gives the ANOVA table

SPSS

- Choose **Compare Means** from the **Analyze** menu.
- Choose **One-way ANOVA** from the **Compare Means** submenu.
- In the One-Way ANOVA dialogue, select the Y-variable and move it to the dependent target. Then move the X-variable to the independent target.
- Click the **OK** button.

COMMENTS

SPSS expects data in stacked format. The **Contrasts** and **Post Hoc** buttons offer ways to test contrasts and perform multiple comparisons. See your SPSS manual for details.

STATCRUNCH

To perform a one-way ANOVA:

- Enter the factor in one column and the response variable in a second
- Click on **Stat**
- Choose **ANOVA > One Way**
- Under **Compare**, choose **Selected columns** for unstacked data, or **Values in a single column** for stacked data. Fill in the correct columns, then click **Compute!**

COMMENTS

In addition to the ANOVA output, STATCRUNCH also displays group means.

In the dialogue box, under **Options**, instead of the Bonferroni multiple comparisons procedure, a better one is offered, called **Tukey HSD**.

Chapter Exercises

Ⓣ indicates data is available in a saved file at the Companion Website at www.pearsoncanada.ca/deveaux

13. Cat hair A Toronto veterinarian is interested to see if the amount of hair a cat will shed is related to the type of food consumed by the cat. The vet randomly divides 60 cats from a kennel into three groups and gives one group wet food and another group dry food. The third group acts as a control and receives no food that day. She measures the weight of cat hair remaining in the cage at the end of the day, and performs an ANOVA, obtaining an *F*-ratio of 2.58.
a) What are the null and alternative hypotheses?

b) How many degrees of freedom does the treatment sum of squares have? How about the error sum of squares?
c) Assuming that the conditions are satisfied, what is the P-value? What would you conclude?
d) Do you have any concerns about this experiment design?

14. Consulting A researcher from another department wants to know if he used the correct analysis. He collected resting heart rates and diastolic blood pressure for

100 students from five different majors, and performed an ANOVA. Since heart rate and diastolic blood pressure have a similar mean and variance, he decided to use a Bonferroni test to compare the heart rate of Biology majors to the blood pressure of Physical Education majors, and found no significant difference. Do you see any problems with this technique?

15. Yogourt An experiment to determine the effect of several methods of preparing cultures for use in commercial yogourt was conducted by a food science research group. Three batches of yogourt were prepared using each of three methods: traditional, ultrafiltration, and reverse osmosis. A trained expert then tasted each of the nine samples, presented in random order, and judged them on a scale from 1 to 10. A partially complete analysis of variance table of the data is shown below.

An incomplete ANOVA Table for the Yogourt Data

Source	DF	Sum of Squares	Mean Square	F-ratio	P-value
Treatment		17.300			
Residual		0.460			
Total		17.769			

a) Calculate the mean square of the treatments and the mean square of the error.
b) Form the F-statistic by dividing the two mean squares.
c) The P-value of this F-statistic turns out to be 0.000017. What does this say about the null hypothesis of equal means?
d) What assumptions have you made in order to answer part c)?
e) What would you like to see in order to justify the conclusions of the F-test?
f) What is the average size of the error standard deviation in the judge's assessment?

16. Smokestack scrubbers Particulate matter is a serious form of air pollution often arising from industrial production. One way to reduce the pollution is to put a filter, or scrubber, at the end of the smokestack to trap the particulates. An experiment to determine which smokestack scrubber design is best was run by placing four scrubbers of different designs on an industrial stack in random order. Each scrubber was tested five times. For each run, the same material was produced, and the particulate emissions coming out of the scrubber were measured (in parts per billion). A partially complete analysis of variance table of the data is shown below.

An incomplete ANOVA Table for the Smokestack Data

Source	DF	Sum of Squares	Mean Square	F-ratio	P-value
Treatment		81.2			
Residual		30.8			
Total		112.0			

a) Calculate the mean square of the treatments and the mean square of the error.
b) Form the F-statistic by dividing the two mean squares.
c) The P-value of this F-statistic turns out to be 0.00000949. What does this say about the null hypothesis of equal means?
d) What assumptions have you made in order to answer part c)?
e) What would you like to see in order to justify the conclusions of the F-test?
f) What is the average size of the error standard deviation in particulate emissions?

T 17. Eggs A student wants to investigate the effects of real versus substitute eggs on his favourite brownie recipe. He enlists the help of 10 friends and asks them to rank each of eight batches on a scale from 1 to 10. Four of the batches were made with real eggs, four with substitute eggs. The judges tasted the brownies in random order. Here is a boxplot of the data:

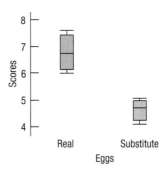

Analysis of Variance

Source	DF	Sum of Squares	Mean Square	F-ratio	P-value
Eggs	1	9.010013	9.01001	31.0712	0.0014
Error	6	1.739875	0.28998		
Total	7	10.749883			

The mean score for the real eggs was 6.78 with a standard deviation of 0.651. The mean score for the substitute eggs was 4.66 with a standard deviation of 0.395.
a) What are the null and alternative hypotheses?
b) What do you conclude from the ANOVA table?
c) Do the assumptions for the test seem to be reasonable?
d) Perform a two-sample pooled t-test of the difference. What P-value do you get? Show that the square of the t-statistic is the same (to rounding error) as the F-ratio.

T 18. Auto noise filters In a statement to a U.S. Senate Public Works Committee, a senior executive of Texaco, Inc., cited a study on the effectiveness of auto filters on reducing noise. Because of concerns about performance, two types of filters were studied, a standard silencer and a new device developed by the Associated Octel Company. Here are the boxplots from the data on noise reduction (in decibels) of the two filters. Type 1 = standard; Type 2 = Octel.

Analysis of Variance

Source	DF	Sum of Squares	Mean Square	F-ratio	P-value
Type	1	6.31	6.31	0.7673	0.3874
Error	33	271.47	8.22		
Total	34	2.77			

Level	n	Mean	StdDev
Standard	18	81.5556	3.22166
Octel	17	80.7059	2.43708

a) What are the null and alternative hypotheses?
b) What do you conclude from the ANOVA table?
c) Do the assumptions for the test seem to be reasonable?
d) Perform a two-sample pooled *t*-test of the difference. What P-value do you get? Show that the square of the *t*-statistic is the same (to rounding error) as the *F*-ratio.

T 19. School system A school district superintendent wants to test a new method of teaching arithmetic in Grade 4 at his 15 schools. He plans to select eight students from each school to take part in the experiment, but to make sure they are roughly of the same ability, he first gives a test to all 120 students. Here are the scores of the test by school:

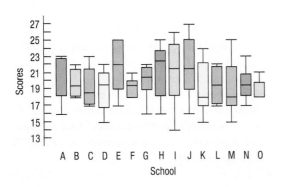

The ANOVA table shows:

Analysis of Variance

Source	DF	Sum of Squares	Mean Square	F-ratio	P-value
School	14	108.800	7.7714	1.0735	0.3899
Error	105	760.125	7.2392		
Total	119	868.925			

a) What are the null and alternative hypotheses?
b) What does the ANOVA table say about the null hypothesis? (Be sure to report this in terms of scores and schools.)
c) An intern reports that he has done *t*-tests of every school against every other school and finds that several of the schools seem to differ in mean score. Does this match your finding in part b)? Give an explanation for the difference, if any, between the two results.

T 20. Fertilizers A biology student is studying the effect of 10 different fertilizers on the growth of mung bean sprouts. She sprouts 12 beans in each of 10 different petri dishes, and adds the same amount of fertilizer to each dish. After one week she measures the heights of the 120 sprouts in millimetres. Here are boxplots and an ANOVA table of the data:

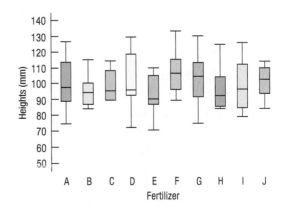

Analysis of Variance

Source	DF	Sum of Squares	Mean Square	F-ratio	P-value
Fertilizer	9	2073.708	230.412	1.1882	0.3097
Error	110	21 331.083	193.919		
Total	119	23 404.791			

a) What are the null and alternative hypotheses?
b) What does the ANOVA table say about the null hypothesis? (Be sure to report this in terms of heights and fertilizers.)
c) Her lab partner looks at the same data. He says that he did *t*-tests of every fertilizer against every other

fertilizer and finds that several of the fertilizers seem to have significantly higher mean heights. Does this match your finding in part b)? Give an explanation for the difference, if any, between the two results.

T 21. Cereals Grocery stores often place similar types of cereal on the same shelf. The same data set we met in the Step-By-Step of Chapter 7 keeps track of the shelf as well as the sugar, sodium, and calorie content of 77 cereals. Does sugar content vary by shelf? Here's a boxplot and an ANOVA table for the 77 cereals:

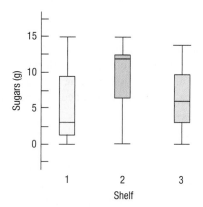

Analysis of Variance

Source	DF	Sum of Squares	Mean Square	F-ratio	P-value
Shelf	2	248.4079	124.204	7.3345	0.0012
Error	74	1253.1246	16.934		
Total	76	1501.5325			

Level	n	Mean	StdDev
1	20	4.80000	4.57223
2	21	9.61905	4.12888
3	36	6.52778	3.83582

a) What are the null and alternative hypotheses?
b) What does the ANOVA table say about the null hypothesis? (Be sure to report this in terms of sugars and shelves.)
c) Can we conclude that cereals on shelf 2 have a higher mean sugar content than cereals on shelf 3? Can we conclude that cereals on shelf 2 have a higher mean sugar content than cereals on shelf 1? What *can* we conclude?
d) To check for significant differences between the shelf means, we can use a Bonferroni test, whose results are shown below. For each pair of shelves, the difference is shown along with its standard error and significance level. What does it say about the questions in part c)?

Dependent Variable: SUGARS

(I) SHELF	(J) SHELF	Mean Difference (I-J)	Std. Error	P-value	95% Confidence Interval Lower Bound	95% Confidence Interval Upper Bound
Bonferroni						
1	2	−4.819(*)	1.2857	0.001	−7.969	−1.670
	3	−1.728	1.1476	0.409	−4.539	1.084
2	1	4.819(*)	1.2857	0.001	1.670	7.969
	3	3.091(*)	1.1299	0.023	0.323	5.859
3	1	1.728	1.1476	0.409	−1.084	4.539
	2	−3.091(*)	1.1299	0.023	−5.859	−0.323

*The mean difference is significant at the 0.05 level.

T 22. Cereals redux We also have data on the protein content of cereals by their shelf number. Here are the boxplot and ANOVA table:

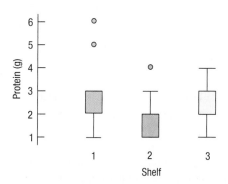

Analysis of Variance

Source	DF	Sum of Squares	Mean Square	F-ratio	P-value
Shelf	2	12.4258	6.2129	5.8445	0.0044
Error	74	78.6650	1.0630		
Total	76	91.0909			

Means and Std Deviations

Level	n	Mean	StdDev
1	20	2.65000	1.46089
2	21	1.90476	0.99523
3	36	2.86111	0.72320

a) What are the null and alternative hypotheses?
b) What does the ANOVA table say about the null hypothesis? (Be sure to report this in terms of protein content and shelves.)
c) Can we conclude that cereals on shelf 2 have a lower mean protein content than cereals on shelf 3? Can we conclude that cereals on shelf 2 have a lower mean protein content than cereals on shelf 1? What *can* we conclude?

d) To check for significant differences between the shelf means, we can use a Bonferroni test, whose results are shown below. For each pair of shelves, the difference is shown along with its standard error and significance level. What does it say about the questions in part c)?

Dependent Variable: PROTEIN

	(I) SHELF	(J) SHELF	Mean Difference (I-J)	Std. Error	P-value	95% Confidence Interval Lower Bound	Upper Bound
Bonferroni	1	2	0.75	0.322	0.070	−0.04	1.53
		3	−0.21	0.288	1.000	−0.92	0.49
	2	1	−0.75	0.322	0.070	−1.53	0.04
		3	−0.96(*)	0.283	0.004	−1.65	−0.26
	3	1	0.21	0.288	1.000	−0.49	0.92
		2	0.96(*)	0.283	0.004	0.26	1.65

*The mean difference is significant at the 0.05 level.

T 23. **Downloading** To see how much of a difference time of day made on the speed at which he could download files, a second-year university student performed an experiment. He placed a file on a remote server and then proceeded to download it at three different time periods of the day. He downloaded the file 48 times in all, 16 times at each time of day, and recorded the time in seconds that the download took.

Early (7 a.m.) Time (sec)	Evening (5 p.m.) Time (sec)	Late night (12 a.m.) Time (sec)
68	299	216
138	367	175
75	331	274
186	257	171
68	260	187
217	269	213
93	252	221
90	200	139
71	296	226
154	204	128
166	190	236
130	240	128
72	350	217
81	256	196
76	282	201
129	320	161

a) State the null and alternative hypotheses, being careful to talk about download time and time of day as well as parameters.
b) Perform an ANOVA on these data. What can you conclude?
c) Check the assumptions and conditions for an ANOVA. Do you have any concerns about the experimental design or the analysis?
d) (Optional) Perform a multiple comparisons test to determine which times of day differ in terms of mean download time.

T 24. **Analgesics** A pharmaceutical company tested three formulations of a pain relief medicine for migraine headache sufferers. For the experiment, 27 volunteers were selected and 9 were randomly assigned to one of three drug formulations. The subjects were instructed to take the drug during their next migraine headache episode and to report their pain on a scale of 1 = no pain to 10 = extreme pain 30 minutes after taking the drug.

Drug	Pain	Drug	Pain	Drug	Pain
A	4	B	6	C	6
A	5	B	8	C	7
A	4	B	4	C	6
A	3	B	5	C	6
A	2	B	4	C	7
A	4	B	6	C	5
A	3	B	5	C	6
A	4	B	8	C	5
A	4	B	6	C	5

a) State the null and alternative hypotheses, being careful to talk about drug and drug pain as well as parameters.
b) Perform an ANOVA on these data. What can you conclude?
c) Check the assumptions and conditions for an ANOVA. Do you have any concerns about the experimental design or the analysis?
d) (Optional) Perform a multiple comparisons test to determine which drugs differ in terms of mean pain level reported.

T 25. **Popsicle sticks** In a bridge-building competition, popsicle sticks are often used to construct joints by gluing two or more sticks together. A high school class performed an experiment to test several different factors, including: four types of *Glue* (hot, white, carpenter's, Gorilla); two types of *Clamping* (clamped, not clamped); three types of *Overlap* distance (S, M, L);

two types of *Colour* (C, B); and three numbers of *Sticks* in the joint (2, 3, 4).

Twelve batches of roughly 30 sticks were tested at various combinations of the factors. One research question of interest was whether coloured sticks broke at a different stress than regular beige sticks. Some analysis is shown below.

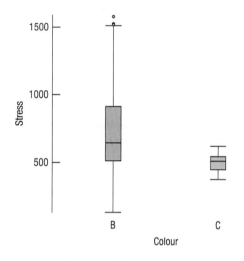

Analysis of Variance

Source	DF	Sum of Squares	Mean Square	F-ratio	P-value
Colour	1	1 303 181	1 303 181	18.01	<0.0001
Error	350	25 320 104	72 343		
Total	351	26 623 285			

Level	n	Mean	StdDev
B	323	722.3	279.72
C	29	501.0	66.79

a) How many factors are present in this experiment? How many possible treatments are there if the design were to be completely balanced?
b) Does it appear from the boxplots that coloured sticks break at a lower stress than beige sticks? Does the ANOVA output support this conclusion?
c) Do you have any concerns about this analysis? What are they?
d) Do you see any problems with the study design? What could you have done differently when designing the study to best answer the research question?

T 26. Popsicle sticks, redux Consider again the popsicle stick data from the previous question. Suppose we want to see which glue type is strongest (the original purpose of the experiment, as we had no idea). Some boxplots are shown below, as well as the ANOVA output.

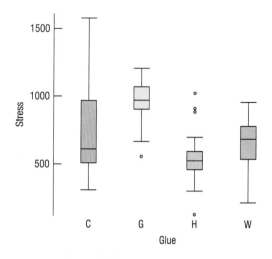

Analysis of Variance

Source	DF	Sum of Squares	Mean Square	F-ratio	P-value
Glue	3	3 686 845	1 228 948	18.65	<0.0001
Error	348	22 936 439	65 909		
Total	351	26 623 284			

Level	n	Mean	StdDev
C	209	737.2	301.53
G	25	957.0	146.53
H	59	535.7	146.85
W	59	648.1	197.35

The global test is very significant, so there are some differences among the glues. It looks like Gorilla may be the best, and Hot the worst. Carpenter's has the second highest mean, but is it significantly worse than Gorilla? Why not do a follow-up test to see if this difference is statistically significant. Perform a two-sample *t*-test. What do you conclude? Do you have any concerns about this method?

27. Popsicle sticks, another taste The students decide to compare the breaking force for 28 joints made with each glue type. One joint could not be tested and was removed. An incomplete ANOVA table is shown below.

An incomplete ANOVA table for the Popsicle stick data

Source	DF	SS	MS	F	P
Glue	3	724.32	241.44		
Error					
Total	110	1237.41			

a) What are the null and alternative hypotheses?
b) How many error degrees of freedom does this experiment have?

c) Fill in the sum of squares and the mean square for the error row.
d) What do you conclude, in the context of this study? What else would you like to see?

28. **Popsicle sticks, dessert** A residual plot and normal probability plot for the popsicle stick experiment is shown below. Comment on the following aspects of the graphs:
a) Equal variance assumption
b) Outliers
c) Normality of residuals
d) Do you feel justified performing inference in this case?

 Just Checking ANSWERS

1. The null hypothesis is that the mean flight distance for all four designs is the same.

2. Yes, it looks as if the variation between the means is greater than the variation within each boxplot.

3. Yes, the *F*-test rejects the null hypothesis with a P-value < 0.0001.

4. No. The alternative hypothesis is that *at least* one mean is different from the other three. Rejecting the null hypothesis does not imply that all four means are different.

MathXL	Go to MathXL at **www.mathxl.com** or MyLab™ Statistics at **www.mystatlab.com**. You can practise exercises
MyLab Statistics	for this chapter as often as you want. The guided solutions will help you find answers step by step. You'll find a personalized study plan available to you too!

Multifactor Analysis of Variance

LEARNING OBJECTIVES

1 Understand the advantages of an experiment in two factors

2 Know how to set up an additive model in two factors

3 Interpret the main effects

4 Know how to add interaction to the model

5 Interpret a two-way ANOVA with interaction

6 Check the appropriate assumptions and conditions

Connections

When production problems forced a computer manufacturer to halt operations, teams of engineers, scientists, production workers, and managers were formed to solve the problem. They found themselves over-whelmed with many potential causes. But experiments that test one factor at a time would have been inefficient. And they might have missed important combinations of factors. In this chapter, we'll see how to look at two factors at a time, and we'll discover why two-factor designs are so much more interesting than one-factor experiments.

We first discussed designing experiments in Chapter 10. In Chapter 24, we saw that the one-way ANOVA generalized the pooled *t*-test to more than two levels of one factor. Now we'll add a second factor.

Without randomization, your inferences haven't got a chance.

—*George Cobb*

How accurately can you throw a dart? It probably depends on which hand you use and on how far from the target you're standing. A student designed an experiment to test the effects of both factors. He used three levels of the factor *Distance: Near, Middle,* and *Far*—and the two natural levels *Left* and *Right* for *Hand.* (He was right-handed.) In random order, he threw six darts at a dartboard under each of the six treatment conditions, and measured the distance each dart landed from the bull's-eye in inches.[1]

It makes sense that dart throws would be more accurate from nearer distances, especially when using his regular throwing hand. Side-by-side boxplots (Figure 25.1) of the distance from the bull's-eye by each factor seem to support our intuition:

Figure 25.1

Boxplots of *Accuracy* (distance in inches from the bull's-eye) plotted for each experimental factor, *Distance* and *Hand*.

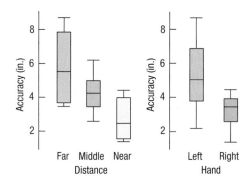

25.1 A Two-Factor ANOVA Model

It appears that both the distance and the hand he used affected the accuracy of his throw, and by about the same amount. How can we measure the effect of each variable? And how can we test whether the observed differences are larger than we'd expect just from sampling variation?

Two Factors at Once?!

Unlike the experiments we've seen so far, this experiment has not one but two factors. You might expect this to confuse things, but the surprising fact is that it actually improves the experiment and its analysis. It also lets us see something we'd never be able to see if we ran two different one-factor experiments.

---

[1]He called the response variable *Accuracy,* although it measures inches away from the target. That means that bigger values correspond to *less* accuracy, not more.

Who	Attempts at hitting a bull's-eye on a dartboard
What	Distance from the centre of the target
Units	Inches
How	Two-factor experiment, randomized order of runs

True, with two factors, we have two hypothesis tests. Each of those hypotheses asks whether the mean of the response variable is the same for each of the factor's levels. In comparing means, we know that any excess variation just gets in the way. Back in Chapter 10, we considered ways to remove or avoid extra variation in designing experiments.

A two-factor experiment does just that. If both factors do in fact have an effect on the response (as we'd expect *Hand* choice and *Distance* to do for dart throwing accuracy), then removing the effects of one factor should make it easier to see and assess the effects of the other. We'll have removed variability due to an identifiable cause.

For Example HOW FAR CAN A PAPER AIRPLANE FLY?

Some curious students set up an experiment to investigate factors that determine how far a paper airplane will fly. Among the factors are whether the plane is constructed of copier paper or heavier construction paper, whether it is a "dart" or a glider, and whether the ends of the wings are folded up into "flaps." We'll combine the paper and shape factors into a single *Design* factor with four levels: *Dart/Copier*, *Dart/Construction*, *Glider/Copier*, and *Glider/Construction*. The second factor, *Flaps*, has two levels: *Flaps* and *None*.

Four planes were constructed of each of the four designs. For each design, two planes were given flaps and two were not. Each plane was flown 4 times and the flight distances were carefully measured. (Data in **Paper_planes**)

QUESTION: What are the details of the experiment design?

ANSWER: We have two factors: *Design* and *Flaps*. *Design* has 4 levels, Dart/Copier, Dart/Construction, Glider/Copier and Glider/Construction. *Flaps* has 2: Flaps and None. There are thus 4 × 2 = 8 treatment groups. Each treatment group has 2 planes and each plane was flown 4 times for a total of 8 observations. The response variable is the distance flown.

A Simple Additive Model

The model for one-way ANOVA represented each observation as the sum of simple components. It broke each observation into the sum of three effects: the grand mean, the treatment effect, and an error:

$$y_{ij} = \mu + \tau_j + \varepsilon_{ij}$$

Now we have two factors, *Hand* and *Distance*. So each treatment is a combination of a *Hand* assignment and a *Distance*. For example, throwing with the left hand from a near distance would be one treatment. Our **two-way ANOVA model** will reflect the effects of both factors. We've already used τ for the treatment effects for the first factor, so for the second factor, we'll denote the effects by γ. Now, for each observation, we write:

$$y_{ijk} = \mu + \tau_j + \gamma_k + \varepsilon_{ijk}$$

Here we see the i-th observation at level j of factor A and level k of factor B. Why the i subscript? We're told that the student recorded six trials at each treatment, so the subscript i denotes the different observations at each combination of the two factors. The error has three subscripts because we associate a possibly different error with each observation. (If there were only one observation at each treatment, we wouldn't need the extra subscript.)

When you write out a model, you'll usually find it clearer to name the factors rather than using Greek letters. So the model will look like this:

$$y_{ijk} = \mu + \textit{Hand effect}_j + \textit{Distance effect}_k + \textit{Error}_{ijk}$$

To estimate the effect due to each level of a factor, we take the *difference* between the mean response at that level of the factor and the grand mean, $\bar{\bar{y}}$. So, to find the effect of throwing with the left hand, we'd find the mean accuracy of *all* runs using the left hand and subtract the grand average from that. This effect is also referred to as the factor's *main effect* since we are averaging over all levels of the other factor.

In general, we'd like to know whether the mean dart accuracy changes with changes to the levels of either factor. The model helps us to think about this question. Our null hypothesis on each factor is that the effects of that factor are all zero. So we can write two null hypotheses we'd like to test as

$$H_0: \tau_1 = \tau_2 \quad \text{and} \quad H_0: \gamma_1 = \gamma_2 = \gamma_3$$

where the τ's (taus) denote the effects due to the two levels of *Hand* and the γ's (gammas) denote the effects due to the three levels of *Distance*. We omit the superfluous '= 0' in each hypothesis.

The alternative hypotheses are that the factor's effects are not all equal. The details of the analysis are pretty much the same as for a one-factor ANOVA. We want to compare the differences in mean across the levels of the factor with the underlying variability within the treatments. But now with two factors, that underlying variability has the effect of both factors removed from it. So the error term holds the variability that's left after removing the effects of both factors. Here's the ANOVA table for the dart experiment:

Analysis of Variance Table for *Accuracy*

Source	DF	Sum of Squares	Mean Square	F-Ratio	P-value
Distance	2	51.041	25.521	28.561	<0.0001
Hand	1	36.693	36.693	44.415	<0.0001
Error	32	28.595	0.894		
Total	35	116.329			

This analysis splits up the total variation into three sources: the variation due to changes in *Distance*, the variation due to changes in *Hand*, and the error variation. The square root of the Error Mean Square, $\sqrt{0.894} = 0.946$ inches, estimates the standard deviation we'd expect if we made repeated observations at any treatment condition. Both *F*-ratios are large. Looking back at the boxplots, we shouldn't be surprised. It was clear that there were differences in accuracy among *near*, *middle*, and *far* even without accounting for the fact that each level contained both hands. The *F*-ratio for *Distance* is 28.56. The probability of an *F*-ratio that big occurring by chance is very small (< 0.0001). So we can reject the hypothesis that the mean accuracy at *near*, *middle*, and *far* distances is the same.

The means of *left* and *right* don't look the same either. The *F*-ratio of 44.42 with a P-value of < 0.0001 leads us to reject the null hypothesis about *Hand* effects as well.

Of course, there are assumptions and conditions to check. The good news is that most are the same as for one-way ANOVA.

Section 25.1 Exercises

1. **Student TV time** In the previous chapter we considered TV watching by male and female student athletes. In that example, we categorized the students into four groups, but now we have seen that these data could be analyzed with two factors, *Sex* and *Athlete*. Write the ANOVA model for the design in this form.

2. **Cookies** A student performed an experiment to compare chocolate chip cookie recipes. He baked batches of cookies with different amounts of *Sugar*: 0.5, 0.375, and 0.25 cups, and with different kinds of *Chips*: milk, semisweet, and dark chocolate. Cookie quality was scored on a 0 to 10 scale by other students in the class, with the mean score being the result for each recipe. Write the ANOVA model for the design of this experiment.

25.2 Assumptions and Conditions

Even before checking any assumptions, you should have already plotted the data. Box-plots give a first look that draws our attention to the aspects tested by the ANOVA.

Plot the Data . . .

You could start by examining side-by-side boxplots of the data across levels of *each factor* as we did earlier, looking for outliers within any of the groups and correcting them if they are errors, or consider omitting them if we can't correct them.[2] (There don't seem to be any.) We'd like to get an idea of whether the groups have similar spreads (as we'll want) and whether the centres seem to be alike (as the null hypotheses claim) or different.

The problem of just looking at the boxplots is that the responses at each level of one factor contain *all levels of the other factor*. For example, the responses for the *left Hand* were measured at all three distances. As a result, the boxplots show more variability than they need to. Our model will deal with both factors so it will take care of this extra variability, but the simple side-by-side boxplots can't do that.

On the other hand, if you do see a difference here, that's a difference *in spite of this extra variation*, so there's a good chance that the *F*-test will reject the null hypothesis for that factor. Just be careful not to prematurely rule out the possibility of differences that aren't apparent.

A better alternative might be to make boxplots for each factor level *after removing the effects of the other factor*. How would you do that? You could compute a one-way ANOVA on one factor and find the residuals. Then make boxplots of those *residuals* for each level of the other factor. We might call such a display a **partial boxplot**.[3] We will also add the overall mean to all of these residuals; then we can call them 'adjusted responses,' being responses for levels of one factor, while effectively *holding constant* the other factor. By removing the variability due to the other factor, you should see differences in the responses among the levels of the factor you are graphing more readily. For example, Figure 25.2 shows the boxplots of dart toss *Accuracy* plotted for the two levels of *Hand*, both in the original data and after removing the effects of *Distance*:

Figure 25.2

Ordinary boxplots of *Accuracy* by *Hand*, on the left, show the effect of changing *Hand* less clearly than the corresponding partial boxplots on the right, which show the effects of *Hand* after the effect of *Distance* has been removed. The effect of changing hands is much easier to see without the unwanted variation caused by changing the other factor, *Distance*.

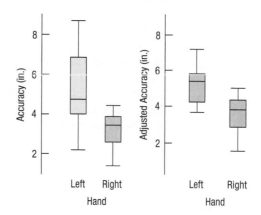

[2]Unlike regression coefficients, which can be changed drastically by outliers, ANOVA tests tend to fail on the "safe side." That is, outliers generally inflate the Error Mean Square, which reduces the *F*-statistics and increases the P-values. That makes it harder to reject the null hypotheses, so we are more likely to make Type II errors. That's generally thought to be safer than increasing the Type I errors. For example, we are not likely to be mistakenly led to approve a drug that isn't really effective. But it also means that we might fail to approve a drug that really is effective. Generally, it is best to remove outliers and report them separately.

[3]This is not a standard term. You are not likely to find it in a statistics package. But it is a natural term because we've removed *part* of the variability.

Removing the variation due to *Distance* makes the differences between the two hands in the partial boxplots much clearer. We'll want to do this for each factor, removing the effects of the other factor by computing the one-way ANOVA and finding the residuals. Figure 25.3 shows the usual boxplots for *Distance* and the partial boxplots for *Distance*, removing the variability due to *Hand:*

Figure 25.3

Partial boxplots after removing the effect of *Hand* (on the right) show the differences in *Accuracy* among the three *Distances* more clearly than the boxplots of *Accuracy* by *Distance* (on the left).

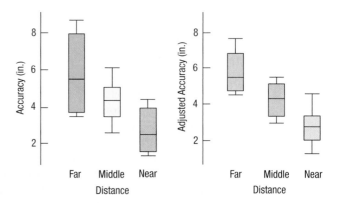

We'll use the partial boxplots for diagnostic purposes here, just as we used regular boxplots in a one-way analysis. If the partial boxplots were all skewed in the same direction, or if their spreads changed systematically from level to level across a factor, we would have considered re-expressing the response variable to make them more symmetric or more similar in spread (often achieving both). Re-expression would be likely to make the analysis more powerful and more correct.

Additivity Assumption

Each observation is assigned a treatment that combines two factor levels, one from each factor. Our model assumes that we can just *add* the effects of these two factor levels together:

$$y_{ijk} = \mu + \textit{Hand effect}_j + \textit{Distance effect}_k + \textit{Error}_{ijk}$$

That's a nice simple model, but it may not be a good one for our data. Just like linearity, **additivity** is an assumption. Like all assumptions, it helps us by making the model simpler, but it may be wrong for our data. Of course, we can't know for sure, but we can check the **Additive Enough Condition**. You won't be surprised that we check it with displays.

For the effects of *Hand* and *Distance* to be additive, changing hands must make the *same* difference in accuracy no matter what distance you throw from (and *vice versa*). Is this reasonable for this experiment? Alternatively, we could ask whether moving away from the target makes a greater difference to accuracy when throwing with the weaker hand. That's a *conditional* question, so we look at how *Accuracy* changes by *Distance* for each *Hand* separately (Figure 25.4).

Figure 25.4

Boxplots of *Accuracy* by *Distance* conditional on each *Hand*. Changing the *Distance* seems to have a greater effect on left *Hand* accuracies.

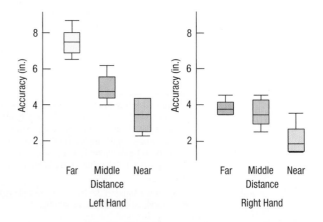

The changes in accuracy due to *Distance* do look larger for the *left* hand, especially at the *far* distance.

When the effects of one factor *change* for different levels of another factor, we say that there's an **interaction**. To show the interaction, we often plot the treatment means in a single display, called an **interaction plot** (also known as the *means plot*). Figure 25.5 shows the average of the observations at *each* level of one factor broken up by the levels of the other factor.

Figure 25.5

An interaction plot showing the mean *Accuracy* at the six combinations of *Distance* and *Hand*. Connecting lines for right and left hands shows the greater effect of the far *Distance* on left-hand *Accuracy* than on right-hand *Accuracy*.

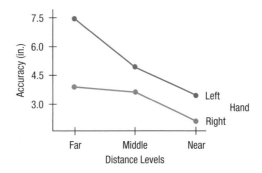

Here it's easier to see that throwing from the Far distance has more of an effect for the Left hand. If the effect of *Distance* were constant, the lines in this plot should be pretty much parallel. If they aren't, that's evidence that the effects are not additive. But how parallel is enough? Are these lines sufficiently far from parallel for us to doubt whether our data satisfy the **Additive Enough Condition**? A graphical check like this is usually good enough to proceed. Later in the chapter, we'll see a way to include a term for non-additivity in our model and test for it.

Independence Assumptions

The **Independence Assumptions** for the two-factor model are the same as in one-way ANOVA. That is, the observed responses *within each treatment group* must be independent of each other, and the groups must be independent. As usual, no test can verify these assumptions. Ideally, the data should be generated by a randomized comparative experiment or be drawn independently, at random from the population of interest. Without randomization of some sort, our P-values are meaningless. We have no sampling distribution models and no valid inference.

Check the **Randomization Condition**. Were the data collected with suitable randomization? For surveys, are the data for each group a representative random sample of the population of interest? For experiments, was the experiment properly randomized? (But note that some more complicated but well randomized designs may require a more complex ANOVA model.)

We were told that the order of treatment conditions in the dart experiment was randomized. This is especially crucial for an experiment like this because of the possibility of either learning or tiring. Your accuracy might increase after a few throws as you warm up, but after an hour or two, your concentration—and your aim—might start to wander.

If we are satisfied with the Additivity and Independence Assumptions, then we can look at the ANOVA table. For the dart data, we might have some doubts about the additivity, but we'll go ahead for now.

Source	DF	Sum of Squares	Mean Square	F-ratio	P-value
Distance	2	51.041	25.521	28.561	<0.0001
Hand	1	36.693	36.693	44.415	<0.0001
Error	32	28.595	0.894		
Total	35	116.329			

These may look like small P-values, but inference requires additional assumptions. Before we believe the *F*-tests and use the P-values to make decisions, we have other assumptions to think about.

Equal Variance Assumption

Like the one-way ANOVA, the two-factor ANOVA requires that the variances of all treatment groups be equal. It's the residuals after fitting *both* effects that we'll pool for the Error Mean Square, so it's their variance whose stability we care about most. We check this assumption by checking the **Similar Variance Condition**.

In one-way ANOVA, we looked at the side-by-side boxplots of the groups to see whether they had roughly the same spread. But here, we need to check for equal spread across all treatment groups. The easiest way to check this is to plot the residuals from the two-way ANOVA model.

■ Look at the residuals plotted against the predicted values. If the variance is constant, the plot should be patternless. If the plot thickens (to one side or the other), it's a sign that the variance is changing *systematically*. Then it's usually a good idea to consider re-expressing the response variable.

Figure 25.6

A scatterplot of residuals versus predicted values from the two-way ANOVA model for *Accuracy* shows a U-shaped pattern. This suggests that some condition is violated.

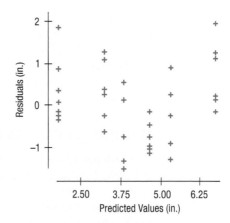

Figure 25.6 doesn't seem to show changing variance, but it's certainly not patternless. Residuals at both ends of the predicted values are higher than those in the middle. This could mean that something is missing from our model. Could that nagging suspicion we had about the additivity be resurfacing? Stay tuned.

■ We can also plot the residuals grouped by each factor (or by treatment combination, but the smaller the collection of residuals, the less we can learn). Figure 25.7 is a boxplot of residuals by factor level for each factor in the dart experiment:

Figure 25.7

Boxplots of the residuals from the two-way ANOVA of the dart accuracy experiment show roughly equal variability when plotted for each factor.

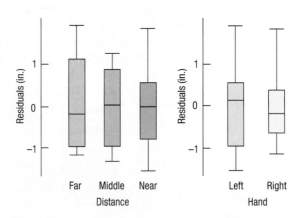

Remember, we can't use the original boxplots to check changing spread. From those, it seemed that the spread of the *near* level was smaller than that of the *far* level. But much of that difference was due to changes in the way the factor *Hand* behaves at different distances rather than a change in the underlying variance. By contrast, these boxplots of residuals look similar enough in spread.

Normal Error Assumption

As with one-way ANOVA, the *F*-tests require that the underlying errors follow a Normal model. We'll check a corresponding **Nearly Normal Condition** with a Normal probability plot or histogram of the residuals (Figure 25.8).

Figure 25.8

A Normal probability plot of the residuals from the two-way ANOVA model for dart accuracy seems reasonably straight.

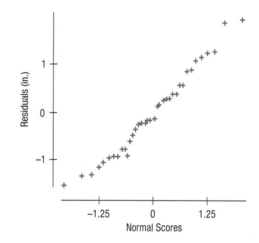

This condition appears to be satisfied.

What Have We Learned so Far?

The ANOVA table confirmed what we saw in the boxplots. Both *Distance* and *Hand* affect accuracy of throwing a dart, at least at the levels chosen in this experiment. The small P-values give us confidence that the differences we see are due to the factors and not to chance. We did worry that the Additive Enough Condition might not be satisfied, and there seems to be something amiss in the plot of residuals versus predicted values. Before we examine that more, let's look at a different example with two factors, Step-By-Step.

Step-by-Step Example   TWO-FACTOR ANALYSIS OF VARIANCE

Another student, who prefers the great outdoors to damp pub basements, wonders whether leaving her tennis balls in the trunk of her car for several days after the can is opened affects their performance, especially in the winter when it can get quite cold. She also wonders if the more expensive brand might retain its bounce better. To investigate, she performed a two-factor experiment on *Brand* and *Temperature*, using two *Brands* and three levels of *Temperature*. She bounced three balls under each of the six treatment conditions by first randomly selecting a brand and, for that ball, randomly selecting whether to leave it at room temperature or to put it in the refrigerator or the freezer. After 8 hours, she dropped the balls from a height of 1 metre to a concrete floor, recording the *Height* of the bounce in centimetres. (Data in **TennisBalls**)

This is a completely randomized design in two factors with three observations (replications) per treatment. The factor *Brand* has two levels: *Premium* and *Standard*. The factor *Temperature* has three levels: *Room, Fridge,* and *Freezer*. The null hypotheses are that *Brand* has no effect and that *Temperature* has no effect.

Brand	Temperature	Bounce Height (cm)
Standard	Freezer	37
Standard	Fridge	59
Standard	Room	59
Premium	Freezer	45
Premium	Fridge	60
Premium	Room	63
Standard	Freezer	37
Standard	Fridge	58
Standard	Room	60
Premium	Freezer	39
Premium	Fridge	64
Premium	Room	62
Standard	Freezer	41
Standard	Fridge	60
Standard	Room	61
Premium	Freezer	37
Premium	Fridge	63
Premium	Room	61

Who   Tennis balls of two *Brands* allocated to three *Temperatures*.

What   *Height* of the ball after it bounces from an initial height of 100 centimetres

Units   Centimetres

How   Two-factor experiment, randomized order of runs

THINK ➡ State what we want to know and the null hypotheses we wish to test. For two-factor ANOVA, the null hypotheses are that each factor has the same mean at each of its levels. The alternatives are that the means are not all equal.

I want to know whether the storage *Temperature* affects the mean bounce *Height* of tennis balls. I also want to know whether the two *Brands* have different mean bounce *Heights*. Writing τ_j for the effect of *Brand*$_j$, the null hypothesis is that the two *Brand* effects are the same:

$$H_0: \tau_P = \tau_S$$

Writing γ_k for the effect of *Temperature* level k, then the other null hypothesis is

$$H_0: \gamma_{room} = \gamma_{fridge} = \gamma_{freezer}$$

The alternative for the first hypothesis is that the effects of the two *Brands* on the bounce *Height* are different. The alternative for the second is that the effects of the *Temperature* levels are not all equal.

Plot   Examine the side-by-side partial box-plots of the data.

It seems pretty clear that *Temperature* has an effect, but we can't be sure about the effect of *Brand* just from the plots.

✓ **Independence Assumption**: Because the order was randomized, I can assume the observations are independent. However, I might want to check to see if the balls of each brand came from the same box. It might be better to have selected balls of the same brand from different boxes.

✓ **Randomization Condition**: The experimenter performed the trials in random order.

✓ **Additive Enough Condition**: Here is the plot of the mean bounce *Heights* at each treatment condition:

The lines look reasonably parallel. Now, I can fit the additive model in two factors:

$$y_{ijk} = \mu + Brand\ effect_j + Temp\ effect_k + Error_{ijk}.$$

Under these conditions, it's appropriate to fit the two-way Analysis of Variance model.

Plan Think about the assumptions and check the appropriate conditions.

Show the ANOVA table.

			Analysis of Variance for Height		
Source	**DF**	**Sum of Squares**	**Mean Square**	**F-ratio**	**P-value**
Temp	2	1849.33	924.665	209.55	<0.0001
Brand	1	26.8889	26.8889	6.0935	0.0271
Error	14	61.7778	4.41270		
Total	17	1938.00			

The Error Mean Square is 4.41, which means that the common error standard deviation is estimated to be $\sqrt{4.41}$, a bit more than 2 centimetres. This seems like a reasonable accuracy for measuring bounce *Heights* of tennis balls dropped from a height of 1 metre.

Before testing the hypotheses, I need to check some more conditions:

✓ **Similar Variance Condition:** A plot of residuals versus predicted values shows some increased spread on the left, but not a huge difference and with only five points it's hard to say. (Unless the spread in one group is clearly more than twice the spread in another group, our results should still be trustworthy)

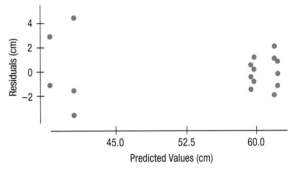

✓ **Nearly Normal Condition, Outlier Condition:** A Normal probability plot of the residuals looks okay, too:

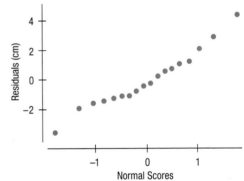

Under these conditions, it is appropriate to perform and interpret the *F*-test.

SHOW ➡ Mechanics Discuss the tests.

Display the main effects for each level of the significant factors. Each represents the average difference in response between that level of the factor and the overall mean for the data. Remember, significance does not guarantee that the differences are meaningful from a practical standpoint.

The *F*-ratio for *Brand* is 6.09, with a P-value of 0.0271. The *F*-ratio for *Temperature* is 209.55, with a P-value less than 0.0001.

The overall mean bounce *Height* is 53.67 cm.

The *Brand* effects are estimated as

Premium	1.22
Standard	−1.22

The *Temperature* effects are estimated as

Room	7.33
Fridge	7.00
Freezer	−14.33

TELL ➡ **Interpretation** Tell what the *F*-tests mean.

P-values this small would be very unlikely if the factors had no effect. I conclude that the means of the two *Brand* levels are not equal. I also conclude that the means of the three *Temperature* levels are not all equal.

Overall, the premium brand bounces about 2.44 cm (1.22 − (−1.22)) higher than the standard brand, averaging across the temperature ranges. Generally, there seems to be little difference between the average bounce height for room and fridge temperatures, but tennis balls from the freezer bounce almost 22 cm less, on average.

Conclusions: It appears that the premium brand outperforms the standard brand overall. It looks from the plots like our student needn't be too worried about playing tennis on a cold day, but if the temperature drops below freezing, she should try to warm the tennis balls first.[4]

For Example

RECAP: Two factors were tested in an experiment to investigate the effects of paper airplane design on flight distance.

Below are boxplots of distance by each of the factors:

QUESTION: What can we learn from the boxplots?

ANSWER: It appears that the darts travel farther than the gliders. There does not appear to be much of an effect from using flaps. The spreads of all the groups are comparable. There are two observations plotted as outliers for the glider/copier design, but they don't seem to be very extreme and are probably not a problem.[5]

---

[4]It's quite obvious from the plots and the tables of effects that there's a difference between *Freezer* and the other temperatures, but little difference between *Room* and *Fridge*. A Bonferroni test would confirm this.

[5]But is independence a problem? Note that 2 planes were used for each treatment, with 4 flights per plane. The planes are not identical so there may be some differences on average between them, hence the 8 observations per treatment are not mutually independent, in violation of the independence assumption. It is possible to enter the 'plane' effect into our model, but this is a bit more complicated than we can deal with here.

For Example

RECAP: An experiment on paper airplanes considered two factors: *Design* and *Flaps*.

Here is the two-factor ANOVA table:

Source	DF	Sum of Squares	Mean Square	F-ratio	P-value
Design	3	1549.50	516.499	56.610	<0.0001
Flap	1	27.0140	27.0140	2.9608	0.0905
Error	59	538.303	9.12378		
Total	63	2114.81			

QUESTION: State and test the null hypotheses. What can you conclude?

ANSWER: The first null hypothesis is that the mean distance flown will be the same for each design, ignoring the presence or absence of flaps. That hypothesis is rejected. As we saw in the boxplots, the darts fly farther.

The second null hypothesis is that the mean distance will be the same whether or not flaps are present, averaging over all designs. The P-value for that factor appears to be too large to allow us to safely reject the null hypothesis.

Blocks

Back in Chapter 10, we thought about a *randomized block design.*[6] We imagined comparing two nail polish colours by randomly assigning two fingernails on each hand to each of the nail polish colours. We can view the choice of *Hand* as a factor in this design. We now have a two-factor model, with *Hand* and nail polish *Colour* as factors. The ANOVA looks just the same, but now we're not likely to be interested in testing the *Hand* effect. We *assumed* that a tendency for the polish on her fingers to chip might possibly differ between her two hands—that's why we used a blocked design.

The two-way ANOVA takes advantage of the blocking to remove the unwanted variation due to *Hand*. As in a completely randomized two-factor design, removing the unwanted variation makes the effects of the other factor easier to see.

We wouldn't call a blocked design *completely randomized* because we can't assign her fingernails at random to her hands. We only randomize *within* the blocks when we assign her fingernails randomly to choices of nail polish *Colour*.

*How ANOVA Works—the Gory Details

You'll probably never carry out a two-way ANOVA by hand. But to understand the model it can help to see exactly how it decomposes each observation into components. Our tennis ball model says

$$y_{ijk} = \mu + \textit{Brand effect}_j + \textit{Temp effect}_k + \textit{Error}_{ijk}.$$

To apply this model to data, we need to estimate each of these quantities:

$$y_{ijk} = \bar{\bar{y}} + \widehat{\textit{Brand effect}_j} + \widehat{\textit{Temp effect}_k} + \textit{Residual}_{ijk}.$$

---

[6]You may have noticed in the tennis ball example that balls couldn't be randomly assigned to a brand. *Brand* may be viewed as a blocking factor, and the experiment is a randomized block design. But it doesn't change the analysis at all.

Here are the data again, displayed a little differently:

		Temperature		
		Room	**Fridge**	**Freezer**
Brand	**Premium**	63	60	45
		62	64	39
		61	63	37
	Standard	59	59	37
		60	58	37
		61	60	41

The first term in the model is μ, the overall mean. It's reasonable to estimate that with the grand mean, $\bar{\bar{y}}$, which is just the mean of all the observations. For the tennis balls, the average bounce was 53.67 cm. The grand mean has the same value for every observation, which we can lay out as a table like this:

		Temperature		
		Room	**Fridge**	**Freezer**
Brand	**Premium**	53.67	53.67	53.67
		53.67	53.67	53.67
		53.67	53.67	53.67
	Standard	53.67	53.67	53.67
		53.67	53.67	53.67
		53.67	53.67	53.67

Now, we turn to the *Brand* effects. To estimate the effect of, say, the *premium Brand*, we find the mean of all *premium* balls and subtract the grand mean. That's 54.89 cm − 53.67 cm = 1.22 cm. In other words, the *premium Brand* balls bounced an average of 1.22 cm higher than the overall average. The *standard Brand* balls bounced an average of 52.444 cm, so their effect is 52.444 cm − 53.667 cm = −1.22 cm. Of course, with only two levels, the fact that these two effects are equal and opposite is no coincidence. Effects *must sum to* 0.

We can put the *Brand* effects into a table like this:

		Temperature		
		Room	**Fridge**	**Freezer**
Brand	**Premium**	1.22	1.22	1.22
		1.22	1.22	1.22
		1.22	1.22	1.22
	Standard	−1.22	−1.22	−1.22
		−1.22	−1.22	−1.22
		−1.22	−1.22	−1.22

We can find a *Temperature* effect table in a similar way. We take the treatment average at each level of *Temperature* and subtract the grand mean. Here's the table:

		Temperature		
		Room	Fridge	Freezer
Brand	Premium	7.33	7.00	−14.33
		7.33	7.00	−14.33
		7.33	7.00	−14.33
	Standard	7.33	7.00	−14.33
		7.33	7.00	−14.33
		7.33	7.00	−14.33

We're missing only one table, the *Residuals*. Finding the residuals is easy, too. The additive model represents each observation as the sum of these three components. As always, the residuals are the differences between the observed values and those given by the model. Look at the first ball in the table. It was a *premium Brand* ball at *room Temperature*. From the effects, we'd predict it to bounce 53.67 (grand mean) +1.22 (from being premium) +7.33 (from being at room temperature) = 62.22 cm. Because it actually bounced 63 cm, its residual is 63 − 62.22 = 0.78 cm. If we compute all the residuals this way (or have the computer do it), we'd find the residual table to be

		Temperature		
		Room	Fridge	Freezer
Brand	Premium	0.78	−1.89	4.44
		−0.22	2.11	−1.56
		−1.22	1.11	−3.56
	Standard	−0.78	−0.44	−1.11
		0.22	−1.44	−1.11
		1.22	0.56	2.89

What we have done with all these tables is to decompose each original bounce height into the three estimated components plus a residual:

Bounce Height = mean Bounce Height + Brand effect + Temperature effect + Residual.

These tables actually display the workings of the ANOVA. Each term in the ANOVA model corresponds to one of these tables. Because the overall mean doesn't vary, we usually leave it out of the ANOVA table. Each Sum of Squares term in the ANOVA is simply the sum of the squares of all the values in its table.

The degrees of freedom are easy to find too. Look at the grand mean table. Because it has the same value for all observations, it has 1 degree of freedom. The *Brand* effect table seems to have 2 degrees of freedom, but we know that the two effects always sum to 0. So it really has only 1. In general, the degrees of freedom for a factor is one less than its number of levels. So *Temperature* has 2 degrees of freedom.

The original table has 18 observations and so 18 degrees of freedom. But what about the residuals? Remarkably enough, the degrees of freedom from the tables add. That means that we can find the degrees of freedom for the residuals by subtraction. They must have 18 − (1 + 1 + 2) = 14 degrees of freedom.

The mean squares are just the sum of squares for each effect divided by the degrees of freedom. The *F*-ratios are the ratios of each factor's Mean Square to the Error Mean Square. You can check the degrees of freedom and the mean squares in the step-by-step

ANOVA table. The original table has 18 observations and so 18 degrees of freedom, but most programs choose not to display the overall mean and subtract out the degree of freedom associated with it to give what's sometimes called a "corrected" total, but usually just called "total." That's why our ANOVA table has only 17 degrees of freedom in "total."

Section 25.2 Exercises

3. **Student TV assumptions** The TV watching study of Exercise 1 was collected as a survey of students at a small college. Do the assumptions of ANOVA appear to be met? Here are some displays to help in your decision:

4. **Cookie assumptions** The student performing the chocolate chip cookie experiment of Exercise 2 planned to analyze his results with an Analysis of Variance on two factors. Here are some displays. Do you think the Assumptions for ANOVA are satisfied?

25.3 Adding Interaction to the Model

Until now, we have assumed that whatever effects the two factors have on the response can be modelled by simply adding the separate (main) effects together. What if that's not good enough? In the dart-throwing experiment, we worried that the experimenter's accuracy when throwing from farther away may deteriorate *even more* when he uses his nondominant hand. Looking at the interaction plot added evidence to that suspicion.

From Figure 25.9 (next page), it doesn't look like the effect of *Distance* is constant for the two hands. In particular, moving from *middle* to *far* has a much larger effect when using the *left Hand*. It's not unusual for the effect of a factor's level to depend on the level of the other factor. Unfortunately, our model doesn't take this into account. So, we need a new model.

Figure 25.9

Interaction plot of *Accuracy* by *Hand* and *Distance*. The effect of *Distance* appears to be greater for the left hand. Equivalently, we see that the effect of changing from right to left hand is bigger at the far distance.

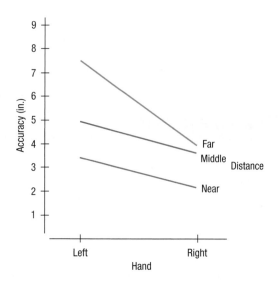

When the effect of one factor *changes* depending on the levels of the other factor, the factors are said to *interact*. We can model this interaction by adding a term to our model and testing whether the adjustment is worthwhile with an *F*-test.

We include a new term to our model, adding an interaction effect at every *combination* of the levels of the other two factors. The model can now be written:

$$y_{ijk} = \mu + \tau_j + \gamma_k + (\tau\gamma)_{jk} + \varepsilon_{ijk}$$

The new term, $(\tau\gamma)_{jk}$, represents the *interaction* effect at level j of factor 1 and level k of factor 2. This is the **Two-way ANOVA Model with Interaction**.

We add a corresponding line to the ANOVA table. Of course, we'll have to spend some degrees of freedom on this new term. We'll have to take them from the error term where we parked all the leftover degrees of freedom after fitting the two factors. We find the Interaction degrees of freedom simply by taking the *product* of each factor's degrees of freedom. For the darts, this is only 2 degrees of freedom (1 from *Hand* × 2 from *Distance*), but for factors with more levels, this could get expensive.

Here's the new ANOVA table for the dart experiment:

ANOVA Table for Accuracy					
Source	DF	Sum of Squares	Mean Square	F-ratio	P-value
Distance	2	51.044	25.522	41.977	<0.0001
Hand	1	39.690	39.690	65.28	<0.0001
Distance × Hand	2	10.355	5.178	8.516	0.0012
Error	30	18.240	0.608		
Total	35	119.329			

What has changed? Nearly everything. (Of course, the Total Sum of Squares remains the same because it was calculated from the observations minus the overall mean, and the overall mean is still the same.) The interaction term is significant. Because we've now removed a significant source of variability, our Error Mean Square is smaller in spite of the loss of the 2 degrees of freedom devoted to estimating the interaction.

The *F*-test for the interaction leads us to reject the null hypothesis of no interaction effect and indicates that the effect of *Distance* really does change depending on which *Hand* is used. (Or, equivalently, the effect of *Hand* changes depending on the *Distance*.)

What about the main effects? When a significant interaction term is present, it's difficult to *Tell* about the main effect of each factor. How much does switching from *right* hand to *left* hand matter? It depends! It depends on the *Distance*. It no longer makes sense to talk about the main effects of each factor separately. The main effects are just averages of the effects over all the levels of the other factor. As we saw with Simpson's paradox (in Chapter 2), it can be misleading to average over very different situations.

When a significant interaction is present, the best way to display the results is with an interaction plot.

Now that we have the interaction term in our model, we recalculate the residuals. They pick up the variation in the data left over after accounting for the overall mean, the effect of each factor, and the interaction effect. Figure 25.10 is a plot of residuals versus predicted values for the two-factor model with interaction:

Figure 25.10

Residuals for the two-way ANOVA of dart accuracy with an interaction term included. Now, unlike in Figure 25.6, there is no U-shaped pattern.

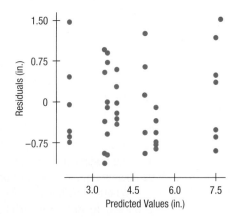

How has the interaction term helped? These residuals have less structure. Fitting the interaction term succeeded in removing structure from the error. This new model seems to satisfy the assumptions more successfully, and so our inferences are likely to be closer to the truth.

Why Not Always Start by Including an Interaction Term?

With two factors in the model, there's always the possibility that they interact. We checked the Additive Enough Condition with an interaction plot. Now we know we can include the interaction term and use the *F*-test for added information. If the interaction plot shows parallel lines and the *F*-test for the interaction term is not significant, you can comfortably proceed with a two-way ANOVA model without interaction. This will put back into the error term the degrees of freedom you used for the interaction.

On the other hand, if the interaction is significant, the model *must* contain more than just the main effects. In this case, the interpretation of the analysis depends crucially on the interaction term, and the interaction plot shows most of what to *Tell* about the data. When there's a significant interaction term, the effect of each factor depends on the level of the other factor, so it often does not make sense to talk about the effects of a factor by itself, though sometimes there is some sense to be made of the main effects. For example, we can see that the *right* hand is better *on average* than the *left* at any distance, and the closer you are to the target the better the accuracy, no matter which hand you use. So both main effects are significant, and we can discuss them. A significant interaction effect tells us more. It says that *how much* better the performance is at near distances *depends* on which hand you use.

If the lines in the interaction plot cross, you may need to be careful. Figure 25.11 is an interaction plot from an experiment on fuels and alternative engine designs:

Figure 25.11

An interaction plot of gas mileage by *Car Type* and *Gas Additive*. If the lines cross, as they do here, be careful of interpreting the main effects.

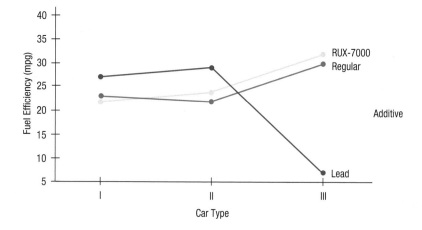

It looks like the fuel with lead is great for two of the engine designs, but a disaster for the third. Suppose someone asked you to summarize the effect of adding lead. Nothing simpler than saying it *depends* on the engine type makes sense. *On average*, it lowers gas mileage, but is that relevant? The real message is that sometimes it improves mileage and sometimes it hurts. You won't want to *Tell anything* about the main effect. Its *F*-test is irrelevant. Crossing of lines in this plot is just one type of non-parallelism that is easy to spot and may reveal an important interaction. But these lines can cross and still be very nearly parallel, suggesting little interaction! And lines that don't cross can be severely non-parallel, suggesting interaction.

Whether the lines cross or not, including an interaction term in our model is always a good way to start. But sometimes we just can't. Suppose we had run the dart experiment with only one dart at each treatment. A design like that is called an **unreplicated two-factor design**. Imagine that the throw at the combination of *left* hand and *far* distance missed the bull's-eye by a mile. Would this be evidence of an interaction between *Distance* and *Hand*, or was it just a bad throw? How could we tell? The obvious thing to do would be to *repeat* the treatment. If accuracy was consistently poor at this combination of factor levels, we'd want to model the interaction. If the other throws were no worse than we'd expect from the two main effects, then we might set the outlier aside. We can't distinguish an interaction effect from error unless we replicate the experiment. In fact, the design tells us that. Without replication, if we try to fit an interaction term, there are exactly 0 degrees of freedom left for the error. That makes any inference or testing impossible. If we are willing to *assume* that the additive model is adequate, then we can reduce the number of runs required and use an unreplicated design. In that case, although the residuals are indistinguishable from the interaction effects, the residual plot may help reveal whether the assumption made sense. And when one of the factors is actually a blocking variable, running an unreplicated design is often appropriate and adequate for inference (though for reasons some of which are beyond the level of our discussion here).

When you do replicate, it's best to replicate *all* treatment conditions equally. (Unbalanced and other more complicated designs can be analyzed, but they are beyond the scope of this book.)

Just Checking

The student conducting the paper airplane design decides to do a follow-up experiment to investigate two other factors in more detail. Using his favourite design, he uses one of three different-sized weights and places the weight either on the nose or on the tail of the plane. At each combination, he performs four test flights (randomizing the order of the flights). An interaction plot looks like this:

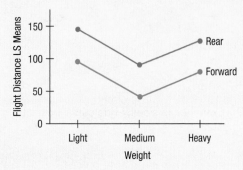

The ANOVA table shows:

Analysis of Variance					
Source	DF	Sum of Squares	Mean Square	F-ratio	Prob > F
Position	1	13 371.9314	13 371.9314	53.386	<0.0001
Weight	2	12 625.5486	6 312.7743	25.203	<0.0001
Position × Weight	2	10.6208	5.3104	0.021	0.9790
Error	18	4 508.501	250.47		
Total	23	30 516.602			

1. What are the null hypotheses?
2. From the interaction plot, do the effects appear to be additive?
3. Does the *F*-test in the ANOVA table support your conclusion in (2)?
4. Does the position of the weight seem to matter?
5. Does the amount of weight appear to affect flight distance?
6. What would you recommend to the student to increase flight distance?

Step-by-Step Example   TWO-FACTOR ANOVA WITH INTERACTION

In Chapter 24, we looked at how much TV four groups of students watched on average. Let's look at their grade point averages (GPAs). Back in that chapter, we treated the four groups (male athletes, female athletes, male non-athletes, and female non-athletes) as four levels of the factor *Group*. Now we recognize that there are really two factors: the factor *Sex* with levels *male* and *female* and the factor *Varsity* with levels *yes* and *no*. Let's analyze the GPA data with a two-factor ANOVA.

As with many Social Science studies, this isn't an experiment. (Some fans might object if we assigned students at random to varsity teams.[7]) Instead, it is an application of ANOVA to an observational study.

THINK ➡ State what we want to know and the null hypotheses we wish to test. For two-factor ANOVA with interaction, the null hypotheses are that all the levels of a factor have the same mean—for each factor—and that interaction is not present. The alternatives are that at least one effect is not 0.

I want to know whether the mean *GPA* is the same for men and women, whether the mean *GPA* is the same for athletes and non-athletes, and whether the factors interact. Writing τ_j for the effect of Sex level j, the null hypothesis is that the effects are the same:

$$H_0: \tau_M = \tau_F$$

Writing γ_k for the effect of Varsity level k, the other null hypothesis is

$$H_0: \gamma_Y = \gamma_N$$

---

[7]And randomly assigning the *other* factor would be problematic.

Finally, the null hypothesis for interaction is that the interaction terms are all 0.

The alternative for the first hypothesis is that the mean *GPA* is different for men and women. The alternative for the second is that the mean *GPA* is different for varsity athletes than for other students. The alternative for the last is that there is some interaction effect.

Plot Examine the side-by-side partial boxplots of the data.

It appears that there may be no difference in mean *GPA* between men and women (after controlling for varsity status), but there may be a difference between varsity and nonvarsity athletes (after controlling for sex).

Plan Think about the assumptions and check the conditions.

The study is a random sample, so students' GPAs should be independent of each other.

✓ **Independence Assumption, Randomization Condition:** The study was based on a random sample of students.

✓ **Additive Enough Condition:** Here's the interaction plot of the means at each treatment condition:

The lines are not close to parallel, so I'll add an interaction term to the model and fit the two-way ANOVA with interaction:

$$y_{ijk} = \mu + Sex\ effect_j + Varsity\ effect_k$$
$$+ Interaction_{jk} + Error_{ijk}.$$

Show the ANOVA table.

The SSs do not add up to the TotalSS here, because the four sample sizes are not equal. Unlike one-way ANOVA, for two-way ANOVA, they only add up when the design is balanced. Computations are much more complicated, but interpretation is still straightforward.

Analysis of Variance for GPA

Source	DF	Sum of Squares	Mean Square	F-ratio	P-value
Sex	1	0.3040	0.3040	1.7681	0.1852
Varsity	1	2.4345	2.4345	14.1871	0.0002
Sex × Varsity	1	1.0678	1.0678	6.2226	0.0134
Error	196	33.6397	0.1716		
Total	199	37.4898			

✓ **Similar Variance Condition**: A plot of residuals versus predicted shows no thickening or other patterns:

✓ **Nearly Normal Condition, Outlier Condition**: The Normal probability plot of the residuals (below) shows only mild non-Normality. No problem.

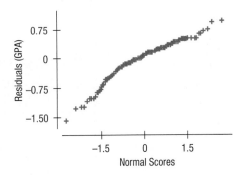

Under these conditions it is appropriate to interpret the F-tests and their P-values.

SHOW ➡ Mechanics

The Error Mean Square is 0.172 points, which means that the common error standard deviation is estimated to be $\sqrt{0.172} = 0.415$ points (nearly half a letter grade).

The F-ratio for the interaction term is 6.223, with a P-value of 0.0134. Based on this, I reject the hypothesis of no interaction. Because there is interaction, I am cautious in testing the main effects, but will examine the means further to understand the factor effects.

Overall mean *GPA* = 3.32.

Level of Varsity	Effect
N	0.1281
Y	−0.1281

Show the table of means.

The main effect of *Varsity* is significant. Varsity athletes have lower *GPA* by about a quarter of a grade point averaged over all, but this effect is much greater for men than for women. Here are the four group mean GPAs:

Group	Mean GPA
Female, nonvarsity	3.35
Female, varsity	3.26
Male, nonvarsity	3.43
Male, varsity	3.00

TELL ➡ Interpretation Tell what the *F*-tests mean.

The interaction term is significant, and it appears that most of the effect we see is due to a difference of 0.43 in mean *GPA* between male athletes and male non-athletes, whereas the similar difference between female athletes and non-athletes is quite slight and probably insignificant.

For Example

RECAP: The ANOVA for paper airplane distance on two factors, *Design* and *Flaps*, showed a significant effect for *Design*, but not for *Flaps*. But maybe there is an interaction. Here is the ANOVA with the interaction term, as well as an interaction plot.

Source	DF	Sum of Squares	Mean Square	F-ratio	P-value
Design	3	1549.50	516.499	81.065	0.0001
Flap	1	27.014	27.0140	4.2399	0.0441
Design × Flap	3	181.503	60.5009	9.4956	0.0001
Error	56	356.800	6.37144		
Total	63	2114.81			

QUESTION: Interpret the interactions. What do they say about paper airplane design?

ANSWER: Including the interaction term reveals that the *Flaps* factor is significant. But it matters in different ways for different designs. Flaps help improve the distance of both kinds of gliders, but they hinder the performance of darts made of lighter copier paper (although they help darts made of construction paper).

What Next?

This chapter introduced some of the issues that arise in advanced experimental design and analysis. We have scratched the surface of this large and important area by showing the analysis of a balanced two-way design with interaction.

Of course, there's no reason to stop with two factors. And if we have more than two, there are many ways in which they may interact. As you can imagine, real-life experiments and observational studies may pose many interesting challenges. Anyone who is serious about analyzing or designing multifactor studies should consult a professional statistician. You now know the vocabulary and principles, so you are well equipped to do that.

Section 25.3 Exercises

5. Student TV interactions Here is an ANOVA table with an interaction term and the corresponding interaction plot for the TV watching data of Exercise 1. What does the interaction term mean here?

Analysis of Variance For ⎯⎯⎯⎯ **TVhours**
No Selector

Source	DF	Sums of Squares	Mean Square	F-ratio	P-value
Intercept	1	2635.13	2635.13	134.51	<0.0001
Sex	1	464.839	464.839	23.728	<0.0001
Ath	1	374.634	374.634	19.123	<0.0001
Sex x Ath	1	312.877	312.877	15.971	<0.0001
Error	193	3780.97	19.5905		
Total	196	4765.12			

6. Have ANOVA cookie Here is the ANOVA table for the cookie experiment of Exercise 2 along with an

interaction plot. What does the interaction term say about the cookie recipes?

Analysis of Variance For Score

Source	DF	Sums of Squares	Mean Square	F-ratio	P-value
Intercept	1	2515.16	2515.16	677.74	<0.0001
Ch	2	27.8990	13.9495	3.7588	0.0270
Sgr	2	27.6566	13.8283	3.7262	0.0279
Ch x Sgr	4	38.2828	9.57071	2.5789	0.0426
Error	90	334.000	3.71111		
Total	98	427.838			

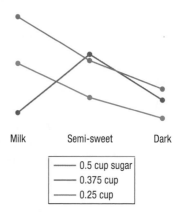

WHAT CAN GO WRONG?

■ **Beware of unreplicated designs unless you are sure there is no interaction.** Without replicating the experiment for each treatment combination, there is no way to distinguish the interaction terms from the residuals. If you are designing a two-factor experiment, you must be willing to assume that there is no interaction if you choose not to replicate. In such a case, you can fit an additive model only in the two factors. If there is an interaction, it will show up in a malformed error term. You should examine a residual plot to help reveal a possible interaction effect. You must be prepared to defend the assumption of no interaction and the decision not to replicate.

(continued)

- **Don't attempt to fit an interaction term to an unreplicated two-factor design.** If you have an unreplicated two-factor experiment or observational study, you'll find that if you try to fit an interaction term, you'll get a strange ANOVA table. The design exhausts the degrees of freedom for error, so fitting the interactions leaves no degrees of freedom for residuals. That wipes out the mean square errors, F-ratios, and P-values, which may appear in the computer output as dots, dashes, or some other indication of things gone wrong (like a simple error message). Remove the interaction term from the model and try again.

- **Be sure to fit an interaction term when it exists.** When the design is replicated, it is always a good idea to fit an interaction term. If it turns out not to be statistically significant, you can then choose to fit a simpler two-factor main effects model instead.

- **When the interaction effect is significant, don't interpret the main effects.** Main effects can be very misleading in the presence of interaction. Look at Figure 25.12 below.

Figure 25.12

An interaction plot of *Yield* by *Temperature* and *Pressure*. The main effects are misleading. There is no main (average) effect of *Pressure* because the average *Yield* at the two pressures is the same. That doesn't mean that *Pressure* has no effect on *Yield*. In the presence of an interaction effect, be careful when interpreting the main effects.

The experiment was run at two temperatures and two pressure levels. High amounts of material were produced at high pressure with high temperature and at low pressure with low *Temperature*. What's the effect of *Temperature?* Of *Pressure?* Both main effects are 0, but it would be silly (and wrong) to say that neither *Temperature* nor *Pressure* was important. The *real story* is in the interaction.

- **Always check for outliers.** As in any analysis, outliers can distort your conclusions. An outlier can inflate the Error Mean Square so much that it may be hard to discern any effects that may be present. Use residual plots and boxplots to search for outliers. Consider setting outliers aside and re-analyzing the results. An outlier can make an interaction term appear significant. For example, a single male varsity athlete with a very low GPA could account for the results we saw.

- **Check for skewness.** If the underlying data distributions are skewed, you should consider a transformation (re-expression) to make them more symmetric.

- **Beware of unbalanced designs and designs with empty cells.** We've been assuming that the data are *balanced* over the design—that we have equal numbers of observations in each cell. There are methods that will easily compensate for small amounts of imbalance, but empty cells and other more serious violations of balance require different methods and additional assumptions.

What Have We Learned?

In Chapter 24, we learned that the Analysis of Variance is a natural way to extend the *t*-test for testing the means of two groups to compare several groups. Often, those groups are different levels of a single factor. Now we've learned that we can extend the Analysis of Variance to designs with more than one factor.

- Understand the advantages of an experiment in two factors.
- Know how to set up an additive model in two factors.
- Be able to use a statistics package to compute a two-way ANOVA.

We've learned that partial boxplots are a good way to graphically examine the effect of each factor on the response. We can carry out separate tests of hypotheses for each of the factors, testing whether the factor has any discernible effect on the response (averaged over levels of the other factor and called the *main effect* of the factor).

- Know how to interpret partial boxplots.
- Be able to interpret main effects in a two-way ANOVA.

We've also learned that sometimes factors can interact with each other. When we have at least two observations at each combination of factor levels, we can add an interaction term to our model to account for the possible interaction.

- Know how to examine the Additivity Assumption and when to consider an interaction term.
- Know how to make an interaction plot for replicated data.
- Be able to use an interaction plot to explain an interaction effect.
- Be able to distinguish when a discussion of main effects is appropriate in the presence of a significant interaction.

Of course, we still need to check the appropriate assumptions and conditions as we did for simple ANOVA.

- **Equal Variance Assumption:** Check the Similar Variance Condition by looking at side-by-side partial boxplots of the groups and a plot of residuals against predicted values. A common problem is *increasing* spread with increasing predicted values—the *plot thickens*.
- **Independence Assumptions:** The groups and the observations within each group must be independent of each other. Think about how the data were collected and the **Randomization Condition**. Check a residual plot.
- **Normal Error Assumption:** Check the **Nearly Normal Condition** by *looking at a normal probability plot of the residuals.*
- **Additivity Assumption:** Sometimes, we assume that we can *add* the effects of the two factors together. Check the **Additive Enough Condition** that the effects of one factor are about the same across all levels of the other factor. An interaction plot should show roughly parallel lines. *Of course, this assumption (and associated condition) is not required when an interaction term is included in the model.*

All the methods we learned for displaying data in the early chapters are still important. And we've begun to see the deep connection between the design concepts we learned in Chapter 10, the exploratory methods we learned in the early chapters, and the inference methods of the latter part of the book.

Review of Terms

Partial boxplot Plots the response at each of the levels of one factor with the effects of the second factor removed. A simple way to generate partial boxplots of factor A is to first save the residuals from a one-way ANOVA of the response on factor B. Add the overall mean to the residuals and make boxplots of the resulting data against factor A. This gives you a good visual impression of factor A's effects and variability, while effectively holding constant factor B. (p. 829).

Interaction When the effects of the levels of one factor change depending on the level of the other factor, the two factors are said to interact. When interaction terms are present, it can be misleading to talk about the main effect of one factor because how large it is *depends* on the level of the other factor (p. 831).

Interaction plot A plot that shows the means at each treatment combination, highlighting the factors' effects and their behaviour at all the combinations (p. 831).

Two-way ANOVA model (main effects model) The simplest model for a two-way (one response, two factors) ANOVA is

$$y_{ijk} = \mu + \tau_j + \gamma_k + \varepsilon_{ijk}$$

where τ_j represents the (main) effect of level j of factor one and γ_k represents the (main) effect of level k of factor two. The subscript i designates the ith replication of the treatment combination. This is also called an *additive* model (p. 827).

Additivity, additive model	A model in two factors is additive if it simply adds the effects of the two factors and doesn't include an interaction term (p. 830).
Two-way ANOVA model with interaction	For a replicated two-way design, an interaction term can also be fit. The resulting model is

$$y_{ijk} = \mu + \tau_j + \gamma_k + (\tau\gamma)_{jk} + \varepsilon_{ijk}$$

where τ_j represents the (main) effect of level j of factor one and γ_k represents the (main) effect of level k of factor two. Now $(\tau\gamma)_{jk}$ represents the effect of the interaction at levels j of factor one and k of factor two. The subscript i designates the ith replication of the treatment combination (p. 842).

On the Computer TWO-WAY ANOVA

Some statistics packages distinguish between models with one factor and those with two or more factors. You must be alert to these differences when analyzing a two-factor ANOVA. It's not unusual to find ANOVA models in several different places in the same package.

Sometimes, you must specify the interaction term yourself. That's because these features of statistics packages typically are designed for models with three or more factors where the number of interactions can explode unless the data analyst carefully selects the interactions to be included in the model.

EXCEL

- The Excel Data Analysis Add-in offers a two-way ANOVA "with and without replication."

COMMENTS

Excel requires that the data be in a special format and cannot deal with unbalanced data.

JMP

- From the **Analyze** menu, select **Fit Model**.
- Select variables: and **Add** them to the **Construct Model Effects** box.
- To specify an interaction, select both factors and press the **Cross** button.
- Click **Run Model**.
- JMP opens a **Fit Least Squares** window.
- Click on the red triangle beside each effect to see the means plots for that factor. For the interaction term, this is the interaction plot.
- Consult the JMP documentation for information about other features.

COMMENTS

JMP expects data in "stacked" format with one continuous response and two nominal factor variables.

MINITAB

- Choose **ANOVA** from the **Stat** menu.
- Choose **Balanced ANOVA.** . . . from the **ANOVA** submenu.
- Under **Model**, enter the columns containing the factors, say C1 and C2, and if you wish to include interaction, also write C1*C2. Enter the response variable under **Responses.**
- In the ANOVA-Graphs dialogue, under **Residual plots**, check **Four in one**, for a useful set of four residual plots.

COMMENTS

Minitab expects "stacked" format data for two-way designs.

R

To perform an analysis of variance of a variable y on factors $x1$ and $x2$:

- myaov = aov($y \sim x1 + x2$)
- anova(myaov) # gives the ANOVA table

To include the interaction of $x1$ and $x2$:

- myaov = aov($y \sim x1 + x2 + x1{:}x2$) or equivalently,

- myaov = aov($y \sim x1{*}x2$)
- summary(myaov) # shows the individual effects
- anova(myaov) # gives the ANOVA table
- interaction.plot(x1, x2, y) # provides an interaction plot

SPSS

- Choose **Analyze > General Linear Model > Univariate.**
- Assign the response variable to the **Dependent Variable** box.
- Assign the two factors to the **Fixed Factor(s)** box. This will fit the model with interactions by default.

- To omit interactions, click on **Model**. Select **Custom**. Highlight the factors. Select **Main Effects** under the **Build Terms** arrow and click the arrow.
- Click **Continue** and **OK** to compute the model.

STATCRUNCH

To find an ANOVA of response variable y on two factors, x_1 and x_2, from the **Stat** menu,

- Choose **ANOVA >Two Way**
- In the dialog, specify the response variable.
- Specify the factors. It really doesn't matter which you specify as the "row factor" and which as the "column factor."

StatCrunch automatically includes an interaction term in your model. Choose the "Fit Additive Model" option to omit the interaction term.

A table of cell means is available.

TI-83/84 PLUS

COMMENTS

You need a special program to compute a two-way ANOVA on the TI-83.

Chapter Exercises

T indicates data is available in a saved file at the Companion Website at www.pearsoncanada.ca/deveaux

7. **Popcorn revisited** A student runs a two-factor experiment to test how microwave power and temperature affect popping. She chooses three levels of *Power* (low, medium, and high) and three *Times* (three minutes, four minutes, and five minutes), running one bag at each condition. She counts the number of uncooked kernels as the response variable.
   a) What are the null and alternative hypotheses for the main effects?
   b) How many degrees of freedom does each factor sum of squares have? How about the error sum of squares?
   c) Should she consider fitting an interaction term to the model? Why or why not?

8. **Gas mileage revisited** A student runs an experiment to study the effect of *Tire Pressure* and *Acceleration* on gas mileage. He devises a system so that his Jeep Wagoneer uses gasoline from a one-litre container. He uses three levels of tire pressure (low, medium, and full) and two levels of acceleration, either holding the pedal steady or pumping it every few seconds. He randomizes the trials, performing four runs under each treatment condition, carefully recording the number of kilometres he can go in his Jeep Wagoneer on one litre of gas.
   a) What are the null and alternative hypotheses for the main effects?
   b) How many degrees of freedom does each treatment sum of squares have? How about the error sum of squares?
   c) Should he consider fitting an interaction term to the model? Why might it be a good idea?
   d) If he fits an interaction term, how many degrees of freedom would it have?

9. **Popcorn again** Refer to the experiment in Exercise 7. After collecting her data and analyzing the results, the student reports that the *F*-ratio for *Power* is 13.56 and the *F*-ratio for *Time* is 9.36.
   a) What are the P-values?
   b) What would you conclude?
   c) What else about the data would you like to see in order to check the assumptions and conditions?

10. **Gas mileage again** Refer to the experiment in Exercise 8. After analyzing his data, the student reports that the *F*-ratio for *Tire Pressure* is 4.29 with a P-value of 0.030, the *F*-ratio for *Acceleration* is 2.35 with a P-value of 0.143, and the *F*-ratio for the interaction effect is 1.54 with a P-value of 0.241.
    a) What would you conclude?
    b) What else about the data would you like to see in order to check the assumptions and conditions?

c) If your conclusion about the *Acceleration* factor in part a) is wrong, what type of error have you made?

11. **Crash analysis** The U.S. National Highway Transportation Safety Administration runs crash tests in which stock automobiles are crashed into a wall at 56.33 km/hr with dummies in both the passenger's and the driver's seats. The THOR Alpha crash dummy is capable of recording 134 channels of data on the impact of the crash at various sites on the dummy. In this test, 335 cars are crashed. The response variable is a measure of head injury. Researchers want to know if which seat the dummy is in affects head injury severity, as well as whether the type of car affects severity.

Here are partial boxplots for the two different *Seat* (driver, passenger) and the six different *Size* classifications (compact, light, medium, mini, pickup, van):

An interaction plot shows:

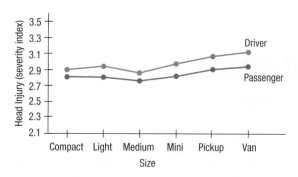

A scatterplot of residuals versus predicted values shows:

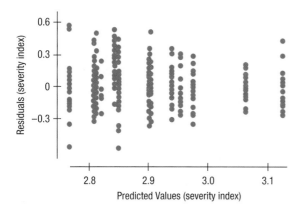

The ANOVA table follows:

Analysis of Variance for Head Severity: 293 cases

Source	DF	Sum of Squares	Mean Square	F-ratio	P-value
Seat	1	0.88713	0.88713	25.501	<0.0001
Size	5	1.49253	0.29851	8.581	<0.0001
Seat × Size	5	0.07224	0.01445	0.415	0.838
Error	282	9.8101	0.03479		
Total	293	12.3853			

a) State the hypotheses about the main effects (both numerically and in words).
b) Are the conditions for a two-way ANOVA met?
c) If so, perform the hypothesis tests and state your conclusion. Be sure to state it in terms of head injury severity, seats, and vehicle types.

12. Sprouts An experiment on mung beans was performed to investigate the environmental effects of salinity and water temperature on sprouting. Forty beans were randomly allocated to each of 36 petri dishes that were subject to one of four levels of *Salinity* (0, 4, 8, and 12 ppm) and one of three *Temperatures* (32°, 34°, or 36°C). After 48 hours, the biomass of the sprouts was measured.

Here are partial boxplots of *Biomass* on *Salinity* and *Temperature*:

The interaction plot shows:

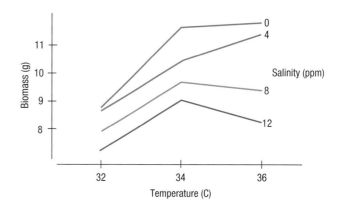

A two-way ANOVA model is fit, and the following ANOVA table results:

Analysis of Variance for Biomass (g)

Source	DF	Sum of Squares	Mean Square	F-ratio	P-value
Salinity	3	36.4701	12.1567	16.981	<0.0001
Temp	2	34.7168	17.3584	24.247	<0.0001
Salinity × Temp	6	5.2972	0.8829	1.233	0.3244
Error	24	17.1816	0.7159		
Total	35	93.6656			

The plot of residuals versus predicted values shows:

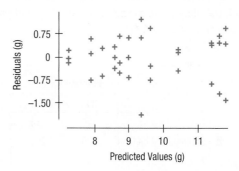

a) State the hypotheses about the factors (both numerically and in words).
b) Are the conditions for a two-way ANOVA met?
c) Perform the hypothesis tests and state your conclusions. Be sure to state your conclusions in terms of biomass, salinity, and water temperature.

13. Baldness and heart disease A retrospective study examined the link between baldness and the incidence of heart disease. In the study, 1435 middle-aged men were selected at random and examined to see whether they showed signs of *Heart Disease* (or not) and what amount of *Baldness* they exhibited (none, little, some, or much). A student runs a two-factor ANOVA on these data (given below) and finds the following ANOVA table:

Source	DF	Sum of Squares	Mean Square	F-ratio	P-value
Baldness	3	62 441.375	20 813.792	17.956	0.020
Heart disease	1	1 485.125	1 485.125	1.281	0.340
Error	3	3 477.375	1 159.125		

a) Comment on her analysis. What problems, if any, do you find?
b) What sort of analysis might you do instead?

Baldness	Heart Disease	Number of Men
None	Yes	251
None	No	331
Little	Yes	165
Little	No	221
Some	Yes	195
Some	No	185
Much	Yes	52
Much	No	35

14. **Fish and prostate** In the Chapter 8 Step-By-Step, we looked at a Swedish study that asked 6272 men how much fish they ate and whether or not they had prostate cancer. Here are the data:

		Prostate Cancer?	
		No	Yes
Eat Fish?	Never/seldom	110	14
	Small part of diet	2420	201
	Moderate part	2769	209
	Large part	507	42

Armed with the methods of this chapter, a student performs a two-way ANOVA on the data. Here is her ANOVA table:

Source	DF	Sum of Squares	Mean Square	F-ratio	P-value
Fish	3	3 110 203.0	1 036 734.3	1.3599	0.4033
Prostate cancer	1	3 564 450.0	3 564 450.0	4.6756	0.1193
Error	3	2 287 051.0	762 350.0		

a) Comment on her analysis. What problems, if any, do you find?
b) What sort of analysis might you do instead?

15. **Baldness and heart disease again** Refer back to Exercise 13. Perform your own analysis of the data to see if baldness and heart disease are related. Do your conclusions support the claim that baldness is a cause of heart disease? Explain.

16. **Prostate and fish** Refer back to Exercise 14. Perform your own analysis of the data to see if eating fish and contracting prostate cancer are related.

T 17. Basketball shots A student performed an experiment to see if her favourite sneakers and the time of day might affect her free-throw percentage. She tried shooting with and without her favourite sneakers and in the early morning and at night. For each treatment combination, she shot 50 baskets on four different occasions, recording the number of shots made each time. She randomized the treatment conditions by drawing numbers out of a hat. Here are her data:

Time of Day	Shoes	Shots Made
Morning	Others	25
Morning	Others	26
Night	Others	27
Night	Others	27
Morning	Favourite	32
Morning	Favourite	22
Night	Favourite	30
Night	Favourite	34
Morning	Others	35
Morning	Others	34
Night	Others	33
Night	Others	30
Morning	Favourite	33
Morning	Favourite	37
Night	Favourite	36
Night	Favourite	38

a) What are the null and alternative hypotheses?
b) Write a short report on your findings, being sure to include diagnostic analysis as well as practical conclusions.

T 18. Washing For his final project, Jonathan examined the effects of two factors on how well stains are removed when washing clothes. On each of 16 new white handkerchiefs, he spread a teaspoon of dirty motor oil (obtained from a local garage). He chose four *Temperature* settings (each of which is a combination of wash and rinse: cold-cold, cold-warm, warm-hot, and hot-hot) and four *Cycle* lengths (short, med short, med long, and long). After its washing, each handkerchief was dried in a dryer for 20 minutes and hung up. He rounded up 10 family members to judge the handkerchiefs for cleanliness on a scale of 1 to 10 and used the average score as his response. Here are the data:

Temp	Cycle	Score
Cold-cold	Med long	3.7
Warm-hot	Med long	6.5
Cold-warm	Med long	4.9
Hot-hot	Med long	6.5
Cold-cold	Long	4.6
Warm-hot	Long	8.3
Cold-warm	Long	4.7
Hot-hot	Long	9.1
Cold-cold	Short	3.4
Warm-hot	Short	5.6
Cold-warm	Short	3.8
Hot-hot	Short	7.1
Cold-cold	Med short	3.1
Warm-hot	Med short	6.3
Cold-warm	Med short	5
Hot-hot	Med short	6.1

You may assume, as Jonathan did, that interactions between *Temperature* and *Cycle* are negligible. Write a report showing what you found about washing factors and stain removal.

19. Sprouts again The students running the sprouts experiment (Exercise 12) also kept track of the number of beans sprouted (out of 40) for each of the 36 dishes. The partial boxplots of *Sprouts* plotted against *Salinity* and *Temperature* are shown below.

An ANOVA table shows:

Source	DF	Sum of Squares	Mean Square	F-ratio	P-value
Salinity	3	2014.750	671.583	23.657	<0.0001
Temp	2	57.556	28.778	1.014	0.3779
Salinity × Temp	6	96.000	16.000	0.564	0.7549
Error	24	681.333	28.389		
Total	35	2849.639			

An interaction plot shows:

A plot of residuals vs. predicted values shows:

a) State the hypotheses about the factors (both numerically and in words).
b) Perform the hypothesis tests and state your conclusions. Be sure to check conditions.

T 20. Containers revisited Building on the cup experiment of the Chapter 3 Step-By-Step, a student selects one type of container and designs an experiment to see whether the type of *Liquid* stored and the outside *Environment* affect the ability of a cup to maintain temperature. He randomly chooses an experimental condition and runs each twice:

Liquid	Environment	Change in Temperature (F)
Water	Room	13
Water	Room	14
Water	Outside	31
Water	Outside	31
Coffee	Room	11
Coffee	Room	11
Coffee	Outside	27
Coffee	Outside	29

After fitting a two-way ANOVA model, he obtains the following interaction plot, ANOVA table, and effects table:

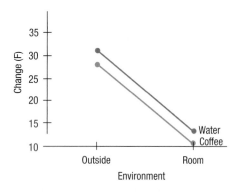

Source	DF	Sum of Squares	Mean Square	F-ratio	P-value
Liquid	1	15.125	15.125	24.2	0.0079
Envirn	1	595.125	595.125	952.2	<0.0001
Interaction	1	0.125	0.125	0.2	0.6779
Error	4	2.500	0.625		
Total	7	612.875			

Term	Estimate
Overall mean	20.875
Liquid II [Coffee]	−1.375
Liquid II [Water]	1.375
Environment II [Outside]	8.625
Environment II [Room]	−8.625

a) State the null and alternative hypotheses.
b) Test the hypotheses at $\alpha = 0.05$.
c) Perform a residual analysis.
d) Summarize your findings.

21. **Gas additives** An experiment to test a new gasoline *Additive*, Gasplus, was performed on three different car *Types*: a sports car, a minivan, and a hybrid. Each car was tested with both Gasplus and regular gas on 10 different occasions and their gas mileage was recorded. Here are the partial boxplots:

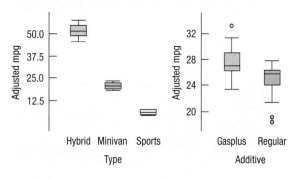

A two-way ANOVA with interaction model was run, and the following ANOVA table resulted:

Source	DF	Sum of Squares	Mean Square	F-ratio	P-value
Type	2	23 175.4	11 587.7	2 712.2	<0.0001
Additive	1	92.1568	92.1568	21.57	<0.0001
Type × Additive	2	51.8976	25.9488	6.0736	0.0042
Error	54	230.711	4.27242		
Total	59	23 550.2			

A plot of the residuals versus predicted values showed:

What conclusions about the *Additive* and car *Types* do you draw? Do you see any potential problems with the analysis?

22. **Chromatography** A gas chromatograph is an instrument that measures the amounts of various compounds contained in a sample by separating the constituents. Because different components are flushed through the system at different rates, chromatographers are able to both measure and distinguish the constituents of the sample. A counter is placed somewhere along the instrument that records how much material is passing at various times. By looking at the counts at various times, the chemist is able to reconstruct the amounts of various compounds present. The total count is proportional to the amount of the compound present.

An experiment was performed to see whether slowing down the flow rate would increase total counts. A mixture was produced with three different *Concentration* levels: low, medium, and high. The two *Flow Rates* used were slow and fast. Each mixture was run five times and the total count recorded each time. Partial boxplots for *Concentration* and *Flow Rates* show:

A two-way ANOVA with interaction was run, and the following ANOVA table resulted:

Source	DF	Sum of Squares	Mean Square	F-ratio	P-value
Conc	2	483 655E5	241 828E5	1969.44	<0.0001
Flow rate	1	364 008E3	364 008E3	29.65	<0.0001
Interaction	2	203 032E3	101 516E3	8.27	0.0019
Error	24	294 698E3	122 791E3		
Total	29	492 272E5			

A plot of residuals versus predicted values showed:

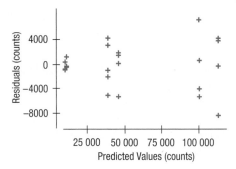

What conclusions about the effect of flow rate do you draw? Do you see any potential problems with the analysis?

23. Gas additives again Refer back to the experiment in Exercise 21. Instead of *mpg*, redo the analysis using log(*mpg*) as the response. Do your conclusions change? How? Are the assumptions of the model better satisfied?

24. Chromatography again Refer back to the experiment in Exercise 22. Instead of *total counts,* the analysis was redone using log(*total counts*) as the response. Do your conclusions change? How? Are the assumptions of the model better satisfied?

Analysis of Variance

Source	DF	Sum of Squares	Mean Square	F-ratio	P-value
Type	2	10.1254	5.06268	5923.1	<0.0001
Additive	1	0.026092	0.0260915	30.526	<0.0001
Interaction	2	7.57E-05	3.78E-05	0.044265	0.9567
Error	54	0.046156	8.55E-04		
Total	59	10.1977			

25. Batteries A student experiment was run to test the performance of four *Brands* of batteries under two different *Environments* (room temperature and cold). For each of the eight treatments, two batteries of a particular brand were put into a flashlight. The flashlight was then turned on and allowed to run until the light went out. The number of minutes the flashlight stayed on was recorded. Each treatment condition was run twice.
Partial boxplots showed:

An interaction plot showed:

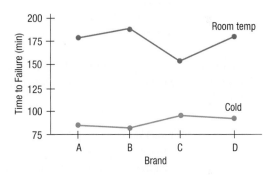

An ANOVA table showed:

Source	DF	Sum of Squares	Mean Square	F-ratio	P-value
Envir	1	30 363.1	30 363.1	789.93	<0.0001
Brand	3	338.187	112.729	2.9328	0.0994
Interaction	3	1 278.19	426.063	11.085	0.0032
Error	8	307.5	38.4375		
Total	15	32 286.9			

a) What are the main effect null and alternative hypotheses?
b) From the partial boxplots, do you think that the *Brand* has an effect on the time the batteries last? How about the condition?
c) Do the conclusions of the ANOVA table match your intuition?
d) What does the interaction plot say about the performance of the brands?
e) Why might you be uncomfortable with a recommendation to go with the cheapest battery (brand C)?

26. Peas In an experiment on growing sweet peas, a team of students selected two factors at four levels each and recorded *Weight*, *Stem Length*, and *Root Length* after six days of growth. They grew plants using various amounts of *Water* and *Quickgrow* solution, a fertilizer designed to help plants grow faster. Each factor was run at four levels: little, some, moderate, and full. They grew two plants under each of the 16 conditions. The response variable was *Weight* in mg.

An interaction plot shows:

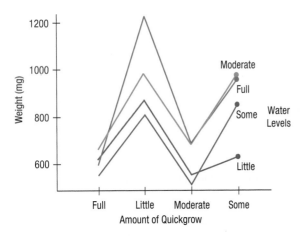

Because of this, a two-way ANOVA with interaction model was fit to the data, resulting in the following ANOVA table:

Analysis of Variance for Weight

Source	DF	Sum of Squares	Mean Square	F-ratio	P-value
Water	3	246 255	82 084.8	2.5195	0.0948
QS	3	827 552	275 851	8.4669	0.0013
Water × QS	8	176 441	22 055.1	0.6770	0.7130
Error	16	521 275	32 579.7		
Total	31	1 771 523			

Residuals plots showed no violations of the variance or Normality conditions. A table of effects for Quickgrow shows:

Level of Quickgrow	Effect
Little	213.8
Some	97.1
Moderate	−155.5
Full	−155.4

A table of effects for water shows:

Level of Water	Effect
Little	−91.1
Some	−81.9
Moderate	66.4
Full	106.6

Write a report summarizing what the students have learned about the effects of *Water* and *Quickgrow* solution on the early stages of sweet pea growth as measured by weight.

27. Batteries again Another student analyzed the battery data from Exercise 25, using a one-way ANOVA. He considered the experimental factor to be an eight-level factor consisting of the eight possible combinations of *Brand* and *Environment*. Here are the boxplots for the eight treatments and a one-way ANOVA:

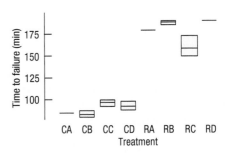

Source	DF	Sum of Squares	Mean Square	F-ratio	P-value
Treatment	7	31 895.8	4 556.54	93.189	<0.0001
Error	8	391.167	48.8958		
Total	15	32 286.9			

Compare this analysis with the one performed in Exercise 25. Which one provides a better understanding of the data? Explain.

28. Containers, one more time Another student performs a one-way ANOVA on the container data of Exercise 20, using the four treatments *water room*, *water outside*, *coffee room*, and *coffee outside*. Perform this analysis and comment on the differences between this analysis and the one in Exercise 20.

Just Checking ANSWERS

1. The null hypotheses are that the mean flight distance for all three weights is the same, that the mean flight distance for both positions is the same, and that the interaction effects are all 0.

2. Yes, the effects appear to be additive. The lines are nearly parallel.

3. Yes, the F-test provides no evidence to reject the null hypothesis with a P-value of ≥ 0.9.

4. Yes, we reject the null hypothesis with a P-value ≤ 0.0001.

5. Yes, we reject the null hypothesis with a P-value ≤ 0.0001.

6. Because both factors are significant, it appears that using the light weight in the rear position may result in the longest mean flight distance.

MathXL

MyLab Statistics

Go to MathXL at **www.mathxl.com** or MyLab™ Statistics at **www.mystatlab.com**. You can practise exercises for this chapter as often as you want. The guided solutions will help you find answers step by step. You'll find a personalized study plan available to you too!

Multiple Regression

Connections

We discussed simple regression in Chapter 7, and related inferences in Chapter 23. In Chapter 7, we found that the top wind speed in a hurricane depends on the central barometric pressure. But what about the sea surface temperature? Can we include other variables in our model? For more realism, we often need models with *more than one explanatory variable.*

Regression inference is closely connected to other

Old stories are called teaching, new ones research.

—*Professor Edward Chamberlin*

In Chapter 23, we tried to predict the percent body fat of male subjects from their waist size, and we did pretty well. The R^2 of 67.8% says that we accounted for almost 68% of the variability in *%Body fat* by knowing only *Waist* size. We completed the analysis by performing a hypothesis test on the coefficient of *Waist* size (or slope) and looking at the residuals.

But that remaining 32% of the variance has been bugging us. Couldn't we do a better job of accounting for *%Body fat* if we weren't limited to a single predictor? In the full data set, there were 15 other measurements on the 250 men. We might be able to use other predictor variables to help us account for the leftover variation that wasn't accounted for by *Waist* size.

What about *Height*? Does *Height* help to predict *%Body fat*? Men with the same waist size can vary from short and corpulent to tall and emaciated. Knowing a man has a 50-inch waist tells us that he's likely to carry a lot of body fat. If we found out that he was 7 feet tall, that might change our impression of his body type. Knowing his *Height* as well as his *Waist* size might help us make a more accurate prediction.

26.1 What is Multiple Regression?

Does a regression with *two* predictors even make sense? It does—and that's fortunate, because the world is too complex a place for linear regression to model it with a single predictor. A regression with two or more predictor variables is called a **multiple regression**. (When we need to note the difference, a regression on a single predictor is called a *simple* regression.) We'd never try to find a regression by hand, and even calculators aren't really up to the task. This is a job for a statistics program on a computer. If you know how to find the regression of *%Body fat* on *Waist* size with a statistics package, you can usually just add *Height* to the list of predictors without having to think hard about how to do it.

For simple regression, we found the **least squares** solution, the one whose coefficients made the sum of the squared residuals as small as possible. For multiple regression, we'll do the same thing but this time with more coefficients. Remarkably enough, we can still solve this problem. Even better, a statistics package can find the coefficients of the least squares model easily.

Here's a typical example of a multiple regression table (Data in **Body fat complete**):

Dependent variable is: %Body fat
R-squared = 71.3% R-squared (adjusted) = 71.1%
s = 4.460 with 250 − 3 = 247 degrees of freedom

Variable	Coefficient	SE(Coeff)	t-ratio	P-value
Intercept	−3.10088	7.686	−0.403	0.6870
Waist	1.77309	0.0716	24.8	≤ 0.0001
Height	−0.60154	0.1099	−5.47	≤ 0.0001

You should recognize most of the numbers in this table. Most of them mean what you expect them to.

R^2 has the same interpretation in multiple regression as in simple regression (though it is now the square of the correlation between the observed responses and the predicted responses). Here, for example, R^2 gives the fraction of the variability of *%Body fat* accounted for by the *multiple* regression model. (With *Waist* size alone predicting *%Body fat*, the R^2 was 67.8%.) The multiple regression model accounts for 71.3% of the variability in *%Body fat*. We shouldn't be surprised that R^2 has gone up. It was the hope of accounting for some of that leftover variability that led us to try a second predictor ('R^2 adjusted' in the output will be discussed later in this chapter).

The standard deviation of the residuals is still denoted s (or sometimes s_e to distinguish it from the standard deviation of y).

The degrees of freedom calculation follows our rule of thumb: The degrees of freedom for s and the t-ratio is the number of observations (250) minus one for each coefficient estimated—for this model, three.

For each predictor, we have a coefficient, its standard error, a t-ratio, and the corresponding P-value. As with simple regression, the t-ratio measures how many standard errors the coefficient is away from 0. So, we can find a P-value from a Student's t-model to test the null hypothesis that the true value of the coefficient is 0.

Using the coefficients from this table, we can write the fitted regression model:

$$\widehat{\%Body\,fat} = -3.10 + 1.77\,Waist - 0.60\,Height.$$

As before, we define the residuals as

$$Residuals = \%Body\,fat - \widehat{\%Body\,fat}.$$

We've fit this model with the same least squares principle: The sum of the squared residuals is as small as possible for any choice of coefficients.

So, What's Different?

With so much of the multiple regression looking just like simple regression, why devote an entire chapter (or two) to the subject?

There are several answers to this question. First—and most important—the *meaning* of the coefficients in the regression model has changed in a subtle but important way. Because that change is not obvious, multiple regression coefficients are often misinterpreted. We'll show some examples to help make the meaning clear.

Second, multiple regression is an extraordinarily versatile calculation, underlying many widely used Statistics methods. A sound understanding of the multiple regression model will help you to understand these other applications.

Third, multiple regression offers our first glimpse into statistical models that use more than two quantitative variables. The real world is complex. Simple models of the kind we've seen so far are a great start, but often they're just not detailed enough to be useful for understanding, predicting, and decision making. Models that use several variables can be a big step toward realistic and useful modelling of complex phenomena and relationships.

For Example REAL ESTATE

As a class project, students in a large Statistics class collected publicly available information on recent home sales in their home towns. We have 894 properties. These are not a random sample, but they may be representative of home sales during a short period of time, nationwide. (Data in **Real estate project**)

Variables available include the price paid, the size of the living area (sq. ft.), the number of bedrooms, the number of bathrooms, the year of construction, the lot size (acres), and a coding of the location as urban, suburban, or rural made by the student who collected the data.

---

inference methods for measurement data, with similar assumptions, like independent Normal errors with constant variance, which we check by pooling and plotting residuals.

We'll see more ANOVA tables here. This is not coincidental. Multiple Regression, one-way and two-way ANOVA, are all part of a more general statistical model known as the General Linear Model.

Who	250 male subjects
What	Body fat, waist size, and height
Units	%body fat, inches, and inches
When	1990s
Where	United States
Why	Scientific research

A NOTE ON TERMINOLOGY

When we have two or more predictors and fit a linear model by the least squares method, we are formally said to fit a least squares linear multiple regression. Most folks just call it "multiple regression." You may also see the abbreviation OLS used with this kind of analysis. It stands for "Ordinary Least Squares."

Here's a regression to model the sale price from the living area (sq. ft.) and the number of bedrooms.

Dependent variable is: Price
R squared = 14.6% R squared (adjusted) = 14.4%
s = 266 899 with 894 − 3 = 891 degrees of freedom

Variable	Coefficient	SE(Coeff)	t-ratio	P-value
Intercept	308 100	41 148	7.49	< 0.0001
Living area	135.089	11.48	11.8	< 0.0001
Bedrooms	−43 346.8	12 844	−3.37	0.0008

QUESTION: How should we interpret the regression output?

ANSWER: The model is

$$\widehat{Price} = 308\,100 + 135\,LivingArea - 43\,347\,Bedrooms$$

The R-squared says that this model accounts for only 14.6% of the variation in *Price*. The value of *s* also leads us to doubt that this model would provide very good predictions because the standard deviation of the residuals is more than $266 000. Nevertheless, we may be able to learn about home prices because the P-values of the coefficients are all very small, so we can be quite confident that none of them is really zero.

Section 26.1 Exercises

1. House prices The following regression model was found for houses in an area of upstate New York:

$$\widehat{Price} = 20\,986.09 - 7\,483.10\,Bedrooms + 93.84\,Living\;Area.$$

a) Find the predicted price of a 2 bedroom, 1000-sq-ft house from this model.
b) The house sold for $135 000. Find the residual corresponding to this house.
c) What does that residual say about this transaction?

2. Candy sales A candy maker surveyed chocolate bars available in a local supermarket and found the following least squares regression model:

$$\widehat{Calories} = 28.4 + 11.37\,Fat(g) + 2.91\,Sugar(g).$$

a) The hand-crafted chocolate she makes has 15g of fat and 20g of sugar. How many calories does the model predict for a serving?
b) In fact, a laboratory test shows that her candy has 227 calories per serving. Find the residual corresponding to this candy. (Be sure to include the units.)
c) What does that residual say about her candy?

26.2 Interpreting Multiple Regression Coefficients

We said that height might be important in predicting body fat in men. What is the relationship between *%Body fat* and *Height* in men? We know how to approach this question; we follow the three rules. Below (Figure 26.1) is the scatterplot.

It doesn't look like *Height* tells us much about *%Body fat*. You just can't tell much about a man's *%Body fat* from his *Height*. Or can you? Remember, in the multiple regression

Figure 26.1

The scatterplot of *%Body fat* against *Height* seems to say that there is little relationship between these variables.

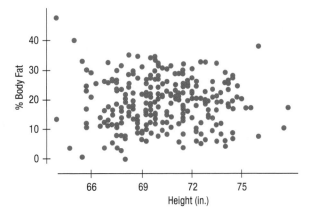

model, the coefficient of *Height* was −0.60, had a *t*-ratio of −5.47, and had a very small P-value. So it *did* contribute to the *multiple* regression model. How could that be?

The answer is that the multiple regression coefficient of *Height* takes account of the other predictor, *Waist* size, in the regression model.

To understand the difference, let's think about all men whose waist size is about 37 inches—right in the middle of our sample. If we think only about *these* men, what do we expect the relationship between *Height* and *%Body fat* to be? Now a negative association makes sense because taller men probably have less body fat than shorter men *who have the same waist size*. Let's look at the plot:

Figure 26.2

When we restrict our attention to men with waist sizes between 36 and 38 inches (points in blue), we can see a relationship between *%Body fat* and *Height*.

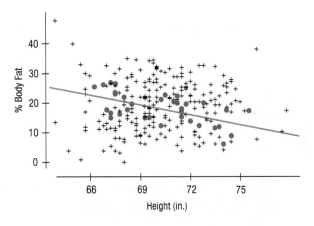

In Figure 26.2 we've highlighted the men with waist sizes between 36 and 38 inches. Overall, there's little relationship between *%Body fat* and *Height*, as we can see from the full set of points. But when we focus on *particular* waist sizes, there *is* a relationship between body fat and height. This relationship is *conditional* because we've restricted our set to only those men within a certain range of waist sizes. For men with that waist size, an extra inch of height is associated with a decrease of about 0.60% in body fat. If that relationship is consistent for each *Waist* size, then the multiple regression coefficient will estimate it. The simple regression coefficient simply couldn't see it.

We've picked one particular *Waist* size to highlight. How could we look at the relationship between *%Body fat* and *Height* conditioned on *every individual waist size at the same time*? Once again, residuals come to the rescue.

We plot the residuals of *%Body fat* after a regression on *Waist* size against the residuals of *Height* after regressing *it* on *Waist* size. This display is called a *partial regression plot*. Figure 26.3 shows us just what we asked for: the relationship of *%Body fat* to *Height* after removing the linear effects of *Waist* size.

> **Residuals**
> As their name reminds us, residuals are what's left over after we fit a model. That lets us remove the effects of some variables. The residuals are what's left.

Figure 26.3

A partial regression plot for the coefficient of *Height* in the regression model has a slope equal to the coefficient value in the multiple regression model.

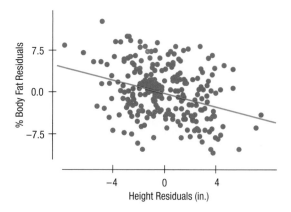

A **partial regression plot** (also known as added variable plot, adjusted variable plot, and individual coefficient plot) for a particular predictor has a slope that is the same as the *multiple* regression coefficient for that predictor. Here, it's −0.60. It also has the same residuals as the full multiple regression,[1] so you can spot any outliers or influential points and tell whether they are likely to have affected the estimation of this particular coefficient.

Many modern statistics packages offer partial regression plots as an option for any coefficient of a multiple regression. For the same reasons that we always look at a scatterplot before interpreting a simple regression coefficient, it's a good idea to make a partial regression plot for any multiple regression coefficient that you hope to understand or interpret.

Section 26.2 Exercises

3. Movie profits What can predict how much a motion picture will make? We have data on a number of recent releases that includes the *USGross* (in $M), the *Budget* ($M), the *Run Time* (minutes), and the average number of *Stars* awarded by reviewers. The first several entries in the data table look like this:

Movie	USGross ($M)	Budget ($M)	Run Time (minutes)	Stars
White Noise	56.094360	30	101	2
Coach Carter	67.264877	45	136	3
Elektra	24.409722	65	100	2
Racing Stripes	49.772522	30	110	3
Assault on Precinct 13	20.040895	30	109	3
Are We There Yet?	82.674398	20	94	2
Alone in the Dark	5.178569	20	96	1.5
Indigo	51.100486	25	105	3.5

We want a regression model to predict *USGross*. Parts of the regression output computed in Excel look like this:

Dependent variable is: USGross($)
R squared = 47.4%   R squared (adjusted) = 46.0%
s = 46.41 with 120 − 4 = 116 degrees of freedom

Variable	Coefficient	SE(Coeff)	t-ratio	P-value
Intercept	−22.9898	25.70	−0.895	0.3729
Budget($)	1.13442	0.1297	8.75	<0.0001
Stars	24.9724	5.884	4.24	<0.0001
Run Time	−0.403296	0.2513	−1.60	0.1113

a) Write the multiple regression equation.
b) What is the interpretation of the coefficient of *Budget* in this regression model?

4. Movie profits again A middle manager at an entertainment company, upon seeing the analysis of Exercise 3, concludes that longer movies make less money. He argues that his company's films should all be cut by 30 minutes to improve their gross. Explain the flaw in his interpretation of this model.

---

[1] If this reminds you of the partial boxplots you saw in Chapter 25, you're on the right track. Here, we've removed the effects of all the other predictors by plotting their residuals against the predictor of interest.

26.3 Model Assumptions and Conditions

We can write a multiple regression model like this, numbering the predictors arbitrarily (we don't care which one is x_1), writing β's for the model coefficients (which we will estimate from the data), and including the errors in the model:

$$y = \beta_0 + \beta_1 x_1 + \beta_2 x_2 + \varepsilon$$

Of course, the multiple regression model is not limited to two predictor variables, and regression model equations are often written to indicate summing any number (a typical letter to use is k) of predictors. That doesn't really change anything, so we'll often stick with the two-predictor version just for simplicity. But don't forget that we can have many predictors. As in simple regression, the symbol b_j will indicate the least-squares estimate of β_j.

The assumptions and conditions for the multiple regression model sound nearly the same as for simple regression, but with more variables in the model, we'll have to make a few changes.

Linearity Assumption

We are fitting a *linear model*.[2] For that to be the right kind of model, we need an underlying linear relationship. But now we're thinking about several predictors. To see whether the assumption is reasonable, we'll check the Straight Enough Condition for *each* of the predictors.

Straight Enough Condition: Scatterplots of y against each of the predictors are reasonably straight. As we have seen with *Height* in the body fat example, the scatterplots need not show a strong (or any!) slope; we just check that there isn't a bend or other nonlinearity. For the body fat data, the scatterplot is beautifully linear in *Waist*, as we saw in Chapter 23. For *Height*, we saw no relationship at all, but at least there was no bend.

As we did in simple regression, it's a good idea to check the residuals for linearity after we fit the model. It's good practice to plot the residuals against the predicted values and check for patterns, especially for bends or other nonlinearities. (We'll watch for other things in this plot as well.) We can also plot the residuals against each predictor to gain some more insight into the relationship between the response, y, and the particular predictor, x_i.

> **CHECK THE RESIDUAL PLOT (PART 1A)**
>
> The residuals should appear to have no pattern when plotted against the predicted values.
>
> **(PART 1B)**
>
> The residuals should appear to have no pattern when plotted against each of the predictors.

If we're willing to assume that the multiple regression model is reasonable, we can fit the regression model by least squares. But we must check the other assumptions and conditions before we can interpret the model or test any hypotheses.

Independence Assumption

As with simple regression, the errors in the true underlying regression model must be independent of each other. As usual, there's no way to be sure that the Independence Assumption is true. Fortunately, even though there can be many predictor variables, there is only one response variable and only one set of errors. The Independence Assumption concerns the errors, so we check the corresponding conditions on the residuals.

Randomization Condition: The data ideally should arise from a random sample or randomized experiment. Randomization assures us that the data are representative of some identifiable population. If you can't identify the population, you can interpret the regression model only as a description of the data you have, and it may not be possible to interpret the hypothesis tests in a meaningful way because they are about a regression model for that population. Regression methods are often applied to data that were not collected with randomization. Regression models fit to such data may still do a good job of modelling the data at hand, but

---

[2]By *linear*, we mean that each x appears simply multiplied by its coefficient and added to the model. No x appears in an exponent or some other more complicated function. That means that as we move along any x-variable, our prediction for y will change at a constant rate (given by the coefficient) if nothing else changes. When we have two predictors, this is just the equation of a plane, which is linear in the sense of being flat in all directions. With more predictors, the geometry is harder to visualize, but the simple structure of the model is consistent; the predicted values change consistently with equal size changes in any predictor.

CHECK THE RESIDUAL PLOT (PART 2)

The residuals should appear to be randomly scattered and show no patterns or clumps when plotted against the predicted values, or against *time* if data taken at different time points.

CHECK THE RESIDUAL PLOT (PART 3)

The spread of the residuals should be fairly uniform when plotted against any of the *x*'s or against the predicted values.

---

without some reason to believe that the data are representative of a particular population, you should be reluctant to believe that the model generalizes to other situations.

We also check displays of the regression residuals for evidence of patterns, trends, or clumping, any of which would suggest a failure of independence. As discussed in Chapter 23, in the special case when observations are taken in some time order, plotting the residuals against *time* will allow us to see whether the errors exhibit time dependence, a special form of violating the Independence Assumption.

The body fat data were collected on a sample of men. The men were not related in any way, so their measurements should be pretty much independent.

Equal Variance Assumption

The variability of the errors should be about the same for all values of *each* predictor. To see if this is reasonable, we look at scatterplots.

Does the Plot Thicken? Condition: Scatterplots of the regression residuals against each *x* variable or against the predicted values, $\hat{y}$, offer a visual check. The spread around the line should be nearly constant. Be alert for a "fan" shape or other tendency for the variability to grow or shrink in one part of the scatterplot.

Figure 26.4 shows the residuals plotted against *Waist* size and *Height*. Neither plot shows patterns that might indicate a problem.

Figure 26.4

Residuals plotted against each predictor show no pattern. That's a good indication that the Straight Enough Condition and the Does the Plot Thicken? Condition are satisfied.

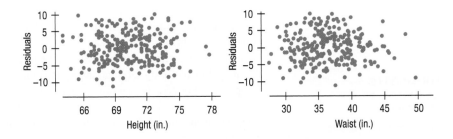

If residual plots show no pattern, if the residuals are plausibly independent, and if the plots don't thicken, we can feel good about interpreting the regression model. Before we test hypotheses, however, we must check one final assumption.

Normality Assumption

We assume that the errors around the idealized regression model at any specified values of the *x*-variables follow a Normal model. We need this assumption so that we can use a Student's *t*-model for inference. As with other times when we've used Student's *t*, we'll settle for the residuals satisfying the Nearly Normal Condition.

Nearly Normal Condition: Because we have only one set of residuals, this is the same set of conditions we had for simple regression. Look at a histogram or Normal probability plot of the residuals. The histogram of residuals in the body fat regression certainly looks nearly Normal, and the Normal probability plot is fairly straight (Figure 26.5). And, as we have said before, the Normality Assumption becomes less important as the sample size grows.

Let's summarize all the checks of conditions that we've made and the order that we've made them:

1. Check the Straight Enough Condition with scatterplots of the *y*-variable against each *x*-variable.
2. If the scatterplots are straight enough (that is, if it looks like the regression model is plausible), fit a multiple regression model to the data. (Otherwise, stop or consider re-expressing an *x*-variable or the *y*-variable. Another possibility is to try fitting a model with squared or cubic powers of some of the predictors.)
3. Find the residuals and predicted values.

Figure 26.5

Check a histogram of the residuals. The distribution of the residuals should be unimodal and symmetric. Better yet, check a Normal probability plot to see whether it is straight.

4. Make a scatterplot of the residuals against the predicted values.[3] This plot should look patternless. Check in particular for any bend (which would suggest that the data weren't all that straight after all) and for any thickening. If there's a bend, and especially if the plot thickens, consider re-expressing the *y*-variable and starting over. You should also check for extraordinary cases here. If you find any, investigate them for errors, or consider doing the analysis with and without them. (More details about outliers, leverage, and influence are in the next chapter.)

5. Consider also making scatterplots of the residuals against each predictor, particularly if you see patterns in the plot in 4. Check in particular for any bend, which would suggest the *y*-variable and *x*-variable do not have a straight line relationship. If this is the case, consider re-expressing that *x*-variable and starting over. You can look for unusual observations in this step as well.

6. Think about how the data were collected. Was suitable randomization used? Are the data representative of some identifiable population? If the data were collected in some time order, plot the residuals against time to check for non-random patterns that might suggest a lack of independence.

7. If the conditions check out this far, feel free to interpret the regression model and use it for prediction. If you want to investigate a particular coefficient, a partial regression plot for that coefficient may be useful.

8. If you wish to test hypotheses about the coefficients or about the overall regression (or construct confidence intervals for the coefficients or for predictions), then make a histogram and Normal probability plot of the residuals to check the Nearly Normal Condition.

Step-by-Step Example MULTIPLE REGRESSION

Question: How should we model %*Body Fat* in terms of *Height* and *Waist* size?

THINK ➡ **Variables** Name the variables, report the *W*'s, and specify the questions of interest.	I have quantitative body measurements on 250 adult males from the BYU Human Performance Research Center. I want to understand the relationship between %*Body fat*, *Height*, and *Waist* size.
Plan Think about the assumptions and check the conditions.	✓ **Straight Enough Condition:** There is no obvious bend in the scatterplots of %*Body Fat* against either *x*-variable. The scatterplot of residuals against predicted values a bit farther below shows no pattern that would suggest nonlinearity.
	✓ **Independence Assumption:** These data are not collected over time, and there's no reason to think that the %*Body fat* of one man influences that of another. I don't know whether the men measured were sampled randomly, but the data are presented as being representative of the male population of the United States.
Now we can find the regression and examine the residuals.	✓ **Does the Plot Thicken? Condition:** The scatterplot of residuals against predicted values below shows no obvious changes in the spread about the line.

---

[3]In Chapter 23, we noted that a scatterplot of residuals against the predicted values looked just like the plot of residuals against *x*. But for a multiple regression, there are several *x*'s. Now the predicted or $\hat{y}$ values are a combination of the *x*'s—in fact, they're the combination given by the regression equation we have computed. So they combine the effects of all the *x*'s in a way that makes sense for our particular regression model. That makes them a good choice to plot against.

(Actually, we need the Nearly Normal Condition only if we want to do inference.)

✓ **Nearly Normal Condition, Outlier Condition:** A histogram of the residuals is unimodal and symmetric.

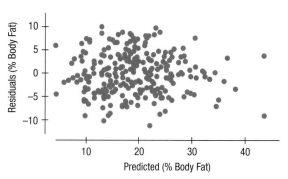

The Normal probability plot of the residuals is reasonably straight:

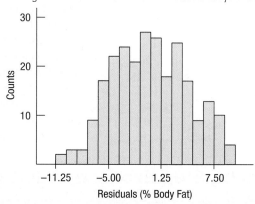

Choose your method.

Under these conditions a full multiple regression analysis is appropriate.

SHOW ➡ Mechanics

Here is the computer output for the regression:

Dependent variable is: %Body fat
R-squared = 71.3% R-squared (adjusted) = 71.1%
s = 4.460 with 250 − 3 = 247 degrees of freedom

Source	Sum of Squares	DF	Mean Square	F-ratio	P-value
Regression	12 216.6	2	6 108.28	307	<0.0001
Residual	4 912.26	247	19.8877		

Variable	Coefficient	SE(Coeff)	t-ratio	P-value
Intercept	−3.10088	7.686	−0.403	0.6870
Waist	1.77309	0.0716	24.8	<0.0001
Height	−0.60154	0.1099	−5.47	<0.0001

The estimated regression equation is

$$\widehat{\%Body\ fat} = -23.10 + 1.77\ Waist$$
$$- 0.60\ Height$$

TELL ➡ Interpretation

The R^2 for the regression is 71.3%. *Waist* size and *Height* together account for about 71% of the variation in *%Body fat* among men. The regression equation indicates that each additional inch in *Waist* size is associated with about a 1.77 increase in *%Body fat* among men who are of a particular height. Each inch of *Height* is associated with a decrease in *%Body fat* of about 0.60 among men with a particular waist size.

More Interpretation

The standard errors for the slopes of 0.07 (*Waist*) and 0.11 (*Height*) are both small compared with the slopes themselves, so it looks like the coefficient estimates are fairly precise. The residuals have a standard deviation of 4.46%, which gives some indication of how precisely we can predict *%Body fat* with this model.

Section 26.3 Exercises

5. More movies For the movies examined in Exercise 3, here is a scatterplot of *USGross* vs. *Budget*:

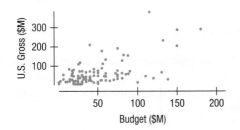

What (if anything) does this scatterplot tell us about the following Assumptions and Conditions for the regression?
a) Linearity condition
b) Equal Spread condition
c) Normality assumption

6. Movie residuals For the movies regression, here is a histogram of the residuals. What does it tell us about these Assumptions and Conditions?

a) Linearity condition
b) Nearly Normal condition
c) Equal Spread condition

26.4 Multiple Regression Inference

I Thought I Saw an ANOVA Table . . .

There are several hypothesis tests in the multiple regression output, but all of them talk about the same thing. Each is concerned with whether the underlying model parameters are actually zero.

The first of these hypotheses is one we skipped over for simple regression (other than in an optional section), for reasons that will be clear in a minute. Now that we've looked at some ANOVAs (in Chapters 24 and 25),[4] we can recognize the ANOVA table sitting in the middle of the regression output. Where'd that come from?

The answer is that now that we have more than one predictor, there's an *overall test* we should consider before we do more inference on the coefficients. We ask the global question, "Is this multiple regression model any good at all?" That is, would we do as well using just $\bar{y}$ to model y? Well, if all the coefficients (except the intercept) were zero, we'd have

$$y = \beta_0 + 0x_1 + \ldots + 0x_k + \varepsilon$$

and the least-squares estimate for β_0 turns out to be $b_0 = \bar{y}$ or $\hat{y} = \bar{y}$.

To address this overall question, we'll test

$$H_0: \beta_1 = \beta_2 = \ldots = \beta_k = 0.$$

(That null hypothesis looks very much like the null hypothesis we tested with an F-test in the one-way Analysis of Variance in Chapter 24.) This hypothesis test is often referred to as the "overall F-test" or "global F-test" in the multiple regression context.

We can test this hypothesis with a statistic that is labelled with the letter F (in honour of Sir Ronald Fisher, the developer of Analysis of Variance). In our example, the F-value is 307 on 2 and 247 degrees of freedom. The alternative hypothesis is just that the slope coefficients aren't all equal to zero, and only bigger F-values contradict the null hypothesis and lead to smaller P-values. If the null hypothesis were true, the F-statistic would be in the general vicinity of 1. The F-statistic here is quite large, so we can easily reject the null hypothesis and conclude that the multiple regression model is better than just using the mean.[5] (On page 759 you can see an F density curve and an F-table).

Why didn't we emphasize this for simple regression? Because the null hypothesis would have just been that the lone model slope coefficient was zero, and we were already testing that with the t-statistic for the slope. In fact, the *square* of that t-statistic is equal to the ANOVA F-statistic for the simple regression, so it really was the identical test.

Testing Individual Coefficients

Once we check the F-test and reject the null hypothesis—and, if we are being careful, *only* if we reject that hypothesis—we can move on to checking the test statistics for the individual coefficients. Those tests look like what we did for the slope of a simple regression in Chapter 23. For each coefficient we test

$$H_0: \beta_j = 0$$

against the (typically two-sided) alternative that it isn't zero. The regression table gives a standard error for each estimated coefficient, b_j, and the ratio of the estimated coefficient to its standard error. If the assumptions and conditions are met (and now we need the Nearly Normal Condition), these ratios follow a Student's t-distribution.

$$t = \frac{b_j - 0}{SE(b_j)}$$

But how many degrees of freedom does this t-distribution have? We have a rule of thumb that works here. The degrees of freedom is the number of data values minus the

> **PREDICTION INTERVALS**
>
> Just as with simple regression, we can also use the model to predict a response or estimate the mean response at specified values of the predictors, along with a SE and confidence interval. But just as with the SE(b_1), these SE formulas are messy. Just look for the "Prediction" option for your software. The only difference is that you have to specify the values of all of your multiple predictors. The interpretation of these intervals is the same as for simple regression.

---

[4]If you skipped over Chapters 24 and 25, you can just take our word on this and read on.

[5]There are F tables in the back of the book, and they work pretty much as you'd expect. Most regression tables include a P-value for the F-statistic, but often there's no need to perform this particular test in a multiple regression. Usually we just glance at the F-statistic to see that it's reasonably far from 1.0, the value it would have if the true coefficients were really all zero.

number of coefficients estimated (including the intercept). For our regression on two ($k = 2$) predictors, that's $n - 3$. You shouldn't have to look up the *t*-values. Almost every regression report includes the corresponding P-values.

We can build a confidence interval in the usual way, as an estimate ± a margin of error. As always, the margin of error is just the product of the standard error and a critical value. Here the critical value comes from the *t*-distribution on $n - k - 1$ degrees of freedom. So a confidence interval for β_1 is

$$b_1 \pm t_{n-k-1}^{*} \, SE(b_1)$$

The tricky parts of these tests are that the standard errors of the coefficients now require harder calculations (which we leave to technology), and the meaning of a coefficient, as we have seen, depends on all the *other* predictors in the multiple regression model.

That last bit is important. The null hypothesis for a multiple regression coefficient does **not** say that the corresponding predictor variable has no linear relationship to the response *y*. It says that the corresponding predictor contributes nothing to modelling *y* *after allowing for all the other predictors.*

Be Careful Interpreting the Coefficients

This last point bears repeating. The multiple regression model looks so simple and straightforward:

$$y = \beta_0 + \beta_1 x_1 + \cdots + \beta_k x_k + \varepsilon$$

It *looks* like each β_j tells us the effect of its associated predictor, x_j, on the response variable, *y*. But that is not so. This is, without a doubt, the most common error that people make with multiple regression:

■ It is possible for there to be no simple relationship between *y* and x_j, and yet b_j in a *multiple* regression can be very significantly different from 0. We saw this happen for the coefficient of *Height* in our example.

■ It is also possible for there to be a strong two-variable relationship between *y* and x_j, and yet b_j in a multiple regression can be almost 0 with a large P-value, so that we cannot reject the null hypothesis that the true coefficient is zero. For example, if we're trying to model the horsepower of a car using both its weight and its engine size, it may turn out that the coefficient for *Engine Size* is nearly 0. That *doesn't* mean that engine size isn't important for understanding horsepower. It simply means that after allowing for the weight of the car, the engine size doesn't give much *additional* information.

■ It is even possible for there to be a significant linear relationship between *y* and x_j in one direction, and yet b_j can be of the *opposite* sign and strongly significant in a multiple regression. More expensive cars tend to be bigger, and since bigger cars have worse fuel efficiency, the price of a car has a slightly negative association with fuel efficiency. But in a multiple regression of *Fuel Efficiency* on *Weight* and *Price*, the coefficient of *Price* may be positive. If so, it means that *among cars of the same weight*, more expensive cars have better fuel efficiency. The simple regression on *Price*, though, has the opposite direction, because, *overall*, more expensive cars are bigger. This switch in sign may seem a little strange at first, but it's not really a contradiction at all. It's due to the change in the *meaning* of the coefficient of *Price* when it is in a multiple regression rather than in a simple regression.

So we'll say it once more: The coefficient of x_j in a multiple regression depends as much on the *other* predictors as it does on x_j. Remember that when you interpret a multiple regression model.

For Example   INTERPRETING COEFFICIENTS

We looked at a multiple regression to predict the price of a house from its living area and the number of bedrooms. We found the model

$$\widehat{Price} = 308\,100 + 135\ LivingArea - 43\,347\ Bedrooms$$

However, common sense says that houses with more bedrooms are usually worth more. And, in fact, the simple regression of *Price* on *Bedrooms* finds the model

$$\widehat{Price} = 33\,897 + 40\,234\ Bedrooms$$

and the P-value for the slope coefficient is 0.0005.

QUESTION: How should we understand the coefficient of *Bedrooms* in the multiple regression?

ANSWER: The coefficient of *Bedrooms* in the multiple regression does not mean that houses with more bedrooms are generally worth less. It must be interpreted taking into account the other predictor (*LivingArea*) in the regression. If we consider houses with a given amount of living area, those that devote more of that area to bedrooms either must have smaller bedrooms or less living area for other parts of the house. Those differences could result in reducing the home's value.

 # Just Checking

Recall the regression example in Chapter 7 to predict hurricane maximum wind speed from central barometric pressure. Another researcher, interested in the possibility that global warming was causing hurricanes to become stronger, added the variable *Year* as a predictor and obtained the following regression:

Dependent variable is: Max. Winds (km)
275 total cases of which 113 are missing

R squared = 77.9% R squared (adjusted) = 77.6%
s = 7.727 with 162 − 3 = 159 degrees of freedom

Source	Sum of Squares	DF	Mean Square	F-ratio
Regression	33 446.2	2	16 723.1	280
Residual	9 493.45	159	59.7072	

Variable	Coefficient	s.e. of Coeff	t-ratio	prob
Intercept	1 009.99	46.53	21.7	≤0.0001
Central Pressure	−0.933491	0.0395	−23.6	≤0.0001
Year	−0.010084	0.0123	−0.821	0.4128

1. Interpret the R^2 of this regression.
2. Interpret the coefficient of *Central Pressure*.
3. The researcher concluded that "there has been no change over time in the strength of Atlantic hurricanes." Is this conclusion a sound interpretation from the regression model?

Another Example: Modelling Infant Mortality

Infant Mortality is often used as a general measure of the quality of healthcare for children and mothers. It is reported as the rate of deaths of newborns per 1000 live births. Data recorded for each of the 50 states of the United States may allow us to build regression models to help understand or predict infant mortality. The variables available for our model are *Child Deaths* (deaths per 100 000 children aged 1–14), percent of teens (ages 16–19) who drop out of high school (*HS Drop%*), percent of low-birth weight babies (*Low BW%*), *Teen Births* (births per 100 000 females aged 15–17), and *Teen Deaths* by accident, homicide, and suicide (deaths per 100 000 teens ages 15–19).[6] (Data in **Infant mortality**)

All of these variables were displayed and found to have no outliers.[7] One useful way to check many of our conditions is with a **scatterplot matrix**. Figure 26.6 shows an array of scatterplots set up so that the plots in each row have the same variable on their *y*-axis and those in each column have the same variable on their *x*-axis. This way every pair of variables is graphed. On the diagonal, rather than plotting a variable against itself, you'll usually find either a Normal probability plot or a histogram of the variable to help us assess the **Nearly Normal Condition**.

Who	U.S. states
What	Various measures relating to children and teens
When	1999
Why	Research and policy

Figure 26.6

A scatterplot matrix shows a scatterplot of each pair of variables arrayed so that the vertical and horizontal axes are consistent across rows and down columns. You can tell which variable is plotted on the *x*-axis of any plot by reading all the way down to the diagonal and looking to the vertical axis label on the left. The diagonal cells may hold Normal probability plots (as they do here), histograms, or just the names of the variables. These are a great way to check the Straight Enough Condition and to check for simple outliers.

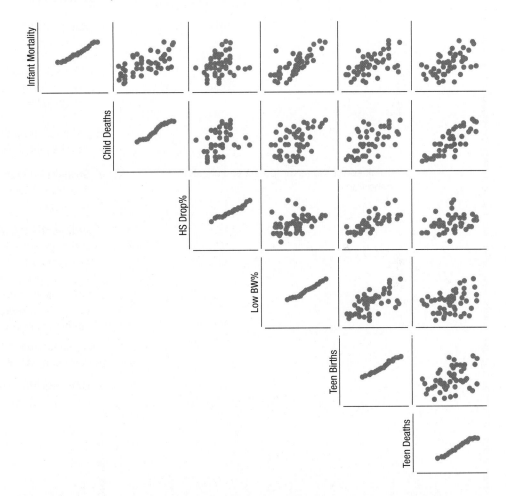

The individual scatterplots show at a glance that each of the relationships is straight enough for regression. There are no obvious bends, clumping, or outliers. And the plots don't thicken. So it looks like we can examine some multiple regression models with inference.

---

[6]The data are available from the Kids Count section of the Annie E. Casey Foundation (http://datacenter.kidscount.org/), and are all for 1999.

[7]In the interest of complete honesty, we should point out that the original data include the District of Columbia, but it proved to be an outlier on several of the variables, so we've restricted attention to the 50 states here.

Step-by-Step Example INFERENCE FOR MULTIPLE REGRESSION

Question: How should we model *Infant Mortality* using the available predictors?

THINK ➡ **Hypotheses** State what we want to know.

Plan State the hypotheses of interest.

(Hypotheses on the intercept are not particularly interesting for these data, as is usually the case.)

Think about the assumptions and check the conditions.

I wonder whether all or some of these predictors contribute to a useful model for *Infant Mortality*.

First, I'll check the overall null hypothesis that asks whether the entire model is better than just modelling the y variable with its mean:

H_0: The model itself contributes nothing useful, and all the slope coefficients are zero:

$$\beta_1 = \beta_2 = \ldots = \beta_k = 0$$

H_A: At least one of the β_j is not 0.

If I reject this global hypothesis, I can then test a null hypothesis for any of the coefficients of the form:

H_0: The *j*-th variable contributes nothing useful, after allowing for the other predictors in the model: $\beta_j = 0$.

H_A: The *j*-th variable makes a useful contribution to the model: $\beta_j \neq 0$.

✓ **Straight Enough Condition, Outlier Condition:** The scatterplot matrix shows no bends, clumping, or outliers.

✓ **Independence/Randomization:** Independence is not really an issue here, since we selected all the states, not a sample of them, but it also makes sense to recognize some inherent random variation in these state reported statistics, as well as their representativeness of general conditions in our southern neighbour.

These conditions are enough to compute the regression model and find residuals.

✓ **Does the Plot Thicken? Condition:** The residual plot shows no obvious trends in the spread:

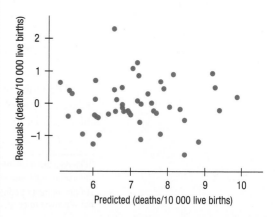

✓ **Nearly Normal Condition**: A histogram of the residuals is unimodal and symmetric.

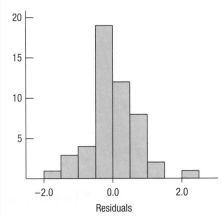

The one possible (though mild) outlier is South Dakota. I may want to repeat the analysis after removing South Dakota to see whether it changes substantially.

Choose your method.

Under these conditions I can continue with the multiple regression analysis.

SHOW ➡ Mechanics Multiple regressions are usually found from a computer program.

Computer output for this regression looks like this:

Dependent variable is: Infant Mort
R-squared = 71.3 % R-squared (adjusted) 68.0 %
s = 0.7520 with 50 − 6 = 44 degrees of freedom

Source	Sum of Squares	DF	Mean Square	F-ratio
Regression	61.7319	5	12.3464	21.8
Residual	24.8843	44	0.565553	

Variable	Coefficient	SE(Coeff)	t-ratio	P-value
Intercept	1.63168	0.9124	1.79	0.0806
Child Deaths	0.03123	0.0139	2.25	0.0292
HS Drop%	−0.09971	0.0610	−1.63	0.1096
Low BW%	0.66103	0.1189	5.56	<0.0001
Teen Births	0.01357	0.0238	0.57	0.5713
Teen Deaths	0.00556	0.0113	0.49	0.6245

The P-values given in the regression output table are from the Student's t-distribution on $(n − 6) = 44$ degrees of freedom. They are appropriate for two-sided alternatives.

Consider the hypothesis tests.

The F-ratio of 21.8 on 5 and 44 degrees of freedom is certainly large enough to reject the global null hypothesis that the regression model is useless. So I will examine the individual coefficients.

Under the assumptions we're willing to accept, and considering the conditions we've checked, the individual coefficient estimates follow Student's t-distributions on 44 degrees of freedom.

Most of these coefficients have relatively small t-ratios, so there is little evidence that their true underlying values are non-zero. Two of the coefficients, *Child Deaths* and *Low BW%*, have P-values less than 5%. So I can be confident that in this model both of these variables are unlikely to really have zero coefficients.

TELL ➡ Interpretation

Overall, the R^2 indicates that more than 71% of the variability in infant mortality can be accounted for with this regression model.

After allowing for the linear effects of the other variables in the model, an increase in child deaths of 1 death per 100 000 is associated with an increase of 0.03 deaths per 1000 live births in the infant mortality rate. And an increase of 1% in the percentage of live births that are low birth weight is associated with an increase of 0.66 deaths per 1000 live births.

*Testing a *Set of Predictors* in a Model

For the multiple regression model,

$$y = \beta_0 + \beta_1 x_1 + \ldots + \beta_g x_g + \beta_{g+1} x_{g+1} + \ldots + \beta_k x_k + \varepsilon$$

we use the F-test to test the null hypothesis that the overall regression is no improvement over just modelling y with its mean:

$$H_0 : \beta_1 = \beta_2 = \ldots = \beta_k = 0$$

And the t-ratios for the coefficients are used to test the significance of each regression coefficient. Besides these two tests, sometimes we might want to test the hypothesis that the overall regression is no improvement over modelling y with only the predictors $x_1, \ldots, x_g$:

$$y = \beta_0 + \beta_1 x_1 + \ldots + \beta_g x_g + \varepsilon$$

That is, we are testing

$$H_0 : \beta_{g+1} = \beta_{g+2} = \ldots = \beta_k = 0$$

which is a statistical way of saying that the set of predictors $x_{g+1}, \ldots, x_k$ are not needed and may be removed from the original model.

Let us revisit the *Infant Mortality* example. We have already determined that

$$InfantMort = \beta_0 + \beta_1 LowBW\% + \beta_2 ChildDeaths + \beta_3 HSDrop\% +$$
$$\beta_4 TeenBirths + \beta_5 TeenDeaths + \varepsilon$$

is a significant model. Individually, we found that *HSDrop%*, *TeenBirths*, and *TeenDeaths* may not make a useful contribution to the model. However, deleting any one of them changes the model and possibly the significance of the other predictors. Rather than examining them individually, we can examine their joint contribution by testing if the above model is no improvement over the simpler model

$$InfantMort = \beta_0 + \beta_1 LowBW\% + \beta_2 ChildDeaths\% + \varepsilon.$$

In other words, we are testing the hypothesis:

$$H_0 : \beta_3 = \beta_4 = \beta_5 = 0.$$

How can we test this hypothesis?

Before we answer the question, we need to introduce some new terminology. The model with all the predictors in it is known as the *Complete Model*. If H_0 is true, the model that deletes the predictors we think may be worthless, as specified in the hypothesis is known as the *Reduced Model*. For the *Infant Mortality* example, the Complete Model is:

$$InfantMort = \beta_0 + \beta_1 LowBW\% + \beta_2 ChildDeaths + \beta_1 HSDrop\% +$$
$$\beta_4 TeenBirths + \beta_5 TeenDeaths + \varepsilon$$

and the Reduced Model

$$InfantMort = \beta_0 + \beta_1 LowBW\% + \beta_2 ChildDeaths\% + \varepsilon$$

If the null hypothesis is true, then the Reduced Model is as good a fit to the data as the Complete Model.

The $SS_{Residuals}$ of a regression model measures the portion of the total variation not explained by the model (or equivalently, $SS_{Regression}$ measures the portion of the total variation explained by the model). Therefore by examining the difference in the $SS_{Residuals}$ between the two models (equivalent to the difference in the $SS_{Regression}$, because SS_{Total} remains unchanged), we have a measure of the extra variation explained by the additional predictors in the model. The Complete Model must have a smaller $SS_{Residuals}$ (larger $SS_{Regression}$), unless you are extremely unlucky, in which case you just might get equality. The difference is known as SS_{Drop}, which can be calculated by

$$SS_{Drop} = SS_{Residuals}(Reduced) - SS_{Residuals}(Complete)$$
$$= SS_{Regression}(Complete) - SS_{Regression}(Reduced)$$

So, what we need to do is to take this difference and convert into an appropriate test statistic. As in Chapter 24, we cannot directly compare SS's since they have different numbers of squared deviations, but mean squared deviations are comparable. By comparing the extra variation explained by the additional predictors, using its Mean Square, with the Mean Squared Error, we can see if this is a significant source of variation. More specifically, the test statistic is an F-statistic which takes the form

$$\frac{MS_{Drop}}{MS_{Residuals}(Complete)}$$

where

$$MS_{Drop} = \frac{SS_{Drop}}{k - g}$$

$$MS_{Residuals}(Complete) = \frac{SS_{Residuals}(Complete)}{n - (k + 1)}.$$

Note that $(k - g)$ is just the number of predictors we are testing. The P-value of the test can then be obtained from the F-distribution with $(k - g)$ and $(n - k - 1)$. Again the smaller the P-value, the more the evidence is against the null hypothesis.

Now back to our example. We have $n = 50$; $k = 5$; $g = 2$; $SS_{Residuals}(Complete) = 24.884$, $SS_{Residuals}(Reduced) = 26.938$. Therefore $SS_{Drop} = 26.938 = 24.884 - 2.054$ and the F-statistic is

$$\frac{2.054/(5 - 3)}{24.884/(50 - 6)} = 1.816.$$

Using a computer, the corresponding P-value from the F-distribution with 2 and 44 degrees of freedom is 0.1746. (If we are using the tables in this text, we can only report that P-value > 0.1.) With a large P-value, we have little evidence that the complete model is an improvement over a reduced model. That is, we have little evidence that these three "teen" variables (*HS drop%*, *Teen Births*, and *Teen Deaths*) would add anything useful to the model including just *Low BW%* and *Child Deaths*.[8]

---

[8]Note that looking at the three non-significant individual *t*-ratios in the earlier output does not assure us of a similar conclusion, since if you drop one predictor from the model, you have to refit the new model before assessing any other predictors.

Section 26.4 Exercises

7. **Movie tests** In the regression output for the movies of Exercise 3,
   a) What is the null hypothesis tested for the coefficient of *Stars* in this table?
   b) What is the *t*-statistic corresponding to this test?
   c) What is the P-value corresponding to this *t*-statistic?
   d) Complete the hypothesis test. Do you reject the null hypothesis?

8. **More movie tests**
   a) What is the null hypothesis tested for the coefficient of *Run Time* in the regression of Exercise 3?
   b) What is the *t*-statistic corresponding to this test?
   c) Why is this *t*-statistic negative?
   d) What is the P-value corresponding to this *t*-statistic?
   e) Complete the hypothesis test. Do you reject the null hypothesis?

26.5 Comparing Multiple Regression Models

There may be even more variables available to model *Infant Mortality*. Moreover, several of those we tried don't seem to contribute to the model. How do we know that some other choice of predictors might not provide a better model? What exactly *would* make an alternative model better?

These are not easy questions. There is no simple measure of the success of a multiple regression model. Many people look at the R^2 value, and certainly we are not likely to be happy with a model that accounts for only a small fraction of the variability of *y*. But that's not enough. You can always drive the R^2 up simply by piling on more and more predictors, but models with many predictors are hard to understand. Keep in mind that the meaning of a regression coefficient depends on all the *other* predictors in the model, so it is best to keep the number of predictors as small as possible.

Regression models should make sense. Predictors that are easy to understand are usually better choices than obscure variables. Similarly, if there is a known mechanism by which a predictor has an effect on the response variable, that predictor is usually a good choice for the regression model.

How can we know whether we have the best possible model? The simple answer is that we can't. There's always the chance that some other predictors might bring an improvement (in higher R^2 or fewer predictors or simpler interpretation).

Adjusted R^2

You may have noticed that the full regression tables shown in this chapter include another statistic we haven't discussed. It is called adjusted R^2 and sometimes appears in computer output as R^2 (adjusted). The **adjusted R^2** statistic is a rough attempt to adjust for the simple fact that when we add another predictor to a multiple regression, the R^2 can't go down and will almost surely go up. Only if we were to add a predictor whose estimated coefficient turned out to be exactly zero would the R^2 remain the same. This fact complicates the comparison of alternative regression models that have different numbers of predictors.

We can write a formula for R^2 using the Sums of Squares in the ANOVA table portion of the regression output table:

$$R^2 = \frac{SS_{Regression}}{SS_{Regression} + SS_{Residual}} = 1 - \frac{SS_{Residual}}{SS_{Total}}$$

Adjusted R^2 simply substitutes the corresponding *Mean Squares* for the SS's:[9]

$$R^2_{adj} = 1 - \frac{MS_{Residual}}{MS_{Total}}$$

---

[9]We learned about Mean Squares in Chapter 24. A Mean Square is just a Sum of Squares divided by its appropriate degrees of freedom. Mean Squares are just special types of variances.

Because the Mean Squares are Sums of Squares divided by degrees of freedom, they are adjusted for the number of predictors in the model. As a result, the adjusted R^2 value won't necessarily increase when a new predictor is added to the multiple regression model. That's fine. But adjusted R^2 no longer simply tells the fraction of variability accounted for by the model and it isn't even bounded by 0 and 100%, so it can be awkward to interpret.

Comparing alternative regression models is a challenge, especially when they have different numbers of predictors. The search for a summary statistic to help us choose among models is the subject of much contemporary research in Statistics. Adjusted R^2 is one common—but not necessarily the best—choice often found in computer regression output tables. It's certainly not the sole decision criterion when you compare different regression models. (The test discussed in the previous optional section is only useful when one set of predictors is a *subset* of the other set of predictors.)

Section 26.5 Exercises

9. Movie R-squared In the regression model of Exercise 3,
   a) What is the R^2 for this regression? What does it mean?
   b) Why is the "Adjusted R Square" in the table different from the "R Square"?

10. Movie output Here is another part of the regression output for the movies in Exercise 3:

Source	Sum of Squares	df	Mean Square	F-ratio
Regression	224 995	3	74 998.4	34.8
Residual	249 799	116	2 153.44	

a) Using the values from the table, show how the value of R^2 could be computed. Don't try to do the calculation, just show what is computed.
b) What is the F-statistic value for this regression?
c) What null hypothesis can you test with it?
d) Would you reject that null hypothesis?

WHAT CAN GO WRONG?

Interpreting Coefficients

- **Don't interpret regression causally.** Regressions are usually applied to observational data. Without deliberately assigned treatments, randomization, and control, we can't draw conclusions about causes and effects. We can never be certain that there are no variables lurking in the background, causing everything we've seen. Don't interpret b_1, the coefficient of x_1 in the multiple regression, by saying, "If we were to change an individual's x_1 by 1 unit (holding the other x's constant), it would change that individual's y by b_1 units." We have no way of knowing what applying a change to an individual would do.

- **Be cautious about interpreting a regression model as predictive.** Yes, we do call the x's predictors, and you can certainly plug in values for each of the x's and find a corresponding *predicted value* $\hat{y}$. But the term "prediction" suggests extrapolation into the future or beyond the data, and we know that we can get into trouble when we use models to estimate $\hat{y}$ values for x's not in the range of the data. Be careful not to extrapolate far from the span of your data. In simple regression, it was easy to tell when you extrapolated. With many predictor variables, it's often harder to know when you are outside the bounds of your original data.[10] We usually think of fitting models to the data more as modelling than as prediction, so that's often a more appropriate term.

(continued)

[10]With several predictors we can wander beyond the data because of the *combination* of values, even when individual values are not extraordinary. For example, both 28-inch waists and 76-inch heights can be found in men in the body fat study, but a single individual with both these measurements would not be at all typical. The model we fit is probably not appropriate for predicting the *%Body fat* for such a tall and skinny individual.

- **Don't claim to "hold everything else constant" for a single individual.** It's often meaningless to say that a regression coefficient says what we expect to happen if all variables but one were held constant for an individual and the predictor in question changed. While it's mathematically correct, it often just doesn't make any sense. We can't gain a year of experience or have another child without getting a year older. Instead, we *can* think about all those who fit given criteria on some predictors and ask about the conditional relationship between *y* and one *x* variable for those individuals. The coefficient −0.60 of height for predicting *%Body fat* says that among men of the same waist size, those who are one inch taller in height tend to be, on average, 0.60% lower in *%Body fat*. The multiple regression coefficient measures that average conditional relationship.

- **The sign of a coefficient is also model-dependent.** Sometimes our primary interest in a predictor is whether it has a positive or negative association with *y*. As we have seen, though, the sign of the coefficient also depends on the other predictors in the model. Don't look at the sign in isolation and conclude that "the direction of the relationship is positive (or negative)." Just like the value of the coefficient, the sign is about the relationship after allowing for the linear effects of the other predictors. The sign of a variable can change depending on which other predictors are in or out of the model. For example, in the regression model for infant mortality, the coefficient of *HS drop%* was negative and its P-value was fairly small, but the simple association between *HS drop%* and *Infant Mortality* is positive. (Check the matrix plot.)

- **If a coefficient's *t*-statistic is not significant, don't interpret it at all.** You can't be sure that the value of the corresponding parameter in the underlying regression model isn't really zero.

WHAT ELSE CAN GO WRONG?

- **Don't fit a linear regression to data that aren't straight.** Linearity is the most fundamental regression assumption. If the relationship between the *x*'s and *y* isn't approximately linear, there's no sense in fitting a linear model to it. What we mean by "linear" is a model of the form we have been writing for the regression. When we have two predictors, this is the equation of a plane, which is linear in the sense of being flat in all directions. With more predictors, the geometry is harder to visualize, but the simple structure of the model is consistent; the predicted values change consistently with equal size changes in any predictor.

  Usually we're satisfied when plots of *y* against each of the *x*'s are straight enough. We'll also check a scatterplot of the residuals against the predicted values for signs of nonlinearity.

- **Watch out for the plot thickening.** The estimate of the error standard deviation shows up in all the inference formulas (which were too complicated to show you, without using vector-matrix notation). But that estimate assumes that the error standard deviation is the same throughout the range of the *x*'s so that we can combine (pool, actually) all the residuals when we estimate it. If s_e changes with any *x*, these estimates won't make sense. The most common check is a plot of the residuals against the predicted values. If plots of residuals against several of the predictors all show a thickening, and especially if they also show a bend, then consider re-expressing *y*. If the scatterplot against only one predictor shows thickening, consider re-expressing that predictor.

- **Make sure the errors are nearly Normal.** All of our inferences require that the true errors be modelled well by a Normal model. Check the histogram and Normal probability plot of the residuals to see whether this assumption looks reasonable.

- **Watch out for outliers and high-influence points.** We always have to be on the lookout for a few points that may have undue influence on our model, and regression is certainly no exception. Partial regression plots are a good place to look for influential points and to understand how they might affect each of the coefficients. Chapter 27 discusses this issue in greater depth.

What Have We Learned?

In Chapter 23 we learned to apply our inference methods to linear regression models. Now we've seen that much of what we know about those models is also true for multiple regression:

- The assumptions and conditions are the same: linearity (checked now with scatterplots of y against each x), independence (think about it), constant variance (checked with the scatterplot of residuals against predicted values), and Nearly Normal residuals (checked with a histogram or probability plot).
- R^2 is still the fraction of the variation in y accounted for by the regression model.
- s_e is still the standard deviation of the residuals—a good indication of the precision of the model.
- The degrees of freedom (in the denominator of s_e and for each of the t-tests) follows the same rule: n minus the *number of parameters estimated*.
- The regression table produced by any statistics package shows a row for each coefficient, giving its estimate, a standard error, a t-statistic, and a P-value.
- If all the conditions are met, we can test each coefficient against the null hypothesis that its parameter value is zero with a Student's t-test.

And we've learned some new things that are useful now that we have multiple predictors:

- We can perform an overall test of whether the multiple regression model provides a better summary for y than its mean by using the F-distribution we saw in the previous chapters.
- We learned that R^2 may not be appropriate for comparing multiple regression models with different numbers of predictors. Adjusted R^2 is one approach to this problem.

Finally, we've learned that multiple regression models extend our ability to model the world to many more situations, but that we must take great care when we interpret its coefficients. To interpret a coefficient of a multiple regression model, remember that it estimates the linear relationship between y and that predictor *after allowing for the linear effects of all the other predictors on both y and that x*.

Review of Terms

Multiple regression
A linear regression with two or more predictors whose coefficients are found to minimize the sum of the squared residuals is a least squares linear multiple regression. But it is usually just called a multiple regression. When the distinction is needed, a least squares linear regression with a single predictor is called a simple regression. The multiple regression model is (p. 862)

$$y = \beta_0 + \beta_1 x_1 + \ldots + \beta_k x_k + \varepsilon$$

Least Squares
We still fit multiple regression models by choosing the coefficients that make the sum of the squared residuals as small as possible. This is called the method of least squares (p. 862).

Partial regression plot
The partial regression plot for a specified coefficient is a display that helps in understanding the meaning of that coefficient in a multiple regression. It has a slope equal to the coefficient value and gives some idea of the influence of each case on that value. Partial regression plots display the residuals when y is regressed on the *other* predictors against the residuals when the specified x-variable is regressed on the other predictors (p. 866).

ANOVA table (for Regression)
The Analysis of Variance table that is ordinarily part of the multiple regression results offers an F-test to test the null hypothesis that the overall regression is no improvement over just modelling the response y with its mean:

$$H_0: \beta_1 = \beta_2 = \ldots = \beta_k = 0$$

If this null hypothesis is not rejected at some reasonable level, you should not proceed to test any of the individual coefficients (p. 872).

t-ratios for the coefficients

The *t*-ratios for the coefficients can be used to test the null hypotheses that the true value of each coefficient is zero against the alternative that it is not. This tests whether the corresponding predictor actually contributes to this particular model. (p. 872):

$$t = \frac{b_j - 0}{SE(b_j)}$$

Scatterplot matrix

A scatterplot matrix displays scatterplots for all pairs of a collection of variables, arranged so that all the plots in a row have the same variable displayed on their *y*-axis and all plots in a column have the same variable on their *x*-axis. Usually, the diagonal holds a display of a single variable such as a histogram or Normal probability plot, and identifies the variable in its row and column (p. 875).

Adjusted R^2

An adjustment to the R^2 statistic that attempts to allow for the number of predictors in the model. It is sometimes used when comparing regression models with different numbers of predictors (p. 880).

$$R^2_{adj} = 1 - \frac{MS_{Residual}}{MS_{Total}}$$

On the Computer REGRESSION ANALYSIS

All statistics packages make a table of results for a regression. If you can read a package's regression output table for simple regression, then you can read its table for a multiple regression. You'll want to look at the ANOVA table, and you'll see information for each of the coefficients, not just for a single slope.

Most packages offer to plot residuals against predicted values. Some will also plot residuals against the individual *x*'s. With some packages you must request plots of the residuals when you request the regression. Others let you find the regression first and then analyze the residuals afterward. Either way, your analysis is not complete if you don't check the residuals with a histogram or Normal probability plot and a scatterplot of the residuals against the *x*'s or the predicted values.

One good way to check assumptions before embarking on a multiple regression analysis is with a scatterplot matrix. Multiple regressions are always found with a computer or programmable calculator. Before computers were available, a full multiple regression analysis could take months of work.

EXCEL

- In Excel 2003 and earlier, select **Data Analysis** from the **Tools** menu.
- In Excel 2007, select **Data Analysis** from the **Analysis Group** on the Data Tab.
- Select Regression from the **Analysis Tools** list.
- Click the **OK** button.
- Enter the data range holding the Y-variable in the box labelled "Y-range."
- Enter the range of cells holding the X-variables in the box labelled "X-range."
- Select the **New Worksheet Ply** option.
- Select **Residuals** options. Click the **OK** button.

COMMENTS

The Y and X ranges do not need to be in the same rows of the spreadsheet, although they must cover the same number of cells. But it is a good idea to arrange your data in parallel columns, as in a data table. The X-variables must be in adjacent columns. No cells in the data range may hold non-numeric values.

Although the dialogue offers a Normal probability plot of the residuals, the data analysis add-in does not make a correct probability plot, so don't use this option.

JMP

- From the **Analyze** menu select **Fit Model**.
- Specify the response, Y. Assign the predictors, X, in the **Construct Model Effects** dialogue box.
- Click on **Run Model**.

COMMENTS

JMP chooses a regression analysis when the response variable is "Continuous." The predictors can be any combination of quantitative or categorical. If you get a different analysis, check the variable types.

MINITAB

- Choose **Regression** from the **Stat** menu.
- Choose **Regression** and then **Fit Regression Model . . .** from the **Regression** submenu.
- In the Regression dialogue, assign the Y-variable to the Response box and assign the X-variables to the Continuous Predictors box.
- Click the **Graphs** button.

- In the Regression-Graphs dialogue, select **Standardized residuals**, and check **Normal plot of residuals** and **Residuals versus fits** or choose **Four in One** for a useful batch of four residual plots.
- Click the **OK** button to return to the Regression dialogue.
- Click the **OK** button to compute the regression.

R

Suppose the response variable y and predictor variables $x_1, \ldots, x_k$ are in a data frame called mydata. To fit a multiple regression of y on x_1 and x_2:

- mylm = lm(y ~ x_1 + x_2, data = mydata)
- summary(mylm) # gives the details of the fit, including the ANOVA table
- plot(mylm) # gives a variety of plots
  To fit the model with *all* the predictors in the data frame,
- mylm = lm(y ~., data = mydata) # The period means use all other variables

COMMENTS

To get confidence or prediction intervals use:
- predict(mylm, interval = "confidence")
or
- predict(mylm, interval = "prediction")

SPSS

- Choose **Regression** from the **Analyze** menu.
- Choose **Linear** from the **Regression** submenu.
- When the Linear Regression dialogue appears, select the Y-variable and move it to the dependent target. Then, move the X-variables to the independent target.
- Click the **Plots** button.

- In the Linear Regression Plots dialogue, choose to plot the *SRESIDs against the *ZPRED values.
- Click the **Continue** button to return to the Linear Regression dialogue.
- Click the **OK** button to compute the regression.

STATCRUNCH

To find a multiple regression of a response variable y on two factors, x_1 and x_2, from the **Stat** menu,
- Choose **Regression** > **Multiple Linear**.
- In the dialog, specify the response in **Y Variable**. Specify the predictors in **X Variables**.
- Click Compute! to perform the regression.

COMMENTS

You can save residuals and predicted values among other quantities with the **save options**. In the dialog box, you can also specify **Interactions**—not discussed in this chapter, but similar to those discussed in Chapter 25; more about this topic in Chapter 27.

TI-83/84

You need a special program to compute a multiple regression on the TI-83.

Chapter Exercises

T indicates data is available in a saved file at the Companion Website at www.pearsoncanada.ca/deveaux

11. Interpretations A regression performed to predict selling price of houses found the equation

$\widehat{Price} = 169\,328 + 35.3\,Area + 0.718\,Lotsize - 6543\,Age$

where price is in dollars, area is in square feet, lotsize is in square feet, and age is in years. The R^2 is 92%. One of the interpretations below is correct. Which is it? Explain what's wrong with the others.

a) Each year a house ages, it is worth $6543 less.
b) Every extra square foot of area is associated with an additional $35.30 in average price, for houses with a given lotsize and age.
c) Every dollar in price means lotsize increases 0.718 square feet.
d) This model fits 92% of the data points exactly.

12. More interpretations A household appliance manufacturer wants to analyze the relationship between total sales and the company's three primary means of advertising (television, magazines, and radio). All values were in millions of dollars. They found the regression equation

$\widehat{Sales} = 250 + 6.75\,TV + 3.5\,Radio + 2.3\,Magazines.$

One of the interpretations below is correct. Which is it? Explain what's wrong with the others.

a) If they did no advertising, their income would be $250 million.
b) Every million dollars spent on radio makes sales increase $3.5 million, all other things being equal.
c) Every million dollars spent on magazines increases television spending by $2.3 million.
d) Sales increase on average about $6.75 million for each million spent on television, after allowing for the effects of the other kinds of advertising.

13. Predicting final exams How well do tests given during the semester predict performance on the final? One class had three tests during the semester. Computer output of the regression gives

Dependent variable is: Final
s = 13.46 R-Sq = 77.7% R-Sq(adj) = 74.1%

Predictor	Coeff	SE(Coeff)	t	P-value
Intercept	−6.72	14.00	−0.48	0.636
Test1	0.2560	0.2274	1.13	0.274
Test2	0.3912	0.2198	1.78	0.091
Test3	0.9015	0.2086	4.32	<0.0001

Analysis of Variance

Source	DF	SS	MS	F	P-value
Regression	3	11 961.8	3 987.3	22.02	<0.0001
Error	19	3 440.8	181.1		
Total	22	15 402.6			

a) Write the equation of the regression model.
b) How much of the variation in final exam scores is accounted for by the regression model?
c) Explain in context what the coefficient of Test3 scores means.
d) A student argues that clearly the first test doesn't help to predict final performance. She suggests that this exam not be given at all. Does Test1 have no effect on the final exam score? Can you tell from this model? (*Hint:* Do you think test scores are related to each other?)

T 14. Scottish hill races 2008 Hill running—races up and down hills—has a written history in Scotland dating back to the year 1040. Races are held throughout the year at different locations around Scotland. A recent compilation of information for 91 races (for which full information was available and omitting two unusual races) includes the *Distance* (km), the *Climb* (m), and the *Record time* (minutes). A regression to predict the men's records as of 2008 looks like this:

Dependent variable is: Men's Time (mins)
R-squared 98.0% R-squared (adjusted) 98.0%
s = 6.623 with 90 − 3 = 87 degrees of freedom

Source	Sum of Squares	DF	Mean Square	F-ratio
Regression	18 9204	2	94 602.1	2157
Residual	3 815.92	87	43.8612	

Variable	Coefficient	SE(Coeff)	t-ratio	P-value
Intercept	−10.3723	1.245	−8.33	<0.0001
Climb (m)	0.034227	0.0022	15.7	<0.0001
Distance (km)	4.04204	0.1448	27.9	<0.0001

a) Write the regression equation. Give a brief report on what it says about men's record times in hill races.
b) Interpret the value of R^2 in this regression.
c) What does the coefficient of *Climb* mean in this regression?

15. Home prices Many variables have an impact on determining the price of a house. A few of these are size of the house (*Sq ft*), *Lotsize*, and number of *Bathrooms*. Information for a random sample of homes for sale in the Statesboro, GA, area was obtained from the Internet. Regression output modelling the asking price with *Square Footage* and number of *Bathrooms* gave the following result:

Dependent Variable is: Asking Price
s = 67 013 R-Sq = 71.1% R-Sq (adj) = 64.6%

Predictor	Coeff	SE(Coeff)	t	P-value
Intercept	−152 037	85 619	−1.78	0.110
Baths	9 530	40 826	0.23	0.821
Sq ft	139.87	46.67	3.00	0.015

Analysis of Variance

Source	DF	SS	MS	F	P-value
Regression	2	99 303 550 067	49 651 775 033	11.06	0.004
Residual	9	40 416 679 100	4 490 742 122		
Total	11	1.39720E+11			

a) Write the regression equation.
b) How much of the variation in home asking prices is accounted for by the model?
c) Explain in context what the coefficient of *Square Footage* means.
d) The owner of a construction firm, upon seeing this model, objects because the model says that the number of bathrooms has no effect on the price of the home. He says that when *he* adds another bathroom, it increases the value. Is it true that the number of bathrooms is unrelated to house price? (*Hint:* Do you think bigger houses have more bathrooms?)

T 16. **More hill races 2008** Here is the regression for the women's records for the same Scottish hill races we considered in Exercise 14:

Dependent variable is: Women's Time (mins)
R-squared 96.7% R-squared (adjusted) 96.7%
s = 10.06 with 90 − 3 = 87 degrees of freedom

Source	Sum of Squares	DF	Mean Square	F-ratio
Regression	261 029	2	130 515	1 288
Residual	8 813.02	87	101.299	

Variable	Coefficient	SE(Coeff)	t-ratio	P-value
Intercept	−11.6545	1.891	−1.16	≤0.0001
Climb (m)	0.045195	0.0033	13.7	≤0.0001
Distance (km)	4.43427	0.2200	20.2	≤0.0001

a) Compare the regression model for the women's records with that found for the men's records in Exercise 14.
Here's a scatterplot of the residuals for this regression:

b) Discuss the residuals and what they say about the assumptions and conditions for this regression.

17. **Predicting finals II** Here are some diagnostic plots for the final exam data from Exercise 13. These were generated by a computer package and may look different from the plots generated by the packages you use. (In particular, note that the axes of the Normal probability plot are swapped relative to the plots we've made in the text. We only care about the pattern of this plot, so it shouldn't affect your interpretation.) Examine these plots and discuss whether the assumptions and conditions for the multiple regression seem reasonable.

18. **Home prices II** Here are some diagnostic plots for the home prices data from Exercise 15. These were generated by a computer package and may look different from

the plots generated by the packages you use. (In particular, note that the axes of the Normal probability plot are swapped relative to the plots we've made in the text. We only care about the pattern of this plot, so it shouldn't affect your interpretation.) Examine these plots and discuss whether the assumptions and conditions for the multiple regression seem reasonable.

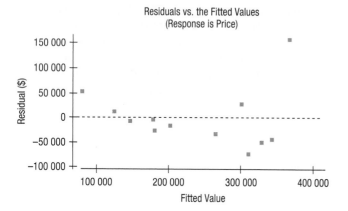

Residuals vs. the Fitted Values
(Response is Price)

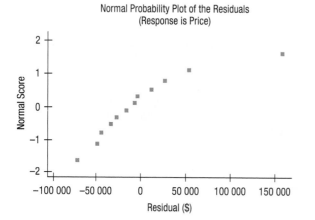

Normal Probability Plot of the Residuals
(Response is Price)

Histogram of the Residuals
(Response is Price)

19. Receptionist performance The American Federation of Labor has undertaken a study of 30 receptionists' yearly salaries (in thousands of dollars). The organization wants to predict *Salaries* from several other variables.

The variables considered to be potential predictors of salary are:

X1 = months of service

X2 = years of education

X3 = score on standardized test

X4 = words per minute (wpm) typing speed

X5 = ability to take dictation in words per minute

A multiple regression model with all five variables was run on a computer package, resulting in the following output:

Variable	Coefficient	Std. Error	*t*-value
Intercept	9.788	0.377	25.960
X1	0.110	0.019	5.178
X2	0.053	0.038	1.369
X3	0.071	0.064	1.119
X4	0.004	0.307	0.013
X5	0.065	0.038	1.734

$s = 0.430$ $R^2 = 0.863$

Assume that the residual plots show no violations of the conditions for using a linear regression model.
a) What is the regression equation?
b) From this model, what is the predicted *Salary* (in thousands of dollars) of a receptionist with 10 years (120 months) of experience, Grade 9 education (nine years of education), a 50 on the standardized test, 60 wpm typing speed, and the ability to take 30 wpm dictation?
c) Test whether the coefficient for words per minute of typing speed (X4) is significantly different from zero at $\alpha = 0.05$.
d) How might this model be improved?
e) A correlation of *Age* with *Salary* finds $r = 0.682$, and the scatterplot shows a moderately strong positive linear association. However, if X6 = *Age* is added to the multiple regression, the estimated coefficient of *Age* turns out to be $b_6 = -0.154$. Explain some possible causes for this apparent change of direction in the relationship between *Age* and *Salary*.

20. GPA and SATs A large section of Stat 200 was asked to fill out a survey on grade point average and SAT scores. A regression was run to find out how well Math and Verbal SAT scores could predict academic performance as measured by GPA. The regression was run on a computer package with the following output:

Response: GPA

	Coefficient	Std Error	*t*-ratio	Prob > |t|
Intercept	0.574968	0.253874	2.26	0.0249
SAT Verbal	0.001394	0.000519	2.69	0.0080
SAT Math	0.001978	0.000526	3.76	0.0002

a) What is the regression equation?
b) From this model, what is the predicted GPA of a student with an SAT Verbal score of 500 and an SAT Math score of 550?

c) What else would you want to know about this regression before writing a report about the relationship between SAT scores and grade point averages? Why would these be important to know?

T 21. Body fat revisited The data set on body fat contains 15 body measurements on 250 men from 22 to 81 years old. Is average *%Body fat* related to *Weight*? Here's a scatterplot:

Weight (lb)

And here's the simple regression:

Dependent variable is: Pct BF
R-squared = 38.1% R-squared (adjusted) = 37.9%
s = 6.538 with 250 − 2 = 248 degrees of freedom

Variable	Coefficient	SE(Coeff)	t-ratio	P-value
Intercept	−14.6931	2.760	−5.32	<0.0001
Weight	0.18937	0.0153	12.4	<0.0001

a) Is the coefficient of *%Body fat* on *Weight* statistically distinguishable from 0? (Perform a hypothesis test.)
b) What does the slope coefficient mean in this regression?

We saw before that the slopes of both *Waist* size and *Height* are statistically significant when entered into a multiple regression equation. What happens if we add *Weight* to that regression? Recall that we've already checked the assumptions and conditions for regression on *Waist* size and *Height* in the chapter. Here is the output from a regression on all three variables:

Dependent variable is: Pct BF
R-squared = 72.5% R-squared (adjusted) = 72.2%
s = 4.376 with 250 − 4 = 246 degrees of freedom

Source	Sum of Squares	DF	Mean Square	F-ratio
Regression	12 418.7	3	4 139.57	216
Residual	4 710.11	246	19.1468	

Variable	Coefficient	SE(Coeff)	t-ratio	P-value
Intercept	−31.4830	11.54	−2.73	0.0068
Waist	2.31848	0.1820	12.7	<0.0001
Height	−0.224932	0.1583	−1.42	0.1567
Weight	−0.100572	0.0310	−3.25	0.0013

c) Interpret the slope for *Weight*. How can the coefficient for *Weight* in this model be negative when its coefficient was positive in the simple regression model?
d) What does the P-value for *Height* mean in this regression? (Perform the hypothesis test.)

T 22. Breakfast cereals We saw in Chapter 7 that the calorie content of a breakfast cereal is linearly associated with its sugar content. Is that the whole story? Here's the output of a regression model that regresses *Calories* for each serving on its *Protein* (g), *Fat* (g), *Fibre* (g), *Carbohydrate* (g), and *Sugars* (g) content.

Dependent variable is: calories
R-squared = 84.5% R-squared (adjusted) = 83.4%
s = 7.947 with 77 − 6 = 71 degrees of freedom

Source	Sum of Squares	DF	Mean Square	F-ratio
Regression	24 367.5	5	4 873.50	77.2
Residual	4 484.45	71	63.1613	

Variable	Coefficient	SE(Coeff)	t-ratio	P-value
Intercept	20.2454	5.984	3.38	0.0012
Protein	5.69540	1.072	5.32	<0.0001
Fat	8.35958	1.033	8.09	<0.0001
Fibre	−1.02018	0.4835	−2.11	0.0384
Carbo	2.93570	0.2601	11.3	<0.0001
Sugars	3.31849	0.2501	13.3	<0.0001

Assuming that the conditions for multiple regression are met,
a) What is the regression equation?
b) Do you think this model would do a reasonably good job at predicting calories? Explain.
c) To check the conditions, what plots of the data might you want to examine?
d) What does the coefficient of *Fat* mean in this model?

T 23. Body fat again Chest size might be a good predictor of body fat. Here's a scatterplot of *%Body fat* versus *Chest size.*

Chest Size (in.)

A regression of *%Body Fat* on *Chest Size* gives the following equation:

Dependent variable is: Pct BF
R-squared = 49.1% R-squared (adjusted) = 48.9%
S = 5.930 with 250 − 2 = 248 degrees of freedom

Variable	Coefficient	SE(Coeff)	t-ratio	P-value
Intercept	−52.7122	4.654	−11.3	<0.0001
Chest Size	0.712720	0.0461	15.5	<0.0001

a) Is the slope of *%Body fat* on *Chest size* statistically distinguishable from 0? (Perform a hypothesis test.)
b) What does the answer in part a) mean about the relationship between *%Body fat* and *Chest size*?

We saw before that the slopes of both *Waist* size and *Height* are statistically significant when entered into a multiple regression equation. What happens if we add *Chest size* to that regression? Here is the output from a regression on all three variables:

Dependent variable is: Pct BF
R-squared = 72.2% R-squared (adjusted) = 71.9%
s = 4.399 with 250 − 4 = 246 degrees of freedom

Source	Sum of Squares	DF	Mean Square	F-ratio	P-value
Regression	12 368.9	3	4 122.98	213	<0.0001
Residual	4 759.87	246	19.3491		

Variable	Coefficient	SE(Coeff)	t-ratio	P-value
Intercept	2.07220	7.802	0.266	0.7908
Waist	2.19939	0.1675	13.1	<0.0001
Height	−0.561058	0.1094	−5.13	<0.0001
Chest Size	−0.233531	0.0832	−2.81	0.0054

c) Interpret the coefficient for *Chest size*.
d) Would you consider removing any of the variables from this regression model? Why or why not?

T **24. Grades** The table below shows the five scores from an Introductory Statistics course. Find a model for predicting the final exam score by trying all possible models with two predictor variables. Which model would you choose? Be sure to check the conditions for multiple regression.

Name	Final	Midterm 1	Midterm 2	Project	Homework
Timothy F.	117	82	30	10.5	61
Karen E.	183	96	68	11.3	72
Verena Z.	124	57	82	11.3	69
Rodrigo L.	169	88	86	10.6	84
Alok M.	164	93	81	10	71
Zygmunt E.	134	90	83	11.3	79

Lele A.	98	83	21	11.2	51
Ashish K.	136	59	62	9.1	58
Justin E.	183	89	57	10.7	79
Zhang E.	171	83	86	11.5	78
Wang B.	173	95	75	8	77
Annie C.	164	81	66	10.7	66
Zheng A.	150	86	63	8	74
Brian J.	153	81	86	9.2	76
Gregory J.	149	81	87	9.2	75
Kristina G.	178	98	96	9.3	84
Timothy B.	75	50	27	10	20
Jason C.	159	91	83	10.6	71
Whitney E.	157	87	89	10.5	85
Alexis P.	158	90	91	11.3	68
Nicholas T.	171	95	82	10.5	68
Amandeep S.	173	91	37	10.6	54
Irena R.	165	93	81	9.3	82
Yvon T.	168	88	66	10.5	82
Sara M.	186	99	90	7.5	77
Yang P.	157	89	92	10.3	68
Benjamin S.	177	87	62	10	72
David W.	170	92	66	11.5	78
Josef H.	78	62	43	9.1	56
Rebecca S.	191	93	87	11.2	80
Joshua D.	169	95	93	9.1	87
Ian M.	170	93	65	9.5	66
Katharine A.	172	92	98	10	77
Emily R.	168	91	95	10.7	83
Brian M.	179	92	80	11.5	82
Li M.	148	61	58	10.5	65
Yang R.	103	55	65	10.3	51
John M.	144	76	88	9.2	67
Zhu J.	155	63	62	7.5	67
Peng G.	141	89	66	8	72
Peter H.	138	91	42	11.5	66
Catherine R.M.	180	90	85	11.2	78
Christina M.	120	75	62	9.1	72
Enrique J.	86	75	46	10.3	72
Sarah K.	151	91	65	9.3	77
Thomas J.	149	84	70	8	70
Sonya P.	163	94	92	10.5	81
Michael B.	153	93	78	10.3	72
Wesley M.	172	91	58	10.5	66
Xiaoling R.	165	91	61	10.5	79
Adam J.	155	89	86	9.1	62

Jared A.	181	98	92	11.2	83
Michael T.	172	96	51	9.1	83
Kathryn D.	177	95	95	10	87
Nicole M.	189	98	89	7.5	77
Wayne E.	161	89	79	9.5	44
Elizabeth S.	146	93	89	10.7	73
Wang R.	147	74	64	9.1	72
Valentin A.	160	97	96	9.1	80
David T.O.	159	94	90	10.6	88
Marc I.	101	81	89	9.5	62
Samuel E.	154	94	85	10.5	76
Brooke S.	183	92	90	9.5	86

25. Fifty states 2009 Here is a data set on various measures of the 50 states in the United States. The *Murder* rate is per 100 000, high school graduation rate is in % (*HS Grad*), *Income* is per capita income in dollars, *Illiteracy* rate is per 1000, and *Life expectancy* is in years. Find a regression model for *Life expectancy* with three predictor variables by trying all four of the possible models.
a) Which model appears to do the best?
b) Would you leave all three predictors in this model?
c) Does this model mean that by changing the levels of the predictors in this equation, we could affect *Life expectancy* in that state? Explain.
d) Be sure to check the conditions for multiple regression. What do you conclude?

State	Life expectancy	Murder rate08	Graduation Rate	Income/cap07	Illiteracy
Alabama	74.4	7.6	62%	27 557	0.15
Alaska	77.1	4.1	67	34 316	0.09
Arizona	77.5	6.3	59	28 088	0.13
Arkansas	75.2	5.7	72	25 563	0.14
California	78.2	5.8	68	35 352	0.23
Colorado	78.2	3.2	68	34 902	0.10
Connecticut	78.7	3.5	75	46 021	0.09
Delaware	76.8	6.5	73	34 533	0.11
Florida	77.5	6.4	59	32 693	0.20
Georgia	75.3	6.6	54	28 452	0.17
Hawaii	80.0	1.9	69	33 369	0.16
Idaho	77.9	1.5	78	26 530	0.11
Illinois	76.4	6.1	78	34 290	0.13
Indiana	76.1	5.1	74	28 587	0.08
Iowa	78.3	2.5	93	29 784	0.07
Kansas	77.3	4	76	31 268	0.08
Kentucky	75.2	4.6	71	26 457	0.12

Louisiana	74.2	11.9	69	29 557	0.16
Maine	77.6	2.4	78	28 677	0.07
Maryland	76.3	8.8	75	39 136	0.11
Massachusetts	78.4	2.6	75	41 740	0.10
Michigan	76.3	5.4	75	29 837	0.08
Minnesota	78.8	2.1	82	34 896	0.06
Mississippi	73.6	8.1	62	24 530	0.16
Missouri	75.9	7.7	75	29 245	0.07
Montana	77.2	2.4	83	27 602	0.09
Nebraska	77.8	3.8	85	31 015	0.07
Nevada	75.8	6.3	58	34 424	0.16
New Hampshire	78.3	1	71	35 302	0.06
New Jersey	77.5	4.3	75	41 835	0.17
New Mexico	77.0	7.2	65	26 766	0.16
New York	77.7	4.3	70	40 296	0.22
North Carolina	75.8	6.5	63	28 604	0.14
North Dakota	78.3	0.5	88	29 633	0.06
Ohio	76.2	4.7	77	29 657	0.09
Oklahoma	75.2	5.8	74	29 044	0.12
Oregon	77.8	2.2	67	29 580	0.10
Pennsylvania	76.7	5.6	82	32 986	0.13
Rhode Island	78.3	2.8	72	33 560	0.08
South Carolina	74.8	6.8	62	26 374	0.15
South Dakota	77.7	3.2	80	28 833	0.07
Tennessee	75.1	6.6	60	28 301	0.13
Texas	76.7	5.6	67	31 624	0.19
Utah	78.7	1.4	81	26 523	0.09
Vermont	78.2	2.7	84	31 184	0.07
Virginia	76.8	4.7	74	35 162	0.12
Washington	78.2	2.9	70	34 368	0.10
West Virginia	75.1	3.3	82	25 118	0.13
Wisconsin	77.9	2.6	85	30 655	0.07
Wyoming	76.7	1.9	81	36 760	0.09

26. Breakfast cereals again We saw in Chapter 7 that the calorie count of a breakfast cereal is linearly associated with its sugar content. Can we predict the calories of a serving from its vitamin and mineral content? Here's a multiple regression model of *Calories* per serving on its *Sodium* (mg), *Potassium* (mg), and *Sugars* (g):

Dependent variable is: Calories
R-squared = 38.4% R-squared (adjusted) = 35.9%
s = 15.60 with $77 - 4 = 73$ degrees of freedom

Source	Sum of Squares	DF	Mean Square	F-ratio	P-value
Regression	11 091.8	3	3 697.28	15.2	<0.0001
Residual	17 760.1	73	243.289		

Variable	Coefficient	SE(Coeff)	t-ratio	P-value
Intercept	83.0469	5.198	16.0	<0.0001
Sodium	−0.05721	0.0215	2.67	0.0094
Potass	0.01933	0.0251	−0.769	0.4441
Sugars	2.38757	0.4066	5.87	<0.0001

Assuming that the conditions for multiple regression are met,
a) What is the regression equation?
b) Do you think this model would do a reasonably good job at predicting calories? Explain.
c) Would you consider removing any of these predictor variables from the model? Why or why not?
d) To check the conditions, what plots of the data might you want to examine?

 27. Burger King revisited Recall the Burger King menu data from Chapter 7. BK's nutrition sheet lists many variables. Here's a multiple regression to predict calories for Burger King foods from *Protein* content (g), *Total fat* (g), *Carbohydrate* (g), and *Sodium* (mg) per serving:

Dependent variable is: Calories
R-squared = 100.0% R-squared (adjusted) = 100.0%
s = 3.140 with 31 − 5 = 26 degrees of freedom

Source	Sum of Squares	DF	Mean Square	F-ratio
Regression	1 419 311	4	354 828	35 994
Residual	256.307	26	9.85796	

Variable	Coefficient	SE(Coeff)	t-ratio	P-value
Intercept	6.53412	2.425	2.69	0.0122
Protein	3.83855	0.0859	44.7	<0.0001
Total fat	9.14121	0.0779	117	<0.0001
Carbs	3.94033	0.0336	117	<0.0001
Na/Serv.	−0.69155	0.2970	−2.33	0.0279

a) Do you think this model would do a good job of predicting calories for a new BK menu item? Why or why not?
b) The mean of *Calories* is 455.5 with a standard deviation of 217.5. Discuss what the value of s in the regression means about how well the model fits the data.
c) Does the R^2 value of 100.0% mean that the residuals are all actually equal to zero?

T **28. Popsicle sticks** Recall the Popsicle stick data from Chapter 24. Suppose we want to fit a multiple regression model to predict the breaking stress from the overlap area and the breaking force, on the medium overlap joints only. Taking logs of all three variables, we get the following output:

Dependent variable is: log(stress)
R-squared = 100% R-squared (adjusted) = 100%
S = 0.0001746 on 287 degrees of freedom

Variable	Coefficient	SE(Coeff)	t-ratio	P-value
Intercept	9.191	0.0008452	10 875	≤0.0001
log(force)	1.000	0.00002631	38 002	≤0.0001
log(area)	−1.000	0.0001666	−6 004	≤0.0001

This output looks too good to be true, with $R^2 = 100\%$ and MSE of virtually zero. Can you explain what happened?

✓ Just Checking ANSWERS

1. 77.9% of the variation in maximum wind speed can be accounted for by multiple regression on *Central Pressure* and *Year*.

2. In any given year, hurricanes with a central pressure that is 1 mb lower can be expected to have, on average, winds that are 0.933 kn faster.

3. First, the researcher is trying to prove his null hypothesis for this coefficient and, as we know, statistical inference won't permit that. Beyond that problem, we can't even be sure we understand the relationship of *Wind Speed* to *Year* from this analysis. For example, both *Central Pressure* and *Wind Speed* might be changing over time, but their relationship might stay the same during any given year.

MathXL
MyLab Statistics
Go to MathXL at **www.mathxl.com** or MyLab™ Statistics at **www.mystatlab.com**. You can practise exercises for this chapter as often as you want. The guided solutions will help you find answers step by step. You'll find a personalized study plan available to you too!

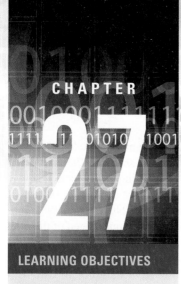

27

Multiple Regression Wisdom

More wisdom when using multiple regression!

1 Use indicator variables to introduce categorical variables into the model.

2 Use interaction terms to allow for different slopes.

3 Identify extraordinary cases that may be influential on the fitted model

4 Recognize when a regression model may suffer from collinearity, and understand its consequences.

5 Know how to build a multiple regression model, selecting predictors from a large collection of potential predictors.

6 Understand that building multiple regression models is art as well as science. There is no *right* model.

Connections

Wisdom in building and interpreting multiple regressions uses all the tools and concepts that we've discussed throughout this book. We use models to help us understand the world with data. This chapter is about building that understanding even when we use powerful, complex methods.

Once we learn how to insert categorical predictors into our model, using indicator variables, we'll see that multiple regression and ANOVA are really the same analysis. Both can be viewed as part of a more

Where the world ceases to be the scene of our personal hopes and wishes, where we face it as free beings admiring, asking and observing, there we enter the realm of Art and Science.

—*Albert Einstein*

Roller coasters are an old thrill that continue to grow in popularity. Engineers and designers compete to make them bigger and faster. Canada's fastest roller coaster (as of July 2013) is the *Leviathan* at Canada's Wonderland in Vaughan, Ontario. With a cost of $28 million to build, it reaches 306 feet tall (93 m) and maxes out at 92 mph (148 km/h). And at 3 minutes and 28 seconds, it's also one of the longest rides around! Can we learn what makes a roller coaster fast? Or how long the ride will last? Here is a small selection from our data file on nearly 200 roller coasters, worldwide. (Data in **Roller_coasters_2014**)

Name	Park	Country	Type	Duration (sec.)	Speed (mph)	Height (ft.)	Drop (ft.)	Length (ft.)	Inversion?
New Mexico Rattler	Cliff's Amusement Park	USA	Wooden	75	47	80	75	2750	No
Fujiyama	Fuji-Q Highlands	Japan	Steel	216	80.8	259.2	229.7	6708.67	No
Goliath	Six Flags Magic Mountain	USA	Steel	180	85	235	255	4500	No
Great American Scream Machine	Six Flags Great Adventure	USA	Steel	140	68	173	155	3800	Yes
Hangman	Wild Adventures	USA	Steel	125	55	115	95	2170	Yes
Hayabusa	Tokyo SummerLand	Japan	Steel	108	60.3	137.8	124.67	2559.1	No
Hercules	Dorney Park	USA	Wooden	135	65	95	151	4000	No
Hurricane	Myrtle Beach Pavilion	USA	Wooden	120	55	101.5	100	3800	No

Table 27.1

A small selection of roller coasters from the larger data set available online at www.pearsoncanada.ca/deveaux.

Here, more precisely, are the variables and their units:

- *Type* indicates what kind of track the roller coaster has. The possible values are "wooden" and "steel." (The frame usually is of the same construction as the track, but doesn't have to be.)
- *Duration* is the duration of the ride in seconds.
- *Speed* is the top speed in miles per hour.
- *Height* is the maximum height above ground level in feet.
- *Drop* is the greatest drop in feet.

general statistical model known as the General Linear Model which accounts for why they seem so connected, with similar concepts and approaches.

Who	Roller coasters
What	See Table 27.1. (For multiple regression, we have to know "What" and the units for each variable.)
Where	Worldwide
When	All were in operation in 2014.
Source	The Roller Coaster DataBase, www.rcdb.com

- *Length* is the total length of the track in feet.
- *Inversions* reports whether riders are turned upside down during the ride. It has the values "yes" or "no."

It's always a good idea to explore the data before starting to build a model. Let's first consider the ride's *Duration*. We have that information for only 136 of the 195 coasters in our data set, but there's no reason to believe that the data are missing in any patterned way so we'll look at those 136 coasters. The average *Duration* for these coasters is 124.5 seconds, but one ride is as short as 28 seconds and another as long as 240 seconds. We might wonder whether the duration of the ride should depend on the maximum speed of the ride.

Figure 27.1 shows the scatterplot of *Duration* against *Speed* and the regression line. Below it is the usual summary with a residual plot in the margin.

Figure 27.1

Duration of the ride appears to be linearly related to the maximum *Speed* of the ride.

Response variable is: Duration
R-squared = 34.5%　R-squared (adjusted) = 34.0%
s = 36.36 with 134 − 2 = 132 degrees of freedom

Source	Sum of Squares	DF	Mean Square	F-ratio	P-value
Regression	91951.7	1	91951.7	69.6	< 0.0001
Residual	174505 1	132	1322.01		

Variable	Coefficient	SE(Coeff)	t-ratio	P-value
Intercept	20.4744	12.93	1.58	0.1156
Speed	1.82262	0.2185	8.34	< 0.0001

The regression conditions seem to be met, and the regression makes sense. We'd expect faster coasters to give longer rides. The duration of the ride increases, on average, by about 1.82 seconds for each additional mile per hour of speed—faster coasters actually have rides that last *longer*.

27.1 Indicator Variables

Of course, there's more to these data. One interesting variable might not be one you'd naturally think of. Many modern coasters have "inversions." That's a nice way of saying that they turn riders upside down, with loops, corkscrews, or other devices. These inversions add excitement, but they must be carefully engineered, and that enforces some speed limits on that portion of the ride.

We'd like to add the information of whether the roller coaster has an inversion to our model. Until now, all our predictor variables have been quantitative. Whether or not a roller coaster has any inversions is a categorical variable ("yes" or "no"). Can we introduce the categorical variable *Inversions* as a predictor in our regression model? What would it mean if we did?

The *Leviathan* at Canada's Wonderland
robyelo357/Fotolia

Let's start with a plot. Figure 27.2 below shows the same scatterplot of *Duration* against *Speed*, but now with the roller coasters that have inversions shown as red *x*'s and a separate regression line drawn for each type of roller coaster.

It's easy to see that, for a given speed, the roller coasters with inversions take a bit longer, and that for each type of roller coaster, the slopes of the relationship between *Duration* and *Speed* are not quite equal but are similar.

We could split the data into two groups—coasters without inversions and those with inversions—and compute the regression for each group. The outputs would look like this:

Figure 27.2

The two lines fit to coasters with inversions and without are roughly parallel.

Response variable is: Duration
Cases selected according to no Inversions
R-squared = 57.0%   R-squared (adjusted) = 56.2%
s = 31.40 with 55 − 2 = 53 degrees of freedom

Variable	Coefficient	SE(Coeff)	t-ratio	P-value
Intercept	33.9448	13.31	2.55	0.0137
Speed	1.79522	0.2142	8.38	< 0.0001

Response variable is: Duration
Cases selected according to Inversions
R-squared = 12.8%   R-squared (adjusted) = 11.7%
s = 37.61 with 79 − 2 = 77 degrees of freedom

Variable	Coefficient	SE(Coeff)	t-ratio	P-value
Intercept	22.7725	27.71	0.822	0.4137
Speed	1.63518	0.4865	3.36	0.0012

As the scatterplot showed, the slopes are very similar, but the intercepts are different.

When we have a situation like this with roughly parallel regressions for each group,[1] there's an easy way to add the group information to a single regression model. We make up a special variable that *indicates* what type of roller coaster we have, giving it the value 1 for roller coasters that have inversions and the value 0 for those that don't. (We could have reversed the coding; it's an arbitrary choice.[2]) Such variables are called **indicator variables** or *indicators* because they indicate which of the categories each case is in.[3]

When we add our new indicator, *Inversions*, to the regression model, the model looks like this:

Response variable is: Duration
R-squared = 39.5%   R-squared (adjusted) = 38.5%
s = 35.09 with 134 − 3 = 131 degrees of freedom

Variable	Coefficient	SE(Coeff)	t-ratio	P-value
Intercept	35.9969	13.34	2.70	0.0079
Speed	1.76038	0.2118	8.31	<0.0001
Inversions	−20.2726	6.187	−3.28	0.0013

This looks like a better model than the simple regression for all the data. The R^2 is larger, the *t*-ratios of both coefficients are large, and the residuals look reasonable. But what does the coefficient for *Inversions* mean?

---

[1]The fact that the individual regression lines are nearly parallel is really part of the Linearity Assumption and Straight Enough Condition. You should check that the lines are nearly parallel before using this method, or read on to see what to do if they are not parallel enough.

[2]Some implementations of indicator variables use −1 and 1 for the levels of the categories.

[3]They are also commonly called *dummies* or *dummy variables*. But this sounds like an insult, so the more politically correct term is *indicator variable*.

Let's see how an indicator variable works when we calculate predicted values for two of the roller coasters given at the start of the chapter:

Name	Park	Country	Type	Duration	Speed	Height	Drop	Length	Inversion?
Hangman	Wild Adventures	USA	Steel	125	55	115	95	2170	Yes
Hayabusa	Tokyo SummerLand	Japan	Steel	108	60.3	137.8	124.67	2559.1	No

The model says that for all coasters, the predicted *Duration* is

$$36 + 1.76 \; Speed - 20.27 \; Inversions$$

For *Hayabusa*, the speed is 55 mph and the value of *Inversions* is 0, so the model predicts a duration of[4]

$$35.9969 + (1.76038 \times 55) - (20.2726 \times 0) = 132.82 \text{ seconds}$$

That's a bit higher than the actual duration of 108 seconds.

For the *Hangman*, the speed is 60.3 mph. It has an inversion, so the value of *Inversions* is 1, and the model predicts a duration of

$$35.9969 + (1.76038 \times 60.3) - (20.2726 \times 1) = 121.88 \text{ seconds}$$

That compares well with the actual duration of 125 seconds.

Notice how the indicator works in the model. When there is an inversion (as in *Hangman*), the value 1 for the indicator causes the amount of the indicator's coefficient, -20.2726, to be added to the prediction. When there is no inversion (as in *Hayabusa*), the indicator is zero, so nothing is added. Looking back at the scatterplot, we can see that this is exactly what we need. The difference between the two lines is a vertical shift of about 20 seconds.

This may seem a bit confusing at first. We usually think of the coefficients in a multiple regression as slopes. For indicator variables, however, they act differently. They're vertical shifts (by category) that keep the slopes for the other variables apart.

For Example USING INDICATOR VARIABLES

As a class project, students in a large Statistics class collected publicly available information on recent home sales in their hometowns. There are 894 properties. These are not a random sample, but they may be representative of home sales during a short period of time, nationwide. In Chapter 26, we looked at these data and constructed a multiple regression model. Let's look further. Among the variables available is an indication of whether the home was in an urban, suburban, or rural setting.

QUESTION: How can we incorporate information such as this in a multiple regression model?

ANSWER: We might suspect that homes in rural communities might differ in price from similar homes in urban or suburban settings. We can define an indicator (dummy) variable to be 1 for homes in rural communities and 0 otherwise. A scatterplot shows that rural homes have, on average, lower prices for a given living area:

---

[4]We round coefficient values when we write the model but calculate with the full precision, rounding at the end of the calculation.

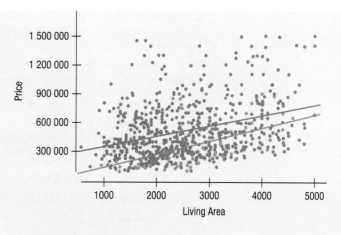

The multiple regression model is:
Dependent variable is: Price
R squared = 18.4%   R squared (adjusted) = 18.2%
s = 260 996 with 894 − 3 = 891 degrees of freedom

Variable	Coefficient	SE(Coeff)	t-ratio	P-value
Intercept	230 945	25 706	8.98	<0.0001
Living area	112.534	9.353	12.0	<0.0001
Rural	−172 359	23 749	−7.26	<0.0001

The coefficient of *Rural* indicates that, for a given living area, rural homes sell for, on average, about $172 000 less than comparable homes in urban or suburban settings.

Adjusting for Different Slopes

What if the lines aren't parallel? An indicator variable that is 0 or 1 can only shift the line up and down. It can't change the slope, so it works only when we have lines with the same slope and different intercepts.

Let's return to the Burger King data we looked at in Chapter 7, and look at how *Calories* are related to *Carbohydrates* (*Carbs* for short). Figure 27.3 in the margin shows the scatterplot.

It's not surprising to see that more *Carbs* goes with more *Calories*, but the plot seems to thicken as we move from left to right. Could there be something else going on?[5]

Burger King foods can be divided into two groups: those with meat (including chicken and fish) and those without. When we colour the plot (red for meat, blue for nonmeat) and look at the regressions for each group, we see a different picture (in Figure 27.4).

Clearly, meat-based dishes contribute more calories for each gram of carbohydrate than do other Burger King foods. But we can't account for the kind of difference we see here by just including an indicator variable in a regression. It isn't just the height of the lines that is different; they have entirely different slopes. How can we deal with that in our regression model?

The trick is to adjust the slopes with another constructed variable. This one is the product of an indicator for one group and the predictor variable. The coefficient of this constructed **interaction term** in a multiple regression gives an adjustment to the slope, b_1, to be made for the individuals in the indicated group.[6] Here we have the indicator variable *Meat*, which is 1 for meat-containing foods and 0 for the others. We then construct an

Figure 27.3

Calories of Burger King foods plotted against *Carbohydrates* seems to fan out.

Figure 27.4

Plotting the meat-based and nonmeat items separately, we see two distinct linear patterns.

---

[5]Would we even ask if there weren't?

[6]Chapter 25 discussed interaction effects in two-way ANOVA. Interaction terms, such as these, are exactly the same idea.

interaction variable, *Carbs*Meat,* which is just the *product of those two variables.* That's right; just multiply them. The resulting variable has the value of *Carbs* for foods containing meat (those coded 1 in the *Meat* indicator) and the value 0 for the others. By including the interaction variable in the model, we can adjust the slope of the line fit to the meat-containing foods. Here's the resulting analysis:

Dependent variable is: Calories
R-squared = 78.1%   R-squared (adjusted) = 75.7%
s = 106.0 with 32 − 4 = 28 degrees of freedom

Source	Sum of Squares	df	Mean Square	F-ratio	P-value
Regression	1 119 979	3	373 326	33.2	< 0.0001
Residual	314 843	28	11 244.4		

Variable	Coefficient	SE(Coeff)	t-ratio	P-value
Intercept	137.395	58.72	2.34	0.0267
Carbs(g)	3.93317	1.113	3.53	0.0014
Meat	−26.1567	98.48	−0.266	0.7925
Carbs*Meat	7.87530	2.179	3.61	0.0012

The regression model is

$$\widehat{Calories} = 137.40 + 3.93\,Carbs - 26.16\,Meat + 7.88\,Carbs*Meat$$

What does the coefficient for the indicator *Meat* mean? It provides a different intercept to separate the meat and nonmeat items at the origin (where *Carbs* = 0). For these data, there is a different slope, but the two lines nearly meet at the origin, so there seems to be no need for an additional adjustment. The estimated difference of 26.16 calories is small; that's why the coefficient for *Meat* has a small *t*-statistic.

By contrast, the coefficient of the interaction term, *Carbs*Meat,* says that the slope relating calories to carbohydrates is steeper by 7.875 calories per carbohydrate gram for meat-containing foods than for meat-free foods. Its small P-value suggests that this difference is real.

Let's consider the regression model more carefully. For nonmeat items (*Meat* = 0), the regression model is

$$\widehat{Calories} = 137.40 + 3.93\,Carbs - 26.16\times0 + 7.88\,Carbs\times0$$

$$= 137.40 + 3.93\,Carbs$$

For meat items (*Meat* = 1), the regression model is

$$\widehat{Calories} = 137.40 + 3.93\,Carbs - 26.16\times1 + 7.88\,Carbs\times1$$

$$= 111.24 + 11.81\,Carbs$$

Figure 27.5 shows these two lines. Let's see how these adjustments work. A BK Whopper has 53 g of *Carbs* and is a meat dish. The model predicts its *Calories* as

$$137.395 + (3.93317 \times 53) - (26.1567 \times 1) + (7.8753 \times 53 \times 1) = 737.1,$$

not far from the measured calorie count of 680. By contrast, the Veggie Burger, with 43 g of *Carbs,* is predicted to have

$$137.395 + (3.93317 \times 43) - (26.1567 \times 0) + (7.87530 \times 43 \times 0) = 306.5\text{ calories,}$$

not far from the 330 measured officially. The last two terms in the equation for the Veggie Burger are zero because the indicator for *Meat* is 0 for the Veggie Burger.

Figure 27.5
The Whopper and Veggie Burger belong to different groups.

Section 27.1 Exercises

1. **Indicators** For each of these potential predictor variables say whether they should be represented in a regression model by indicator variables. If so, then suggest what specific indicators should be used (that is, what values they would have).
   a) In a regression to predict income, the sex of respondents in a survey.
   b) In a regression to predict the square footage available for rent, the number of stories in a commercial building.
   c) In a regression to predict the amount an individual's medical insurance would pay for an operation, whether the individual was over 65 (and eligible for Medicare).

2. **More indicators** For each of these potential predictor variables say whether they should be represented in a regression model by indicator variables. If so, then suggest what specific indicators should be used (that is, what values they would have).
   a) In a regression to predict income, the age of respondents in a survey.
   b) In a regression to predict the square footage available for rent, whether a commercial building has an elevator or not.
   c) In a regression to predict annual medical expenses, whether a person was a child (in pediatric care).

27.2 Diagnosing Regression Models: Looking at the Cases

We often use regression analyses to try to understand the world. By working with the data and creating models, we can learn a great deal about the relationships among variables. As we saw with simple regression, sometimes we can learn as much from the cases that *don't* fit the model as from the bulk of cases that do. Extraordinary cases often tell us more about the world simply by the ways in which they fail to conform and the reasons we can discover for those deviations.

If a case doesn't conform to the others, we should identify it and, if possible, understand why it is different. As in simple regression, a case can be extraordinary by standing away from the model in the y direction or by having unusual values in an x-variable. In multiple regression it can also be extraordinary by having an unusual *combination* of values in the x-variables. Deviations in the y direction show up in the residuals. Deviations in the x variables show up as *leverage*.

Leverage

Recent events have focused attention on airport screening of passengers, but screening has a longer history. The *Sourcebook of Criminal Justice Statistics Online* lists the numbers of various violations found by U.S. airport screeners for each of several types of violations in each year from 1977 to 2000. Here's a regression of the number of long guns (rifles and the like) found versus the number of times false information was discovered.

Response variable is: Long guns
R-squared = 3.8%   R-squared (adjusted) = −0.6%
s = 56.34 with 24 − 2 = 22 degrees of freedom

Variable	Coefficient	SE(Coeff)	t-ratio	P-value
Intercept	87.0071	19.68	4.42	0.0002
false information	0.246762	0.2657	0.929	0.3630

That summary doesn't look like it's a particularly successful regression. The R^2 is only 3.8%, and the P-value for *False Info* is large. But a look at the scatterplot (Figure 27.6) tells us more.

Figure 27.6

A high-leverage point can hide a strong relationship, so that you can't see it in the regression. Make a plot.

The unusual cases are from 1988 and 2000. In 2000, there were nearly 300 long gun confiscations. But because this point is at a typical value for false information, it doesn't have a strong effect on the slope of the regression line. But in 1988, the number of false information reports jumped over 200. The resulting case has high leverage because it is so far from the *x*-values of the other points. It's easy to see the influence of that one high-leverage case if we look at the regression lines with and without that case (Figure 27.7).

Figure 27.7

A single high-leverage point can change the regression slope quite a bit. The (blue) line omitting the point for 1988 is quite different from the line that includes the outlier.

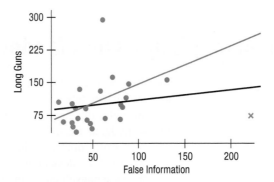

The **leverage** of a case measures its ability to move the regression model all by itself by just moving the case in the *y* direction. In Chapter 8, when we had only one predictor variable, we could *see* high leverage points in a scatterplot because they stood far from the mean of *x*. But now, with several predictors, we can't count on seeing them in our plots.

Fortunately, we can put a number on the leverage. If we keep everything else the same, change the *y*-value of a case by 1.0, and find a new regression, the leverage of that case is the amount by which the *predicted* value at that case would change. Leverage can never be greater than 1.0—we wouldn't expect the line to move *farther* than we move the case, only to try to keep up. Nor can it be less than 0.0—we'd hardly expect the line to move in the *opposite* direction. A point with zero leverage has no effect at all on the regression model, although it does participate in the calculations of R^2, *s*, and the *F*- and *t*-statistics.

For the airport inspections, the leverage of 1988 is 0.63. That's quite high. If there had been even one fewer long gun discovered that year (decreasing the *observed y*-value by 1), the *predicted y*-value for 1988 would have decreased by 0.63, dragging the regression line down still farther. For comparison, the point for 2000 that has an extraordinary value for *Long guns* only has leverage 0.42. We would consider it an outlier because it is far from the other values in the *y* (here, *Long guns*) direction. But because it is near the mean in *x* (*False information*) it doesn't have enough leverage to change the slope of the regression line.

The leverage of a case is a measure of how far that case is from the centre of the *x*'s. As always in Statistics, we expect to measure that distance with a ruler based on a standard deviation—here, the standard deviation of the *x*'s. And that's really all the leverage is: an indication of how far each point is away from the centre of all the *x*-values, measured in standard deviations. Fortunately, there's a less tedious way to calculate leverage than moving each case in turn, but it's beyond the scope of this book and you'd never want to do it by hand anyway. So, just let the computer do the computing and think about what the result *means*. Most statistics programs calculate leverage values, and you should examine them.

A case can have large leverage in two different ways:

- It might be extraordinary in one or more individual variables. For example, the fastest or slowest roller coaster may stand out.
- It may be extraordinary in a *combination* of variables. For example, suppose we were predicting *Drop* using *Duration* and *Speed* as predictors. High leverage points would show up in a plot of one predictor versus the other, like the *plot in the margin* of *Duration* versus *Speed* (actually, based on some older data). Note that one roller coaster stands out in this scatterplot. It isn't extraordinarily fast and others have shorter duration, but the *combination* of high speed and short duration is unusual. Looking at leverage values can be a very effective way to discover cases that are extraordinary on a combination of *x*-variables.

There are no tests for whether the leverage of a case is too large. The average leverage value among all cases in a regression is (k + 1)/n, where k + 1 is the number of betas in the model, but that doesn't give us much of a guide. Sometimes we classify a point as high leverage if it has leverage greater than twice the mean or 2(k + 1)/n. Another common approach is just to make a histogram of the leverages. Any case whose leverage stands out in a histogram of leverages probably deserves special attention. You may decide to leave the case in the regression or to see how the regression model changes when you delete the case, but you should be aware of its potential to influence the regression.

For Example DIAGNOSING A REGRESSION

Here's another regression model for the real estate data we looked at earlier in the chapter.

Dependent variable is: Price
R squared = 23.1%   R squared (adjusted) = 22.8%
s = 253 709 with 893 − 5 = 888 degrees of freedom

Variable	Coefficient	SE(Coeff)	t-ratio	P-value
Intercept	322 470	40 192	8.02	≤0.0001
Living area	92.6272	13.09	7.08	≤0.0001
Bedrooms	−69 720.6	12 764	−5.46	≤0.0001
Bathrooms	82 577.6	13 410	6.16	≤0.0001
Rural	−161 575	23 313	−6.93	≤0.0001

QUESTION: What do diagnostic statistics tell us about these data and this model?

ANSWER: A boxplot of the leverage values shows one extraordinarily large leverage:

Investigation of that case reveals it to be a home that sold for $489 000. It has 8 bedrooms and only 2.5 bathrooms. This is a particularly unusual combination, especially for a home with that value. If we were pursuing this analysis further, we'd want to check the records for this house to be sure that the number of bedrooms and bathrooms were reported accurately. (The boxplot above was produced by software that uses different symbols for less vs. more extreme outliers.)

Residuals and Standardized Residuals

Residuals are not all alike. Consider a point with leverage 1.0. That's the highest a leverage can be, and it means that the line follows the point perfectly. So, a point like that must have a zero residual. And, since we know the residual exactly, that residual has zero standard deviation. This tendency is true in general: The larger the leverage, the *smaller* the standard deviation of its residual.[7]

When we want to compare values that have differing standard deviations, it's a good idea to standardize them.[8] We can do that with the regression residuals, dividing each one by an estimate of its own standard deviation. When we do that in a certain way, the resulting standardized residuals are called **Studentized residuals** and will follow a Student's *t*-distribution[9]. It's a good idea to examine the Studentized residuals (rather than the simple residuals) to assess the Nearly Normal Condition and the **Does the Plot Thicken? Condition**. Any Studentized residual that stands out from the others deserves our attention.

It may occur to you that we've always plotted the *unstandardized* residuals when we made regression models. And we've treated them as if they all had the same standard deviation when we checked the **Nearly Normal Condition**. It turns out that this was a simplification. It didn't matter much for simple regression, but for multiple regression models, it's a better idea to use the Studentized residuals when checking the Nearly Normal Condition. (Of course, Student's *t* isn't exactly Normal either—that's why we say "nearly" Normal.)

> ### IT ALL FITS TOGETHER DEPARTMENT
>
> Make an indicator variable for a single case—that is, construct a variable that is 0 everywhere except that it is 1 just for the case in question. When you include that indicator in the regression model, its *t*-ratio will be what that case's Studentized residual was in the original model without the indicator. That tells us that a Studentized residual can be used to perform a *t*-test of the null hypothesis that a case is *not* an outlier. If we reject that null hypothesis, we can call the point an outlier.[10]

Influential Cases

A case that has *both* high leverage and large Studentized residuals is likely to have changed the regression model substantially all by itself. Such a case is said to be **influential**. An influential case cries out for special attention because removing it is likely to give a very different regression model.

The surest way to tell whether a case is influential is to omit it[11] and see how much the regression model changes. You should call a case "influential" if omitting it changes the regression model by enough to matter for *your* purposes. To identify possibly influential cases, check the leverage and Studentized residuals. Two statistics that combine leverage and Studentized residuals into a single measure of influence, Cook's distance and DFFITs, are offered by many statistics programs. If either measure is unusually large for a case, that case should be checked as a possibly influential point.

[7]Technically, $SD(e_i) = \sigma \sqrt{1 - h_i}$ where h_i is the leverage of the *i*-th case, e_i is its residual, and σ is the standard deviation of the regression model errors.

[8]Be cautious when you encounter the term "standardized residual." It is used in different books and by different statistics packages to mean quite different things. Be sure to check the meaning.

[9]There's more than one way to Studentize residuals according to how you estimate σ. You may find statistics packages referring to *externally Studentized residuals* and *internally Studentized residuals*. It is the *externally* Studentized version that follows a *t*-distribution, so those are the ones we recommend. This Studentized residual is found using a model fitted *without* the case of interest: We compare the response with its prediction from *that* model (hence removing the influence of this case), and divide that difference or residual by its estimated standard deviation.

[10]Finally we have a test to decide whether a case is a regression outlier (i.e. an unusually high or low *y*-value, after fitting a model excluding this case). Up until now, all we've had was our judgment based on how the plots looked. But you must still use your common sense and understanding of the data to decide *why* the case is extraordinary and whether it should be corrected or removed from the analysis. That important decision is *still* a judgment call.

[11]Or, equivalently, include an indicator variable that selects only for that case.

When a regression analysis has cases that have both high leverage and large Studentized residuals, it would be irresponsible to report only the regression on all the data. You should also compute and discuss the regression found with such cases removed, and discuss the extraordinary cases individually if they offer additional insight. If your interest is to understand the world, the extraordinary cases may well tell you more than the rest of the model. If your only interest is in the model (for example, because you hope to use it for prediction), then you'll want to be certain that the model wasn't determined by only a few influential cases, but instead was built on the broader base of the bulk of your data.

Step-by-Step Example DIAGNOSING A MULTIPLE REGRESSION

Let's consider what makes a roller coaster fast and then diagnose the model to understand more. Roller coasters get their *Speed* from gravity (the "coaster" part), so we'd naturally look to such variables as the *Height* and largest *Drop* as predictors. Let's make and diagnose that multiple regression.

THINK ➡ **Variables** Name the variables, report the *W*'s, and specify the questions of interest.

Plot

I have data on 105 roller coasters that give their top *Speed* (mph), maximum *Height*, and largest *Drop* (both in feet).

Plan Think about the assumptions and check the conditions.

✓ **Straight Enough Condition**: The plots look reasonably straight.

✓ **Independence Assumption**: There are only a few manufacturers of roller coasters worldwide. Coasters made by the same company may be similar in some respects, but each roller coaster in our data is individualized for its site, so the coasters are likely to be independent.

Because these conditions are met, I computed the regression model and found the Studentized residuals.

Actually, we need the Nearly Normal Condition only if we want to do inference, but it's hard not to look at the P-values, so we usually check it out. In a multiple regression, it's best to check the Studentized residuals, although the difference is rarely large enough to change our assessment of the Normality.

✓ **Straight Enough Condition (2)**: The values for one roller coaster don't seem to affect the values for the others in any systematic fashion. This makes the Independence Assumption more plausible.

✓ **Does the Plot Thicken? Condition**: The scatterplot of Studentized residuals against predicted values shows no obvious changes in the spread about the line. There do seem to be some large residuals that might be outliers.

✓ **Nearly Normal Condition**: A histogram of the Studentized residuals is unimodal and reasonably symmetric, but has a few moderately high outliers.

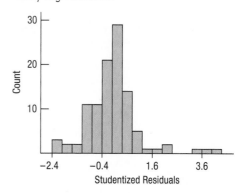

✗ **Outlier Condition**: The histogram suggests that there may be a few large positive residuals. I'll want to look at those.

Under these conditions, the multiple regression model is appropriate as long as we are cautious about the possible outliers.

Choose your method.

SHOW ➡ Mechanics

Here is the computer output for the regression:

Response variable is: Speed
R-squared = 85.7%   R-squared (adjusted) = 85.4%
s = 4.789 with 105 − 3 = 102 degrees of freedom

Source	Sum of Squares	DF	Mean Square	F-ratio	P-value
Regression	14 034.3	2	7 017.15	306	<0.0001
Residual	2 338.88	102	22.9302		

Variable	Coefficient	SE(Coeff)	t-ratio	P-value
Intercept	34.7035	1.288	27.0	<0.0001
Height	0.050380	0.0185	2.72	0.0077
Drop	0.150264	0.0183	8.20	<0.0001

The estimated regression equation is

$$\widehat{Speed} = 34.7 + 0.05\ Height + 0.15\ Drop.$$

TELL ➡ Interpretation

Diagnosis

Leverage Most computer regression programs will calculate leverage. There is a leverage value for each case.

It may or may not be necessary to remove high leverage points from the model, but it's certainly wise to know where they are and, if possible, why they are unusual.

The R^2 for the regression is 85.7%. *Height* and *Drop* account for 86% of the variation in *Speed* in roller coasters like these. Both *Height* and *Drop* contribute significantly to the *Speed* of a roller coaster.

A histogram of the leverages shows one roller coaster with a rather high leverage of more than 0.21.

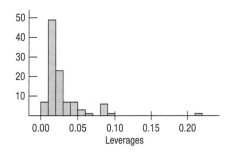

This high-leverage point is the Oblivion coaster in Alton, England. Neither the *Height* nor the *Drop* is extraordinary. To see what's going on, I made a scatterplot of *Drop* against *Height* with *Oblivion's* shown as a red x.

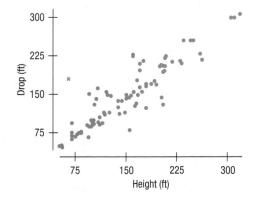

Although *Oblivion's* maximum height is a modest 65 feet, it has a surprisingly long drop of 180 feet. Looking it up, I discovered that the unusual feature of the *Oblivion's* coaster is that it plunges riders down a deep hole below the ground.

Big Residuals At this point, we might consider recomputing the regression model after removing these three coasters. That's what we do in the next section.

The histogram of the Studentized residuals (above) also nominates some cases for special attention. That bar on the right of the histogram holds three roller coasters with large positive residuals: the Xcelerator; Hypersonic XCL; and Volcano, the Blast Coaster. New technologies, such as hydraulics or compressed air, are used to launch all three roller coasters. These three coasters are different in that their speed doesn't depend only on gravity.

Diagnosis Wrapup

What have we learned from diagnosing the regression? We've discovered four roller coasters that may be influencing the model. And for each of them, we've been able to understand why and how they differed from the others. The oddness of *Oblivion* in plunging into a hole in the ground may cause us to prefer *Drop* as a predictor of *Speed* rather than *Height*.

The three cases with large residuals turned out to be different from the other roller coasters because they are "blast coasters" that don't rely solely on gravity for their acceleration. Although we can't count on always discovering why unusual cases are special, diagnosing them raises the question of what about them might be different. Understanding unusual cases can help us understand our data better.

When there are unusual cases that may possibly be influential, we always want to consider the regression model without them, shown here:

Response variable is: Speed
R-squared = 91.5%   R-squared (adjusted) = 91.3%
s = 3.683 with 102 − 3 = 99 degrees of freedom

Variable	Coefficient	SE(Coeff)	t-ratio	P-value
Intercept	34.3993	0.9926	34.7	<0.0001
Drop	0.198348	0.0153	13.0	<0.0001
Height	0.00183	0.0155	0.118	0.9062

Without the three blast coasters, *Height* no longer appears to be important in the model, so we might try omitting it:

Response variable is: Speed
R-squared = 91.5%   R-squared (adjusted) = 91.4%
s = 3.664 with 102 − 2 = 100 degrees of freedom

Variable	Coefficient	SE(Coeff)	t-ratio	P-value
Intercept	34.4285	0.9567	36.0	<0.0001
Drop	0.200004	0.0061	32.7	<0.0001

That looks like a good model. It seems that our diagnosis has led us back to a simple regression.

INDICATORS FOR EXTRAORDINARY CASES: REMOVING THEIR INFLUENCE

One good way to deal with an extraordinary case in a regression is to construct a special indicator variable that is zero for all cases *except* the one we want to isolate. *Including such an indicator in the regression model has the same effect as removing the case from the data*, but it has two special advantages. First, it makes it clear to anyone looking at the regression model that we have treated that case specially. Second, the *t*-statistic for the indicator variable's coefficient can be used as a test of whether the case is a regression outlier (i.e. having a surprisingly high or low *y*-value if we fit a model not influenced by this case). If the P-value is small, then that case really didn't fit well with the rest of the data. Typically, we name such an indicator with the identifier of the case we want to remove. Here's the last roller coaster model in which we have removed the influence of the three blast coasters by constructing indicators for them instead of by removing them from the data. Notice that the coefficients for the other predictors are just the same as the ones we found by omitting the cases.

Response variable is: Speed
R-squared = 91.8%   R-squared (adjusted) = 91.5%
s = 3.664 with 105 − 5 = 100 degrees of freedom

Variable	Coefficient	SE(Coeff)	t-ratio	P-value
Intercept	34.4285	0.9567	36.0	<0.0001
Drop	0.200004	0.0061	32.7	<0.0001
Xcelerator	21.5711	3.684	5.86	<0.0001
HyperSonic	18.9711	3.683	5.15	<0.0001
Volcano	19.5713	3.704	5.28	<0.0001

The P-values for the three indicator variables confirm that each of these roller coasters doesn't fit with the others.

Section 27.2 Exercises

3. Residual, leverage, influence For each of the following cases, would your primary concern about them be that they had a large residual, large leverage, or likely large influence on the regression model? Explain your thinking.
   a) In a regression to predict the construction cost of roller coasters from the length of track, the height of the highest point, and the type of construction (metal or wood), the *Kingda Ka* coaster, which opened in 2005 and, at (456 ft), is quite tall.
   b) In a regression to predict income of graduates of a college five years after graduation, a graduate who created a high-tech start-up company based on his senior thesis, and sold it for several million dollars.

4. Residual, leverage, influence, 2 For each of the following cases, would your primary concern about them be that they had a large residual, large leverage, or likely large influence on the regression model?
   a) In a regression to predict Freshman grade point averages as part of the admissions process, a student whose Math SAT was 750, whose Verbal SAT was 585, and who had a 4.0 GPA at the end of her Freshman year.
   b) In a regression to predict life expectancy in countries of the world from generally available demographic, economic, and health statistics, a country that, due to a high prevalence of HIV, has an unusually low life expectancy.

27.3 Building Multiple Regression Models

> "It is the mark of an educated mind to be able to entertain a thought without accepting it."
>
> —*Aristotle*

When many possible predictors are available, we will naturally want to select only a few of them for a regression model. But which ones? The first and most important thing to realize is that often there is no such thing as the "best" regression model. (After all, all models are wrong.) Several alternative models may be useful or insightful. The "best" for one purpose may not be best for another. And the one with the highest R^2 may not be best for many purposes. There is nothing wrong with continuing to work with several models without choosing among them.

Multiple regressions are subtle. The coefficients often don't mean what they at first appear to mean. The choice of which predictors to use determines almost everything about the regression.

Relationships between predictors can complicate interpretation and understanding. So it is usually best to build a *parsimonious* model, using as few predictors as you can. On the other hand, we don't want to leave out predictors that are theoretically or practically important. Making this tradeoff is the heart of the challenge of selecting a good model. The best regression models, in addition to satisfying the assumptions and conditions of multiple regression, have:

- Relatively few predictors to keep the model simple.
- A relatively high R^2 indicating that much of the variability in the *y*-variable is accounted for by the regression model.
- A relatively small value of *s*, the standard deviation of the residuals, indicating that the magnitude of the errors is small.
- Relatively small P-values for their *F*- and *t*-statistics, showing that the overall model is better than a simple summary with the mean, and that the individual coefficients are reliably different from zero.
- No cases with extraordinarily high leverage that might dominate and alter the model.
- No cases with extraordinarily large residuals, and Studentized residuals that appear to be nearly Normal. Outliers can alter the model and certainly weaken the power of any test statistics. And the Nearly Normal Condition is required for inference.
- Predictors that are reliably measured and relatively unrelated to each other.

The term "relatively" in this list is meant to suggest that you should favour models with these attributes over others that satisfy them less, but, of course, there are many tradeoffs and no absolute rules.

Cases with high leverage or large residuals can be dealt with by introducing indicator variables.

In addition to favouring predictors that can be measured reliably, you may want to favour those that are less expensive to measure, especially if your model is intended for prediction with values not yet measured.

Seeking Multiple Regression Models Automatically

How can we find the best multiple regression model? The list of desirable features we just looked at should make it clear that there is no simple definition of the "best" model. A computer can try all combinations of the predictors to find the regression model with the highest R^2, or optimize some other criterion,[12] but models found that way are not best for all purposes and may not even be particularly good for many purposes.

Another alternative is to have the computer build a regression "stepwise." In a **stepwise regression**, at each step, a predictor is either added to or removed from the model. The predictor chosen to add is the one whose addition increases the R^2 the most (or similarly improves some other measure). The predictor chosen to remove is the one whose removal reduces the R^2 the least (or similarly loses the least on some other measure). The hope is that by following this path, the computer can settle on a good model. The model will gain or lose a predictor only if that change in the model makes a *big enough* change in the performance measure. The changes stop when no more changes pass the criterion.

Stepping in the wrong direction.   Here's an example of how stepwise regression can go astray. We might want to find a regression to model *Horsepower* in a sample of cars from the car's engine size *(Displacement)* and *Weight*. The simple correlations are as follows:

	HP	Disp	Wt
Horsepower	1.000		
Displacement	0.872	1.000	
Weight	0.917	0.951	1.000

Because *Weight* has a slightly higher correlation with *Horsepower*, stepwise regression will choose it first. Then, because *Weight* and engine size *(Displacement)* are so highly correlated, once *Weight* is in the model, *Displacement* won't be added to the model. But *Weight* is, at best, a lurking variable leading to both the need for more horsepower and a larger engine. Don't try to tell an engineer that the best way to increase horsepower is to add weight to the car and that the engine size isn't important! From an engineering standpoint, *Displacement* is a far more appropriate predictor of *Horsepower*, but stepwise regression won't find that model.

Stepwise methods can be valuable when there are hundreds or thousands of potential predictors, as can happen in data mining applications. They can build models that are useful for prediction or as starting points in the search for better models. Because they do each step *automatically*, however, stepwise methods are inevitably affected by influential points and nonlinear relationships. A better strategy might be to mimic the stepwise procedure yourself, but more carefully. You could consider adding or removing a variable yourself with a careful look at the assumptions and conditions each time a variable is considered. That kind of guided stepwise method is still not guaranteed to find a good model, but it may be a sensible way to search among the potential candidates. (Some software will allow pauses in its stepwise procedure, so you can intervene to make changes.)

---

[12]This is literally true. Even for many variables and a moderately large number of cases, it is computationally possible to find the "best subset" of predictors that maximizes R^2. Many statistics programs offer this impressive capability; note that if there are just 10 possible predictors under consideration, there are 1024 different subsets of predictors to compare!

Building Regression Models Sequentially

You can build a regression model by adding variables to a growing regression. This is generally referred to as a *forward selection* method. Each time you add a predictor, you hope to account for a little more of the variation in the response. What's left over are the residuals. At each step, consider the predictors still available to you. Those that are most highly correlated with the current residuals are the ones that are most likely to improve the model.

If you see a variable with a high correlation at this stage and it is *not* among those that you thought were important, stop and think about it. Is it correlated with another predictor or with several other predictors? Don't let a variable that doesn't make sense enter the model just because it has a high correlation, but at the same time, don't exclude a predictor just because you didn't initially think it was important. (That would be a good way to make sure that you never learn anything new.) Finding the balance between these two choices underlies the art of successful model building.

Alternatively, you can start with all available predictors in the model and remove those with small *t*-ratios. This is generally referred to as a *backward elimination* method. At each step, make a plot of the residuals to check for outliers, and check the leverages (say, with a histogram of the leverage values) to help spot any high-leverage points. Influential cases can strongly affect which variables appear to be good or poor predictors in the model. It's also a good idea to check that a predictor doesn't appear to be unimportant in the model only because it's correlated with other predictors in the model. It may (as is true of *Displacement* in the example of predicting *Horsepower*) actually be a more useful or meaningful predictor than some of those in the model.

In either method, adding or removing a predictor will usually change *all* of the coefficients, sometimes by quite a bit.

Step-by-Step Example BUILDING MULTIPLE REGRESSION MODELS

Let's return to the Kids Count infant mortality data from Chapter 26. (Data in **Infant_Mortality**) We fitted a large multiple regression model in which several of the *t*-ratios for coefficients were too small to be discernibly different from zero. Maybe we can build a more parsimonious model. Which model should we build?

The most important thing to do is to think about the data. Regression models can and should make sense. Many factors can influence your choice of a model, including the cost of measuring particular predictors, the reliability or possible biases in some predictors, and even the political costs or advantages to selecting predictors.

THINK ➡ **Variables** Name the available variables, report the *W*'s, and specify the question of interest or the purpose of finding the regression model.	I have data on the 50 states. The available variables are (all for 1999):
	Infant Mortality (deaths per 1000 live births)
	Low Birth Weight (Low BW%—%babies with low birth weight)
	Child Deaths (deaths per 100 000 children ages 1–14)
	%Poverty (percent of children in poverty in the previous year)
	HS Drop% (percent of teens who are high-school dropouts, ages 16–19)
	Teen Births (births per 100 000 females ages 15–17)

Plan Think about the assumptions and check the conditions.

We examined a scatterplot matrix and the regression with all potential predictors in Chapter 26.

Teen Deaths (by accident, homicide, and suicide; deaths per 100 000 teens ages 15–19)

I hope to gain a better understanding of factors that affect infant mortality.

✓ **Straight Enough Condition**: As we observed in the previous chapter, the scatterplot matrix showed no bends, clumping, or outliers in any of the scatterplots.

✓ **Independence Assumption**: See previous chapter.

With this assumption and condition satisfied, let's fit the model with all the predictors and find residuals.

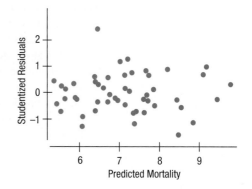

Remember that in a multiple regression, we usually plot Studentized residuals against the predicted values (sometimes supplemented by plots against individual predictors).

✓ **Does the Plot Thicken? Condition**: This scatterplot of Studentized residuals versus predicted values for the full model (all predictors) shows no obvious trends in the spread.

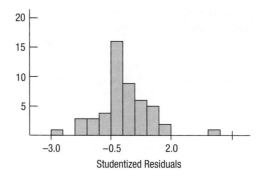

Nearly Normal Condition, Outlier Condition: A histogram of the Studentized residuals from the full model is unimodal and symmetric, but it seems to have an outlier. The unusual state is South Dakota. I'll test whether it really is an outlier by making an indicator variable for South Dakota and including it in the predictors.

SHOW ➡ Mechanics Multiple regressions are always found from a computer program.

I'll start with the full regression and work backward:

Dependent variable is: Infant Mort
R-squared = 78.7% R-squared (adjusted) = 75.2%
s = 0.6627 with 50 − 8 = 42 degrees of freedom

Variable	Coefficient	SE(Coeff)	*t*-ratio	P-value
Intercept	1.31183	0.8639	1.52	0.1364
Low BW%	0.73272	0.1067	6.87	<0.0001
Child Deaths	0.02857	0.0123	2.31	0.0256
%Poverty	−5.3026e-3	0.0332	−0.160	0.8737
HS Drop%	−0.10754	0.0540	−1.99	0.0531
Teen Births	0.02402	0.0234	1.03	0.3111
Teen Deaths	−1.5516e-4	0.0101	−0.015	0.9878
S. Dakota	2.74813	0.7175	3.83	0.0004

For model building, look at the P-values *only as general indicators* of how much a predictor contributes to the model.

The coefficient for the S. Dakota indicator variable has a very small P-value, so that case is an outlier in this regression model (and we can just carry along this indicator, which has the effect of removing any influence of the outlier). *Teen Births*, *Teen Deaths*, and *%Poverty* have large P-values and look like they are less successful predictors in this model.

I'll remove *Teen Deaths* first:

Dependent variable is: Infant Mort
R-squared = 78.7% R-squared (adjusted) = 75.7%
s = 0.6549 with 50 − 7 = 43 degrees of freedom

Variable	Coefficient	SE(Coeff)	*t*-ratio	P-value
Intercept	1.30595	0.7652	1.71	0.0951
Low BW%	0.73283	0.1052	6.97	<0.0001
Child Deaths	0.02844	0.0085	3.34	0.0018
%Poverty	−5.3548e-3	0.0326	−0.164	0.8703
HS Drop%	−0.10749	0.0533	−2.02	0.0501
Teen Births	0.02402	0.0231	1.04	0.3053
S. Dakota	2.74651	0.7014	3.92	0.0003

You shouldn't remove more than one predictor at a time from the model because each predictor can influence how the others contribute to the model. If removing a predictor from the model doesn't change the remaining coefficients very much (or reduce the R^2 by very much), that predictor wasn't contributing very much to the model. When we carry an indicator for a data point, we don't need to check its actual influence as the indicator variable controls for this.

Removing *Teen Births* and *%Poverty*, in turn, gives this model:

Dependent variable is: Infant Mort
R-squared = 78.1% R-squared (adjusted) = 76.2%
s = 0.6489 with 50 − 5 = 45 degrees of freedom

Variable	Coefficient	SE(Coeff)	*t*-ratio	P-value
Intercept	1.03782	0.6512	1.59	0.1180
Low BW%	0.78334	0.0934	8.38	<0.0001
Child Deaths	0.03104	0.0075	4.12	0.0002
HS Drop%	−0.06732	0.0381	−1.77	0.0837
S. Dakota	2.66150	0.6899	3.86	0.0004

Adjusted R^2 can increase when you remove a predictor if that predictor wasn't contributing very much to the regression model.

Compared with the full model, the R^2 has come down only very slightly, and the adjusted R^2 has actually increased. The P-value for *HS drop%* is somewhat bigger than the common .05 level, so if we have no particular practical reasons to keep it, let's take that variable out, and then the model looks like this:

Dependent variable is: Infant Mort
R-squared = 76.6% R-squared (adjusted) = 75.1%
s = 0.6638 with 50 − 4 = 46 degrees of freedom

Variable	Coefficient	SE(Coeff)	*t*-ratio	P-value
Intercept	0.760145	0.6465	1.18	0.2457
Child Deaths	0.026988	0.0073	3.67	0.0006
Low BW%	0.750461	0.0937	8.01	<0.0001
S.Dakota	2.74057	0.7042	3.89	0.0003

This looks like a good model. It has a reasonably high R^2 and small P-values for each of the coefficients.

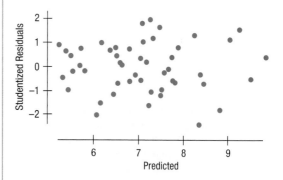

Before deciding that any regression model is a "keeper," remember to check the residuals.

TELL ➡ Summarize the features of this model.

Here's an example of an outlier that might help us learn something about the data or the world. Whatever makes South Dakota's infant mortality rate so much higher than the model predicts, it might be something we could address with new policies or interventions.

The scatterplot of Studentized residuals against predicted values shows no structure, and the histogram of Studentized residuals is nearly Normal. So this looks like a good model for *Infant Mortality*. The coefficient for S. Dakota is still very significant, so I'd prefer to keep South Dakota separate and look into why its *Infant Mortality* rate is so much higher

(2.74 deaths per 1000 live births) than we would otherwise expect from its *Child Death* rate and *Low Birth Weight* percent.

SHOW

Let's try the other way and build a regression model using the forward selection method. If just building a model for prediction, we might start with the variable having the highest correlation with *Infant Mortality*. But let's assume we are very interested in how teen behaviour might be predictive of infant mortality.

The data include several variables that concern teens: *Teen Births*, *Teen Deaths*, and the *HS Drop%*.

Both *Teen Births* and *Teen Deaths* are promising predictors, but births to teens seem more directly relevant. Here's the regression model:

Dependent variable is: Infant Mort
R-squared = 29.3% R-squared (adjusted) = 27.9%
s = 1.129 with 50 − 2 = 48 degrees of freedom

Variable	Coefficient	SE(Coeff)	t-ratio	P-value
Intercept	4.96399	0.5098	9.74	<0.0001
Teen Births	0.081217	0.0182	4.47	<0.0001

One way to select variables to add to a growing regression model is to find the correlation of the residuals of the current state of the model with the potential new predictors. Predictors with higher correlations can be expected to account for more of the remaining residual variation if we include them in the regression model.

The correlations of the residuals with other (teen related) predictors look like this:

	Resids
HS Drop%	−0.188
Teen Deaths	0.333
%Poverty	0.105

Teen Deaths looks like a good choice to add to the model:

Dependent variable is: Infant Mort
R-squared = 39.1% R-squared (adjusted) = 36.5%
s = 1.059 with 50 − 3 = 47 degrees of freedom

Variable	Coefficient	SE(Coeff)	t-ratio	P-value
Intercept	3.98643	0.5960	6.69	<0.0001
Teen Births	0.057880	0.0191	3.04	0.0039
Teen Deaths	0.028228	0.0103	2.75	0.0085

Finally, I'll try adding *HS Drop%* to the model:

Dependent variable is: Infant Mort
R-squared = 44.0% R-squared (adjusted) = 40.4%
s = 1.027 with 50 − 4 = 46 degrees of freedom

Variable	Coefficient	SE(Coeff)	t-ratio	P-value
Intercept	4.51922	0.6358	7.11	<0.0001
Teen Births	0.097855	0.0272	3.60	0.0008
Teen Deaths	0.026844	0.0100	2.69	0.0099
HS Drop%	−0.164347	0.0819	−2.01	0.0506

Notice that adding a predictor that does not contribute to the model can reduce the adjusted R^2.

The regression that models *Infant Mortality* on *Teen Births*, *Teen Deaths*, and *HS Drop%* may be worth keeping as well. But, of course, we're not finished until we check the residuals:

Here is one more step, adding *%Poverty* to the model:

Dependent variable is: Infant Mort
R-squared = 44.0% R-squared (adjusted) = 39.1%
s = 1.038 with 50 − 5 = 45 degrees of freedom

Variable	Coefficient	SE(Coeff)	*t*-ratio	P-value
Intercept	4.49810	0.7314	6.15	<0.0001
Teen Births	0.09690	0.0317	3.06	0.0038
Teen Deaths	0.02664	0.0106	2.50	0.0160
HS Drop%	−0.16397	0.0830	−1.98	0.0544
%Poverty	3.1053e-3	0.0513	0.061	0.9520

The P-value for *%Poverty* is quite high, so I prefer the previous model.

Here are the residuals:

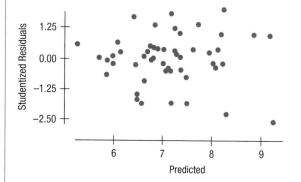

This histogram hints of a low mode holding some large negative residuals, and the scatterplot shows two in particular that trail off at the bottom right corner of the plot, Texas and New Mexico. These states are neighbours and may share some regional attributes. To be careful, I'll try removing them from the model. Or equivalently, I'll construct two indicator variables that are 1 for the named state and 0 for all others:

Dependent variable is: Infant Mort
R-squared = 58.9% R-squared (adjusted) = 54.2%
s = 0.8997 with 50 − 6 = 44 degrees of freedom

Variable	Coefficient	SE(Coeff)	t-ratio	P-value
Intercept	4.15748	0.5673	7.33	<0.0001
Teen Births	0.13823	0.0259	5.33	<0.0001
Teen Deaths	0.02669	0.0090	2.97	0.0048
HS Drop%	−0.22808	0.0735	−3.10	0.0033
New Mexico	−3.01412	0.9755	−3.09	0.0035
Texas	−2.74363	0.9748	−2.81	0.0073

Removing the two outlying states has improved the model noticeably. The indicators for both states have small P-values, so I conclude that they were in fact outliers for this model. The R^2 has improved to 58.9%, and the P-values of all the other coefficients have been reduced.

A final check on the residuals from this model shows that they satisfy the regression conditions:

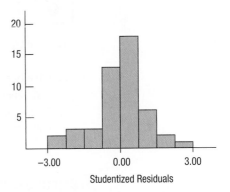

This model is an alternative to the first one I found. It has a smaller R^2 (58.9%) and larger s value, but it might be useful for understanding the relationships between these teen-related variables and infant mortality.

TELL ➡ Compare and contrast the models.	I have found two reasonable regression models for *Infant Mortality*. The first finds that *Infant Mortality* can be modelled by *Child Deaths* and *%Low Birth Weight*, removing the influence of South Dakota:

$$\overline{Infant\ Mortality} = 0.76 + 0.027\ Child\ Deaths$$
$$+ 0.75\ LowBW\%$$

It may be worthwhile to look into why South Dakota is so different from the other states.

For a more complete understanding of infant mortality, we should look into South Dakota's early childhood variables and the teen-related variables in New Mexico and Texas. We might learn as much about infant mortality by understanding why these states stand out—and how they differ from each other—as we would from the regression models themselves.

The other model focused on teen behaviour, modelling *Infant Mortality* by *Teen Births*, *Teen Deaths*, and *HS Drop%*, removing the influence of Texas and New Mexico:

$$\overline{Infant\ Mortality} = 4.16 + 0.138\ Teen\ Births$$
$$+ 0.027\ Teen\ Deaths$$
$$- 0.228\ HS\ Drop\%$$

The coefficient of *HS Drop%* is the opposite sign of the simple relationship between *Infant Mortality* and *HS Drop%*.

Each model has nominated different states as outliers. For a more complete understanding of infant mortality, it might be worthwhile to look into why these states are outliers in these models.

Which model is better? That depends on what you want to know. Remember—all models are wrong. But both may offer useful information and insights about infant mortality and its relationship with other variables, as well as the states that stood out and why they differ from the others.

Just Checking

1. Give two ways that we use histograms to support the construction, inference, and understanding of multiple regression models.
2. Give two ways that we use scatterplots to support the construction, inference, and understanding of multiple regression models.
3. What role does the Normal model play in the construction, inference, and understanding of multiple regression models?

Regression Roles

We build regression models for a number of reasons. One reason is to model how variables are related to each other in the hope of understanding the relationships. Another is to build a model that might be used to predict values for a response variable when given values for the predictor variables.

When we hope to understand, we are often particularly interested in simple, straightforward models in which predictors are as unrelated to each other as possible. We are especially happy when the *t*-statistics are large, indicating that the predictors each contribute to the model. We are likely to want to look at partial regression plots to understand the coefficients and to check that no outliers or influential points are affecting them.

When prediction is our goal, we are more likely to care about the overall R^2. Good prediction occurs when much of the variability in the response *y* is accounted for by the model. We might be willing to keep variables in our model that have relatively small *t*-statistics simply for the stability that having several predictors can provide. We care less whether the predictors are related to each other because we don't intend to interpret the coefficients anyway.

In both roles, we may include some predictors to "get them out of the way." Regression offers a way to approximately control for factors when we have observational data because each coefficient measures effects *after removing the effects* of the other predictors. Of course, it would be better to control for factors in a randomized experiment, but often that's just not possible.

*Indicators for Three or More Levels

It's easy to construct indicators for a categorical variable with two levels; we just assign 0 to one level and 1 to the other. But variables like *Month* or *Class* often have several levels. You can construct indicators for a categorical variable with several levels by constructing a separate indicator for each of these levels. There's just one trick: You have to choose one of the categories as a "baseline" and *leave out* its indicator. Then the coefficients of the other indicators can be interpreted as the amount by which their categories differ from the baseline category, after allowing for the linear effects of the other variables in the model.[13]

Make sure your collection of indicators doesn't exhaust all the categories. One category must be left out to serve as a baseline or the regression model can't be found. For the two-category variable *Inversions*, we used "no inversion" as the baseline and coasters with an inversion got a 1. We needed only one variable for two levels. If we wished to represent *Month* with indicators, we would need 11 of them. We might, for example, define *January* as the baseline, and make indicators for *February*, *March*, …, *November*, and *December*. Each of these indicators would be 0 for all cases except for the ones that had that value for the variable *Month*. So, for example, each case in February has every indicator = 0 except for the February indicator, and when all the indicators = 0 for a case, we know that case falls in January. Why not just a single variable with "1" for *January*, "2" for *February*, and so on? That is quite unlikely to work. It would impose the pretty strict assumption that the responses to the months are ordered and equally spaced—that is, that the change in our response variable from January to February is the same in both direction and amount as the change from July to August. That's a pretty severe restriction and may not be true for many kinds of data. Using 11 indicators releases the model from that restriction, but, of course, at the expense of having 10 fewer degrees of freedom for all of our *t*-tests.

Collinearity

Let's look at the infant mortality data one more time. One useful predictor of *Infant Mortality* is *Teen Deaths*. The scatterplot is in the margin, with regression output below.

Dependent variable is: Infant Mort
R-squared = 27.2% R-squared (adjusted) = 25.7%
s = 1.146 with 50 − 2 = 48 degrees of freedom

Variable	Coefficient	SE(Coeff)	*t*-ratio	P-value
Intercept	4.73979	0.5866	8.08	<0.0001
Teen Deaths	0.042129	0.0100	4.23	0.0001

---

[13]There are alternative coding schemes that compare all the levels with the mean. Make sure you know how the indicators are coded.

Teen Deaths has a positive coefficient (as we might expect) and a very small P-value. Suppose we now add *Child Deaths* to the regression model:

Dependent variable is: Infant Mort
R-squared = 42.6% R-squared (adjusted) = 40.1%
s = 1.029 with 50 − 3 = 47 degrees of freedom

Variable	Coefficient	SE(Coeff)	t-ratio	P-value
Intercept	5.79561	0.6049	9.58	<0.0001
Teen Deaths	−1.86877e−3	0.0153	−0.122	0.9032
Child Deaths	0.059398	0.0168	3.55	0.0009

Suddenly, *Teen Deaths* has a small negative coefficient and a very large P-value. What happened? The problem is that *Teen Deaths* and *Child Deaths* are closely associated (Figure 27.8). The coefficient of *Teen Deaths* now reports how *Infant Mortality* is related to *Teen Deaths* after *allowing for the linear effects of Child Deaths* on both variables.

When we have several predictors, we must think about how the predictors are related to each other. When predictors are unrelated to each other, each provides new information to help account for more of the variation in the response *y*. Just as we need a predictor to have a large enough variability to provide a stable base for simple regression, when we have several predictors, we need for them to vary in different directions for the multiple regression to have a stable base. If you wanted to build a deck on the back of your house, you wouldn't build it with supports placed just along one diagonal. Instead, you'd want the supports spread out in different directions as much as possible to make the deck stable. We're in a similar situation with multiple regression. When predictors are highly correlated, they line up together, which makes the regression they support balance precariously. Even small variations can rock it one way or the other. A more stable model can be built when predictors have low correlation and the points are spread out.

When two or more predictors are linearly related, they are said to be **collinear**. The general problem of predictors with close (but perhaps not perfect) linear relationships is called the problem of **collinearity**.

Fortunately, there's an easy way to assess collinearity. To measure how much one predictor is linearly related to the others, just find the regression of that predictor on the others and look at the R^2. That R^2 gives the fraction of the variability of the predictor in question that is accounted for by the other predictors. So $1 − R^2$ is the amount of the predictor's variance that is left after we allow for the effects of the other predictors. That's what the predictor has left to bring to the regression model. And we know that a predictor with little variance can't do a good job of predicting.[14]

Collinearity can hurt our analysis in yet another way. We've seen that the variance of a predictor plays a role in the standard error of its associated coefficient. Small variance leads to a larger SE. In fact, it's exactly this leftover variance that shows up in the formula for the SE of the coefficient. That's what happened in the infant mortality example.

As a final blow, when a predictor is collinear with the other predictors, it's often difficult to figure out what its coefficient means in the multiple regression. We've blithely talked about "removing the effects of the other predictors," but now when we do that, there may not be much left. What is left is not likely to be about the original predictor, but more about the fractional part of that predictor not associated with the others. In a regression of *Horsepower* on *Weight* and engine *Size*, once we've removed the effect of weight on horsepower, engine size doesn't tell us anything *more* about horsepower. That's certainly not the same as saying that engine size doesn't tell us anything about horsepower. It's just that most cars with big engines also weigh a lot.

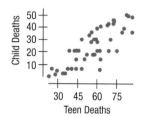

Figure 27.8

Child Deaths and *Teen Deaths* are linearly related.

You may find this problem referred to as "multicollinearity." But there is no such thing as "unicollinearity"—we need at least two predictors for there to be a linear association between them—so there is no need for the extra two syllables.

WHY NOT JUST LOOK AT THE CORRELATIONS?

It's sometimes suggested that we examine the table of correlations of all the predictors to search for collinearity. But this will find only associations among *pairs* of predictors. Collinearity can—and does—occur among *several* predictors working together. You won't find that more subtle collinearity with a correlation table.

---

[14]The statistic $1/(1 − R^2)$ for the R^2 found from the regression of one predictor on the other predictors in the model is also called the *Variance Inflation Factor*, or *VIF*, in some computer programs and books.

When a predictor is collinear with the other predictors in the model, two things can happen:

1. Its coefficient can be surprising, taking on an unanticipated sign or being unexpectedly large or small.
2. The standard error of its coefficient can be large, leading to a smaller *t*-statistic and correspondingly large P-value.

One telltale sign of collinearity is the paradoxical situation in which the overall *F*-test for the multiple regression model is significant, showing that at least one of the coefficients is discernably different from zero, and yet most or all of the individual coefficients have small *t*-values, each in effect, denying that *it* is the significant one.

What should you do about a collinear regression model? The simplest cure is to remove some of the predictors. That both simplifies the model and generally improves the *t*-statistics. And, if several predictors give pretty much the same information, removing some of them won't hurt the model. Which should you remove? Keep the predictors that are most reliably measured, least expensive to find, or even those that are politically important.

> A Mathematics department at a large university built a regression model to help them predict success in graduate study. They were shocked when the coefficient for Mathematics GRE score was not significant. But the Math GRE was collinear with some of the other predictors, such as math course GPA and Verbal GRE, which made its slope not significant. They decided to omit some of the other predictors and retain Math GRE as a predictor because that model seemed more appropriate—even though it predicted no better (and no worse) than others without Math GRE.

Section 27.3 Exercises

5. Significant coefficient? In a regression to predict compensation of employees in a large firm, the predictors in the regression were *Years with the firm*, *Age*, and *Years of Experience*. The coefficient of *Age* is negative and statistically significantly different from zero. Does this mean that the company pays workers less as they get older? Explain.

6. Better model? Joe wants to impress his boss. He builds a regression model to predict sales that has 20 predictors and an R^2 of 80%. Sally builds a competing model with only 5 predictors, but an R^2 of only 78%. Which model is likely to be most useful for understanding the drivers of sales? How could the boss tell? Explain.

WHAT CAN GO WRONG?

In the Oscar-winning movie *The Bridge on the River Kwai* and in the book on which it is based,[15] the character Colonel Green famously says, "As I've told you before, in a job like yours, even when it's finished, there's always one more thing to do." It is wise to keep Colonel Green's advice in mind when building, analyzing, and understanding multiple regression models.

■ **Beware of collinearities.** When the predictors are linearly related to each other, they add little to the regression model *after allowing* for the contributions of the other predictors. Check the R^2 when each predictor is regressed on the others. If these are high, consider omitting some of the predictors.

■ **Don't check for collinearity only by looking at pairwise correlations.** Collinearity is a relationship among any number of the predictors. Pairwise correlations can't always show that. (Of course, a high pairwise correlation between two predictors does indicate collinearity.)

■ **Don't be fooled when high-influence points and collinearity show up together.** A single high-influence point can be the difference between your predictors being collinear and seeming not to be collinear. (Picture that deck supported only along its diagonal and with a single additional post in another corner. Supported in this way, the deck is stable, but the height of that single post completely determines the tilt of the deck, so it may be very influential.) Removing a high-influence point may surprise you with unexpected

(continued)

---

[15]The author of the book, Pierre Boulle, also wrote the book and script for *Planet of the Apes*. The director, David Lean, also directed *Lawrence of Arabia*.

collinearity. Alternatively, a single value that is extreme on several predictors can make them appear to be collinear when in fact they would not be if you removed that point. Removing that point may make apparent collinearities disappear (and would probably result in a more useful regression model).

- **Beware missing data.** Values may be missing or unavailable for any case in any variable. In simple regression, when the cases are missing for reasons that are unrelated to the variable we're trying to predict, that's not a problem. We just analyze the cases for which we have data. But when several variables participate in a multiple regression, any case with data missing on any of the variables will be omitted from the analysis. You can unexpectedly find yourself with a much smaller set of data than you started with. Be especially careful, when comparing regression models with different predictors, that the cases participating in the models are the same.

- **Remember linearity.** The Linearity Assumption (and the Straight Enough Condition) require linear relationships among the variables in a regression model. As you build and compare regression models, be sure to plot the data to check that relationships are straight. Violations of this assumption make everything else about a regression model invalid.

  Actually it is quite simple to add quadratic or cubic terms to your model and fit by least squares—similar to adding an interaction term. Though not linear in the predictor itself, the model is still linear in a bigger sense—we describe it as linear in the betas, called a General Linear Model—and hence easy to fit by least-squares and use for inference. Re-expressing variables though (with logs, etc.) should generally be the first approach when trying to straighten relationships. However, if the relationship goes up, then down, or down, then up, a re-expression won't work.

- **Check for parallel regression lines.** When you introduce an indicator variable for a category, check the underlying assumption that the other coefficients in the model are essentially the same for both groups. If not, consider adding an interaction term.

What Have We Learned?

In Chapter 26, we learned that multiple regression is a natural way to extend what we knew about linear regression models to include several predictors. Now we've learned that multiple regression is both more powerful and more complex than it may appear at first. As with other chapters in this book whose titles spoke of greater "wisdom," this chapter has drawn us deeper into the uses and cautions of multiple regression.

- Know how to incorporate categorical data by using indicator variables, modelling relationships that have parallel slopes but at different levels for different groups.

- Know how to use interaction terms, to allow for different slopes. We can create identifier variables that isolate individual cases to remove their influence from the model while exhibiting how they differ from the other points and testing whether that difference is statistically significant.

- Beware unusual cases. A single unusual case can greatly influence the entire regression. Such cases should be treated specially, possibly by fitting the model both with and without them or by including indicator variables to isolate their influence.

- Be cautious in complex models because one has to be careful in interpreting the coefficients. Associations among the predictors can change the coefficients to values that can be quite different from the coefficient in the simple regression of a predictor and the response, even changing the sign.

- Understand that building multiple regression models is an art that speaks to the central goal of statistical analysis: understanding the world with data. We've learned that there is no right model. We've seen that the same response variable can be modelled with several alternative models, each showing us different aspects of the data and of the relationships among the variables and nominating different cases as special and deserving of our attention.

■ We've also seen that everything we've discussed throughout this book fits together to help us understand the world. The graphical methods are the same ones we learned in the early chapters, and the inference methods are those we originally developed for means. In short, there's been a consistent tale of how we understand data to which we've added more and more detail and richness, but which has been consistent throughout.

Review of Terms

Indicator variable A variable constructed to indicate for each case whether it is in a designated group or not. A common way to assign values to indicator variables is to let them take on the values 0 and 1, where 1 indicates group membership (p. 895).

Interaction term A constructed variable found as the product of a predictor and an indicator variable. An interaction term adjusts the *slope* of the cases identified by the indicator against the predictor (p. 897).

Leverage The leverage of a case measures how far its x-values are from the centre of the x's and, consequently, how much influence it is capable of exerting on the regression model. Points with high leverage can determine a regression model and should, therefore, be examined carefully (p. 900).

Studentized residual When a residual is divided by an independent estimate of its standard deviation (based on a model with this particular case deleted), the result is a Studentized residual. The type of Studentized residual that has a *t*-distribution is termed an *externally* Studentized residual (or Studentized deleted residual) (p. 902).

Influential case A case is *influential* on a multiple regression model if, when it is omitted, the model changes by enough to matter for your purposes. (There is no specific amount of change defined to declare a case influential.) Cases with high leverage and large Studentized residual are likely to be influential (p. 902).

Stepwise regression An automated method of building regression models in which predictors are added to or removed from the model one at a time in an attempt to optimize a measure of the success of the regression. Stepwise methods rarely find the best model and are easily influenced by influential cases, but they can be valuable in winnowing down a large collection of candidate predictors (p. 908).

Collinearity When one (or more) of the predictors can be fit closely by a multiple regression on the other predictors (so it has little new to contribute to a model in addition to what those other predictors contribute), we have collinearity. When collinear predictors are in a regression model, they may have unexpected coefficients and often have inflated standard errors (and correspondingly small *t*-statistics) (p. 918).

On the Computer REGRESSION ANALYSIS

Statistics packages differ in how much information they provide to diagnose a multiple regression. Most packages provide leverage values. Many provide far more, including statistics that we have not discussed. But for all, the principle is the same. We hope to discover any cases that don't behave like the others in the context of the regression model and then to understand why they are special.

Many of the ideas in this chapter rely on the concept of examining a regression model and then finding a new one based on your growing understanding of the model and the data. Regression diagnosis is meant to provide steps along that road. A thorough regression analysis may involve finding and diagnosing several models.

See the previous chapter for how to run a multiple regression using these packages.

EXCEL

Excel does not offer diagnostic statistics with its regression function.

COMMENTS

Although the dialogue offers a Normal probability plot of the residuals, the data analysis add-in does not make a correct probability plot, so don't use this option. The "standardized residuals" are just the residuals divided by their standard deviation (with the wrong df), so they, too, should be ignored.

JMP

- From the **Analyze** menu select **Fit Model**.
- Specify the response, Y. Assign the predictors, X, in the **Construct Model Effects** dialogue box.
- Click on **Run Model**.
- Click on the red triangle in the title of the Model output to find a variety of plots and diagnostics available.

COMMENTS

JMP chooses a regression analysis when the response variable is "Continuous."

MINITAB

- Choose **Regression** from the **Stat** menu.
- Choose **Regression > Fit Regression Model** from the **Regression** submenu.
- In the Regression dialogue, assign the Y variable to the **Responses** box and assign the X-variables to the **Continuous Predictors** box. You even have a box for **Categorical predictors** and Minitab will form the indicator variables for you.
- Click the **Storage** button.
- In the Regression **Storage** dialogue, you can select a variety of diagnostic statistics. They will be stored in columns of your worksheet. 'Deleted residuals' are just the Studentized residuals.

- Click the **OK** button to return to the Regression dialogue.
- To specify displays, click **Graphs**, and check the displays you want. The 'Four in one' batch of residual plots is a good choice.
- Click **OK**, then **OK** again.

COMMENTS

You will probably want to make displays of the stored diagnostic statistics. Use the usual Minitab methods for creating displays.

R

Suppose the response variable *y* and predictor variables *x1,…,xk* are in a data frame called mydata. After fitting a multiple regression of *y* on *x1* and *x2* via:

- mylm = lm(y~x1+x2, data=mydata)
- summary(mylm) #gives the details of the fit, including the ANOVA table
- plot(mylm) #gives a variety of plots
- lm.influence(mylm) #gives a variety of regression diagnostic values

To get partial regression plots (called Added Variable plots in **R**), you need the library car:

- library(car)

Then to get the partial regression plots:

- avPlots(mylm) #one plot for each predictor variable – interactions not permitted

SPSS

- Choose **Regression** from the **Analyze** menu.
- Choose **Linear** from the **Regression** submenu.
- When the Linear Regression dialogue appears, select the Y-variable and move it to the dependent target. Then move the X-variables to the independent target.
- Click the **Save** button.

- In the Linear Regression Save dialogue, choose diagnostic statistics. These will be saved in your worksheet along with your data.
- Click the **Continue** button to return to the Linear Regression dialogue.
- Click the **OK** button to compute the regression.

STATCRUNCH

StatCrunch offers some of the diagnostic statistics discussed in this chapter in the regression dialog. It does not currently make partial regression plots.

COMMENTS

You can also model interactions when entering predictor variables.

TI-83/84 PLUS

COMMENTS

You need a special program to compute a multiple regression on the TI-83.

Chapter Exercises

T indicates data is available in a saved file at the Companion Website at www.pearsoncanada.ca/deveaux

T 7. Climate change 2013 Recent concern with the rise in global temperatures has focused attention on the level of carbon dioxide (CO_2) in the atmosphere. The National Oceanic and Atmospheric Administration (NOAA) records the CO_2 levels in the atmosphere atop the Mauna Loa volcano in Hawaii, far from any industrial contamination, and calculates the annual overall temperature of the atmosphere and the oceans using an established method. (See data.giss.nasa.gov/gistemp/tabledata_v3/GLB.Ts+dSST.txt and ftp://aftp.cmdl.noaa.gov/products/trends/co2/co2_annmean_mlo.txt. We examined these data in Chapter 7 and again in Chapter 23. There we saw a strong relationship between global mean temperature and the level of CO_2 in the atmosphere.) Here is a regression predicting Mean Annual Temperature from annual CO_2 levels (parts per million). We'll examine the data from 1970 to 2013.

Dependent variable is: Global Temperature Anomaly
Response variable is: Global Avg Temp
R-squared = 73.6%   R-squared (adjusted) = 72.3%
s = 0.1331 with 44 − 3 = 41 degrees of freedom

Source	Sum of Squares	DF	Mean Square	F-ratio
Regression	2.02629	2	1.01315	57.2
Residual	0.726571	41	0.017721	

Variable	Coefficient	SE(Coeff)	t-ratio	P-value
Intercept	18.8061	35.03	0.537	0.5942
Year	−4.60135e-3	0.0197	−0.233	0.8169
CO_2	0.013127	0.0121	1.09	0.2830

A histogram of the externally Studentized residuals looks like this:

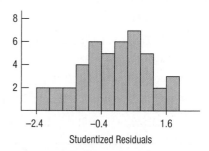

Studentized Residuals

a) Comment on the distribution of the Studentized residuals.
b) It is widely understood that global temperatures have been rising consistently during this period. But the coefficient of *Year* is negative and its *t*-ratio is small. Does this contradict the common wisdom?

T **8. Pizza** Consumers' Union rated frozen pizzas. Their report includes the number of *Calories*, *Fat* content, and *Type* (cheese or pepperoni, represented here as an indicator variable that is 1 for cheese and 0 for pepperoni). Here's a regression model to predict the *Score* awarded each pizza from these variables:

Dependent variable is: Score
R-squared = 28.7% R-squared (adjusted) = 20.2%
s = 19.79 with 29 − 4 = 25 degrees of freedom

Source	Sum of Squares	df	Mean Square	F-ratio
Regression	3947.34	3	1315.78	3.36
Residual	9791.35	25	391.654	

Variable	Coefficient	SE(Coeff)	t-ratio	P-value
Intercept	−148.817	77.99	−1.91	0.0679
Calories	0.743023	0.3066	2.42	0.0229
Fat	−3.89135	2.138	−1.82	0.0807
Type	15.6344	8.103	1.93	0.0651

a) What is the interpretation of the coefficient of cheese in this regression?
b) What displays would you like to see to check assumptions and conditions for this model?

9. Healthy breakfast In Chapter 26, we considered a regression model for the calories in breakfast cereals that looked like this:

Dependent variable is: Calories
R-squared = 84.5% R-squared (adjusted) = 83.4%
s = 7.947 with 77 − 6 = 71 degrees of freedom

Source	Sum of Squares	df	Mean Square	F-ratio
Regression	24 367.5	5	4 873.50	77.2
Residual	4 484.45	71	63.1613	

Variable	Coefficient	SE(Coeff)	t-ratio	P-value
Intercept	20.2454	5.984	3.38	0.0012
Protein	5.69540	1.072	5.32	≤0.0001
Fat	8.35958	1.033	8.09	≤0.0001
Fibre	−1.02018	0.4835	−2.11	0.0384
Carbo	2.93570	0.2601	11.3	≤0.0001
Sugars	3.31849	0.2501	13.3	≤0.0001

Let's take a closer look at *Fibre*. Here's the partial regression plot for *Fibre* in that regression model:

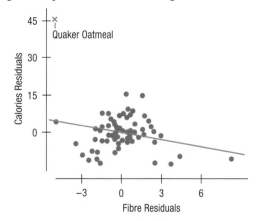

a) The line on the plot is the least squares line fit to this plot. What is its slope? (You may need to look back at the facts about partial regression plots in Chapter 26.)
b) One point is labelled as corresponding to Quaker Oatmeal. What effect does this point have on the slope of the line? (Does it make it larger, smaller, or have no effect at all?)

Here is the same regression with Quaker Oatmeal removed from the data:

Dependent variable is: Calories
77 total cases of which + is missing
R-squared = 93.9% R-squared (adjusted) = 93.5%
s = 5.002 with 76 − 6 = 70 degrees of freedom

Source	Sum of Squares	DF	Mean Square	F-ratio
Regression	27 052.4	5	5 410.49	216
Residual	1 751.51	70	25.0216	

Variable	Coefficient	SE(Coeff)	t-ratio	P-value
Intercept	−1.25891	4.292	−0.293	0.7701
Protein	3.88601	0.6963	5.58	≤0.0001
Fat	8.69834	0.6512	13.4	≤0.0001
Fibre	0.250140	0.3277	0.763	0.4478
Carbo	4.14458	0.2005	20.7	≤0.0001
Sugars	3.96806	0.1692	23.4	≤0.0001

c) Compare this regression with the previous one. In particular, which model is likely to make the best predictions of calories? Which seems to fit the data better?
d) How would you interpret the coefficient of *Fibre* in this model? Does *Fibre* contribute significantly to modelling calories?

T **10. Fifty states** Let's look again at the data from the 50 states in the United States given in Chapter 26. The *Murder* rate is per 100 000, *HS Graduation* rate is in %, *Income* is per capita income in dollars, *Illiteracy* rate is per 1000, and *Life expectancy* is in years. We are trying to find a regression model for *Life expectancy*.

Here's the result of a regression on all the available predictors:

Dependent variable is: Lifeexp
R-squared = 67.0% R-squared (adjusted) = 64.0%
s = 0.8049 with 50 − 5 = 45 degrees of freedom

Source	Sum of Squares	DF	Mean Square	F-ratio
Regression	59.1430	4	14.7858	22.8
Residual	29.1560	45	0.6479	

Variable	Coefficient	SE(Coeff)	t-ratio	P-value
Intercept	69.4833	1.325	52.4	≤0.0001
Murder	−0.261940	0.0445	−5.89	≤0.0001
HS grad	0.046144	0.0218	2.11	0.0403
Income	1.24948e−4	0.0002	0.516	0.6084
Illiteracy	0.276077	0.3105	0.889	0.3787

Here's a histogram of the leverages and a scatterplot of the externally Studentized residuals against the leverages:

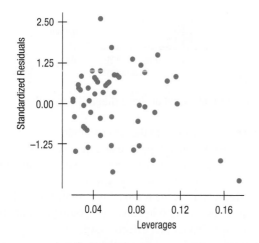

a) The two states with high leverages and large (negative) Studentized residuals are Nevada and Alaska. Do you think they are likely to be influential in the regression? From just the information you have here, why or why not?

Here's the regression with indicator variables for Alaska and Nevada added to the model to remove those states from affecting the model:

Dependent variable is: Lifeexp
R-squared = 74.1% R-squared (adjusted) = 70.4%
s = 0.7299 with 50 − 7 = 43 degrees of freedom

Variable	Coefficient	SE(Coeff)	t-ratio	P-value
Intercept	66.9280	1.442	46.4	≤0.0001
Murder	−0.207019	0.0446	−4.64	≤0.0001
HS grad	0.065474	0.0206	3.18	0.0027
Income	3.91600e−4	0.0002	1.63	0.1105
Illiteracy	0.302803	0.2984	1.01	0.3159
Alaska	−2.57295	0.9039	−2.85	0.0067
Nevada	−1.95392	0.8355	−2.34	0.0241

b) What evidence do you have that Nevada and Alaska are outliers with respect to this model?
Do you think they should continue to be treated specially? Why?

c) Would you consider removing any of the predictors from this model? Why or why not?

T **11. A nutritious breakfast** In Exercise 26 of Chapter 26, we considered a multiple regression model for predicting calories in breakfast cereals. The regression looked like this:

Dependent variable is: Calories
R-squared = 38.4% R-squared (adjusted) = 35.9%
s = 15.60 with 77 − 4 = 73 degrees of freedom

Source	Sum of Squares	DF	Mean Square	F-ratio	P-value
Regression	11 091.8	3	3 697.28	15.2	≤0.0001
Residual	17 760.1	73	243.289		

Variable	Coefficient	SE(Coeff)	t-ratio	P-value
Intercept	83.0469	5.198	16.0	≤0.0001
Sodium	0.057211	0.0215	2.67	0.0094
Potassium	−0.019328	0.0251	−0.769	0.4441
Sugars	2.38757	0.4066	5.87	≤0.0001

Here's a histogram of the leverages and a partial regression plot for potassium in which the three high-leverage points are plotted with red x's. (They are *All-Bran, 100% Bran,* and *All-Bran with Extra Fibre.*)

Source	Sum of Squares	DF	Mean Square	F-ratio
Regression	261 029	2	130 515	1288
Residual	8 813.02	87	101.299	

Variable	Coefficient	SE(Coeff)	t-ratio	P-value
Intercept	−11.6545	1.891	−6.16	≤0.0001
Distance	4.43427	0.2200	20.2	≤0.0001
Climb	0.045195	0.0033	13.7	≤0.0001

Here is the scatterplot of externally Studentized residuals against predicted values, as well as a histogram of leverages for this regression:

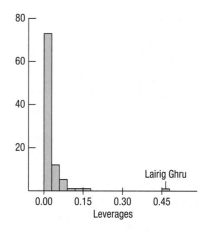

With this additional information, answer the following:

a) How would you interpret the coefficient of *Potassium* in the multiple regression?

b) *Without doing any calculating,* how would you expect the coefficient and *t*-statistic for *Potassium* to change if we were to omit the three high-leverage points?

Here's a histogram of the externally Studentized residuals. The selected bar, holding the two most negative residuals constant, holds the two bran cereals that had the largest leverages.

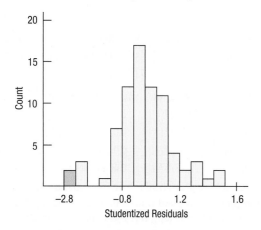

With this additional information, answer the following:

c) What term would you apply to these two cases? Why?

d) Do you think they should be omitted from this analysis? Why or why not? (*Note:* There is no correct choice. What matters is your reasons.)

T **12. Scottish hill races 2008** In Chapter 26, Exercises 14 and 16, we considered data on hill races in Scotland. These are overland races that climb and descend hills—sometimes several hills in the course of one race. Here is a regression analysis to predict the *Women's record* times from the *Distance* and total vertical *Climb* of the races:

Dependent variable is: Womens record
R-squared = 96.7% R-squared (adjusted) = 96.7%
s = 10.06 with 90 − 3 = 87 degrees of freedom

a) Comment on what these diagnostic displays indicate.

b) The two races with the largest Studentized residuals are the Arochar Alps race and the Glenshee 9. Both are relatively new races, having been run only one or two times with relatively few participants. What effects can you be reasonably sure they have had on the regression? What displays would you want to see to investigate other effects? Explain.

c) If you have access to a suitable statistics package, make the diagnostic plots you would like to see and discuss what you find.

T **13. Traffic delays 2011** The Texas Transportation Institute studies traffic delays. Data the institute published for the year 2011 include information on the Cost of Congestion per auto commuter ($) (hours per year spent delayed by traffic), *Congested%* (*Percent of vehicle miles traveled that were congested*), and the *Size* of the city (small, medium, large, very large). The regression model based on these variables looks like this. The variables *Small, Large,* and *Very Large* are indicators constructed to be 1 for cities of the named size and 0 otherwise.

Response variable is: Congestion per auto commuter ($)
R-squared = 63.3% R-squared (adjusted) = 61.8%
s = 152.4 with 101 − 5 = 96 degrees of freedom

Source	Sum of Squares	DF	Mean Square	F-ratio
Regression	3 852 487	4	963 122	41.5
Residual	2 229 912	96	232 28.2	

Variable	Coefficient	SE(Coeff)	t-ratio	P-value
Intercept	501.490	43.27	11.6	<0.0001
Small	−104.239	43.27	−2.41	0.0179
Large	106.026	41.46	2.56	0.0121
Very Large	348.147	59.67	5.83	<0.0001
Congested% ...	3.13481	0.9456	3.32	0.0013

a) Explain how the coefficients of *Small, Large,* and *Very Large* account for the size of the city in the model. Why is there no coefficient for *Medium*?
b) What is the interpretation of the coefficient of *Large* in this regression model?

14. Gourmet pizza Here's a plot of the Studentized residuals against the predicted values for the regression model found in Exercise 8:

The extraordinary cases in the plot of residuals are Reggio's and Michelina's, two gourmet pizzas.
a) Interpret these residuals. What do they say about these two brands of frozen pizza? Be specific—that is, talk about the *Scores* they received and might have been expected to receive.

We can create indicator variables to isolate these cases. Adding them to the model results in the following:

Dependent variable is: Score
R-squared = 65.2% R-squared (adjusted) = 57.7%
s = 14.41 with 29 − 6 = 23 degrees of freedom

Source	Sum of Squares	df	Mean Square	F-ratio
Regression	8964.13	5	1792.83	8.64
Residual	4774.56	23	207.590	

Variable	Coefficient	SE(Coeff)	t-ratio	P-value
Intercept	−363.109	72.15	−5.03	≤0.0001
Calories	1.56772	0.2824	5.55	≤0.0001
Fat	28.82748	1.887	−4.68	0.0001
Cheese	25.1540	6.214	4.05	0.0005
Reggio's	−67.6401	17.86	−3.79	0.0010
Michelina's	−67.0036	16.62	−4.03	0.0005

b) What does the coefficient of *Michelina's* mean in this regression model? Do you think that Michelina's pizza is an outlier for this model for these data? Explain.

15. More traffic Here's a plot of Studentized residuals against *Congested%* for the model of Exercise 13. The plot is coloured according to *City Size,* and regression lines are fit for each size.

a) The model of Exercise 13 includes indicators for *City Size.* Considering this display, have these indicator variables accomplished what is needed for the regression model? Explain.

We constructed additional indicators as the product of *Small* with *Arterial mph* and the product of *Very Large* with *Arterial mph.* Here's the resulting model:

Response variable is: Congestion per auto commuter ($)
R-squared = 64.9%   R-squared (adjusted) = 63.1%
s = 149.9 with 101 − 6 = 95 degrees of freedom

Source	Sum of Squares	DF	Mean Square	F-ratio
Regression	3 947 593	5	789 519	35.1
Residual	2 134 805	95	22 471.6	

Variable	Coefficient	SE(Coeff)	t-ratio	P-value
Intercept	461.515	46.79	9.86	<0.0001
Small	27.8150	77.02	0.361	0.7188
Large	86.1516	41.90	2.06	0.0425
Very Large	305.853	62.19	4.92	<0.0001
Congested% ...	4.24056	1.074	3.95	0.0002
Sml*C%V	−4.41683	2.147	−2.06	0.0424

b) What does the predictor *Sml*C,V* (Small by Congestion%) do in this model? Interpret the coefficient.
c) Does this appear to be a good regression model? Would you consider removing any predictors? Why or why not?

16. Another slice of pizza A plot of Studentized residuals against predicted values for the regression model found in Exercise 14 now looks like this. It has been coloured according to *Type* of pizza, with separate regression lines fitted for each type:

a) Comment on this diagnostic plot. What does it say about how the regression model deals with cheese and pepperoni pizzas?

Based on this plot, we constructed yet another variable consisting of the indicator *Cheese* multiplied by *Calories*:

Dependent variable is: Score
R-squared = 73.7% R-squared (adjusted) = 66.5%
s = 12.82 with 29 − 7 = 22 degrees of freedom

Source	Sum of Squares	DF	Mean Square	F-ratio
Regression	10 121.4	6	1 686.90	10.3
Residual	3 617.32	22	164.424	

Variable	Coefficient	SE(Coeff)	t-ratio	P-Value
Intercept	−464.498	74.73	−6.22	≤0.0001
Calories	1.92005	0.2842	6.76	≤0.0001
Fat	−10.3847	1.779	−5.84	≤0.0001
Cheese	183.634	59.99	3.06	0.0057
Cheese*cals	−0.461496	0.1740	−2.65	0.0145
Reggio's	−64.4237	15.94	−4.04	0.0005
Michelina's	−51.4966	15.90	−3.24	0.0038

b) Interpret the coefficient of *Cheese*Cals* in this regression model.

c) Would you prefer this regression model to the model of Exercise 14? Explain.

T 17. Influential traffic? Here are histograms of the leverage and Studentized residuals for the regression model of Exercise 15.

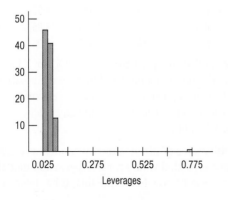

The city with the highest leverage is Laredo, TX. It's highlighted in both displays.

Do you think Laredo is an influential case? Explain your reasoning.

T 18. The final slice Here's the residual plot corresponding to the regression model of Exercise 16:

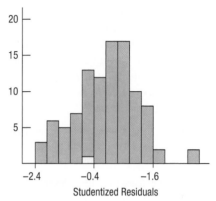

The extreme case this time is Weight Watchers Pepperoni (makes sense, doesn't it?). We can make one more indicator for *Weight Watchers*. Here's the model:

Dependent variable is: Score
R-squared = 77.1% R-squared (adjusted) = 69.4%
s = 12.25 with 29 − 8 = 21 degrees of freedom

Source	Sum of Squares	DF	Mean Square	F-ratio
Regression	10 586.8	7	1 512.41	10.1
Residual	3 151.85	21	150.088	

Variable	Coefficient	SE(Coeff)	t-ratio	P-value
Intercept	2525.063	79.25	−6.63	≤0.0001
Calories	2.10223	0.2906	7.23	≤0.0001
Fat	210.8658	1.721	−6.31	≤0.0001
Cheese	231.335	63.40	3.65	0.0015
Cheese*cals	−0.586007	0.1806	−3.24	0.0039
Reggio's	−66.4706	15.27	−4.35	0.0003
Michelina's	−52.2137	15.20	−3.44	0.0025
Weight W ...	−8.3265	16.09	1.76	0.0928

a) Compare this model with the others we've seen for these data. In what ways does this model seem better or worse than the others?

b) Do you think the indicator for *Weight Watchers* should be in the model? (Consider the effect that including it has had on the other coefficients as well.)

c) What do the Consumers' Union tasters seem to think makes for a really good pizza?

 19. Popsicle sticks Recall the Popsicle stick data from Chapter 24. Suppose we want to fit a multiple regression model to predict the breaking stress from the overlap area, allowing for different slopes depending on the level of overlap (small, medium, and large). Fitting this model, we get the following output and scatter plot.

Dependent variable is: stress
R-squared = 12.4% R-squared (adjusted) = 11.1%
S = 259.7 on 346 degrees of freedom

Variable	Coefficient	SE(Coeff)	t-ratio	P-value
Intercept	253.5732	1231.2640	0.206	0.837
area	0.7773	4.8330	0.161	0.872
overlapM	413.6458	1254.0276	0.330	0.742
overlapS	1646.5704	1314.0899	1.253	0.211
area*overlapM	−0.5074	5.0368	−0.101	0.920
area*overlapS	−10.6376	6.5436	−1.626	0.105

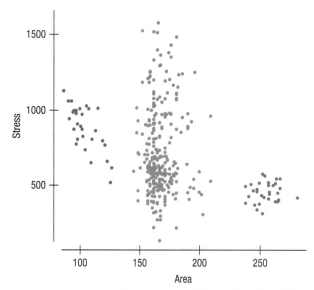

a) What does the coefficient for *Area* mean in this question?

b) It seems like there might be a linear relationship between stress and area for the small (green) overlap joints, but not the other two sizes, judging by the scatter plot. Yet, none of the slope P-values are significant. Explain why that came to be.

Assuming the residuals work out nicely, and holding everything else constant (and ignoring non-significance of the coefficients):

c) How much do we expect the stress to change on large overlap joints with a unit increase in area?

d) How much do we expect the stress to change on small overlap joints with an increase of 10 in area?

![checkmark] **Just Checking** ANSWERS

1. Histograms are used to examine the shapes of distributions of individual variables. We check especially for multiple modes, outliers, and skewness. They are also used to check the shape of the distribution of the residuals for the Nearly Normal Condition.

2. Scatterplots are used to check the Straight Enough Condition in plots of *y* versus any of the *x* variables. They are also used to check plots of the residuals or Studentized residuals against the predicted values, against any of the predictors, or against *time* to check for patterns. Scatterplots are also the display used in partial regression plots, where we check for influential points and unexpected subgroups.

3. The Normal model is needed only when we use inference; it isn't needed for computing a regression model. We check the Nearly Normal Condition on the residuals.

MathXL

MyLab Statistics

Go to MathXL at **www.mathxl.com** or MyLab™ Statistics at **www.mystatlab.com**. You can practise exercises for this chapter as often as you want. The guided solutions will help you find answers step by step. You'll find a personalized study plan available to you too!

Modelling the World at Large

Quick Review

This part introduces inference for the most widely used class of statistical models: 'linear' models. You may have read one, two, or all of the chapters in this part. They have a consistent central theme, and it is one you'll likely see if you study more Statistics.

- Linear models predict a single quantitative variable from one or more other variables. The predictor variables can be quantitative or categorical. Linear models require only that the effects of the predictor variables be added together (or equivalently that they be 'linear in the betas').

- When the predictors are categorical, we may use an analysis of variance (ANOVA) model.

- When the predictors are quantitative, we may use a regression model. With two or more predictors, we have a *multiple* regression model. (And categorical predictors may be included by using indicator variables.)

- The overall fit of a linear model is tested with an F-test. The null hypothesis for this test is that we'd do just as well modelling y with its mean.

- If the overall F-test is significant, we can move on to consider the individual coefficients. For regression, this means considering the t-statistics. For ANOVA, it means first considering the F-statistics for individual factors. We may then move on to consider *comparisons* among different levels of significant factors.

- Models with two or more predictors can fail conditions in more ways than those with only one, and some of these can be quite subtle. These models require care in checking assumptions and conditions, and especially in checking for outliers and high-leverage points.

Remember (have we said this often enough yet?): Never use any inference procedure without first checking the assumptions and conditions. We summarized those in the last unit; have another look below.

Here come more opportunities to practise using these concepts and skills, *but we've also sprinkled in some exercises that call for methods learned in earlier chapters, so stay alert.*

Review Exercises

1. Mutual funds In March 2002, *Consumer Reports* listed the rate of return for several large cap mutual funds over the previous three-year and five-year periods. ("Large cap" refers to companies worth over $10 billion.)
a) Create a 95% confidence interval for the difference in rate of return for the three- and five-year periods covered by these data. Clearly explain what your interval means.
b) It's common for advertisements to carry the disclaimer that "past returns may not be indicative of future performance," but do these data indicate that there was an association between three-year and five-year rates of return?

Annualized Returns (%)		
Fund name	Three-year	Five-year
Ameristock	7.9	17.1
Clipper	14.1	18.2
Credit Suisse Strategic Value	5.5	11.5
Dodge & Cox Stock	15.2	15.7
Excelsior Value	13.1	16.4
Harbor Large Cap Value	6.3	11.5
ICAP Discretionary Equity	6.6	11.4
ICAP Equity	7.6	12.4
Neuberger Berman Focus	9.8	13.2
PBHG Large Cap Value	10.7	18.1
Pelican	7.7	12.1
Price Equity Income	6.1	10.9
USAA Cornerstone Strategy	2.5	4.9
Vanguard Equity Income	3.5	11.3
Vanguard Windsor	11.0	11.0

2. Football A student runs an experiment to test four different grips on his football throwing distance, recording the distance in yards that he can throw a bag of 20 footballs using each grip. He randomizes the grip used each time by drawing numbers out of a hat until each grip has been used five times. After collecting his data and analyzing the results, he reports that the P-value of his test is 0.0032.
a) What kind of test should he have performed?
b) What are the null and alternative hypotheses?
c) Assuming that the conditions required for the test are satisfied, what would you conclude?
d) What else about the data would you like to see to check the assumptions and conditions?
e) What might he want to test next?
f) Suggest one problem with this experiment design that can be fixed, and one that likely cannot.

3. Golf A student runs an experiment to test four different golf clubs on her putting accuracy, recording the distance in centimetres from a small target that she places on the green. She randomizes the club used each time by drawing numbers out of a hat until each club has been used six times. After collecting her data and analyzing the results, she reports that the P-value of her test is 0.0245.
a) What kind of test should she have performed?
b) What are the null and alternative hypotheses?
c) Assuming that the conditions required for the test are satisfied, what would you conclude?
d) What else about the data would you like to see to check the assumptions and conditions?
e) What might she want to test next?

4. Wild horses Large herds of wild horses can become a problem on some U.S. federal lands in the West. Researchers hoping to improve the management of these herds collected data to see if they could predict the number of foals that would be born based on the size of the current herd. Their attempt to model this herd growth is summarized in the output shown.

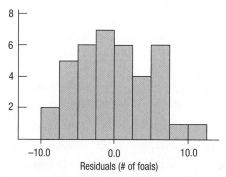

Variable	Count	Mean	StdDev
Adults	38	110.237	71.1809
Foals	38	15.3947	11.9945

Dependent variable is: Foals
R-squared = 83.5%
s = 4.941 with 38 − 2 = 36 degrees of freedom

Variable	Coefficient	SE(Coeff)	t-ratio	P-value
Intercept	−1.57835	1.492	−1.06	0.2970
Adults	0.153969	0.0114	13.5	≤0.0001

a) How many herds of wild horses were studied?
b) Are the conditions necessary for inference satisfied? Explain.
c) Create a 95% confidence interval for the slope of this relationship.
d) Explain in this context what that slope means.
e) Suppose that a new herd with 80 adult horses is located. Estimate, with a 90% prediction interval, the number of foals that may be born.

T 5. North and hard Consider the following data on water hardness and mortality that includes information on whether a given town is north or south of Derby. When that information is coded in a variable as 1 for "north of Derby" and 0 for "south of Derby," and that variable is included in the regression, the resulting table looks like this:

Dependent variable is: Mortality
R-squared = 56.2% R-squared (adjusted) = 54.7%
s = 126.4 with 61 − 3 = 58 degrees of freedom

Variable	Coefficient	SE(Coeff)	t-ratio	P-value
Intercept	1537.50	42.02	36.6	≤0.0001
Calcium	−2.16011	0.4979	−4.34	≤0.0001
North of Derby	158.892	37.87	4.20	≤0.0001

a) What is the name for the kind of variable that is being used for being north of Derby?
b) What does the coefficient of that variable mean in this regression?
c) Do you prefer this regression? Explain why or why not.

T 6. Horses again, a bit less wild In an attempt to control the growth of the herds of wild horses discussed in Exercise 4, managers sterilized some of the stallions. The variable *Sterilized* is coded 0 for herds and years in which no stallions were sterilized and 1 for herds and years in which some stallions were sterilized. The resulting regression looks like this:

Dependent variable is: Foals
R-squared = 84.8% R-squared (adjusted) = 83.9%
s = 4.814 with 38 − 3 = 35 degrees of freedom

Source	Sum of squares	DF	Mean Square	F-ratio
Regression	4511.85	2	2255.92	97.3
Residual	811.229	35	23.1780	

Variable	Coefficient	SE(Coeff)	t-ratio	P-value
Intercept	6.23667	4.801	1.30	0.2024
Adults	0.112271	0.0268	4.18	0.0002
Sterilized	−6.43657	3.769	−1.71	0.0965

a) What is the name for the kind of variable used here to represent sterilizing?
b) What is the interpretation of the coefficient of *Sterilized*?
c) Does sterilizing appear to work? That is, do these data show a statistically significant effect of sterilizing stallions on the number of foals born?

7. Bowling A student runs an experiment to test how different factors affect her bowling performance. She uses three levels for the weight of the ball (low, medium, and high) and two approaches (standing and walking), throwing four balls at each condition and choosing the conditions in random order. She counts the number of pins knocked down as the response variable.
a) What are the null and alternative hypotheses for the main effects?
b) How many degrees of freedom does each factor sum of squares have? How about the error sum of squares?
c) Should she consider fitting an interaction term to the model? Why or why not? How many degrees of freedom would it take?

8. Video racing A student runs an experiment to test how different factors affect his reaction time while playing video games. He uses a specified race course with random hazards and times how long it takes him (in seconds) to finish four laps under a variety of experimental conditions. As factors, he uses three different types of *Mouse* (cordless ergonomic, cordless regular, or corded) and keeps the *Lights* on or off. He measures his time once at each condition. To avoid a learning effect, he runs the race five times before selecting the conditions in random order.
a) What are the null and alternative hypotheses for the main effects?
b) How many degrees of freedom does each factor sum of squares have? How about the error sum of squares?
c) Should he consider fitting an interaction term to the model? Why or why not? How many degrees of freedom would it take?

9. Resumé fraud In 2002, the Veritas Software company found out that its chief financial officer did not actually have the MBA he had listed on his resumé. They fired him, and the value of the company's stock dropped 19%. Kroll, Inc., a firm that specializes in investigating such matters, said that they believe as many as 25% of background checks might reveal false information. How many such random checks would they have to do to estimate the true percentage of people who misrepresent their backgrounds to within ±5% with 98% confidence?

T 10. Paper airplanes In preparation for a regional paper airplane competition, a student tried out her latest design. The distances her plane travelled (in feet) in 11 trial flights are given here. (The world record is an astounding 193.01 feet!) The data were 62, 52, 68, 23, 34, 45, 27, 42, 83, 56, and 40 feet.

Here are some summaries:

Count	11
Mean	48.3636
Median	45
StdDev	18.0846
StdErr	5.45273
IntQRange	25
25th %tile	35.5000
75th %tile	60.5000

Distance (ft.)

a) Construct a 95% confidence interval for the mean distance.
b) Based on your confidence interval, is it plausible that the true mean distance is 40 feet? Explain.
c) How would a 99% confidence interval for the mean distance differ from your answer in part a)? Explain briefly, without actually calculating a new interval.
d) How large a sample size would the student need to get a confidence interval half as wide as the one you got in part a), at the same confidence level?

T **11.** **Nuclear power** Here are data on 32 light water nuclear power plants. The variables are:
Cost: In $100 000, adjusted to 1976 base.
Date: Date the construction permit was issued in years after 1900. Thus, 68.58 is roughly halfway through 1968.
MWatts: Power plant net capacity in megawatts.

We are interested in the cost of the plants as a function of *Date* and *MWatts*.

Cost	MWatts	Date	Cost	MWatts	Date
345.39	514	67.92	457.12	822	68.42
460.05	687	68.58	690.19	792	68.33
452.99	1065	67.33	350.63	560	68.58
443.22	1065	67.33	402.59	790	68.75
652.32	1065	68.00	412.18	530	68.42
642.23	1065	68.00	495.58	1050	68.92
272.37	822	68.17	394.36	850	68.92
317.21	457	68.42	423.32	778	68.42
712.27	845	69.50	473.64	538	70.42
289.66	530	68.42	697.14	1130	71.08
881.24	1090	69.17	207.51	745	67.25

Cost	MWatts	Date	Cost	MWatts	Date
490.88	1050	68.92	288.48	821	67.17
567.79	913	68.75	284.88	886	67.83
665.99	828	70.92	280.36	886	67.83
621.45	786	69.67	217.38	745	67.25
608.80	821	70.08	270.71	886	67.83

a) Examine the relationships between *Cost* and *MWatts* and between *Cost* and *Date*. Make appropriate displays and interpret them with a sentence or two.
b) Find the regression of *Cost* on *MWatts*. Write a sentence that explains the relationship as described by the regression.
c) Make a scatterplot of residuals versus predicted values and discuss what it shows. Make a Normal probability plot or histogram of the residuals. Discuss the four assumptions needed for regression analysis and indicate whether you think they are satisfied here. Give your reasons.
d) State the standard null hypothesis for the slope coefficient and complete the *t*-test at the 5% level. State your conclusion.
e) Estimate the cost of a 1000-MWatt plant. Show your work.
f) Compute the residuals for this regression. Discuss the meaning of the R^2 in this regression. Plot the residuals against *Date*. Does it appear that *Date* can account for some of the remaining variability?
g) Compute the multiple regression of *Cost* on both *MWatts* and *Date*. Compare the coefficient in this regression with those you have found for each of these predictors.
h) Would you expect *MWatts* and *Date* to be correlated? Why or why not? Examine the relationship between *MWatts* and *Date*. Make a scatterplot and find the correlation coefficient, for example. It's only because of the extraordinary nature of this relationship that the relationships you saw at earlier steps were this simple.

T **12.** **Barbershop music** At a barbershop music singing competition, choruses are judged on three scales: *Music* (quality of the arrangement, etc.), *Performance*, and *Singing*. The scales are supposed to be independent of each other, and each is scored by a different judge, but a friend claims that he can predict a chorus's singing score from the other two scores. He offers the following regression based on the scores of all 34 choruses in a recent competition:

Dependent variable is: Singing
R-squared = 90.9% R-squared (adjusted) = 90.3%
s = 6.483 with 34 − 3 = 31 degrees of freedom

Variable	Coefficient	SE(Coeff)	t-ratio	P-value
Intercept	2.08926	7.973	0.262	0.7950
Performance	0.793407	0.0976	8.13	≤0.0001
Music	0.219100	0.1196	1.83	0.0766

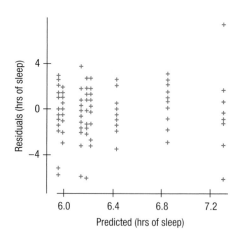

a) What do you think of your friend's claim? Can he predict singing scores? Explain.

b) State the standard null hypothesis for the coefficient of performance and complete the *t*-test at the 5% level. State your conclusion.

c) Complete the analysis. Check assumptions and conditions to the extent you can with the information provided.

13. Sleep Using a simple random sample, a student group asked 450 students about their sleep and study habits and received about 200 responses. The group wanted to know if the average amount of sleep varied by *Sex* (F or M) or by *Year* (first, second, third, or fourth) of the respondent. Partial boxplots of the amount of sleep last night by the two factors are shown below with the interaction plot:

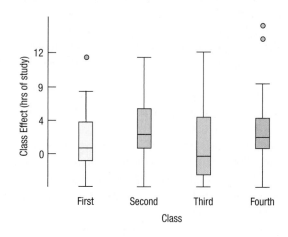

a) What are the null and alternative hypotheses for the main effects?

b) What effects appear to be significant?

c) What, if any, reservations do you have about the conclusions?

14. Study habits The survey in Exercise 13 also asked students about the number of hours they studied. Those doing the survey wanted to know if the average amount of studying varied by *Sex* or by *Class* of the respondent. Partial boxplots of the hours studied last night by the two factors are shown below:

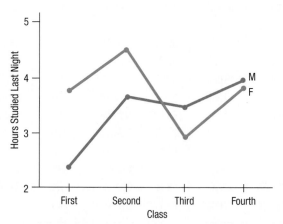

The ANOVA table shows:				
Source	**DF**	**Sum of squares**	**F-ratio**	**P-value**
Sex	1	8.0658	2.2821	0.1325
Year	3	8.8990	0.8393	0.4739
Sex × Year	3	19.4075	1.8303	0.1431
Error	189	670.6457		
Total	196	707.0180		

Source	DF	Sum of squares	F-ratio	P-value
Sex	1	7.1861	1.0744	0.3013
Class	3	34.9922	1.7440	0.1595
Sex × Class	3	27.8188	1.3865	0.2483
Error	189	1259.8049		
Total	196	1329.8020		

A plot of residuals versus predicted shows:

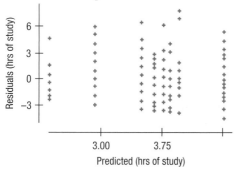

a) What are the null and alternative hypotheses for the main effects?
b) Should those doing the survey consider fitting an interaction term to the model? Why or why not?
c) What effects appear to be significant?
d) What, if any, reservations do you have about the conclusions?

15. Pregnancy In 1998, a San Diego reproductive clinic reported 42 live births to 157 women under the age of 38, but only 7 successes for 89 clients aged 38 and older. Is this evidence of a difference in the effectiveness of the clinic's methods for older women?
a) Test the appropriate hypotheses using the two-proportion z-procedure.
b) Repeat the analysis using an appropriate chi-square procedure.
c) Explain how the two results are equivalent.

16. Lost baggage The Bureau of Transportation Statistics of the U.S. Department of Transportation reports statistics about airline performance. For July 2012, they reported the following number of bags lost per 1000 passengers.

Airline	Lost Bags
Airtran	1.73
Jetblue	2.18
Delta	2.43
US Airways	2.46
American	2.99
Southwest	3.33
United	4.84
Skywest	6.03
Expressjet	6.84

Are the large airlines roughly equal in their baggage performance? Perform a chi-square goodness-of-fit test, or explain why that would not be appropriate.

17. Lefties and music In an experiment to see if left- and right-handed people have different abilities in music, subjects heard a tone and were then asked to identify which of several other tones matched the first. Of 76 right-handed subjects, 38 were successful in completing this test, compared with 33 of 53 lefties. Is this strong evidence of a difference in musical abilities based on handedness?

18. Preemies Do the effects of being born prematurely linger into adulthood? Researchers examined 242 Cleveland area children born prematurely between 1977 and 1979, and compared them with 233 children of normal birth weight; 24 of the "preemies" and 12 of the other children were described as being of "subnormal height" as adults.[1] Is this evidence that babies born with a very low birth weight are more likely to be smaller than normal adults?

T 19. Teen traffic deaths 2007 The Insurance Institute for Highway Safety publishes data on a variety of traffic-related risks. One report gives the numbers of male and female teenagers killed in highway accidents during each year from 1975 to 2002. Here is a regression predicting *Female deaths* by *Year*:

Dependent variable is: Female deaths
R-squared = 57.4% R-squared (adjusted) = 56.1%
s = 182.6 with 33 = 2 = 31 degrees of freedom

Source	Sum of squares	DF	Mean Square	F-ratio
Regression	1 396 025	1	1 396 025	41.9
Residual	1 033 964	31	33 353.7	

Variable	Coefficient	SE(Coeff)	t-ratio	P-value
Intercept	45 074.0	6 648	6.78	≤0.0001
Year	−21.6006	3.339	−6.47	≤0.0001

a) Here's a scatterplot of residuals vs. predicted values for this regression. Discuss the assumptions and conditions of the regression.

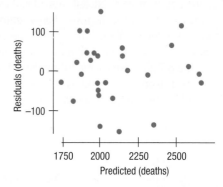

b) What is the meaning of the R^2 value of 52.3% for this regression?
c) What is the regression equation?
d) Give an interpretation of the coefficient of *Year* in this regression.

---

[1]"Outcomes in young adulthood for very-low-birth-weight infants," *New England Journal of Medicine*, 346, no. 3 [January 2002].

T **20.** **More teen traffic 2007** The data discussed in Exercise 21 included the numbers of male teen traffic deaths as well. We can add that as a predictor to obtain the following model:

Dependent variable is: Female deaths
R-squared = 89.5% R-squared (adjusted) = 88.8%
s = 92.5 with 33 − 3 = 30 degrees of freedom

Source	Sum of squares	DF	Mean Square	F-ratio
Regression	2 175 799	2	1 087 899	128
Residual	254 191	30	8 473.02	

Variable	Coefficient	SE(Coeff)	t-ratio	P-value
Intercept	−20 186.2	7 583	−2.66	0.0124
Year	10.5383	3.749	2.81	0.0086
Male deaths	0.271321	0.0283	9.59	≤0.0001
Year	20.3628	4.326	4.71	≤0.0001

a) How does this regression compare with the regression of Exercise 21? Which would you prefer to use? Why?
b) What does the coefficient of *Year* mean in this regression?

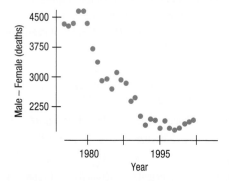

c) Considering both the regression model of Exercise 21 and this one, would you say that female teen traffic fatalities have been increasing or decreasing over time? How does the scatterplot above help to explain what is happening?

21. **Typing** For a class project, Nick designed and carried out an experiment to see if the room *Temperature* and the wearing of *Gloves* affected his typing speed. He ran each combination of hot and cold temperature and gloves on and off eight times, recording the net number of words typed (words typed minus mistakes). Partial boxplots and interaction plots show:

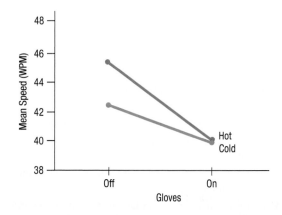

The ANOVA table shows:				
Source	**DF**	**Sum of squares**	**F-ratio**	**P-value**
Gloves	1	120.1250	57.2511	<0.0001
Temperature	1	18.0000	8.5787	0.0067
Gloves*Temp	1	15.1250	7.2085	0.0121
Error	28	58.7496		
Total	31	212.0000		

A scatterplot of residuals versus predicted values shows:

a) What are the null and alternative hypotheses for the main effects?
b) Given the partial boxplots, what do you suspect the *F*-test will say about the two main effects?
c) Should he consider fitting an interaction term to the model? Why or why not?
d) If he does fit an interaction term, do you suspect it will be significant on the basis of the interaction plot? Explain.
e) Which effects appear to be significant?
f) Describe the effects of the factors.
g) What is the size of the estimated standard deviation of the errors? Does this seem reasonable given the partial boxplots? Explain.
h) If Nick wants to increase his typing speed, what recommendations would you give him?
i) What reservations, if any, do you have about the conclusions?

22. Typing again Nick (see Exercise 21) designed a follow-up experiment to see if having *Music* or *Television* on would affect his typing speed. In particular, he'd like to know if he can type just as effectively with the music and/or the television on while he types. He ran each combination of *Music* and *Television* on or off eight times, recording the net number of words typed (words typed minus mistakes). Partial boxplots and interaction plots show:

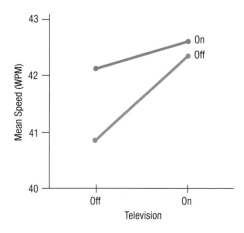

The ANOVA table shows:

Source	DF	Sum of squares	F-ratio	P-value
Music	1	4.5000	0.6380	0.4312
Television	1	8.0000	1.1342	0.2960
Interaction	1	2.0000	0.2835	0.5986
Error	28	197.5000		

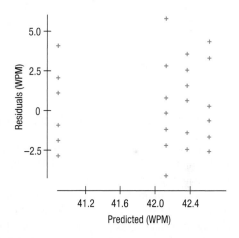

a) What are the null and alternative hypotheses for the main effects?
b) Given the partial boxplots, what do you suspect the *F*-test will say about the two main effects?
c) Should he consider fitting an interaction term to the model? Why or why not?
d) If he does fit an interaction term, do you suspect it will be significant on the basis of the interaction plot? Explain.
e) What effects appear to be significant?
f) Describe the effects of the factors.
g) What is the size of the estimated standard deviation of the errors? Does this seem reasonable given the partial boxplots? Explain.
h) If Nick wants to increase his typing speed, what recommendations would you give him?
i) What reservations, if any, do you have about the conclusions?

23. NY Marathon The *New York Times* reported the results of the 2003 NY Marathon by listing time brackets and the number of racers who finished within that bracket. Because the brackets are of different sizes, we look at the number of racers finishing per minute against the time at the middle of the time bracket. The resulting regression looks like this:

Dependent variable is: #/minute
R-squared = 10.9% R-squared (adjusted) = 9.5%
s = 78.45 with 63 − 2 = 61 degrees of freedom

Variable	Coefficient	SE(Coeff)	*t*-ratio	P-value
Intercept	45.1407	51.74	0.873	0.3863
Mid time …	0.519037	0.1899	2.73	0.0082

a) How would you interpret the coefficient of *Mid time*?
  Here's a scatterplot of the Studentized residuals against the predicted values for this regression:

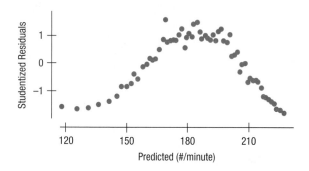

b) Comment on the regression in the light of this plot.

Ⓣ 24. Births The U.S. National Vital Statistics Report provides information on live Births (per 1000 women), according to the age of the woman (in five-year brackets—*Age* used here is the midpoint of the bracket) and the *Year* from 1990 to 1999. The report isolates births to women younger than 20 as a separate category. Looking only at women 20 years old and older, we find the following regression:

Dependent variable is: Births
R-squared = 98.1% R-squared (adjusted) = 98.0%
s = 15.55 with 50 − 3 = 47 degrees of freedom

Variable	Coefficient	SE(Coeff)	*t*-ratio	P-value
Intercept	4422.98	1527	2.90	0.0057
Age	−15.1898	0.3110	−48.8	≤0.0001
Year	−1.89830	0.7656	−2.48	0.0168

a) Write out the regression model
b) How would you interpret the coefficient of *Year* in this regression? What happened to pregnancy rates during the decade of the 1990s?

Here's a scatterplot of the Studentized residuals against the predicted values.

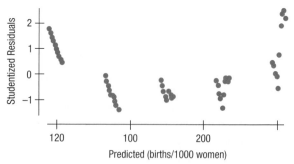

c) What might you do to improve the regression model?

25. Depression and the Internet The September 1998 issue of the *American Psychologist* published an article reporting on an experiment examining "the social and psychological impact of the Internet on 169 people in 73 households during their first 1 to 2 years on-line." In the experiment, a sample of households was offered free Internet access for one or two years in return for allowing their time and activity online to be tracked. The members of the households who participated in the study were also given a battery of tests at the beginning and again at the end of the study. One of the tests measured the subjects' levels of depression on a four-point scale, with higher numbers meaning the person was more depressed. Internet usage was measured in average number of hours per week. The regression analysis examines the association between the subjects' depression levels and the amounts of Internet use. The conditions for inference were satisfied.

Dependent variable is: Depression After
R-squared = 4.6%
s = 0.4563 with 162 − 2 = 160 degrees of freedom

Variable	Coefficient	SE(coeff)	*t*-ratio	P-value
Constant	0.565485	0.0399	14.2	≤0.0001
Intr_use	0.019948	0.0072	2.76	0.0064

a) Do these data indicate that there is an association between Internet use and depression? Test an appropriate hypothesis and state your conclusion clearly.
b) One conclusion of the study was that those who spent more time online tended to be more depressed at the end of the experiment. News headlines said that too much time on the Internet can lead to depression. Does the study support this conclusion? Explain.
c) As noted, the subjects' depression levels were tested at both the beginning and the end of this study; higher scores indicated the person was more depressed. Results are summarized in the table. Is there evidence that the depression level of the subjects changed during this study?

Depression Level
162 subjects

Variable	Mean	StdDev
DeprBfore	0.730370	0.487817
DeprAfter	0.611914	0.461932
Difference	−0.118457	0.552417

26. Learning math Developers of a new math curriculum called Accelerated Math compared performances of students taught by their system with control groups of students in the same schools who were taught using traditional instructional methods and materials. Statistics about pretest and posttest scores are shown in the table.[2]

a) Did the groups differ in average math score at the start of this study?

b) Did the group taught using the Accelerated Math program show a significant improvement in test scores?

c) Did the control group show significant improvement in test scores?

d) Were gains significantly higher for the Accelerated Math group than for the control group?

		Instructional Method	
		Acc. math	Control
Number of students		231	245
Pretest	Mean	560.01	549.65
	St. Dev	84.29	74.68
Posttest	Mean	637.55	588.76
	St. Dev	82.9	83.24
Individual gain	Mean	77.53	39.11
	St. Dev.	78.01	66.25

27. Pesticides A study published in 2002 in the journal *Environmental Health Perspectives* examined the sex ratios of children born to workers exposed to dioxin in Russian pesticide factories. The data covered the years 1961 to 1988 in the city of Ufa, Bashkortostan, Russia. Of 227 children born to workers exposed to dioxin, only 40% were male. Overall in the city of Ufa, the proportion of males was 51.2%. Is this evidence that human exposure to dioxin may results in the birth of more girls? (An interesting note: It appeared that paternal exposure was most critical; 51% of babies born to mothers exposed to the chemical were boys.)

Source: Environ Health Perspect. 2002 Nov; 110(11): A699–A701. PMCID: PMC1241090 Research Article Sex ratios of children of Russian pesticide producers exposed to dioxin. John Jake Ryan, Zarema Amirova, and Gaetan Carrier

Ⓣ 28. Is Old Faithful getting older? The data on Old Faithful eruptions we saw in Exercise 21, Part VI Review, include another variable, the *Day* on which the eruption occurred

[2]J. Ysseldyke and S. Tardrew, *Differentiating Math Instruction*, Renaissance Learning, 2002.

(where 1 is the first day and each successive day just counts one more). The correlation of *Interval* (minutes until the next eruption) with *Day* is −0.004. But when we include *Day* in a multiple regression along with the *Duration* (in minutes) of the previous eruption, we get the following model:

Dependent variable is: Interval
R-squared = 77.6% R-squared (adjusted) = 77.3%
s = 6.092 with 222 − 3 = 219 degrees of freedom

Variable	Coefficient	SE(Coeff)	t-ratio	P-value
Intercept	35.2463	1.509	23.4	≤0.0001
Duration	10.4348	0.3794	27.5	≤0.0001
Day	−0.126316	0.0523	−2.42	0.0166

a) What is the model fit by this regression?

b) Is the *Interval* changing over time? Perform a formal test of the relevant hypothesis.

c) Doesn't the small P-value for *Day* contradict the correlation between *Interval* and *Day* being virtually zero? Explain.

d) Is the amount of change in *Interval* due to *Day* meaningful?

29. Video pinball A student runs an experiment to test how different factors affect his score while playing video pinball. Here are the results of 16 runs of an experiment performed in random order. Factor *Eyes* has two levels: both open and right eye closed. Factor *Tilt* has two levels: tilt on and tilt off. The response is the score of one ball at a combination of factors. Each combination was repeated four times in random order.

a) What are the null and alternative hypotheses for the main effects?

b) Analyze the data and write a short report on your findings. Include appropriate graphics and diagnostic plots.

Eyes	Tilt	Score
Both	On	67 059
Right eye	Off	21 036
Both	On	59 520
Right eye	Off	3 100
Both	Off	61 272
Right eye	On	55 957
Both	Off	72 472
Right eye	On	18 460
Right eye	On	16 556
Both	Off	89 553
Right eye	On	37 950
Both	Off	74 336
Right eye	Off	700
Both	On	79 037
Right eye	Off	36 591
Both	On	74 610

30. **Javelin** Brianna, a member of the track and field team, runs an experiment to test how different factors affect her javelin throw. She wants to know if the more expensive (premium) javelin is worth the extra price and is curious to know how much warming up helps her distance. She tries all four combinations of the two *Javelins* (standard and premium) and *Preparation* (cold—and warm-up), repeating each combination twice in random order. She measures the distance of her throw to the nearest metre. Here are the results of the eight runs.

Javelin	Preparation	Distance (metres)
Premium	Warm-up	46
Standard	Warm-up	39
Premium	Cold	37
Standard	Cold	30
Premium	Warm-up	45
Standard	Warm-up	40
Premium	Cold	35
Standard	Cold	32

a) What are the null and alternative hypotheses for the main effects?

b) Analyze the data and write a short report on your findings. Include appropriate graphics and diagnostic plots and a recommendation for Brianna to optimize her javelin distance.

T 31. Eye and hair colour A survey of 1021 school-age children was conducted by randomly selecting children from several large urban elementary schools. Two of the questions concerned eye and hair colour. In the survey, the following codes were used:

Hair Colour	Eye Colour
1 = Blond	1 = Blue
2 = Brown	2 = Green
3 = Black	3 = Brown
4 = Red	4 = Grey
5 = Other	5 = Other

The Statistics students analyzing the data were asked to study the relationship between eye and hair colour.

a) One group of students produced the output shown below. What kind of analysis is this? What are the null and alternative hypotheses? Is the analysis appropriate? If so, summarize the findings, being sure to include any assumptions you've made and/or limitations to the analysis. If it's not an appropriate analysis, explicitly state why not.

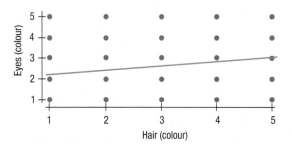

Dependent variable is: Eye Colour
R-squared = 3.7%
s = 1.112 with 1021 − 2 = 1019 degrees of freedom

Variable	Coefficient	SE(Coeff)	t-ratio	P-value
Constant	1.99541	0.08346	23.9	≤0.0001
Hair Colour	0.211809	0.03372	0.28	≤0.0001

b) A second group of students used the same data to produce the output shown below. What kind of analysis is this? What are the null and alternative hypotheses? Is the analysis appropriate? If so, summarize the findings, being sure to include any assumptions you've made and/or limitations to the analysis. If it's not an appropriate analysis, explicitly state why not.

Table contents: Counts
Standardized Residuals

	Eye Colour				
	1	**2**	**3**	**4**	**5**
1	143	30	58	15	12
	7.6754	0.417988	−5.88169	−0.63925	−0.314506
2	90	45	215	30	20
	−2.57141	0.290189	1.72235	0.491885	−0.0824592
3	28	15	190	10	10
	−5.39425	22.3478	6.28154	−1.76376	−0.803818
4	30	15	10	10	5
	2.06116	2.71589	−4.0554	2.37402	0.759931
5	10	5	15	5	5
	−0.521945	0.332621	−0.941918	1.36326	−.07578

Hair Colour (row label)

$$\sum \frac{(observed - Expected)^2}{Expected} = 223.6 \quad \text{P-value} = 0.00001$$

32. Infliximab In an article appearing in the journal *Lancet* in 2002, medical researchers reported on the experimental use of the arthritis drug infliximab in treating Crohn's disease. In a trial, 573 patients were given initial five-mg injections of the drug. Two weeks later, 335 had responded positively. These patients were then randomly assigned to three groups. Group I received continued injections of a placebo, Group II continued with five mg of infliximab, and Group III received 10 mg of the drug. After 30 weeks, 23 of 110 Group I patients were in remission, compared with 44 of 113 Group II and 50 of 112 Group III patients. Do these data indicate that continued treatment with infliximab is of value for Crohn's disease patients who exhibit a positive initial response to the drug?

Source: The basic info is available on https://www.ncbi.nlm.nih.gov/pubmed/12047962 and it has the detail information about the paper, authors, …

(T) 33. LA rainfall The Los Angeles Almanac website reports recent annual rainfall (in inches), as shown in the table.
a) Create a 90% confidence interval for the mean annual rainfall in L.A.
b) If you wanted to estimate the mean annual rainfall with a margin of error of only two inches, how many years' data would you need?
c) Do these data suggest any change in annual rainfall as time passes? Check for an association between rainfall and year.

Year	Rain (in.)	Year	Rain (in.)
1980	8.96	1991	21.00
1981	10.71	1992	27.36
1982	31.28	1993	8.14

Year	Rain (in.)	Year	Rain (in.)
1983	10.43	1994	24.35
1984	12.82	1995	12.46
1985	17.86	1996	12.40
1986	7.66	1997	31.01
1987	12.48	1998	9.09
1988	8.08	1999	11.57
1989	7.35	2000	17.94
1990	11.99	2001	4.42

34. TV and athletics Using a simple random sample, a student asked 200 students questions about their study and workout habits, and received 124 responses. One of the questions asked, "Do you participate in intramural (IM) athletics, varsity athletics, or no athletics?" while the second asked, "How many hours of television did you watch last week?" The student wants to see if participation in athletics is associated with amount of television watched. Here are the boxplots of *Television watching* by *Athletic participation*:

Analysis of Variance					
Source	DF	Sum of squares	Mean Square	F-ratio	P-value
Athl. Part.	2	128.271	64.1353	4.2343	0.0167
Error	121	1832.72	15.1465		
Total	123	1960.99			

a) State the hypothesis about the students (both numerically and in words).
b) Do the conditions for ANOVA appear to be satisfied? What concerns, if any, do you have?
c) Assuming that the assumptions for inference are satisfied, perform the hypothesis test and state your conclusion. Be sure to state it in terms of television watching and athletic participation.
d) An analysis of the data with all the points highlighted as outliers removed was performed. The *F*-test showed a P-value of 0.0049. How does this affect your answer to part b)?

35. Weight and athletics Using the same survey as in Exercise 36, the student examined the relationship between *Athletic Participation* and *Weight*. Here are the boxplots of *Weight* by *Athletic* participation:

Analysis of Variance					
Source	DF	Sum of squares	Mean Square	F-ratio	P-value
Athl. Part.	2	9 025.55	4 512.78	5.7368	0.0042
Error	121	95 183.2	786.638		
Total	123	10 4209			

a) State the null hypothesis about the students (both numerically and in words).

b) Do the conditions for ANOVA appear to be satisfied? What concerns, if any, do you have?

c) Assuming that the assumptions for inference are satisfied, perform the hypothesis test and state your conclusion. Be sure to state it in terms of *Weight* and *Athletic participation*. What might explain the apparent relationship?

d) An analysis of the data with the highlighted outlier point(s) removed was performed. The *F*-test showed a P-value of 0.0030. How does this affect your answer to part b)?

T 36. Weight loss A weight loss clinic advertises that its program of diet and exercise will allow clients to lose 10 pounds in one month. A local reporter investigating weight reduction gets permission to interview a randomly selected sample of clients who report the given weight losses during their first month in this program. Create a confidence interval to test the clinic's claim that typical weight loss is 10 pounds.

Pounds Lost	
9.5	9.5
13	9
9	8
10	7.5
11	10
9	7
5	8
9	10.5
12.5	10.5
6	9

T 37. Cramming Students in two basic Spanish classes were required to learn 50 new vocabulary words. One group of 45 students received the list on Monday and studied the words all week. Statistics summarizing this group's scores on Friday's quiz are given. The other group of 25 students did not get the vocabulary list until Thursday. They also took the quiz on Friday after "cramming" Thursday night. Then, when they returned to class the following Monday they were retested—without advance warning. Both sets of test scores for these students are shown.

> **Group 1**
> **Fri.**
> Number of students = 45
> Mean = 43.2 (of 50)
> StDev = 3.4
> Students passing (score ≥ 40) = 33%

a) Did the week-long study group have a mean score significantly higher than that of the overnight crammers?

b) Was there a significant difference in the percentages of students who passed the quiz on Friday?

c) Is there any evidence that when students cram for a test their "learning" does not last for three days?

d) Use a 95% confidence interval to estimate the mean number of words that might be forgotten by crammers.

e) Is there any evidence that how much students forget depends on how much they "learned" to begin with?

Group 2			
Fri.	Mon.	Fri.	Mon.
42	36	50	47
44	44	34	34
45	46	38	31
48	38	43	40
44	40	39	41
43	38	46	32
41	37	37	36
35	31	40	31
43	32	41	32
48	37	48	39
43	41	37	31
45	32	36	41
47	44		

T 38. Airport screening Concern with terrorism leads us to look at the records of airport screening for the years between 1977 and 1999 as provided by *the Sourcebook of Criminal Justice Statistics Online*. We find the following regression predicting the incidence of false information from other problems discovered while screening passengers:

Dependent variable is: False info
R-squared = 8.7% R-squared (adjusted) = −9.5%
s = 48.41 with 19 − 4 = 15 degrees of freedom

Source	Sum of squares	DF	Mean Square	F-ratio
Regression	3 364.40	3	1 121.47	0.479
Residual	35 150.2	15	2 343.35	

Variable	Coefficient	SE(Coeff)	t-ratio	P-value
Intercept	27.8181	65.24	0.426	0.6759
Long guns	0.545680	0.6010	0.908	0.3783
Handguns	−4.25650e-3	0.0357	−0.119	0.9066
Explosives	−0.093442	0.0952	−0.982	0.3417

a) Does this appear to be a successful model for incidence of false information? Why or why not?
Here's a scatterplot of the Studentized residuals versus the predicted values:

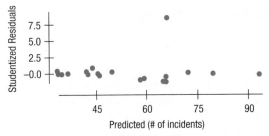

The outlying value is the year 1988. We created an indicator variable for 1988 and included it in the model with the following result:

Dependent variable is: False info
R-squared = 86.8% R-squared (adjusted) = 83.0%
s = 19.05 with 19 − 5 = 14 degrees of freedom

Source	Sum of squares	DF	Mean Square	F-ratio
Regression	33 436.2	4	8 359.05	23.0
Residual	5 078.44	14	362.746	

Variable	Coefficient	SE(Coeff)	t-ratio	P-value
Intercept	47.2152	25.76	1.83	0.0882
Long guns	0.750819	0.2375	3.16	0.0069
Handguns	−0.023583	0.0142	−1.66	0.1190
Explosives	−0.089368	0.0374	−2.39	0.0316
1988	181.310	19.91	9.10	≤0.0001

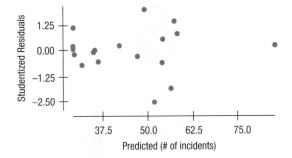

b) Complete the analysis. Check the assumptions and conditions, so far as you can with the information provided. Summarize the model. Discuss any concerns you may still have about the data or the model.

c) Would you remove any of the predictors from the model? If so, which one and why?

Nonparametric Tests

Connections

The most common and powerful approaches to statistical analysis of quantitative data assume that the data are random samples from Normal distributions (with common variance). We then estimate the Normal curve parameters, μ and σ, that will give a good fit to the data. Such an approach is said to be "parametric," since we estimate the parameters of a density curve model. If

I have not failed. I have just found 10,000 ways that won't work.

—Thomas Edison

In the course Statistics 220, offered at the University of Toronto in the summer of 2006, there were six students who were Biology majors and five who were Psychology majors. Their final course grades were (Data in **CourseGrades**):

Biology Majors	Psychology Majors
C+	B
B−	C
A−	D−
D+	B+
A+	C+
B+	

We might ask if there is evidence of a difference in achievement levels between the two types of students. The grades appear higher for the Biology majors, but are they significantly higher? Though this was not a random sample, many students each year take Statistics 220, so it may be reasonable to view these students as similar to a random sample of the many Biology and Psychology students who take this course.

If the numerical grades were available to us, or if we were to convert the letter grades into corresponding numerical grades, we might be able to proceed using a *t*-statistic—after checking the data for possible violations of the assumption of Normality. Though the two-sample *t*-test is rather robust against violations of Normality, particularly for larger samples, the samples in this case are quite small, so noticeable departures from Normality would be of concern.

There are times when an alternative approach may be necessary or desirable, one that is less restrictive in its assumptions. Might there be some other way of attacking this problem? Surely, there must be more than one way to *scan a cat!*[1]

28.1 Wilcoxon Rank Sum Test

Let's start by ranking all 11 observations from low to high, so that the lowest grade receives a rank of 1, the next lowest a rank of 2, and so on, until we reach the highest grade, which receives a rank of 11. How should we deal with grades that are identical? While we might consider just flipping a coin to decide which student gets which rank, the

---

[1]Indeed www.ncbi.nlm.nih.gov/pubmed/24389047 describes an experiment comparing different ways to *scan* a cat. Hope you appreciate our political correctness.

the Normal model is a poor fit, we might try a transformation (or work with some other parametric model for the data).

But usually when we need an alternative, we look for a *distribution-free* method (e.g. the Sign Test—in Chapter 20). Such methods don't impose any particular density curve on the data, so there are no parameters to estimate; hence they are also called *nonparametric* methods. We'll see some examples in this chapter.

Basic concepts of hypothesis testing stay the same: Find a test statistic with a known distribution (under the null hypothesis), calculate the *P*-value for the sample, and interpret it the usual way.

Who	Students taking STA220
What	Course (letter) grades in STA220
Why	To compare performance of students of different majors
When	Summer 2006
Where	University of Toronto

A common element of the methods in this chapter is the replacement of observations by their ranks (so, in some sense, we are employing a transformation approach, but using a very special sort, a "rank-transformation"). This frees us from potentially risky parametric assumptions and leads to a rather simple form of analysis, since the rank values of 1, 2, 3, . . . never change (while raw data vary from sample to sample). And, surprisingly, little loss of information or power occurs when we replace the raw data with their ranks.

usual approach is to assign the same rank value to each of the observations that are tied, where the rank value is the average of the ranks that the observations would have been assigned if not tied. So, if the third, fourth, and fifth biggest values in the data are identical, each one's rank will be recorded as $(3 + 4 + 5)/3 = 4$.

Below are the ranks of our 11 observations. We rank them *all together*, as if this were one big sample of 11 numbers. Let's also compute the sum of ranks for each group, denoted T_1 and T_2. We will need them for making inferences.

Biology Majors	Rank	Psychology Majors	Rank
C+	4.5	B	7
B−	6	C	3
A−	10	D−	1
D+	2	B+	8.5
A+	11	C+	4.5
B+	8.5		
Rank Sum:	$42 = T_1$		$24 = T_2$

Alternatively, we could put all the grades in order, as below, while keeping track of which grades come from which sample (**bold** to indicate **Psychology majors**), and then add up the ranks in bold (or those not in bold), to compute the **rank sum** for Psychology (or Biology) majors. Either way, we get rank sums of 24 for the Psychology majors and 42 for the Biology majors.

D−	D+	**C**	**C+**	C+	B−	**B**	**B+**	B+	A−	A+
1	2	**3**	**4.5**	4.5	6	**7**	**8.5**	8.5	10	11

We want to know if the Biology majors generally tend to perform better or worse than the Psychology majors. Lacking a parametric representation, we think of this as a systematic shift higher or lower for the values in one distribution when compared with the values in the other distribution. (We will define *systematic shift* more precisely later in the chapter.) In a nonparametric setting, our null hypothesis simply states that the populations are identical. We test:

H_0: The two distributions of grades are identical.

vs.

H_A: The values in one distribution are systematically higher or lower than the values in the other distribution.

If one group is superior to the other, it should show up in generally higher ranks, so maybe we should compare the rank averages for the two groups. But instead we'll do something even simpler. We have already calculated the two rank sums, denoted T_1 and T_2. The sum of all of the ranks (1 to 11) equals $1 + 2 + 3 + \ldots + 11 = 66$. And, in general, the sum of the first n integers 1, 2, 3, . . . and so on up to n is equal to $n(n + 1)/2$. So if we know one rank sum, the other is completely determined. Since the rank sum of the first sample is 42, we can predict, without even counting, that the other rank sum must equal the grand sum of 66 minus 42, or 24. Hence a test can be based on either of the two rank sums. We do not need to compare two rank sums or averages with each other (as in a two-sample *t*-test). We pick one rank sum, say $T_2 = 24$, and then determine whether it is unusually big or small.

What do we know about T_2? It could be as small as $1 + 2 + 3 + 4 + 5 = 15$, or as big as $11 + 10 + 9 + 8 + 7 = 45$. It seems possible that our expectation under H_0

might be right in the middle of these numbers, or equal to 30, and indeed this is true. Under the null hypothesis, the rankings in any sample are just a random selection of all rankings, so each sample should, on average, receive its fair share, which would be a rank sum proportionate to its size. Sample two has 5 of the 11 grades, so its expected rank sum is 5/11 of the sum of all ranks, or $(5/11) \times 66 = 30$.

Under the null, T_2 can range from 15 to 45, with a mean of 30. Here it equals 24, somewhat below its mean. But how unusual is this? In other words, what is the P-value?

If we had chosen to work with the other rank sum, T_1, a similar analysis would tell us that it could be anywhere between 21 and 51, with a mean of 36. The actual rank sum is 42, and comparing this high number with its expectation of 36 is exactly the same as comparing the second sample's low rank sum of 24 with its expectation of 30. So either rank sum will do. But when using tables of critical values, the convention is to use the rank sum corresponding to the smaller sample (with fewer observations). Here this is sample two.

We calculated $T_2 = 24$. We need to find the probability of getting such a small rank sum. A deck of cards and some time (well, lots of time) would suffice. Remove every card except the 1, 2, . . . , 10, and Jack, of hearts (think of the Jack as 11). Under the null hypothesis, the ranks for sample two are a random draw of five cards from this deck. Deal out five cards, and count the rank sum. You now have one observation, by simulation, on T_2. Repeat, repeat, repeat . . . , thousands of times, to approximate the distribution of T_2. Find the proportion of times that the rank sum was 24 or below, and you'll have the P-value (but double it for a two-sided alternative). This is a good way to think about the P-value, but not exactly practical (though simulating by computer would be easy enough).

Some basic knowledge of combinations is really all that's needed. We have to count all the ways that any particular rank sum can occur (for example, $T_2 = 17$ only for rankings of 1, 2, 3, 4, 7 or 1, 2, 3, 5, 6, so in 2 ways), and divide that count by the total number of possible combinations or $_{11}C_5$ (which, as in Chapter 13, denotes the number of ways of choosing 5 items from 11 items, and equals 462) to compute the probability of obtaining this rank sum. If we did this for every possible rank sum, we could then find any P-value by summing the appropriate individual probabilities.

Fortunately, this dreary combinatorial work has already been done for you in tables of critical values for the **Wilcoxon Rank Sum test**. The corresponding table in Appendix C gives some tail probabilities for the rank sum corresponding to the smaller sized sample (for equal sample sizes, use either rank sum). In our example, the rank sum for the smaller sample is $T_2 = 24$. From the table, for $n_1/n_2 = 5/6$, the closest critical value is 20, and the probability of a rank sum ≤ 20 is approximately 0.05. So the P-value for our rank sum of 24, or $P(T_2 \leq 24) \times 2$, is bigger than 0.05×2, after doubling the lower tail area for a two-tailed test. At the 10% level of significance, we do not have evidence of a difference in course grades between the two types of majors in our study.

If a table is unavailable, or if your sample sizes do not appear in the table, what to do? Recall that the Central Limit Theorem tells us that averaging or summing many observations will often produce Normality,[2] and indeed it is the case here that if both sample sizes exceed 10, the rank sums will be approximately Normally distributed. Then all you need is the parameters of the Normal model. It can be shown that

$$E(T_i) = n_i(n_1 + n_2 + 1)/2$$

and

$$Var(T_i) = n_1 n_2(n_1 + n_2 + 1)/12$$

> **The Null Model** For the Normal curve parametric approach, our null model draws each sample from that well-shuffled deck of cards used by Gossett (in Chapter 18), where all the numbers in the deck, piled up, produce a bell-shaped population. Here, our null model shuffles 11 cards numbered 1 to 11, and draws 5 of them.

---

[2]The Central Limit Theorem discussed in Chapter 14 has been generalized in various ways over the years. Under appropriate conditions, a sum or average of random variables—even when they do not all have the same distribution and are not independent—will have an approximately Normal distribution, as long as you sum enough of them. The conditions essentially ensure that one variable is not very dominant in its effects over the others. As long as they all make small contributions to the sum, Normality often results.

for $i = 1$ or 2, so either rank sum can be standardized using the above formula, and then the P-value can be approximated from the standard Normal distribution. It is possible to improve slightly on this formula using a continuity correction in the top and a correction for ties in the bottom, but we will omit these corrections here. Unless there are a great number of ties relative to the sample size, the correction for ties makes little difference.

Assumptions and Conditions

Since this is a distribution-free procedure, there are no distributional assumptions such as Normality to check, but pay attention to the other common design issues mentioned in Chapter 19, such as:

Independence Assumptions

Independence Groups Assumption: The groups must be independent of each other. No test can verify this, so think about how the survey or experiment was conducted.

Independence Assumption: The individual responses within each group must be independent as well. Check the **Randomization Condition.** For surveys, were the data drawn using simple random samples, or can the data be viewed as independent observations reasonably representative of their populations if collected in a less controlled observational setting? For an experiment, were the experimental units assigned at random to the treatments?

Ordinal or Quantitative Data Condition

Ordinal Data Condition: The data in our example were *ordinal* in nature—we knew which grades were higher or lower, but not the exact amount of the difference between two grades. If we had the students' actual numerical grades, which are on a precise quantitative scale, we still could have proceeded exactly as before, since numerical data can easily be converted to ranks. *Quantitative data can always be reduced to an ordinal scale.* So the Wilcoxon test can be used for quantitative data or for purely ordinal data.

As another example, suppose that we wanted to compare opinions in Ontario and Alberta about a proposed greenhouse emissions bill, where the possible responses in a survey were "strongly disagree," "mildly disagree," "neutral," "mildly agree," or "strongly agree." These responses are on an ordinal scale, so we can rank all the responses from each province, though we would have a large number of ties—which the Normal approximation with correction for ties can handle well enough. (Warning: Be wary of assigning numbers to ordinal data and treating the results as if they were actual quantitative data.)

> Computer software may report a confidence interval for the difference in population medians. This interval is only valid for two quantitative populations possessing similarly shaped distributions.

> **WILCOXON RANK SUM TEST**[3]
>
> When the assumptions and conditions for the Wilcoxon Rank Sum test are met, we can test:
>
> H_0: The two distributions are identical.
>
> vs.
>
> H_A: The values in one distribution are systematically higher or lower than the values in the other distribution.
>
> Calculate the rank sum for the smaller sample. Then estimate the P-value using tables of critical values (or the Normal approximation, if either sample size is off the table).
>
> Due to the nonparametric nature of the test, we cannot compare distributions using parameters of density curves. A general definition of *systematically higher* would say that the probability massed to the right of any particular value is always greater (or sometimes the same) in one of the distributions. But for two similarly shaped quantitative populations, this test may be viewed as comparing medians, and software will typically provide a confidence interval for the difference in the medians.

---

[3]There is an alternative but equivalent form of this test known as the Mann-Whitney test.

For Example   THE WILCOXON RANK SUM TEST WITH QUANTITATIVE DATA

Consider the Chapter 19 opening example, comparing the lifetimes of generic versus brand-name batteries. The battery lifetime measurements in minutes are shown in two columns below (Data in **Batteries**). The 12 lifetimes were ordered, from low to high, with each assigned a rank, displayed to the right of the lifetime.

Brand Name	Rank	Generic	Rank
194.0	5	190.7	4
205.5	9	203.5	7.5
199.2	6	203.5	7.5
172.4	2	206.5	10
184.0	3	222.5	12
169.5	1	209.4	11

QUESTION: Does the Wilcoxon test provide evidence of a difference between the two brands?

ANSWER: The sum of the ranks for brand-name batteries, denoted T_1, is 26.0. We find from our tables ($n_1 = 6$, $n_2 = 6$) that $P(T_1 \leq 26) \leq 0.025$, so the P-value is double this, or ≤ 0.05, which provides some evidence of a difference. In Chapter 19, the 95% confidence interval barely missed zero, indicating that the P-value for a test of no difference would be slightly less than 0.05. In fact, a precise calculation with a two-sample t-test gives $P = 0.032$, so the parametric test sees the difference (assuming it exists) slightly more clearly. Of course, this is assuming that the parametric calculation is trustworthy, and with so little data, it's hard to tell. However, the similarity between the two P-values should be somewhat reassuring.

If the samples were larger, we could use the Normal approximation to estimate the P-value. Let's check it out here anyway:

$$E(T_1) = (6 \times 13)/2 = 39$$

and

$$\text{Var}(T_1) = (6 \times 6 \times 13)/12 = 39 \quad \text{or} \quad \sigma = 6.245, \text{ so we get:}$$

$z = \dfrac{(26 - 39)}{6.245} = -2.08$, yielding a P-value $= 0.038$, which is a pretty decent approximation to the P-value from the exact Wilcoxon test, which, more accurately calculated (using software), would be 0.042.

This Wilcoxon test has the advantage of a greater range of applicability compared with the t-test, primarily due to less restrictive assumptions about the data. It may give you trustworthy P-values when those calculated from the t-test are deemed suspect.

This raises another interesting question: How does the Wilcoxon test stack up against the t-test when the data are in fact Normally distributed? The issue is one of power. We would expect the Wilcoxon test to be less powerful than the t-test since it does not use as much of the information present in the data. We have thrown out the numbers in favour of their ranks, so some loss of power should result (just as power would decrease if we lost some observations), but how much loss?

It turns out that the power of the two-sample t-test for samples of size 95 is approximately the same as the power of the Wilcoxon test for samples of size 100 (for Normally distributed data).

Quite surprising! The loss of power in using the Wilcoxon test rather than the t-test is equivalent to a loss of 5% of sample size, so those ranks do a pretty good job of holding on to the relevant information present in the data. And when severe skewness or extreme outliers are present, the Wilcoxon test will usually have greater power to discern a genuine difference than the t-test.

Just Checking

1. If we compare data in two samples, one of size three and one of size four,
   a. What will be the sum of all the ranks?
   b. What are the biggest and smallest rank sums possible for the larger sample?
   c. What would be the expected or average value for the rank sum of the larger sample if there were no actual difference between the two populations?

Section 28.1 Exercises

1. For which one of the following situations would a Wilcoxon Rank Sum test be appropriate?
   a) Judging whether the height of water at high tide at a marina increases more or less consistently as the phase of the moon changes from new to full. We'll take measurements over several lunar (29½ day) months, during each of which the moon goes through all of its eight phases.
   b) Estimating the association between the value of homes and their age in a community known to have a number of very expensive mansions.
   c) Comparing the sweetness of a diet drink (rated from 1 to 10) as judged by a panel of teenaged tasters and a panel of adult tasters.

2. For which one of the following situations would a Wilcoxon Rank Sum test be appropriate?

a) The Richter scale measures the damage potential of earthquakes. Some scientists think it's best used for judging the relative size of earthquakes and not their absolute size. Are earthquakes around the Pacific "Ring of Fire" stronger in general than those around the Atlantic? Compare earthquakes over a recent five-year period.
b) Each member of a group of professional chefs rates the quality of a basket of organic fruits on a seven-point scale. The same chefs then rate the quality of a basket of non-organic fruits on the same scale. Are the organic fruits rated higher?
c) A dog's intelligence can be judged roughly by how many times a new command must be repeated until it is learned. A researcher trained several dogs from four different breeds. He taught the dogs three new commands. Are the breeds equally "intelligent"?

28.2 Kruskal-Wallis Test

Suppose now that we have more than two independent samples of measurements to compare. We need a generalization of the Wilcoxon Rank Sum test. Below are bacteria count data for four different hand-washing methods, as examined in a Chapter 24 example (Data in **HandWashing**):

Method:	Alcohol	AB Soap	Soap	Water
	51	70	84	74
	5	164	51	135
	19	88	110	102
	18	111	67	124
	58	73	119	105
	50	119	108	139
	82	20	207	170
	17	95	102	87

Transform the data to ranks, ranking them as one big group from 1 to 32 (using tied ranks when necessary). Ranks are shown in parentheses below. As a check, note that the sum of all the ranks should equal $n(n + 1)/2 = 32(33)/2 = 528$.

Method:	Alcohol	AB Soap	Soap	Water
	51(7.5)	70(11)	84(15)	74(13)
	5(1)	164(30)	51(7.5)	135(28)
	19(4)	88(17)	110(23)	102(19.5)
	18(3)	111(24)	67(10)	124(27)
	58(9)	73(12)	119(25.5)	105(21)
	50(6)	119(25.5)	108(22)	139(29)
	82(14)	20(5)	207(32)	170(31)
	17(2)	95(18)	102(19.5)	87(16)
Rank Sums:	**46.5**	**142.5**	**154.5**	**184.5**

Unlike the Wilcoxon test, we cannot simply focus on one rank sum for these data (again, any one of the rank sums would be redundant, but that still leaves three free to vary). We need a measure of the variation among all the rank sums (adjusted for differences in sample size), and this is provided by the **Kruskal-Wallis statistic**:

$$H = \left(\frac{12}{N(N + 1)} \sum \frac{T_i^2}{n_i} \right) - 3(N + 1)$$

where T_i are the rank sums, and $N =$ total sample size.

The Kruskal-Wallis statistic measures variation among the rank sums and grows bigger as the differences among them increase. So, a very large value for H provides evidence of a genuine difference among the populations. To assess this, we need an appropriate sampling model. If the samples are not too small, the chi-square distribution does the job nicely.[4] The number of degrees of freedom in this distribution is the number of squared terms in the summation (or the number of samples) minus 1.

This chi-square model is actually a large-sample approximation, so for very small samples, it would be best to hunt down a table of exact critical values (or similar software). Again, we omit a correction for ties that could be inserted into the formula above. When there are just two samples, this test is identical to the Wilcoxon test, and the chi-square value will be the square of the z value for the Wilcoxon test.

Assumptions and Conditions

The assumptions and conditions for the Kruskal-Wallis test are the same as for the Wilcoxon Rank Sum test, but the Kruskal-Wallis test is suitable for *comparing any number of independent groups*, whereas the Wilcoxon test can only be used to compare two independent groups.

KRUSKAL-WALLIS TEST

When the assumptions and conditions for the Kruskal-Wallis test are met, we may assess:
  H_0: The distributions are all identical.
                                vs.
  H_A: At least two of the distributions differ (systematically higher or lower) from each other.

Calculate the Kruskal-Wallis test statistic. Then, estimate the P-value using the upper tail (only) of the chi-square model with $t - 1$ degrees of freedom, where $t =$ the number of samples.

---

[4]The operation of squaring and summing often produces a quantity with approximately a chi-square distribution, after some doodling, whereas a ratio of two independent sums of squares, as in a one-way ANOVA, can often be worked into a random variable with an F-distribution, after some doodling (dividing each by df).

For Example

Continuing with our example, let's test the null hypothesis that the hand-washing method has no effect on bacteria count. We take the rank sums calculated earlier and plug into the Kruskal-Wallis statistic:

$$H = \left(\frac{12}{32(32+1)} \times \frac{46.5^2 + 142.5^2 + 154.5^2 + 184.5^2}{8} \right) - 3(32+1) = 15.17$$

The P-value $= P(\chi^2_{3df} > 15.17) = 0.002$ (from software), so we have very strong evidence that the washing method does indeed affect the bacteria count.

Comparing with the parametric analysis in Chapter 24, we see that the P-value here is less impressive, though still quite small. This mostly reflects the fact that the ANOVA paid much more attention to the big difference between the alcohol spray and the other methods, whereas this test takes note that the alcohol spray values were lower, without paying attention to how much lower.

Just Checking

2. If there are four samples with three responses in each, what is the biggest value possible for the Kruskal-Wallis statistic? And what would be the estimated P-value in that case?

Section 28.2 Exercises

3. For which one of the following situations would a Kruskal-Wallis test be appropriate?
   a) Testing whether the distributions of the ratings of a new distance-learning course were the same as rated by first-, second-, third-, and fourth-year students who had taken the course.
   b) Comparing the ratings of a new product on a seven-point scale by a panel of consumers with their ratings of a competitor's product on the same scale.
   c) Comparing the sweetness of a diet drink (rated from 1 to 10) as judged by a panel of teenaged tasters and a panel of adult tasters.

4. For which one of the following situations would a Kruskal-Wallis test be appropriate?
   a) The Richter scale measures the damage potential of earthquakes. Some scientists think it's best used for judging the relative size of earthquakes and not their absolute size. Are earthquakes around the Pacific "Ring of Fire" stronger in general than those around the Atlantic? Compare earthquakes over a recent five-year period.
   b) The Mohs scale rates the hardness of minerals. If one mineral can scratch another, it's judged to be harder. Is hardness related to the weight of a one-by-one-by-one cm cube of the mineral?
   c) A dog's intelligence can be judged roughly by how many times a new command must be repeated until it is learned. A researcher trained several dogs from four different breeds. He taught the dogs three new commands. Are the breeds equally "intelligent"?

28.3 Wilcoxon Signed Rank Test for Paired Data

Suppose we want to compare two similarly priced pinot noir (red) wines newly marketed in Ontario: a Russian River (California) pinot noir and a Niagara region pinot noir. Some experienced wine tasters are available for the study. Should we proceed using a completely randomized design or a randomized block design? Randomly dividing tasters into two groups—one to try the Russian River and the other to try the Niagara wine—makes no sense, since we would be confounding differences in tasters with wine differences. Clearly, *blocking* or *pairing* to filter out (see Chapter 10) tasters' differences is the way to go. Each taster should taste each wine, in a random order. Below are the results from a tasting (wines abbreviated as RR and NG). The tasters rated both wines on a scale of 0 to 10, with 10 being best (Data in **WineTasting**):

Taster#	1	2	3	4	5	6	7	8	9	10
RR	10	8	10	9	7	8	10	8	8	6
NG	8	8	7	5	6	8	4	5	3	7

To test the null hypothesis of equivalence of the wines, we might try the parametric approach, calculating the 10 differences, assuming them to be (hopefully) Normal, and performing a *t*-test. But how else might we proceed?

In Chapter 20, we discussed the nonparametric *sign test* approach. Note that we have seven who prefer the Russian River, one who prefers the Niagara, and two ties. Ignoring the ties, we have a simple model under the hypothesis of equivalence:

If X = number who prefer the RR (number of "+" differences), then X is Binomially distributed with $n = 8$, $p = 0.5$.

The P-value equals $2 \times P(X \geq 7) = .07$ by direct calculation, using the Binomial distribution (or Binomial tables or software).

But the sign test is quite wasteful of information, since it throws away the numbers while only retaining their signs. Maybe, as with the first Wilcoxon test discussed in this chapter, we can gain by replacing the numerical differences with their ranks. First though, take the *absolute value* of each difference (denoted $|d|$), and then rank those absolute differences from low to high (ignoring zero differences):

Taster#	1	2	3	4	5	6	7	8	9	10
RR	10	8	10	9	7	8	10	8	8	6
NG	8	8	7	5	6	8	4	5	3	7
difference:	+2	0	+3	+4	+1	0	+6	+3	+5	−1
$|d|$	2		3	4	1		6	3	5	1
rank $|d|$	3		4.5	6	1.5		8	4.5	7	1.5

Let $T+$ = the sum of the ranks corresponding to the positive differences, which equals 34.5.

Let $T-$ = the sum of the ranks corresponding to the negative differences, which equals 1.5.

Consider the null hypothesis:

H_0: There is no true systematic difference in ratings between the two wines.

The alternative is simply 'Not H_0.' If the null hypothesis is true, then we expect $T+$ and $T-$ to be close in value. But if the null is false, we expect them to be far apart; in other words, we expect a rather large or rather small $T+$ (or $T-$), keeping in mind that these two numbers always add up to the same thing: $n(n + 1)/2$, which equals $8(9)/2 = 36$, in this case. The expected value of either sum, under the null here is, rather obviously, just half of 36 or 18.

Is $T+$ significantly large? This is the same as asking whether $T-$ is significantly small. The standard tables for this test provide critical values for the *numerically* smaller of the two rank sums.

So, for convenience, when using tables, we'll likewise define our test statistic to be the smaller of $(T+, T-)$, which we'll denote as T. Here, $T+ = 34.5$ and $T- = 1.5$, so $T = 1.5$.

For any specified value of n (the number of non-zero differences), we can refer to the table of critical values for the **Wilcoxon Signed Rank test** in Appendix C to find critical values for a two-sided or one-sided alternative.

In our wine-tasting example, we may assume a two-sided alternative. In this case, either a small $T+$ or a small $T-$ would be evidence of a difference between the wines. From the table, we find the critical $T^* = 3$ at $\alpha = 0.05$ for a two-sided alternative. We calculated $T = 1.5$, which is < 3, and hence has a P-value < 0.05, so we may reject the null hypothesis of equivalence of the wines at the 5% level.[5]

If the alternative were one-sided, stating that the RR has higher ratings than the NG, we would have proof if $T+$ is big, or T− is small, and from the table we see that a $T-$ smaller than 3 has a P-value below 0.025. So we would be able to reject H_0 at the 2.5% level.

On the other hand, if the alternative hypothesis stated that the NG has ratings higher than the RR, our conclusion would be immediate. Such a hypothesis could only be supported if there were many minuses and a large $T-$ (or small $T+$). But here $T-$ is quite small (T+ large), so the P-value must be very large indeed (in fact, $> 1 - 0.025$)!

If $n \geq 25$, then under H_0, $T+$ (or $T-$) has an approximately Normal distribution with

$$E(T+ \text{ or } T-) = n(n + 1)/4 \quad \text{and} \quad \text{Var}(T+ \text{ or } T-) = n(n + 1)(2n + 1)/24$$

Even in our wine-tasting example, with much smaller n, the Normal approximation produces a result similar to the tables:

$$z = \frac{1.5 - 18}{\sqrt{(8 \times 9 \times 17)/24}} = -2.31$$

and from the standard Normal table, the P-value $= 2(.01) = 0.02$. More detailed Wilcoxon tables would give a P-value equal to 0.024 for $T = 2$. If instead we chose to use $T+$ in this calculation, we would get $z = +2.31$ and the same P-value.

Again, we omit corrections for ties and for continuity from the z-statistic formula. The correction for ties can be very useful, since throwing out and ignoring a substantial number of ties (as in the exact test) can bias the test against the null hypothesis (ties clearly support the null hypothesis). Software will always apply a correction for ties in the z-statistic.

> *For one-sided alternatives, be careful to check that the data actually point in the direction of and support the alternative*, and not in the exact opposite direction, particularly when using tables of critical values. More generally, always plot and inspect the data right at the start, and then make sure your final conclusions bear some resemblance to the story that your plots tell.

Assumptions and Conditions

The responses must be quantitative. Other than that, there are no distributional assumptions to check, but pay attention to other common design issues for paired data studies (discussed in Chapter 20), such as:

Paired Data Assumption: There must be a clear linkage between responses in the two samples (for example, from the same person, or from the same pair of siblings, or from two individuals paired together *a priori* to improve the precision of the treatment comparison). Do not try to create some linkage yourself with the data values. It must be part of the study design.

Independence Assumption: The groups are not independent due to pairing, but the *differences* must be independent of each other. Independence must be ensured by the study design, as no test can verify this assumption (though plotting the differences versus any lurking variable, such as time order, may help point out possible problems), so think about how the data were collected. Check the **Randomization Condition:** Were the data

---

[5]We are assuming that the tasters were well trained and able to apply the same rating scale *consistently*. A difference of three for one taster must be genuinely bigger than a difference of two for another taster, or the rankings used in our test would not make much sense.

Computer software may report a confidence interval for the median difference. This interval is only valid if we make the additional assumption that the numerical differences have a symmetrically shaped distribution. The Wilcoxon Signed Rank test can also be used with a single sample to test for a hypothesized median, but this likewise requires an assumption of symmetry.

collected with suitable randomization? For surveys (or before–after studies), were the pairs drawn using a proper random sampling scheme, or, if collected in a less controlled observational setting, can they be viewed as reasonably representative of some population? In experiments, either the treatments should be randomly assigned to one member of each pair or the order of the (same-subject) treatments should be randomized, depending on the nature of the study.

WILCOXON SIGNED RANK TEST

When the appropriate assumptions and conditions are met, we may test:

H_0: There is no systematic difference within pairs.

vs.

H_A: The values are systematically higher or lower for one of the distributions or treatments.

Calculate $T+$ = the sum of ranks of the positive differences, and $T-$ = the sum of ranks of the negative differences. Then estimate the P-value using a table of critical values for T = minimum $\{T-, T+\}$, or, if the sample size is off the table, using the Normal approximation for either rank sum.

Step-by-Step Example   ANALYZING PAIRED DATA WITH THE WILCOXON SIGNED RANK TEST

"Hormones may boost profits," stated the headline of science reporter Anne McIlroy's April 15, 2008, article in the *Globe and Mail*. To examine this theory, back in June 2005, Canadian researcher John Coates took saliva samples twice a day for eight days from 17 men working on a London, England mid-size trading floor, trading a wide range of assets, with largest exposure to German interest rate futures. He classified each trader according to whether his testosterone level was high or low on that day (compared with the trader's median over the period). High testosterone days differed from trader to trader, and high days differed from low days by 25%, on average, in testosterone level. He also recorded the daily profits or losses (P&L) in pounds sterling of each trader (from 11 a.m. to 4 p.m.). The *Globe and Mail* also published boxplots comparing P&L on high testosterone days with profits on low testosteroned days, similar to the plots shown here:[6]

| THINK ➡ Plan State what we want to know. Identify variables and check the *W*'s. | I want to determine if there is evidence of a testosterone effect on daily profits for traders similar to those examined in Coates's study. For each trader, we have recorded profit and loss averages for high testosterone days (testosterone level above each trader's own median) during an eight-day period in June 2005, and another profit average for low testosterone days during the same eight-day period. |

Hypotheses State the null and alternative hypotheses.

We want to know if profits shift higher or lower according to testosterone level. We don't want to make assumptions about the direction of the difference, so we'll go with the two-sided alternative.

H_0: There is no difference in the distribution of profits between high testosterone days and low testosterone days.

H_A: Profits tend to be either generally higher or generally lower on a trader's high testosterone days than on his low testosterone days.

[6]"Hormones may boost profits," by Anne McIlroy. *Globe and Mail*, April 15, 2008, L1; J.M. Coates and J. Herbert, Proceedings of the National Academy of Sciences of the 0USA (*PNAS*) 2008; Vol. 105, pp. 6167–6172, available at www.pnas.org/content/105/15.toc. Copyright 2008 National Academy of Sciences. USA.

REALITY CHECK There appears to be a difference from the boxplots (previous page), but for a proper analysis we must recognize the paired structure of the data (each trader appears once in each of the two boxplots), which is not visible.

Model Think about the assumptions and check the conditions.

Are the data paired? How?

Paired data condition: We took an average on both high and low testosterone days for the same trader, so each trader provided paired responses.

Was randomization used?

Randomization: This was not a random sample of traders, but we believe them to be representative, of futures traders. And there was nothing special about this time period or trading floor.

Are responses independent?

Independence: We'll assume that each trader acts independently of the others.

Check any distributional assumptions of the intended analysis.

Quantitative data: All that we require for the Wilcoxon test is quantitative data. But let's check for Normality, too, since for Normal data, a t-test would be more powerful.

Make a picture of the *differences*, which form the actual data in our analysis—perhaps a histogram or Normal probability plot. Professor Vukov contacted Dr. Coates and asked for the data (even if we could read the points in the boxplots, we wouldn't know how to pair them!).

Dr. Coates kindly provided the following data from his study (Data in **Traders_Profits**)

Trader	P&L Low 11 a.m. Testosterone	P&L High 11 a.m. Testosterone
1	127	−63
2	113	725
3	−2	−5
4	−292	1780
5	−98	262
6	−1308	1401
7	88	2092
8	361	887
9	−700	2500
10	143	82
11	1448	1547
12	1625	7000
13	−127	38
14	−2066	1941
15	183	268
16	483	513
17	−115	−04

There is a high outlier in one of the boxplots, which would be a concern if we were performing, say, an independent samples t-test, but here, it's the differences that matter. They are, in order by trader *(High–Low):* −190, 612, −3, 2072, 360, 2709, 2004, 526, 3200, −61, 99, 5375, 165, 4007, 85, 30, 111, and plotted below.

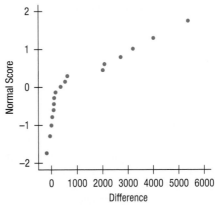

Choose the method of analysis.

The data are highly skewed. A paired (one-sample) *t*-test on the differences is risky. If we try it, we get $t = 3.30$ with $P = 0.008$. But it would be rather hard to trust this P-value. The sign test gives us a P-value of 0.0127 based on 14 positive differences out of 17 differences. But let's extract more information about the differences by applying the **Wilcoxon Signed Rank test**.

SHOW ➡ **Mechanics** Compute the difference for each trader. Throw out any zero differences.

List the differences in order of magnitude. Sum the ranks for the positive differences and for the negative differences. Use the sum that is smaller as test statistic, as required by the table of critical values.

Test your rank sum for significance by using the table of critical values, software or the Normal approximation (if appropriate).

Here are the differences, in order by magnitude, with ranks shown below.

Diff:		−3	30	−61	85	99	111	165	−190	360
Rank|Diffl|:	1	2	3	4	5	6	7	8	9

Diff:		526	612	2004	2072	2709	3200	4007	5375
Rank|Diffl|:	10	11	12	13	14	15	16	17

The smaller rank sum is clearly for the three negative differences: $T— = 1 + 3 + 8 = 12$. According to the table of critical values, we

	find that the P-value corresponding to this sum is < 0.002, since $P \le 0.002$ for the bigger rank sum of 14 displayed in the table. The other rank sum must be $n(n + 1)/2 - 12 = 17(18)/2 - 12 = 141$, either of which could be used if working with the Normal approximation. The Normal approximation would yield a standard Normal statistic: $z = \frac{(141 - 76.5)}{21.12} = 3.05$, giving a similar P-value estimate of 0.002.
TELL ➡ **Conclusion** Link the P-value to your decision about the null hypothesis, and state your conclusion in context.	The P-value is tiny, so we have strong evidence of a difference, and we can conclude that higher profits are associated with higher testosterone levels.

Section 28.3 Exercises

5. For which one of the following situations would a Wilcoxon Signed Rank test be appropriate?
   a) Estimating the association between the value of homes and their age in a community known to have a number of very expensive mansions.
   b) Comparing the ratings of a new product on a seven-point scale by a panel of consumers with their ratings of a competitor's product on the same scale.
   c) Judging whether the height of water at high tide at a marina increases more or less consistently as the phase of the moon changes from new to full. We'll take measurements over several lunar months, during each of which the moon goes through all of its eight phases.

6. For which one of the following situations would a Wilcoxon Signed Rank test be appropriate?

   a) During a smartphone game's initial release, is there an association between the number of downloads and the number of hours from the time it is released?
   b) Each member of a group of professional chefs rates the quality of a basket of organic fruits on a seven-point scale. The same chefs then rate the quality of a basket of non-organic fruits on the same scale. Are the organic fruits rated higher?
   c) The Richter scale measures the damage potential of earthquakes. Some scientists think it's best used for judging the relative size of earthquakes and not their absolute size. Are earthquakes around the Pacific "Ring of Fire" stronger in general than those around the Atlantic? Compare earthquakes over a recent five-year period.

28.4 Friedman Test for a Randomized Block Design

In Chapter 25, we mentioned that a two-way layout of data may arise when one of the factors is a blocking variable (see Chapter 10 to review the concept of blocking). In this case, we often have just one observation at each combination of block and treatment, and then proceed after making the (often reasonable) assumption of an additive model. If we have a randomized block design with just two treatments, the Wilcoxon Signed Rank test can be used to compare the treatments. For more than two treatments, it turns out that a simple twist on the Kruskal-Wallis procedure works very well.

The table displays data showing weight gain, in grams, for nine young rats, three from each of three litters. Three different diets (the treatments) were assigned at random to the three siblings in litter one (block 1). This was repeated for the other two litters (blocks 2 and 3).

Treatment:	1	2	3
Block 1:	90	95	80
Block 2:	77	90	72
Block 3:	75	84	64

We need to convert to ranks, but now we *rank separately within each block*, which gives the following ranks (in parentheses):

Sample:	1	2	3
Block 1:	90(2)	95(3)	80(1)
Block 2:	77(2)	90(3)	72(1)
Block 3:	75(2)	84(3)	64(1)
Rank Sums:	6	9	3

To test for significant variation among the treatment rank sums, we can use the **Friedman statistic**:

$$S = \left(\frac{12}{bt(t+1)} \sum T_i^2 \right) - 3b(t+1),$$

where T_i are the rank sums for the various treatments, b = the number of blocks, and t = the number of treatments.

The Friedman statistic measures variation among the three rank sums, and gets bigger as the sums get further apart from each other. It equals zero when the three sums are identical. So we should reject the null hypothesis that all treatments are identical if the Friedman statistic is *unusually big*. For moderate sample sizes, we can get a good approximation to the sampling distribution using the chi-square model with $t-1$ degrees of freedom. Tables of exact critical values though are recommended for very small sample sizes (like in this example!).

Assumptions and Conditions

Our assumptions and conditions are generally the same as for the Wilcoxon Signed Rank test, except that we may compare any number of treatments where the responses are classified according to treatment and block (in a randomized block design). The data may be on either an ordinal or quantitative scale.

FRIEDMAN TEST

When the appropriate assumptions and conditions are met, we may test:

H_0: The treatments are all identical.

vs.

H_A: At least one of the treatments has responses systematically higher or lower than some other treatment.

Calculate the Friedman test statistic. Estimate the P-value using the upper tail of the chi-square model with $t-1$ degrees of freedom, where t = number of treatments being compared. Or use a table of exact critical values for very small samples.

For Example

Continuing with our rat weight gain example, if we apply the Friedman statistic, we get:

$$S = \left[\frac{12}{3 \times 3(3 + 1)} (6^2 + 3^2 + 9^2) \right] - 3 \times 3(3 + 1) = 6.0$$

which gives a P-value = 0.05 using the χ^2 distribution with 2 df. We have some evidence that weight gain depends on diet, or that superior weight gains are achievable by proper choice of diet.

Note that the evidence here was as strong as it could possibly be for treatment differences, since we had all the "1" rankings in one diet, all the "2" rankings in another diet, and all the "3" rankings in the remaining diet. This gives the maximal spread among the rank totals. So for such small samples, with only three responses in each of three treatments, the best that we can ever hope to attain is 5% significance (and just barely). Considering the small sample size in this example, it would also be advisable to search for software or critical value tables for the exact test, in order to verify the accuracy of our P-value.

Section 28.4 Exercises

7. For which one of the following situations would a Friedman test be appropriate?
  a) Comparing the sweetness of a diet drink (rated from 1 to 10) as judged by a panel of teenaged tasters and a panel of adult tasters.
  b) Testing whether the distributions of the ratings of a new distance-learning course were the same as rated by first-, second-, third-, and fourth-year students who had taken the course.
  c) Judging whether the height of water at high tide at a marina increases more or less consistently as the phase of the moon changes from new to full. We'll take measurements over several lunar months, during each of which the moon goes through all of its eight phases.

8. For which one of the following situations would a Friedman test be appropriate?
  a) The Mohs scale rates the hardness of minerals. If one mineral can scratch another, it's judged to be harder. Is hardness related to the weight of a one-by-one-by-one cm cube of the mineral?
  b) During a smartphone game's initial release, is there an association between the number of downloads and the number of hours from the time it is released?
  c) A dog's intelligence can be judged roughly by how many times a new command must be repeated until it is learned. A researcher trained several dogs from four different breeds. He taught the dogs three new commands. Are the breeds equally "intelligent"?

28.5 Rank Correlation

With bivariate data, we sometimes wish to define a measure of correlation that is valid and useful under less restrictive assumptions than the correlation coefficient defined in Chapter 6, where we first discussed bivariate quantitative data. Maybe converting observations to ranks will do the trick. The correlation coefficient defined in Chapter 6 is commonly referred to as *Pearson's product moment correlation*. One formula used to calculate it is:

$$r = \frac{\sum (x_i - \bar{x})(y_i - \bar{y})}{\sum (x_i - \bar{x})^2}$$

Recall that this correlation coefficient is a measure of the strength of linear (straight-line type) association between two variables x and y. We'll generalize the notion of linear relation to the notion of **monotonic relation**. Consider the following graphs:

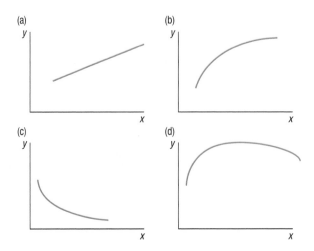

The relations in graphs (a), (b), and (c) are all perfect monotonic relations. In other words,

■   as x increases, y increases (a positive monotonic relation); or
■   as x increases, y decreases (a negative monotonic relation)

Only in graph (a) is this monotonic relation also a linear one. A linear relation is monotonic, but a monotonic relation need not be linear. In graph (d), the relation is not monotonic, since y neither goes up consistently nor goes down consistently with increases in x. About all we can say is that there is a positive monotonic relation over the first half of the range of x values, and a negative monotonic relation over the second half. However, the overall relationship is not monotonic.

Just as Pearson's r measures the strength of linear association, we can define a measure of the strength or extent of monotonic association. Replace the x values by their ranks, from 1 to n, where n is the number of pairs. Replace the y values by their ranks, also from 1 to n. Compute Pearson's r for these two columns of rankings rather than for the original observations. This produces a measure of the linear association of the ranks rather than the linear association of the actual data. The resulting value is **Spearman's correlation**, or r_s.

Consider the following simple data:

x:	1	2	3	4	5
y:	1	2	4	8	16

Plotted, these data look like:

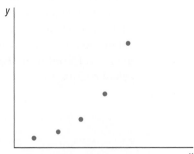

Pearson's correlation r is equal to 0.933 in this example. This number *partly* captures the strong rise in y with increases in x, but is less than 1 because we do not have a perfect linear relation. However, the data are a perfect monotonic relation, so r_s should capture this fact by producing a value of 1. Let's check:

Converting to ranks (x's and y's ranked separately), we find:

x rank:	1	2	3	4	5
y rank:	1	2	3	4	5

So, the ranks, unlike the actual data, exhibit a perfect linear association. Hence, r_s will equal 1, since Pearson's r applied to a perfect linear association must equal 1.

But there is a simpler way to calculate r_s. Instead of ranking the data and then plugging the rankings into the messy formula for Pearson's r, it can be shown that, in the absence of ties, we can reduce that formula to the following (applied to the ranks, not the raw data):

$$r_s = 1 - \frac{6 \sum d_i^2}{n(n^2 - 1)}$$

where d_i is the difference between the ranks for the ith observation on x and y.

Even if there are some ties, this simple formula will provide a good enough approximation to the more precise calculation applying Pearson's r to the ranks.

So r_s measures the strength of the linear relation between the rankings of x and the rankings of y. Like r, this measure varies in value from -1 to 1, with the values of -1 or $+1$ indicating a perfect monotonic relationship between x and y and a value like $r_s = 0.7$, for example, indicating a moderate degree of positive monotonicity, in other words, a moderately strong increasing—but not necessarily linear—relation.

When compared with Pearson's r, there are a few possible advantages to using Spearman's nonparametric measure of correlation:

- Greater resistance to the effects of a small number of observations (such as outliers) on the calculated correlation
- Less restrictive assumptions regarding the form of the relationship (so monotonic as well as linear associations are appropriate)
- Less restrictive distributional assumptions when testing the significance of the observed association (in other words, Normality of x or y is not required)

Sometimes, we may wish to test the null hypothesis that there is no relationship between the variables x and y

$$H_0: \rho_s = 0 \text{ where } \rho_s \text{ is the population equivalent of } r_s$$

For small samples, we can readily find tables of critical values to complete the test, but if n is not too small (say $n > 10$), we can plug r_s into the following t-statistic (which also works when testing the significance of the Pearson correlation, though under stronger parametric assumptions):

$$t = \frac{r\sqrt{n - 2}}{\sqrt{1 - r^2}}$$

and evaluate significance using the t_{n-2} distribution.

Assumptions and Conditions

Assumptions are rather minimal. Looking back at the assumptions for Pearson's correlation in Chapter 6, we see that the concerns expressed about nonlinearity and outlier effects have vanished. The **Quantitative Variables Condition** may be relaxed somewhat, as we could also apply Spearman's correlation in situations with ordinal responses—for example, if we wanted to correlate the opinions of adult Canadians about two different House of Commons bills, where the possible responses were "strongly disagree," "mildly disagree," "neutral," "mildly agree," or "strongly agree." We could still rank all the ordinal responses for each bill, though with a large number of ties.

In Chapter 23, we presented a test of significance for Pearson's correlation (equivalent to testing the regression slope). With Spearman's correlation we only need to be concerned with the **Independence Assumption** (of the pairs), and can ignore the other (distributional and linearity) assumptions mentioned in that chapter. Check the **Randomization Condition** that the individuals are or may be considered to be a representative sample from the population to which we want our conclusions to apply.

SPEARMAN'S RANK CORRELATION

For two quantitative or ordinal variables, we can measure the strength of monotonic association between them using Spearman's correlation.

To compute Spearman's correlation, rank the responses separately for each variable, and then compute Pearson's correlation for the two columns of rankings, or use the shortcut formula:

$$r_s = 1 - \frac{6 \sum d_i^2}{n(n^2 - 1)}$$

where each d_i is the different between the ranks for the ith observation on x and y.

Assuming the data are a simple random sample from some population, we can test:

H_0: No monotonic association in the population ($\rho_s = 0$).

To estimate the P-value, use tables of critical values for r_s or plug into the t-statistic discussed earlier (using $n - 2$ df):

$$t = \frac{r_s \sqrt{n - 2}}{1 - r_s^2} \text{ (if } n > 10)$$

For Example

Below is a scatterplot for a sample of 12 Toronto Blue Jay baseball players, where x = the number of home runs scored in 2009 and y = salary in 2009 in $10 000's (Data in **HomeRuns**):[7]

QUESTION: How would you characterize the relationship between home runs and salary? Below are the actual data.

Home Runs	Salary	Home Runs	Salary
36	259	13	240
15	469	7	85
35	41	9	40
12	110	2	50
19	250	8	243
16	795	0	41

Pearson's correlation for the data above is equal to 0.183. If we replace the data with their ranks, we get:

Home Run Rank	Salary Rank	Home Run Rank	Salary Rank
12	10	7	7
8	11	3	5
11	2.5	5	1
6	6	2	4
10	9	4	8
9	12	1	2.5

Using software to calculate Pearson's correlation for the two columns of ranks, we find Spearman's correlation = 0.518. Alternatively, one could plug into the formula for r_s to get

$$r_s = 1 - \frac{6 \sum d_i^2}{n(n^2 - 1)}$$

$$= 1 - \frac{6(2^2 + 3^2 + 8.5^2 + 0^2 + 1^2 + 3^2 + 0^2 + 2^2 + 4^2 + 2^2 + 4^2 + 1.5^2)}{12(12^2 - 1)}$$

$$= 1 - 0.481 = 0.519$$

We see that Pearson's $r = 0.18$, whereas Spearman's $r_S = 0.52$. There is some indication that the relation might not be very linear, and there are a few unusual observations. This limits the applicability of Pearson's r, which is noticeably smaller in value than Spearman's r. The latter is less affected by the unusual observations and pays no attention to any curvature present, but picks up on the moderately monotone positive relation (an increase in home runs is accompanied by an increase in salary most of the time—what a shock, eh?). The bigger value for r_S relays this message more clearly. (However, a test of significance only produces a P-value of 0.08, due to the small sample size.)

 Just Checking

3. For the (column-wise) paired data below

x:	1	2	3	4	5	6	7
y:	1	4	9	16	9	4	1

which correlation measure is more useful?

Section 28.5 Exercises

9. For which one of the following situations would a rank correlation be appropriate?
a) Judging whether the height of water at high tide at a marina increases more or less consistently as the phase of the moon changes from new to full. We'll take measurements over several lunar months, during each of which the moon goes through all of its eight phases.
b) Estimating the association between the value of homes and their age in a community known to have a number of very expensive mansions.
c) Testing whether the distributions of the ratings of a new distance-learning course were the same as rated by first-, second-, third-, and fourth-year students who had taken the course.

10. For which one of the following situations would a rank correlation be needed?
a) During a smartphone game's initial release, is there an association between the number of downloads and the number of hours from the time it is released?
b) The Richter scale measures the damage potential of earthquakes. Some scientists think it's best used for judging the relative size of earthquakes and not their absolute size. Are earthquakes around the Pacific "Ring of Fire" stronger in general than those around the Atlantic? Compare earthquakes over a recent five-year period.
c) The Mohs scale rates the hardness of minerals. If one mineral can scratch another, it's judged to be harder. Is hardness related to the weight of a one-by-one-by-one cm cube of the mineral?

WHAT CAN GO WRONG?

■ **Make sure that the data values are independent and from an appropriately randomized sample.** Though we avoid making unnecessary distributional assumptions in these tests, for valid inferences, we still need observations that are independent of each other (and from independent groups for the Wilcoxon Rank Sum or Kruskal-Wallis tests), with proper random sampling or randomization. Selecting survey data in clusters, for example, violates independence (though analysis is still possible, just more complicated), and convenience samples may be hard to generalize.

■ **Beware of small samples with low power.** With distribution-free procedures, power and sample size calculations are not possible, but we should still be aware that, as usual, tests based on small samples may let important treatment effects slip by undetected due to lack of power. Power calculations using a similar parametric test may provide some rough guidance on sample size.

■ **Be cautious about interpreting confidence intervals provided by statistical software for these procedures.** We have not discussed the confidence intervals for medians, or for differences in medians, that are often provided by statistical software in the output for these procedures. There are additional assumptions required for such confidence intervals to be valid (for example, similarly shaped distributions for the Wilcoxon Rank Sum test). Be sure to learn more about the particular procedure before attempting to interpret such an interval.

What Have We Learned?

There is no need to restrict ourselves solely to the well-known and oft-used "classical" or Normal distribution-based methods when making inferences. There are other means for extracting relevant evidence about a hypothesis from the data. If concerned about being able to use one of the common parametric inferential methods after checking the necessary assumptions or conditions, seek out another approach, with different or fewer assumptions—a nonparametric procedure.

- We can use rank transformations to replace parametric modelling of data in many common types of studies, and then perform a rank-based test of significance.
- A P-value—*is a P-value*—**is a P-value**. It is always interpreted the same way, as the weight of evidence against the null hypothesis, regardless of the test procedure.

Review of Terms

Rank sum	Sum of the ranks in a sample after replacing the responses with their respective rankings (p. 949).
Wilcoxon Rank Sum test	Nonparametric rank-based test that compares two distributions to assess whether one distribution has systematically larger values than the other, in an independent samples study. Also known as the Mann-Whitney test. (p. 950).
Kruskal-Wallis statistic	Used to test for systematic differences among any number of distributions in an independent samples study, after converting the data to ranks (p. 954).
Wilcoxon Signed Rank test	Nonparametric rank-based test for a systematic difference within pairs in a paired data study (p. 957).
Friedman statistic	Used to test for systematic differences among the treatments in a randomized block design, after converting the data to ranks within each block (p. 962).
Spearman's correlation	A measure of monotonic association between two quantitative or ordinal variables based on the ranks of the data values (p. 964).
Monotonic relationship	An association where one variable increases whenever the other variable increases, or decreases whenever the other variable increases (p. 964)

On the Computer NONPARAMETRIC METHODS

Statistical packages generally provide an assortment of nonparametric procedures. Often, they are listed separately under the heading "Nonparametrics" or something similar, or they may pop up as available options when you execute more common parametric procedures. Computer output is similar to the output for parametric tests, showing P-values and often confidence intervals (for medians or differences in medians). However, if you want to use the confidence interval, be sure to read up on the procedure, as additional data assumptions are required.

EXCEL

Nonparametric procedures are not included in Excel's standard data analysis toolkit, though there is a RANK function (or RANK.AVG in Excel 2010) available.

JMP

JMP includes nonparametric procedures as additional options (under the **red triangle** or in the dialogue box) once you have run the corresponding parametric procedure:

- For two independent samples, run **Analyze>Fit y by x**, as in Chapter 19. Click the red triangle to find **Nonparametric** options, including the **Wilcoxon (Rank Sum) Test**.

- To compare several independent groups, run **Analyze>Fit y by x**, as in Chapter 24, then click the red triangle for **Nonparametric** options. Choose **Wilcoxon Test** (this produces the Kruskal-Wallis generalization of the Wilcoxon test).

- For a paired data comparison, select **Analyze>Matched Pairs**, as in Chapter 20. Click the red triangle at the top of the analysis and select **Wilcoxon Signed Rank**. The Friedman test for blocked data is not available.

- For correlation analysis, choose **Analyze>Multivariate Methods>Multivariate**, enter the variables, hit **OK**, then click the red triangle and choose **Nonparametric Correlations>Spearman's p**.

MINITAB

- From the **Stat** menu, choose the **Nonparametrics** submenu. The **Mann-Whitney** test is the same as the Wilcoxon Rank Sum test.
- For the Wilcoxon Signed Rank test, first calculate the differences and store them in a column, then input this column to **1-Sample Wilcoxon**.
- Minitab does not list Spearman's rank correlation under **Nonparametrics**, so to calculate this, use **Data>Rank** and convert each data column to corresponding ranks. Then use **Stat>Basic Statistics>Correlation** to correlate the ranks, which will also provide a P-value (valid if n > 10).

COMMENTS

Stat>Tables>Cross Tabulation and Chi-Square>Other Stats>Correlation coefficients for ordinal categories will also calculate Spearman's correlation, but no P-value is reported.

R

- For a Wilcoxon test, use wilcox.test(y1, y2, paired=FALSE); change to paired=TRUE for the signed rank test.
- To do the Kruskal-Wallis test, use kruskal.test(y~A) similar to an ANOVA model.

SPSS

- Select **Analyze>Nonparametric Tests** from the menu, then choose your option according to your data structure (either **Independent Samples** or **Related Samples**).

COMMENTS

For independent samples, make sure your data are stacked. Note that in the output, SPSS refers to the Wilcoxon Rank Sum test by its alternate name, the Mann-Whitney test.

STATCRUNCH

Several nonparametric procedures are available under **Stat > Nonparametrics**

TI-83/84 PLUS

These procedures are not automated, though they can be executed via a series of steps described in the TI-83/84 Statistics Handbook.

Chapter Exercises

T indicates data is available in a saved file at the Companion Website at www.pearsoncanada.ca/deveaux

11. Course grades again Though the samples in the chapter-opening Statistics course grades example were small, use the Normal approximation to estimate the P-value for these data, and compare with the previously reported P-value. If the grade of A– were changed to A+, how would this affect the Wilcoxon test? How would it affect the *t*-test (after replacing letter grades with numerical values)?

12. Course grades amended Add two additional Biology students, both with an A grade, to the chapter-opening Statistics grade data. How do you think this will affect the P-value? Redo the test and report the P-value again using the table of exact critical values. Though the samples were small, use the Normal approximation to estimate the P-value for these data, and compare with the exact distribution P-value.

13. Music students Fourteen music students from two different schools gave performances and were ranked as follows (best first, worst last). Students from Lawrence School of Music are in bold.
Joe Bini **Jack Bob Shuang** John Sally **Mary** Mo **Celina** Zheng **Aditi** Rodrigo Astrud
Test for a significant difference in performance between students from the two schools.

14. Art students Thirteen art students from two different classes painted portraits, which were ranked as follows (best first, worst last). Artists from Andrey's class are in bold.
John Ben **Jen Bo Sharon** Joe Sam **Matt** May Zee **Ashish** Rod Amy
Test for a significant difference in performance between students from the two classes.

T 15. Buying from a friend In Chapter 19, we examined the effect of friendship on transactions where prices offered were compared for buying from a friend versus buying from a stranger. Use an appropriate nonparametric technique to compare the two conditions. Compare results with those from the parametric test executed in Chapter 19.[8] Here are the prices offered for a used camera in good condition:

Price Offered for a Used Camera ($)	
Buying from a Friend	Buying from a Stranger
275	260
300	250
260	175
300	130

[8]J. J. Halpern, "The Transaction Index: A method for standardizing comparisons of transaction characteristics across different contexts," *Group Decision and Negotiation*, 6: 557–572.

255	200
275	225
290	240
300	

T 16. Ontario and U.K. students, again Look back at Exercise 40 in Chapter 19, where we tested for a difference in mean travel time to school between Ontario and U.K. secondary school students. Use an appropriate nonparametric procedure to test for a difference, and compare your results with the parametric approach.

U.K students travel times (minutes)
45 5 4 15 50 20 20 20 20 20 25 35 15 30 20
10 45 10 3 60 25 20 5 15 5 15 17 30 40 20
10 30 10 15 20 10 15 17 10 25

Ontario students travel times (minutes)
30 10 8 30 5 8 7 15 10 35 15 10 25 22 20
25 30 10 25 8 15 18 25 15 10 25 5 2 5 25
20 15 47 20 20 13 20 5 15 12

T 17. Downloading again In Exercise 23 of Chapter 24, we examined how much of a difference time of day makes on the speed at which a university student can download files. Analyze the same data using an appropriate nonparametric procedure, and compare your results with the results from the one-way ANOVA.

Early (7 a.m.) Time (sec)	Evening (5 p.m.) Time (sec)	Late night (12 a.m.) Time (sec)
68	299	216
138	367	175
75	331	274
186	257	171
68	260	187
217	269	213
93	252	221
90	200	139
71	296	226
154	204	128
166	190	236
130	240	128
72	350	217
81	256	196
76	282	201
129	320	161

T **18. Analgesics again** In Exercise 24 of Chapter 24, we compared the effect on migraine pain of three different drug formulations. Analyze the data using an appropriate nonparametric procedure, and compare your results with the results from the one-way ANOVA.

Drug	Pain	Drug	Pain	Drug	Pain
A	4	B	6	C	6
A	5	B	8	C	7
A	4	B	4	C	6
A	3	B	5	C	6
A	2	B	4	C	7
A	4	B	6	C	5
A	3	B	5	C	6
A	4	B	8	C	5
A	4	B	6	C	5

19. Extreme values Considering the information in Exercise 18, if drug A were even more effective—say, if all the data values there were reduced by two (to 2, 3, 2, . . .)—how would the one-way ANOVA and the Kruskal-Wallis test be affected? In which case would the effect be greater, and why?

T **20. Extreme values again** Consider again the hand-washing data in this chapter and our Kruskal-Wallis test. Would doubling the values in the sample for hand-washing with water have a bigger effect on the Kruskal-Wallis test or on the ANOVA *F*-test? Explain. If we doubled the values in this sample once more, how would it affect the tests?

T **21. Washing again** In Exercise 18 of Chapter 25, we examined the effect of temperature setting and cycle length on cleanliness. Assume that temperature is the factor of interest, and that cycle length is of minor interest, similar to a blocking variable. Analyze the temperature effect using Friedman's test, and compare with results from the ANOVA in Chapter 25.

Temp	Cycle	Score
Cold-cold	Med long	3.7
Warm-hot	Med long	6.5
Cold-warm	Med long	4.9
Hot-hot	Med long	6.5
Cold-cold	Long	4.6
Warm-hot	Long	8.3
Cold-warm	Long	4.7
Hot-hot	Long	9.1
Cold-cold	Short	3.4
Warm-hot	Short	5.6
Cold-warm	Short	3.8
Hot-hot	Short	7.1
Cold-cold	Med short	3.1
Warm-hot	Med short	6.3
Cold-warm	Med short	5
Hot-hot	Med short	6.1

22. Competition Suppose we want to compare four different competitive conditions on the success of the grass species *Agropyron*. These conditions are different ways of introducing a competing species into the plots. We use four plots in one region, four in another region, and four in a third region. Conditions are assigned randomly in each region, and the regions may vary slightly in fertility level. The table displays (coded) biomass measurements for the 12 plots. Analyze the results with a parametric procedure and with a nonparametric procedure. If the P-values differ greatly, explain why.

	Condition 1	Condition 2	Condition 3	Condition 4
Region 1	20	11	25	4
Region 2	42	25	39	16
Region 3	53	32	58	19

T **23. Women's 1500-m skate** Consider the speed skating example discussed in Chapter 20.

Inner Lane		Outer Lane	
Name	**Time**	**Name**	**Time**
OLTEAN Daniela	129.24	(no competitor)	
ZHANG Xiaolei	125.75	NEMOTO Nami	122.34
ABRAMOVA Yekaterina	121.63	LAMB Maria	122.12
REMPEL Shannon	122.24	NOH Seon Yeong	123.35
LEE Ju-Youn	120.85	TIMMER Marianne	120.45
ROKITA Anna Natalia	122.19	MARRA Adelia	123.07
YAKSHINA Valentina	122.15	OPITZ Lucille	122.75
BJELKEVIK Hedvig	122.16	HAUGLI Maren	121.22
ISHINO Eriko	121.85	WOJCICKA Katarzyna	119.96
RANEY Catherine	121.17	BJELKEVIK Annette	121.03
OTSU Hiromi	124.77	LOBYSHEVA Yekaterina	118.87
SIMIONATO Chiara	118.76	JI Jia	121.85
ANSCHUETZ THOMS Daniela	119.74	WANG Fei	120.13
BARYSHEVA Varvara	121.60	van DEUTEKOM Paulien	120.15
GROENEWOLD Renate	119.33	GROVES Kristina	116.74
RODRIGUEZ Jennifer	119.30	NESBITT Christine	119.15
FRIESINGER Anni	117.31	KLASSEN Cindy	115.27
WUST Ireen	116.90	TABATA Maki	120.77

a) Apply the sign test to these data.
b) Apply the Wilcoxon Signed Rank test to these data.

c) Compare results from a) and b) with the *t*-test results in Chapter 20. Compare the P-values among these three procedures and, if possible, explain some of the differences.

T 24. Wine tasting For the data presented in this chapter about comparing two wines, add two additional tasters to the data, with one giving the Russian River and Niagara wines ratings of 8 and 6, respectively, and the other giving them ratings of 9 and 5, respectively.
   a) Execute the *t*-test for these data. Is the P-value here trustworthy?
   b) Execute the sign test and the Wilcoxon Signed Rank test.
   c) Compare P-values among the three procedures, and explain the differences.

T 25. Gasoline again Using an appropriate nonparametric procedure, tackle the hypothesis posed in Exercise 34a of Chapter 20, where we studied whether premium gas produces superior mileage over regular gas. Compare your results with the results from the parametric approach.

Here are the results (litres per 100 km):

Car #	1	2	3	4	5	6	7	8	9	10
Regular	14.7	11.8	11.5	10.7	10.2	10.7	8.7	9.4	8.7	8.4
Premium	12.4	10.7	9.8	9.8	9.4	9.4	9.0	9.0	8.4	7.4

T 26. Job satisfaction again Using an appropriate nonparametric procedure, tackle the hypothesis posed in Exercise 31b of Chapter 20 about the effects of an exercise program on job satisfaction, and compare your results with the results from the parametric approach.

Worker Number	Job Satisfaction Index Before	After
1	34	33
2	28	36
3	29	50
4	45	41
5	26	37
6	27	41
7	24	39
8	15	21
9	15	20
10	27	37

T 27. Life expectancy 2013, again Consider again Exercise 45 in Chapter 8, where we displayed births per woman and life expectancy for 26 countries:

Country	Birth Rate (births/ 1000 population)	Life Expectancy
Argentina	17.12	76.1
Barbados	12.1	77
Belize	25.58	76.3
Canada	10.28	81.1
Chile	14.12	79.3
Colombia	16.98	73.9
Costa Rica	16.25	79.4
Cuba	9.92	79.3
Dominican Republic	19.21	73.6
Ecuador	19.23	75.8
El Salvador	17.12	72.4
Guatemala	25.99	71.4
Guyana	16.31	70.2
Haiti	23.35	62.4
Honduras	24.16	73.4
Jamaica	18.65	73.3
Mexico	18.61	77.1
Nicaragua	18.77	74.3
Panama	18.91	76.3
Paraguay	16.95	72.7
Saint Kitts and Nevis	13.79	73.3
Saint Lucia	14.19	74.8
Saint Vincent and the Grenadines	14.12	72.5
Trinidad and Tobago	14.07	70.3
Uruguay	13.28	77.2

   a) Calculate Spearman's correlation and Pearson's correlation for all 26 countries. Remove the obvious outlier, and recompute the two correlations.
   b) Explain any differences observed between Pearson's correlation and Spearman's correlation in part a).

T 28. Gestation, again Look back at Exercise 39 in Chapter 8, which displays gestation periods and life expectancies for 18 mammal species.
   a) Calculate Spearman's correlation and Pearson's correlation.
   b) Remove the humans, and recompute the two correlations.
   c) Compare the two correlations in a) and b), and explain any big differences between them.
   d) Now remove the elephants too. If you were to recompute the correlations, which do you predict will change the most (and why)? Recompute both to check your prediction.

T **29. Antidepressants, again** Consider our example in Exercise 22, Chapter 6, in which we discussed drug improvement scores versus placebo improvement scores. Would you expect Spearman's or Pearson's correlation to be bigger? Why? Pearson's correlation was given as 0.898. Compute Spearman's correlation.

30. Pace of course Survey researchers often ask people to rank their opinions or attitudes on a scale that lacks units. (Common scales run from 1 to 5, 1 to 7, or 1 to 10.) For example, we might ask:

"How would you assess the pace of your Statistics course so far?"

1 = Way too slow 2 = A little too slow 3 = About right
4 = A little too fast 5 = Way too fast

Scales of this sort that attempt to measure attitudes numerically are called Likert scales, after their developer, Rensis Likert. Likert scales have order. The higher the number, the faster you think the course has gone, but a 4 doesn't necessarily mean *twice* as fast as 2. Suppose we take a sample of students from this course and record the GPA, major field of study, and opinion about pace for each student.

a) What measure would you use to estimate the strength of the association between perceived pace and GPA? Explain briefly.

b) How would you test for an association between perceived pace and major field of study?

T **31. Egyptians again** Consider the data from Exercise 24 in Chapter 19 comparing skull breadths of ancient Egyptians from two different eras. Use an appropriate nonparametric technique to test for a difference between the two eras. Compare your results with those from the appropriate parametric test.

Maximum Skull Breadth (mm)			
4000 B.C.E.	**4000 B.C.E.**	**200 B.C.E.**	**200 B.C.E.**
131	131	141	131
125	135	141	129
131	132	135	136
119	139	133	131
136	132	131	139
138	126	140	144
139	135	139	141
125	134	140	130
131	128	138	133
134	130	132	138
129	138	134	131
134	128	135	136
126	127	133	132
132	131	136	135
141	124	134	141

T **32. Cuckoos again** In Exercise 50 of Chapter 19, we looked at cuckoo egg lengths for three species. Use an appropriate nonparametric approach to test for a difference in egg lengths between each pair of species. Compare your results with results from the appropriate parametric test for the sparrows versus robins comparison.

Cuckoo Egg Length (mm)		
Foster Parent Species		
Sparrow	**Robin**	**Wagtail**
20.85	21.05	21.05
21.65	21.85	21.85
22.05	22.05	21.85
22.85	22.05	21.85
23.05	22.05	22.05
23.05	22.25	22.45
23.05	22.45	22.65
23.05	22.45	23.05
23.45	22.65	23.05
23.85	23.05	23.25
23.85	23.05	23.45
23.85	23.05	24.05
24.05	23.05	24.05
25.05	23.05	24.05
	23.25	24.85
	23.85	

33. Using cards Using a deck of cards, how could you estimate the P-value for the Wilcoxon Rank Sum test where the two rank sums are 11 and 34 for two independent samples of size four and five, respectively?

34. Marooned If you were marooned on a desert island with no books but with plenty of time, how would you estimate the P-value for the results we obtained in our chapter-opening grades example, using no math (but pretend there are no ties)?

35. Exact P-values How many distinct hands are possible if you were to deal yourself five cards from a deck of 11 distinct cards (denoted $_{11}C_5$, or "11 choose 5" and defined in Chapter 13 for the Binomial model as the number of ways of choosing the trials in which to place the successes)? What would be the exact P-value if the rank sum for the smaller sample in this chapter's course grades example were equal to 15 (which could only happen if this sample had the ranks 1, 2, 3, 4, and 5)? What would be the P-value for a rank sum of 16 (count the ways that the ranks will sum to 15 or 16)? For a rank sum of 17?

36. More P-values How many distinct hands are possible if you were to deal yourself four cards from a deck of nine distinct cards (denoted $_9C_4$ and defined in Chapter 13 for the Binomial model)? What would be the exact P-value if the rank sums for two independent samples of size four and size five were 10 and 35, respectively?

Just Checking ANSWERS

1. **a.** Sum of all ranks $= 7(8)/2 = 28$.

   b. Biggest $= 7 + 6 + 5 + 4 = 22$, smallest $= 1 + 2 + 3 + 4 = 10$

   c. Expected value $= (4/7) \times 28 = 16$

2. We would have to have ranks 1, 2, and 3 in one sample, 4, 5, and 6 in another sample, etc., to produce rank sums of 6, 15, 24, and 33.

   So $H = \dfrac{12}{12 \times 13} \times (6^2/3 + 15^2/3 + 24^2/3 + 33^2/3) - 3(13) = 10.38$ and $0.01 < P < .025$ from the chi-square distribution with 3 df.

3. Neither, since the graph goes up then down, which is nonlinear and not monotonic to any degree. Both correlations are zero. Neither picks up the strong (quadratic) relationship.

The Bootstrap

Bootstrapping is like breathing your own carbon dioxide.

—*Chairman George*

"**P**ulling oneself up by the bootstraps" may be a physical impossibility in the real world, but in the statistical world we can pull ourselves up from the data itself to inference about the larger population without using mathematical power (though we do need to tie our bootstraps to a computer to help pull us up). Statistics Canada and many other researchers increasingly rely on *bootstrapping* to produce standard errors, confidence limits, and P-values (the basic tools of inference), particularly for situations in which no textbook answer can be found. In the late 1980s, as cheap computer power was growing quickly, Bradley Efron and others[1] saw a way to harness the power of computer simulation to "bootstrap" answers. While there are various refinements possible, we will just introduce the basic idea, with application to estimating standard errors and confidence intervals. Unfortunately, "canned" bootstrapping capabilities vary greatly among statistical software packages at this time.

Connections

In the previous chapter, we learned about alternative approaches to inference which use rank transformation of the data. In this chapter, we'll learn another nonparametric approach to inference, called the *Bootstrap*, a data-based computer intensive simulation method that substitutes computing power for the power of mathematical derivation (which often falls short when we encounter complex studies, estimators, and models).

The idea harks back to Chapter 14 where we studied sampling distributions by drawing repeated samples from hypothetical populations. In practice though, we cannot repeatedly sample the population, so we'll do next best—we'll re-sample (with replacement) the sample—which serves as a proxy for the population—over and over, until voila, we see the *bootstrapped* sampling distribution! And as you've learned, the sampling distribution provides a basis for inference.

The full chapter is available in the etext in MyStatLab.

[1]Including Efron's PhD student and former University of Toronto student and professor Rob Tibshirani.

972

APPENDIX A

Answers

Here are the answers to the odd-numbered exercises for the chapters and the reviews. As we said in Chapter 1, the answers provide outlines of the complete solution. Your solution should follow the model of the Step-By-Step examples, where appropriate. You should explain the context, show your reasoning and calculations, and draw conclusions. For some problems, what you decide to include in an argument may differ somewhat from the answers here. However, make sure the numerical part of your answer matches the numbers in the answers shown, aside from small differences due to rounding errors or differences in the definitions of quartiles (which may also affect boxplots).

Chapter 1

1. Retailers, and suppliers to whom they sell the information, will use the information about what products consumers buy to target their advertisements to customers more likely to buy their products.

3. The individual games.

5. a) If we compare the percentage of first-graders who can tie their own shoes to the percentage of second-graders who can tie their own shoes, grade-level is treated as categorical.
   b) If studying the relationship between grade-level and height, grade-level will be quantitative.

7. Categorical.

9. Quantitative.

11. Answers will vary.

13. *Who:* 40 undergraduate women.
    What: Whether or not the women could identify the sexual orientation of men based on a picture.
    Why: To see if ovulation affects a woman's ability to identify sexual orientation of a male.
    How: Showing very similar photos to the women, with half gay.
    Variables: Categorical variable: "He's gay" or "He's not gay."

15. *Who:* 48 China/India/Chindia funds listed at globeinvestor.com.
    What: 1 month, 1 year, and 5 year returns for each fund.
    When: The most recent periods of time.
    Where: globeinvestor.com website.
    Why: To compare investment returns, for future investment decisions.
    How: globeinvestor.com uses reports from the fund companies.
    Variables (Q=quantitative, C=categorical): 1 month return (Q, percentage), 1 year return (Q, percentage), 5 year return, annualized (Q, percentage).

17. *Who:* All airline flights in Canada.
    What: Type of aircraft, number of passengers, whether departures and arrivals were on schedule, and mechanical problems.
    When: This information is currently reported.
    Where: Canada.
    Why: This information is required by Transport Canada and the Canadian Transportation Agency.
    How: Data is collected from airline flight information.
    Variables: Type of aircraft, departure and arrival timeliness, and mechanical problems are categorical variables, and number of passengers is a quantitative variable.

19. *Who:* Automobiles.
    What: Make, country of origin, type of vehicle, and age of vehicle (probably in years).
    When: Not specified.
    Where: A large university.

Why: Not specified.
How: A survey was taken in campus parking lots.
Variables: Make, country of origin, and type of vehicle are categorical variables, and age of vehicle is a quantitative variable.

21. *Who:* Workers who buy coffee in an office.
    What: Amount of money contributed to the collection tray.
    Where: Newcastle.
    Why: To see if people behave more honestly when feeling watched.
    How: Counting money in the tray each week.
    Variables: Amount contributed (pounds) is a quantitative variable.

23. *Who:* 54 bears.
    What: Weight, neck size, length (no specified units), and sex.
    When: Not specified.
    Where: Not specified.
    Why: Since bears are difficult to weigh, the researchers hope to use the relationships between weight, neck size, length, and sex of bears to estimate the weight of bears, given the other, more observable features of the bear.
    How: Researchers collected data on 54 bears they were able to catch.
    Variables: Weight, neck size and length are quantitative variables, and sex is a categorical variable. No units are specified for the quantitative variables.
    Concerns: The researchers are (obviously!) only able to collect data from bears they were able to catch. This method is a good one, as long as the researchers believe the bears caught are representative of all bears, in regard to the relationships between weight, neck size, length, and sex.

25. *Who:* Doughnut types for sale at Tim Hortons.
    What: Various nutritional characteristics (see variables below).
    When: Not stated, but presumably the measurements were taken recently.
    Where: Tim Hortons website.
    Why: To help customers make good nutritional choices.
    How: Not specified, but presumably at some specialized food analysis type lab.
    Variables: All eight variables are quantitative: Number of calories (kcal/s), amounts of trans fat (g), total fat (g), sodium (mg), sugar (g), protein (g), % daily value of iron (percentage), and % daily value of calcium (percentage). Units (found at the website) are all per serving.

27. *Who:* 882 births.
    What: Mother's age (in years), length of pregnancy (in weeks), type of birth (Caesarean, induced, or natural), level of prenatal care (none, minimal, or adequate), birth weight of baby (unit of measurement not specified, but probably pounds and ounces), gender of baby (male or female), and baby's health problems (none, minor, major).
    When: 1998–2000.
    Where: Large city hospital.

Why: Researchers were investigating the impact of prenatal care on newborn health.
How: It appears that they kept track of all births in the form of hospital records, although it is not specifically stated.
Variables: There are three quantitative variables: mother's age, length of pregnancy, and birth weight of baby. There are four categorical variables: type of birth, level of prenatal care, gender of baby, and baby's health problems.

29. *Who:* Experiment volunteers.
What: Herbal cold remedy or sugar solution, and cold severity.
When: Not specified.
Where: Major pharmaceutical firm.
Why: Scientists were testing the efficacy of an herbal compound on the severity of the common cold.
How: The scientists set up a controlled experiment.
Variables: Type of treatment (herbal or sugar solution) is categorical, and severity rating is quantitative.
Concerns: The severity of a cold seems subjective and difficult to quantify. Also, the scientists may feel pressure to report negative findings about the herbal product.

31. *Who:* Streams.
What: Name of stream, substrate of the stream (limestone, shale, or mixed), acidity of the water (measured in pH), temperature (in degrees Celsius), and BCI (unknown units).
When: Not specified.
Where: Upstate New York.
Why: Research is conducted for an ecology class.
How: Not specified.
Variables: Name and substrate of the stream are categorical variables, and acidity, temperature, and BCI are quantitative variables.

33. *Who:* 41 refrigerators.
What: Brand, cost (probably in dollars), size (in cu. ft.), type, estimated annual energy cost (probably in dollars), overall rating, and repair history (in percent requiring repair over the past five years).
When: 2002–2006.
Where: United States.
Why: The information was compiled to provide information to the readers of *Consumer Reports*.
How: Not specified.
Variables: Brand, type, and overall rating are categorical variables. Cost, size, estimated energy cost, and repair history are quantitative variables.

35. *Who:* Kentucky Derby races.
What: Year, winner, jockey, trainer, owner, and time (in minutes, seconds, and hundredths of a second).
When: 1875–2015.
Where: Churchill Downs, Louisville, Kentucky.
Why: It is interesting to examine the trends in the Kentucky Derby.
How: Official statistics are kept for the race each year.
Variables: Date, winner, jockey, trainer, and owner are categorical variables. Duration is a quantitative variable.

Chapter 2

1. Subcompact and mini 0.1128
Compact 0.3163
Intermediate 0.3380
Full 0.2193
Unknown 0.0137

3. 1996: B, 2006: C, 2008: A, 2012: D

5. a) 170/348 = 48.9%
b) 41/348 = 11.8%
c) 41/170 = 24.1%
d) 41/90 = 45.6%

7. Answers will vary.

9. Answers will vary.

11. The relative frequency distribution of the forest fires is shown below:

Cause of fire	Percentage
Lightning	46.1%
Human activities	51.4%
Unknown	2.5%

Most forest fires are caused by human activities and lightning. Only 2.5% of forest fires are due to unknown causes.

13. According to the Monitoring the Future study, teen smoking brand preferences differ somewhat by region. Although Marlboro is the most popular brand in each region, with about 58% of teen smokers preferring this brand in each region, teen smokers from the South prefer Newports at a higher percentage than teen smokers from the West, with 22.5% of teen smokers preferring this brand, compared to only 10.1% in the South. Camels are more popular in the West, with 9.5% of teen smokers preferring this brand, compared to only 3.3% in the South. Teen smokers in the West are also more likely to have no particular brand than teen smokers in the South. 12.9% of teen smokers in the West have no particular brand, compared to only 6.7% in the South. Both regions have about 9% of teen smokers that prefer one of over 20 other brands.

15. a) The bar chart shows that grounding and collision are the most frequent causes of oil spills, whereas hull or equipment failures and fire & explosions are the least likely causes of oil spills. Relatively few oil spills have unknown causes.
b) A pie chart seems appropriate as well.
c) Much easier to see from the bar chart.

17. There's no title, the percentages total only 93%, and the three-dimensional display distorts the sizes of the regions.

19. a) A bar chart would be appropriate. A pie chart would not because these are counts and not fractions of a whole.
b) The *Who* for these data is athletic trainers who used cryotherapy, which should be a cause for concern. A trainer who treated many patients with cryotherapy would be more likely to have seen complications than one who used cryotherapy rarely. We would prefer a study in which the *Who* referred to patients so we could assess the risks of each complication.

21. a) The relative frequency distribution of quadrant location is given below. Not all proportions are equal. In particular, the relative frequency for Quadrant 4 is approximately twice the other frequencies.

Quadrant	Quadrant 1	Quadrant 2	Quadrant 3	Quadrant 4
Relative Frequency	0.18	0.21	0.22	0.39

b) The relative frequency distribution of quadrant location is given below. There seems to be some similarity with the relative frequency distribution of quadrant location given in part a). For example, Quadrant 4 has the highest relative frequency and Quadrant 1 has the lowest. However the values don't look very similar.

Quadrant	Quadrant 1	Quadrant 2	Quadrant 3	Quadrant 4
Relative Frequency	0.12	0.24	0.28	0.36

23. a) The females in this course were 45.5% Liberal, 46.8% Moderate, and 7.8% Conservative.
b) The males in this course were 43.5% Liberal, 38.3% Moderate, and 18.3% Conservative.
c) A segmented bar chart comparing the distributions is below.

d) Politics and sex do not appear to be independent in this course. Although the percentage of Liberals was roughly the same for each sex, females had a greater percentage of Moderates and a lower percentage of Conservatives than males.

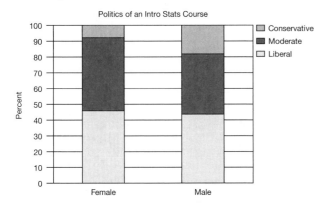

Politics of an Intro Stats Course

25. a) 68.1%   b) 30.1%   c) 94.4%   d) 74.1%

e) If language knowledge were independent of Province, we would expect the percentage of French-speaking residents of Quebec to be the same as the overall percentage of Canadians who speak French. Since 30.1% of all Canadians speak French while 94.4% of residents of Quebec speak French, there is evidence of an association between language knowledge and province.

27. a) 9.3%   b) 24.7%   c) 80.8%

d) No, there appears to be no association between weather and ability to forecast weather. On rainy days, his forecast was correct 79.4% of the time. When there was no rain, his forecast was correct 81.0% of the time.

29. a) 4.9%   b) 51.5%

c) In each age group, there are more male drivers than female drivers. The age group in which the percentage of male drivers is closest to 50% is 35–44 (50.7%). The age group in which the percentage of male drivers is furthest from 50% is under 16 (54.2%), followed by 65 and over (53.1%).

d) There appears to be a slight association between age and sex of drivers. Younger and older drivers are slightly more likely to be male.

31. These data provide no evidence that Prozac might be helpful in treating anorexia. About 71% of the patients who took Prozac were diagnosed as "Healthy," while about 73% of the patients who took a placebo were diagnosed as "Healthy." Even though the percentage was higher for the placebo patients, this does not mean that Prozac is hurting patients. The difference between 71% and 73% is not likely to be statistically significant.

33. a) The marginal distribution of genotype is given below:

Genotype	Marginal percentage
GG	42.71
GT	45.08
TT	45.08

b) The conditional distributions of genotype for the four categories of smokers are given in columns 2–5 of the table below.

Genotype	Cigarettes per day				
	1–10	11–20	21–30	31 and more	All
GG	48.06	42.60	38.32	36.75	42.71
GT	42.96	44.75	47.39	48.28	45.08
TT	8.99	12.65	14.29	14.98	12.21
All	100.00	100.00	100.00	100.00	100.00

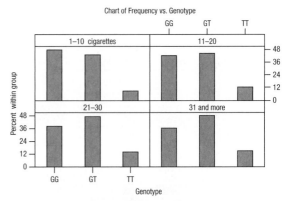

Chart of Frequency vs. Genotype

c) Though not a very noticeable difference, the percentages of smokers with genotype GT (also TT) are slightly higher among heavy smokers. However, this is only an observed association. This does not prove that presence of the T allele increases susceptibility to nicotine addiction. We cannot conclude that this increase was caused by the presence of the T allele. There can be many factors associated with the presence of the T allele, and some of these factors might be the reason for the increase in susceptibility to nicotine addiction.

35. Yes, the risk of bone fractures is about twice as high (10%) among the people who were taking SSRIs than among those who were not (5%).

37. a) The two-way table and the conditional distributions (percentages) of 'car accident' (crash or non-crash) for cell phone users and non-cell phone users are given below. The proportion of crashes is higher for cell phone users.

	Cell phone user	Non-cell phone user	All
Crash	20	10	30
Non-crash	58	92	150
All	78	102	180

Cell Contents: Count

	Cell phone user	Non-cell phone user	All
Crash	25.64	9.80	16.67
Non-crash	74.36	90.20	83.33
All	100.00	100.00	100.00

Cell Contents: % of Column

b) On the basis of this study, we cannot conclude that the use of a cell phone increases the risk of a car accident. This is only an observed association between cell phone usage and the risk of car accidents. We cannot conclude that the higher proportion of accidents was caused by the use of a cell phone. There can be other factors common to cell phone users that contribute to causing accidents.

39. a)

Blood Pressure	Total	Percent
Low	95	20.0%
Normal	232	48.9%
High	147	31.0%

b)

Blood Pressure	Under 30	Percent
Low	27	27.6%
Normal	48	49.0%
High	23	23.5%

Blood Pressure	30–49	Percent
Low	37	20.7%
Normal	91	50.8%
High	51	28.5%

Blood Pressure	Over 50	Percent
Low	31	15.7%
Normal	93	47.2%
High	73	37.1%

c)

d) As age increases, the percent of employees with high blood pressure increases. On the other hand, the percent of employees with low blood pressure decreases.

e) No, but it gives an indication that it might. There might be additional reasons that explain the differences in blood pressures.

41. a) The second column includes some individuals in the 3rd, 4th, and the 5th columns, so it is not a standard contingency table.

b) Use "Indigenous population not included in columns 3, 4 and 5" (or call them "other Indigenous populations").

c) 0.08% (approx.)   d) 1.93% (approx.)   e) 4.24%   f) 1.95%

g) 85.39%   h) 98.94%   i) 45.54%   j) 3.63%

k) A table of percentages of total provincial population for each Indigenous identity group (Inuit, Metis, N.A. Indian) for NL, Ontario, Saskatchewan, and Alberta is given below. The second table is a bit easier to make using MINITAB. The side-by-side bar charts below show that Saskatchewan has the highest proportion of N.A. Indian and Metis proportions. Ontario, Saskatchewan and Alberta have very small proportions of Inuit.

Region	Percent N.A. Indian	Percent Metis	Percent Inuit
Newfoundland and Labrador	3.81%	1.51%	1.24%
Ontario	1.59%	0.68%	0.03%
Saskatchewan	10.23%	5.20%	0.03%
Alberta	3.27%	2.71%	0.06%

Group	Newfoundland and Labrador	Ontario	Saskatchewan	Alberta
N.A. Indian	3.81%	1.59%	10.23%	3.27%
Metis	1.51%	0.68%	5.20%	2.71%
Inuit	1.24%	0.03%	0.03%	0.06%

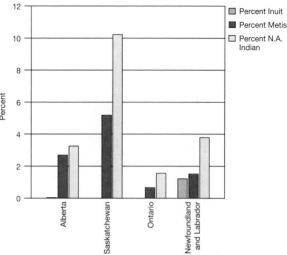

Percent Indigenous Group by Province

43. a) The marginal totals have been added to the table:

		Discharge delayed		
		Large Hospital	**Small Hospital**	**Total**
Procedure	Major surgery	120 of 800	10 of 50	130 of 850
	Minor surgery	10 of 200	20 of 250	30 of 450
	Total	130 of 1000	30 of 300	160 of 1300

160 of 1300, or about 12.3% of the patients had a delayed discharge.

b) Yes. Major surgery patients were delayed 130 of 850 times, or about 15.3% of the time. Minor Surgery patients were delayed 30 of 450 times, or about 6.7% of the time.

c) Large Hospital had a delay rate of 130 of 1000, or 13%. Small Hospital had a delay rate of 30 of 300, or 10%. The small hospital has the lower overall rate of delayed discharge.

d) Large Hospital: Major Surgery 15% delayed and Minor Surgery 5% delayed. Small Hospital: Major Surgery 20% delayed and Minor Surgery 8% delayed. Even though small hospital had the lower overall rate of delayed discharge, the large hospital had a lower rate of delayed discharge for each type of surgery.

e) No. While the overall rate of delayed discharge is lower for the small hospital, the large hospital did better with *both* major surgery and minor surgery.

f) The small hospital performs a higher percentage of minor surgeries than major surgeries. 250 of 300 surgeries at the small hospital were minor (83%). Only 200 of the large hospital's 1000 surgeries were minor (20%). Minor surgery had a lower delay rate than major surgery (6.7% to 15.3%), so the small hospital's overall rate was artificially inflated. Simply put, it is a mistake to look at the overall percentages. The real truth is found by looking at the rates after the information is broken down by type of surgery, since the delay rates for each type of surgery are so different. The larger hospital is the better hospital when comparing discharge delay rates.

45. a) 42.6%

b) A higher percentage of males than females were admitted: males: 47.2% to females: 30.9%.

c) Program 1: Males 61.9%, Females 82.4%
Program 2: Males 62.9%, Females 68.0%
Program 3: Males 33.7%, Females 35.2%
Program 4: Males 5.9%, Females 7.0%

d) The comparisons in part c) show that males have a *lower* admittance rate in every program, even though the overall rate shows males with a higher rate of admittance. This is an example of Simpson's paradox.

Chapter 3

1. a) Bar chart is for categorical data, whereas histogram is for quantitative data.

Bar chart uses gaps to separate categories and a gap (or gaps) in histogram indicate that there is a bin (or bins) with no values.

b) A stem-and-leaf plot is like a histogram but it shows the individual values.

It is easier to make by hand for data sets with small number of data.

3. Probably it will be unimodal and skewed to the high end.

The reason is that the higher the individual income, the less number of individuals will be able to achieve it.

5. 80.5

7. 11

9. The five number summary is: {14, 22, 26, 29, 34}. The boxplot is:

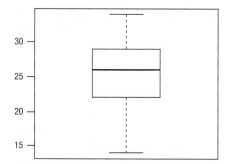

11. The mean will be higher. The distribution is unimodal and skewed to the right, so the mean will be pulled by the tail toward the higher values.

13. The IQR will not be affected at all. The standard deviation will increase.

15. Answers will vary.

17. Answers will vary.

19. a) Unimodal (near 0) and skewed. Many seniors will have 0 or 1 speeding ticket(s). Some may have several, and a few have more than that.

b) Probably unimodal and slightly skewed to the right. It is easier to score 15 strokes over the mean than 15 strokes under the mean.

c) Probably unimodal and symmetric. Weights may be equally likely to be over or under the average.

d) Probably bimodal. Men's and women's distributions may have different modes. It may also be skewed to the right, since it is possible to have very long hair, but hair length can't be negative.

21. a) Bimodal. Looks like two groups. Modes are near 6% and 46%. No real outliers.

b) Looks like two groups of cereals, a low-sugar and a high-sugar group.

23. a) 22.7% b) 47.3%

c) The distribution is symmetric. The centre (the median) is around 32. The scores range from 0–60, but the few scores close to zero may be outliers (with a gap of just one bin we might not be able to conclude they are clear outliers, but they are somewhat unusual compared to the rest of the scores).

25. a) The distribution is right skewed and so it is logical to expect the mean to be greater than the median. The median is in the interval 0.45–0.55.

b) The distribution is right skewed and could be bimodal. The median is between 0.45 and 0.55. The values of the percentage of rejected ballots range from 0–2.5 (approx.). The largest values (after 1.8) could be outliers, so might warrant further investigation.

27. a) $1001.50 b) $1025, $850, $1200 c) $835, $350

29. a) Median will be unaffected. The mean will be larger.

b) The range and standard deviation will increase; IQR will be unaffected.

31. a) The standard deviation will be larger for set 2, since the values are more spread out. SD(set 1) = 2.2, SD(set 2) = 3.2.

b) The standard deviation will be larger for set 2, since 11 and 19 are farther from 15 than are 14 and 16. Other numbers are the same. SD(set 1) = 3.6, SD(set 2) = 4.5.

c) The standard deviation will be the same for both sets, since the values in the second data set are just the values in the first data set + 80. The spread has not changed. SD(set 1) = 4.2. SD(set 2) = 4.2.

33. a) Mean $525, median $450.

b) Two employees earn more than the mean.

c) The median because of the outlier.

d) The IQR will be least sensitive to the outlier of $1200, so it would be the best to report. But really no one simple measure of spread here can provide as good a summary of spread as the 5-number summary.

35. The stem and leaf plot, a dotplot, and the five-number summary (plus mean) for these data are given below. The distribution looks slightly right-skewed (mean larger than median). The median number of slot machines is 428. The interquartile range is 731 – 299 = 432.

Stem and leaf plot:

```
 1 | 7
 2 | 0035
 3 | 0033
 4 | 0224
 5 | 3
 6 | 000
 7 | 0678
 8 | 66
 9 |
10 | 0
```

(6 | 0 means 600 slot machines)

Dotplot:

Numbers of Slot Machines

5-number summary: numbers of slot machines

Min.	1st Qu.	Median	3rd Qu.	Max.
170	299	428	731	1000

37. The histogram shows some low outliers in the distribution of height estimates. These are probably poor estimates and will pull the mean down. The median is likely to give a better estimate of the professor's true height.

39. The distribution of the number of home runs hit by Joe Carter during the 1983–1998 seasons is skewed to the left, with a typical number of home runs per season in the high 20s to low 30s. The season in which Joe hit no home runs looks to be an outlier. With the exception of this no home run season, Joe's total number of home runs per season was between 13 and 35. The median is 27 home runs.

41. a) A dotplot of the number of hurricanes each year from 1944 through 2015 is displayed. Each dot represents a year in which there were that many hurricanes.

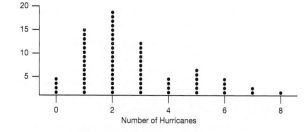

Number of Hurricanes

b) The distribution of the number of hurricanes per year is unimodal and skewed to the right, with centre around 2 hurricanes per year. The number of hurricanes per year ranges from 0 to 8. There are no outliers. There may be a second mode at 5 hurricanes per year, but since there were only 6 years in which 5 hurricanes occurred, this may simply be natural variability.

43. a) This is not a histogram. The horizontal axis should the number of home runs per year, split into bins of a convenient width. The vertical axis should show the frequency; that is, the number of years in which Carter hit a number of home runs within the interval of each bin. The display shown is a bar chart/time plot hybrid that simply displays the data table visually. It is of no use in describing the shape, centre, spread, or unusual features of the distribution of home runs hit per year by Carter.

b) The histogram is shown below.

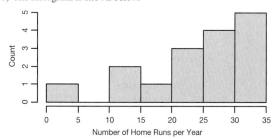

45. The distribution of the pH readings of water samples in Allegheny County, Pennsylvania, is bimodal. A roughly uniform cluster is centred on a pH of 4.4. This cluster ranges from pH of 4.1 to 4.9. Another smaller, tightly packed cluster is centred on a pH of 5.6. The two readings in the middle seem to belong to neither cluster.

Acidity of Water Samples in Allegheny County, Penn.

47. a)

```
−0 | 421
 0 | 000000001222244
 0 | 789
```

Percent Increase (rounded)

(0 | 9 means 9 percent)

b) Minimum = −4.0. Q1 = 0.0. Median = 0. Q3 = 3.0. Maximum = 9.0

c) The mean of the data is approximately 1.6. The mean is larger than the median because the distribution is slightly right-skewed.

d) The percentage increases in price range from −4% to 9%. The median increase is 0% (actually, this could be anywhere between 0 and 1% due to the truncation in the display). The distribution is slightly right-skewed.

e) The largest increases are in the Greater Toronto Area and Niagara regions of Ontario, and in British Columbia.

49. a) mean = 6.4 median = 7   10% trimmed mean = 6.875

b) 5% trimmed mean = 4.5.

c) The data in part (a) is left skewed. As a result, the median is bigger than the trimmed mean, which is bigger than the mean. The data in part (b) is right skewed, so the median is smaller than the trimmed mean, which is smaller than the mean.

51. Histogram bins are too wide to be useful.

53. Zip codes are categorical data, not quantitative. The summary statistics are meaningless. A bar chart would be more appropriate.

55. a)

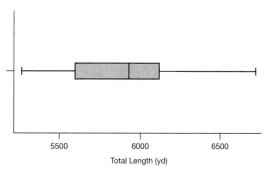

b) Between 5585.75 yd and 6131 yd.

c) Mean and SD, since this is roughly unimodal and symmetric.

d) The distribution of the lengths of all the golf courses in Vermont is roughly unimodal and symmetric. The mean length of the golf courses is approximately 5900 yards. Vermont has golf courses anywhere from 5185 yards to 6796 yards long. There are no outliers in the distribution.

e) Close to empirical rule for bell-shaped distributions.

57. a) The 5-number summary of the national averages is given below:

Min.	1st Qu.	Median	3rd Qu.	Max.
419	487	496.5	514	546

The IQR, mean, and the standard deviation of the national averages is given below:

Variable	Mean	StDev	IQR
Avg. Score	495.7	29.66	27

The distribution of average scores is left-skewed. The long left tail pulls the mean down toward the smaller values (in fact, three of these small values are smaller than Q1 − 1.5 × IQR, and are thus outliers). The outliers attract the mean toward them, whereas the median is resistant to outliers.

b) Since there are outliers in the data set, the 5-number summary is better than the mean & standard deviation. The values in the 5-number summary are resistant to outliers. The mean and the standard deviation are not resistant measures.

c) Thirty-four countries participated in the program. The highest national average is 546 and the lowest is 418. The median national average is 496.3. The interquartile range is 27.2. Twenty-five percent (i.e., 9 countries) of the participating countries had a national average

487.1 or below, and at least 25% of the countries had a national average of 514.34 or above. Canada's national average (which is 526.8)
is in the top 25% of all participating countries, more specifically, the
5th highest of all participating countries.

d) About two-thirds of students scored between 439 and 615.
Only about 5% scored less than 351 or more than 703.
Only a real math genius could have scored above 791.

59. a) A histogram or stemplot (below) is appropriate. The distribution is
right skewed. This means the portion of large registry groups is relatively small. There are some (one or two) outliers. The median size is
717. The interquartile range is 1220.

(7 | 4 means 7400)
The 5-number summary is:

Minimum	Q1	Median	Q3	Maximum
41	452	717	1672	11 202

b) If we calculate the band sizes, most band sizes will be same as the
registry group sizes since only Six Nations of the Grand River in
Ontario consists of more than one registry group. This very large
band, consisting of 13 registry groups, will increase considerably the
mean and standard deviation, while having a much smaller effect on
the more resistant median and IQR. The histogram will have a bigger
gap due to this very large band.

61. a) The histogram of the proportion of city residents who are bilingual
is shown below. The highest proportion of bilinguals is in Montreal
with more than 50% bilinguals, and the lowest proportion is in
Brantford with less than 5% bilinguals. The median is about 8%. The
distribution of the proportion of city residents who are bilingual is
right skewed. This means a relatively smaller number of cities with
a high proportion of bilinguals and more cities with a low proportion
of bilinguals. The distribution of proportion of bilinguals appears
bimodal—it looks like the distribution of a few cities with a large
proportion of bilinguals has a distinct shape compared to the other
cities. Summary statistics are given below. About 25% of the cities
have less than 7% bilinguals.

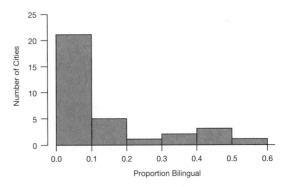

Descriptive Statistics: Proportion Bilingual

Mean	StDev	Min.	Q1	Median	Q3	Max.
0.153	0.145	0.042	0.066	0.077	0.148	0.539

b) The dotplot below shows the cities in Quebec and New Brunswick
on top. They have relatively high proportions of bilinguals.

c) The histogram of the proportion of city residents who speak neither English nor French is shown below. The highest proportion
of residents who speak neither English nor French is in Vancouver
with more than 5%, and the lowest proportion is in Saguenay with
about 0.03%. The median is about 0.69%. The distribution of the
proportion of city residents who speak neither English nor French
is right skewed. This means there is a relatively smaller number of
cities with a high proportion of residents who speak neither English
nor French and more cities with a low proportion. Three cities
(Vancouver, Abbotsford, and Toronto, the bars above 0.031 on the
histogram) appear to be outliers.

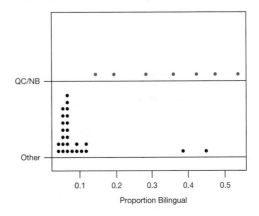

Descriptive Statistics: Proportion Ni-lingual

Mean	StDev	Minimum	Q1	Median	Q3	Maximum
0.0111	0.0135	0.0003	0.0031	0.0069	0.0144	0.0560

63. a) The closeness of the mean and median suggests that the distribution of the percentage of voter turnout in the 2011 election is approximately symmetric.

b) The distribution of the percentage of voter turnout in the 2011 election is approximately symmetric, with mean 61.068 and standard deviation 5.772.

c) 72.612 and 49.524

d) The overall percentage won't be exactly the same as the mean of the percentages as the number of eligible voters is not exactly the same in each electoral district. However, since the number of eligible voters in each riding is pretty close, we should expect the overall percentage to be fairly close to 61.07%.

65. a) The mean will be smaller because it depends on the sum of all of the values. The standard deviation will be smaller because we have truncated the distribution at 32. The range will be smaller for the same reason. The statistics based on quantiles of the data (median, Q1, Q3, IQR) won't change because the distribution was truncated to the right of Q3.

b) The spread of the distribution will be relatively smaller because we have added observations near the middle. Thus, we expect the standard deviation and IQR to be smaller. We should only see a small change to the mean and median because the added observations are near the middle of the distribution. The range will not change.

c) The spread of the distribution will be relatively larger because we have added observations at about 2 minutes left and right of the mean. Thus, we expect the standard deviation and IQR to be bigger. Because the additional observations were added symmetrically to either side of the mean, there should be little change to the mean or median, and the range will remain unchanged.

d) Subtracting one minute from each time shifts the entire distribution to the left by one minute. The mean and median should decrease by one minute. However, there has been no change to the shape of the distribution, so the standard deviation, IQR, and range will all remain the same.

e) The values added at 35 are further from the mean than the values at 29.5, so the mean will increase. The overall spread of the distribution has increased, so the standard deviation will increase. The added observations at 35 minutes lie above Q3, so the IQR will increase. Because the same number of observations was added on either side of the median, its value won't change. The minimum and maximum values have not changed, so the range remains unchanged as well.

f) We have moved some values outside of the Q1 to Q3 interval further away from the centre of the distribution. Q1 and Q3 remain unchanged, so the IQR remains unchanged. Because the adjustment is symmetric, the mean and median are unchanged. The range remains the same because the minimum and maximum are unchanged. The standard deviation increases because the spread of the distribution has increased.

67. a) **Descriptive Statistics: Height (class midpoint)**

Variable	N	Mean	StDev
Height (class mid)	130	67.069	3.996

Descriptive Statistics: Height (actual)

Variable	N	Mean	StDev
Height (actual)	130	67.115	3.792

The mean and the standard deviation of the actual data are very close to those calculated using the midpoints. For grouped data, we assume that all the values in a class are equal to the midpoint. This assumption is usually reasonable unless the class width is big.

b) Mean $= \dfrac{\sum_i f_i m_i}{\sum_i f_i}$, Variance $= \dfrac{\sum_i f_i (m_i - \overline{X})}{\sum_i f_i - 1}$

69. a) The last wide bin now has a rather big rectangle above it, with height equal to about 13. The number of run times exceeding 32.5 appears to have grown!

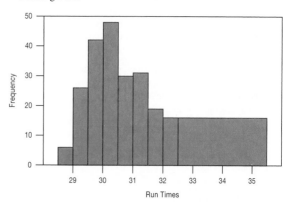

b) Now the rectangle height over the last bin equals the average of the 6 rectangle heights in the original histogram, i.e., the area of this rectangle equals the total area of the 6 rectangles it replaced. We have been true to the Area Principle, and kept area proportional to frequency. (Since area is height multiplied by width, the rectangle height must be proportional to frequency divided by bin width.)

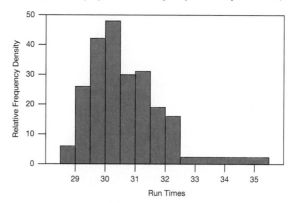

71. a) The histogram and boxplot of the distribution of ages both show that a typical crowd crush victim was approximately 18–20 years of age, that the range of ages is 36 years, and that there are two outliers, one at age 36–38 and another at age 46–48.

b) This histogram shows that there may have been two modes in the distribution of ages, one at 18–20 years of age and another at 22–24 years of age, but this is not shown in the boxplot.

c) Median is the better measure of centre, since the distribution of ages has outliers.

d) IQR is a better measure of spread, since the distribution of ages has outliers.

Chapter 4

1. Both distributions are unimodal and skewed to the left. The lowest international value may be an outlier. Because the distributions are skewed, we choose to compare medians and IQRs. The medians are very similar. The IQRs show that the domestic load factors vary a bit more.

3. Load factors are generally highest and least variable in the summer months (June–August). They are lower and more variable in the winter and spring.

5. Air travel immediately after the events of 9/11 was not typical of air travel in general. If we want to analyze monthly patterns, it might be best to set these months aside.

7. a) After a period of little change in 2000–2001, load factors have been increasing steadily.

b) No. These are percentages, so they can't exceed 100%. Moreover, we have no reason to believe this trend will continue.

9. Data with a distribution this skewed are difficult to summarize. The extremely large values will dominate any summary or description.

11. Answers will vary.

13. The data are a time series, so it makes sense to plot them as a time series. A time series plot is shown below. The data show an increase in the percentage of Ontario residents aged 65 years and older over the time period, especially after the year 1971.

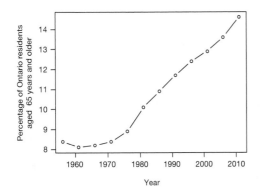

15. a) Roughly symmetric. The mean and median are reasonably close.
b) No, all data are within the fences.
c)

d) The 48 universities graduate, on average, about 68% of freshman "on time," with percents ranging from 43% to 87%. The middle 50% of these universities graduate between 59% and 75% of their freshman in 4 years.

17.

19. a) They should be put on the same scale, from 0 to 20 days.
b) Men have a mode at 1 day, then tapering off from there. Women have a mode near 5 days, with a sharp drop afterward.
c) A possible reason is childbirth.

21. a) Both women have a median score of about 17 points per game, but Scyrine is much more consistent. Her IQR is about 2 points, while Alexandra's is over 10.
b) If the coach wants a consistent performer, she should take Scyrine. She'll almost certainly deliver somewhere between 15 and 20 points. But, if she wants to take a chance and needs a "big game," she should take Alexandra. Alexandra scores over 24 points about a quarter of the time. (On the other hand, she scores under 11 points about as often.)

23. Class A is Class 1. The median is 60, but has less spread than Class B, which is Class 2. Class C is Class 3, since it's median is higher, which corresponds to the skew to the left.

25. a) *Who:* 45 volunteers. *What:* Level of caffeine consumption and memory test score. *When:* Not specified. *Where:* Not specified. *Why:* The student researchers want to see the possible effects of caffeine

on memory. *How:* It appears that the researchers imposed the treatment of level of caffeine consumption in an experiment. However, this point is not clear.
b) Caffeine level: categorical; Test score: quantitative.
c)

d) The participants scored about the same with no caffeine and low caffeine. The medians for both were 21 points, with slightly more variation for the low caffeine group. The group consuming high caffeine had lower memory scores in general, with a median score of about 19. No one in the high caffeine group scored above 24, but 25% of each of the other groups scored above 24.

27. a) Comparative boxplots are below.
b) The distribution of population growth in NE/MW states is unimodal, symmetric and tightly clustered around 5% growth. The distribution of population growth in S/W states is much more spread out, with most states having population growth between 5% and 25%.

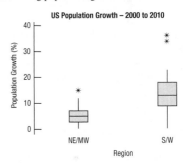

A typical state had about 15% growth. There were two outliers, with 34% and 35% growth, respectively. Generally, the growth rates in the S/W states were higher and more variable than the rates in the NE/MW states.

29. a) About 36 mph.
b) Q1 about 35 mph and Q3 about 37 mph.
c) The range appears to be about 7 mph, from about 31 to 38 mph. The IQR is about 2 mph.
d) We can't know exactly, but the boxplot may look something like this:

e) The median winning speed has been about 36 mph, with a max of about 38 and a min of about 31 mph. Half have run between about 35 and 37 mph, for an IQR of 2 mph.

31. a) Boys   b) Boys   c) Girls
d) The boys appeared to have more skew, as their scores were less symmetric between quartiles. The girls' quartiles are the same distance

from the median, although the left tail stretches a bit farther to the left.

e) Girls. Their median and upper quartiles are larger. The lower quartile is slightly lower, but close.

f) 14 (4.2) + 11 (4.6)/25 = 4.38

33. a) The median for males is slightly higher than that for females. There are two low outliers in each boxplot. The median of the differences between the mean scores for males and females is about 12. All but one country has the mean for males greater than that for females. Only one country has mean score for females greater than that for males and that is identified as an outlier.

b) In the boxplots below, Canada, the U.S., and Mexico are indicated by the symbols x, U, and M, respectively. In the boxplots for males, Israel, Turkey, Chile, and Mexico are outliers because their mean scores are unusually low. In the boxplot for females, Mexico and Chile are outliers because their mean scores are unusually low.

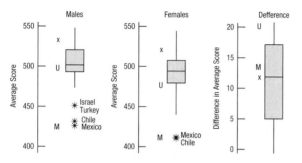

c) The 5-number summaries for the male scores, female scores, and differences are given below:

Column	Min.	Q1	Median	Q3	Max.
Male	425.4	492.9	501.3	520.4	547.8
Female	410.4	479.2	494.1	507.5	544.5
Difference	−1.75	5.00	11.8	17.1	21.8

The average female score for Mexico is 411.8, which is less than $Q1 - 1.5 * IQR = 436.8$ and is thus an outlier.

d) The separate boxplots for males and females show that some countries have particularly low average scores (i.e, Israel, Chile, Turkey, and Mexico) compared with that of other countries. The boxplot for differences shows that although the average female scores are lower, the difference between male and female scores is not independent of country. The graph for differences is more useful for learning about gender differences in OECD countries.

35. There appears to be an outlier. That distance should be removed before looking at a plot of the drilling distances. With the outlier removed, we can determine that the slow drilling process is more accurate.

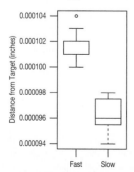

37. a) The side by side boxplots for Canada and the U.S. are given below. Homicide, aggravated assaults and robberies are higher in the U.S. Break and enter is higher in Canada. Theft and motor vehicle theft

is about the same in Canada and the U.S. The pattern appears to be similar in the two countries, for example homicide is the least frequent among the six types in both countries and theft is the most frequent.

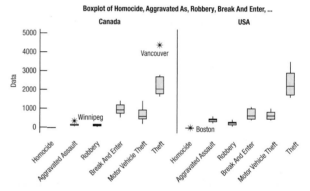

b) Winnipeg has aggravated assaults in the U.S. range and Boston has homicides in the Canadian range. Vancouver is a high outlier in Theft.

c) Population in the city and poverty can be lurking variables.

39. a) Side-by-side boxplots for % voter turnout are given below. PEI has the highest voter turnout (highest median, highest first quartile, and the highest third quartile). Newfoundland has the lowest % voter turnout (lowest median, first quartile, and the third quartile, other than the territories).

b) They are outliers compared to the observations (ridings) within those territories, but they are not outliers compared to all Canadian ridings.

c) The boxplot Canadian ridings (overall) is given below. There are outliers: Cardigan in PEI is the only positive outlier. The negative outliers are Calgary East, Nunavut, Random-Burin-St. George's,

Bonavista-Gander-Grand Falls-Windsor, Churchill, and Fort McMurray-Athabasca. Fort McMurray-Athabasca is the only outlier in the combined boxplot that was also an outlier within its province/territory. Cardigan is a country-level outlier but is not an outlier within PEI because of the overall high voter turnout in PEI. With the exception of Calgary East, the negative outliers all seem to be in relatively remote areas of the country, which could account for the low turnout.

d) The mean must be smaller than the median in Manitoba, while in BC, the mean must be larger than the median.

41. a) The side-by-side boxplots of homicide rates are shown below. Large cities have the highest median homicide rate. Medium-sized cities have the least variable homicide rate. There are two large outliers.

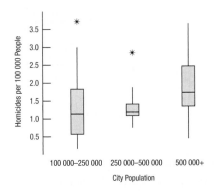

b) City size is a quantitative variable. In our analysis in a) above we used the variable city size to create a categorical variable (with three categories). This helps us get some idea about how the homicide rates are related to the city size, but the actual city size values have more information about the city size and forming categories combining cities with sizes within certain ranges (even though the actual sizes are different) leads to some loss of information in the data. A plot of homicide rate versus population size (a scatterplot) will show the relation between the two variables.

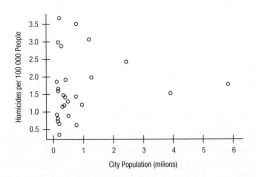

c) The side-by-side boxplots of homicide rates are shown below. Quebec has the lowest homicide rate and the West has the highest.

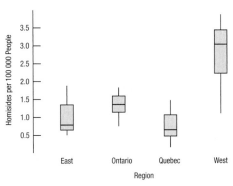

d) If we have the annual data for the 11-year period, we could examine the variation in homicide rate over time.

43. a)

Variable	Minimum	Q1	Median	Q3	Maximum
points scored	48.0	98.3	145.5	192.8	215.0

b)

c) $48 > Q1 - 1.5 \, IQR = -43.45$ and so 48 is not an outlier.

d) The plot of the total points versus year is shown below. The total points decreases with time. The boxplot does not show the relationship with time.

45.

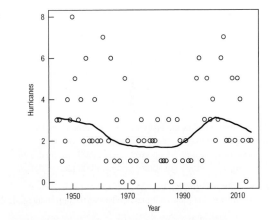

It appears that the number of hurricanes per year had a lull during the period from the mid-1960s to about 1990, increased fairly rapidly until the early 2000's, and has been declining since.

47. a) The distribution is unimodal and slightly left skewed.

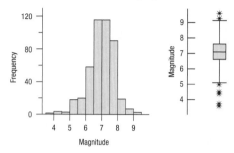

b) It doesn't appear that magnitude depends on time. That is, over time it doesn't appear that the distribution has really changed all that much.

c) There are several countries with only 1 observation and thus only appear as a single point. There are also too many countries. We could group countries together in similar regions.

d) If we go by the middle 50% or the median, the quakes seemed to be the biggest in Peru. The smallest ones seemed to occur in Greece. The three countries with the biggest quakes were Chile (9.5 in 1960), the U.S. (9.2 in 1964), and Indonesia (9.1 in 2004). The smallest were in Greece in 1962 and in USA in 1952, both at 4.5 (though there was also a 4.4 quake in a US territory in 1989).

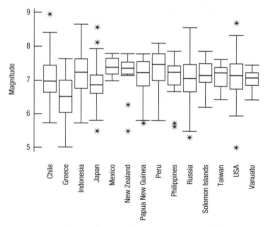

49. a) The logarithm makes the histogram more unimodal and symmetric. It is easy to see that the centre is around 3.5 in log assets.
b) The company's assets are approximately 2,500 million dollars.
c) The company's assets are approximately 1,000 million dollars.

51. a) Fusion time and group.
b) Fusion time is quantitative (units = seconds). Group is categorical.

c) Both distributions are skewed to the right with high outliers. The boxplot indicates that visual information may reduce fusion time. The median for the Verbal/Visual group seems to be about the same as the lower quartile of the No/Verbal group.

53. a) 24 000, 7500. The median band size is slightly higher in Saskatchewan but it is hard to read the difference. Boxplots do not show the mean. Both of the means must be greater than the medians.
b) Add gridlines.
c) Right skewed. Many suspect outliers may just be part of the long upper tail of this skewed distribution, rather than genuine outliers.
d) Median and IQR do not change.
e) Because the distribution of band size is right skewed in both provinces.
f) At least 1500. Between 1500 and 2500 in size.
g) The distribution of the transformed sizes is much less skewed in both provinces.
h) Six Nations is still an outlier but it is much less extreme (less unusual) that it was before the transformation. There are additional outliers, most notably the Lucky Man band (size = 110) in North Central Saskatchewan.

Chapter 5

1. 65

3. a) 72 oz., 40 oz.   b) 4.5 lb, 2.5 lb

5. a)

b) 18.6 to 31.0 mpg   c) 16%   d) 13.5%   e) Less than 12.4 mpg

7. a) 12.2%   b) 71.6%   c) 23.3%

9. a) No. The plot is not straight.
b) It is skewed to the right.

11. a) Skewed to the right; mean is higher than median.
b) $350 and $950
c) Minimum $350. Mean $750. Median $550. Range $1200. IQR $600. Q1 $400. SD $400.
d) Minimum $330. Mean $770. Median $550. Range $1320. IQR $660. Q1 $385. SD $440.

13. Lowest score = 910. Mean = 1230. SD = 120. Q3 = 1350. Median = 1270. IQR = 240.

15. In January, a high temperature of 13°C is less than 2 standard deviations above the mean. In July, a high temperature of 13° is more than two standard deviations below the mean. So it is less likely to happen in July.

17. a) Megan   b) Anna

19. a) About 1.81 standard deviations below the mean.
b) 1000 ($z = 21.81$) is more unusual than 1250 ($z = 1.17$).

21. a) Mean = $1152 - 1000 = 152$ pounds; SD is unchanged at 84 pounds.
b) Mean = $0.40(1152) = \$460.80$; SD = $0.40(84) = \$33.60$.

23. Min = $0.40(980) - 20 = \$372$; median = $0.40(1140) - 20 = \$436$; SD = $0.40(84) = \$33.60$; IQR = $0.40(102) = \$40.80$.

25. College professors can have between 0 and maybe 40 (or possibly 50) years of experience. A standard deviation of 1/2 year is impossible, because many professors would be 10 or 20 SDs away from the mean, whatever it is. A SD of 16 years would mean that 2 SDs on either side of the mean is plus or minus 32, for a range of 64 years. That's too high. So, the SD must be 6 years.

27. a)

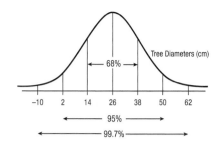

b) Between 2 cm and 50 cm.   c) 2.5%   d) 34%   e) 16%

29. Since the histogram is not unimodal and symmetric, it is not wise to have faith in numbers from the Normal model.

31. a) 16%   b) 13.56%

c) The percentages in parts a) and b) do not agree because the Normal model is not appropriate in this situation.

d) The histogram of 2010 Winter Olympic Downhill times is skewed to the right. The Normal model is not appropriate for the distribution of times, because the distribution is not symmetric.

33. a) 16%

b) 16% of the students should be watching less than 1.27 hours a week, but no one can watch less than 0 hours, so the model doesn't fit well.

c) Data are strongly skewed to the right, not symmetric.

35. a) 6.68%   b) 98.78%   c) 71.63%   d) 38.29%

37. a) 0.842   b) −0.674   c) −1.88   d) −1.645 to 1.645

39. a) 11.75%   b) 71.31%   c) 23.10%

41. a) 261.95 days.   b) 303.22 days.   c) 21.58 days.

43. a)

b) 30.85%   c) 17.00%   d) 32.4 points   e) 212.9 points

45. a) 11.1 %   b) (90.72, 103.28) cm   c) 103.28 cm

47. About 1% of Greek men are expected to be taller than the average Dutch man. He wouldn't stand out radically, but he'd probably have a hard time keeping a "low profile"!

49. a) The smallest size is 50.672 cm and the biggest is 61.328 cm.

b) Heads in the range 54.488 cm and 57.512 cm.

c) z-values for 51 cm and 54 cm are −2.78 and −1.11 respectively. The proportion is 13.08%.

d) 57.872 cm   e) Less than or equal to 52.31 cm.

51. a) 16 %

b) If the distribution is normal, the 10th and the 90th percentile are 463.2 and 616.8, which are not very different from the given values.

c) For a normal distribution with mean = 540 and standard deviation 60, IQR = 80.4. This is not that far from the given value.

d) 1.13 %

53. a) 5.3 grams   b) 6.4 grams   c) Younger because SD is smaller.

d) Mean 62.7 grams, SD 6.2 grams

55. a) No. The Normal probability plot does not look straight.

b) The distribution of delay times is strongly skewed to the right.

57. The histogram, boxplot, and NPP are shown below. All three plots show several low outliers. Apart from these outliers, the histogram looks close to symmetric (maybe slightly left skewed). The boxplot has similar information. The normal probability plot shows a slight curvature (to the right). The many low outliers make the distribution look slightly left skewed.

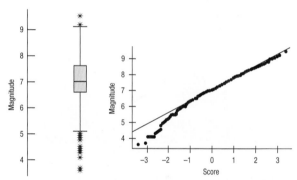

59. a) The histogram is close to a bell shape and the normal probability plot is close to a straight line. Usually it is easier to judge straightness. It can be difficult to judge *bell-shapedness* with small samples.

b) The histograms and the normal probability plots for the four simulated samples from the normal (0, 1) are given below. The normal probability plots are close to straight lines. We expect this because the data were simulated from a normal distribution. The shapes of the histograms vary from sample to sample even though the data are from a normal distribution.

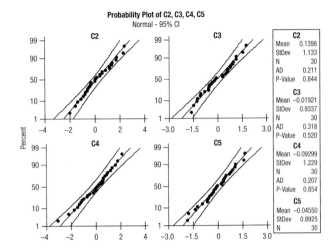

Probability Plot of C2, C3, C4, C5
Normal - 95% CI

61. a) The histogram is close to a bell shape and the normal probability plot is close to a straight line. Usually is easier to judge straightness. It can be difficult to judge *bell-shapedness* with small samples.

b) The normal probability plots are close to straight lines. We expect this because the data were simulated from a normal distribution. The shapes of the histograms vary from sample to sample even though the data are from a normal distribution.

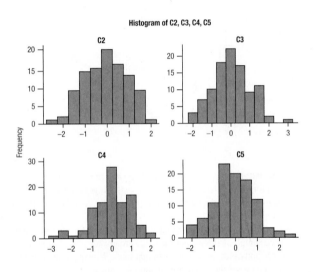

Histogram of C2, C3, C4, C5

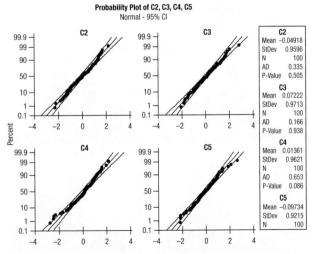

Probability Plot of C2, C3, C4, C5
Normal - 95% CI

Part I Review

1. a) A histogram for the homicide rates is shown below:

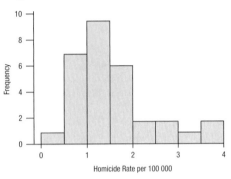

The homicide rates have median 1.36, quartiles Q1 = 0.92 and Q3 = 1.92, and interquartile range 1.00. The distribution is skewed to the right.

b) It appears that there are substantial differences between regions in Canada. The Prairies have the highest homicide rates, and Quebec has the lowest homicide rates.

3. a) If enough sopranos have a height of 65 inches, this can happen.

b) The distribution of heights for each voice part is roughly symmetric. The basses are slightly taller than the tenors. The sopranos and altos have about the same median height. Heights of basses and sopranos are more consistent than those of altos and tenors.

5. a) It means their heights are also more variable.

b) The *z*-score for women to qualify is 2.40, compared with 1.75 for men, so it is harder for women to qualify.

7. a) *Who:* Local residents near a Canadian university.
What: Age, whether or not the respondent attended college or university, and whether or not the respondent had a favourable opinion of the university.
When: Not specified.
Where: Region around a Canadian university.
Why: The information will be included in a report to the university's directors.
How: 850 local residents were surveyed by phone.

b) There is one quantitative variable, age, probably measured in years. There are two categorical variables, college or university attendance (yes or no), and opinion of this university (favourable or unfavourable).

c) There are several problems with the design of the survey. No mention is made of a random selection of residents. Furthermore, there may be a non-response bias present. People with an unfavourable opinion of the university may hang up as soon as the staff member identifies himself or herself. Also, response bias may be introduced by the interviewer. The responses of the residents may be influenced by the fact that employees of the university are asking the questions. There may be a greater percentage of favourable responses to the survey than truly exist.

9. a) These are categorical data, so mean and standard deviation are meaningless.
   b) Not appropriate. Even if it fits well, the Normal model is meaningless for categorical data.

11. a)

   b) The scores on Friday were higher by about 5 points. This is a drop of more than 10% of the average score and shows that students fared worse on Monday after preparing for the test on Friday. The spreads are about the same, but the scores on Monday are a bit skewed to the right.

   c)

   d) The changes (Friday – Monday) are unimodal and centred near 4 points, with a spread of about 5 (SD). They are fairly symmetric, but slightly skewed to the right. Only three students did better on Monday (had a negative difference).

13. a) There is a higher proportion living on reserves in the 0–24 age group and a lower proportion living on reserves in the 35–64 age group.

   b) Side by side barplots for the four conditional residence distributions are shown below. For the youngest age group, there are many more living on reserve. As the age groups increase in age, the percentage living on reserve declines.

   c) The residence variable is categorical. The age variable is quantitative, but has been made categorical for this display.
   d) Yes. The following mean ages will occur if only 1/8 of the males live on reserves, but 3/4 of the females live on reserves.

Residency	Male	Female
On	30	40
Off	50	60
Overall	47.5	45

15. a) Annual mortality rate for males (quantitative) in deaths per 100 000 and water hardness (quantitative) in parts per million.
   b) Calcium is skewed right, possibly bimodal. There looks to be a mode down near 12 ppm that is the centre of a fairly tight symmetric distribution and another mode near 62.5 ppm that is the centre of a much more spread out, symmetric (almost uniform) distribution. Mortality, however, appears unimodal and symmetric with the mode near 1500 deaths per 100 000.

17. a) They are on different scales.
   b) January's values are lower and more spread out.
   c) Roughly symmetric but slightly skewed to the left. There are more low outliers than high ones. Centre is around 40 degrees with an IQR of around 7.5 degrees.

19. a) Bimodal with modes near 2 and 4.5 minutes. Fairly symmetric around each mode.
   b) Because there are two modes, which probably correspond to two different groups of eruptions, an average might not make sense.
   c) The intervals between eruptions are longer for long eruptions. There is very little overlap. More than 75% of the short eruptions had intervals less than about an hour (62.5 minutes), while more than 75% of the long eruptions had intervals longer than about 75 minutes. Perhaps the interval could even be used to predict whether the next eruption will be long or short.

21. a) The distribution of the ratio of arm length to nose length of 18 girls in a statistics class is unimodal and roughly symmetric, with centre around 15. There is one low outlier, a ratio of 11.8. A boxplot is shown at the right. A histogram or stemplot is also an appropriate display.

   b) In the presence of an outlier, the 5-number summary is the appropriate choice for summary statistics. The 5-number summary is 11.8, 14.4, 15.25, 15.7, 16.9. The IQR is 1.3.
   c) The ratio of 9.3 for the Statue of Liberty is very low, well below the lowest ratio in the statistics class, 11.8, which is already a low outlier. Compared to the girls in the statistics class, the Statue of Liberty's nose is very long in relation to her arm.

23. If we look only at the overall statistics, it appears that the follow-up group is insured at a much lower rate than those not traced (11.1% of the time as compared with 16.6%). But most of the follow-up group were black, who have a lower rate of being insured. When broken down by race, the follow-up group actually has a higher rate of being insured for both blacks and whites. So the overall statistic is misleading and is attributable to the difference in race makeup of the two groups.

25. a)

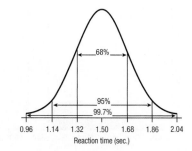

b) According to the model, reaction times are symmetric with centre at 1.5 seconds. About 95% of all reaction times are between 1.14 and 1.86 seconds.

c) 8.2%   d) 24.1%

e) Quartiles are 1.38 and 1.62 seconds, so the IQR is 0.24 seconds.

f) The slowest 1/3 of all drivers have reaction times of 1.58 seconds or more.

27. a)

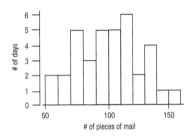

b) Mean $100.25, SD $25.54.

c) The distribution is somewhat symmetric and unimodal, but the centre is rather flat, almost uniform.

d) 64%. The Normal model seems to work reasonably well, since it predicts 68%.

29. a) *Who:* 100 health food store customers.
What: Have you taken a cold remedy?, and Effectiveness (scale 1 to 10).
When: Not stated.
Where: Not stated.
Why: Promotion of herbal medicine.
How: In-person interviews.

b) Have you taken a cold remedy?—categorical. Effectiveness—categorical or ordinal.

c) No. Customers are not necessarily representative, and the Council had an interest in promoting the herbal remedy.

31. a) 38 cars

b) Possibly because the distribution is skewed to the right.

c) Centre—median is 148.5 cubic inches. Spread—IQR is 126 cubic inches.

d) No. It's bigger than average, but smaller than more than 25% of cars. The upper quartile is at 231 inches.

e) No. 1.5 IQR is 189, and 105–189 is negative, so there can't be any low outliers. $231 + 189 = 420$. There aren't any cars with engines bigger than this, since the maximum has to be at most Q1 + Range = $105 + 275 = 380$ cubic inches.

f) Because the distribution is skewed to the right, this is probably not a good approximation.

g) Mean, median, range, quartiles, IQR, and SD all get multiplied by 16.4.

33.

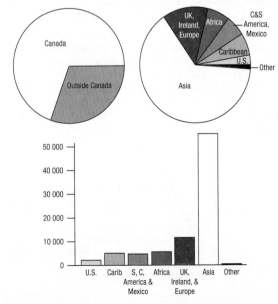

35. a) Side by side boxplots for the two teams are shown below.

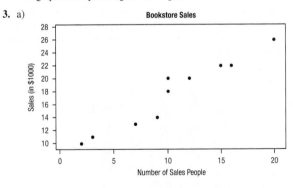

The median salaries for the two teams are similar, $1 675 000 for the Blue Jays and $1 725 000 for the Maple Leafs. The Blue Jay salaries are more spread out, with IQR $3 888 500 versus $2 500 000 for the Maple Leafs. There are many more very high salaries for the Blue Jays.

b) As a superstar I would prefer playing for the Blue Jays, where the very high salaries occur. As a new player I would prefer the Maple Leafs, as the minimum is higher ($500 000 versus $392 200). As an average player I would prefer the Blue Jays as the mean salaries are higher ($3 417 367 versus $2 099 545).

c) As an owner, I would prefer the salary structure of the Maple Leafs as the total salary budget is much smaller ($46 190 000 versus $82 016 800).

37. a) 0.43 hour   b) 1.4 hours   c) 0.89 hour (or 53.4 minutes)

d) Survey results vary, and the mean and the SD may have changed.

Chapter 6

1. a) Weight in ounces: explanatory; Weight in grams: response. (Could be other way around.) To predict the weight in grams based on ounces. Scatterplot: positive, straight, strong (perfectly linear relationship).

b) Circumference: explanatory. Weight: response. To predict the weight based on the circumference. Scatterplot: positive, linear, moderately strong.

c) Shoe size: explanatory; GPA: response. To try to predict GPA from shoe size. Scatterplot: no direction, no form, very weak.

d) Miles driven: explanatory; Gallons remaining: response. To predict the gallons remaining in the tank based on the miles driven since filling up. Scatterplot: negative, straight, moderate.

3. a)

b) Positive   c) Linear   d) Strong   e) No

5. a) True.

b) False. The correlation will remain the same.

c) False. Correlation has no units.

7. Correlation does not demonstrate causation. The analyst's argument is that sales staff cause sales. However, the data may reflect the store hiring more people as sales increase, so any causation would run the other way.

9. 4, 5, and 6. Yes, it will be very close to a straight line.

11. a) Altitude: explanatory; Temperature: response. (Other way around possible as well.) To predict the temperature based on the altitude. Scatterplot: negative, possibly straight, weak to moderate.

b) Ice cream cone sales: explanatory. Air conditioner sales: response—although the other direction would work as well. To predict one from the other. Scatterplot: positive, straight, moderate.

c) Age: explanatory; Grip strength: response. To predict the grip strength based on age. Scatterplot: curved down, moderate. Very young and elderly would have grip strength less than that of adults.

d) Reaction time: explanatory; Blood alcohol level: response. To predict blood alcohol level from reaction time test. (Other way around is possible.) Scatterplot: positive, nonlinear, moderately strong.

13. a) None   b) 3 and 4   c) 2, 3, and 4   d) 2
e) 3 and possibly 1

15. There seems to be a very weak—or possibly no—relation between brain size and performance IQ.

17. a)

b) Unimodal, skewed to the right. The skew.

c) The positive, somewhat linear relation between batch number and broken pieces.

d) Answers may vary.

19. a) 0.006   b) 0.777   c) −0.923   d) −0.487

21. a) Yes. It shows a linear form and no outliers.

b) There is a strong, positive, linear association between drop and speed; the greater the coaster's initial drop, the higher the top speed.

23. a) −0.649

b) The correlation will not change. Correlation does not change with change of units.

c) There is a moderately strong negative linear association between number of calories consumed and time at the table for toddlers. Longer times are associated with fewer calories.

d) There appears to be a relation between time at the table and calories consumed, but there is no reason to believe increased time *causes* decreased calorie consumption. There may be a lurking variable.

25. a)

b) Negative, linear, strong, with one high outlier.

c) −0.848

d) There is a fairly strong linear relation in a negative direction between engine size and gas mileage. Lower fuel efficiency is generally associated with larger engines.

27.

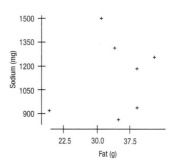

(Plot could have explanatory and predictor variables swapped.) Correlation is 0.199. There does not appear to be a relation between sodium and fat content in burgers, especially without the low-fat low-sodium item. The correlation of 0.199 shows a weak relationship, even with the outlier included.

29. a) Yes, although the association is weak, the form does seem to be somewhat linear.

b) As the number of runs increases, the attendance also increases.

c) There is a slight positive association, but even if it were stronger it does not prove that more fans would come if the number of runs increased. Association does not indicate causality.

31. There may be an association but not a correlation unless the variables are quantitative. There could be a correlation between average number of hours of TV watched per week per person and the number of crimes committed per year. Even if there is a relationship, it doesn't mean one causes the other.

33. a) Actually, yes, taller children will tend to have higher reading scores, but this doesn't imply causation.

b) Older children are generally both taller and are better readers. Age is the lurking variable.

35. The scatterplot is not linear; correlation is not appropriate.

37. a) Assuming the relation is linear, a correlation of −0.772 shows a strong relation in a negative direction.

b) Continent is a categorical variable. Correlation does not apply.

39. This is categorical data even though it is represented by numbers. The correlation is meaningless.

41. The scatterplot below shows that the association between duration and length is straight, positive, and moderate, with no outliers. Generally, rides on coasters with a greater length tend to last longer. The correlation between length and duration is 0.698, indicating a moderate association.

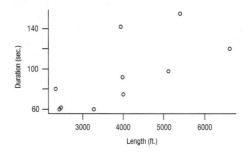

43. a) We would expect that as one variable (say length of ride) increases, the rank will improve, which means it will decrease.

b) Drop has the strongest correlation ($r = -0.193$), but even that correlation is very weak. The scatterplot shows no apparent association. The number one ranked coaster, Bizarro, has a fairly typical drop. There appear to be other factors that influence the rank of coaster more than any of the ones measured in this data set.

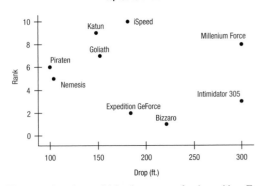

Top Ten Roller Coasters

c) There may be other variables that account for the ranking. For example, other quantitative variables such as number of loops and number of corkscrews, categorical variables such as whether the coaster is made of wood or steel, and whether or not there are tunnels may all have an affect on the rank.

45. a) Correlations: Year, Bust size (in.), Waist size (in.), Hips size (in.)

	Year	**Bust size**	**Waist size**
Bust size	1.000		
Waist size	0.992	0.992	
Hips size	0.936	0.936	0.929

b) All these relationships are linear, strong and positive in direction, with no clear outliers. The relationship between bust size and year is the strongest.

Matrix Plot of Year, Bust Size (in.), Waist Size (in.), Hips Size (in.)

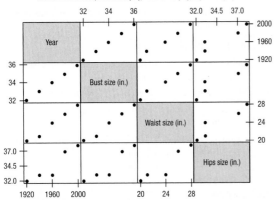

c) The correlations between body measures are not quite surprising as we expect some proportionality between them (otherwise women's body shape is changing over time). The positive correlation with time can happen for example if we are getting bigger (increasing weight) over time due to various reasons (maybe lack of activity because most of the work we did in the good old days are taken over by machines).

47. a) Armspan and height have the highest correlation at 0.781.
b) Linear, positive, moderately strong relationship.
c) Relationship looks similar for both genders, but slightly stronger for females than for males.

49. a)

The relation between position and distance is very strong, positive, and curved. There is very little scatter from the trend.
b) The relation is not linear.
c)

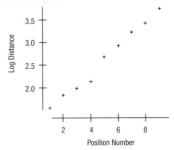

The relation between position number and log of distance appears to be roughly linear.

51. a) 0.995. This is very close to 1.
b) Most points are concentrated to small region and just two points far away from them but in the same direction. This can make the correlation very high and correlation is not an appropriate measure in situations like this.

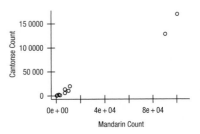

c) Yes, now there is concentration of points to smaller communities. The correlation is 0.985. The correlation is now an appropriate measure.

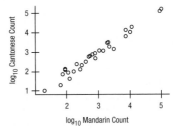

d) In larger cities it is reasonable to expect relatively larger counts for both these groups (as well as for any other group) and thinking this way, the larger Mandarin counts might not be because of the larger Cantonese counts. It is even possible that these two counts are not even related. In other words what we are suggesting is that the population in those cities is a possible lurking variable. If we think population is a possible lurking variable, we can use proportions of Cantonese and Mandarin speaker counts as the variables.

Chapter 7

The calculations in this chapter are particularly sensitive to rounding. If your answers differ slightly from those given here, they may still be correct in the sense that you calculated them by a correct method, but you may have used values that were rounded differently than those used to find these answers. In general, we use the original data and do no intermediate rounding. Don't be concerned about minor differences—what is important is your interpretations of the results.

1. a) False. The line usually touches none of the points. We minimize the sum of the squared errors.
b) True.

c) False. It is the sum of the squares of all the residuals that is minimized.

3. The weight of a newborn boy can be predicted as −5.94 kg plus 0.1875 kg per cm of length. This is a model fit to data. No particular baby should be expected to fit this model exactly.

5. a) $b_1 = 0.914$ if found by hand. $b_1 = 0.913$ if found by technology. (Difference is due to rounding error.)
   b) It means that an additional $0.914 ($1000) or $914 of sales is associated with each additional sales person working.
   c) $b_0 = 8.10$
   d) It would mean that, on average, we expect sales of $8.10 ($1000) or $8100 with 0 sales people working. Doesn't really make sense in this context.
   e) $\widehat{sales} = 8.10 + 0.913\ Number\ of\ Sales\ People\ Working$
   f) $24.54 ($1000) or $24 540. ($24 530 if using the technology solution.)
   g) $460 ($470, using technology).
   h) Underestimated.

7. The winners may be suffering from regression to the mean. Perhaps they weren't really better than other rookie executives, but just happened to have a lucky year.

9. a) Thousands of dollars.
   b) 2.77 (the largest residual in magnitude).
   c) 0.07 (the smallest residual in magnitude).

11. $R^2 = 93.12\%$. About 93% of the variance in *Sales* can be accounted for by the regression of *Sales* on *Number of Sales Workers*.

13. a) Linearity assumption.
    b) Outlier condition.
    c) Equal spread condition.

15.

	$\bar{x}$	s_x	$\bar{y}$	s_y	r	$\hat{y} = b_0 + b_1 x$
a)	10	2	20	3	0.5	$\hat{y} = 12.5 + .75x$
b)	2	0.06	7.2	1.2	−0.4	$\hat{y} = 23.2 − 8x$
c)	12	6	152	30	−0.8	$\hat{y} = 200 − 4x$
d)	2.5	1.2	25	100	0.6	$\hat{y} = −100 + 50x$

17. a) Model is appropriate.
    b) Model is not appropriate. Relationship is nonlinear.
    c) Model may not be appropriate. Spread is changing.

19. The line $\hat{y} = 7 + 1.1x$ minimizes the sum of squared vertical distances from the points to the line.

21. a) The response variable is the asking price (in thousands of dollars) and the explanatory variable is the size (in square feet).
    b) Millions of dollars per square feet.
    c) The price usually increases as the size increases and so the slope must be positive.

23. 70.2% of the variability in Price is explained by the linear regression of Price on size.

25. a) 0.84
    b) Price should be 0.84 standard deviations above the average price.
    c) Price should be 1.68 standard deviations below the average price.

27. a) For every square foot increase in size, the price is expected to increase by 0.37 thousand dollars.
    b) 419.3 thousand dollars.
    c) The predicted price is 493 300 dollars so the asking price is 487 300 dollars. The value $6000 (actually − $6000) is the residual for this condo.

29. a) Probably. The residuals show some initially low points, but there is no clear curvature.

b) The linear model on Tar content accounts for 92.4% of the variability in Nicotine.

31. a) $r = \sqrt{R^2} = \sqrt{0.924} = 0.961$
    b) Nicotine should be 1.922 SDs below average.
    c) Tar should be 0.961 SDs above average.

33. a) $\widehat{Nicotine} = 0.15403 + 0.065052\ Tar$
    b) 0.414 mg
    c) Nicotine content increases by 0.065 mg of nicotine per milligram of tar.
    d) We'd expect a cigarette with no tar to have 0.154 mg of nicotine.
    e) 0.1094 mg

35. 300 pounds/foot. If a "typical" car is 15 feet long, all of 3, 30, and 3000 would give ridiculous weights.

37. a) R^2 does not tell whether the model is appropriate, but measures the strength of the linear relationship. High R^2 could also be due to an outlier.
    b) Predictions based on a regression line are for average values of *y* for a given *x*. The actual wingspan will vary around the prediction.

39. a) Probably not. Your score is better than about 97.5% of people, assuming scores follow the Normal model. Your next score is likely to be closer to the mean.
    b) The friend probably should retake the test. His score is better than only about 16% of people. His score is likely to be closer to the mean.

41. a) Moderately strong, fairly straight, and positive. Possibly some outliers (higher than expected math scores).
    b) The student with 500 verbal and 800 math.
    c) Positive, fairly strong linear relationship. 46.9% of variation in math scores is explained by verbal scores.
    d) $\widehat{Math} = 217.692 + 0.662\ Verbal$
    e) Every point of verbal score adds an average 0.662 points to the predicted math score.
    f) 548.7 points   g) 53.0 points

43. a) 0.685
    b) $\widehat{Verbal} = 162.106 + 0.709\ Math$
    c) The observed verbal score is higher than predicted from the math score.
    d) 516.6 points   e) 559.7 points
    e) Regression to the mean. Someone whose math score is below average is predicted to have a verbal score below average, but not as far (in SDs). So if we use that verbal score to predict math, they will be even closer to the mean in predicted math score than their observed math score. If we kept cycling back and forth, eventually we would predict the mean of each and stay there.

45. a)

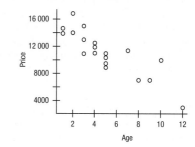

b) There is a strong, negative, linear association between price and age of used Toyota Corollas. A few observations are somewhat unusual, but not very unusual.
c) The scatterplot provides evidence that the relationship is Straight Enough. A linear model will likely be appropriate.
d) $r = −0.865$ e) 74.8%
f) The relationship is not perfect. Other factors, such as options, condition, and mileage, explain the rest of the variability in price.

47. a) The scatterplot from Exercise 45 shows that the relationship is straight, so the linear model is appropriate. The regression equation to predict the price of a used Toyota Corolla from its age is: Price = $15,729 − $935.9 × Age.

The regression output is below:

Coefficients:

	Estimate	Std. Error	t value	$Pr(>|t|)$
(Intercept)	15729.0	723.2	21.749	2.26e − 14 ***
Age	−935.9	127.9	−7.315	8.56e − 07 ***

b) According to the model, for each additional year in age, the car is expected to drop $936 in price.

c) The model predicts that a new Toyota Corolla (0 years old) will cost $15 729.

d) $10 114

e) Buy the car with the negative residual. Its actual price is lower than predicted.

f) −$2050

g) No, predicting a price after 17 years would be an extrapolation. In fact, the predicted price using the model would be negative.

49. a)

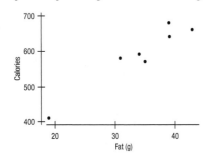

b) 92.3% of the variation in calories can be accounted for by the fat content.

c) $\widehat{Calories} = 210.954 + 11.0555\ Fat$

d)

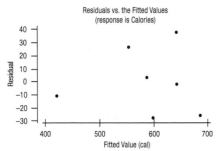

Residuals show no clear pattern, so the model seems appropriate.

e) Could say a fat-free burger still has 211.0 calories, but this is extrapolation (no data close to 0).

f) Every gram of fat adds 11.06 calories, on average.

g) 553.5 calories

51. a) The regression was for predicting calories from fat, not the other way around.

b) $\widehat{Fat} = −14.9622 + 0.083471\ Calories$ Predict 35.1 grams of fat.

53. a) This model predicts that every bridge will be deficient. The intercept is less than 5 and the slope is negative, so all predicted conditions are less than 5.

b) This model says bridges in New York City are decreasing in condition at only 0.004 per year—less rapidly than bridges in Tompkins County.

c) The R^2 for this model is only 2.6%. I don't think it has much predictive value.

55. a) 0.858

b) CO_2 levels account for 73.6% of the variation in mean temperature.

c) $\widehat{Mean\ Temp} = 10.64 + 0.00103\ CO_2$

d) The predicted mean temperature has been increasing at an average rate of 0.0103 degrees Celsius/ppm of CO_2.

e) One could say that with no CO_2 in the atmosphere, there would be a temperature of 10.64 degrees Celsius, but this is extrapolation to a nonsensical point.

f) No.

g) Predicted 14.76 degrees Celsius

h) No. The model indicates only that CO_2 and Temperature are associated. It doesn't imply any causal relationship.

57. a) $\widehat{\%Fat} = −27.3763 + 0.249874\ Weight$

b) Residuals look randomly scattered around 0, so conditions are satisfied.

c) % Body Fat increases on average by 0.25% per pound of Weight.

d) Reliable is relative. R^2 is 48.5%, but residuals have a standard deviation of 7%, so variation around the line is large.

e) 0.9%

59. a) $\widehat{HighJump} = 2.42 − 0.004690(800\ m\ time)$. According to the model, the predicted high jump decreases by an average of 0.0047 metres for each additional second in 800 metre time.

b) 6.7%

c) Yes, good high jumpers tend to be fast runners. The slope of the association is negative. Faster runners tend to jump higher, as well.

d) The residuals plot is fairly patternless. The scatterplot shows a slight tendency for less variation in high jump height among the slower runners than the faster runners. Overall, the linear model is appropriate.

e) The linear model is not particularly useful for predicting high jump performance. First of all, 6.7% of the variability in high jump height is accounted for by the variability in 800 metre time, leaving 93.3% of the variability unaccounted for. Secondly, the residual standard deviation is 0.07288 metres, which is not much smaller than the standard deviation of all high jumps, 0.074 metres. Predictions are not likely to be accurate.

61. a) As calcium levels increase, mortality rate decreases. Relationship is fairly strong, negative, and linear.

b) $Mortality = 1676 − 3.23\ Calcium$

c) Mortality decreases 3.23 deaths per 100 000, on average people, for each part per million of calcium. The intercept indicates a baseline mortality of 1676 deaths per 100 000 people with no calcium, but this is extrapolation.

d) Exeter has 348.6 fewer deaths per 100 000 people than the model predicts.

e) 1353 deaths per 100 000 people.

f) Calcium concentration explains 43.0% of the variation in death rate per 100 000 people.

63. a) The association between speed and stopping distance is strong, positive, and appears straight. Higher speeds are generally associated with greater stopping distances. The linear regression model, with equation $\widehat{Stopping\ Distance} = −20.667 + 1.151\ Speed$. The model explains 96.7% of the variability in stopping distance. However, the residuals plot has a curved pattern. The linear model is not appropriate. A model using re-expressed variables should be used.

b) Stopping distances appear to be relatively higher for higher speeds. This increase in the rate of change might be able to be straightened by taking the square root of the response variable, stopping distance.

c) The model for the re-expressed data is

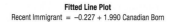
$\sqrt{Stopping\ distance} = 1.777 + 0.08226\ Speed$. The residuals plot is better than the linear fit before re-expressing the data, and shows no pattern. 98.4% of the variability in the square root of the stopping distance can be explained by the model.

d) According to the model, a car traveling 90 km/h is expected to require approximately 84.3 metres to come to a stop.

65. a)

Scatterplot of Recent Immigrant vs Canadian Born

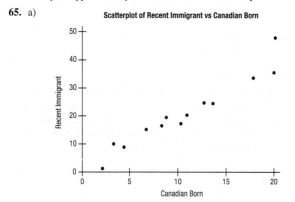

There is a strong positive linear relationship between the immigrant percentage and the Canadian born percentage. No outliers.

b) The fitted line is given below.

Fitted Line Plot
Recent Immigrant = −0.227 + 1.990 Canadian Born

c) 1.99% (This is the slope.)

d) Montreal has a large residual compared to other observations. The predicted value for Montreal is 39.8 but the actual value is 47.8.

e) The plot of residuals vs. the *Canadian born* % is given below. It doesn't look quite random. For example, most of the points after 10% Canadian born are negative. The model overestimates the immigrant percentage for cities with a Canadian born % of 10% or more. The plot of residuals vs. *Canadian born* % without Montreal is given below. This problem is much less, but the variability of residuals seems to decrease with Canadian born %.

Scatterplot of Residuals vs Canadian Born

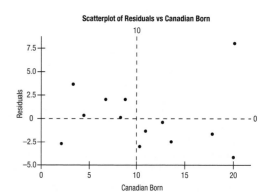

Residuals Versus Canadian Born (without Montreal)
(Response is Recent Immigrant)

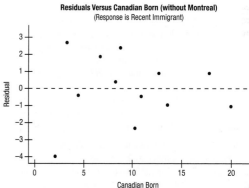

67. a) 79   b) 75.4

c) For a test score of 85, you predict an exam score of 79, but if the exam score is 79, you predict a test score of 75.4, not 85. Predictions are not reversible in direction. The regression model changes when the response does, and our prediction is always nearer the mean.

Chapter 8

1. The different segments are not scattered at random throughout the scatterplot. Each segment may have a different relationship.

3. Yes, it is clear that the relationship between January and December spending is not the same for all five segments. Using one overall model to predict January spending would be very misleading.

5. This observation was influential. After it was removed, the R^2 value and the slope of the regression line both changed by a large amount.

7. Your friend is extrapolating. It is impossible to know if a trend like this will continue so far into the future.

9. a) $354 472
b) An extrapolation this far from the data is unreliable.

11. No. In warm weather, more children will go outside and play.

13. Individual student scores will vary greatly. The class averages will have much less variability and may disguise important trends.

15. a) The trend appears to be somewhat linear up to about 1940, but from 1940 to about 1970 the trend appears to be nonlinear with a large drop from 1940 to 1950. From 1975 or so to the present, the trend appears to be linear.

b) Relatively strong for certain periods.

c) No, as a whole the graph is clearly nonlinear. Within certain periods (e.g., 1975 to the present) the correlation is high.

d) Overall, no. You could fit a linear model to the period from 1975 to 2003, but why? You don't need to interpolate, since every year is reported, and extrapolation seems dangerous.

17. a) Fairly linear, negative, strong.

b) Gas mileage decreases an average 7.652 mpg for each thousand pounds of weight.

c) No. Residuals show a curved pattern.

19. a) No. We need to see the scatterplot and residual plot first to see if the conditions are satisfied.

b) With such a high R^2, it is likely that we can make very accurate predictions—unless data plots show big problems like extreme outliers.

21. a) Millions of dollars per minute of run time.

b) Costs for longer movies increase at the same rate.

c) Regardless of run time, dramas cost less by about $20 million.

23. a) There are several features to comment on in this plot. There is a strong monthly pattern around the general trend. From 1997 to 2008, passengers increased fairly steadily with a notable exception of September 2001, probably due to the attack on the twin towers. Then sometime in late 2008, departures dropped dramatically, possibly due to the economic crisis. Recently, they have been recovering, but not at the same rate as their previous increase.

b) The trend was fairly linear until late 2008 then passengers dropped suddenly.

c) The trend since 2009 has been linear (overall, ignoring monthly oscillations). If the increase continues to be linear, the predictions should be reasonable for the short term.

25. a) 1) The point has high leverage and a small residual.

2) The point is not influential. It has the *potential* to be influential, because its position far from the mean of the explanatory variable gives it high leverage. However, the point is not *exerting* much influence, because it reinforces the association.

3) If the point were removed, the correlation would become weaker. The point heavily reinforces the positive association. Removing it would weaken the association.

4) The slope would remain roughly the same, since the point is not influential.

b) 1) The point has high leverage and probably has a small residual.

2) The point is influential. The point alone gives the scatterplot the appearance of an overall negative direction, when the points are actually fairly scattered.

3) If the point were removed, the correlation would become weaker. Without the point, there would be very little evidence of linear association.

4) The slope would increase, from a negative slope to a slope near 0. Without the point, the slope of the regression line would be nearly flat.

c) 1) The point has moderate leverage and a large residual.

2) The point is somewhat influential. It is well away from the mean of the explanatory variable, and has enough leverage to change the slope of the regression line, but only slightly.

3) If the point were removed, the correlation would become stronger. Without the point, the positive association would be reinforced.

4) The slope would increase slightly, becoming steeper after the removal of the point. The regression line would follow the general cloud of points more closely.

d) 1) The point has little leverage and a large residual.

2) The point is not influential. It is very close to the mean of the explanatory variable, and the regression line is anchored at the point $(\bar{x}, \bar{y})$, and would only pivot if it were possible to minimize the sum of the squared residuals. No amount of pivoting will reduce the residual for the stray point, so the slope would not change.

3) If the point were removed, the correlation would become slightly stronger, decreasing to become more negative. The point detracts from the overall pattern, and its removal would reinforce the association.

4) The slope would remain roughly the same. Since the point is not influential, its removal would not affect the slope.

27. 1) Point e is very influential. Its addition will give the appearance of a strong, negative correlation like $r = -0.90$.

2) Point d is influential (but not as influential as point e). Its addition will give the appearance of a weaker, negative correlation like $r = -0.40$.

3) Point c is directly below the middle of the group of points. Its position is directly below the mean of the explanatory variable. It has no influence. Its addition will leave the correlation the same, $r = 0.00$.

4) Point b is almost in the centre of the group of points, but not quite. Its addition will give the appearance of a very slight positive correlation like $r = 0.05$.

5) Point a is very influential. Its addition will give the appearance of a strong, positive correlation like $r = 0.75$.

29. Perhaps high blood pressure causes high body fat, high body fat causes high blood pressure, or both could be caused by a lurking variable such as a genetic or lifestyle issue.

31. a) The graph shows that, on average, students progress at about one reading level per year. This graph shows averages for each grade. The linear trend has been enhanced by using averages.

b) Very close to 1.

c) The individual data points would show much more scatter, and the correlation would be lower.

d) A slope of 1 would indicate that for each 1-year grade level increase, the average reading level is increasing by 1 year.

33. a) *Cost* decreases by $2.13 per degree of average daily *Temp*. So warmer temperatures indicate lower costs.

b) For an avg. monthly temperature of 0°F, the cost is predicted to be $133.

c) Too high; the residuals (observed – predicted) around 32°F are negative, showing that the model overestimates the costs.

d) $111.70   e) About $105.70

f) No, the residuals show a definite curved pattern. The data are probably not linear.

g) No, there would be no difference. The relationship does not depend on the units.

35. a) 0.88

b) Interest rates during this period grew at about 0.25% per year, starting from an interest rate of about 0.61%.

c) Substituting 50 in the model yields a predicted rate of about 13%.

d) Not really. Extrapolating 20 years beyond the end of these data would be dangerous and unlikely to be accurate.

37. a) Interest rates peaked about 1980 and decreased afterward. This regression has a negative slope and high intercept.

b) This model predicts the interest rate in 2000 to be 3.71%, much lower than the other model predicts.

c) Separating the data into parts will make the linear model predict better on each part, but the model is still not appropriate. There is no way of knowing if the linear trend will continue.

d) The first predicts 17.96%, the second −2.35%.

e) Both predictions are absurd. Neither model can extrapolate that far. The first would be historically high and the second is negative.

39. a) Stronger. Both slope and correlation would increase.
   b) Restricting the study to nonhuman animals would justify it.
   c) Moderately strong.
   d) For every year increase in life expectancy, the gestation period increases by about 15.5 days, on average.
   e) About 270.5 days.

41. a) Removing hippos would make the association stronger, since hippos are more of a departure from the pattern.
   b) Increase.
   c) No, there must be a good reason for removing data points.
   d) Yes, removing it lowered the slope from 15.5 to 11.6 days per year.

43. a) Answers may vary. Using the data for 1955–2010 results in a scatterplot that is relatively linear with some curvature. You might use the data after 1955 only to predict 2015, but that would still call for extrapolation and would not be safe. The prediction is 27.19.
   b) Not much, since the data are not truly linear and 2020 is 10 years from the last data point (extrapolating is risky).
   c) No, extrapolation of more than 50 years would be absurd. There's no reason to believe the trend from 1955 to 2005 will continue.

45. a)

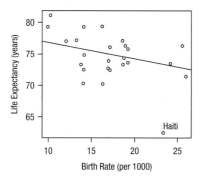

There is a moderate, negative, linear relationship between birth rate and life expectancy.
   b) Haiti is an outlier with a much lower than predicted life expectancy.
   c) $\widehat{lifeexp} = 79.504 - 0.264\ Birth\ Rate$.
   d) The residual plot is reasonably scattered.
   e) Yes.
   f) 14.1% of the variation in life expectancy is accounted for by the regression on birth rate.
   g) While there is an association, there is no reason to expect causality. Lurking variables may be involved.

47. a) The scatterplot of the percent earning over $150 000 versus year is given below. The percent earning over $150 000 increases over time, but not quite linearly. In the period 1986–1992, the relationship has a different form than that in the period 1993–2011. The relationship in the period 1993–2011 is strong, and roughly linear.

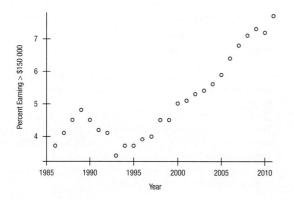

   b) The regression equation is: $\widehat{Percent > \$150\ 000} = -288 + 0.14667\ Year$. 76.03% of the variation in the percentage earning over $150 000 is explained by the straight line relationship (this is R^2).
   c) The predicted percent for 1997 is $-288 + 0.14667(1997) = 4.87\%$. The actual value (in the data set) is 4.0%. The predicted value is a bit too high—the residual is –0.87%.
   d) Dropping the first seven years (from 1986 to 1992), we obtain the following model: $\widehat{Percent > \$15\ 0000} = -481 + 0.2428\ Year$. 98.4% of the variation in the percentage earning over $150 000 is explained by the linear relationship (this is R^2). This model fits the later data better, and is more appropriate because the trend after 1992 appears to be close to linear.

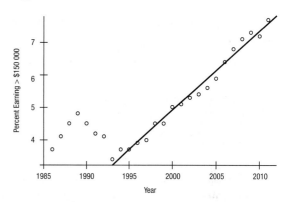

The predicted percent for 2013 is $-481 + 0.2428(2013) = 8.066\%$. The predicted percent for 2040 is $-481 + 0.2428(2040) = 14.62\%$. Both 2013 and 2040 are out of the period for which we had information (to estimate the model) and so both these are extrapolations. 2040 is way too far from this period (1986–2001) and so the predicted value is not very reliable. We should not be very confident that the linear increase will continue until 2040 given the earlier data from 1986–1992, but the prediction for 2013 might be closer to accurate.

49. a) The scatterplot of the difference in median earnings versus year is given below. The difference decreases with year, but not quite linearly. The relationship is strong with no outliers.

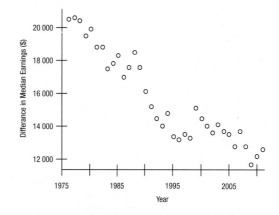

   b) The linear regression model (the straight line fit) is:
   Difference in Median earnings = 497806 − 241.84 Year.
   The R^2 for the model is 87.9%, i.e. 87.9% of the variation in the differences in medians is explained by this model. The plot of residuals versus year is given below. The curving pattern on this plot indicates that a higher order model (e.g., a quadratic model) or some other curving pattern is needed to describe the relationship.

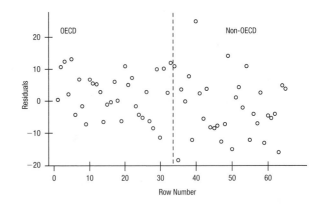

c) The predicted difference for 2070 (assuming that the form of the relationship will remain the same after 2011, and that the linear model is appropriate) is $497\,806 - 241.84 \times 2070 = -2804$. The difference is likely to be higher than this as we can see that the decrease in the difference reduces with time.

d) The plot of residuals for the quadratic model is much more random (free of patterns) than that for the linear model and so this is a more appropriate model than the linear model.

51. a) The plot of male mean score versus female mean score with the regression is shown below.

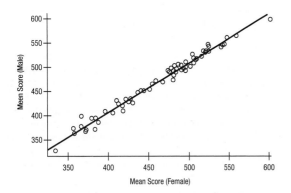

b) The plot of residuals versus the predicted values is shown below. The plot looks random (free of patterns) and so a straight line model seems appropriate. There is one large residual whose magnitude is greater than 20 points (Colombia).

c) The histogram looks roughly bell shaped and the normal probability plot is close to straight, both suggesting that the residuals are approximately normally distributed.

d) The plot of residuals versus row number is shown below. It seems that the OECD countries are less variable than the non-OECD countries.

e) Shanghai-China and Kyrgyzstan have the greatest leverage on the fit. Their x-values (i.e., the means for females) are far from the average of all x-values. The average female score in Shanghai-China is 601 and the average female score in Kyrgyzstan is 334.

f) Slope close to 1 implies when a country's average for females is higher than another country's female average by a certain amount, the country average for males also is higher by approximately the same amount, on average.
When the slope is 1, the estimate of the Y-intercept is given by $\bar{Y} - \bar{X} =$ Average for males – Average for females $= 471.98 - 463.18 = 8.80$, so the estimated prediction equation is: Mean score (Male) $= 8.80 +$ Mean score (Female).

g) The correlation will decrease as we can see from the scatter plot that the points with both male and female averages greater than 500 look less linear. The other points following the same pattern on the scatterplot make it look more linear and so add strength to the linear relationship. Restricting the range usually decreases the correlation.

53. a) Correlations: Year, Bust size (in.), Waist size (in.), Hips size (in.)

	Year	Bust size	Waist size
Bust size	1.000		
Waist size	0.992	0.992	
Hips size	0.936	0.936	0.929

b) These values are averages. The correlations with averages are usually high since they do not show the variability in individual observations.

c) The prediction equation is Waist size (in.) $= -182 + 0.105$ Year. The predicted waist size for 1950 is 22.75 and for 2080 is 36.40 inches. The year 2080 is way outside the period for which we have information and so the prediction for 2080 is not very reliable (extrapolation).

d) The prediction equation is Waist size (in.) $= -47.6 + 2.10$ Bust size (in.). The predicted waist size for UK women if their average bust size gets up to 44 is 44.8. The average bust size is outside the range of data on bust sizes and so this prediction is not very reliable.

55. a) *Arm span* is the best predictor of *Middle finger*. 32% of the variation in *Middle finger* measurements can be explained by a linear association *Arm span*.

b) *Height* is the best predictor of *Arm span*. 61% of the variation in *Arm span* measurements can be explained by a linear association *Height*.

c) The prediction equation is *Arm span* $= 16.8 + 0.897$ *Height*. The slope of this regression equation is 0.987. That means for every unit in increase in *Height*, the mean *Arm span* increases by 0.987 units.

d) The histogram is close to a bell-shape and the normal scores plot is close to straight, both indicating that there are no serious violations of the assumption of normality of residuals.

e) The plot of residuals against height is shown below. It looks pretty random, indicating no serious violations of the assumption of independents of error terms.

Residuals versus Height
(Response is armspan)

f) The individual with height 164 and arm span 187 (i.e., observation number 75) has the most unusual arm span (3.13 units longer than the average arm span for individuals with height = 164).

g) The plot of residuals versus gender (with GenderCode = 0 for girls and 1 for boys) is shown below.

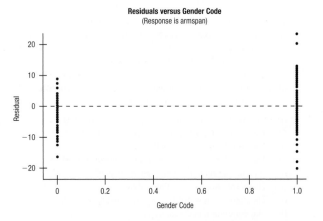

Residuals versus Gender Code
(Response is armspan)

The residuals are mostly negative for girls indicating that this regression model often overestimates the arm span of girls.

57. a) The association between distance from the sun and planet year is strong, positive, and curved concave upward. Generally, planets farther from the Sun have longer years than closer planets.

b) The rate of change in length of year per unit distance appears to be increasing, but not exponentially. Re-expressing with the logarithm of each variable may straighten a plot such as this:

Planetary Distance and Length of Year

The scatterplot and residuals plot for the linear model relating log(Distance) and log(Length of Year) appear below.

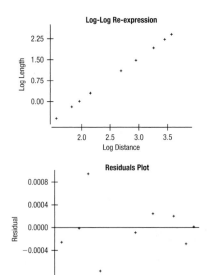

Log-Log Re-expression

Residuals Plot

The regression model for the log-log re-expression is:
$$\widehat{\log (Length)} = -2.95 + 1.5 \log(Distance).$$

c) The model explains 100% of the variability in the log of the length of the planetary year, at least according to the accuracy of the statistical software. The residuals plot is scattered, and the residuals are all extremely small. This is a very accurate model.

59. a) The value of R^2 is 60.6%. That is, 60.6% of the variation in the neonatal brain weight is explained by the linear regression model. The plot of residuals versus predicted values shows a curving pattern indicating that a straight line model is not appropriate. This can also be seen on the scatterplot of the neonatal brain weight versus gestation, but the curvature is more prominent in the plot of residual versus predicted values.

Fitted Line Plot
Brain Weight (g) = −78.46 + 0.9176 Gestation (days)

S	50.9455
R-Sq	60.6%
R-Sq(adj)	58.7%

Residuals vs. the Fitted Values
(Response is Brain Weight(g))

b) The value of R^2 is 71.6%. That is, 71.6% of the variation in the logarithms of the neonatal brain weight is explained by this regression model. The plot of residuals versus predicted values looks random (free of patterns), indicating that this model is appropriate for describing the relationship between the two variables.

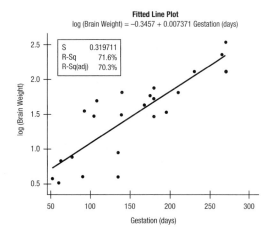

Fitted Line Plot
log (Brain Weight) = −0.3457 + 0.007371 Gestation (days)

S	0.319711
R-Sq	71.6%
R-Sq(adj)	70.3%

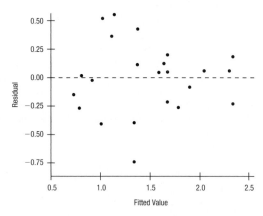

Residuals versus the Fitted Values
(Response is log (Brain Weight))

c) The plot of the square roots of the neonatal brain weight versus gestation with the regression line is shown below. The value of R^2 for this model is 74.09%. That is, 74.09% of the variation in the square roots of the neonatal brain weight is explained by this regression model. The plot of residuals versus predicted values is also given below. It looks like the variability of residuals increases with the fitted value (and also the plot of residuals versus predicted values for the log transformation looks a bit more random than that for the square root model) and so the log transformation is a bit better (even though the R^2 is a bit higher for the square root transformation).

Fitted Line Plot
sqrt (Brain Weight) = −1.650 + 0.05443 Gestation (days)

S	2.22144
R-Sq	74.0%
R-Sq(adj)	72.8%

Residuals versus the Fitted Values
(Response is sqrt (Brain Weight))

d) The predicted log brain weight for a 200-day gestation period is 1.82 and so the predicted brain weight is 66.1g.
e) The predicted square root brain weight for a 200-day gestation period is 9.23 and so the predicted brain weight is 85.19g.

Part II Review

1. % over 50: $r = 0.69$
    % under 20: $r = -0.71$
    % Full-time Fac.: $r = 0.09$
    % Gr. on time: $r = -0.51$

3. a) There does not appear to be a linear relationship.
    b) Nothing, there is no reason to believe that the results for the Finger Lakes region are representative of the vineyards of the world.
    c) $\overline{CasePrice} = 92.765 + 0.567284\ Years$
    d) Only 2.7% of the variation in case price is accounted for by the ages of vineyards. Most of that is due to two outliers. We are better off using the mean price rather than this model.

5. a) (i)   The correlation is negative, approximately $r = -0.7$.
        (ii) The correlation is negative, approximately $r = -0.85$.
        (iii) The correlation is essentially zero.
        (iv) The correlation is negative, approximately $r = -0.75$.
    b) (i)   The correlation is positive, approximately $r = 1$.
        (ii) The correlation is essentially zero.
        (iii) The correlation is positive, approximately $r = 0.8$.
        (iv) The correlation is essentially zero.

7. a) −0.52
    b) Negative, moderate, linear, but with more variation as pH increases.
    c) The BCI would also be average.
    d) The predicted BCI will be 1.56 SDs of BCI below the mean BCI.

9. a) $\overline{Manatee\ Deaths} = -51.87 + 0.1429\ Powerboat\ Registrations$ (in 1000s)
    b) According to the model, for each increase of 10 000 powerboat registrations, the number of manatees killed increases by approximately 1.43 on average.
    c) If there were 0 powerboat registrations, the number of manatee deaths would be −51.87. This is obviously a silly extrapolation.
    d) The predicted number is 83.72 deaths. The actual number of deaths was 97. The residual is 97 − 83.72 = 13.28. The model underestimated the number of deaths by 13.28 manatees.
    e) Negative residuals would suggest that the actual number of deaths was lower than the predicted number.
    f) As the number of powerboat registrations has increased the number of manatees killed has increased. We don't know whether powerboat registrations will continue to increase. And even if they do, we don't know whether the trend will continue. Even though these manatees were killed by powerboats, we have no way of knowing if the relationship will continue as it is.

11. a) −0.984   b) 96.9%   c) 32.95 mph   d) 1.66 mph
    e) Slope will increase.

f) Correlation will weaken (become less negative).

g) Correlation is the same, regardless of units.

13. a) A scatterplot of the 2011 versus 2001 census counts is shown below. The correlation appears to be very close to 1. In fact, $r = 0.994$.

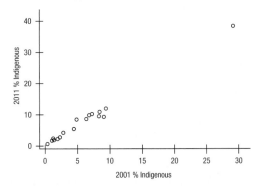

b) The fitted least squares line $\widehat{2011Percent} = 0.19 + 1.29(2001Percent)$.

c) The slope tells us that for each increase of 1% in the percentage of Indigenous people in 2001 there is a 1.29% increase in the percentage of Indigenous people in 2011.

d) Prince Albert has the highest percentage of Indigenous people. It also has the highest leverage of all census areas. However, if you set Prince Albert aside, there appear to be two large outliers, or perhaps even a pattern in the residuals – possible curvature and/or increasing spread.

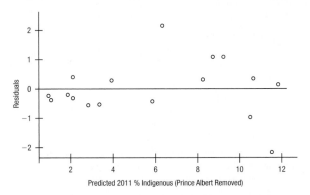

e) Taking logs seems to have resolved the issue. The new model is $\log_{10}(\widehat{2011Percent}) = 0.21 + 0.89 \log_{10}(2001Percent)$.

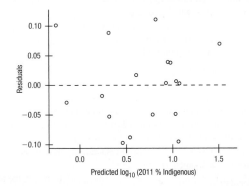

f) 12.6% (using the log transformed model)

15. a) $\widehat{Horsepower} = 3.50 + 34.31\ Weight$

b) The weight is measured in thousands of pounds. The slope of the model predicts an increase of about 34.3 horsepower for each additional unit of weight. 34.3 horsepower for each additional thousand pounds makes more sense than 34.3 horsepower for each additional pound.

c) Since the residuals plot shows no pattern, the linear model is appropriate for predicting horsepower from weight.

d) According to the model, a car weighing 2595 pounds is expected to have 92.54 horsepower. The actual horsepower of the car is 115.0 horsepower.

17. a) The scatterplot shows a fairly strong, linear and positive relation. There seem to be two distinct clusters of data.

b) $\widehat{Interval} = 33.97 + 10.36\ Duration$

c) The time between eruptions increases by about 10.4 minutes per minute of *Duration* on average.

d) 77% of the variation in *Interval* is accounted for by *Duration*. The prediction is fairly accurate but not precise.

e) 75.4 minutes

f) A residual is the observed value minus the predicted value. So the residual $= 79 - 75.4 = 3.6$ minutes, indicating that the model underestimated the interval in this case.

19. a) $r = 0.888$. Although r is high, you must look at the scatterplot and verify that the relation is linear in form.

b)

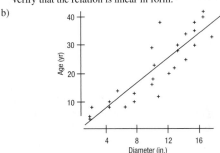

The association between diameter and age appears to be strong, somewhat linear, and positive.

c) $\widehat{Age} = -0.97 + 2.21 \times Diameter$

d)

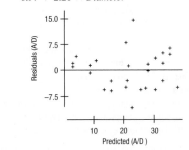

The residuals show a curved pattern (and two outliers).

e) The residuals for five of the seven largest trees (15 in. or larger) are positive, indicating that the predicted values underestimate the age.

21. TV screen sizes might vary from 19 to 70 inches. A TV with a screen that was 10 inches larger would be predicted to cost $10(0.03) = +0.3$, $10(0.3) = +3$, $10(3) = +30$, or $10(30) = +300$. Only $300 seems reasonable, so the slope must be 0.3.

23. The best one was the reciprocal square root re-expression, resulting in the equation $\dfrac{1}{\sqrt{DrainTime}} = 0.00243 + 0.219\ Diameter$.

25. There is a strong, roughly linear, negative association between mean January temperature and latitude. U.S. cities with higher latitudes generally have lower mean January temperatures. There are two outliers, cities with higher mean January temperatures than the pattern would suggest.

27. a) 71.9%

b) The negative correlation indicates that as the latitude increases, the average January temperature generally decreases.

c) $\widehat{JanTemp} = 108.796 - 2.111\ Latitude$

d) For each additional degree of latitude, the model predicts a decrease of approximately 2.1°F in average January temperature.

e) The model predicts that the mean January temperature will be approximately 108.8 °F when the latitude is 0°. This is an extrapolation, and may not be meaningful.

f) 24.4°F

g) In this context, a positive residual means that the actual average temperature in the city was higher than the temperature predicted by the model. In other words, the model underestimated the average January temperature.

h) Temperature in degrees Celsius are related to temperature in degrees Fahrenheit as

$$C = \frac{5}{9}(F - 32)$$

(i) Correlation is unchanged by linear transformations in either or both variables.

(ii) The equation of the line in Celsius is:

$$C = \frac{5}{9}(F - 32)$$

$$\approx \frac{5}{9}(108.796 - 2.111\ Latitude - 32)$$

$$\approx 42.664 - 1.173\ Latitude$$

(iii) The predicted temperature at Denver is $42.664 - 1.173(40)$ $= -4.256$ °C. Note that this is the same as the previous prediction, once it is converted from °F to °C, i.e.,

$$\frac{5}{9}(24.4 - 32) = -4.22 \text{ apart from round-off error.}$$

29. a) The association between Olympic long jump distances and high jump heights is strong, linear, and positive. Years with longer long jumps tended to have higher high jumps. There is one departure from the pattern. The year in which the Olympic gold medal long jump was the longest had a shorter gold medal high jump than we might have predicted.

b) There is an association between long jump and high jump performance, but it is likely that training and technique have improved over time and affected both jump performances.

c) The correlation would be the same.

d) In a year when the high jumper jumped one standard deviation better than the average jump, the long jumper would be predicted to jump $r = 0.912$ standard deviations above the average long jump.

31. a) No relation; the correlation would probably be close to 0.

b) The relation would have a positive direction and the correlation would be strong, assuming students were studying French in each grade level. Otherwise, no correlation.

c) No relation; correlation close to 0.

d) The relation would have a positive direction and the correlation would be strong, since vocabulary would increase with each grade level.

33. $\widehat{Calories} = 560.7 - 3.08\ Time$

Each extra minute at the table corresponds to in 3.08 fewer calories consumed on average. Perhaps hungrier children eat faster and eat more.

35. There seems to be a strong, positive, linear relationship with one high-leverage point (Northern Ireland) that makes the overall R^2 quite low. Without that point, the R^2 increases to 61.5%. Of course, these data are averaged across thousands of households, and so the correlation appears to be higher than individuals would be. Any conclusions about individuals would be suspect.

37. a) 30 818 pounds

b) 1302 pounds

c) 31 187.6 pounds

d) $R^2 = 93\%$, so the model explains 93% of the variability in weight, but some of the residuals are 1000 pounds or more.

e) Negative residuals will be more of a problem, as the predicted weight would overestimate the weight of the truck; trucking companies might be inclined to take the ticket to court.

Chapter 9

1. The University of British Columbia administrators should take a survey. They should sample a part of the student body, selecting respondents with a randomization method. They should be sure to draw a sufficiently large sample.

3. The proportion in the sample is a statistic. The proportion of all students is the parameter of interest. The statistic estimates that parameter, but is not likely to be exactly the same.

5. This is not an SRS. Although each student may have an equal chance to be in the survey, groups of friends who choose to sit together will either all be in or out of the sample, so the selection is not independent.

7. a) Cluster sample.

b) Stratified sample.

c) Systematic sample.

9. Several terms are poorly defined. The survey needs to specify the meaning of "family" for this purpose and the meaning of "higher education." The term "seek" may also be poorly defined (for example, would applying to college but not being admitted qualify for seeking more education?)

11. a) This would suffer from voluntary response bias.

b) This would be a convenience sample.

13. a) No. It would be nearly impossible to get exactly 500 males and 500 females from every country by random chance.

b) A stratified sample, stratified by whether the respondent is male or female.

15. a) Voluntary response.

b) We have no confidence at all in estimates from such studies.

17. a) The population of interest is all adults in the United States aged 18 and older.

b) The sampling frame is U.S. adults with landline telephones.

c) Some members of the population (e.g., many college students) don't have landline phones, which could create a bias.

19. a) Population—Human resources directors of Fortune 500 companies.

b) Parameter—Proportion who don't feel surveys intruded on their work day.

c) Sampling Frame—List of HR directors at Fortune 500 companies.

d) Sample—23% who responded.

e) Method—Questionnaire mailed to all (nonrandom).

f) Bias—Nonresponse. Hard to generalize because who responds is related to the question itself.

21. a) Population—All U.S. adults.

b) Parameter—Proportion who have used and benefited from alternative medicine.

c) Sampling Frame—All Consumers Union subscribers.

d) Sample—Those who responded (random).

e) Method—Questionnaire to all (nonrandom).

f) Bias—Nonresponse. Those who respond may have strong feelings one way or another.

23. a) Population—Adults.

b) Parameter—Proportion who think drinking and driving is a serious problem.

c) Sampling Frame—Bar patrons.

d) Sample—Every 10th person leaving the bar.

e) Method—Systematic sampling.

f) Bias—Those interviewed had just left a bar. They probably think drinking and driving is less of a problem than do adults in general.

25. a) Population—Soil around a former waste dump.
    b) Parameter—Concentrations of toxic chemicals.
    c) Sampling Frame—Accessible soil around the dump.
    d) Sample—16 soil samples.
    e) Method—Not clear.
    f) Bias—Don't know if soil samples were randomly chosen. If not, may be biased toward more or less polluted soil.

27. a) Population—Snack food bags.
    b) Parameter—Weight of bags, proportion passing inspection.
    c) Sampling Frame—All bags produced each day.
    d) Sample—Bags in 10 randomly selected cases, 1 bag from each case for inspection.
    e) Method—Multistage sampling.
    f) Bias—Should be unbiased.

29. Bias. Only people watching the news will respond, and their preference may differ from that of other voters. The sampling method may systematically produce samples that don't represent the population of interest.

31. a) Voluntary response. Only those who both see the ad *and* feel strongly enough will respond.
    b) Cluster sampling. One school may not be typical of all.
    c) Attempted census. Will have nonresponse bias.
    d) Stratified sampling with follow-up. Should be unbiased.

33. a) This is a systematic sample.
    b) This sample is likely to be representative of those waiting in line for the roller coaster, especially if those people at the front of the line (after their long wait) respond differently from those at the end of the line.
    c) The sampling frame is patrons willing to wait for the roller coaster on that day at that time. It should be representative of the people in line, but not of all people at the amusement park.

35. a) This is a voluntary response sample.
    b) We have absolutely no confidence in estimates made from voluntary response samples. Voluntary response bias exists. Additionally, undercoverage limits the scope of any conclusions to the population of young adults, since only those that visited gamefaqs.com could have been chosen to be in the sample.

37. a) Answers will definitely differ. Question 1 will probably get many "No" answers, while Question 2 will get many "Yes" answers. This is wording bias.
    b) "Do you think standardized tests are appropriate for deciding whether a student should be promoted to the next grade?" (Other answers will vary.)

39. Only those who think it worth the wait are likely to be in line. Those who don't like roller coasters are unlikely to be in the sampling frame, so the poll won't get a fair picture of whether park patrons overall would favour still more roller coasters.

41. a) Biased toward yes because of "pollute." "Should companies be responsible for any costs of environmental cleanup?"
    b) Biased toward no because of "old enough to serve in the military." "Do you think the drinking age should be lowered from 21?"

43. a) Not everyone has an equal chance. People with unlisted numbers, people without phones, and those at work cannot be reached.
    b) Generate random numbers and call at random times.
    c) Under the original plan, those families in which one person stays home are more likely to be included. Under the second plan, many more are included. People without phones are still excluded.
    d) It improves the chance of selected households being included.
    e) This takes care of phone numbers. Time of day may be an issue. People without phones are still excluded.

45. a) Answers will vary.
    b) The parameter estimated by 10 measurements is the true length of your arm. The population is all possible measurements of your arm length.
    c) The population is now the arm lengths of your friends. The average now estimates the mean of the arm lengths of your friends.
    d) These 10 arm lengths are unlikely to be representative of the community, or the country. Your friends are likely to be of the same age, and not very diverse.

47. a) Assign numbers 001 to 120 to each order. Use random numbers to select 10 transactions to examine.
    b) Sample proportionately within each type. (Do a stratified random sample.)

49. a) Select three cases at random; then select one jar randomly from each case.
    b) Use random numbers to choose 3 cases from numbers 61 through 80; then use random numbers between 1 and 12 to select the jar from each case.
    c) No. Multistage sampling.

51. a) Depends on the Yellow Page listings used. If from regular (line) listings, this is fair if all doctors are listed. If from ads, probably not, as those doctors may not be typical.
    b) Not appropriate. This cluster sample will probably contain listings for only one or two business types.

Chapter 10

1. This is an observational study because the sports writer is not randomly assigning players to take steroids or not take steroids; the writer is merely observing home run totals between two eras. It would be unwise to conclude steroids caused any increases in home runs because we need to consider other factors besides steroids—factors possibly leading to more home runs include better equipment, players training more in the offseason, smaller ballparks, better scouting techniques, etc.

3. Each of the 40 deliveries is an experimental unit. He has randomized the experiment by flipping a coin to decide whether or not to phone.

5. The factor is calling, and the levels are whether or not he calls the customer. The response variable is the tip percentage for each delivery.

7. By calling some customers but not others during the same run, the driver has controlled many variables, such as day of the week, season, and weather. The experiment was randomized because he flipped a coin to determine whether or not to phone and it was replicated because he did this for 40 deliveries.

9. Because customers don't know about the experiment, those that are called don't know that others are not, and vice versa. Thus, the customers are blind. That would make this a single-blind study. It can't be double-blind because the delivery driver must know whether or not he phones.

11. Yes. Driver is now a block. The experiment is randomized within each block. This is a good idea because some drivers might generally get higher tips than others, but the goal of the experiment is to study the effect of phone calls. Blocking on driver eliminates the variability in tips inherent to the driver.

13. Answers will vary. The cost or size of a delivery may confound his results. Larger orders may generally tip a higher or lower percentage of the bill.

15. a) No, this is not an experiment. There are no imposed treatments. This is an observational study.
    b) There can be other factors related to Facebook usage that are common among Facebook users and some of those factors might be the reason for the changes in the differences in average graduating grades.

17. a) This is a retrospective observational study.
b) That's appropriate because MS is a relatively rare disease.
c) The subjects were U.S. military personnel, some of whom had developed MS.
d) The variables were the vitamin D blood levels and whether or not the subject developed MS.

19. a) Randomized, comparative, placebo-controlled experiment.
b) Yes, such an experiment is the right way to determine whether black cohosh is an effective treatment for hot flashes.
c) The subjects were 351 women, aged 45 to 55, who reported at least two hot flashes a day.
d) The treatments were black cohosh, a multi-herb supplement, plus advice to consume more soy foods, estrogen, and a placebo. The response was the women's self-reported symptoms, presumably the frequency of hot flashes.

21. a) Experiment.   b) 30 patients with bipolar disorder.
c) The experiment has 1 factor (omega-3 fats from fish oil) at 2 levels (high dose of omega-3 fats from fish oil and no omega-3 fats from fish oil).
d) One factor at 2 levels gives a total of 2 treatments.
e) The response variable is "improvement," but there is no indication of how the response variable was measured.
f) There is no information about the design of the experiment.
g) The experiment is blinded, since the use of a placebo keeps the patients from knowing whether or not they received the omega-3 fats from fish oils. It is not stated whether or not the evaluators of the "improvement" were blind to the treatment, which would make the experiment double-blind.
h) Although it needs to be replicated, the experiment can determine whether or not omega-3 fats from fish oils cause improvements in patients with bipolar disorder, at least over the short term. The experiment design would be stronger if it were double-blind.

23. a) Observational study.
b) Prospective.
c) Men and women with moderately high blood pressure and normal blood pressure, unknown selection process.
d) Memory and reaction time.
e) As there is no random assignment, there is no way to know that high blood pressure caused subjects to do worse on memory and reaction time tests. A lurking variable may also be the cause.

25. a) Observational study.   b) Retrospective.
c) Swedish men, unknown selection process.
d) Occurrence of kidney cancer.
e) As there is no random assignment, there is no way to know that being overweight or high blood pressure caused the higher risk for kidney cancer.

27. a) This is an experiment, since treatments were imposed on randomly assigned groups.
b) Twenty-four post-menopausal women were the subjects in this experiment.
c) There is 1 factor (type of drink) at 2 levels (alcoholic and non-alcoholic). (Supplemental estrogen is not a factor in the experiment, but rather a blocking variable. The subjects were not given estrogen supplements as part of the experiment.).
d) One factor, with 2 levels, is 2 treatments.
e) The response variable is an increase in estrogen level.
f) This experiment uses a blocked design. The subjects were blocked by whether or not they used supplemental estrogen. This design reduces variability in the response variable of estrogen level that may be associated with the use of supplemental estrogen.
g) This experiment does not use blinding.

29. a) Experiment.   b) Locations in a garden.
c) 1 factor: traps (2 levels).
d) 2 treatments.   e) Number of bugs in the trap.
f) Blocked by location.   g) Not blind.

h) One type of trap is more effective than the other.

31. a) Observational study.
b) Retrospective.
c) Women in Finland, unknown selection process with data from church records.
d) Mothers' lifespans.
e) As there is no random assignment, there is no way to know that having sons or daughters shortens or lengthens the lifespan of mothers.

33. a) Observational study.   b) Prospective.
c) People with or without depression, unknown selection process.
d) Frequency of crying in response to sad situations.
e) There is no apparent difference in crying response (to sad movies) for depressed and nondepressed groups.

35. a) Experiment.   b) Rats.
c) 1 factor: sleep deprivation; four levels.
d) 4 treatments.
e) Glycogen content in the brain.
f) No discussion of randomness.
g) Blinding is not discussed.
h) The conclusion could be that rats deprived of sleep have significantly lower glycogen levels and may need sleep to restore that brain energy fuel. Extrapolating to humans would be very speculative.

37. a) This is an experiment. Subjects were randomly assigned to treatments.
b) The subjects were the people in this experiment experiencing migraines.
c) There are 2 factors (pain reliever and water temperature). The pain reliever factor has 2 levels (pain reliever or placebo), and the water temperature factor has 2 levels (ice water and regular water).
d) Two factors, at 2 levels each, results in 4 treatments.
e) The response variable is the level of pain relief.
f) The experiment is completely randomized.
g) The subjects are blinded to the pain reliever factor through the use of a placebo. The subjects are not blinded to the water factor. They will know whether they are drinking ice water or regular water.
h) The experiment may indicate whether pain reliever alone or in combination with ice water give pain relief, but patients are not blinded to ice water, so the placebo effect may also be the cause of any relief seen due to ice water.

39. a) This is an experiment. Athletes were randomly assigned to one of two exercise programs.
b) The subjects are athletes suffering hamstring injuries.
c) There is one factor (type of exercise) at 2 levels (static stretching, and agility and trunk stabilization).
d) One factor, at 2 levels, results in 2 treatments.
e) The response variable is the time before the athletes were able to return to sports.
f) The experiment is completely randomized.
g) The experiment employs no blinding. The subjects know what kind of exercise they do.
h) Assuming that the athletes actually followed the exercise program, this experiment can help determine which of the two exercise programs is more effective at rehabilitating hamstring injuries.

41. The experimenters need to compare omega-3 results to something. Perhaps bipolarity is seasonal and would have improved during the experiment anyway.

43. a) Subjects' responses might be related to many other factors (diet, exercise, genetics, etc.). Randomization should equalize the two groups with respect to unknown factors.
b) More subjects would minimize the impact of individual variability in the responses, but the experiment would become more costly and time consuming.

45. People who engage in regular exercise might respond differently to the omega-3 fats, and that additional variability could obscure the effectiveness of this treatment.

47. Answers may vary. Number the tomatoes plants 1 to 24. Use a random number generator to randomly select 24 numbers from 01 to 24 without replication. Assign the tomato plants matching the first 8 numbers to the first group, the second 8 numbers to the second group, and the third group of 8 numbers to the third group.

49. a) First, they are using athletes who have a vested interest in the success of the shoe by virtue of their sponsorship. They should choose other athletes. Second, they should randomize the order of the runs, not run all the races with their shoes second. They should blind the athletes by disguising the shoes if possible, so they don't know which is which. The timers shouldn't know which athletes are running with which shoes, either. Finally, they should replicate several times, since times will vary under both shoe conditions.
b) Because of the problems in (a), the results they obtain may be biased in favour of their shoes. In addition, the results obtained for Olympic athletes may not be the same as for runners in general.

51. a) Allowing athletes to self-select treatments could confound the results. Other issues such as severity of injury, diet, age, etc., could also affect time to heal, and randomization should equalize the two treatment groups with respect to any such variables.
b) A control group could have revealed whether either exercise program was better (or worse) than just letting the injury heal.
c) Doctors who evaluated the athletes to approve their return to sports should not know which treatment the subject had engaged in.
d) It's hard to tell. The difference of 15 days seems large, but the standard deviations indicate that there was a great deal of variability in the times.

53. a) The differences among the Mozart and quiet groups were more than would have been expected from ordinary sampling variation.
b)

c) The Mozart group seems to have the smallest median difference and thus the *least* improvement, but there does not appear to be a significant difference.
d) No, if anything, there is less improvement, but the difference does not seem significant compared with the usual variation.

55. a) Observational. Randomly select a group of children, ages 10 to 13, have them taste the cereal, and ask if they like the cereal.
b) Answers may vary. Get volunteers ages 10 to 13. Each volunteer will taste both cereals, randomizing the order in which they taste them. Compare the percentage of favourable ratings for each cereal.

c) Answers may vary. From the volunteers, identify the children who watch Frump and identify the children who do not watch Frump. Use a blocked design to reduce variation in cereal preference that may be associated with watching the Frump cartoon.

57. a) Observational, prospective study.
b) The supposed relation between health and wine consumption might be explained by the confounding variables of income and education.
c) None of these. While the variables have a relation, there is no causality indicated for the relation.

59. a) Arrange the 20 containers in 20 separate locations. Use a random-number generator to identify the 10 containers that should be filled with water.
b) Guessing, the dowser should be correct about 50% of the time. A record of 60% (12 out of 20) does not appear to be significantly different.
c) Answers may vary. You would need to see a high level of success—say, 90% to 100%, that is, 18 to 20 correct.

61. Randomly assign half the reading teachers in the district to use each method. Students should be randomly assigned to teachers as well. Make sure to block both by school and grade (or control grade by using only one grade). Construct an appropriate reading test to be used at the end of the year, and compare scores.

63. a) They mean that the difference is higher than they would expect from normal sampling variability.
b) An observational study.
c) No. Perhaps the differences are attributable to some confounding variable (e.g., people are more likely to engage in riskier behaviours on the weekend) rather than the day of admission.
d) Perhaps people have more serious accidents and traumas on weekends and are thus more likely to die as a result.

65. Answers may vary. This experiment has 1 factor (pesticide), at 3 levels (pesticide A, pesticide B, no pesticide), resulting in 3 treatments. The response variable is the number of beetle larvae found on each plant. Randomly select a third of the plots to be sprayed with pesticide A, a third with pesticide B, and a third with no pesticide (since the researcher also wants to know whether the pesticides even work at all). To control the experiment, the plots of land should be as similar as possible with regard to amount of sunlight, water, proximity to other plants, etc. If not, plots with similar characteristics should be blocked together. If possible, use some inert substance as a placebo pesticide on the control group, and do not tell the counters of the beetle larvae which plants have been treated with pesticides. After a given period of time, count the number of beetle larvae on each plant and compare the results.

67. Answers may vary. Find a group of volunteers. Each volunteer will be required to shut off the machine with his or her left hand and right hand. Randomly assign the left or right hand to be used first. Complete the first attempt for the whole group. Now repeat the experiment with the alternate hand. Check the differences in time for the left and right hands.

69. a) Jumping with or without a parachute.
b) Volunteer skydivers (the dimwitted ones).
c) A parachute that looks real but doesn't work.
d) A good parachute and a placebo parachute.
e) Whether parachutist survives the jump (or extent of injuries).
f) All should jump from the same altitude in similar weather conditions and land on similar surfaces.
g) Randomly assign people the parachutes.
h) The skydivers (and the people involved in distributing the parachute packs) shouldn't know who got a working chute. And the people evaluating the subjects after the jumps should not be told who had a real parachute either!

Part III Review

1. Observational prospective study. Indications of behaviour differences can be seen in the two groups. May show a link between premature birth and behaviour, but there may be lurking variables involved.

3. Experiment, matched by gender and weight. The experiment had one factor (diet), at two levels (allowing the dogs to eat as much as they want, or restricted diet), resulting in two treatments. The response variable was length of life. On average, dogs with a lower-calorie diet live longer.

5. Completely randomized experiment, with the treatment being receiving folic acid or not (one factor, two levels). The response variable is the number of precancerous growths, or simply the occurrence of additional precancerous growths. Neither blocking nor matching is mentioned, but in a study such as this one, it is likely that researchers and patients are blinded. Since treatments were randomized, it seems reasonable to generalize results to all people with precancerous polyps, though caution is warranted since these results contradict a previous study.

7. Sampling. No information is given about the sampling procedure, so hopefully the tested fireworks are selected randomly. It would probably be a good idea to test a few of each type of firework, so stratification by type seems likely. The population is all fireworks produced each day, and the parameter of interest is the proportion of duds. With a random sample, the manufacturers can make inferences about the proportion of duds in the entire day's production, and use this information to decide whether or not the day's production is suitable for sale.

9. Observational retrospective study. Researcher can conclude that for anyone's lunch, even when packed with ice, food temperatures are rising to unsafe levels.

11. Experiment, with a control group being the genetically engineered mice who received no antidepressant and the treatment group being the mice who received the drug. The response variable is the amount of plaque in their brains after one dose and after four months. There is no mention of blinding or matching. Conclusions can be drawn to the general population of mice and we should assume treatments were randomized. To conclude the same for humans would be risky, but researchers might propose an experiment on humans based on this study.

13. Experiment. Factor is gene therapy. Hamsters were randomized to treatments. Treatments were gene therapy or not. Response variable is heart muscle condition. Can conclude that gene therapy is beneficial (at least in hamsters).

15. Sampling. Population is all oranges on the truck. Parameter is proportion of unsuitable oranges. Procedure is probably simple random sampling. Can conclude whether or not to accept the truckload.

17. Observational prospective study. Physically fit men may have a lower risk of death from cancer.

19. a) The waiters performed an experiment, since treatments were imposed on randomly assigned groups. This experiment has one factor (candy), at two levels (candy or no candy), resulting in two treatments. The response variable is the percentage of the bill given as a tip.
    b) If the decision whether to give candy or not was made before the people were served, the server may have subconsciously introduced bias by treating the customers better. If the decision was made just before the check was delivered, then it is reasonable to conclude that the candy was the cause of the increase in the percentage of the bill given as a tip.
    c) "Statistically significant" means that the difference in the percentage of tips between the candy and no candy groups was more than expected due to sampling variability.

21. Everyone responding to the poll is 'online', and probably younger than the average voter. It may be the case that younger Canadians vote Liberal at a higher proportion than older Canadians. Older citizens also tend to show up to vote on election day in a higher proportion than young Canadians.

23. a) Yes.
    b) No. Residences without phones are excluded. Residences with more than one phone had a higher chance.
    c) No. People who respond to the survey may be of age but may not be registered voters.
    d) No. Households who answered the phone may be more likely to have someone at home when the phone call was generated. These may not be representative of all households.

25. a) Does not prove it. There may be other confounding variables. Only way to prove this would be to do a controlled experiment.
    b) Alzheimer's usually shows up late in life. Perhaps smokers have died of other causes before Alzheimer's can be seen.
    c) An experiment would be unethical. One could design a prospective study in which groups of smokers and nonsmokers are followed for many years and the incidence of Alzheimer's is tracked.

27.

Numerous subjects will be randomly assigned to see shows with violent, sexual, or neutral content. They will see the same commercials. After the show, they will be interviewed for their recall of brand names in the commercials.

29. a) May have been a simple random sample, but given the relative equality in age groups, may have been stratified.
    b) 38.2%
    c) We don't know. If data were from precincts that are primarily Democratic or Republican, that would bias the results. Also, party preference and actual party registration are not exactly the same thing. Probably OK, though.
    d) Do party affiliations differ for different age groups?

31. The factor in the experiment will be type of bird control. I will have three treatments: scarecrow, netting, and no control. I will randomly assign several different areas in the vineyard to one of the treatments, taking care that there is sufficient separation that the possible effect of the scarecrow will not be confounded. At the end of the season, the response variable will be the proportion of bird-damaged grapes.

33. a) We want all subjects treated as alike as possible. If there were no "placebo surgery," subjects would know this and perhaps behave differently.
    b) The experiment was intended to see if there was a difference in the effectiveness of the two treatments. (If we wanted to generalize, we would need to assume that the results for these volunteers are the same as on all patients who might need this operation.)
    c) Statistically significant means a difference in results so large it cannot be explained by natural sampling variability.

35. a) Use stratified sampling to select 2 first-class passengers and 12 from coach.
    b) Number passengers alphabetically, 01 = Bergman to 20 = Testut. Read in blocks of two, ignoring any numbers more than 20. This gives 65, 43, 67, 11 (selects Fontana), 27, 04 (selects Castillo).
    c) Number passengers alphabetically 001 to 120. Use the random number table to find three-digit numbers in this range until 12 have been selected.

Chapter 11

1. In the long run, a fair coin will generate 50% heads and 50% tails, approximately. But for each flip we cannot predict the outcome.

3. There is no law of averages for the short run. The first five flips do not affect the sixth flip.

5. a) 0.30   b) 0.80

7. P(blue) = 5/10
   P(black) = 2/10
   Since a shirt can either be blue colour or black colour, and cannot be both, the Addition Rule can be applied. Therefore, P(blue *or* black) = 5/10 + 2/10 = 7/10

9. 0.42

11. a) S = {HH, HT, TH, TT}, equally likely
    b) S = {0, 1, 2, 3}, not equally likely
    c) S = {H, TH, TTH, TTT}, not equally likely
    d) S = {1, 2, 3, 4, 5, 6}, not equally likely

13. In this context "truly random" should mean that every number is equally likely to occur.

15. There is no "Law of Averages." She would be wrong to think they are "due" for a harsh winter.

17. There is no "Law of Averages." If at bats are independent, his chance for a hit does not change based on recent successes or failures.

19. a) There is some chance you would have to pay out much more than the $300.
    b) Many customers pay for insurance. The small risk for any one customer is spread among all.

21. a) Legitimate.
    b) Legitimate.
    c) Not legitimate (sum more than 1).
    d) Legitimate.
    e) Not legitimate (can't have negatives).

23. a) 0.72   b) 0.89   c) 0.28

25. a) 0.32   b) 0.408

27. a) 0.69   b) 0.03   c) 0.50   d) 0.31

29. a) 0.68   b) 0.32   c) 0.04

31. a) 0.09   b) 0.91   c) 0.09

33. a) 0.4095   b) 0.2001   c) 0.0904   d) 0.5192

35. 1/9

37. If the cards are selected at random (simple random sampling) from the deck of 52 cards, all sets of five cards have the same chance to be the selected set.

39. Answers will vary. Author simulations estimate the probability to be about 90%.

Chapter 12

1. 0.675

3. 0.135

5. No, $P(S) = 0.323$ but $P(S \mid FC) = 0.625$. These are not the same.

7. Let A = mattress 1 will be delivered on time and B = mattress 2 will be delivered on time.
   a) P(A *and* B) = P(A)P(B) = 0.995(0.995) = 0.990025
   b) $P(A^C$ *and* $B^C) = P(A^C)P(B^C) = (1 - 0.995)(1 - 0.995) = 0.000025$

9. We have joint and marginal probabilities, not conditional probabilities, so a table is better.

11. A tree is better because we have conditional and marginal probabilities. The joint probabilities are found at the end of the branches.

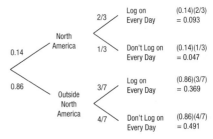

13. P(North America | Log on every day) = P(North America and Log on every day) / P(Log on every day) = 0.093/ (0.093 + 0.369) = 0.201. It has increased.

15. a) 0.0198   b) 0.328
    c) We assume that every group of three people in this group of 500 people has the same chance to be the group selected. (i.e., a simple random sample of size 3).
    d) We assume that random sampling here will ensure it.

17. a) 0.027   b) 0.128   c) 0.512   d) 0.271

19. a) Disjoint.   b) Independent.
    c) No. If two events are disjoint, whenever one event occurs, the one cannot occur and so they are not independent.

21. a) (1/6)(1/6)(1/6) = 1/216 = .0046
    b) (3/6)(3/6)(3/6) = 27/216 = 0.125
    c) (4/6)(4/6)(4/6) = 64/216 = 0.2963
    d) 1 − (5/6)(5/6)(5/6) = 1 − 125/216 = 91/216 = 0.4213
    e) 1 − (1/6)(1/6)(1/6) = 1 − 1/216 = 215/216 = 0.9954

23. a) 0.027   b) 0.063   c) 0.973   d) 0.013841287201

25. a) 0.438976   b) 0.729   c) 0.295031

27. 0.0776

29. a) 0.676   b) 0.0954   c) 0.324   d) 0.213

31. a) 1/2   b) 1   c) 1/13   d) 1/3

33. a) 0.11   b) 0.27   c) 0.407   d) 0.344

35. a) 0.452   b) 0.221   c) 0.552   d) 0.401

37. Having a fever and having a sore throat are not independent events, so: P(fever and sore throat) = 0.21.

39. a) 0.145   b) 0.118   c) 0.414   d) 0.217

41. a) 0.318   b) 0.955   c) 0.071   d) 0.009

43. Yes. The overall probability of getting an ace is 1/13. If you consider just one suit, there is only 1 ace out of 13 cards, so the probability of getting an ace given that the card is a diamond, for instance, is 1/13.

45. a) Yes.   b) No.   c) No.   d) No.

47. High blood pressure and high cholesterol are not independent events. 28.8% of men with OK blood pressure have high cholesterol, while 40.7% of men with high blood pressure have high cholesterol.

49. a) 0.54   b) Not independent.

51. a) No, the flight leaving on time and the luggage making the connection are not independent events. The probability is 0.95 if the flight is on time and only 0.65 if it is not on time.
    b) 0.695

53. 0.975

55. a) No. The rate of absenteeism for the night shift is 2%, while the rate for the day shift is only 1%.
    b) 1.4%

57. 57.1%

59. a) 0.2   b) 0.272   c) 0.353   d) 0.033

61. 0.563

Chapter 13

1. 33 oranges.

3. 4.58 oranges.

5. $1700, $141

7. 0.0596

9. 0.140

11. 0.055

13. 0.011

15. a) 19   b) 4.2

17. a) $50   b) $50

19. a)

Win	$0	$5	$10	$30
P(amount won)	$\frac{26}{52}$	$\frac{13}{52}$	$\frac{12}{52}$	$\frac{1}{52}$

b) $4.13   c) $4 or less (answers may vary).

21. a)

Children	1	2	3
P(Children)	0.5	0.25	0.25

b) 1.75 children.

c) 0.875 boys.

Boys	0	1	2	3
P(Boys)	0.5	0.25	0.125	0.125

23. $27 000

25. a) 7   b) 1.89

27. $5.44

29. 0.83

31. a) 1.7   b) 0.9

33. $\mu = 0.64, \sigma = 0.93$

35. a) $50   b) $100

37. a) No. The probability of winning the second depends on the outcome of the first.

b) 0.42   c) 0.08

d)

Games won	0	1	2
P(Games won)	0.42	0.50	0.08

e) $\mu = 0.66, \sigma = 0.62$

39. a)

Number good	0	1	2
P(Number good)	0.067	0.467	0.467

b) 1.40   c) 0.61

41. a) 30; 6   b) 26; 5   c) 30; 5.39   d) −10; 5.39   e) 20; 2.83

43. a) 240; 12.8   b) 140; 24   c) 720; 34.18

d) 60; 39.40   e) 1200; 32

45. a) 1.8

b) 0.87

c) Cartons are independent of each other.

47. 13.6; 2.55 (assuming the hours are independent of each other).

49. a) 23.4; 2.97

b) We are assuming that trucks are ticketed independently.

c) An unusually bad month might be one in which the company got 30 or more tickets.

51. a) There will be many gains of $150 with a few large losses.

b) $\mu = \$300, \sigma = \$8485.28$

c) $\mu = \$1\,500\,000, \sigma = \$600\,000$

d) Yes. $0 is 2.5 SDs below the mean for 10 000 policies.

e) Losses are independent of each other. A major catastrophe with many policies in an area would violate the assumption.

53. a) No. More than two outcomes are possible.

b) Yes, assuming the people are unrelated to each other.

c) No. The chance of a heart changes as cards are dealt.

d) No, 500 is more than 10% of 3000.

e) If packages in a case are independent of each other, yes; otherwise, no.

55. a) Use single random digits. Let 0, 1 = De Grasse. Examine random digits in groups of five, counting the number of 0's and 1's.

b) Results will vary.

c) Results will vary.

d) X = number of drivers wearing seatbelts in five cars.

x	0	1	2	3	4	5
P(x)	0.33	0.41	0.20	0.05	0.01	0.0

57. Departures from the same airport during a 2-hour interval may not be independent. All could be delayed by weather, for example.

59. a) 50   b) E(heads) = np = 100(0.5) = 50 heads.

61. a) $\mu = 10.44, \sigma = 1.16$ b) 0.812   c) 0.475

d) 0.00193   e) 0.998

63. $\mu = 20.28, \sigma = 4.22$

65. a) 0.118   b) 0.324   c) 0.744   d) 0.580

67. a) $\mu = 56, \sigma = 4.10$

b) Yes, $np = 56 \geq 10, nq = 24 \geq 10$, serves are independent.

c) In a match with 80 serves, approximately 68% of the time she will have between 51.9 and 60.1 good serves, approximately 95% of the time she will have between 47.8 and 64.2 good serves, and approximately 99.7% of the time she will have between 43.7 and 68.3 good serves.

d) Normal, approx.: 0.0141; Binomial, exact: 0.0161.

69. a) Assuming apples fall and become blemished independently of each other, Binom(300, 0.06) is appropriate. Since $np \geq 10$ and $nq \geq 10$, $N(18, 4.11)$ is also appropriate.

b) Normal, approx.: 0.072; Binomial, exact: 0.085

c) No, 50 is 7.8 SDs above the mean.

71. Normal, approx.: 0.053; Binomial, exact: 0.061

73. The mean number of sales should be 24 with SD 4.60. Ten sales is more than 3.0 SDs below the mean. He was probably misled.

75. a) 0.0707   b) 0.0265

77. a) 4 cases.   b) 0.9817

79. a) 0.107   b) $\mu = 24, \sigma = 2.19$

c) Normal, approx.: 0.819; Binomial, exact: 0.848

81. $\mu = 20, \sigma = 4$. I'd want *at least* 32 (3 SDs above the mean). (Answers will vary.)

83. Probably not. There's a more than 9% chance that he could hit 4 shots in a row, so he can expect this to happen nearly once in every 10 sets of 4 shots he takes. That does not seem unusual.

85. Yes. We'd expect him to make 22 shots, with a standard deviation of 3.15 shots. 32 shots is more than 3 standard deviations above the expected value, an unusually high rate of success.

87. a) 1/360   b) 0.25

89. a) 1 oz   b) 0.5 oz   c) 0.023   d) $\mu = 4$ oz, $\mu = 0.5$ oz
   e) 0.159   f) $\mu = 12.3$ oz, $\sigma = 0.54$ oz

91. a) 12.2 oz   b) 0.51 oz   c) 0.058

93. a) $\mu = 200.57$ sec., $\sigma = 0.46$ sec.
   b) No, $z = -2.36$. There is only 0.009 probability of swimming that fast or faster.

95. a) A = price of a pound of apples; P = price of a pound of potatoes; $Profit = 100A + 50P - 2$.
   b) $63.00   c) $20.62
   d) Mean—no; SD—yes (independent sales prices).

97. a) Yes.
   b) No, since we are counting the number of events in a fixed number of trials.
   c) 0.03125   d) 0.125
   e) Very skewed since as you increase x, the power of q increases, and since q is less than 1, this part of the distribution formula decreases (geometrically), so the probabilities always decrease as x increases.

99. a) 2.0   b) 1.0   c) 3.0   d) 4.0
   e) The smallest variance occurs when the investments are negatively correlated.

Part IV Review

1. a) 0.34   b) 0.27   c) 0.069
   d) No, 2% of cars have both types of defects.
   e) Of all cars with cosmetic defects, 6.9% have functional defects. Overall, 7.0% of cars have functional defects. The probabilities here are estimates, so these are probably close enough to say the defects are independent.

3. a) C = Price to China; F = Price to France; Total $= 3C + 5F$.
   b) $\mu = \$8100$, $\sigma = \$672.68$
   c) $\mu = \$300$, $\sigma = \$180.28$
   d) Means—no. Standard deviations—yes; ticket prices must be independent of each other for different countries, but all tickets to the same country are at the same price.

5. a) $\mu = -\$0.20$, $\sigma = \$1.89$
   b) $\mu = -\$0.40$, $\sigma = \$2.67$

7. a) 0.999   b) 0.944   c) 0.993

9. a) 0.237   b) 0.015   c) 0.896

11. a) $\mu = 118.5$, $\sigma = 5.44$
   b) Yes. $np \geq 10$ and $nq \geq 10$.
   c) Normal, approx.: 0.059; Binomial, exact: 0.073

13. a) 0.0173
   b) 0.590
   c) $E(\text{left}) = 960$ people, $E(\text{right}) = 120$ people, $E(\text{two} - \text{sided}) = 120$ people
   d) Let R = the number of people with right-brain language control. $E(R) = 120$ people and $SD(R) \approx 10.39$ people.
   e) Since $np_R = 120$ and $nq_R = 1080$ are both greater than 10, the Normal model, $N(120, 10.39)$, may be used to approximate $Binom$ (1200, 0.10). According to the Normal model, about 68% of randomly selected groups of 1200 people could be expected to have between 109.61 and 130.39 people with right-brain language control. About 95% of randomly selected groups of 1200 people could be expected to have between 99.22 and 140.78 people with right-brain language control. About 99.7% of randomly selected groups of 1200

people could be expected to have between 88.83 and 151.17 people with right-brain language control.

15. a) Men's heights are more variable than women's.
   b) Men.
   c) M = Man's height; W = Woman's height; $M = W$ is how much taller the man is.
   d) 5.1"   e) 3.75"   f) 0.913
   g) If independent, it should be about 91.3%. We are told 92%. This difference seems small and may be due to natural sampling variability.

17. a) No. The chance is 0.00000016.   b) 0.952

19. −$2080.00

21. a) 0.717   b) 0.588

23. a) $\mu = 100$, $\sigma = 8$ b) $\mu = 1000$, $\sigma = 60$
   c) $\mu = 100$, $\sigma \approx 8.54$ d) $\mu = -50$, $\sigma = 10$
   e) $\mu = 100$, $\sigma \approx 11.31$

25. a) Many do both, so the two categories can total more than 100%.
   b) No. They can't be disjoint. If they were, the total would be 100% or less.
   c) No. Probabilities are different for boys and girls.

27. a) 21 days.   b) 1649.73 soms.
   c) Bring an extra 3300 soms. This gives you a cushion of about 157 soms for each day.

29. You'd expect 549.4 homeowners, with an SD of 13.46. 523 is about 2 SDs below the mean; somewhat unusual.

31. a) 0.005   b) 0.4746   c) 0.17

33. a) 15   b) 0.402

35. a) 38%   b) 41%   c) 19.6%
   d) 19.6% of calculator users were computer users. In classes where calculators were not used, 22.4% of the classes used computers. Since the percentages are different, there is evidence of an association between computer use and calculator use. Calculator classes are less likely to use computers.

37. a) A Poisson model.   b) 0.1 failures.   c) 0.090   d) 0.095

39. a) 1/11   b) 7/22   c) 5/11   d) 0   e) 19/66

41. a) Expected number of stars with planets.
   b) Expected number of planets with intelligent life.
   c) Probability of a planet with a suitable environment having intelligent life.
   d) f_l: If a planet has a suitable environment, the probability that life develops.
   f_i: If a planet develops life, the probability that the life evolves intelligence.
   f_c: If a planet has intelligent life, the probability that it develops radio communication.

43. 0.991

Chapter 14

1. a) Roughly Normal.   b) 0.36   c) 0.034

3. All the histograms are centered near 0.05. As n gets larger, the histograms (possibly approaching a Normal model) while the variability in the sample proportions decreases.

5. Check conditions: The data come from a random sample, so the Randomization condition is met. We don't know the exact value of p, but if the sample is anywhere near correct, $np \approx 1500 * 0.27 = 405$, and $nq \approx 1500 * 0.73 = 1095$. So there are well over 10 successes and 10 failures, meeting the Success/Failure condition. Since there are more than 10 * 1500 = 15 000 adults in the United States, the 10%

condition is met. A Normal model is appropriate for the sampling distribution of the sample proportion.

7. a) The histogram is unimodal and symmetric for $n = 200$.
b) The Success/Failure condition says that np and nq should both be at least 10, which is not satisfied until $n = 200$ for $p = 0.05$. The theory predicted my choice.

9. a) A Normal model is not appropriate since the histogram is strongly left skewed.
b) No. The 95% rule is based on the Normal model, and the Normal model is not appropriate here.

11. a) Can't use a Normal model to estimate probabilities. The distribution is skewed right—not Normal.
b) 4 is probably not a large enough sample to say the average follows the Normal model.
c) No. This is 3.16 SDs above the mean.

13. a) 0.0003. Model is $N(384, 34.15)$.
b) $427.77 or more ($427.71 using tables).

15. a) Symmetric, because probability of heads and tails is equal.
b) 0.5
c) 0.125
d) $np = 8 < 10$

17. a) About 68% should have proportions between 0.4 and 0.6, about 95% between 0.3 and 0.7, and about 99.7% between 0.2 and 0.8.
b) $np = 12.5, nq = 12.5$; both are ≥ 10.
c)

0.3125 0.3750 0.4375 0.5000 0.5625 0.6250 0.6875
Proportion

$np = nq = 32$; both are ≥ 10.
d) Becomes narrower (less spread around 0.5).

19. This is a fairly unusual result: about 2.26 SDs below the mean. The probability of that is about 0.012. So, in a class of 100 this is certainly a reasonable possibility.

21. a)

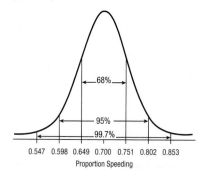

0.547 0.598 0.649 0.700 0.751 0.802 0.853
Proportion Speeding

b) Both $np = 56$ and $nq = 24 \geq 10$. Drivers *may* be independent of each other, but if flow of traffic is very fast, they may not be. Or weather conditions may affect all drivers. In these cases they may get more or fewer speeders than they expect.

23. a) Assume that these children are typical of the population. They represent fewer than 10% of all children. We expect 20.4 nearsighted and 149.6 not; both are at least 10.

b)

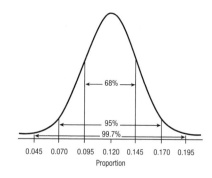

0.045 0.070 0.095 0.120 0.145 0.170 0.195
Proportion

c) Probably between 12 and 29.

25. a) $\mu = 7\%, \sigma = 1.8\%$
b) Assume that clients pay independently of each other, that we have a random sample of all possible clients, and that these represent less than 10% of all possible clients. $np = 14$ and $nq = 186$ are both at least 10.
c) 0.048

27. 68% of the time within 0.014 (one SD), 95% of the time within 0.028 (two SD), and 99.7% of the time within 0.042 (three SD).

29. It is unlikely that rural families compose a SRS. However, proceeding as if they do, we find $\sigma(\hat{p}) = 1.10\%$ and so 16% would be a very unusual result, as it is outside the $3\sigma(\hat{p})$ range by the 68-95-99.7 rule.

31. 0.801 using $N(0.92, 0.0215)$ model.

33. 0.088 using $N(0.08, 0.022)$ model.

35. Answers will vary. Using $\mu + 3\sigma$ for "very sure," the restaurant should have 89 nonsmoking seats. Assumes customers at any time are independent of each other, a random sample, and represent less than 10% of all potential customers. $np = 72, nq = 48$, so the Normal model is reasonable ($\mu = 0.60, \sigma = 0.045$).

37. a) Normal, centre at μ, standard deviation $\dfrac{\sigma}{\sqrt{n}}$
b) Standard deviation will be smaller. Centre will remain the same.

39. a) The histogram is unimodal and slightly skewed to the right, centred at 36 inches, with a standard deviation near 4 inches.
b) All the histograms are centred near 36 inches. As n gets larger, the histograms approach the Normal shape and the variability in the sample means decreases. The histograms are fairly normal by the time the sample reaches size 5.

41. a)

n	Observed mean	Theoretical mean	Observed st. dev.	Theoretical st. dev.
2	36.314	36.33	2.855	2.842
5	36.314	36.33	1.805	1.797
10	36.341	36.33	1.276	1.271
20	36.339	36.33	0.895	0.899

b) They are all very close to what we would expect.
c) For samples as small as 5, the sampling distribution of sample means is unimodal and very symmetric.
d) The distribution of the original data is nearly unimodal and symmetric, so it doesn't take a very large sample size for the distribution of sample means to be approximately Normal.

43. Normal, $\mu = 73, \sigma = 1.4$. We assume that the students are randomly assigned to the seminars and represent less than 10% of all possible students, and that the individual's GPAs are independent of one another.

45. a) The mean of 100 coins varies less from sample to sample than the individual coins.

b) $SD(\bar{y}) = \dfrac{\sigma}{\sqrt{n}} = \dfrac{0.09}{\sqrt{100}} = 0.009$

47. a) 21.1%   b) 276.8 days or more.
c) $N(266, 2.07)$ d) 0.002

49. a) There are more premature births than very long pregnancies. Modern practice of medicine stops pregnancies at about 2 weeks past normal due date.
b) Parts (a) and (b)—yes—we can't use the Normal model if it's very skewed. Part (c)—no—CLT guarantees a Normal model for this large sample size.

51. a) $\mu = \$2.00, \sigma = \$3.61$
b) $\mu = \$4.00, \sigma = \$5.10$
c) 0.191. Model is $N(80, 22.83)$.

53. a) $\mu = 2.442, \sigma = 1.325$.
b) They would not be expected to follow a Normal Distribution. They would follow the frequency distribution given in the table.
c) Approximately $N(2.442, 0.210)$.

55. For $n = 36$, using $N(2.442, 0.293)$, $z = 1.91$ and $P(z > 1.91) = 0.0281$, so farm families may be larger on average.

57. a) $N(2.9, 0.045)$ b) 0.0131   c) 2.97 gm/mi

59. a) 0.3085   b) 0.3783   c) 0.0031   d) 0.0455

61. Answers may vary.

63. Answers may vary.

65. Answers may vary.

Chapter 15

1. a) This means that 49% of the 799 teens in the sample said they have misrepresented their age online. This is our best estimate of p, the proportion of all U.S. teens who would say they have done so.
b) $SE(\hat{p}) = \sqrt{\dfrac{(0.49)(0.51)}{799}} \approx 0.018$
c) Because we don't know p, we use $\hat{p}$ to estimate the standard deviation of the sampling distribution. So the standard error is our estimate of the amount of variation in the sample proportion we expect to see from sample to sample when we ask 799 teens whether they've misrepresented their age online.

3. a) We are 95% confident that, if we were to ask all U.S. teens whether they have misrepresented their age online, between 45.6% and 52.5% of them would say they have.
b) If we were to collect many random samples of 799 teens, about 95% of the confidence intervals we construct would contain the proportion of all U.S. teens who admit to misrepresenting their age online.

5. a) ME = 0.01.   b) Larger.

7. a) Yes, all conditions for the confidence interval are met.
b) At least 412 voters.

9. He believes the true proportion is within 4% of his estimate, with some (probably 95%) degree of confidence.

11. a) Population—all cars; sample—134 cars actually stopped at the checkpoint; p—proportion of all cars with safety problems; $\hat{p}$—proportion actually seen with safety problems (10.4%); if sample (a cluster sample) is representative, then the methods of this chapter will apply.
b) Population—general public; sample—602 viewers who logged onto the website; p—population proportion of those who favour prayer in school; $\hat{p}$—proportion of those who voted in the poll who favoured prayer in school (81.1%); can't use methods of this chapter—sample is biased and nonrandom.
c) Population—parents at the school; sample—380 parents who returned the questionnaire; p—proportion of all parents who favour uniforms; $\hat{p}$—proportion of respondents who favour uniforms (60%); should not use methods of this chapter, since not SRS (possible non-response bias).

d) Population—students at the college; sample—the 1632 students who entered that year; p—proportion of all students who will graduate on time; $\hat{p}$—proportion of that year's students who graduate on time (85.0%); can use methods of this chapter if that year's students (a cluster sample) are viewed as a representative sample of all possible students at the school.

13. a) Not correct. This implies certainty.
b) Not correct. Different samples will give different results. Many fewer than 95% will have 88% on-time orders.
c) Not correct. The interval is about the population proportion, not the sample proportion in different samples.
d) Not correct. In this sample, we *know* 88% arrived on time.
e) Not correct. The interval is about the parameter, not about the days.

15. a) False.   b) True.   c) True.   d) False.

17. On the basis of this sample, we are 90% confident that the proportion of Japanese cars is between 29.9% and 47.0%.

19. a) (0.162, 0.280)
b) We are 95% confident that between 16.2% and 28.0% of all seafood packages sold in these three states are mislabelled.
c) The size of the population is irrelevant. If *Consumer Reports* had a random sample, 95% of intervals generated by studies like this are expected to capture the true proportion of seafood packages that are mislabelled.

21. a) 0.026
b) We are 90% sure that the true proportion of Canadians who are proud of health care is within ±2.6% of 58%.
c) Larger. To be more certain, we must be less precise.
d) 0.041   e) Less confidence.
f) Our new interval (55.4%, 60.6%) does not include the old measurement of 50%. Even with the wider margins associated with the 99% CI, we still don't include 50%. Thus, we have strong evidence that the real proportion has changed.

23. a) (0.0465, 0.0491). The assumptions and conditions for constructing a confidence interval are satisfied.
b) The confidence interval gives the set of plausible values (with 95% confidence). Since 0.05 is outside the interval, that seems to be a bit too optimistic.

25. a) (12.7%, 18.6%)
b) We are 95% confident, based on this sample, that the proportion of all auto accidents that involve teenage drivers is between 12.7% and 18.6%.
c) About 95% of all random samples will produce confidence intervals that contain the true population proportion.
d) Contradicts. The interval is completely below 20%.

27. Probably nothing. Those who bothered to fill out the survey may be a biased sample.

29. a) Leading question bias.
b) (0.48, 0.54)
c) Smaller—the sample size is larger.

31. a) (18.2%, 21.8%)
b) We are 98% confident, based on the sample, that between 18.2% and 21.8% of English children are deficient in vitamin D.
c) About 98% of all random samples will produce a confidence interval that contains the true proportion.

33. CI = (0.11, 0.19). The method used assumes independence of observations. This is unlikely to be true when observing individuals with a block. The method we used (assuming independence) underestimates the variability of the sample proportion and so the SE of the sample proportion.

35. a) (15.5%, 26.3%)   b) 612
c) Sample may not be random or representative. Deer that may legally be hunted may not be representative of all sexes and ages.

37. a) 141   b) 318   c) 564

39. 1801

41. 384 total, using $\hat{p} = 0.15$.

43. 90%

45. a) 99% m.e. = 0.03 and so a 99% CI = (0.49, 0.55).
b) The margin of error increases as the sample size decreases. The sample size for BC must be smaller than the sample size for Canada and so, we should expect an increase in margin of error. The margin of error of the confidence interval for population proportions also depends on the sample proportion $\hat{p}$. The value of $\hat{p}(1 - \hat{p})$ takes its maximum possible value when $\hat{p} = 0.5$ and decreases as $\hat{p}$ moves away from 0.5 (either increases or decreases). It is most likely that the increase in the margin of error due to the decrease in sample size will dominate the decrease due to $\hat{p}$ moving away from 0.5 (since it only moves from 0.52 to 0.58, but the decrease in the sample size can be substantial).
c) 0.028   d) 8004

47. a) Answers may vary. (0.690, 0.830)
b) Answers may vary. 92% of my confidence intervals contained the true population proportion. For 100 simulated samples, if X = number of CIs that contain 0.8, the X should have a binomial distribution with $n = 100$ and $p = 0.9$.

49. a) $n\hat{p} = 5 < 10$ and so the sample size is not large enough for the distribution of the sample proportion to be approximately Normal.
b) (0.05, 0.25). We are 95% confident that between 5% to 25% of all puppies in this population has early hip dysplasia.

51. a) Since there is no time trend, and also the timeplot shows little relationship between one month and the next month, the monthly on-time departure rates appear to be nearly independent. This is not a random sample, but should be representative of future flights. These months represent fewer than 10% of all possible months. The histogram looks unimodal and slightly skewed to the left. Since the sample size is 201, the sample mean should be approximately Normal, by the CLT, and we can plug in s for σ in the confidence interval.
b) (80.22, 81.29).
c) On-time departures average somewhere between 80.22% and 81.29% of flights per month (with 90% confidence).

53. a) 7.9   b) (832.0, 872.8) km/sec.
c) Michelson is 99% sure that this interval captures the true speed of light.
d) You would need to check that the data was distributed normally, as your use of the z^* critical values depends on this assumption. Also, the measurements would need to be independent and unbiased for our analysis to hold.
e) This value is well outside his 99% confidence interval. He was either very unlucky, or some of our assumptions about the normality and unbiasedness of his data are not correct. However, the result is still remarkably close for 1897!

55. To achieve a reduction of 50%: $n = 174$; for a ME of 10.0: $n = 415$.

Chapter 16

1. a) The new drug is not more effective than aspirin, and reduces the risk of heart attack by 44% ($p = 0.44$).
b) The new drug is more effective than aspirin, and reduces the risk of heart attack by more than 44% ($p > 0.44$).

3. a) Since the P-value of 0.28 is large, there is insufficient evidence to conclude that the new drug is better than aspirin.
b) Since the P-value of 0.004 is very small, there is evidence that the new drug is more effective than aspirin.

5. a) $H_0: p = 0.043$, $H_A: p \neq 0.043$.
b) The conditions are met.

c) $z = 8.3$; P-value < 0.001
d) Since the P-value < 0.001, we reject H_0. There is evidence that the Indigenous population in this province differs from the nation as a whole.

7. The alternative hypothesis would be one-sided, because the only evidence that would support the friend's claim is guessing more than 25% of the suits correctly.

9. a) The drug may not be approved for use, and people would miss out on a beneficial product.
b) The drug will go into production and people will suffer the side effect.

11. a) $H_0: p = 0.30$; $H_A: p < 0.30$
b) $H_0: p = 0.50$; $H_A: p \neq 0.50$
c) $H_0: p = 0.20$; $H_A: p > 0.20$

13. Statement d is correct.

15. No, we can say only that there is a 27% chance of seeing the observed effectiveness just from natural sampling variation. There is no *evidence* that the new formula is more effective, but we can't conclude that they are equally effective.

17. a) No. There's a 25% chance of losing twice in a row. That's not unusual.
b) 0.125
c) No, we expect that to happen 1 time in 8.
d) Maybe 5? The chance of 5 losses in a row is only 1 in 32, which seems unusual.

19. 1. Use p, not $\hat{p}$, in hypotheses.
2. The question was about failing to meet the goal, so H_A should be $p < 0.96$.
3. Did not check 0.04(200) = 8. Since $nq < 10$, the Success/Failure condition is violated. Didn't check 10% condition.
4. 188/200 = 0.94; $SD(\hat{p}) \approx 0.014$.
5. z is incorrect; should be $z = -1.43$.
6. $P = P(z < -1.43) = 0.076$
7. There is only weak evidence that the new instructions do not work.

21. a) $H_0: p = 0.30$; $H_A: p > 0.30$
b) Possibly an SRS; we don't know if the sample is less than 10% of his customers, but it could be viewed as less than 10% of all possible customers; $(0.3)(80) \geq 10$ and $(0.7)(80) \geq 10$. Wells are independent only if customers don't have farms on the same underground springs.
c) $z = 0.73$; P-value = 0.232
d) If his dowsing is no different from standard methods, there is more than a 23% chance of seeing results as good as those of the dowser's, or better, by natural sampling variation.
e) These data provide no evidence that the dowser's chance of finding water is any better than normal drilling.

23. There is no need to conduct any hypothesis testing since the figures (i.e., 64.1% in 2013 and 60.7% in 2006) are already the parameters (recall that it is a census).

25. a) $H_0: p = 0.05$ vs. $H_A: p < 0.05$
b) We assume the whole mailing list is at least 10 times bigger than 100 000 names. This is a random sample, and we expect 5000 successes and 95 000 failures.
c) $z = -3.178$; P-value = 0.00074, so we reject H_0; there is strong evidence that the donation rate would be below 5%.

27. a) $H_0: p = 0.296$, $H_A: p > 0.296$
b) The sample is representative. 240 < 10% of all law school applicants. We expect 240(0.296) = 71.04 to be admitted and 240(0.704) = 168.96 not to be, both at least 10. $z = 1.69$; P-value = 0.046.
c) Although the evidence is weak, there is some indication that the program may be successful. Candidates should decide whether they can afford the time and expense.

29. H_0: $p = 0.20$; H_A: $p > 0.20$. SRS (not clear from information provided); 22 is more than 10% of the population of 150; $(0.20)(22) < 10$. Do not proceed with a test.

31. $H_0 = p = 0.03$; H_A: $p \neq 0.03$; $\hat{p} = 0.015$. One mother having twins will not affect another, so observations are independent; not an SRS; sample is less than 10% of all births. However, the mothers at this hospital may not be representative of all teenagers; $(0.03)(469) = 14.07 \geq 10$; $(0.97)(469) \geq 10$. $z = -1.91$. P-value = 0.0556. With a P-value this low, reject H_A. These data show some evidence that the rate of twins born to teenage girls at this hospital is less than the national rate of 3%. It is not clear whether this can be generalized to all teenagers.

33. H_0: $p = 0.25$; H_A: $p > 0.25$. SRS; sample is less than 10% of all potential subscribers; $(0.25)(500) \geq 10$; $(0.75)(500) \geq 10$. $z = 1.24$; P-value = 0.1076. The P-value is high, so do not reject H_0. These data do not show that more than 25% of current readers would subscribe; the company should not go ahead with the WebZine on the basis of these data.

35. H_0: $p = 0.40$; H_A: $p < 0.40$. Data are for all executives in this company and may not be able to be generalized to all companies; $(0.40)(43) \geq 10$; $(0.60)(43) \geq 10$. $z = -1.31$; P-value = 0.0955. Because the P-value is high, we fail to reject H_0. These data do not show that the proportion of women executives is less than the 40% of women in the company in general.

37. a) H_0: $p = 0.5$ vs. H_A: $p \neq 0.5$. $z = -3.34$. P-value ≈ 0.0008. The sample does provide evidence to conclude that an injured Atom player is not equally likely to be from either quartile.
   b) $z = -6.48$. P-value ≈ 0. The sample does provide evidence to conclude that an injured youth is not equally likely to be from either quartile.
   c) We really should be looking at proportions injured of those in each age group, since there are more older players actually selected to play for the teams. We could try to obtain ages or age quintiles for all players, then compare the proportion injured in one age group with the proportion injured in the other age group.

39. H_0: $p = 0.90$; H_A: $p < 0.90$. $\hat{p} = 0.844$; $z = -2.05$; P-value = 0.0201. Because the P-value is so low, we reject H_0. There is strong evidence that the actual rate at which passengers with lost luggage are reunited with it within 24 hours is less than the 90% claimed by the airline.

41. a) Yes; assuming this sample to be a typical group of people, $P = 0.0008$. This cancer rate is very unusual.
   b) No, this group of people may be atypical for reasons that have nothing to do with the radiation.

43. H_0: $p = 0.5$ vs. H_A: $p \neq 0.5$. The test is two-sided (given). $\hat{p} = 0.4468$, $z = -2.31$, P-value = 0.0208. We have moderately strong evidence that more Canadians would choose Obama as their most admired figure.

Chapter 17

1. a) Let p = probability of winning on the slot machine.
   H_0: $p = 0.01$ vs. H_A: $p \neq 0.01$
   b) Let μ = mean spending per customer this year.
   H_0: $\mu = \$35.32$ vs. H_A: $\mu \neq \$35.32$
   c) Let p = proportion of patients cured by the new drug.
   H_0: $p = 0.3$ vs. H_A: $p \neq 0.3$
   d) Let p = proportion of clients now using the website.
   H_0: $p = 0.4$ vs. H_A: $p \neq 0.4$

3. a) False. A high P-value shows that the data are consistent with the null hypothesis, but provides no evidence for rejecting the null hypothesis.
   b) False. It results in rejecting the null hypothesis, but does not prove that it is false.

   c) False. A high P-value shows that the data are consistent with the null hypothesis, but does not prove that the null hypothesis is true.
   d) False. Whether a P-value provides enough evidence to reject the null hypothesis depends on the risk of a Type I error that one is willing to assume (the a level).

5. a) H_0: $p = 0.40$ vs. H_A: $p \neq 0.40$. Two-sided.
   b) H_0: $p = 0.42$ vs. H_A: $p > 0.42$. One-sided.
   c) H_0: $p = 0.50$ vs. H_A: $p > 0.50$. One-sided.

7. a) True.
   b) False. The alpha level is set independently and does not depend on the sample size.
   c) False. The P-value would have to be less than 0.01 to reject the null hypothesis.
   d) False. It simply means we do not have enough evidence at that alpha level to reject the null hypothesis.

9. a) $z = \pm 1.96$ b) $z = 1.645$ c) $z = 2.33$

11. a) Type I error. The actual value is not greater than 0.3, but they rejected the null hypothesis.
   b) No error. The actual value is 0.50, which was not rejected.
   c) Type II error. The null hypothesis was not rejected, but it was false. The true relief rate was greater than 0.25.

13. Power = 0.543

15. a) Two sided. Let p be the percentage of students who prefer Diet Pepsi. H_0: $p = 0.50$ vs. H_A: $p \neq 0.50$
   b) One sided. Let p be the percentage of teenagers who prefer the new formulation. H_0: $p = 0.50$ vs. H_A: $p > 0.50$
   c) One sided. Let p be the percentage of people who intend to vote for the override. H_0: $p = 2/3$ vs. H_A: $p > 2/3$
   d) Two sided. Let p be the percentage of days that the market goes up. H_0: $p = 0.50$ vs. H_A: $p \neq 0.50$

17. If there is no difference in effectiveness, the chance of seeing an observed difference this large or larger is 4.7% by natural sampling variation.

19. $\alpha = 0.05$. Yes. The P-value is less than 0.05, so it's less than 0.10. But to reject H_0 at $\alpha = 0.01$, the P-value must be below 0.01, which isn't necessarily the case.

21. a) There is only a 1.1% chance of seeing a sample proportion as low as 89.4% vaccinated by natural sampling variation if 90% have really been vaccinated.
   b) We conclude that p is below 0.9, but a 95% confidence interval would show that the true proportion is between (0.889, 0.899). Most likely, a decrease from 90% to 89.9% would not be considered important. On the other hand, with 1 000 000 children a year vaccinated, even 0.1% represents about 1000 kids—so this may very well be important.

23. a) (48.6%, 53.4%)
   b) Because 45% is not in the interval, there is strong evidence that fewer than 45% of all men use work as their primary measure of success.
   c) $\alpha = 0.02$; it's a two-tailed test based on a 98% confidence interval.

25. a) (0.065520, 0.079994).
   b) Since 0.13 is not in the confidence interval, we reject the null hypothesis that $p = 0.13$. Since all values in the 95% CI lie below 0.13, we have some reasons to suspect a possible lack of rural applicants.

27. a) Type II error.   b) Type I error.
   c) By making it easier to get the loan, the bank has reduced the alpha level.
   d) The risk of a Type I error is decreased and the risk of a Type II error is increased.

29. a) Power is the probability that the bank denies a loan that would not have been repaid.
   b) Raise the cutoff score.
   c) A larger number of trustworthy people would be denied credit, and the bank would miss the opportunity to collect interest on those loans.

31. a) The null hypothesis is that the obesity rate in First Nations youth has not decreased and the alternative hypothesis is that the obesity rate in First Nations youth has decreased.

b) Type I error in this situation is the error in concluding that the obesity rate in First Nations youth has decreased when in fact it has not decreased.

c) Type II error in this situation is the error in concluding that the obesity rate in First Nations youth has not decreased when in fact it has decreased.

d) If we make a Type I error, this harms the federal and provincial governments. If we make a Type II error, this harms the First Nations children as they are losing a program effective in reducing the obesity rate.

e) The Power of the test represents the probability of concluding that the obesity rate in First Nations youth has decreased when in fact it has decreased.

33. a) It is decided that the shop is not meeting standards when it is.

b) The shop is certified as meeting standards when it is not.

c) Type I.   d) Type II.

35. a) The probability of detecting a shop that is not meeting standards.

b) 40 cars. Larger n.

c) 10%. More chance to reject H_0.

d) A lot. Larger differences are easier to detect.

37. a) One-tailed. The company wouldn't be sued if "too many" minorities were hired.

b) Deciding the company is discriminating when it is not.

c) Deciding the company is not discriminating when it is.

d) The probability of correctly detecting discrimination when it exists.

e) Increases Power.

f) Lower, since n is smaller.

39. a) One-tailed. Software is supposed to decrease the dropout rate.

b) $H_0: p = 0.13$; $H_A: p < 0.13$

c) He buys the software when it doesn't help students.

d) He doesn't buy the software when it does help students.

e) The probability of correctly deciding the software is helpful.

41. a) $z = -3.21$, $p = 0.007$. The change is statistically significant. A 95% confidence interval is (2.3%, 8.5%). This is clearly lower than 13%. If the cost of the software justifies it, the professor should consider buying the software.

b) The chance of observing 11 or fewer dropouts in a class of 203 is only 0.07% if the dropout rate is really 13%.

43. a) $H_A: p = 0.30$, where p is the probability of heads.

b) Reject the null hypothesis if the coin comes up tails—otherwise fail to reject.

c) P(tails given the null hypothesis) $= 0.1 = \alpha$

d) P(tails given the alternative hypothesis) $=$ Power $= 0.70$

e) Spin the coin more than once and base the decision on the sample proportion of heads.

45. a) 0.0464   b) Type I.   c) 37.6%

d) Increase the number of shots. Or keep the number of shots at 10, but increase alpha by declaring that 8, 9, or 10 will be deemed as having improved.

47. a) $\hat{p} > 0.6901003607$ b) 0.8925   c) $n = 147$

49. a) Answers may vary. $\hat{p} = 0.826667, n = 75, z = 0.58$, P-value $= 2 \times P(Z > 0.58) = 0.564 > 0.05$ and so there is no evidence to reject the null hypothesis $p = 0.8$ against the two-sided alternative.

b) Answers may vary. Ninety-four of the tests do not reject the null hypothesis at the 5% level and 6 of them reject the null hypothesis. We would expect 95% percent of tests to produce a correct decision if we repeated these simulations many times. If $X =$ number wrong decisions, then X has a binomial distribution with $n = 100$ and $p = 0.05$.

51. a) Answers may vary. $\hat{p} = 0.906667, n = 75, z = 3.91$, P-value $= 2 \times P(Z > 3.91) = 0.00 < 0.05$ and so we have evidence to reject the null hypothesis $p = 0.7$ against the two-sided alternative.

b) Answers may vary. Forty-seven of the tests reject the null hypothesis at the 5% percent level and 53 of them failed to reject the null hypothesis. The probability of a Type II error $= 0.53$ and the Power $= 0.47$. An estimate of Power from a simulation based on 1500 samples (each of size 75) was 0.465.

c) Exact Power $= 0.468$

53. Let $X =$ the number of significant tests. Then X is Binomial $(10, 0.05)$, so $P(X \geq 1) = 0.40$. A single significant result in 10 trials is easily explained by chance, so we should not reject the null hypothesis.

Chapter 18

1. a) SD's are 1 lb, 0.5 lb, and 0.2 lb respectively.

b) The distribution of pallets. The CLT tells us that the Normal model is approached in the limit regardless of the underlying distribution. As samples get larger the approximation gets better.

3. a) 1.74   b) 2.37

5. The shape is unimodal, symmetric, and bell-shaped. It becomes closer to Normal. The centre is at 0. The standard deviation becomes smaller.

7. a) Houses are independent; randomly sampled. Should check Nearly Normal condition by making a histogram. Sample size: 36 should be big enough.

b) ($9052.50, $10 067.50)

c) We are 95% confident that the interval ($9052.50, $10 067.50) contains the true mean loss in home value.

9. a) 0.0524   b) 0.0889

11. $t = -2.56$ with 35 df, $P = 0.0075$. We reject the null hypothesis because the P-value is small, and conclude that the loss of home values in this community does appear to be unusual.

13. 27 using z, 31 using t.

15. a) The confidence interval is for the population mean, not the individual cows in the study.

b) The confidence interval is not for individual cows.

c) We *know* the average gain in this study was 56 pounds!

d) The average weight gain of all cows does not vary. It's what we're trying to estimate.

e) No. There is not a 95% chance for another sample to have an average weight gain between 45 and 67 pounds. There is a 95% chance that another sample will have its average weight gain within two standard errors of the true mean.

17. a) No. A confidence interval is not about individuals in the population.

b) No. It's not about individuals in the sample, either.

c) No. We know the mean cost for students in the sample was $1467.

d) No. A confidence interval is not about other sample means.

e) Yes. A confidence interval estimates a population parameter.

19. a) Based on this sample, we can say, with 95% confidence, that the mean pulse rate of adults is between 70.9 and 74.5 beats per minute.

b) 1.8 beats per minute.

c) Larger.

21. The assumptions and conditions for a t-interval are not met. The distribution is highly skewed to the right and there is a large outlier.

23. a) Yes. Randomly selected group; the histogram is not unimodal and symmetric, but it is not highly skewed and there are no outliers, so with a sample size of 52, there is no problem.

b) (36.70, 36.96)

c) We are 98% confident that the interval 36.7°C to 36.96°C contains the true mean body temperature for adults.

d) 98% of all random samples of size 52 will produce intervals that contain the true mean body temperature of adults.

e) Since the interval is completely below the body temperature of 37°C, there is strong evidence that the true mean body temperature of adults is lower than 37°C.

25. a) Narrower. A smaller margin of error, so less confident.
   b) Advantage: more chance of including the true value. Disadvantage: wider interval.
   c) Narrower; due to the larger sample, the SE will be smaller.
   d) About 313.

27. a) (709.9, 802.5)
   b) We are 95% confident that the interval 299 709.9 to 299 802.5 km/sec. contains the speed of light.
   c) Assumed that the measurements are independent of each other and that the distribution of the population of all possible measurements is Normal.

29. a) Given no time trend, the monthly on-time departure rates should be independent. Though not a random sample, these months should be representative. The histogram looks unimodal, but slightly left skewed; not a concern with this large sample.
   b) (77.48%, 78.56%)
   c) We can be 90% confident that the interval from 77.48% to 78.56% holds the true mean monthly percentage of on-time flight departures.

31. The 95% confidence interval lies entirely above the 0.08 ppm limit, evidence that mirex contamination is too high and consistent with rejecting the null. We used an upper-tail test, so the P-value should therefore be smaller than $\frac{1}{2}(1 - 0.95) = 0.025$, and it was.

33. If in fact the mean cholesterol of pizza eaters does not indicate a health risk, then only 7 of every 100 samples would have mean cholesterol levels as high (or higher) as observed in this sample.

35. a) Upper-tail. We want to show it will hold 500 pounds (or more) easily.
   b) They will decide the stands are safe when they're not.
   c) They will decide the stands are unsafe when they are in fact safe.

37. a) Decrease α. This means a smaller chance of declaring the stands safe if they are not.
   b) The probability of correctly detecting that the stands are capable of holding more than 500 pounds.
   c) Decrease the standard deviation—probably costly. Increase the sample size—takes more time for testing and is costly. Increase α—more Type I errors. Increase the "design load" to be well above 500 pounds—again, costly.

39. a) $H_0: \mu = 22.6$; $H_A: \mu > 22.6$
   b) We have a random sample of the population. Population may not be normally distributed, as it would be easier to have a few much older women at their first marriage than some very young women. However, with a sample size of 40, it should be safe to proceed.
   c) The standard deviation of the population, $\sigma(\bar{y})$ will be estimated by $SE(\bar{y}) = \dfrac{s}{\sqrt{n}}$, and we will use a Student's t model, with $40 - 1 = 39$ degrees of freedom.
   d) The P-value is <0.00001.
   e) If the mean age at first marriage is still 22.6 years, there is a near-zero chance of getting a sample mean of 27.2 years or older simply from natural sampling variation.
   f) Since the P-value is low, we reject the null hypothesis.

41. a) Probably a representative sample; the Nearly Normal condition seems reasonable from a Normal probability plot. The histogram is nearly uniform, with no outliers or skewness.
   b) $\bar{y} \approx 28.78$ grams, $s \approx 0.40$ grams
   c) (28.36, 29.21) grams
   d) Based on this sample, we are 95% confident the average weight of the content of Ruffles bags is between 28.36 and 29.21 grams.
   e) The company is erring on the safe side, as it appears that, on average, it is putting in slightly more chips than stated.

43. a) Type I; he mistakenly rejected the null hypothesis that $p = 0.10$ (or worse).
   b) Yes. These are a random sample of bags and the Nearly Normal condition is met; $t = -2.51$ with 7 df for a one-sided P-value of 0.0203.

45. a) Random sample; the Nearly Normal condition seems reasonable from a Normal probability plot. The histogram is roughly unimodal and symmetric with no outliers.
   b) (1187.9, 1288.4) chips
   c) Based on this sample, the mean number of chips in an 18-ounce bag is between 1187.9 and 1288.4, with 95% confidence. The *mean* number of chips is clearly greater than 1000. However, if the claim is about individual bags, then it's not necessarily true. If the mean is 1188 and the SD deviation is near 94, then 2.5% of the bags will have fewer than 1000 chips, using the Normal model. If in fact the mean is 1288, the proportion below 1000 will be less than 0.1%, but the claim is still false.

47. a) The Normal probability plot is relatively straight, with one outlier at 93.8 sec. Without the outlier, the conditions seem to be met. The histogram is roughly unimodal and symmetric with no other outliers.
   b) $t = -2.63$, P-value $= 0.0160$. With the outlier included, we might conclude that the mean completion time for the maze is not 60 seconds; in fact, it is less.
   c) $t = -4.46$, P-value $= 0.0003$. Because the P-value is so small, we reject H_0. Without the outlier, we see strong evidence that the average completion time for the maze is less than 60 seconds. The outlier here did not change the conclusion.
   d) The maze does not meet the "one-minute average" requirement. Both tests rejected a null hypothesis of a mean of 60 seconds.

49. a) (289.9, 292.3)
   b) These data are not a random sample of golfers. The top professionals are not representative and were not selected at random. We might consider the 2011 data to represent the population of all professional golfers, past, present, and future.
   c) The data are means for each golfer, so they are less variable than if we looked at all the separate drives.
   d) Independence assumption violated.

51. a) (75.35, 142.63)
   b) The plots suggest that the distribution of the data is close enough to Normal but there is one outlier (Red Deer). When there are outliers in the data set, the sample mean and the standard deviation are not good summary statistics of the data set and so the 95% confidence level quoted in a) may not be trustworthy. The summary statistics and the confidence intervals calculated from the data set with and without the outlier are shown below. The mean is approximately 10 percentage points higher with the outlier and the standard deviation is about 15 percentage points higher.

Histogram of % Change

Probability Plot of % Change

	n	Mean	StDev	SE Mean	95% CI
With the outlier	20	108.990	71.871	16.071	(75.353, 142.627)
Without the outlier	19	98.7105	56.7602	13.0217	(71.3530, 126.0681)

c) No, because our interval is for an average change per urban area, not an average per capita.

53. a) Answers may vary. 90% CI is (48.42, 55.74). This interval includes the mean (i.e., 50).
  b) Answers may vary. 91 of intervals contain 50. We expect 90% percent of intervals to contain 50, if we repeated these simulations many times. X has a binomial distribution with $n = 100$ and $p = 0.95$.

55. a) Answers may vary. It is right skewed.

Histogram of x

  90% CI = (0.6408, 1.7987); This interval includes the mean (i.e., 1.0).
  b) Answers may vary. 82 of the intervals contain the mean 1.0. The sample mean of a sample of size 15 from an exponential distribution will not have a Normal distribution and so this interval will not have the requested confidence level (i.e., 90%). If we change the sample size to 100, then the distribution of the sample mean will become approximately Normal and so approximately 90% of the CIs will contain 1.0.

57. a) Answers may vary. Mean = 51.6308, Standard deviation = 5.4848, Standard error of the mean = 2.4529, t-statistic = 0.66, P-value = 0.543 > 0.10, and so we do not reject the null hypothesis. Passed a good batch. Since the null hypothesis is true, the error of rejecting is the possible error here. That is Type I error.
  b) Answers may vary. We expect to reject 10% of the tests. Note that this is the Type I error and the probability of a Type I error is the level of the test. The distribution of X is binomial with ($n = 100$, $p = 0.10$).

59. a) $z^* = 1.645$ for $\alpha = 0.05$ and so the criterion is reject the null hypothesis if $\dfrac{\bar{y} - 0.08}{s/\sqrt{150}} > 1.645$.
  b) Reject the null hypothesis if $\bar{y} > 0.08671568438$.
  c) The probability of rejecting the null hypothesis is 0.7881.
  d) Using software (MINITAB), the power at sample sizes 150, 75, and 38 with a difference 0.01 (i.e., 0.09–0.08) and $\alpha = 0.05$ is 0.786, 0.528, and 0.332 respectively. Note that the Power decreases as the sample size decreases.

Part V Review

1. H_0: The probability of choosing whole wheat bread is 50% ($p_{wheat} = 0.5$)
  H_A: The probability of choosing whole wheat bread is not 50% ($p_{wheat} \neq 0.5$)

3. a) 10.29
  b) Not really. The z-score is −1.11. Not any evidence to suggest that the proportion for Monday is low.
  c) Yes. The z-score is 2.26 with a P-value of 0.024 (two-sided).
  d) Some births are scheduled for the convenience of the doctor and/or the mother.

5. a) H_0: $p = 0.40$; H_A: $p < 0.40$
  b) Random sample; less than 10% of all California gas stations, $0.4(27) = 10.8$, $0.6(27) = 16.2$. Assumptions and conditions are met.
  c) $z = -1.49$, P-value = 0.0677
  d) With a P-value this high, we fail to reject H_0. These data do not provide evidence that the proportion of leaking gas tanks is less than 40% (or that the new program is effective in decreasing the proportion).
  e) Yes, Type II.
  f) Increase α, increase the sample size.
  g) Increasing α—increases Power, lowers chance of Type II error, but increases chance of Type I error. Increasing sample size—increases Power, costs more time and money.

7. a) The researcher believes that the true proportion of "A's" is within 10% of the estimated 54%, namely, between 44% and 64%.
  b) Small sample.
  c) No, 63% is contained in the interval.

9. a) Pew uses a 95% confidence level. We can be 95% confident that the true proportion is within 2% of 13%—that is, it is between 11% and 15%.
  b) The cell phone group would have the larger ME because its sample size is smaller.
  c) CI = (78.5%, 85.5%)
  d) The ME is 0.035, which is larger than the 2% ME in part a), largely because of the smaller sample size. It is larger than the ME in part b), mostly because 0.82 is smaller than 0.87 and proportions closer to 0.50 have larger MEs.

11. a) Bimodal.
  b) μ, the population mean. Sample size does not matter.
  c) $\dfrac{\sigma}{\sqrt{n}}$; sample size does matter.
  d) It becomes closer to a Normal model and narrower as the sample size increases.

13. a) H_0: The mean weight of bags of Fritos is 35.4 grams. ($\mu = 35.4$)
  H_A: The mean weight of bags of Fritos is less than 35.4 grams. ($\mu < 35.4$)
  b) It is reasonable to think that the 6 bags are representative of all bags of Fritos.
  The histogram of bags weights shows one unusually heavy bag. Although not technically an outlier, it probably should be excluded for the purposes of the test. (We will leave it in for the preliminary test, then remove it and test again.)
  c) We will perform a one-sample t-test. $t = 0.726$. Since the P-value = 0.7497 is high, we fail to reject the null hypothesis. There is no evidence to suggest that the mean weight of bags of Fritos is less than 35.4 grams.
  d) $t = -0.53$. Since the P-value = 0.3107 is high, we fail to reject the null hypothesis. There is no evidence to suggest that the mean weight of bags of Fritos is less than 35.4 grams.
  e) Neither test provides evidence that the mean weight of bags of Fritos is less than 35.4 grams. It is reasonable to believe that the mean weight of the bags is the same as the stated weight. However, the sample sizes are very small, and the tests have very little power to detect lower mean weights. It would be a good idea to weigh more bags.

15. a) Different samples have different means. Since this is a fairly small sample, the difference may be due to natural sampling variation. Also, we have no idea how to quantify "a lot less" without considering the variation as measured by the standard deviation.
  b) H_0: The mean life of a battery is 100 hours. ($\mu = 100$)
  H_A: The mean life of a battery is less than 100 hours. ($\mu < 100$)
  c) It is reasonable to think that these 16 batteries are representative of all batteries of this type.
  Since we don't have the actual data, we can't check a graphical display, and the sample is not large. Assume that the population of battery lifetimes is Normal.

d) We will perform a one-sample t-test. $t = -1$. Since the P-value = 0.1666 is greater than $\alpha = 0.05$, we fail to reject the null hypothesis. There is no evidence to suggest that the mean battery life is less than 100 hours.

e) If the mean life of the company's batteries is only 98 hours, then the mean life is less than 100, and the null hypothesis is false. We failed to reject a false null hypothesis, making a Type II error.

17. a) 0.0201, or about 2%.

b) We cannot do this without knowing the sample size for Quebec. If we are told Quebec's population is 23% of Canada's, we could base a calculation on an approximate sample size for Quebec of 0.23(1000) = 230, which gives ME $\approx$ 0.0474, or about 4.74%.

c) The margin of error is inversely proportional to the square root of the sample size. Increasing the sample size by a factor f reduces the ME

for both all of Canada and just Quebec by $\dfrac{1}{\sqrt{f}}$, so we want $\dfrac{1}{\sqrt{f}} = 1/4$

or $f = 16$. The sample size for Canada would need to increase from 1000 to $1000 \times 16 = 16\,000$.

19. a) $H_0: p = 0.126$. $H_A: p < 0.126$. $z = -0.748$, P-value = 0.227. Because the P-value is high, we fail to reject H_0. There is little evidence that the fatal accident rate is lower for teenage girls than for teens in general.

b) If the proportion is really 12.6%, we will see the observed proportion (11.34%) or lower in about 22.7% of samples of size 388 due to sampling variation.

21. a) One would expect many small fish, with a few large ones.

b) We don't know the exact distribution, but we know it's not Normal.

c) Probably not. With a skewed distribution, a sample size of five is not a large enough sample to say the sampling model for the mean is approximately Normal.

d) 0.961

23. a) Yes. (60)(0.80) = 48, (60)(0.20) = 12. Both are $\geq$ 10.

b) 0.834

c) Higher. Bigger sample means smaller standard deviation for $\hat{p}$.

d) Answers will vary. For $n = 500$, the probability is 0.997.

25. a) 54.4% to 62.5%.

b) Based on this study, with 95% confidence the proportion of Crohn's disease patients who will respond favourably to infliximab is between 54.4% and 62.5%.

c) 95% of all such random samples will produce confidence intervals that contain the true proportion of patients who respond favourably.

27. At least 423, assuming that p is near 50%.

29. **Bread**

a) Since the histogram shows that the distribution of the number of loaves sold per day is skewed strongly to the right, we can't use the Normal model to estimate the number of loaves sold on the busiest 10% of days.

b) **Randomization Condition**: Assume that these days are representative of all days.

Nearly Normal Condition: The histogram is skewed strongly to the right.

However, since the sample size is large, the Central Limit Theorem guarantees that the distribution of averages will be approximately Normal.

The days in the sample had a mean of 103 loaves sold and a standard deviation of 9 loaves sold. Since the conditions are satisfied, the sampling distribution of the mean can be modelled by a Student's t-model, with $100 - 1 = 99$ degrees of freedom. We will use a one-sample t-interval with 95% confidence for the mean number of loaves sold. (By hand, use $t^*_{50} \approx 2.403$ from the table.)

c) (101.2, 104.8) We are 95% confident that the mean number of loaves sold per day at the Tobermory Bakery is between 101.2 and 104.8.

d) We know that in order to cut the margin of error in half, we need a sample four times as large. If we allow a margin of error that is twice as wide, that would require a sample only one-fourth the size. In this case, our original sample is 100 loaves; so 25 loaves would be a sufficient number to estimate the mean with a margin of error twice as wide.

e) Since the interval is completely above 100 loaves, there is strong evidence that the estimate was incorrect. The evidence suggests that the mean number of loaves sold per day is greater than 100. This difference is statistically significant, but may not be practically significant. It seems like the owners made a pretty good estimate!

31. a) H_0: The computer is undamaged. H_A: The computer is damaged.

b) 20% of good PCs will be classified as damaged (bad), while all damaged PCs will be detected (good).

c) 3 or more.

d) 20%

e) By switching to two or more as the rejection criterion, 7% of the good PCs will be misclassified, but only 10% of the bad ones will be misclassified, increasing the Power from 20% to 90%.

33. a) The company is interested only in confirming that the athlete is well-known.

b) Type I: the company concludes that the athlete is well-known, but that's not true. It offers an endorsement contract to someone who lacks name recognition. Type II: the company overlooks a well-known athlete, missing the opportunity to sign a potentially effective spokesperson.

c) Type I would be more likely, but Type II less likely.

35. a) There are $n = 100$ fires, with mean $\bar{y} = 6056$ and standard deviation $s = 10811.98$. The 90% confidence interval is $6065 \pm 1.65(10811.98)/\sqrt{100}$ or (4260.35, 7850.77).

b) The interval does contain the actual value of 6398. If we took 200 independent random samples from the population and counted the number of intervals which contained the actual mean 6398, this number would follow a binomial distribution with index $n = 200$ and probability $p = 0.10$.

c) The distribution is extremely skewed, as shown in the histogram below. However, the sample size is quite large, so the confidence interval should be approximately valid.

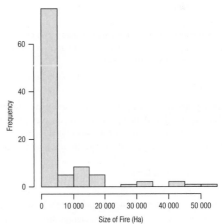

Histogram of Fires

37. a) $H_0: \mu = 22$; $H_A: \mu > 22$. $t = 5.985$. P-value < 0.00001. There is strong evidence that the injected rats drink more water than the non-injected rats.

b) (26.6, 35.4) cubic centimetres.

39. a) Based on a random sample of 1443, the margin of error should be ME = 0.0258 or 2.6%. The published ME is larger at 4.7%. This is not a random sample, but a stratified sample where random samples

are taken within each province. In addition, the results are post-stratified by age to focus on the 15–24 age group. The published ME takes these factors into account.

b) A 99% confidence interval requires adjustment of the confidence level from 0.95 to 0.99 in the margin of error, using $\hat{p} \pm 2.58\,ME/1.96$. Here $\hat{p} = 0.529$, and ME = 0.047, so the confidence interval is $0.529 \pm 2.58(0.047)/1.96$ or (0.4671, 0.5909). An interval constructed in this way will contain the true proportion 0.99 of the time.

Chapter 19

1. We must assume the samples were random or otherwise independent of each other. We also assume that the distributions are roughly Normal, so it would be a good idea to check a histogram to make sure there isn't strong skewness or outliers.

3. We are 95% confident that the mean purchase amount at Walmart is between $1.85 and $14.15 less than the mean purchase amount at Target.

5. The difference is −$8 with an SE of 3.115, so the *t*-statistic is −2.569. With 162.75 (or 163) df, the P-value is 0.011 which is less than 0.05. Reject the null hypothesis that the means are equal.

7. The *t*-statistic is still −2.561 using the pooled estimate of the standard deviation, 20.06. There are 163 df so the P-value is still 0.011. Same conclusion as before. Because the sample standard deviations are nearly the same and the sample sizes are large, the pooled test is essentially the same as the two-sample *t*-test.

9. We need a sample size of 153.

11. Yes. The high P-value means that we lack evidence of a difference, so 0 is a possible value for $\mu_{Meat} - \mu_{Beef}$.

13. a) Plausible values of $\mu_{Meat} - \mu_{Beef}$ are all negative, so the mean fat content is probably higher for beef hot dogs.
    b) The difference is significant.
    c) 10%

15. a) False; the confidence interval is about means, not about individual hot dogs.
    b) False; the confidence interval is about means, not about individual hot dogs.
    c) True.
    d) False; CIs based on other samples will also try to estimate the true difference in population means; there's no reason to expect other samples to conform to this result.
    e) True.

17. a) 2.927   b) Larger.
    c) Based on this sample, we are 95% confident that students who learn Math using the CPMP method will score, on average, between 5.57 and 11.43 points better on a test solving applied Algebra problems with a calculator than students who learn by traditional methods.
    d) Yes; 0 is not in the interval.

19. a) $H_0: \mu_C - \mu_T = 0$ vs. $H_A: \mu_C - \mu_T \neq 0$
    b) Yes. Groups are independent, though we don't know if students were randomly assigned to the programs. Sample sizes are large.
    c) If the means for the two programs are really equal, there is less than a 1 in 10 000 chance of seeing a difference as large as or larger than the observed difference just from natural sampling variation.
    d) On average, students who learn with the CPMP method do significantly worse on Algebra tests that do not allow them to use calculators than students who learn by traditional methods.

21. a) (1.36, 4.64)
    b) No; 5 minutes is beyond the high end of the interval.

23.

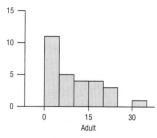

Independent group assumption—percentage of sugar in children's cereals is unrelated to the percentage of sugar in adults' cereals; Randomization condition—reasonable to assume that the cereals are representative of all adults and children's cereals in regard to sugar content; Nearly Normal condition—histogram of adult cereal sugar content is skewed to the right, but the sample sizes are of reasonable size, allowing us to proceed. Based on these samples, with 95% confidence, children's cereals average between 32.49% and 40.80% more sugar content than adults' cereals.

25.

There are several concerns here. First, we don't have a random sample. We have to assume that the actual number of hurricanes in a given year is a random sample of the hurricanes that might occur under similar weather conditions. Also, the data for 1944–1969 are not symmetric and have four outliers. The outliers will tend to make the average for the period 1944–1969 larger. These data are not appropriate for inference. The boxplots provide little evidence of a change in the mean number of hurricanes in the two periods. Moreover, a *t*-test of the two periods fails to show evidence of a difference in means with or without the outliers.

27. The histograms of the scores are unimodal and symmetric; $t = 2.076$ and P-value = 0.0228. There is evidence that the students taught using the new activities have a higher mean score on the reading comprehension test than the students taught using traditional methods.

29.

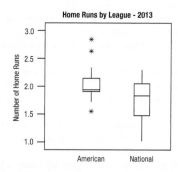

a) Both are reasonably symmetric. There are two possible high outliers and one possible low outlier in the American League (depending on how quartiles are calculated).
b) Based on these data, the average number of home runs hit per game in an American League stadium is between 1.89 and 2.25 (or 1.90 to 2.10 with all 3 outliers removed) with 95% confidence.
c) No. The boxplot indicates that there are no National League outliers (and Coors is not even the highest).

d) We want to work directly with the average difference. The two separate confidence intervals do not answer questions about the difference. The difference has a different standard deviation, found by adding variances.

e) (0.035, 0.573). Without the 3 outliers it is (0.008, 0.461).

f) Based on these data, with 95% confidence, American League stadiums average between 0.035 and 0.573 more home runs per game than National League stadiums.

g) Yes, 0 is not in the interval with or without the outliers.

31. These are not two independent samples. These are before and after scores for the same individuals.

33. a) These data do not provide evidence of a difference in ad recall between shows with sexual content and violent content.

b) $H_0: \mu_S - \mu_N = 0$ vs. $H_A: \mu_S - \mu_N \neq 0$. $t = -6.08$, df = 213.99, P-value = 5.5×10^{-9}. Because the P-value is low, we reject H_0. These data suggest that ad recall between shows with sexual and neutral content is different; those who saw shows with neutral content had higher average recall.

c) P-value = 0.136. The pooled t-test makes sense since this is an experiment and the sample variances are similar.

35. a) $H_0: \mu_V - \mu_N = 0$ vs. $H_A: \mu_V - \mu_N \neq 0$. $t = -7.21$, df = 201.96, P-value = 1.1×10^{-11}. Because of the very small P-value, we reject H_A. There is a significant difference in mean ad recall between shows with violent content and neutral content; viewers of shows with neutral content remember more brand names, on average.

b) With 95% confidence, the average number of brand names remembered 24 hours later is between 1.45 and 2.41 higher for viewers of neutral content shows than for viewers of sexual content shows, based on these data.

37. $H_0: \mu_{big} - \mu_{small} = 0$ vs. $H_A: \mu_{big} - \mu_{small} \neq 0$; bowl size was assigned randomly; amount scooped by individuals and by the two groups should be independent. With 34 df, $t = 2.104$ and P-value = 0.0428. The low P-value leads us to reject the null hypothesis and conclude that there is a difference in the average amount of ice cream that people scoop when given a bigger bowl.

39. The data for males has an extreme outlier; we should remove it before using t-tests.

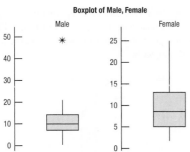

Boxplot of Male, Female

The boxplots and some summary statistics after removing the extreme outlier are shown below. We see that the spreads in the boxplots (and also the standard deviations in the summary statistics) are very similar and so we can use the pooled t-test.

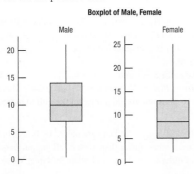

Boxplot of Male, Female

Descriptive Statistics: Male, Female				
Variable	N	Mean	SE Mean	StDev
Male	45	10.322	0.820	5.498
Female	47	9.309	0.777	5.326

Test $H_0: \mu_M - \mu_F = 0$ vs. $H_A: \mu_M - \mu_F > 0$. $S_{pooled} = 5.4107$, $SE_{pooled} = 1.1285$. $t = 0.90$. P-value = 0.186. Even if the population means were equal, there is a high chance for $\overline{Y}_M - \overline{Y}_F$ to be greater than or equal to 1.013 and so the data do not provide evidence against the null hypothesis. We do not reject the null hypothesis. The data do not provide sufficient evidence to conclude that women are better at parallel parking.

41. a) We can be 95% confident that the interval −216.74 to 154.64 minutes includes the true difference in mean crossing times between men and women. Because the interval includes zero, we cannot be confident that there is any difference at all.

b) **Independence Assumption:** There is no reason to believe that the swims are not independent or that the two groups are not independent of each other.

Randomization Condition: The swimmers are not a random sample from any identifiable population, but they may be representative of swimmers who tackle challenges such as this.

Nearly Normal Condition: The boxplots show two high outliers for the men and some skewness for both. The histograms are unimodal; the histogram for men is somewhat skewed to the right.

43. Independent Groups Assumption: The runners are different women, so the groups are independent. The Randomization Condition is satisfied since the runners are selected at random for these heats.

400-m Run Times

Nearly Normal Condition: The boxplots show an outlier, but we will proceed and then redo the analysis with the outlier deleted. When we include the outlier, $t = 0.048$ with a two-sided P-value of 0.96. With the outlier deleted, $t = -1.10$, with P = 0.3001. Either P-value is so large that we fail to reject the null hypothesis of equal means and conclude that there is no evidence of a difference in the mean times for runners in unseeded heats.

45. With $t \approx 4.57$ and a very low P-value of 0.0013, we reject the null hypothesis of equal mean velocities. There is strong evidence that golf balls hit off Stinger tees will have a higher mean initial velocity.

47. The 90% confidence interval for the difference in mean number of months to publication between *Applied Statistics* and *The American Statistician* is (8.43, 11.57) months. We assume that the articles are a random sample of the kinds of articles typically submitted to these two journals and that they represent fewer than 10% of all articles that could have been submitted. Both samples are large.

49. a) $H_0: \mu_R - \mu_N = 0$ vs. $H_A: \mu_R - \mu_N < 0$. $t = -1.36$, df = 20.00, P-value = 0.0944. Because the P-value is large, we fail to reject H_0. These data show no evidence of a difference in mean number of objects recalled between listening to rap music or no music at all.

b) Didn't conclude any difference.

51. $n = 69$

53. Using MINITAB, the required sample size is 27 for each sample.

Power and Sample Size

2-Sample t Test

Testing mean 1 = mean 2 (versus not =)

Calculating power for mean 1 = mean 2 + difference

Alpha = 0.05 Assumed standard deviation = 3

Sample Target			
Difference	Size	Power	Actual Power
3	27	0.95	0.950077

Chapter 20

1. a) Paired.   b) Not paired.   c) Paired.

3. a) No. The vehicles have no natural pairing.
  b) Possibly. The data are quantitative and paired by vehicle.
  c) The sample size is large, but there is at least one extreme outlier that should be investigated before applying these methods.

5. (7.34, 7.76). We are 95% confident that the interval from 7.34 mpg to 7.76 mpg captures the true improvement in highway gas mileage compared to city gas mileage.

7. We need a sample of size 139.

9. The difference between fuel efficiency of cars and that of trucks can be large, but isn't relevant to the question asked about highway vs. city driving. Pairing places each vehicle in its own block to remove that variation from consideration.

11. H_0: The median difference between highway and city fuel efficiency is 0. H_A: The median difference between highway and city fuel efficiency is different than 0. The test statistic is 0.9844. The z-score is 23.27, which corresponds to a P-value of 0.

13. a) Randomly assign 50 hens to each of the two kinds of feed. Compare production at the end of the month.
  b) Give all 100 hens the new feed for 2 weeks and the old food for 2 weeks, randomly selecting which feed the hens get first. Analyze the differences in production for all 100 hens.
  c) Matched pairs. Because hens vary in egg production, the matched-pairs design will control for that.

15. a) Show the same people ads with and without sexual images, and record how many products they remember in each group. Randomly decide which ads a person sees first. Examine the differences for each person.
  b) Randomly divide volunteers into two groups. Show one group ads with sexual images and the other group ads without. Compare how many products each group remembers.

17. a) The paired *t*-test is appropriate here. The weight measurement before and after surgical treatment is paired by subject. H_0: There is no difference in weight due to the surgical treatment. Since the P-value is very low, we conclude that the mean weight is lower after surgical treatment than before.
  b) The two-sample *t*-test is appropriate here. The subjects are not paired across the control and experimental groups. H_0: There is no difference in weight loss between the two treatment groups. Since the P-value is very low, we conclude that the mean weight loss of the group receiving surgical treatment is higher than the group receiving conventional treatment.

19. a) The paired *t*-test is appropriate since we have pairs of Fridays in 5 different months. Data from adjacent Fridays within a month may be more similar than data from randomly chosen Fridays.
  b) We conclude that there is evidence (P-value of 0.0212) that the mean number of cars found on the M25 motorway on Friday the 13th is less than on the previous Friday.

  c) We don't know if these Friday pairs were selected at random. Obviously, if these are the Fridays with the largest differences, this will affect our conclusion. The Nearly Normal condition appears to be met by the differences, but the sample size is small.

21. Adding variances requires that the variables be independent. These price quotes are for the same cars, so they are paired. Drivers quoted high insurance premiums by the local company will be likely to get a high rate from the online company, too.

23. a) The histogram—we care about differences in price.
  b) Insurance cost is based on risk, so drivers are likely to see similar quotes from each company, making the differences relatively smaller.
  c) The price quotes are paired; they were for a random sample of fewer than 10% of the agent's customers; the histogram of differences looks approximately Normal.

25. H_0: $\mu_{Local-Online} = 0$ vs. H_A: $\mu_{Local-Online} > 0$; with 9 df, t = 0.83. With a high P-value of 0.215, we don't reject the null hypothesis. These data don't provide evidence that online premiums are lower, on average.

27.

July – January

Data are paired for each city; cities are independent of each other; histogram shows the temperature differences are roughly unimodel and symmetric. This is probably not a random sample, so we might be wary of inferring that this difference applies to all European cities. Based on these data, we are 90% confident that the average temperature in European cities in July is between 32.3°F and 41.3°F higher than in January.

29. Based on these data, we are 90% confident that boys, on average, can do between 1.6 and 13.0 more push-ups than girls (independent samples—not paired).

31. a) Paired sample test. Data are before/after for the same workers; workers randomly selected; assume fewer than 10% of all this company's workers; histogram of differences shows them to be roughly unimodel and symmetric.

After – Before

  b) H_0: $\mu_d = 0$ vs. H_A: $\mu_d > 0$. $t \approx 3.60$, P-value = 0.0029. Because $P < 0.01$, reject H_0. These data show that average job satisfaction has increased after implementation of the exercise program.
  c) Type I error.
  d) The P-value for 8 +'s and 2 −'s is
  $P(X = 0) + P(X = 1) + P(X = 2) = 0.0547$, where X is the number of −'s. Thus, we would not reject the null hypothesis, unlike in part b).

33. H_0: $\mu_D = 0$ *vs.* H_A: $\mu_D \neq 0$. Data are paired by brand; brands are independent of each other; fewer than 10% of all yogourts (questionable); histogram of differences shows an outlier for Great Value:

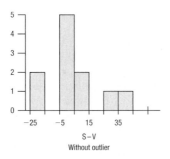

With the outlier included, the mean difference (Strawberry – Vanilla) is 12.5 calories with a *t*-statistic of 1.332, with 11 df, for a P-value of 0.2098. Deleting the outlier, the difference is even smaller, 4.55 calories with a *t*-statistic of only 0.833 and a P-value of 0.4241. With P-values so large, we do not reject H_0. We conclude that the data do not provide evidence of a difference in mean calories.

35. a) Not a simple random sample, but most likely representative; stops most likely independent of each other; histogram is roughly unimodel and symmetric.

Based on these data, with 95% confidence, the average braking distance for these tires on dry pavement is between 133.6 and 145.2 feet.
b) Not simple random samples, but most likely representative; stops most likely independent of each other; less than 10% of all possible wet stops; Normal probability plots are relatively straight. Based on these data, with 95% confidence, the average increase in distance for these tires on wet pavement is between 51.4 and 74.6 feet.

37. a) These students are not random, having been selected from writers of the PISA, but they are probably representative of all students. They are less than 10% of all students. A histogram shows a roughly normal distribution with some left skew. Using paired *t*-test. 90% CI = (5.851, 14.524).
b) The confidence interval does not include zero, we have strong evidence that there is a non-zero difference between genders in math scores. The confidence interval includes 10, we cannot conclude that the mean difference is at least 10 points.
c) It is an estimate of average difference in MATH scores, not scores.

39. a) 60% is 30 strikes; $H_0: \mu_A = 30$ vs. $H_A: \mu_A > 30$. $t = 6.06$, P-value $= 3.92 \times 10^{-6}$. With a very small P-value, we reject H_0. There is very strong evidence that players can throw more than 60% strikes after training, based on this sample.
b) $H_0: \mu_d = 0$ vs. $H_A: \mu_d > 0$. $t \approx 0.135$, P-value $= 0.4472$. With such a high P-value, we do not reject H_0. These data provide no evidence that the program has improved pitching in these Little League players.
c) H_0: The median difference in number of strikes thrown before and after training is zero. H_A: The median difference in number of strikes thrown before and after training is greater than zero. Taking differences in number of strikes thrown as After – Before, we get the following:
$+, +, +, -, -, -, 0, +, +, +, 0, +, -, -, -, +, -, -, -$
(two ties). The P-value for 8 +'s and 10 −'s is
$P(X = 0) + P(X = 1) + \cdots + P(X = 10) = 0.760$, where X is the number of −'s. Thus, we would not reject the null hypothesis, just like in part b).

41. a) The data are clearly paired. Both races may have improved over time, but the pairwise differences are likely to be independent. We don't have any information to check the Nearly Normal condition.
b) With 95% confidence we can say the mean time difference is between −16.63 minutes and 6.87 minutes.
c) The interval contains 0, so we would not reject the null hypothesis of no mean difference at a significance level of 0.05. We are unable to discern a difference between the female wheelchair times and the male running times.

43. a) Same cows before and after injection; the cows should be representative of others of their breed; cows are independent of each other; don't know about Nearly Normal differences.
b) (12.66, 15.34)
c) Based on this sample, with 95% confidence, the average increase in milk production for Ayrshire cows given BST is between 12.66 and 15.34 pounds per day.
d) 0.25(47) = 11.75. The average increase is much more than this, so we would recommend he assume the extra expense.

45. $n = 80$. This sample size is large enough for the use of z^* in place of t^*_{n-1}.

47. Using MINITAB, we need a sample of size 57.
Power and Sample Size
1-Sample t Test
Testing mean = null (versus not = null)
Calculating power for mean = null + difference
Alpha = 0.05
Assumed standard deviation = 1139.6

Sample Target			
Difference	Size	Power	Actual Power
500	57	0.9	0.902376

The probability that the test will conclude that the mean difference is non-zero when the true mean difference is 500 is 0.90.

Chapter 21

1. 0.0161

3. We must assume the data were collected randomly and that the Americans selected are independent of the Canadians selected. Both assumptions should be met. Also, for both groups, we have at least 10 national-born and foreign-born citizens and the sample sizes are less than 10% of the population sizes. All conditions for inference are met.

5. We are 95% confident that, based on these data, the proportion of foreign-born Canadians is between 3.24% and 9.56% more than the proportion of foreign-born Americans.

7. If we were to take repeated samples of these sizes of Canadians and Americans, and compute two-proportion confidence intervals, we would expect 95% of the intervals to contain the true difference in the proportion of foreign-born citizens.

9. a) 0.064 b) 0.1638 c) 0.0159 d) 4.030
e) P-value is < 0.001. Very strong evidence, so reject the null hypothesis that the proportions are the same in the two countries.

11. It is very unlikely that samples would show an observed difference this large if, in fact, there was no real difference in the proportions of teens that have Facebook accounts and teens that have MySpace accounts.

13. Without knowing anything about how the polls were taken, we should be cautious about drawing conclusions from this study. For example, the second poll could have been taken at a support rally for this candidate, causing a large source of bias. If the same group was polled twice, we do not satisfy the Independence assumption. Assuming proper polling

techniques, a P-value this small would give us good evidence that the campaign ads are working.

15. The responses are not from two independent groups, but are from the same individuals.

17. a) Stratified.   b) 6% higher among males.   c) 4%
d)

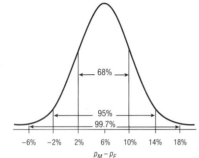

68%

95%

99.7%

−6%   −2%   2%   6%   10%   14%   18%
$p_M - p_F$

e) Yes. A poll result showing little difference is only 1–2 standard deviations below the expected outcome.

19. a) Yes. Random sample; less than 10% of the population; samples are independent; more than 10 successes and failures in each sample.
b) (0.055, 0.140)
c) We are 95% confident, based on these samples, that the proportion of American women age 65 and older who suffer from arthritis is between 5.5% and 14.0% more than the proportion of American men of the same age who suffer from arthritis.
d) Yes. The entire interval lies above 0.

21. a) 0.035   b) (0.356, 0.495)
c) We are 95% confident, based on these data, that the proportion of pets with a malignant lymphoma in homes where herbicides are used is between 35.6% and 49.5% higher than the proportion of pets with lymphoma in homes where no pesticides are used.

23. a) Yes. Subjects were randomly divided into independent groups, and more than 10 successes and failures were observed in each group.
b) (4.7%, 8.9%)
c) Yes, we're 95% confident that the rate of infection is 5–9 percentage points lower. That's a meaningful reduction, considering the 20% infection rate among the unvaccinated kids.

25. a) $H_0: p_V - p_{NV} = 0$, $H_A: p_V = p_{NV} < 0$
b) Because 0 is not in the confidence interval, reject the null. There's evidence that the vaccine reduces the rate of ear infections.
c) Type I.
d) Babies would be given ineffective vaccinations.

27. a) Prospective study.
b) $H_0: p_1 - p_2 = 0$; $H_A: p_1 - p_2 \neq 0$ where p_1 is the proportion of students whose parents disapproved of smoking who became smokers and p_2 is the proportion of students whose parents are lenient about smoking who became smokers.
c) Yes. We assume the students were randomly selected; they are less than 10% of the population; samples are independent; at least 10 successes and failures in each sample.
d) $z = -1.17$, P-value = 0.1211. These samples do not show evidence that parental attitudes influence teens' decisions to smoke.
e) If there is no difference in the proportions, there is about a 12% chance of seeing the observed difference or larger by natural sampling variation.
f) Type II.

29. a) (−0.221, 0.065)
b) We are 95% confident that the proportion of teens whose parents disapprove of smoking who will eventually smoke is between 22.1% less and 6.5% more than for teens with parents who are lenient about smoking.

c) 95% of all random samples will produce intervals that contain the true difference.

31. a) No. This is observational data.
b) $H_0: p_1 - p_2 = 0$; $H_A: p_1 - p_2 \neq 0$. $z = 3.56$, P-value = 0.0004. With a P-value this low, we reject H_0. There is a significant difference in the clinic's effectiveness. Younger mothers have a higher birth rate than older mothers. Note that the Success/Failure condition is met based on the pooled estimate of p.
c) We are 95% confident, based on these data, that the proportion of successful live births at the clinic is between 10.0% and 27.8% higher for mothers under 38 than in those 38 and older. However, the Success/Failure condition is not met for the older women, so we should be cautious when using this interval.

33. a) $H_0: p_1 - p_2 = 0$; $H_A: p_1 - p_2 > 0$. $z = 1.18$, P-value = 0.118. With a P-value this high, we fail to reject H_0. These data do not show evidence of a decrease in the voter support for the candidate.
b) Type II error.
c) It is estimated with 95% confidence that the change in support is anywhere from an 8% decrease to a 2% increase. We could not conclude, with 95% confidence, that there was a decrease in support.

35. a) We are 95% confident, based on this study, that between 67.0% and 83.0% of patients with joint pain will find medication A effective.
b) We are 95% confident, based on this study, that between 51.9% and 70.3% of patients with joint pain will find medication B effective.
c) Yes, they overlap. This might indicate no difference in the effectiveness of the medications, although this is not a proper test.
d) We are 95% confident that the proportion of patients with joint pain who will find medication A effective is between 1.7% and 26.1% higher than the proportion who will find medication B effective.
e) No. There is a difference in the effectiveness of the medications.
f) To estimate the variability in the difference of proportions, we must add variances. The two one-sample intervals do not. The two-sample method is the correct approach.

37. The conditions are satisfied to test $H_0: p_{young} = p_{old}$ against $H_A: p_{young} > p_{old}$. The one-sided P-value is 0.0619, so we may reject the null hypothesis. Although the evidence is not strong, *Time* may be justified in saying that younger men are more comfortable discussing personal problems.

39. Yes. With a low P-value of 0.003, reject the null hypothesis of no difference. There's evidence of an increase in the proportion of parents checking the websites visited by their teens.

41. Yes. With a low P-value of 0.0014, reject the null hypothesis of no difference. There is strong evidence to suggest that the percentages are different for the two groups: People from urban areas are more likely to agree with the statement than those from rural areas.

43. In order to use the methods discussed in this chapter, the two samples must be independent. Since we have the same 44 species-size combinations at two different times, they may not be independent. If this is the case, the methods we have discussed in this chapter cannot be used to carry out a test for comparing the two proportions.

45. a) $H_0: p_M - p_F = 0$ vs. $H_A: p_M - p_F \neq 0$. $z = 1.41$, P-value = 0.1586. The samples observed do not provide evidence against the null hypothesis at the 5% level of significance. That is, the samples do not provide sufficient evidence of a difference in the proportion of males and females who read newspapers on the subway.
b) $H_0: p_M - p_F = 0$ vs. $H_A: p_M - p_F \neq 0$. $z = 1.96$, P-value = 0.05. The samples provide some evidence of a difference in the proportion of males and females who read novels on the subway.
c) Yes, the probability of one or more erroneous rejections of the null hypothesis (known as the family error rate) in the series increases as the number of tests in the series increases. For a 5% level test, the

probability of Type I error is 0.05. If we perform 40 tests, then the probability that all 40 tests will fail to reject when all of them are true is less than or equal to 0.95^{40} and the probability of rejecting the null hypothesis in one or more of these 40 tests when all of them are true is greater than or equal to 0.8714878434, which is much higher than the acceptable level (0.05).

47. a) The table shows the conditional distributions as percentages.

Age	Male victim	Female victim
0–11	3.24	6.79
12–29	46.76	27.78
≥30	50.00	65.43
Total	100	100

b) $z = -0.3362$ and $P < .001$; very strong evidence against the null hypothesis.

c) $z = -1.9187$ and $P = 0.0550$; weak evidence against the null hypothesis.

d) $z = 4.1803$ and $P < .001$; very strong evidence against the null hypothesis.

e) There are proportionally fewer female than male victims in the 12–29 age group and more in the over 30 age group. An important problem with doing multiple tests is that the overall error rate, or probability of concluding at least one of the null hypotheses is incorrect, is inflated.

f) $z = 0.60$ and P-value $= 0.55$; no evidence against the null hypothesis of no gender effect.

49. a) 1538   b) 388

Chapter 22

1. a) (30, 30, 30, 30), 30 for each season.
b) 1.933
c) 3

3. a) 3
b) No. It's smaller than the mean.
c) It would say that there is not enough evidence to reject the null hypothesis that births are distributed uniformly across the seasons.
d) 7.815
e) Do not reject the null hypothesis. As in part c), there is not enough evidence of a seasonal effect on human births.

5. a) The age distributions of customers at the two branches are the same.
b) Chi-square test of homogeneity.
c)

	Age			
	Less than 30	30–55	56 or older	Total
In-Town Branch	25	45	30	100
Mall Branch	25	45	30	100
Total	50	90	60	200

d) 9.778
e) 2
f) 0.0075
g) Reject the null hypothesis and conclude that the age distributions at the two branches are not the same.

7. a) (−0.913, 0.913, 0.365, −0.365)
b) No. They are quite small for z-values.
c) Because we did not reject the null hypothesis, we should expect the standardized residuals to be relatively small.

9. a)

Did you use the Internet yesterday?			
	Yes	No	Total
White	2506.38	895.62	3402
Black	338.90	121.10	460
Hispanic/Other	445.73	159.27	605
Total	3291	1176	4467

b) 11.176   c) 2   d) 0.0037
e) Reject the null hypothesis and conclude that Race and Internet use are not independent.

11. a) Chi-square test of independence. We have one sample and two variables. We want to see if the variable *Account Type* is independent of the variable *Trade Type*.
b) Other test. The variable *Account Size* is quantitative, not counts.
c) Chi-square test of homogeneity. We want to see if the distribution of one variable, *Courses*, is the same for two groups (resident and nonresident students).

13. a) 10   b) Goodness-of-fit.
c) H_0: The die is fair (all faces have $p = 1/6$). H_A: The die is not fair.
d) Count data; rolls are random and independent; expected frequencies are all bigger than 5.
e) 5   f) $\chi^2 = 5.6$, P-value $= 0.3471$
g) Because the P-value is high, do not reject H_0. The data show no evidence that the die is unfair.

15. a) Weights are quantitative, not counts.
b) Count the number of each kind of nut, assuming the company's percentages are based on counts rather than weights.

17. H_0: The distribution of blood types of residents on Bongainville Island is the same as the distribution of blood types of the general population; H_A: The distribution of blood types of residents on Bongainville Island differs from the distribution of blood types of the general population. One cell has expected count less than 5, the chi-square procedures are not appropriate. Cells would have to be combined in order to proceed. Combine Blood type B with blood type AB. $\chi^2 \approx 38.39$, df $= 2$, P-value $= 0.0000$. Because the P-value is so low, we reject H_0. There is strong evidence that the distribution of blood types of residents on Bongainville Island differs from the distribution of blood types of the general population.

19. a) $\chi^2 \approx 5.671$, df $= 3$, P-value $= 0.1288$. With a P-value this high, we fail to reject H_0. Yes, these data are consistent with those predicted by genetic theory.
b) $\chi^2 \approx 11.342$, df $= 3$, P-value $= 0.0100$. Because of the low P-value, we reject H_0. These data provide evidence that the distribution is not as specified by genetic theory.
c) With small samples, many more data sets will be consistent with the null hypothesis. With larger samples, small discrepancies will show evidence against the null hypothesis.

21. a) 6   b) Goodness of Fit.
c) H_0: The number of large hurricanes remains constant over decades. H_A: The number of large hurricanes has changed.
d) 15   e) 0.63
f) The very high P-value means these data offer no evidence that the numbers of large hurricanes has changed.
g) The final period is only 6 years rather than 10 and already 7 large hurricanes have been observed. Perhaps this decade will have an unusually large number of such hurricanes.

23. a) Independence.
b) H_0: Breastfeeding success is independent of having an epidural. H_A: There's an association between breastfeeding success and having an epidural.

25. a) 1   b) 159.34
  c) Breastfeeding behaviour should be independent for these babies. They are fewer than 10% of all babies; we assume they are representative. We have counts, and all the expected counts are at least 5.

27. a) 5.90   b) P-value < 0.0001
  c) The P-value is very low, so reject the null. There's strong evidence of an association between having an epidural and subsequent success in breastfeeding.

29. a) $c = \dfrac{Obs - Exp}{\sqrt{Exp}} = \dfrac{190 - 159.34}{\sqrt{159.34}} = 2.43$
  b) It appears that babies whose mothers had epidurals during childbirth are much less likely to be breastfeeding 6 months later.

31. These factors would not be mutually exclusive. There would be yes or no responses for every baby for each.

33. a) 40.2%   b) 8.1%   c) 62.2%   d) 285.48
  e) H_0: Survival was independent of status on the ship. H_A: Survival depended on the status.
  f) 3
  g) We reject the null hypothesis. Survival depended on status. We can see that first-class passengers were more likely to survive than passengers of any other class.

35. First class passengers were most likely to survive, while third class passengers and crew were under-represented among the survivors.

37. a) Independence.
  b) H_0: *Discipline* is independent of *Birth Order*. H_A: There is an association between *Birth Order* and *Discipline*.
  c) Counted Data condition—these are counts of students; Randomization condition—a class is not a random sample, but there's little reason to think this particular group of students isn't representative; Expected Cell Frequency condition—the expected frequencies are low for both the Social Science and Professional disciplines, for third and fourth or higher birth orders (check residuals later).
  d) 9
  e) With a P-value this low, we reject the null hypothesis. There is some evidence of an association between *Birth Order* and *Discipline*.
  f) Unfortunately, 3 of the 4 largest standardized residuals are in cells with expected counts less than 5. We should be very wary of drawing conclusions from this chi-square test.

39. a) Experiment—actively imposed treatments (different drinks).
  b) Homogeneity.
  c) H_0: The rate of urinary tract infection is the same for all three groups. H_A: The rate of urinary tract infection is different among the groups.
  d) Count data; random assignment to treatments; all expected frequencies larger than 5.
  e) 2
  f) $\chi^2 = 7.776$, P-value ≈ 0.020
  g) With a P-value this low, we reject H_0. These data provide reasonably strong evidence that there is a difference in urinary tract infection rates between cranberry juice drinkers, lactobacillus drinkers, and the control group.
  h) The standardized residuals are

	Cranberry	Lactobacillus	Control
Infection	−1.87276	1.191759	0.681005
No infection	1.245505	−0.79259	−0.45291

From the standardized residuals (and the sign of the residuals), it appears those who drank cranberry juice were less likely to develop urinary tract infections and those who drank lactobacillus were more likely to have infections.

41. a) Independence.
  b) H_0: delivery time is independent from computer-assisted ordering. H_A: There is an association between delivery time and computer-assisted ordering.

c) Counted data, probably a random sample; one cell has expected count less than 5, the chi-square procedures are not appropriate. Cells would have to be combined in order to proceed. Combine last two columns.
  d) $\chi^2 \approx 8.86$, P-value ≈ 0.003
  e) Since the *P*-value is small, we reject H_0. There is evidence of an association between delivery time and computer-assisted ordering.

43. a) Homogeneity.
  b) H_0: The distribution of grades is the same for the two professors. H_A: The distribution of grades is different for the two professors.
  c) The expected counts are:

	Prof. Alpha	Prof. Beta
A	6.667	5.333
B	12.778	10.222
C	12.222	9.778
D	6.111	4.889
F	2.222	1.778

Since three cells have expected counts less than 5, the chi-square procedures are not appropriate. Cells would have to be combined in order to proceed.

45. a)

	Prof. Alpha	Prof. Beta
A	6.667	5.333
B	12.778	10.222
C	12.222	9.778
Below C	8.333	6.667

All expected frequencies are now larger than 5.
  b) Decreased from 4 to 3.
  c) $\chi^2 = 9.306$, P-value ≈ 0.0255. Because the P-value is so low, we reject H_0. The grade distributions for the two professors are different. Professor Alpha gives fewer A's and more grades below C than Professor Beta.

47. a) Answers will vary. Stacked bar charts or pie charts would work well.
  b) Chi-square test of independence. One person's response should not influence another's. We have counts, an SRS of less than 10% of the population, and all expected cell frequencies are much larger than 5. With 8 df, $\chi^2 \approx 190.96$. Because the P-value < 0.001, reject the null. There is strong evidence to suggest that responses are not independent of age.

49. $\chi^2 = 14.058$, df $= 1$, P-value $= 0.0002$. With a P-value this low, we reject H_0. There is evidence of racial steering. Blacks are much less likely to rent in Section A than Section B.

51. a) $z \approx 3.74936$, $z^2 \approx 14.058$
  b) The resulting P-values were both ≈ 0.0002. The two tests are equivalent.

53. $\chi^2 \approx 2815.968$, df $= 9$, P-value < 0.0001. Because the P-value is so low, we reject H_0. There are definite differences in education levels attained by the groups. The largest component indicates that Hispanics are more likely not to have high-school diplomas. The next-largest residual indicates that fewer Whites than expected do not have a high-school diploma.

55. $\chi^2 \approx 0.870$, df $= 2$, P-value ≈ 0.6471. Because the P-value is so large, we do not reject H_0. These data show no evidence of a difference in higher education opportunities between Blacks and Hispanics.

57. A goodness-of-fit test with 2 df yields $\chi^2 = 12.26$, with P-value $= 0.002$. Compared to the top 500 universities, the top 100 seem to be more heavily concentrated in the Americas and sparser in the Asia, Africa, and Pacific regions.

Chapter 23

1. The scatterplot shows a linear relationship with equal spread about a regression line throughout the range of *Acceptance Rates*. The residual plot has no structure, and there are not any striking outliers. The histogram of residuals is symmetric and bell-shaped. All conditions are satisfied to proceed with the regression analysis.

3. $s = 0.0254$ or 2.54 percentage points. Since the range of *Graduation Rates* is only about 17 percentage points, this number is fairly small and helps us to understand the amount of spread about the regression model.

5. The standard error for the slope tells us how much the slope of the regression equation would vary from sample to sample.

7. The administrators can conclude that there is a relationship between *Graduation Rates* and *Admission Rates* for these prestigious schools. It seems the lower the *Admission Rate*, the higher the *Graduation Rate*.

9. The administrators can conclude, with 95% confidence, that the schools with *Admission Rates* that are lower by 1 percentage point will have, on average, *Graduation Rates* that are higher by between 0.238 and 0.332 percentage points.

11. The standard error for the mean will be smaller. We can be much more precise about the prediction for the mean graduation rate of all colleges compared to the prediction for just one.

13. The administrators can expect, with 95% confidence, that schools that admit 33% of applicants will have average graduation rates between 87.35% and 88.88%.

15. The administrators can be 95% confident that if their *Admission Rate* is 33%, their *Graduation Rate* would be between 82.96% and 93.28%.

17. Equivalently, we test whether the true population correlation is 0. The test statistic is $t = -12.07$, under the t_{48} distribution. The P-value is less than 0.0001, so we are confident that a linear relationship exists between graduation rate and acceptance rate.

19. 75.3%

21. a) $\widehat{Error} = -454.448 - 8.479\ Year$; according to the model, the error made in predicting a hurricane's path was about 455 nautical miles, on average, in 1970. It has been declining at a rate of about 8.48 nautical miles per year.
    b) $H_0: \beta_1 = 0$; there has been no change in prediction accuracy. $H_A: \beta_1 \neq 0$; there has been a change in prediction accuracy.
    c) With a P-value < 0.001, I reject the null hypothesis and conclude that prediction accuracies have in fact been changing during this period.
    d) 71.1% of the variation in hurricane prediction accuracy is accounted for by this linear model on time.

23. a) $\widehat{Budget} = -63.9981 + 1.02648\ RunTime$. The model suggests that each additional minute of run time for a movie costs about $1 026 000.
    b) A negative intercept makes no sense, but a P-value of 0.07 indicates that we can't discern a difference between our estimated value and zero. It makes sense that a movie of zero length should cost $0.
    c) Amounts by which movie costs differ from predictions made by this model vary, with a standard deviation of about $33 million.
    d) The standard error of the slope is 0.1541 million dollars per minute.
    e) If we constructed other models based on different samples of movies, we'd expect the slopes of the regression lines to vary, with a standard deviation of about $154 000 per minute.

25. a) The scatterplot is straight enough, and the residuals plot looks unpatterned.
    b) I'm 95% confident that the cost of making longer movies increases at a rate of between 0.41 and 1.12 million dollars per minute.

27. a) $H_0: \beta_1 = 0$; there's no association between calories and sodium content in all-beef hot dogs. $H_A: \beta_1 \neq 0$: there is an association.
    b) Based on the low P-value (0.0018), I reject the null. There is evidence of an association between the number of calories in all-beef hot dogs and their sodium contents.

29. a) Among all-beef hot dogs with the same number of calories, the sodium content varies, with a standard deviation of about 60 mg.
    b) 0.561 mg/cal
    c) If we tested many other samples of all-beef hot dogs, the slopes of the resulting regression lines would be expected to vary, with a standard deviation of about 0.56 mg of sodium per calorie.

31. I'm 95% confident that for every additional calorie, all-beef hot dogs have, on average, between 1.03 and 3.57 mg more sodium.

33. a) H_0: Difference in age between men and women at first marriage has not been decreasing since 1921. ($\beta_1 = 0$). H_A: Difference in age between men and women at first marriage has been decreasing since 1921. ($\beta_1 < 0$).
    b) Residual plot shows a curved pattern; histogram is not particularly unimodal and symmetric, and it is skewed to the right.
    c) $t = -44.02$, P-value < 0.0001. With such a low P-value, we reject H_0. These data show evidence that difference in age between men and women at first marriage appears to be decreasing over time.

35. Based on these data, we are 95% confident that the mean difference in age between men and women at first marriage decreases by between 0.020 and 0.022 years in age per year.

37. a) $H_0: \beta_1 = 0$; $H_A: \beta_1 \neq 0$. P-value < 0.0001, indicating strong evidence of an association.
    b) Yes, the conditions seem satisfied. Histogram of residuals is unimodal and symmetric; residual plot looks OK, but some "thickening" of the plot with increasing values.
    c) $t = -12.2$, P-value < 0.0001 These data show evidence that *Fuel Economy* decreases with the *Weight* of the car.

39. a) $(-9.57, -6.86)$ mpg per 1000 pounds.
    b) Based on these data, we are 95% confident that *Fuel Efficiency* decreases between 6.86 and 9.57 miles per gallon, on average, for each additional 1000 pounds of *Weight*.

41. a) Based on this regression, we are 95% confident that 2500-pound cars will average between 27.34 and 29.07 miles per gallon.
    b) Based on the regression, a 3450-pound car will get between 15.44 and 25.37 miles per gallon, with 95% confidence.

43. a) Yes. $t = 2.73$, P-value $= 0.0079$. With a P-value so low, we reject H_0. There is a positive relationship between *Calories* and *Sodium* content.
    b) No. $R^2 = 9\%$ and s appears to be large, although without seeing the data, it is a bit hard to tell.

45. Plot of *Calories* against *Fibre* does not look linear; this is borne out in the residuals plot, which also shows increasing variance as predicted values get large. The histogram of residuals is right skewed.

47. a) $H_0: \beta_1 = 0$; $H_A: \beta_1 \neq 0$.
    b) $t = -7.73$ with 161 df; P-value < 0.0001
    c) With such a small P-value, this means that the association we see in the data is unlikely to occur by chance. We reject the null hypothesis, and conclude that there is strong evidence of a linear relationship between *BCI* and *pH*. Streams with higher *pH* tend to have lower *BCI*.

49. a) $H_0: \beta_1 = 0$; $H_A: \beta_1 > 0$. $t = 3.48$, P-value $= 0.0018$. With a P-value so low, we reject H_0. These data show evidence that *Ozone* does increase with *Population*.
    b) Yes, *Population* accounts for 84% of the variability in *Ozone* level, and s is just over 5 parts per million.

51. a) Based on this regression, each additional million residents corresponds to an increase in average ozone level of between 3.29 and 10.01 ppm, with 90% confidence.
    b) Based on the regression, the mean *Ozone* level for cities with 600 000 people is between 18.47 and 27.29 ppm, with 90% confidence.

53. a) 33 batteries.

b) The scatterplot of the data is roughly linear with lots of scatter; plot of residuals vs. predicted values shows no overt patterns; Normal probability plot of residuals is reasonably straight.

c) $H_0: \beta_1 = 0$; $H_A: \beta_1 > 0$. $t = 3.23$; P-value $0.0029/2 = 0.0015$. With a P-value so low, we reject H_0. These data provide evidence that more expensive batteries do have more cranking amps.

d) No. $R^2 = 25.2\%$ and s = 116 amps. Considering that the range of amperage is only about 400 amps, an s of 116 is not very useful.

e) $\widehat{Power} = 384.594 + 4.14649 \, Cost$

f) (1.97, 6.32) cold cranking amps per dollar.

g) *Power* increases, on average, between 1.97 and 6.32 per dollar of battery *Cost* increase, with 90% confidence.

55. a) 34 tablets.

b) Yes. The scatterplot is roughly linear with lots of scatter; plot of residuals vs. predicted values shows no overt patterns; Normal probability plot of residuals is reasonably straight.

c) $H_0: \beta_1 = 0$; $H_A: \beta_1 < 0$. $t = 2.02$; one-sided P-value $= 0.0522$. This is not a particularly small P-value. There may be some evidence of a relationship, but it is not strong.

d) Not particularly. $R^2 = 11.3$, and s = 2.128 hours.

e) $Hours = 5.387 + 0.009 \, Screen \, Brightness$

f) (0.00142, 0.0167) hours per cd/m² units

g) *Battery life* is longer, on average, between 0.00142 to 0.0167 hours per one cd/m² units, with 90% confidence.

57. a) $H_0: \beta_1 = 0$; $H_A: \beta_1 \neq 0$. $t \approx 8.1$; P-value < 0.0001. %*Body Fat* seems to increase with *Waist* size.

b) With 95% confidence, mean %*Body Fat* for people with 40-inch waists is between 23.58 and 29.03, based on this regression.

59. a) The regression model is $\widehat{Midterm_2} = 12.005 + 0.72 \, Midterm_1$ with output:

Dependent variable is: Midterm 2
No Selector
R squared = 19.9% R squared (adjusted) = 18.6%
s = 16.78 with 64 − 2 = 62 degrees of freedom

Source	Sum of Squares	Df	Mean Square	F-ratio
Regression	4337.14	1	4337.14	15.4
Residual	17459.5	62	281.604	

Variable	Coefficient	s.o. of Coeff	t-ratio	prob
Constant	12.0054	15.96	0.752	0.4546
Midterm 1	0.720990	0.1837	3.92	0.0002

b) The scatterplot shows a weak, somewhat linear, positive relationship. There are several outlying points, but removing them only makes the relationship slightly stronger. There is no obvious pattern in the residual plot. The regression model appears appropriate. The small P-value for the slope shows that the slope is statistically distinguishable from 0 even though the R^2 value of 0.199 suggests that the overall relationship is weak.

c) No. The R^2 value is only 0.199 and the value of s of 16.8 points indicates that she would not be able to predict performance on *Midterm2* very accurately.

61. a) We want to know whether the times to swim across Lake Ontario have changed over time. We know the time of each swim in minutes and the year of the swim. $H_0: \beta_1 = 0$; $H_A: \beta_1 \neq 0$.

b) The Straight Enough condition fails because there is an outlier.

c) We cannot continue this analysis.

d) With the outlier removed, the relationship looks straight enough. Swims are separate, but some swimmers performed the feat more than once, so swims may not be entirely independent. There is no thickening of the plot. The normal probability plot of the residuals looks straight. The conditions appear to be met.

e) The analysis shows that swim *Times* have been increasing at about 14.9 minutes per *Year*. The P-value of 0.0158 is small enough to reject the null hypothesis, so this looks like a real increase.

f) The SE depends on three things: the standard deviation of the residuals around the line, σ, the standard deviation of x, and the number of data values. Only the first of these was changed substantially by removing the outlier. Indeed, we can see that s_e was 447.3 with the outlier present and only 277.2 when it was removed. This accounts for most of the change in the SE and therefore in the P-value. Because the outlier was at a value of x near the middle of the x range, it didn't affect the slope very much.

63. a) Data plot looks linear; no overt pattern in residuals; histogram of residuals roughly symmetric and unimodal.

b) $H_0: \beta_1 = 0$; $H_A: \beta_1 \neq 0$. $t = -6.24$; P-value < 0.001. There is evidence that cities in which the mean education level is higher also tend to have a lower mortality rate.

c) No. Data are on cities, not individuals. Also, these are observational data. We cannot predict causal consequences from them.

d) (−65.95, −33.89) deaths per 100 000 people.

e) We are 95% confident that the mean number of deaths per 100 000 people decreases by between 33.89 and 65.95 deaths for an increase of one year in average education level.

f) r = −0.64.
We test whether the true population correlation, $\rho = 0$. $t = -6.233$. P-value < 0.0001 (using the t_{58} distribution), so we are confident that a linear relationship exists between mortality and education level.

*g) The ANOVA table is:

Source	DF	SS	MS	F-value	P-value
Regression	1	89433	89433	38.942	0.0000
Residual	56	128609	2297		
Total	57	218042			

h) (874.2, 914.2)
The total variation explained by the regression model is 89433.

65. a) $\widehat{January} = 120.73 + 0.6995 \, December$ (We are told this is an SRS. One cardholder's spending should not affect another's. These are quantitative data with no apparent bend in the scatterplot. The residual plot shows some increased spread for larger values of January charges. A histogram of residuals is unimodal and slightly skewed to the right with several high outliers.)

b) $1519.73

c) ($1330.24, $1709.24)

d) ($290.76, $669.76)

e) The residuals show increasing spread, so the confidence intervals may not be valid. I would be skeptical of interpreting them too literally.

67. a) Yes, three points that are outliers and have high leverage.

b) Residuals plot show some increased spread for larger values of fitted values. Normal probability plot of residuals is reasonably straight.

c) ($170.05, $300.60)

d) ($5.56, $133.79)

e) The squared standard error of b1 is so small compared to s_e^2.

*69. Cost of higher education

a) Logit (Type) = −13.1461 + 0.08455 *Top*10% + 0.000259 \$ /*student*

b) Statistically significant, since the P-value of 0.033 is less than $\alpha = 0.05$.

c) Statistically significant, since the P-value of 0.003 is less than $\alpha = 0.05$.

Part VI Review

1. a) $H_0: \mu_{Jan} - \mu_{Jul} = 0$; $H_A: \mu_{Jan} - \mu_{Jul} \neq 0$. $t = -1.94$, df $= 43.0$, P-value $= 0.0590$. Since P-value is fairly low, reject the null. These data show some evidence of a difference in mean *Age* to crawl between January and July babies.

   b) $H_0: \mu_{Apr} - \mu_{Oct} = 0$; $H_A: \mu_{Apr} - \mu_{Oct} \neq 0$. $t = -0.92$; df $= 59.40$; P-value $= 0.3610$. Since P-value is high, do not reject the null; these data do not provide evidence of a significant difference between April and October with regard to the mean age at which crawling begins.

   c) These results are not consistent with the claim.

3. a) For First Nations, the confidence interval is (3638.6, 3651.4). For Immigrants of Chinese origin, the interval is (3386.7, 3399.3).

   b) The value from the BC population, 3558, is not in either confidence interval, so both groups differ significantly from the BC population.

   c) (243.01, 260.99)

5. We are 95% confident that the difference in the mean percentage of aluminum oxide content of the pottery at the two sites is between –3.37% and 1.65%. Since 0 is in the interval, there is no evidence that the aluminum oxide content at the two sites is different. It would be reasonable for the archaeologists to think that the same ancient people inhabited the sites.

7. a) The observed difference between the proportions is $0.38 - 0.26 = 0.12$. Since the P-value $= 0.0060$ is very low, we reject the null hypothesis. There is strong evidence that the proportion of ALS patients who are athletes is greater than the proportion of patients with other disorders who are athletes.

   b) This was a retrospective observational study. In order to make the inference, we must assume that the patients studied are representative of all patients with neurological disorders.

9. H_0: The proportions of traits are as specified by the ratio 1:3:3:9. H_A: The proportions of traits are not as specified. $\chi^2 = 5.01$, P-value $= 0.1711$. Since the P-value is high, we fail to reject the null hypothesis. There is no evidence that the proportions of traits are anything other than 1:3:3:9.

11. a) H_0: The proportion of Asian students with perfect pitch is the same as the proportion of non-Asians with perfect pitch.
    H_A: The proportion of Asian students with perfect pitch is different than the proportion of non-Asians with perfect pitch.

    b) There is strong evidence of a difference in the proportion of Asians with perfect pitch and the proportion of non-Asians with perfect pitch. There is evidence that Asians are more likely to have perfect pitch.

    c) If there is no difference in the proportion of students with perfect pitch, we would expect the observed difference of 25% to be seen simply due to sampling variation in less than one out of every 10 000 samples of 2700 students.

    d) The data do not prove anything about genetic differences causing differences in perfect pitch.

13. a) If there is no difference in the average fish sizes, the chance of observing a difference this large, or larger, just by natural sampling variation is 0.1%.

    b) There is evidence that largemouth bass that are fed a natural diet are larger. The researchers would advise people who raise largemouth bass to feed them a natural diet.

    c) If the advice is incorrect, the researchers have committed a Type I error.

15. $\chi^2 = 6.14$; P-value ≈ 0.1887. Since the P-value ≈ 0.1887 is high, we fail to reject the null hypothesis. There is no evidence of an association between duration of pregnancy and level of prenatal care in twin births.

17. a) We are 95% confident that the mean age of patients with cardiac disease is between 3.39 and 5.01 years higher than the mean age of patients without cardiac disease.

    b) Older patients are at greater risk for a variety of health problems, thus if an older patient does not survive a heart attack, the researchers will not know to what extent depression was involved because there will be a variety of other possible variables influencing the death rate. Additionally, older patients may be more (or less) likely to be depressed than younger ones.

19. a) It is unlikely that an equal number of boys and girls were contacted strictly by chance. It is likely that this was a stratified random sample, stratified by gender.

    b) We are 95% confident that the proportion of computer gamers is between 7.0% and 17.0% higher for boys than for girls.

    c) Since the interval lies entirely above 0, there is evidence that a greater percentage of boys play computer games than girls.

21. a) Positive direction, generally linear trend; moderate scatter.

    b) $H_0: \beta_1 = 0$. $H_A: \beta_1 \neq 0$.

    c) Yes; histogram is unimodal and roughly symmetric; residuals plot shows random scatter.

    d) $t \approx 27.1$; P-value $= 0.001$. With such a small P-value, we reject H_0. There is a significant positive linear relationship between *Duration* and time to next eruption of Old Faithful.

    e) Based on this regression, the average time *Interval* to next eruption after a 2-minute eruption is between 53.24 and 56.12 minutes, with 95% confidence.

    f) Based on this regression, we will have to wait between 63.23 and 87.57 minutes after a 4-minute eruption, with 95% confidence.

23. a) Assume a random sample; certainly less than 10% of all preemies and normal babies; more than 10 failures and successes in each group. 1.7% to 16.3% greater for children of normal birth weight.

    b) Since 0 is not in the interval, there is evidence that preemies have a lower high school graduation rate than children of normal birth weight.

    c) Type I, since we rejected the null hypothesis.

25. $H_0: \mu_C - \mu_U = 0$, $H_A: \mu_C - \mu_U > 0$, $t = 1.57$, df $= 86$, P-value $= 0.0598$. The P-value is just greater than 0.05. Although there may be some indication that students of certified teachers achieve higher mean reading scores than students of uncertified teachers, we cannot reject the null hypothesis at the 5% level.

27. Based on these data we are 95% confident that the proportions of streams in the Adirondacks with shale substrates is between 32.8% and 47.4%.

29. a) The students scores are dependent on what school they went to, so the normal standard deviation calculations do not apply. Assuming the standard errors provided in the table account for the dependence, the confidence interval for overall score is (522.296, 531.704).

    b) Again, a dependence issue with school. The confidence interval is (27.49, 36.51), indicating that females score higher.

31. a) $H_0: \mu_d = 0$, $H_A: \mu_d \neq 0$, $t \approx 0.381$, df $= 19$, P-value $= 0.7078$. With a high P-value, we do not reject H_0. There is no evidence that the tests differ in mean difficulty.

    b) $H_0: \mu_d = 0$, $H_A: \mu_d \neq 0$. The boxplot of the differences shows three outliers. With these outliers removed, the histogram is roughly unimodal and symmetric. A paired *t*-test with them included shows $t = -0.2531$ with 19 df and a P-value of 0.8029, showing no evidence of a difference in mean scores. With the three outliers removed, the *t*-test shows $t = -1.1793$ with 16 df and a P-value of 0.2555— still no evidence of a difference in means. We conclude the testing environment does not affect mean score.

33. a) Parallel boxplots of the distributions of petal lengths for the two species of flower are shown below. No units are specified, but millimetres seems like a reasonable guess.

b) *Versicolor* generally has longer petals than *virginica*.

c) (10.14, 14.46)

d) We are 95% confident, based on the information given, that the average petal length for *versicolor* irises is between 10.14 and 14.46 mm longer than that of *virginica* irises.

e) Yes. The entire interval is above 0.

35. $\chi^2 = 10.146$, P-value $= 0.0174$. We reject the null hypothesis that teams are evenly matched at $\alpha = 0.05$. If they were, we would expect to see fewer 4-game series.

37. a) 59 products.

b) 84.5%

c) (2.21, 2.78) dollars per minute.

d) Based on this regression, average Price increases between $2.21 and $2.78 for each minute of polishing Time, with 95% confidence.

39. a) $373.50

b) They are 95% confident that the average *Loss* in a home burglary is between $1644 and $2391, based on their sample.

c) 95% of all random samples will produce confidence intervals that contain the true mean *Loss*.

41. a) $H_0: \beta_1 = 0$ vs. $H_A: \beta_1 \neq 0$

With conditions for inference satisfied (given), an appropriate test statistic is $t = 4.16$, modelled by a t-distribution with 140 df. P-value < 0.0001. We reject the null hypothesis, and conclude that there is strong evidence of a linear relationship between grades and the number of meals eaten as a family.

b) This relationship would not be particularly useful for predicting a student's grade point average. Only 11% of the variation in GPA can be explained by the number of meals eaten together per week.

c) These conclusions are not contradictory. There is strong evidence that the slope is not zero, and that means strong evidence of a linear relationship. This does not mean that the relationship itself is strong, or useful for predictions.

43. a) Based on the data, we are 95% confident that the difference in recruiting is between −4.2% and 8.2%.

b) No. The interval contains 0 and negative values—it may actually be the case that the new strategy lowers acceptance rates.

Chapter 24

1. a) $H_0: \mu_1 = \mu_2 = \mu_3 = \mu_4$ vs. the alternative that not all means are equal. Here μ_k refers to the mean number of popcorn kernels left unpopped for brand k.

b) MS_T has 3 df; MS_E has 12.

c) The data provide very strong evidence that the means are not all equal. The brands do not produce the same mean number of unpopped kernels.

3. a) The null hypothesis is that the mean mileage driven on one litre of gas using each muffler is the same. The alternative is that not all means are equal.

b) MS_T has 2 df; MS_E has 21 df.

c) We would not reject the null hypothesis of equal means with a P-value of 0.1199. The data from this experiment provide no evidence that the muffler type affects gas mileage (assuming any necessary conditions are met).

d) Type II error.

5. a) $H_0: \mu_1 = \mu_2 = \mu_3 = \mu_4$ vs. the alternative that not all means are equal. Here μ_k refers to the mean activation time using recipe k.

b) The data provide strong evidence with a P-value of 0.0001 to reject the null hypothesis and conclude that the means are not all equal. This experiment provides strong evidence that the mean activation times differ among the recipes.

7. a) $H_0: \mu_4 = \mu_5 = \mu_6 = \mu_7$ vs. the alternative that not all means are equal. Here, μ_k refers to the mean mileage using k cylinders.

b) The spreads of the four groups look very different. The Similar Variance condition is not met. The response variable might need to be re-expressed before proceeding with an Analysis of Variance.

9. a) $H_0: \mu_1 = \mu_2 = \mu_3 = \mu_4 = \mu_5 = \mu_6$ vs. the alternative that not all means are equal. Here, μ_k refers to the mean *Time* it takes to serve a customer by *Teller k*.

b) P-value $= 0.1914$. The data do not provide evidence that the mean *Time* it takes to serve a customer differs by *Teller*.

c) No. We do not reject the null hypothesis of equal means, so we cannot perform multiple comparisons.

11. An ANOVA is not appropriate because *Eye Colour* is a categorical variable. The students should consider analyzing the data with a χ^2 test of independence between eye and hair colour.

13. a) $H_0: \mu_1 = \mu_2 = \mu_3$ vs. the alternative that not all means are the same. Here μ_k refers to the mean weight of cat hair remaining in cage at the end of the day for cats on food type k.

b) MS_T has 2 df; MS_E has 57 df.

c) P-value $= 0.0846$. We fail to reject the null hypothesis and conclude that there is no evidence that the mean weights of cat hair for the three groups are not all the same.

d) The runs were randomized. We may need to be cautious in generalizing the results to all cats since each breed could react differently.

15. a) $MS_T = 8.65$; $MS_E = 0.0767$

b) $F = 112.826$

c) P-value $= 0.000017$. The data provide very strong evidence that the means are not equal.

d) We have assumed that the experimental runs were performed in random order, that the variances of the treatment groups are similar, and that the residuals are nearly Normal.

e) A boxplot of the *Scores* by *Method*, a plot of residuals vs. predicted values, a Normal probability plot, and a histogram of the residuals.

f) $s_p = \sqrt{0.0767}$ points $= 0.277$ points.

17. a) The null hypothesis is that the mean *Scores* are the same for both *Types* of eggs. The alternative is that they are different.

b) P-value $= 0.0014$. The data provide strong evidence that the means are different. The real eggs have a higher mean score.

c) Yes. The conditions look reasonable. It would be good to examine the residuals as well.

d) The pooled estimate of the variance is equal to MS_E. The *t*-ratio is 5.5727 and $5.5727^2 = 31.0550$, which is approximately the *F*-ratio. The two-sided P-value for the *t* with 6 df is 0.0014, which agrees with the P-value for the *F*-statistic.

19. a) The null hypothesis is that the mean *Test Scores* from all the schools are equal. The alternative is that not all the means are equal.

b) P-value = 0.3899 The data provide insufficient evidence that the mean *Test Scores* differ.

c) The intern's Type I error rate is higher because he is performing many (over 100) *t*-tests rather than an Analysis of Variance or a multiple comparisons method.

21. a) The null hypothesis is that the mean *Sugar Content* is the same for the cereals on each *Shelf*. The alternative is that not all the means are equal.

b) The P-value of 0.0012 provides strong evidence that the mean *Sugar Content* is not the same for each *Shelf*.

c) We cannot conclude that cereals on *Shelf* 2 have a higher mean *Sugar Content* than cereals on *Shelf* 3 or that cereals on *Shelf* 2 have a higher mean *Sugar Content* than cereals on *Shelf* 1. We can conclude only that the means are not all equal.

d) The Bonferroni rejects the null hypothesis of equal means at $\alpha = 0.05$. Now we can conclude that the mean *Sugar Content* of cereals on *Shelf* 2 is *not* equal to the mean *Sugar Content* on *Shelf* 1 and is *not* equal to the mean *Sugar Content* on *Shelf* 3. In other words, we can conclude from this test what we wanted to conclude in part c).

23. a) The null hypothesis states that the mean *Download Time* (DT) is the same at each *Time of Day*. The alternative is that not all the means are equal.

b)

Analysis of Variance for Time (in sec.)					
Source	DF	Sum of Squares	Mean Square	F-Ratio	P-Value
DT	2	204641	102320	46.035	<0.0001
Error	45	100020	2222.67		
Total	47	304661			

The data provide very strong evidence that mean *Download Time* is not the same at each *Time of Day*.

c) The Similar Variance condition seems to be met. The residual plots below show no evidence of violating the Nearly Normal or Similar Variance condition. The runs were not randomized because they were in specific time periods. It would be good to check a timeplot of residuals by run order to see if there is dependence from one run to the next.

d) A Bonferroni test shows that all three pairs can be distinguished from each other at $\alpha = 0.05$.

25. a) There are five factors. The number of possible treatments is $4 \times 2 \times 3 \times 2 \times 3 = 144$.

b) It appears from the boxplots that coloured sticks break at a lower stress than beige sticks. The ANOVA output supports this with $p < 0.0001$.

c) The SD of the stress of the coloured group is way lower than that of the beige group, which violates our assumption of equal variance.

d) Since we only have 12 batches to assign to 144 possible treatment combinations, we have clearly not controlled the other factors well. It could be that the coloured group also has the worst glue type, in which case glue type is confounded with colour. Also, the sample sizes of the two colour groups should be closer for higher efficiency.

27. a) H_0: The mean breaking force is the same for each of the glue types. H_A: The mean breaking forces are not all the same for the four glue types.

b) MSE has 107 df.

c) The completed ANOVA table:

Source	DF	SS	MS	F	P
glue	3	724.32	241.44	50.35	<0.0001
Error	107	513.09	4.80		
Total	110	1237.41			

d) We reject the null hypothesis and conclude that there is strong evidence to suggest that there is at least one pair of glue types with different mean breaking force. We would also like to check the Equal Variance Assumption by looking at a residual plot.

Chapter 25

1. $TV_{ijk} = \mu + Sex_j + Athlete_k + \varepsilon_{ijk}$

3. The boxplots show different distributions for Women and Men, so the Additivity assumption is probably not satisfied. The plots show that, at least for men, the Equal Variance assumption seems unlikely. And both plots show outliers, making the Normality assumption questionable. If the survey was randomized, the responses may be mutually independent, but we don't have enough information to tell.

5. The interaction term reflects the fact that there is very little difference in TV watching among women between athletes and non-athletes, but there is a difference for men with athletes watching more TV.

7. a) $H_0: \gamma_1 = \gamma_2 = \gamma_3$ where γ represents the effect of the *Power* level vs. H_A: Not all of the *Power* levels have the same effect on the response. $H_0: \tau_1 = \tau_2 = \tau_3$ where τ represents the effect of *Time* level vs. H_A: Not all of the *Time* levels have the same effect on the response.

9. a) The P-value for *Power* is 0.0165. The P-value for *Time* is 0.0310.

b) We reject the null hypothesis that *Power* has no effect and conclude that the mean number of uncooked kernels is not equal across all 3 *Power* levels. We also reject the null hypothesis that *Time* has no effect and conclude that the mean number of uncooked kernels is not the same across all 3 *Times*.

c) Partial boxplots, scatterplots of residuals vs. predicted values, and a Normal probability plot of the residuals to check the assumptions.

11. a) $H_0: \gamma_1 = \gamma_2$ where γ represents the effect of the *Seat* level vs. H_A: One of the *Seat* levels has a different effect. The null hypothesis states that *seat* choice (driver vs. passenger) has no effect on mean head injury sustained. The alternative states that it does. $H_0: \tau_1 = \tau_2 = \tau_3 = \tau_4 = \tau_5 = \tau_6$ where τ represents the effect of the *Size* level vs. H_A: The *Size* level effects are not all equal. The null hypothesis states that *Size* of vehicle (the 6 represented) has no effect on mean *Head Injury* sustained. The alternative states that it does.

b) The conditions appear to be met. The effects are additive enough, the data we assume were collected independently, the boxplots show that the variance is roughly constant, and there are no patterns in the scatterplot of residuals vs. predicted values.

c) There is no significant interaction. The P-values for both *Seat* and *Size* are 0.0001. Thus, we reject the null hypotheses and conclude that both the *Seat* and the *Size* of car affect the severity of head injury. From the partial boxplots we see that the mean *Head Injury* level is higher for the driver's side. The effect of driver's seat seems to be roughly the same for all 6 car sizes.

13. a) A two-factor ANOVA must have a quantitative response variable. Here the response is whether they exhibited baldness or not, which is a categorical variable. A two-factor ANOVA is not appropriate.

b) We could use a chi-square analysis to test whether *Baldness* and *Heart Disease* are independent.

15. a) A chi-square test of independence gives a chi-square statistic of 14.510 with a P-value of 0.0023. We reject the hypothesis that *Baldness* and *Heart Disease* are independent.

b) No, the fact that these are not independent does NOT mean that one causes the other. There could be a lurking variable (such as age) that influences both.

17. a) $H_0: \gamma_1 = \gamma_2$ where γ represents the effect of the *Time of Day* on *Shots Made*. H_A: The means of *Shots Made* differ across the two *Time of Day* levels. The null hypothesis states that *Time of Day* has no effect on *Shots Made*. The alternative states that it does. $H_0: \tau_1 = \tau_2$ where τ represents the kind of *Shoes* worn vs. H_A: The means of *Shots Made* differ across the two types of *Shoes*. The null hypothesis states that type of *Shoes* worn has no effect on *Shots Made*. The alternative states that it does.

b) Partial boxplots show little effect of either *Time of Day* or *Shoes* on *Shots Made*.

Neither the scatterplot of residuals vs. predicted values nor the Normal probability plots of residuals show any conditions that aren't met. We assume the number of shots made were independent from one treatment condition to the next.

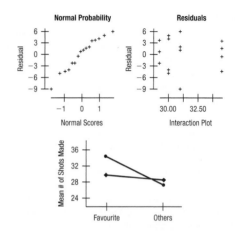

	Analysis of Variance				
Source	**DF**	**Sum of Squares**	**Mean Square**	**F-ratio**	**P-value**
Type	3	33.2519	11.084	23.468	0.0001
Cycle	3	7.19688	2.39896	5.0794	0.0250
Error	9	4.25062	0.472292		
Total	15	44.6994			

An interaction plot shows a possible interaction effect. It looks as though the favourite shoes may make more of a difference at night. However, the *F*-test shows that the null hypothesis of no interaction effect is not rejected. None of the effects appears to be significant. It looks as though she cannot conclude that either *Shoes* or *Time of Day* affect her mean *Shots Made*.

19. a) $H_0: \gamma_1 = \gamma_2 = \gamma_3$ where γ represents the effect of the *Temperature* level vs. H_A: Not all the γ are equal. The null hypothesis states that *Temperature* level (32°, 34°, and 36°C) has no effect on the *Number of Sprouts*. The alternative states that it does. $H_0: \tau_1 = \tau_2 = \tau_3 = \tau_4$ where τ represents the effect of the *Salinity* level vs. H_A: Not all the τ are equal. The null hypothesis states that *Salinity* level (0, 4, 8, and 12 ppm) has no effect on the *Number of Sprouts*. The alternative states that it does.

b) The partial boxplots show that *Salinity* appears to have an effect on the *Number of Sprouts,* but *Temperature* does not. The ANOVA supports this. The P-value for *Salinity* is < .0001, but the P-value for *Temperature* is 0.3779, providing no evidence of a temperature effect over this range of temperatures. There appears to be no interaction as well. There appears to be more spread in *Number of Sprouts* for the lower *Salinity* levels (where the response is higher). This is evident in both the partial boxplot and the residual vs. predicted values. This is cause for some concern, but most likely does not affect the conclusions.

21. The ANOVA table shows that both *Car Type* and *Additive* affect *Gas Mileage,* with P-values 0.0001. There is a significant interaction effect as well that makes interpretation of the main effects problematic. However, the residual plot shows a strong increase in variance, which makes the whole analysis suspect. The Similar Variance condition appears to be violated.

23. The ANOVA table now shows only main effects to be significant:

Source	DF	Sum of Squares	Mean Square	F-ratio	P-value
Type	2	10.1254	5.06268	5923.1	< 0.0001
Additive	1	0.026092	0.0260915	30.526	< 0.0001
Interaction	2	7.57E-05	3.78E-05	0.044265	0.9567
Error	54	0.046156	8.55E-04		
Total	59	10.1977			

The residuals now show *no* violation of the conditions:

We can conclude that both the *Car Type* and *Additive* have an effect on mileage and that the effects are constant (in log10 *Mpg*) over the values of the various levels of the other factor.

25. a) $H_0: \gamma_1 = \gamma_2$ where γ represents the effect of the *Environment* level vs. H_A: The mean *Time* the battery lasted is not the same for the two conditions. $H_0: \tau_1 = \tau_2 = \tau_3 = \tau_4$ where τ represents the effect of the *Brand* vs. H_A: Not all the τ are equal.

b) From the partial boxplot, it is not clear that *Brand* has an effect. The condition clearly has an effect.

c) Yes, the *Brand* effect has a P-value of 0.099, while the *Environment* effect is clearly significant with a P-value 0.0001.

d) There is also an interaction, however, which makes the statement about *Brands* problematic. Not all brands are affected by the environment in the same way. Brand C, which works best in the cold, performs worst at room temperature.

e) Because it performs the worst of the four at room temperature.

27. In this one-way ANOVA, we can see that the means vary across treatments. (However, boxplots with only 2 observations are not appropriate.) By looking closely, it seems obvious that the four flashlights at room temperature lasted much longer than the ones in the cold. But it is much harder to see whether the means of the four brands are different, or whether they differ by the same amounts across both environmental conditions. The two-way ANOVA with interaction makes these distinctions clear.

Chapter 26

1. a) $99\,859.89   b) $35\,140.11$
c) The house sold for about $35\,100 more than our estimate.

3. a) $\widehat{USGross} = -22.9898 + 1.12442\,Budget + 24.9724\,Stars - 0.403296\,RunTime$
b) After allowing for the effects of *RunTime* and *Stars*, each million dollars spent making a film yields about 1.13 million dollars in gross revenue.

5. a) Linearity: The plot is reasonably linear with no bends.
b) Equal Spread: The plot fails the Equal Spread condition. It is much more spread out to the right than on the left.
c) Normality: A scatterplot of two of the variables doesn't tell us anything about the distribution of the residuals.

7. a) $H_0: \beta_{Stars} = 0$
b) $t = 4.24$
c) $P < 0.0001$
d) Reject the null hypothesis and conclude that the coefficient of *Stars* is not zero.

9. a) $R^2 = 0.474$
About 47.4% of the variation in *USGross* is accounted for by the least squares regression on *Budget*, *RunTime*, and *Stars*.
b) Adjusted R^2 accounts for the number of predictors, so it differs from R^2, which does not.

11. a) Doesn't mention the other predictors.
b) This is correct.
c) Can't predict from *y* to *x*.
d) R^2 is about the fraction of variability accounted for by the regression model, not the fraction of data values.

13. a) $\widehat{Final} = -6.72 + 0.2560\,Test1 + 0.3912\,Test2 + 0.9015\,Test3$
b) R^2 is 77.7%, so 77.7% of the variation in final grade is accounted for by the regression model.
c) After allowing for the linear effects of the other predictors, each point on *Test3* is associated with an average increase of 0.9015 points on the final.
d) Test scores are probably collinear. If all we are concerned about is predicting final exam score, *Test1* may not add much to the regression. However, we'd expect it to be associated with final exam score.

15. a) $\widehat{Price} = -152037 + 9530\,Baths + 139.87\,Sqft$
b) 71.1% of the variation in asking price is accounted for by this regression model.
c) For homes with the same number of bathrooms, asking price increases, on average, by about $139.87 per square foot.
d) The number of bathrooms is probably correlated with the size of the house (even after considering the square footage of the bathroom itself). That correlation may account for the coefficient of *Baths* not being discernibly different from zero. Moreover, the regression model does not predict what will happen when a house is modified (for example, by converting existing space into a bathroom).

17. The plot of residuals vs. predicted values looks bent, rising in the middle and falling at both ends. This violates the Straight Enough condition. The Normal probability plot and the histogram of the residuals suggest that the highest five residuals (which we know are in the middle of the predicted value range) are extraordinarily high. These data may benefit from a re-expression.

19. a) $\widehat{Salary} = 9.788 + 0.11\,Service + 0.053\,Education + 0.071\,Score + 0.004\,Speed + 0.065\,Dictation$
b) 29.2 thousand dollars
c) Testing $H_0: \beta_4 = 0$ vs. $H_A: \beta_4 \ne 0$ has a *t*-statistic value of 0.013. That is not significant at any reasonable alpha level.
d) Omitting typing *Speed* would simplify the model and might result in one that was almost as good. Other predictors might also be omitted, but we can't tell that from what we know here.
e) *Age* may be collinear with other predictors in the model. In particular, it is likely to be highly associated with months of *Service*.

21. a) $H_0: \beta = 0; H_A: \beta \ne 0; t = 12.4$ on 248 df; $P < 0.0001$, so we reject the null hypothesis. The coefficient of *Weight* is statistically discernible from zero.
b) Each pound of weight is associated, on average, with an increase of 0.189 in *%Body Fat*.
c) After removing the linear effects of *Waist* and *Height*, each pound of *Weight* is associated on average with a decrease of 0.10 in *%Body Fat*. Alternatively, for men with the same *Waist* and *Height*, each pound of *Weight* is associated, on average, with a decrease of 0.10 in *%Body Fat*. The change in coefficient and sign is a result of including the other predictors. We expect *Weight* to be correlated with both *Waist* and *Height*. It may be collinear with them.
d) The P-value of 0.1567 says that if the coefficient of *Height* in this model is truly zero, we could expect to observe a sample regression coefficient as far from zero as the one we have here about 15.6% of the time.

23. a) P-value is less than 0.0001, so we can reject the null hypothesis that the true slope is zero.

b) It says that there is a clear relationship between *%Body Fat* and *Chest Size*. In fact, *%Body Fat* grows, on average, by 0.71272% per inch of *Chest Size*.

c) After allowing for the linear effects of *Waist* and *Height*, each inch of *Chest Size* is associated, on average, with a decrease of 0.233531 in *%Body Fat*. This coefficient is statistically discernible from zero (P = 0.0054).

d) Each of the variables appears to contribute to the model. There appears to be no advantage in removing one.

25. a) The only model that seems to do poorly is the one that omits *Murder*. The other three are hard to choose among.

b) Each of the models has at least one coefficient with a large P-value. This predictor variable could be omitted to simplify the model without degrading it much.

c) No. Regression models can't be interpreted in this way.

d) Plots of the residuals highlight some states as possible outliers. You may want to consider setting them aside to see if the model changes.

27. a) With an R^2 of 100%, the model should make excellent predictions.

b) The value of *s*, 3.140 calories, is very small compared with the initial standard deviation of *Calories*. This indicates that the model fits the data quite well, leaving very little variation unaccounted for.

c) No, the residuals are not all zero. Indeed, we know that their standard deviation, *s*, is 3.140. But they are very small compared with the original values. The true value of R^2 was rounded up to 100%.

Chapter 27

1. a) Use an indicator. Code Male = 0, Female = 1 (or the reverse).

b) Treat it as a quantitative predictor.

c) Use an indicator. Code Older than 65 = 1, Younger than 65 = 0.

3. a) Likely high leverage.

b) Likely outlier.

5. No. The predictors are almost certainly collinear, which would affect the coefficient of *Age*.

7. a) They are unimodal and fairly symmetric. Looks OK for regression.

b) That would not be a correct interpretation. Because CO_2 levels have risen steadily over recent years, CO_2 and *Year* are likely to be collinear.

9. a) The slope of a partial regression plot is the coefficient of the corresponding predictor in the multiple regression—in this case, −1.020.

b) Quaker oatmeal makes the slope more strongly negative. It appears to have substantial influence on this slope.

c) Not surprisingly, omitting Quaker oatmeal changes the coefficient of *Fibre*. It is now positive (although not significantly different from 0). This second regression model has a higher R^2, suggesting that it fits the data better. Without the influential point, the second regression is probably the better model.

d) The coefficient of *Fibre* is not discernibly different from zero. We have no evidence that it contributes significantly to *Calories* after allowing for the other predictors in the model.

11. a) After allowing for the effects of *Sodium* and *Sugars*, each gram of *Potassium* is associated with a 0.019 calorie decrease in *Calories*.

b) Those points pull the slope of the relationship down. Omitting them should increase the value of the coefficient of *Potassium*. It would likely become positive, since the remaining points show a positive slope in the partial regression plot.

c) These appear to be influential points. They have both high leverage and large residual, and the partial regression plot shows their influence.

d) If our goal is to understand the relationships among these variables, then it might be best to omit these cereals because they seem to behave as if they are from a separate subgroup.

13. a) This set of indicators uses *medium* size as its base. The coefficients of *Small*, *Large*, and *Very Large* estimate the average change in amount of *Delay/Person* for each of these three city sizes compared to the amount of *Delay/Person* for *medium* size cities. If there were an indicator variable for *medium* as well, the four indicators would be collinear, so the coefficients could not be estimated.

b) The costs of congestion per commuter in Large cities are, on average, about $106 greater than in *medium* size cities, after allowing for the effects of *Congested%*.

15. a) An assumption required for indicator variables to be useful is that the regression model's fit for the different groups identified by the indicators be parallel. These lines are not parallel.

b) The coefficient of SML*C%V adjusts the slope of the regression model fit for the small cities. We would say that the slope of *Congestion Cost per person* on *Congested% of motor vehicle miles* for *small* cities (after allowing for the linear effects of the other variables in the model) is $27.815 − $4.42 = $23.395.

c) The regression model seems to do a good job. The R^2 shows that 65% of the variability in *Congestion cost per person* is accounted for by the model. Most of the P-values for the coefficients are small.

17. Laredo has high leverage, but it does not have a particularly high Studentized residual. It appears that the point has influence but doesn't exert it. Removing this case from the regression may not result in a large change in the model, so the case is probably not influential.

19. a) It is the expected change in stress for a unit change in area, for a large overlap joint, holding everything else constant.

b) There are way more medium joints than small joints. Since the relationship is very weak for medium joints, the P-value is "swamped" by them.

c) 0.7773   d) −98.603

Part VII Review

1. a) Based on these data, with 95% confidence 5-year yields are between 3.15% and 5.93% higher than 3-year yields on average (paired data).

b) Yes (at least for this data set). The regression line is 5-year = 6.93 + 0.719 3-year. H_0: $\beta_1 = 0$ against H_A: $\beta_1 \neq 0$ has $t \approx 4.27$; P-value = 0.0009. Since P is so small, we reject H_0. There is evidence of an association. (But we don't know that this was an SRS or even a representative sample of large cap funds.)

3. a) An ANOVA and an *F*-test.

b) The null hypothesis is that the mean *Distance* from the target is the same for all 4 *Clubs*. The alternative is that it is not.

c) Conclude that the mean *Distance* from the target is not the same for all the *Clubs*.

d) Boxplots of *Distance* by *Club*, residual vs. predicted value scatterplot, and a Normal probability plot of the residuals.

e) A multiple comparison test to see which club is best.

5. a) Indicator variable.

b) The *Mortality* rate north of Derby is, on average, 158.9 deaths per 100 000 higher than south of Derby, after allowing for the linear effects of the water hardness.

c) This appears to be a better regression. The indicator has a strongly significant coefficient, and the R^2 for the overall model has increased from 43% to 56.2%.

7. a) The null hypotheses are that the mean number of *Pins* is the same for all three *Weights* and for both *Approaches*. The alternatives are that the means are not all the same for the three *Weights* or for the two *Approaches*.

b) *Weight* has 2 df; *Approach* has 1. The error has 20 df.

c) If the interaction plot shows any evidence of not being parallel, she should fit an interaction term, using 2 df to fit it.

9. 406 checks.

11. a) Both are straight enough. Larger plants are more expensive. Plants got more expensive over time.

b) *Cost* of nuclear plants increased by \$42 383 on average for each *Megawatt* of power.

c) Relationship is straight enough; there is no reason to doubt independence. Residual plot shows reasonably constant variance. Residuals are nearly Normal.

d) $t = 2.93$ on 30 df, $P = 0.0064$

e) Cost estimate $= 111.7412 + 432.83 = 535.57$, or \$535 570 000.

f) The R^2 means that 22.3% of the variation in *Cost* of nuclear plants can be accounted for by the linear relationship between *Cost* and the size of the plant, measured in megawatts (*Mwatts*). A scatterplot of residuals against *Date* shows a strong linear pattern, so it looks like *Date* could account for more of the variation.

g) The coefficient of *Mwatts* changed very little.

h) Because the coefficient changed little when we added *Date* to the model, we can expect that *Date* and *Mwatts* are relatively uncorrelated. In fact, their correlation is 0.02.

13. a) Null hypotheses are that mean hours slept are the same for both sexes and that the mean hours slept are the same for all four classes. Alternatives are that the mean hours slept are not the same for both sexes and that not all four classes have the same mean hours slept.

b) With all three P-values so large, none of the effects appear to be significant.

c) There are a few outliers. The Similar Variance condition appears to be met, but we do not have Normal probability plots of residuals. The main concern is that we may not have enough power to detect differences between groups.

15. a) H_0: $p_{<38} - p_{\geq38} = 0$; H_A: $p_{<38} - p_{\geq38} \neq 0$.
$z = 3.56$; P-value $= 0.0004$. With such a small P-value, we reject H_0. We conclude there is evidence of a difference in effectiveness; it appears the methods are not as good for older women.

b) $\chi^2 = 12.70$; P-value $= 0.0004$. Same conclusion.

c) The P-values are the same; $z^2 = (3.563944)^2 = 12.70 = \chi^2$.

17. H_0: The proportion of right-handed people who can match the tone is the same as the proportion of left-handed people who can match the tone.
($p_L = p_R$ or $p_L - p_R = 0$).
H_A: The proportion of right-handed people who can match the tone is different from the proportion of left-handed people who can match the tone. ($p_L \neq p_R$ or $p_L - p_R \neq 0$).
Conditions have been satisfied; $z = 1.38$. P-value $= 0.1683$ is high, we fail to reject the null hypothesis. There is no evidence that the proportion of people able to match the tone differs between right-handed and left-handed people.

19. a) **Straight Enough condition:** The scatterplots of the response versus each predicted value should be reasonably straight.
Independence: These data are measured over time. We should plot the data and the residuals against time to look for failures of independence.
Nearly Normal condition: The Normal probability plot should be straight, and the histogram of residual should be unimodal and symmetric.
Similar Variance condition: The spread of the residuals plot is constant.

b) $R^2 = 57.4\%$, so 57.4% of the variation in female teen traffic deaths is accounted for by the linear model.

c) $\widehat{deaths} = 45074.0 - 21.6006\,year$

d) The model predicts that female teen traffic deaths have been declining at a rate of approximately 21.6 per year.

21. a) H_0: Typing speed is the same whether the gloves are on or off.
($\gamma_{On} = \gamma_{Off}$)
H_A: Typing speeds are not the same when the gloves are on and off.
H_0: Typing speed is the same at hot or cold temperatures. ($\tau_H = \tau_C$)
H_A: Typing speeds are not the same at hot and cold temperatures.

b) Based on the boxplots, it appears that both the effects of *Temperature* and *Gloves* affect typing speed.

c) The interaction plot is not parallel, so the Additive Enough condition is not met. Therefore, an interaction term should be fit.

d) Yes, It will be significant because a difference in typing speed due to temperature seems to be significant only when he is not wearing the gloves.

e) The P-values for all three effects: *Gloves*, *Temperature*, and the interaction term, are all very small, so all three effects are significant.

f) Gloves decreased typing speed at both temperatures. A hot temperature increases typing speed with the gloves off, but has little effect with the gloves on.

g) $s = \sqrt{\dfrac{58.75}{28}} = 1.45$ words per minute. Yes, this seems consistent with the size of the variation shown on the partial boxplots.

h) Tell him to type in a warm room without wearing his gloves.

i) The Normal probability plot of residuals should be straight. The residuals plot shows constant spread across the predicted values. The Similar Variance condition appears to be met. For the levels of the factors that he used, it seems clear that a warm room and not wearing gloves are beneficial for his typing.

23. a) The number of finishers per minute appears to increase by about 0.519037 finishers per minute.

b) There is a definite pattern in the residuals. This is a clear violation of the Linearity assumption. While there does seem to be a strong association between time and number of finishers, it is not linear.

25. a) H_0: There is no linear relationship between depression and Internet usage. ($\beta_1 = 0$).
H_A: There is a linear relationship between depression and Internet usage. ($\beta_1 \neq 0$).
Conditions for a regression slope t-test are satisfied.
$t \approx 2.76$ (df = 160). P-value = 0.0064. We reject the null hypothesis, and conclude that there is strong evidence of a linear relationship between depression and Internet usage. Those with high levels of Internet usage tend to have high levels of depression.

b) The study says nothing about causality, merely association. Furthermore, there are almost certainly other factors involved. In fact, if 4.6% of the variation in depression level is related to Internet usage, the other 95.4% of the variation must be related to something else!

c) H_0: The mean difference in depression before and after the experiment is zero.
H_A: The mean difference in depression before and after the experiment is not equal to zero. We use a paired t-test, with $\bar{d} = -0.118457$. $t = -2.73$ (df = 161).
Since the P-value = 0.007 is close to 0, we reject the null hypothesis. There is evidence that the mean depression level changed over the course of the experiment.

27. H_0: $p = 0.512$
H_A: $p < 0.512$
$z = -3.35$
P-value = 0.0004. We reject the null hypothesis. There is strong evidence that the percentage of males born to workers is less than 51.2%. This provides evidence that human exposure to dioxin may result in the birth of more girls.

29. a) H_0: Pinball score is the same whether the tilt is on or off ($\gamma_{On} = \gamma_{Off}$).
H_A: Pinball scores are different when the tilt is on and when it is off.
H_0: Pinball score is the same whether both eyes are open or the right is closed ($\tau_B = \tau_R$).
H_A: Pinball scores are different when both eyes are open or when the right is closed.

b) The partial boxplots show that the *Eye* effect is strong, but there appears to be little or no effect due to *Tilt*.

Randomization condition: The experiment was performed in random order.
Similar Variance condition: The side-by-side boxplots have similar spreads. The residuals plot shows no pattern, and no systematic change in spread.
Nearly Normal condition: The Normal probability plot of the residuals is reasonably straight.

According to the ANOVA, only the *Eyes* effect is significant with a P-value less than 0.0001. Both the *Tilt* and interaction effects are not significant. We conclude that keeping both eyes open improves score but that using the tilt does not.

	Df	Sum Sq	Mean Sq	F value	Pr (>F)
Eyes	1	9.385e + 09	9.385e + 09	44.937	2.19e − 05 ***
Tilt	1	1.568e + 08	1.568e + 08	0.751	0.403
Eyes: Tilt	1	4.505e + 08	4.505e + 08	2.157	0.168
Residuals	12	2.506e + 09	2.089e + 08		

31. a) This is an attempt at linear regression. Regression inference is meaningless here, since eye and hair colour are categorical variables.

b) This is an analysis based upon a chi-square test for independence.
H_0: Eye colour and hair colour are independent.
H_A: There is an association between eye colour and hair colour.
Since we have two categorical variables, this analysis seems appropriate. However, if you check the expected counts, you will find that four of them are less than 5. We would have to combine several cells in order to perform the analysis. (Always check the conditions!)
Since the value of chi-square is so high, it is likely that we would find an association between eye and hair colour, even after the cells were combined. There are many cells of interest, but some of the most striking differences that would not be affected by cell combination involve people with fair hair. Blonds are likely to have blue eyes, and not likely to have brown eyes. Those with red hair are not likely to have brown eyes. Additionally, those with black hair are much more likely to have brown eyes than blue.

33. a) (11.65,17.39)
b) You would need about 44 years of data.
c) H_0: There is no linear relationship between year and annual LA rainfall ($\beta_1 = 0$).
H_A: There is a linear relationship between year and annual LA rainfall ($\beta_1 \neq 0$).
Straight Enough condition: The scatterplot is straight enough to try linear regression, although there is no apparent pattern.
Independence assumption: The residuals plot shows no pattern.
Does the Plot Thicken? condition: The spread of the residuals is consistent.
Nearly Normal condition: The Normal probability plot of residuals is not straight, and the histogram of the residuals is skewed to the right, but a sample of 22 years is large enough to proceed.
Conditions for inference are satisfied.
The value of $t \approx 0.124$ (df = 20). P-value = 0.92029. We fail to reject the null hypothesis, and conclude that there is no evidence of a linear relationship between the annual rainfall in LA and the year.

35. a) H_0: Mean weight is the same for all three groups ($\mu_1 = \mu_2 = \mu_3$).
H_A: Mean weights are not the same for all three groups.

b) According to the boxplots the spread appears constant for all three groups. There is one outlier. We do not have a Normal probability plot, but we suspect that the data may be Normal enough.

c) With a P-value equal to 0.0042, we reject the null hypothesis; there is evidence that the mean weights are not the same for all three groups. It may be that more men are involved with athletics, which might explain the weight differences. On the other hand, it may simply be that those who weigh more are more likely to be involved with sports.

d) It seems that differences are evident even when the outlier is removed. It seems that the conclusion is valid.

37. a) H_0: The mean score of week-long study group students is the same as the mean score of overnight cramming students ($\mu_1 = \mu_2$ or $\mu_1 - \mu_2 = 0$).
   H_A: The mean score of week-long study group students is greater than the mean score of overnight cramming students ($\mu_1 > \mu_2$ or $\mu_1 - \mu_2 > 0$).
   We perform a two-sample t-test. $t = 0.90$ (df = 39.94). Since the P-value = 0.1864 is high, we fail to reject the null hypothesis. There is no evidence that students with a week to study have a higher mean score than students who cram the night before.

b) H_0: The proportion of study group students who will pass is the same as the proportion of crammers who will pass ($p_1 = p_2$ or $p_1 - p_2 = 0$).
   H_A: The proportion of study group students who will pass is different from the proportion of crammers who will pass ($p_1 \neq p_2$ or $p_1 - p_2 \neq 0$).
   $z = -3.11$. P-value = 0.0019. We reject the null hypothesis. There is strong evidence to suggest a difference in the proportion of passing grades for study group participants and overnight crammers. The crammers generally did better.

c) H_0: There is no mean difference in the scores of students who cram, after three days ($\mu_d = 0$).
   H_A: The scores of students who cram decreases, on average, after three days ($\mu_d > 0$).
   We use a paired t-test; $t = 5.17$.
   Since the P-value < 0.0001, we reject the null hypothesis. There is strong evidence that the mean difference is greater than zero. Students who cram seem to forget a significant amount after three days.

d) (3.03, 7.05) We are 95% confident that students who cram will forget an average of 3.03 to 7.05 words in three days.

e) H_0: There is no linear relationship between Friday score and Monday score ($\beta_1 = 0$).
   H_A: There is a linear relationship between Friday score and Monday score ($\beta_1 \neq 0$).

Straight Enough condition: The scatterplot is straight enough to try linear regression.
Independence assumption: The residuals plot shows no pattern.
Does the Plot Thicken? condition: The spread of the residuals is consistent.
Nearly Normal condition: The Normal probability plot of residuals is reasonably straight, and the histogram of the residuals is roughly unimodal and symmetric.

Since the conditions for inference are satisfied, the sampling distribution of the regression slope can be modelled by a Student's t-model with $(25 - 2) = 23$ degrees of freedom. We will use a regression slope t-test. $t \approx 2.57$. P-value = 0.0170. We reject the null hypothesis, and conclude that there is strong evidence of a linear relationship between Friday score and Monday score. Students who do better in the first place tend to do better after three days.
However, since R^2 is only 22.4%, Friday score is not a very good predictor of Monday score.

Chapter 28

1. c

3. a

5. b

7. c

9. b

11. If we assume that the rank sum has a normal distribution, its mean = 30 and the variance = 30. $z = -1.10$ and P-value = 0.2714 (for a non-directional test).

13. P-value for observed sum of 43 is greater than 0.05×2, so the difference is still not significant at the 5% level.

15. Both Wilcoxon's rank sum test (Mann-Whitney test) and two sample t-test show a significant difference between the populations (P-values 0.0026, or 0.0025 adjusted for ties, for the Wilcoxon's rank sum test and 0.007 for the two sample t-test).

17. The test statistic H = 31.64, $\chi^2_{2,0.05} = 5.991$, $H > \chi^2_{2,0.05}$ and so reject H_0: all populations have the same distributions. The results from the one-way ANOVA are copied below and based on that we reject the null hypothesis of equal population means.

One-way ANOVA: Time(sec.) versus Time of day

Source	DF	SS	MS	F	P
Time of day	2	204641	102320	46.03	0.000
Error	45	100020	2223		
Total	47	304661			

19. The one-way ANOVA would show a greater difference, since the means move much further apart, whereas the ranks don't change much.

21. The test statistic S = 10.88, $\chi^2_{3,0.05} = 7.815$, $S > \chi^2_{2,0.05}$ and so reject H_0: all populations have the same distributions.
   The results from the ANOVA are copied below and based on that we also (and even more convincingly) reject the null hypothesis of equal population means.

Two-way ANOVA: Score versus Temp, Cycle

Source	DF	SS	MS	F	P
Temp	3	33.2519	11.0840	23.47	0.000
Cycle	3	7.1969	2.3990	5.08	0.025
Error	9	4.2506	0.4723		
Total	15	44.6994			

23. a) There are 7 negative differences and 10 positive differences, and the P-value of the sign test is 0.6291. The P-value is large so there's insufficient evidence to declare any difference.

b) $W = 93.0$ and the P-value is 0.449. The P-value is large so there's insufficient evidence to declare any difference.

c) The P-values for the paired *t*-test, Wilcoxon's signed rank test, and sign test are 0.391, 0.449 and 0.6291.

25. Wilcoxon's signed rank test shows a significant improvement (P-value = 0.005). Paired *t*-test similarly shows a significant improvement (P-value = 0.001).

27. a) **Correlations with outlier:**
   - Pearson: −0.467
   - Spearman: −0.359

   Correlations without outlier;
   - Pearson: −0.375
   - Spearman: −0.301

   b) Removing the outlier (Haiti) changed the Pearson correlation (−0.467 to −0.375) more than the Spearman rank correlation (−0.359 to −0.301). Rank correlation is less sensitive to outliers, so this makes some sense.

29. The curvature in the scatterplot will be less for the ranked data, so Spearman's correlation should be greater. The Spearman's correlation = 0.94.

31. Both Wilcoxon's rank sum test (Mann-Whitney test) and two sample *t*-test show a significant difference between the populations (P-values 0.0016 for the Wilcoxon's rank sum test and 0.001 for the two sample *t*-test).

33. Choose 1–10 and Jack from Spades. Mix them up well and then draw 4 of them. Note if the total is ≤ 11. Repeat thousands of times. Divide # time ≤ 11 by # repetitions. This estimated the probability for ≤ 11. Double for two sided test.

35. There are 462 distinct possible hands.
   P-value for sum of 15 = $1/462 \times 2$
   ≤ 16: P-value = $2/462 \times 2$
   ≤ 17: P-value = $4/462 \times 2$

Tables

Row	Table of Random Digits									
1	96299	07196	98642	20639	23185	56282	69929	14125	38872	94168
2	71622	35940	81807	59225	18192	08710	80777	84395	69563	86280
3	03272	41230	81739	74797	70406	18564	69273	72532	78340	36699
4	46376	58596	14365	63685	56555	42974	72944	96463	63533	24152
5	47352	42853	42903	97504	56655	70355	88606	61406	38757	70657
6	20064	04266	74017	79319	70170	96572	08523	56025	89077	57678
7	73184	95907	05179	51002	83374	52297	07769	99792	78365	93487
8	72753	36216	07230	35793	71907	65571	66784	25548	91861	15725
9	03939	30763	06138	80062	02537	23561	93136	61260	77935	93159
10	75998	37203	07959	38264	78120	77525	86481	54986	33042	70648
11	94435	97441	90998	25104	49761	14967	70724	67030	53887	81293
12	04362	40989	69167	38894	00172	02999	97377	33305	60782	29810
13	89059	43528	10547	40115	82234	86902	04121	83889	76208	31076
14	87736	04666	75145	49175	76754	07884	92564	80793	22573	67902
15	76488	88899	15860	07370	13431	84041	69202	18912	83173	11983
16	36460	53772	66634	25045	79007	78518	73580	14191	50353	32064
17	13205	69237	21820	20952	16635	58867	97650	82983	64865	93298
18	51242	12215	90739	36812	00436	31609	80333	96606	30430	31803
19	67819	00354	91439	91073	49258	15992	41277	75111	67496	68430
20	09875	08990	27656	15871	23637	00952	97818	64234	50199	05715
21	18192	95308	72975	01191	29958	09275	89141	19558	50524	32041
22	02763	33701	66188	50226	35813	72951	11638	01876	93664	37001
23	13349	46328	01856	29935	80563	03742	49470	67749	08578	21956
24	69238	92878	80067	80807	45096	22936	64325	19265	37755	69794
25	92207	63527	59398	29818	24789	94309	88380	57000	50171	17891
26	66679	99100	37072	30593	29665	84286	44458	60180	81451	58273
27	31087	42430	60322	34765	15757	53300	97392	98035	05228	68970
28	84432	04916	52949	78533	31666	62350	20584	56367	19701	60584
29	72042	12287	21081	48426	44321	58765	41760	43304	13399	02043
30	94534	73559	82135	70260	87936	85162	11937	18263	54138	69564
31	63971	97198	40974	45301	60177	35604	21580	68107	25184	42810
32	11227	58474	17272	37619	69517	62964	67962	34510	12607	52255
33	28541	02029	08068	96656	17795	21484	57722	76511	27849	61738
34	11282	43632	49531	78981	81980	08530	08629	32279	29478	50228
35	42907	15137	21918	13248	39129	49559	94540	24070	88151	36782
36	47119	76651	21732	32364	58545	50277	57558	30390	18771	72703
37	11232	99884	05087	76839	65142	19994	91397	29350	83852	04905
38	64725	06719	86262	53356	57999	50193	79936	97230	52073	94467
39	77007	26962	55466	12521	48125	12280	54985	26239	76044	54398
40	18375	19310	59796	89832	59417	18553	17238	05474	33259	50595

	Second decimal place in z									
0.09	0.08	0.07	0.06	0.05	0.04	0.03	0.02	0.01	0.00	z
									0.0000[†]	−3.9
0.0001	0.0001	0.0001	0.0001	0.0001	0.0001	0.0001	0.0001	0.0001	0.0001	−3.8
0.0001	0.0001	0.0001	0.0001	0.0001	0.0001	0.0001	0.0001	0.0001	0.0001	−3.7
0.0001	0.0001	0.0001	0.0001	0.0001	0.0001	0.0001	0.0001	0.0002	0.0002	−3.6
0.0002	0.0002	0.0002	0.0002	0.0002	0.0002	0.0002	0.0002	0.0002	0.0002	−3.5
0.0002	0.0003	0.0003	0.0003	0.0003	0.0003	0.0003	0.0003	0.0003	0.0003	−3.4
0.0003	0.0004	0.0004	0.0004	0.0004	0.0004	0.0004	0.0005	0.0005	0.0005	−3.3
0.0005	0.0005	0.0005	0.0006	0.0006	0.0006	0.0006	0.0006	0.0007	0.0007	−3.2
0.0007	0.0007	0.0008	0.0008	0.0008	0.0008	0.0009	0.0009	0.0009	0.0010	−3.1
0.0010	0.0010	0.0011	0.0011	0.0011	0.0012	0.0012	0.0013	0.0013	0.0013	−3.0
0.0014	0.0014	0.0015	0.0015	0.0016	0.0016	0.0017	0.0018	0.0018	0.0019	−2.9
0.0019	0.0020	0.0021	0.0021	0.0022	0.0023	0.0023	0.0024	0.0025	0.0026	−2.8
0.0026	0.0027	0.0028	0.0029	0.0030	0.0031	0.0032	0.0033	0.0034	0.0035	−2.7
0.0036	0.0037	0.0038	0.0039	0.0040	0.0041	0.0043	0.0044	0.0045	0.0047	−2.6
0.0048	0.0049	0.0051	0.0052	0.0054	0.0055	0.0057	0.0059	0.0060	0.0062	−2.5
0.0064	0.0066	0.0068	0.0069	0.0071	0.0073	0.0075	0.0078	0.0080	0.0082	−2.4
0.0084	0.0087	0.0089	0.0091	0.0094	0.0096	0.0099	0.0102	0.0104	0.0107	−2.3
0.0110	0.0113	0.0116	0.0119	0.0122	0.0125	0.0129	0.0132	0.0136	0.0139	−2.2
0.0143	0.0146	0.0150	0.0154	0.0158	0.0162	0.0166	0.0170	0.0174	0.0179	−2.1
0.0183	0.0188	0.0192	0.0197	0.0202	0.0207	0.0212	0.0217	0.0222	0.0228	−2.0
0.0233	0.0239	0.0244	0.0250	0.0256	0.0262	0.0268	0.0274	0.0281	0.0287	−1.9
0.0294	0.0301	0.0307	0.0314	0.0322	0.0329	0.0336	0.0344	0.0351	0.0359	−1.8
0.0367	0.0375	0.0384	0.0392	0.0401	0.0409	0.0418	0.0427	0.0436	0.0446	−1.7
0.0455	0.0465	0.0475	0.0485	0.0495	0.0505	0.0516	0.0526	0.0537	0.0548	−1.6
0.0559	0.0571	0.0582	0.0594	0.0606	0.0618	0.0630	0.0643	0.0655	0.0668	−1.5
0.0681	0.0694	0.0708	0.0721	0.0735	0.0749	0.0764	0.0778	0.0793	0.0808	−1.4
0.0823	0.0838	0.0853	0.0869	0.0885	0.0901	0.0918	0.0934	0.0951	0.0968	−1.3
0.0985	0.1003	0.1020	0.1038	0.1056	0.1075	0.1093	0.1112	0.1131	0.1151	−1.2
0.1170	0.1190	0.1210	0.1230	0.1251	0.1271	0.1292	0.1314	0.1335	0.1357	−1.1
0.1379	0.1401	0.1423	0.1446	0.1469	0.1492	0.1515	0.1539	0.1562	0.1587	−1.0
0.1611	0.1635	0.1660	0.1685	0.1711	0.1736	0.1762	0.1788	0.1814	0.1841	−0.9
0.1867	0.1894	0.1922	0.1949	0.1977	0.2005	0.2033	0.2061	0.2090	0.2119	−0.8
0.2148	0.2177	0.2206	0.2236	0.2266	0.2296	0.2327	0.2358	0.2389	0.2420	−0.7
0.2451	0.2483	0.2514	0.2546	0.2578	0.2611	0.2643	0.2676	0.2709	0.2743	−0.6
0.2776	0.2810	0.2843	0.2877	0.2912	0.2946	0.2981	0.3015	0.3050	0.3085	−0.5
0.3121	0.3156	0.3192	0.3228	0.3264	0.3300	0.3336	0.3372	0.3409	0.3446	−0.4
0.3483	0.3520	0.3557	0.3594	0.3632	0.3669	0.3707	0.3745	0.3783	0.3821	−0.3
0.3859	0.3897	0.3936	0.3974	0.4013	0.4052	0.4090	0.4129	0.4168	0.4207	−0.2
0.4247	0.4286	0.4325	0.4364	0.4404	0.4443	0.4483	0.4522	0.4562	0.4602	−0.1
0.4641	0.4681	0.4721	0.4761	0.4801	0.4840	0.4880	0.4920	0.4960	0.5000	−0.0

Table Z — Areas under the standard Normal curve

[†]For $z \leq -3.90$, the areas are 0.0000 to four decimal places.

Table Z (cont.)
Areas under the
standard Normal curve

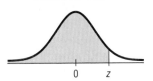

z	0.00	0.01	0.02	0.03	0.04	0.05	0.06	0.07	0.08	0.09
					Second decimal place in z					
0.0	0.5000	0.5040	0.5080	0.5120	0.5160	0.5199	0.5239	0.5279	0.5319	0.5359
0.1	0.5398	0.5438	0.5478	0.5517	0.5557	0.5596	0.5636	0.5675	0.5714	0.5753
0.2	0.5793	0.5832	0.5871	0.5910	0.5948	0.5987	0.6026	0.6064	0.6103	0.6141
0.3	0.6179	0.6217	0.6255	0.6293	0.6331	0.6368	0.6406	0.6443	0.6480	0.6517
0.4	0.6554	0.6591	0.6628	0.6664	0.6700	0.6736	0.6772	0.6808	0.6844	0.6879
0.5	0.6915	0.6950	0.6985	0.7019	0.7054	0.7088	0.7123	0.7157	0.7190	0.7224
0.6	0.7257	0.7291	0.7324	0.7357	0.7389	0.7422	0.7454	0.7486	0.7517	0.7549
0.7	0.7580	0.7611	0.7642	0.7673	0.7704	0.7734	0.7764	0.7794	0.7823	0.7852
0.8	0.7881	0.7910	0.7939	0.7967	0.7995	0.8023	0.8051	0.8078	0.8106	0.8133
0.9	0.8159	0.8186	0.8212	0.8238	0.8264	0.8289	0.8315	0.8340	0.8365	0.8389
1.0	0.8413	0.8438	0.8461	0.8485	0.8508	0.8531	0.8554	0.8577	0.8599	0.8621
1.1	0.8643	0.8665	0.8686	0.8708	0.8729	0.8749	0.8770	0.8790	0.8810	0.8830
1.2	0.8849	0.8869	0.8888	0.8907	0.8925	0.8944	0.8962	0.8980	0.8997	0.9015
1.3	0.9032	0.9049	0.9066	0.9082	0.9099	0.9115	0.9131	0.9147	0.9162	0.9177
1.4	0.9192	0.9207	0.9222	0.9236	0.9251	0.9265	0.9279	0.9292	0.9306	0.9319
1.5	0.9332	0.9345	0.9357	0.9370	0.9382	0.9394	0.9406	0.9418	0.9429	0.9441
1.6	0.9452	0.9463	0.9474	0.9484	0.9495	0.9505	0.9515	0.9525	0.9535	0.9545
1.7	0.9554	0.9564	0.9573	0.9582	0.9591	0.9599	0.9608	0.9616	0.9625	0.9633
1.8	0.9641	0.9649	0.9656	0.9664	0.9671	0.9678	0.9686	0.9693	0.9699	0.9706
1.9	0.9713	0.9719	0.9726	0.9732	0.9738	0.9744	0.9750	0.9756	0.9761	0.9767
2.0	0.9772	0.9778	0.9783	0.9788	0.9793	0.9798	0.9803	0.9808	0.9812	0.9817
2.1	0.9821	0.9826	0.9830	0.9834	0.9838	0.9842	0.9846	0.9850	0.9854	0.9857
2.2	0.9861	0.9864	0.9868	0.9871	0.9875	0.9878	0.9881	0.9884	0.9887	0.9890
2.3	0.9893	0.9896	0.9898	0.9901	0.9904	0.9906	0.9909	0.9911	0.9913	0.9916
2.4	0.9918	0.9920	0.9922	0.9925	0.9927	0.9929	0.9931	0.9932	0.9934	0.9936
2.5	0.9938	0.9940	0.9941	0.9943	0.9945	0.9946	0.9948	0.9949	0.9951	0.9952
2.6	0.9953	0.9955	0.9956	0.9957	0.9959	0.9960	0.9961	0.9962	0.9963	0.9964
2.7	0.9965	0.9966	0.9967	0.9968	0.9969	0.9970	0.9971	0.9972	0.9973	0.9974
2.8	0.9974	0.9975	0.9976	0.9977	0.9977	0.9978	0.9979	0.9979	0.9980	0.9981
2.9	0.9981	0.9982	0.9982	0.9983	0.9984	0.9984	0.9985	0.9985	0.9986	0.9986
3.0	0.9987	0.9987	0.9987	0.9988	0.9988	0.9989	0.9989	0.9989	0.9990	0.9990
3.1	0.9990	0.9991	0.9991	0.9991	0.9992	0.9992	0.9992	0.9992	0.9993	0.9993
3.2	0.9993	0.9993	0.9994	0.9994	0.9994	0.9994	0.9994	0.9995	0.9995	0.9995
3.3	0.9995	0.9995	0.9995	0.9996	0.9996	0.9996	0.9996	0.9996	0.9996	0.9997
3.4	0.9997	0.9997	0.9997	0.9997	0.9997	0.9997	0.9997	0.9997	0.9997	0.9998
3.5	0.9998	0.9998	0.9998	0.9998	0.9998	0.9998	0.9998	0.9998	0.9998	0.9998
3.6	0.9998	0.9998	0.9999	0.9999	0.9999	0.9999	0.9999	0.9999	0.9999	0.9999
3.7	0.9999	0.9999	0.9999	0.9999	0.9999	0.9999	0.9999	0.9999	0.9999	0.9999
3.8	0.9999	0.9999	0.9999	0.9999	0.9999	0.9999	0.9999	0.9999	0.9999	0.9999
3.9	1.0000[†]									

[†]For $z \geq 3.90$, the areas are 1.0000 to four decimal places.

Two-tail probability		0.20	0.10	0.05	0.02	0.01	
One-tail probability		0.10	0.05	0.025	0.01	0.005	
Table T	df						df
Value of t_α	1	3.078	6.314	12.706	31.821	63.657	1
	2	1.886	2.920	4.303	6.965	9.925	2
	3	1.638	2.353	3.182	4.541	5.841	3
	4	1.533	2.132	2.776	3.747	4.604	4
	5	1.476	2.015	2.571	3.365	4.032	5
	6	1.440	1.943	2.447	3.143	3.707	6
	7	1.415	1.895	2.365	2.998	3.499	7
	8	1.397	1.860	2.306	2.896	3.355	8
	9	1.383	1.833	2.262	2.821	3.250	9
	10	1.372	1.812	2.228	2.764	3.169	10
	11	1.363	1.796	2.201	2.718	3.106	11
	12	1.356	1.782	2.179	2.681	3.055	12
	13	1.350	1.771	2.160	2.650	3.012	13
	14	1.345	1.761	2.145	2.624	2.977	14
	15	1.341	1.753	2.131	2.602	2.947	15
	16	1.337	1.746	2.120	2.583	2.921	16
	17	1.333	1.740	2.110	2.567	2.898	17
	18	1.330	1.734	2.101	2.552	2.878	18
	19	1.328	1.729	2.093	2.539	2.861	19
	20	1.325	1.725	2.086	2.528	2.845	20
	21	1.323	1.721	2.080	2.518	2.831	21
	22	1.321	1.717	2.074	2.508	2.819	22
	23	1.319	1.714	2.069	2.500	2.807	23
	24	1.318	1.711	2.064	2.492	2.797	24
	25	1.316	1.708	2.060	2.485	2.787	25
	26	1.315	1.706	2.056	2.479	2.779	26
	27	1.314	1.703	2.052	2.473	2.771	27
	28	1.313	1.701	2.048	2.467	2.763	28
	29	1.311	1.699	2.045	2.462	2.756	29
	30	1.310	1.697	2.042	2.457	2.750	30
	32	1.309	1.694	2.037	2.449	2.738	32
	35	1.306	1.690	2.030	2.438	2.725	35
	40	1.303	1.684	2.021	2.423	2.704	40
	45	1.301	1.679	2.014	2.412	2.690	45
	50	1.299	1.676	2.009	2.403	2.678	50
	60	1.296	1.671	2.000	2.390	2.660	60
	75	1.293	1.665	1.992	2.377	2.643	75
	100	1.290	1.660	1.984	2.364	2.626	100
	120	1.289	1.658	1.980	2.358	2.617	120
	140	1.288	1.656	1.977	2.353	2.611	140
	180	1.286	1.653	1.973	2.347	2.603	180
	250	1.285	1.651	1.969	2.341	2.596	250
	400	1.284	1.649	1.966	2.336	2.588	400
	1000	1.282	1.646	1.962	2.330	2.581	1000
	∞	1.282	1.645	1.960	2.326	2.576	∞
Confidence levels		80%	90%	95%	98	99%	

Right-tail probability		0.10	0.05	0.025	0.01	0.005
Table χ	df					
Values of χ^2_α	1	2.706	3.841	5.024	6.635	7.879
	2	4.605	5.991	7.378	9.210	10.597
	3	6.251	7.815	9.348	11.345	12.838
	4	7.779	9.488	11.143	13.277	14.860
	5	9.236	11.070	12.833	15.086	16.750
	6	10.645	12.592	14.449	16.812	18.548
	7	12.017	14.067	16.013	18.475	20.278
	8	13.362	15.507	17.535	20.090	21.955
	9	14.684	16.919	19.023	21.666	23.589
	10	15.987	18.307	20.483	23.209	25.188
	11	17.275	19.675	21.920	24.725	26.757
	12	18.549	21.026	23.337	26.217	28.300
	13	19.812	22.362	24.736	27.688	29.819
	14	21.064	23.685	26.119	29.141	31.319
	15	22.307	24.996	27.488	30.578	32.801
	16	23.542	26.296	28.845	32.000	34.267
	17	24.769	27.587	30.191	33.409	35.718
	18	25.989	28.869	31.526	34.805	37.156
	19	27.204	30.143	32.852	36.191	38.582
	20	28.412	31.410	34.170	37.566	39.997
	21	29.615	32.671	35.479	38.932	41.401
	22	30.813	33.924	36.781	40.290	42.796
	23	32.007	35.172	38.076	41.638	44.181
	24	33.196	36.415	39.364	42.980	45.559
	25	34.382	37.653	40.647	44.314	46.928
	26	35.563	38.885	41.923	45.642	48.290
	27	36.741	40.113	43.195	46.963	49.645
	28	37.916	41.337	44.461	48.278	50.994
	29	39.087	42.557	45.722	59.588	52.336
	30	40.256	43.773	46.979	50.892	53.672
	40	51.805	55.759	59.342	63.691	66.767
	50	63.167	67.505	71.420	76.154	79.490
	60	74.397	79.082	83.298	88.381	91.955
	70	85.527	90.531	95.023	100.424	104.213
	80	96.578	101.879	106.628	112.328	116.320
	90	107.565	113.145	118.135	124.115	128.296
	100	118.499	124.343	129.563	135.811	140.177

Table F

Numerator df

α = .01	1	2	3	4	5	6	7	8	9	10	11	12	13	14	15	16	17	18	19	20	21	22
1	4052.2	4999.3	5403.5	5624.3	5764.0	5859.0	5928.3	5981.0	6022.4	6055.9	6083.4	6106.7	6125.8	6143.0	6157.0	6170.0	6181.2	6191.4	6200.7	6208.7	6216.1	6223.1
2	98.50	99.00	99.16	99.25	99.30	99.33	99.36	99.38	99.39	99.40	99.41	99.42	99.42	99.43	99.43	99.44	99.44	99.44	99.45	99.45	99.45	99.46
3	34.12	30.82	29.46	28.71	28.24	27.91	27.67	27.49	27.34	27.23	27.13	27.05	26.98	26.92	26.87	26.83	26.79	26.75	26.72	26.69	26.66	26.64
4	21.20	18.00	16.69	15.98	15.52	15.21	14.98	14.80	14.66	14.55	14.45	14.37	14.31	14.25	14.20	14.15	14.11	14.08	14.05	14.02	13.99	13.97
5	16.26	13.27	12.06	11.39	10.97	10.67	10.46	10.29	10.16	10.05	9.96	9.89	9.82	9.77	9.72	9.68	9.64	9.61	9.58	9.55	9.53	9.51
6	13.75	10.92	9.78	9.15	8.75	8.47	8.26	8.10	7.98	7.87	7.79	7.72	7.66	7.60	7.56	7.52	7.48	7.45	7.42	7.40	7.37	7.35
7	12.25	9.55	8.45	7.85	7.46	7.19	6.99	6.84	6.72	6.62	6.54	6.47	6.41	6.36	6.31	6.28	6.24	6.21	6.18	6.16	6.13	6.11
8	11.26	8.65	7.59	7.01	6.63	6.37	6.18	6.03	5.91	5.81	5.73	5.67	5.61	5.56	5.52	5.48	5.44	5.41	5.38	5.36	5.34	5.32
9	10.56	8.02	6.99	6.42	6.06	5.80	5.61	5.47	5.35	5.26	5.18	5.11	5.05	5.01	4.96	4.92	4.89	4.86	4.83	4.81	4.79	4.77
10	10.04	7.56	6.55	5.99	5.64	5.39	5.20	5.06	4.94	4.85	4.77	4.71	4.65	4.60	4.56	4.52	4.49	4.46	4.43	4.41	4.38	4.36
11	9.65	7.21	6.22	5.67	5.32	5.07	4.89	4.74	4.63	4.54	4.46	4.40	4.34	4.29	4.25	4.21	4.18	4.15	4.12	4.10	4.08	4.06
12	9.33	6.93	5.95	5.41	5.06	4.82	4.64	4.50	4.39	4.30	4.22	4.16	4.10	4.05	4.01	3.97	3.94	3.91	3.88	3.86	3.84	3.82
13	9.07	6.70	5.74	5.21	4.86	4.62	4.44	4.30	4.19	4.10	4.02	3.96	3.91	3.86	3.82	3.78	3.75	3.72	3.69	3.66	3.64	3.62
14	8.86	6.51	5.56	5.04	4.69	4.46	4.28	4.14	4.03	3.94	3.86	3.80	3.75	3.70	3.66	3.62	3.59	3.56	3.53	3.51	3.48	3.46
15	8.68	6.36	5.42	4.89	4.56	4.32	4.14	4.00	3.89	3.80	3.73	3.67	3.61	3.56	3.52	3.49	3.45	3.42	3.40	3.37	3.35	3.33
16	8.53	6.23	5.29	4.77	4.44	4.20	4.03	3.89	3.78	3.69	3.62	3.55	3.50	3.45	3.41	3.37	3.34	3.31	3.28	3.26	3.24	3.22
17	8.40	6.11	5.19	4.67	4.34	4.10	3.93	3.79	3.68	3.59	3.52	3.46	3.40	3.35	3.31	3.27	3.24	3.21	3.19	3.16	3.14	3.12
18	8.29	6.01	5.09	4.58	4.25	4.01	3.84	3.71	3.60	3.51	3.43	3.37	3.32	3.27	3.23	3.19	3.16	3.13	3.10	3.08	3.05	3.03
19	8.18	5.93	5.01	4.50	4.17	3.94	3.77	3.63	3.52	3.43	3.36	3.30	3.24	3.19	3.15	3.12	3.08	3.05	3.03	3.00	2.98	2.96
20	8.10	5.85	4.94	4.43	4.10	3.87	3.70	3.56	3.46	3.37	3.29	3.23	3.18	3.13	3.09	3.05	3.02	2.99	2.96	2.94	2.92	2.90
21	8.02	5.78	4.87	4.37	4.04	3.81	3.64	3.51	3.40	3.31	3.24	3.17	3.12	3.07	3.03	2.99	2.96	2.93	2.90	2.88	2.86	2.84
22	7.95	5.72	4.82	4.31	3.99	3.76	3.59	3.45	3.35	3.26	3.18	3.12	3.07	3.02	2.98	2.94	2.91	2.88	2.85	2.83	2.81	2.78
23	7.88	5.66	4.76	4.26	3.94	3.71	3.54	3.41	3.30	3.21	3.14	3.07	3.02	2.97	2.93	2.89	2.86	2.83	2.80	2.78	2.76	2.74
24	7.82	5.61	4.72	4.22	3.90	3.67	3.50	3.36	3.26	3.17	3.09	3.03	2.98	2.93	2.89	2.85	2.82	2.79	2.76	2.74	2.72	2.70
25	7.77	5.57	4.68	4.18	3.85	3.63	3.46	3.32	3.22	3.13	3.06	2.99	2.94	2.89	2.85	2.81	2.78	2.75	2.72	2.70	2.68	2.66
26	7.72	5.53	4.64	4.14	3.82	3.59	3.42	3.29	3.18	3.09	3.02	2.96	2.90	2.86	2.81	2.78	2.75	2.72	2.69	2.66	2.64	2.62
27	7.68	5.49	4.60	4.11	3.78	3.56	3.39	3.26	3.15	3.06	2.99	2.93	2.87	2.82	2.78	2.75	2.71	2.68	2.66	2.63	2.61	2.59
28	7.64	5.45	4.57	4.07	3.75	3.53	3.36	3.23	3.12	3.03	2.96	2.90	2.84	2.79	2.75	2.72	2.68	2.65	2.63	2.60	2.58	2.56
29	7.60	5.42	4.54	4.04	3.73	3.50	3.33	3.20	3.09	3.00	2.93	2.87	2.81	2.77	2.73	2.69	2.66	2.63	2.60	2.57	2.55	2.53
30	7.56	5.39	4.51	4.02	3.70	3.47	3.30	3.17	3.07	2.98	2.91	2.84	2.79	2.74	2.70	2.66	2.63	2.60	2.57	2.55	2.53	2.51
32	7.50	5.34	4.46	3.97	3.65	3.43	3.26	3.13	3.02	2.93	2.86	2.80	2.74	2.70	2.65	2.62	2.58	2.55	2.53	2.50	2.48	2.46
35	7.42	5.27	4.40	3.91	3.59	3.37	3.20	3.07	2.96	2.88	2.80	2.74	2.69	2.64	2.60	2.56	2.53	2.50	2.47	2.44	2.42	2.40
40	7.31	5.18	4.31	3.83	3.51	3.29	3.12	2.99	2.89	2.80	2.73	2.66	2.61	2.56	2.52	2.48	2.45	2.42	2.39	2.37	2.35	2.33
45	7.23	5.11	4.25	3.77	3.45	3.23	3.07	2.94	2.83	2.74	2.67	2.61	2.55	2.51	2.46	2.43	2.39	2.36	2.34	2.31	2.29	2.27
50	7.17	5.06	4.20	3.72	3.41	3.19	3.02	2.89	2.78	2.70	2.63	2.56	2.51	2.46	2.42	2.38	2.35	2.32	2.29	2.27	2.24	2.22
60	7.08	4.98	4.13	3.65	3.34	3.12	2.95	2.82	2.72	2.63	2.56	2.50	2.44	2.39	2.35	2.31	2.28	2.25	2.22	2.20	2.17	2.15
75	6.99	4.90	4.05	3.58	3.27	3.05	2.89	2.76	2.65	2.57	2.49	2.43	2.38	2.33	2.29	2.25	2.22	2.18	2.16	2.13	2.11	2.09
100	6.90	4.82	3.98	3.51	3.21	2.99	2.82	2.69	2.59	2.50	2.43	2.37	2.31	2.27	2.22	2.19	2.15	2.12	2.09	2.07	2.04	2.02
120	6.85	4.79	3.95	3.48	3.17	2.96	2.79	2.66	2.56	2.47	2.40	2.34	2.28	2.23	2.19	2.15	2.12	2.09	2.06	2.03	2.01	1.99
140	6.82	4.76	3.92	3.46	3.15	2.93	2.77	2.64	2.54	2.45	2.38	2.31	2.26	2.21	2.17	2.13	2.10	2.07	2.04	2.01	1.99	1.97
180	6.78	4.73	3.89	3.43	3.12	2.90	2.74	2.61	2.51	2.42	2.35	2.28	2.23	2.18	2.14	2.10	2.07	2.04	2.01	1.98	1.96	1.94
250	6.74	4.69	3.86	3.40	3.09	2.87	2.71	2.58	2.48	2.39	2.32	2.26	2.20	2.15	2.11	2.07	2.04	2.01	1.98	1.95	1.93	1.91
400	6.70	4.66	3.83	3.37	3.06	2.85	2.68	2.56	2.45	2.37	2.29	2.23	2.17	2.13	2.08	2.05	2.01	1.98	1.95	1.92	1.90	1.88
1000	6.66	4.63	3.80	3.34	3.04	2.82	2.66	2.53	2.43	2.34	2.27	2.20	2.15	2.10	2.06	2.02	1.98	1.95	1.92	1.90	1.87	1.85

Denominator df

Table F (cont.)

α = .01

Numerator df

Denominator df	23	24	25	26	27	28	29	30	32	35	40	45	50	60	75	100	120	140	180	250	400	1000
1	6228.7	6234.3	6239.9	6244.5	6249.2	6252.9	6257.1	6260.4	6266.9	6275.3	6286.4	6295.7	6302.3	6313.0	6323.7	6333.9	6339.5	6343.2	6347.9	6353.5	6358.1	6362.8
2	99.46	99.46	99.46	99.46	99.46	99.46	99.46	99.47	99.47	99.47	99.48	99.48	99.48	99.48	99.48	99.49	99.49	99.49	99.49	99.50	99.50	99.50
3	26.62	26.60	26.58	26.56	26.55	26.53	26.52	26.50	26.48	26.45	26.41	26.38	26.35	26.32	26.28	26.24	26.22	26.21	26.19	26.17	26.15	26.14
4	13.95	13.93	13.91	13.89	13.88	13.86	13.85	13.84	13.81	13.79	13.75	13.71	13.69	13.65	13.61	13.58	13.56	13.54	13.53	13.51	13.49	13.47
5	9.49	9.47	9.45	9.43	9.42	9.40	9.39	9.38	9.36	9.33	9.29	9.26	9.24	9.20	9.17	9.13	9.11	9.10	9.08	9.06	9.05	9.03
6	7.33	7.31	7.30	7.28	7.27	7.25	7.24	7.23	7.21	7.18	7.14	7.11	7.09	7.06	7.02	6.99	6.97	6.96	6.94	6.92	6.91	6.89
7	6.09	6.07	6.06	6.04	6.03	6.02	6.00	5.99	5.97	5.94	5.91	5.88	5.86	5.82	5.79	5.75	5.74	5.72	5.71	5.69	5.68	5.66
8	5.30	5.28	5.26	5.25	5.23	5.22	5.21	5.20	5.18	5.15	5.12	5.09	5.07	5.03	5.00	4.96	4.95	4.93	4.92	4.90	4.89	4.87
9	4.75	4.73	4.71	4.70	4.68	4.67	4.66	4.65	4.63	4.60	4.57	4.54	4.52	4.48	4.45	4.41	4.40	4.39	4.37	4.35	4.34	4.32
10	4.34	4.33	4.31	4.30	4.28	4.27	4.26	4.25	4.23	4.20	4.17	4.14	4.12	4.08	4.05	4.01	4.00	3.98	3.97	3.95	3.94	3.92
11	4.04	4.02	4.01	3.99	3.98	3.96	3.95	3.94	3.92	3.89	3.86	3.83	3.81	3.78	3.74	3.71	3.69	3.68	3.66	3.64	3.63	3.61
12	3.80	3.78	3.76	3.75	3.74	3.72	3.71	3.70	3.68	3.65	3.62	3.59	3.57	3.54	3.50	3.47	3.45	3.44	3.42	3.40	3.39	3.37
13	3.60	3.59	3.57	3.56	3.54	3.53	3.52	3.51	3.49	3.46	3.43	3.40	3.38	3.34	3.31	3.27	3.25	3.24	3.23	3.21	3.19	3.18
14	3.44	3.43	3.41	3.40	3.38	3.37	3.36	3.35	3.33	3.30	3.27	3.24	3.22	3.18	3.15	3.11	3.09	3.08	3.06	3.05	3.03	3.02
15	3.31	3.29	3.28	3.26	3.25	3.24	3.23	3.21	3.19	3.17	3.13	3.10	3.08	3.05	3.01	2.98	2.96	2.95	2.93	2.91	2.90	2.88
16	3.20	3.18	3.16	3.15	3.14	3.12	3.11	3.10	3.08	3.05	3.02	2.99	2.97	2.93	2.90	2.86	2.84	2.83	2.81	2.80	2.78	2.76
17	3.10	3.08	3.07	3.05	3.04	3.03	3.01	3.00	2.98	2.96	2.92	2.89	2.87	2.83	2.80	2.76	2.75	2.73	2.72	2.70	2.68	2.66
18	3.02	3.00	2.98	2.97	2.95	2.94	2.93	2.92	2.90	2.87	2.84	2.81	2.78	2.75	2.71	2.68	2.66	2.65	2.63	2.61	2.59	2.58
19	2.94	2.92	2.91	2.89	2.88	2.87	2.86	2.84	2.82	2.80	2.76	2.73	2.71	2.67	2.64	2.60	2.58	2.57	2.55	2.54	2.52	2.50
20	2.88	2.86	2.84	2.83	2.81	2.80	2.79	2.78	2.76	2.73	2.69	2.67	2.64	2.61	2.57	2.54	2.52	2.50	2.49	2.47	2.45	2.43
21	2.82	2.80	2.79	2.77	2.76	2.74	2.73	2.72	2.70	2.67	2.64	2.61	2.58	2.55	2.51	2.48	2.46	2.44	2.43	2.41	2.39	2.37
22	2.77	2.75	2.73	2.72	2.70	2.69	2.68	2.67	2.65	2.62	2.58	2.55	2.53	2.50	2.46	2.42	2.40	2.39	2.37	2.35	2.34	2.32
23	2.72	2.70	2.69	2.67	2.66	2.64	2.63	2.62	2.60	2.57	2.54	2.51	2.48	2.45	2.41	2.37	2.35	2.34	2.32	2.30	2.29	2.27
24	2.68	2.66	2.64	2.63	2.61	2.60	2.59	2.58	2.56	2.53	2.49	2.46	2.44	2.40	2.37	2.33	2.31	2.30	2.28	2.26	2.24	2.22
25	2.64	2.62	2.60	2.59	2.58	2.56	2.55	2.54	2.52	2.49	2.45	2.42	2.40	2.36	2.33	2.29	2.27	2.26	2.24	2.22	2.20	2.18
26	2.60	2.58	2.57	2.55	2.54	2.53	2.51	2.50	2.48	2.45	2.42	2.39	2.36	2.33	2.29	2.25	2.23	2.22	2.20	2.18	2.16	2.14
27	2.57	2.55	2.54	2.52	2.51	2.49	2.48	2.47	2.45	2.42	2.38	2.35	2.33	2.29	2.26	2.22	2.20	2.18	2.17	2.15	2.13	2.11
28	2.54	2.52	2.51	2.49	2.48	2.46	2.45	2.44	2.42	2.39	2.35	2.32	2.30	2.26	2.23	2.19	2.17	2.15	2.13	2.11	2.10	2.08
29	2.51	2.49	2.48	2.46	2.45	2.44	2.42	2.41	2.39	2.36	2.33	2.30	2.27	2.23	2.20	2.16	2.14	2.12	2.10	2.08	2.07	2.05
30	2.49	2.47	2.45	2.44	2.42	2.41	2.40	2.39	2.36	2.34	2.30	2.27	2.25	2.21	2.17	2.13	2.11	2.10	2.08	2.06	2.04	2.02
32	2.44	2.42	2.41	2.39	2.38	2.36	2.35	2.34	2.32	2.29	2.25	2.22	2.20	2.16	2.12	2.08	2.06	2.05	2.03	2.01	1.99	1.97
35	2.38	2.36	2.35	2.33	2.32	2.30	2.29	2.28	2.26	2.23	2.19	2.16	2.14	2.10	2.06	2.02	2.00	1.98	1.96	1.94	1.92	1.90
40	2.31	2.29	2.27	2.26	2.24	2.23	2.22	2.20	2.18	2.15	2.11	2.08	2.06	2.02	1.98	1.94	1.92	1.90	1.88	1.86	1.84	1.82
45	2.25	2.23	2.21	2.20	2.18	2.17	2.16	2.14	2.12	2.09	2.05	2.02	2.00	1.96	1.92	1.88	1.85	1.84	1.82	1.79	1.77	1.75
50	2.20	2.18	2.17	2.15	2.14	2.12	2.11	2.10	2.08	2.05	2.01	1.97	1.95	1.91	1.87	1.82	1.80	1.79	1.76	1.74	1.72	1.70
60	2.13	2.12	2.10	2.08	2.07	2.05	2.04	2.03	2.01	1.98	1.94	1.90	1.88	1.84	1.79	1.75	1.73	1.71	1.69	1.66	1.64	1.62
75	2.07	2.05	2.03	2.02	2.00	1.99	1.97	1.96	1.94	1.91	1.87	1.83	1.81	1.76	1.72	1.67	1.65	1.63	1.61	1.58	1.56	1.53
100	2.00	1.98	1.97	1.95	1.93	1.92	1.91	1.89	1.87	1.84	1.80	1.76	1.74	1.69	1.65	1.60	1.57	1.55	1.53	1.50	1.47	1.45
120	1.97	1.95	1.93	1.92	1.90	1.89	1.87	1.86	1.84	1.81	1.76	1.73	1.70	1.66	1.61	1.56	1.53	1.51	1.49	1.46	1.43	1.40
140	1.95	1.93	1.91	1.89	1.88	1.86	1.85	1.84	1.81	1.78	1.74	1.70	1.67	1.63	1.58	1.53	1.51	1.48	1.46	1.43	1.40	1.37
180	1.92	1.90	1.88	1.86	1.85	1.83	1.82	1.81	1.78	1.75	1.71	1.67	1.64	1.60	1.55	1.49	1.47	1.45	1.42	1.39	1.35	1.32
250	1.89	1.87	1.85	1.83	1.82	1.80	1.79	1.77	1.75	1.72	1.67	1.64	1.61	1.56	1.51	1.46	1.43	1.41	1.38	1.34	1.31	1.27
400	1.86	1.84	1.82	1.80	1.79	1.77	1.76	1.75	1.72	1.69	1.64	1.61	1.58	1.53	1.48	1.42	1.39	1.37	1.33	1.30	1.26	1.22
1000	1.83	1.81	1.79	1.77	1.76	1.74	1.73	1.72	1.69	1.66	1.61	1.58	1.54	1.50	1.44	1.38	1.35	1.33	1.29	1.25	1.21	1.16

Denominator df

Table F (cont.)

Numerator df

$\alpha = .05$	1	2	3	4	5	6	7	8	9	10	11	12	13	14	15	16	17	18	19	20	21	22
1	161.4	199.5	215.7	224.6	230.2	234.0	236.8	238.9	240.5	241.9	243.0	243.9	244.7	245.4	245.9	246.5	246.9	247.3	247.7	248.0	248.3	248.6
2	18.51	19.00	19.16	19.25	19.30	19.33	19.35	19.37	19.38	19.40	19.40	19.41	19.42	19.42	19.43	19.43	19.44	19.44	19.44	19.45	19.45	19.45
3	10.13	9.55	9.28	9.12	9.01	8.94	8.89	8.85	8.81	8.79	8.76	8.74	8.73	8.71	8.70	8.69	8.68	8.67	8.67	8.66	8.65	8.65
4	7.71	6.94	6.59	6.39	6.26	6.16	6.09	6.04	6.00	5.96	5.94	5.91	5.89	5.87	5.86	5.84	5.83	5.82	5.81	5.80	5.79	5.79
5	6.61	5.79	5.41	5.19	5.05	4.95	4.88	4.82	4.77	4.74	4.70	4.68	4.66	4.64	4.62	4.60	4.59	4.58	4.57	4.56	4.55	4.54
6	5.99	5.14	4.76	4.53	4.39	4.28	4.21	4.15	4.10	4.06	4.03	4.00	3.98	3.96	3.94	3.92	3.91	3.90	3.88	3.87	3.86	3.86
7	5.59	4.74	4.35	4.12	3.97	3.87	3.79	3.73	3.68	3.64	3.60	3.57	3.55	3.53	3.51	3.49	3.48	3.47	3.46	3.44	3.43	3.43
8	5.32	4.46	4.07	3.84	3.69	3.58	3.50	3.44	3.39	3.35	3.31	3.28	3.26	3.24	3.22	3.20	3.19	3.17	3.16	3.15	3.14	3.13
9	5.12	4.26	3.86	3.63	3.48	3.37	3.29	3.23	3.18	3.14	3.10	3.07	3.05	3.03	3.01	2.99	2.97	2.96	2.95	2.94	2.93	2.92
10	4.96	4.10	3.71	3.48	3.33	3.22	3.14	3.07	3.02	2.98	2.94	2.91	2.89	2.86	2.85	2.83	2.81	2.80	2.79	2.77	2.76	2.75
11	4.84	3.98	3.59	3.36	3.20	3.09	3.01	2.95	2.90	2.85	2.82	2.79	2.76	2.74	2.72	2.70	2.69	2.67	2.66	2.65	2.64	2.63
12	4.75	3.89	3.49	3.26	3.11	3.00	2.91	2.85	2.80	2.75	2.72	2.69	2.66	2.64	2.62	2.60	2.58	2.57	2.56	2.54	2.53	2.52
13	4.67	3.81	3.41	3.18	3.03	2.92	2.83	2.77	2.71	2.67	2.63	2.60	2.58	2.55	2.53	2.51	2.50	2.48	2.47	2.46	2.45	2.44
14	4.60	3.74	3.34	3.11	2.96	2.85	2.76	2.70	2.65	2.60	2.57	2.53	2.51	2.48	2.46	2.44	2.43	2.41	2.40	2.39	2.38	2.37
15	4.54	3.68	3.29	3.06	2.90	2.79	2.71	2.64	2.59	2.54	2.51	2.48	2.45	2.42	2.40	2.38	2.37	2.35	2.34	2.33	2.32	2.31
16	4.49	3.63	3.24	3.01	2.85	2.74	2.66	2.59	2.54	2.49	2.46	2.42	2.40	2.37	2.35	2.33	2.32	2.30	2.29	2.28	2.26	2.25
17	4.45	3.59	3.20	2.96	2.81	2.70	2.61	2.55	2.49	2.45	2.41	2.38	2.35	2.33	2.31	2.29	2.27	2.26	2.24	2.23	2.22	2.21
18	4.41	3.55	3.16	2.93	2.77	2.66	2.58	2.51	2.46	2.41	2.37	2.34	2.31	2.29	2.27	2.25	2.23	2.22	2.20	2.19	2.18	2.17
19	4.38	3.52	3.13	2.90	2.74	2.63	2.54	2.48	2.42	2.38	2.34	2.31	2.28	2.26	2.23	2.21	2.20	2.18	2.17	2.16	2.14	2.13
20	4.35	3.49	3.10	2.87	2.71	2.60	2.51	2.45	2.39	2.35	2.31	2.28	2.25	2.22	2.20	2.18	2.17	2.15	2.14	2.12	2.11	2.10
21	4.32	3.47	3.07	2.84	2.68	2.57	2.49	2.42	2.37	2.32	2.28	2.25	2.22	2.20	2.18	2.16	2.14	2.12	2.11	2.10	2.08	2.07
22	4.30	3.44	3.05	2.82	2.66	2.55	2.46	2.40	2.34	2.30	2.26	2.23	2.20	2.17	2.15	2.13	2.11	2.10	2.08	2.07	2.06	2.05
23	4.28	3.42	3.03	2.80	2.64	2.53	2.44	2.37	2.32	2.27	2.24	2.20	2.18	2.15	2.13	2.11	2.09	2.08	2.06	2.05	2.04	2.02
24	4.26	3.40	3.01	2.78	2.62	2.51	2.42	2.36	2.30	2.25	2.22	2.18	2.15	2.13	2.11	2.09	2.07	2.05	2.04	2.03	2.01	2.00
25	4.24	3.39	2.99	2.76	2.60	2.49	2.40	2.34	2.28	2.24	2.20	2.16	2.14	2.11	2.09	2.07	2.05	2.04	2.02	2.01	2.00	1.98
26	4.23	3.37	2.98	2.74	2.59	2.47	2.39	2.32	2.27	2.22	2.18	2.15	2.12	2.09	2.07	2.05	2.03	2.02	2.00	1.99	1.98	1.97
27	4.21	3.35	2.96	2.73	2.57	2.46	2.37	2.31	2.25	2.20	2.17	2.13	2.10	2.08	2.06	2.04	2.02	2.00	1.99	1.97	1.96	1.95
28	4.20	3.34	2.95	2.71	2.56	2.45	2.36	2.29	2.24	2.19	2.15	2.12	2.09	2.06	2.04	2.02	2.00	1.99	1.97	1.96	1.95	1.93
29	4.18	3.33	2.93	2.70	2.55	2.43	2.35	2.28	2.22	2.18	2.14	2.10	2.08	2.05	2.03	2.01	1.99	1.97	1.96	1.94	1.93	1.92
30	4.17	3.32	2.92	2.69	2.53	2.42	2.33	2.27	2.21	2.16	2.13	2.09	2.06	2.04	2.01	1.99	1.98	1.96	1.95	1.93	1.92	1.91
32	4.15	3.29	2.90	2.67	2.51	2.40	2.31	2.24	2.19	2.14	2.10	2.07	2.04	2.01	1.99	1.97	1.95	1.94	1.92	1.91	1.90	1.88
35	4.12	3.27	2.87	2.64	2.49	2.37	2.29	2.22	2.16	2.11	2.07	2.04	2.01	1.99	1.96	1.94	1.92	1.91	1.89	1.88	1.87	1.85
40	4.08	3.23	2.84	2.61	2.45	2.34	2.25	2.18	2.12	2.08	2.04	2.00	1.97	1.95	1.92	1.90	1.89	1.87	1.85	1.84	1.83	1.81
45	4.06	3.20	2.81	2.58	2.42	2.31	2.22	2.15	2.10	2.05	2.01	1.97	1.94	1.92	1.89	1.87	1.86	1.84	1.82	1.81	1.80	1.78
50	4.03	3.18	2.79	2.56	2.40	2.29	2.20	2.13	2.07	2.03	1.99	1.95	1.92	1.89	1.87	1.85	1.83	1.81	1.80	1.78	1.77	1.76
60	4.00	3.15	2.76	2.53	2.37	2.25	2.17	2.10	2.04	1.99	1.95	1.92	1.89	1.86	1.84	1.82	1.80	1.78	1.76	1.75	1.73	1.72
75	3.97	3.12	2.73	2.49	2.34	2.22	2.13	2.06	2.01	1.96	1.92	1.88	1.85	1.83	1.80	1.78	1.76	1.74	1.73	1.71	1.70	1.69
100	3.94	3.09	2.70	2.46	2.31	2.19	2.10	2.03	1.97	1.93	1.89	1.85	1.82	1.79	1.77	1.75	1.73	1.71	1.69	1.68	1.66	1.65
120	3.92	3.07	2.68	2.45	2.29	2.18	2.09	2.02	1.96	1.91	1.87	1.83	1.80	1.78	1.75	1.73	1.71	1.69	1.67	1.66	1.64	1.63
140	3.91	3.06	2.67	2.44	2.28	2.16	2.08	2.01	1.95	1.90	1.86	1.82	1.79	1.76	1.74	1.72	1.70	1.68	1.66	1.65	1.63	1.62
180	3.89	3.05	2.65	2.42	2.26	2.15	2.06	1.99	1.93	1.88	1.84	1.81	1.77	1.75	1.72	1.70	1.68	1.66	1.64	1.63	1.61	1.60
250	3.88	3.03	2.64	2.41	2.25	2.13	2.05	1.98	1.92	1.87	1.83	1.79	1.76	1.74	1.71	1.69	1.66	1.65	1.63	1.61	1.60	1.58
400	3.86	3.02	2.63	2.39	2.24	2.12	2.03	1.96	1.90	1.85	1.81	1.78	1.74	1.72	1.69	1.67	1.65	1.63	1.61	1.60	1.58	1.57
1000	3.85	3.00	2.61	2.38	2.22	2.11	2.02	1.95	1.89	1.84	1.80	1.76	1.73	1.70	1.68	1.65	1.63	1.61	1.60	1.58	1.57	1.55

Denominator df

Table F (cont.)

$\alpha = .05$	23	24	25	26	27	28	29	30	32	35	40	45	50	60	75	100	120	140	180	250	400	1000
1	248.8	249.1	249.3	249.5	249.6	249.8	250.0	250.1	250.4	250.7	251.1	251.5	251.8	252.2	252.6	253.0	253.3	253.4	253.6	253.8	254.0	254.2
2	19.45	19.45	19.46	19.46	19.46	19.46	19.46	19.46	19.46	19.47	19.47	19.47	19.48	19.48	19.48	19.49	19.49	19.49	19.49	19.49	19.49	19.49
3	8.64	8.64	8.63	8.63	8.63	8.62	8.62	8.62	8.61	8.60	8.59	8.59	8.58	8.57	8.56	8.55	8.55	8.55	8.54	8.54	8.53	8.53
4	5.78	5.77	5.77	5.76	5.76	5.75	5.75	5.75	5.74	5.73	5.72	5.71	5.70	5.69	5.68	5.66	5.66	5.65	5.65	5.64	5.64	5.63
5	4.53	4.53	4.52	4.52	4.51	4.50	4.50	4.50	4.49	4.48	4.46	4.45	4.44	4.43	4.42	4.41	4.40	4.39	4.39	4.38	4.38	4.37
6	3.85	3.84	3.83	3.83	3.82	3.82	3.81	3.81	3.80	3.79	3.77	3.76	3.75	3.74	3.73	3.71	3.70	3.70	3.69	3.69	3.68	3.67
7	3.42	3.41	3.40	3.40	3.39	3.39	3.38	3.38	3.37	3.36	3.34	3.33	3.32	3.30	3.29	3.27	3.27	3.26	3.25	3.25	3.24	3.23
8	3.12	3.12	3.11	3.10	3.10	3.09	3.08	3.08	3.07	3.06	3.04	3.03	3.02	3.01	2.99	2.97	2.97	2.96	2.95	2.95	2.94	2.93
9	2.91	2.90	2.89	2.89	2.88	2.87	2.87	2.86	2.85	2.84	2.83	2.81	2.80	2.79	2.77	2.76	2.75	2.74	2.73	2.73	2.72	2.71
10	2.75	2.74	2.73	2.72	2.72	2.71	2.70	2.70	2.69	2.68	2.66	2.65	2.64	2.62	2.60	2.59	2.58	2.57	2.57	2.56	2.55	2.54
11	2.62	2.61	2.60	2.59	2.59	2.58	2.58	2.57	2.56	2.55	2.53	2.52	2.51	2.49	2.47	2.46	2.45	2.44	2.43	2.43	2.42	2.41
12	2.51	2.51	2.50	2.49	2.48	2.48	2.47	2.47	2.46	2.44	2.43	2.41	2.40	2.38	2.37	2.35	2.34	2.33	2.33	2.32	2.31	2.30
13	2.43	2.42	2.41	2.41	2.40	2.39	2.39	2.38	2.37	2.36	2.34	2.33	2.31	2.30	2.28	2.26	2.25	2.25	2.24	2.23	2.22	2.21
14	2.36	2.35	2.34	2.33	2.33	2.32	2.31	2.31	2.30	2.28	2.27	2.25	2.24	2.22	2.21	2.19	2.18	2.17	2.16	2.15	2.15	2.14
15	2.30	2.29	2.28	2.27	2.27	2.26	2.25	2.25	2.24	2.22	2.20	2.19	2.18	2.16	2.14	2.12	2.11	2.11	2.10	2.09	2.08	2.07
16	2.24	2.24	2.23	2.22	2.21	2.21	2.20	2.19	2.18	2.17	2.15	2.14	2.12	2.11	2.09	2.07	2.06	2.05	2.04	2.03	2.02	2.02
17	2.20	2.19	2.18	2.17	2.17	2.16	2.15	2.15	2.14	2.12	2.10	2.09	2.08	2.06	2.04	2.02	2.01	2.00	1.99	1.98	1.98	1.97
18	2.16	2.15	2.14	2.13	2.13	2.12	2.11	2.11	2.10	2.08	2.06	2.05	2.04	2.02	2.00	1.98	1.97	1.96	1.95	1.94	1.93	1.92
19	2.12	2.11	2.11	2.10	2.09	2.08	2.08	2.07	2.06	2.05	2.03	2.01	2.00	1.98	1.96	1.94	1.93	1.92	1.91	1.90	1.89	1.88
20	2.09	2.08	2.07	2.07	2.06	2.05	2.05	2.04	2.03	2.01	1.99	1.98	1.97	1.95	1.93	1.91	1.90	1.89	1.88	1.87	1.86	1.85
21	2.06	2.05	2.05	2.04	2.03	2.02	2.02	2.01	2.00	1.98	1.96	1.95	1.94	1.92	1.90	1.88	1.87	1.86	1.85	1.84	1.83	1.82
22	2.04	2.03	2.02	2.01	2.00	2.00	1.99	1.98	1.97	1.96	1.94	1.92	1.91	1.89	1.87	1.85	1.84	1.83	1.82	1.81	1.80	1.79
23	2.01	2.01	2.00	1.99	1.98	1.97	1.97	1.96	1.95	1.93	1.91	1.90	1.88	1.86	1.84	1.82	1.81	1.81	1.79	1.78	1.77	1.76
24	1.99	1.98	1.97	1.97	1.96	1.95	1.95	1.94	1.93	1.91	1.89	1.88	1.86	1.84	1.82	1.80	1.79	1.78	1.77	1.76	1.75	1.74
25	1.97	1.96	1.96	1.95	1.94	1.93	1.93	1.92	1.91	1.89	1.87	1.86	1.84	1.82	1.80	1.78	1.77	1.76	1.75	1.74	1.73	1.72
26	1.96	1.95	1.94	1.93	1.92	1.91	1.91	1.90	1.89	1.87	1.85	1.84	1.82	1.80	1.78	1.76	1.75	1.74	1.73	1.72	1.71	1.70
27	1.94	1.93	1.92	1.91	1.90	1.90	1.89	1.88	1.87	1.86	1.84	1.82	1.81	1.79	1.76	1.74	1.73	1.72	1.71	1.70	1.69	1.68
28	1.92	1.91	1.91	1.90	1.89	1.88	1.88	1.87	1.86	1.84	1.82	1.80	1.79	1.77	1.75	1.73	1.71	1.71	1.69	1.68	1.67	1.66
29	1.91	1.90	1.89	1.88	1.88	1.87	1.86	1.85	1.84	1.83	1.81	1.79	1.77	1.75	1.73	1.71	1.70	1.69	1.68	1.67	1.66	1.65
30	1.90	1.89	1.88	1.87	1.86	1.85	1.85	1.84	1.83	1.81	1.79	1.77	1.76	1.74	1.72	1.70	1.68	1.68	1.66	1.65	1.64	1.63
32	1.87	1.86	1.85	1.84	1.84	1.83	1.82	1.82	1.80	1.79	1.77	1.75	1.74	1.71	1.69	1.67	1.66	1.65	1.64	1.63	1.61	1.60
35	1.84	1.83	1.82	1.81	1.81	1.80	1.79	1.79	1.77	1.76	1.74	1.72	1.70	1.68	1.66	1.63	1.62	1.61	1.60	1.59	1.58	1.57
40	1.80	1.79	1.78	1.77	1.77	1.76	1.75	1.74	1.73	1.72	1.69	1.67	1.66	1.64	1.61	1.59	1.58	1.57	1.55	1.54	1.53	1.52
45	1.77	1.76	1.75	1.74	1.73	1.73	1.72	1.71	1.70	1.68	1.66	1.64	1.63	1.60	1.58	1.55	1.54	1.53	1.52	1.51	1.49	1.48
50	1.75	1.74	1.73	1.72	1.71	1.70	1.69	1.69	1.67	1.66	1.63	1.61	1.60	1.58	1.55	1.52	1.51	1.50	1.49	1.47	1.46	1.45
60	1.71	1.70	1.69	1.68	1.67	1.66	1.66	1.65	1.64	1.62	1.59	1.57	1.56	1.53	1.51	1.48	1.47	1.46	1.44	1.43	1.41	1.40
75	1.67	1.66	1.65	1.64	1.63	1.63	1.62	1.61	1.60	1.58	1.55	1.53	1.52	1.49	1.47	1.44	1.42	1.41	1.40	1.38	1.37	1.35
100	1.64	1.63	1.62	1.61	1.60	1.59	1.58	1.57	1.56	1.54	1.52	1.49	1.48	1.45	1.42	1.39	1.38	1.36	1.35	1.33	1.31	1.30
120	1.62	1.61	1.60	1.59	1.58	1.57	1.56	1.55	1.54	1.52	1.50	1.47	1.46	1.43	1.40	1.37	1.35	1.34	1.32	1.30	1.29	1.27
140	1.61	1.60	1.58	1.57	1.56	1.56	1.55	1.54	1.53	1.51	1.48	1.46	1.44	1.41	1.38	1.35	1.33	1.32	1.30	1.29	1.27	1.25
180	1.59	1.58	1.57	1.56	1.55	1.54	1.53	1.52	1.51	1.49	1.46	1.44	1.42	1.39	1.36	1.33	1.31	1.30	1.28	1.26	1.24	1.22
250	1.57	1.56	1.55	1.54	1.53	1.52	1.51	1.50	1.49	1.47	1.44	1.42	1.40	1.37	1.34	1.31	1.29	1.27	1.25	1.23	1.21	1.18
400	1.56	1.54	1.53	1.52	1.51	1.50	1.50	1.49	1.47	1.45	1.42	1.40	1.38	1.35	1.32	1.28	1.26	1.25	1.23	1.20	1.18	1.15
1000	1.54	1.53	1.52	1.51	1.50	1.49	1.48	1.47	1.46	1.43	1.41	1.38	1.36	1.33	1.30	1.26	1.24	1.22	1.20	1.17	1.14	1.11

Denominator df

Table F (cont.)

Numerator df

$\alpha = .1$	1	2	3	4	5	6	7	8	9	10	11	12	13	14	15	16	17	18	19	20	21	22
1	39.9	49.5	53.6	55.8	57.2	58.2	58.9	59.4	59.9	60.2	60.5	60.7	60.9	61.1	61.2	61.3	61.5	61.6	61.7	61.7	61.8	61.9
2	8.53	9.00	9.16	9.24	9.29	9.33	9.35	9.37	9.38	9.39	9.40	9.41	9.41	9.42	9.42	9.43	9.43	9.44	9.44	9.44	9.44	9.45
3	5.54	5.46	5.39	5.34	5.31	5.28	5.27	5.25	5.24	5.23	5.22	5.22	5.21	5.20	5.20	5.20	5.19	5.19	5.19	5.18	5.18	5.18
4	4.54	4.32	4.19	4.11	4.05	4.01	3.98	3.95	3.94	3.92	3.91	3.90	3.89	3.88	3.87	3.86	3.86	3.85	3.85	3.84	3.84	3.84
5	4.06	3.78	3.62	3.52	3.45	3.40	3.37	3.34	3.32	3.30	3.28	3.27	3.26	3.25	3.24	3.23	3.22	3.22	3.21	3.21	3.20	3.20
6	3.78	3.46	3.29	3.18	3.11	3.05	3.01	2.98	2.96	2.94	2.92	2.90	2.89	2.88	2.87	2.86	2.85	2.85	2.84	2.84	2.83	2.83
7	3.59	3.26	3.07	2.96	2.88	2.83	2.78	2.75	2.72	2.70	2.68	2.67	2.65	2.64	2.63	2.62	2.61	2.61	2.60	2.59	2.59	2.58
8	3.46	3.11	2.92	2.81	2.73	2.67	2.62	2.59	2.56	2.54	2.52	2.50	2.49	2.48	2.46	2.45	2.45	2.44	2.43	2.42	2.42	2.41
9	3.36	3.01	2.81	2.69	2.61	2.55	2.51	2.47	2.44	2.42	2.40	2.38	2.36	2.35	2.34	2.33	2.32	2.31	2.30	2.30	2.29	2.29
10	3.29	2.92	2.73	2.61	2.52	2.46	2.41	2.38	2.35	2.32	2.30	2.28	2.27	2.26	2.24	2.23	2.22	2.22	2.21	2.20	2.19	2.19
11	3.23	2.86	2.66	2.54	2.45	2.39	2.34	2.30	2.27	2.25	2.23	2.21	2.19	2.18	2.17	2.16	2.15	2.14	2.13	2.12	2.12	2.11
12	3.18	2.81	2.61	2.48	2.39	2.33	2.28	2.24	2.21	2.19	2.17	2.15	2.13	2.12	2.10	2.09	2.08	2.08	2.07	2.06	2.05	2.05
13	3.14	2.76	2.56	2.43	2.35	2.28	2.23	2.20	2.16	2.14	2.12	2.10	2.08	2.07	2.05	2.04	2.03	2.02	2.01	2.01	2.00	1.99
14	3.10	2.73	2.52	2.39	2.31	2.24	2.19	2.15	2.12	2.10	2.07	2.05	2.04	2.02	2.01	2.00	1.99	1.98	1.97	1.96	1.96	1.95
15	3.07	2.70	2.49	2.36	2.27	2.21	2.16	2.12	2.09	2.06	2.04	2.02	2.00	1.99	1.97	1.96	1.95	1.94	1.93	1.92	1.92	1.91
16	3.05	2.67	2.46	2.33	2.24	2.18	2.13	2.09	2.06	2.03	2.01	1.99	1.97	1.95	1.94	1.93	1.92	1.91	1.90	1.89	1.88	1.88
17	3.03	2.64	2.44	2.31	2.22	2.15	2.10	2.06	2.03	2.00	1.98	1.96	1.94	1.93	1.91	1.90	1.89	1.88	1.87	1.86	1.86	1.85
18	3.01	2.62	2.42	2.29	2.20	2.13	2.08	2.04	2.00	1.98	1.95	1.93	1.92	1.90	1.89	1.87	1.86	1.85	1.84	1.84	1.83	1.82
19	2.99	2.61	2.40	2.27	2.18	2.11	2.06	2.02	1.98	1.96	1.93	1.91	1.89	1.88	1.86	1.85	1.84	1.83	1.82	1.81	1.81	1.80
20	2.97	2.59	2.38	2.25	2.16	2.09	2.04	2.00	1.96	1.94	1.91	1.89	1.87	1.86	1.84	1.83	1.82	1.81	1.80	1.79	1.79	1.78
21	2.96	2.57	2.36	2.23	2.14	2.08	2.02	1.98	1.95	1.92	1.90	1.87	1.86	1.84	1.83	1.81	1.80	1.79	1.78	1.78	1.77	1.76
22	2.95	2.56	2.35	2.22	2.13	2.06	2.01	1.97	1.93	1.90	1.88	1.86	1.84	1.83	1.81	1.80	1.79	1.78	1.77	1.76	1.75	1.74
23	2.94	2.55	2.34	2.21	2.11	2.05	1.99	1.95	1.92	1.89	1.87	1.84	1.83	1.81	1.80	1.78	1.77	1.76	1.75	1.74	1.74	1.73
24	2.93	2.54	2.33	2.19	2.10	2.04	1.98	1.94	1.91	1.88	1.85	1.83	1.81	1.80	1.78	1.77	1.76	1.75	1.74	1.73	1.72	1.71
25	2.92	2.53	2.32	2.18	2.09	2.02	1.97	1.93	1.89	1.87	1.84	1.82	1.80	1.79	1.77	1.76	1.75	1.74	1.73	1.72	1.71	1.70
26	2.91	2.52	2.31	2.17	2.08	2.01	1.96	1.92	1.88	1.86	1.83	1.81	1.79	1.77	1.76	1.75	1.73	1.72	1.71	1.71	1.70	1.69
27	2.90	2.51	2.30	2.17	2.07	2.00	1.95	1.91	1.87	1.85	1.82	1.80	1.78	1.76	1.75	1.74	1.72	1.71	1.70	1.70	1.69	1.68
28	2.89	2.50	2.29	2.16	2.06	2.00	1.94	1.90	1.87	1.84	1.81	1.79	1.77	1.75	1.74	1.73	1.71	1.70	1.69	1.69	1.68	1.67
29	2.89	2.50	2.28	2.15	2.06	1.99	1.93	1.89	1.86	1.83	1.80	1.78	1.76	1.75	1.73	1.72	1.71	1.69	1.68	1.68	1.67	1.66
30	2.88	2.49	2.28	2.14	2.05	1.98	1.93	1.88	1.85	1.82	1.79	1.77	1.75	1.74	1.72	1.71	1.70	1.69	1.68	1.67	1.66	1.65
32	2.87	2.48	2.26	2.13	2.04	1.97	1.91	1.87	1.83	1.81	1.78	1.76	1.74	1.72	1.71	1.69	1.68	1.67	1.66	1.65	1.64	1.64
35	2.85	2.46	2.25	2.11	2.02	1.95	1.90	1.85	1.82	1.79	1.76	1.74	1.72	1.70	1.69	1.67	1.66	1.65	1.64	1.63	1.62	1.62
40	2.84	2.44	2.23	2.09	2.00	1.93	1.87	1.83	1.79	1.76	1.74	1.71	1.70	1.68	1.66	1.65	1.64	1.62	1.61	1.61	1.60	1.59
45	2.82	2.42	2.21	2.07	1.98	1.91	1.85	1.81	1.77	1.74	1.72	1.70	1.68	1.66	1.64	1.63	1.62	1.60	1.59	1.58	1.58	1.57
50	2.81	2.41	2.20	2.06	1.97	1.90	1.84	1.80	1.76	1.73	1.70	1.68	1.66	1.64	1.63	1.61	1.60	1.59	1.58	1.57	1.56	1.55
60	2.79	2.39	2.18	2.04	1.95	1.87	1.82	1.77	1.74	1.71	1.68	1.66	1.64	1.62	1.60	1.59	1.58	1.56	1.55	1.54	1.53	1.53
75	2.77	2.37	2.16	2.02	1.93	1.85	1.80	1.75	1.72	1.69	1.66	1.63	1.61	1.60	1.58	1.57	1.55	1.54	1.53	1.52	1.51	1.50
100	2.76	2.36	2.14	2.00	1.91	1.83	1.78	1.73	1.69	1.66	1.64	1.61	1.59	1.57	1.56	1.54	1.53	1.52	1.50	1.49	1.48	1.48
120	2.75	2.35	2.13	1.99	1.90	1.82	1.77	1.72	1.68	1.65	1.63	1.60	1.58	1.56	1.55	1.53	1.52	1.50	1.49	1.48	1.47	1.46
140	2.74	2.34	2.12	1.99	1.89	1.82	1.76	1.71	1.68	1.64	1.62	1.59	1.57	1.55	1.54	1.52	1.51	1.50	1.48	1.47	1.46	1.45
180	2.73	2.33	2.11	1.98	1.88	1.81	1.75	1.70	1.67	1.63	1.61	1.58	1.56	1.54	1.53	1.51	1.50	1.48	1.47	1.46	1.45	1.44
250	2.73	2.32	2.11	1.97	1.87	1.80	1.74	1.69	1.66	1.62	1.60	1.57	1.55	1.53	1.51	1.50	1.49	1.47	1.46	1.45	1.44	1.43
400	2.72	2.32	2.10	1.96	1.86	1.79	1.73	1.69	1.65	1.61	1.59	1.56	1.54	1.52	1.50	1.49	1.47	1.46	1.45	1.44	1.43	1.42
1000	2.71	2.31	2.09	1.95	1.85	1.78	1.72	1.68	1.64	1.61	1.58	1.55	1.53	1.51	1.49	1.48	1.46	1.45	1.44	1.43	1.42	1.41

Denominator df

Table F (cont.)

Numerator df

$\alpha = .1$

Denominator df	23	24	25	26	27	28	29	30	32	35	40	45	50	60	75	100	120	140	180	250	400	1000
1	61.9	62.0	62.1	62.1	62.1	62.2	62.2	62.3	62.3	62.4	62.5	62.6	62.7	62.8	62.9	63.0	63.1	63.1	63.1	63.2	63.2	63.3
2	9.45	9.45	9.45	9.45	9.45	9.46	9.46	9.46	9.46	9.46	9.47	9.47	9.47	9.47	9.48	9.48	9.48	9.48	9.49	9.49	9.49	9.49
3	5.18	5.18	5.17	5.17	5.17	5.17	5.17	5.17	5.17	5.16	5.16	5.16	5.15	5.15	5.15	5.14	5.14	5.14	5.14	5.14	5.14	5.1
4	3.83	3.83	3.83	3.83	3.82	3.82	3.82	3.82	3.81	3.81	3.80	3.80	3.80	3.79	3.78	3.78	3.78	3.77	3.77	3.77	3.77	3.76
5	3.19	3.19	3.19	3.18	3.18	3.18	3.18	3.17	3.17	3.16	3.16	3.15	3.15	3.14	3.13	3.13	3.12	3.12	3.12	3.11	3.11	3.11
6	2.82	2.82	2.81	2.81	2.81	2.81	2.80	2.80	2.80	2.79	2.78	2.77	2.77	2.76	2.75	2.75	2.74	2.74	2.74	2.73	2.73	2.72
7	2.58	2.58	2.57	2.57	2.56	2.56	2.56	2.56	2.55	2.54	2.54	2.53	2.52	2.51	2.51	2.50	2.49	2.49	2.49	2.48	2.48	2.47
8	2.41	2.40	2.40	2.40	2.39	2.39	2.39	2.38	2.38	2.37	2.36	2.35	2.35	2.34	2.33	2.32	2.32	2.31	2.31	2.30	2.30	2.30
9	2.28	2.28	2.27	2.27	2.26	2.26	2.26	2.25	2.25	2.24	2.23	2.22	2.22	2.21	2.20	2.19	2.18	2.18	2.18	2.17	2.17	2.16
10	2.18	2.18	2.17	2.17	2.17	2.16	2.16	2.16	2.15	2.14	2.13	2.12	2.12	2.11	2.10	2.09	2.08	2.08	2.07	2.07	2.06	2.06
11	2.11	2.10	2.10	2.09	2.09	2.08	2.08	2.08	2.07	2.06	2.05	2.04	2.04	2.03	2.02	2.01	2.00	2.00	1.99	1.99	1.98	1.98
12	2.04	2.04	2.03	2.03	2.02	2.02	2.01	2.01	2.01	2.00	1.99	1.98	1.97	1.96	1.95	1.94	1.93	1.93	1.92	1.92	1.91	1.91
13	1.99	1.98	1.98	1.97	1.97	1.96	1.96	1.96	1.95	1.94	1.93	1.92	1.92	1.90	1.89	1.88	1.88	1.87	1.87	1.86	1.86	1.85
14	1.94	1.94	1.93	1.93	1.92	1.92	1.92	1.91	1.91	1.90	1.89	1.88	1.87	1.86	1.85	1.83	1.83	1.82	1.82	1.81	1.81	1.80
15	1.90	1.90	1.89	1.89	1.88	1.88	1.88	1.87	1.87	1.86	1.85	1.84	1.83	1.82	1.80	1.79	1.79	1.78	1.78	1.77	1.76	1.76
16	1.87	1.87	1.86	1.86	1.85	1.85	1.84	1.84	1.83	1.82	1.81	1.80	1.79	1.78	1.77	1.76	1.75	1.75	1.74	1.73	1.73	1.72
17	1.84	1.84	1.83	1.83	1.82	1.82	1.81	1.81	1.80	1.79	1.78	1.77	1.76	1.75	1.74	1.73	1.72	1.71	1.71	1.70	1.70	1.69
18	1.82	1.81	1.80	1.80	1.80	1.79	1.79	1.78	1.78	1.77	1.75	1.74	1.74	1.72	1.71	1.70	1.69	1.69	1.68	1.67	1.67	1.66
19	1.79	1.79	1.78	1.78	1.77	1.77	1.76	1.76	1.75	1.74	1.73	1.72	1.71	1.70	1.69	1.67	1.67	1.66	1.65	1.65	1.64	1.64
20	1.77	1.77	1.76	1.76	1.75	1.75	1.74	1.74	1.73	1.72	1.71	1.70	1.69	1.68	1.66	1.65	1.64	1.64	1.63	1.62	1.62	1.61
21	1.75	1.75	1.74	1.74	1.73	1.73	1.72	1.72	1.71	1.70	1.69	1.68	1.67	1.66	1.64	1.63	1.62	1.62	1.61	1.60	1.60	1.59
22	1.74	1.73	1.73	1.72	1.72	1.71	1.71	1.70	1.69	1.68	1.67	1.66	1.65	1.64	1.63	1.61	1.60	1.60	1.59	1.59	1.58	1.57
23	1.72	1.72	1.71	1.70	1.70	1.69	1.69	1.69	1.68	1.67	1.66	1.64	1.64	1.62	1.61	1.59	1.59	1.58	1.57	1.57	1.56	1.55
24	1.71	1.70	1.70	1.69	1.69	1.68	1.68	1.67	1.66	1.65	1.64	1.63	1.62	1.61	1.59	1.58	1.57	1.57	1.56	1.55	1.54	1.54
25	1.70	1.69	1.68	1.68	1.67	1.67	1.66	1.66	1.65	1.64	1.63	1.62	1.61	1.59	1.58	1.56	1.56	1.55	1.54	1.54	1.53	1.52
26	1.68	1.68	1.67	1.67	1.66	1.66	1.65	1.65	1.64	1.63	1.61	1.60	1.59	1.58	1.57	1.55	1.54	1.54	1.53	1.52	1.52	1.51
27	1.67	1.67	1.66	1.65	1.65	1.64	1.64	1.64	1.63	1.62	1.60	1.59	1.58	1.57	1.55	1.54	1.53	1.53	1.52	1.51	1.50	1.50
28	1.66	1.66	1.65	1.64	1.64	1.63	1.63	1.63	1.62	1.61	1.59	1.58	1.57	1.56	1.54	1.53	1.52	1.51	1.51	1.50	1.49	1.48
29	1.65	1.65	1.64	1.63	1.63	1.62	1.62	1.62	1.61	1.60	1.58	1.57	1.56	1.55	1.53	1.52	1.51	1.50	1.50	1.49	1.48	1.47
30	1.64	1.64	1.63	1.63	1.62	1.62	1.61	1.61	1.60	1.59	1.57	1.56	1.55	1.54	1.52	1.51	1.50	1.49	1.49	1.48	1.47	1.46
32	1.63	1.62	1.62	1.61	1.60	1.60	1.59	1.59	1.58	1.57	1.56	1.54	1.53	1.52	1.50	1.49	1.48	1.47	1.47	1.46	1.45	1.44
35	1.61	1.60	1.60	1.59	1.58	1.58	1.57	1.57	1.56	1.55	1.53	1.52	1.51	1.50	1.48	1.47	1.46	1.45	1.44	1.43	1.43	1.42
40	1.58	1.57	1.57	1.56	1.56	1.55	1.55	1.54	1.53	1.52	1.51	1.49	1.48	1.47	1.45	1.43	1.42	1.42	1.41	1.40	1.39	1.38
45	1.56	1.55	1.55	1.54	1.53	1.53	1.52	1.52	1.51	1.50	1.48	1.47	1.46	1.44	1.43	1.41	1.40	1.39	1.38	1.37	1.37	1.36
50	1.54	1.54	1.53	1.52	1.52	1.51	1.51	1.50	1.49	1.48	1.46	1.45	1.44	1.42	1.41	1.39	1.38	1.37	1.36	1.35	1.34	1.33
60	1.52	1.51	1.50	1.50	1.49	1.49	1.48	1.48	1.47	1.45	1.44	1.42	1.41	1.40	1.38	1.36	1.35	1.34	1.33	1.32	1.31	1.30
75	1.49	1.49	1.48	1.47	1.47	1.46	1.45	1.45	1.44	1.43	1.41	1.40	1.38	1.37	1.35	1.33	1.32	1.31	1.30	1.29	1.27	1.26
100	1.47	1.46	1.45	1.45	1.44	1.43	1.43	1.42	1.41	1.40	1.38	1.37	1.35	1.34	1.32	1.29	1.28	1.27	1.26	1.25	1.24	1.22
120	1.46	1.45	1.44	1.43	1.43	1.42	1.41	1.41	1.40	1.39	1.37	1.35	1.34	1.32	1.30	1.28	1.26	1.26	1.25	1.23	1.22	1.20
140	1.45	1.44	1.43	1.42	1.42	1.41	1.41	1.40	1.39	1.38	1.36	1.34	1.33	1.31	1.29	1.26	1.25	1.24	1.24	1.22	1.20	1.19
180	1.43	1.43	1.42	1.41	1.40	1.40	1.39	1.39	1.38	1.36	1.34	1.33	1.32	1.29	1.27	1.25	1.23	1.23	1.21	1.20	1.18	1.16
250	1.42	1.41	1.41	1.40	1.39	1.39	1.38	1.37	1.36	1.35	1.33	1.31	1.30	1.28	1.26	1.23	1.22	1.22	1.19	1.18	1.16	1.14
400	1.41	1.40	1.39	1.39	1.38	1.37	1.37	1.36	1.35	1.34	1.32	1.30	1.29	1.26	1.24	1.21	1.20	1.19	1.17	1.16	1.14	1.12
1000	1.40	1.39	1.38	1.38	1.37	1.36	1.36	1.35	1.34	1.32	1.30	1.29	1.27	1.25	1.23	1.20	1.18	1.17	1.15	1.13	1.11	1.08

Wilcoxon Rank Sum Test

The table gives some lower and upper tail probabilities, under H_0, for $T =$ sum of the ranks for the smaller of two independent samples (or for either sample if $n_1 = n_2$). Probabilities shown at the top are either $P(T \leq \textit{critical value})$ or $P(T \geq \textit{critical value})$, for the critical values displayed inside the table.

For some sample sizes, for practical reasons related to the discreteness of the distribution, we have chosen to display critical values that have a slightly higher probability (shown in parentheses) than what is written at the top of the table, but which are as close as possible. Otherwise, these probabilities are accurate (within a small rounding factor) or are conservative (high).

| n_1/n_2 | Lower tail probability | | | // | Upper tail probability | | |
	.01	.025	.05	//	.05	.025	.01
3/3			6		15		
3/4		6(.029)	7(.057)		17(.057)	18(.029)	
3/5	6(.018)	7(.036)	8(.07)		19(.07)	20(.036)	21(.018)
3/6	6	7	8		22	23	24
3/7	6	8(.033)	9(.058)		24(.058)	25(.033)	27
3/8	7	8	9		27	28	29
3/9	7	8	10		29	31	32
3/10	8(.014)	9	11(.056)		31(.056)	33	34(.014)
4/4	10(.014)	11(.029)	12(.057)		24(.057)	25(.029)	26(.014)
4/5	10	11	13(.056)		27(.056)	29	30
4/6	11	12	14(.057)		30(.057)	32	33
4/7	12	13	15(.055)		33(.055)	35	36
4/8	12	14	16(.055)		36(.055)	38	40
4/9	13	15	17(.053)		39(.053)	41	43
4/10	14	16(.027)	18(.053)		42(.053)	44(.027)	46
5/5	16	18(.028)	19		36	37(.028)	39
5/6	17	19	20		40	41	43
5/7	18	20	22(.053)		43(.053)	45	47
5/8	19	21	23		47	49	51
5/9	20	22	25(.056)		50(.056)	53	55
5/10	21	24(.028)	26		54	56(.028)	59
6/6	24	26	28		50	52	54
6/7	26	28	30		54	56	58
6/8	27	29	32(.054)		58(.054)	61	63
6/9	28	31	34(.057)		62(.057)	65	68
6/10	30	33(.028)	35		67	69(.028)	72
7/7	34	37(.027)	39		66	68(.027)	71
7/8	36	39(.027)	41		71	73(.027)	76
7/9	38	41(.027)	43		76	78(.027)	81
7/10	39	43(.028)	46(.054)		80(.054)	83(.028)	87
8/8	46	49	52		84	87	90
8/9	48	51	55(.057)		89(.057)	93	96
8/10	50	54(.027)	57		95	98(.027)	102
9/9	59	63	66		105	108	112
9/10	62	66(.027)	69		111	114(.027)	118
10/10	74	79	83(.053)		127(.053)	131	136

If sample sizes are off the table, use the normal approximation for the distribution of $T =$ rank sum of sample i, where $E(T) = n_i(n_1 + n_2 + 1)/2$ and $V(T) = n_1 n_2 (n_1 + n_2 + 1)/12$. (In the presence of many ties, this approach will be conservative.)

Adapted from Table A.6, Hollander & Wolf, *Nonparametric Statistical Methods*, Wiley (1973).

Tables for the Wilcoxon Signed Rank Test

Below are critical values for the two-sided signed rank test. Reject H_0 at indicated level if $T=$ smaller of $\{T+, T-\} \leq$ critical value. Divide the given P-value by 2 to obtain P-values for one-sided alternatives, using either $T+$ or $T-$ as appropriate to your alternative (since a small value for only one of these, and not the other, will support your H_a).

n	$P \leq .1$	$P \leq .05$	$P \leq .02$	$P \leq .01$	$P \leq .002$
5	0				
6	2	0			
7	3	2	0		
8	5	3	1	0	
9	8	5	3	1	
10	10	8	5	3	0
11	13	10	7	5	1
12	17	13	9	7	2
13	21	17	12	9	4
14	25	21	15	12	6
15	30	25	19	15	8
16	35	29	23	19	11
17	41	34	27	23	14
18	47	40	32	27	18
19	53	46	37	32	21
20	60	52	43	37	26
21	67	58	49	42	30
22	75	65	55	48	35
23	83	73	62	54	40
24	91	81	69	61	45
25	100	89	76	68	51

If off the table, use the normal approximation for the distribution of the rank sum T, with $E(T) = n(n + 1)/4$ and $V(T) = n(n + 1)(2n + 1)/24$

Adapted from Table A.6, Hollander & Wolf, *Nonparametric Statistical Methods*, Wiley (1973).